PIPING DATABOOK

Mohinder L. Nayyar, P. E.
ASME Fellow

McGRAW-HILL
New York Chicago San Francisco Lisbon London
Madrid Mexico City Milan New Delhi San Juan Seoul
Singapore Sydney Toronto

Cataloging-in-Publication Data is on file with the Library of Congress.

McGraw-Hill

A Division of The ***McGraw-Hill*** *Companies*

2 3 4 5 6 7 8 9 0 PBT/PBT 0 8 7 6 5 4 3

ISBN 0-07-136451-x

The sponsoring editor for this book was Larry Hager, the editing supervisor was Caroline Levine, and the production supervisor was Sherri Souffrance. This book was set in Times Roman by SNP Best-set Typesetter Ltd., Hong Kong.

Printed and bound by Custom Graphics.

This book is printed on recycled, acid-free paper containing a minimum of 50% recycled, de-inked fiber.

To those whose endeavors add to knowledge
and improve life

ABOUT THE AUTHOR

Mohinder L. Nayyar, P.E., ASME Fellow, is Senior Engineering Specialist for piping and valves in the engineering department of the Bechtel Power Corporation in Frederick, Maryland, and the Editor-in-Chief and an author of the sixth and seventh editions of the field-leading *Piping Handbook*. As a Bechtel consultant, he has assisted many leading companies, and has supervised and supported both nuclear- and fossil-fueled projects around the world. He is Chairman of the Materials Technical Committee of the ASME B31, Pressure Piping Code, and of the Materials Group of ASME B31.1, Power Piping Code; Vice Chairman and member of ASME B16 Main Committee; member of ASME Section II, Materials Subcommittee; and several other standards committees. He lives in Brookeville, Maryland.

CONTENTS

Section 2 Piping Materials

Section 3 Metallic Piping

Section 7 Pipe Supports

Section 8 Gaskets

Section 9 Resource List

HONORS LIST

CONTRIBUTORS

Chapter number	Chapter title	Contributor
1	Terminology	*Mohinder L. Nayyar, P.E.*, ASME Fellow, Bechtel Power Corporation, 5275 Westview Drive, Frederick, MD 21703
2	Systems of Units and Conversion Tables	*Mohinder L. Nayyar, P.E.*, ASME Fellow, Bechtel Power Corporation, 5275 Westview Drive, Frederick, MD 21703 *Ervin L. Geiger, P.E.*, Engineering Supervisor, Bechtel Power Corporation, 5275 Westview Drive, Frederick, MD 21703 *Kenneth R. Collier*, Systems Engineer, Pittsburgh Corning, 800 Presque Isle Drive, Pittsburgh, PA 15239 *Kathleen Posteraro*, Systems Engineer, Pittsburgh Corning, 800 Presque Isle Drive, Pittsburgh, PA 15239
3	Materials for Metallic Piping	*Shriman Gurram*, Mechanical Engineer, Bechtel Power Corporation, 5275 Westview Drive, Frederick, MD 21703 *R. Peter Deubler, P.E.*, Technical Director, Fronek Company, Inc., 15 Engle Street, Englewood, NJ 07631 *Jill M. Gillette,* Mechanical Engineer, Bechtel Power Corporation, 5275 Westview Drive, Frederick, MD 21703 *James M. Tanzosh*, Supervisor, Materials Engineering, Babcock & Wilcox Co., 20 S. Van Buren Ave., Barberton, OH 44203 *Akhil Prakash, P.E.*, Supervisor Engineer, 12741 King Street, Overland Park, KS 66213 *Mohinder L. Nayyar, P.E.*, ASME Fellow, Bechtel Power Corporation, 5275 Westview Drive, Frederick, MD 21703
4	Nonmetallic Piping Materials	*Mohinder L. Nayyar, P.E.*, ASME Fellow, Bechtel Power Corporation, 5275 Westview Drive, Frederick, MD 21703 *Shriman Gurram*, Mechanical Engineer, Bechtel Power Corporation, 5275 Westview Drive, Frederick, MD 21703 *Dr. Timothy J. McGrath*, Principal, Simpson, Gumpertz & Heger, Inc., 297 Broadway, Arlington, MA 02174-5310 *Stanley A. Mruk*, Executive Director, Plastics Pipe Institute, Wayne Interchange Plaza II, 155 Route 46 West, Wayne, NJ 07470 *Carl E. Martin*, Director of Marketing, Fibercast Company, P. O. Box 968, Sand Springs, Oklahoma 74063-0968

Chapter number	Chapter title	Contributor
5	ASME Wrought Pipe, Fittings, Flanges, and Joints	*Dr. C. Basavarajru*, Senior Engineering Specialist, Bechtel Power Corporation, 5275 Westview Drive, Frederick, MD 21703 *Sandeep Singh*, Piping Engineer, Bechtel India Limited, 249A Udyog Vihar, Phase IV, Gurgaon-122015, Haryana, India *Rohit Goel*, Piping Engineer, Bechtel India Limited, 249A Udyog Vihar, Phase IV, Gurgaon-122015, Haryana, India *Dheeraj Modawel*, Piping Engineer, Bechtel India Limited, 249A Udyog Vihar, Phase IV, Gurgaon-122015, Haryana, India *Shriman Gurram*, Mechanical Engineer, Bechtel Power Corporation, 5275 Westview Drive, Frederick, MD 21703 *Mohinder L. Nayyar, P.E.*, ASME Fellow, Bechtel Power Corporation, 5275 Westview Drive, Frederick. MD 21703
6	API Wrought Pipe, Flanges, Fittings, and Joints	*Mohinder L. Nayyar, P.E.*, ASME Fellow, Bechtel Power Corporation, 5275 Westview Drive, Frederick, MD 21703 *Dheeraj Modawel*, Piping Engineer, Bechtel India Limited, 249A Udyog Vihar, Phase IV, Gurgaon-122015, Haryana, India
7	AWWA Wrought Pipe, Fittings, Flanges, and Joints	*Mohinder L. Nayyar, P.E.*, ASME Fellow, Bechtel Power Corporation, 5275 Westview Drive, Frederick, MD 21703
8	MSS Fittings and Flanges	*Mohinder L. Nayyar, P.E.*, ASME Fellow, Bechtel Power Corporation, 5275 Westview Drive, Frederick, MD 21703 *Sandeep Singh*, Piping Engineer, Bechtel India Limited, 249A Udyog Vihar, Phase IV, Gurgaon-122015, Haryana, India
9	Cast Iron Pipe, Fittings, Flanges, and Joints	*Soami D. Suri, P.E.*, Senior Mechanical Engineer, Bechtel Power Corporation, 5275 Westview Drive, Frederick, MD 21703 *Mohinder L. Nayyar, P.E.*, ASME Fellow, Bechtel Power Corporation, 5275 Westview Drive, Frederick, MD 21703
10	Ductile Iron Pipe, Fittings, Flanges, and Joints	*Ramesh Tiwari*, Senior Mechanical Engineer, Bechtel Power Corporation, 5275 Westview Drive, Frederick, MD 21703 *Soami D. Suri, P.E.*, Senior Mechanical Engineer, Bechtel Power Corporation, 5275 Westview Drive, Frederick, MD 21703
11	Malleable Iron and Gray Iron Fittings	*Mohinder L. Nayyar, P.E.*, ASME Fellow, Bechtel Power Corporation, 5275 Westview Drive, Frederick, MD 21703
12	Copper and Copper Alloy Pipe, Fittings, Flanges, and Joints	*Ramesh Tiwari*, Senior Mechanical Engineer, Bechtel Power Corporation, 5275 Westview Drive, Frederick, MD 21703
13	Aluminum and Aluminum Alloy Pipe and Tube, Fittings, Flanges, and Joints	*Mohinder L. Nayyar, P.E.*, ASME Fellow, Bechtel Power Corporation, 5275 Westview Drive, Frederick, MD 21703
14	Titanium and Titanium Alloy Pipe and Tube, Zirconium and Zirconium Alloy Pipe and Tube	*Mohinder L. Nayyar, P.E.*, ASME Fellow, Bechtel Power Corporation, 5275 Westview Drive, Frederick, MD 21703
15	Piping Joints	*Shriman Gurram*, Mechanical Engineer, Bechtel Power Corporation, 5275 Westview Drive, Frederick, MD 21703

Chapter number	Chapter title	Contributor
		Mohinder L. Nayyar, P.E., ASME Fellow, Bechtel Power Corporation, 5275 Westview Drive, Frederick, MD 21703
16	Fabrication of Piping	*Mohinder L. Nayyar, P.E.*, ASME Fellow, Bechtel Power Corporation, 5275 Westview Drive, Frederick, MD 21703
17	PVC Plastic Pipe, Fittings, Flanges, and Joints	*Ramesh Tiwari*, Senior Mechanical Engineer, Bechtel Power Corporation, 5275 Westview Drive, Frederick, MD 21703 *Mohinder L. Nayyar, P.E.*, ASME Fellow, Bechtel Power Corporation, 5275 Westview Drive, Frederick, MD 21703
18	CPVC Pipe, Tube, Fittings, Flanges, and Joints	*Ramesh Tiwari*, Senior Mechanical Engineer, Bechtel Power Corporation, 5275 Westview Drive, Frederick, MD 21703 *Mohinder L. Nayyar, P.E.*, ASME Fellow, Bechtel Power Corporation, 5275 Westview Drive, Frederick, MD 21703
19	Polyethylene Pipe, Fittings, Flanges, and Joints	*Ratna Tewari*, Civil and Structural Engineer, Piping Engineer, Bechtel India Limited, 249A Udyog Vihar, Phase IV, Gurgaon-122015, Haryana, India *Ved Prakash Tewari*, Piping Designer, Piping Engineer, Bechtel India Limited, 249A Udyog Vihar, Phase IV, Gurgaon-122015, Haryana, India
20	Acrylonitrile-Butadiene-Styrene (ABS) Pipe, Fittings, and Joints	*Mohinder L. Nayyar, P.E.*, ASME Fellow, Bechtel Power Corporation, 5275 Westview Drive, Frederick, MD 21703
21	Fiberglass Pipe, Fittings, Flanges, and Joints	*Mohinder L. Nayyar, P.E.*, ASME Fellow, Bechtel Power Corporation, 5275 Westview Drive, Frederick, MD 21703 *Carl E. Martin*, Director of Marketing, Fibercast Company, P. O. Box 968, Sand Springs, Oklahoma 74063-0968
22	Concrete and Reinforced Concrete Piping	*Mohinder L. Nayyar, P.E.*, ASME Fellow, Bechtel Power Corporation, 5275 Westview Drive, Frederick, MD 21703 *Richard E. Deremiah, P.E.*, Project Manager, Price Brothers Company, 367 West Second Avenue, Dayton, OH 45402
23	Glass Piping, Fittings, and Joints	*Mohinder L. Nayyar, P.E.*, ASME Fellow, Bechtel Power Corporation, 5275 Westview Drive, Frederick, MD 21703
24	Vitrified Clay Pipe, Fittings, and Joints	*Mohinder L. Nayyar, P.E.*, ASME Fellow, Bechtel Power Corporation, 5275 Westview Drive, Frederick, MD 21703
25	Lined Piping	*Mohinder L. Nayyar, P.E.*, ASME Fellow, Bechtel Power Corporation, 5275 Westview Drive, Frederick, MD 21703 *Richard E. Deremiah, P.E.*, Project Manager, Price Brothers Company, 367 West Second Avenue, Dayton, OH 45402 *J. Alan Kehr*, Technical Marketing Manager, 3M Company, 3M Austin Center, Building A 147-4N-02, 6801 River Place Boulevard, Austin, TX 78726-9000 *Richard K. Lewis*, Executive Vice President, Blair Rubber Company, 1252 Mina Avenue, Akron, OH 44321 *David Jentzsch*, General Manager, Blair Rubber Company, 1252 Mina Avenue, Akron, OH 44321

Chapter number	Chapter title	Contributor
		Michael B. Ferg, Marketing Engineer, Resistoflex Company, One Quality Way, Marion, NC 28752 *John M. Kalnins*, Dow Plastic-Lined Piping Products, 4675 E. Wilder Road, Bay City, MI 48706
26	Pipe Supports and Hangers for Metallic Piping	*Lorenzo DiGiacomo, Jr.*, Senior Engineer, Bechtel Power Corporation, 5275 Westview Drive, Frederick, MD 21703
27	Pipe Supports and Hangers for Nonmetallic Piping	*Lorenzo DiGiacomo, Jr.*, Senior Engineer, Bechtel Power Corporation, 5275 Westview Drive, Frederick, MD 21703 *Dr. Timothy J. McGrath*, Principal, Simpson, Gumpertz & Heger, Inc., 297 Broadway, Arlington, MA 02174-5310 *Stanley A. Mruk*, Executive Director, Plastics Pipe Institute, Wayne Interchange Plaza II, 155 Route 46 West, Wayne, NJ 07470 *Carl E. Martin*, Director of Marketing, Fibercast Company, P. O. Box 968, Sand Springs, Oklahoma 74063-0968
28	Gaskets: Selection, Application, and Installation	Walter M. Stephan, Engineering Manager, Flexitallic, Inc., 1300 Route 73, Suite 311, Mt. Laurel, NJ 08054 *Mohinder L. Nayyar, P.E.*, ASME Fellow, Bechtel Power Corporation, 5275 Westview Drive, Frederick, MD 21703
29	Manufacturers, Suppliers, and Fabricators of Piping and Specialty Items	*Soami D. Suri, P.E.*, Senior Mechanical Engineer, Bechtel Power Corporation, 5275 Westview Drive, Frederick, MD 21703 *James Kenyon White*, Piping and Valve Technician, Bechtel Power Corporation, 5275 Westview Drive, Frederick, MD 21703 *Mohinder L. Nayyar, P.E.*, ASME Fellow, Bechtel Power Corporation, 5275 Westview Drive, Frederick, MD 21703

TECHNICAL AND ADMINISTRATIVE SUPPORT

Shriman Gurram, Mechanical Engineer, Bechtel Power Corporation, 5275 Westview Drive, Frederick, MD 21703

Dheeraj Modawel, Piping Engineer, Bechtel India Limited, 249A Udyog Vihar, Phase IV, Gurgaon-122015, Haryana, India

Sandeep Singh, Piping Engineer, Bechtel India Limited, 249A Udyog Vihar, Phase IV, Gurgaon-122015, Haryana, India

Troy J. Skillen, Mechanical Engineer, Bechtel Power Corporation, 5275 Westview Drive, Frederick, MD 21703

Ramesh Tiwari, Senior Mechanical Engineer, Bechtel Power Corporation, 5275 Westview Drive, Frederick, MD 21703

James Kenyon White, Piping and Valve Technician, Bechtel Power Corporation, 5275 Westview Drive, Frederick, MD 21703

PREFACE

It has been my privilege to work as a piping engineer. During the course of my work there have been numerous instances in which the piping-related data required to perform tasks at hand were not readily available. Like others, I had to look beyond the limits of my office and the reference library. At times, it has been frustrating and disappointing not to find the standards, specifications, reference books, catalogs, and other sources of information to assist me in my work. The urgency of the schedule-constrained demands of my profession did not go away. Thus I feel honored to share the *Piping Databook*, a resource that brings together information from many areas, with my fellow pipers.

The primary objective of this effort is to make commonly used piping-related data available in an easy-to-locate format in one source book. Users must find it time saving, convenient, and rewarding to see information from many standards, specifications, codes, reference books, vendor catalogs, and other similar publications. It would be unusual, if not impossible, for a piping professional to have all the resources available at his/her disposal in one location. Therefore it has been the objective of myself and my associates to make this book a source for engineers, designers, fabricators, constructors, installers, inspectors, suppliers, and manufacturers, maintenance, and operation personnel involved with piping of civic and industrial facilities. Others engaged in the fields of education, research, and planning associated with piping systems should find it handy in accomplishing their tasks.

Piping Databook contains definitions of commonly used terms for metallic and nonmetallic pipe and pipe lined with nonmetallics, and fittings, flanges, piping joints, fabrication, installation, testing, and inspection. It lists dimensions and tolerances for pipe, fittings, and flanges conforming to national and international codes and standards, in addition to other accepted industry standards used all over the world. Tabulation of domestic and international material specifications for pipe, fittings, and flanges makes it very helpful to specify the acceptable materials and determine their compatibility. Hard-to-locate but commonly required values of various material properties enhance the value of this book.

A chapter on units and conversion factors includes fast-lookup tables for ease of conversion from one type of unit to another. An introduction to different types of unit systems used around the globe provides marked understanding of differences and clarifies any confusion between metric and SI units.

Since flanged joints are an integral part of most piping systems, we realized that relevant data for selection and application of gaskets are needed by users to make leak-tight joints. We have made an attempt to satisfy this need by adding a chapter on gaskets, covering different types of gaskets, materials of construction, dimensions, and tolerances, and capability, suitability, compatibility, and limitation of gaskets in relation to flow media, pressure-temperature ratings, surface finish, and types of flange facings. In addition, the *Databook* includes tables for selection of proper bolts and nuts. Two chapters on pipe supports for metallic and nonmetallic piping systems define and illustrate types of pipe supports and hangers, and contain tables for selection of support spans.

Knowing the daily need of contacting pipe, valve, fitting, flange, and other piping specialty item manufacturers, in addition to suppliers, fabricators, installers, and other companies, we have furnished a source list with addresses, telephone and fax numbers, and email addresses, where available. I ask for the reader's understanding while using this list, since companies merge and go through other changes with the changing economic realities of business.

The contents of *Piping Databook* and their extent have been limited by the limit on the number of pages; however, we hope to include additional data later. We ask users to take the time to offer their opinions, suggestions, and constructive criticism.

As always I am a firm believer in team effort. I rely on a team approach to utilize the best ideas, collective effort, and synergetic energy focused on a common objective. The *Piping Databook* is an example of such a team effort. I have been fortunate to enjoy the cooperation of several professionals who have contributed partial or complete chapters. I thank them for their efforts. I owe my gratitude especially to Dheeraj Modawel, Sandeep Singh, Shriman Gurram, Ramesh Tiwari, Troy Skillen, Kenny White, and Soami Suri. They are the backbone of this team. They sacrificed personal time and made efforts beyond commitment to accomplish the publication of *Piping Databook*. No words of appreciation can express my respect for their professionalism and gratitude for their dedicated efforts and cooperation. I salute them and wish them success.

Each and every professional listed as a contributor in the Honors Roll has been key in making this databook a reality. Larry Hager, Caroline Levine, and others at McGraw-Hill have been instrumental in moving the publication process forward. I consider myself fortunate to associate with the well-known and recognized contributors, and goal-oriented professionals at McGraw-Hill. I offer my heart-felt appreciation for their contributions and cooperation.

As in past efforts, I have enjoyed full, silent, and solid support of my wife Prabha while devoting weekends and weekdays toward this team effort. Formal expressions of thanks in so many words cannot express my gratitude for her support. I am proud to recognize the efforts of my daughters, Mukta and Mahak, and my son, Manav, who have helped me in researching and collecting data, preparing manuscript, reviewing proof pages, and performing other tasks.

Finally, I ask the users of *Piping Databook* to identify opportunities for improving and enhancing its contents. I look forward to your input. I take full responsibility for any shortcoming and commit myself to improving the contents and their utility. Hoping to hear from you.

Mohinder L. Nayyar

S · E · C · T · I · O · N · 1

TERMINOLOGY AND UNITS

CHAPTER 1
TERMINOLOGY

See p. 36 for a list of tables in Chap. 1.

PIPING

Piping includes pipe, flanges, fittings, bolting, gaskets, valves, and the pressure-containing portions of other piping components. It also includes pipe hangers and supports and other items necessary to prevent over-pressurization and overstressing of the pressure-containing components. It is evident that pipe is only one element or a part of piping. Therefore, pipe sections when joined with fittings, valves, and other mechanical equipment and properly supported by hangers and supports, are called *piping*.

Pipe

Pipe is a tube with round cross section conforming to the dimensional requirements of the following ASME standards.

- ASME B36.10M — Welded and Seamless Wrought Steel Pipe
- ASME B36.19M — Stainless Steel Pipe

Pipe Size

Initially a system known as *iron pipe size* (IPS) was established to designate the pipe size. The size represented the approximate inside diameter of the pipe in inches. An IPS 6 pipe is one whose inside diameter is approximately 6 inches (in). Users started to refer to the pipe as 2-in, 4-in, 6-in pipe and so on. To begin with, each pipe size was produced to have one thickness, which later was termed as *standard* (STD) or *standard weight* (STD WT). The outside diameter of the pipe was standardized.

As the industrial requirements demanded the handling of fluids with higher pressure, pipes were produced having thicker walls, which came to be known as *extra strong* (XS) or *extra heavy* (XH). The higher pressure requirements further increased, requiring thicker wall pipes. Accordingly, pipes were manufactured with *double extra strong* (XXS) or *double extra heavy* (XXH) walls while the standardized outside diameters remained unchanged.

With the development of stronger and corrosion-resistant piping materials, the need for thinner wall pipe resulted in a new method of specifying pipe size and wall thickness. The designation known as *nominal pipe size* (NPS) replaced IPS, and the term *schedule* (SCH) was invented to specify the nominal wall thickness of pipe.

Nominal pipe size (NPS) is a dimensionless designator of pipe size. It indicates standard pipe size when followed by the specific size designation number without an inch symbol. The NPS 12 and smaller pipe has an outside diameter greater than the size designator (say, 2, 4, 6, . . .). For example, NPS 2 indicates a pipe whose outside diameter is 2.375 in. However, the outside diameter of NPS 14 and larger pipe is the same as the size designator in inches. For example, NPS 14 pipe has an outside diameter equal to 14 in. The inside diameter will depend upon the pipe wall thickness specified by the schedule number. Refer to ASME B36.10M or ASME B36.19M.

Diameter nominal (DN) is also a dimensionless designator of pipe size in the metric unit system, developed by the International Standards Organization (ISO). It indicates standard pipe size when followed by the specific size designation number without a millimeter symbol. For example, DN 50 is the equivalent designation of NPS 2. Refer to Table 1.1 for NPS and DN pipe size equivalents.

TABLE 1.1 Pipe Size Designators: NPS and DN

NPS	DN	NPS	DN	NPS	DN	NPS	DN
1/8	6	3½	90	22	550	44	1100
1/4	8	4	100	24	600	48	1200
3/4	10	5	125	26	650	52	1300
1/2	15	6	150	28	700	56	1400
3/4	20	8	200	30	750	60	1500
1	25	10	250	32	800	64	1600
1¼	32	12	300	34	850	68	1700
1½	40	14	350	36	900	72	1800
2	50	16	400	38	950	76	1900
2½	65	18	450	40	1000	80	2000
3	80	20	500	42	1050	—	—

1. For sizes larger than NPS 80, determine the DN equivalent by multiplying NPS size designation number by 25.

Pipe Wall Thickness

Schedule is expressed in numbers (5, 5S, 10, 10S, 20, 20S, 30, 40, 40S, 60, 80, 80S, 100, 120, 140, 160). A schedule number indicates the approximate value of the expression 1000 *P/S*, where *P* is the service pressure and *S* is the allowable stress, both expressed in pounds per square inch (psi). The higher the schedule number, the thicker the pipe is. The outside diameter of each pipe size is standardized. Therefore, a particular nominal pipe size will have a different inside diameter depending upon the schedule number specified.

Note that the original pipe wall thickness designations of STD, XS, and XXS have been retained; however, they correspond to a certain schedule number depending upon the nominal pipe size. The nominal wall thickness of NPS 10 and smaller schedule 40 pipe is the same as that of STD WT pipe. Also, NPS 8 and smaller schedule 80 pipe have the same wall thickness as XS pipe.

The schedule numbers followed by the letter S are specified in ASME B36.19M, and they are primarily intended for use with stainless steel pipe; however, they are also being used with pipe other than stainless steel pipe. The pipe wall thickness specified by a schedule number followed by the letter S may or may not be the same as that specified by a schedule number without the letter S. Refer to Chap. 5 or ASME B36.19M and ASME B36.10M.

ASME B36.19M does not cover all pipe sizes. Therefore, the dimensional requirements of ASME B36.10M apply to stainless steel pipe of the sizes and schedules not covered by ASME B36.19M.

PIPING CLASSIFICATION

It is usual industry practice to classify pipe in accordance with the pressure-temperature rating system used for classifying flanges. However, it is not essential that piping be classified as Class 150, 300, 400, 600, 900, 1500, and 2500. *The piping rating must be governed by the pressure-temperature rating of the weakest pressure-containing item in the piping.* The weakest item in a piping system may be a fitting made of weaker material or rated lower because of design and other considerations. Table 1.2 lists the standard pipe class ratings based on ASME B16.5 along with corresponding *pression nominal* (PN) rating designators. *Pression nominal* is the French equivalent of pressure nominal.

In addition, the piping may be classified by class ratings covered by other ASME standards, such as ASME B16.1, B16.3, B16.24, and B16.42. A piping system may be rated for a unique set of pressures and temperatures not covered by any standard.

TABLE 1.2 Piping Class Ratings Based on ASME B16.5 and Corresponding PN Designators

Class	150	300	400	600	900	1500	2500
PN	20	50	68	110	150	260	420

Notes:
1. Pressure-temperature ratings of different classes vary with the temperature and the material of construction.
2. For pressure-temperature ratings, refer to tables in ASME B16.5, or ASME B16.34.

Pression nominal is the rating designator followed by a designation number, which indicates the approximate pressure rating in *bars*. The bar is a unit of pressure, and 1 bar is equal to 14.5 psi or 100 kilopascals (kPa). Table 1.2 provides a cross-reference of the ASME class ratings to PN rating designators. It is evident that the PN ratings do not provide a proportional relationship between different PN numbers, whereas the class numbers do. Therefore, it is recommended that class numbers be used to designate the ratings.

OTHER PIPE RATINGS

Manufacturer's Rating

A manufacturer may assign a pressure-temperature rating, based upon a unique or proprietary design of a pipe, fitting, or joint that may form the design basis for the piping system.

In no case shall the manufacturer's rating be exceeded. In addition, the manufacturer may impose limitations that must also be adhered to.

NFPA Ratings

The piping systems within the jurisdiction of the National Fire Protection Association (NFPA) requirements are required to be designed and tested to certain pressures. These systems are usually rated for 175 psi (1207.5 kPa), 200 psi (1380 kPa), or as specified.

AWWA Ratings

The American Water Works Association (AWWA) publishes standards and specifications, which are used to design and install water pipelines and distribution system piping. The rating used may be in accordance with the flange ratings of AWWA C207, Steel Pipe Flanges, or the rating could be based upon the rating of the joints used in the piping.

Specific or Unique Rating

When the design pressure and temperature conditions of a piping system do not fall within the pressure-temperature ratings of any of the above-described rating systems, the designer may assign a specific rating to the piping system. Examples of such applications include main steam or hot reheat piping of a power plant, whose design pressure and design temperature may exceed the pressure-temperature rating of ASME B16.5 Class 2500 flanges. It is normal to assign a specific rating to the piping. This rating must be equal to or higher than the design conditions. The rating of all pressure-containing components in the piping system must meet or exceed the specific rating assigned by the designer.

Dual Ratings

Sometimes a piping system may be subjected to full-vacuum conditions or submerged in water and thus experience external pressure, in addition to withstanding the internal pressure of the flow medium. Such piping systems must be rated for both internal and external pressures at the given temperatures. In addition, a piping system may handle more than one flow medium during its different modes of operation. Therefore, such a piping system may be assigned a dual rating for two different flow media. For example, a piping system may have condensate flowing through it at some lower temperature during one mode of operation while steam may flow through it at some higher temperature during another mode of operation. It may be assigned two pressure ratings at two different temperatures.

GENERAL DEFINITIONS

Absolute Viscosity. Absolute viscosity or the coefficient of absolute viscosity is a measure of the internal resistance. In the centimeter, gram, second (cgs) or metric system, the unit of absolute viscosity is the poise (abbreviated P), which is equal to 100 centipoise (cP). The English units used to measure or express viscosity are slugs per foot-second or pound force seconds per square foot. Sometimes, the English units are also expressed as pound mass per foot-second or poundal seconds per square foot.

Adhesive Joint. A joint made in plastic piping by the use of an adhesive substance which forms a continuous bond between the mating surfaces without dissolving either one of them.

Air-Hardened Steel. A steel that hardens during cooling in air from a temperature above its transformation range.

Alloy Steel. A steel that owes its distinctive properties to elements other than carbon. Steel is considered to be alloy steel when the maximum of the range given for the content of alloying elements exceeds one or more of the following limits:

Manganese	1.65 percent
Silicon	0.60 percent
Copper	0.60 percent

or a definite range or a definite minimum quantity of any of the following elements is specified or required within the limits of the recognized field of constructional alloy steels:

Aluminum	Nickel
Boron	Titanium
Chromium (up to 3.99 percent)	Tungsten
Cobalt	Vanadium
Columbium	Zirconium
Molybdenum	

or any other alloying element added to obtain a desired alloying effect.

Small quantities of certain elements are unavoidably present in alloy steels. In many applications, these are not considered to be important and are not specified or required. When not specified or required, they should not exceed the following amounts:

Copper	0.35 percent
Chromium	0.20 percent
Nickel	0.25 percent
Molybdenum	0.06 percent

Ambient Temperature. The temperature of the surrounding medium, usually used to refer to the temperature of the air in which a structure is situated or a device operates.

Anchor. A rigid restraint providing substantially full fixation, permitting neither translatory nor rotational displacement of the pipe.

Annealing. Heating a metal to a temperature above a critical temperature and holding above that range for a proper period of time, followed by cooling at a suitable rate to below that range for such purposes as reducing hardness, improving machinability, facilitating cold working, producing a desired microstructure, or obtaining desired mechanical, physical, or other properties. (A softening treatment is often carried out just below the critical range and is referred to as a subcritical annealing.)

Arc Cutting. A group of cutting processes in which the severing or removing of metals is effected by melting with the heat of an arc between an electrode and the base metal (includes carbon, metal, gas metal, gas tungsten, plasma, and air carbon arc cutting). *See also* Oxygen Cutting.

Arc Welding. A group of welding processes in which coalescence is produced by heating with an electric arc or arcs, with or without the application of pressure and with or without the use of filler metal.

Assembly. The joining together of two or more piping components by bolting, welding, caulking, brazing, soldering, cementing, or threading into their installed location as specified by the engineering design.

Automatic Welding. Welding with equipment that performs the entire welding operation without constant observation and adjustment of the controls by an operator. The equipment may or may not perform the loading and unloading of the work.

Backing Ring. Backing in the form of a ring that can be used in the welding of piping to prevent weld spatter from entering a pipe and to ensure full penetration of the weld to the inside of the pipe wall.

Ball Joint. A component that permits universal rotational movement in a piping system.

Base Metal. The metal to be welded, brazed, soldered, or cut. It is also referred to as *parent* metal.

Bell-Welded Pipe. Furnace-welded pipe produced in individual lengths from cut-length skelp, having its longitudinal butt joint forge-welded by the mechanical pressure developed in drawing the furnace-heating skelp through a cone-shaped die (commonly known as a *welding bell*), which serves as a combined forming and welding die.

Bevel. A type of edge or end preparation.

Bevel Angle. The angle formed between the prepared edge of a member and a plane perpendicular to the surface of the member. See Fig. 1.1.

Blank Flange. A flange that is not drilled but is otherwise complete.

Blind Flange. A flange used to close the end of a pipe. It produces a blind end which is also known as a *dead end.*

Bond. The junction of the weld metal and the base metal, or the junction of the base metal parts when weld metal is not present. See Fig. 1.2.

Branch Connection. The attachment of a branch pipe to the run of a main pipe with or without the use of fittings.

Braze Welding. A method of welding whereby a groove, fillet, plug, or slot weld is made using a nonferrous filler metal having a melting point below that of the base metals, but above 800°F (427°C). The filler metal is not distributed in the joint by capillary action. (*Bronze* welding, the term formerly used, is a misnomer.)

Brazing. A metal joining process in which coalescence is produced by use of a nonferrous filler metal having a melting point above 800°F (427°C) but lower than that of the base metals joined. The filler metal is distributed between the closely fitted surfaces of the joint by capillary action.

Butt Joint. A joint between two members lying approximately in the same plane.

Butt Weld. Weld along a seam that is butted edge to edge. See Fig. 1.3.

Bypass. A small passage around a large valve for warming up a line. An emergency connection around a reducing valve, trap, etc., to use in case it is out of commission.

Carbon Steel. A steel that owes its distinctive properties chiefly to the carbon (as distinguished from the other elements) which it contains. Steel is considered to be carbon steel when no minimum content is specified or required for aluminum, boron, chromium, cobalt, columbium, molybdenum, nickel, titanium, tungsten, vanadium, or zirconium or for any other element added to obtain a desired alloying effect; when the

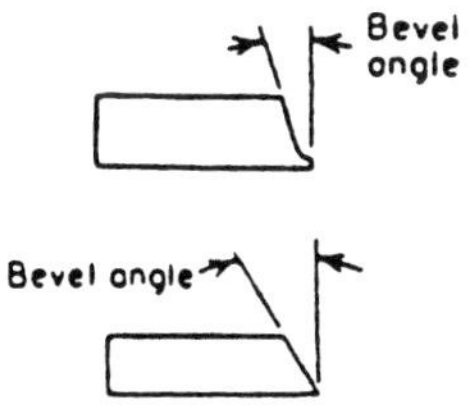

FIGURE 1.1 Bevel angle.

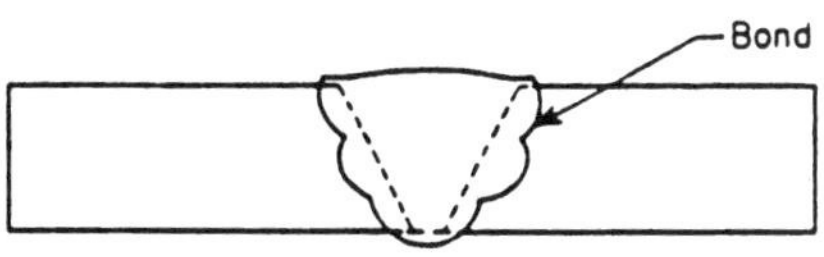

FIGURE 1.2 Bond between base metal and weld metal.

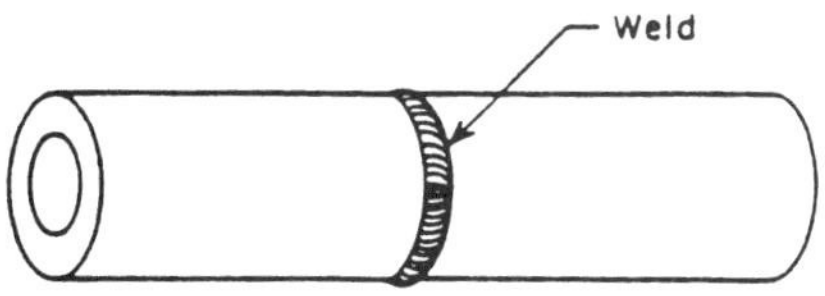

FIGURE 1.3 A circumferential butt-welded joint.

specified minimum for copper does not exceed 0.40 percent; or when the maximum content specified for any of the following elements does not exceed the percentages noted: manganese, 1.65 percent; silicon, 0.60 percent; copper, 0.60 percent.

Cast Iron. A generic term for the family of high carbon-silicon-iron casting alloys including gray, white, malleable, and ductile iron.

Centrifugally Cast Pipe. Pipe formed from the solidification of molten metal in a rotating mold. Both metal and sand molds are used. After casting, if required the pipe is machined, to sound metal, on the internal and external diameters to the surface roughness and dimensional requirements of the applicable material specification.

Certificate of Compliance. A written statement that the materials, equipment, or services are in accordance with the specified requirements. It may have to be supported by documented evidence.

Certified Material Test Report (CMTR). A document attesting that the material is in accordance with specified requirements, including the actual results of all required chemical analyses, tests, and examinations.

Chamfering. The preparation of a contour, other than for a square groove weld, on the edge of a member for welding.

Cold Bending. The bending of pipe to a predetermined radius at any temperature below some specified phase change or transformation temperature but usually at or near room temperature.

Cold Working. Deformation of a metal plastically. Although ordinarily done at room temperature, cold working may be done at the temperature and rate at which strain hardening occurs. Bending of steel piping at 1300°F (704°C) would be considered a cold-working operation.

Companion Flange. A pipe flange suited to connect with another flange or with a flanged valve or fitting. A loose flange that is attached to a pipe by threading, van stoning, welding, or similar method as distinguished from a flange that is cast integrally with a fitting or pipe.

Consumable Insert. Preplaced filler metal that is completely fused into the root of the joint and becomes part of the weld. See Fig. 1.4.

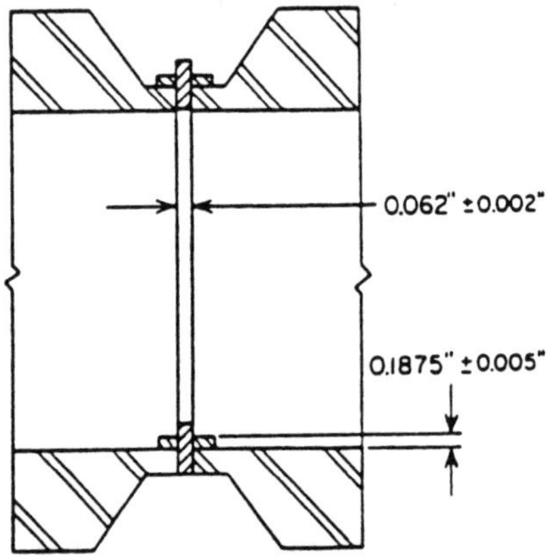

FIGURE 1.4 Consumable insert ring inserted in pipe joint eccentrically for welding in horizontal position.

Continuous-Welded Pipe. Furnace-welded pipe produced in continuous lengths from coiled skelp and subsequently cut into individual lengths, having its longitudinal butt joint forge-welded by the mechanical pressure developed in rolling the hot-formed skelp through a set of round pass welding rolls.

Contractor. The entity responsible for furnishing materials and services for fabrication and installation of piping and associated equipment.

Control Piping. All piping, valves, and fittings used to interconnect air, gas, or hydraulically operated control apparatus or instrument transmitters and receivers.

Controlled Cooling. A process of cooling from an elevated temperature in a predetermined manner to avoid hardening, cracking, or internal damage or to produce a desired metallurgical microstructure. This cooling usually follows the final hot-forming or postheating operation.

Corner Joint. A joint between two members located approximately at right angles to each other in the form of an L. See Fig. 1.5.

Coupling. A threaded sleeve used to connect two pipes. Commercial couplings have internal threads to fit external threads on pipe.

Covered Electrode. A filler metal electrode, used in arc welding, consisting of a metal core wire with a relatively thick covering which provides protection for the molten metal from the atmosphere, improves the properties of the weld metal, and stabilizes the arc. Covered electrodes are extensively used in shop fabrication and field erection of piping of carbon, alloy, and stainless steels.

Crack. A fracture-type imperfection characterized by a sharp tip and high ratio of length and depth to opening displacement.

Creep or Plastic Flow of Metals. At sufficiently high temperatures, all metals flow under stress. The higher the temperature and stress, the greater the tendency to plastic flow for any given metal.

Cutting Torch. A device used in oxygen, air, or powder cutting for controlling and directing the gases used for preheating and the oxygen or powder used for cutting the metal.

Defect. A flaw or an imperfection of such size, shape, orientation, location, or properties as to be rejectable according to the applicable minimum acceptance standards.

Density. The density of a substance is the mass of the substance per unit volume. It may be expressed in a variety of units.

Deposited Metal. Filler metal that has been added during a welding operation.

Depth of Fusion. The distance that fusion extends into the base metal from the surface melted during welding. See Fig. 1.6.

Designer. Responsible for ensuring that the engineering design of piping complies with the requirements of the applicable code and standard and any additional requirements established by the owner.

Dew Point. The temperature at which the vapor condenses when it is cooled at constant pressure.

Dilatant Liquid. If the viscosity of a liquid increases as agitation is increased at constant temperature, the liquid is termed *dilatant*. Examples are clay slurries and candy compounds.

Discontinuity. A lack of continuity or cohesion; an interruption in the normal physical structure of material or a product.

Double Submerged Arc-Welded Pipe. Pipe having a longitudinal butt joint produced by at least two passes, one of which is on the inside of the pipe. Coalescence is produced by heating with an electric arc or arcs between the bare metal electrode or electrodes and the work. The welding is shielded by a blanket of granular, fusible material on the work. Pressure is not used, and filler metal for the inside and outside welds is obtained from the electrode or electrodes.

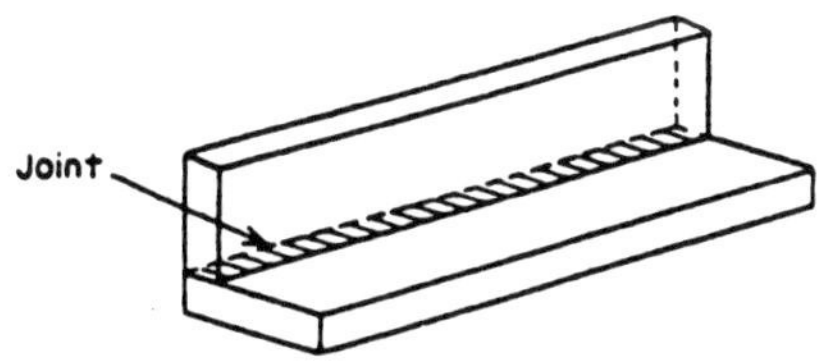

FIGURE 1.5 Corner joint.

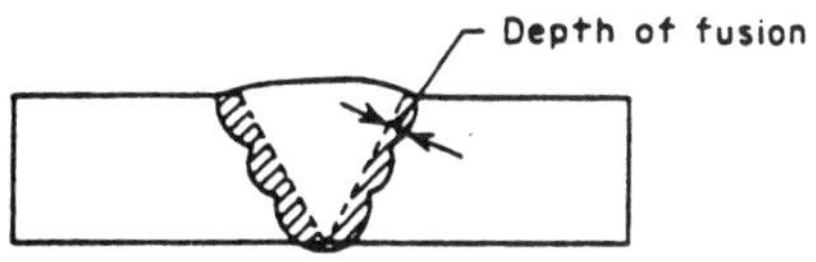

FIGURE 1.6 Depth of fusion.

Ductile Iron. A cast ferrous material in which the free graphite is in a spheroidal form rather than a fluke form. The desirable properties of ductile iron are achieved by means of chemistry and a ferritizing heat treatment of the castings.

Eddy Current Testing. This is a nondestructive testing method in which eddy current flow is induced in the test object. Changes in the flow caused by variations in the object are reflected into a nearby coil or coils for subsequent analysis by suitable instrumentation and techniques.

Edge Joint. A joint between the edges of two or more parallel or nearly parallel members.

Edge Preparation. The contour prepared on the edge of a member for welding. See Fig. 1.7.

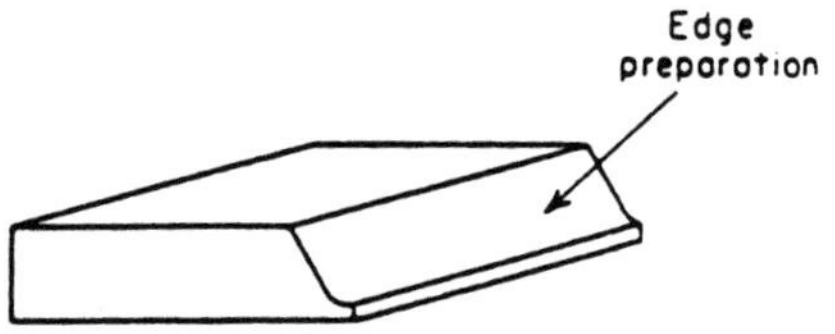

FIGURE 1.7 Edge preparation.

Electric Flash-Welded Pipe. Pipe having a longitudinal butt joint in which coalescence is produced simultaneously over the entire area of abutting surfaces by the heat obtained from resistance to the flow of electric current between the two surfaces and by the application of pressure after heating is substantially completed. Flashing and upsetting are accompanied by expulsion of metal from the joint.

Electric Fusion-Welded Pipe. Pipe having a longitudinal or spiral butt joint in which coalescence is produced in the performed tube by manual or automatic electric arc welding. The weld may be single or double and may be made with or without the use of filler metal.

Electric Resistance-Welded Pipe. Pipe produced in individual lengths or in continuous lengths from coiled skelp and subsequently cut into individual lengths having a longitudinal butt joint in which coalescence is produced by the heat obtained from resistance of the pipe to the flow of electric current in a circuit of which the pipe is a part and by the application of pressure.

Electrode. *See* Covered Electrode.

End Preparation. The contour prepared on the end of a pipe, fitting, or nozzle for welding. The particular preparation is prescribed by the governing code.

Engineering Design. The detailed design developed from process requirements and conforming to established design criteria, including all necessary drawings and specifications, governing a piping installation.

Equipment Connection. An integral part of such equipment as pressure vessels, heat exchangers, pumps, etc., designed for attachment of pipe or piping components.

Erection. The complete installation of a piping system, including any field assembly, fabrication, testing, and inspection of the system.

Erosion. Destruction of materials by the abrasive action of moving fluids, usually accelerated by the presence of solid particles.

Examination. The procedures for all visual observation and nondestructive testing.

Expansion Joint. A flexible piping component that absorbs thermal and/or terminal movement.

Extruded Nozzles. The forming of nozzle (tee) outlets in pipe by pulling hemispherically or conically shaped dies through a circular hole from the inside of the pipe. Although some cold extruding is done, it is generally performed on steel after the area to be shaped has been heated to temperatures between 2000 and 1600°F (1093 and 871°C).

Extruded Pipe. Pipe produced from hollow or solid round forgings, usually in a hydraulic extrusion press. In this process the forging is contained in a cylindrical die. Initially a punch at the end of the extrusion plunger pierces the forging. The extrusion plunger then forces the contained billet between the cylindrical die and the punch to form the pipe, the latter acting as a mandrel.

One variation of this process utilizes autofrettage (hydraulic expansion) and heat treatment, above the recrystallization temperature of the material, to produce a wrought structure.

Fabrication. Primarily, the joining of piping components into integral pieces ready for assembly. It includes bending, forming, threading, welding, or other operations upon these components, if not part of assembly. It may be done in a shop or in the field.

Face of Weld. The exposed surface of a weld on the side from which the welding was done.

Filler Metal. Metal to be added in welding, soldering, brazing, or braze welding.

Fillet Weld. A weld of an approximately triangular cross section joining two surfaces approximately at right angles to each other in a lap joint, tee joint, corner joint, or socket weld. See Fig. 1.8.

Fire Hazard. Situation in which a material of more than average combustibility or explodibility exists in the presence of a potential ignition source.

Flat-Land Bevel. A square extended root face preparation extensively used in inert-gas, root-pass welding of piping. See Fig. 1.9.

Flat Position. The position of welding that is performed from the upper side of the joint, while the face of the weld is approximately horizontal. See Fig. 1.10.

Flaw. An imperfection or unintentional discontinuity that is detectable by a nondestructive examination.

Flux. Material used to dissolve, prevent accumulation of, or facilitate removal of oxides and other undesirable substances during welding, brazing, or soldering.

Flux-Cored Arc Welding (FCAW). An arc welding process that employs a continuous tubular filler metal (consumable) electrode having a core of flux for shielding. Added shielding may or may not be obtained from an externally supplied gas or gas mixture.

Forge Weld. A method of manufacture similar to hammer welding. The term *forge welded* is applied more particularly to headers and large drums, while *hammer welded* usually refers to pipe.

Forged and Bored Pipe. Pipe produced by boring or trepanning of a forged billet.

Full-Fillet Weld. A fillet weld whose size is equal to the thickness of the thinner member joined.

Fusion. The melting together of filler and base metal, or of base metal only, which results in coalescence.

Fusion Zone. The area of base metal melted as determined on the cross section of a weld. See Fig. 1.11.

Galvanizing. A process by which the surface of iron or steel is covered with a layer of zinc.

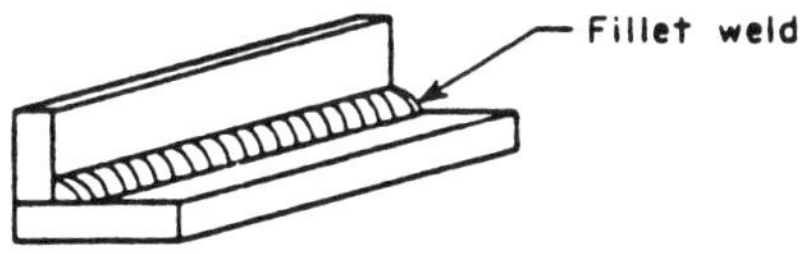

FIGURE 1.8 Fillet weld.

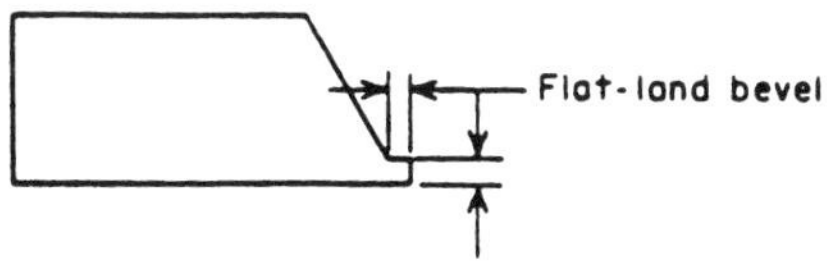

FIGURE 1.9 Flat-land bevel.

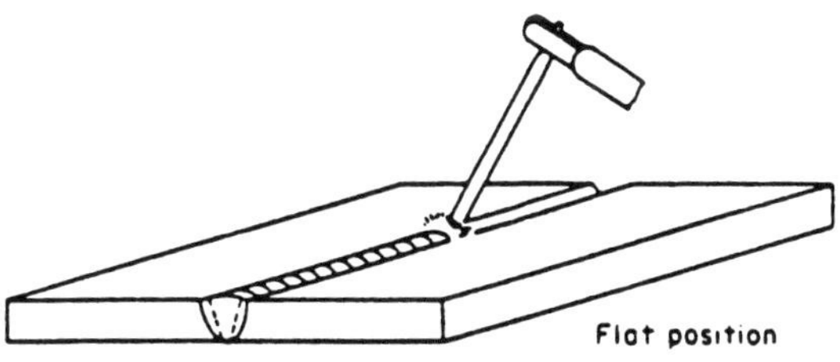

FIGURE 1.10 Welding in the flat position.

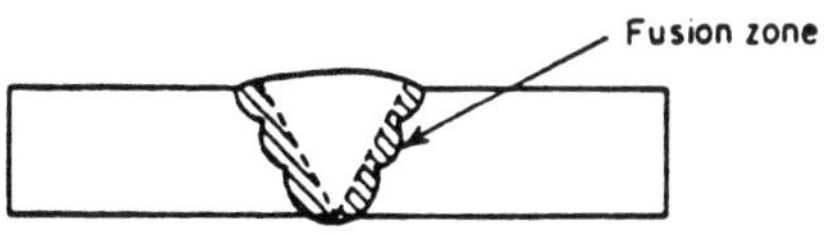

FIGURE 1.11 Fusion zone is the section of the parent metal that melts during the welding process.

Gas Metal Arc Welding (GMAW). An arc welding process that employs a continuous solid filler metal (consumable) electrode. Shielding is obtained entirely from an externally supplied gas or gas mixture. (Some methods of this process have been called *MIG* or CO_2 welding.)

Gas Tungsten Arc Welding (GTAW). An arc welding process that employs a tungsten (nonconsumable) electrode. Shielding is obtained from a gas or gas mixture. Pressure may or may not be used, and filler metal may or may not be used. (This process has sometimes been called *TIG* welding.) When shielding is obtained by the use of an inert gas such as helium or argon, this process is called *inert-gas tungsten arc welding*.

Gas Welding. Welding process in which coalescence is produced by heating with a gas flame or flames, with or without the application of pressure and with or without the use of filler metal.

Groove. The opening provided for a groove weld.

Groove Angle. The total included angle of the groove between parts to be joined by a groove weld. See Fig. 1.12.

Groove Face. That surface of a member included in the groove. See Fig. 1.13.

Groove Radius. The radius of a J or U groove. See Fig. 1.14.

Groove Weld. A weld made in the groove between two members to be joined. The standard types of groove welds are square, single-V, single-bevel, single-U, single-J, double-V, double-U, double-bevel, double-J, flat-land single, and double-V groove welds. See Fig. 1.15 for a typical groove weld.

Hammer Weld. Method of manufacturing large pipe (usually NPS 20 or DN 500 and larger) by bending a plate into circular form, heating the overlapped edges to a welding temperature, and welding the longitudinal seam with a power hammer applied to the outside of the weld while the inner side is supported on an overhung anvil.

Hangers and Supports. Hangers and supports include elements that transfer the load from the pipe or structural attachment to the supporting structure or equipment. They include hanging-type fixtures such as hanger rods, spring hangers, sway braces, counterweights, turnbuckles, struts, chains, guides, and anchors and bearing-type fixtures such as saddles, bases, rollers, brackets, and sliding supports. Refer to Chaps. 26 and 27.

Header. A pipe or fitting to which a number of branch pipes are connected.

Heat-Affected Zone. That portion of the base metal that has not been melted but whose mechanical properties or microstructure have been altered by the heat of welding or cutting. See Fig. 1.16.

Heat Fusion Joint. A joint made in thermoplastic piping by heating the parts sufficiently to permit fusion of the materials when the parts are pressed together.

Horizontal Fixed Position. In pipe welding, the position of a pipe joint in which the axis of the pipe is approximately horizontal and the pipe is not rotated during the operation.

Horizontal-Position Fillet Weld. Welding is performed on the upper side of an approximately horizontal surface and against an approximately vertical surface. See Fig. 1.17.

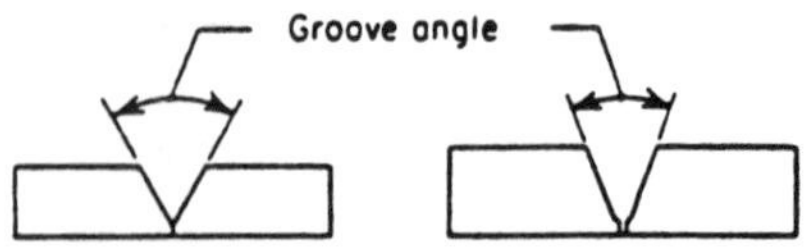

FIGURE 1.12 The groove angle is twice the bevel angle.

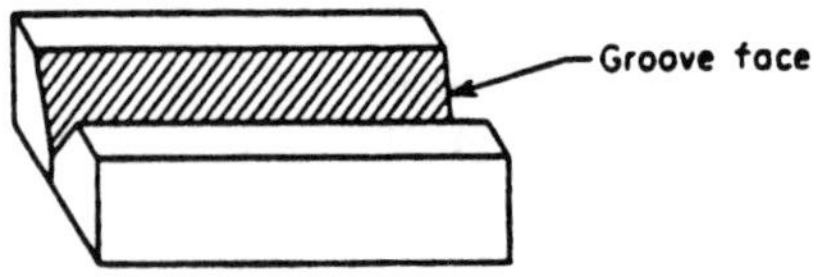

FIGURE 1.13 A groove face.

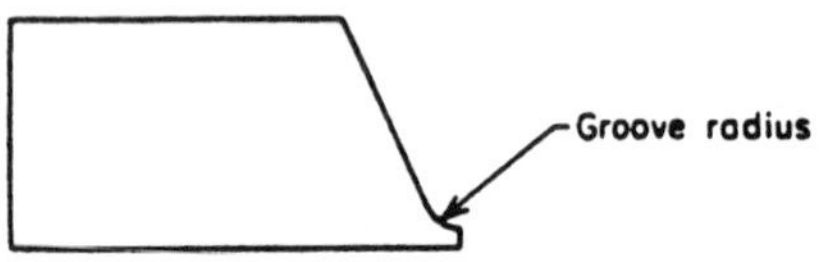

FIGURE 1.14 A groove radius.

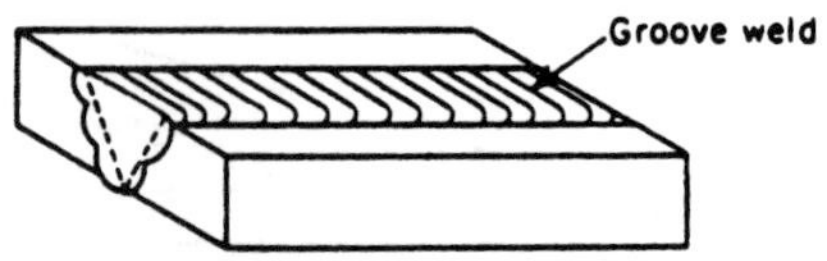

FIGURE 1.15 Groove weld.

Horizontal-Position Groove Weld. The position of welding in which the weld axis lies in an approximately horizontal plane and the face of the weld lies in an approximately vertical plane. See Fig. 1.18.

Horizontal Rolled Position. The position of a pipe joint in which welding is performed in the flat position by rotating the pipe. See Fig. 1.19.

Hot Bending. Bending of piping to a predetermined radius after heating to a suitably high temperature for hot working. On many pipe sizes, the pipe is firmly packed with sand to avoid wrinkling and excessive out-of-roundness.

Hot Taps. Branch piping connections made to operating pipelines, mains, or other facilities while they are in operation.

Hot Working. The plastic deformation of metal at such a temperature and rate that strain hardening does not occur. Extruding or swaging of chrome-moly piping at temperatures between 2000 and 1600°F (1093 and 871°C) would be considered hot-forming or hot-working operations.

Hydraulic Radius. The ratio of area of flowing fluid to the wetted perimeter.

$$\text{Hydraulic radius} = \frac{\text{area of flowing fluid}}{\text{wetted perimeter}}$$

Impact Test. A test to determine the behavior of materials when subjected to high rates of loading, usually in bending, tension, or torsion. The quantity measured is the energy absorbed in breaking the specimen by a single blow, as in Charpy or Izod tests.

Imperfection. A condition of being imperfect; a departure of a quality characteristic from its intended condition.

Incomplete Fusion. Fusion that is less than complete and that does not result in melting completely through the thickness of the joint.

Indication. The response or evidence from the application of a nondestructive examination.

Induction Heating. Heat treatment of completed welds in piping by means of placing induction coils around the piping. This type of heating is usually performed during field erection in those cases where stress relief of carbon- and alloy-steel field welds is required by the applicable code.

Inspection. Activities performed by an authorized inspector to verify whether an item or activity conforms to specified requirements.

Instrument Piping. All piping, valves, and fittings used to connect instruments to main piping, to other instruments and apparatus, or to measuring equipment.

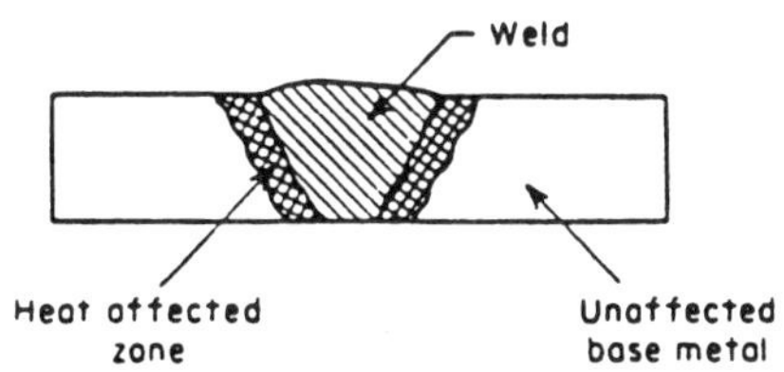

FIGURE 1.16 Welding zones.

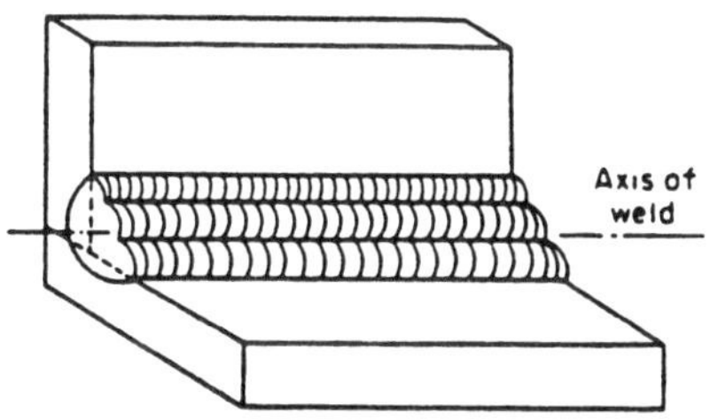

FIGURE 1.17 Horizontal position fillet weld.

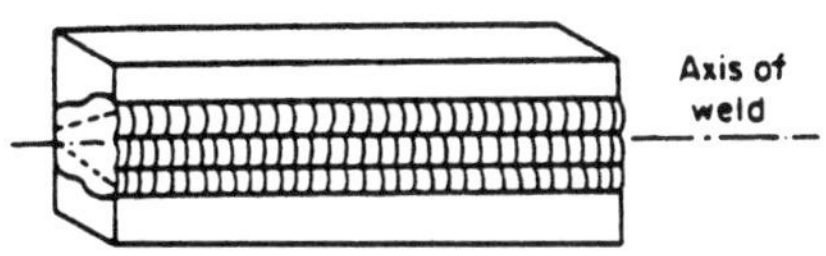

FIGURE 1.18 Horizontal position groove weld.

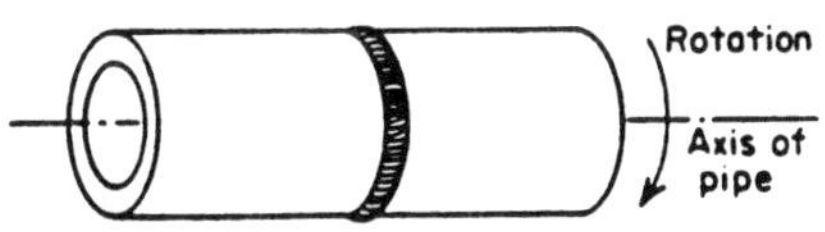

FIGURE 1.19 Horizontal rolled position.

Interpass Temperature. In a multiple-pass weld, the minimum or maximum temperature of the deposited weld metal before the next pass is started.

Interrupted Welding. Interruption of welding and preheat by allowing the weld area to cool to room temperature as generally permitted on carbon-steel and on chrome-moly alloy-steel piping after sufficient weld passes equal to at least one-third of the pipe wall thickness or two weld layers, whichever is greater, have been deposited.

Joint. A connection between two lengths of pipe or between a length of pipe and a fitting.

Joint Penetration. The minimum depth a groove weld extends from its face into a joint, exclusive of reinforcement. See Fig. 1.20.

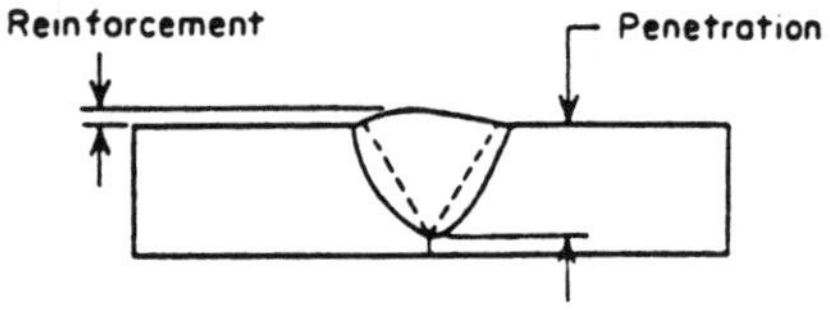

FIGURE 1.20 Weld joint penetration.

Kinematic Viscosity. The ratio of the absolute viscosity to the mass density. In the metric system, kinematic viscosity is measured in strokes or square centimeters per second.

Laminar Flow. Fluid flow in a pipe is usually considered laminar if the Reynolds number is less than 2000. Depending upon many possible varying conditions, the flow may be laminar at a Reynolds number as low as 1200 or as high as 40,000; however, such conditions are not experienced in normal practice.

Lap Weld. Weld along a longitudinal seam in which one part is overlapped by the other. A term used to designate pipe made by this process.

Lapped Joint. A type of pipe joint made by using loose flanges on lengths of pipe whose ends are lapped over to give a bearing surface for a gasket or metal-to-metal joint.

Liquid Penetrant Examination or Inspection. This is a nondestructive examination method for finding discontinuities that are open to the surface of solid and essentially nonporous materials. This method is based on capillary action or capillary attraction by which the surface of a liquid in contact with a solid is elevated or depressed. A liquid penetrant, usually a red dye, is applied to the clean surface of the specimen. Time is allowed for the penetrant to seep into the opening. The excess penetrant is removed from the surface. A developer, normally white, is applied to aid in drawing the penetrant up or out to the surface. The red penetrant is drawn out of the discontinuity, which is located by the contrast and distinct appearance of the red penetrant against the white background of the developer.

Local Preheating. Preheating of a specific portion of a structure.

Local Stress-Relief Heat Treatment. Stress-relief heat treatment of a specific portion of a weldment. This is done extensively with induction coils, resistance coils, or propane torches in the field erection of steel piping.

Machine Welding. Welding with equipment that performs the welding operation under the observation and control of an operator. The equipment may or may not perform the loading and unloading of the work.

Magnetic Particle Examination or Inspection. This is a nondestructive examination method to locate surface and subsurface discontinuities in ferromagnetic materials. The presence of discontinuities is detected by the use of finely divided ferromagnetic particles applied over the surface. Some of these magnetic particles are gathered and held by the magnetic leakage field created by the discontinuity. The particles gathered at the surface form an outline of the discontinuity and generally indicate its location, size, shape, and extent.

Malleable Iron. Cast iron that has been heat-treated in an oven to relieve its brittleness. The process somewhat improves the tensile strength and enables the material to stretch to a limited extent without breaking.

Manual Welding. Welding wherein the entire welding operation is performed and controlled by hand.

Mean Velocity of Flow. Under steady state of flow, the mean velocity of flow at a given cross section of pipe is equal to the rate of flow Q divided by the area of cross section A. It is expressed in feet per second or meters per second.

$$\upsilon = \frac{Q}{A} \tag{1.1}$$

where υ = mean velocity of flow, in feet per second, ft/s (meters per second, m/s)
Q = rate of flow, in cubic feet per second, ft^3/s (cubic meters per second, m^3/s)
A = area of cross section, in square feet, ft^2 (square meters, m^2)

Mechanical Joint. A joint for the purpose of mechanical strength or leak resistance or both, where the mechanical strength is developed by threaded, grooved, rolled, flared, or flanged pipe ends or by bolts, pins, and compounds, gaskets, rolled ends, caulking, or machined and mated surfaces. These joints have particular application where ease of disassembly is desired.

Mill Length. Also known as *random length*. The usual run-of-mill pipe is 16 to 20 ft (5 to 6 m) in length. Line pipe and pipe for power plant use are sometimes made in double lengths of 30 to 35 ft (10 to 12 m).

Miter. Two or more straight sections of pipe matched and joined on a line bisecting the angle of junction so as to produce a change in direction. See Fig. 1.21.

Newtonian Liquid. A liquid is called *newtonian* if its viscosity is unaffected by the kind and magnitude of motion or agitation to which it may be subjected, as long as the temperature remains constant. Water and mineral oil are examples of newtonian liquids.

Nipple. A piece of pipe less than 12 in (0.3 m) long that may be threaded on both ends or on one end and provided with ends suitable for welding or a mechanical joint. Pipe over 12 in (0.3 m) long is regarded as cut pipe. Common types of nipples are close nipple, about twice the length of a standard pipe thread and without any shoulder; shoulder nipple, of any length and having a shoulder between the pipe threads; short nipple,

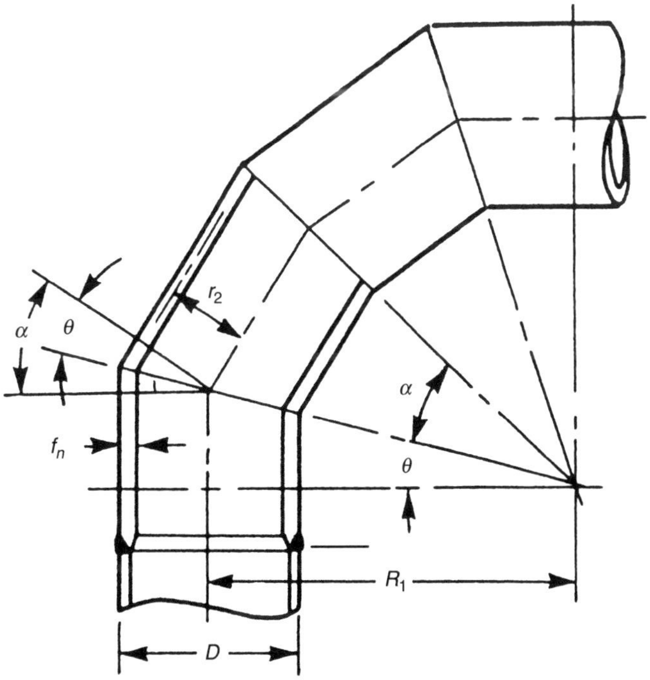

FIGURE 1.21 Miter.
D = outside diameter of pipe
f_n = nominal thickness of pipe or as measured
R_1 = effective radius of miter bend
r_2 = mean radius of pipe
θ = angle of miter cut
α = angle of change in direction = 2θ

a shoulder nipple slightly longer than a close nipple and of a definite length for each pipe size which conforms to manufacturer's standard; long nipple, a shoulder nipple longer than a short nipple which is cut to a specific length.

Nominal Diameter (DN). A dimensionless designator of pipe in metric system. It indicates standard pipe size when followed by the specific size designation number without the millimeter symbol (for example, DN 40, DN 300).

Nominal Pipe Size (NPS). A dimensionless designator of pipe. It indicates standard pipe size when followed by the specific size designation number without an inch symbol (for example, NPS $1\frac{1}{2}$, NPS 12).

Nominal Thickness. The thickness given in the product material specification or standard to which manufacturing tolerances are applied.

Nondestructive Examination or Inspection. Inspection by methods that do not destroy the item, part, or component to determine its suitability for use.

Normalizing. A process in which a ferrous metal is heated to a suitable temperature above the transformation range and is subsequently cooled in still air at room temperature.

Nozzle. As applied to piping, this term usually refers to a flanged connection on a boiler, tank, or manifold consisting of a pipe flange, a short neck, and a welded attachment to the boiler or other vessel. A short length of pipe, one end of which is welded to the vessel with the other end chamfered for butt welding, is also referred to as a *welding nozzle*.

Overhead Position. The position of welding performed from the underside of the joint.

Oxidizing Flame. An oxyfuel gas flame having an oxidizing effect caused by excess oxygen.

Oxyacetylene Cutting. An oxygen-cutting process in which metals are severed by the chemical reaction of oxygen with the base metal at elevated temperatures. The necessary temperature is maintained by means of gas flames obtained from the combustion of acetylene with oxygen.

Oxyacetylene Welding. A gas welding process in which coalescence is produced by heating with a gas flame or flames obtained from the combustion of acetylene with oxygen, with or without the addition of filler metal.

Oxyfuel Gas Welding (OFGW). A group of welding processes in which coalescence is produced by heating with a flame or flames obtained from the combustion of fuel gas with oxygen, with or without the application of pressure and with or without the use of filler metal.

Oxygen Cutting (OC). A group of cutting processes used to sever or remove metals by means of the reaction of oxygen with the base metal at elevated temperatures. In the case of oxidation-resistant metals, the reaction is facilitated by use of a chemical flux or metal powder.

Oxygen Gouging. An application of oxygen cutting in which a chamfer or groove is formed.

Pass. A single progression of a welding or surfacing operation along a joint, weld deposit, or substrate. The result of a pass is a weld bead, layer, or spray deposit.

Peel Test. A destructive method of examination that mechanically separates a lap joint by peeling.

Peening. The mechanical working of metals by means of hammer blows.

Pickle. The chemical or electrochemical removal of surface oxides. Following welding operations, piping is frequently *pickled* in order to remove mill scale, oxides formed during storage, and the weld discolorations.

Pipe. A tube with a round cross section conforming to the dimensional requirements for nominal pipe size as tabulated in ASME B36.10M and ASME B36.19M. For special pipe having a diameter not listed in the above-mentioned standards, the nominal diameter corresponds to the outside diameter.

Pipe Alignment Guide. A restraint in the form of a sleeve or frame that permits the pipeline to move freely only along the axis of the pipe.

Pipe Supporting Fixtures. Elements that transfer the load from the pipe or structural attachment to the support structure or equipment.

Pipeline or Transmission Line. A pipe installed for the purpose of transmitting gases, liquids, slurries, etc., from a source or sources of supply to one or more distribution centers or to one or more large-volume customers; a pipe installed to interconnect source or sources of supply to one or more distribution centers or to one or more large-volume customers; or a pipe installed to interconnect sources of supply.

Piping System. Interconnected piping subject to the same set or sets of design conditions.

Plasma Cutting. A group of cutting processes in which the severing or removal of metals is effected by melting with a stream of hot ionized gas.

Polarity. The direction of flow of current with respect to the welding electrode and workpiece.

Porosity. Presence of gas pockets or voids in metal.

Positioning Weld. A weld made in a joint that has been so placed as to facilitate the making of the weld.

Postheating. The application of heat to a fabricated or welded section subsequent to a fabrication, welding, or cutting operation. Postheating may be done locally, as by induction heating, or the entire assembly may be postheated in a furnace.

Postweld Heat Treatment. Any heat treatment subsequent to welding.

Preheating. The application of heat to a base metal immediately prior to a welding or cutting operation.

Pressure. The force per unit that is acting on a real or imaginary surface within a fluid is the pressure or intensity of pressure. It is expressed in pounds per square inch:

$$p = \frac{w \cdot h}{144} + p_a \tag{1.2}$$

$$p = \frac{w \cdot h}{10^4} + p_a \tag{1.2M}$$

where p = absolute pressure at a point, psi (kg/cm^2)
w = specific weight, lb/ft^3 (kg/m^3)
h = height of fluid column above the point, ft (m)
p_a = atmospheric pressure, psi (kg/cm^2)

The gauge pressure at a point is obtained by designating atmospheric pressure as zero:

$$p = \frac{w \cdot h}{144} \tag{1.3}$$

$$p = \frac{w \cdot h}{10^4} \quad \text{(metric)} \tag{1.3M}$$

where p = gauge pressure. To obtain absolute pressure from gauge pressure, add the atmospheric pressure to the gauge pressure.

Pressure Head. From the definition of pressure, the expression p/w is the pressure head. It can be defined as the height of the fluid above a point, and it is normally measured in feet.

Purging. The displacement during welding, by an inert or neutral gas, of the air inside the piping underneath the weld area in order to avoid oxidation or contamination of the underside of the weld. Gases most commonly used are argon, helium, and nitrogen (the last is principally limited to austenitic stainless steel). Purging can be done within a complete pipe section or by means of purging fixtures on a small area underneath the pipe weld.

Quenching. Rapid cooling of a heated metal.

Radiographic Examination or Inspection. Radiography is a nondestructive test method that makes use of short-wavelength radiation, such as x-rays or gamma rays, to penetrate objects for detecting the presence and nature of macroscopic detects or other structural discontinuities. The shadow image of defects or discontinuities is recorded either on a fluorescent screen or on photographic film.

Reinforcement. In branch connections, reinforcement is material around a branch opening that serves to strengthen it. The material is either integral in the branch components or added in the form of weld metal, a pad, a saddle, or a sleeve. In welding, reinforcement is weld metal in excess of the specified weld size.

Reinforcement Weld. Weld metal on the face of a groove weld in excess of the metal necessary for the specified weld size.

Repair. The process of physically restoring a nonconformance to a condition such that an item complies with the applicable requirements, including the code requirements.

Resistance Weld. Method of manufacturing pipe by bending a plate into circular form and passing electric current through the material to obtain a welding temperature.

Restraint. A structural attachment, device, or mechanism that limits movement of the pipe in one or more directions.

Reverse Polarity. The arrangement of direct-current arc welding leads with the work as the negative pole and the electrode as the positive pole of the welding arc; a synonym for direct-current electrode positive.

Reynolds Number. A dimensionless number. It is defined as the ratio of the dynamic forces of mass flow to the shear stress due to viscosity. It is expressed as

$$R = \frac{D\upsilon\rho}{\mu} \tag{1.4}$$

where R = Reynolds number
υ = mean velocity of flow, ft/s (m/s)
ρ = weight density of fluid, lb/ft^3 (kg/m^3)
D = internal diameter of pipe, ft (m)
μ = absolute viscosity, in pound mass per foot second [lbm/(ft·s)] or poundal seconds per square foot (centipoise)

Rolled Pipe. Pipe produced from a forged billet that is pierced by a conical mandrel between two diametrically opposed rolls. The pierced shell is subsequently rolled and expanded over mandrels of increasingly large diameter. Where closer dimensional tolerances are desired, the rolled pipe is cold- or hot-drawn through dies and then machined. One variation of this process produces the hollow shell by extrusion of the forged billet over a mandrel in a vertical hydraulic piercing press.

Root Edge. A root face of zero width.

Root Face. That portion of the groove face adjacent to the root of the joint. This portion is also referred to as the *root land*. See Fig. 1.22.

Root of Joint. That portion of a joint to be welded where the members to be joined come closest to each other. In cross section, the root of a joint may be a point, a line, or an area. See Fig. 1.22.

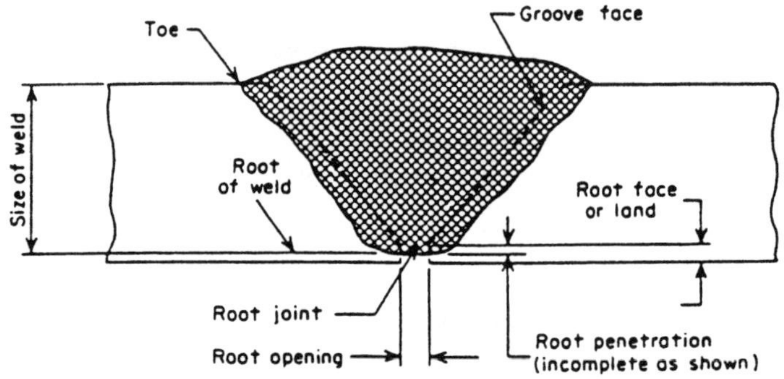

FIGURE 1.22 Nomenclature at joint of groove weld.

Root Opening. The separation, between the members to be joined, at the root of the joint. See Fig. 1.22.

Root Penetration. The depth that a groove weld extends into the root of a joint, as measured on the centerline of the root cross section. Sometimes welds are considered unacceptable if they show incomplete penetration. See Fig. 1.22.

Root Reinforcement. Weld reinforcement at the side other than that from which the welding was done.

Root Surface. The exposed surface of a weld on the side other than that from which the welding was done.

Run. The portion of a fitting having its end in line, or nearly so, as distinguished from branch connections, side outlets, etc.

Saddle Flange. Also known as *tank flange* or *boiler flange*. A curved flange shaped to fit a boiler, tank, or other vessel and to receive a threaded pipe. A saddle flange is usually riveted or welded to the vessel.

Sample Piping. All piping, valves, and fittings used for the collection of samples of gas, steam, water, oil, etc.

Sargol. A special type of joint in which a lip is provided for welding to make the joint fluid-tight, while mechanical strength is provided by bolted flanges. The Sargol joint is used with both Van Stone pipe and fittings.

Sarlun. An improved type of Sargol joint.

Schedule Numbers. Approximate values of the expression 1000*P*/*S*, where *P* is the service pressure and *S* is the allowable stress, both expressed in pounds per square inch.

Seal Weld. A fillet weld used on a pipe joint primarily to obtain fluid tightness as opposed to mechanical strength; usually used in conjunction with a threaded joint.

Seamless Pipe. A wrought tubular product made without a welded seam. It is manufactured by hot-working steel or, if necessary, by subsequently cold-finishing the hot-worked tubular product to produce the desired shape, dimensions, and properties.

Semiautomatic Arc Welding. Arc welding with equipment which controls only the filler metal feed. The advance of the welding is manually controlled.

Semisteel. A high grade of cast iron made by the addition of steel scrap to pig iron in a cupola or electric furnace. More correctly described as *high-strength gray iron*.

Service Fitting. A street ell or street tee having a male thread at one end.

Shielded Metal Arc Welding (SMAW). An arc welding process in which coalescence is produced by heating with an electric arc between a covered metal electrode and the work. Shielding is obtained from decomposition of the electrode covering. Pressure is not used, and filler metal is obtained from the electrode.

Shot Blasting. Mechanical removal of surface oxides and scale on the pipe inner and outer surfaces by the abrasive impingement of small steel pellets.

Single-Bevel-, Single-J-, Single-U-, Single-V-Groove Welds. All are specific types of groove welds and are illustrated in Fig. 1.23.

Single-Welded Butt Joint. A butt joint welded from one side only.

Size of Weld. For a groove weld, it is the joint penetration, which is the depth of chamfering plus the root penetration. See Fig. 1.22. For fillet welds, the leg length of the largest isosceles right triangle which can be inscribed within the fillet-weld cross section. See Fig. 1.24.

Skelp. A piece of plate prepared by forming and bending, ready for welding into pipe. Flat plates when used for butt-welded pipe are called *skelp*.

Slag Inclusion. Nonmetallic solid material entrapped in weld metal or between weld metal.

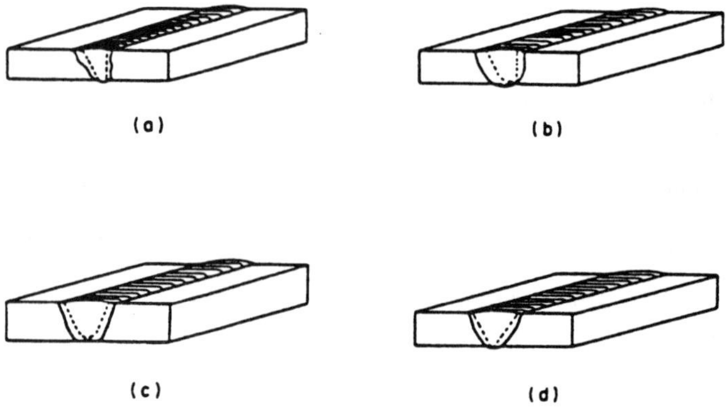

FIGURE 1.23 Groove welds. (*a*) Single-bevel; (*b*) single-J; (*c*) double-U; (*d*) double-V.

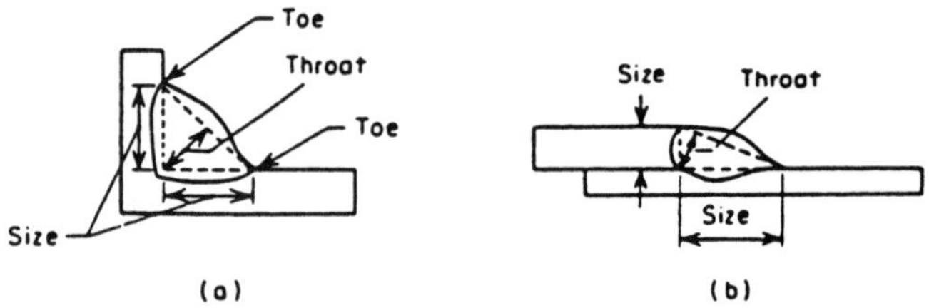

FIGURE 1.24 Size of weld (*a*) in fillet weld of equal legs and (*b*) in fillet weld of unequal legs.

Slurry. A two-phase mixture of solid particles in an aqueous phase.

Socket Weld. Fillet-type seal weld used to join pipe to valves and fittings or to other sections of pipe. Generally used for piping whose nominal diameter is NPS 2 (DN 50) or smaller.

Soldering. A metal-joining process in which coalescence is produced by heating to a suitable temperature and by using a nonferrous alloy fusible at temperatures below that of the base metals being joined. The filler metal is distributed between closely fitted surfaces of the joint by capillary action.

Solution Heat Treatment. Heating an alloy to a suitable temperature, holding at that temperature long enough to allow one or more constituents to enter into solid solution, and then cooling rapidly enough to hold the constituents in solution.

Source Nipple. A short length of heavy-walled pipe between high-pressure mains and the first valve of bypass, drain, or instrument connections.

Spatter. In arc and gas welding, the metal particles expelled during welding that do not form part of the weld.

Spatter Loss. Difference in weight between the amount of electrode consumed and the amount of electrode deposited.

Specific Gravity. The ratio of a material's weight to the weight of an equal volume of water at standard conditions.

Specific Volume. The volume of a unit mass of a fluid is its specific volume, and it is measured in cubic feet per pound mass (ft^3/lbm). In the metric system, the units are cubic meters per kilogram (m^3/kg).

Specific Weight. The weight of a unit volume of a fluid is its specific weight. In English units, it is expressed in pounds per cubic foot (lb/ft^3). In the metric system, the units are kilograms per cubic meter (kg/m^3).

Spiral-Riveted. A method of manufacturing pipe by coiling a plate into a helix and riveting together the overlapped edges.

Spiral-Welded. A method of manufacturing pipe by coiling a plate into a helix and fusion-welding the overlapped or abutted edges.

Spiral-Welded Pipe. Pipe made by the electric-fusion-welded process with a butt joint, a lap joint, or a lock-seam joint.

Square-Groove Weld. A groove weld in which the pipe ends are not chamfered. Square-groove welds are generally used on piping and tubing of wall thickness no greater than $^1/_8$ in (3 mm).

Stainless Steel. An alloy steel having unusual corrosion-resisting properties, usually imparted by nickel and chromium.

Statically Cast Pipe. Pipe formed by the solidification of molten metal in a sand mold.

Straight Polarity. The arrangement of direct-current arc welding leads in which the work is the positive pole and the electrode is the negative pole of the welding arc; a synonym for *direct-current electrode negative*.

Stress Relieving. Uniform heating of a structure or portion thereof to a sufficient temperature to relieve the major portion of the residual stresses, followed by uniform cooling.

Stringer Bead. A type of weld bead made by moving the electrode in a direction essentially parallel to the axis of the bead. There is no appreciable transverse oscillation of the electrode. The deposition of a number of string beads is known as *string beading* and is used extensively in the welding of austenitic stainless-steel materials. *See also* Weave Bead.

Structural Attachments. Brackets, clips, lugs, or other elements welded, bolted, or clamped to the pipe support structures, such as stanchions, towers, building frames, and foundation. Equipment such as vessels, exchangers, and pumps is not considered to be pipe-supporting elements.

Submerged Arc Welding (SAW). An arc welding process that produces coalescence of metals by heating them with an arc or arcs drawn between a bare metal electrode or electrodes and the base metals. The arc is shielded by a blanket of granular fusible material. Pressure is not used, and filler metal is obtained from the electrode and sometimes from a supplementary welding rod, flux, or metal granules.

Supplemental Steel. Structural members that frame between existing building framing steel members and are significantly smaller than the existing steel.

Swaging. Reducing the ends of pipe and tube sections with rotating dies that are pressed intermittently against the pipe or tube end.

Swivel Joint. A joint that permits single-plane rotational movement in a piping system.

Tack Weld. A small weld made to hold parts of a weldment in proper alignment until the final welds are made.

Tee Joint. A welded joint between two members located approximately at right angles to each other in the form of a **T**.

Tempering. A process of heating a normalized or quench-hardened steel to a temperature below the transformation range and, from there, cooling at any rate desired. This operation is also frequently called *stress relieving*.

Testing. An element of verification for the determination of the capability of an item to meet specified requirements by subjecting the item to a set of physical, chemical, environmental, or operating conditions.

Thixotropic Liquid. If the viscosity of a liquid decreases as agitation is increased at constant temperature, the liquid is called *thixotropic*. Examples include glues, greases, and paints.

Throat of a Weld. A term applied to fillet welds. It is the perpendicular distance from the beginning of the root of a joint to the hypotenuse of the largest right triangle that can be inscribed within the fillet-weld cross section. See Fig. 1.24.

Toe of a Weld. The junction between the face of a weld and the base metal. See Fig. 1.22.

Transformation Range. A temperature range in which a phase change is initiated and completed.

Transformation Temperature. A temperature at which a phase change occurs.

Trepanning. The removal by destructive means of a small section of piping (usually containing a weld) for an evaluation of weld and base-metal soundness. The operation is frequently performed with a hole saw.

Tube. A hollow product of round or any other cross section having a continuous periphery. Round tube size may be specified with respect to any two, but not all three, of the following: outside diameter, inside diameter, and wall thickness. Dimensions and permissible variations (tolerances) are specified in the appropriate ASTM or ASME specifications.

Turbinizing. Mechanical removal of scale from the inside of the pipe by means of air-driven centrifugal rotating cleaners. The operation is performed on steel pipe bends after hot bending to remove loose scale and sand.

Turbulent Flow. Fluid flow in a pipe is usually considered turbulent if the Reynolds number is greater than 4000. Fluid flow with a Reynolds number between 2000 and 4000 is considered to be in "transition."

Ultrasonic Examination or Inspection. A nondestructive method in which beams of high-frequency sound waves that are introduced into the material being inspected are used to detect surface and subsurface flaws. The sound waves travel through the material with some attendant loss of energy and are reflected at interfaces. The reflected beam is detected and analyzed to define the presence and location of flaws.

Underbead Crack. A crack in the heat-affected zone or in previously deposited weld metal paralleling the underside contour of the deposited weld bead and usually not extending to the surface.

Undercut. A groove melted into the base material adjacent to the toe or root of a weld and left unfilled by weld material.

Van Stoning. Hot upsetting of lapping pipe ends to form integral lap flanges, the lap generally being of the same diameter as the raised face of standard flanges.

Vapor Pressure. The pressure exerted by the gaseous form, or vapor, of liquid. When the pressure above a liquid equals its vapor pressure, boiling occurs. If the pressure at any point in the flow of a liquid falls below the vapor pressure or becomes equal to the vapor pressure, the liquid flashes into vapor. This is called *cavitation*. The vapor thus formed travels with the liquid and collapses where the pressure is greater than vapor pressure. This could cause damage to piping and other components.

Vertical Position. With respect to pipe welding, the position in which the axis of the pipe is vertical, with the welding being performed in the horizontal position. The pipe may or may not be rotated.

Viscosity. In flowing liquids, the internal friction or the internal resistance to relative motion of the fluid particles with respect to one another.

Weave Bead. A type of weld bead made with oscillation of the electrode transverse to the axis of the weld. Contrast to string bead.

Weld. A localized coalescence of material produced either by heating to suitable temperatures, with or without the application of pressure, or by application of pressure alone, with or without the use of filler material.

Weld Bead. A weld deposit resulting from a pass.

Weld Metal. That portion of a weld that has been melted during welding. The portion may be the filler metal, base metal, or both.

Weld Metal Area. The area of the weld metal as measured on the cross section of a weld.

Weld Penetration. *See* Joint Penetration *and* Root Penetration.

Weld-Prober Sawing. Removal of a boat-shaped sample from a pipe weld for examination of the weld and its adjacent base-metal area. This operation is usually performed in graphitization studies.

Weld Reinforcement. Weld material in excess of the specified weld size.

Weldability. The ability of a metal to be welded under the fabrication conditions imposed into a specific, suitably designed structure and to perform satisfactorily in the intended service.

Welded Joint. A localized union of two or more members produced by the application of a welding process.

Welder. One who is capable of performing a manual or semiautomatic welding operation.

Welder Performance Qualification. Demonstration of a welder's ability to produce welds in a manner described in a welding procedure specification that meets prescribed standards.

Welding Current. The current that flows through the electric welding circuit during the making of a weld.

Welding Fittings. Wrought- or forged-steel elbows, tees, reducers, and similar pieces for connection by welding to one another or to pipe. In small sizes, these fittings are available with counterbored ends for connection to pipe by fillet welding and are known as *socket-weld* fittings. In large sizes, the fittings are supplied with ends chamfered for connection to pipe by means of butt welding and are known as *butt-welding* fittings.

Welding Generator. The electric generator used for supplying welding current.

Welding Machine. Equipment used to perform the welding operation.

Welding Operator. One who operates a welding machine or automatic welding equipment.

Welding Procedure. The detailed methods and practices involved in the production of a weldment.

Welding Procedure Qualification Record. Record of welding data and test results of the welding procedure qualifications, including essential variables of the process and the test results.

Welding Procedure Specification (WPS). The document that lists the parameters to be used in construction of weldments in accordance with the applicable code requirements.

Welding Rod. Filler metal, in wire or rod form, used in gas welding and brazing procedures and those arc welding processes where the electrode does not furnish the filler metal.

Welding Sequence. The order of making the welds in a weldment.

Weldment. An assembly whose component parts are to be joined by welding.

Wrought Iron. Iron refined in a plastic state in a puddling furnace. It is characterized by the presence of about 3 percent of slag irregularly mixed with pure iron and about 0.5 percent carbon and other elements in solution.

Wrought Pipe. The term *wrought pipe* refers to both wrought steel and wrought iron. Wrought in this sense means *worked*, as in the process of forming furnace-welded pipe from skelp or seamless pipe from plates or billets. The expression *wrought pipe* is thus used as a distinction from cast pipe. Wrought pipe in this sense should not be confused with *wrought-iron pipe*, which is only one variety of wrought pipe. When wrought-iron pipe is referred to, it should be designated by its complete name.

TERMINOLOGY RELATED TO PLASTIC PIPING

Acrylonitrile-Butadiene-Styrene (ABS). Plastics containing polymers, blends of polymers, or both, having the following composition:

Butadiene	6% minimum
Acrylonitrile	15% minimum
Styrene or substitute styrene, or both	15% minimum
All other monomers	5% maximum
Plus lubricants, stabilizers, and colorants	

Adhesive. A substance capable of holding materials together by surface attachment.

Aging. The effect on materials of exposure to an environment for an interval of time.

Alloy. Two or more immiscible polymers united, usually by another component, to form a plastic vein having enhanced properties.

Antioxidant. A compounding ingredient used to retard deterioration caused by oxidation.

Apparent Density. The weight per unit volume of a material, including voids inherent in the material.

Backfill. Material used to fill the trench from bedding to finished surface.

Backfill, Final. Material used to fill the trench from top of the initial backfill to finished surface.

Backfill, Initial. Materials used to fill the trench from top of bedding to a designated height over the pipe.

Bedding. Materials that provide support for buried pipe.

Bell End. The enlarged portion of a pipe that resembles the socket portion of a fitting and that is intended to be used to make a joint.

Beveled Pipe. A pipe with a chamfered end to mate or adjust to another surface or to assist in assembly.

Binder. In a reinforced plastic, the continuous phase that holds together the reinforcement.

Blister. Rounded elevation of the surface of a plastic, with boundaries that may be more or less sharply defined.

Brittle Failure. A failure mode that exhibits no visible material deformation in the area of the break.

Burst Strength. The internal pressure required to cause a pipe or fitting to fail.

Butylene Plastics. Plastics based on resins made by the polymerization of butene or copolymerization of butene with one or more unsaturated compounds, the butene being the greatest amount by weight.

Catalyst. The chemical added to vinyl ester resins that causes them to harden, usually methyl ethyl ketone peroxide (MEKP).

Cell. A small partially or completely enclosed cavity. A cell totally enclosed by its walls is called a *closed cell*. A cell not totally enclosed by its walls is called an *open cell*.

Cellular Plastic. A plastic containing numerous cells, intentionally introduced, interconnecting or not, distributed throughout the mass.

Cellulose Acetate Butyrate (CAB). Plastic made by compounding a cellulose acetate butyrate ester with plasticizers and other ingredients.

Centrifugal Casting. A process for making pipe in which resin and fiberglass are placed into the interior of spinning rotary mold, forming the pipe through centrifugal force.

Chemical Resistance. Ability to resist chemical attack. Its severity is measured by determining the changes in physical properties. Time, temperature, stress, and reagent may all be factors that affect the chemical resistance of a material.

Chlorinated Poly(Vinyl Chloride) Plastics (CPVC). Plastics based on chlorinated poly (vinyl chloride) in which the chlorinated poly (vinyl chloride) is the greatest amount by weight.

Chopped Fiber. Continuous glass fibers cut into short (0.125 to 2.0 in or 3 to 50 mm) lengths.

Chopped Strand Mat. Coarse fabric sheets made from chopped strands randomly placed and held together by resin binders.

Collapse. It is defined as:

- Inadvertent densification of cellular material during manufacture resulting from breakdown of cell structure
- The buckling of the inner liner of composite piping
- The buckling or flattening of a plastic rehabilitation liner

- The buckling or crushing of a plastic pipe from external forces such as earth loads or external hydrostatic load

Compaction, Soil. An act of packing soil with mechanical force to increase its density.

Compatible. Components of a plastic piping system or different specific plastic materials, or both, are termed *compatible* if they can be joined together for satisfactory joints.

Composite Pipe. A pipe consisting of two or more different materials arranged with specific functional purpose to serve as pipe.

Compound. A mixture of a polymer with other ingredients such as fillers, stabilizers, catalysts, processing aids, lubricants, modifiers, pigments, or curing agents.

Contamination. Presence of a substance not intentionally incorporated in a product.

Crack. Any narrow opening or fissure in the surface that is visible to the naked eye.

Crater. A small, shallow surface imperfection.

Crazing. Apparent fine cracks at or under the surface of a plastic.

Creep. The time-dependent part of strain resulting from stress.

Cross Laminate. A laminate in which some of the layers of material are oriented approximately at right angles to the remaining layers with respect to the grain or strongest direction in tension.

Cross-Linking. The formation of a three-dimensional polymer by means of interchain reactions resulting in changes in physical properties.

Cure. To change the properties of a polymeric system into a more stable, usable condition by the use of heat, radiation, or reaction with chemical additives.

Curing Agent. Any of a number of chemicals added to epoxy resin to cause it to harden. Aromatic amine curing agents are commonly used for high-temperature corrosion service. Anhydride-cured epoxy resins are typically used for less stringent applications.

Cure Time. The time it takes for a resin system to reach full strength.

Density of Plastics. The weight per unit volume of material at 23°C expressed as D 23c, g/cm^3 (kg/m^3) or lb/ft^3.

Diffusion. The movement of a material such as gas or liquid in the body of a plastic.

Elastomer. A macromolecular material that returns rapidly to approximately its initial dimensions and shape after substantial deformation by a weak stress and release of the stress.

Elastomeric Seal. A material or device that uses an elastomer to effect a seal between separable piping components.

Elevated Temperature Testing. Tests on plastics at temperatures above 23°C.

Embedment. The placement of materials completely around the pipe to provide support.

Engineering Plastics. Those plastics and polymeric compositions for which well-defined properties are available.

Epoxy Resin. A thermosetting resin used in caustic, solvent, salt, and some acid solutions.

Ethylene Plastics. Plastics based on polymers of ethylene or copolymers of ethylene with other monomers, ethylene being in the greatest amount by mass.

Exfiltration, Pipe. The passage of fluid from a pipe section through small holes or leaks.

Fabmat. A combination of woven roving and chopped strand mat held together with resin binders. Usually used for making contact-molded fittings and butt weld joints.

Filament. A single fiber of glass (i.e., a monofilament).

Filler. A relatively inert material added to plastic to modify its strength, permeance, working properties, or other qualities or to lower costs.

Flow Rate. Rate of extrusion, weight per unit of time, g/10 min (kg/s), of molten resins through a die of specified length and diameter, under prescribed conditions of temperature, load, and piston position in the barrel as the timed measurement is being made.

Fluorocarbon Plastics. Plastics based on polymers made with perfluoromonomers.

Fungus Resistance. The ability of plastic pipe to withstand fungus growth or the metabolic products of fungi, or both, under normal conditions of service or laboratory tests simulating such conditions.

Gel Time. The time it takes for a resin system to harden so flow will not occur.

Hardener. Any of a number of chemicals added to resin that cause hardening to occur.

Heat Joining. Making a joint by heating the mating surfaces of the pipe components to be joined and pressing them together so that they fuse and become essentially one piece.

High-Density Polyethylene (HDPE) Plastics. Those linear polyethylene plastics having a standard density of 0.941 g/cm^3 or greater.

Hydrostatic Design Basis (HDB). The recommended maximum hoop stress that can be applied continuously with a high degree of certainty that failure of pipe will not occur.

Incasement. A rigid structure or pipe surrounding a buried pipe to provide additional support or protection.

Inhibitor. A substance used in low concentration that suppresses chemical reaction.

ISO Equation. An equation showing relationships between stress, pressure, and dimensions in pipe:

$$S = P(\mathrm{ID} + t)/2t \qquad \text{or} \qquad S = P(\mathrm{OD} - t)/2t$$

where S = hoop stress
P = pressure
ID = average inside diameter
OD = average outside diameter
t = minimum wall thickness

Joint, Adhesive-Bonded. A joint made by using an adhesive to bond the piping components.

Joint, Bell and Spigot. A connection between piping components consisting of a bell on one component, an elastomeric gasket between the components, and a spigot end on the other component.

Joint, Butt-Fused. A joint in which the prepared ends of the joint components are heated and then placed in contact to form the joint.

Joint, Clamped Insert-Fitting. A mechanical joint using external metal clamps or other mechanical devices to form a pressure seal between an inset fitting and the outside surface of the pipe.

Joint, Compression Fitting. A mechanical joint made by compressing the components either externally, internally, or axially to form a pressure seal between the fitting and the surface of the pipe.

Joint, Compression Gasket. A mechanical joint that utilizes a compression nut or a gland nut against a gasket to develop a pressure seal.

Joint, Flare. A mechanical compression connection between flared-end plastic pipe and a fitting specifically designed to accept flared-end plastic pipe.

Joint, Heat-Fused. A joint made using heat and pressure only.

Joint, Mechanical. A joint between piping components employing physical force to develop a seal or produce alignment.

Joint, Push-on. A joint in which an elastomeric ring gasket is compressed in the annular space between a bell end or socket and a spigot end of pipe.

Joint, Saddle-Fused. A joint in which the curved base of the saddle fitting and a corresponding area of the pipe surface are heated and then placed together to form the joint.

Joint, Socket-fused or Insert-Fused. A joint in which the joining surfaces of the components are heated, and the joint is made by inserting one component into the other.

Joint, Solvent-Cemented. A joint made using a solvent cement to unite the components.

Joint, Threaded. A mechanical joint that utilizes threaded pipe and fittings.

Laminate. A product made by bonding together two or more layers of material or materials.

Liner. The resin-rich interior surface of the pipe or fitting. The liner provides the corrosion resistance for chemical service.

Live Load. Portion of load transmitted to the pipe from wheel or traveling loads or other surcharged loads. The weight of the flow medium in the pipe is also termed *live load*.

Long-Term Hydrostatic Strength (LTHS). The hoop stress that when applied continuously will cause failure of the pipe at 100,000 h (11.43 years).

Low-Density Polyethylene (LDPE) Plastics. Those branched polyethylene plastics having a standard density of 0.910 to 0.925 g/cm^3.

Lubricant. A material used to reduce the friction between two mating surfaces that are being joined by sliding contact. An additive that is added to a plastic compound to lower the viscosity or otherwise improve the processing or product characteristics.

Medium-Density Polyethylene (MDPE) Plastics. Those branched polyethylene plastics having a standard density of 0.926 to 0.940 g/cm^3.

Monomer. A relatively simple compound that can react to form a polymer.

Necking. The localized reduction in cross section that may occur in a material under tensile stress.

Nexus. Porous surfacing mat of synthetic used to provide a resin-rich layer or liner.

Nonpressure Pipe. Pipe designed for a gravity-conveyed medium that must resist only intermittent static pressures and does not have a pressure rating.

Novolac Resin. A premium epoxy vinyl ester resin used in a broad range of corrosive applications.

Pimple. A small, sharp, or conical elevation on the surface of a plastic.

Pipe Spigot. Portion of a pipe or fitting that fits into a bell or socket of a preceding pipe or fitting.

Pit. A small crater in the surface of the plastic, with its width approximately the same order of magnitude as its depth.

Plastic. A material that contains as an essential ingredient one or more organic polymeric substances of large molecular weight, is solid in its finished state, and, at some stage in its manufacture or processing into finished articles, can be shaped by flow.

Plasticizer. A substance incorporated in a material to increase its workability, flexibility, or distensibility.

Plastic Pipe. A hollow cylinder of a plastic material in which the wall thickness is usually small compared to the diameter and in which the inside and outside walls are essentially concentric.

Plastic Tube. A particular size of smooth-wall plastic pipe in which the outside diameter is essentially the same as the corresponding size of copper tube.

Polybutylene. A polymer prepared by the polymerization of butene as the sole monomer.

Polyethylene. A polymer prepared by the polymerization of ethylene as the sole monomer.

Polymer. A substance consisting of molecules characterized by the repetition of one or more types of monomeric units.

Polypropylene. A polymer prepared by the polymerization of propylene as the sole monomer.

Polystyrene. A polymer prepared by the polymerization of styrene as the sole monomer.

Poly(Vinyl Chloride) (PVC). A polymer prepared by the polymerization of vinyl chloride as the sole monomer.

Pressure Rating. The estimated maximum pressure that the medium in the pipe can exert continuously with a high degree of certainty that failure of pipe will not occur.

Primer. An organic solvent that enhances adhesion, applied to plastic pipe and fittings prior to application of a solvent cement.

Reinforcement. Typically, fibers of glass used to provide strength and stiffness to a composite material.

Reinforced Plastic. A plastic with high-strength fillers embedded in the composition, resulting in some mechanical properties superior to those of the base resin.

Release Agent. A material used to keep a molding material from adhering to the mold.

Resin. A solid or pseudosolid organic material, often of high molecular weight, that exhibits a tendency to flow when subjected to stress, usually has a softening or melting range, and usually fractures conhoidally.

Roving. A collection of one or more strands of glass filaments. The typical form of glass fiber used in the manufacture of filament-wound pipe.

Rubber. A material that is capable of recovering from large deformations quickly and forcibly, and can be, or already is, modified to a state in which it is essentially insoluble in boiling solvent, such as benzene, methyl ethyl ketone, and ethanol toluene azeotrope. A rubber in its modified state, free of diluents, retracts within 1 min to less than 1.5 times its original length after being stretched at room temperature to twice its length and held for 1 min before release.

Service Factor. A factor used to reduce a strength value to obtain an engineering design stress. The factor may vary depending on the service conditions, the hazard, the length of service desired, and the properties of the pipe.

Slit Failure. A form of brittle failure that exhibits only a very small crack through the wall of the pipe with no visible material deformation in the area of the break.

Solvent Cement. An adhesive made by dissolving a plastic resin or compound in a suitable solvent or mixture of solvents. The solvent cement dissolves the surfaces of pipe and fittings to form a bond between the mating surfaces provided the proper cement is used for the particular materials and proper techniques are followed.

Solvent Joint. A joint made by using a solvent cement to unite the components.

Stabilizer. An ingredient added to a plastic to retard possible degradation. Usually added for heat protection, environmental protection, or both.

Standard Dimension Ratio (SDR). The ratio of the average specified outside diameter to the minimum specified wall thickness for outside diameter–controlled pipe. The SDR value is derived by adding 1 to the pertinent number selected from the ANSI Preferred Number Series 10. See the table below.

ANSI Preferred Number Series 10	SDR
5.0	6.0
6.3	7.3
8.0	9.0
10.0	11.0
12.5	13.5
16.0	17.0
20.0	21.0
25.0	26.0
31.5	32.5
40.0	41.0
50.0	51.0
63.0	64.0

Standard Inside Diameter Ratio (SIDR). The ratio of the specified average inside diameter to the specified minimum wall thickness for inside diameter–controlled plastic pipe. The SIDR value is derived by subtracting 1 from the pertinent number selected from the ANSI Preferred Number Series 10. See the table below.

ANSI Preferred Number Series 10	SIDR
5.0	4.0
6.3	5.3
8.0	7.0
10.0	9.0
12.5	11.5
16.0	15.0
20.0	19.0
25.0	24.0
31.5	30.5
40.0	39.0
50.0	49.0
63.0	62.0

Styrene-Rubber (SR) Plastics. Plastics containing at least 50 percent styrene plastics combined with rubber and other compounding materials, but not more than *15* percent acrylonitrile.

Sustained Pressure Test. A constant internal pressure test for an extended period of time.

Thermoplastic. A plastic that can be repeatedly softened by heating and hardened by cooling through a temperature range characteristic of the plastic, and that in the softened state can be shaped by flow into articles by molding or extrusion.

Thermoset. A plastic that, when cured by application of heat or chemical means, changes into a substantially infusible and insoluble product.

Thermoset Resin. A resin cured by heat or chemical additives. Once cured, a thermoset resin cannot be remelted.

Thermosetting. Capable of being changed into a substantially infusible or insoluble product when cured by heat or other means.

Toe-in. A small reduction of the outside diameter at the cut end of a length of thermoplastic pipe.

Veil. Surfacing mat of porous fabric made from filaments. Used to provide a resin-rich layer or liner.

Vinyl Ester Resin. A thermosetting resin used in strong acids, chlorine, and oxidizing agents.

Void. An unfilled space of such a size, in a solid plastic, that it scatters radiant energy such as light. Also, it is defined as a cavity unintentionally formed in a cellular material, and substantially larger than the characteristic individual cells.

Vulcanization. An irreversible process during which a rubber compound, through a change in its chemical structure, becomes less plastic and more resistant to swelling by organic liquids and elastic properties are conferred, improved, or extended over a range of temperature.

Woven Roving. Coarse cloth-like material made by weaving fiberglass roving.

TERMINOLOGY RELATED TO PRESTRESSED CONCRETE CYLINDER PIPE (PCCP)

Bell Ring. Receiving portion of a PCCP joint. See Fig. 1.25.

External Dead Load. A load applied to a pipe that is generally constant, such as earth weight and foundation load.

External Live Load. A load applied to a pipe that is transient in nature and usually of short duration, such as that from motor vehicles.

Field Test Pressure. An internal pressure applied to a pipeline or a portion of a pipeline to test its structural and watertight performance.

Laying Length. A measure of a pipe or fitting's length along its axis for purpose of advancing the length of a pipeline.

O-Ring Gasket. Rubber ring of a circular cross section that, when compressed into the spigot-ring groove by the bell ring, provides a watertight seal.

Spigot Ring. The protruding end of a PCCP joint that contains a shaped groove to retain the O-ring rubber gasket. See Fig. 1.26.

Transient Pressure. The incremental change in internal pressure in a pipeline, usually of short duration, that is caused by a relatively sudden change in flow velocity.

Working Pressure. The long-term, steady-state internal pressure.

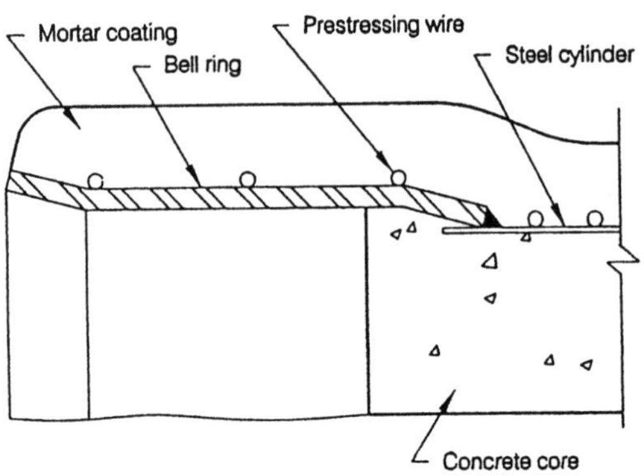

FIGURE 1.25 Bell ring.

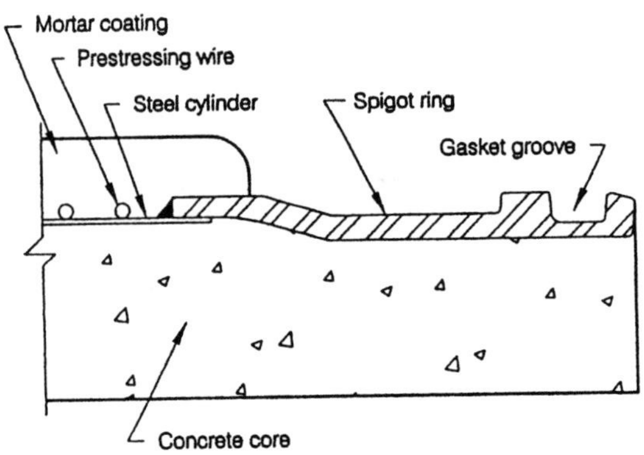

FIGURE 1.26 Spigot ring.

TERMINOLOGY RELATED TO LINED PIPING

Terminology Related to Cement Mortar- and Concrete-Lined Piping

Centrifugal Casting. A method of placing a concrete or mortar lining by centrifugal force inside a revolving cylinder.

Concrete. A mixture of portland cement, fine aggregate (sand), coarse aggregate (gravel), and water.

Mortar. A mixture of portland cement, fine aggregate (sand), and water.

Passivation. The formation of a tightly adhering and protective chemical oxide film on the metal surface due to the presence of highly alkaline portland cement mortar or concrete.

Pneumatic Placement. A method of applying a mortar lining to a surface using compressed air.

Radial Compaction. A vertical method of placing a pipe's concrete lining by the use of a special revolving packerhead.

Tuberculation. The formation of localized corrosion products scattered over the metal surface in the form knob-like mounds.

Vertical Casting. A method of placing the concrete for a pipe involving the use of a steel inside form and outside form.

Terminology Related to Epoxy Lined and Coated Piping

Cathodic Disbondment. Loss of coating adhesion around a holiday (hole in the coating) when the coated pipe is provided with cathodic protection. The phenomenon of cathodic disbondment happens with all organic coatings.

Cathodic Protection. When external pipeline corrosion takes place, there are anodic and cathodic areas on the surface. In the anodic areas (the corroding sites), current flows from the pipeline steel into the surrounding electrolyte (water in the soil). Where the current flows from the electrolyte to the pipeline, the area is cathodic and does not corrode. An external source of current flowing to the pipe will make it cathodic and will prevent corrosion. This method of corrosion prevention is called *cathodic protection*.

Cathodic Shielding. Cathodic shielding can take place when an electrically insulating barrier prevents current flow from reaching the pipe. An example is a disbonded coating that allows water to come in contact with the pipe, but does not allow cathodic protection current to passivate the pipe surface. Current flow under the coating will depend on the resistance of the water there and, after a short distance, will be insufficient to protect the pipe from corrosion.

Coating. Application of FBE on an internal or external surface of a pipe or other metallic equipment. When applied on an internal surface, it is referred to as a *lining*, and when applied to an external surface, it is called an *external coating*.

Cure Time. The time required to complete the heat-induced cross-linking reaction to the point where the coating provides the acceptable level of performance.

DGEBA. Diglycidyl ether of bisphenol A is a widely used solid epoxy resin used in the manufacture of fusion-bonded epoxy pipe coatings.

FBE. Fusion-bonded epoxy.

Fusion-Bonded Epoxy. A one-part, heat-curable, thermosetting epoxy-resin powdered-coating material normally composed of an epoxy and a curing agent, plus other materials to provide a tough, well-adhered coating. The powder is sprayed onto a hot pipe where it melts, flows, and cures.

Gel Time. The time interval during which a coating material converts from a molten liquid to a solid.

Holiday. A discontinuity, or hole, in a coating that allows an electric charge to pass at a specific voltage.

Nested. Rows of pipe that rest on each other in storage are said to be *nested*.

NSF. National Sanitation Foundation. U.S. certification agency for potable drinking water systems.

Powder Coating. See Fusion-Bonded Epoxy.

Soil Stress. Damage to a pipeline coating caused by wet/dry cycling of the soil and/or pipe movement caused by temperature change.

Stacked. Rows of pipe placed on each other in storage with an intervening layer of stripping, normally wood, are said to be *stacked*.

Stress-Corrosion Cracking. In the presence of stress, corrosion cracks can initiate. The incidence of failures on pipelines has been in cathodically protected areas of disbonded coatings where liquids at a pH of 9 to 11 and carbonate/bicarbonate solutions have collected between the coating and the surface of the steel.

Three-Layer External Pipe Coating. It is composed of an epoxy primary corrosion-protection layer, a tie layer of modified-polyolefin adhesive, and an overcasting of polyolefin.

Terminology Related to Rubber-Lined Piping

Abrasion Resistance. The resistance of a material to loss of surface particles due to frictional forces.

Acid Resistance. The ability to resist the action of identified acids within specified limits of concentration and temperature.

Adhesion. The state in which two surfaces are held together by interfacial forces, which may consist of molecular forces or interlocking action, or both. Adhesion values for tank linings are often determined by the procedures described in ASTM D429, Test Methods for Rubber Property–Adhesion to Rigid Substrates.

Adhesion Failure. The separation of two materials at the surface interface rather than within one of the materials itself.

Adhesive. A substance capable of holding materials together by surface attachment.

Age Resistance. The resistance to deterioration by oxygen, heat, light, or ozone, alone or in combination, during storage or use.

Antioxidant. Same as age resistor; a chemical compounding material used to retard deterioration caused by oxygen.

ASTM. American Society for Testing and Materials.

Autoclave. A pressure vessel used for the curing or vulcanization of rubber parts by means of steam under pressure.

Blasting. Surface cleaning and preparation of a substrate with abrasives such as airborne sand, grit, or shot.

Blemish. A superficial mark or impression on the surface of green or cured rubber lining.

Blister. A cavity or sac deforming the surface of a material, usually due to expansion of an entrapped liquid or gas. Permeation failures of tank linings in service are sometimes evidenced by blister formation.

Bond. The union of materials by use of adhesives, usually used in relation to parts vulcanized after being attached or assembled together.

Bromobutyl. Used in a general sense to mean a bromobutyl tank lining construction. The ASTM designation is BIIR.

Butt Seam. A seam made by placing the two pieces to be joined edge to edge.

Butt Splice. A joint made in a rubber part before or after vulcanization by placing the two pieces to be joined edge to edge.

Butyl Rubber. A copolymer of isobutylene and isoprene rubber; ASTM designation IIR; chlorobutyl or bromobutyl rubber is the common name for such materials used in lining.

Calender. A machine equipped with two or more heavy, internally heated or cooled rolls, used for the continuous sheeting or plying up of rubber compounds.

Calender Blister. Trapped air between calender plies of a multi-ply rubber buildup.

Cement. A dispersion or solution of an elastomer or compound in a solvent for use as an adhesive or coating.

Chlorobutyl. Used in a general sense to mean a chlorobutyl tank lining construction. ASTM designation CIIR.

Chlorosulfonated Polyethylene. Generic name of an elastomeric material sold as Hypalon™. ASTM designation CSM.

Closed Skive. A reverse angle cut along the edge of a rubber panel. This enables the installer to stitch down the cut edge so that the tie gum is protected from exposure to the commodity contained in a tank.

Cure. Similar to cross-linking, while cure covers all types (sulfur, peroxide, radiation etc.).

Delamination. Separation or splitting, either between plies in laminated goods or occasionally within the homogeneous part itself.

Durometer. An instrument for measuring the hardness of rubber and plastics. The A durometer scale is used for flexible materials and the D for rigids.

Durometer Hardness. A value that indicates the indentation, or resistance to indentation, by the indentor point of a durometer. High values indicate harder materials. See ASTM D2280, Test Method for Rubber Property–Durometer Hardness.

EPDM. ASTM abbreviation for a terpolymer of ethylene, propylene, and a diene with the residual unsaturated portion of the diene in the side chain.

Ebonite. A term for hard rubber, made by vulcanization of rubber with high levels (greater than 30 parts) of sulfur, where the high hardness is due to the sulfur content.

Elastomer. A polymeric material which, at room temperature, is capable of recovering substantially in shape and size after removal of a deforming force.

Elongation. Extension produced by tensile stress, usually expressed as a percentage of original unit length.

Face. The commodity-contacting surface in tank lining construction.

Face Stock. The commodity-contacting stock in a multicomponent lining.

Fish Eye. A thin elongated void in a calendered sheet. This slight cosmetic blemish does not affect the service life of the rubber lining.

Gasket. A deformable material clamped between essentially stationary faces to prevent the escape of matter through an opening or joint.

Gauge (gage). Refers to a dimension, generally the thickness of a product, as measured by a gauge of some type.

Hardness. The measured resistance to indentation of a material.

Heat Resistance. The property or ability of rubber articles to resist the deteriorating effects of elevated temperatures.

Holiday. A small uncovered or noncoated area in a substrate; usually refers to pin holes in thin coatings of rubber sheeting.

I.D. or ID. Inside diameter.

Immersion Testing. Commonly used to determine the resistance of tank lining compounds to various chemicals. See ASTM D-471, Test Method for Rubber Property–Effect of Liquids.

Lap Seam. A seam made by extending the flat edge of one piece of material flat over the edge of a second piece of material.

Light Aging. Deterioration of compounds when exposed to light (direct or indirect, natural or artificial).

NACE. National Association of Corrosion Engineers.

Natural Rubber. Rubber formed in a living plant or tree, usually referring to *Hevea brasiliensis*; ASTM designation NR.

NBR. ASTM designation for copolymers of acrylonitrile and butadiene.

Neoprene. Originally the trade name, now the generic name of polymers and copolymers based on chloroprene; ASTM designation CR.

Nerve. The elastic resistance of raw rubber or compounds to permanent deformation during processing. A nervy tank lining will be difficult to lay around tight bends or in corners because of spring-back.

Nitrile Rubber. Copolymers of acrylonitrile and butadiene; same as NBR or Buna-N.

O.D. or OD. Outside diameter.

Off Gauge. Not conforming to specified tolerance on thickness.

Open Seam. A seam in which edges do not meet, forming a void.

Optimum Cure. The time and temperature of cure necessary to develop a desired combination of properties. Several laboratory procedures are available to help determine this point.

Overlap Splice. The edge of a panel that overlaps the end of an adjoining panel. The splice can be open or closed.

Overlay. To add another layer of lining over an in-place tank lining construction. Typical overlays are cap strips and dome ends of tank cars.

Oxidation. The reaction of oxygen with a rubber product, usually accompanied by a change in feel, appearance of surface, or (usually adverse) in physical properties.

pH. The measure, on a logarithmic scale of 1 to 14, of the relative acidity or alkalinity of an aqueous solution. Neutral pH (pure water) is 7. Hydrochloric acid would be approximately 1 and sodium hydroxide approximately 13.

Plasticity. The tendency of a material to remain deformed after reduction of the deforming stress to or below its yield stress.

Ply Adhesion. The force required to separate two adjoining plies in a specified width of a rubber product.

Ply Separation. A condition that occurs because of a loss of adhesion between plies.

Polymer. A macromolecular material formed by the chemical combination of monomers having either the same or different chemical composition.

Polymer Chain. The chain of elements that form the basis of the structure of a polymer. The elements may be all carbon atoms or carbon and oxygen, silicon, nitrogen, etc.

Pressure Cure. Vulcanization under pressure.

Profile. Surface profile is a measure of the roughness of a surface that results from abrasive blast cleaning. The height of the profile produced on the surface is measured from the bottom of the lowest valley to the top of the highest peak.

psi. Pounds per square inch.

Qualification Inspection or Test. The examination of samples from a typical production run of products to determine adherence to a given specification for approval to become a supplier.

Quality Conformance Inspection or Test. The examination of samples from a production run of products to determine adherence to a given specification for acceptance of that production run.

Rate of Cure. The relative time required to reach a predetermined state of vulcanization under specified conditions.

Recovery. The degree to which a rubber product returns to its normal dimensions after being distorted.

Reversion. The softening of vulcanized rubber when it is heated too long or exposed to elevated temperatures. It is a deterioration in physical properties. (Extreme reversion may result in tackiness.) This most commonly affects natural rubber linings.

Rubber. An elastomer, generally implying natural rubber, but used loosely to mean any elastomer, vulcanized and unvulcanized. By definition, a material that is capable of recovering from large deformations quickly and forcibly.

Sand Blast. To clean with sand or steel grit in a high-velocity stream of air.

Seam. A line formed by joining material to form a single ply or layer. A splice.

Semihard Rubber. A term used for hard rubber which when fully cured is flexible and can be bent without shattering.

Shelf Aging. The natural deterioration of rubber articles kept in storage or "on the shelf" under normal atmospheric conditions. This slow deterioration is due primarily to oxygen and ozone attack.

Shelf Life. An expression describing the time a material can be stored without losing any of its properties.

Shore Hardness. A term denoting a hardness value derived from an instrument developed by the Shore Instrument & Manufacturing Co., Inc.

Skive. A cut made on an angle to the surface to produce a tapered or feathered edge.

Skive Butt Seam. Two skived edges joined edge to edge to form a smooth surface.

Spark Tester. A high-voltage test unit used to detect breaks or holes in a lining.

Specific Gravity. The ratio of the mass of a unit volume of a material to that of the same volume of water at a specified temperature. Specific gravity is:

$$\frac{\text{Grams/cm}^3\text{ of material}}{\text{Grams/cm}^3\text{ of water (1)}} \quad \text{or} \quad \frac{\text{pounds/ft}^3\text{ of material}}{\text{pounds/ft}^3\text{ of water (62.4)}}$$

Specific Volume. The reciprocal of specific gravity (1/specific gravity). Also expressed as the ratio of the volume of one pound of water to the volume of one pound of material.

SSPC. Steel Structures Painting Council.

Static Spark. A high-voltage discharge of static electricity.

Steam Cure (Open). A method of vulcanizing rubber parts by exposing them directly to steam.

Stitching. A method of joining two pieces of uncured rubber compound together by means of a stitching roller, a hand-held tool composed of a wheel with a narrow edge, which is often serrated. It is commonly used to make tank lining splices.

Substrate. The surface on which a coating or lining is applied.

Surface Preparation. The preparation of a substrate prior to applying tank lining: welding, grinding, blasting, and cleaning.

Swelling. An increase in volume or linear dimension of a specimen immersed in liquid or exposed to a vapor.

Tack. The property of a polymer, compound, or adhesive that causes two layers to stick together on application of mild pressure. Tacky polymers or compounds do not necessarily stick to other surfaces.

Tack Cement. A formulated rubber/cement mixture that can be rolled or brushed on surfaces and will hold a rubber panel in place until cure takes place. Normally considered a part of the adhesive system.

Tensile Strength. The maximum tensile stress applied during stretching of a specimen to rupture, per unit area of the specimen, i.e., psi or MPa.

Tie Gum. An intermediate layer of rubber employed to promote bonding of two surfaces; usually a soft rubber compound.

Vapor Phase. Vapor above the liquid in a tank car or closed storage tank; often the most severe condition for rubber tank lining.

Vulcanization. (1) An irreversible process during which a rubber compound, through a change in its chemical structure (cross-linking), becomes less plastic and more elastic. Elastic properties are conserved, improved, or extended over a greater range of temperature. (2) The reaction of rubber specifically with sulfur; *curing* covers other methods of cross-linking. Both terms are often used interchangeably.

Weathering. A surface deterioration of a rubber article during outdoor exposure.

White Metal Blast. To sand or shot blast a steel substrate to an SSPC No. 5 finish, as specified by the Steel Structures Painting Council (NACE 1).

Tables in Chapter 1

CHAPTER 2
SYSTEMS OF UNITS AND CONVERSION TABLES

See p. 68 for a list of tables in Chap. 2.

INTRODUCTION

This section provides a brief introduction to commonly used systems of units. However, the discussion is limited to the units and their conversion factors in relation to piping and tubing.

UNITED STATES CUSTOMARY SYSTEM OF UNITS

The system of units used in the United States is referred to as the U.S. Customary System (USCS); it is also called the *inch-pound* system. It utilizes the inch to measure length and the pound to measure weight. Seconds are used to measure time. The USCS is primarily identical to the English units with the exception of differences in units used to measure capacity; these are evident from the following examples:

1 U.S. gallon = 231 cubic inches = 3.7853 liters = 0.00378 cubic meters

1 Imperial gallon = 277.42 cubic inches = 4.5459 liters = 0.00455 cubic meters

Units for Length

Table 2.1 lists the USCS units for length that are used for piping and tubing.

Units for Area

Table 2.2 provides a listing of USCS units for area.

Units for Volume

Table 2.3 lists the commonly used USCS units for volume.

Units for Weight

Table 2.4 lists the units for weight.

Units for Circular Measures

Table 2.5 lists the units for circular measures.

TABLE 2.1 USCS Units for Length

12 inches	1 foot
3 feet	1 yard
220 yards (= 40 poles)	1 furlong
8 furlongs (1760 yards = 5280 feet)	1 mile
7.92 inches	1 link
100 links (= 66 feet = 4 rods)	1 chain
80 chains (= 5280 feet)	1 mile

TABLE 2.2 USCS Units for Area

144 square inches	1 square foot
9 square feet	1 square yard
4840 square yards	1 acre
10 square chains	1 acre
43, 560 square feet	1 acre
640 acres	1 square mile

TABLE 2.3 USCS Units for Volume

1728 cubic inches	1 cubic foot
231 cubic inches	1 U.S. gallon
27 cubic feet	1 cubic yard
1 cubic foot	7.4805 U.S. gallons
4 gills	1 pint (liquid or fluid)
2 pints	1 quart (liquid or fluid)
4 quarts	1 U.S. gallon (liquid or fluid)
8 drams	1 liquid ounce
16 ounces	1 pint
8 quarts	1 peck (dry measure)
4 pecks	1 bushel
1 register ton	100 cubic feet
1 U.S. shipping ton	40 cubic feet
1 barrel (oil)*	42 U.S. gallons

* The capacity of a barrel varies with the industry as follows:
1 barrel of beer = 31 U.S. gallons
1 barrel of wine = 31.5 U.S. gallons
1 barrel of whiskey = 45 U.S. gallons

TABLE 2.4 USCS Units for Weight

16 drams (437.5 grains)	1 ounce
16 ounces (= 7000 grains)	1 pound
100 pounds	1 cental
2000 pounds	1 short ton
2240 pounds	1 long ton
24 grains	1 pennyweight (troy weights)
20 pennyweights (= 480 grains)	1 ounce (troy weights)
12 ounces (= 5760 grains)	1 pound

TABLE 2.5 USCS Units for Circular Measures

60 seconds	1 minute
60 minutes	1 degree
90 degrees	1 quadrant
360 degrees	Circumference
57.2957705 degrees	1 radian*

* Angle having arc of length equal to radius.

METRIC SYSTEM OF UNITS

The metric system of units was developed in France. In the metric system, the base units for length, force, and time are metre, kilogram, and second. It is commonly referred to as the MKS system of units. The following provides a brief introduction to the metric system of units.

Units for Length

The unit for length is called the metre and it is equal to 1/10,000,000 of a quarter meridian, North Pole to equator.

Units for Volume

The volume of a cube with each side equal to one-tenth of a metre is termed a litre, the unit for volume.

Units for Mass

The mass of a litre of water is called a kilogram, the unit for mass.

Units for Time

The metric unit for time is the second, as in the USCS.

Units for Force

The metric unit for force is the kilogram, which is also the unit for mass. As such, it creates confusion.

BRITISH GRAVITATIONAL SYSTEM OF UNITS

The British gravitational system of units is based on the base units for length, force, and time as follows.

Units for Length

The unit for length is called the foot. Refer to the USCS system, which is based on the British system.

Units for Volume

The volume of a cube with each side equal to one foot is termed one cubic foot, the unit for volume. Refer to USCS of units.

Units for Mass

The unit for mass is called the pound.

Units for Time

The British system also uses the second as the unit for time.

Units for Force

The British unit for force is the pound, which is also the unit for mass. It creates confusion.

CGS SYSTEM OF UNITS

The CGS system of units is an abbreviated version of centimetre, gram, and second. This system can be defined as follows.

Units for Length

The CGS unit for length is called the centimetre, which is one-hundredth of a metre.

Units for Mass

The CGS unit for mass is called the gram, which is one-thousandth of a kilogram.

Units for Time

The CGS system also uses the second as the unit of time.

Units for Force

The CGS unit for force is the dyne. The dyne is the force that, when applied to the standard gram body, would give the body an acceleration of one centimetre per second per second ($1\,cm/s^2$).

THE INTERNATIONAL SYSTEM OF UNITS (SI)

In the British system of units, the pound is a unit for mass and force. Similarly, in the metric system of units, kilogram is a unit used for mass and force. Due to these and other reasons, a new system of units, the International System of Units, was developed. This system of units is called le Systeme International (SI) d'Unités in French.

SI units are divided into three categories of units, as follows:

- Base units
- Supplementary units
- Derived units

SI units also include prefixes, which are listed in Table 2.6. These prefixes are used to form decimal multiples and submultiples of SI units.

In accordance with the U.S. National Institute of Standards and Technology (NIST), we shall use dot, or period, as the decimal marker rather than comma, used by European countries; the spellings *meter*, *liter*, and *deka* rather than *metre*, *litre*, and *deca*; and the name *metric ton* rather than *tonne*. Henceforth, we will follow this approach, although we have done otherwise up to this point.

Base Units

The following provides a definition of the seven base SI units.

Meter. The meter is defined as the length of path traveled by light in a vacuum during a time interval of 1/299,792,458 of a second.

Kilogram. The kilogram is the unit of mass; it is equal to the mass of the international prototype of the kilogram.

Second. The second is the duration of 9,192,631,770 periods of the radiation corresponding to the transition between the two hyperfine levels of the ground state of the cesium 133 atom.

Ampere. The ampere is the constant current that, if maintained in two straight parallel conductors of infinite length, of negligible cross section, and placed 1 meter apart in vacuum, would produce between these conductors a force equal to 2×10^{-7} newton per meter of length.

Kelvin. The kelvin, the unit of thermodynamic temperature, is the fraction 1/273.16 of the thermodynamic temperature of the triple point of water.

Mole. The mole is the amount of substance of a system that contains as many elementary entities as there are atoms in 0.012 kilogram of carbon 12. When the mole is used, the elementary entities must be specified and may be atoms, molecules, ions, electrons, other particles, or specified groups of such particles.

Candela. The candela is the luminous intensity, in a given direction, of a source that emits monochromatic radiation of frequency 540×10^{12} hertz and that has a radiant intensity in that direction of 1/683 watt per steradian. Table 2.7 lists the SI base units and their symbols.

TABLE 2.6 SI Prefixes

Factor	Prefix	Symbol	Factor	Prefix	Symbol
$10^{24} = (10^3)^8$	yotta	Y	10^{-1}	deci	d
$10^{21} = (10^3)^7$	zetta	Z	10^{-2}	centi	c
$10^{18} = (10^3)^6$	exa	E	$10^{-3} = (10^3)^{-1}$	milli	m
$10^{15} = (10^3)^5$	peta	P	$10^{-6} = (10^3)^{-2}$	micro	μ
$10^{12} = (10^3)^4$	tera	T	$10^{-9} = (10^3)^{-3}$	nano	n
$10^{9} = (10^3)^3$	giga	G	$10^{-12} = (10^3)^{-4}$	pico	p
$10^{6} = (10^3)^2$	mega	M	$10^{-15} = (10^3)^{-5}$	femto	f
$10^{3} = (10^3)^1$	kilo	k	$10^{-18} = (10^3)^{-6}$	atto	a
10^{2}	hecto	h	$10^{-21} = (10^3)^{-7}$	zepto	z
10^{1}	deka	da	$10^{-24} = (10^3)^{-8}$	yocto	y

Supplementary Units

There are two supplementary SI units: radian and steradian. Radian is the unit for the quantity plane angle. Steradian is the unit for the quantity solid angle.

Radian. The radian is the plane angle between two radii of a circle that cut off on the circumference an arc equal in length to the radius.

Steradian. The steradian is the solid angle that, having its vertex in the center of a sphere, cuts off an area of the surface of the sphere equal to that of a square with sides of length equal to the radius of the sphere. Refer to Table 2.8 for a listing and symbols of supplementary SI units.

TABLE 2.7 SI Base Units

	SI base unit	
Base quantity	Name	Symbol
Length	meter	m
Mass	kilogram	kg
Time	second	s
Electric current	ampere	A
Thermodynamic temperature	kelvin	K
Amount of substance	mole	mol
Luminous intensity	candela	cd

TABLE 2.8 Supplementary SI Units

Quantity	Unit	Symbol
Plane angle	radian	rad
Solid angle	steradian	sr

Derived Units

Derived units are expressed algebraically in terms of base units, supplementary units, or other derived units. The symbols for derived units are obtained by mathematical operations of multiplication and division. As an example, the derived unit for area is obtained by multiplication of length by length ($m \times m = m^2$). Some commonly used derived units are listed in Table 2.9.

There are many more derived SI units, which are used for variety of tasks. Refer to Table 2.10 for listing of SI units in alphabetical order with their symbols and conversion factors.

TABLE 2.9 Examples of SI Derived Units Expressed in Terms of SI Base Units

	SI derived unit	
Derived quantity	Name	Symbol
Area	square meter	m^2
Volume	cubic meter	m^3
Speed, velocity	meter per second	m/s
Acceleration	meter per second squared	m/s^2
Wave number	reciprocal meter	m^{-1}
Mass density (density)	kilogram per cubic meter	kg/m^3
Specific volume	cubic meter per kilogram	m^3/kg
Current density	ampere per square meter	A/m^2
Magnetic field strength	ampere per meter	A/m
Amount of substance concentration (concentration)	mole per cubic meter	mol/m^3
Luminance	candela per square meter	cd/m^2

TABLE 2.10 Conversion Factors to Convert to SI Units*

To convert from	To	Multiply by
	Acceleration	
acceleration of free fall, standard (g_n)	meter per second squared (m/s^2)	**9.806 65**
foot per second squared (ft/s^2)	meter per second squared (m/s^2)	**3.048×10^{-1}**
gal (Gal)	meter per second squared (m/s^2)	**1.0×10^{-2}**
inch per second squared (in/s^2)	meter per second squared (m/s^2)	**2.54×10^{-2}**
	Angle	
degree (°)	radian (rad)	1.745329×10^{-2}
gon (also called grade) (gon)	radian (rad)	1.570796×10^{-2}
gon (also called grade) (gon)	degree (°)	**9.0×10^{-1}**
mil	radian (rad)	9.817477×10^{-4}
mil	degree (°)	**5.625×10^{-2}**
minute (′)	radian (rad)	2.908882×10^{-4}
revolution (r)	radian (rad)	6.283 185
second (″)	radian (rad)	4.848137×10^{-6}
	Area and sccond moment of area	
acre (based on U.S. survey foot)	square meter (m^2)	4.046873×10^{3}
are (a)	square meter (m^2)	**1.0×10^{2}**
barn (b)	square meter (m^2)	**1.0×10^{-28}**
circular mil	square meter (m^2)	5.067075×10^{-10}
circular mil	square millimeter (mm^2)	5.067075×10^{-4}
foot to the fourth power (ft^4)	meter to the fourth power (m^4)	8.630975×10^{-3}
hectare (ha)	square meter (m^2)	**1.0×10^{4}**
inch to the fourth power (in^4)	meter to the fourth power (m^4)	4.162314×10^{-7}
square foot (ft^2)	square meter (m^2)	**9.290304×10^{-2}**
square inch (in^2)	square meter (m^2)	**6.4516×10^{-4}**
square inch (in^2)	square centimeter (cm^2)	**6.4516**
square mile (mi^2)	square meter (m^2)	2.589998×10^{6}
square mile (mi^2)	square kilometer (km^2)	2.589 998
square mile (based on U.S. survey foot (mi^2)	square meter (m^2)	2.589998×10^{6}
square mile (based on U.S. survey foot) (mi^2)	square kilometer (km^2)	2.589 998
square yard (yd^2)	square meter (m^2)	8.361274×10^{-1}
	Electricity and magnetism	
abampere	ampere (A)	**1.0×10**
abcoulomb	coulomb (C)	**1.0×10**
abfarad	farad (F)	**1.0×10^{9}**
abhenry	henry (H)	**1.0×10^{-9}**
abmho	siemens (S)	**1.0×10^{9}**
abohm	ohm (Ω)	**1.0×10^{-9}**
abvolt	volt (V)	**1.0×10^{-8}**
ampere hour (A·h)	coulomb (C)	**3.6×10^{3}**
biot (Bi)	ampere (A)	**1.0×10**
EMU* of capacitance (abfarad)	farad (F)	**1.0×10^{9}**
EMU of current (abampere)	ampere (A)	**1.0×10**
EMU of electric potential (abvolt)	volt (V)	**1.0×10^{-8}**
EMU of inductance (abhenry)	henry (H)	**1.0×10^{-9}**
EMU of resistance (abohm)	ohm (Ω)	**1.0×10^{-9}**
ESU† of capacitance (statfarad)	farad (F)	1.112650×10^{-12}
ESU of current (statampere)	ampere (A)	3.335641×10^{-10}
ESU of electric potential (statvolt)	volt (V)	2.997925×10^{2}
ESU of inductance (stathenry)	henry (H)	8.987552×10^{11}
ESU of resistance (statohm)	ohm (Ω)	8.987552×10^{11}
faraday (based on carbon 12)	coulomb (C)	9.648531×10^{4}
franklin (Fr)	coulomb (C)	3.335641×10^{-10}

TABLE 2.10 Conversion Factors to Convert to SI Units* (Continued)

To convert from	To	Multiply by
gamma (γ)	tesla (T)	**1.0×10^{-9}**
gauss (Gs, G)	tesla (T)	**1.0×10^{-4}**
gilbert (Gi)	ampere (A)	7.957747×10^{-1}
maxwell (Mx)	weber (Wb)	**1.0×10^{-8}**
mho	siemens (S)	**1.0**
oersted (Oe)	ampere per meter (A/m)	7.957747×10
ohm centimeter (Ω·cm)	ohm meter (Ω·m)	**1.0×10^{-2}**
ohm circular-mil per foot	ohm meter (Ω·m)	1.662426×10^{-9}
ohm circular-mil per foot	ohm square millimeter per meter (Ω·mm^2/m)	1.662426×10^{-3}
statampere	ampere (A)	3.335641×10^{-10}
statcoulomb	coulomb (C)	3.335641×10^{-10}
statfarad	farad (F)	1.112650×10^{-12}
stathenry	henry (H)	8.987552×10^{11}
statmho	siemens (S)	1.112650×10^{-12}
statohm	ohm (Ω)	8.987552×10^{11}
statvolt	volt (V)	2.997925×10^{2}
unit pole	weber (Wb)	1.256637×10^{-7}
	Energy (includes work)	
British thermal unit$_{IT}$ (Btu$_{IT}$)‡	joule (J)	1.055056×10^{3}
British thermal unit$_{th}$ (Btu$_{th}$)	joule (J)	1.054350×10^{3}
British thermal unit (mean) (Btu)	joule (J)	1.05587×10^{3}
British thermal unit (39°F) (Btu)	joule (J)	1.05967×10^{3}
British thermal unit (59°F) (Btu)	joule (J)	1.05480×10^{3}
British thermal unit (60°F) (Btu)	joule (J)	1.05468×10^{3}
calorie$_{IT}$ (cal$_{IT}$)	joule (J)	**4.1868**
calorie$_{th}$ (cal$_{th}$)	joule (J)	**4.184**
calorie (mean) (cal)	joule (J)	4.19002
calorie (15°C) (cal$_{15}$)	joule (J)	4.18580
calorie (20°C) (cal$_{20}$)	joule (J)	4.18190
calorie$_{IT}$, kilogram (nutrition)	joule (J)	**4.1868×10^{3}**
calorie$_{th}$, kilogram (nutrition)	joule (J)	**4.184×10^{3}**
calorie (mean), kilogram (nutrition)	joule (J)	4.19002×10^{3}
electronvolt (eV)	joule (J)	1.602177×10^{-19}
erg (erg)	joule (J)	**1.0×10^{-7}**
foot poundal	joule (J)	4.214011×10^{-2}
foot pound-force (ft·lbf)	joule (J)	1.355818
kilocalorie$_{IT}$ (kcal$_{IT}$)	joule (J)	**4.1868×10^{3}**
kilocalorie$_{th}$ (kcal$_{th}$)	joule (J)	**4.184×10^{3}**
kilocalorie (mean) (kcal)	joule (J)	4.19002×10^{3}
kilowatt hour (kW·h)	joule (J)	**3.6×10^{6}**
kilowatt hour (kW·h)	megajoule (MJ)	**3.6**
quad (10^{15} Btu$_{IT}$)	joule (J)	1.055056×10^{18}
therm (EC)	joule (J)	**1.05506×10^{8}**
therm (U.S.)	joule (J)	**1.054804×10^{8}**
ton of TNT (energy equivalent)	joule (J)	**4.184×10^{9}**
watt hour (W·h)	joule (J)	**3.6×10^{3}**
watt second (W·s)	joule (J)	**1.0**
	Energy divided by area time	
erg per square centimeter second [erg/(cm^2·s)]	watt per square meter (W/m^2)	**1.0×10^{-3}**
watt per square centimeter (W/cm^2)	watt per square meter (W/m^2)	**1.0×10^{4}**
watt per square inch (W/in^2)	watt per square meter (W/m^2)	1.550003×10^{3}
	Force	
dyne (dyn)	newton (N)	**1.0×10^{-5}**
kilogram-force (kgf)	newton (N)	**9.80665**
kilopond (kilogram-force) (kp)	newton (N)	**9.80665**
kip (1 kip = 1000 lbf)	newton (N)	4.448222×10^{3}

TABLE 2.10 Conversion Factors to Convert to SI Units* (Continued)

To convert from	To	Multiply by
kip (1 kip = 1000 lbf)	kilonewton (kN)	4.448 222
ounce (avoirdupois)-force (ozf)	newton (N)	$2.780\,139 \times 10^{-1}$
poundal	newton (N)	$1.382\,550 \times 10^{-1}$
pound-force (lbf)	newton (N)	4.448 222
pound-force per pound (lbf/lb) (thrust to mass ratio)	newton per kilogram (N/kg)	**9.806 65**
ton-force (2000 lbf)	newton (N)	$8.896\,443 \times 10^{3}$
ton-force (2000 lbf)	kilonewton (kN)	8.896 443
	Force divided by length	
pound-force per foot (lbf/ft)	newton per meter (N/m)	$1.459\,390 \times 10$
pound-force per inch (lbf/in)	netwon per meter (N/m)	$1.751\,268 \times 10^{2}$
	Heat	
Available energy:		
British thermal unit$_{IT}$ per cubic foot (Btu_{IT}/ft^3)	joule per cubic meter (J/m^3)	$3.725\,895 \times 10^{4}$
British thermal unit$_{th}$ per cubic foot (Btu_{th}/ft^3)	joule per cubic meter (J/m^3)	$3.723\,403 \times 10^{4}$
British thermal unit$_{IT}$ per pound (Btu_{IT}/lb)	joule per kilogram (J/kg)	**2.326×10^{3}**
British thermal unit$_{th}$ per pound (Btu_{th}/lb)	joule per kilogram (J/kg)	$2.324\,444 \times 10^{3}$
calorie$_{IT}$ per gram (cal_{IT}/g)	joule per kilogram (J/kg)	**4.1868×10^{3}**
calorie$_{th}$ per gram (cal_{th}/g)	joule per kilogram (J/kg)	**4.184×10^{3}**
Coefficient of heat transfer:		
British thermal unit$_{IT}$ per hour square foot degree Fahrenheit [$Btu_{IT}/(h \cdot ft^2 \cdot °F)$]	watt per square meter kelvin [$W/(m^2 \cdot K)$]	5.678 263
British thermal unit$_{th}$ per hour square foot degree Fahrenheit [$Btu_{th}/(h \cdot ft^2 \cdot °F)$]	watt per square meter kelvin [$W/(m^2 \cdot K)$]	5.674 466
British thermal unit$_{IT}$ per second square foot degree Fahrenheit [$Btu_{IT}/(s \cdot ft^2 \cdot °F)$]	watt per square meter kelvin [$W/(m^2 \cdot K)$]	$2.044\,175 \times 10^{4}$
British thermal unit$_{th}$ per second square foot degree Fahrenheit [$Btu_{th}/(s \cdot ft^2 \cdot °F)$]	watt per square meter kelvin [$W/(m^2 \cdot K)$]	$2.042\,808 \times 10^{4}$
Density of heat:		
British thermal unit$_{IT}$ per square foot (Btu_{IT}/ft^2)	joule per square meter (J/m^2)	$1.135\,653 \times 10^{4}$
British thermal unit$_{th}$ per square foot (Btu_{th}/ft^2)	joule per square meter (J/m^2)	$1.134\,893 \times 10^{4}$
calorie$_{th}$ per square centimeter (cal_{th}/cm^2)	joule per square meter (J/m^2)	**4.184×10^{4}**
langley (cal_{th}/cm^2)	joule per square meter (J/m^2)	**4.184×10^{4}**
Density of heat flow rate:		
British thermal unit$_{IT}$ per square foot hour [$Btu_{IT}/(ft^2 \cdot h)$]	watt per square meter (W/m^2)	3.154 591
British thermal unit$_{th}$ per square foot hour [$Btu_{th}/(ft^2 \cdot h)$]	watt per square meter (W/m^2)	3.152 481
British thermal unit$_{th}$ per square foot minute [$Btu_{th}/(ft^2 \cdot min)$]	watt per square meter (W/m^2)	$1.891\,489 \times 10^{2}$
British thermal unit$_{IT}$ per square foot second [$Btu_{th}/(ft^2 \cdot s)$]	watt per square meter (W/m^2)	$1.135\,653 \times 10^{4}$
British thermal unit$_{th}$ per square foot second [$Btu_{th}/(ft^2 \cdot s)$]	watt per square meter (W/m^2)	$1.134\,893 \times 10^{4}$
British thermal unti$_{th}$ per square inch second [$Btu_{th}/(in^2 \cdot s)$]	watt per square meter (W/m^2)	$1.634\,246 \times 10^{6}$
calorie$_{th}$ per square centimeter minute [$cal_{th}/(cm^2 \cdot min)$]	watt per square meter (W/m^2)	$6.973\,333 \times 10^{2}$
calorie$_{th}$ per square centimeter second [$cal_{th}/(cm^2 \cdot s)$]	watt per square meter (W/m^2)	**4.184×10^{4}**
Fuel consumption:		
gallon (U.S.) per horsepower hour [$gal/(hp \cdot h)$]	cubic meter per joule (m^3/J)	$1.410\,089 \times 10^{-9}$

TABLE 2.10 Conversion Factors to Convert to SI Units* (Continued)

To convert from	To	Multiply by
gallon (U.S.) per horsepower hour [gal/(hp·h)]	liter per joule (L/J)	1.410089×10^{-6}
mile per gallon (U.S.) (mpg) (mi/gal)	meter per cubic meter (m/m^3)	4.251437×10^{5}
mile per gallon (U.S.) (mpg) (mi/gal)	kilometer per liter (km/L)	4.251437×10^{-1}
mile per gallon (U.S.) (mpg) (mi/gal)	liter per 100 kilometer (L/100 km)	Divide 235.215 by number of miles per gallon
pound per horsepower hour [lb/(hp·h)]	kilogram per joule (kg/J)	1.689659×10^{7}
Heat capacity and entropy:		
British thermal unit$_{IT}$ per degree Fahrenheit (Btu$_{IT}$/°F)	joule per kelvin (J/K)	1.899101×10^{3}
British thermal unit$_{th}$ per degree Fahrenheit (Btu$_{th}$/°F)	joule per kelvin (J/K)	1.897830×10^{3}
British thermal unit$_{IT}$ per degree Rankine (Btu$_{IT}$/°R)	joule per kelvin (J/K)	1.899101×10^{3}
British thermal unit$_{th}$ per degree Rankine (Btu$_{th}$/°R)	joule per kelvin (J/K)	1.897830×10^{3}
Heat flow rate:		
British thermal unit$_{IT}$ per hour (Btu$_{IT}$/h)	watt (W)	2.930711×10^{-1}
British thermal unit$_{th}$ per hour (Btu$_{th}$/h)	watt (W)	2.928751×10^{-1}
British thermal unit$_{th}$ per minute (Btu$_{th}$/min)	watt (W)	1.757250×10
British thermal unit$_{IT}$ per second (Btu$_{IT}$/s)	watt (W)	1.055056×10^{3}
British thermal unit$_{th}$ per second (Btu$_{th}$/s)	watt (W)	1.054350×10^{3}
calorie$_{th}$ per minute (cal$_{th}$/min)	watt (W)	6.973333×10^{-2}
calorie$_{th}$ per second (cal$_{th}$/s)	watt (W)	**4.184**
kilocalorie$_{th}$ per minute (kcal$_{th}$/min)	watt (W)	6.973333×10
kilocalorie$_{th}$ per second (kcal$_{th}$/s)	watt (W)	**4.184×10^{3}**
ton of refrigeration (12,000 Btu$_{IT}$/h)	watt (W)	3.516853×10^{3}
Specific heat capacity and specific entropy:		
British thermal unit$_{IT}$ per pound degree Fahrenheit [Btu$_{IT}$/(lb·°F)]	joule per kilogram kelvin [J/(kg·K)]	**4.1868×10^{3}**
British thermal unit$_{th}$ per pound degree Fahrenheit [Btu$_{th}$/(lb·°F)]	joule per kilogram kelvin [J/(kg·K)]	**4.184×10^{3}**
British thermal unit$_{IT}$ per pound degree Rankine [Btu$_{IT}$/(lb·°R)]	joule per kilogram kelvin [J/(kg·K)]	**4.1868×10^{3}**
British thermal unit$_{th}$ per pound degree Rankine [Btu$_{th}$/(lb·°R)]	joule per kilogram kelvin [J/(kg·K)]	**4.184×10^{3}**
calorie$_{IT}$ per gram degree Celsius [cal$_{IT}$/(g·°C)]	joule per kilogram kelvin [J/(kg·K)]	**4.1868×10^{3}**
calorie$_{th}$ per gram degree Celsius [cal$_{th}$/(g·°C)]	joule per kilogram kelvin [J/(kg·K)]	**4.184×10^{3}**
calorie$_{IT}$ per gram kelvin [cal$_{IT}$/(g·K)]	joule per kilogram kelvin [J/(kg·K)]	**4.1868×10^{3}**
calorie$_{th}$ per gram kelvin [cal$_{th}$/(g·K)]	joule per kilogram kelvin [J/(kg·K)]	**4.184×10^{3}**
Thermal conductivity:		
British thermal unit$_{IT}$ foot per hour square foot degree Fahrenheit [Btu$_{IT}$·ft/(h·ft^2·°F)]	watt per meter kelvin [W/(m·K)]	1.730735
British thermal unit$_{th}$ foot per hour square foot degree Fahrenheit [Btu$_{th}$·ft/(h·ft^2·°F)]	watt per meter kelvin [W/(m·K)]	1.729577
British thermal unit$_{IT}$ inch per hour square foot degree Fahrenheit [Btu$_{IT}$·in/(h·ft^2·°F)]	watt per meter kelvin [W/(m·K)]	1.442279×10
British thermal unit$_{th}$ inch per hour square foot degree Fahrenheit [Btu$_{th}$·in/(h·ft^2·°F)]	watt per meter kelvin [W/(m·K)]	1.441314×10
British thermal unit$_{IT}$ inch per second square foot degree Fahrenheit [Btu$_{IT}$·in/(s·ft^2·°F)]	watt per meter kelvin [W/(m·K)]	5.192204×10^{2}
British thermal unit$_{th}$ inch per second square foot degree Fahrenheit [Btu$_{th}$·in/(s·ft^2·°F)]	watt per meter kelvin [W/(m·K)]	5.188732×10^{2}
calorie$_{th}$ per centimeter second degree Celsius [cal$_{th}$/(cm·s·°C)]	watt per meter kelvin [W/(m·K)]	**4.184×10^{2}**
Thermal diffusivity:		
square foot per hour (ft^2/h)	square meter per second (m^2/s)	**2.58064×10^{-5}**
Thermal insulance:		
clo	square meter kelvin per watt (m^2·K/W)	1.55×10^{-1}

TABLE 2.10 Conversion Factors to Convert to SI Units* (Continued)

To convert from	To	Multiply by
degree Fahrenheit hour square foot per British thermal unit$_{IT}$ (°F·h·ft^2/Btu$_{IT}$)	square meter kelvin per watt (m^2·K/W)	1.761102×10^{-1}
degree Fahrenheit hour square foot per British thermal unit$_{th}$ (°F·h·ft^2/Btu$_{th}$)	square meter kelvin per watt (m^2·K/W)	1.762280×10^{-1}
Thermal resistance:		
degree Fahrenheit hour per British thermal unit$_{IT}$ (°F·h/Btu$_{IT}$)	kelvin per watt (K/W)	1.895634
degree Fahrenheit hour per British thermal unit$_{th}$ (°F·h/Btu$_{th}$)	kelvin per watt (K/W)	1.896903
degree Fahrenheit second per British thermal unit$_{IT}$ (°F·s/Btu$_{IT}$)	kelvin per watt (K/W)	5.265651×10^{-4}
degree Fahrenheit second per British thermal unit$_{th}$ (°F·s/Btu$_{th}$)	kelvin per watt (K/W)	5.269175×10^{-4}
Thermal resistivity:		
degree Fahrenheit hour square foot per British thermal unit$_{IT}$ inch [°F·h·ft^2/(Btu$_{IT}$·in)	meter kelvin per watt (m·K/W)	6.933472
degree Fahrenheit hour square foot per British thermal unit$_{th}$ inch [°F·h·ft^2/(Btu$_{th}$·in)]	meter kelvin per watt (m·K/W)	6.938112×10^{4}
	Length	
ångström (Å)	meter (m)	**1.0×10^{-10}**
ångström (Å)	nanometer (nm)	**1.0×10^{-1}**
astronomical unit (AU)	meter (m)	1.495979×10^{11}
chain (based on U.S. survey foot) (ch)	meter (m)	2.011684×10
fathom (based on U.S. survey foot)	meter (m)	1.828804
fermi	meter (m)	**1.0×10^{-15}**
fermi	femtometer (fm)	**1.0**
foot (ft)	meter (m)	**3.048×10^{-1}**
foot (U.S. survey) (ft)	meter (m)	3.048006×10^{-1}
inch (in)	meter (m)	**2.54×10^{-2}**
inch (in)	centimeter (cm)	**2.54**
kayser (K)	reciprocal meter (m^{-1})	**1×10^{2}**
light year (l.y.)	meter (m)	9.46073×10^{15}
microinch	meter (m)	**2.54×10^{-8}**
microinch	micrometer (μm)	**2.54×10^{-2}**
micron (μ)	meter (m)	**1.0×10^{-6}**
micron (μ)	micrometer (μm)	**1.0**
mil (0.001 in)	meter (m)	**2.54×10^{-5}**
mil (0.001 in)	millimeter (mm)	**2.54×10^{-2}**
mile (mi)	meter (m)	**1.609344×10^{3}**
mile (mi)	kilometer (km)	**1.609344**
mile (based on U.S. survey foot) (mi)	meter (m)	1.609347×10^{3}
mile (based on U.S. survey foot) (mi)	kilometer (km)	1.609347
mile, nautical	meter (m)	**1.852×10^{3}**
parsec (pc)	meter (m)	3.085678×10^{16}
pica (computer) (1/6 in)	meter (m)	4.233333×10^{-3}
pica (computer) (1/6 in)	millimeter (mm)	4.233333
pica (printer's)	meter (m)	4.217518×10^{-3}
pica (printer's)	millimeter (mm)	4.217518
point (computer) (1/72 in)	meter (m)	3.527778×10^{-4}
point (computer) (1/72 in)	millimeter (mm)	3.527778×10^{-1}
point (printer's)	meter (m)	3.514598×10^{-4}
point (printer's)	millimeter (mm)	3.514598×10^{-1}
rod (based on U.S. survey foot) (rd)	meter (m)	5.029210
yard (yd)	meter (m)	**9.144×10^{-1}**
	Light	
candela per square inch (cd/in^2)	candela per square meter (cd/m^2)	1.550003×10^{3}
footcandle	lux (lx)	1.076391×10
footlambert	candela per square meter (cd/m^2)	3.426259
lambert	candela per square meter (cd/m^2)	3.183099×10^{3}

TABLE 2.10 Conversion Factors to Convert to SI Units* (Continued)

To convert from	To	Multiply by
lumen per square foot (lm/ft^2)	lux (lx)	1.076391×10
phot (ph)	lux (lx)	$\mathbf{1.0 \times 10^4}$
stilb (sb)	candela per square meter (cd/m^2)	$\mathbf{1.0 \times 10^4}$
	Mass and moment of inertia	
carat, metric	kilogram (kg)	$\mathbf{2.0 \times 10^{-4}}$
carat, metric	gram (g)	$\mathbf{2.0 \times 10^{-1}}$
grain (gr)	kilogram (kg)	$\mathbf{6.479891 \times 10^{-5}}$
grain (gr)	milligram (mg)	$\mathbf{6.479891 \times 10}$
hundredweight (long, 112 lb)	kilogram (kg)	5.080235×10
hundredweight (short, 100 lb)	kilogram (kg)	4.535924×10
kilogram-force second squared per meter ($kgf{\cdot}s^2/m$)	kilogram (kg)	**9.80665**
ounce (avoirdupois) (oz)	kilogram (kg)	2.834952×10^{-2}
ounce (avoirdupois) (oz)	gram (g)	2.834952×10
ounce (troy or apothecary) (oz)	kilogram (kg)	3.110348×10^{-2}
ounce (troy or apothecary) (oz)	gram (g)	3.110348×10
pennyweight (dwt)	kilogram (kg)	1.555174×10^{-3}
pennyweight (dwt)	gram (g)	1.555174
pound (avoirdupois) (lb)	kilogram (kg)	4.535924×10^{-1}
pound (troy or apothecary) (lb)	kilogram (kg)	3.732417×10^{-1}
pound foot squared ($lb{\cdot}ft^2$)	kilogram meter squared ($kg{\cdot}m^2$)	4.214011×10^{-2}
pound inch squared ($lb{\cdot}in^2$)	kilogram meter squared ($kg{\cdot}m^2$)	2.926397×10^{-4}
slug (slug)	kilogram (kg)	1.459390×10
ton, assay (AT)	kilogram (kg)	2.916667×10^{-2}
ton, assay (AT)	gram (g)	2.916667×10
ton, long (2240 lb)	kilogram (kg)	1.016047×10^3
ton, metric (t)	kilogram (kg)	$\mathbf{1.0 \times 10^3}$
tonne (called "metric ton" in U.S.) (t)	kilogram (kg)	$\mathbf{1.0 \times 10^3}$
ton, short (2000 lb)	kilogram (kg)	9.071847×10^2
	Mass divided by area	
ounce (avoirdupois) per square foot (oz/ft^2)	kilogram per square meter (kg/m^2)	3.051517×10^{-1}
ounce (avoirdupois) per square inch (oz/in^2)	kilogram per square meter (kg/m^2)	4.394185×10
ounce (avoirdupois) per square yard (oz/yd^2)	kilogram per square meter (kg/m^2)	3.390575×10^{-2}
pound per square foot (lb/ft^2)	kilogram per square meter (kg/m^2)	4.882428
pound per square inch (*not* pound force) (lb/in^2)	kilogram per square meter (kg/m^2)	7.030696×10^2
	Mass divided by length	
denier	kilogram per meter (kg/m)	1.111111×10^{-7}
denier	gram per meter (g/m)	1.111111×10^{-4}
pound per foot (lb/ft)	kilogram per meter (kg/m)	1.488164
pound per inch (lb/in)	kilogram per meter (kg/m)	1.785797×10
pound per yard (lb/yd)	kilogram per meter (kg/m)	4.960546×10^{-1}
tex	kilogram per meter (kg/m)	$\mathbf{1.0 \times 10^{-6}}$
	Mass divided by time (includes flow)	
pound per hour (lb/h)	kilogram per second (kg/s)	1.259979×10^{-4}
pound per minute (lb/min)	kilogram per second (kg/s)	7.559873×10^{-3}
pound per second (lb/s)	kilogram per second (kg/s)	4.535924×10^{-1}
ton, short, per hour	kilogram per second (kg/s)	2.519958×10^{-1}
	Mass divided by volume (includes mass density and mass concentration)	
grain per gallon (U.S.) (gr/gal)	kilogram per cubic meter (kg/m^3)	1.711806×10^{-2}
grain per gallon (U.S.) (gr/gal)	milligram per liter (mg/L)	1.711806×10
gram per cubic centimeter (g/cm^3)	kilogram per cubic meter (kg/m^3)	$\mathbf{1.0 \times 10^3}$
ounce (avoirdupois) per cubic inch (oz/in^3)	kilogram per cubic meter (kg/m^3)	1.729994×10^3
ounce (avoirdupois) per gallon [Canadian and U.K. (Imperial)] (oz/gal)	kilogram per cubic meter (kg/m^3)	6.236023

TABLE 2.10 Conversion Factors to Convert to SI Units* (Continued)

To convert from	To	Multiply by
ounce (avoirdupois) per gallon [Canadian and U.K. (Imperial)] (oz/gal)	gram per liter (g/L)	6.236023
ounce (avoirdupois) per gallon (U.S.) (oz/gal)	kilogram per cubic meter (kg/m^3)	7.489152
ounce (avoirdupois) per gallon (U.S.) (oz/gal)	gram per liter (g/L)	7.489152
pound per cubic foot (lb/ft^3)	kilogram per cubic meter (kg/m^3)	1.601846×10
pound per cubic inch (lb/in^3)	kilogram per cubic meter (kg/m^3)	2.767990×10^4
pound per cubic yard (lb/yd^3)	kilogram per cubic meter (kg/m^3)	5.932764×10^{-1}
pound per gallon [Canadian and U.K. (Imperial)] (lb/gal)	kilogram per cubic meter (kg/m^3)	9.977637×10
pound per gallon [Canadian and U.K. (Imperial)] (lb/gal)	kilogram per liter (kg/L)	9.977637×10^{-2}
pound per gallon (U.S.) (lb/gal)	kilogram per cubic meter (kg/m^3)	1.198264×10^2
pound per gallon (U.S.) (lb/gal)	kilogram per liter (kg/L)	1.198264×10^{-1}
slug per cubic foot ($slug/ft^3$)	kilogram per cubic meter (kg/m^3)	5.153788×10^2
ton, long, per cubic yard	kilogram per cubic meter (kg/m^3)	1.328939×10^3
ton, short, per cubic yard	kilogram per cubic meter (kg/m^3)	1.186553×10^3
	Moment of force or torque	
dyne centimeter (dyn·cm)	newton meter (N·m)	**1.0×10^{-7}**
kilogram-force meter (kgf·m)	newton meter (N·m)	**9.80665**
ounce (avoirdupois)-force inch (ozf·in)	newton meter (N·m)	7.061552×10^{-3}
ounce (avoirdupois)-force inch (ozf·in)	millinewton meter (mN·m)	7.061552
pound-force foot (lbf·ft)	newton meter (N·m)	1.355818
pound-force inch (lbf·in)	newton meter (N·m)	1.129848×10^{-1}
	Moment of force or torque, divided by length	
pound-force foot per inch (lbf·ft/in)	newton meter per meter (N·m/m)	5.337866×10
pound-force inch per inch (lbf·in/in)	newton meter per meter (N·m/m)	4.448222
	Permeability	
darcy	meter squared (m^2)	9.869233×10^{-13}
perm (0°C)	kilogram per pascal second square meter [$kg/(Pa \cdot s \cdot m^2)$]	5.72135×10^{-11}
perm (23°C)	kilogram per pascal second square meter [$kg/(Pa \cdot s \cdot m^2)$]	5.74525×10^{-11}
perm inch (0°C)	kilogram per pascal second meter [kg/(Pa·s·m)]	1.45322×10^{-12}
perm inch (23°C)	kilogram per pascal second meter [kg/(Pa·s·m)]	1.45929×10^{-12}
	Power	
erg per second (erg/s)	watt (W)	**1.0×10^{-7}**
foot pound-force per hour (ft·lbf/h)	watt (W)	3.766161×10^{-4}
foot pound-force per minute (ft·lbf/min)	watt (W)	2.259697×10^{-2}
foot pound-force per second (ft·lbf/s)	watt (W)	1.355818
horsepower (550 ft·lbf/s)	watt (W)	7.456999×10^2
horsepower (boiler)	watt (W)	9.80950×10^3
horsepower (electric)	watt (W)	**7.46×10^2**
horsepower (metric)	watt (W)	7.354988×10^2
horsepower (U.K.)	watt (W)	7.4570×10^2
horsepower (water)	watt (W)	7.46043×10^2
	Pressure or stress (force divided by area)	
atmosphere, standard (atm)	pascal (Pa)	**1.01325×10^5**
atmosphere, standard (atm)	kilopascal (kPa)	**1.01325×10^2**
atmosphere, technical (at)	pascal (Pa)	**9.80665×10^4**
atmosphere, technical (at)	kilopascal (kPa)	**9.80665×10**
bar (bar)	pascal (Pa)	**1.0×10^5**
bar (bar)	kilopascal (kPa)	**1.0×10^2**
centimeter of mercury (0°C)	pascal (Pa)	1.33322×10^3

TABLE 2.10 Conversion Factors to Convert to SI Units* (Continued)

To convert from	To	Multiply by
centimeter of mercury (0°C)	kilopascal (kPa)	1.33322
centimeter of mercury, conventional (cmHg)	pascal (Pa)	1.333224×10^3
centimeter of mercury, conventional (cmHg)	kilopascal (kPa)	1.333224
centimeter of water (4°C)	pascal (Pa)	9.80638×10
centimeter of water, conventional (cmH_2O)	pascal (Pa)	$\mathbf{9.80665 \times 10}$
dyne per square centimeter (dyn/cm^2)	pascal (Pa)	$\mathbf{1.0 \times 10^{-1}}$
foot of mercury, conventional (ftHg)	pascal (Pa)	4.063666×10^4
foot of mercury, conventional (ftHg)	kilopascal (kPa)	4.063666×10
foot of water (39.2°F)	pascal (Pa)	2.98898×10^3
foot of water (39.2°F)	kilopascal (kPa)	2.98898
foot of water, conventional (ftH_2O)	pascal (Pa)	2.989067×10^3
foot of water, conventional (ftH_2O)	kilopascal (kPa)	2.989067
gram-force per square centimeter (gf/cm^2)	pascal (Pa)	$\mathbf{9.80665 \times 10}$
inch of mercury (32°F)	pascal (Pa)	3.38638×10^3
inch of mercury (32°F)	kilopascal (kPa)	3.38638
inch of mercury (60°F)	pascal (Pa)	3.37685×10^3
inch of mercury (60°F)	kilopascal (kPa)	3.37685
inch of mercury, conventional (inHg)	pascal (Pa)	3.386389×10^3
inch of mercury, conventional (inHg)	kilopascal (kPa)	3.386389
inch of water (39.2°F)	pascal (Pa)	2.49082×10^2
inch of water (60°F)	pascal (Pa)	2.4884×10^2
inch of water, conventional (inH_2O)	pascal (Pa)	2.490889×10^2
kilogram-force per square centimeter (kgf/cm^2)	pascal (Pa)	$\mathbf{9.80665 \times 10^4}$
kilogram-force per square centimeter (kgf/cm^2)	kilopascal (kPa)	$\mathbf{9.80665 \times 10}$
kilogram-force per square meter (kgf/m^2)	pascal (Pa)	**9.80665**
kilogram-force per square millimeter (kgf/mm^2)	pascal (Pa)	$\mathbf{9.80665 \times 10^6}$
kilogram-force per square millimeter (kgf/mm^2)	megapascal (MPa)	**9.80665**
kip per square inch (ksi) (kip/in^2)	pascal (Pa)	6.894757×10^6
kip per square inch (ksi) (kip/in^2)	kilopascal (kPa)	6.894757×10^3
millibar (mbar)	pascal (Pa)	$\mathbf{1.0 \times 10^2}$
millibar (mbar)	kilopascal (kPa)	$\mathbf{1.0 \times 10^{-1}}$
millimeter of mercury, conventional (mmHg)	pascal (Pa)	1.333224×10^2
millimeter of water, conventional (mmH_2O)	pascal (Pa)	**9.80665**
poundal per square foot	pascal (Pa)	1.488164
pound-force per square foot (lbf/ft^2)	pascal (Pa)	4.788026×10
pound-force per square inch (psi) (lbf/in^2)	pascal (Pa)	6.894757×10^3
pound-force per square inch (psi) (lbf/in^2)	kilopascal (kPa)	6.894757
psi (pound-force per square inch) (lbf/in^2)	pascal (Pa)	6.894757×10^3
psi (pound-force per square inch) (lbf/in^2)	kilopascal (kPa)	6.894757
torr (torr)	pascal (Pa)	1.333224×10^2
	Radiology	
curie (Ci)	becquerel (Bq)	$\mathbf{3.7 \times 10}$
rad (absorbed dose) (rad)	gray (Gy)	$\mathbf{1.0 \times 10^{-2}}$
rem (rem)	sievert (Sv)	$\mathbf{1.0 \times 10^{-2}}$
roentgen (R)	coulomb per kilogram (C/kg)	$\mathbf{2.58 \times 10^{-4}}$
	Temperature	
degree Celsius (°C)	kelvin (K)	$T/\text{K} = t/°\text{C} + \mathbf{273.15}$
degree centigrade	degree Celsius (°C)	t (°C) ≈ t (deg. cent.)
degree Fahrenheit (°F)	degree Celsius (°C)	$t/°\text{C} = (t/°\text{F} - \mathbf{32})/\mathbf{1.8}$
degree Fahrenheit (°F)	kelvin (K)	$T/\text{K} = (t/°\text{F} + \mathbf{459.67})/\mathbf{1.8}$
degree Rankine (°R)	kelvin (K)	$T/\text{K} = (T/°\text{R})/\mathbf{1.8}$
kelvin (K)	degree Celsius (°C)	$t/°\text{C} = T/\text{K} - \mathbf{273.15}$

TABLE 2.10 Conversion Factors to Convert to SI Units* (Continued)

To convert from	To	Multiply by
	Temperature interval	
degree Celsius (°C)	kelvin (K)	**1.0**
degree centigrade	degree Celsius (°C)	1.0
degree Fahrenheit (°F)	degree Celsius (°C)	5.555556×10^{-1}
degree Fahrenheit (°F)	kelvin (K)	5.555556×10^{-1}
degree Rankine (°R)	kelvin (K)	5.555556×10^{-1}
	Time	
day (d)	second (s)	$\mathbf{8.64 \times 10^4}$
day (sidereal)	second (s)	8.616409×10^4
hour (h)	second (s)	$\mathbf{3.6 \times 10^3}$
hour (sidereal)	second (s)	3.590170×10^3
minute (min)	second (s)	$\mathbf{6.0 \times 10}$
minute (sidereal)	second (s)	5.983617×10
second (sidereal)	second (s)	9.972696×10^{-1}
shake	second (s)	$\mathbf{1.0 \times 10^{-8}}$
shake	nanosecond (ns)	$\mathbf{1.0 \times 10}$
year (365 days)	second (s)	$\mathbf{3.1536 \times 10^7}$
year (sidereal)	second (s)	3.155815×10^7
year (tropical)	second (s)	3.155693×10^7
	Velocity (includes speed)	
foot per hour (ft/h)	meter per second (m/s)	8.466667×10^{-5}
foot per minute (ft/min)	meter per second (m/s)	$\mathbf{5.08 \times 10^{-3}}$
foot per second (ft/s)	meter per second (m/s)	$\mathbf{3.048 \times 10^{-1}}$
inch per second (in/s)	meter per second (m/s)	$\mathbf{2.54 \times 10^{-2}}$
kilometer per hour (km/h)	meter per second (m/s)	2.777778×10
knot (nautical mile per hour)	meter per second (m/s)	5.144444×10
mile per hour (mi/h)	meter per second (m/s)	$\mathbf{4.4704 \times 10}$
mile per hour (mi/h)	kilometer per hour (km/h)	**1.609344**
mile per minute (mi/min)	meter per second (m/s)	$\mathbf{2.68224 \times 10}$
mile per second (mi/s)	meter per second (m/s)	$\mathbf{1.609344 \times 10^3}$
revolution per minute (rpm) (r/min)	radian per second (rad/s)	1.047198×10^{-1}
rpm (revolution per minute) (r/min)	radian per second (rad/s)	1.047198×10^{-1}
	Viscosity, dynamic	
centipoise (cP)	pascal second (Pa·s)	$\mathbf{1.0 \times 10^{-3}}$
poise (P)	pascal second (Pa·s)	$\mathbf{1.0 \times 10^{-1}}$
poundal second per square foot	pascal second (Pa·s)	1.488164
pound-force second per square foot ($lbf{\cdot}s/ft^2$)	pascal second (Pa·s)	4.788026×10
pound-force second per square inch ($lbf{\cdot}s/in^2$)	pascal second (Pa·s)	6.894757×10^3
pound per foot hour [lb/(ft·h)]	pascal second (Pa·s)	4.133789×10^{-4}
pound per foot second [lb/(ft·s)]	pascal second (Pa·s)	1.488164
rhe	reciprocal pascal second [$(Pa{\cdot}s)^{-1}$]	$\mathbf{1.0 \times 10}$
slug per foot second [slug/(ft·s)]	pascal second (Pa·s)	4.788026×10
	Viscosity, kinematic	
centistokes (cSt)	meter squared per second (m^2/s)	$\mathbf{1.0 \times 10^{-6}}$
square foot per second (ft^2/s)	meter squared per second (m^2/s)	$\mathbf{9.290304 \times 10^{-2}}$
stokes (St)	meter squared per second (m^2/s)	$\mathbf{1.0 \times 10^{-4}}$
	Volume (includes capacity)	
acre-foot (based on U.S. survey foot)	cubic meter (m^3)	1.233489×10^3
barrel [for petroleum, 42 gallons (U.S.)] (bbl)	cubic meter (m^3)	1.589873×10^{-1}
barrel [for petroleum, 42 gallons (U.S.)] (bbl)	liter (L)	1.589873×10^2
bushel (U.S.) (bu)	cubic meter (m^3)	3.523907×10^{-2}
bushel (U.S.) (bu)	liter (L)	3.523907×10
cord ($128 ft^3$)	cubic meter (m^3)	3.624556

TABLE 2.10 Conversion Factors to Convert to SI Units* (Continued)

To convert from	To	Multiply by
cubic foot (ft^3)	cubic meter (m^3)	2.831685×10^{-2}
cubic inch (in^3)	cubic meter (m^3)	1.638706×10^{-5}
cubic mile (mi^3)	cubic meter (m^3)	4.168182×10^{9}
cubic yard (yd^3)	cubic meter (m^3)	7.645549×10^{-1}
cup (U.S.)	cubic meter (m^3)	2.365882×10^{-4}
cup (U.S.)	liter (L)	2.365882×10^{-1}
cup (U.S.)	milliliter (mL)	2.365882×10^{2}
fluid ounce (U.S.) (fl oz)	cubic meter (m^3)	2.957353×10^{-5}
fluid ounce (U.S.) (fl oz)	milliliter (mL)	2.957353×10
gallon [Canadian and U.K. (Imperial)] (gal)	cubic meter (m^3)	**4.54609×10^{-5}**
gallon [Canadian and U.K. (Imperial)] (gal)	liter (L)	**4.54609**
gallon (U.S.) (gal)	cubic meter (m^3)	3.785412×10^{-3}
gallon (U.S.) (gal)	liter (L)	3.785412
gill [Canadian and U.K. (Imperial)] (gi)	cubic meter (m^3)	1.420653×10^{-4}
gill [Canadian and U.K. (Imperial)] (gi)	liter (L)	1.420653×10^{-1}
gill (U.S.) (gi)	cubic meter (m^3)	1.182941×10^{-4}
gill (U.S.) (gi)	liter (L)	1.182941×10^{-1}
liter (L)	cubic meter (m^3)	**1.0×10^{-3}**
ounce [Canadian and U.K. fluid (Imperial)] (fl oz)	cubic meter (m^3)	2.841306×10^{-5}
ounce [Canadian and U.K. fluid (Imperial)] (fl oz)	milliliter (mL)	2.841306×10
ounce (U.S. fluid) (fl oz)	cubic meter (m^3)	2.957353×10^{-5}
ounce (U.S. fluid) (fl oz)	milliliter (mL)	2.957353×10
peck (U.S.) (pk)	cubic meter (m^3)	8.809768×10^{-3}
peck (U.S.) (pk)	liter (L)	8.809768
pint (U.S. dry) (dry pt)	cubic meter (m^3)	5.506105×10^{-4}
pint (U.S. dry) (dry pt)	liter (L)	5.506105×10^{-1}
pint (U.S. liquid) (liq pt)	cubic meter (m^3)	4.731765×10^{4}
pint (U.S. liquid) (liq pt)	liter (L)	4.731765×10^{-1}
quart (U.S. dry) (dry qt)	cubic meter (m^3)	1.101221×10^{-3}
quart (U.S. dry) (dry qt)	liter (L)	1.101221
quart (U.S. liquid) (liq qt)	cubic meter (m^3)	9.463529×10^{-4}
quart (U.S. liquid) (liq qt)	liter (L)	9.463529×10^{-1}
stere (st)	cubic meter (m^3)	**1.0**
tablespoon	cubic meter (m^3)	1.478676×10^{-5}
tablespoon	milliliter (mL)	1.478676×10
teaspoon	cubic meter (m^3)	4.928922×10^{-6}
teaspoon	milliliter (mL)	4.928922
ton, register	cubic meter (m^3)	2.831685
	Volume divided by time (includes flow)	
cubic foot per minute (ft^3/min)	cubic meter per second (m^3/s)	4.719474×10^{-4}
cubic foot per minute (ft^3/min)	liter per second (L/s)	4.719474×10^{-1}
cubic foot per second (ft^3/s)	cubic meter per second (m^3/s)	2.831685×10^{-2}
cubic inch per minute (in^3/min)	cubic meter per second (m^3/s)	2.731177×10^{-7}
cubic yard per minute (yd^3/min)	cubic meter per second (m^3/s)	1.274258×10^{-2}
gallon (U.S.) per day (gal/d)	cubic meter per second (m^3/s)	4.381264×10^{-8}
gallon (U.S.) per day (gal/d)	liter per second (L/s)	4.381264×10^{-5}
gallon (U.S.) per minute (gpm) (gal/min)	cubic meter per second (m^3/s)	6.309020×10^{-5}
gallon (U.S.) per minute (gpm) (gal/min)	liter per second (L/s)	6.309020×10^{-2}

* EMU = electromagnetic units.
† ESU = electrostatic units.
‡ IT: International Table.
th: Thermochemical.

SUMMARY OF SYSTEMS OF UNITS

Table 2.11 provides a summary of systems of units that are in use. The International System of Units (SI) has been adopted by most countries and it is gaining momentum in the United States. It is expected that in time all units in the United States will be in SI form.

TABLE 2.11 System of Units

Quantity	British "gravitational system"	Metric "gravitational system"	CGS system	SI
Length	1 ft	1 m	1 cm	1 m
Mass	1 slug		1 g	1 kg
Force	1 lb	1 kg	1 dyne	1 N
Time	1 s	1 s	1 s	1 s

Units of Temperature

The relative "hotness" or "coldness" of a body is denoted by the term *temperature*. The most commonly used scales for measuring temperature are the Fahrenheit and Celsius (formerly centigrade). These thermometer scales have as their base the melting and boiling points of water, both at atmospheric pressure. The relation of the Fahrenheit and Celsius scales is as follows:

Temperature scale	Absolute zero	Freezing point of water	Boiling point of water
Fahrenheit	–459.6°	32°	212°
Celsius	–273.16°	0°	100°

The relationship between the two scales is:

$$\text{Degrees C} = \tfrac{5}{9}(\text{degrees F} - 32) \qquad \text{and} \qquad \text{Degrees F} = \tfrac{9}{5}(\text{degrees C}) + 32$$

where C is the reading on the Celsius scale and F is the reading on the Fahrenheit scale.

In certain calculations, it is necessary to express the temperature in "absolute" units. The absolute temperature associated with the Fahrenheit scale is called the Rankine temperature, denoted by degrees Rankine (°R), and it has the following association:

$$\text{Degrees R} = \text{degrees F} + 459.6$$

At times 459.6 is rounded to 460 when the accuracy of the results is not of serious concern.

The absolute temperature associated with the Celsius scale is termed degrees Kelvin (K). This relationship is expressed as follows:

$$\text{Kelvins} = \text{degrees C} + 273.16$$

For sake of approximation, 273.16 is usually rounded off to 273.

Units of Terrestrial Gravity

Standard acceleration of gravity, g^0, is equal to 9.80665 meters per second per second, or 32.1740 feet per second per second. This value of g^0 is assumed to be the value of acceleration due to gravity at sea level and latitude 45°.

Unit Conversion Factors

The American Society for Testing and Materials (ASTM) standard E 380, Standard Practice for Use of the International System of Units (SI) (the modernized metric system), provides conversion factors for converting from various systems of units to SI units. The conversion factors are listed as a number greater than 1 and less than 10 with six or fewer decimal places. The number is followed by the power of 10 by which the number must be multiplied to obtain the correct value. For example:

EXAMPLE 1 To convert from atmospheres (normal) to pascals (Pa), multiply by the conversion factor of 1.01325×10^5.

$$\begin{aligned} 1 \text{ atmosphere (normal)} &= 1 \times 1.01325 \times 10^5 \text{ Pa} \\ &= 101{,}325 \text{ Pa} \end{aligned}$$

EXAMPLE 2 To convert 1 bar to pascals (Pa), multiply by 1.000000×10^5.

$$\begin{aligned} 1 \text{ bar} &= 1 \times 1.000000 \times 10^5 \text{ Pa} \\ &= 10{,}000 \text{ Pa} \end{aligned}$$

Refer to Table 2.10 for SI conversion factors.

Conversion Factors for Commonly Used Units for Piping

Tables 2.12 through 2.22 provide conversion factors for various quantities associated with piping. Also reference may be made to Table 2.10 for conversion factors for SI units.

Frequently Used USCS Units to SI Units. Refer to Table 2.12 for conversion factors for frequently used USCS units to SI units.

TABLE 2.12 Conversion Factors—Frequently Used U.S. Customary Units to SI Standard Units

To convert	To	Multiply by	To convert	To	Multiply by
Acceleration			Power		
feet per second sq	meters per second sq	0.3048	Btu per second	watt	1054.350
Area					
square feet	square meters	0.09290304	foot pounds per second	watt	1.355818
Energy			horsepower	watt	746.
Btu	joule	1055.06	Pressure		
calorie	joule	4.19002	atmosphere	newtons per sq meter	101325.0
foot pound	joule	1.355818	bar	newtons per sq meter	100000.
watthour	joule	3600.355818	kilogram per sq cm	newtons per sq meter	98066.50
Force			pounds per sq in.	newtons per sq meter	6894.757
dyne	newton	0.00001	torr (mm Hg 0°C)	newtons per sq meter	133.322
kilogram	newton	9.80665			
pound	newton	4.448222	Viscosity		
Length			centipoise	newton second per meter	0.001
Foot	meter	0.3048000			
mile (U.S. statute)	meter	1609.344	pounds per foot second	newton second per meter	1.488164
Mass			Volume		
pound	kilogram	0.4535924	cubic foot	cubic meter	0.02831685
slug	kilogram	14.59390	gallon (U.S. liquid)	cubic meter	0.003785412
ton (2000 lb)	kilogram	907.1847			

Mass Equivalents. Refer to Table 2.13 for conversion factors for units of mass.

Length Equivalents. Refer to Table 2.14 for conversion factors for units of length.

Area Equivalents. Refer to Table 2.15 for conversion factors for units of area.

TABLE 2.13 Mass Equivalents

To obtain →, multiply ↓, by ↘	pound (avdp)	ounce (avdp)	gram	kilogram	slug	ton (short)*
pound (avdp)	1	16	453.6	0.4536	0.0310	0.0005
ounce (avdp)	0.0625	1	28.3495		0.0019	
gram	0.0022	0.0353	1	0.001	68.5×10^{-5}	
kilogram	2.2046	35.274	1000	1	0.0685	0.0011
slug	32.174	514.784	$1.459 \times 10^{+4}$	14.5939	1	
ton (short)*	2000			907.185		1

* 1 short ton = 0.8926 long tons; 1 short ton = 0.9072 metric tons.

TABLE 2.14 Length Equivalents

To obtain →, multiply ↓, by ↘	inch	foot	miles (statute)	millimeter	centimeter	meter	kilometer
inch	1	0.08333		25.4	2.54	0.0254	
foot	12	1		304.8	30.48	0.3048	
mile (statute)	63360	5280	1			1609.344	1.609344
millimeter	0.03937	0.003281		1	0.1	0.001	
centimeter	0.3937	0.032808		10	1	0.01	
meter	39.3701	3.28084		1000	100	1	0.001
kilometer	39,370	3280.8	0.62137		100,000	1000	1

TABLE 2.15 Area Equivalents

To obtain →, multiply ↓, by ↘	square inch	square foot	acre	square millimeter	square centimeter	square meter
square inch	1	0.006944		645.16	6.4516	0.00064516
square foot	144	1	2.2956×10^{-5}	92903.04	929.0304	0.09290
acre		43560	1			4046.8564
square millimeter	0.00155			1	0.01	1×10^{-6}
square centimeter	0.1550	0.001076	2.5×10^{-8}	100	1	0.0001
square meter	1550.0031	10.76391	0.000247	1×10^{6}	10,000	1

Volume Equivalents. Refer to Table 2.16 for conversion factors for units of volume.

Volumetric Flow Rate Equivalents. Refer to Table 2.17 for conversion factors for units of flow rate.

TABLE 2.16 Volume Equivalents

To obtain →, multiply ↓, by ↘	U.S. gallon	Imperial gallon	cubic inch	cubic foot	liter	cubic meter	barrel (oil)*
U.S. gallon	1	0.83267	231	0.13368	3.7853	0.00378	0.02381
Imperial gallon	1.2009	1	277.42	0.16054	4.5459	0.00455	0.02859
cubic inch	0.004329	0.003604	1	0.000579	0.0164	0.000016	0.00010
cubic foot	7.4805	6.2288	1728	1	28.316	0.02832	0.17813
liter	0.26418	0.21997	61.025	0.0353	1	0.001	0.00629
cubic meter	264.17	219.97	61023.74	35.3147	1000	1	6.2899
barrel (oil)*	42	34.977	9702.00	5.614	158.983	0.15876	1

* The capacity of a barrel varies with industry as follows:
1 barrel of beer = 31 U.S. gallons
1 barrel of wine = 31.5 U.S. gallons
1 barrel of oil = 42 U.S. gallons
1 barrel of whiskey = 45 U.S. gallons

TABLE 2.17 Volumetric Flow Rate Equivalents

To obtain →, multiply ↓, by ↘	U.S. gallons per minute	Imperial gallons per minute	U.S. million gallons per day	cubic feet per second	cubic meters per hour	liter per second	barrel (42 gallons) per minute	barrel (42 gallons) per day
U.S. gallons per minute	1	0.8327	0.00144	0.00223	0.2271	0.0631	0.0238	34.286
Imperial gallons per minute	1.201	1	0.00173	0.002676	0.2727	0.0758	0.02859	41.176
U.S. million gallons per day	694.4	578.25	1	1.547	157.7	43.8	16.53	23810
cubic feet per second	448.83	373.7	0.646	1	101.9	28.32	10.686	15388
cubic meters per second	15850	13199	22.83	35.315	3600	1000	377.4	543447
cubic meters per minute	264.2	220	0.3804	0.5886	60.0	16.667	6.290	9058
cubic meters per hour	4.403	3.67	0.00634	0.00982	1	0.2778	0.1048	151
liter per second	15.85	13.20	0.0228	0.0353	3.60	1	0.3773	543.3
liter per minute	0.2642	0.220	0.000380	0.000589	0.060	0.0167	0.00629	9.055
barrel (42 gallons) per minute	42	34.97	0.0605	0.09357	9.5256	2.65	1	1440
barrel (42 gallons) per day	0.0292	0.0243	0.000042	0.000065	0.00662	0.00184	0.00069	1

Density Equivalents. Refer to Table 2.18 for conversion factors for units of density.

Pressure Equivalents. Refer to Table 2.19 for conversion factors for units of pressure.

TABLE 2.18 Density Equivalents

To obtain →, multiply ↓, by ↘	grams per cubic centimeter	pounds per cubic inch	pounds per cubic foot	slugs per cubic foot	kilograms per cubic meter
grams per cubic centimeter	1	0.03613	62.42806	1.9403	1000
pounds per cubic inch	27.67991	1	1728	53.708	27679.905
pounds per cubic foot	0.01602	0.0005787	1	0.31081	16.01846
slugs per cubic foot	0.51538	0.0186	32.17	1	515.379
kilograms per cubic meter	0.001	3.6128×10^{-5}	0.06243	0.00194	1

TABLE 2.19 Pressure Equivalents

To obtain →, multiply ↓, by ↘	pounds per square inch	pounds per square foot	atmosphere	kilograms per square centimeter	kilograms per square meter	inches water (68°F)	foot water (68°F)	inches mercury (32°F)	millimeters mercury (32°F)	bar	mega-pascal (MPa)
pounds per square inch	1	144	0.068046	0.070307	703.070	27.7276	2.310636	2.03602	51.71497	0.068947	0.006895
pounds per square foot	0.0069444	1	0.000473	0.000488	4.88242	0.1926	0.01605	0.014139	0.35913	0.000479	0.0000479
atmosphere	14.696	2116.22	1	1.03323	10332.27	407.484	33.9570	29.9213	760	1.01325	0.101325
kilograms per square centimeter	14.2233	2048.155	0.96784	1	10000.	394.38	32.8650	28.959	735.559	0.98067	0.098067
kilograms per square meter	0.001422	0.204768	0.0000968	0.0001	1	0.03944	0.003287	0.002896	0.073556	0.000098	0.0000098
inches water (68°F)	0.036065	5.1972	0.002454	0.00253	25.375	1	0.08333	0.073430	1.8651	0.002466	0.000249
foot water (68°F)	0.432781	62.3205	0.029449	0.03043	304.275	12	1	0.88115	22.3813	0.029839	0.0029839
inches mercury (32°F)	0.491154	70.7262	0.033421	0.03453	345.316	13.6185	1.1349	1	25.4	0.033864	0.0033864
millimeters mercury (32°F)	0.0193368	2.78450	0.0013158	0.0013595	13.59509	0.53616	0.044680	0.03937	1	0.001333	0.0001333
bar	14.5038	2088.55	0.98692	1.01972	10197.2	402.156	33.5130	29.5300	750.062	1	0.10
MPa	145.038	20885.5	9.8692	10.1972	101972.0	4021.56	335.130	295.300	7500.62	10.0	1

Energy Equivalents. Refer to Table 2.20 for conversion factors for units of energy.

Power Equivalents. Refer to Table 2.21 for conversion factors for units of power.

Thermal Conductivity Equivalents. Refer to Table 2.22 for conversion factors for units of thermal conductivity.

Dynamic Viscosity Equivalents. Refer to Table 2.23 for conversion factors for units of dynamic viscosity.

TABLE 2.20 Energy Equivalents

To obtain →, multiply ↓, by ↘	Btu*	calorie (gram)†	foot pound	horsepower hour	joule	kilowatt hour	kilogram meter
Btu	1	251.8	778.169	0.000393	1055.06	0.000293	107.586
calorie (gram)†	0.00397	1	3.08596	1.56×10^{-6}	4.184	1.16×10^{-6}	0.426649
foot pound	0.00129	0.32405	1	5.05×10^{-7}	1.35582	3.77×10^{-7}	0.13825
horsepower hour	2544.5	641616	1.98×10^{6}	1	2.68×10^{6}	0.7457	273745
joule	0.00095	0.2390	0.73756	3.72×10^{-7}	1	2.77×10^{-7}	0.102
kilowatt hour	3412.97	860421	2.65×10^{6}	1.34102	3.6×10^{6}	1	367098
kilogram meter	0.00929	2.344	7.233	3.65×10^{-6}	9.807	2.72×10^{-6}	1

* Based on 1 Btu = 778.169 ft · lb.
† Based on 1 cal = 4.1840 joules.

TABLE 2.21 Power Equivalents*

To obtain →, multiply ↓, by ↘	horsepower	foot pound per second	watt	Btu per hour
horsepower	1	550	745.7	2544
foot pound per second	0.00182	1	1.3558	4.626
watt	0.00134	0.7376	1	3.412
Btu per hour	0.00039	0.2161	0.2931	1

* Based on 1 Btu = 778.169 ft · lb.

TABLE 2.22 Conversion Factors for Thermal Conductivity, *k*

To obtain →, multiply ↓, by ↘	watt/ (m·K)	cal/ (s·cm·K)	Btu/ (h·ft·°R)	Btu·in/ (h·ft²·°R)
W/m·K	1	0.0022901	0.57818	6.9381
W/cm·K	100	0.23901	57.818	693.81
W/ft·°R	5.9055	0.014114	3.4144	40.973
W/in·°R	70.866	0.16937	40.973	491.68
cal/(s·cm·K)	418.4	1	241.91	2902.9
cal/(s·in·°R)	296.50	0.70866	171.43	2057.2
kcal/(h·m·K)	1.1622	0.0027778	0.67197	8.0636
Btu/(s·ft·°R)	6226.5	14.882	3600	43200
Btu/(s·in·°R)	74717×10^{4}	178.58	43200	5.184×10^{5}
Btu·in/(s·ft²·°R)	518.87	1.2401	300	3600
Btu/(h·ft·°R)	1.7296	0.0041338	1	12
Btu/(h·in·°R)	20.755	0.049605	12	144
Btu in/(h·ft²·°R)	0.14413	0.00034448	0.083333	1
lb·ft/(h·ft·°R)	0.0022241	5.3157×10^{-6}	0.0012859	0.015431

TABLE 2.23 Conversion Factors for (Dynamic) Viscosity

To obtain → / multiply, by ↓	$\text{Pa} \times \text{s}$	$\frac{\text{lbf} \times \text{s}}{\text{ft}^2}$	$\frac{\text{lbm}}{\text{ft} \times \text{s}}$	$\frac{\text{lbm}}{\text{h} \times \text{ft}}$	$\frac{\text{g}}{\text{cm} \times \text{s}}$ (poise)	$\frac{\text{kg}}{\text{m} \times \text{s}}$
$\text{Pa} \times \text{s}$	1	$\frac{(0.3048)^2}{9.80665 \times 0.45359237} = 0.0208854342$	$\frac{0.3048}{0.45359237} = 0.671968975$	$\frac{0.3048 \times 3600}{0.45359237} = 2419.08831$	10	1
$\frac{\text{lbf} \cdot \text{s}}{\text{ft}^2}$	$\frac{9.80665 \times 0.45359237}{(0.3048)^2} = 47.8802590$	1	$\frac{980.665}{30.48} = 32.1740486$	$\frac{980.665 \times 3600}{30.48} = 115826.575$	$\frac{980.665 \times 453.59237}{30.48^2} = 478.802590$	$\frac{980.665 \times 453.59237}{10 \times 30.48^2} = 47.8802590$
$\frac{\text{lbm}}{\text{ft} \times \text{s}}$	$\frac{0.45359237}{0.3048} = 1.48816394$	$\frac{30.48}{980.665} = 0.0310809502$	1	3600	$\frac{453.59237}{30.48} = 14.8816394$	$\frac{453.59237}{304.8} = 1.48816394$
$\frac{\text{lbm}}{\text{h} \times \text{ft}}$	$\frac{0.45359237}{0.3048 + 3600} = 0.413378873 \times 10^{-3}$	$\frac{30.48}{980.665 \times 3600} = 8.63359727 \times 10^{-6}$	$\frac{1.0}{3600} = 0.000277777778$	1	$\frac{453.59237}{30.48 \times 3600} = 0.00413378873$	$\frac{453.59237}{304.8 \times 3600} = 0.000413378873$
$\frac{\text{g}}{\text{cm} \times \text{s}}$ (poise)	0.1	$\frac{30.48^2}{980.665 \times 453.59237} = 0.00208854342$	$\frac{30.48}{453.59237} = 0.0671968975$	$\frac{3600 \times 30.48}{453.59237} = 241.908831$	1	0.1
$\frac{\text{kg}}{\text{m} \times \text{s}}$	1	$\frac{10 \times 30.48^2}{980.665 \times 453.59237} = 0.0208854342$	$\frac{304.8}{453.59237} = 0.671968975$	$\frac{10 \times 3600 \times 30.48}{453.59237} = 2.41908831$	10	1

Kinematic Viscosity Equivalents. Refer to Tables 2.24 and 2.25 for conversion factors for units of kinematic viscosity.

TABLE 2.24 Conversion Factors for Kinematic Viscosity

To obtain → multiply, by ↓ ↘	$\frac{m^2}{s}$	$\frac{ft^2}{s}$	$\frac{cm^2}{s}$ (stoke)	$\frac{cm^2}{h}$	$\frac{m^2}{h}$
$\frac{m^2}{s}$	1	$\frac{1.0}{(0.3048)^2}$ = 10000	$(100)^2$ = 10000	$(100)^2 \times 3600$ = 36×10^6	3600
$\frac{ft^2}{s}$	$(0.3048)^2$ = 0.09290304	1	$(30.48)^2$ = 929.0304	$(30.48)^2 \times 3600$ = 3334509.440000	$(0.3048)^2 \times 3600$ = 334.450944000
$\frac{cm^2}{s}$ (stoke)	$\frac{1.0}{(100)^2}$ = 10^{-4}	$\frac{1.0}{(30.48)^2}$ = 0.00107639104	1	3600	$\frac{3600}{(100)^2}$ = 0.36
$\frac{cm^2}{h}$	$\frac{1.0}{(100)^2 \times 3600}$ = $27.7777778 \times 10^{-9}$	$\frac{1.0}{(30.48)^2 \times 3600}$ = $0.298997512 \times 10^{-6}$	$\frac{1.0}{3600}$ = 0.000277777778	1	$\frac{1.0}{(100)^2}$ = 0.0001
$\frac{m^2}{h}$	$\frac{1.0}{3600}$ = $277.777778 \times 10^{-6}$	$\frac{1.0}{(0.3048)^2 \times 3600}$ = $2.98997512 \times 10^{-3}$	$\frac{(100)^2}{3600}$ = 2.77777778	$(100)^2$ = 10000	1

TABLE 2.25 Kinematic Viscosity Conversion Table (centistokes to Engler, Saybolt, and Redwood units)

centistokes	Engler degrees	Saybolt seconds at 130°F	Redwood seconds at 140°F	centistokes	Engler degrees	Saybolt seconds at 130°F	Redwood seconds at 140°F	centistokes	Engler degrees	Saybolt seconds at 130°F	Redwood seconds at 140°F
2.0	1.140	32.66	30.95	18.0	2.644	89.37	78.45	41.0	5.465	190.6	168.3
2.5	1.182	34.46	32.20	19.0	2.755	93.48	82.10	42.0	5.590	195.1	172.3
3.0	1.224	36.07	33.45	20.0	2.870	97.69	85.75	43.0	5.720	199.6	176.4
3.5	1.266	37.67	34.70	21.0	2.984	101.9	89.50	44.0	5.845	204.2	180.4
4.0	1.308	39.17	35.95	22.0	3.100	106.2	93.25	45.0	5.975	208.8	184.5
4.5	1.350	40.78	37.20	23.0	3.215	110.5	97.05	46.0	6.105	213.4	188.5
5.0	1.400	42.38	38.45	24.0	3.335	114.8	100.9	47.0	6.235	218.0	192.6
5.5	1.441	43.98	39.80	25.0	3.455	119.1	104.7	48.0	6.365	222.6	196.6
6.0	1.481	45.59	41.05	26.0	3.575	123.5	108.6	49.0	6.495	227.2	200.7
6.5	1.521	47.19	42.40	27.0	3.695	127.9	112.5	50.0	6.630	231.8	204.7
7.0	1.563	48.79	43.70	28.0	3.820	132.4	116.5	52.0	6.890	241.1	212.8
7.5	1.605	50.44	45.00	29.0	3.945	136.8	120.4	54.0	7.106	250.3	221.0
8.0	1.653	52.10	46.35	30.0	4.070	141.2	124.4	56.0	7.370	259.5	229.1
8.5	1.700	53.80	47.75	31.0	4.195	145.6	128.3	58.0	7.633	268.7	237.2
9.0	1.746	55.51	49.10	32.0	4.320	150.0	132.3	60.0	7.896	277.9	245.3
9.5	1.791	57.21	50.55	33.0	4.445	154.5	136.3	62.0	8.159	287.2	253.5
10.0	1.837	58.91	52.00	34.0	4.570	159.0	140.2	64.0	8.422	296.4	261.6
11.0	1.928	62.42	55.00	35.0	4.695	163.5	144.2	66.0	8.686	305.6	269.8
12.0	2.020	66.03	58.10	36.0	4.825	168.0	148.2	68.0	8.949	314.8	277.9
13.0	2.120	69.73	61.30	37.0	4.955	172.5	152.2	70.0	9.212	324.0	286.0
14.0	2.219	73.54	64.55	38.0	5.080	177.0	156.2	72.0	9.475	333.3	294.1
15.0	2.323	77.35	67.95	39.0	5.205	181.5	160.3	74.0	9.738	342.5	302.2
16.0	2.434	81.25	71.40	40.0	5.335	186.0	164.3	75.0	9.870	347.2	306.3
17.0	2.540	85.26	74.85								

Supplementary kinematic viscosity conversion table

centistokes	2	6	10	20	30	40	50	60	70
Saybolt at 100°F	32.60	45.50	58.80	97.50	40.9	185.7	231.4	277.4	323.4
Saybolt at 210°F	32.83	45.82	59.21	98.18	141.9	187.0	233.0	279.3	325.7
Redwood at 70°F	30.20	40.50	51.70	85.40	123.7	163.2	203.3	243.5	283.9
Redwood at 200°F	31.20	41.50	52.55	86.90	126.0	166.7	208.3	250.0	291.7

Temperature conversion: $t_c = \frac{5}{9}(t_F - 32)$

70°F = 21.1°C; 140°F = 60.0°C
100°F = 37.8°C; 200°F = 93.3°C
130°F = 54.5°C; 210°F = 98.9°C

Insulation Properties. Refer to Tables 2.26 through 2.33 for tables of insulation properties.

TABLE 2.26 Properties and Limitations of Insulation Materials

Insulation material	Compressive strength (pH)	Maximum temperature (°F)	Minimum temperature (°F)	Permeability (perm·in)	Conductivity [Btu·in/(b·ft²·°F)] at 50°F, 200°F	Spread/smoke density index
Calcium silicate	100	1000	250	NA	NA, 0.43	0/0
Cellular glass*	87	900	–450	0.005	0.28, 0.37	5/0
Elastomeric foam	NA	200	–40	0.03	0.29, NA	25/50
Fiberglass	2.5 at 10%†	850	42	75	0.29, 0.39	25/50
Mineral wool	10 at 10%†	1200	42	150		
Expanded perlite	70	1000	250	18	NA, 0.52	25/50
Phenolic foam	22 at 10%†	300	75	6–7	0.15, 0.25	25/50
Polystyrene foam	25	165	–65	1–5	0.23, NA	NA
Polyurethane and polyisocyanurate	30	250	–200	2–4	0.17, NA	25/50

* For flat surfaces cupped per ASTM C240 (box asphalt, 15$ felt).

† While compressive strength technically measures the stress at which a material fails under load, deformation monitors a material's structural distortion *with or without* actual failure. So, with compressible insulations, failure is recorded as the point at which an insulation's deformation reaches a percentage of its thickness—usually between 5 and 25 percent—for various densities.

TABLE 2.27 ASTM Specification Reference

Material	ASTM Specification
Calcium silicate	C533–95
Cellular glass	C552–91
Elastomeric foam	C534–94
Fiberglass and mineral wool	C553–92, C547–95, C612–93
Expanded perlite	C610–95
Phenolic foam	C1126–96
Polystyrene foam	C578–95
Polyurethane modified polyisocyanurate foam	C591–94
Aluminum jacketing and banding	C921–96
Stainless steel jacketing and banding	C921–96
FRP	C921–96
ASJ	C921–96
Asphalt cutback mastic	C647–95
Acrylic intex mastic	C647–95
Hypalon mastic	C647–95
6 × 6 mesh fabric	NA
10 × 10 glass scrim	D1668–85

TABLE 2.28 Calcium Silicate Insulation *k* Value Table
Thermal conductivity as a function of temperature for calcium silicate (calsil)

English units		Non-SI metric units		SI units	
At uniform temperature (°F)	Thermal conductivity [Btu·in/(h·ft²·°F)]	At uniform temperature (°C)	Thermal conductivity [kcal/(h·m²·°C)]	At uniform temperature (K)	Thermal conductivity [W/(m·K)]
0.0	0.373	−17.8	0.0462	255.4	0.0537
50.0	0.388	10.0	0.0481	283.1	0.0559
100.0	0.403	37.8	0.0499	310.9	0.0580
150.0	0.418	65.6	0.0518	338.7	0.0602
200.0	0.433	93.3	0.0537	366.5	0.0624
250.0	0.449	121.1	0.0557	394.3	0.0647
300.0	0.466	148.9	0.0577	422.0	0.0671
350.0	0.483	176.7	0.0599	449.8	0.0696
400.0	0.501	204.4	0.0622	477.6	0.0722
450.0	0.521	232.2	0.0646	505.4	0.0750
500.0	0.541	260.0	0.0671	533.2	0.0780
550.0	0.563	287.8	0.0698	560.9	0.0812
600.0	0.587	315.6	0.0728	588.7	0.0846
650.0	0.612	343.3	0.0759	616.5	0.0882
700.0	0.639	371.1	0.0793	644.3	0.0921
750.0	0.668	398.9	0.0829	672.0	0.0963
800.0	0.700	426.7	0.0868	699.8	0.1008
850.0	0.733	454.4	0.0909	727.6	0.1057
900.0	0.769	482.2	0.0954	755.4	0.1109
950.0	0.808	510.0	0.1002	783.2	0.1165
1000.0	0.849	537.8	0.1053	810.9	0.1224

The values of the thermal conductivity of the insulation shown above were determined by evaluating a polynomial at the insulation temperature. This polynomial, giving the thermal conductivity in Btu·in/(h·ft²·°F) as a function of temperature in °F, is $K(T) = 0.3728 + 2.98 \times 10^{-4}(T) - 2.3 \times 10^{-8}(T^2) + 2.02 \times 10^{-10}(T^3)$.

This curve is based on an insulation density of 16.0 lb/ft³ and may be subject to decreasing reliability outside the temperature range from 250.0 to 1000.0°F.

TABLE 2.29 Super_k™ FOAMGLAS® Insulation *K* Value Table
Thermal conductivity as a function of temperature for Super_k™ FOAMGLAS® insulation[a]

English units		Non-SI metric units		SI units	
At uniform temperature (°F)	Thermal conductivity [Btu·in/(h·ft²·°F)]	At uniform temperature (°C)	Thermal conductivity [kcal/(h·m²·°C)]	At uniform temperature (K)	Thermal conductivity [W/(m·K)]
–300.0	0.109	–184.4	0.0135	88.7	0.0156
–250.0	0.126	–156.7	0.0156	116.5	0.0181
–200.0	0.146	–128.9	0.0181	144.3	0.0210
–150.0	0.168	–101.1	0.0208	172.0	0.0242
–100.0	0.193	–73.3	0.0239	199.8	0.0278
–50.0	0.219	–45.6	0.0272	227.6	0.0316
0.0	0.247	–17.8	0.0307	255.4	0.0356
50.0	0.277	10.0	0.0344	283.1	0.0399
100.0	0.309	37.8	0.0383	310.9	0.0455
150.0	0.342	65.6	0.0425	338.7	0.0494
200.0	0.378	93.3	0.0469	356.5	0.0545
250.0	0.476	121.1	0.0516	394.3	0.0600
300.0	0.457	148.9	0.0566	422.0	0.0658
350.0	0.500	176.7	0.0621	449.8	0.0721
400.0	0.548	204.4	0.0679	477.6	0.0789
450.0	0.599	232.2	0.0742	505.4	0.0863
500.0	0.654	260.0	0.0811	533.2	0.0943
550.0	0.715	287.8	0.0887	560.9	0.1031
600.0	0.782	315.6	0.0969	588.7	0.1127
650.0	0.855	343.3	0.1060	616.5	0.1232
700.0	0.935	371.1	0.1159	644.3	0.1347
750.0	1.023	398.9	0.1268	672.0	0.1474
800.0	1.119	426.7	0.1388	699.8	0.1613
850.0	1.225	454.4	0.1520	727.6	0.1766
900.0	1.342	482.2	0.1654	755.4	0.1934

[a] Super_k™ FOAMGLAS® insulation is a registered trademark of Pittsburgh Corning Corporation.

The values of the thermal conductivity of the insulation shown above were determined by evaluating a polynomial at the insulation temperature. This polynomial, giving the thermal conductivity in Btu·in/(h·ft²·°F) as a function of temperature in °F, is $K(T) = 0.2472 + 5.811 \times 10^{-4}(T) + 3.4561 \times 10^{-7}(T^2) + 3.2 \times 10^{-13}(T^3) + 5.3092 \times 10^{-13}(T^4) - 9.64 \times 10^{-17}(T^5)$.

This curve is based on an insulation density of 7.5 lb/ft³ and may be subject to decreasing reliability outside the temperature range from –250.0 to 800.0°F.

TABLE 2.30 Fiberglas Insulation K Value Table

Thermal conductivity as a function of temperature for fiberglass

English units		Non-SI metric units		SI units	
At uniform temperature (°F)	Thermal conductivity [Btu·in/(h·ft²·°F)]	At uniform temperature (°C)	Thermal conductivity [kcal/(h·m²·°C)]	At uniform temperature (K)	Thermal conductivity [W/(m·K)]
0.0	0.195	−17.8	0.0241	255.4	0.0280
50.0	0.216	10.0	0.0268	283.1	0.0311
100.0	0.237	37.8	0.0294	310.9	0.0342
150.0	0.258	65.6	0.0320	338.7	0.0372
200.0	0.280	93.3	0.0347	366.5	0.0403
250.0	0.301	121.1	0.0373	394.3	0.0433
300.0	0.322	148.9	0.0399	422.0	0.0464
350.0	0.343	176.7	0.0426	449.8	0.0495
400.0	0.364	204.4	0.0452	477.6	0.0525
450.0	0.386	232.2	0.0478	505.4	0.0556
500.0	0.407	260.0	0.0505	533.2	0.0587
550.0	0.478	287.8	0.0531	560.9	0.0617
600.0	0.449	315.6	0.0557	588.7	0.0648
650.0	0.471	343.3	0.0584	616.5	0.0678
700.0	0.492	371.1	0.0610	644.3	0.0709
750.0	0.513	398.9	0.0636	672.0	0.0740
800.0	0.534	426.7	0.0663	699.8	0.0770
850.0	0.556	454.4	0.0689	727.6	0.0801

The values of the thermal conductivity of the insulation shown above were determined by evaluating a polynomial at the insulation temperature. This polynomial, giving the thermal conductivity in Btu·in/(h·ft²·°F) as a function of temperature in °F, is $K(T) = 0.195 + 4.25 \times 10^{-4}(T) + 0.0000(T^2)$.

This curve is based on an insulation density of 5.63 lb/ft³ and may be subject to decreasing reliability outside the temperature range from 42.0 to 800.0°F.

TABLE 2.31 Mineral Wool Insulation K Value Table

Thermal conductivity as a function of temperature for mineral wool

English units		Non-SI metric units		SI units	
At uniform temperature (°F)	Thermal conductivity [Btu·in/(h·ft²·°F)]	At uniform temperature (°C)	Thermal conductivity [kcal/(h·m²·°C)]	At uniform temperature (K)	Thermal conductivity [W/(m·K)]
0.0	0.228	−17.8	0.0283	255.4	0.0329
50.0	0.248	10.0	0.0307	283.1	0.0357
100.0	0.271	37.8	0.0336	310.9	0.0391
150.0	0.297	65.6	0.0368	338.7	0.0428
200.0	0.326	93.3	0.0405	366.5	0.0470
250.0	0.358	121.1	0.0444	354.3	0.0516
300.0	0.393	148.9	0.0488	422.0	0.0567
350.0	0.432	176.7	0.0535	449.8	0.0622
400.0	0.473	204.4	0.0586	477.6	0.0681
450.0	0.517	232.2	0.0641	505.4	0.0745
500.0	0.564	260.0	0.0699	533.2	0.0813
550.0	0.614	287.8	0.0761	560.9	0.0885
600.0	0.667	315.6	0.0827	588.7	0.0961
650.0	0.723	343.3	0.0897	616.5	0.1042
700.0	0.782	371.1	0.0970	644.3	0.1127
750.0	0.844	398.9	0.1047	672.0	0.1217
800.0	0.909	426.7	0.1128	699.8	0.1311
850.0	0.978	454.4	0.1212	727.6	0.1409
900.0	1.049	482.2	0.1300	755.4	0.1511
950.0	1.123	510.0	0.1392	783.2	0.1618
1000.0	1.200	537.8	0.1488	810.9	0.1729
1050.0	1.280	565.6	0.1587	838.7	0.1845
1100.0	1.363	593.3	0.1690	866.5	0.1965
1150.0	1.499	621.1	0.1797	894.3	0.2089
1200.0	1.538	648.9	0.1908	922.2	0.2217

The values of the thermal conductivity of the insulation shown above were determined by evaluating a polynomial at the insulation temperature. This polynomial, giving the thermal conductivity in Btu·in/(h·ft²·°F) as a function of temperature in °F, is $K(T) = 0.228 + 3.72 \times 10^{-4}(T) + 6.0 \times 10^{-7}(T^2)$.

This curve is based on an insulation density of 11.7 lb/ft³ and may be subject to decreasing reliability outside the temperature range from 42.0 to 1100.0°F.

TABLE 2.32 Perlite Insulation K Value Table
Thermal conductivity as a function of temperature for perlite

English units		Non-SI metric units		SI units	
At uniform temperature (°F)	Thermal conductivity [Btu·in/(h·ft²·°F)]	At uniform temperature (°C)	Thermal conductivity [kcal/(h·m²·°C)]	At uniform temperature (K)	Thermal conductivity [W/(m·K)]
0.0	0.403	−17.8	0.0500	255.4	0.0581
50.0	0.434	10.0	0.0538	283.1	0.0626
100.0	0.464	37.8	0.0575	310.9	0.0668
150.0	0.492	65.6	0.0610	338.7	0.0709
200.0	0.519	93.3	0.0644	366.5	0.0748
250.0	0.546	121.1	0.0677	394.3	0.0787
300.0	0.572	148.9	0.0709	422.0	0.0824
350.0	0.598	176.7	0.0741	449.8	0.0862
400.0	0.624	204.4	0.0774	477.6	0.0899
450.0	0.650	232.2	0.0806	505.4	0.0937
500.0	0.677	260.0	0.0840	533.2	0.0976
550.0	0.705	287.8	0.8740	560.9	0.1016
600.0	0.734	315.6	0.0910	588.7	0.1058
650.0	0.764	343.3	0.0948	616.5	0.1101
700.0	0.796	371.1	0.0987	644.3	0.1148
750.0	0.830	398.9	0.1029	672.0	0.1196
800.0	0.866	426.7	0.1074	698.8	0.1248
850.0	0.905	454.4	0.1122	727.6	0.1304
900.0	0.946	482.2	0.1173	755.4	0.1363
950.0	0.990	510.0	0.1228	783.2	0.1427
1000.0	1.038	537.8	0.1287	810.9	0.1496

The values of the thermal conductivity of the insulation shown above were determined by evaluating a polynomial at the insulation temperature. This polynomial, giving the thermal conductivity in Btu·in/(h·ft²·°F) as a function of temperature in °F, is $K(T) = 0.4030 + 6.38 \times 10^{-4}(T) - 3.56 \times 10^{-7}(T^2) + 3.53 \times 10^{-10}(T^3)$.

This curve is based on an insulation density of 12.0 lb/ft³ and may be subject to decreasing reliability outside the temperature range from 250.0 to 1000.0°F.

TABLE 2.33 Polyurethane or Polyisocyanurate Foam Insulation K Value Table
Thermal conductivity as a function of temperature for urethane

English units		Non-SI metric units		SI units	
At uniform temperature (°F)	Thermal conductivity [Btu·in/(h·ft²·°F)]	At uniform temperature (°C)	Thermal conductivity [kcal/(h·m²·°C)]	At uniform temperature (K)	Thermal conductivity [W/(m·K)]
−100.0	0.179	−73.3	0.0222	199.8	0.0258
−50.0	0.179	−45.6	0.0223	227.6	0.0259
0.0	0.174	−17.8	0.0215	255.4	0.0250
50.0	0.166	10.0	0.0206	283.1	0.0239
100.0	0.165	37.8	0.0204	310.9	0.0238
150.0	0.180	65.6	0.0223	338.7	0.0260
200.0	0.225	93.3	0.0279	366.5	0.0324
250.0	0.316	121.1	0.0391	394.3	0.0455
300.0	0.470	148.9	0.0583	422.0	0.0677
350.0	0.710	176.7	0.0880	449.8	0.1023

The values of the thermal conductivity of the insulation shown above were determined by evaluating a polynomial at the insulation temperature. This polynomial, giving the thermal conductivity in Btu·in/(h·ft²·°F) as a function of temperature in °F, is $K(T) = 0.1735 - 1.549 \times 10^{-4}(T) - 3.389 \times 10^{-7}(T^2) + 8.377 \times 10^{-9}(T^3) + 1.819 \times 10^{-11}(T^4)$.

This curve is based on an insulation density of 2.0 lb/ft³ and may be subject to decreasing reliability outside the temperature range from −50.0 to 250.0°F.

Tables in Chapter 2

S · E · C · T · I · O · N · 2

PIPING MATERIALS

CHAPTER 3
METALLIC PIPING MATERIALS

See p. 102 for a list of tables in Chap. 3.

Metals are divided into two types: ferrous, which includes iron and iron-base alloys; and nonferrous, covering other metals and alloys.

FERROUS METALS

Metallic iron, one of the most common of metals, is very rarely found in nature in its pure form. It occurs in the form of mineral oxides (Fe_2O_3 or Fe_3O_4), and as such it constitutes about 6 percent of the earth's crust. The product removed from the blast furnace is called *pig iron* and is an impure form of iron containing about 4 percent carbon by weight. Liquid pig iron cast from the blast furnace is sometimes used directly for metal castings. More often, however, the iron is remelted in a cupola, or furnace, to further refine it and adjust its composition.

CAST IRON

Pig iron that has been remelted is known as *cast iron*, a term applicable to iron possessing carbon in excess of 2 weight percent. Compared with steel, cast iron is inferior in malleability, strength, toughness, and ductility. On the other hand, cast iron has better fluidity in the molten state and can be cast satisfactorily into complicated shapes. It is also less costly than steel. The most important types of cast iron are white and gray cast irons.

White cast iron is so known because of the silvery appearance of its fracture surface when broken. In this alloy, the carbon is present in the form of iron carbide (Fe_3C), also known as *cementite*. This carbide is chiefly responsible for the high hardness, brittleness, and poor machinability characteristic of white cast iron. Chilled iron, a form of white cast iron, is cast against metal chills that cause rapid cooling, promoting the formation of cementite. Consequently, a structure is obtained that possesses high wear and abrasion resistance, the principal attribute of the material, but retains white cast iron's characteristic brittleness.

Malleable cast iron is the name given to white cast iron that has been heat treated to change its cementite into nodules of graphite. The iron becomes more malleable because, in this condition, the carbon as carbide no longer exists continuously through the metal matrix.

Gray iron is a widely used type of cast iron. In this alloy, the carbon predominantly exists in the form of graphite flakes. The typical appearance of a fracture of this iron is gray since the graphite flakes are exposed. The strength of gray iron depends on the size of the graphite particles and the amount of cementite formed together with the graphite. The strength of the iron increases as the graphite crystal size decreases and the amount of cementite increases. This material is easily machined. A wide range of tensile strengths can be achieved by alloying gray iron with elements such as nickel, chromium, and molybdenum.

Another member of the cast-iron family is so-called *ductile iron*. It is a high-carbon magnesium-treated product containing graphite in the form of spheroids. Ductile iron is similar to gray cast iron in machinability, but it possesses superior mechanical properties. This alloy is especially suited for pressure castings.

STEEL

Steel is defined as an alloy of iron with not more that 2.0 weight percent carbon.

Alloying additions are made, if required, to the molten steel while it is in the reducing furnace as already noted, in the ladle into which the steel is put, or in the ingot into which steel is poured from the ladle.

Steel of the killed type is deoxidized almost completely; that is, sufficient deoxidizing agent is added to the molten pool to combine with all the excess entrained oxygen. The result is a large number of tiny oxides in the melt. The lack of gas in the molten pool gives the effect of "killing" any visible bubbling activity of the steel, thus the name. Killed steel has a more uniform composition than any of the other types, and usually possesses the best formability at room temperature. A fine-grained structure results from this practice because the many oxides formed act as initiation sites of new grains upon solidification and subsequent recrystallization. This fine-grained character offers toughness superior to the other types of steel.

Rimmed steel employs no purposeful addition of deoxidizing agents, and is characterized by relatively violent bubbling and stirring action in the ingot mold. This type exhibits a marked variation in composition across and from top to bottom of the ingot. The outer rim or outer edge of the solidified ingot is relatively pure and ductile material. The amounts of carbon, phosphorus, sulfur, and nonmetallic inclusions in this rim are lower than the average composition of the whole ingot. The amount of these constituents in the inner portion or core is higher than the ingot average. This type of steel costs less to make than the other types and is widely used for structural applications, where good surface appearance of the final product is desired.

Semikilled steel is only partially deoxidized with silicon, aluminum, or both, taking advantage of the positive attributes of killed and rimmed steel.

ALLOYING OF STEEL

The alloying of carbon steel with other elements to obtain a wide range of desired properties is a mature science. In general, when used in combination, alloying elements may complement each other and give greater overall benefits than when used singly in much larger quantities. The following summarizes the known effects of adding certain elements to steel.

Carbon. In general, increasing the carbon content of steel alloys produces higher ultimate strength and hardness but may lower ductility and toughness. Carbon also increases air-hardening tendencies and weld hardness. In low-alloy steel for high-temperature applications, the carbon content is usually restricted to a maximum of about 0.15 percent in order to assure optimum ductility for welding, expanding, and bending operations. An increasing carbon content lessens the thermal and electrical conductivities of steel.

Phosphorus. High phosphorus content has an undesirable effect on the properties of carbon steel, notably on shock resistance and ductility. Phosphorus is effective, however, in improving machinability. In steels, it is normally controlled to less than 0.04 weight percent.

Silicon. Used as a deoxidizing agent, silicon increases the tensile strength of steel without increasing brittleness when limited to less than about 2 percent. Silicon increases resistance to oxidation, increases electrical resistivity, and decreases hysteresis losses. Thus it is used for electrical applications. Adding silicon may reduce creep rupture strength.

Manganese. Manganese is normally present in all commercial steels. The manganese combines with sulfur, thus improving hot-working characteristics. In alloy steels, manganese decreases the critical cooling rate to cause a hardened or martensitic structure and thus contributes to deep hardening.

Nickel. As an alloying element in alloy steels, nickel is a ferrite strengthener and toughener and is soluble in all proportions. Nickel steels are easily hardened because nickel lowers the critical cooling rate necessary to produce hardening on quenching. In heat-treated steel, nickel increases the strength and toughness. In combination with chromium, nickel produces alloy steels possessing higher impact and fatigue resistance than can be obtained with straight carbon steels.

Chromium. As an alloying element in steel, chromium is miscible in iron as a solid solution, and forms a complex series of carbide compounds. Chromium is essentially a hardening element and is frequently used with a toughening element such as nickel to produce superior mechanical properties. At higher temperatures, chromium contributes increased strength and is ordinarily used in conjunction with molybdenum. Additions of chromium significantly improve the elevated-temperature oxidation resistance of steels.

Molybdenum. In steel, molybdenum can form a solid solution with the iron, and depending on the molybdenum and carbon content, can also form a carbide. A deeper-hardening steel results. The molybdenum carbide is very stable and is responsible for matrix strengthening in long-term creep service.

Vanadium. This element is one of the strong carbide formers. It dissolves to some degree in ferrite, imparting strength and toughness. Vanadium steels show a much finer grain structure than steels of a similar composition without vanadium.

Boron. Boron is usually added to steel to improve hardenability, that is, to increase the depth of hardening during quenching.

Aluminum. Aluminum is widely used as a deoxidizer in molten steel and for controlling grain size. When added to steel in controlled amounts, it produces a fine grain size.

Sulfur. Present to some degree in all steel (less than 0.04 weight percent), sulfur forms a nonmetallic impurity that, in large amounts, results in cracking during forming at high temperatures (hot shortness). Combining it with manganese forms an MnS compound that is relatively harmless.

Copper. Copper dissolves in steel and strengthens the iron as a substitutional element. The use of copper in certain alloys increases resistance to atmospheric corrosion and increases yield strength. However, excessive amounts of copper (usually above 0.3 percent) are harmful to elevated-temperature performance, since the lower melting point element segregates to grain boundaries and locally melts (liquates), causing intergranular separation under applied stress.

CLASSIFICATION OF STEELS

There are literally hundreds of wrought grades of steel that range in composition with the variation of the many major and minor alloying elements. The simplest of these classes is known as *plain carbon steel*, with carbon varying between approximately 0.05 and 1.0 weight percent. Within this broad range fall three general groups according to carbon content; they are defined as follows:

1. Low-carbon steels contain 0.05 to 0.25 percent carbon
2. Medium-carbon steels contain 0.25 to 0.50 percent carbon
3. High-carbon steels contain 0.50 percent and greater carbon content

Alloy steels are generally considered to be steels to which one or more alloying elements, other than carbon, have been added to give them special properties that are different from those of straight carbon steels. From the standpoint of composition, steel is considered to be an alloy steel when amounts of manganese, silicon, or copper exceed the maximum limits for the carbon steels, or when minimum quantities of other alloying elements are intentionally added. These could be chromium, molybdenum, nickel, copper, cobalt, niobium, vanadium, or others.

The next higher class of alloyed steel useful to the piping industry is ferritic and martensitic stainless steels. These are steels alloyed with chromium contents above about 12 percent. Because of the chromium, these materials possess good corrosion resistance. They retain a ferritic (BCC) crystal structure, allowing the grades to be hardened by heat treatment.

When sufficient nickel is added to iron-chromium alloys, an austenitic (FCC) structure is retained at room temperature. Austenitic stainless steels possess an excellent combination of strength, ductility, and corrosion resistance. These steels cannot be hardened by quenching, since the austenite does not transform to martensite.

A stronger type of stainless steel has been developed that takes advantage of precipitation reactions within the metal matrix made possible by addition of elements such as aluminum, titanium, copper, and nitrogen. These materials are referred to as precipitation-hardenable stainless steels. Both martensitic and austenitic stainless steels can be enhanced in this manner.

As annealed, these materials are soft and readily formed. When fully hardened through aging heat treatments, they attain their full strength potential.

ACCEPTABILITY OF SPECIFICATIONS

It is noted that not all the specifications are acceptable for use within all sections of ASME B31, Pressure Piping Code, and the ASME Boiler and Pressure Vessel Code. Tables 3.1 through 3.6 list the ASTM material specifications that are acceptable for use within the scope of the following codes:

- ASME B31.1, Power Piping
- ASME B31.3, Process Piping
- ASME B31.4, Pipeline Transportation Systems for Liquid Hydrocarbons and Other Liquids
- ASME B31.5, Refrigeration Piping

- B31.8, Gas Transmission and Distribution Piping Systems
- B31.9, Building Services Piping
- B31.11, Slurry Transportation Piping Systems
- ASME Section I, Power Boilers
- ASME Section III, Nuclear Power Plant Components

In the tables that follow, "Yes" against a specification indicates that the material is acceptable for construction of pipe, tube, or fittings, as applicable, within the scope of the code for which "Yes" is listed. A "No" indicates that the material specification is not allowed for use within the jurisdiction of the code.

MATERIAL SPECIFICATIONS

Carbon Steel Pipe, Tube, and Fittings

Table 3.1 lists ASTM material specifications for carbon steel pipe, tube, and fittings.

Alloy Steel Pipe, Tube, and Fittings

Table 3.2 lists ASTM material specifications for alloy steel pipe, tube, and fittings.

Stainless Steel Pipe, Tube, and Fittings

Table 3.3 lists ASTM material specifications for stainless steel pipe, tube, and fittings.

Nickel and Nickel Alloy Pipe, Tube, and Fittings

Table 3.4 lists ASTM material specifications for nickel and nickel alloy pipe, tube, and fittings.

Copper and Copper Alloy Pipe, Tube, and Fittings

Table 3.5 lists ASTM material specifications for copper and copper alloy pipe, tube, and fittings.

Aluminum and Aluminum Alloy Pipe, Tube, and Fittings

Table 3.6 lists ASTM material specifications for aluminum and aluminum alloy pipe, tube, and fittings.

Pipe, Tube, and Fitting Material Specifications

Tables 3.7 through 3.36 list ASTM material specifications for seamless and welded pipe, tube, and fittings made from carbon steel, alloy steel, stainless steel, copper and copper alloys, aluminum and aluminum alloys, and titanium and titanium alloys. These specifications are limited to those referenced in different sections of ASME B31, Pressure Piping Code. Refer to the corresponding ASME Section II material specifications starting with suffix SA for ferrous materials and SB for nonferrous materials for use within the jurisdiction of different sections of the ASME Boiler and Pressure Vessel Code.

INTERNATIONAL MATERIAL SPECIFICATIONS

Tables 3.37 through 3.41 provide a cross reference for material specifications from different countries: United States, Britain, Germany, Belgium, France, Italy, and Japan. The specifications listed from various countries are closely related in terms of technical requirements; however, the extent of equivalency must be established prior to using one in lieu of the other.

TABLE 3.1 Carbon Steel Pipe, Tube, and Fittings[1,2]

Material spec	Description	Notes	B31.1	B31.3	B31.4	B31.5	B31.8	B31.9	B31.11	ASME I	ASME III
A53	Seamless pipe	(3) (4)	Yes	Yes	Yes	Yes	Yes	Yes	Yes	Yes	Yes
A106	Seamless pipe	(3)	Yes	Yes	Yes	Yes	Yes	Yes	Yes	Yes	Yes
A333	Seamless pipe	(3)	Yes	Yes	No	Yes	Yes	No	No	No	Yes
A369	Seamless pipe	(3)	Yes	Yes	No	No	No	No	No	Yes	Yes
API-5L	Seamless pipe	(3)	Yes	Yes	Yes	Yes	Yes	Yes	Yes	No	No
A524	Seamless pipe/tube	—	No	Yes	Yes	No	No	No	Yes	No	Yes
A179	Seamless tube	—	Yes	Yes	No	Yes	No	No	Yes	No	No
A192	Seamless tube	—	Yes	No	No	Yes	No	No	No	Yes	No
A210	Seamless tube	(3)	Yes	No	No	Yes	No	No	No	Yes	Yes
A334	Seamless tube	—	No	Yes	No	Yes	No	No	No	No	Yes
A53	Seamwelded pipe	(4)	Yes	Yes	Yes	Yes	No	Yes	Yes	Yes	Yes
A134	Seamwelded pipe	(3) (6)	Yes	Yes	Yes	Yes	Yes	No	Yes	No	No
A135	Seamwelded pipe	—	Yes	Yes	Yes	Yes	Yes	Yes	Yes	No	No
A139	Seamwelded pipe	(3) (5)	Yes	Yes	Yes	Yes	Yes	No	Yes	No	No
A211	Seamwelded pipe	(3)	Yes	Yes	No	Yes	Yes	Yes	Yes	No	No
A333	Seamwelded pipe	—	Yes	Yes	No	Yes	Yes	No	No	No	Yes
A381	Seamwelded pipe	—	No	Yes	Yes	No	Yes	No	Yes	No	No
A587	Seamwelded pipe	—	Yes	Yes	No	Yes	No	No	No	No	Yes
A671	Seamwelded pipe	(3)	Yes	Yes	Yes	No	Yes	No	Yes	No	Yes
A672	Seamwelded pipe	(3)	Yes	Yes	Yes	No	Yes	No	Yes	No	Yes
A691	Seamwelded pipe	(3)	Yes	Yes	No	No	No	No	Yes	No	Yes
API-5L	Seamwelded pipe	(3)	Yes	Yes	Yes	Yes	Yes	Yes	Yes	No	No
API-5LU	Seamwelded pipe	—	No	No	Yes	No	No	No	Yes	No	No
A178	Seamwelded tube	—	Yes	No	No	Yes	No	No	No	Yes	Yes
A214	Seamwelded tube	—	Yes	No	No	Yes	No	No	No	No	No
A226	Seamwelded tube	—	Yes	No	No	Yes	No	No	No	Yes	No
A105	Fittings	(3)	No	Yes	Yes	Yes	Yes	Yes	Yes	Yes	Yes
A181	Fittings	(3)	No	Yes	Yes	Yes	No	Yes	Yes	Yes	Yes
A234	Fittings	(3)	Yes	Yes	Yes	Yes	No	Yes	Yes	Yes	Yes
A350	Fittings	(3)	No	Yes	Yes	Yes	No	No	Yes	No	Yes
A420	Fittings	(3)	No	Yes	No	Yes	No	No	Yes	No	Yes
A694	Fittings	(3)	No	No	Yes	No	No	No	Yes	No	No

Notes:

(1) For materials acceptable for use in ASME I and/or ASME III applications, use ASME specifications (e.g., SA 234) in lieu of ASTM specifications (e.g., A 234). The ASME specifications for ferrous materials are prefixed by SA, whereas ASTM specifications are prefixed by A.

(2) For limitations or additional requirements on use of these material specifications, refer to the applicable code.

(3) Refer to code for acceptable grades, class, temper, and/or UNS number.

(4) Excluding grade F.

(5) For sizes NPS 4 and over.

(6) For sizes NPS 16 and over.

TABLE 3.2 Alloy Steel Pipe, Tube, and Fittings[(1),(2)]

Material spec	Description	Notes	B31.1	B31.3	B31.4	B31.5	B31.8	B31.9	B31.11	ASME I	ASME III
A333	Seamless pipe	(3)	Yes	Yes	No	Yes	Yes	No	No	No	Yes
A335	Seamless pipe	(3)	Yes	Yes	No	No	No	No	No	Yes	Yes
A369	Seamless pipe	(3)	Yes	Yes	No	No	No	No	No	Yes	Yes
A714	Seamless pipe		Yes	No	No	No	No	No	No	No	No
A199	Seamless tube	(3)	Yes	No	No	No	No	No	No	No	No
A213	Seamless tube	(3)	Yes	No	No	Yes	No	No	No	Yes	Yes
A334	Seamless tube		No	Yes	No	Yes	No	No	No	No	Yes
A333	Seamwelded pipe	(3)	Yes	Yes	No	Yes	Yes	No	No	No	Yes
A671	Seamwelded pipe		No	Yes	Yes	No	Yes	No	Yes	No	Yes
A672	Seamwelded pipe	(3)	Yes	Yes	Yes	No	Yes	No	Yes	No	Yes
A691	Seamwelded pipe	(3)	Yes	Yes	No	No	No	No	Yes	No	Yes
A714	Seamwelded pipe		Yes	No	No	No	No	No	No	No	No
A182	Fittings	(3)	No	Yes	Yes	Yes	No	No	Yes	Yes	Yes
A234	Fittings	(3)	Yes	Yes	Yes	Yes	No	Yes	Yes	Yes	Yes
A350	Fittings		No	Yes	Yes	Yes	No	No	Yes	No	Yes
A420	Fittings	(3)	No	Yes	No	Yes	No	No	Yes	No	Yes
A694	Fittings		No	No	Yes	No	No	No	Yes	No	No

Notes:

(1) For materials acceptable for use in ASME I and/or ASME III applications, use ASME specifications (e.g., SA 234) in lieu of ASTM specifications (e.g., A 234). The ASME specifications for ferrous materials are prefixed by SA, whereas ASTM specifications are prefixed by A.

(2) For limitations or additional requirements on use of these material specifications, refer to the applicable code.

(3) Refer to code for acceptable grades, class, temper, and/or UNS number.

PROPERTIES OF MATERIALS

Modulus of Elasticity

Table 3.42 lists the modulus of elasticity of various piping materials at different temperatures.

Density of Piping Materials

Refer to Table 3.43.

Specific Heat of Piping Materials

Table 3.43 lists the specific heat of piping materials at the mean temperature included in parentheses next to the value of specific heat.

TABLE 3.3 Stainless Steel Pipe, Tube, and Fittings[(1),(2)]

Material spec	Description	Notes	B31.1	B31.3	B31.4	B31.5	B31.8	B31.9	B31.11	ASME I	ASME III
A312	Seamless pipe	(3)	Yes	Yes	No	Yes	No	Yes	No	Yes	Yes
A376	Seamless pipe	(3)	Yes	Yes	No	Yes	No	No	No	Yes	Yes
A430	Seamless pipe	(3)	Yes	Yes	No	No	No	No	No	Yes	Yes
A731	Seamless pipe	(3)	Yes	No	No	No	No	No	No	No	No
A790	Seamless pipe		No	Yes	No	No	No	No	No	No	No
A213	Seamless tube	(3)	Yes	No	No	No	No	No	No	Yes	Yes
A268	Seamless tube	(3)	Yes	Yes	No	No	No	No	No	Yes	Yes
A269	Seamless tube		No	Yes	No	Yes	No	No	No	No	No
A789	Seamless tube		No	Yes	No	No	No	No	No	No	No
A312	Seamwelded pipe		Yes	Yes	No	Yes	No	No	No	Yes	Yes
A358	Seamwelded pipe	(3)	Yes	Yes	No	Yes	No	No	No	No	Yes
A409	Seamwelded pipe		No	Yes	No	Yes	No	No	No	No	Yes
A731	Seamwelded pipe		Yes	No	No	No	No	No	No	No	No
A790	Seamwelded pipe		No	Yes	No	No	No	No	No	No	No
A249	Seamwelded tube		Yes	No	No	Yes	No	No	No	Yes	Yes
A268	Seamwelded tube		Yes	Yes	No	No	No	No	No	Yes	Yes
A269	Seamwelded tube		No	Yes	No	Yes	No	No	No	No	No
A789	Seamwelded tube		No	Yes	No	No	No	No	No	No	No
A182	Fittings	(3)	No	Yes	No	No	No	No	No	Yes	Yes
A403	Fittings	(3)	Yes	Yes	No	Yes	No	Yes	No	No	Yes

Notes:

(1) For materials acceptable for use in ASME I and/or ASME III applications, use ASME specifications (e.g., SA 182) in lieu of ASTM specifications (e.g., A 182). The ASME specifications for ferrous materials are prefixed by SA, whereas ASTM specifications are prefixed by A.

(2) For limitations or additional requirements on use of these material specifications, refer to the applicable code.

(3) Refer to code for acceptable grades, class, temper, and/or UNS number.

TABLE 3.4 Nickel and Nickel Alloy Pipe, Tube, and Fittings[(1),(2)]

Material spec	Description	Notes	B31.1	B31.3	B31.4	B31.5	B31.8	B31.9	B31.11	ASME I	ASME III
B165	Seamless pipe/tube	(3)	Yes	Yes	No	Yes	No	No	No	No	No
B167	Seamless pipe/tube	(3)	Yes	Yes	No	No	No	No	No	No	No
B407	Seamless pipe/tube	(3)	Yes	Yes	No	No	No	No	No	No	No
B423	Seamless pipe/tube	(3)	Yes	No	No	No	No	No	No	No	No
B444	Seamless pipe/tube		No	Yes	No	No	No	No	No	No	No
B622	Seamless pipe/tube		No	Yes	No	No	No	No	No	No	No
B690	Seamless pipe/tube	(3)	Yes	Yes	No	No	No	No	No	No	No
B729	Seamless pipe/tube	(3)	Yes	Yes	No	No	No	No	No	No	No
B161	Seamless tube	(3)	Yes	Yes	No	No	No	No	No	No	No
B163	Seamless tube	(3)	Yes	No	No	No	No	No	No	No	No
B464	Seamwelded pipe	(3)	Yes	Yes	No	No	No	No	No	No	No
B514	Seamwelded pipe		No	Yes	No	No	No	No	No	No	N

TABLE 3.4 Nickel and Nickel Alloy Pipe, Tube, and Fittings[(1),(2)] (Continued)

Material spec	Description	Notes	B31.1	B31.3	B31.4	B31.5	B31.8	B31.9	B31.11	ASME I	ASME III
B517	Seamwelded pipe		No	Yes	No	No	No	No	No	No	No
B619	Seamwelded pipe		No	Yes	No	No	No	No	No	No	No
B675	Seamwelded pipe	(3)	Yes	Yes	No	No	No	No	No	No	No
B705	Seamwelded pipe		No	Yes	No	No	No	No	No	No	No
B804	Seamwelded pipe	(3)	Yes	Yes	No	No	No	No	No	No	No
B468	Seamwelded tube	(3)	Yes	No	No	No	No	No	No	No	No
B676	Seamwelded tube	(3)	Yes	No	No	No	No	No	No	No	No
B160	Fittings	(3)	No	Yes	No	No	No	No	No	No	Yes
B164	Fittings	(3)	No	Yes	No	No	No	No	No	No	Yes
B166	Fittings	(3)	No	Yes	No	No	No	No	No	No	Yes
B366	Fittings	(3)	Yes	Yes	No	No	No	No	No	No	No
B462	Fittings	(3)	Yes	Yes	No	No	No	No	No	No	Yes
B564	Fittings	(3)	No	Yes	No	No	No	No	No	No	Yes

Notes:
(1) For materials acceptable for use in ASME I and/or ASME III applications, use ASME specifications (e.g., SB 160) in lieu of ASTM specifications (e.g., B 160). The ASME specifications for nonferrous materials are prefixed by SB, whereas ASTM specifications are prefixed by B.
(2) For limitations or additional requirements on use of these material specifications, refer to the applicable code.
(3) Refer to code for acceptable grades, class, temper, and/or UNS number.

TABLE 3.5 Copper and Copper Alloy Pipe, Tube, and Fittings[(1),(2)]

Material spec	Description	Notes	B31.1	B31.3	B31.4	B31.5	B31.8	B31.9	B31.11	ASME I	ASME III
B42	Seamless pipe	(3)	Yes	Yes	No	Yes	Yes	Yes	No	No	No
B43	Seamless pipe		Yes	Yes	No	Yes	Yes	Yes	No	No	No
B302	Seamless pipe	(3)	Yes	No	No	Yes	No	Yes	No	No	No
B315	Seamless pipe/tube	(3)	Yes	No	No	Yes	No	No	No	No	No
B466	Seamless pipe/tube	(3)	Yes	Yes	No	Yes	No	No	No	No	No
B68	Seamless tube	(3)	Yes	Yes	No	Yes	Yes	Yes	No	No	No
B75	Seamless tube	(3)	Yes	Yes	No	Yes	Yes	Yes	No	No	No
B88	Seamless tube	(3)	Yes	Yes	No	Yes	Yes	Yes	No	No	No
B111	Seamless tube	(3)	Yes	No	No	Yes	No	No	No	No	No
B135	Seamless tube		No	No	No	No	No	Yes	No	No	No
B280	Seamless tube		No	Yes	No	Yes	No	Yes	No	No	No
B467	Seamwelded pipe	(3)	Yes	Yes	No	Yes	No	No	No	No	No
B608	Seamwelded pipe	(3)	Yes	No	No	No	No	No	No	No	No
B21	Fittings		No	Yes	No	Yes	No	No	No	No	No
B98	Fittings	(3)	No	Yes	No	Yes	No	No	No	No	Yes
B133	Fittings		No	Yes	No	No	No	No	No	No	Yes
B150	Fittings	(3)	Yes	Yes	No	No	No	No	No	No	Yes

Notes:
(1) For materials acceptable for use in ASME I and/or ASME III applications, use ASME specifications (e.g., SB 98) in lieu of ASTM specifications (e.g., B 98). The ASME specifications for nonferrous materials are prefixed by SB, whereas ASTM specifications are prefixed by B.
(2) For limitations or additional requirements on use of these material specifications, refer to the applicable code.
(3) Refer to code for acceptable grades, class, temper, and/or UNS number.

TABLE 3.6 Aluminum and Aluminum Alloy Pipe, Tube, and Fittings[(1),(2)]

Material spec	Description	Notes	B31.1	B31.3	B31.4	B31.5	B31.8	B31.9	B31.11	ASME I	ASME III
B210	Seamless tube		Yes	Yes	No	Yes	No	Yes	No	No	No
B234	Seamless tube		No	No	No	Yes	No	No	No	No	No
B241	Seamless tube	(3)	Yes	Yes	No	Yes	No	Yes	No	No	No
B345	Seamless pipe/tube		No	Yes	No	Yes	No	No	No	No	No
B247	Fittings	(3)	No	Yes	No	Yes	No	Yes	No	No	Yes
B361	Fttings		No	Yes	No	Yes	No	Yes	No	No	No

Notes:
(1) For materials acceptable for use in ASME I and/or ASME III applications, use ASME specifications (e.g., SB 247) in lieu of ASTM specifications (e.g., B 247). The ASME specifications for nonferrous materials are prefixed by SB, whereas ASTM specifications are prefixed by B.
(2) For limitations or additional requirements on use of these material specifications, refer to the applicable code.
(3) Refer to code for acceptable grades, class, temper, and/or UNS number.

TABLE 3.7 ASTM Material Specifications for Carbon Steel Seamless Pipe

ASTM Spec	Title
A53	Pipe, Steel, Black and Hot-Dipped, Zinc Coated, Welded and Seamless
A106	Seamless Carbon Steel Pipe for High-Temperature Service
A333	Seamless and Welded Steel Pipe for Low-Temperature Service
A369	Carbon and Ferritic Alloy Steel Forged and Bored Pipe for High-Temperature Service
API-5L	Line Pipe
A524	Seamless Carbon Steel Pipe for Atmospheric and Lower Temperatures

TABLE 3.8 ASTM Material Specifications for Carbon Steel Seamwelded Pipe

ASTM Spec	Title
A53	Pipe, Steel, Black and Hot-Dipped, Zinc Coated, Welded and Seamless
A134	Pipe, Steel, Electric-Fusion (Arc)-Welded (Sizes NPS 16 and Over)
A135	Electric-Resistance-Welded Steel Pipe
A139	Electric-Fusion (ArC)-Welded Steel Pipe (Sizes NPS 4 and Over)
A211	Spiral-Welded Steel or Iron Pipe
A333	Seamless and Welded Steel Pipe for Low-Temperature Service
A381	Metal-Arc-Welded Steel Pipe for Use with High-Pressure Transmission Systems
A587	Electric-Welded Low-Carbon Steel Pipe for the Chemical Industry
A671	Electric-Fusion-Welded Steel Pipe for Atmospheric and Lower Temperatures
A672	Electric-Fusion-Welded Steel Pipe for High-Pressure Service at Moderate Temperatures
A691	Carbon and Alloy Steel Pipe, Electric Fusion-Welded for High-Pressure Service at High Temperatures
API-5L	Line Pipe

TABLE 3.9 ASTM Material Specifications for Carbon Steel Seamless Tube

ASTM Spec	Title
A179	Seamless Cold-Drawn Low-Carbon Steel Heat-Exchanger and Condenser Tubes
A192	Seamless Carbon Steel Boiler Tubes for Pressure Service
A210	Seamless Medium-Carbon Steel Boiler and Superheater Tubes
A334	Seamless and Welded Carbon and Alloy-Steel Tubes for Low-Temperature Service

TABLE 3.10 ASTM Material Specifications for Carbon Steel Seamwelded Tube

ASTM Spec	Title
A178	Electric-Resistance-Welded Carbon Steel and Carbon-Manganese Steel Boiler and Superheater Tubes
A214	Electric-Resistance-Welded Carbon Steel Heat-Exchanger and Condenser Tubes
A226	Electric-Resistance-Welded Carbon Steel Boiler and Superheater Tubes for High-Pressure Service

TABLE 3.11 ASTM Material Specifications for Carbon Steel Pipe Fittings

ASTM Spec	Title
A105	Forgings, Carbon Steel, for Piping Components
A181	Forgings, Carbon Steel for General Purpose Piping
A234	Piping Fittings of Wrought Carbon Steel and Alloy Steel for Moderate and Elevated Temperatures
A350	Forgings, Carbon and Low-Alloy Steel Requiring Notch Toughness Testing for Piping Components
A420	Piping Fittings of Wrought Carbon Steel and Alloy Steel for Low-Temperature Service
A694	Forgings, Carbon and Alloy Steel, for Pipe Flanges, Fittings, Valves, and Parts for High-Pressure Transmission Service

TABLE 3.12 ASTM Material Specifications for Alloy Steel Seamless Pipe

ASTM Spec	Title
A333	Seamless and Welded Steel Pipe for Low-Temperature Service
A335	Seamless Ferritic Alloy Steel Pipe for High-Temperature Service
A369	Carbon Steel and Ferritic Alloy Steel Forged and Bored Pipe for High-Temperature Service
A714	High-Strength Low-Alloy Welded and Seamless Steel Pipe

TABLE 3.13 ASTM Material Specifications for Alloy Steel Seamwelded Pipe

ASTM Spec	Title
A333	Seamless and Welded Steel Pipe for Low-Temperature Service
A671	Electric-Fusion-Welded Steel Pipe for Atmospheric and Lower Temperatures
A672	Electric-Fusion-Welded Steel Pipe for High-Pressure Service at Moderate Temperatures
A691	Carbon and Alloy Steel Pipe, Electric Fusion-Welded for High-Pressure Service at High Temperatures
A714	High-Strength Low-Alloy Welded and Seamless Steel Pipe

TABLE 3.14 ASTM Material Specifications for Alloy Steel Seamless Tube

ASTM Spec	Title
A199	Seamless Cold-Drawn Intermediate Alloy Steel Heat Exchanger and Condenser Tubes
A213	Seamless Ferritic and Austenitic Alloy Steel Boiler, Superheater, and Heat Exchanger Tubes
A334	Seamless and Welded Carbon and Alloy Steel Tubes for Low-Temperature Service

TABLE 3.15 ASTM Material Specifications for Alloy Steel Pipe Fittings

ASTM Spec	Title
A182	Forged or Rolled Alloy-Steel Pipe Flanges, Forged Fittings, and Valves and Parts for High-Temperature Service
A234	Piping Fittings of Wrought Carbon Steel and Alloy Steel for Moderate and Elevated Temperatures
A350	Forgings, Carbon and Low-Alloy Steel Requiring Notch Toughness Testing for Piping Components
A420	Piping Fittings of Wrought Carbon Steel and Alloy Steel for Low-Temperature Service
A694	Forgings, Carbon and Alloy Steel, for Pipe Flanges, Fittings, Valves, and Parts for High-Pressure Transmission Service

TABLE 3.16 ASTM Material Specifications for Stainless Steel Seamless Pipe

ASTM Spec	Title
A312	Seamless and Welded Austenitic Stainless Steel Pipe
A376	Seamless Austenitic Steel Pipe for High-Temperature Central-Station Service
A430	Austenitic Steel Forged and Bored Pipe for High-Temperature Service
A731	Seamless, Welded Ferritic, and Martensitic Stainless Steel Pipe
A790	Seamless and Welded Ferritic/Austenitic Stainless Steel Pipe

TABLE 3.17 ASTM Material Specifications for Stainless Steel Seamwelded Pipe

ASTM Spec	Title
A312	Seamless and Welded Austenitic Stainless Steel Pipe
A358	Electric-Fusion-Welded Austenitic Chromium-Nickel Alloy Steel Pipe for High-Temperature Service
A409	Welded Large Diameter Austenitic Steel Pipe for Corrosive or High-Temperature Service
A731	Seamless, Welded Ferritic, and Martensitic Stainless Steel Pipe
A790	Seamless and Welded Ferritic/Austenitic Stainless Pipe

TABLE 3.18 ASTM Material Specifications for Stainless Steel Seamless tube

ASTM Spec	Title
A213	Seamless Ferritic and Austenitic Alloy Steel Boiler, Superheater, and Heat Exchanger Tubes
A268	Seamless and Welded Ferritic Stainless Steel Tubing for General Service
A269	Seamless and Welded Austenitic Stainless Steel Tubing for General Service
A789	Seamless and Welded Ferritic/Austenitic Stainless Steel Tubing for General Service

TABLE 3.19 ASTM Material Specifications for Stainless Steel Seamwelded tube

ASTM Spec	Title
A249	Welded Austenitic Steel Boiler, Superheater, Heat Exchanger, and Condenser Tubes
A268	Seamless and Welded Ferritic and Martensitic Stainless Steel Tubing for General Service
A269	Seamless and Welded Austenitic Stainless Steel Tubing for General Service
A789	Seamless and Welded Ferritic/Austenitic Stainless Steel Tubing for General Service

TABLE 3.20 ASTM Material Specifications for Stainless Steel Pipe Fittings

ASTM Spec	Title
A182	Forged or Rolled Alloy-Steel Pipe Flanges, Forged Fittings, and Valves and Parts for High-Temperature Service
A403	Wrought Austenitic Stainless Steel Piping Fittings

TABLE 3.21 ASTM Material Specifications for Nickel and Nickel Alloy Seamless Pipe/Tube

ASTM Spec	Title
B165	Nickel-Copper Alloy (UNS N04400) Seamless Pipe and Tube
B167	Nickel-Chromium-Iron Alloys (UNS N06600, N06601, N06690, N06025, and N06045) Seamless Pipe and Tube
B407	Nickel-Iron-Chromium Alloy Seamless Pipe and Tube
B423	Nickel-Iron-Chromium-Molybdenum-Copper Alloy (UNS N08825 and N08221) Seamless Pipe and Tube
B444	Nickel-Chromium-Molybdenum-Columbium Alloys (UNS N06625) Pipe and Tube
B622	Seamless Nickel and Nickel-Cobalt Alloy Pipe and Tube
B690	Iron-Nickel-Chromium-Molybdenum Alloys (UNS N08366 and UNS N08367) Seamless Pipe and Tube
B729	Seamless UNS N08020, UNS N08026, and UNS N08024 Nickel-Alloy Pipe and Tube

TABLE 3.22 ASTM Material Specifications for Nickel and Nickel Alloy Seamwelded Pipe

ASTM Spec	Title
B464	Welded UNS N08020, UNS N08024, and UNS N08026 Alloy Pipe
B514	Welded Nickel-Iron-Chromium Alloy Pipe
B517	Welded Nickel-Chromium-Iron Alloy (UNS N06600), UNS N06025, and UNS N06045 Pipe
B619	Welded Nickel and Nickel-Cobalt Alloy Pipe
B675	UNS N08366 and UNS N08367 Welded Pipe
B705	Nickel-Alloy (UNS N06625 and UNS N08825) Welded Pipe
B804	UNS N08367 and UNS N08926 Welded Pipe

TABLE 3.23 ASTM Material Specifications for Nickel and Nickel Alloy Seamless Tube

ASTM Spec	Title
B161	Nickel Seamless Pipe and Tube
B163	Seamless Nickel and Nickel Alloy Condenser and Heat Exchanger Tubes

TABLE 3.24 ASTM Material Specifications for Nickel and Nickel Alloy Seamwelded Tube

ASTM Spec	Title
B468	Welded UNS N08020, UNS N08024, and UNS N08026 Alloy Tubes
B676	UNS N08366 and UNS N08367 Welded Tube

TABLE 3.25 ASTM Material Specifications for Nickel and Nickel Alloy Pipe Fittings

ASTM Spec	Title
B160	Nickel Rod and Bar
B164	Nickel-Copper Alloy Rod, Bar, and Wire
B166	Nickel-Chromium-Iron Alloys (UNS N06600, N06601, N06690, N06025, and N06045) and Nickel-Chromium-Cobalt-Molybdenum Alloy (UNS N06617) Rod, Bar and Wire
B366	Factory-Made Wrought Nickel and Nickel Alloy Fittings
B462	Forged or Rolled UNS N08020, UNS N08024, UNS N08026, and UNS N08367 Alloy Pipe Flanges, Forged Fittings, and Valves and Parts for Corrosive High-Temperature Service
B564	Nickel Alloy Forgings

TABLE 3.26 ASTM Material Specifications for Copper and Copper Alloy Seamless Pipe

ASTM Spec	Title
B42	Seamless Copper Pipe, Standard Sizes
B43	Seamless Red Brass Pipe, Standard Sizes
B302	Threadless Copper Pipe

TABLE 3.27 ASTM Material Specifications for Copper and Copper Alloy Seamwelded Pipe

ASTM Spec	Title
B467	Welded Copper-Nickel Pipe
B608	Welded Copper Alloy Pipe

TABLE 3.28 ASTM Material Specifications for Copper and Copper Alloy Seamless Tube

ASTM Spec	Title
B68	Seamless Copper Tube, Bright Annealed
B75	Seamless Copper Tube
B88	Seamless Copper Water Tube
B111	Copper and Copper-Alloy Seamless Condenser Tubes and Ferrule Stock
B135	Seamless Brass Tube
B280	Seamless Copper Tube for Air Conditioning and Refrigeration Fluid Service

TABLE 3.29 ASTM Material Specifications for Copper and Copper Alloy Seamless Pipe/Tube

ASTM Spec	Title
B315	Seamless Copper Alloy Pipe and Tube
B466	Seamless Copper-Nickel Pipe and Tube

TABLE 3.30 ASTM Material Specifications for Copper and Copper Alloy Pipe Fittings

ASTM Spec	Title
B21	Naval Brass Rod, Bar, and Shapes
B98	Copper-Silicon Alloy Rod, Bar and Shapes
B133	Copper Rod, Bar and Shapes
B150	Aluminum-Bronze Rod, Bar and Shapes

TABLE 3.31 ASTM Material Specifications for Aluminum and Aluminum Alloy Seamless Tube

ASTM Spec	Title
B210	Aluminum and Aluminum-Alloy Drawn Seamless Tubes
B234	Aluminum and Aluminum-Alloy Drawn Seamless Tubes for Condensers and Heat Exchangers
B241	Aluminum-Alloy Seamless Pipe and Seamless Extruded Tube

TABLE 3.32 ASTM Material Specifications for Aluminum and Aluminum Alloy Seamless Pipe/Tube

ASTM Spec	Title
B345	Aluminum-Alloy Seamless Extruded Tube and Seamless Pipe for Gas and Oil Transmission and Distribution Piping Systems

TABLE 3.33 ASTM Material Specifications for Aluminum and Aluminum Alloy Pipe Fittings

ASTM Spec	Title
B247	Aluminum-Alloy Die, Hand and Rolled Ring Forgings
B361	Factory-Made Wrought Aluminum and Aluminum-Alloy Welding Fittings

TABLE 3.34 ASTM Material Specifications for Titanium and Titanium Alloy Seamless Pipe and Tube

ASTM Spec	Title
B337	Seamless and Welded Titanium and Titanium-Alloy Pipe
B338	Seamless and Welded Titanium and Titanium Alloy Tubes for Condensers and Heat Exchangers

TABLE 3.35 ASTM Material Specifications for Titanium and Titanium Alloy Seamwelded Pipe and Tube

ASTM Spec	Title
B337	Seamless and Welded Titanium and Titanium-Alloy Pipe
B338	Seamless and Welded Titanium and Titanium Alloy Tubes for Condensers and Heat Exchangers

TABLE 3.36 ASTM Material Specifications for Titanium and Titanium Alloy Pipe Fittings

ASTM Spec	Title
B381	Titanium and Titanium Alloy Forgings

Melting Temperature of Piping Materials

Refer to Table 3.43.

Thermal Conductivity of Piping Materials

Table 3.44 lists thermal conductivity of various piping materials.

Linear Coefficient of Thermal Expansion

Refer to Table 3.44. It lists the average coefficient of linear thermal expansions within the temperature range indicated.

TABLE 3.37 Cross Reference for International Materials—Carbon Steel Pipe, Tube, Forgings, Fittings, and Castings

American				British			German			Belgian		French		Italian		Japanese	
Nominal composition	Spec	Grade	Type or class	BS number	BS grade	BS process	DIN type	DIN number	DIN material number	NBN type	NBN number	AFNOR type	NF number	UNI type	UNI number	JIS symbol	JIS number
Seamless pipe and tube																	
C	A53	A	S	3601	360		St 35	1629	1.0308	E 37-2	630			Fe 35-1	663	STPG 370	G3454
C-Mn	A53	B	S	3601	430		St 45	1629	1.0408	E 42-2	630					STPG 410	G3454
C-Si	A106	A		3602	360	HFS	St 35.8	17175	1.0305	D 37.2	629	TU 37C	A49-213	C 14	5462	STPT 370	G3456
C-Si	A106	B		3602	430	HFS	St 45.8	17175	1.0405	D 42.2	629	TU 42C	A49-213	C 18	5462	STPT 410	G3456
C-Si	A106	C					17 Mn 4	17175								STPT 480	G3456
C	A179			3606	320	CFS	St 35.4 St 35.8	1629 17175	1.0309 1.0305	D 37.2	629	TU 37C	A49-215	C 14	5462	STB 340	G3461
C-Si	A192			3059	320		St 35.8	17175	1.0305	D45	837	TU 37C	A49-213	C 14	5462	STB 340	G3461
C-Si	A210	A1		3059	320		St 45.8	17175	1.0405	D45	837	TU 42C	A49-213	C 18	5462	STB 410	G3461
C-Mn-Si	A210	C															
C-Mn	A333	1					TT St 41N		1.0437			TU 42 BT	A49-230	C 15	5949	STPL 380	G3460
C-Mn-Si	A333	6		3603	410	HFS	TT St 45N		1.0456			TU 42 BT	A49-230	C 20	5949	STPL 690	G3460
C-Si	A369	FPA															
C-Mn	A369	FPB															
C	API-5L	A		3601	360												
C-Mn	API-5L	B		3601	430												
Furnace butt-welded pipe																	
C	A53		F														
C	API-5L	A25	I & II	1387													
Electric resistance-welded pipe																	
C	A53	A	E	3601	320		St 35	1629									
C-Mn	A53	B	E	3601	410		St 45	1629									
C	A135	A															
C-Mn	A135	B															
C	A178	A		3059	320	ERW	St 37.8	17177	1.0315							STB 340	G3461
C		C					St 42.8	17177	1.0498							STB 410	G3461
C	A214			3606	30	ERW	St 37.2	1626	1.0036	D37.1	629	TS 34e	A49-245			STB 340	G3461
C-Si	A226																

C-Mn	A333	1					TT St 35N		1.0437		TU 42 BT	A49-230	C 15	5949	STPL 380	G3460
C-Mn-Si	A333	6		3603	410	HFS	TT St 44N		1.5637		TU 42 BT	A49-230	C 20	5949	STPL 690	G3460
C	API-5L	A25	I & II	1387			St 33	1626								
C	API-5L	A		3601	360		St 34.7	17172								
C-Mn	API-5L	B		3601	410		St 38.7	17172								
C	A587	C														
							Electric fusion-welded pipe (filler metal added)									
C	A134	A283A		3602	410	SAW									STPY 400	G3457
C	A134	A283B													STPY 400	G3457
C	A134	A283C													STPY 400	G3457
C	A134	A283D													STPY 400	G3457
C	A134	A285A													STPY 400	G3457
C	A134	A285B													STPY 400	G3457
C	A134	A285C													STPY 400	G3457
C	A139	A														
C-Mn	A139	B		3602	410	SAW										
C	API-5L	A		3601	360		St 34.7	17172								
C-Mn	API-5L	B		3601	410		St 38.7	17172								
C	A211	A570-30														
C	A211	A570-33														
C	A211	A570-340														
C-Mn-Si	A671	CD70														
C-Mn-Si	A671	CD80														
C	A672	A45					H I	1626								
C	A672	A50					H I	1626								
C	A672	A55					H I	1626								
C	A672	B55														
C	A672	B60														

TABLE 3.37 Cross Reference for International Materials—Carbon Steel Pipe, Tube, Forgings, Fittings, and Castings (Continued)

American				British			German			Belgian		French		Italian		Japanese	
Nominal composition	Spec	Grade	Type or class	BS number	BS grade	BS process	DIN type	DIN number	DIN material number	NBN type	NBN number	AFNOR type	NF number	UNI type	UNI number	JIS symbol	JIS number
C	A672	B65					H II	1626									
C	A672	B70					17 Mn 4	1626									
C	A672	C55															
C	A672	C60															
C	A672	C65															
C	A672	C70															
C-Mn-Si	A672	D70															
C-Mn-Si	A672	D80															
C-Mn-Si	A672	N75															
C-Mn-Si	A691	CMSH70															
C-Mn-Si	A691	CMSH80															
C-Mn-Si	A691	CMS75															
Forgings																	
C-Si	A105			1503	221-490		19 Mn 5	17155	1.0482			AF 48N	M87-508	Fe 510B	7746	SFVC2A	G3202
C-Si	A181	60		1503	221-430		St 44-2	17100	1.0044	E 37-1	630	AF 42-2	A33-101	Fe 410B	7746	SFVC1	G3202
C-Si	A181	70		1503	221-490		C22	17200	1.0402	E 52-1	630	AF 48N	M87-508	Fe 510B	7746	SFVC2A	G3202
Fittings																	
C-Si	A234	WPB		1503	221-460		St 45.8	17175	1.0405			TU 42 C	A49-213			STPT 42	G3456
C-Si	A234	WPC		1503	221-510									C 18	5462	STPT 49	G3456
Castings																	
C-Si	A216	WCA		1540-161	221-430		GS-C25	17245	1.0519			A42C-M	A32-055			SCPH 1	G5151
C-Si	A216	WCB		1540-161	221-480							A42C-M	A32-055			SCPH 2	G5151
C-Mn-Si	A216	WCC															

TABLE 3.38 Cross Reference for International Materials—Alloy Steel Pipe, Tube, Forgings, Fittings, and Castings

American				British			German			Belgian		French		Italian		Japanese	
Nominal composition	Spec	Grade	Type or class	BS number	BS grade	BS process	DIN type	DIN number	DIN material number	NBN type	NBN number	AFNOR type	NF number	UNI type	UNI number	JIS symbol	JIS number
Seamless pipe and tube																	
5Cr-½Mo	A 199	T5		3606	625	CSF	12 Cr Mo 19 5		1.7362	X 12 Cr Mo	837	Z 12 CD 5.05				STBA25	G3462
9Cr-1Mo	A 199	T9														STBA26	G3462
1¼Cr-½Mo	A 199	T11		3606	621	CFS	13 Cr Mo 44	17175	1.7335			10 CD 5.05				STBA23	G3462
3Cr-1Mo	A 199	T21															
2¼Cr-1Mo	A 199	T22		3606	622	CFS	10 Cr Mo 9 10	17175	1.738	10 Cr Mo 9	837	10 CD 9.10		12 Cr Mo 9	5462	STBA24	G3462
½Cr-½Mo	A 213	T2														STBA20	G3462
2Cr-½Mo	A 213	T3b															
5Cr-½Mo	A 213	T5		3606	625		12 Cr Mo 195		1.7362			TU Z 12 CD	A49-213			STBA25	G3462
5Cr-½Mo-1½Si	A 213	T5b															
5Cr-½Mo-Ti	A 213	T5c															
9Cr-1Mo	A 213	T9														STBA26	G3462
1¼Cr-½Mo	A 213	T11		3606	621		13 Cr Mo 44	17175	1.7335			TU 10 CD 5.05				STBA23	G3462
1Cr-½Mo	A 213	T12		3606	620		13 Cr Mo 44	17175	1.7335							STBA22	G3462
3Cr-1Mo	A 213	T21															
2¼Cr-1Mo	A 213	T22		3606	622		10 Cr Mo 9 10	17175	1.738			TU 10 CD 9.10		12 Cr Mo 9	5462	STBA24	G3462
9Cr-1Mo-V	A 213	T91															
3½Ni	A 333	3		3603	503	HFS	10 Ni 14		1.5637			TU 16 N 14	A49-230	18 Ni 14	5949	STPL450	G3460
Ni-Cr-Cu-Al	A 333	4															
2½Ni	A 333	7															
2Ni-1Cu	A 333	9															
C-½Mo	A 335	P1					16 Mo 5		1.5423			TU 15 D 3	A49-213	16 Mo 5	5462	STPA12	G3458
½Cr-½Mo	A 335	P2														STPA20	G3458
5Cr-½Mo	A 335	P5		3604	625	HFS	12 Cr Mo 19 5		1.7362			TU Z 12 CD	A49-213			STPA25	G3458
5Cr-½Mo-1½Si	A 335	P5b															
5Cr-½Mo-Ti	A 335	P5c															
9Cr-1Mo	A 335	P9			629	HFS						TU Z 10 CD	A49-213			STPA26	G3458
1¼Cr-½Mo-Si	A 335	P11		3604	621	HFS	13 Cr Mo 44	17175	1.7335			TU 10 CD 5	A49-213			STPA23	G3458
1Cr-½Mo	A 335	P12		3604	620	HFS	13 Cr Mo 44	17175	1.7335					14 Cr Mo 3	5462	STPA22	G3458
3Cr-1Mo	A 335	P21															
2¼Cr-1Mo	A 335	P22		3604	622	HFS	10 Cr Mo 9 10	17175	1.738			TU 10 CD 5	A49-213	12 Cr Mo 9	5462	STPA24	G3458
9Cr-1Mo-V	A 335	P91															
C-½Mo	A 369	FP1															
½Cr-½Mo	A 369	FP2															
2Cr-½Mo	A 369	FP3b															
5Cr-½Mo	A 369	FP5															
9Cr-1Mo	A 369	FP9															
1¼Cr-½Mo-Si	A 369	FP11															
1Cr-½Mo	A 369	FP12															
3Cr-1Mo	A 369	FP21															
2¼Cr-1Mo	A 369	FP22															
2Ni-1Cu	A 714	V															
Centrifugally cast pipe																	
C-½Mo	A 426	CP1														SCPH-11C	G5202
½Cr-½Mo	A 426	CP2															
5Cr-½Mo	A 426	CP5															
5Cr-½Mo-Si	A 426	CP5B															
9Cr-1Mo	A 426	CP9															
1¼Cr-½Mo	A 426	CP11															
1Cr-½Mo	A 426	CP12					GS-17 CrMo55	17245	1.7357							SCPH-21C	G5202
3Cr-1Mo	A 426	CP21															
2¼Cr-1Mo	A 426	CP22					GS-18 CrMo9.10		1.7379							SCPH-32C	G5202

TABLE 3.38 Cross Reference for International Materials—Alloy Steel Pipe, Tube, Forgings, Fittings, and Castings (Continued)

American				British			German			Belgian		French		Italian		Japanese	
Nominal composition	Spec	Grade	Type or class	BS number	BS grade	BS process	DIN type	DIN number	DIN material number	NBN type	NBN number	AFNOR type	NF number	UNI type	UNI number	JIS symbol	JIS number
Electric resistance-welded pipe																	
3½Ni	A 333	3		3603	503	HFS	10 Ni 14		1.5637			TU 16 N 14	A49-230	18 Ni 14	5949	STPL450	G3460
2½Ni	A 333	7															
2Ni	A 333	9															
2Ni-Cu	A 714	V	E														
Electric fusion-welded pipe—filler metal added																	
C-½Mo	A 672	L65	20, 23, 30, 33, 40, 43														
C-½Mo	A 672	L70	20, 23, 30, 33, 40, 43														
C-½Mo	A 672	L75	20, 23, 30, 33, 40, 43														
C-½Mo	A 691	CM65	20, 23, 30, 33, 40, 43														
C-½Mo	A 691	CM70	20, 23, 30, 33, 40, 43														
C-½Mo	A 691	CM75	20, 23, 30, 33, 40, 43														
½Cr-½Mo	A 691	½Cr	20, 23														
1Cr-½Mo	A 691	1Cr	20, 23														
1¼Cr-½Mo-Si	A 691	1¼Cr	20, 23														
2¼Cr-1Mo	A 691	2¼Cr	20, 23														
3Cr-1Mo	A 691	3Cr	20, 23														
5Cr-½Mo	A 691	5Cr	20, 23														
Forgings																	
Cr-1/2 Mo	A182	F1		1503	245-420		16Mo5		1.5432			25 D5	M87-508			SFVA F1	G3203
1/2 Cr-1/2 Mo	A182	F2															
5Cr-1/2 Mo	A182	F5		1503	625-520		12CrMo19 5		1.7362			Z 10 CD 5-	M87-508			SFVA F5B	G3203
5Cr-1/2 Mo	A182	F5a															
9Cr-1Mo	A182	F9															
9Cr-1Mo-V	A182	F91															
1 1/4 Cr-1/4 Mo-Si	A182	F11	Class 2	1503	621-460		13CrMo44	17175	1.7335			15 SD 5-03	M87-508	25CO4		SFVA F11	G3203
1Cr-½Mo	A182	F12	Class 2	1503	620-540		13CrMo44	17175	1.7335			15 CD 4-05	M87-508	13CO4		SFVA F12	G3203
3Cr-1Mo	A182	F21															
2 1/4 Cr-1Mo	A182	F22	Class 3	1503	622-560		10CrMo9 10	17175	1.738			10 CD 9-10	M87-508			SFVA F221	G3203
3 1/2 Ni	A350	LF3		1640	WPL-3		10 Ni 14		1.5637			AFN 14	M87-508			SFL 3	G3205
3/4Cr-3/4Ni-Cu-Al	A350	LF4															
Mn-1 1/2 Ni	A350	LF5	Class 1														
Mn-1 1/2 Ni	A350	LF5	Class 2														
2Ni-1Cu	A350	LF9															
Fittings																	
Cr-1/2 Mo	A234	WP1		1503	245-420		16Mo5		1.5423					16 Mo 5	5462	STPA 12	G3458
5Cr-1/2 Mo	A234	WP5		1503	625-520		12CrMo19 5		1.7362			TU Z 12 CD	A49-213			STPA 25	G3458
1 1/4 Cr-1/2 Mo	A234	WP11		1503	621-460		13CrMo44	17175	1.7335			TU 6 CD 5	A49-213	14 Cr Mo 3	5462	STPA 23	G3458
1Cr-1/2 Mo	A234	WP12		1503	620-440		13CrMo44	17175	1.7335			TU 6 CD 5	A49-213	14 Cr Mo 3	5462	STPA 22	G3458
2 1/4 Cr-1Mo	A234	WP22															
9Cr-1Mo-V	A234	WP91		1503	622-490		10CrMo9 10	17175	1.738			TU 10 CD 5	A49-213	12 Cr Mo 9	5462	STPA 24	G3458
Castings																	
Cr-1/2 Mo	A217	WC1		1504-240			GS-22 Mo 4	17245	1.5419			20 D5-M	A32-055			SCPH 11	G5151
Ni-Cr-1/2 Mo	A217	WC4															
Ni-Cr-1Mo	A217	WC5															
1 1/4Cr-1/2 Mo	A217	WC6		1504-621			GS-17 CrMo 5 5	17245	1.7357			15 CD 5 05	A32-055			SCPH 21	G5151
2 1/4Cr-1Mo	A217	WC9										15 CD 5 10	A32-055				
5Cr-1/2 Mo	A217	C5		1504-625			GS-12 CrMo 19 5		1.7363			Z 15 CD 5	A32-055			SCPH 61	G5151
9Cr-1Mo	A217	C12		1504-629			G-X-12 CrMo 10 1		1.7389								
9Cr-1Mo-V	A217	C12A															

TABLE 3.39 Cross Reference for International Materials—Stainless Steel (Austenitic) Pipe and Tube

American				British			German			Belgian		French		Italian		Japanese	
Nominal composition	Spec	Grade	Type or class	BS number	BS grade	BS process	DIN type	DIN number	DIN material number	NBN type	NBN number	AFNOR type	NF number	UNI type	UNI number	JIS symbol	JIS number
Seamless pipe and tube																	
18Cr-8Ni	A213	TP304		3606	304S25		X 5 CrNi 18 9	17440	1.4301			Z 6 CN 18-09	A49-218	X 5 Cr Ni 18 10	6904	SUS304TB	G3463
18Cr-8Ni	A213	TP304H														SUS304HTB	G3463
18Cr-8Ni	A213	TP304L		3606	304S22		X 2 CrNi 18 9	17440	1.4306			Z 2 CN 18-09	A49-218	X 2 Cr Ni 18 11	6904	SUS304LTB	G3463
18Cr-8Ni-N	A213	TP304N															
23Cr-12Ni	A213	TP309H															
25Cr-20Ni	A213	TP310H															
16Cr-12Ni-2Mo	A213	TP316		3606	316S30		X 5 CrNiMo 18 10	17440	1.4401			Z 6 CND 17-12	A49-218	X 5 Cr Ni Mo 17 12	6904	SUS316TB	G3463
16Cr-12Ni-2Mo	A213	TP316H														SUS316HTB	G3463
16Cr-12Ni-2Mo	A213	TP316L		3606	316S29		X 2 CrNiMo 18 10	17440	1.4404			Z 2 CND 17-13	A49-218	X 2 Cr Ni Mo 17 12	6904	SUS316LTB	G3463
16Cr-12Ni-2Mo-N	A213	TP316N															
18Cr-10Ni-Ti	A213	TP321		3606	321S22		X 10 CrNiTi 18 9	17440	1.4541			Z 6 CNT 18-11	A49-218	X 6 Cr Ni Ti 18 11	6904	SUS321TB	G3463
18Cr-10Ni-Ti	A213	TP321H															
18Cr-10Ni-Cb	A213	TP347		3606	347S17		X 10 CrNiNb 18 9	17440	1.455			Z 6 CNNb 18-11	A49-218	X 6 Cr Ni Nb 18 11	6904	SUS347TB	G3463
18Cr-10Ni-Cb	A213	TP347H														SUS347HTB	G3463
18Cr-10Ni-Cb	A213	TP348															
18Cr-10Ni-Cb	A213	TP348H															
18Cr-8Ni	A312	TP304		3605	304S18		X 5 CrNi 18 9	17440	1.4301			TU Z 6 CN 19-10	A49-230	X 5 Cr Ni 18 10	6904	SUS304TB	G3459
18Cr-8Ni	A312	TP304H		3605	394S59							TU Z 6 CN 19-10	A49-214	X 8 Cr Ni 19 10	6904	SUS304HTP	G3459
18Cr-8Ni	A312	TP304L		3605	304S14		X 2 CrNi 18 9	17440	1.4306			TU Z 2 CN 19-11	A49-230	X 2 Cr Ni 18 11	6904	SUS304LTP	G3459
18Cr-8Ni-N	A312	TP304N															
23Cr-12Ni	A312	TP309H															
25Cr-20Ni	A312	TP310H															
16Cr-12Ni-2Mo	A312	TP316		3605	316S18		X 5 CrNiMo 18 10	17440	1.4401			Z 6 CND 17-11	A35-573	X 5 Cr Ni Mo 17 12	6904	SUS316TP	G3459
16Cr-12Ni-2Mo	A312	TP316H		3605	316S59							TU Z 6 CND 17-12	A49-230	X 8 Cr Ni Nb 17 12	6904	SUS316HTP	G3459
16Cr-12Ni-2Mo	A312	TP316L		3605	316S14		X 2 CrNiMo 18 10	17440	1.4404			TU Z 2 CND 17-12	A49-230	X 2 Cr Ni Nb 17 12	6904	SUS316LTP	G3459
16Cr-12Ni-2Mo-N	A312	TP316N															
18Cr-13Ni-3Mo	A312	TP317															
18Cr-10Ni-Ti	A312	TP321		3605	321S18		X 10 CrNiTi 18 9	17440	1.4541			Z 6 CNT 18-11	A49-230	X 6 Cr Ni Ti 18 11	6904	SUS321TP	G3459
18Cr-10Ni-Ti	A312	TP321H														SUS321HTP	G3459
18Cr-10Ni-Cb	A312	TP347		3605	347S18		X 10 CrNiNb 18 9		1.4543			Z 6 CNNb 18-11	A35-573	X 6 Cr Ni Nb 18 11	6904	SUS347TP	G3459
18Cr-10Ni-Cb	A312	TP347H												X 6 Cr Ni Nb 18 11	6904	SUS347HTP	G3459
18Cr-10Ni-Cb	A312	TP348															
18Cr-10Ni-Cb	A312	TP348H															
18Cr-18Ni-2Si	A312	TPXM15															
22Cr-13Ni-5Mn	A312	TPXM19															
20Cr-18Ni-6Mo	A312	. . .															
18Cr-8Ni	A376	TP304															
18Cr-8Ni	A376	TP304H															
18Cr-8Ni-N	A376	TP304N															
16Cr-12Ni-2Mo	A376	TP316															
16Cr-12Ni-2Mo	A376	TP316H															
16Cr-12Ni-2Mo-N	A376	TP316N															
18Cr-10Ni-Ti	A376	TP321															
18Cr-10Ni-Ti	A376	TP321H															
18Cr-10Ni-Cb	A376	TP347															
18Cr-10Ni-Cb	A376	TP347H															
18Cr-10Ni-Cb	A376	TP348															
18Cr-8Ni	A430	FP304															
18Cr-8Ni	A430	FP304H															
18Cr-8Ni-N	A430	FP304N															
16Cr-12Ni-2Mo	A430	FP316															
16Cr-12Ni-2Mo	A430	FP316H															
16Cr-12Ni-2Mo-N	A430	FP316N															
18Cr-10Ni-Ti	A430	FP321															

TABLE 3.39 Cross Reference for International Materials—Stainless Steel (Austenitic) Pipe and Tube (Continued)

American				British			German			Belgian		French		Italian		Japanese	
Nominal composition	Spec	Grade	Type or class	BS number	BS grade	BS process	DIN type	DIN number	DIN material number	NBN type	NBN number	AFNOR type	NF number	UNI type	UNI number	JIS symbol	JIS number
18Cr-10Ni-Ti	A430	FP321H															
18Cr-10Ni-Cb	A430	FP347															
18Cr-10Ni-Cb	A430	FP347H															
Centrifugally cast pipe																	
18Cr-8Ni	A451	CPF8		4534			G-X 6 CrNi 18 9	17445	1.4308								
18Cr-10Ni-Cb	A451	CPF8C					G-X 7 CrNiNb 18 9	17445	1.4552								
18Cr-9Ni-2Mo	A451	CPF8M		4534			G-X 6 CrNiMo 18 9	17445	1.4408								
25Cr-12Ni	A451	CPH8															
25Cr-12Ni	A451	CPH10															
25Cr-12Ni	A451	CPH20															
25Cr-20Ni	A451	CPK20		4534			G-X 15 CrNi 25 20		1.484								
18Cr-8Ni	A452	TP304H															
16Cr-12Ni-2Mo	A452	TP316H															
18Cr-10Ni-Cb	A452	TP347H															
Welded pipe and tube (without filler metal)																	
18Cr-8Ni	A249	TP304															
18Cr-8Ni	A249	TP304H															
18Cr-8Ni	A249	TP304L															
18Cr-8Ni-N	A249	TP304N															
23Cr-12Ni	A249	TP309H															
16Cr-12Ni-2Mo	A249	TP316															
16Cr-12Ni-2Mo	A249	TP316H															
16Cr-12Ni-2Mo	A249	TP316L															
16Cr-12Ni-2Mo-N	A249	TP316N															
19Cr-13Ni-3Mo	A249	TP317															
18Cr-10Ni-Ti	A249	TP321															
18Cr-10Ni-Ti	A249	TP321H															
18Cr-10Ni-Cb	A249	TP347															
18Cr-10Ni-Cb	A249	TP347H															
18Cr-10Ni-Cb	A249	TP348															
18Cr-10Ni-Cb	A249	TP348H															
20Cr-18Ni-6Mo	A249	. . .															
18Cr-8Ni	A312	TP304		3605	304S18		X 5 CrNi 18 9	17440	1.4301			TU Z 6 CN 19-10	A49-230	X 5 Cr Ni 18 10	6904	SUS304TP	G3459
18Cr-8Ni	A312	TP304H		3605	394S59							TU Z 6 CN 19-10	A49-214	X 8 Cr Ni 19 10	6904	SUS304HTP	G3459
18Cr-8Ni	A312	TP304L		3605	304S14		X 2 CrNi 18 9	17440	1.4306			TU Z 2 CN 19-11	A49-230	X 2 Cr Ni 18 11	6904	SUS304LTP	G3459
18Cr-8Ni-N	A312	TP304N															
23Cr-12Ni	A312	TP309H															
25Cr-20Ni	A312	TP310H															
16Cr-12Ni-2Mo	A312	TP316		3605	316S18		X 5 CrNiMo 18 10	17440	1.4401			Z 6 CND 17-11	A35-573	X 5 Cr Ni Mo 17 12	6904	SUS316TP	G3459
16Cr-12Ni-2Mo	A312	TP316H		3605	316S59							TU Z 6 CND 17-12	A49-230	X 8 Cr Ni Nb 17 12	6904	SUS316HTP	G3459
16Cr-12Ni-2Mo	A312	TP316L		3605	316S14		X 2 CrNiMo 18 10	17440	1.4404			TU Z 2 CND 17-12	A49-230	X 2 Cr Ni Nb 17 12	6904	SUS316LTP	G3459
16Cr-12Ni-2Mo-N	A312	TP316N															
19Cr-13Ni-3Mo	A312	TP317															
18Cr-10Ni-Ti	A312	TP321		3605	321S18		X 10 CrNiTi 18 9	17440	1.4541			Z 6 CNT 18-11	A49-230	X 6 Cr Ni Ti 18 11	6904	SUS321TP	G3459
18Cr-10Ni-Ti	A312	TP321H														SUS321HTP	G3459

18Cr-10Ni-Cb	A312	TP347		3605	347S18	X 10 CrNiNb 18 9		1.4543		Z 6 CNNb 18-11	A35-573	X 6 Cr Ni Nb 18 11	6904	SUS347TP	G3459
18Cr-10Ni-Cb	A312	TP347H										X 6 Cr Ni Nb 18 11	6904	SUS347HTP	G3459
18Cr-10Ni-Cb	A312	TP348													
18Cr-10Ni-Cb	A312	TP348H													
18Cr-18Ni-2Si	A312	TPXM15													
20Cr-18Ni-6Mo	A312	. . .													
Welded pipe—filler metal added															
18Cr-8Ni	A358	304	1	3605	304S25	X 5 CrNi 18 9	17440	1.4301		Z 6 CN 18.09	A36-209	X 6 Cr Ni 18 10	8317	SUS304 TPY	G3468
18Cr-8Ni	A358	304	2	3605	304S25	X 5 CrNi 18 9	17440	1.4301		Z 6 CN 18.09	A36-209	X 6 Cr Ni 18 10	8317	SUS304 TPY	G3468
18Cr-8Ni	A358	304	3	3605	304S25	X 5 CrNi 18 9	17440	1.4301		Z 6 CN 18.09	A36-209	X 6 Cr Ni 18 10	8317	SUS304 TPY	G3468
18Cr-8Ni	A358	304L	1												
18Cr-8Ni	A358	304L	2												
18Cr-8Ni	A358	304L	3												
18Cr-8Ni-N	A358	304N	1												
18Cr-8Ni-N	A358	304N	2												
18Cr-8Ni-N	A358	304N	3												
23Cr-12Ni	A358	309	1												
23Cr-12Ni	A358	309	2												
23Cr-12Ni	A358	309	3												
25Cr-20Ni	A358	310	1			X 12 CrNi 25 21		1.4845			A36-209	X 22 Cr Ni 25 20	5900	SUS310S TPY	G3468
25Cr-20Ni	A358	310	2			X 12 CrNi 25 21		1.4845			A36-209	X 22 Cr Ni 25 20	5900	SUS310S TPY	G3468
25Cr-20Ni	A358	310	3			X 12 CrNi 25 21		1.4845			A36-209	X 22 Cr Ni 25 20	5900	SUS310S TPY	G3468
16Cr-12Ni-2Mo	A358	316	1	3605	361S26	X 5 CrNiMo 18 10	17440	1.4401		Z 6 CND 17.11	A36-209	X 6 Cr Ni Mo 17 12 2	8317	SUS316 TPY	G3468
16Cr-12Ni-2Mo	A358	316	2	3605	316S26	X 5 CrNiMo 18 10	17440	1.4401		Z 6 CND 17.11	A36-209	X 6 Cr Ni Mo 17 12 2	8317	SUS316 TPY	G3468
16Cr-12Ni-2Mo	A358	316	3	3605	316S26	X 5 CrNiMo 18 10	17440	1.4401		Z 6 CND 17.11	A36-209	X 6 Cr Ni Mo 17 12 2	8317	SUS316 TPY	G3468
16Cr-12Ni-2Mo	A358	316L	1												
16Cr-12Ni-2Mo	A358	316L	2												
16Cr-12Ni-2Mo	A358	316L	3												
16Cr-12Ni-2Mo-N	A358	316N	1												
16Cr-12Ni-2Mo-N	A358	316N	2												
16Cr-12Ni-2Mo-N	A358	316N	3												
18Cr-10Ni-Ti	A358	321	1	3605	321S22	X 10 CrNiTi 18 9	17440	1.4541		Z 6 CNT 18.10	A36.209	X 6 Cr Ni Ti 18 10	8317	SUS321 TPY	G3468
18Cr-10Ni-Ti	A358	321	2	3605	321S22	X 10 CrNiTi 18 9	17440	1.4541		Z 6 CNT 18.10	A36.209	X 6 Cr Ni Ti 18 10	8317	SUS321 TPY	G3468
18Cr-10Ni-Ti	A358	321	3	3605	321S22	X 10 CrNiTi 18 9	17440	1.4541		Z 6 CNT 18.10	A36.209	X 6 Cr Ni Ti 18 10	8317	SUS321 TPY	G3468
18Cr-10Ni-Cb	A358	347	1	3605	347S17	X 10 CrNiNb 18 9	17440	1.455		Z 6 CnNb 18.10	A36-209	X 6 Cr Ni Nb 18 10	8317	SUS347 TPY	G3468
18Cr-10Ni-Cb	A358	347	2	3605	347S17	X 10 CrNiNb 18 9	17440	1.455		Z 6 CnNb 18.10	A36-209	X 6 Cr Ni Nb 18 10	8317	SUS347 TPY	G3468
18Cr-10Ni-Cb	A358	347	3	3605	347S17	X 10 CrNiNb 18 9	17440	1.455		Z 6 CnNb 18.10	A36-209	X 6 Cr Ni Nb 18 10	8317	SUS347 TPY	G3468
18Cr-10Ni-Cb	A358	348	1												
18Cr-10Ni-Cb	A358	348	2												
18Cr-10Ni-Cb	A358	348	3												
20Cr-18Ni-6Mo	A358	. . .	1												
20Cr-18Ni-6Mo	A358	. . .	2												
20Cr-18Ni-6Mo	A358	. . .	3												

TABLE 3.40 Cross Reference for International Materials—Stainless Steel (Ferritic/Martensitic) Pipe and Tube

American				British			German			Belgian		French		Italian		Japanese	
Nominal composition	Spec	Grade	Type or class	BS number	BS grade	BS process	DIN type	DIN number	DIN material number	NBN type	NBN number	AFNOR type	NF number	UNI type	UNI number	JIS symbol	JIS number
Seamless pipe and tube																	
12Cr-A1	A268	TP405					X 7 CrAl	17440	1.4002								
13Cr	A268	TP410					X 10 Cr 13	17440	1.4006							SUS405TB	G3463
15Cr	A268	TP429															
17Cr	A268	TP430															
26Cr-1Mo	A268	TPXM27															
27Cr	A268	TP446-1															
27Cr-1Mo-Ti	A268	TPXM33															
27Cr-1Mo	A731	TPXM27															
27Cr-1Mo-Ti	A731	TPXM33															
Welded pipes and tubes—without filler metal																	
12Cr-Al	A268	TP405					X 7 CrAl	17440	1.4002								
13Cr	A268	TP410					X 10 Cr 13	17440	1.4006							SUS405TB	G3463
15Cr	A268	TP429															
17Cr	A268	TP430															
27Cr	A268	TP446-1															
26Cr-1Mo	A268	TPXM27															
27Cr-1Mo-Ti	A268	TPXM33															
27Cr-1Mo	A731	TPXM27															
27Cr-1Mo-Ti	A731	TPXM33															

TABLE 3.41 Cross Reference for International Materials—Stainless Steel Forgings, Fittings, and Castings

American				British			German			Belgian		French		Italian		Japanese	
Nominal composition	Spec	Grade	Type or class	BS number	BS grade	BS process	DIN type	DIN number	DIN material number	NBN type	NBN number	AFNOR type	NF number	UNI type	UNI number	JIS symbol	JIS number
Forgings																	
20Cr-18Ni-6Mo	A182	F44															
18Cr-8Ni	A182	F304		1503	304S31		X 5 CrNi 18 9	17440	1.4301			Z 5 CN 18-09	M87-508	X 5 Cr Ni 18 10	6901	SUSF 304	G3214
18Cr-8Ni	A182	F304H		1503	304S51									X 8 Cr Ni 19 10	6901	SUSF 304H	G3214
18Cr-8Ni	A182	F304L		1503	304S11		X 2 CrNi 18 9	17440	1.4306			Z 2 CN 18-10	M87-508	X 2 Cr Ni 18 11	6901	SUSF 304L	G3214
18Cr-8Ni-N	A182	F304N															
25Cr-20Ni	A182	F310		1503	310S31									X 22 Cr Ni 25 20	6901	SUSF 310S	G3214
16Cr-12Ni-2Mo	A182	F316		1503	316S31		X 5 CrNiMo 18 10	17440	1.4401			Z 6 CND 17.11	M87-508	X 5 Cr Ni Mo 17 12	6901	SUSF 316	G3214
16Cr-12Ni-2Mo	A182	F316H		1503	316S51									X 8 Cr Ni Mo 17 12	6901	SUSF 316H	G3214
16Cr-12Ni-2Mo	A182	F316L		1503	316S11		X 2 CrNiMo 18 10	17440	1.4404			Z 2 CND 17.11	M87-508	X 2 Cr Ni Mo 17 12	6901	SUSF 316L	G3214
16Cr-12Ni-2Mo-N	A182	F316N															
18Cr-10Ni-Ti	A182	F321		1503	321S31		X 10 CrNiTi 18 9	17440	1.4541			Z 6 CNT 18.11	M87-508	X 6 Cr Ni Ti 18 11	6901	SUSF 321	G3214
18Cr-10Ni-Ti	A182	F321H															
18Cr-10Ni-Cb	A182	F347		1503	347S31		X 10 CrNiNb 18 9	17440	1.455			Z 6 CNNb 18.11	M87-508	X 6 Cr Ni Nb 18 11	6901	SUSF 347	G3214
18Cr-10Ni-Cb	A182	F347H															
18Cr-10Ni-Cb	A182	F348															
18Cr-10Ni-Cb	A182	F348H															
27Cr-1Mo	A182	FXM-27b															
27Cr-1Mo	A182	FXM-27Cb															
Fittings																	
18Cr-8Ni	A403	WP304															
18Cr-8Ni	A403	WP304H															
18Cr-8Ni	A403	WP304L															
18Cr-8Ni-N	A403	WP304N															
23Cr-12Ni	A403	WP309															
23Cr-20Ni	A403	WP310															
16Cr-12Ni-2Mo	A403	WP316															
16Cr-12Ni-2Mo	A403	WP316H															
16Cr-12Ni-2Mo	A403	WP316L															
16Cr-12Ni-2Mo-N	A403	WP316N															
18Cr-13Ni-3Mo	A403	WP317															
18Cr-10Ni-Ti	A403	WP321															
18Cr-10Ni-Ti	A403	WP321H															
18Cr-10Ni-Cb	A403	WP347															
18Cr-10Ni-Cb	A403	WP348															
18Cr-10Ni-Cb	A403	WP348H															
Castings																	
18Cr-8Ni	A351	CF3															
18Cr-8Ni	A351	CF3A															
18Cr-12Ni-2Mo	A351	CF3M															
18Cr-8Ni	A351	CF8		1504-304	C15		G-X 6 CrNi 18 9	17445	1.4308			Z 6 CN 20.11	A35-578	X 6 Cr Ni 18 10		SCS 13 A	G5121
18Cr-10Ni-Cb	A351	CF8C															
16Cr-12Ni-2Mo	A351	CF8M		1504-316	C16		G-X 6 CrNiMo 18 10	17445	1.4408					X 6 Cr Ni Mo 17 12		SCS 14A	G5121
25Cr-12Ni	A351	CH8															
25Cr-12Ni	A351	CH20															
25Cr-20Ni	A351	CK20		1504-310	C40		G-X 15 CrNiSi 25 20		1.4849			Z 12 CNS 25.20	A35-578			SCS 18	G5121
13Cr-1/2 Mo	A217	CA15															

TABLE 3.42 Modulus of Elasticity in U.S. Units for Metals*

Material	E = modulus of elasticity, msi (millions of psi),† at temperature °F‡																				
	−425	−400	−350	−325	−200	−100	70	200	300	400	500	600	700	800	900	1000	1100	1200	1300	1400	1500
Ferrous metals																					
Gray cast iron	...	...	...	...	...	...	13.4	13.2	12.9	12.6	12.2	11.7	11.0	10.2	...	...	...	...	...	...	...
Carbon steels, C $\leq$ 0.3%	31.9	...	...	31.4	30.8	30.2	29.5	28.8	28.3	27.7	27.3	26.7	25.5	24.2	22.4	20.4	18.0	...	...	...	...
Carbon steels, C > 0.3%	31.7	...	...	31.2	30.6	30.0	29.3	28.6	28.1	27.5	27.1	26.5	25.3	24.0	22.2	20.2	17.9	15.4	...	...	...
Carbon-moly steels	31.7	...	...	31.1	30.5	29.9	29.2	28.5	28.0	27.4	27.0	26.4	25.3	23.9	22.2	20.1	17.8	15.3	...	...	...
Nickel steels, Ni 2%–9%	30.1	...	...	29.6	29.1	28.5	27.8	27.1	26.7	26.1	25.7	25.2	24.6	23.0	...	...	...	...	...	...	...
Cr-Mo steels, Cr ½%–2%	32.1	...	...	31.6	31.0	30.4	29.7	29.0	28.5	27.9	27.5	26.9	26.3	25.5	24.8	23.9	23.0	21.8	20.5	18.9	...
Cr-Mo steels, Cr 2¼%–3%	33.1	...	...	32.6	32.0	31.4	30.6	29.8	29.4	28.8	28.3	27.7	27.1	26.3	25.6	24.6	23.7	22.5	21.1	19.4	...
Cr-Mo steels, Cr 5%–9%	33.4	...	...	32.9	32.3	31.7	30.9	30.1	29.7	29.0	28.6	28.0	27.3	26.1	24.7	22.7	20.4	18.2	15.5	12.7	...
Chromium steels, Cr 12%, 17%, 27%	31.8	...	...	31.2	30.7	30.1	29.2	28.5	27.9	27.3	26.7	26.1	25.6	24.7	23.2	21.5	19.1	16.6	...	...	...
Austenitic steels (TP304, 310, 316, 321, 347)	30.8	...	...	30.3	29.7	29.1	28.3	27.6	27.0	26.5	25.8	25.3	24.8	24.1	23.5	22.8	22.1	21.2	20.2	19.2	18.1
Copper and copper alloys																					
Comp. and leaded-Sn bronze (C83600, C92200)	...	...	...	14.8	14.6	14.4	14.0	13.7	13.4	13.2	12.9	12.5	12.0	...	...	...	...	...	...	...	...
Naval brass, Si- & Al-bronze (C46400, C65500, C95200, C95400)	...	...	...	15.9	15.6	15.4	15.0	14.6	14.4	14.1	13.8	13.4	12.8	...	...	...	...	...	...	...	...
Copper (C11000)	...	...	...	16.9	16.6	16.5	16.0	15.6	15.4	15.0	14.7	14.2	13.7	...	...	...	...	...	...	...	...
Copper, red brass, Al-bronze (C10200, C12000, C12200, C12500, C14200, C23000, C61400)	...	...	...	18.0	17.7	17.5	17.0	16.6	16.3	16.0	15.6	15.1	14.5	...	...	...	...	...	...	...	...
90Cu-10Ni (C70600)	...	...	...	19.0	18.7	18.5	18.0	17.6	17.3	16.9	16.6	16.0	15.4	...	...	...	...	...	...	...	...
Leaded Ni-bronze	...	...	...	20.1	19.8	19.6	19.0	18.5	18.2	17.9	17.5	16.9	16.2	...	...	...	...	...	...	...	...
80Cu-20Ni (C71000)	...	...	...	21.2	20.8	20.6	20.0	19.5	19.2	18.8	18.4	17.8	17.1	...	...	...	...	...	...	...	...
70Cu-30Ni (C71500)	...	...	...	23.3	22.9	22.7	22.0	21.5	21.1	20.7	20.2	19.6	18.8	...	...	...	...	...	...	...	...

Nickel and nickel alloys																					
Monel 400 (N04400)	28.3	...	...	27.8	27.3	26.8	26.0	25.4	25.0	24.7	24.3	24.1	23.7	23.1	22.6	22.1	21.7	21.2	...	...	...
Alloys G, G1, 20 Mod. (N06007, N08320)	30.3	...	...	29.5	29.2	28.6	27.8	27.1	26.7	26.4	26.0	25.7	25.3	24.7	24.2	23.6	23.2	22.7	...	...	...
Alloys 800, 800H, × (N08800, N08810, N06002)	31.1	...	...	30.5	29.9	29.4	28.5	27.8	27.4	27.1	26.6	26.4	25.9	25.4	24.8	24.2	23.8	23.2	...	...	...
Alloys C-4, C276 (N06455, N10276)	32.5	...	...	31.6	31.3	30.6	29.8	29.1	28.6	28.3	27.9	27.6	27.1	26.5	25.9	25.3	24.9	24.3	...	...	...
Nickel 200, 201, Alloy 625 (N02200, N02201, N06625)	32.7	...	...	32.1	31.5	30.9	30.0	29.3	28.8	28.5	28.1	27.8	27.3	26.7	26.1	25.5	25.1	24.5	...	...	...
Alloy 600 (N06600)	33.8	...	...	33.2	32.6	31.9	31.0	30.2	29.9	29.5	29.0	28.7	28.2	27.6	27.0	26.4	25.9	25.3	...	...	...
Alloy B (N10001)	33.9	...	...	33.3	32.7	32.0	31.1	30.3	29.9	29.5	29.1	28.8	28.3	27.7	27.1	26.4	26.0	25.3	...	...	...
Alloy B-2 (N10665)	34.2	...	...	33.3	33.0	32.3	31.4	30.6	30.1	29.8	29.4	29.0	28.6	27.9	27.3	26.7	26.2	25.6	...	...	...
Unalloyed titanium																					
Grades 1, 2, 3, and 7	...	...	...	...	...	...	15.5	15.0	14.6	14.0	13.3	12.6	11.9	11.2	...	...	...	...	...	...	...
Aluminum and aluminum alloys																					
Grades 443, 1060, 1100, 3003, 3004, 6061, 6063 (A24430, A91060, A91100, A93003, A93004, A96061, A96063)	11.4	...	...	11.1	10.8	10.5	10.0	9.6	9.2	8.7											
Grades 5052, 5154, 5454, 5652 (A95052, A95154, A95454, A95652)	11.6	...	...	11.3	11.0	10.7	10.2	9.7	9.4	8.9											
Grades 356, 5083, 5086, 5456 (A03560, A95083, A95086, A95456)	11.7	...	...	11.4	11.1	10.8	10.3	9.8	9.5	9.0											

* These data are for information, and it is not to be implied that materials are suitable for all the temperatures shown. Data are taken from Code for Pressure Piping, ASME B31.1-1995.

† To convert psi into kPa, multiply the tabulated values by 6.895×10^6.

‡ To convert °F into °C, divide (tabulated °F − 32) by 1.8.

TABLE 3.43 Some Physical Properties of Piping Materials

Material	Density, lb/in^3*	Specific heat, mean (temperature, °F)	Melting temperature,† °F
Pure iron	0.2845	0.112 (122–212) 0.170 (1562–1652)	2781–2799
Gray cast iron	0.251–0.265	...	2150–2360
Malleable cast iron ferritic	0.260–0.265	0.11 (at 70) 0.165 (at 800)	2750
Malleable cast iron pearlitic	0.264	0.11 (at 70) 0.165 (at 800)	2750
Nodular iron	0.257	...	2050–3150
Wrought iron	0.28	...	2750
0.06 C, 0.38 Mn	0.2844	0.115 (122–212) 0.264 (1292–1382)	2600
0.23 C, 0.35 Mn	0.2839	0.116 (122–212) 0.342 (1292–1382)	2600
0.43 C, 0.69 Mn	0.2834	0.116 (122–212) 0.227 (1292–1472)	2600
1.22 C, 0.35 Mn	0.2839	0.116 (122–212) 0.499 (1292–1382)	2600
Carbon-½Mo	0.28	...	2600–2800
1¼ Cr-½Mo	0.283	0.114 (122–212)	2600–2800
2¼ Cr-1Mo	0.283	0.11	2600–2800
5 Cr-½Mo	0.28	0.11	2700–2800
9 Cr-1Mo	0.28	0.11	2700–2800
3½% Ni steel	0.28	0.115 (212)	...
Type 304 wrought	0.29	0.12 (32–212)	2550–2650
CF-8 cast	0.28	0.12	2600
Type 316 wrought	0.29	0.12 (32–212)	2500–2550
CF-8M cast	0.28	0.12	2550
Type 321 wrought	0.29	0.12 (32–212)	2550–2600
Type 347 wrought	0.29	0.12 (32–212)	2550–2660
CF-8C cast	0.28	0.12	2550–2600
405 wrought	0.28	0.11	2700–2790
CA15 cast	0.275	0.11	2750
410 wrought	0.28	0.11	2700–2790
446 wrought	0.273	0.144	2550–2750
CC50 cast	0.272	0.12	2725
Aluminum 1100	0.098	0.23 (212)	1190–1215
Aluminum 6061	0.098	0.23 (212)	1080–1200
Aluminum 4043	0.097	0.23 (212)	1065–1170
Aluminum 356	0.097	0.23 (212)	1035–1135
Copper (DHP)	0.323	0.092	1981
Red brass (wrought)	0.316	0.09	1810–1880
Yellow brass (wrought)	0.306	0.09	1660–1710
Admiralty brass (wrought)	0.308	0.09	1650–1720
Manganese bronze (wrought)	0.302	0.09	1590–1630
Cupronickel (70–30) (wrought)	0.323	0.09	2140–2260
Aluminum bronze (3) (wrought)	0.281	0.09	1910–1940
Beryllium copper (wrought)	0.297	0.10 (86–212)	1587–1750
Chemical lead	0.4097	0.0309	618
50/50 Sn Pb solder	0.321	0.046	361–421
Nickel (A) (wrought)	0.321	0.13	2615–2635
Monel (70Ni–30Cu) (wrought)	0.319	0.127	2370–2460
Inconel (wrought)	0.307	0.11	2540–2600
Incoloy (wrought)	0.290	0.12	2540–2600
Hastelloy B (wrought)	0.334	0.091	2410–2460
Hastelloy C (wrought)	0.323	0.092	2320–2380
Tin	0.26	0.0534	449.4
Titanium (99.0%)	0.163	0.125	3002–3038
Tantalum	0.600	0.034	5425

* $lb/in^3 = 2.76798 \times 10^{-2}\ kg/cm^3$.
† °C = (°F − 32)/1.8.
Source: Reprinted with permission from SAE J 933 ©1989. Society of Automotive Engineers, Inc.

TABLE 3.44 Thermal Conductivity and Expansion of Piping Material

Material	Thermal conductivity*		Linear thermal expansion average	
	Btu/(hr·°F·ft^2/ft)	Temperature, °F[†]	Microinch/(in·°F)	Temperature range, °F
Pure iron	43	70	6.83	68–212
	28	752	8.97	932–1112
Gray cast iron	27	70	5.83	32–212
	23.7	752		
Malleable cast iron ferritic	. . .	. . .	6.6	70–750
Malleable cast iron pearlitic	. . .	. . .	6.6	70–750
Nodular iron	18 (pearlitic)	212	6.46	68–212
	20 (ferritic)	212	6.97	68–600
			7.49	68–1000
			7.69	68–1400
Wrought iron	34	212		
	26	752		
Wrought carbon steel:	34	32	7.83	70–800
0.06 C, 0.38 Mn			7.95	70–900
			8.02	70–1000
			8.21	70–1100
			8.36	70–1200
0.23 C, 0.635 Mn	30	70		
	25	752	6.50	68–212
	17	2192		
0.435 C, 0.69 Mn	. . .	. . .	6.44	68–212
			8.39	68–1292
1.22 C, 0.35 Mn	26	70		
	22	752	5.89	122–212
	16	2192	9.33	932–1832
Carbon-½Mo	25.8	212	7.70	68–800
			7.85	68–1000
			7.95	68–1100
			8.07	68–1200
1¼Cr-½Mo	17.9	212	7.32	70–800
			7.44	70–900
			7.56	70–1000
			7.63	70–1100
			7.74	70–1200
			7.82	70–1300
2¼Cr-1Mo	16.3	212	7.49	70–800
			7.65	70–900
			7.72	70–1000
			7.78	70–1100
			7.84	70–1200
			7.88	70–1300
5Cr-½Mo	21.2	212	6.44	0–212
	20.8	392	6.91	70–800
	20.4	572	7.02	70–900
	19.8	752	7.10	70–1000
	19.5	932	7.19	70–1100
			7.31	70–1200
			7.35	70–1300
9Cr-1Mo	. . .	. . .	6.28	70–300
			6.60	70–800
			6.75	70–900
			6.81	70–1000
			6.95	70–1100
			7.07	70–1200
			7.13	70–1300

TABLE 3.44 Thermal Conductivity and Expansion of Piping Material (Continued)

Material	Thermal conductivity*		Linear thermal expansion average	
	Btu/(hr·°F·ft²/ft)	Temperature, °F†	Microinch/(in·°F)	Temperature range, °F
3½%Ni steel	21	212		
	14	1472		
Type 304 wrought	9.4	212	9.6	32–212
	10.3	392	9.9	32–600
	11.0	572	10.2	32–1000
	11.8	752	10.4	32–1200
	12.5	932	11.2	32–1800
CF-8 cast	9.2	212		
	12.1	1000		
Type 316 wrought	9.0	212	8.9	32–212
	12.1	932	9.0	32–600
			9.7	32–1000
			10.3	32–1200
			11.1	32–1500
CF-8M cast	9.4	212	8.9	68–212
	12.3	1000	9.7	68–1000
Type 321 wrought	9.3	212	9.3	32–212
	10.2	392	9.5	32–600
	11.1	752	10.3	32–1000
	11.9	932	10.7	32–1200
	12.8		11.2	32–1500
Type 347 wrought	9.3	212	9.3	32–212
	10.2	392	9.5	32–600
	11.1	572	10.3	32–1000
	11.9	752	10.6	32–1200
	12.8	932	11.1	32–1500
CF-8C cast	9.3	212	9.3	68–212
	12.8	1000	10.3	68–1000
405 wrought	...	...	6.0	32–212
			6.4	32–600
			6.7	32–1000
			7.5	32–1200
CA15 cast	14.5	212	5.5	68–212
	16.7	1000	6.4	68–1000
			6.7	68–1300
410 wrought	14.4	212	6.1	32–212
	16	752	7.2	32–1000
			7.6	32–1832
446 wrought	12.1	212	5.9	32–212
	14.1	932	6.3	32–1000
			7.6	32–1832
CC50 cast	12.6	212		
	17.9	1000	5.9	68–212
	20.3	1500	6.4	68–1000
	24.2	2000		
Aluminum 1100	128	70	12.2	–58 to +68
			13.1	68–212
			13.7	68–392
			14.2	68–572
Aluminum 6061	99 (0 temper)	70	12.1	–76 to +68
	90 (T4 temper)	70	13.0	68–212
	90 (T6 temper)	70	13.5	68–392
			14.1	68–572

TABLE 3.44 Thermal Conductivity and Expansion of Piping Material (Continued)

Material	Thermal conductivity*		Linear thermal expansion average	
	Btu/(hr·°F·ft²/ft)	Temperature, °F†	Microinch/(in·°F)	Temperature range, °F
Aluminum 43	82 (as cast) 94 (annealed)	70 70	12.2 12.8 13.3	68–212 68–392 68–572
Aluminum 356	97 sand cast T51 88 sand cast T6	70 70	11.9 12.8 13.0	68–212 68–392 68–572
Copper (DHP)	196	68	9.8	68–572
Red brass	92	68	10.4 cold rolled	68–572
Yellow brass	67	68	11.3	68–572
Admiralty brass	64	68	11.2	68–572
Manganese bronze	61	68	11.8	68–572
Cupronickel (70-30)	17	68	9.0	68–572
Aluminum bronze (3)	44	68	9.0	68–572
Beryllium copper	33–41 cold worked 48–68 precipitation hardened	68	9.3 9.4 9.9	68–212 68–392 68–572
Chemical lead	. . .	. . .	16.3 14.7	65–212 –130 to +66
50/40 SnPb solder	27	129	13.0	60–230
Nickel (A) wrought	35	32–212	7.4	77–212
Monel (70Ni-30Cu) (wrought)	15	32–212	7.8	32–212
Inconel	8.4	70–212	6.4 8.3	68–212 70–1000
Incoloy	6.8	32–212	8.0	32–212
Hastelloy B	6	. . .	5.3 7.8	70–200 70–1600
Hastelloy C	5	70	6.6 8.2	70–200 70–1600
Tin	36	32	12.8	32–212
Titanium (99.0%)	9.0–11.5 12.4	68 1500	4.8 5.6 5.7	68–200 68–1200 58–1600
Tantalum	31	68	3.6	

* Btu/(hr °F·ft²/ft) = 0.5780101 watt/meter²/°C/meter (W/m²/°C/m).
† °C = (°F − 32)/1.8.

Tables in Chapter 3

CHAPTER 4
NONMETALLIC PIPING MATERIALS

See p. 132 for a list of tables in Chap. 4.

Plastic piping is made from either of two basic groups of synthetic materials, thermoplastic and thermosetting. Thermoplastics can be softened and reshaped repeatedly by the application of heat. In contrast, thermosetting materials are irreversibly *set*, or *cured*, or *hardened* into a permanent shape during factory manufacture. Once hardened into their final shape, thermosetting products cannot be softened and therefore, they may not be reshaped by heating.

ADDITIVES FOR THERMOPLASTICS

Additives are incorporated into a thermoplastic composition to achieve specific purposes during fabrication or service. The main kinds of additives that may be used in thermoplastic piping compositions include the following:

Heat stabilizers. To protect the plastic against thermal degradation, particularly during processing.

Antioxidants. To protect against oxidation during processing and when in service.

Ultraviolet screens, or stabilizers. To protect against ultraviolet radiation in sunlight during outdoor storage and weather-exposed service.

Lubricants. To facilitate and improve fabrication by reducing viscosity and lessening frictional drag through dies and other surfaces.

Pigments. To give the product a distinctive color.

Processing aids. To facilitate material mixing and fusion during processing and, thereby, optimize the homogeneity of the material and its properties.

Property modifiers. To enhance a particular property such as impact strength or flexibility.

Fillers. Most often used to reduce volume cost; however, fillers may also be used to increase stiffness or to modify processing characteristics.

The precise nature and quantities of additives that can be used for piping compositions are delimited by their effect on engineering properties, such as rigidity, impact strength, chemical resistance, creep resistance, rupture strength under long-term loading, and fatigue endurance.

THERMOPLASTIC MATERIALS

Table 4.1 lists the approximate values of physical properties of commercially used thermoplastic piping materials.

Table 4.2 lists the approximate ratio of creep modulus to short-term modulus as a function of duration of continuous loading at 73°F (23°C).

Table 4.3 provides a chemical resistance guide for thermoplastic piping materials at ambient temperatures.

Table 4.4 lists approximate values of temperature derating factors for various thermoplastic materials. They are based on the effect of temperature on strength and stiffness of thermoplastic pipe.

TABLE 4.1 Approximate Values of Physical Properties of Thermoplastic Piping Materials

Property*	ASTM Test Method	Approximate value at 75°F						
		ABS	PVC	CPVC	PE	PEX	PB	PVDF
Specific gravity	D792	1.08	1.4	1.54	0.95	0.94	0.92	1.76
Tensile strength, 10^3 psi	D638	7.0	8.0	8.0	3.2	2.8	4.2	7.0
Tensile modulus, 10^3 psi	D638	340	410	420	120	9–	55	220
Impact strength, Izod, ft-lb/in of notch	D256	4	1	1.5	>10	>10	>10	3.8
Coeff. of linear expansion, 10^{-6} in/(in·°F)	D696	60	30	35	90	90	72	70
Thermal conductivity, Btu·in/(h·ft²·°F)	C177	1.35	1.1	1.0	3.2	3.2	1.5	1.5
Specific heat, Btu/(lb·°F)	—	0.34	0.25	0.20	0.55	0.55	0.45	0.29
Approx. operating temp. limits	—							
Nonpressure, °F		180	150	210	160	210	210	300
(°C)		(80)	(65)	(100)	(70)	(100)	(100)	(150)
Pressure, °F		160	130	180	140	200	180	280
(°C)		(70)	(55)	(80)	(60)	(95)	(80)	(140)

* Conversion factors to commonly used metric equivalents are as follows:
1. psi (pounds-force per square inch) to Mpa (megapascals)—multiply by 6.90 x 10^3.
2. ft·lb/in of notch (foot-pounds force/inch of notch) to J/m of notch (joules/meter)—multiply by 53.4.
3. in/(in·°F) to m/(m·°C)—multiply by 1.8.
4. Btu·in/(h·ft²·°F) to cal·cm/(s·cm²·°C)—multiply by 344 x 10^6.
5. Btu/(lb·°F) to cal/(g·°C)—multiply by 1.0.

TABLE 4.2 Approximate Ratio of Creep Modulus to Short-Term Modulus as a Function of Duration of Continuous Loading, for 73°F (23°C)

Duration of uninterrupted loading, h	Approximate ratio of creep to short-term modulus*	
	High-Density PE	Type I PVC
1	0.80	0.84
100	0.52	0.60
10,000	0.28	0.40
438,000 (50 years)	0.22	0.34

* Exact ratio depends on the specific commercial grade of material, stress field, fabrication residual stresses, and other factors.

POLY(VINYL CHLORIDE) (PVC)

In its virgin state PVC is a translucent colorless rigid polymer. Of the commonly available thermoplastics, rigid PVC offers the highest strength and stiffness at the least volume cost. Major uses include: water mains; irrigation; drain, waste, and vent (DWV); sewage and drainage; well casing; electric conduit; and power and communications duct.

Table 4.5 lists the principal PVC piping standards.

Table 4.6 lists the cell classification limits for PVC materials in accordance with ASTM D1784.

Table 4.7 lists PVC compositions with suffix designation for chemical resistance.

Table 4.8 lists the classifications of commercial types and grades of rigid PVC piping materials: comparison of former to newer designations.

Table 4.9 lists the suggested maximum sustained working pressures for water at 73°F (23°C) for Schedule 40 and Schedule 80 PVC fittings.

TABLE 4.3 Thermoplastic Piping Materials: Chemical Resistance Guide for Ambient Temperatures*

Chemical	ABS	PVC Type I	PVC Type II	CPVC	PE	PEX	PB	PP	PVDF
Inorganic									
Acids, dilute	G	G	L	G	G	G	G	G	G
Acids, concentrated (<80%)	L	L	L	G	L	L	L	L	G
Acids, oxidizing	L	P	P	L	P	L	P	P	G
Alkalies, dilute	G	G	G	G	G	G	G	G	G
Alkalies, concentrated (<80%)	L	G	L	G	G	G	G	G	G
Gases, HCl and HF, dry	L	L	L	L	L	L	L	L	G
Gases, HCl and HF, wet	L	G	L	G	L	L	L	L	G
Gases, ammonia, dry	L	G	L	G	G	G	G	G	G
Gases, halogen, dry	L	L	L	L	L	L	L	P	G
Gases, sulfur containing, dry	P	G	L	G	G	G	L	P	G
Salts, acidic	G	G	G	G	G	G	G	G	G
Salts, basic	G	G	G	G	G	G	G	G	G
Salts, neutral	G	G	G	G	G	G	G	G	G
Salts, oxidizing	L	L	L	L	G	G	G	G	G
Organic									
Acids	G	G	G	L	G	G	G	G	G
Acid anhydrides	L	L	L	P	L	L	L	L	L
Alcohols, glycols	L	G	L	G	L[†]	G	G	G	G
Esters, ethers, ketones	P	P	P	P	L	L	L	L	L
Hydrocarbons, aliphatic	L	L	L	G	L	L	L	L	G
Hydrocarbons, aromatic	P	P	P	L	P	L	P	P	G
Hydrocarbons, halogenated	L	L	L	L	P	L	P	P	L
Natural gas (fuel)	G	G	G	G	G	G	G	G	G
Mineral oil	G[†]	G	G	G	L[†]	G	G	G	G
Oils, animal and vegetable	G[†]	G	G	G	L[†]	G	G	G	G
Synthetic gas (fuel)	L	L	L	L	L	L	L	L	G

* G denotes good, L limited resistance, and P poor. These ratings are only for general guidance. For determination of suitability of chemical resistance under the actual anticipated end use conditions more detailed information should be consulted.

† Stress crack–resistant grade of plastic material should be used.

TABLE 4.4 Effect of Temperature on Strength and Stiffness of Thermoplastic Pipe: Approximate Temperature Derating Factors*

Temperature, °F	PE	PEX	PB	PVC Type I	CPVC	PVDF
70	1.0	1.0	1.0	1.0	1.0	1.0
80	0.95	0.95	0.97	0.88		0.93
90	0.90	0.91	0.92	0.75		0.87
100	0.80	0.87	0.86	0.62	0.78	0.82
110	0.75	0.83	0.82	0.50		0.76
120	0.70	0.79	0.77	0.40	0.65	0.71
130	0.50	0.76	0.72	0.30		0.65
140	0.40	0.73	0.68	0.22	0.50	0.61
150	0.20	0.69	0.69	NR[†]		0.57
160	NR	0.66	0.58	NR	0.40	0.54
180	NR	0.63	0.48	NR	0.25	0.47
200	NR	0.50	0.40	NR	0.20	0.41
220	NR	NR	NR	NR	NR	0.38
250	NR	NR	NR	NR	NR	0.35
280	NR	NR	NR	NR	NR	0.28

* Check with pipe manufacturer for recommendations for derating factors for the specific composition under consideration.

† NR—not resistant for continuous service at indicated temperature.

TABLE 4.5 Principal PVC Piping Standards

Piping material and product standard	Subject product, or abbreviated title of standard	Nominal size range, inches	Principal applications
ASTM D3311	DWV fittings patterns	1¼–8	Drain, waste, and vent
ASTM D2680	ABS and PVC sewer pipe of composite wall construction	4–15	Sewer and drain
ASTM F409	Accessible and replaceable tube and fittings	1¼–1½	Drain, waste, and vent
ASTM F480	Thermoplastic water well casing	2–16	Water-well casing
ASTM F1499	Coextruded, composite DWV pipe	1¼–8	Drain, waste, and vent
CSA B181.5*	Coextruded ABS/PVC DWV pipe		Drain, waste, and vent
ASTM F1488	Coextruded composite pipe	2–12	Drain, waste, and vent; sewer and drain; electrical and communications conduit
ASTM D2513	Thermoplastic gas pressure pipe and fittings	¼–24	Natural gas and LPG distribution
AASHTO M278*	PVC Drainage pipe		Drainage
ASTM D1785	PVC Pipe, Schedules 40, 80 and 120	½–12	Cold water; industrial
ASTM D2241	PVC Pipe, Dimension ratio series	⅛–36	Cold water; industrial
ASTM D2466	PVC Socket fittings, Schedule 40	⅛–8	Cold water; industrial
ASTM D2467	PVC Socket fittings, Schedule 80	⅛–8	Cold water; industrial
ASTM D2665	PVC DWV Pipe and fittings	1¼–12	Drain, waste, and vent
ASTM D2672	Specification for solvent cement type PVC pipe joints	⅛–12	Cold water; industrial
ASTM D2729	PVC Drain pipe and fittings	2–6	Drain; leaching fields
ASTM D2949	PVC Thin walled DWV pipe	3	Drain, waste, and vent
ASTM D3034	PVC Sewer pipe and fittings	4–15	Sewer and drain
ASTM F512	PVC Conduit for buried applications	1–6	Electrical duct
ASTM F679	PVC Sewer pipe and fittings, larger diameter	18–36	Sewer and drain
ASTM F758	PVC Underdrain piping	4–8	Drain
ASTM F789	PVC Sewer pipe, 46 psi stiffness	4–15	Sewer and drain
ASTM F794	PVC Sewer pipe, ribbed wall	4–48	Sewer and drain
ASTM F891	Coextruded PVC pipe with a cellular core	2–18	Drain, waste, and vent; sewer and drain
ASTM F949	PVC Sewer pipe, corrugated wall, smooth interior surface	4–10	Sewer and drain
ASTM F1336	PVC Gasketed sewer fittings	4–27	Sewer and drain
ASTM F1483	Oriented PVC pipe	4–16	Water distribution and transmission
ASTM F1504	Folded PVC pipe for pipeline rehabilitation	4–15	Rehabilitation of existing sewers and conduits
ASTM F1732	PVC Sewer and drain pipe containing recycled material	2–6	Sewer and drain
ASTM F1760	Coextruded PVC non-pressure pipe containing recycled material	4–15	Drain, waste, and vent; sewer and drain; electrical and communications duct
ASTM F1803	PVC Closed profile gravity pipe made to controlled inside diameters	18–60	Sewer and drain
AWWA C900*	PVC Pipe for water distribution	4–12	Water distribution
AWWA C905*	PVC Pipe for water distribution	14–36	Water distribution and transmission
AWWA C907*	PVC Gasketed fittings for water	4–8	Water distribution
CSA B137.2*	PVC Gasketed fittings for water		Water distribution
CSA B 137.3*	PVC Pipe for pressure applications		Cold water; industrial
CSA B181.2*	PVC DWV Pipe and fittings		Drain, waste, and vent
CSA B182.1*	Drain and sewer pipe and fittings		Drain and sewer piping
CSA B182.2*	PVC Sewer pipe and fittings		Sewer collection
CSA B182.4*	Profile wall PVC sewer pipe and fittings		Sewer collection
CSA B182.7*	Multilayer PVC Sewer pipe having reprocessed-recycled content		Sewer collection
CSA B196.3*	PVC Underground telecommunications cable ducting and fittings		Communications ducting
NEMA TC-2*	PVC Electrical conduit	½–6	Electrical conduit
NEMA TC-3*	PVC Conduit fittings	½–6	Electrical conduit
UL 514*	PVC Electrical outlet boxes and fittings	½–6	Electrical conduit
UL 651*	PVC Rigid non-metallic conduit	½–6	Electrical conduit

* AASHTO: American Association of State Highway and Transportation Officials.
AWWA: American Water Works Association.
CSA: Canadian Standards Association.
NEMA: National Electrical Manufacturers Association.
UL: Underwriters Laboratories, Inc.

TABLE 4.6 Primary Properties: Cell Classification Limits for PVC Materials in Accordance with ASTM D1784

Property and test method[†]	Property cell limits*								
	Cell 0	Cell 1	Cell 2	Cell 3	Cell 4	Cell 5	Cell 6	Cell 7	Cell 8
Base resin	Unspecified	Poly (vinyl) homopolymer	Chlorinated poly(vinyl) chloride	Vinyl copolymer					
Min. impact strength (Izod), ft·lb/in of notch, ASTM D256	Unspecified	<0.65	0.65	1.5	5.0	10.0	15.0		
Min. tensile strength, psi, ASTM D638	Unspecified	<5000	5000	6000	7000	8000			
Min. modulus of elasticity, psi, ASTM D638	Unspecified	<280,000	280,000	320,000	360,000	400,000	440,000		
Min. deflection temperature under load (264 psi), °F (°C)	Unspecified	<131 (55)	131 (55)	140 (60)	158 (70)	176 (80)	194 (90)	212 (100)	230 (110)
Flamability[‡]									

* The minimum property value determines the cell number, even though the maximum measured value may fall within a higher cell.

[†] Conversion factors to common metric units are as follows:
psi to MPa—multiply by 6.90×10^{-3}.
ft·lb/in of notch to J/m of notch—multiply by 53.4.

[‡] All compounds covered by this specification when tested in accordance with ASTM method D635 shall yield not less than the following result: average extent of burning <25 mm; average time of burning <10 s.

TABLE 4.7 PVC Compositions with Suffix Designations for Chemical Resistance

Test solution	Suffix			
	A	B	C	D
H_2SO_4 (93%)—14 days immersion at 55 ± 2°C				
Change in weight:				
Increase, max., %	1.0*	5.0*	25.0	NA[†]
Decrease, max., %	0.1*	0.1*	0.1	NA
Change in flexural yield strength:				
Increase, max., %	5.0*	5.0*	5.0	NA
Decrease, max., %	5.0*	25.0*	50.0	NA
H_2SO_4 (80%)—30 days immersion at 60 ± 2°C				
Change in weight:				
Increase, max., %	NA	NA	5.0	15.0
Decrease, max., %	NA	NA	5.0	0.1
Change in flexural yield strength				
Increase, max., %	NA	NA	15.0	25.0
Decrease, max., %	NA	NA	15.0	25.0
ASTM Oil No. 3—30 days immersion at 23°C				
Change in weight:				
Increase, max., %	0.5	1.0	1.0	10.0
Decrease, max., %	0.5	1.0	1.0	0.1

* Specimens washed in running water and dried by air blast or other mechanical means shall show no sweating within 2 hours after removal from the acid bath.

[†] NA—not applicable.

TABLE 4.8 Classification of Commercial Types and Grades of Rigid Poly(Vinyl Chloride) Piping Materials and Comparison of Old to New Designations

Type and grade designation based on older edition of ASTM D1784	Corresponding cell classification based on current edition of ASTM D1784, Tables D1.4 and D1.5
Rigid PVC materials	
Type I, Grade 1	12454-B
Type I, Grade 2	12454-C
Type I, Grade 3	11443-B
Type II, Grade 1	14333-D
Type III, Grade 1	13233
CPVC materials	
Type IV, Grade 1	23447-B

TABLE 4.9 Suggested Maximum Sustained Working Pressures for Water for 73°F (23°C) for Schedule 40 and Schedule 80 PVC Fittings*

	Schedule 40		Schedule 80	
Nominal size, in	Required minimum burst pressure by ASTM D2466, psig	Maximum suggested working pressure, psig	Required minimum burst pressure by ASTM D2467, psig	Maximum suggested working pressure, psig
1/2	1910	358	2720	509
3/4	1540	289	2200	413
1	1440	270	2020	378
1 1/4	1180	221	1660	312
1 1/2	1060	198	1510	282
2	890	166	1290	243
2 1/2	970	182	1360	255
3	840	158	1200	225
3 1/2	770	144	1110	207
4	710	133	1040	194
5	620	117	930	173
6	560	106	890	167
8	500	93	790	148
10	450	84	600	140
12	420	79	580	137

* This table is intended only as a general guide. Appropriate maximum working pressures may vary widely depending on specific fitting design and field conditions, particularly when repetitive surge pressures are present, as these may lower the long-term strength because of fatigue effects. The fitting manufacturer should be consulted for recommendations. Refer to ASTM D2466 and ASTM D2467.

CHLORINATED POLY(VINYL CHLORIDE) (CPVC)

As implied by its name, CPVC is a chemical modification of PVC. It is very similar to PVC in many properties, including strength and stiffness at ambient temperature. However, the extra chlorine in CPVC chemical structure increases this material's maximum operating temperature limit by about 50°F (28°C) above that for PVC. Principal uses for CPVC are domestic hot and cold water piping, residential fire-sprinkling piping, and many industrial applications that take advantage of its elevated-temperature capabilities and superior chemical resistance.

Table 4.10 lists the principal CPVC piping standards.

TABLE 4.10 Principal CPVC Piping Standards

Piping material and product standard	Subject product, or abbreviated title of standard	Nominal size range, inches	Principal applications
ASTM F1488	Coextruded composite pipe	2–12	Drain, waste, and vent; sewer and drain; electrical and communications conduit
ASTM F1803	PVC Closed profile gravity pipe made to controlled inside diameters	18–60	Sewer and drain
AWWA C900	PVC Pipe for water distribution	4–12	Water distribution
AWWA C905	PVC Pipe for water distribution	14–36	Water distribution and transmission
AWWA C907	PVC Gasketed fittings for water	4–8	Water distribution
CSA B137.2	PVC Gasketed fittings for water		Water distribution
CSA B137.3	PVC Pipe for pressure applications		Cold water, industrial
CSA B181.2	PVC DWV Pipe and fittings		Drain, waste, and vent
CSA B182.1	Drain and sewer pipe and fittings		Drain and sewer piping
CSA B182.2	PVC Sewer pipe and fittings		Sewer collection
CSA B182.4	Profile wall PVC sewer pipe and fittings		Sewer collection
CSA B182.7	Multilayer PVC Sewer pipe having reprocessed-recycled content		Sewer collection
CSA B196.3	PVC Underground telecommunications cable ducting and fittings		Communications ducting
NEMA TC-2	PVC Electrical conduit	$^1/_2$–6	Electrical conduit
NEMA TC-3	PVC Conduit fittings	$^1/_2$–6	Electrical conduit
UL 514	PVC Electrical outlet boxes and fittings	$^1/_2$–6	Electrical conduit
UL 651	PVC Rigid non-metallic conduit	$^1/_2$–6	Electrical conduit
ASTM D2846	CPVC hot & cold water piping	$^3/_8$–2	Hot and cold water
ASTM F437	CPVC fittings, threaded, Schedule 80	$^1/_4$–6	Hot and cold water, industrial
ASTM F438	CPVC fittings, socket, Schedule 40	$^1/_4$–6	Hot and cold water, industrial
ASTM F439	CPVC fittings, socket, Schedule 80	$^1/_4$–6	Hot and cold water, industrial
ASTM F441	CPVC pipe, Schedules 40 and 80	$^1/_4$–12	Hot and cold water, industrial
ASTM F442	CPVC PIPE, Dimension ratio series	$^1/_4$–12	Hot and cold water, industrial
CSA B137.6	CPVC Pipe, tubing and fittings for hot & cold water distributing systems		Hot and cold water, plumbing

AWWA: American Water Works Association.
CSA: Canadian Standards Association.
NEMA: National Electrical Manufacturers Association.
UL: Underwriters Laboratories, Inc.

POLYETHYLENE (PE)

Polyethylene (PE) is possibly the best-known member of the polyolefin family (materials derived from the polymerization of olefin gases including ethylene, propylene, and butylene) because it has penetrated so widely into everyday household uses. PE in its virgin form is a translucent and tough substance with a waxy feel and appearance. Like the other polyolefins, PE is a partly crystalline and partly amorphous material.

Table 4.11 lists the cell classification limits for PE piping materials in accordance with ASTM D3350.

Table 4.12 lists the classifications of commercial polyethylene piping materials and comparison of former to newer designations.

Table 4.13 lists the principal PE piping standards.

POLYBUTYLENE (PB)

Polybutylene is a polyolefin with a stiffness resembling that of low-density polyethylene, but with a long-term strength that is greater than that of high-density polyethylene (PE). Its most distinctive feature, however, is that the long-term strength is less affected by increasing temperature than that of PE. While most PEs have an upper temperature limit of around 140°F (60°C), the limit for PB is nearly 200°F (93°C).

Major applications for PB pipe and tubing are for water service lines and uses that take advantage of its improved elevated-temperature strength.

TABLE 4.11 Primary Properties: Cell Classification Limits for PE Piping Materials in Accordance with ASTM D3350

Property and Test Method*	Property cell limits									
	Cell 0	Cell 1	Cell 2	Cell 3	Cell 4	Cell 5	Cell 6	Cell 7	Cell 8	Cell 9
Density, ASTM D1505, g/cm^3	Not specified	0.910 to 0.925	0.926 to 0.940	0.941 to 0.955	>0.955					Specify value
Melt Index ASTM, D1238, g/10 min	Not specified	>1.0	0.4 to 1.0	0.15 to 0.4	<0.15	†	‡			Specify value
Flexural modulus, ASTM D790, psi	Not specified	<20,000	20,000 to 40,000	40,000 to 80,000	80,000 to 110,000	110,000 to 160,000	>160,000			Specify value
Yield strength, ASTM D638, psi	Not specified	<2,200	2,200 to 2,600	2,600 to 3,000	3,000 to 3,500	3,500 to 4,000	>4,000			Specify value
Environmental stress crack resistance	Not specified									Specify value
Method A, ASTM F1473, min. hours§						0.15	3	10	30	
Method B, ASTM D1693										
Test condition		A	B	C	C					
Test duration, h		48	24	192	600					
Failure, max., %		50	50	20	20					
Hydrostatic design basis, ASTM D2837, h	Not specified¶	800	1,000	1,250	1,600					Specify value

* To convert psi to MPa multiply by 6.90×10^{-3}.

† Materials with MI less than Cell 4 but which have a flow rate of <4.0 g/10 min when tested in accordance with D1238, Condition F.

‡ Materials with MI less than Cell 4 but which have flow rate of <0.30 g/10 min when tested in accordance with ASTM D1238, but at a temperature of 590°F (310°C) and under a total load of 12,480 g.

§ Test is conducted on compression-molded samples, notched to a depth in accordance with Table 1 of F1473, at a load of 350 psi (2.4 MPa) and at a temperature of 176°F (80°C).

¶ Compositions for which an HDB has not been established are designated as NPR (non-pressure-rated).

TABLE 4.12 Classification of Commercial Polyethylene Piping Materials: Comparison of Old to New Designations

PE material designation based on type and grade in accordance with former ASTM D1248, plus a code for material's maximum recommended hydrostatic design stress for water, for 73°F	Corresponding minimum cell classification in accordance with cell classification system of ASTM D3350*
PE 2406	PE 213333
PE 3406	PE 324433
PE 3408	PE 334434

* For cells 1 and 6 the resultant values must fall within the limits shown for the particular property cell. For the other properties, material values may fall within or be above the designated cell limits. Refer to a list of ranges of property values for each property cell.

TABLE 4.13 Principal PE Piping Standards

Piping material and product standard	Subject product, or abbreviated title of standard	Nominal size range, inches	Principal applications
ASTM D2513	Thermoplastic gas pressure pipe and fittings	¼–24	Natural gas and LPG distribution
CSA B181.3	Polyolefin drainage systems		Industrial
AASHTO M252	Corrugated PE drainage tubing	3–15	Subsurface drainage
AASHTO M294	Corrugated PE pipe	12–24	Subsurface drainage
API 15LE	PE line pipe	½–12	Oil and gas production; cold water
ASTM D2104	PE pipe, Schedule 40, ID based	½–6	Cold water; industrial
ASTM D2239	PE pipe, dimension ratio series, ID based	½–6	Cold water; industrial
ASTM D2447	PE pipe, Schedules 40 and 80	½–12	Cold water; industrial
ASTM D2609	Plastic insert fittings for PE pipe	½–6	Cold water
ASTM D2683	PE fittings, socket fusion type	½–4	Cold water; natural gas; industrial
ASTM D2737	PE tubing	½–2	Cold water
ASTM D3035	PE pipe, dimension ratio series	½–6	Cold water; industrial
ASTM D3261	PE fittings, butt fusion type	½–48	Cold water; natural gas; industrial
ASTM F405	PE corrugated tubing and fittings	3–6	Drainage; leaching fields
ASTM F667	PE corrugated tubing, larger diameter	8–24	Drainage; leaching fields
ASTM F714	PE pipe, large diameter	3–48	Cold water; sewer and drain; industrial
ASTM F771	PE pipe for irrigation	½–6	Irrigation
ASTM F810	PE pipe for drainage and waste disposal	3–6	Drainage; leaching fields
ASTM F892	PE corrugated pipe with smooth interior	4	Sewer; drain; conduit
ASTM F894	PE profile wall pipe, large diameter	18–120	Sewer; drain; industrial
ASTM F1055	PE electrofusion fittings	½–12	Water; natural gas; industrial
ASTM F1533	Deformed PE liner for pipeline rehabilitation	3–18	Rehabilitation of existing pressure and nonpressure pipelines
ASTM F1759	PE manholes for subsurface applications		Sewer and drain; industrial
AWWA C901	PE pipe and tubing for water service	½–3	Cold water service
AWWA C906	PE pipe for water distribution	4–63	Water distribution and transmission
CSA B137.1	PE pipe, tubing, and fittings for cold water		Cold water; industrial
CSA B137.4	PE piping for gas service		Natural gas distribution
CSA B137.4.1	Electrofusion-type PE fittings for gas service		Natural gas distribution
CSA B182.6	Profile PE pipe and fittings		Drainage; leaching fields
CSA B137.9	Polyethylene/aluminum/polyethylene Composite pressure piping		Cold water service; industrial

AASHTO: American Association of State Highway and Transportation Officials.
AWWA: American Water Works Association.
CSA: Canadian Standards Association.

Table 4.14 lists the principal PB piping standards.

CROSS-LINKED POLYETHYLENE (PEX)

Cross-linked polyethylene is, as its name implies, actually a thermoset. PEX pipe and tubing are made from PE, a thermoplastic, by essentially the same extrusion process used to manufacture all thermoplastic pipes. The only difference is that in the case of PEX there is a cross-linking of the polymer chains that occurs during, or soon after, pipe extrusion.

Table 4.15 lists the principal PEX piping standards.

POLYPROPYLENE (PP)

Polypropylene is a polyolefin similar in properties to high-density PE but is somewhat harder, more temperature resistant, and lighter in weight, but less tough. It is also similar to PE in its chemical resistance and its heat fusibility. Like PE, PP can be joined to itself by socket fusion, butt fusion, and electrofusion.

Because of its greater stiffness and better tolerance to elevated temperature, PP is sometimes chosen over PE where these qualities are advantageous—e.g., for aboveground piping and for the conveying of hot fluids. A principal application is for corrosive drainage piping, for which PP offers better solvent resistance than either ABS or PVC.

Table 4.16 lists the principal PP piping standards.

TABLE 4.14 Principal PB Piping Standards

Piping material and product standard	Subject product, or abbreviated title of standard	Nominal size range, inches	Principal applications
ASTM D2662	PB pipe, dimension ratio series, ID based	½–6	Hot and cold water; industrial
ASTM D2666	PB tubing	½–2	Hot and cold water; industrial
ASTM D3000	PB pipe, dimension ratio series	½–6	Hot and cold water; industrial
ASTM D3309	PB tubing for hot and cold water distribution	⅛–2	Hot and cold water
ASTM F809	Larger diameter PB pipe	3–42	Hot and cold water; industrial
ASTM F845	Plastic insert fittings for PB pipe and tubing	⅜–¼	Hot and cold water; water service
ASTM F878	PB drip irrigation tubing	½	Drip irrigation
ASTM F1380	Metal insert fittings for PB tubing	⅜–1	Hot and cold water
CSA B137.7*	Pipe for cold water distribution systems		Hot and cold water; water service
CSA B137.8*	PB piping for pressure applications		Hot and cold water; industrial

* CSA: Canadian Standards Association.

TABLE 4.15 Principal PEX Piping Standards

Piping material and product standard	Subject product, or abbreviated title of standard	Nominal size range, inches	Principal applications
ASTM F876	Cross-linked PE tubing	¼–2	Hydronic heating
ASTM F877	Cross-linked PE tubing	¼–2	Hot and cold water distributing
ASTM F1807	Metal insert fittings utilizing a copper crimp ring for cross-linked PE tubing	⅜–1	Hot and cold water
CSA B137.5*	Cross-linked PE tubing		Hot and cold water
CSA B 137.10*	Cross-linked PE/aluminum/cross-linked PE composite pressure piping		Hot and cold water; hydronic heating

* CSA: Canadian Standards Association.

TABLE 4.16 Principal PP Piping Standards

Piping material and product standard	Subject product, or abbreviated title of standard	Nominal size range, inches	Principal applications
ASTM D3311	DWV fittings patterns	$1^1/_4$–8	Drain, waste, and vent
CSA B181.3*	Polyolefin drainage systems		Industrial
CSA B137.11	Polypropylene pipe and fittings for pressure applications		Industrial

* CSA: Canadian Standards Association.

ACRYLONITRILE-BUTADIENE-STYRENE (ABS)

ABS plastics are made by combining styrene-acrylonitrile copolymers with copolymers formed by reacting styrene-acrylonitrile with butadiene. The butadiene copolymers impart toughness, while the acrylonitrile copolymers contribute strength, rigidity, and hardness. The result is a tough, relatively strong plastic that is easy to mold and extrude.

Table 4.17 lists the principal ABS piping standards.

FLUOROPLASTICS

Fluoroplastics designate a broad family of paraffinic polymers that have some or all of the hydrogen replaced by fluorine. The fully fluorinated fluorocarbons include perfluoroalkoxy (PFA), polytrafluoroethylene (PTFE), and fluorinated ethylene propylene (FEP). The partially fluorinated fluoroplastics include ethylene tetrafluoroethylene (ETFE), polychlorotrifluoroethylene (CTFE), ethylene-chlorotrifluoroethylene (ECTFE), and polyvinylidene fluoride (PVDF).

Fluorinated polymers have outstanding resistance to chemicals and excellent resistance to solvents. They also offer improved elevated-temperature properties and are very stable and durable. Fluorinated plastics also have outstanding resistance to weathering and electromagnetic radiation.

Table 4.18 lists the principal PVDF piping standards.

OTHER STANDARDS

Table 4.19 lists the principal PA piping standards.

THERMOPLASTIC VALVES

Table 4.20 lists the size range of readily available thermoplastic valves and other appurtenances.

THERMOPLASTIC PIPING JOINTS

Table 4.21 lists the common methods for joining thermoplastics piping.

The following notation has been used to denote different thermoplastic piping materials (notation list continues on p. 115):

ABS	Acrylonitrile-butadiene-styrene
CPVC	Chlorinated polyvinyl chloride
PA	Polyamide (nylon)
PB	Polybutylene
PE	Polyethylene

TABLE 4.17 Principal ABS Piping Standards

Piping material and product standard	Subject product, or abbreviated title of standard	Nominal size range, inches	Principal applications
ASTM D3311	DWV fittings patterns	$1\frac{1}{4}$–8	Drain, waste, and vent
ASTM D2680	ABS and PVC sewer pipe of composite wall construction	4–15	Sewer and drain
ASTM F409	Accessible and replaceable tube and fittings	$1\frac{1}{4}$–$1\frac{1}{2}$	Drain, waste, and vent
ASTM F480	Thermoplastic water well casing	2–16	Water-well casing
ASTM F1499	Coextruded, composite DWV pipe	$1\frac{1}{4}$–8	Drain, waste, and vent
CSA B181.5*	Coextruded ABS/PVC DWV pipe		Drain, waste, and vent
ASTM F1488	Coextruded composite pipe	2–12	Drain, waste, and vent; sewer and drain; electrical and communications conduit
ASTM D1527	ABS pipe, Schedules 40 and 80	$\frac{1}{8}$–12	Cold water; industrial
ASTM D2282	ABS pipe, dimension ratio series	$\frac{1}{8}$–12	Cold water; industrial
ASTM D2468	ABS socket fittings, Schedule 40	$\frac{1}{8}$–8	Cold water; industrial
ASTM D2661	ABS DWV pipe and fittings	$1\frac{1}{4}$–6	Drain, waste, and vent
ASTM D2750	ABS utility conduit and fittings	1–6	Electrical duct
ASTM D2751	ABS sewer pipe and fittings	3–12	Sewer and drain
ASTM F 628	ABS foam core DWV	$1\frac{1}{4}$–6	Drain, waste, and vent
CSA B181.1	ABS DWV pipe and fittings		Drain, waste, and vent
ASTM D3311	DWV fittings patterns	$1\frac{1}{4}$–8	Drain, waste, and vent

* CSA: Canadian Standards Association.

TABLE 4.18 Principal PVDF Piping Standards

Piping material and product standard	Subject product, or abbreviated title of standard	Nominal size range, inches	Principal applications
ASTM F1673	PVDF corrosive waste drainage piping	$1\frac{1}{4}$–12	Industrial

TABLE 4.19 Principal PA Piping Standards

Piping material and product standard	Subject product, or abbreviated title of standard	Nominal size range, inches	Principal applications
ASTM F1733	Butt heat fusion fittings for polyamide pipe and tubing	$\frac{1}{2}$–48	Gas distribution
CSA B137.12*	Polyamide piping systems for gas service		Gas distribution
ASTM D2513	Thermoplastic gas pressure pipe and fittings	$\frac{1}{4}$–24	Natural gas and LPG distribution

* CSA: Canadian Standards Association.

TABLE 4.20 Size Range of Readily Available Thermoplastic Valves and Other Appurtenances

Nominal pipe sizes expressed in inches

Item	PVC	CPVC	PE	PP	PVDF
Valves					
Needle	1/4–1/2	1/4–1/2	—	—	—
Ball	1/4–6	1/4–4	—	1/4–4	1/4–6
Gate	1/2–6	1/2–6	—	—	—
Butterfly	4–24	—	—	—	—
Check	1/4–16	1/2–4	—	1/2–4	1/2–4
Globe	1/8–4	—	—	—	—
Diaphragm	1/2–10	1/2–10	—	—	1/2–6
Foot	1/2–4	1/2–4	—	1/2	—
Solenoid	1/8–3	1/8–3	—	—	—
Gas service	—	—	1/4–6	—	—
Strainers	1/8–3	1/8–3	—	—	—
Saddles, tapping	2–8	2–12	—	—	—
Expansion joints	1/2–12	1/2–6	—	1/2–4	1/2–4
Flange adapters	2–12	2–12	2–42	2–6	2–6

TABLE 4.21 Common Methods for Joining Thermoplastic Piping

Joining method	ABS	PVC	CPVC	PE	PEX	PB	PP	PVDF
Solvent cementing	x	x	x					
Heat fusion				x		x	x	x
Threading*	x	x	x	x			x	x
Flanged connectors†	x	x	x	x		x	x	x
Grooved joints‡	x	x	x	x			x	x
Mechanical compression§	x	x	x	x	x	x	x	x
Elastomeric seal	x	x	x	x	x	x	x	x
Flaring¶				x		x		

* Generally limited to pipes of wall thickness not less than that of Schedule 80.
† Flanged adapters can be fastened to plastic pipe by means of heat fusion, solvent cementing, or threading.
‡ Grooving requires a minimum pipe wall thickness, which depends on the material. The pipe manufacturer should be consulted.
§ In most cases internal stiffeners are required to permanently support the pipe against compressive forces generated by these fittings.
¶ Not all commercial grades may be flared. Consult the pipe manufacturer for specific recommendations.

PEX	Cross-linked polyethylene
PE/AL/PE	Polyethylene/aluminum/polyethylene composite
PEX/AL/PEX	Cross-linked polyethylene/aluminium/cross-linked polyethylene composite
PP	Polypropylene
PVC	Poly(vinyl chloride)
PVDF	Poly(vinylidene fluoride)

MATERIALS FOR FIBERGLASS PIPING

Terminology

Catalyst. The chemical added to vinyl ester resins that causes them to harden. Usually methyl ethyl ketone peroxide (MEKP) is used.

Chopped Fiber. Continuous glass fibers cut into short lengths (0.125 to 2.0 in or 3 to 50 mm).

Chopped Strand Mat. Coarse fabric sheets made from chopped strands randomly placed and held together by resin binders.

Cure. The hardening of a thermoset resin system by the action of heat or chemical action.

Cure Time. The time it takes for a resin system to reach full strength.

Curing Agent. Any of a number of chemicals added to epoxy resin to cause it to harden. Aromatic amine curing agents are commonly used for high-temperature corrosion service. Anhydride-cured epoxy resins are typically used for less stringent applications.

Epoxy Resin. A thermosetting resin used in caustic, solvent, salt, and some acid solutions.

Fabmat. A combination of woven roving and chopped strand mat held together with resin binders. Usually, it is used for making contact molded fittings and butt weld joints.

Filament. A single fiber of glass.

Gel Time. The time it takes for a resin system to harden so flow will not occur.

Hardener. Any of a number of chemicals added to resin that cause hardening to occur.

Liner. The resin-rich interior surface of the pipe or fitting. The liner provides the corrosion resistance to the chemical service.

Nexus. Porous surfacing mat of synthetic used to provide a resin-rich layer or liner.

Novolac Resin. A premium epoxy vinyl ester resin used in a broad range of corrosive applications.

Reinforcement. Typically, fibers of glass used to provide strength and stiffness to a composite material.

Resin. The polymer or plastic material used to bind the glass fibers together in fiberglass pipe and fittings.

Roving. A collection of one or more strands of glass filaments. The typical form of glass fiber used in the manufacture of filament-wound pipe.

Thermoset Resin. A resin cured by heat or chemical additives. Once cured, a thermoset resin cannot be remelted.

Veil. Surfacing mat of porous fabric made from filaments. It is used to provide a resin-rich layer or liner.

Vinyl Ester Resin. A thermosetting resin used in strong acids, chlorine, and oxidizing agents.

Woven Roving. Coarse cloth-like material made by weaving fiberglass roving.

Resins

The following types of resins are used for manufacturing fiberglass pipe:

Epoxy Resins. Bisphenol-A epoxies and epoxy novolacs are two types of epoxy resins. Bisphenol epoxies cured with aromatic amines and bisphenol epoxies cured with aromatic anhydrides are the two most widely used resin systems for manufacture of fiberglass reinforced pipe (FRP). Pipe made from these resins has an upper temperature limit of 250°F (121°C), and it is resistant to salt solutions and rather severe alkaline and solvent exposures.

NPS 2 through 16 (DN 50 through 400) FRP, used for oil field and water handling applications, is manufactured from bisphenol epoxies cured with aromatic anhydrides. These pipes have an upper temperature limit of 150°F (66°C).

Epoxy novolacs are used where increased temperature resistance and/or better solvent resistance are required.

Epoxy resins are not resistant to strong mineral acids or strong oxidizers.

Polyester Resins

Vinyl Ester Resins. These resins have excellent resistance to oxidizers and strong mineral acids and good resistance to alkaline environments.

Standard vinyl ester resins are limited to 200 to 225°F (93 to 107°C). High-performance vinyl esters are suitable for general use up to 250°F (121°C), and have been used in some special applications in temperatures as high as 350°F (177°C).

Bisphenol-A Fumarate Polyester Resins. These chemical-resistant polyester resins are equivalent to vinyl ester resins at temperatures up to 250°F (121°C), but they are more rigid than vinyl ester resins. As such, these resins are unsuitable for manufacturing small-diameter pipe on mass-production equipment. The rigidity of this resin does not pose any difficulty in manufacture of large-diameter pipe or vessels.

Chlorendic Polyester Resins. These resins have a chlorinated backbone in their molecular structure that makes them particularly well suited for elevated-temperature applications, up to 350°F (177°C). These resins are more rigid than bisphenol-A fumarate resins. These resins offer:

- Excellent resistance to concentrated mineral acids and highly oxidizing environments
- Very good solvent resistance compared to other polyester resins
- Poor resistance to alkaline solutions

Isophthalic Polyester Resins. These are the least expensive and least chemical resistant of the corrosion-resistant polyester resins. For temperatures up to 180°F (82°C), they offer:

- Good resistance to water, dilute acids, and very weak alkaline solutions
- Good resistance to petroleum solvents, such as gasoline and oil

Other Resins. Furon resins are limited to applications that require a unique blend of superior acid, alkali, and solvent resistance at temperatures up to 300°F (177°C). They are not suitable for handling oxidizing services.

Phenolic resins require special processing techniques, which limits their economical application to systems that require superior acid and solvent resistance at temperatures up to 300°F (177°C). These resins are often used in flame-resistant duct applications.

Maximum Temperature Rating Range and Suitability of Resins

Table 4.22 provides the maximum temperature rating ranges for the resins discussed above with regard to their suitability for various service fluids. Within the temperature rating range listed, the resin is considered suitable for use with the service fluid listed.

This table is a preliminary guideline only, based on the data published by several manufacturers. The user must consult the pipe manufacturer for additional input prior to making a decision.

Fiberglass Pipe and Fitting Standard Product Specifications

Table 4.23 lists the ASTM, API, AWWA, ISO, and British standard product specifications used for fiberglass pipe and fittings. It also provides the nominal pipe size (NPS and DN) of the pipe and/or the fittings covered by the specification.

Table 4.24 lists the codes and other standards that contain requirements for fiberglass piping.

ASTM, DIN, and ISO Standards for Testing

Table 4.25 lists the ASTM, DIN, and ISO standards for testing of various properties of fiberglass pipe and fittings.

Physical Properties of Fiberglass Pipe

Table 4.26 lists typical values of ultimate physical properties of fiberglass pipe made from epoxy, vinyl ester, and isophthalic resins. It lists values for the following properties:

TABLE 4.22 Maximum Temperature Rating Range by Chemical and Resin Class

Service/fluid	Temp. scale	Epoxy	Vinyl ester	Novolac VE	Furan	Isophthalic
Acetic acid	°F	75–100	100–210	150–215	225	170
	°C	24–93	38–99	66–102	107	77
Acetone	°F	75–180	125–150	100–200	AMB	NR
	°C	24–82	52–66	38–93	AMB	NR
Acrylic acid	°F	100–120	75–100	100–100	NT	NR
	°C	38–49	24–38	38–38	NT	NR
Adipic acid	°F	225–250	75–200	100–210	NT	NL
	°C	107–121	24–93	38–99	NT	NL
Air	°F	210–300	200–360	225–450	NL	NL
	°C	99–149	93–182	107–232	NL	NL
Alcohol, ethyl	°F	120–180	75–150	80–150	NT	NL
	°C	49–82	24–66	27–66	NT	NL
Alcohol, isopropyl	°F	100–180	80–150	100–150	NT	NL
	°C	38–82	27–66	38–66	NT	NL
Alcohol, methyl	°F	75–150	100–150	100–150	NT	NL
	°C	24–66	38–66	38–66	NT	NL
Alcohol-methyl isobutyl	°F	100–150	100–150	120–150	NT	NL
	°C	38–66	38–66	49–66	NT	NL
Alcohol, secondary butyl	°F	100–175	120–150	120–150	NT	NL
	°C	38–79	49–66	49–66	NT	NL
Allyl chloride	°F	75–120	80	75–80	NL	NR
	°C	24–49	27	24–27	NL	NR
Aluminum chloride	°F	150–300	180–210	210–250	250	170
	°C	66–149	82–99	99–121	121	77
Aluminum fluoride	°F	75–150	70–85	70–85	225	NR
	°C	24–66	21–29	21–29	107	NR
Aluminum hydroxide	°F	150–200	80–180	175–180	225	NL
	°C	66–93	27–82	79–82	107	NL
Aluminum nitrate	°F	150–250	125–200	160–225	NT	140
	°C	66–121	52–93	71–107	NT	60
Aluminum sulfate	°F	210–300	160–210	210–250	250	170
	°C	99–149	71–99	99–121	121	77
Alums	°F	225–300	210	210–250	NT	170
	°C	107–149	99	99–121	NT	77
Ammonia gas, dry	°F	150–225	80–100	100	250	NL
	°C	66–107	27–38	38	121	NL
Ammonia gas, wet	°F	125–150	100	100	NT	NR
	°C	52–66	38	38	NT	NR

TABLE 4.22 Maximum Temperature Rating Range by Chemical and Resin Class (Continued)

Service/fluid	Temp. scale	Epoxy	Vinyl ester	Novolac VE	Furan	Isophthalic
Ammonium carbonate	°F	150–250	75–150	150	225	80
	°C	66–96	24–66	66	107	27
Ammonium chloride	°F	150–270	180–210	210–224	220	170
	°C	66–132	82–99	99–107	104	77
Ammonium fluoride	°F	75–150	85–150	150	225	NL
	°C	24–66	29–66	66	107	NL
Ammonium hydroxide	°F	100–200	100–180	100–150	180	NR
	°C	38–93	38–82	38–66	82	NR
Ammonium nitrate	°F	200–250	150–200	200–250	220	140
	°C	93–121	66–93	93–121	104	60
Ammonium persulfate	°F	100–250	170	180	180	NR
	°C	38–121	77	82	82	NR
Ammonium phosphate	°F	150–200	75–200	210	190	140
	°C	66–93	24–93	99	88	60
Ammonium sulfate	°F	200–300	120–210	210–250	220	170
	°C	93–149	49–99	99–121	104	77
Amyl acetate	°F	75–120	NL	80–120	200	NR
	°C	24–49	NL	27–49	93	NR
Amyl chloride	°F	100	120	120	NR	NR
	°C	38	49	49	NR	NR
Aniline	°F	75–100	100–120	70	250	NR
	°C	24–38	38–49	21	121	NR
Antimony chloride	°F	NR	200–210	NR	225	NR
	°C	NR	93–99	NR	107	NR
Antimony trichloride	°F	150–220	NR	200–200	NR	NR
	°C	66–104	NR	93–93	NR	NR
Barium carbonate	°F	200–250	180–210	210–250	NT	80
	°C	93–121	82–99	99–121	NT	27
Barium chloride	°F	210–250	180–210	210	200	170
	°C	99–121	82–99	99	93	77
Barium hydroxide	°F	180–220	100–200	150–200	200	NR
	°C	82–104	38–93	66–93	93	NR
Barium sulfate	°F	210–250	200–210	210–250	NT	170
	°C	99–121	93–99	99–121	NT	77
Barium sulfide	°F	210–300	150–200	180–215	150	NR
	°C	99–149	66–93	82–102	66	NR
Beer	°F	200–225	90–200	120–200	NT	80
	°C	93–107	32–83	49–93	NT	27

TABLE 4.22 Maximum Temperature Rating Range by Chemical and Resin Class (Continued)

Service/fluid	Temp. scale	Epoxy	Vinyl ester	Novolac VE	Furan	Isophthalic
Benzene	°F	100–150	75	100	NT	NR
	°C	38–66	24	38	NT	NR
Benzene sulfonic acid	°F	100–220	125–200	125–215	200	NR
	°C	38–104	52–93	52–102	93	NR
Benzoic acid	°F	100–220	150–210	210	250	170
	°C	38–104	66–99	99	121	77
Black liquor (pulp mill)	°F	100–225	150–180	180–225	NL	NR
	°C	38–107	66–82	82–107	NL	NR
Borax	°F	225–250	200–210	210	NL	170
	°C	107–121	93–99	99	NL	77
Boric acid	°F	200–225	180–210	200–210	200	170
	°C	93–107	82–99	93–99	93	77
Bromic acid	°F	150	NL	150	NT	NL
	°C	66	NL	66	NT	NL
Bromine water	°F	100–150	100	100–190	NT	NL
	°C	38–66	38	38–88	NT	NL
Butadiene, gas	°F	100–150	100	100	NT	NL
	°C	38–66	38	38	NT	NL
Butane, gas	°F	75–150	100	100	NL	NL
	°C	24–66	38	38	NL	NL
Butyl acetate	°F	AMB-150	NL	80	220	NR
	°C	AMB-66	NL	27	104	NR
Butyl cellosolve	°F	150	100	100	NT	NR
	°C	66	38	38	NT	NR
Butyric acid	°F	150–200	80–210	120–210	200	80
	°C	66–93	27–99	49–99	93	27
Calcium bisulfite	°F	200–270	180–200	180 180 180	NL	140
	°C	93–132	82–93	82 82 82	NL	60
Calcium carbonate	°F	150–300	100–200	180–200	NT	160
	°C	66–149	38–93	82–93	NT	71
Calcium chlorate	°F	150–200	100–210	210–250	NT	150
	°C	66–93	38–99	99–121	NT	66
Calcium chloride	°F	210–300	200–210	215–250	250	170
	°C	99–149	93–99	102–121	121	77
Calcium hydroxide	°F	150–200	100–180	150–210	225	NL
	°C	66–93	38–82	66–99	107	NL

TABLE 4.22 Maximum Temperature Rating Range by Chemical and Resin Class (Continued)

Service/fluid	Temp. scale	Epoxy	Vinyl ester	Novolac VE	Furan	Isophthalic
Calcium hypochlorite	°F	100–150	100–200	150–180	NT	NR
	°C	38–66	38–93	66–82	NT	NR
Calcium nitrate	°F	150–250	125–210	210–215	220	170
	°C	66–121	52–99	99–102	104	77
Calcium sulfate	°F	200–250	200–210	210–250	250	170
	°C	93–121	93–99	99–121	121	77
Carbon dioxide gas	°F	225–250	200–210	200–350	NT	NL
	°C	107–121	93–99	93–177	NT	NL
Carbon tetrachloride	°F	100–150	90–180	125–200	225	NR
	°C	38–66	32–82	52–93	107	NR
Carbonic acid	°F	150–180	175	NL	NL	NL
	°C	66–82	79	NL	NL	NL
Castor oil	°F	200–225	150	160–210	NT	NL
	°C	93–107	66	71–99	NT	NL
Chlorine dioxide gas	°F	75	80–200	180–210	150	NR
	°C	24	27–93	82–99	66	NR
Chlorine gas	°F	125	200–210	210–250	225	NR
	°C	52	93–99	99–121	107	NR
Chloroacetic acid	°F	100–200	75–200	100–135	NT	NR
	°C	38–93	24–93	38–57	NT	NR
Chromic acid	°F	75–150	80–200	90–150	NT	NL
	°C	24–66	27–93	32–66	NT	NL
Chromic fluoride	°F	75	75	75	NL	NL
	°C	24	24	24	NL	NL
Citric acid	°F	200–250	200–210	210–215	190	160
	°C	93–121	93–99	99–102	88	71
Copper chloride	°F	150–250	150–210	210–250	250	170
	°C	66–121	66–99	99–121	121	77
Copper fluoride	°F	200–250	175–200	175–210	NT	NR
	°C	93–121	79–93	79–99	NT	NR
Copper nitrate	°F	150–250	100–210	210–215	220	170
	°C	66–121	38–99	99–102	104	77
Copper sulfate	°F	150–250	150–210	210–250	250	170
	°C	66–121	66–99	99–121	121	77
Crude oil	°F	210–250	200–210	210–250	NL	170
	°C	99–121	93–99	99–121	NL	77
Deionized water	°F	212	180	180–210	NL	170
	°C	100	82	82–99	NL	77

TABLE 4.22 Maximum Temperature Rating Range by Chemical and Resin Class (Continued)

Service/fluid	Temp. scale	Epoxy	Vinyl ester	Novolac VE	Furan	Isophthalic
Dichlorobenzene (ortho)	°F	100–150	NL	120	220	NR
	°C	38–66	NL	49	104	NR
Diesel fuel	°F	210–250	150–180	190–225	NL	140
	°C	99–121	66–82	88–107	NL	60
Ethyl acetate	°F	75–150	NL	70	125	NR
	°C	24–66	NL	21	52	NR
Ethyl cellosolve	°F	100–150	NL	NR	NT	NL
	°C	38–66	NL	NR	NT	NL
Ethyl chloride	°F	75–80	NL	80	220	NR
	°C	24–27	NL	27	104	NR
Ethylene glycol	°F	200–270	200–210	210–225	250	170
	°C	93–132	93–99	99–107	121	77
Fatty acids	°F	200–225	200–210	210–250	250	170
	°C	93–107	93–99	99–121	121	77
Ferric chloride	°F	150–300	150–210	210–225	250	170
	°C	66–149	66–99	99–107	121	77
Ferric nitrate	°F	150–250	150–200	210–215	250	170
	°C	66–121	66–93	99–102	121	77
Ferric sulfate	°F	200–225	200–210	210–215	225	170
	°C	93–107	93–99	99–102	107	77
Ferrous chloride	°F	175–250	165–210	210–225	NT	170
	°C	79–121	74–99	99–107	NT	77
Ferrous sulfate	°F	200–225	200–210	210–215	NT	170
	°C	93–107	93–99	99–102	NT	77
Fluorine gas	°F	75	75–80	75–80	265	NL
	°C	24	24–27	24–27	129	NL
Fluoboric acid	°F	200	180–210	190–210	200	150
	°C	93	82–99	88–99	93	66
Fluosilicic acid	°F	100–200	80–200	80–210	200	NR
	°C	38–93	27–93	27–99	93	NR
Formaldehyde	°F	75–150	75–150	75–260	225	NR
	°C	24–66	24–66	24–127	107	NR
Formic acid	°F	100–140	75–180	100–180	225	100
	°C	38–60	24–82	38–82	107	38
Freon	°F	75–150	75–80	75–100	225	NL
	°C	24–66	24–27	24–38	107	NL
Gas, natural	°F	200–225	180–200	200–210	260	NL
	°C	93–107	82–93	93–99	127	NL

TABLE 4.22 Maximum Temperature Rating Range by Chemical and Resin Class (Continued)

Service/fluid	Temp. scale	Epoxy	Vinyl ester	Novolac VE	Furan	Isophthalic
Gasoline	°F	150–225	100–180	100–180	NL	110
	°C	66–107	38–82	38–82	NL	43
Glucose	°F	200–250	200–210	210–250	NL	110
	°C	93–121	93–99	99–121	NL	43
Glycerine	°F	210–300	200–210	210	NL	170
	°C	99–149	93–99	99	NL	77
Glycol ethylene	°F	200	200	210	NL	NL
	°C	93	93	99	NL	NL
Heptane	°F	150–200	150–210	200–210	250	140
	°C	66–93	66–99	93–99	121	60
Hexane	°F	75–150	100–160	150–160	150	NL
	°C	24–66	38–71	66–71	66	NL
Hexylene glycol alcohol	°F	150	150	150	NL	NL
	°C	66	66	66	NL	NL
Hydraulic fluid	°F	200–225	100–180	180–195	NT	NR
	°C	93–107	38–82	82–91	NT	NR
Hydrobromic acid	°F	100–150	100–180	100–180	NT	NL
	°C	38–66	38–82	38–82	NT	NL
Hydrochloric acid	°F	75–200	75–200	165–230	150	110
	°C	24–93	24–93	74–110	66	43
Hydrocyanic acid	°F	120	150–210	210	200	80
	°C	49	66–99	99	93	27
Hydrofluoric acid	°F	75	100–150	100–150	NT	NR
	°C	24	38–66	38–66	NT	NR
Hydrogen peroxide	°F	80–150	100–150	75–150	NT	NR
	°C	27–66	38–66	24–66	NT	NR
Hydrogen sulfide	°F	150–250	150–180	175–210	NL	140
	°C	66–121	66–82	79–99	NL	60
Hypochlorous acid	°F	120–200	150–180	135–180	NT	NL
	°C	49–93	66–82	57–82	NT	NL
Jet fuel	°F	150–250	120–180	180	NT	140
	°C	66–121	49–82	82	NT	60
Kerosene	°F	150–250	150–200	175–180	NT	140
	°C	66–121	66–93	79–82	NT	60
Lactic acid	°F	200–225	125–210	210–215	225	130
	°C	93–107	52–99	99–102	107	54
Lauric acid	°F	200–225	125–200	175–215	NT	NL
	°C	93–107	52–93	79–102	NT	NL

TABLE 4.22 Maximum Temperature Rating Range by Chemical and Resin Class (Continued)

Service/fluid	Temp. scale	Epoxy	Vinyl ester	Novolac VE	Furan	Isophthalic
Lead acetate	°F	150–250	150–210	210–230	NT	110
	°C	66–121	66–99	99–110	NT	43
Levulinic acid	°F	200–250	200–210	215–230	225	NL
	°C	93–121	93–99	102–110	107	NL
Magnesium carbonate	°F	150–250	150–200	180–200	NL	130
	°C	66–121	66–93	82–93	NL	54
Magnesium chloride	°F	210–270	200–210	225–250	NL	140
	°C	99–132	93–99	107–121	NL	60
Magnesium hydroxide	°F	120–270	125–200	210–215	NL	NL
	°C	49–132	52–93	99–102	NL	NL
Magnesium nitrate	°F	210–300	200–210	210–225	NT	170
	°C	99–149	93–99	99–107	NT	77
Magnesium sulfate	°F	150–270	200–210	210–250	200	150
	°C	66–132	93–99	99–121	93	66
Maleic acid	°F	150–220	200–210	200–250	200	140
	°C	66–104	93–99	93–121	93	60
Mercury	°F	225–270	200–210	210–250	NT	170
	°C	107–132	93–99	99–121	NT	77
Methyl ethyl ketone	°F	75–150	NL	75–125	150	NR
	°C	24–66	NL	24–52	66	NR
Methyl isobutyl carbitol	°F	100–120	NL	NL	NT	NL
	°C	38–49	NL	NL	NT	NL
Methyl isobutyl ketone	°F	100–150	NL	NL	NL	NL
	°C	38–66	NL	NL	NL	NL
Mineral oil	°F	210–270	120	NL	NL	170
	°C	99–132	49	NL	NL	77
Monochlorobenzene	°F	100–150	NL	NL	NL	NR
	°C	38–66	NL	NL	NL	NR
Naphtha	°F	200–250	125–180	120–210	NT	110
	°C	93–121	52–82	49–99	NT	43
Naphthalene	°F	150–200	100–210	190–210	NL	130
	°C	66–93	38–99	88–99	NL	54
Nickel chloride	°F	210–300	200–210	210–215	220	140
	°C	99–149	93–99	99–102	104	60
Nickel nitrate	°F	200–250	200–210	210–215	220	140
	°C	93–121	93–99	99–102	104	60
Nitric acid	°F	75–120	75–200	120–215	NT	150
	°C	24–49	24–93	49–102	NT	66

TABLE 4.22 Maximum Temperature Rating Range by Chemical and Resin Class (Continued)

Service/fluid	Temp. scale	Epoxy	Vinyl ester	Novolac VE	Furan	Isophthalic
Oleic acid	°F	200–225	100–210	200–210	NL	170
	°C	93–107	38–99	93–99	NL	77
Oxalic acid	°F	150–250	120–200	120–215	150	170
	°C	66–121	49–93	49–102	66	77
Perchloric acid	°F	75	75–150	75–150	NT	NR
	°C	24	24–66	24–66	NT	NR
Phenol	°F	75–150	100–150	75–150	NL	NR
	°C	24–66	38–66	24–66	NL	NR
Phosphoric acid	°F	75–225	100–210	210–225	250	170
	°C	24–107	38–99	99–107	121	77
Phosphoric pentoxide	°F	100–200	200	210	NL	NL
	°C	38–93	93	99	NL	NL
Picric acid	°F	75–100	100	100–205	165	NR
	°C	24–38	38	38–96	74	NR
Potassium bicarbonate	°F	150–270	100–180	100–180	NL	NL
	°C	66–132	38–82	38–82	NL	NL
Potassium bromide	°F	200–225	100–200	120–215	200	150
	°C	93–107	38–93	49–102	93	66
Potassium carbonate	°F	100–250	110–180	120–180	200	80
	°C	38–121	43–82	49–82	93	27
Potassium chloride	°F	210–300	200–210	210–215	250	170
	°C	99–149	93–99	99–102	121	77
Potassium dichromate	°F	200–250	200–210	210	NT	170
	°C	93–121	93–99	99	NT	77
Potassium hydroxide	°F	100–200	100–180	150–180	220	NR
	°C	38–93	38–82	66–82	104	NR
Potassium nitrate	°F	200–300	200–210	210–215	250	170
	°C	93–149	93–99	99–102	121	77
Potassium sulfate	°F	210–270	180–210	210–215	250	170
	°C	99–132	82–99	99–102	121	77
Propane	°F	75–150	100–200	200	NL	NL
	°C	24–66	38–93	93	NL	NL
Silicic acid	°F	200–250	125–200	200–210	NT	170
	°C	93–121	52–93	93–99	NT	77
Silver nitrate	°F	150–250	150–210	210	NL	170
	°C	66–121	66–99	99	NL	77
Soaps	°F	200–250	150–200	210	NL	NL
	°C	93–121	66–93	99	NL	NL

TABLE 4.22 Maximum Temperature Rating Range by Chemical and Resin Class (Continued)

Service/fluid	Temp. scale	Epoxy	Vinyl ester	Novolac VE	Furan	Isophthalic
Sodium acetate	°F	150–250	150–210	210–215	225	170
	°C	66–121	66–99	99–102	107	77
Sodium bicarbonate	°F	200–275	100–180	170–180	225	100
	°C	93–135	38–82	77–82	107	38
Sodium bisulfate	°F	150–250	150–210	210–215	225	170
	°C	66–121	66–99	99–102	107	77
Sodium bromide	°F	200–250	200–210	210–215	NL	170
	°C	93–121	93–99	99–102	NL	77
Sodium carbonate	°F	100–205	100–180	180	NL	NR
	°C	38–96	38–82	82	NL	NR
Sodium chlorate	°F	180–250	180–210	210–215	NT	NR
	°C	82–121	82–99	99–102	NT	NR
Sodium chloride	°F	210–300	200–210	210–250	250	130
	°C	99–149	93–99	99–121	121	54
Sodium cyanide	°F	225–250	150–210	210	225	80
	°C	107–121	66–99	99	107	27
Sodium dichromate	°F	200–250	200–210	210	NT	140
	°C	93–121	93–99	99	NT	60
Sodium ferrocyanide	°F	200–300	150–210	210–215	NL	170
	°C	93–149	66–99	99–102	NL	77
Sodium fluoride	°F	150–250	150–180	150–180	250	80
	°C	66–121	66–82	66–82	121	27
Sodium hydroxide	°F	100–200	100–210	150–180	212	NR
	°C	38–93	38–99	66–82	100	NR
Sodium hypochlorite	°F	75	75–180	110–180	NT	NR
	°C	24	24–82	43–82	NT	NR
Sodium nitrate	°F	200–300	200–210	210–215	225	170
	°C	93–149	93–99	99–102	107	77
Sodium phosphate	°F	200	200–210	210	250	NL
	°C	93	93–99	99	121	NL
Sodium silicate	°F	150–220	160–210	210–215	250	80
	°C	66–104	71–99	99–102	121	27
Sodium sulfate	°F	200–300	200–210	210–215	220	170
	°C	93–149	93–99	99–102	104	77
Sodium sulfite	°F	200	200–210	210–215	NL	80
	°C	93	93–99	99–102	NL	27
Sodium thiosulfate	°F	150	150–200	180	NL	140
	°C	66	66–93	82	NL	60

TABLE 4.22 Maximum Temperature Rating Range by Chemical and Resin Class (Continued)

Service/fluid	Temp. scale	Epoxy	Vinyl ester	Novolac VE	Furan	Isophthalic
Stannic chloride	°F	150–270	150–210	210	NL	170
	°C	66–132	66–99	99	NL	77
Stearic acid	°F	150–225	150–210	210–215	NL	170
	°C	66–107	66–99	99–102	NL	77
Sulfamic acid	°F	100–150	100–210	150–210	NL	110
	°C	38–66	38–99	66–99	NL	43
Sulfite liquor	°F	150	200	200	225	NL
	°C	66	93	93	107	NL
Sulfur dioxide	°F	150–250	120–210	225–250	250	NL
	°C	66–121	49–99	107–121	121	NL
Sulfuric acid	°F	75–205	75–210	120–215	225	170
	°C	24–96	24–99	49–102	107	77
Sulfurous acid	°F	75–200	80–150	100–150	200	NL
	°C	24–93	27–66	38–66	93	NL
Tannic acid	°F	200–225	100–210	100–210	NT	170
	°C	93–107	38–99	38–99	NT	77
Tartaric acid	°F	150–250	200–210	210	250	170
	°C	66–121	93–99	99	121	77
Toluene	°F	100–150	75–80	75–120	225	NR
	°C	38–66	24–27	24–49	107	NR
Trichloroethylene	°F	120–150	NL	NL	NL	NR
	°C	49–66	NL	NL	NL	NR
Triethanolamine	°F	100–150	100–120	120	NL	110
	°C	38–66	38–49	49	NL	43
Trisodium phosphate	°F	100–200	100–210	210–250	NL	120
	°C	38–93	38–99	99–121	NL	49
Tung oil	°F	200	100	200	NL	NL
	°C	93	38	93	NL	NL
Turpentine	°F	100–150	150	100–210	NL	NR
	°C	38–66	66	38–99	NL	NR
Urea	°F	150–200	150	150	225	120
	°C	66–93	66	66	107	49
Vinegar	°F	150–200	NL	210	NL	150
	°C	66–93	NL	99	NL	66
Vinyl acetate	°F	75–150	150–210	NR	NL	NR
	°C	24–66	66–99	NR	NL	NR
Water	°F	200–250	175–200	180–210	NL	160
	°C	93–121	79–93	82–99	NL	71

TABLE 4.22 Maximum Temperature Rating Range by Chemical and Resin Class (Continued)

Service/fluid	Temp. scale	Epoxy	Vinyl ester	Novolac VE	Furan	Isophthalic
Water, sea	°F	210–270	175–200	210	NL	170
	°C	99–132	79–93	99	NL	77
Xylene	°F	125–150	75–80	75–120	225	NR
	°C	52–66	24–27	24–49	107	NR
Zinc chloride	°F	210–250	200	200–310	NL	170
	°C	99–121	93	93–154	NL	77
Zinc sulfate	°F	200–250	200–210	210–350	250	170
	°C	93–121	93–99	99–177	121	77

NL = Not listed.
NR = Not recommended for the service fluid.
NT = Not tested for the service fluid.

TABLE 4.23 Fiberglass Pipe and Fittings Standard Product Specifications

Product description	Standard	Pipe	Fittings	Size,* NPS (DN)	Pressure, psig (bar)
Fittings					
Contact molded	ASTM D 6041	No	Yes	All	0–150 (0–10)
Dimensions, nominal	ISO 7370	Yes	Yes	1–144 (25–3600)	NA
Flanges, contact molded	ASTM D 5421	No	Yes	1–96 (25–2400)	25–150 (2–10)
Flanges other than contact molded	ASTM D 4024	No	Yes	All	50–500 (3–34)
Gravity flow	ASTM D 3840	No	Yes	8–144 (200–3600)	Gravity
Line pipe, low pressure	API 15LR	Yes	Yes	1–16 (25–400)	up to 1000 (68)
Pressure	ASTM D 5685	No	Yes	1–16 (25–400)	25–1000 (2–68)
Jet fuel lines, belowground	ASTM D 5677	Yes	Yes	All	up to 150 (10)
Joints					
Bell and spigot gasket joints	ASTM D 4161	Yes	Yes	8–144 (200–3600)	up to 250 (17)
Marine pipe and fittings	ASTM F 1173	Yes	Yes	1–48 (25–1200)	All
Laminates, contact molded	ASTM C 582	Yes	Yes	All	All
Pipe					
Casing and tubing	API 15AR	Yes	No	1–10 (25–250)	
Centrifugally cast	ASTM D 2997	Yes	No	All	All
Contact molded	ASTM C 582	Yes	Yes	All	All
Dimensions, nominal	ISO 7370	Yes	Yes	1–144 (10–3600)	NA
Filament wound	ASTM D 2996	Yes	No	1–16 (25–400)	All
Line pipe, high pressure	API 15HR	Yes	No	1–8 (25–200)	500–1000 (34–68)
Line pipe, low pressure	API 15LR	Yes	Yes	1–16 (25–400)	up to 1000 (68)

* For sizes other than those listed, contact the manufacturer.

TABLE 4.23 Fiberglass Pipe and Fittings Standard Product Specifications (Continued)

Product description	Standard	Pipe	Fittings	Size, NPS (DN)	Pressure, psig (bar)
Machine made classification	ASTM D 2310	Yes	No	NA	NA
Sewer	ASTM D 3262	Yes	No	8–144 (200–3600)	Gravity
Water	ASTM D 3517	Yes	No	8–144 (200–3600)	up to 250 (17)
Industrial wastes and corrosive fluids	ASTM D 3754	Yes	No	8–144 (200–3600)	up to 250 (17)
Process plant piping	BS 6464	Yes	Yes	1–36 (25–1000)	up to 940 (64)
Water supply or sewerage piping	BS 5480	Yes	Yes		
Water systems	AWWA C-950 M45 design manual	Yes	Yes	1–144 (25–3600)	50–250 (3–17)

TABLE 4.24 Fiberglass Pipe and Fittings, Other Standards

Description	Standard
Pressure piping code	
Power piping	ASME B31.1
Process piping	ASME B31.3
British standard, individual plants or sites	BS 7159
Underground installation	ASTM D 6041 ISO TR 10465-1
Terminology and definitions	ASTM F 412 ISO 8572
Building services piping	ASME B31.9

TABLE 4.25 Fiberglass Pipe and Fittings Standard Test Methods

Property tested	ASTM standards			DIN standards			ISO standards		
	Number	Pipe	Fittings	Number	Pipe	Fittings	Number	Pipe	Fittings
Beam bending stress and modulus	D790 modified	Yes	No				ISO 178	Yes	No
Beam deflection, full bore flow	D2925	Yes	No						
Chemical resistance									
Laminates	C581	Yes	Yes	53 393	Yes	Yes			
Molding compounds	D3615	No	Yes						
Pipe, deflected	D3681	Yes	No						
Circumferential flexural modulus									
Short-term	D2412	Yes	No	53 769-3	Yes	No	10466	Yes	No
Long-term creep				EN 761	Yes	No	7684	Yes	No
Circumferential tensile strength	D2290	Yes	No	EN 1393	Yes	No			
Compressive stress and modulus	D695	Yes	No						

TABLE 4.25 Fiberglass Pipe and Fittings Standard Test Methods (Continued)

Property tested	ASTM standards			DIN standards			ISO standards		
	Number	Pipe	Fittings	Number	Pipe	Fittings	Number	Pipe	Fittings
Constituents; % resin, glass, aggregate, filler	D2584	Yes	No	EN 637	Yes	Yes	7510	Yes	Yes
Dimensions	D3567	Yes	Yes						
Density	D792	Yes	Yes	53 479	Yes	Yes			
Dielectric strength	D149	Yes	Yes						
Electrical resistance, dc	D257	Yes	Yes						
Indentation hardness, barchol impressor	D2583	Yes	Yes	EN 59					
Impact resistance	D2444	Yes	Yes						
Joints, pressure and bending as applicable									
Cemented socket and spigot				EN 1449	Yes	Yes			
Bolted flanges				EN 1450	Yes	Yes			
Socket and spigot with elastomeric seals				EN 1448	Yes	Yes			
Pipe stiffness									
Short-term	D2412	Yes	No	53 769-3	Yes	No	10466	Yes	No
Long-term creep				EN 761	Yes	No	7684	Yes	No
Pressure, external	D2924	Yes	No						
Pressure, internal									
Cyclic	D2143	Yes	No						
Hydrostatic design basis	D2992	Yes	Yes	53 769-2					
Short-time hydraulic failure	D1599	Yes	Yes	53 758					
Time to failure, constant pressure	D1598	Yes	No	EN 1447	Yes	No			
Regression analysis	D2992	Yes	Yes	53 768	Yes	Yes	10928	Yes	Yes
				EN 705	Yes	Yes			
Ring-bending strain	D5365	Yes	No						
Shear strength				53 769-1	No	Yes			
Specific gravity	D792	Yes	Yes						
Stiffness factor									
Short-term	D2412	Yes	No	53 769-3	Yes	No	10466	Yes	No
Long-term creep				EN 761	Yes	No	7684	Yes	No
Tensile elongation ultimate	D2105	Yes	No	EN 1393	Yes	No			
Tensile stress and modulus									
Hoop	D1599	Yes	Yes						
Laminate	D638	Yes	Yes				527-4	Yes	Yes
Longitudinal	D2105	Yes	No						
Thermal conductivity	C177	Yes	Yes						
Thermal expansion, linear coefficient									
Between –30°C and 30°C	D696	Yes	No						
Other temperatures	E228	Yes	No						

TABLE 4.26 Typical Ultimate Physical Properties for Fiberglass Pipe

Property at 75°F (24°C)	Filament-wound epoxy pipe	Centrifugally cast epoxy pipe	Filament-wound vinyl ester pipe	Centrifugally cast vinyl ester pipe	Filament-wound isophthalic pipe
Hydrostatic burst, ASTM D-1599					
Ultimate stress, psi (kg/cm^2)*	38,000 (2700)	27,000 (1900)	40,000 (2800)	34,500 (2400)	42,000 (3000)
Modulus of elasticity, 10^6 psi (10^5 kg/cm^2)*	3.4 (2.4)	2.6 (1.8)	3.2 (2.2)	2.3 (1.6)	3.2 (2.2)
Hydrostatic design stress, ASTM D-2992, psi (kg/cm^2)*	11,000 (770)	11,000 (770)	7000 (490)	8000 (560)	7500 (530)
Axial tensile, ASTM D-2105					
Ultimate stress, psi (kg/cm^2)*	15,000 (1100)	35,000 (2500)	9000 (630)	36,000 (2500)	9400 (660)
Modulus of elasticity, 10^6 psi (10^5 kg/cm^2)*	1.8 (1.3)	2.8 (2.0)	1.5 (1.1)	2.6 (1.8)	1.9 (1.3)
Beam bending, ASTM D-2925					
Ultimate stress, psi (kg/cm^2)*	25,000 (1800)	41,000 (2900)	20,000 (1400)	38,000 (2700)	
Modulus of elasticity, 10^6 psi (10^5 kg/cm^2)*	2.0 (1.4)	2.6 (1.8)	1.8 (1.3)	2.4 (1.7)	1.3 (0.91)
Axial compressive, ASTM D-695					
Ultimate stress, psi (kg/cm^2)*	30,000 (2100)	38,000 (2700)	16,000 (1100)	26,000 (1800)	23,000 (1600)
Modulus of elasticity, 10^6 psi (10^5 kg/cm^2)*	2.0 (1.4)	2.5 (1.8)	1.6 (1.1)	2.5 (1.8)	8.5 (0.60)
Hoop bending, ASTM D-2412 modulus of elasticity, 10^6 psi (10^5 kg/cm^2)*	2.5 (1.8)	3.4 (2.4)	2.3 (1.6)	3.1 (2.2)	2.2 (1.5)
Coefficient of thermal expansion ASTM D-696, 10^{-5} in/in/°F (10^{-5} mm/mm/°C)†	1.0 (1.8)	1.1 (2.0)	1.2 (2.2)	0.87 (1.6)	1.4 (2.5)
Coefficient of thermal conductivity ASTM D-177, Btu/(ft^2·h·°F·in), (w/(m·k))†	2.4 (0.35)	0.87 (0.13)	1.7 (0.25)	0.87 (0.13)	1.7 (0.25)
Specific gravity, ASTM D-792†	1.8	1.45	1.8	1.55	2.1
Poisson's ratio, ASTM D-2105, ratio axial strain-to-hoop strain*	0.26	0.15	0.3	0.15	
Flow factor					
Hazen-Williams C	150	150	150	150	150
Manning's n	0.009	0.009	0.009	0.009	0.009
Absolute roughness, 10^{-5} ft	1.7	1.7	1.7	1.7	1.7

* Based on reinforced wall thickness.
† Based on total wall thickness.

TABLE 4.27 Piping System Thermal Expansion, Uninsulated Pipe, in/1000 ft

Temperature change, °F	PVC	CPVC	Carbon steel	Stainless steel	Typical fiberglass pipe
25	0.90	1.14	0.18	0.27	0.31
50	1.80	2.28	0.36	0.54	0.61
75	2.70	3.42	0.54	0.82	0.92
100	3.60	4.56	0.72	1.09	1.23
125	4.50	5.70	0.90	1.36	1.53
150	5.40	6.84	1.08	1.63	1.84

Temperature change, °F = °C/1.8

- Hydrostatic burst stress
- Hydrostatic design stress
- Axial tensile strength
- Beam bending ultimate stress and modulus of elasticity
- Axial compressive strength
- Hoop bending modulus of elasticity
- Coefficient of thermal expansion
- Coefficient of thermal conductivity
- Specific gravity
- Poisson's ratio
- Flow factors (Hazen-Williams C; Manning's n; absolute roughness, ft)

Thermal Expansion

Table 4.27 provides a comparison of thermal expansion, inches per 100 ft of pipe length, for pipes made from PVC, CPVC, carbon steel, stainless steel, and fiberglass for changes in temperature ranging from 25°F to 150°F.

Tables in Chapter 4

S • E • C • T • I • O • N • 3

METALLIC PIPING

CHAPTER 5
ASME WROUGHT PIPE, FITTINGS, FLANGES, AND JOINTS

See pp. 270–271 for a list of tables in Chap. 5.

DIMENSIONS AND OTHER PROPERTIES OF WELDED AND SEAMLESS PIPE

Tables 5.1 and 5.1M list the dimensions of welded and seamless steel and stainless pipe conforming to the following ASME standards for pipe:

- ASME B36.10M—Welded and Seamless Wrought Steel Pipe
- ASME B36.19M—Stainless Steel Pipe

DIMENSIONS OF STAINLESS STEEL PIPE PER ASME B36.19M

ASME B36.19M covers dimensions of welded and seamless stainless steel pipe in Schedules 5S, 10S, 40S, and 80S. The wall thickness of stainless steel pipe may or may not be the same as of wrought steel pipe of corresponding size and schedule without suffix S.

Tables 5.1 and 5.1M cover all sizes and schedules of wrought steel pipe, including stainless steel pipe, covered by Tables 5.2 and 5.2M. Tables 5.2 and 5.2M cover nominal pipe size, outside diameter, and wall thickness of welded and seamless wrought stainless steel pipe.

TOLERANCES FOR WELDED AND SEAMLESS STEEL OR STAINLESS WROUGHT PIPE

Tolerances for welded or seamless wrought steel or stainless steel pipe are covered by ASTM A530 or ASME SA530. The ASTM A530 lists the pipe specifications for which pipe tolerances are covered. For pipe manufactured to a specification not listed in ASTM A530, refer to the pipe specification.

Table 5.3 lists tolerances for outside diameter, inside diameter, length, and weight.

DIMENSIONS OF WROUGHT STEEL BUTT WELDING FITTINGS PER ASME B16.9

Dimensions of Straight and Reducing 90° and 45° Long-Radius Elbows

See Tables 5.4, 5.4M, 5.5, 5.5M, 5.6, and 5.6M.

Dimensions of Long-Radius 180° Return Elbows

See Tables 5.7 and 5.7M.

TABLE 5.1 Principal Properties of Commercial Pipe (in U.S. Customary Units)*

Nominal dia. NPS	Schedule[†]		D, outside dia., in	t, wall thick., in	d, inside dia., in	A_i, inside area, in²	A_M, metal area, in²	A_{so}, outside surf., in²/ft	A_{si}, inside surf., in²/ft	w_p, pipe wt., lb/ft	w_w, water wt., lb/ft	I, mom. of inert, in⁴	z_e, elast sec. mod., in³	R_g, rad. of gyr., in	z_p, plast sec. mod., in³
1/8	10S		0.406	0.049	0.308	0.074	0.055	0.106	0.081	0.186	0.032	0.001	0.004	0.127	0.006
	Std	40	0.406	0.068	0.270	0.057	0.072	0.106	0.071	0.245	0.025	0.001	0.005	0.122	0.008
	XS	80	0.406	0.095	0.216	0.036	0.093	0.106	0.056	0.315	0.016	0.001	0.006	0.115	0.009
1/4	10S		0.539	0.065	0.409	0.132	0.097	0.141	0.107	0.329	0.057	0.003	0.010	0.169	0.015
	Std	40	0.539	0.088	0.363	0.104	0.125	0.141	0.095	0.424	0.045	0.003	0.012	0.163	0.018
	XS	80	0.539	0.119	0.301	0.071	0.157	0.141	0.079	0.534	0.031	0.004	0.014	0.154	0.022
3/8	10S		0.675	0.065	0.545	0.233	0.125	0.177	0.143	0.423	0.101	0.006	0.017	0.217	0.024
	Std	40	0.675	0.091	0.493	0.191	0.167	0.177	0.129	0.567	0.083	0.007	0.022	0.209	0.031
	XS	80	0.675	0.126	0.423	0.141	0.217	0.177	0.111	0.738	0.061	0.009	0.026	0.199	0.039
1/2	5S		0.84	0.065	0.710	0.396	0.158	0.220	0.186	0.537	0.172	0.012	0.028	0.275	0.039
	10S		0.84	0.083	0.674	0.357	0.197	0.220	0.176	0.670	0.155	0.014	0.034	0.269	0.048
	Std	40	0.84	0.109	0.622	0.304	0.250	0.220	0.163	0.850	0.132	0.017	0.041	0.261	0.059
	XS	80	0.84	0.147	0.546	0.234	0.320	0.220	0.143	1.087	0.101	0.020	0.048	0.250	0.072
		160	0.84	0.187	0.466	0.171	0.384	0.220	0.122	1.303	0.074	0.022	0.053	0.240	0.082
	XXS		0.84	0.294	0.252	0.050	0.504	0.220	0.066	1.713	0.022	0.024	0.058	0.219	0.096
3/4	5S		1.05	0.065	0.920	0.665	0.201	0.275	0.241	0.683	0.288	0.025	0.047	0.349	0.063
	10S		1.05	0.083	0.884	0.614	0.252	0.275	0.231	0.856	0.266	0.030	0.057	0.343	0.078
	Std	40	1.05	0.113	0.824	0.533	0.333	0.275	0.216	1.130	0.231	0.037	0.071	0.334	0.100
	XS	80	1.05	0.154	0.742	0.432	0.434	0.275	0.194	1.472	0.187	0.045	0.085	0.321	0.125
		160	1.05	0.218	0.614	0.296	0.570	0.275	0.161	1.935	0.128	0.053	0.100	0.304	0.154
	XXS		1.05	0.308	0.434	0.148	0.718	0.275	0.114	2.438	0.064	0.058	0.110	0.284	0.179
1	5S		1.315	0.065	1.185	1.103	0.255	0.344	0.310	0.867	0.478	0.050	0.076	0.443	0.102
	10S		1.315	0.109	1.097	0.945	0.413	0.344	0.287	1.403	0.409	0.076	0.115	0.428	0.159
	Std	40	1.315	0.133	1.049	0.864	0.494	0.344	0.275	1.677	0.374	0.087	0.133	0.421	0.187
	XS	80	1.315	0.179	0.957	0.719	0.639	0.344	0.251	2.170	0.312	0.106	0.161	0.407	0.233

		160	1.315	0.250	0.815	0.522	0.836	0.344	0.213	2.841	0.226	0.125	0.190	0.387	0.289
	XXS		1.315	0.358	0.599	0.282	1.076	0.344	0.157	3.655	0.122	0.140	0.214	0.361	0.343
$1^1/_4$	5S		1.66	0.065	1.530	1.839	0.326	0.435	0.401	1.106	0.797	0.104	0.125	0.564	0.165
	10S		1.66	0.109	1.442	1.633	0.531	0.435	0.378	1.804	0.708	0.161	0.193	0.550	0.263
	Std	40	1.66	0.140	1.380	1.496	0.669	0.435	0.361	2.270	0.648	0.195	0.235	0.540	0.324
	XS	80	1.66	0.191	1.278	1.283	0.881	0.435	0.335	2.993	0.556	0.242	0.291	0.524	0.415
		160	1.66	0.250	1.160	1.057	1.107	0.435	0.304	3.761	0.458	0.284	0.342	0.506	0.502
	XXS		1.66	0.382	0.896	0.631	1.534	0.435	0.235	5.209	0.273	0.341	0.411	0.472	0.643
$1^1/_2$	5S		1.9	0.065	1.770	2.461	0.375	0.497	0.463	1.273	1.066	0.158	0.166	0.649	0.219
	10S		1.9	0.109	1.682	2.222	0.613	0.497	0.440	2.083	0.963	0.247	0.260	0.634	0.350
	Std	40	1.9	0.145	1.610	2.036	0.799	0.497	0.421	2.715	0.882	0.310	0.326	0.623	0.448
	XS	80	1.9	0.200	1.500	1.767	1.068	0.497	0.393	3.627	0.766	0.391	0.412	0.605	0.581
		160	1.9	0.281	1.338	1.406	1.429	0.497	0.350	4.853	0.609	0.482	0.508	0.581	0.744
	XXS		1.9	0.400	1.100	0.950	1.885	0.497	0.288	6.401	0.412	0.568	0.598	0.549	0.922
			1.9	0.525	0.850	0.567	2.268	0.497	0.223	7.702	0.246	0.614	0.647	0.520	1.041
			1.9	0.625	0.650	0.332	2.503	0.497	0.170	8.502	0.144	0.631	0.664	0.502	1.098
2	5S		2.375	0.065	2.245	3.958	0.472	0.622	0.588	1.602	1.715	0.315	0.265	0.817	0.347
	10S		2.375	0.109	2.157	3.654	0.776	0.622	0.565	2.635	1.583	0.499	0.421	0.802	0.560
	Std	40	2.375	0.154	2.067	3.356	1.075	0.622	0.541	3.649	1.454	0.666	0.561	0.787	0.761
	XS	80	2.375	0.218	1.939	2.953	1.477	0.622	0.508	5.017	1.279	0.868	0.731	0.767	1.018
		160	2.375	0.343	1.689	2.241	2.190	0.622	0.442	7.436	0.971	1.163	0.979	0.729	1.430
	XXS		2.375	0.436	1.503	1.774	2.656	0.622	0.393	9.019	0.769	1.312	1.105	0.703	1.667
			2.375	0.562	1.251	1.229	3.201	0.622	0.328	10.871	0.533	1.442	1.214	0.671	1.907
			2.375	0.687	1.001	0.787	3.643	0.622	0.262	12.372	0.341	1.513	1.274	0.644	2.066

TABLE 5.1 Principal Properties of Commercial Pipe (in U.S. Customary Units)* (Continued)

Nominal dia. NPS	Schedule[†]		D, outside dia., in	t, wall thick., in	d, inside dia., in	A_i, inside area, in^2	A_M, metal area, in^2	A_{so}, outside surf., in^2/ft	A_{si}, inside surf., in^2/ft	w_p, pipe wt., lb/ft	w_w, water wt., lb/ft	I, mom. of inert, in^4	z_e, elast sec. mod., in^3	R_g, rad. of gyr., in	z_p, plast sec. mod., in^3
$2^1/_2$	5S		2.875	0.083	2.709	5.764	0.728	0.753	0.709	2.472	2.497	0.710	0.494	0.988	0.647
	10S		2.875	0.120	2.635	5.453	1.039	0.753	0.690	3.527	2.363	0.988	0.687	0.975	0.912
	Std	40	2.875	0.203	2.469	4.788	1.704	0.753	0.646	5.787	2.074	1.530	1.064	0.947	1.452
	XS	80	2.875	0.276	2.323	4.238	2.253	0.753	0.608	7.653	1.836	1.925	1.339	0.924	1.872
		160	2.875	0.375	2.125	3.547	2.945	0.753	0.556	10.002	1.537	2.353	1.637	0.894	2.362
	XXS		2.875	0.552	1.771	2.463	4.029	0.753	0.464	13.681	1.067	2.872	1.998	0.844	3.035
			2.875	0.675	1.525	1.827	4.665	0.753	0.399	15.843	0.791	3.089	2.149	0.814	3.370
			2.875	0.800	1.275	1.277	5.215	0.753	0.334	17.710	0.553	3.225	2.243	0.786	3.616
3	5S		3.5	0.083	3.334	8.730	0.891	0.916	0.873	3.026	3.783	1.301	0.744	1.208	0.969
	10S		3.5	0.120	3.260	8.347	1.274	0.916	0.853	4.327	3.617	1.822	1.041	1.196	1.372
	Std	40	3.5	0.216	3.068	7.393	2.228	0.916	0.803	7.567	3.203	3.018	1.724	1.164	2.333
	XS	80	3.5	0.300	2.900	6.605	3.016	0.916	0.759	10.242	2.862	3.895	2.226	1.136	3.082
		160	3.5	0.437	2.626	5.416	4.205	0.916	0.687	14.281	2.347	5.033	2.876	1.094	4.129
	XXS		3.5	0.600	2.300	4.155	5.466	0.916	0.602	18.564	1.800	5.994	3.425	1.047	5.119
			3.5	0.725	2.050	3.301	6.321	0.916	0.537	21.464	1.430	6.501	3.715	1.014	5.711
			3.5	0.850	1.800	2.545	7.076	0.916	0.471	24.031	1.103	6.853	3.916	0.984	6.175
$3^1/_2$	5S		4	0.083	3.834	11.545	1.021	1.047	1.004	3.468	5.002	1.960	0.980	1.385	1.274
	10S		4	0.120	3.760	11.104	1.463	1.047	0.984	4.967	4.811	2.756	1.378	1.372	1.807
	Std	40	4	0.226	3.548	9.887	2.679	1.047	0.929	9.099	4.284	4.789	2.394	1.337	3.223
	XS	80	4	0.318	3.364	8.888	3.678	1.047	0.881	12.492	3.851	6.282	3.141	1.307	4.323
	XXS		4	0.636	2.728	5.845	6.721	1.047	0.714	22.825	2.533	9.850	4.925	1.210	7.284
4	5S		4.5	0.083	4.334	14.753	1.152	1.178	1.135	3.911	6.392	2.810	1.249	1.562	1.620
	10S		4.5	0.120	4.260	14.253	1.651	1.178	1.115	5.608	6.176	3.964	1.762	1.549	2.303
			4.5	0.188	4.124	13.358	2.547	1.178	1.080	8.648	5.788	5.932	2.636	1.526	3.498
	Std	40	4.5	0.237	4.026	12.730	3.174	1.178	1.054	10.779	5.516	7.235	3.215	1.510	4.312

	XS	80	4.5	0.337	3.826	11.497	4.408	1.178	1.002	14.968	4.981	9.613	4.273	1.477	5.854
		120	4.5	0.437	3.626	10.326	5.578	1.178	0.949	18.943	4.474	11.646	5.176	1.445	7.243
			4.5	0.500	3.500	9.621	6.283	1.178	0.916	21.338	4.169	12.766	5.674	1.425	8.043
		160	4.5	0.531	3.438	9.283	6.621	1.178	0.900	22.484	4.022	13.274	5.900	1.416	8.416
	XXS		4.5	0.674	3.152	7.803	8.101	1.178	0.825	27.512	3.381	15.288	6.795	1.374	9.970
			4.5	0.800	2.900	6.605	9.299	1.178	0.759	31.580	2.862	16.661	7.405	1.338	11.125
			4.5	0.925	2.650	5.515	10.389	1.178	0.694	35.280	2.390	17.713	7.872	1.306	12.088
5	5S		5.563	0.109	5.345	22.438	1.868	1.456	1.399	6.343	9.722	6.950	2.499	1.929	3.244
	10S		5.563	0.134	5.295	22.020	2.286	1.456	1.386	7.762	9.541	8.428	3.030	1.920	3.952
	Std	40	5.563	0.258	5.047	20.006	4.300	1.456	1.321	14.602	8.668	15.166	5.452	1.878	7.268
	XS	80	5.563	0.375	4.813	18.194	6.112	1.456	1.260	20.756	7.883	20.676	7.433	1.839	10.113
		120	5.563	0.500	4.563	16.353	7.953	1.456	1.195	27.008	7.085	25.738	9.253	1.799	12.861
		160	5.563	0.625	4.313	14.610	9.696	1.456	1.129	32.927	6.330	30.033	10.798	1.760	15.324
	XXS		5.563	0.750	4.063	12.965	11.340	1.456	1.064	38.512	5.618	33.643	12.095	1.722	17.518
			5.563	0.875	3.813	11.419	12.887	1.456	0.998	43.763	4.948	36.645	13.174	1.686	19.457
			5.563	1.000	3.563	9.971	14.335	1.456	0.933	48.682	4.320	39.111	14.061	1.652	21.158
6	5S		6.625	0.109	6.407	32.240	2.232	1.734	1.677	7.579	13.969	11.850	3.577	2.304	4.630
	10S		6.625	0.134	6.357	31.739	2.733	1.734	1.664	9.281	13.752	14.403	4.348	2.295	5.648
			6.625	0.219	6.187	30.064	4.408	1.734	1.620	14.968	13.026	22.642	6.835	2.266	8.993
	Std	40	6.625	0.280	6.065	28.890	5.581	1.734	1.588	18.954	12.518	28.149	8.498	2.245	11.282
	XS	80	6.625	0.432	5.761	26.067	8.405	1.734	1.508	28.544	11.294	40.502	12.227	2.195	16.599
		120	6.625	0.562	5.501	23.767	10.705	1.734	1.440	36.353	10.298	49.624	14.981	2.153	20.723
		160	6.625	0.718	5.189	21.148	13.324	1.734	1.358	45.248	9.163	58.988	17.808	2.104	25.181
	XXS		6.625	0.864	4.897	18.834	15.638	1.734	1.282	53.105	8.160	66.350	20.030	2.060	28.897
			6.625	1.000	4.625	16.800	17.672	1.734	1.211	60.012	7.279	72.119	21.772	2.020	31.980
			6.625	1.125	4.375	15.033	19.439	1.734	1.145	66.013	6.514	76.597	23.124	1.985	34.513

TABLE 5.1 Principal Properties of Commercial Pipe (in U.S. Customary Units)* (Continued)

Nominal dia. NPS	Schedule†		D, outside dia., in	t, wall thick., in	d, inside dia., in	A_i, inside area, in^2	A_M, metal area, in^2	A_{so}, outside surf., in^2/ft	A_{si}, inside surf., in^2/ft	w_p, pipe wt., lb/ft	w_w, water wt., lb/ft	I, mom. of inert., in^4	z_e, elast sec. mod., in^3	R_g, rad. of gyr., in	z_p, plast sec. mod., in^3
8	5S		8.625	0.109	8.407	55.510	2.917	2.258	2.201	9.905	24.051	26.451	6.133	3.011	7.908
	10S		8.625	0.148	8.329	54.485	3.941	2.258	2.181	13.384	23.607	35.422	8.214	2.998	10.638
			8.625	0.219	8.187	52.643	5.784	2.258	2.143	19.642	22.809	51.134	11.857	2.973	15.482
		20	8.625	0.250	8.125	51.849	6.578	2.258	2.127	22.338	22.465	57.737	13.388	2.962	17.544
		30	8.625	0.277	8.071	51.162	7.265	2.258	2.113	24.671	22.167	63.371	14.695	2.953	19.315
	Std	40	8.625	0.322	7.981	50.027	8.399	2.258	2.089	28.524	21.676	72.509	16.814	2.938	22.215
		60	8.625	0.406	7.813	47.944	10.483	2.258	2.045	35.600	20.773	88.756	20.581	2.909	27.453
	XS	80	8.625	0.500	7.625	45.664	12.763	2.258	1.996	43.342	19.785	105.743	24.520	2.878	33.056
		100	8.625	0.593	7.439	43.463	14.963	2.258	1.948	50.815	18.832	121.354	28.140	2.847	38.333
		120	8.625	0.718	7.189	40.591	17.835	2.258	1.882	60.569	17.587	140.570	32.596	2.807	45.022
		140	8.625	0.812	7.001	38.495	19.931	2.258	1.833	67.685	16.679	153.762	35.655	2.777	49.756
		160	8.625	0.906	6.813	36.456	21.970	2.258	1.784	74.610	15.796	165.928	38.476	2.748	54.240
			8.625	1.000	6.625	34.472	23.955	2.258	1.734	81.350	14.936	177.132	41.074	2.719	58.486
			8.625	1.125	6.375	31.919	26.507	2.258	1.669	90.018	13.830	190.621	44.202	2.681	63.769
10	5S		10.75	0.134	10.482	86.293	4.470	2.814	2.744	15.179	37.389	62.991	11.719	3.754	15.107
	10S		10.75	0.165	10.420	85.276	5.487	2.814	2.728	18.633	36.949	76.884	14.304	3.743	18.492
			10.75	0.219	10.312	83.517	7.246	2.814	2.700	24.607	36.186	100.517	18.701	3.724	24.298
		20	10.75	0.250	10.250	82.516	8.247	2.814	2.683	28.006	35.753	113.743	21.162	3.713	27.573
		30	10.75	0.307	10.136	80.691	10.072	2.814	2.654	34.205	34.962	137.458	25.574	3.694	33.497
	Std	40	10.75	0.365	10.020	78.854	11.908	2.814	2.623	40.440	34.166	160.776	29.912	3.674	39.389
	XS	60	10.75	0.500	9.750	74.662	16.101	2.814	2.553	54.678	32.350	212.005	39.443	3.628	52.583
		80	10.75	0.593	9.564	71.841	18.922	2.814	2.504	64.258	31.127	244.904	45.564	3.597	61.258
		100	10.75	0.718	9.314	68.134	22.629	2.814	2.438	76.846	29.521	286.203	53.247	3.556	72.397
		120	10.75	0.843	9.064	64.526	26.237	2.814	2.373	89.101	27.958	324.307	60.336	3.515	82.955
			10.75	0.875	9.000	63.617	27.145	2.814	2.356	92.185	27.564	333.570	62.060	3.505	85.567

	140	10.75	1.000	8.750	60.132	30.631	2.814	2.291	104.021	26.054	367.900	68.447	3.465	95.415
	160	10.75	1.125	8.500	56.745	34.018	2.814	2.225	115.523	24.587	399.410	74.309	3.426	104.716
		10.75	1.250	8.250	53.456	37.307	2.814	2.160	126.692	23.162	428.259	79.676	3.388	113.486
		10.75	1.500	7.750	47.173	43.590	2.814	2.029	148.030	20.439	478.587	89.039	3.313	129.495
12	5S	12.75	0.156	12.438	121.505	6.172	3.338	3.256	20.959	52.646	122.408	19.201	4.453	24.747
	10S	12.75	0.180	12.390	120.568	7.108	3.338	3.244	24.139	52.240	140.456	22.032	4.445	28.449
	20	12.75	0.250	12.250	117.859	9.818	3.338	3.207	33.340	51.066	191.874	30.098	4.420	39.076
	30	12.75	0.330	12.090	114.800	12.876	3.338	3.165	43.727	49.741	248.517	38.983	4.393	50.927
	Std	12.75	0.375	12.000	113.098	14.579	3.338	3.142	49.510	49.003	279.407	43.829	4.377	57.457
	40	12.75	0.406	11.938	111.933	15.744	3.338	3.125	53.466	48.498	300.275	47.102	4.367	61.896
	XS	12.75	0.500	11.750	108.434	19.242	3.338	3.076	65.346	46.983	361.637	56.727	4.335	75.088
	60	12.75	0.562	11.626	106.157	21.519	3.338	3.044	73.079	45.996	400.528	62.828	4.314	83.561
	80	12.75	0.687	11.376	101.641	26.036	3.338	2.978	88.416	44.039	475.231	74.546	4.272	100.099
		12.75	0.750	11.250	99.402	28.274	3.338	2.945	96.019	43.069	511.057	80.166	4.251	108.162
	100	12.75	0.843	11.064	96.143	31.534	3.338	2.897	107.088	41.657	561.790	88.124	4.220	119.740
		12.75	0.875	11.000	95.033	32.643	3.338	2.880	110.856	41.176	578.671	90.772	4.210	123.637
	120	12.75	1.000	10.750	90.763	36.914	3.338	2.814	125.358	39.326	641.829	100.679	4.169	138.424
	140	12.75	1.125	10.500	86.590	41.086	3.338	2.749	139.528	37.518	700.731	109.919	4.129	152.538
		12.75	1.250	10.250	82.516	45.161	3.338	2.683	153.364	35.753	755.572	118.521	4.090	165.997
	160	12.75	1.312	10.126	80.531	47.145	3.338	2.651	160.104	34.893	781.330	122.562	4.070	172.434
14	5S	14	0.156	13.688	147.154	6.784	3.665	3.584	23.039	63.759	162.590	23.227	4.895	29.903
	10S	14	0.188	13.624	145.781	8.157	3.665	3.567	27.702	63.164	194.609	27.801	4.884	35.873
		14	0.210	13.580	144.841	9.098	3.665	3.555	30.896	62.757	216.364	30.909	4.876	39.946
		14	0.219	13.562	144.456	9.482	3.665	3.551	32.201	62.590	225.215	32.174	4.873	41.606
	10	14	0.250	13.500	143.139	10.799	3.665	3.534	36.674	62.020	255.366	36.481	4.862	47.280
		14	0.281	13.438	141.828	12.110	3.665	3.518	41.126	61.452	285.105	40.729	4.851	52.902

TABLE 5.1 Principal Properties of Commercial Pipe (in U.S. Customary Units)* (Continued)

Nominal dia. NPS	Schedule[†]	D, outside dia., in	t, wall thick., in	d, inside dia., in	A_i, inside area, in^2	A_M, metal area, in^2	A_{so}, outside surf., in^2/ft	A_{si}, inside surf., in^2/ft	w_p, pipe wt., lb/ft	w_w, water wt., lb/ft	I, mom. of inert, in^4	z_e, elast sec. mod., in^3	R_g, rad. of gyr., in	z_p, plast sec. mod., in^3
	20	14	0.312	13.376	140.521	13.417	3.665	3.502	45.564	60.885	314.472	44.925	4.841	58.480
		14	0.344	13.312	139.180	14.759	3.665	3.485	50.121	60.304	344.345	49.192	4.830	64.180
	Std 30	14	0.375	13.250	137.887	16.052	3.665	3.469	54.511	59.744	372.856	53.265	4.819	69.647
	40	14	0.437	13.126	135.318	18.621	3.665	3.436	63.235	58.631	428.724	61.246	4.798	80.434
		14	0.469	13.062	134.001	19.937	3.665	3.420	67.707	58.060	456.950	65.279	4.787	85.923
	XS	14	0.500	13.000	132.733	21.206	3.665	3.403	72.014	57.511	483.881	69.126	4.776	91.185
	60	14	0.593	12.814	128.962	24.977	3.665	3.355	84.820	55.877	562.425	80.346	4.745	106.680
		14	0.625	12.750	127.677	26.262	3.665	3.338	89.185	55.320	588.681	84.097	4.734	111.910
	80	14	0.750	12.500	122.719	31.220	3.665	3.273	106.021	53.172	687.495	98.214	4.692	131.839
	100	14	0.937	12.126	115.485	38.454	3.665	3.175	130.587	50.038	824.654	117.808	4.630	160.199
	120	14	1.093	11.814	109.619	44.319	3.665	3.093	150.508	47.496	929.755	132.822	4.580	182.554
	140	14	1.250	11.500	103.869	50.069	3.665	3.011	170.034	45.005	1027.464	146.781	4.529	203.895
	160	14	1.406	11.188	98.310	55.628	3.665	2.929	188.912	42.596	1116.925	159.561	4.480	223.973
16	5S	16	0.165	15.670	192.854	8.208	4.189	4.102	27.875	83.560	257.369	32.171	5.599	41.383
	10S	16	0.188	15.624	191.724	9.339	4.189	4.090	31.713	83.071	291.967	36.496	5.591	47.013
	10	16	0.250	15.500	188.692	12.370	4.189	4.058	42.008	81.757	383.763	47.970	5.569	62.033
	20	16	0.312	15.376	185.685	15.377	4.189	4.025	52.221	80.454	473.381	59.173	5.548	76.815
	Std 30	16	0.375	15.250	182.655	18.408	4.189	3.992	62.513	79.141	562.229	70.279	5.526	91.589
	XS 40	16	0.500	15.000	176.715	24.347	4.189	3.927	82.683	76.568	732.130	91.516	5.483	120.191
	60	16	0.656	14.688	169.441	31.622	4.189	3.845	107.386	73.416	932.556	116.570	5.430	154.569
	80	16	0.844	14.312	160.870	40.193	4.189	3.747	136.493	69.702	1157.907	144.738	5.367	194.139
	100	16	1.031	13.938	152.579	48.484	4.189	3.649	164.650	66.110	1364.767	170.596	5.305	231.426
	120	16	1.218	13.564	144.500	56.563	4.189	3.551	192.085	62.609	1555.804	194.475	5.244	266.796
	140	16	1.437	13.126	135.318	65.745	4.189	3.436	223.267	58.631	1760.316	220.040	5.174	305.812
	160	16	1.593	12.814	128.962	72.101	4.189	3.355	244.852	55.877	1894.017	236.752	5.125	332.058

18	5S		18	0.165	17.670	245.225	9.245	4.712	4.626	31.396	106.251	367.715	40.857	6.306	52.496
	10S		18	0.188	17.624	243.950	10.520	4.712	4.614	35.725	105.699	417.348	46.372	6.298	59.658
		10	18	0.250	17.500	240.529	13.941	4.712	4.582	47.343	104.217	549.279	61.031	6.276	78.787
		20	18	0.312	17.376	237.132	17.338	4.712	4.549	58.879	102.745	678.434	75.382	6.255	97.646
	Std		18	0.375	17.250	233.706	20.764	4.712	4.516	70.514	101.261	806.839	89.649	6.233	116.531
		30	18	0.437	17.126	230.357	24.112	4.712	4.484	81.885	99.810	930.519	103.391	6.211	134.854
	XS		18	0.500	17.000	226.981	27.489	4.712	4.451	93.352	98.347	1053.441	117.049	6.190	153.197
		40	18	0.562	16.876	223.681	30.789	4.712	4.418	104.557	96.917	1171.802	130.200	6.168	170.991
		60	18	0.750	16.500	213.825	40.644	4.712	4.320	138.028	92.647	1515.027	168.336	6.105	223.357
		80	18	0.937	16.126	204.241	50.228	4.712	4.222	170.574	88.494	1833.952	203.772	6.042	273.135
		100	18	1.156	15.688	193.298	61.171	4.712	4.107	207.737	83.753	2180.223	242.247	5.969	328.557
		120	18	1.375	15.250	182.655	71.815	4.712	3.992	243.882	79.141	2498.733	277.637	5.898	380.980
		140	18	1.562	14.876	173.805	80.665	4.712	3.895	273.935	75.307	2749.824	305.536	5.838	423.421
		160	18	1.781	14.438	163.722	90.748	4.712	3.780	308.177	70.938	3020.719	335.635	5.769	470.477
20	5S		20	0.188	19.624	302.459	11.701	5.236	5.138	39.736	131.050	574.296	57.430	7.005	73.807
	10S		20	0.218	19.564	300.612	13.548	5.236	5.122	46.007	130.250	662.943	66.294	6.994	85.327
		10	20	0.250	19.500	298.648	15.512	5.236	5.105	52.677	129.399	756.628	75.663	6.983	97.540
	Std	20	20	0.375	19.250	291.040	23.120	5.236	5.040	78.516	126.102	1113.757	111.376	6.940	144.474
	XS	30	20	0.500	19.000	283.529	30.631	5.236	4.974	104.021	122.848	1457.239	145.724	6.897	190.205
		40	20	0.593	18.814	278.006	36.154	5.236	4.926	122.779	120.455	1704.125	170.412	6.865	223.454
		60	20	0.812	18.376	265.211	48.949	5.236	4.811	166.228	114.911	2257.342	225.734	6.790	299.203
			20	0.875	18.250	261.587	52.573	5.236	4.778	178.536	113.341	2409.306	240.931	6.769	320.332
		80	20	1.031	17.938	252.720	61.440	5.236	4.696	208.647	109.499	2772.292	277.229	6.716	371.412
		100	20	1.281	17.438	238.828	75.332	5.236	4.565	255.825	103.480	3315.838	331.584	6.634	449.649
		120	20	1.500	17.000	226.981	87.179	5.236	4.451	296.059	98.347	3755.119	375.512	6.562	514.603

TABLE 5.1 Principal Properties of Commercial Pipe (in U.S. Customary Units)* (Continued)

Nominal dia. NPS	Schedule†	D, outside dia., in	t, wall thick., in	d, inside dia., in	A_i, inside area, in²	A_M, metal area, in²	A_{so}, outside surf., in²/ft	A_{si}, inside surf., in²/ft	w_p, pipe wt., lb/ft	w_w, water wt., lb/ft	I, mom. of inert., in⁴	z_e, elast sec. mod., in³	R_g, rad. of gyr., in	z_p, plast sec. mod., in³
	140	20	1.750	16.500	213.825	100.335	5.236	4.320	340.735	92.647	4216.705	421.670	6.482	584.763
	160	20	1.968	16.064	202.675	111.485	5.236	4.206	378.602	87.815	4586.375	458.637	6.413	642.568
22	5S	22	0.188	21.624	367.252	12.882	5.760	5.661	43.747	159.124	766.356	69.669	7.712	89.460
	10S	22	0.218	21.564	365.216	14.917	5.760	5.645	50.659	158.242	885.013	80.456	7.701	103.452
	10	22	0.250	21.500	363.051	17.082	5.760	5.629	58.012	157.304	1010.524	91.866	7.690	118.294
	Std 20	22	0.375	21.250	354.657	25.476	5.760	5.563	86.517	153.667	1490.053	135.459	7.647	175.418
	XS 30	22	0.500	21.000	346.361	33.772	5.760	5.498	114.690	150.072	1952.953	177.541	7.603	231.213
		22	0.625	20.750	338.164	41.970	5.760	5.432	142.529	146.520	2399.612	218.147	7.560	285.695
		22	0.750	20.500	330.064	50.069	5.760	5.367	170.034	143.011	2830.416	257.311	7.518	338.880
	60	22	0.875	20.250	322.063	58.071	5.760	5.301	197.206	139.544	3245.743	295.068	7.475	390.784
	80	22	1.125	19.750	306.355	73.779	5.760	5.171	250.550	132.738	4031.466	366.497	7.391	490.809
	100	22	1.375	19.250	291.040	89.094	5.760	5.040	302.561	126.102	4759.726	432.702	7.308	585.896
	120	22	1.625	18.750	276.117	104.016	5.760	4.909	353.237	119.637	5433.397	493.945	7.227	676.169
	140	22	1.875	18.250	261.587	118.546	5.760	4.778	402.581	113.341	6055.276	550.480	7.146	761.754
	160	22	2.125	17.750	247.450	132.684	5.760	4.647	450.590	107.216	6628.089	602.554	7.067	842.775
24	10	24	0.250	23.500	433.737	18.653	6.283	6.152	63.346	187.931	1315.681	109.640	8.397	141.049
	Std 20	24	0.375	23.250	424.558	27.833	6.283	6.087	94.519	183.953	1942.798	161.900	8.354	209.362
	XS	24	0.500	23.000	415.477	36.914	6.283	6.021	125.358	180.019	2550.009	212.501	8.310	276.222
	30	24	0.562	22.876	411.008	41.382	6.283	5.989	140.533	178.083	2843.968	236.997	8.289	308.854
		24	0.625	22.750	406.494	45.897	6.283	5.956	155.865	176.127	3137.738	261.478	8.267	341.644
	40	24	0.687	22.626	402.074	50.317	6.283	5.923	170.874	174.212	3422.191	285.183	8.246	373.569
		24	0.750	22.500	397.609	54.782	6.283	5.891	186.037	172.277	3706.409	308.867	8.224	405.644
	5S	24	0.218	23.564	436.103	16.287	6.283	6.169	55.310	188.956	1151.846	95.987	8.409	123.321
		24	0.875	22.250	388.822	63.568	6.283	5.825	215.877	168.470	4256.436	354.703	8.182	468.237
	60	24	0.968	22.064	382.349	70.041	6.283	5.776	237.859	165.665	4653.773	387.814	8.150	513.899
	80	24	1.218	21.564	365.216	87.174	6.283	5.645	296.041	158.242	5673.245	472.770	8.066	632.891

		100	24	1.531	20.938	344.320	108.070	6.283	5.482	367.004	149.188	6853.399	571.117	7.962	774.279
		120	24	1.812	20.376	326.083	126.307	6.283	5.334	428.937	141.286	7826.587	652.216	7.871	894.226
		140	24	2.062	19.876	310.276	142.114	6.283	5.204	482.617	134.437	8627.250	718.938	7.790	995.515
		160	24	2.343	19.314	292.979	159.412	6.283	5.056	541.358	126.943	9457.825	788.152	7.702	1103.432
26			26	0.250	25.500	510.706	20.224	6.807	6.676	68.680	221.280	1676.811	128.985	9.104	165.804
		10	26	0.312	25.376	505.751	25.179	6.807	6.643	85.509	219.133	2077.742	159.826	9.083	205.937
	Std		26	0.375	25.250	500.742	30.189	6.807	6.610	102.521	216.963	2479.062	190.697	9.061	246.307
	XS	20	26	0.500	25.000	490.875	40.055	6.807	6.545	136.027	212.688	3257.834	250.603	9.017	325.232
			26	0.625	24.750	481.107	49.824	6.807	6.480	169.201	208.455	4013.590	308.738	8.974	402.594
			26	0.750	24.500	471.436	59.494	6.807	6.414	202.040	204.265	4746.789	365.138	8.931	478.408
			26	0.875	24.250	461.864	69.066	6.807	6.349	234.547	200.118	5457.882	419.837	8.888	552.691
			26	1.000	24.000	452.390	78.540	6.807	6.283	266.720	196.013	6147.320	472.871	8.846	625.458
			26	1.125	23.750	443.015	87.916	6.807	6.218	298.560	191.950	6815.546	524.273	8.804	696.725
28			28	0.250	27.500	593.959	21.795	7.330	7.200	74.015	257.352	2098.629	149.902	9.812	192.559
		10	28	0.312	27.376	588.614	27.140	7.330	7.167	92.166	255.036	2601.756	185.840	9.790	239.251
	Std		28	0.375	27.250	583.209	32.545	7.330	7.134	110.522	252.694	3105.915	221.851	9.768	286.253
	XS	20	28	0.500	27.000	572.557	43.197	7.330	7.069	146.696	248.079	4085.857	291.847	9.724	378.242
		30	28	0.625	26.750	562.003	53.751	7.330	7.003	182.537	243.506	5038.952	359.925	9.681	468.544
			28	0.750	26.500	551.547	64.206	7.330	6.938	218.044	238.976	5965.696	426.121	9.638	557.174
			28	0.875	26.250	541.190	74.564	7.330	6.872	253.217	234.488	6866.580	490.470	9.595	644.147
			28	1.000	26.000	530.930	84.823	7.330	6.807	288.058	230.043	7742.088	553.006	9.552	729.479
			28	1.125	25.750	520.769	94.984	7.330	6.741	322.565	225.640	8592.702	613.764	9.510	813.186
30	5S		30	0.250	29.500	683.494	23.366	7.854	7.723	79.349	296.146	2585.848	172.390	10.519	221.315
		10	30	0.312	29.376	677.760	29.100	7.854	7.691	98.824	293.661	3207.211	213.814	10.497	275.062
	Std		30	0.375	29.250	671.959	34.901	7.854	7.658	118.524	291.148	3830.429	255.362	10.475	329.199

TABLE 5.1 Principal Properties of Commercial Pipe (in U.S. Customary Units)* (Continued)

Nominal dia. NPS	Schedule[†]	D, outside dia., in	t, wall thick., in	d, inside dia., in	A_i, inside area, in^2	A_M, metal area, in^2	A_{so}, outside surf., in^2/ft	A_{si}, inside surf., in^2/ft	w_p, pipe wt., lb/ft	w_w, water wt., lb/ft	I, mom. of inert., in^4	z_e, elast sec. mod., in^3	R_g, rad. of gyr., in	z_p, plast sec. mod., in^3
	XS 20	30	0.500	29.000	660.521	46.339	7.854	7.592	157.365	286.192	5043.503	336.234	10.431	435.254
	30	30	0.625	28.750	649.182	57.678	7.854	7.527	195.873	281.279	6225.607	415.040	10.388	539.496
	40	30	0.750	28.500	637.941	68.919	7.854	7.461	234.047	276.409	7377.272	491.818	10.345	641.941
		30	0.875	28.250	626.798	80.062	7.854	7.396	271.888	271.581	8499.026	566.602	10.302	742.604
		30	1.000	28.000	615.754	91.106	7.854	7.330	309.395	266.795	9591.390	639.426	10.259	841.502
		30	1.125	27.750	604.807	102.053	7.854	7.265	346.569	262.052	10654.884	710.326	10.217	938.649
32		32	0.250	31.500	779.313	24.936	8.378	8.247	84.684	337.663	3143.183	196.449	11.226	252.071
	10	32	0.312	31.376	773.189	31.061	8.378	8.214	105.481	335.009	3899.988	243.749	11.204	313.369
	Std	32	0.375	31.250	766.992	37.257	8.378	8.181	126.525	332.324	4659.673	291.230	11.182	375.145
	XS 20	32	0.500	31.000	754.769	49.480	8.378	8.116	168.034	327.028	6140.201	383.763	11.138	496.266
	30	32	0.625	30.750	742.645	61.605	8.378	8.050	209.209	321.775	7585.339	474.084	11.095	615.449
	40	32	0.688	30.624	736.572	67.678	8.378	8.017	229.832	319.144	8300.367	518.773	11.073	674.780
		32	0.750	30.500	730.618	73.631	8.378	7.985	250.050	316.564	8995.657	562.229	11.052	732.709
		32	0.875	30.250	718.690	85.560	8.378	7.919	290.558	311.396	10371.718	648.232	11.009	848.063
		32	1.000	30.000	706.860	97.390	8.378	7.854	330.733	306.270	11714.082	732.130	10.966	961.526
		32	1.125	29.750	695.128	109.122	8.378	7.789	370.574	301.187	13023.302	813.956	10.923	1073.113
34		34	0.250	33.500	881.415	26.507	8.901	8.770	90.018	381.902	3775.345	222.079	11.933	284.828
	10	34	0.312	33.376	874.901	33.021	8.901	8.738	112.139	379.079	4685.972	275.645	11.911	354.173
	Std	34	0.375	33.250	868.309	39.614	8.901	8.705	134.527	376.223	5600.717	329.454	11.889	424.093
	XS 20	34	0.500	33.000	855.301	52.622	8.901	8.639	178.702	370.587	7385.377	434.434	11.845	561.279
	30	34	0.625	32.750	842.391	65.532	8.901	8.574	222.545	364.993	9129.933	537.055	11.802	696.402
	40	34	0.688	32.624	835.922	72.000	8.901	8.541	244.512	362.190	9994.075	587.887	11.780	763.719
		34	0.750	32.500	829.579	78.344	8.901	8.509	266.053	359.442	10834.992	637.352	11.759	829.478
		34	0.875	32.250	816.865	91.057	8.901	8.443	309.229	353.933	12501.154	735.362	11.716	960.523
		34	1.000	32.000	804.250	103.673	8.901	8.378	352.070	348.467	14129.016	831.119	11.673	1089.551
		34	1.125	31.750	791.732	116.190	8.901	8.312	394.579	343.044	15719.168	924.657	11.630	1216.579

36			36	0.250	35.500	989.800	28.078	9.425	9.294	95.352	428.863	4487.050	249.281	12.640	319.585
		10	36	0.312	35.376	982.897	34.982	9.425	9.261	118.796	425.872	5571.044	309.502	12.618	397.473
	Std		36	0.375	35.250	975.909	41.970	9.425	9.228	142.529	422.844	6660.633	370.035	12.596	476.041
	XS	20	36	0.500	35.000	962.115	55.763	9.425	9.163	189.371	416.868	8788.458	488.248	12.552	630.293
		30	36	0.625	34.750	948.420	69.459	9.425	9.098	235.881	410.934	10871.173	603.954	12.509	782.357
		40	36	0.750	34.500	934.822	83.056	9.425	9.032	282.056	405.042	12909.418	717.190	12.466	932.249
			36	0.875	34.250	921.323	96.555	9.425	8.967	327.899	399.193	14903.832	827.991	12.422	1079.984
			36	1.000	34.000	907.922	109.956	9.425	8.901	373.408	393.387	16855.048	936.392	12.379	1225.578
			36	1.125	33.750	894.620	123.259	9.425	8.836	418.584	387.623	18763.694	1042.427	12.337	1369.047
42			42	0.250	41.500	1352.655	32.790	10.996	10.865	111.356	586.082	7146.551	340.312	14.761	435.858
	Std		42	0.375	41.250	1336.407	49.038	10.996	10.799	166.533	579.042	10624.310	505.920	14.717	649.888
	XS	20	42	0.500	41.000	1320.257	65.188	10.996	10.734	221.378	572.044	14039.409	668.543	14.674	861.339
		30	42	0.625	40.750	1304.206	81.240	10.996	10.668	275.889	565.089	17392.604	828.219	14.630	1070.227
		40	42	0.750	40.500	1288.252	97.193	10.996	10.603	330.066	558.177	20684.649	984.983	14.586	1276.568
			42	1.000	40.000	1256.640	128.806	10.996	10.472	437.421	544.480	27088.274	1289.918	14.500	1681.670
			42	1.250	39.500	1225.420	160.025	10.996	10.341	543.442	530.953	33256.213	1583.629	14.414	2076.769
			42	1.500	39.000	1194.593	190.852	10.996	10.210	648.130	517.596	39194.321	1866.396	14.329	2461.992

Notes: The following formulas were used in the computation of the values shown in the table:

Weight of pipe per foot (pounds) $= 10.6802t(D - t)$
Weight of water per foot (pounds) $= 0.3405d^2$
Square feet outside surface per foot $= 0.2618D$
Square feet inside surface per foot $= 0.2618d$
Inside area (square inches) $= 0.785d^2$
Area of metal (square inches) $= 0.785(D^2 - d^2)$
Moment of inertia (inches⁴) $= 0.0491(D^4 - d^4)$
$= A_M R_g^2$
Elastic section modulus (inches³) $= 0.0982(D^4 - d^4)/D$
Plastic section modulus (inches³) $= (D^3 - d^3)/6$
Radius of gyration (inches) $= 0.25\sqrt{(D^2 + d^2)}$
where A_M = area of metal (square inches)
d = inside diameter (inches)
D = outside diameter (inches)
R_g = radius of gyration (inches)
t = pipe wall thickness (inches)

* The ferritic stainless steels may be about 5 percent less and the austenitic stainless steels about 2 percent greater than the values shown in this table, which are based on weights for carbon steel.

† Schedule numbers: Standard weight pipe schedule and Schedule 40 are the same in all sizes through NPS 10, and from NPS 12 through 24, standard weight pipe has a wall thickness of 3/8 in. Extra-strong weight pipe and Schedule 80 are the same in all sizes through NPS 8; from NPS 8 through 24 in, extra-strong weight pipe has a wall thickness of 1/2 in. Double extra-strong weight pipe has no corresponding schedule number.

1. ASME B36.10—Welded and Seamless Wrought Steel Pipe.
2. ASME B36.19—Stainless Steel Pipe

TABLE 5.1M Principal Properties of Commercial Pipe (in Metric Units)

D_N, Nom. dia.	Schedule	D, outside dia., mm	t, wall thick., mm	d, inside dia., mm	A_i, inside area, cm^2	A_M, metal area, cm^2	A_{so}, outside surf., m^2/m	A_{si}, inside surf., m^2/m	w_p, pipe wt., kg/m	w_w, water wt., kg/m	I, mom of inert., cm^4	z_e, elast. sec. mod., cm^3	R_g, rad. of gyr., cm	z_p, plast. sec. mod., cm^3
3	10S	10.3	1.2446	7.811	0.479	0.354	0.032	0.025	0.277	0.048	0.037	0.072	0.323	0.103
	Std 40	10.3	1.7272	6.846	0.368	0.465	0.032	0.022	0.364	0.037	0.044	0.086	0.309	0.129
	XS 80	10.3	2.413	5.474	0.235	0.598	0.032	0.017	0.468	0.024	0.051	0.099	0.292	0.155
6	10S	13.7	1.651	10.398	0.849	0.625	0.043	0.033	0.489	0.085	0.116	0.169	0.430	0.241
	Std 40	13.7	2.235	9.230	0.669	0.805	0.043	0.029	0.630	0.067	0.137	0.200	0.413	0.298
	XS 80	13.7	3.023	7.654	0.460	1.014	0.043	0.024	0.794	0.046	0.156	0.228	0.392	0.354
10	10S	17.145	1.651	13.843	1.505	0.804	0.054	0.043	0.629	0.151	0.244	0.285	0.551	0.398
	Std 40	17.145	2.311	12.523	1.232	1.077	0.054	0.039	0.843	0.123	0.303	0.354	0.531	0.513
	XS 80	17.145	3.2	10.745	0.907	1.402	0.054	0.034	1.098	0.091	0.359	0.418	0.506	0.633
15	5S	21.336	1.651	18.034	2.554	1.021	0.067	0.057	0.799	0.255	0.498	0.467	0.698	0.641
	10S	21.336	2.108	17.120	2.302	1.273	0.067	0.054	0.997	0.230	0.596	0.558	0.684	0.782
	Std 40	21.336	2.769	15.798	1.960	1.615	0.067	0.050	1.265	0.196	0.711	0.667	0.664	0.962
	XS 80	21.336	3.734	13.868	1.510	2.065	0.067	0.044	1.617	0.151	0.836	0.783	0.636	1.174
	160	21.336	4.75	11.836	1.100	2.475	0.067	0.037	1.938	0.110	0.921	0.863	0.610	1.342
	XXS	21.336	7.468	6.400	0.322	3.254	0.067	0.020	2.547	0.032	1.009	0.946	0.557	1.575
20	5S	26.67	1.651	23.368	4.289	1.298	0.084	0.073	1.016	0.429	1.020	0.765	0.886	1.035
	10S	26.67	2.108	22.454	3.960	1.627	0.084	0.071	1.273	0.396	1.236	0.927	0.872	1.275
	Std 40	26.67	2.87	20.930	3.441	2.146	0.084	0.066	1.680	0.344	1.541	1.156	0.848	1.634
	XS 80	26.67	3.912	18.846	2.790	2.797	0.084	0.059	2.190	0.279	1.864	1.398	0.816	2.046
	160	26.67	5.537	15.596	1.910	3.676	0.084	0.049	2.878	0.191	2.193	1.645	0.772	2.529
	XXS	26.67	7.823	11.024	0.954	4.632	0.084	0.035	3.626	0.095	2.411	1.808	0.721	2.938
25	5S	33.401	1.651	30.099	7.115	1.647	0.105	0.095	1.289	0.712	2.081	1.246	1.124	1.666
	10S	33.401	2.769	27.863	6.097	2.665	0.105	0.088	2.086	0.610	3.151	1.887	1.087	2.605
	Std 40	33.401	3.378	26.645	5.576	3.186	0.105	0.084	2.494	0.558	3.635	2.177	1.068	3.058
	XS 80	33.401	4.547	24.307	4.640	4.122	0.105	0.076	3.227	0.464	4.396	2.632	1.033	3.817
	160	33.401	6.35	20.701	3.366	5.396	0.105	0.065	4.225	0.337	5.208	3.119	0.982	4.732
	XXS	33.401	9.093	15.215	1.818	6.944	0.105	0.048	5.436	0.182	5.846	3.501	0.918	5.624
32	5S	42.164	1.651	38.862	11.862	2.101	0.132	0.122	1.645	1.186	4.318	2.048	1.434	2.711
	10S	42.164	2.769	36.626	10.536	3.427	0.132	0.115	2.683	1.054	6.681	3.169	1.396	4.305
	Std 40	42.164	3.556	35.052	9.650	4.313	0.132	0.110	3.377	0.965	8.104	3.844	1.371	5.316
	XS 80	42.164	4.851	32.462	8.276	5.686	0.132	0.102	4.452	0.828	10.063	4.773	1.330	6.792
	160	42.164	6.35	29.464	6.818	7.145	0.132	0.093	5.594	0.682	11.815	5.604	1.286	8.230
	XXS	42.164	9.703	22.758	4.068	9.895	0.132	0.071	7.747	0.407	14.198	6.734	1.198	10.529
40	5S	48.26	1.651	44.958	15.875	2.418	0.152	0.141	1.893	1.587	6.573	2.724	1.649	3.588
	10S	48.26	2.769	42.722	14.335	3.957	0.152	0.134	3.098	1.433	10.275	4.258	1.611	5.737
	Std 40	48.26	3.683	40.894	13.134	5.158	0.152	0.128	4.038	1.313	12.899	5.345	1.581	7.335
	XS 80	48.26	5.08	38.100	11.401	6.891	0.152	0.120	5.395	1.140	16.283	6.748	1.537	9.516
	160	48.26	7.137	33.986	9.072	9.220	0.152	0.107	7.219	0.907	20.078	8.321	1.476	12.191
	XXS	48.26	10.16	27.940	6.131	12.161	0.152	0.088	9.521	0.613	23.635	9.795	1.394	15.098
		48.26	13.335	21.590	3.661	14.631	0.152	0.068	11.455	0.366	25.560	10.593	1.322	17.056
		48.26	15.875	16.510	2.141	16.151	0.152	0.052	12.645	0.214	26.262	10.884	1.275	17.983
50	5S	60.325	1.651	57.023	25.538	3.043	0.190	0.179	2.383	2.554	13.106	4.345	2.075	5.685
	10S	60.325	2.769	54.787	23.575	5.007	0.190	0.172	3.920	2.357	20.780	6.890	2.037	9.180

	Std	40	60.325	3.912	52.501	21.648	6.933	0.190	0.165	5.428	2.165	27.713	9.188	1.999	12.470
	XS	80	60.325	5.537	49.251	19.051	9.530	0.190	0.155	7.461	1.905	36.124	11.977	1.947	16.677
		160	60.325	8.712	42.901	14.455	14.126	0.190	0.135	11.059	1.446	48.379	16.039	1.851	23.429
	XXS		60.325	11.074	38.177	11.447	17.134	0.190	0.120	13.415	1.145	54.579	18.095	1.785	27.315
			60.325	14.275	31.775	7.930	20.652	0.190	0.100	16.168	0.793	60.002	19.893	1.705	31.242
			60.325	17.45	25.425	5.077	23.504	0.190	0.080	18.402	0.508	62.955	20.872	1.637	33.850
65	5S		73.025	2.108	68.809	37.186	4.696	0.229	0.216	3.677	3.719	29.550	8.093	2.508	10.605
	10S		73.025	3.048	66.929	35.182	6.701	0.229	0.210	5.246	3.518	41.092	11.254	2.476	14.935
	Std	40	73.025	5.156	62.713	30.889	10.993	0.229	0.197	8.607	3.089	63.662	17.436	2.406	23.796
	XS	80	73.025	7.01	59.005	27.344	14.538	0.229	0.185	11.382	2.734	80.089	21.935	2.347	30.665
		160	73.025	9.525	53.975	22.881	19.002	0.229	0.170	14.876	2.288	97.928	26.820	2.270	38.696
	XXS		73.025	14.021	44.983	15.892	25.990	0.229	0.141	20.348	1.589	119.491	32.726	2.144	49.734
			73.025	17.145	38.735	11.784	30.098	0.229	0.122	23.564	1.178	128.539	35.204	2.067	55.218
			73.025	20.32	32.385	8.237	33.645	0.229	0.102	26.341	0.824	134.190	36.752	1.997	59.243
80	5S		88.9	2.108	84.684	56.324	5.748	0.279	0.266	4.500	5.632	54.153	12.183	3.069	15.883
	10S		88.9	3.048	82.804	53.851	8.221	0.279	0.260	6.436	5.385	75.835	17.061	3.037	22.475
	Std	40	88.9	5.486	77.928	47.696	14.376	0.279	0.245	11.255	4.770	125.575	28.251	2.956	38.227
	XS	80	88.9	7.62	73.660	42.614	19.458	0.279	0.231	15.233	4.261	162.092	36.466	2.886	50.490
		160	88.9	11.1	66.700	34.942	27.130	0.279	0.210	21.240	3.494	209.445	47.119	2.779	67.644
	XXS		88.9	15.24	58.420	26.805	35.267	0.279	0.184	27.610	2.680	249.425	56.114	2.659	83.871
			88.9	18.415	52.070	21.294	40.777	0.279	0.164	31.925	2.129	270.517	60.859	2.576	93.572
			88.9	21.59	45.720	16.417	45.654	0.279	0.144	35.743	1.642	285.153	64.151	2.499	101.173
90	5S		101.6	2.108	97.384	74.485	6.589	0.319	0.306	5.158	7.448	81.562	16.055	3.518	20.870
	10S		101.6	3.048	95.504	71.636	9.437	0.319	0.300	7.388	7.164	114.679	22.575	3.486	29.614
	Std	40	101.6	5.74	90.120	63.787	17.286	0.319	0.283	13.533	6.379	199.267	39.226	3.395	52.810
	XS	80	101.6	8.077	85.446	57.342	23.731	0.319	0.268	18.579	5.734	261.390	51.455	3.319	70.823
	XXS		101.6	16.154	69.292	37.710	43.363	0.319	0.218	33.949	3.771	409.886	80.686	3.074	119.348
100	5S		114.3	2.108	110.084	95.179	7.430	0.359	0.346	5.817	9.518	116.941	20.462	3.967	26.537
	10S		114.3	3.048	108.204	91.955	10.653	0.359	0.340	8.340	9.196	164.938	28.861	3.935	37.735
			114.3	4.775	104.750	86.179	16.430	0.359	0.329	12.863	8.618	246.828	43.189	3.876	57.317
	Std	40	114.3	6.02	102.260	82.130	20.478	0.359	0.321	16.033	8.213	301.050	52.677	3.834	70.656
	XS	80	114.3	8.56	97.180	74.173	28.436	0.359	0.305	22.262	7.417	400.023	69.995	3.751	95.920
		120	114.3	11.1	92.100	66.621	35.988	0.359	0.289	28.175	6.662	484.634	84.800	3.670	118.676
			114.3	12.7	88.900	62.072	40.537	0.359	0.279	31.736	6.207	531.220	92.952	3.620	131.782
		160	114.3	13.487	87.326	59.893	42.715	0.359	0.274	33.442	5.989	552.364	96.652	3.596	137.892
	XXS		114.3	17.12	80.060	50.341	52.267	0.359	0.252	40.920	5.034	636.157	111.314	3.489	163.356
			114.3	20.32	73.660	42.614	59.994	0.359	0.231	46.970	4.261	693.313	121.315	3.399	182.271
			114.3	23.495	67.310	35.584	67.025	0.359	0.211	52.474	3.558	737.062	128.970	3.316	198.056
125	5S		141.3	2.769	135.762	144.760	12.051	0.444	0.427	9.435	14.476	289.197	40.934	4.899	53.148
	10S		141.3	3.404	134.492	142.064	14.747	0.444	0.423	11.545	14.206	350.724	49.642	4.877	64.743
	Std	40	141.3	6.553	128.194	129.070	27.740	0.444	0.403	21.718	12.907	631.073	89.324	4.770	119.077
	XS	80	141.3	9.525	122.250	117.379	39.432	0.444	0.384	30.871	11.738	860.367	121.779	4.671	165.690
		120	141.3	12.7	115.900	105.501	51.309	0.444	0.364	40.170	10.550	1071.021	151.595	4.569	210.719
		160	141.3	15.875	109.550	94.257	62.553	0.444	0.344	48.973	9.426	1249.755	176.894	4.470	251.075
	XXS		141.3	19.05	103.200	83.647	73.164	0.444	0.324	57.280	8.365	1399.967	198.155	4.374	287.014
			141.3	22.225	96.850	73.670	83.141	0.444	0.304	65.091	7.367	1524.867	215.834	4.283	318.791
			141.3	25.4	90.500	64.326	92.484	0.444	0.284	72.406	6.433	1627.473	230.357	4.195	346.663
150	5S		168.275	2.769	162.737	208.000	14.398	0.529	0.511	11.272	20.800	493.108	58.607	5.852	75.858
	10S		168.275	3.404	161.467	204.766	17.631	0.529	0.507	13.804	20.477	599.327	71.232	5.830	92.544
			168.275	5.563	157.149	193.961	28.437	0.529	0.494	22.263	19.396	942.175	111.980	5.756	147.342

TABLE 5.1M Principal Properties of Commercial Pipe (in Metric Units) (Continued)

D_N, Nom. dia.	Schedule	D, outside dia., mm	t, wall thick., mm	d, inside dia., mm	A_i, inside area, cm^2	A_M, metal area, cm^2	A_{so}, outside surf., m^2/m	A_{si}, inside surf., m^2/m	w_p, pipe wt., kg/m	w_w, water wt., kg/m	I, mom of inert., cm^4	z_e, elast. sec. mod., cm^3	R_g, rad. of gyr., cm	z_p, plast. sec. mod., cm^3
	Std 40	168.275	7.112	154.051	186.389	36.009	0.529	0.484	28.191	18.639	1171.357	139.219	5.704	184.847
	XS 80	168.275	10.973	146.329	168.171	54.226	0.529	0.460	42.454	16.817	1685.360	200.310	5.575	271.961
	120	168.275	14.275	139.725	153.334	69.063	0.529	0.439	54.070	15.333	2064.954	245.426	5.468	339.522
	160	168.275	18.237	131.801	136.436	85.962	0.529	0.414	67.300	13.644	2454.613	291.738	5.344	412.570
	XXS	168.275	21.946	124.383	121.510	100.887	0.529	0.391	78.985	12.151	2760.981	328.151	5.231	473.444
		168.275	25.4	117.475	108.388	114.009	0.529	0.369	89.258	10.839	3001.044	356.683	5.131	523.970
		168.275	28.575	111.125	96.987	125.410	0.529	0.349	98.184	9.699	3187.370	378.829	5.041	565.461
200	5S	219.075	2.769	213.537	358.127	18.817	0.688	0.671	14.732	35.813	1100.668	100.483	7.648	129.566
	10S	219.075	3.759	211.557	351.516	25.427	0.688	0.665	19.907	35.152	1473.975	134.563	7.614	174.292
		219.075	5.563	207.949	339.629	37.315	0.688	0.653	29.214	33.963	2127.782	194.251	7.551	253.665
	20	219.075	6.35	206.375	334.507	42.437	0.688	0.648	33.224	33.451	2402.551	219.336	7.524	287.441
	30	219.075	7.036	205.003	330.074	46.870	0.688	0.644	36.694	33.007	2636.984	240.738	7.501	316.465
	Std 40	219.075	8.179	202.717	322.754	54.190	0.688	0.637	42.425	32.275	3017.272	275.456	7.462	363.968
	60	219.075	10.312	198.451	309.313	67.631	0.688	0.623	52.949	30.931	3693.332	337.175	7.390	449.792
	XS 80	219.075	12.7	193.675	294.604	82.340	0.688	0.608	64.464	29.460	4400.206	401.708	7.310	541.595
	100	219.075	15.062	188.951	280.407	96.536	0.688	0.594	75.578	28.041	5049.791	461.010	7.233	628.052
	120	219.075	18.237	182.601	261.877	115.067	0.688	0.574	90.086	26.188	5849.434	534.012	7.130	737.642
	140	219.075	20.625	177.825	248.357	128.587	0.688	0.559	100.671	24.836	6398.374	584.126	7.054	815.203
	160	219.075	23.012	173.051	235.201	141.743	0.688	0.544	110.970	23.520	6904.619	630.343	6.979	888.677
		219.075	25.4	168.275	222.398	154.546	0.688	0.529	120.994	22.240	7370.842	672.906	6.906	958.236
		219.075	28.575	161.925	205.930	171.014	0.688	0.509	133.887	20.593	7932.151	724.149	6.811	1044.792
250	5S	273.05	3.404	266.242	556.729	28.836	0.858	0.836	22.576	55.673	2621.182	191.993	9.534	247.519
	10S	273.05	4.191	264.668	550.166	35.399	0.858	0.831	27.714	55.017	3199.288	234.337	9.507	302.978
		273.05	5.563	261.924	538.817	46.748	0.858	0.823	36.599	53.882	4182.745	306.372	9.459	398.094
	20	273.05	6.35	260.350	532.361	53.204	0.858	0.818	41.654	53.236	4733.097	346.684	9.432	451.763
	30	273.05	7.798	257.454	520.583	64.982	0.858	0.809	50.874	52.058	5719.932	418.966	9.382	548.826
	Std 40	273.05	9.271	254.508	508.738	76.828	0.858	0.800	60.148	50.874	6690.212	490.036	9.332	645.349
	XS 60	273.05	12.7	247.650	481.690	103.875	0.858	0.778	81.324	48.169	8821.961	646.179	9.216	861.533
	80	273.05	15.062	242.926	463.488	122.077	0.858	0.763	95.574	46.349	10190.968	746.454	9.137	1003.653
	100	273.05	18.237	236.576	439.574	145.991	0.858	0.743	114.296	43.957	11909.500	872.331	9.032	1186.168
	120	273.05	21.412	230.226	416.294	169.272	0.858	0.723	132.523	41.629	13495.082	988.470	8.929	1359.143
		273.05	22.225	228.600	410.434	175.131	0.858	0.718	137.110	41.043	13880.566	1016.705	8.903	1401.933
	140	273.05	25.4	222.250	387.949	197.616	0.858	0.698	154.714	38.795	15309.107	1121.341	8.802	1563.289
	160	273.05	28.575	215.900	366.097	219.468	0.858	0.678	171.822	36.610	16620.309	1217.382	8.702	1715.683
		273.05	31.75	209.550	344.879	240.687	0.858	0.658	188.434	34.488	17820.783	1305.313	8.605	1859.372
		273.05	38.1	196.850	304.342	281.223	0.858	0.618	220.170	30.434	19915.017	1458.708	8.415	2121.655

300	5S		323.85	3.962	315.926	783.902	39.817	1.017	0.993	31.172	78.390	5093.677	314.570	11.311	405.454
	10S		323.85	4.572	314.706	777.859	45.859	1.017	0.989	35.903	77.786	5844.653	360.948	11.289	466.104
		20	323.85	6.35	311.150	760.380	63.339	1.017	0.978	49.588	76.038	7984.267	493.084	11.228	640.218
		30	323.85	8.382	307.086	740.646	83.072	1.017	0.965	65.037	74.065	10341.334	638.650	11.157	834.390
	Std		323.85	9.525	304.800	729.660	94.058	1.017	0.958	73.638	72.966	11626.710	718.031	11.118	941.379
		40	323.85	10.312	303.226	722.144	101.574	1.017	0.953	79.523	72.214	12495.092	771.659	11.091	1014.118
	XS		323.85	12.7	298.450	699.575	124.144	1.017	0.938	97.192	69.957	15048.477	929.349	11.010	1230.249
		60	323.85	14.275	295.300	684.885	138.833	1.017	0.928	108.692	68.489	16666.818	1029.292	10.957	1369.066
		80	323.85	17.45	288.950	655.747	167.971	1.017	0.908	131.505	65.575	19775.367	1221.267	10.850	1640.027
			323.85	19.05	285.750	641.303	182.415	1.017	0.898	142.813	64.130	21266.181	1313.335	10.797	1772.143
		100	323.85	21.412	281.026	620.274	203.444	1.017	0.883	159.276	62.027	23377.291	1443.711	10.720	1961.840
			323.85	22.225	279.400	613.117	210.601	1.017	0.878	164.879	61.312	24079.741	1487.092	10.693	2025.678
		120	323.85	25.4	273.050	585.565	238.153	1.017	0.858	186.450	58.557	26707.869	1649.398	10.590	2267.947
		140	323.85	28.575	266.700	558.646	265.072	1.017	0.838	207.525	55.865	29158.897	1800.766	10.488	2499.205
			323.85	31.75	260.350	532.361	291.357	1.017	0.818	228.104	53.236	31440.966	1941.699	10.388	2719.710
		160	323.85	33.325	257.200	519.557	304.162	1.017	0.808	238.128	51.956	32512.782	2007.891	10.339	2825.182
350	5S		355.6	3.962	347.676	949.381	43.768	1.117	1.092	34.266	94.938	6765.712	380.524	12.433	489.929
	10S		355.6	4.775	346.050	940.521	52.628	1.117	1.087	41.202	94.052	8098.077	455.460	12.405	587.746
			355.6	5.334	344.932	934.454	58.695	1.117	1.084	45.952	93.445	9003.350	506.375	12.385	654.472
			355.6	5.563	344.474	931.974	61.175	1.117	1.082	47.894	93.197	9371.672	527.091	12.377	681.683
		10	355.6	6.35	342.900	923.477	69.672	1.117	1.077	54.547	92.348	10626.322	597.656	12.350	774.646
			355.6	7.137	341.326	915.018	78.131	1.117	1.072	61.169	91.502	11863.814	667.256	12.323	866.759
		20	355.6	7.925	339.750	906.588	86.561	1.117	1.067	67.769	90.659	13085.844	735.987	12.295	958.142
			355.6	8.738	338.124	897.931	95.218	1.117	1.062	74.546	89.793	14328.945	805.902	12.267	1051.541
	Std	30	355.6	9.525	336.550	889.590	103.559	1.117	1.057	81.076	88.959	15515.330	872.628	12.240	1141.100
		40	355.6	11.1	333.400	873.016	120.133	1.117	1.047	94.052	87.302	17840.120	1003.381	12.186	1317.833
			355.6	11.913	331.774	864.521	128.628	1.117	1.042	100.703	86.452	19014.654	1069.441	12.158	1407.764
	XS		355.6	12.7	330.200	856.338	136.811	1.117	1.037	107.110	85.634	20135.294	1132.469	12.132	1493.984
		60	355.6	15.062	325.476	832.011	161.138	1.117	1.023	126.155	83.201	23403.706	1316.294	12.052	1747.856
			355.6	15.875	323.850	823.718	169.431	1.117	1.017	132.647	82.372	24496.274	1377.743	12.024	1833.553
		80	355.6	19.05	317.500	791.732	201.417	1.117	0.997	157.689	79.173	28608.131	1609.006	11.918	2160.063
		100	355.6	23.8	308.000	745.062	248.087	1.117	0.968	194.228	74.506	34315.583	1930.010	11.761	2624.718
		120	355.6	27.762	300.076	707.218	285.931	1.117	0.943	223.855	70.722	38689.074	2175.988	11.632	2990.990
		140	355.6	31.75	292.100	670.122	323.027	1.117	0.918	252.898	67.012	42754.938	2404.665	11.505	3340.638
		160	355.6	35.712	284.176	634.258	358.891	1.117	0.893	280.976	63.426	46477.609	2614.039	11.380	3669.605
400	5S		406.4	4.191	398.018	1244.218	52.957	1.277	1.250	41.460	124.422	10709.684	527.051	14.221	678.025
	10S		406.4	4.775	396.850	1236.926	60.248	1.277	1.247	47.168	123.693	12149.361	597.902	14.201	770.272
		10	406.4	6.35	393.700	1217.368	79.807	1.277	1.237	62.481	121.737	15969.173	785.885	14.146	1016.360
		20	406.4	7.925	390.550	1197.965	99.209	1.277	1.227	77.671	119.797	19698.390	969.409	14.091	1258.541
	Std	30	406.4	9.525	387.350	1178.414	118.760	1.277	1.217	92.977	117.841	23395.524	1151.355	14.036	1500.599
	XS	40	406.4	12.7	381.000	1140.094	157.080	1.277	1.197	122.978	114.009	30465.487	1499.286	13.927	1969.218
		60	406.4	16.662	373.076	1093.165	204.010	1.277	1.172	159.719	109.316	38805.645	1909.727	13.792	2532.479
		80	406.4	21.4412	363.518	1037.867	259.307	1.277	1.142	203.012	103.787	48182.963	2371.209	13.631	3180.791
		100	406.4	26.187	354.026	984.376	312.798	1.277	1.112	244.889	98.438	56790.855	2794.826	13.474	3791.705
		120	406.4	30.937	344.526	932.255	364.919	1.277	1.082	285.695	93.226	64740.297	3186.038	13.320	4371.222
		140	406.4	36.5	333.400	873.016	424.158	1.277	1.047	332.074	87.302	73250.489	3604.847	13.141	5010.459
		160	406.4	40.462	325.476	832.011	465.164	1.277	1.023	364.177	83.201	78814.075	3878.645	13.017	5440.481

TABLE 5.1M Principal Properties of Commercial Pipe (in Metric Units) (Continued)

D_N, Nom. dia.	Schedule	D, outside dia., mm	t, wall thick., mm	d, inside dia., mm	A_i, inside area, cm²	A_M, metal area, cm²	A_{so}, outside surf., m²/m	A_{si}, inside surf., m²/m	w_p, pipe wt., kg/m	w_w, water wt., kg/m	I, mom of inert., cm⁴	z_e, elast. sec. mod., cm³	R_g, rad. of gyr., cm	z_p, plast. sec. mod., cm³
450	5S	457.2	4.191	448.818	1582.091	59.645	1.436	1.410	46.696	158.209	15301.422	669.354	16.017	860.107
	10S	457.2	4.775	447.650	1573.867	67.869	1.436	1.406	53.135	157.387	17366.728	759.699	15.997	977.443
	10	457.2	6.35	444.500	1551.795	89.941	1.436	1.396	70.415	155.180	22856.650	999.853	15.942	1290.849
	20	457.2	7.925	441.350	1529.879	111.857	1.436	1.387	87.573	152.988	28231.088	1234.956	15.887	1599.844
	Std	457.2	9.525	438.150	1507.775	133.961	1.436	1.376	104.878	150.777	33574.268	1468.691	15.831	1909.259
	30	457.2	11.1	435.000	1486.173	155.563	1.436	1.367	121.790	148.617	38720.854	1693.826	15.777	2209.458
	XS	457.2	12.7	431.800	1464.388	177.348	1.436	1.357	138.846	146.439	43835.895	1917.581	15.722	2510.002
	40	457.2	14.275	428.650	1443.100	198.636	1.436	1.347	155.512	144.310	48761.162	2133.034	15.668	2801.532
	60	457.2	19.05	419.100	1379.514	262.222	1.436	1.317	205.294	137.951	63043.469	2757.807	15.506	3659.509
	80	457.2	23.8	409.600	1317.683	324.054	1.436	1.287	253.702	131.768	76314.638	3338.348	15.346	4475.070
	100	457.2	29.362	398.476	1247.083	394.653	1.436	1.252	308.974	124.708	90723.723	3968.667	15.162	5383.123
	120	457.2	34.925	387.350	1178.414	463.322	1.436	1.217	362.735	117.841	103977.565	4548.450	14.981	6242.017
	140	457.2	39.675	377.850	1121.320	520.416	1.436	1.187	407.434	112.132	114426.016	5005.513	14.828	6937.385
	160	457.2	45.237	366.726	1056.268	585.468	1.436	1.152	458.363	105.627	125698.549	5498.624	14.653	7708.342
500	5S	508	4.775	498.450	1951.345	75.489	1.596	1.566	59.101	195.135	23897.671	940.853	17.792	1209.260
	10S	508	5.537	496.926	1939.431	87.404	1.596	1.561	68.428	193.943	27586.461	1086.081	17.766	1398.006
	10	508	6.35	495.300	1926.760	100.075	1.596	1.556	78.349	192.676	31484.894	1239.563	17.737	1598.112
	Std 20	508	9.525	488.950	1877.672	149.162	1.596	1.536	116.779	187.767	46345.775	1824.637	17.627	2367.082
	XS 30	508	12.7	482.600	1829.218	197.616	1.596	1.516	154.714	182.922	60638.803	2387.354	17.517	3116.336
	40	508	15.062	477.876	1793.582	233.252	1.596	1.501	182.613	179.358	70912.256	2791.821	17.436	3661.096
	60	508	20.625	466.750	1711.038	315.797	1.596	1.466	247.238	171.104	93932.795	3698.142	17.247	4902.169
		508	22.225	463.550	1687.657	339.178	1.596	1.456	265.543	168.766	100256.341	3947.100	17.193	5248.361
	80	508	26.187	455.626	1630.452	396.383	1.596	1.431	310.329	163.045	115360.942	4541.769	17.060	6085.257
	100	508	32.537	442.926	1540.825	486.010	1.596	1.391	380.497	154.082	137979.067	5432.247	16.849	7367.108
	120	508	38.1	431.800	1464.388	562.447	1.596	1.357	440.340	146.439	156258.466	6151.908	16.668	8431.313
	140	508	44.45	419.100	1379.514	647.320	1.596	1.317	506.787	137.951	175466.041	6908.112	16.464	9580.820
	160	508	49.987	408.026	1307.575	719.260	1.596	1.282	563.109	130.757	190848.787	7513.732	16.289	10527.913
550	5S	558.8	4.775	549.250	2369.360	83.110	1.756	1.726	65.067	236.936	31889.683	1141.363	19.588	1465.722
	10S	558.8	5.537	547.726	2356.230	96.240	1.756	1.721	75.347	235.623	36827.258	1318.084	19.562	1694.966
	10	558.8	6.35	546.100	2342.261	110.209	1.756	1.716	86.283	234.226	42050.049	1505.013	19.533	1938.150
	Std 20	558.8	9.525	539.750	2288.106	164.364	1.756	1.696	128.680	228.811	62004.261	2219.193	19.423	2874.067
	XS 30	558.8	12.7	533.400	2234.585	217.885	1.756	1.676	170.582	223.459	81266.498	2908.608	19.313	3788.219
		558.8	15.875	527.050	2181.697	270.772	1.756	1.656	211.988	218.170	99852.945	3573.835	19.204	4680.862
		558.8	19.05	520.700	2129.443	323.027	1.756	1.636	252.898	212.944	117779.597	4215.447	19.095	5552.253
	60	558.8	22.225	514.350	2077.822	374.648	1.756	1.616	293.312	207.782	135062.256	4834.011	18.987	6402.648
	80	558.8	28.575	501.650	1976.480	475.989	1.756	1.576	372.652	197.648	167757.847	6004.218	18.773	8041.472
	100	558.8	34.925	488.950	1877.672	574.798	1.756	1.536	450.009	187.767	198062.308	7088.844	18.563	9599.384
	120	558.8	41.275	476.250	1781.398	671.072	1.756	1.496	525.383	178.140	226095.163	8092.168	18.355	11078.432
	140	558.8	47.625	463.550	1687.657	764.813	1.756	1.456	598.773	168.766	251972.873	9018.356	18.151	12480.663
	160	558.8	53.975	450.850	1596.449	856.021	1.756	1.416	670.179	159.645	275808.832	9871.469	17.950	13808.127

600	10		609.6	6.35	596.900	2798.299	120.343	1.915	1.875	94.217	279.830	54748.259	1796.203	21.329	2310.964
	Std	20	609.6	9.525	590.550	2739.077	179.565	1.915	1.855	140.581	273.908	80843.940	2652.360	21.219	3430.214
	XS		609.6	12.7	584.200	2680.489	238.153	1.915	1.835	186.450	268.049	106111.265	3481.341	21.108	4525.651
		30	609.6	14.275	581.050	2651.660	266.981	1.915	1.825	209.020	265.166	118343.549	3882.662	21.054	5060.300
			609.6	15.875	577.850	2622.534	296.108	1.915	1.815	231.823	262.253	130567.953	4283.725	20.999	5597.532
		40	609.6	17.45	574.700	2594.020	324.622	1.915	1.805	254.147	259.402	142404.625	4672.068	20.945	6120.590
			609.6	19.05	571.500	2565.213	353.429	1.915	1.795	276.700	256.521	154231.529	5060.090	20.890	6646.112
	5S		609.6	5.537	598.526	2813.565	105.077	1.915	1.880	82.265	281.356	47930.747	1572.531	21.358	2020.505
			609.6	22.225	565.150	2508.525	410.117	1.915	1.775	321.081	250.852	177119.328	5811.002	20.782	7671.646
		60	609.6	24.587	560.426	2466.763	451.879	1.915	1.761	353.776	246.676	193653.350	6353.456	20.702	8419.783
		80	609.6	30.937	547.726	2356.230	562.412	1.915	1.721	440.313	235.623	236075.753	7745.267	20.488	10369.359
		100	609.6	38.887	531.826	2221.417	697.225	1.915	1.671	545.858	222.142	285184.461	9356.446	20.225	12685.870
		120	609.6	46.025	517.550	2103.757	814.885	1.915	1.626	637.974	210.376	325680.896	10685.069	19.992	14651.104
		140	609.6	52.375	504.850	2001.777	916.865	1.915	1.586	717.814	200.178	358998.192	11778.156	19.788	16310.639
		160	609.6	59.512	490.576	1890.181	1028.460	1.915	1.541	805.182	189.018	393560.156	12912.079	19.562	18078.760
650			660.4	6.35	647.700	3294.873	130.477	2.075	2.035	102.151	329.487	69775.665	2113.133	23.125	2716.552
		10	660.4	7.925	644.550	3262.903	162.448	2.075	2.025	127.181	326.290	86459.246	2618.390	23.070	3374.093
	Std		660.4	9.525	641.350	3230.584	194.766	2.075	2.015	152.482	323.058	103159.026	3124.138	23.014	4035.523
	XS	20	660.4	12.7	635.000	3166.929	258.421	2.075	1.995	202.318	316.693	135565.391	4105.554	22.904	5328.634
			660.4	15.875	628.650	3103.907	321.443	2.075	1.975	251.658	310.391	167014.011	5057.965	22.794	6596.139
			660.4	19.05	622.300	3041.519	383.832	2.075	1.955	300.502	304.152	197523.944	5981.949	22.685	7838.294
			660.4	22.225	615.950	2979.764	445.587	2.075	1.935	348.850	297.976	227114.059	6878.076	22.577	9055.357
			660.4	25.4	609.600	2918.642	506.709	2.075	1.915	396.702	291.864	255803.031	7746.912	22.469	10247.582
			660.4	28.575	603.250	2858.154	567.197	2.075	1.895	444.059	285.815	283609.344	8589.017	22.361	11415.227
700			711.2	6.35	698.500	3831.984	140.612	2.234	2.194	110.085	383.198	87328.412	2455.805	24.921	3154.914
		10	711.2	7.925	695.350	3797.500	175.096	2.234	2.185	137.082	379.750	108264.589	3044.561	24.866	3919.915
	Std		711.2	9.525	692.150	3762.629	209.967	2.234	2.174	164.384	376.263	129243.733	3634.526	24.810	4689.995
	XS	20	711.2	12.7	685.800	3693.906	278.690	2.234	2.155	218.186	369.391	170021.160	4781.247	24.700	6197.165
		30	711.2	15.875	679.450	3625.817	346.779	2.234	2.135	271.493	362.582	209681.476	5896.554	24.590	7676.682
			711.2	19.05	673.100	3558.362	414.234	2.234	2.115	324.304	355.836	248245.272	6981.026	24.480	9128.801
			711.2	22.225	666.750	3491.539	481.057	2.234	2.095	376.619	349.154	285732.948	8035.235	24.372	10553.779
			711.2	25.4	660.400	3425.351	547.245	2.234	2.075	428.439	342.535	322164.712	9059.750	24.263	11951.871
			711.2	28.575	654.050	3359.795	612.801	2.234	2.055	479.762	335.980	357560.580	10055.134	24.156	13323.334
750	5S		762	6.35	749.300	4409.632	150.746	2.394	2.354	118.019	440.963	107602.641	2824.216	26.717	3626.052
		10	762	7.925	746.150	4372.635	187.743	2.394	2.344	146.984	437.263	133458.839	3502.857	26.662	4506.642
	Std		762	9.525	742.950	4335.209	225.169	2.394	2.334	176.285	433.521	159392.275	4183.524	26.606	5393.628
	XS	20	762	12.7	736.600	4261.420	298.958	2.394	2.314	234.054	426.142	209870.859	5508.421	26.496	7131.247
		30	762	15.875	730.250	4188.264	372.114	2.394	2.294	291.328	418.826	259060.706	6799.494	26.385	8839.163
		40	762	19.05	723.900	4115.741	444.637	2.394	2.274	348.106	411.574	306983.941	8057.321	26.276	10517.633
			762	22.225	717.550	4043.852	516.526	2.394	2.254	404.389	404.385	353662.495	9282.480	26.167	12166.913
			762	25.4	711.200	3972.596	587.782	2.394	2.234	460.175	397.260	399118.109	10475.541	26.058	13787.259
			762	28.575	704.850	3901.973	658.405	2.394	2.214	515.465	390.197	443372.332	11637.069	25.950	15378.927

TABLE 5.1M Principal Properties of Commercial Pipe (in Metric Units) (Continued)

D_N, Nom. dia.	Schedule	D, outside dia., mm	t, wall thick., mm	d, inside dia., mm	A_i, inside area, cm^2	A_M, metal area, cm^2	A_{so}, outside surf., m^2/m	A_{si}, inside surf., m^2/m	w_p, pipe wt., kg/m	w_w, water wt., kg/m	I, mom of inert., cm^4	z_e, elast. sec. mod., cm^3	R_g, rad. of gyr., cm	z_p, plast. sec. mod., cm^3
800		812.8	6.35	800.100	5027.817	160.880	2.553	2.514	125.953	502.782	130794.496	3218.368	28.513	4129.964
	10	812.8	7.925	796.950	4988.306	200.391	2.553	2.504	156.886	498.831	162286.788	3993.277	28.458	5134.272
	Std	812.8	9.525	793.750	4948.327	240.370	2.553	2.494	188.186	494.833	193898.868	4771.134	28.402	6146.424
	XS 20	812.8	12.7	787.400	4869.470	319.226	2.553	2.474	249.923	486.947	255506.773	6287.076	28.291	8130.878
	30	812.8	15.875	781.050	4791.247	397.450	2.553	2.454	311.163	479.125	315642.058	7766.783	28.181	10083.580
	40	812.8	17.475	777.850	4752.068	436.629	2.553	2.444	341.837	475.207	345395.902	8498.915	28.126	11055.669
		812.8	19.05	774.700	4713.657	475.039	2.553	2.434	371.909	471.366	374328.379	9210.836	28.071	12004.789
		812.8	22.225	768.350	4636.701	551.996	2.553	2.414	432.158	463.670	431589.201	10619.813	27.962	13894.759
		812.8	25.4	762.000	4560.378	628.319	2.553	2.394	491.911	456.038	487447.795	11994.286	27.853	15753.746
		812.8	28.575	755.650	4484.688	704.008	2.553	2.374	551.168	448.469	541927.243	13334.824	27.745	17582.007
850		863.6	6.35	850.900	5686.538	171.014	2.713	2.673	133.887	568.654	157100.119	3638.261	30.309	4666.651
	10	863.6	7.925	847.750	5644.513	213.039	2.713	2.663	166.788	564.451	194993.229	4515.823	30.254	5802.806
	Std	863.6	9.525	844.550	5601.981	255.571	2.713	2.653	200.087	560.198	233057.725	5397.354	30.198	6948.382
	XS 20	863.6	12.7	838.200	5518.057	339.495	2.713	2.633	265.791	551.806	307321.189	7117.211	30.087	9196.058
	30	863.6	15.875	831.850	5434.767	422.785	2.713	2.613	330.999	543.477	379915.890	8798.423	29.977	11409.935
	40	863.6	17.475	828.650	5393.034	464.518	2.713	2.603	363.671	539.303	415874.652	9631.187	29.921	12512.862
		863.6	19.05	825.500	5352.110	505.442	2.713	2.593	395.711	535.211	450867.016	10441.571	29.867	13590.269
		863.6	22.225	819.150	5270.087	587.465	2.713	2.573	459.927	527.009	520199.565	12047.234	29.757	15737.316
		863.6	25.4	812.800	5188.697	668.855	2.713	2.553	523.647	518.870	587938.340	13615.987	29.648	17851.332
		863.6	28.575	806.450	5107.940	749.612	2.713	2.534	586.872	510.794	654107.956	15148.401	29.540	19932.573
900		914.4	6.35	901.700	6385.796	181.148	2.873	2.833	141.821	638.580	186715.653	4083.894	32.105	5236.113
	10	914.4	7.925	898.550	6341.258	225.687	2.873	2.823	176.690	634.126	231822.954	5070.493	32.050	6512.244
	Std	914.4	9.525	895.350	6296.172	270.772	2.873	2.813	211.988	629.617	277163.061	6062.184	31.994	7799.502
	XS 20	914.4	12.7	889.000	6207.181	359.763	2.873	2.793	281.659	620.718	365706.392	7998.827	31.883	10326.788
	30	914.4	15.875	882.650	6118.824	448.120	2.873	2.773	350.834	611.882	452372.558	9894.413	31.773	12818.226
	40	914.4	19.05	876.300	6031.100	535.844	2.873	2.753	419.513	603.110	537188.280	11749.525	31.663	15274.073
		914.4	22.225	869.950	5944.009	622.935	2.873	2.733	487.696	594.401	620180.087	13564.744	31.553	17694.585
		914.4	25.4	863.600	5857.552	709.392	2.873	2.713	555.383	585.755	701374.317	15340.646	31.444	20080.017
		914.4	28.575	857.250	5771.728	795.216	2.873	2.693	622.575	577.173	780797.115	17077.802	31.335	22430.627
1050		1066.8	6.35	1054.100	8726.790	211.551	3.351	3.312	165.623	872.679	297383.155	5575.237	37.493	7141.147
	Std	1066.8	9.525	1047.750	8621.965	316.376	3.351	3.292	247.691	862.196	442100.085	8288.341	37.382	10647.836
	XS 20	1066.8	12.7	1041.400	8517.773	420.568	3.351	3.272	329.263	851.777	584209.582	10952.561	37.271	14112.276
	30	1066.8	15.875	1035.050	8414.214	524.127	3.351	3.252	410.339	841.421	723743.155	13568.488	37.160	17534.723
	40	1066.8	19.05	1028.700	8311.289	627.052	3.351	3.232	490.919	831.129	860732.122	16136.710	37.050	20915.433
		1066.8	25.4	1016.000	8107.339	831.002	3.351	3.192	650.592	810.734	1127200.551	21132.369	36.830	27552.668
		1066.8	31.75	1003.300	7905.922	1032.419	3.351	3.152	808.281	790.592	1383861.579	25944.162	36.612	34026.029
		1066.8	38.1	990.600	7707.039	1231.302	3.351	3.112	963.987	770.704	1630958.853	30576.656	36.395	40337.565

TABLE 5.2 Dimensions of Welded and Seamless Stainless Steel Pipe (U.S. Customary Units)

Nominal pipe size, in	Outside diameter, in	Nominal wall thickness, in			
		Schedule 5S*	Schedule 10S†	Schedule 40S	Schedule 80S
1/8	0.405	...	0.049	0.068	0.095
1/4	0.540	...	0.065	0.088	0.119
3/8	0.675	...	0.065	0.091	0.126
1/2	0.840	0.065	0.083	0.109	0.147
3/4	1.050	0.065	0.083	0.113	0.154
1	1.315	0.065	0.109	0.133	0.179
1 1/4	1.660	0.065	0.109	0.140	0.191
1 1/2	1.900	0.065	0.109	0.145	0.200
2	2.375	0.065	0.109	0.154	0.218
2 1/2	2.875	0.083	0.120	0.203	0.276
3	3.500	0.083	0.120	0.216	0.300
3 1/2	4.000	0.083	0.120	0.226	0.318
4	4.500	0.083	0.120	0.237	0.337
5	5.563	0.109	0.134	0.258	0.375
6	6.625	0.109	0.134	0.280	0.432
8	8.625	0.109	0.148	0.322	0.500
10	10.750	0.134	0.165	0.365	0.500†
12	12.750	0.156	0.180	0.375†	0.500†
14	14.000	0.156	0.188†	...	...
16	16.000	0.165	0.188†	...	...
18	18.000	0.165	0.188†	...	...
20	20.000	0.188	0.218†	...	...
22	22.000	0.188	0.218†	...	...
24	24.000	0.218	0.250	...	...
30	30.000	0.250	0.312	...	...

* Threading of Schedule 5S and 10S is not permitted, in accordance with ASME B1.20.1.
† Wall thickness does not conform to ASME B36.10M.

TABLE 5.2M Dimensions of Welded and Seamless Wrought Stainless Steel Pipe (SI Units)

Nominal pipe size, in	Outside diameter, mm	Nominal wall thickness, mm: Schedule 5S*	Schedule 10S*	Schedule 40S	Schedule 80S
1/8	10.3	...	1.24	1.73	2.41
1/4	13.7	...	1.65	2.24	3.02
3/8	17.1	...	1.65	2.31	3.20
1/2	21.3	1.65	2.11	2.77	3.73
3/4	26.7	1.65	2.11	2.87	3.91
1	33.4	1.65	2.77	3.38	4.55
1 1/4	42.2	1.65	2.77	3.56	4.85
1 1/2	48.3	1.65	2.77	3.68	5.08
2	60.3	1.65	2.77	3.91	5.54
2 1/2	73.0	2.11	3.05	5.16	7.01
3	88.9	2.11	3.05	5.49	7.62
3 1/2	101.6	2.11	3.05	5.74	8.08
4	114.3	2.11	3.05	6.02	8.56
5	141.3	2.77	3.40	6.55	9.53
6	168.3	2.77	3.40	7.11	10.97
8	219.1	2.77	3.76	8.18	12.70
10	273.1	3.40	4.19	9.27	12.70†
12	323.9	3.96	4.57	9.53†	12.70†
14	355.6	3.96	4.78†	...	...
16	406.4	4.19	4.78†	...	...
18	457	4.19	4.78†	...	...
20	508	4.78	5.54†	...	...
22	559	4.78	5.54†	...	...
24	610	5.54	6.35	...	...
30	762	6.35	7.92	...	...

* Schedule 5S and 10S pipe is not permitted to be threaded, in accordance with ASME B1.20.1.

† Wall thickness differs from the wall thickness per ASME 36.10.

TABLE 5.3 Tolerances for Pipe Outside Diameter, Inside Diameter, Length, and Weight

Permissible variations in pipe inside diameter*			
Pipe material	Over	Under	
Forged and bored	0	1/16-in.	
Cast	0	1/16-in.	
Permissible variations in length			
Pipe material	Specification	Over	Under
Forged and bored	Various ASTM	1/4 in	0
Cast	Various ASTM	1/8 in	0
Cast cold wrought	Various ASTM	1/8 in	0
Seamless and welded	Various ASTM	1/8 in	0
Permissible variations in outside diameter			
	Permissible variations in outside diameter		
NPS designator	Over, in		Under, in
1/8 to 1 1/2, incl.	1/64		1/32
Over 1 1/2 to 4, incl.	1/32		1/32
Over 4 to 8, incl.	1/16		1/32
Over 8 to 18, incl.	3/32		1/32
Over 18 to 26, incl.	1/8		1/32
Over 26 to 34, incl.	5/32		1/32
Over 34 to 48, incl.	3/16		1/32
Permissible variations in outside diameter			
	Permissible variations in outside diameter		
NPS designator	Over, mm		Under, mm
1/8 to 1 1/2, incl.	0.4		0.8
Over 1 1/2 to 4, incl.	0.8		0.8
Over 4 to 8, incl.	1.6		0.8
Over 8 to 18, incl.	2.4		0.8
Over 18 to 26, incl.	3.2		0.8
Over 26 to 34, incl.	4.0		0.8
Over 34 to 48, incl.	4.8		0.8
Permissible variations in weight			
	Permissible variations in weight		
NPS	Over		Under
≤12	10%		3.50%
>12	10%		5%

* For pipe fabricated from plate, sheet, or strip, refer to the piping material specification.

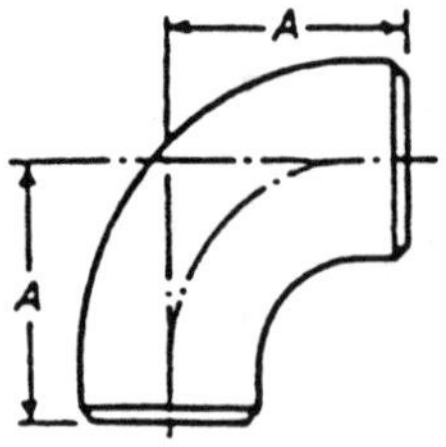

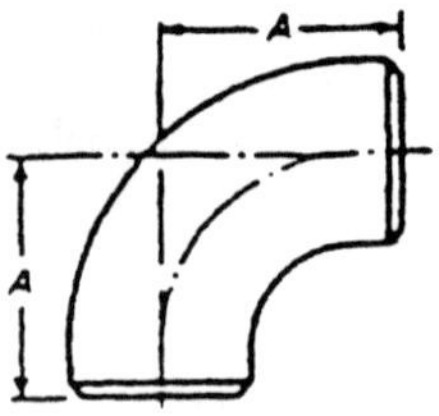

TABLE 5.4 Dimensions of 90° Long-Radius Elbows (U.S. Customary Units)*

Nominal pipe size (NPS)	Outside diameter at bevel, in	Center-to-end, 90° elbows, A, in
½	0.84	1.50
¾†	1.05	1.50
1	1.32	1.50
1¼	1.66	1.88
1½	1.90	2.25
2	2.38	3.00
2½	2.88	3.75
3	3.50	4.50
3½	4.00	5.25
4	4.50	6.00
5	5.56	7.50
6	6.62	9.00
8	8.62	12.00
10	10.75	15.00
12	12.75	18.00
14	14.00	21.00
16	16.00	24.00
18	18.00	27.00
20	20.00	30.00
22	22.00	33.00
24	24.00	36.00
26	26.00	39.00
28	28.00	42.00
30	30.00	45.00
32	32.00	48.00
34	34.00	51.00
36	36.00	54.00
38	38.00	57.00
40	40.00	60.00
42	42.00	63.00
44	44.00	66.00
46	46.00	69.00
48	48.00	72.00

* Dimensions per ASME B16.9. For tolerances, refer to Table 5.15.

† A dimensions of 1.12 in may be furnished for NPS ¾ at the manufacturer's option.

TABLE 5.4M Dimensions of 90° Long Radius Elbows (Metric Units)*

NPS	DN	Outside diameter at bevel, mm	Center-to-end, 90° elbows, A, mm
½	15	21.3	38
¾	20 (1)	26.7	38
1	25	33.4	38
1¼	32	42.2	48
1½	40	48.3	57
2	50	60.3	76
2½	65	73.0	95
3	80	88.9	114
3½	90	101.6	133
4	100	114.3	152
5	125	141.3	190
6	150	168.3	229
8	200	219.1	305
10	250	273.0	381
12	300	323.8	457
14	350	355.6	533
16	400	406.4	610
18	450	457	686
20	500	508	762
22	550	559	838
24	600	610	914
26	650	680	991
28	700	711	1067
30	750	762	1143
32	800	813	1219
34	850	864	1295
36	900	914	1372
38	950	965	1448
40	1000	1016	1524
42	1050	1067	1600
44	1100	1118	1676
46	1150	1168	1753
48	1200	1219	1829

Notes:
* Dimensions per ASME B16.9.
1. Manufacturers may furnish A dimensions of 29 mm for NPS ¾ (DN 20), 90° long radius elbows.
2. For tolerances, refer to Table 5.15.

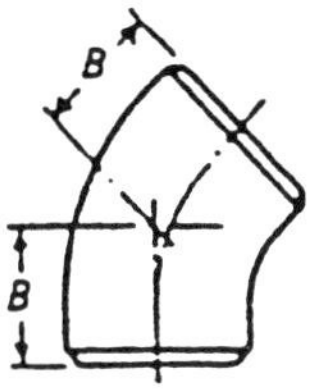

TABLE 5.5 Dimensions of 45° Long-Radius Elbows (U.S. Customary Units)*

Nominal pipe size (NPS)	Outside diameter at bevel, in	Center-to-end, 45° elbows, B, in
½	0.84	0.62
¾†	1.05	0.75
1	1.32	0.88
1¼	1.66	1.00
1½	1.90	1.12
2	2.38	1.38
2½	2.88	1.75
3	3.50	2.00
3½	4.00	2.25
4	4.50	2.50
5	5.56	3.12
6	6.62	3.75
8	8.62	5.00
10	10.75	6.25
12	12.75	7.50
14	14.00	8.75
16	16.00	10.00
18	18.00	11.25
20	20.00	12.50
22	22.00	13.50
24	24.00	15.00
26	26.00	16.00
28	28.00	17.25
30	30.00	18.50
32	32.00	19.75
34	34.00	21.00
36	36.00	22.25
38	38.00	23.62
40	40.00	24.88
42	42.00	26.00
44	44.00	27.38
46	46.00	28.62
48	48.00	29.88

* Dimensions per ASME B16.9. For tolerances, refer to Table 5.15.
† B dimensions of 0.44 in may be furnished for NPS ¾ at the manufacturer's option.

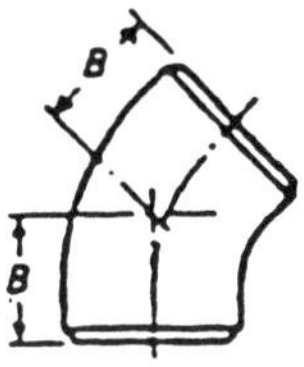

TABLE 5.5M Dimensions of 45° Long-Radius Elbows (Metric Units)*

NPS	DN	Outside diameter at bevel, mm	Center-to-end, 45° elbows, B, mm
½	15	21.3	16
¾	20†	26.7	19
1	25	33.4	22
1¼	32	42.2	25
1½	40	48.3	29
2	50	60.3	35
2½	65	73.0	44
3	80	88.9	51
3½	90	101.6	57
4	100	114.3	64
5	125	141.3	79
6	150	168.3	95
8	200	219.1	127
10	250	273.0	159
12	300	323.8	190
14	350	355.6	222
16	400	406.4	254
18	450	457	286
20	500	508	318
22	550	559	343
24	600	610	381
26	650	660	406
28	700	711	438
30	750	762	470
32	800	813	502
34	850	864	533
36	900	914	565
38	950	965	600
40	1000	1016	632
42	1050	1067	660
44	1100	1118	695
46	1150	1168	727
48	1200	1219	759

* Dimensions per ASME B16.9. For tolerances, refer to Table 5.15.
† B dimension of 11 mm may be furnished for NPS ¾ (DN 20) 45° long-radius elbows.

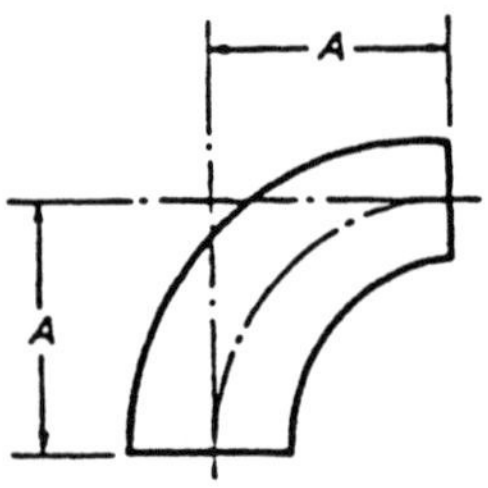

TABLE 5.6 Dimensions of 90° Long-Radius Reducing Elbows (U.S. Customary Units)*

Nominal pipe size (NPS)	Outside diameter at bevel		Center-to-end, *A*	Nominal pipe size (NPS)	Outside diameter at bevel		Center-to-end, *A*
	Large end	Small end			Large end	Small end	
2 × 1½	2.38	1.90	3.00	10 × 8	10.75	8.62	15.00
2 × 1¼	2.38	1.66	3.00	10 × 6	10.75	6.62	15.00
2 × 1	2.38	1.32	3.00	10 × 5	10.75	5.56	15.00
2½ × 2	2.88	2.38	3.75	12 × 10	12.75	10.75	18.00
2½ × 1½	2.88	1.90	3.75	12 × 8	12.75	8.62	18.00
2½ × 1¼	2.88	1.66	3.75	12 × 6	12.75	6.62	18.00
3 × 2½	3.50	2.88	4.50	14 × 12	14.00	12.75	21.00
3 × 2	3.50	2.38	4.50	14 × 10	14.00	10.75	21.00
3 × 1½	3.50	1.90	4.50	14 × 8	14.00	8.62	21.00
3½ × 3	4.00	3.50	5.25	16 × 14	16.00	14.00	24.00
3½ × 2½	4.00	2.88	5.25	16 × 12	16.00	12.75	24.00
3½ × 2	4.00	2.38	5.25	16 × 10	16.00	10.75	24.00
4 × 3½	4.50	4.00	6.00	18 × 16	18.00	16.00	27.00
4 × 3	4.50	3.50	6.00	18 × 14	18.00	14.00	27.00
4 × 2½	4.50	2.88	6.00	18 × 12	18.00	12.75	27.00
4 × 2	4.50	2.38	6.00	18 × 10	18.00	10.75	27.00
5 × 4	5.56	4.50	7.50	20 × 18	20.00	18.00	30.00
5 × 3½	5.56	4.00	7.50	20 × 16	20.00	16.00	30.00
5 × 3	5.56	3.50	7.50	20 × 14	20.00	14.00	30.00
5 × 2½	5.56	2.88	7.50	20 × 12	20.00	12.75	30.00
6 × 5	6.62	5.56	9.00	20 × 10	20.00	10.75	30.00
6 × 4	6.62	4.50	9.00	24 × 22	24.00	22.00	36.00
6 × 3½	6.62	4.00	9.00	24 × 20	24.00	20.00	36.00
6 × 3	6.62	3.50	9.00	24 × 18	24.00	18.00	36.00
8 × 6	8.62	6.62	12.00	24 × 16	24.00	16.00	36.00
8 × 5	8.62	5.56	12.00	24 × 14	24.00	14.00	36.00
8 × 4	8.62	4.50	12.00	24 × 12	24.00	12.75	36.00

* Dimensions in inches. For tolerances, refer to Table 5.15.

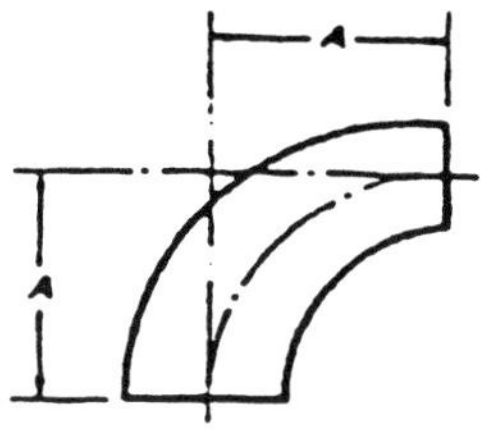

TABLE 5.6M Dimensions of 90° Long-Radius Reducing Elbows (Metric Units)*

NPS	DN	Outside diameter at bevel: Large end, mm	Outside diameter at bevel: Small end, mm	Center-to-end, *A*, mm	NPS	DN	Outside diameter at bevel: Large end, mm	Outside diameter at bevel: Small end, mm	Center-to-end, *A*, mm
2 × 1½	50 × 40	60.3	48.3	76	10 × 8	250 × 200	273.0	219.1	381
2 × 1¼	50 × 32	60.3	42.2	76	10 × 6	250 × 150	273.0	168.3	381
2 × 1	50 × 25	60.3	33.4	76	10 × 5	250 × 125	273.0	141.3	381
2½ × 2	65 × 50	73.0	60.3	95	12 × 10	300 × 250	323.8	273.0	457
2½ × 1½	65 × 40	73.0	48.3	95	12 × 8	300 × 200	323.8	219.1	457
2½ × 1¼	65 × 32	73.0	42.2	95	12 × 6	300 × 150	323.8	168.3	457
3 × 2½	80 × 65	88.9	73.0	114	14 × 12	350 × 300	355.6	323.8	533
3 × 2	80 × 50	88.9	60.3	114	14 × 10	350 × 250	355.6	273.0	533
3 × 1½	80 × 40	88.9	48.3	114	14 × 8	350 × 200	355.6	219.1	533
3½ × 3	90 × 80	101.6	88.9	133	16 × 14	400 × 350	406.4	355.6	610
3½ × 2½	90 × 65	101.6	73.0	133	16 × 12	400 × 300	406.4	323.8	610
3½ × 2	90 × 50	101.6	60.3	133	16 × 10	400 × 250	406.4	273.0	610
4 × 3½	100 × 90	114.3	101.6	152	13 × 16	450 × 400	457	406.4	686
4 × 3	100 × 80	114.3	88.9	152	13 × 14	450 × 350	457	355.6	686
4 × 2½	100 × 65	114.3	73.0	152	13 × 12	450 × 300	457	323.8	686
4 × 2	100 × 50	114.3	60.3	152	13 × 10	450 × 250	457	273.0	686
5 × 4	125 × 100	141.3	114.3	190	20 × 18	500 × 450	508	457	762
5 × 3½	125 × 90	141.3	101.6	190	20 × 16	500 × 400	508	406.4	762
5 × 3	125 × 80	141.3	88.9	190	20 × 14	500 × 350	508	355.6	762
5 × 2½	125 × 65	141.3	73.0	190	20 × 12	500 × 300	508	323.8	762
6 × 5	150 × 125	168.3	141.3	229	20 × 10	500 × 250	610	273.0	762
6 × 4	150 × 100	168.3	114.3	229	24 × 22	600 × 550	610	559	914
6 × 3½	150 × 90	168.3	101.6	229	24 × 20	600 × 500	610	508	914
6 × 3	150 × 80	168.3	88.9	229	24 × 18	600 × 450	610	457	914
8 × 6	200 × 150	219.1	168.3	305	24 × 16	600 × 400	610	406.4	914
8 × 5	200 × 125	219.1	141.3	305	24 × 14	600 × 350	610	355.6	914
8 × 4	200 × 100	219.1	114.3	305	24 × 12	600 × 300	610	323.8	914

* For tolerances, refer to Table 5.15.

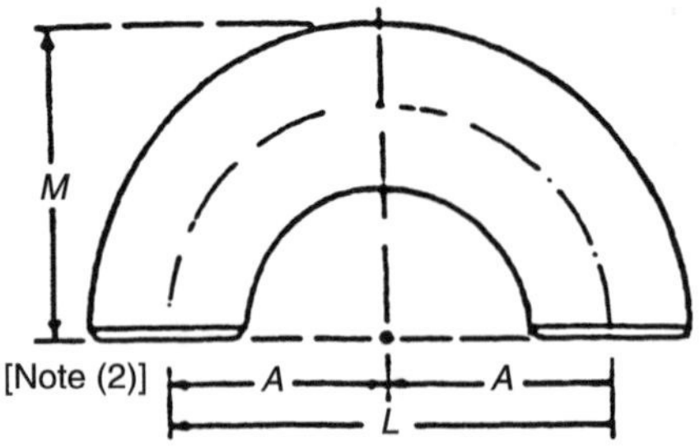

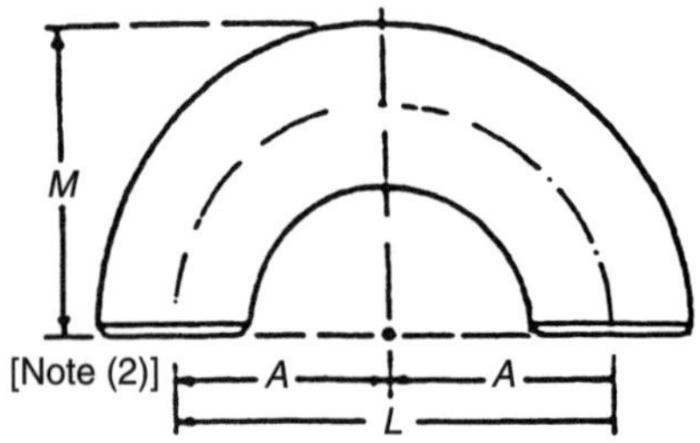

TABLE 5.7 Dimensions of 180° Long-Radius Returns (U.S. Customary Units)*

Nominal pipe size (NPS)	Center-to-center,[†] *L*	Back-to-face, *M*
½	3.00	1.88
¾	3.00[‡]	2.00[‡]
1	3.00	2.19
1¼	3.75	2.75
1½	4.50	3.25
2	6.00	4.19
2½	7.50	5.19
3	9.00	6.25
3½	10.50	7.25
4	12.00	8.25
5	15.00	10.31
6	18.00	12.31
8	24.00	16.31
10	30.00	20.38
12	36.00	24.38
14	42.00	28.00
16	48.00	32.00
18	54.00	36.00
20	60.00	40.00
22	66.00	44.00
24	72.00	48.00

* Dimensions in inches. For tolerances, refer to Table 5.16.
[†] Dimension *A* is equal to one-half of dimension *L*.
[‡] *L* and *M* dimensions of 2.25 in and 1.69 in, respectively, may be furnished for NPS ²⁄₄ at the manufacturer's option.

TABLE 5.7M Dimensions of 180° Long-Radius Returns (Metric Units)*

NPS	DN	Center-to-center, *L*,[†] mm	Back-to-face *M*,[†] mm
½	15	76	48
¾	20[‡]	76	51[‡]
1	25	76	56
1¼	32	95	70
1½	40	114	83
2	50	152	106
2½	65	190	132
3	80	229	159
3½	90	267	184
4	100	305	210
5	125	381	262
6	150	457	313
8	200	610	414
10	250	762	518
12	300	914	619
14	350	1067	711
16	400	1219	813
18	450	1372	914
20	500	1524	1016
22	550	1676	1118
24	600	1829	1219

* Dimensions in millimeters. For tolerances, refer to Table 5.16.
[†] Dimension *A* is equal to one-half of dimension *L*.
[‡] *L* and *M* dimensions of 57 and 43 mm, respectively, may be furnished for NPS ¾ (DN 20) at the manufacturer's option

DIMENSIONS OF SHORT-RADIUS BUTT WELDING FITTINGS

Dimensions of Short-Radius 90° Elbows

See Tables 5.8 and 5.8M.

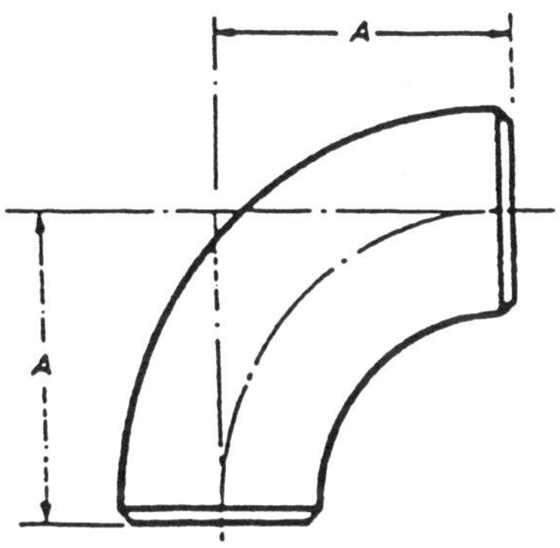

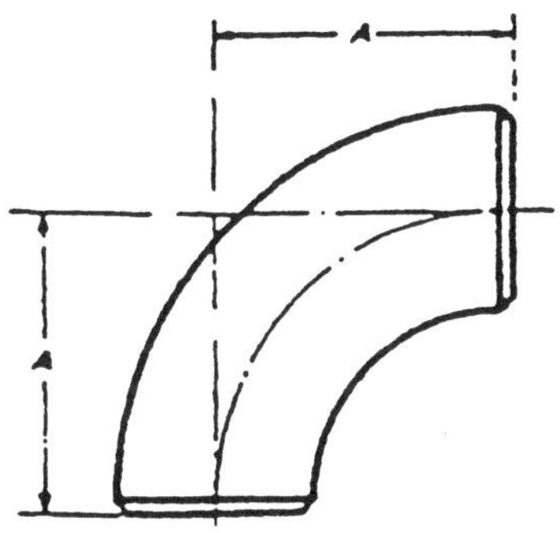

TABLE 5.8 Dimensions of 90° Short-Radius Elbows (U.S. Customary Units)*

Nominal pipe size	Outside diameter at bevel	Center-to-end, A
1	1.32	1.00
$1\frac{1}{4}$	1.66	1.25
$1\frac{1}{2}$	1.90	1.50
2	2.38	2.00
$2\frac{1}{2}$	2.88	2.50
3	3.50	3.00
$3\frac{1}{2}$	4.00	3.50
4	4.50	4.00
5	5.56	5.00
6	6.62	6.00
8	8.62	8.00
10	10.75	10.00
12	12.75	12.00
14	14.00	14.00
16	16.00	16.00
18	18.00	18.00
20	20.00	20.00
22	22.00	22.00
24	24.00	24.00

* Dimensions in inches. For tolerances, refer to Table 5.15.

TABLE 5.8M Dimensions of 90° Short-Radius Elbows (Metric Units)*

NPS	DN	Outside diameter at bevel	Center-to-end, A
1	25	33.4	25
$1\frac{1}{4}$	32	42.2	32
$1\frac{1}{2}$	40	48.3	38
2	50	60.3	51
$2\frac{1}{2}$	65	73.0	64
3	80	88.9	76
$3\frac{1}{2}$	90	101.6	89
4	100	114.3	102
5	125	141.3	127
6	150	168.3	152
8	200	219.1	203
10	250	273.0	254
12	300	323.8	305
14	350	355.6	356
16	400	406.4	406
18	450	457	457
20	500	508	508
22	550	559	559
24	600	610	610

* Dimensions in millimeters. For tolerances, refer to Table 5.15.

Dimensions of Short-Radius 180° Returns

See Tables 5.9 and 5.9M.

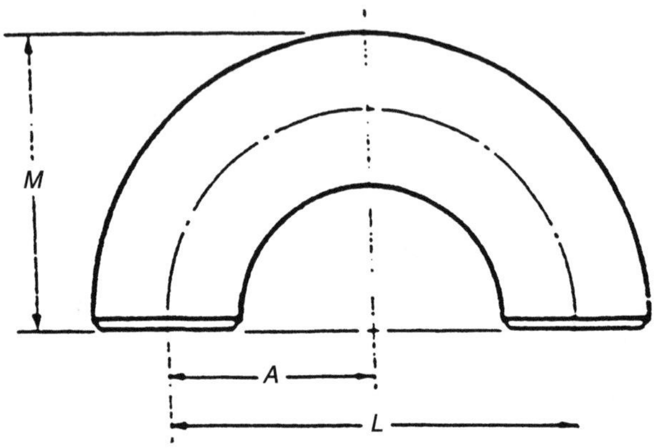

TABLE 5.9 Dimensions of Short-Radius 180° Returns (U.S. Customary Units)*

Nominal pipe size	Center-to-center, $L^{\dagger}$	Back-to-face, M
1	2.00	1.62
$1\frac{1}{4}$	2.50	2.06
$1\frac{1}{2}$	3.00	2.44
2	4.00	3.19
$2\frac{1}{2}$	5.00	3.94
3	6.00	4.75
$3\frac{1}{2}$	7.00	5.50
4	8.00	6.25
5	10.00	7.75
6	12.00	9.31
8	16.00	12.31
10	20.00	15.38
12	24.00	18.38
14	28.00	21.00
16	32.00	24.00
18	36.00	27.00
20	40.00	30.00
22	44.00	33.00
24	48.00	36.00

* Dimensions in inches. For tolerances, refer to Table 6.16.

† Dimension A is equal to one-half of dimension L.

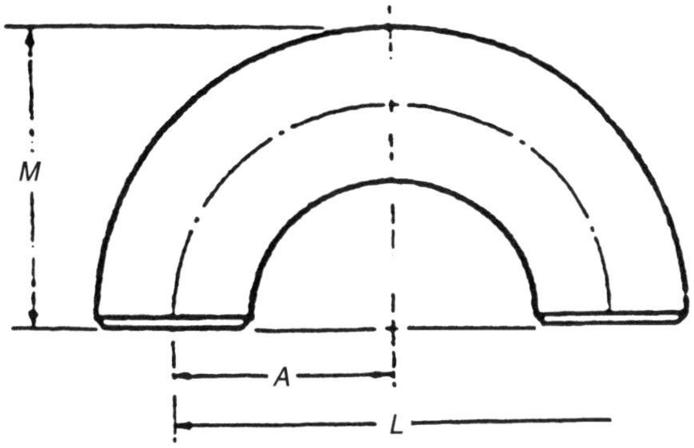

TABLE 5.9M Dimensions of Short-Radius 180° Returns (Metric Units)*

NPS	DN	Outside diameter at bevel	Center-to-center, $L^{\dagger}$	Back-to-face, M
1	25	33.4	51	41
1¼	32	42.2	64	52
1½	40	48.3	76	62
2	50	60.3	102	81
2½	65	73.0	127	100
3	80	88.9	152	121
3½	90	101.6	178	140
4	100	114.3	203	159
5	125	141.3	254	197
6	150	168.3	305	237
8	200	219.1	406	313
10	250	273.0	508	391
12	300	323.8	610	467
14	350	355.6	711	533
16	400	406.4	813	610
18	450	457	914	686
20	500	508	1016	762
22	550	559	1118	838
24	600	610	1219	914

* Dimensions in millimeters. For tolerances, refer to Table 5.16.
† Dimension A is equal to one-half of dimension L.

Dimensions of Butt Welding Straight Tees and Crosses

See Tables 5.10 and 5.10M.

Dimensions of Butt Welding Reducing Outlet Tees and Reducing Outlet Crosses

See Tables 5.11 and 5.11M.

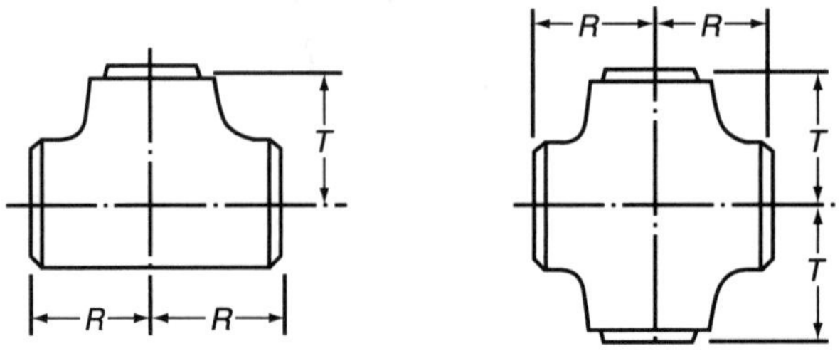

TABLE 5.10 Dimensions of Straight Tees and Crosses (U.S. Customary Units)*

Nominal pipe size (NPS)	Outside diameter at bevel	Center-to-end Run, R	Center-to-end Outlet, T
½	0.84	1.00	1.00
¾	1.05	1.12	1.12
1	1.32	1.50	1.50
1¼	1.66	1.88	1.88
1½	1.90	2.25	2.25
2	2.38	2.50	2.50
2½	2.88	3.00	3.00
3	3.50	3.38	3.38
3½	4.00	3.75	3.75
4	4.50	4.12	4.12
5	5.56	4.88	4.88
6	6.62	5.62	5.62
8	8.62	7.00	7.00
10	10.75	8.50	8.50
12	12.75	10.00	10.00
14	14.00	11.00	11.00
16	16.00	12.00	12.00
18	18.00	13.50	13.50
20	20.00	15.00	15.00
22	22.00	16.50	16.50
24	24.00	17.00	17.00
26	26.00	19.50	19.50
28	28.00	20.50	20.50
30	30.00	22.00	22.00
32	32.00	23.50	23.50
34	34.00	25.00	25.00
36	36.00	26.50	26.50
38	38.00	28.00	28.00
40	40.00	29.50	29.50
42	42.00	30.00	28.00
44	44.00	32.00	30.00
46	46.00	33.50	31.50
48	48.00	35.00	33.00

* Dimensions in inches.

Notes:

1. Dimension T for NPS 26 and larger is recommended but not required.
2. Dimensions apply to crosses NPS 24 and smaller.
3. Refer to Table 5.15 for tolerances.

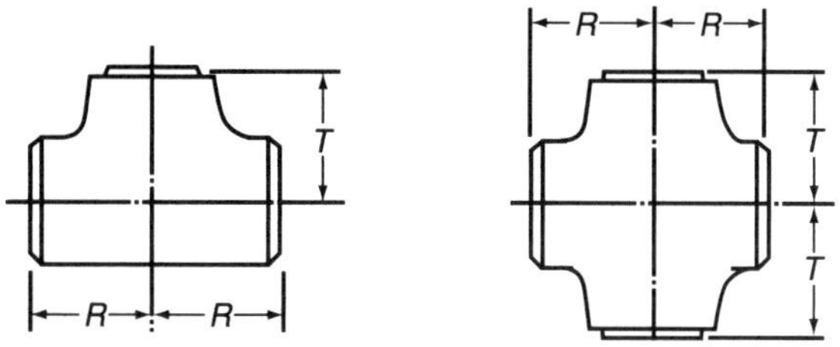

TABLE 5.10M Dimensions of Straight Tees and Crosses (Metric Units)*

NPS	DN	Outside diameter at bevel	Center-to-end	
			Run, *R*	Outlet, *T*
½	15	21.3	25	25
¾	20	26.7	29	29
1	25	33.4	38	38
1¼	32	42.2	48	48
1½	40	48.3	57	57
2	50	60.3	64	64
2½	65	73.0	76	76
3	80	88.9	86	86
3½	90	101.6	95	95
4	100	114.3	105	105
5	125	141.3	124	124
6	150	168.3	143	143
8	200	219.1	178	178
10	250	273.0	216	216
12	300	323.8	254	254
14	350	355.6	279	279
16	400	406.4	305	305
18	450	457	343	343
20	500	508	381	381
22	550	559	419	419
24	600	610	432	432
26	650	660	495	495
28	700	711	521	521
30	750	762	559	559
32	800	813	597	597
34	850	864	635	635
36	900	914	673	673
38	950	965	711	711
40	1000	1016	749	749
42	1050	1067	762	711
44	1100	1118	813	762
46	1150	1168	851	800
48	1200	1219	889	838

* Dimensions in millimeters. Dimensions per ASME B16.9.

Notes:

1. Dimension *T* for NPS 26 (DN 650) and larger is recommended but not required.
2. Dimension applicable to crosses NPS 24 (DN 600) and smaller.
3. For tolerances, refer to Table 5.15.

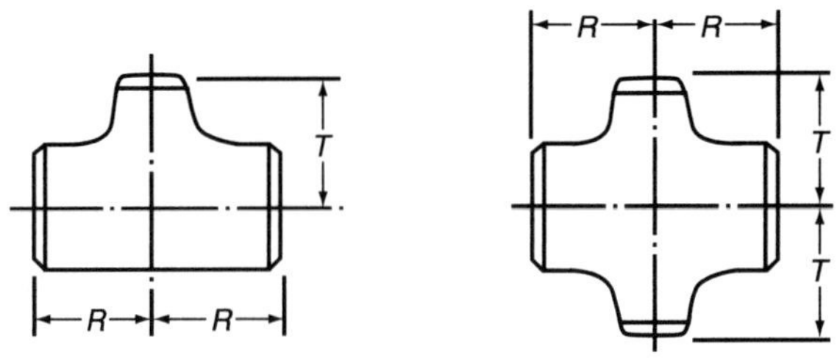

TABLE 5.11 Dimensions of Reducing-Outlet Tees and Reducing-Outlet Crosses (U.S. Customary Units)*

Nominal pipe size (NPS)	Outside diameter at bevel		Center-to-end	
	Run, in	Outlet, in	Run *R*, in	Outlet *T*, in†
½ × ½ × ⅜	0.84	0.68	1.00	1.00
½ × ½ × ¼	0.84	0.54	1.00	1.00
¾ × ¾ × ½	1.05	0.84	1.12	1.12
¾ × ¾ × ⅜	1.05	0.68	1.12	1.12
1 × 1 × ¾	1.32	1.05	1.50	1.50
1 × 1 × ½	1.32	0.84	1.50	1.50
1¼ × 1¼ × 1	1.66	1.32	1.88	1.88
1¼ × 1¼ × ¾	1.66	1.05	1.88	1.88
1¼ × 1¼ × ½	1.66	0.84	1.88	1.88
1½ × 1½ × 1¼	1.90	1.66	2.25	2.25
1½ × 1½ × 1	1.90	1.32	2.25	2.25
1½ × 1½ × ¾	1.90	1.05	2.25	2.25
1½ × 1½ × ½	1.90	0.84	2.25	2.25
2 × 2 × 1½	2.38	1.90	2.50	2.38
2 × 2 × 1¼	2.38	1.66	2.50	2.25
2 × 2 × 1	2.38	1.32	2.50	2.00
2 × 2 × ¾	2.38	1.05	2.50	1.75
2½ × 2½ × 2	2.88	2.38	3.00	2.75
2½ × 2½ × 1½	2.88	1.90	3.00	2.62
2½ × 2½ × 1¼	2.88	1.66	3.00	2.50
2½ × 2½ × 1	2.88	1.32	3.00	2.25
3 × 3 × 2½	3.50	2.88	3.38	3.25
3 × 3 × 2	3.50	2.38	3.38	3.00
3 × 3 × 1½	3.50	1.90	3.38	2.88
3 × 3 × 1¼	3.50	1.66	3.38	2.75
3½ × 3½ × 3	4.00	3.50	3.75	3.62
3½ × 3½ × 2½	4.00	2.88	3.75	3.50
3½ × 3½ × 2	4.00	2.38	3.75	3.25
3½ × 3½ × 1½	4.00	1.90	3.75	3.12
4 × 4 × 3½	4.50	4.00	4.12	4.00
4 × 4 × 3	4.50	3.50	4.12	3.88
4 × 4 × 2½	4.50	2.88	4.12	3.75
4 × 4 × 2	4.50	2.38	4.12	3.50
4 × 4 × 1½	4.50	1.90	4.12	3.38
5 × 5 × 4	5.56	4.50	4.88	4.62
5 × 5 × 3½	5.56	4.00	4.88	4.50
5 × 5 × 3	5.56	3.50	4.88	4.38
5 × 5 × 2½	5.56	2.88	4.88	4.25
5 × 5 × 2	5.56	2.38	4.88	4.12
6 × 6 × 5	6.62	5.56	5.62	5.38
6 × 6 × 4	6.62	4.50	5.62	5.12
6 × 6 × 3½	6.62	4.00	5.62	5.00
6 × 6 × 3	6.62	3.50	5.62	4.88
6 × 6 × 2½	6.62	2.88	5.62	4.75
8 × 8 × 6	8.62	6.62	7.00	6.62
8 × 8 × 5	8.62	5.56	7.00	6.38
8 × 8 × 4	8.62	4.50	7.00	6.12
8 × 8 × 3½	8.62	4.00	7.00	6.00
10 × 10 × 8	10.75	8.62	8.50	8.00
10 × 10 × 6	10.75	6.62	8.50	7.62
10 × 10 × 5	10.75	5.56	8.50	7.50
10 × 10 × 4	10.75	4.50	8.50	7.25
12 × 12 × 10	12.75	10.75	10.00	9.50
12 × 12 × 8	12.75	8.62	10.00	9.00
12 × 12 × 6	12.75	6.62	10.00	8.62
12 × 12 × 5	12.75	5.56	10.00	8.50
14 × 14 × 12	14.00	12.75	11.00	10.62
14 × 14 × 10	14.00	10.75	11.00	10.12
14 × 14 × 8	14.00	8.62	11.00	9.75
14 × 14 × 6	14.00	6.62	11.00	9.38
16 × 16 × 14	16.00	14.00	12.00	12.00
16 × 16 × 12	16.00	12.75	12.00	11.62
16 × 16 × 10	16.00	10.75	12.00	11.12
16 × 16 × 8	16.00	8.62	12.00	10.75
16 × 16 × 6	16.00	6.62	12.00	10.38
18 × 18 × 16	18.00	16.00	13.50	13.00
18 × 18 × 14	18.00	14.00	13.50	13.00
18 × 18 × 12	18.00	12.75	13.50	12.62
18 × 18 × 10	18.00	10.75	13.50	12.12
18 × 18 × 8	18.00	8.62	13.50	11.75
20 × 20 × 18	20.00	18.00	15.00	14.50
20 × 20 × 16	20.00	16.00	15.00	14.00
20 × 20 × 14	20.00	14.00	15.00	14.00
20 × 20 × 12	20.00	12.75	15.00	13.62
20 × 20 × 10	20.00	10.75	15.00	13.12
20 × 20 × 8	20.00	8.62	15.00	12.75
22 × 22 × 20	22.00	20.00	16.50	16.00
22 × 22 × 18	22.00	18.00	16.50	15.50
22 × 22 × 16	22.00	16.00	16.50	15.00
22 × 22 × 14	22.00	14.00	16.50	15.00
22 × 22 × 12	22.00	12.75	16.50	14.62
22 × 22 × 10	22.00	10.75	16.50	14.12
24 × 24 × 22	24.00	22.00	17.00	17.00
24 × 24 × 20	24.00	20.00	17.00	17.00
24 × 24 × 18	24.00	18.00	17.00	16.50
24 × 24 × 16	24.00	16.00	17.00	16.00
24 × 24 × 14	24.00	14.00	17.00	16.00
24 × 24 × 12	24.00	12.75	17.00	15.62
24 × 24 × 10	24.00	10.75	17.00	15.12
26 × 26 × 24	26.00	24.00	19.50	19.00
26 × 26 × 22	26.00	22.00	19.50	18.50
26 × 26 × 20	26.00	20.00	19.50	18.00
26 × 26 × 18	26.00	18.00	19.50	17.50
26 × 26 × 16	26.00	16.00	19.50	17.00
26 × 26 × 14	26.00	14.00	19.50	17.00
26 × 26 × 12	26.00	12.75	19.50	16.62
28 × 28 × 26	28.00	26.00	20.50	20.50
28 × 28 × 24	28.00	24.00	20.50	20.00
28 × 28 × 22	28.00	22.00	20.50	19.50
28 × 28 × 20	28.00	20.00	20.50	19.00
28 × 28 × 18	28.00	18.00	20.50	18.50
28 × 28 × 16	28.00	16.00	20.50	18.00
28 × 28 × 14	28.00	14.00	20.50	18.00
28 × 28 × 12	28.00	12.75	20.50	17.62
30 × 30 × 28	30.00	28.00	22.00	21.50

TABLE 5.11 Dimensions of Reducing-Outlet Tees and Reducing-Outlet Crosses (U.S. Customary Units)* (Continued)

Nominal pipe size (NPS)	Outside Diameter at Bevel		Center-to-end	
	Run, in	Outlet, in	Run R, in	Outlet T, in
30 × 30 × 26	30.00	26.00	22.00	21.50
30 × 30 × 24	30.00	24.00	22.00	21.00
30 × 30 × 22	30.00	22.00	22.00	20.50
30 × 30 × 20	30.00	20.00	22.00	20.00
30 × 30 × 18	30.00	18.00	22.00	19.50
30 × 30 × 16	30.00	16.00	22.00	19.00
30 × 30 × 14	30.00	14.00	22.00	19.00
30 × 30 × 12	30.00	12.75	22.00	18.62
30 × 30 × 10	30.00	10.75	22.00	18.12
32 × 32 × 30	32.00	30.00	23.50	23.00
32 × 32 × 28	32.00	28.00	23.50	22.50
32 × 32 × 26	32.00	26.00	23.50	22.50
32 × 32 × 24	32.00	24.00	23.50	22.00
32 × 32 × 22	32.00	22.00	23.50	21.50
32 × 32 × 20	32.00	20.00	23.50	21.00
32 × 32 × 18	32.00	18.00	23.50	20.50
32 × 32 × 16	32.00	16.00	23.50	20.00
32 × 32 × 14	32.00	14.00	23.50	20.00
34 × 34 × 32	34.00	32.00	25.00	24.50
34 × 34 × 30	34.00	30.00	25.00	24.00
34 × 34 × 28	34.00	28.00	25.00	23.50
34 × 34 × 26	34.00	26.00	25.00	23.50
34 × 34 × 24	34.00	24.00	25.00	23.00
34 × 34 × 22	34.00	22.00	25.00	22.50
34 × 34 × 20	34.00	20.00	25.00	22.00
34 × 34 × 18	34.00	18.00	25.00	21.50
34 × 34 × 16	34.00	16.00	25.00	21.00
36 × 36 × 34	36.00	34.00	26.50	26.00
36 × 36 × 32	36.00	32.00	26.50	25.50
36 × 36 × 30	36.00	30.00	26.50	25.00
36 × 36 × 28	36.00	28.00	26.50	24.50
36 × 36 × 26	36.00	26.00	26.50	24.50
36 × 36 × 24	36.00	24.00	26.50	24.00
36 × 36 × 22	36.00	22.00	26.50	23.50
36 × 36 × 20	36.00	20.00	26.50	23.00
36 × 36 × 18	36.00	18.00	26.50	22.50
36 × 36 × 16	36.00	16.00	26.50	22.00
38 × 38 × 36	38.00	36.00	28.00	28.00
38 × 38 × 34	38.00	34.00	28.00	27.50
38 × 38 × 32	38.00	32.00	28.00	27.00
38 × 38 × 30	38.00	30.00	28.00	26.50
38 × 38 × 28	38.00	28.00	28.00	25.50
38 × 38 × 26	38.00	26.00	28.00	25.50
38 × 38 × 24	38.00	24.00	28.00	25.00
38 × 38 × 22	38.00	22.00	28.00	24.50
38 × 38 × 20	38.00	20.00	28.00	24.00
38 × 38 × 18	38.00	18.00	28.00	23.50
40 × 40 × 38	40.00	38.00	29.50	29.50
40 × 40 × 36	40.00	36.00	29.50	29.00
40 × 40 × 34	40.00	34.00	29.50	28.50
40 × 40 × 32	40.00	32.00	29.50	28.00
40 × 40 × 30	40.00	30.00	29.50	27.50
40 × 40 × 28	40.00	28.00	29.50	26.50
40 × 40 × 26	40.00	26.00	29.50	26.50
40 × 40 × 24	40.00	24.00	29.50	26.00

Nominal pipe size (NPS)	Outside diameter at bevel		Center-to-end	
	Run, in	Outlet, in	Run R, in	Outlet T, in
40 × 40 × 22	40.00	22.00	29.50	25.50
40 × 40 × 20	40.00	20.00	29.50	25.00
40 × 40 × 18	40.00	18.00	29.50	24.50
42 × 42 × 40	42.00	40.00	30.00	28.00
42 × 42 × 38	42.00	38.00	30.00	28.00
42 × 42 × 36	42.00	36.00	30.00	28.00
42 × 42 × 34	42.00	34.00	30.00	28.00
42 × 42 × 32	42.00	32.00	30.00	28.00
42 × 42 × 30	42.00	30.00	30.00	28.00
42 × 42 × 28	42.00	28.00	30.00	27.50
42 × 42 × 26	42.00	26.00	30.00	27.50
42 × 42 × 24	42.00	24.00	30.00	26.00
42 × 42 × 22	42.00	22.00	30.00	26.00
42 × 42 × 20	42.00	20.00	30.00	26.00
42 × 42 × 18	42.00	18.00	30.00	25.50
42 × 42 × 16	42.00	16.00	30.00	25.00
44 × 44 × 42	44.00	42.00	32.00	30.00
44 × 44 × 40	44.00	40.00	32.00	29.50
44 × 44 × 38	44.00	38.00	32.00	29.00
44 × 44 × 36	44.00	36.00	32.00	28.50
44 × 44 × 34	44.00	34.00	32.00	28.50
44 × 44 × 32	44.00	32.00	32.00	28.00
44 × 44 × 30	44.00	30.00	32.00	28.00
44 × 44 × 28	44.00	28.00	32.00	27.50
44 × 44 × 26	44.00	26.00	32.00	27.50
44 × 44 × 24	44.00	24.00	32.00	27.50
44 × 44 × 22	44.00	22.00	32.00	27.00
44 × 44 × 20	44.00	20.00	32.00	27.00
46 × 46 × 44	46.00	44.00	33.50	31.50
46 × 46 × 42	46.00	42.00	33.50	31.00
46 × 46 × 40	46.00	40.00	33.50	30.50
46 × 46 × 38	46.00	38.00	33.50	30.00
46 × 46 × 36	46.00	36.00	33.50	30.00
46 × 46 × 34	46.00	34.00	33.50	29.50
46 × 46 × 32	46.00	32.00	33.50	29.50
46 × 46 × 30	46.00	30.00	33.50	29.00
46 × 46 × 28	46.00	28.00	33.50	29.00
46 × 46 × 26	46.00	26.00	33.50	29.00
46 × 46 × 24	46.00	24.00	33.50	28.50
46 × 46 × 22	46.00	22.00	33.50	28.50
48 × 48 × 46	48.00	46.00	35.00	33.00
48 × 48 × 44	48.00	44.00	35.00	33.00
48 × 48 × 42	48.00	42.00	35.00	32.00
48 × 48 × 40	48.00	40.00	35.00	32.00
48 × 48 × 38	48.00	38.00	35.00	32.00
48 × 48 × 36	48.00	36.00	35.00	31.00
48 × 48 × 34	48.00	34.00	35.00	31.00
48 × 48 × 32	48.00	32.00	35.00	31.00
48 × 48 × 30	48.00	30.00	35.00	30.00
48 × 48 × 28	48.00	28.00	35.00	30.00
48 × 48 × 26	48.00	26.00	35.00	30.00
48 × 48 × 24	48.00	24.00	35.00	29.00
48 × 48 × 22	48.00	22.00	35.00	29.00

* Dimensions in inches. For tolerances, refer to Table 5.15.

† Center-to-end outlet dimension T, listed for run sizes NPS 14 and larger, is a recommended dimension.

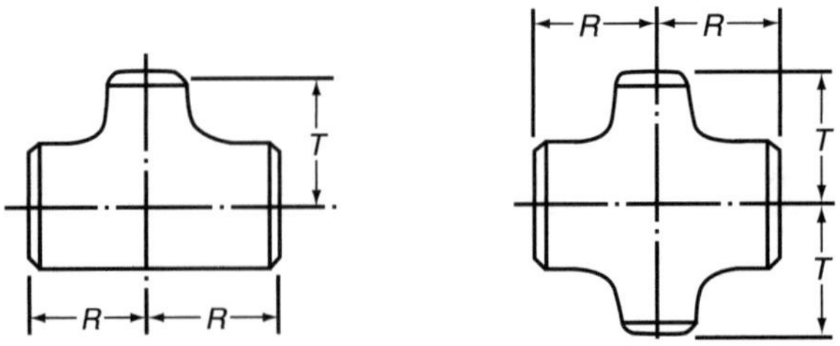

TABLE 5.11M Dimensions of Reducing-Outlet Tees and Reducing-Outlet Crosses (Metric Units)*

NPS	DN	Outside diameter at bevel		Center-to-end	
		Run, mm	Outlet, mm	Run *R*, mm	Outlet *T*, mm†
½ × ½ × ⅜	15 × 15 × 10	21.3	17.3	25	25
½ × ½ × ¼	15 × 15 × 8	21.3	13.7	25	25
¾ × ¾ × ½	20 × 20 × 15	26.7	21.3	29	29
¾ × ¾ × ⅜	20 × 20 × 10	26.7	17.3	29	29
1 × 1 × ¾	25 × 25 × 20	33.4	26.7	38	38
1 × 1 × ½	25 × 25 × 15	33.4	21.3	38	38
1¼ × 1¼ × 1	32 × 32 × 25	42.2	33.4	48	48
1¼ × 1¼ × ¾	32 × 32 × 20	42.2	26.7	48	48
1¼ × 1¼ × ½	32 × 32 × 15	42.2	21.3	48	48
1½ × 1½ × 1¼	40 × 40 × 32	48.3	42.2	57	57
1½ × 1½ × 1	40 × 40 × 25	48.3	33.4	57	57
1½ × 1½ × ¾	40 × 40 × 20	48.3	26.7	57	57
1½ × 1½ × ½	40 × 40 × 15	48.3	21.3	57	57
2 × 2 × 1½	50 × 50 × 40	60.3	48.3	64	60
2 × 2 × 1¼	50 × 50 × 32	60.3	42.2	64	57
2 × 2 × 1	50 × 50 × 25	60.3	33.4	64	51
2 × 2 × ¾	50 × 50 × 20	60.3	26.7	64	44
2½ × 2½ × 2	65 × 65 × 50	73.0	60.3	76	70
2½ × 2½ × 1½	65 × 65 × 40	73.0	48.3	76	67
2½ × 2½ × 1¼	65 × 65 × 32	73.0	42.2	76	64
2½ × 2½ × 1	65 × 65 × 25	73.0	33.4	76	57
3 × 3 × 2½	80 × 80 × 65	88.9	73.0	86	83
3 × 3 × 2	80 × 80 × 50	88.9	60.3	86	76
3 × 3 × 1½	80 × 80 × 40	88.9	48.3	86	73
3 × 3 × 1¼	80 × 80 × 32	88.9	42.2	86	70
3½ × 3½ × 3	90 × 90 × 80	101.6	88.9	95	92
3½ × 3½ × 2½	90 × 90 × 65	101.6	73.0	95	89
3½ × 3½ × 2	90 × 90 × 50	101.6	60.3	95	83
3½ × 3½ × 1½	90 × 90 × 40	101.6	48.3	95	79
4 × 4 × 3½	100 × 100 × 90	114.3	101.6	105	102
4 × 4 × 3	100 × 100 × 80	114.3	88.9	105	98
4 × 4 × 2½	100 × 100 × 65	113.3	73.0	105	95
4 × 4 × 2	100 × 100 × 50	114.3	60.3	105	89
4 × 4 × 1½	100 × 100 × 40	114.3	48.3	105	86
5 × 5 × 4	125 × 125 × 100	141.3	114.3	124	117
5 × 5 × 3½	125 × 125 × 90	141.3	101.6	124	114
5 × 5 × 3	125 × 125 × 80	141.3	88.9	124	111
5 × 5 × 2½	125 × 125 × 65	141.3	73.0	124	108
5 × 5 × 2	125 × 125 × 50	141.3	60.3	124	105
6 × 6 × 5	150 × 150 × 125	168.3	141.3	143	137
6 × 6 × 4	150 × 150 × 100	168.3	114.3	143	130
6 × 6 × 3½	150 × 150 × 90	168.3	101.6	143	127
6 × 6 × 3	150 × 150 × 80	168.3	88.9	143	124
6 × 6 × 2½	150 × 150 × 65	168.3	73.0	143	121
8 × 8 × 6	200 × 200 × 150	219.1	168.3	178	168
8 × 8 × 5	200 × 200 × 125	219.1	141.3	178	162
8 × 8 × 4	200 × 200 × 100	219.1	114.3	178	156
8 × 8 × 3½	200 × 200 × 90	219.1	101.6	178	152

TABLE 5.11M Dimensions of Reducing-Outlet Tees and Reducing-Outlet Crosses (Metric Units)* (Continued)

NPS	DN	Outside diameter at bevel		Center-to-end	
		Run, mm	Outlet, mm	Run *R*, mm	Outlet *T*, mm†
10 × 10 × 8	250 × 250 × 200	273.0	219.1	216	203
10 × 10 × 6	250 × 250 × 150	273.0	168.3	216	194
10 × 10 × 5	250 × 250 × 125	273.0	141.3	216	191
10 × 10 × 4	250 × 250 × 100	273.0	114.3	216	184
12 × 12 × 10	300 × 300 × 250	323.8	273.0	254	241
12 × 12 × 8	300 × 300 × 200	323.8	219.1	254	229
12 × 12 × 6	300 × 300 × 150	323.8	168.3	254	219
12 × 12 × 5	300 × 300 × 125	323.8	141.3	254	216
14 × 14 × 12	350 × 350 × 300	355.6	323.8	279	270
14 × 14 × 10	350 × 350 × 250	355.6	273.0	279	257
14 × 14 × 8	350 × 350 × 200	355.6	219.1	279	248
14 × 14 × 6	350 × 350 × 150	355.6	168.3	279	238
16 × 16 × 14	400 × 400 × 350	406.4	355.6	305	305
16 × 16 × 12	400 × 400 × 300	406.4	323.8	305	295
16 × 16 × 10	400 × 400 × 250	406.4	273.0	305	283
16 × 16 × 8	400 × 400 × 200	406.4	219.1	305	273
16 × 16 × 6	400 × 400 × 150	406.4	168.3	305	264
18 × 18 × 16	450 × 450 × 400	457	406.4	343	330
18 × 18 × 14	450 × 450 × 350	457	355.6	343	330
18 × 18 × 12	450 × 450 × 300	457	323.8	343	321
18 × 18 × 10	450 × 450 × 250	457	273.0	343	308
18 × 18 × 8	450 × 450 × 200	457	219.1	343	298
20 × 20 × 18	500 × 500 × 450	508	457	381	368
20 × 20 × 16	500 × 500 × 400	508	406.4	381	356
20 × 20 × 14	500 × 500 × 350	508	355.6	381	356
20 × 20 × 12	500 × 500 × 300	508	323.8	381	346
20 × 20 × 10	500 × 500 × 250	508	273.0	381	333
20 × 20 × 8	500 × 500 × 200	508	219.1	381	324
22 × 22 × 20	550 × 550 × 500	559	508	419	406
22 × 22 × 18	550 × 550 × 450	559	457	419	394
22 × 22 × 16	550 × 550 × 400	559	406.4	419	381
22 × 22 × 14	550 × 550 × 350	559	355.6	419	381
22 × 22 × 12	550 × 550 × 300	559	323.8	419	371
22 × 22 × 10	550 × 550 × 250	559	273.0	419	359
24 × 24 × 22	600 × 600 × 550	610	559	432	432
24 × 24 × 20	600 × 600 × 500	610	508	432	432
24 × 24 × 18	600 × 600 × 450	610	457	432	419
24 × 24 × 16	600 × 600 × 400	610	406.4	432	406
24 × 24 × 14	600 × 600 × 350	610	355.6	432	406
24 × 24 × 12	600 × 600 × 300	610	323.8	432	397
24 × 24 × 10	600 × 600 × 250	610	273.0	432	384
26 × 26 × 24	650 × 650 × 600	660	610	495	483
26 × 26 × 22	650 × 650 × 550	660	559	495	470
26 × 26 × 20	650 × 650 × 500	660	508	495	457
26 × 26 × 18	650 × 650 × 450	660	457	495	444
26 × 26 × 16	650 × 650 × 400	660	406.4	495	432
26 × 26 × 14	650 × 650 × 350	660	355.6	495	432
26 × 26 × 12	650 × 650 × 300	660	323.8	495	422
28 × 28 × 26	700 × 700 × 650	711	660	521	521
28 × 28 × 24	700 × 700 × 600	711	610	521	508
28 × 28 × 22	700 × 700 × 550	711	559	521	495
28 × 28 × 20	700 × 700 × 500	711	508	521	483
28 × 28 × 18	700 × 700 × 450	711	457	521	470
28 × 28 × 16	700 × 700 × 400	711	406.4	521	457
28 × 28 × 14	700 × 700 × 350	711	355.6	521	457
28 × 28 × 12	700 × 700 × 300	711	323.8	521	448

TABLE 5.11M Dimensions of Reducing-Outlet Tees and Reducing-Outlet Crosses (Metric Units)* (Continued)

NPS	DN	Outside diameter at bevel		Center-to-end	
		Run, mm	Outlet, mm	Run *R*, mm	Outlet *T*, mm[†]
30 × 30 × 28	750 × 750 × 700	762	711	559	546
30 × 30 × 26	750 × 750 × 650	762	660	559	546
30 × 30 × 24	750 × 750 × 600	762	610	559	533
30 × 30 × 22	750 × 750 × 550	762	559	559	521
30 × 30 × 20	750 × 750 × 500	762	508	559	508
30 × 30 × 18	750 × 750 × 450	762	457	559	495
30 × 30 × 16	750 × 750 × 400	762	406.4	559	483
30 × 30 × 14	750 × 750 × 350	762	355.6	559	483
30 × 30 × 12	750 × 750 × 300	762	323.8	559	473
30 × 30 × 10	750 × 750 × 250	762	273.0	559	460
32 × 32 × 30	800 × 800 × 750	813	762	597	584
32 × 32 × 28	800 × 800 × 700	813	711	597	572
32 × 32 × 26	800 × 800 × 650	813	660	597	572
32 × 32 × 24	800 × 800 × 600	813	610	597	559
32 × 32 × 22	800 × 800 × 550	813	559	597	546
32 × 32 × 20	800 × 800 × 500	813	508	597	533
32 × 32 × 18	800 × 800 × 450	813	457	597	521
32 × 32 × 16	800 × 800 × 400	813	406.4	597	508
32 × 32 × 14	800 × 800 × 350	813	355.6	597	508
34 × 34 × 32	850 × 850 × 800	864	813	635	622
34 × 34 × 30	850 × 850 × 750	864	762	635	610
34 × 34 × 28	850 × 850 × 700	864	711	635	597
34 × 34 × 26	850 × 850 × 650	864	660	635	597
34 × 34 × 24	850 × 850 × 600	864	610	635	584
34 × 34 × 22	850 × 850 × 550	864	559	635	572
34 × 34 × 20	850 × 850 × 500	864	508	635	559
34 × 34 × 18	850 × 850 × 450	864	457	635	546
34 × 34 × 16	850 × 850 × 400	864	406.4	635	533
36 × 36 × 34	900 × 900 × 850	914	864	673	660
36 × 36 × 32	900 × 900 × 800	914	813	673	648
36 × 36 × 30	900 × 900 × 750	914	762	673	635
36 × 36 × 28	900 × 900 × 700	914	711	673	622
36 × 36 × 26	900 × 900 × 650	914	660	673	622
36 × 36 × 24	900 × 900 × 600	914	610	673	610
36 × 36 × 22	900 × 900 × 550	914	559	673	597
36 × 36 × 20	900 × 900 × 500	914	508	673	584
36 × 36 × 18	900 × 900 × 450	914	457	673	572
36 × 36 × 16	900 × 900 × 400	914	406.4	673	559
38 × 38 × 36	950 × 950 × 900	965	914	711	711
38 × 38 × 34	950 × 950 × 850	965	864	711	698
38 × 38 × 32	950 × 950 × 800	965	813	711	686
38 × 38 × 30	950 × 950 × 750	965	762	711	673
38 × 38 × 28	950 × 950 × 700	965	711	711	648
38 × 38 × 26	950 × 950 × 650	965	660	711	648
38 × 38 × 24	950 × 950 × 600	965	610	711	635
38 × 38 × 22	950 × 950 × 550	965	559	711	622
38 × 38 × 20	950 × 950 × 500	965	508	711	610
38 × 38 × 18	950 × 950 × 450	965	457	711	597
40 × 40 × 38	1000 × 1000 × 950	1016	965	749	749
40 × 40 × 36	1000 × 1000 × 900	1016	914	749	737
40 × 40 × 34	1000 × 1000 × 850	1016	864	749	724
40 × 40 × 32	1000 × 1000 × 800	1016	813	749	711
40 × 40 × 30	1000 × 1000 × 750	1016	762	749	698
40 × 40 × 28	1000 × 1000 × 700	1016	711	749	673
40 × 40 × 26	1000 × 1000 × 650	1016	560	749	673

TABLE 5.11M Dimensions of Reducing-Outlet Tees and Reducing-Outlet Crosses (Metric Units)* (Continued)

NPS	DN	Outside diameter at bevel		Center-to-end	
		Run, mm	Outlet, mm	Run R, mm	Outlet T, mm†
40 × 40 × 24	1000 × 1000 × 600	1016	610	749	660
40 × 40 × 22	1000 × 1000 × 550	1016	559	749	648
40 × 40 × 20	1000 × 1000 × 500	1016	508	749	635
40 × 40 × 18	1000 × 1000 × 450	1016	457	749	622
42 × 42 × 40	1050 × 1050 × 1000	1067	1016	762	711
42 × 42 × 38	1050 × 1050 × 950	1067	968	762	711
42 × 42 × 36	1050 × 1050 × 900	1067	914	762	711
42 × 42 × 34	1050 × 1050 × 850	1067	864	762	711
42 × 42 × 32	1050 × 1050 × 800	1067	813	762	711
42 × 42 × 30	1050 × 1050 × 750	1067	762	762	711
42 × 42 × 28	1050 × 1050 × 700	1067	711	762	698
42 × 42 × 26	1050 × 1050 × 650	1067	660	762	698
42 × 42 × 24	1050 × 1050 × 600	1067	610	762	660
42 × 42 × 22	1050 × 1050 × 550	1067	559	762	660
42 × 42 × 20	1050 × 1050 × 500	1067	508	762	660
42 × 42 × 18	1050 × 1050 × 450	1067	457	762	648
42 × 42 × 16	1050 × 1050 × 400	1067	406.4	762	635
44 × 44 × 42	1100 × 1100 × 1050	1118	1067	813	762
44 × 44 × 40	1100 × 1100 × 1000	1118	1016	813	749
44 × 44 × 38	1100 × 1100 × 950	1118	965	813	737
44 × 44 × 36	1100 × 1100 × 900	1118	914	813	724
44 × 44 × 34	1100 × 1100 × 850	1118	864	813	724
44 × 44 × 32	1100 × 1100 × 800	1118	813	813	711
44 × 44 × 30	1100 × 1100 × 750	1118	762	813	711
44 × 44 × 28	1100 × 1100 × 700	1118	711	813	698
44 × 44 × 26	1100 × 1100 × 650	1118	660	813	698
44 × 44 × 24	1100 × 1100 × 600	1118	610	813	698
44 × 44 × 22	1100 × 1100 × 550	1118	559	813	686
44 × 44 × 20	1100 × 1100 × 500	1118	508	813	686
46 × 46 × 44	1150 × 1150 × 1100	1168	1118	851	800
46 × 46 × 42	1150 × 1150 × 1050	1168	1067	851	787
46 × 46 × 40	1150 × 1150 × 1000	1168	1016	851	775
46 × 46 × 38	1150 × 1150 × 950	1168	965	851	762
46 × 46 × 36	1150 × 1150 × 900	1168	914	851	762
46 × 46 × 34	1150 × 1150 × 850	1168	864	851	749
46 × 46 × 32	1150 × 1150 × 800	1168	813	851	749
46 × 46 × 30	1150 × 1150 × 750	1168	762	851	737
46 × 46 × 28	1150 × 1150 × 700	1168	711	851	737
46 × 46 × 26	1150 × 1150 × 650	1168	660	851	737
46 × 46 × 24	1150 × 1150 × 600	1168	610	851	724
46 × 46 × 22	1150 × 1150 × 550	1168	559	851	724
48 × 48 × 46	1200 × 1200 × 1150	1219	1168	889	838
48 × 48 × 44	1200 × 1200 × 1100	1219	1118	889	838
48 × 48 × 42	1200 × 1200 × 1050	1219	1067	889	813
48 × 48 × 40	1200 × 1200 × 1000	1219	1016	889	813
48 × 48 × 38	1200 × 1200 × 950	1219	965	889	813
48 × 48 × 36	1200 × 1200 × 900	1219	914	889	787
48 × 48 × 34	1200 × 1200 × 850	1219	864	889	787
48 × 48 × 32	1200 × 1200 × 800	1219	813	889	787
48 × 48 × 30	1200 × 1200 × 750	1219	762	889	762
48 × 48 × 28	1200 × 1200 × 700	1219	711	889	762
48 × 48 × 26	1200 × 1200 × 650	1219	660	889	762
48 × 48 × 24	1200 × 1200 × 600	1219	610	889	737
48 × 48 × 22	1200 × 1200 × 550	1219	559	889	737

* Dimensions in millimeters. For tolerances, refer to Table 5.15.

† Center-to-end outlet dimension T, listed for run sizes NPS 14 and larger, is a recommended dimension.

Dimensions of Butt Welding Lap Joint Stub Ends

See Tables 5.12 and 5.12M.

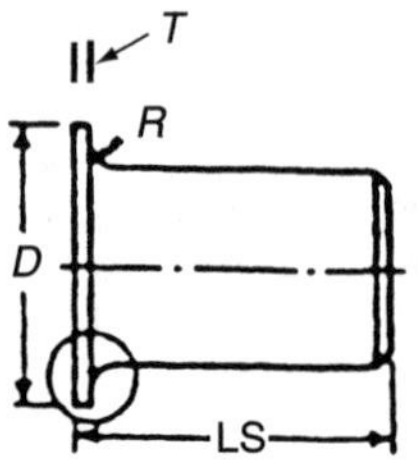

TABLE 5.12 Dimensions of Lap Joint Stub Ends (U.S. Customary Units)

Nominal pipe size (NPS)	Outside diameter of barrel		Long pattern length, LS	Short pattern length, LS	Diameter of lap, *D*
	Max.	Min.			
1/2	0.896	0.809	3.00	2.00	1.38
3/4	1.106	1.019	3.00	2.00	1.69
1	1.376	1.284	4.00	2.00	2.00
1 1/4	1.716	1.629	4.00	2.00	2.50
1 1/2	1.965	1.869	4.00	2.00	2.88
2	2.456	2.344	6.00	2.50	3.62
2 1/2	2.966	2.844	6.00	2.50	4.12
3	3.596	3.469	6.00	2.50	5.00
3 1/2	4.096	3.969	6.00	3.00	5.50
4	4.593	4.469	6.00	3.00	6.19
5	5.683	5.532	8.00	3.00	7.31
6	6.743	6.594	8.00	3.50	8.50
8	8.743	8.594	8.00	4.00	10.62
10	10.913	10.719	10.00	5.00	12.75
12	12.913	12.719	10.00	6.00	15.00
14	14.170	13.969	12.00	6.00	16.25
16	16.180	15.969	12.00	6.00	18.50
18	18.190	17.969	12.00	6.00	21.00
20	20.240	19.969	12.00	6.00	23.00
22	22.240	21.969	12.00	6.00	25.25
24	24.240	23.969	12.00	6.00	27.25

Notes:
1. Dimensions in inches. For tolerances, refer to Table 5.17.
2. *D* dimension conforms to standard machine facings shown in ASME B16.5.

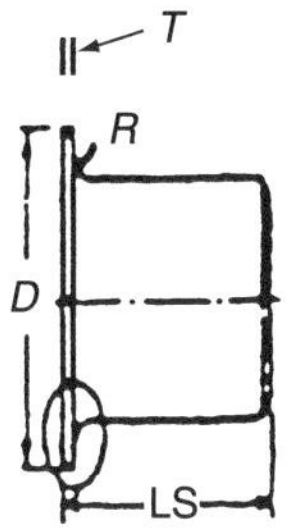

TABLE 5.12M Dimensions of Lap Joint Stub Ends (Metric Units)

NPS	DN	Outside diameter of barrel, mm		Long pattern length, LS	Short pattern length, LS	Diameter of lap, *D*
		Max.	Min.			
½	15	22.8	20.5	76	51	35
¾	20	28.1	25.9	76	51	43
1	25	35.0	32.6	102	51	51
1¼	32	43.6	41.4	102	51	64
1½	40	49.9	47.5	102	51	73
2	50	62.4	59.5	152	64	92
2½	65	75.3	72.2	152	64	106
3	80	91.3	88.1	152	64	127
3½	90	104.0	100.8	152	78	140
4	100	116.7	113.5	152	78	157
5	125	144.3	140.5	203	78	188
6	150	171.3	167.5	203	89	218
8	200	222.1	218.3	203	102	270
10	250	277.2	272.3	264	127	324
12	300	328.0	323.1	264	152	381
14	350	359.9	354.8	305	152	413
16	400	411.0	405.6	305	152	470
18	450	462	456	305	152	533
20	500	514	507	305	152	584
22	550	565	558	305	152	641
24	600	616	609	305	152	682

Notes:
1. Dimensions in millimeters. For tolerances, refer to Table 5.17.
2. *D* dimension conforms to standard machine facings shown in ASME B16.5.

Dimensions of Butt Welding Caps

See Tables 5.13 and 5.13M.

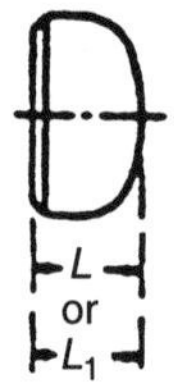

TABLE 5.13 Dimensions of Butt Welding Caps (U.S. Customary Units)*

Nominal pipe size (NPS)	Outside diameter at bevel, in	Length L[†]	Limiting wall thickness for length L	Length L_1[‡]
1/2	0.84	1.00	0.18	1.00
3/4	1.05	1.00	0.15	1.00
1	1.32	1.50	0.18	1.50
1 1/4	1.66	1.50	0.19	1.50
1 1/2	1.90	1.50	0.20	1.50
2	2.38	1.50	0.22	1.75
2 1/2	2.88	1.50	0.28	2.00
3	3.50	2.00	0.30	2.50
3 1/2	4.00	2.50	0.32	3.00
4	4.50	2.50	0.34	3.00
5	5.56	3.00	0.38	3.50
6	6.62	3.50	0.43	4.00
8	8.62	4.00	0.50	5.00
10	10.75	5.00	0.50	6.00
12	12.75	6.00	0.50	7.00
14	14.00	6.50	0.50	7.50
16	16.00	7.00	0.50	8.00
18	18.00	8.00	0.50	9.00
20	20.00	9.00	0.50	10.00
22	22.00	10.00	0.50	10.00
24	24.00	10.50	0.50	12.00
26	26.00	10.50	...	...
28	28.00	10.50	...	...
30	30.00	10.50	...	...
32	32.00	10.50	...	...
34	34.00	10.50	...	...
36	36.00	10.50	...	...
38	38.00	12.00	...	...
40	40.00	12.00	...	...
42	42.00	12.00	...	...
44	44.00	13.50	...	...
46	46.00	13.50	...	...
48	48.00	13.50	...	...

* Dimensions in inches. For tolerances, refer to Table 5.16.

† Length L applies for thickness not exceeding that given in column "Limiting wall thickness for length L."

‡ Length L_1 applies for thickness greater than that given in column "Limiting wall thickness" for NPS 24 (DN 600) and smaller. For NPS 26 (DN 650) and larger, length L_1 shall be per agreement between the purchaser and manufacturer.

TABLE 5.13M Dimensions of Butt Welding Caps (Metric Units)

NPS	DN	Outside diameter at bevel	Length L	Limiting wall thickness for length L	Length L_1
1/2	15	21.3	25	4.57	25
3/4	20	26.7	25	3.81	25
1	25	33.4	38	4.57	38
1 1/4	32	42.2	38	4.83	38
1 1/2	40	48.3	38	5.08	38
2	50	60.3	38	5.59	44
2 1/2	65	73.0	38	7.11	51
3	80	88.9	51	7.62	64
3 1/2	90	101.6	64	8.13	76
4	100	114.3	64	8.64	76
5	125	141.3	76		89
6	150	168.3	89	9.65	102
8	200	219.1	102	10.92	127
10	250	273.0	127	12.70	152
12	300	323.8	152	12.70	178
14	350	355.6	165	12.70	191
16	400	408.4	178	12.70	203
18	450	457	203	12.70	229
20	500	508	229	12.70	254
22	550	559	254	12.70	254
24	600	610	267	12.70	305
26	650	680	267		
28	700	711	267		
30	750	762	267		
32	800	813	267		
34	850	864	267		
36	900	914	267		
38	950	985	305		
40	1000	1016	305		
42	1050	1067	305		
44	1100	1118	343		
46	1150	1168	343		
48	1200	1219	343		

1. Dimensions per ASME B16.9.
2. Dimensions in millimeters.
3. For tolerances, refer to Table 5.16.
4. Length L applies for thickness not exceeding that given in column "Limiting wall thickness for length L."
5. Length L_1 applies for thickness greater than that given in column "Limiting wall thickness" for NPS 24 (DN 600) and smaller. For NPS 26(DN 650) and larger, length L_1 shall be per agreement between the purchaser and manufacturer.

Dimensions of Butt Welding Reducers

See Tables 5.14 and 5.14M.

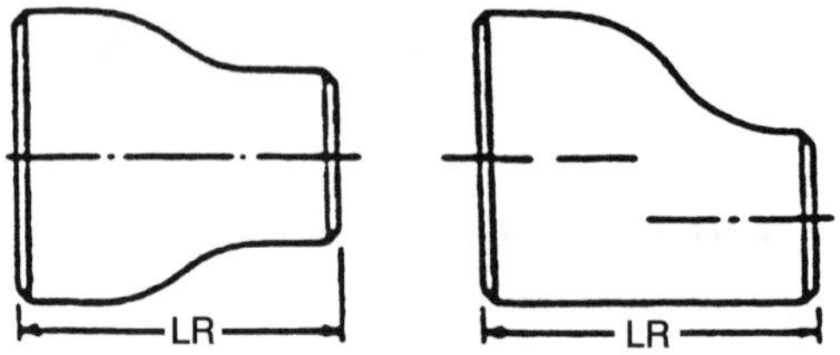

TABLE 5.14 Dimensions of butt welding reducers (U.S. Customary Units)

Nominal pipe size (NPS)	Outside diameter at bevel, in		End-to-end, LR	Nominal pipe size (NPS)	Outside diameter at bevel, in		End-to-end, LR
	Large end	Small end			Large end	Small end	
3/4 × 1/2	1.05	0.84	1.50	8 × 6	8.62	6.62	6.00
3/4 × 3/8	1.05	0.68	1.50	8 × 5	8.62	5.56	6.00
1 × 3/4	1.32	1.05	2.00	8 × 4	8.62	4.50	6.00
1 × 1/2	1.32	0.84	2.00	8 × 3 1/2	8.62	4.00	6.00
1 1/4 × 1	1.66	1.32	2.00	10 × 8	10.75	8.62	7.00
1 1/4 × 3/4	1.66	1.05	2.00	10 × 6	10.75	6.62	7.00
1 1/4 × 1/2	1.66	0.84	2.00	10 × 5	10.75	5.56	7.00
1 1/2 × 1 1/4	1.90	1.66	2.50	10 × 4	10.75	4.50	7.00
1 1/2 × 1	1.90	1.32	2.50	12 × 10	12.75	10.75	8.00
1 1/2 × 3/4	1.90	1.05	2.50	12 × 8	12.75	8.62	8.00
1 1/2 × 1/2	1.90	0.84	2.50	12 × 6	12.75	6.62	8.00
2 × 1 1/2	2.38	1.90	3.00	12 × 5	12.75	5.56	8.00
2 × 1 1/4	2.38	1.66	3.00	14 × 12	14.00	12.75	13.00
2 × 1	2.38	1.32	3.00	14 × 10	14.00	10.75	13.00
2 × 3/4	2.38	1.05	3.00	14 × 8	14.00	8.62	13.00
2 1/2 × 2	2.88	2.38	3.50	14 × 6	14.00	6.62	13.00
2 1/2 × 1 1/2	2.88	1.90	3.50	16 × 14	16.00	14.00	14.00
2 1/2 × 1 1/4	2.88	1.66	3.50	16 × 12	16.00	12.75	14.00
2 1/2 × 1	2.88	1.32	3.50	16 × 10	16.00	10.75	14.00
3 × 2 1/2	3.50	2.88	3.50	16 × 8	16.00	8.62	14.00
3 × 2	3.50	2.38	3.50	18 × 16	18.00	16.00	15.00
3 × 1 1/2	3.50	1.90	3.50	18 × 14	18.00	14.00	15.00
3 × 1 1/4	3.50	1.66	3.50	18 × 12	18.00	12.75	15.00
3 1/2 × 3	4.00	3.50	4.00	18 × 10	18.00	10.75	15.00
3 1/2 × 2 1/2	4.00	2.88	4.00	20 × 18	20.00	18.00	20.00
3 1/2 × 2	4.00	2.38	4.00	20 × 16	20.00	16.00	20.00
3 1/2 × 1 1/2	4.00	1.90	4.00	20 × 14	20.00	14.00	20.00
3 1/2 × 1 1/4	4.00	1.66	4.00	20 × 12	20.00	12.75	20.00
4 × 3 1/2	4.50	4.00	4.00	22 × 20	22.00	20.00	20.00
4 × 3	4.50	3.50	4.00	22 × 18	22.00	18.00	20.00
4 × 2 1/2	4.50	2.88	4.00	22 × 16	22.00	16.00	20.00
4 × 2	4.50	2.38	4.00	22 × 14	22.00	14.00	20.00
4 × 1 1/2	4.50	1.90	4.00	24 × 22	24.00	22.00	20.00
5 × 4	5.56	4.50	5.00	24 × 20	24.00	20.00	20.00
5 × 3 1/2	5.56	4.00	5.00	24 × 18	24.00	18.00	20.00
5 × 3	5.56	3.50	5.00	24 × 16	24.00	16.00	20.00
5 × 2 1/2	5.56	2.88	5.00	26 × 24	26.00	24.00	24.00
5 × 2	5.56	2.38	5.00	26 × 22	26.00	22.00	24.00
6 × 5	6.62	5.56	5.50	26 × 20	26.00	20.00	24.00
6 × 4	6.62	4.50	5.50	26 × 18	26.00	18.00	24.00
6 × 3 1/2	6.62	4.00	5.50	28 × 26	28.00	26.00	24.00
6 × 3	6.62	3.50	5.50	28 × 24	28.00	24.00	24.00
6 × 2 1/2	6.62	2.88	5.50	28 × 20	28.00	20.00	24.00
				28 × 18	28.00	18.00	24.00

TABLE 5.14 Dimensions of butt welding reducers (U.S. Customary Units) (Continued)

Nominal pipe size (NPS)	Outside diameter at bevel, in		End-to-end, LR	Nominal pipe size (NPS)	Outside diameter at bevel, in		End-to-end, LR
	Large end	Small end			Large end	Small end	
30 × 28	30.00	28.00	24.00	40 × 38	40.00	38.00	24.00
30 × 26	30.00	26.00	24.00	40 × 36	40.00	36.00	24.00
30 × 24	30.00	24.00	24.00	40 × 34	40.00	34.00	24.00
30 × 20	30.00	20.00	24.00	40 × 32	40.00	32.00	24.00
32 × 30	32.00	30.00	24.00	40 × 30	40.00	30.00	24.00
32 × 28	32.00	28.00	24.00	42 × 40	42.00	40.00	24.00
32 × 26	32.00	26.00	24.00	42 × 38	42.00	38.00	24.00
32 × 24	32.00	24.00	24.00	42 × 36	42.00	36.00	24.00
34 × 32	34.00	32.00	24.00	42 × 34	42.00	34.00	24.00
34 × 30	34.00	30.00	24.00	42 × 32	42.00	32.00	24.00
34 × 26	34.00	26.00	24.00	42 × 30	42.00	30.00	24.00
34 × 24	34.00	24.00	24.00	44 × 42	44.00	42.00	24.00
36 × 34	36.00	34.00	24.00	44 × 40	44.00	40.00	24.00
36 × 32	36.00	32.00	24.00	44 × 38	44.00	38.00	24.00
36 × 30	36.00	30.00	24.00	44 × 36	44.00	36.00	24.00
36 × 26	36.00	26.00	24.00	46 × 44	46.00	44.00	28.00
36 × 24	36.00	24.00	24.00	46 × 42	46.00	42.00	28.00
38 × 36	38.00	36.00	24.00	46 × 40	46.00	40.00	28.00
38 × 34	38.00	34.00	24.00	46 × 38	46.00	38.00	28.00
38 × 32	38.00	32.00	24.00	48 × 46	48.00	46.00	28.00
38 × 30	38.00	30.00	24.00	48 × 44	48.00	44.00	28.00
38 × 28	38.00	28.00	24.00	48 × 42	48.00	42.00	28.00
38 × 26	38.00	26.00	24.00	48 × 40	48.00	40.00	28.00

Notes:
1. Dimensions in inches.
2. For tolerances, refer to Table 5.17.
3. Reducers may be bell shaped, as shown, or conical.

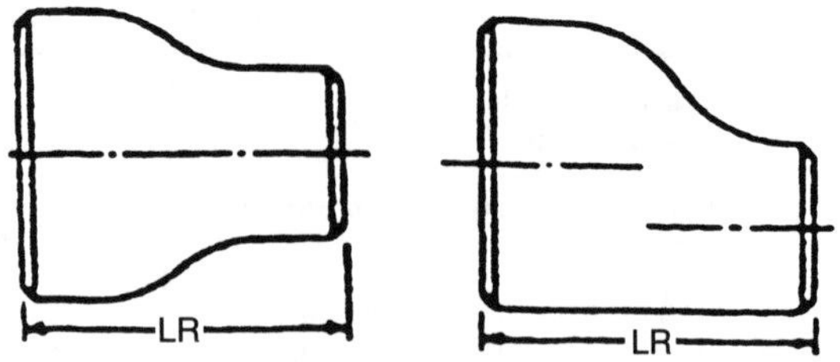

TABLE 5.14M Dimensions of butt welding reducers (Metric Units)

NPS	DN	Outside diameter at bevel, mm		End-to-end, LR, mm
		Large end	Small end	
¾ × ½	20 × 15	26.7	21.3	38
¾ × ⅜	20 × 10	26.7	17.3	38
1 × ¾	25 × 20	33.4	26.7	51
1 × ½	25 × 15	33.4	21.3	51
1¼ × 1	32 × 25	42.2	33.4	51
1¼ × ¾	32 × 20	42.2	26.7	51
1¼ × ½	32 × 15	42.2	21.3	51
1½ × 1¼	40 × 32	48.3	42.2	64
1½ × 1	40 × 25	48.3	33.4	64
1½ × ¾	40 × 20	48.3	26.7	64
1½ × ½	40 × 15	48.3	21.3	64
2 × 1½	50 × 40	60.3	48.3	76
2 × 1¼	50 × 32	60.3	42.2	76
2 × 1	50 × 25	60.3	33.4	76
2 × ¾	50 × 20	60.3	26.7	76
2½ × 2	65 × 50	73.0	60.3	89
2½ × 1½	65 × 40	73.0	48.3	89
2½ × 1¼	65 × 32	73.0	42.2	89
2½ × 1	65 × 25	73.0	33.4	89
3 × 2½	80 × 65	88.9	73.0	89
3 × 2	80 × 50	88.9	60.3	89
3 × 1½	80 × 40	88.9	48.3	89
3 × 1¼	80 × 32	88.9	42.2	89
3½ × 3	90 × 80	101.6	88.9	102
3½ × 2½	90 × 65	101.6	73.0	102
3½ × 2	90 × 50	101.6	60.3	102
3½ × 1½	90 × 40	101.6	48.3	102
3½ × 1¼	90 × 32	101.6	42.2	102
4 × 3½	100 × 90	114.3	101.6	102
4 × 3	100 × 80	114.3	88.9	102
4 × 2½	100 × 65	114.3	73.0	102
4 × 2	100 × 50	114.3	60.3	102
4 × 1½	100 × 40	114.3	48.3	102
5 × 4	125 × 100	141.3	114.3	127
5 × 3½	125 × 90	141.3	101.6	127
5 × 3	125 × 80	141.3	88.9	127
5 × 2½	125 × 65	141.3	73.0	127
5 × 2	125 × 50	141.3	60.3	127
6 × 5	150 × 125	168.3	141.3	140
6 × 4	150 × 100	168.3	114.3	140
6 × 3½	150 × 90	168.3	101.6	140
6 × 3	150 × 80	168.3	88.9	140
6 × 2½	150 × 65	168.3	73.0	140

TABLE 5.14M Dimensions of butt welding reducers (Metric Units) (Continued)

NPS	DN	Outside diameter at bevel, mm		End-to-end, LR, mm
		Large end	Small end	
8 × 6	200 × 150	219.1	168.3	152
8 × 5	200 × 125	219.1	141.3	152
8 × 4	200 × 100	219.1	114.3	152
8 × 3½	200 × 90	219.1	101.6	152
10 × 8	250 × 200	273.0	219.1	178
10 × 6	250 × 150	273.0	168.3	178
10 × 5	250 × 125	273.0	141.3	178
10 × 4	250 × 100	273.0	114.3	178
12 × 10	300 × 250	323.8	273.0	203
12 × 8	300 × 200	323.8	219.1	203
12 × 6	300 × 150	323.8	168.3	203
12 × 5	300 × 125	323.8	141.3	203
14 × 12	350 × 300	355.6	323.8	330
14 × 10	350 × 250	355.6	273.0	330
14 × 8	350 × 200	355.6	219.1	330
14 × 6	350 × 150	355.6	168.3	330
16 × 14	400 × 350	406.4	355.6	356
16 × 12	400 × 300	406.4	323.8	356
16 × 10	400 × 250	406.4	273.0	356
16 × 8	400 × 200	406.4	219.1	356
18 × 16	450 × 400	457	406.4	381
18 × 14	450 × 350	457	355.6	381
18 × 12	450 × 300	457	323.8	381
18 × 10	450 × 250	457	273.0	381
20 × 18	500 × 450	508	457	508
20 × 16	500 × 400	508	406.4	508
20 × 14	500 × 350	508	355.6	508
20 × 12	500 × 300	508	323.8	508
22 × 20	550 × 500	559	508	508
22 × 18	550 × 450	559	457	508
22 × 16	550 × 400	559	406.4	508
22 × 14	550 × 350	559	355.4	508
24 × 22	600 × 550	610	559	508
24 × 20	600 × 500	610	508	508
24 × 18	600 × 450	610	457	508
24 × 16	600 × 400	610	406.4	508
26 × 24	650 × 600	660	610	610
26 × 22	650 × 550	660	559	610
26 × 20	650 × 500	660	508	610
26 × 18	650 × 450	660	457	610
28 × 26	700 × 650	711	660	610
28 × 24	700 × 600	711	610	610
28 × 22	700 × 550	711	508	610
28 × 20	700 × 500	711	457	610
30 × 28	750 × 700	762	711	610
30 × 26	750 × 650	762	660	610
30 × 24	750 × 600	762	610	610
30 × 22	750 × 550	762	508	610
32 × 30	800 × 750	813	762	610
32 × 28	800 × 700	813	711	610
32 × 26	800 × 650	813	660	610
32 × 24	800 × 600	813	610	610

TABLE 5.14M Dimensions of butt welding reducers (Metric Units) (Continued)

NPS	DN	Outside diameter at bevel, mm		End-to-end, LR, mm
		Large end	Small end	
34 × 32	850 × 800	864	813	610
34 × 30	850 × 750	864	762	610
34 × 28	850 × 700	864	660	610
34 × 26	850 × 650	864	610	610
36 × 34	900 × 850	914	864	610
36 × 32	900 × 800	914	813	610
36 × 30	900 × 750	914	762	610
36 × 28	900 × 700	914	660	610
36 × 26	900 × 650	914	610	610
38 × 36	950 × 900	965	914	610
38 × 34	950 × 850	965	864	610
38 × 32	950 × 800	965	813	610
38 × 30	950 × 750	965	762	610
38 × 28	950 × 700	965	711	610
38 × 26	950 × 650	965	660	610
40 × 38	1000 × 950	1016	965	610
40 × 36	1000 × 900	1016	914	610
40 × 34	1000 × 850	1016	864	610
40 × 32	1000 × 800	1016	813	610
40 × 30	1000 × 750	1016	762	610
42 × 40	1050 × 1000	1067	1016	610
42 × 38	1050 × 950	1067	965	610
42 × 36	1050 × 900	1067	914	610
42 × 34	1050 × 850	1067	864	610
42 × 32	1050 × 800	1067	813	610
42 × 30	1050 × 750	1067	762	610
44 × 42	1100 × 1050	1118	1067	610
44 × 40	1100 × 1000	1118	1016	610
44 × 38	1100 × 950	1118	965	610
44 × 36	1100 × 900	1118	914	610
46 × 44	1150 × 1100	1168	1118	711
46 × 42	1150 × 1050	1168	1067	711
46 × 40	1150 × 1000	1168	1016	711
46 × 38	1150 × 950	1168	965	711
48 × 46	1200 × 1150	1219	1168	711
48 × 44	1200 × 1100	1219	1118	711
48 × 42	1200 × 1050	1219	1067	711
48 × 40	1200 × 1000	1219	1016	711

Notes:
1. Dimensions in millimeters. Dimensions per ASME B16.9.
2. Reducers may be bell shaped, as shown, or conical.
3. For tolerances, refer to Table 5.17.

TOLERANCES FOR BUTT WELDING FITTINGS CONFORMING TO ASME B16.9

Tolerances for Butt Welding Elbows, Tees, and Crosses

See Table 5.15.

Tolerances for Butt Welding Caps and 180° Returns

See Table 5.16.

TABLE 5.15 Tolerances for Butt Welding Elbows and Tees (U.S. Customary Units)*

	Elbows, tees, and crosses			90° and 45° elbows, tees, and crosses: center-to-end dimension
Nominal pipe size (NPS)	Outside diameter at bevel	Inside diameter at end	Wall thickness t	A, B, R, T
½ to 2½	+0.06 −0.03	0.03	Not less than 87.5% of nominal thickness	0.06
3 to 3½	0.06	0.06		0.06
4	0.06	0.06		0.06
5 to 8	+0.09 −0.06	0.06		0.06
10 to 18	+0.16 −0.12	0.12		0.09
20 to 24	+0.26 −0.19	0.19		0.09
26 to 30	+0.26 0.19	0.19		0.12
32 to 48	+0.26 0.19	0.19		0.19

* All dimensions are in inches. Multiply inch dimension by 25.4 to get millimeters.

TABLE 5.16 Tolerances for Butt Welding Caps and 180° Returns (U.S. Customary Units)*

	Caps and 180° returns				180° returns		
Nominal pipe size (NPS)	Outside diameter at bevel	Inside diameter at end	Wall thickness t	Caps, overall length E	Center-to-center dimension L	Back-to-face dimension K	Alignment of ends U
½ to 2½	+0.06 −0.03	0.03	Not less than 87.5% of nominal thickness	0.12	0.25	0.25	0.03
3 to 3½	0.06	0.06		0.12	0.25	0.25	0.03
4	0.06	0.06		0.12	0.25	0.25	0.03
5 to 8	+0.09 −0.06	0.06		0.25	0.25	0.25	0.03
10 to 18	+0.16 −0.12	0.12		0.25	0.38	0.25	0.06
20 to 24	+0.25 −0.19	0.19		0.25	0.38	0.25	0.06
26 to 30	+0.25 0.19	0.19		0.38			
32 to 48	+0.25 0.19	0.19		0.38			

* All dimensions are in inches. Multiply inch dimension by 25.4 to get millimeters.

Tolerances for Butt Welding Reducers and Lap Joint Stub Ends

See Table 5.17.

FORGED SOCKET WELDING AND THREADING FITTINGS

Class 3000, 6000, and 9000 forged socket welding and Class 2000, 3000, and 6000 threaded fittings are manufactured in accordance with ASME B16.11. Table 5.18 provides the correlation between the pipe schedule/wall thickness designation and the fitting rating class.

Size Range of Socket Welding Fittings per ASME B16.11

See Tables 5.19 and 5.19M.

Size Range of Threaded Fittings per ASME B16.11

See Tables 5.20 and 5.20M.

TABLE 5.17 Tolerances for Butt Welding Reducers and Lap Joint Stub Ends*

	Reducers and lap joint stub ends			Reducers and lap joint stub ends, overall length	Lap joint stub ends		
Nominal pipe size (NPS)	Outside diameter at bevel	Inside diameter at end	Wall thickness *t*	LS, LR	Outside diameter of lap *D*	Fillet radius of lap *R*	Lap thickness *t*
½ to 2½	+0.06 −0.03	0.03	Not less than 87.5% of nominal thickness	0.06	+0 −0.03	+0 −0.03	+0.06 −0
3 to 3½	0.06	0.06		0.06	+0 −0.03	+0 −0.03	+0.06 −0
4	0.06	0.06		0.06	+0 −0.03	+0 −0.06	+0.06 −0
5 to 8	+0.09 −0.06	0.06		0.06	+0 −0.03	+0 −0.06	+0.06 −0
10 to 18	+0.16 −0.12	0.12		0.09	+0 −0.06	+0 −0.06	+0.12 −0
20 to 24	+0.25 −0.19	0.19		0.09	+0 −0.06	+0 −0.06	+0.12 −0
26 to 30	+0.25 0.19	0.19		0.19			
32 to 48	+0.25 0.19	0.19		0.19			

* All dimensions are in inches. Multiply inch dimension by 25.4 to get millimeters.

TABLE 5.18 Correlation of Class Designation of Forged, Socket Welding, and Threaded Fittings with Pipe Schedule or Wall Designation

Pipe schedule and designation	Threaded class	Socket-welded class
80/XS or less	2000	3000
160	3000	6000
XXS	6000	9000

Dimensions of Class 3000 Socket Welding 90° and 45° Elbows

See Tables 5.21 and 5.21M.

TABLE 5.19 Size Range of Socket Welding Fittings per ASME B16.11

	Socket welding class designation		
Description*	3000	6000	9000
45°, 90° elbows	1/8–4	1/8–2	1/2–2
Tees, crosses	1/8–4	1/8–2	1/2–2
Coupling, half-coupling	1/8–4	1/8–2	1/2–2
Cap	1/8–4	1/8–4	1/2–2

* Plugs and bushings are neither identified by class designation nor are they manufactured as socket welding fittings.

TABLE 5.19M Size Range (DN) of Socket Welding Fittings per ASME B16.11

	Socket welding class designation		
Description*	3000	6000	9000
45°, 90° elbows	DN6-DN100	DN6-DN50	DN15-DN50
Tees, crosses	DN6-DN100	DN6-DN50	DN15-DN50
Coupling, half-coupling	DN6-DN100	DN6-DN50	DN15-DN50
Cap	DN6-DN100	DN6-DN100	DN15-DN50

* Plugs and bushings are neither identified by class designation nor are they manufactured as socket welding fittings.

TABLE 5.20 Size Range (NPS) of Threaded Fittings per ASME B16.11

	Threaded class designation		
Description	2000	3000	6000
45°, 90° elbows	1/8–4	1/8–4	1/8–4
Tees, crosses	1/8–4	1/8–4	1/8–4
Coupling, half-coupling		1/8–4	1/8–4
Cap		1/8–4	1/8–4
Square, hex, round plug		1/8–4	
Hex and flush bushing*		1/8–4	

* Plugs and bushings are not rated by class designation, but they can be used up to Class 6000 designation.

TABLE 5.20M Size Range (DN) of Threaded Fittings per ASME B16.11

	Threaded class designation		
Description	2000	3000	6000
45°, 90° elbows	DN6-DN100	DN6-DN100	DN6-DN100
Tees, crosses	DN6-DN100	DN6-DN100	DN6-DN100
Coupling, half-coupling		DN6-DN100	DN6-DN100
Cap		DN6-DN100	DN6-DN100
Square, hex, round plug		DN6-DN100	
Hex and flush bushing*		DN6-DN100	

* Plugs and bushings are not identified by class designation, but they can be used up to Class 6000 designation.

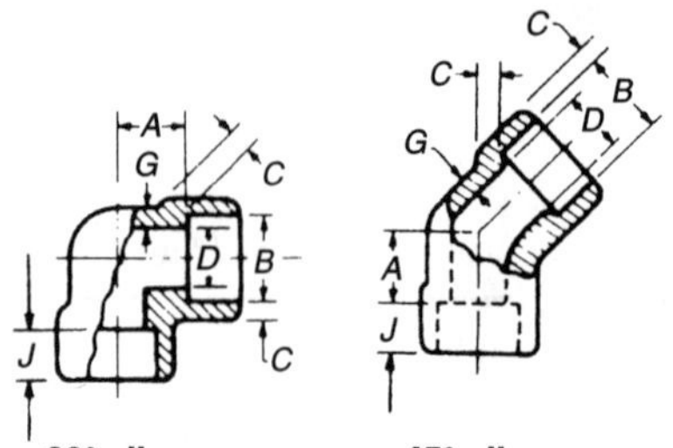

90° elbow **45° elbow**

TABLE 5.21 Class 3000 Forged Socket Welding 90° and 45° Elbows (U.S. Customary Units)*

Nom. pipe size	Socket bore dia., *B*	Bore diameter of fittings, *D*	Socket wall thickness, *C*		Body wall, *G*, min.	Depth of socket, *J*, min.	Center to bottom of socket, *A*		
			Ave.	Min.			90° elbow	45° elbow	Tolerance ±
1/8	0.440 0.420	0.299 0.239	0.125	0.125	0.095	0.38	0.44	0.31	0.03
1/4	0.575 0.555	0.394 0.334	0.149	0.130	0.119	0.38	0.44	0.31	0.03
3/8	0.710 0.690	0.523 0.463	0.158	0.138	0.126	0.38	0.53	0.31	0.06
1/2	0.875 0.855	0.652 0.592	0.184	0.161	0.147	0.38	0.62	0.44	0.06
3/4	1.085 1.065	0.854 0.794	0.193	0.168	0.154	0.50	0.75	0.50	0.06
1	1.350 1.330	1.079 1.019	0.224	0.196	0.179	0.50	0.88	0.56	0.08
1 1/4	1.695 1.675	1.410 1.350	0.239	0.208	0.191	0.50	1.06	0.69	0.08
1 1/2	1.935 1.915	1.640 1.580	0.250	0.218	0.200	0.50	1.25	0.81	0.08
2	2.426 2.406	2.097 2.037	0.273	0.238	0.218	0.62	1.50	1.00	0.08
2 1/2	2.931 2.906	2.529 2.409	0.345	0.302	0.276	0.62	1.62	1.12	0.10
3	3.560 3.535	3.128 3.008	0.375	0.327	0.300	0.62	2.25	1.25	0.10
4	4.570 4.545	4.086 3.966	0.421	0.368	0.337	0.75	2.62	1.62	0.10

* Dimensions in inches. Dimensions per ASME B16.11.

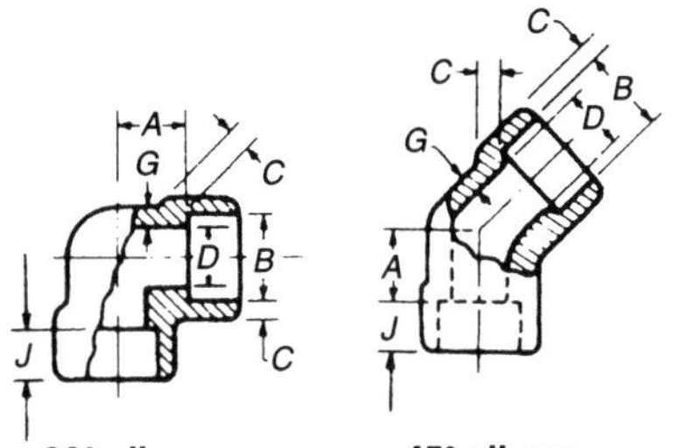

TABLE 5.21M Class 3000 Forged Socket Welding 90° and 45° Elbows (Metric Units)*

DN	Nom. pipe size	Socket bore diameter, B		Bore diameter of fitting, D		Socket wall thickness, C		Body wall, G, min.	Depth of socketk, J, min.	Center to bottom of socket, A		
		Max.	Min.	Max.	Min.	Ave.	Min.			90° elbow	45° elbow	Tolerance ±
6	1/8	11.2	10.8	7.6	6.1	3.18	3.18	2.41	9.5	11.0	8.0	1.0
8	1/4	14.6	14.2	10.0	8.5	3.78	3.30	3.02	9.5	11.0	8.0	1.0
10	3/8	18.0	17.6	13.3	11.8	4.01	3.50	3.20	9.5	13.5	8.0	1.5
15	1/2	22.2	21.8	16.6	15.0	4.67	4.09	3.73	9.5	15.5	11.0	1.5
20	3/4	27.6	27.2	21.7	20.2	4.90	4.27	3.91	12.5	19.0	13.0	1.5
25	1	34.3	33.9	27.4	25.9	5.69	4.98	4.55	12.5	22.5	14.0	2.0
32	1 1/4	43.1	42.7	35.8	34.3	6.07	5.28	4.85	12.5	27.0	17.5	2.0
40	1 1/2	49.2	48.8	41.6	40.1	6.35	5.54	5.08	12.5	32.0	20.5	2.0
50	2	61.7	61.2	53.3	51.7	6.93	6.04	5.54	16.0	38.0	25.5	2.0
65	2 1/2	74.4	73.9	64.2	61.2	8.76	7.67	7.01	16.0	41.0	28.5	2.5
80	3	90.3	89.8	79.4	76.4	9.52	8.30	7.62	16.0	57.0	32.0	2.5
100	4	115.7	115.2	103.8	100.7	10.69	9.35	8.56	19.0	66.5	41.0	2.5

* Dimensions in millimeters.

Dimensions of Class 6000 Socket Welding 90° and 45° Elbows

See Tables 5.22 and 5.22M.

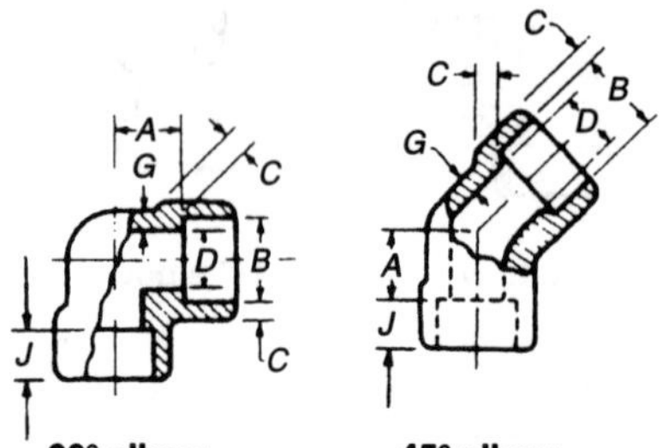

90° elbow **45° elbow**

TABLE 5.22 Class 6000 Forged Socket Welding 90° and 45° Elbows (U.S. Customary Units)*

Nom. pipe size	Socket bore dia. (2), *B*	Bore diameter of fittings (2), *D*	Socket wall thickness (1), *C*		Body wall, *G*, min.	Depth of socket, *J*, min.	Center to bottom of socket, *A*		
			Ave.	Min.			90° elbow	45°elbow	Tolerance ±
1/8	0.440 0.420	0.189 0.126	0.156	0.135	0.124	0.38	0.44	0.31	0.03
1/4	0.575 0.555	0.280 0.220	0.181	0.158	0.145	0.38	0.53	0.31	0.03
3/8	0.710 0.690	0.389 0.329	0.198	0.172	0.158	0.38	0.62	0.44	0.06
1/2	0.875 0.855	0.494 0.434	0.235	0.204	0.188	0.38	0.75	0.50	0.06
3/4	1.085 1.065	0.642 0.582	0.274	0.238	0.219	0.50	0.88	0.56	0.06
1	1.350 1.330	0.845 0.785	0.312	0.273	0.250	0.50	1.06	0.69	0.08
1 1/4	1.695 1.675	1.190 1.130	0.312	0.273	0.250	0.50	1.25	0.81	0.08
1 1/2	1.935 1.915	1.368 1.308	0.351	0.307	0.281	0.50	1.50	1.00	0.08
2	2.426 2.406	1.717 1.657	0.430	0.374	0.344	0.62	1.62	1.12	0.08

* Dimensions in inches. Dimensions per ASME B16.11.

Notes:

(1) Average of socket wall thickness around periphery shall be no less than listed values. The minimum values are permitted in localized areas.

(2) Upper and lower values for each size are the respective maximum and minimum dimensions.

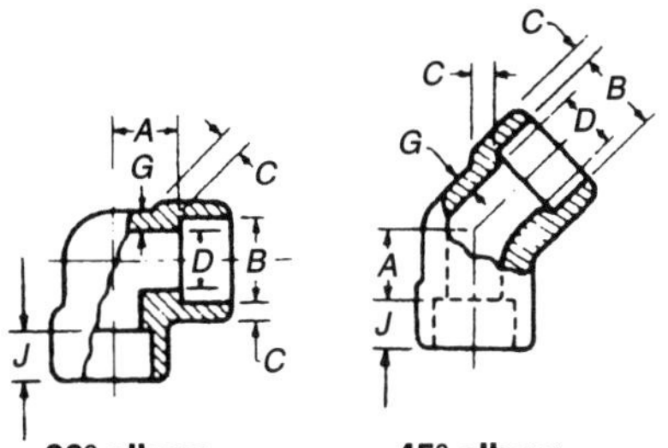

TABLE 5.22M Class 6000 Forged Socket Welding 90° and 45° Elbows (Metric Units)*

DN	Nom. pipe size	Socket bore diameter, B		Bore diameter of fitting, D		Socket wall thickness, C		Body wall, G, min.	Depth of socket, J, min.	Center to bottom of socket, A		
		Max.	Min.	Max.	Min.	Ave.	Min.			90° elbow	45° elbow	Tolerance ±
6	1/8	11.2	10.8	4.8	3.2	3.96	3.43	3.15	9.5	11.0	8.0	1.0
8	1/4	14.6	14.2	7.1	5.6	4.60	4.01	3.68	9.5	13.5	8.0	1.0
10	3/8	18.0	17.6	9.9	8.4	5.03	4.37	4.01	9.5	15.5	11.0	1.5
15	1/2	22.2	21.8	12.5	11.0	5.97	5.18	4.78	9.5	19.0	12.5	1.5
20	3/4	27.6	27.2	16.3	14.8	6.96	6.04	5.56	12.5	22.5	14.0	1.5
25	1	34.3	33.9	21.5	19.9	7.92	6.93	6.35	12.5	27.0	17.5	2.0
32	1 1/4	43.1	42.7	30.2	28.7	7.92	6.93	6.35	12.5	32.0	20.5	2.0
40	1 1/2	49.2	48.8	34.7	33.2	8.92	7.80	7.14	12.5	38.0	25.5	2.0
50	2	61.7	61.2	43.6	42.1	10.92	9.50	8.74	16.0	41.0	28.8	2.0

* Dimensions in millimeters.

Dimensions of Class 9000 Socket Welding 90° and 45° Elbows

See Tables 5.23 and 5.23M.

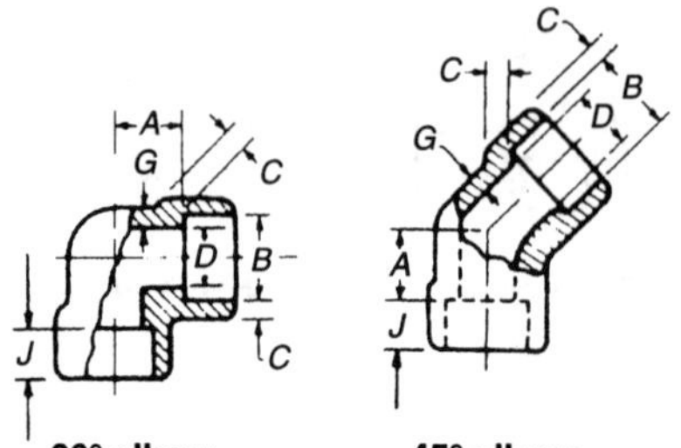

90° elbow 45° elbow

TABLE 5.23 Class 9000 Forged Socket Welding 90° and 45° Elbows (U.S. Customary Units)*

Nom. pipe size	Socket bore dia. (2), *B*	Bore diameter of fittings (2), *D*	Socket wall thickness (1), *C*		Body wall, *G*, min.	Depth of socket, *J*, min.	Center to bottom of socket, *A*		
			Ave.	Min.			90° elbow	45°elbow	Tolerance ±
½	0.875 0.855	0.282 0.222	0.368	0.322	0.294	0.38	1.00	0.62	0.06
¾	1.085 1.065	0.464 0.404	0.385	0.337	0.308	0.50	1.12	0.75	0.06
1	1.350 1.330	0.629 0.569	0.448	0.392	0.358	0.50	1.25	0.81	0.08
1¼	1.695 1.675	0.926 0.866	0.478	0.418	0.382	0.50	1.38	0.88	0.08
1½	1.935 1.915	1.130 1.070	0.500	0.438	0.400	0.50	1.50	1.00	0.08
2	2.426 2.406	1.533 1.473	0.545	0.477	0.436	0.62	2.12	1.12	0.08

* Dimensions in inches. Dimensions per ASME B16.11.

Notes:

(1) Average of socket wall thickness around periphery shall be no less than listed values. The minimum values are permitted in localized areas.

(2) Upper and lower values for each size are the respective maximum and minimum dimensions.

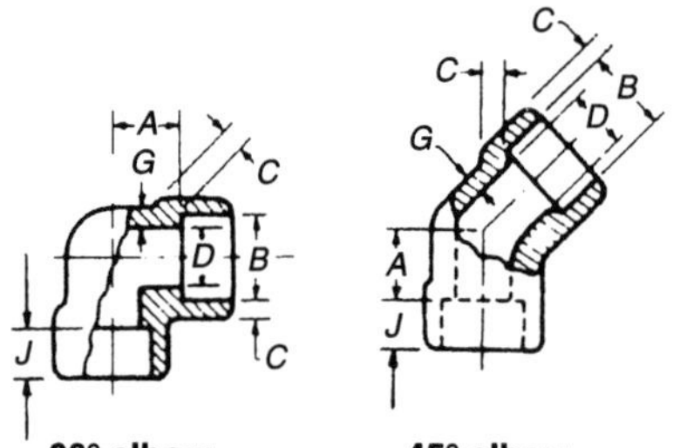

90° elbow 45° elbow

TABLE 5.23M Class 9000 Forged Socket Welding 90° and 45° Elbows (Metric Units)*

DN	Nom. pipe size	Socket bore diameter, *B*		Bore diameter of fitting, *D*		Socket wall thickness, *C*		Body wall, *G*, min.	Depth of socket, *J*, min.	Center to bottom of socket, *A*		
		Max.	Min.	Max.	Min.	Ave.	Min.			90° elbow	45° elbow	Tolerance ±
15	½	22.2	21.8	7.2	5.6	9.35	8.18	7.47	9.5	25.5	15.5	1.5
20	¾	27.6	27.2	11.8	10.3	9.78	8.56	7.82	12.5	28.5	19.0	1.5
25	1	34.3	33.9	16.0	14.4	11.38	9.96	9.09	12.5	32.0	20.5	2.0
32	1¼	43.1	42.7	23.5	22.0	12.14	10.62	9.70	12.5	35.0	22.5	2.0
40	1½	49.2	48.8	28.7	27.2	12.70	11.12	10.15	12.5	38.0	25.5	2.0
50	2	61.7	61.2	38.9	37.4	13.84	12.12	11.07	16.0	54.0	28.5	2.0

* Dimensions in millimeters.

Dimensions of Class 3000 Socket Welding Tees and Crosses

See Tables 5.24 and 5.24M.

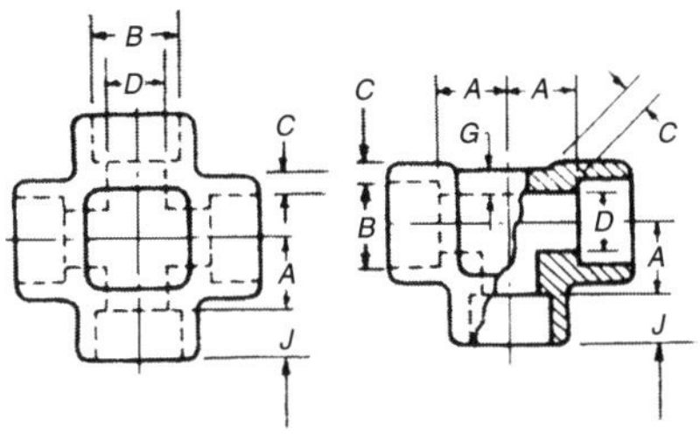

Cross Tee

TABLE 5.24 Class 3000 Forged Socket Welding Tees and Crosses (U.S. Customary Units)*

Nom. pipe size	Socket bore dia., *B*	Bore diameter of fittings, *D*	Socket wall thickness, *C*		Body wall, *G*	Depth of socket, *J*, min.	Center to bottom of socket, *A*	*A* tolerance ±
			Ave.	Min.				
1/8	0.440 0.420	0.299 0.239	0.125	0.125	0.095	0.38	0.44	0.03
1/4	0.575 0.555	0.394 0.334	0.149	0.130	0.119	0.38	0.44	0.03
3/8	0.710 0.690	0.523 0.463	0.158	0.138	0.126	0.38	0.53	0.06
1/2	0.875 0.855	0.652 0.592	0.184	0.161	0.147	0.38	0.62	0.06
3/4	1.085 1.065	0.854 0.794	0.193	0.168	0.154	0.50	0.75	0.06
1	1.350 1.330	1.079 1.019	0.224	0.196	0.179	0.50	0.88	0.08
1 1/4	1.695 1.675	1.410 1.350	0.239	0.208	0.191	0.50	1.06	0.08
1 1/2	1.935 1.915	1.640 1.580	0.250	0.218	0.200	0.50	1.25	0.08
2	2.426 2.406	2.097 2.037	0.273	0.238	0.218	0.62	1.50	0.08
2 1/2	2.931 2.906	2.529 2.409	0.345	0.302	0.276	0.62	1.62	0.10
3	3.560 3.535	3.128 3.008	0.375	0.327	0.300	0.62	2.25	0.10
4	4.570 4.545	4.086 3.966	0.421	0.368	0.337	0.75	2.62	0.10

* Dimensions in inches. Dimensions per ASME B16.11.

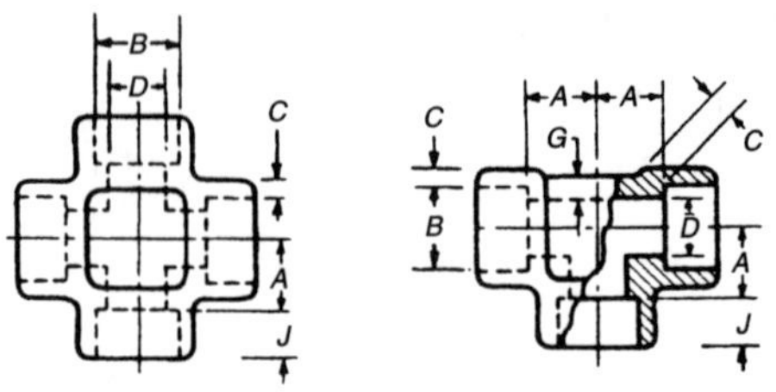

Cross Tee

TABLE 5.24M Class 3000 Forged Socket Welding Tees and Crosses (Metric)*

DN	Nom. pipe size	Socket bore dia., *B*	Bore diameter of fittings, *D*	Socket wall thickness, *C* Ave.	Socket wall thickness, *C* Min.	Body wall, *G*, min.	Depth of socket, *J*, min.	Center to bottom of socket, *A*	*A* tolerance ±
6	1/8	11.2 10.8	7.6 6.1	3.18	3.18	2.41	9.5	11.0	1.0
8	1/4	14.6 14.2	10.0 8.5	3.78	3.30	3.02	9.5	11.0	1.0
10	3/8	18.0 17.6	13.3 11.8	4.01	3.50	3.20	9.5	13.5	1.5
15	1/2	22.2 21.8	16.6 15.0	4.67	4.09	3.73	9.5	15.5	1.5
20	3/4	27.6 27.2	21.7 20.2	4.90	4.27	3.91	12.5	19.0	1.5
25	1	34.3 33.9	27.4 25.9	5.69	4.98	4.55	12.5	22.5	2.0
32	1 1/4	43.1 42.7	35.8 34.3	6.07	5.28	4.85	12.5	27.0	2.0
40	1 1/2	49.2 48.8	41.6 40.1	6.35	5.54	5.08	12.5	32.0	2.0
50	2	61.7 61.2	53.3 51.7	6.93	6.04	5.54	16.0	38.0	2.0
65	2 1/2	74.4 73.9	64.2 61.2	8.76	7.67	7.01	16.0	41.0	2.5
80	3	90.3 89.8	79.4 76.4	9.52	8.30	7.62	16.0	57.0	2.5
100	4	115.7 115.2	103.8 100.7	10.69	9.35	8.56	19.0	66.5	2.5

* Dimensions in millimeters.

Dimensions of Class 6000 Socket Welding Tees and Crosses

See Tables 5.25 and 5.25M.

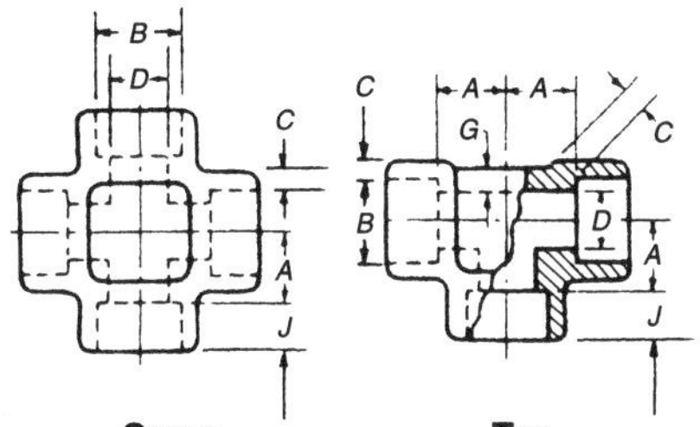

Cross Tee

TABLE 5.25 Class 6000 Forged Socket Welding Tees and Crosses (U.S. Customary Units)*

Nom. pipe size	Socket bore dia. (2), *B*	Bore diameter of fittings (2), *D*	Socket wall thickness (1), *C*		Body wall, *G*, min.	Depth of socket, *J*, min.	Center to bottom of socket, *A*	*A* tolerance ±
			Ave.	Min.				
1/8	0.440 0.420	0.189 0.126	0.156	0.135	0.124	0.38	0.44	0.03
1/4	0.575 0.555	0.280 0.220	0.181	0.158	0.145	0.38	0.53	0.03
3/8	0.710 0.690	0.389 0.329	0.198	0.172	0.158	0.38	0.62	0.06
1/2	0.875 0.855	0.494 0.434	0.235	0.204	0.188	0.38	0.75	0.06
3/4	1.085 1.065	0.642 0.582	0.274	0.238	0.219	0.50	0.88	0.06
1	1.350 1.330	0.845 0.785	0.312	0.273	0.250	0.50	1.06	0.08
1 1/4	1.695 1.675	1.190 1.130	0.312	0.273	0.250	0.50	1.25	0.08
1 1/2	1.935 1.915	1.368 1.308	0.351	0.307	0.281	0.50	1.50	0.08
2	2.426 2.406	1.717 1.657	0.430	0.374	0.344	0.62	1.62	0.08

* Dimensions in inches. Dimensions per ASME B16.11.

Notes:

(1) Average of socket wall thickness around periphery shall be no less than listed values. The minimum values are permitted in localized areas.

(2) Upper and lower values for each size are the respective maximum and minimum dimensions.

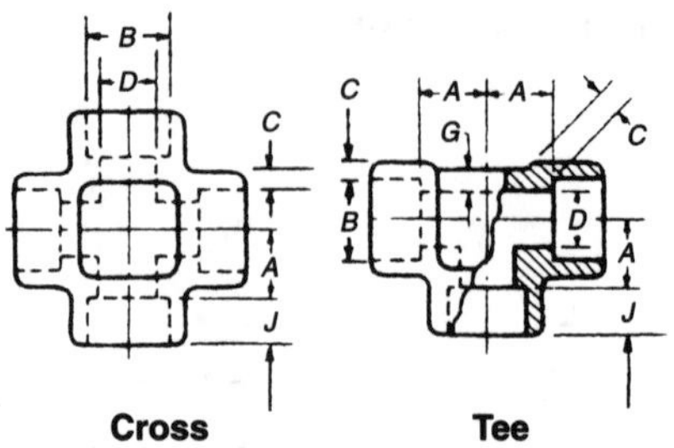

TABLE 5.25M Class 6000 Forged Socket Welding Tees and Crosses (Metric)*

NPS	Socket bore diameter, *B*		Bore diameter of fitting, *D*		Socket wall thickness, *C*		Body wall, *G*, min.	Depth of socket, *J*, min.	Center to bottom of socket, *A*	*A* tolerances ±
	Max.	Min.	Max.	Min.	Ave.	Min.				
1/8	0.440	0.420	0.189	0.126	0.156	0.135	0.124	0.380	0.440	0.030
1/4	0.575	0.555	0.280	0.220	0.181	0.158	0.145	0.380	0.530	0.030
3/8	0.710	0.690	0.389	0.329	0.198	0.172	0.158	0.380	0.620	0.060
1/2	0.875	0.855	0.494	0.434	0.235	0.204	0.188	0.380	0.750	0.060
3/4	1.085	1.065	0.642	0.582	0.274	0.238	0.219	0.500	0.880	0.060
1	1.350	1.330	0.845	0.785	0.312	0.273	0.250	0.500	1.060	0.080
1 1/4	1.695	1.675	1.190	1.130	0.312	0.273	0.250	0.500	1.250	0.080
1 1/2	1.935	1.915	1.368	1.308	0.351	0.307	0.281	0.500	1.500	0.080
2	2.426	2.406	1.717	1.657	0.430	0.374	0.344	0.620	1.620	0.080

* Dimensions in millimeters.

Dimensions of Class 9000 Socket Welding Tees and Crosses

See Tables 5.26 and 5.26M.

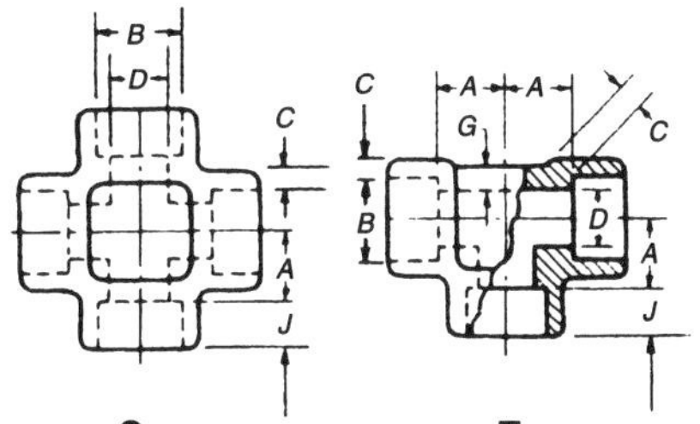

TABLE 5.26 Class 9000 Forged Socket Welding Tees and Crosses (U.S. Customary Units)*

Nom. pipe size	Socket bore dia. (2), *B*	Bore diameter of fittings (2), *D*	Socket wall thickness (1), *C* Ave.	Socket wall thickness (1), *C* Min.	Body wall, *G*, min.	Depth of socket, *J*, min.	Center to bottom of socket, *A*	*A* tolerance ±
½	0.875 0.855	0.282 0.222	0.368	0.322	0.294	0.38	1.00	0.06
¾	1.085 1.065	0.464 0.404	0.385	0.337	0.308	0.50	1.12	0.06
1	1.350 1.330	0.629 0.569	0.448	0.392	0.358	0.50	1.25	0.08
1¼	1.695 1.675	0.926 0.866	0.478	0.418	0.382	0.50	1.38	0.08
1½	1.935 1.915	1.130 1.070	0.500	0.438	0.400	0.50	1.50	0.08
2	2.426 2.406	1.533 1.473	0.545	0.477	0.436	0.62	2.12	0.08

* Dimensions in inches. Dimensions per ASME B16.11.

Notes:

(1) Average of socket wall thickness around periphery shall be no less than listed values. The minimum values are permitted in localized areas.

(2) Upper and lower values for each size are the respective maximum and minimum dimensions.

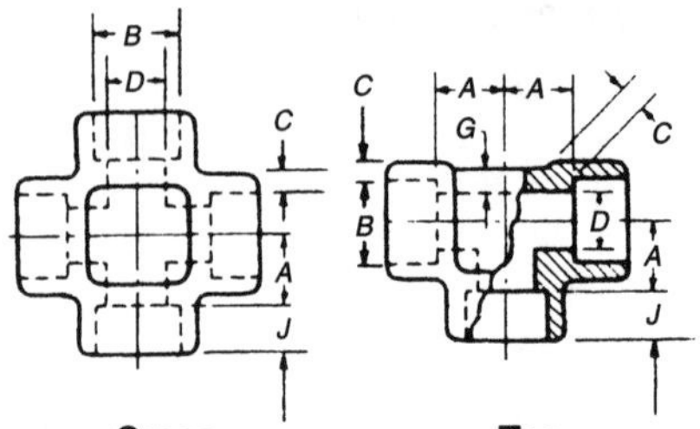

Cross Tee

TABLE 5.26M Class 9000 Forged Socket Welding Tees and Crosses (Metric)*

NPS	Socket bore diameter, *B*		Bore diameter of fitting, *D*		Socket wall thickness, *C*		Body wall, *G*, min.	Depth of socket, *J*, min.	Center to bottom of socket, *A*	*A* tolerance ±
	Max.	Min.	Max.	Min.	Ave.	Min.				
½	0.875	0.855	0.282	0.222	0.368	0.322	0.294	0.38	1.00	0.06
¾	1.085	1.065	0.464	0.404	0.385	0.337	0.308	0.50	1.12	0.06
1	1.350	1.330	0.629	0.569	0.448	0.392	0.358	0.50	1.25	0.08
1¼	1.695	1.675	0.926	0.866	0.478	0.418	0.382	0.50	1.38	0.08
1½	1.935	1.915	1.130	1.070	0.500	0.438	0.400	0.50	1.50	0.08
2	2.426	2.406	1.533	1.473	0.545	0.477	0.436	0.62	2.12	0.08

* Dimensions in millimeters.

Dimensions of Class 3000 Socket Welding Couplings, Half-Couplings, and Caps

See Tables 5.27 and 5.27M.

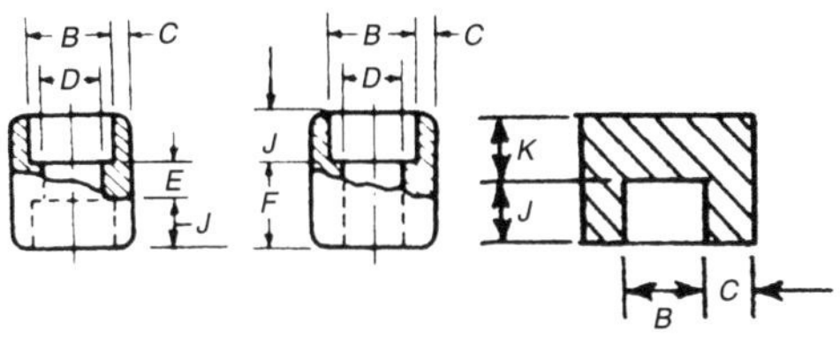

Coupling **Half-coupling** **Cap**

TABLE 5.27 Class 3000 Forged Socket Welding Couplings, Half-Couplings, and Caps (U.S. Customary Units)*

Nom. pipe size	Socket bore dia., B	Bore diameter of fittings, D	Socket wall thickness, C		Depth of socket, J, min.	Laying lengths		Tolerances ±		End wall thickness, K_{min}
			Ave.	Min.		Couplings, E	Half-couplings, F	E	F	
1/8	0.440 0.420	0.299 0.239	0.125	0.125	0.38	0.25	0.62	0.06	0.03	0.19
1/4	0.575 0.555	0.394 0.334	0.149	0.130	0.38	0.25	0.62	0.06	0.03	0.19
3/8	0.710 0.690	0.523 0.463	0.158	0.138	0.38	0.25	0.69	0.12	0.06	0.19
1/2	0.875 0.855	0.652 0.592	0.184	0.161	0.38	0.38	0.88	0.12	0.06	0.25
3/4	1.085 1.065	0.854 0.794	0.193	0.168	0.50	0.38	0.94	0.12	0.06	0.25
1	1.350 1.330	1.079 1.019	0.224	0.196	0.50	0.50	1.12	0.16	0.08	0.38
1 1/4	1.695 1.675	1.410 1.350	0.239	0.208	0.50	0.50	1.19	0.16	0.08	0.38
1 1/2	1.935 1.915	1.640 1.580	0.250	0.218	0.50	0.50	1.25	0.16	0.08	0.44
2	2.426 2.406	2.097 2.037	0.273	0.238	0.62	0.75	1.62	0.16	0.08	0.50
2 1/2	2.931 2.906	2.529 2.409	0.345	0.302	0.62	0.75	1.69	0.20	0.10	0.62
3	3.560 3.535	3.128 3.008	0.375	0.327	0.62	0.75	1.75	0.20	0.10	0.75
4	4.570 4.545	4.086 3.966	0.421	0.368	0.75	0.75	1.88	0.20	0.10	0.88

* Dimensions in inches. Dimensions per ASME B16.11.

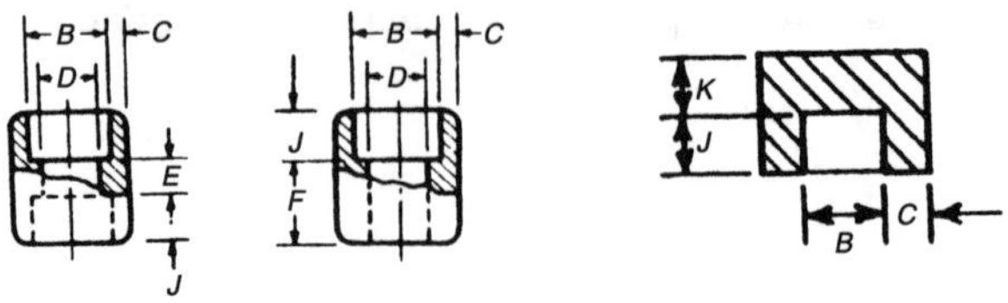

Coupling **Half-coupling** **Cap**

TABLE 5.27M Class 3000 Socket Welding Couplings, Half-Couplings, and Caps (Metric Units)*

DN	Nom. pipe size	Socket bore dia., *B*	Bore diameter of fittings, *D*	Socket wall thickness, *C*		Depth of socket, *J*, min.	Laying lengths		Tolerances ±		End wall thickness, K_{min}
				Ave.	Min.		Couplings, *E*	Half-couplings, *F*	*E*	*F*	
6	1/8	11.2 10.8	7.6 6.1	3.18	3.18	9.5	6.5	16.0	1.5	1.0	4.8
8	1/4	14.6 14.2	10.0 8.5	3.78	3.30	9.5	6.5	16.0	1.5	1.0	4.8
10	3/8	18.0 17.6	13.3 11.8	4.01	3.50	9.5	6.5	17.5	3.0	1.5	4.8
15	1/2	22.2 21.8	16.6 15.0	4.67	4.09	9.5	9.5	22.5	3.0	1.5	6.4
20	3/4	27.6 27.2	21.7 20.2	4.90	4.27	12.5	9.5	24.0	3.0	1.5	6.4
25	1	34.3 33.9	27.4 25.9	5.69	4.98	12.5	12.5	28.5	4.0	2.0	9.6
32	1 1/4	43.1 42.7	35.8 34.3	6.07	5.28	12.5	12.5	30.0	4.0	2.0	9.6
40	1 1/2	49.2 48.8	41.6 40.1	6.35	5.54	12.5	12.5	32.0	4.0	2.0	11.2
50	2	61.7 61.2	53.3 51.7	6.93	6.04	16.0	19.0	41.0	4.0	2.0	12.7
65	2 1/2	74.4 73.9	64.2 61.2	8.76	7.67	16.0	19.0	43.0	5.0	2.5	15.7
80	3	90.3 89.8	79.4 76.4	9.52	8.30	16.0	19.0	44.5	5.0	2.5	19.0
100	4	115.7 115.2	103.8 100.7	10.69	9.35	19.0	19.0	48.0	5.0	2.5	22.4

* Dimensions are in millimeters. Dimensions Conform to ASME B16.11.

Dimensions of Class 6000 Socket Welding Couplings, Half-Couplings, and Caps

See Tables 5.28 and 5.28M.

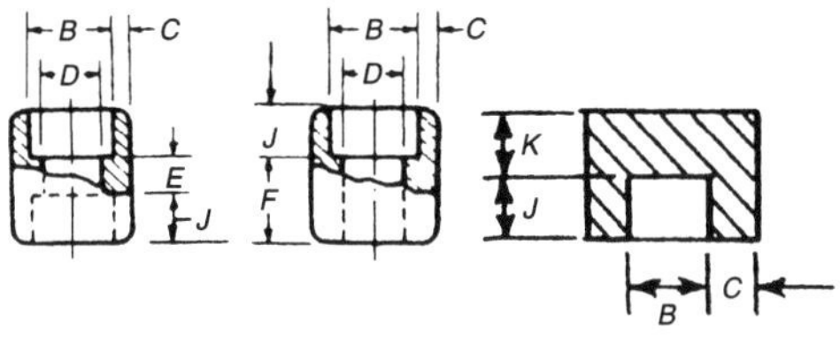

Coupling Half-coupling Cap

TABLE 5.28 Class 6000 Forged Socket Welding Couplings, Half-Couplings, and Caps (U.S. Customary Units)*

Nom. pipe size	Socket bore dia., *B*	Bore diameter of fittings, *D*	Socket wall thickness, *C*		Depth of socket, *J*, min.	Laying lengths		Tolerances ±		End wall thickness, K_{min}
			Ave.	Min.		Couplings, *E*	Half-couplings, *F*	*E*	*F*	
1/8	0.440 0.420	0.189 0.126	0.156	0.135	0.38	0.25	0.62	0.06	0.03	0.25
1/4	0.575 0.555	0.280 0.220	0.181	0.158	0.38	0.25	0.62	0.06	0.03	0.25
3/8	0.710 0.690	0.389 0.329	0.198	0.172	0.38	0.25	0.69	0.12	0.06	0.25
1/2	0.875 0.855	0.494 0.434	0.235	0.204	0.38	0.38	0.88	0.12	0.06	0.31
3/4	1.085 1.065	0.642 0.582	0.274	0.238	0.50	0.38	0.94	0.12	0.06	0.31
1	1.350 1.330	0.845 0.785	0.312	0.273	0.50	0.50	1.12	0.16	0.08	0.44
1 1/4	1.695 1.675	1.190 1.130	0.312	0.273	0.50	0.50	1.19	0.16	0.08	0.44
1 1/2	1.935 1.915	1.368 1.308	0.351	0.307	0.50	0.50	1.25	0.16	0.08	0.50
2	2.426 2.406	1.717 1.657	0.430	0.374	0.62	0.75	1.62	0.16	0.08	0.62

* Dimensions in inches. Dimensions per ASME B16.11.

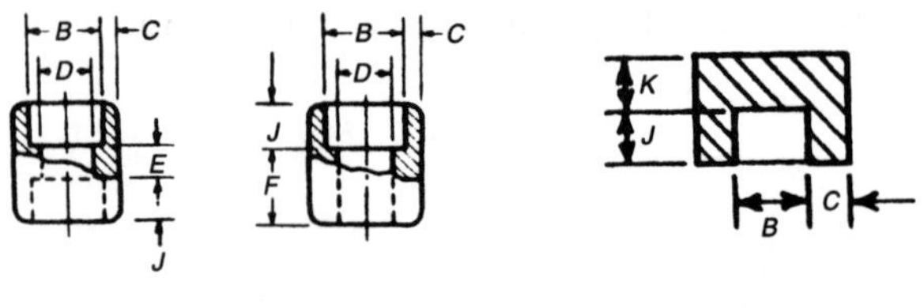

TABLE 5.28M Class 6000 Socket Welding Couplings, Half-Couplings, and Caps (Metric Units)*

DN	Nom. pipe size	Socket bore dia., *B*	Bore diameter of fittings, *D*	Socket wall thickness, *C*		Depth of socket, *J*, min.	Laying lengths		Tolerances ±		End wall thickness, K_{min}
				Ave.	Min.		Couplings, *E*	Half-couplings, *F*	*E*	*F*	
6	1/8	11.2 10.8	4.8 3.2	3.96	3.43	9.5	6.5	16.0	1.5	1.0	6.4
8	1/4	14.6 14.2	7.1 5.6	4.60	4.01	9.5	6.5	16.0	1.5	1.0	6.4
10	3/8	18.0 17.6	9.9 8.4	5.03	4.37	9.5	6.5	17.5	3.0	1.5	6.4
15	1/2	22.2 21.8	12.5 11.0	5.97	5.18	9.5	9.5	22.5	3.0	1.5	7.9
20	3/4	27.6 27.2	16.3 14.8	6.96	6.04	12.5	9.5	24.0	3.0	1.5	7.9
25	1	34.3 33.9	21.5 19.9	7.92	6.93	12.5	12.5	28.5	4.0	2.0	11.2
32	1 1/4	43.1 42.7	30.2 28.7	7.92	6.93	12.5	12.5	30.0	4.0	2.0	11.2
40	1 1/2	49.2 48.8	34.7 33.2	8.92	7.80	12.5	12.5	32.0	4.0	2.0	12.7
50	2	61.7 61.2	43.6 42.1	10.92	9.50	16.0	19.0	41.0	4.0	2.0	15.7

* Dimensions are in millimeters. Dimensions conform to ASME B16.11.

Dimensions of Class 9000 Socket Welding Couplings, Half-Couplings, and Caps

See Tables 5.29 and 5.29M.

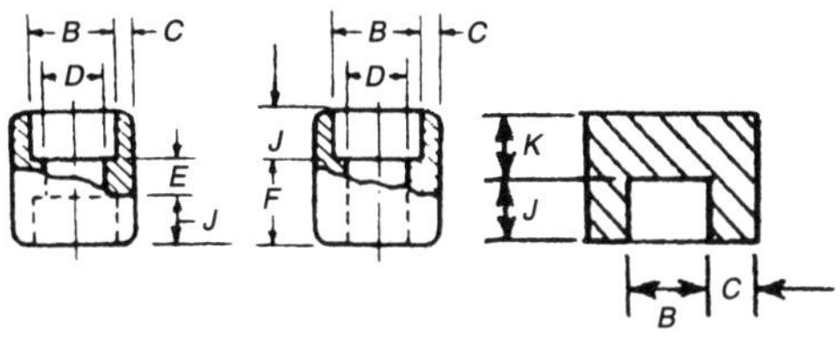

Coupling **Half-coupling** **Cap**

TABLE 5.29 Class 9000 Forged Socket Welding Couplings, Half-Couplings, and Caps (U.S. Customary Units)*

Nom. pipe size	Socket bore dia., *B*	Bore diameter of fittings, *D*	Socket wall thickness, *C*, Min.	Depth of socket, *J*, min.	Laying lengths: Couplings, *E*	Laying lengths: Half-couplings, *F*	Tolerances ±: *E*	Tolerances ±: *F*	End wall thickness, $K_{min.}$
½	0.875 0.855	0.282 0.222	0.322	0.38	0.38	0.88	0.12	0.06	0.44
¾	1.085 1.065	0.464 0.404	0.337	0.50	0.38	0.94	0.12	0.06	0.50
1	1.350 1.330	0.629 0.569	0.392	0.50	0.50	1.12	0.16	0.08	0.56
1¼	1.695 1.675	0.926 0.866	0.418	0.50	0.50	1.19	0.16	0.08	0.56
1½	1.935 1.915	1.130 1.070	0.438	0.50	0.50	1.25	0.16	0.08	0.62
2	2.426 2.406	1.533 1.473	0.477	0.62	0.75	1.62	0.16	0.08	0.75

* Dimensions in inches. Dimensions per ASME B16.11.

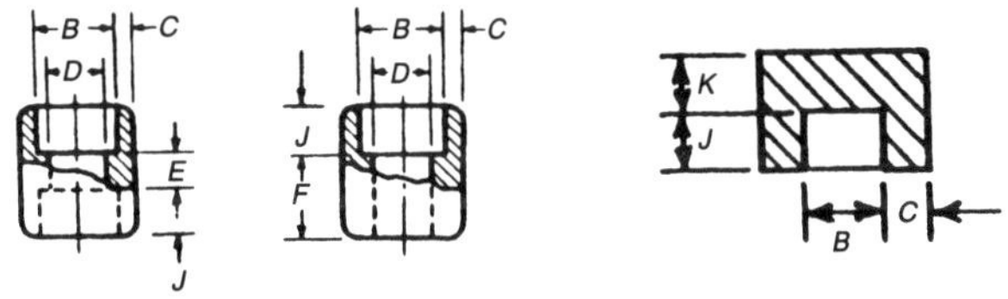

Coupling **Half-coupling** **Cap**

TABLE 5.29M Class 9000 Socket Welding Couplings, Half-Couplings, and Caps (Metric Units)*

DN	Nom. pipe size	Socket bore dia., *B*	Bore diameter of fittings, *D*	Socket wall thickness, *C*: Ave.	Socket wall thickness, *C*: Min.	Depth of socket, *J*, min.	Laying lengths: Couplings, *E*	Laying lengths: Half-couplings, *F*	Tolerances ±: *E*	Tolerances ±: *F*	End wall thickness, $K_{min.}$
15	½	22.2 21.8	7.2 5.6	9.35	8.18	9.5	9.5	22.5	3.0	1.5	11.2
20	¾	27.6 27.2	11.8 10.3	9.78	8.56	12.5	9.5	24.0	3.0	1.5	12.7
25	1	34.3 33.9	16.0 14.4	11.38	9.96	12.5	12.5	28.5	4.0	2.0	14.2
32	1¼	43.1 42.7	23.5 22.0	12.14	10.62	12.5	12.5	30.0	4.0	2.0	14.2
40	1½	49.2 48.8	28.7 27.2	12.70	11.12	12.5	12.5	32.0	4.0	2.0	15.7
50	2	61.7 61.2	38.9 37.4	13.84	12.12	16.0	19.0	41.0	4.0	2.0	19.0

* Dimensions are in millimeters. Dimensions conform to ASME B16.11.

Dimensions of Class 2000 Threaded 90° and 45° Elbows

See Tables 5.30 and 5.30M.

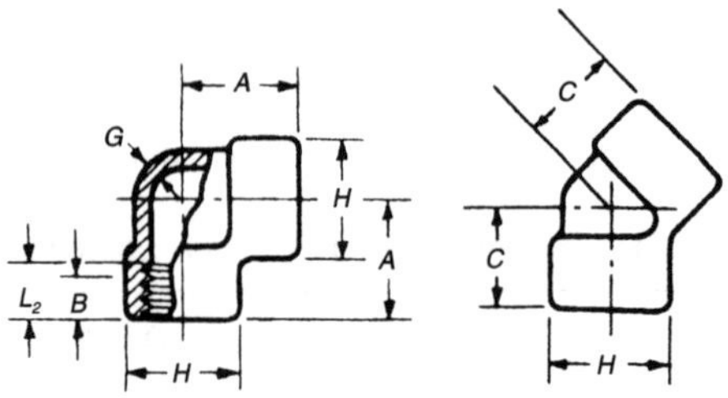

90° elbow **45° elbow**

TABLE 5.30 Class 2000 Forged Threaded 90° and 45° Elbows (U.S. Customary Units)*

Nom. pipe size	Center to end, 90° elbows, *A*	Center to end, 45° elbow, *C*	Outside diameter of band, *H*	Minimum wall thickness, *G*	Length of thread, min.	
					L†	EL‡
1/8	0.81	0.69	0.88	0.125	0.25	0.2639
1/4	0.81	0.69	0.88	0.125	0.32	0.4018
3/8	0.97	0.75	1.00	0.125	0.36	0.4078
1/2	1.12	0.88	1.31	0.125	0.43	0.5337
3/4	1.31	1.00	1.50	0.125	0.50	0.5457
1	1.50	1.12	1.81	0.145	0.58	0.6828
1 1/4	1.75	1.31	2.19	0.153	0.67	0.7068
1 1/2	2.00	1.38	2.44	0.158	0.70	0.7235
2	2.38	1.69	2.97	0.168	0.75	0.7565
2 1/2	3.00	2.06	3.62	0.221	0.93	1.138
3	3.38	2.50	4.31	0.236	1.02	1.200
4	4.19	3.12	5.75	0.258	1.09	1.300

* Dimensions in inches. Threads to conform to ASME B1.20.1.
† *L* = minimum length of perfect thread.
‡ EL = effective length of external thread.

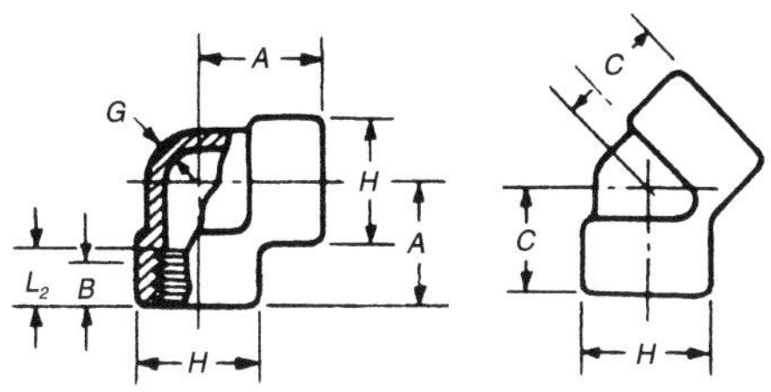

90° elbow **45° elbow**

TABLE 5.30M Class 2000 Forged Threaded 90° and 45° Elbows (Metric Units)*

DN	Nom. pipe size	Center to end, *A*	Center to end, 45° elbow, *C*	Outside diameter of band, *H*	Minimum wall thickness, *G*	Length of thread, min.	
						L†	EL‡
6	1/8	21	17	22	3.18	6.4	6.7
8	1/4	21	17	22	3.18	8.1	10.2
10	3/8	25	19	25	3.18	9.1	10.4
15	1/2	28	22	33	3.18	10.9	13.6
20	3/4	33	25	38	3.18	12.7	13.9
25	1	38	28	46	3.68	14.7	17.3
32	1 1/4	44	33	56	3.89	17.0	18.0
40	1 1/2	51	35	62	4.01	17.8	18.4
50	2	60	43	75	4.27	19.0	19.2
65	2 1/2	76	52	92	5.61	23.6	28.9
80	3	86	64	109	5.99	25.9	30.5
100	4	106	79	146	6.55	27.7	33.0

* Dimensions in millimeters. Threads to conform to ASME B1.20.1.
† *L* = minimum length of external thread.
‡ EL = effective length of external thread.

Dimensions of Class 3000 Threaded 90° and 45° Elbows

See Tables 5.31 and 5.31M.

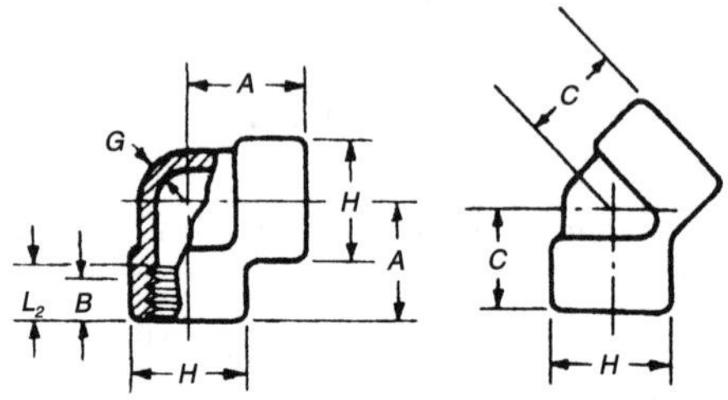

90° elbow 45° elbow

TABLE 5.31 Class 3000 Forged Threaded 90° and 45° Elbows (U.S. Customary Units)*

Nom. pipe size	Center to end, *A*	Center to end, 45° elbow, *C*	Outside diameter of band, *H*	Minimum wall thickness, *G*	Length of thread, min.	
					L†	EL‡
1/8	0.81	0.69	0.88	0.125	0.25	0.2639
1/4	0.97	0.75	1.00	0.130	0.32	0.4018
3/8	1.12	0.88	1.31	0.138	0.36	0.4078
1/2	1.31	1.00	1.50	0.161	0.43	0.5337
3/4	1.50	1.12	1.81	0.170	0.50	0.5457
1	1.75	1.31	2.19	0.196	0.58	0.6828
1 1/4	2.00	1.38	2.44	0.208	0.67	0.7068
1 1/2	2.38	1.69	2.97	0.219	0.70	0.7235
2	2.50	1.72	3.31	0.281	0.75	0.7565
2 1/2	3.25	2.06	4.00	0.301	0.93	1.138
3	3.75	2.50	4.75	0.348	1.02	1.200
4	4.50	3.12	6.00	0.440	1.09	1.300

* Dimensions in inches. Threads to conform to ASME B1.20.1.
† *L* = minimum length of external thread.
‡ EL = effective length of external thread.

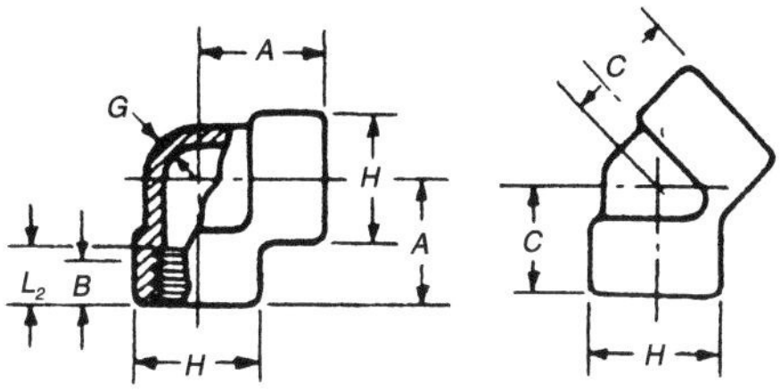

90° elbow **45° elbow**

TABLE 5.31M Class 3000 Forged Threaded 90° and 45° Elbows (Metric Units)*

DN	Nom. pipe size	Center to end, *A*	Center to end, 45° elbow, *C*	Outside diameter of band, *H*	Minimum wall thickness, *G*	Length of thread, min.	
						L†	EL‡
6	1/8	21	17	22	3.18	6.4	6.7
8	1/4	25	19	25	3.30	8.1	10.2
10	3/8	28	22	33	3.51	9.1	10.4
15	1/2	33	25	38	4.09	10.9	13.6
20	3/4	38	28	46	4.32	12.7	13.9
25	1	44	33	56	4.98	14.7	17.3
32	1 1/4	51	35	62	5.28	17.0	18.0
40	1 1/2	60	43	75	5.56	17.8	18.4
50	2	64	44	84	7.14	19.0	19.2
65	2 1/2	83	52	102	7.65	23.6	28.9
80	3	95	64	121	8.84	25.9	30.5
100	4	114	79	152	11.18	27.7	33.0

* Dimensions in millimeters. Threads to conform to ASME B1.20.1.
† *L* = minimum length of external thread.
‡ EL = effective length of external thread.

Dimensions of Class 6000 Threaded 90° and 45° Elbows

See Tables 5.32 and 5.32M.

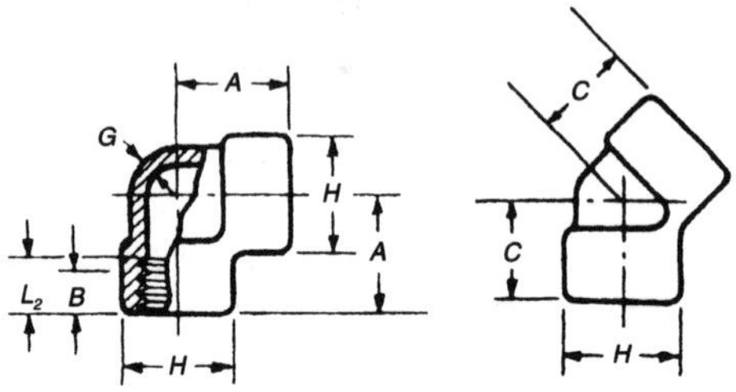

90° elbow **45° elbow**

TABLE 5.32 Class 6000 Forged Threaded 90° and 45° Elbows (U.S. Customary Units)*

Nom. pipe size	Center to end, *A*	Center to end, 45° elbow, *C*	Outside diameter of band, *H*	Minimum wall thickness, *G*	Length of thread, min.	
					L†	EL‡
1/8	0.97	0.75	1.00	0.250	0.25	0.2639
1/4	1.12	0.88	1.31	0.260	0.32	0.4018
3/8	1.31	1.00	1.50	0.275	0.36	0.4078
1/2	1.50	1.12	1.81	0.321	0.43	0.5337
3/4	1.75	1.31	2.19	0.336	0.50	0.5457
1	2.00	1.38	2.44	0.391	0.58	0.6828
1 1/4	2.38	1.69	2.97	0.417	0.67	0.7068
1 1/2	2.50	1.72	3.31	0.436	0.70	0.7235
2	3.25	2.06	4.00	0.476	0.75	0.7565
2 1/2	3.75	2.50	4.75	0.602	0.93	1.138
3	4.19	3.12	5.75	0.655	1.02	1.200
4	4.50	3.12	6.00	0.735	1.09	1.300

* Dimensions in inches. Threads to conform to ASME B1.20.1.
† *L* = minimum length of external thread.
‡ EL = effective length of external thread.

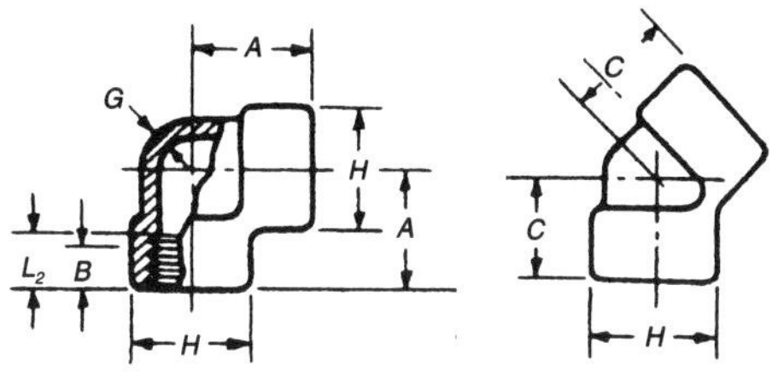

90° elbow 45° elbow

TABLE 5.32M Class 6000 Forged Threaded 90° and 45° Elbows (Metric Units)*

DN	Nom. pipe size	Center to end, *A*	Center to end, 45° elbow, *C*	Outside diameter of band, *H*	Minimum wall thickness, *G*	Length of thread, min.	
						L†	EL‡
6	1/8	25	19	25	6.35	6.4	6.7
8	1/4	28	22	33	6.60	8.1	10.2
10	3/8	33	25	38	6.98	9.1	10.4
15	1/2	38	28	46	8.15	10.9	13.6
20	3/4	44	33	56	8.53	12.7	13.9
25	1	51	35	62	9.93	14.7	17.3
32	1 1/4	60	43	75	10.59	17.0	18.0
40	1 1/2	64	44	84	11.07	17.8	18.4
50	2	83	52	102	12.09	19.0	19.2
65	2 1/2	95	64	121	15.29	23.6	28.9
80	3	106	79	146	16.64	25.9	30.5
100	4	114	79	152	18.67	27.7	33.0

* Dimensions in millimeters. Threads to conform to ASME B1.20.1.
† *L* = minimum length of external thread.
‡ EL = effective length of external thread.

Dimensions of Class 2000 Threaded Tees and Crosses

See Tables 5.33 and 5.33M.

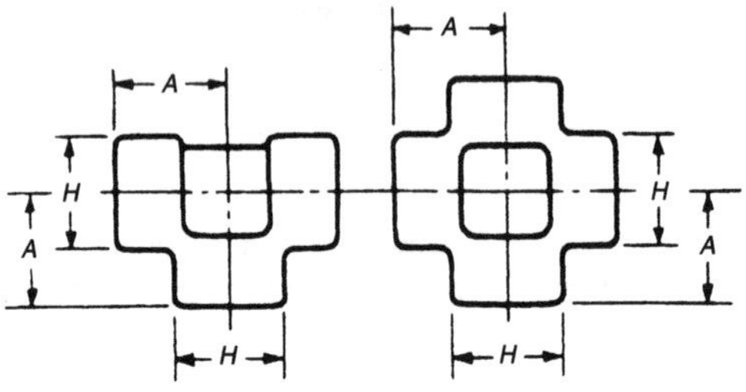

TABLE 5.33 Class 2000 Forged Threaded Tees and Crosses (U.S. Customary Units)*

Nom. pipe size	Center to end, *A*	Outside diameter of band, *H*	Minimum wall thickness, *G*	Length of thread, min.	
				L†	EL‡
1/8	0.81	0.88	0.125	0.25	0.2639
1/4	0.81	0.88	0.125	0.32	0.4018
3/8	0.97	1.00	0.125	0.36	0.4078
1/2	1.12	1.31	0.125	0.43	0.5337
3/4	1.31	1.50	0.125	0.50	0.5457
1	1.50	1.81	0.145	0.58	0.6828
1 1/4	1.75	2.19	0.153	0.67	0.7068
1 1/2	2.00	2.44	0.158	0.70	0.7235
2	2.38	2.97	0.168	0.75	0.7565
2 1/2	3.00	3.62	0.221	0.93	1.138
3	3.38	4.31	0.236	1.02	1.200
4	4.19	5.75	0.258	1.09	1.300

* Dimensions in millimeters. Threads to conform to ASME B1.20.1.
† L = minimum length of perfect thread.
‡ EL = effective length of external thread.

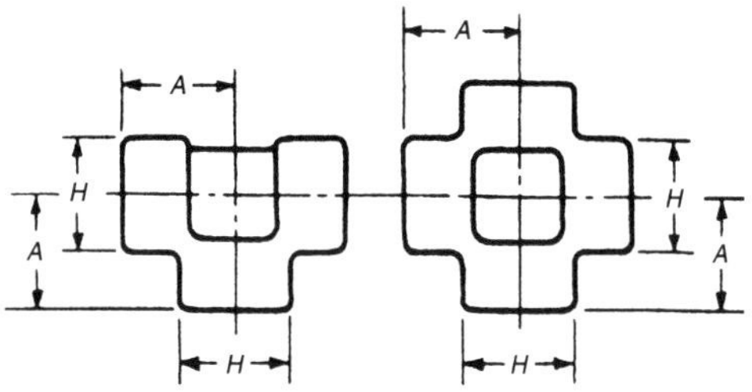

Tee Cross

TABLE 5.33M Class 2000 Forged Threaded Tees and Crosses (Metric Units)*

DN	Nom. pipe size	Center to end, *A*	Outside diameter of band, *H*	Minimum wall thickness, *G*	Length of thread, min. $L^{\dagger}$	$EL^{\ddagger}$
6	1/8	21	22	3.18	6.4	6.7
8	1/4	21	22	3.18	8.1	10.2
10	3/8	25	25	3.18	9.1	10.4
15	1/2	28	33	3.18	10.9	13.6
20	3/4	33	38	3.18	12.7	13.9
25	1	38	46	3.68	14.7	17.3
32	1 1/4	44	56	3.89	17.0	18.0
40	1 1/2	51	62	4.01	17.8	18.4
50	2	60	75	4.27	19.0	19.2
65	2 1/2	76	92	5.61	23.6	28.9
80	3	86	109	5.99	25.9	30.5
100	4	106	146	6.55	27.7	33.0

* Dimensions in millimeters. Threads to conform to ASME B1.20.1.
† *L* = minimum length of perfect thread.
‡ EL = effective length of external thread.

Dimensions of Class 3000 Threaded Tees and Crosses

See Tables 5.34 and 5.34M.

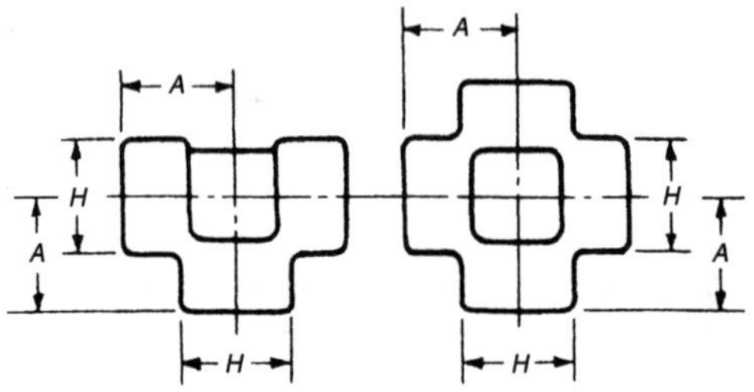

Tee Cross

TABLE 5.34 Class 3000 Forged Threaded Tees and Crosses (U.S. Customary Units)*

Nom. pipe size	Center to end, *A*	Outside diameter of band, *H*	Minimum wall thickness, *G*	Length of thread, min. *L*†	Length of thread, min. EL‡
1/8	0.81	0.88	0.125	0.25	0.2639
1/4	0.97	1.00	0.130	0.32	0.4018
3/8	1.12	1.31	0.138	0.36	0.4078
1/2	1.31	1.50	0.161	0.43	0.5337
3/4	1.50	1.81	0.170	0.50	0.5457
1	1.75	2.19	0.196	0.58	0.6828
1 1/4	2.00	2.44	0.208	0.67	0.7068
1 1/2	2.38	2.97	0.219	0.70	0.7235
2	2.50	3.31	0.281	0.75	0.7565
2 1/2	3.25	4.00	0.301	0.93	1.138
3	3.75	4.75	0.348	1.02	1.200
4	4.50	6.00	0.440	1.09	1.300

* Dimensions in inches. Threads to conform to ASME B1.20.1.
† *L* = minimum length of perfect thread.
‡ EL = effective length of external thread.

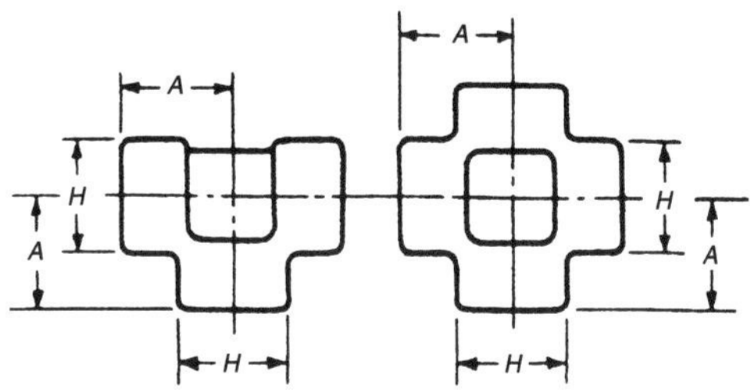

Tee Cross

TABLE 5.34M Class 3000 Forged Threaded Tees and Crosses (Metric Units)*

DN	Nom. pipe size	Center to end, *A*	Outside diameter of band, *H*	Minimum wall thickness, *G*	Length of thread, min.	
					L†	EL‡
6	1/8	21	22	3.18	6.4	6.7
8	1/4	25	25	3.30	8.1	10.2
10	3/8	28	33	3.51	9.1	10.4
15	1/2	33	38	4.09	10.9	13.6
20	3/4	38	46	4.32	12.7	13.9
25	1	44	56	4.98	14.7	17.3
32	1 1/4	51	62	5.28	17.0	18.0
40	1 1/2	60	75	5.56	17.8	18.4
50	2	64	84	7.14	19.0	19.2
65	2 1/2	83	102	7.65	23.6	28.9
80	3	95	121	8.84	25.9	30.5
100	4	114	152	11.18	27.7	33.0

* Dimensions in millimeters. Threads to conform to ASME B1.20.1.
† L = minimum length of perfect thread.
‡ EL = effective length of external thread.

Dimensions of Class 6000 Threaded Tees and Crosses

See Tables 5.35 and 5.35M.

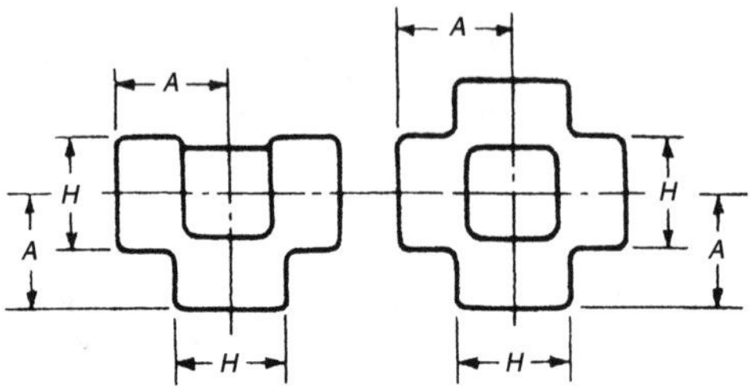

Tee **Cross**

TABLE 5.35 Class 6000 Forged Threaded Tees and Crosses (U.S. Customary Units)*

Nom. pipe size	Center to end, *A*	Outside diameter of band, *H*	Minimum wall thickness, *G*	Length of thread, min. *L*†	EL‡
1/8	0.97	1.00	0.250	0.25	0.2639
1/4	1.12	1.31	0.260	0.32	0.4018
3/8	1.31	1.50	0.275	0.36	0.4078
1/2	1.50	1.81	0.321	0.43	0.5337
3/4	1.75	2.19	0.336	0.50	0.5457
1	2.00	2.44	0.391	0.58	0.6828
1 1/4	2.38	2.97	0.417	0.67	0.7068
1 1/2	2.50	3.31	0.436	0.70	0.7235
2	3.25	4.00	0.476	0.75	0.7565
2 1/2	3.75	4.75	0.602	0.93	1.138
3	4.19	5.75	0.655	1.02	1.200
4	4.50	6.00	0.735	1.09	1.300

* Dimensions in inches. Threads to conform to ASME B1.20.1.
† *L* = minimum length of perfect thread.
‡ EL = effective length of external thread.

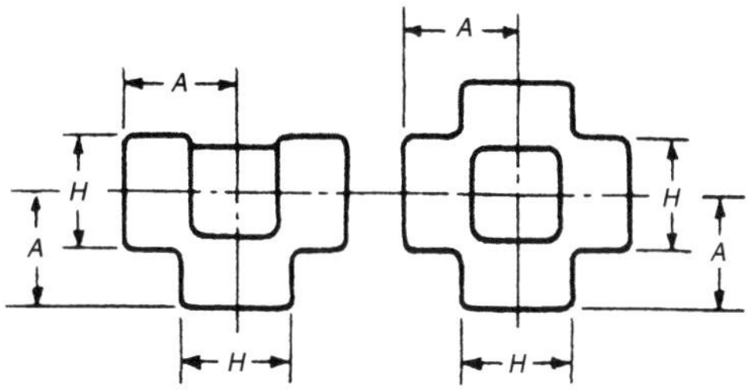

TABLE 5.35M Class 6000 Forged Threaded Tees and Crosses (Metric Units)*

DN	Nom. pipe size	Center to end, *A*	Outside diameter of band, *H*	Minimum wall thickness, *G*	Length of thread, min. *L*[†]	Length of thread, min. EL[‡]
6	1/8	25	25	6.35	6.4	6.7
8	1/4	28	33	6.60	8.1	10.2
10	3/8	33	38	6.98	9.1	10.4
15	1/2	38	46	8.15	10.9	13.6
20	3/4	44	56	8.53	12.7	13.9
25	1	51	62	9.93	14.7	17.3
32	1 1/4	60	75	10.59	17.0	18.0
40	1 1/2	64	84	11.07	17.8	18.4
50	2	83	102	12.09	19.0	19.2
65	2 1/2	95	121	15.29	23.6	28.9
80	3	106	146	16.64	25.9	30.5
100	4	114	152	18.67	27.7	33.0

* Dimensions in millimeters. Threads to conform to ASME B1.20.1.
† *L* = minimum length of perfect thread.
‡ EL = effective length of external thread.

Dimensions of Class 3000 Threaded Couplings and Half-Couplings

See Tables 5.36 and 5.36M.

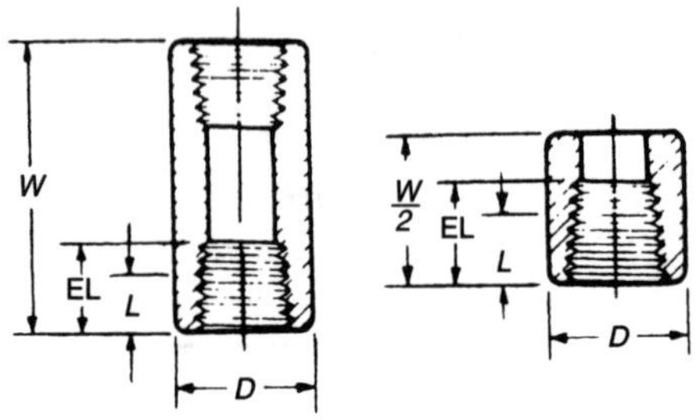

Coupling **Half-coupling**

TABLE 5.36 Class 3000 Forged Threaded Couplings and Half-Couplings (U.S. Customary Units)*

Nom. pipe size	End to end, *W*	Outside diameter, *D*	Length of thread, min.	
			L†	EL‡
1/8	1.25	0.62	0.25	0.2639
1/4	1.38	0.75	0.32	0.4018
3/8	1.50	0.88	0.36	0.4078
1/2	1.88	1.12	0.43	0.5337
3/4	2.00	1.38	0.50	0.5457
1	2.38	1.75	0.58	0.6828
1 1/4	2.62	2.25	0.67	0.7068
1 1/2	3.12	2.50	0.70	0.7235
2	3.38	3.00	0.75	0.7565
2 1/2	3.62	3.62	0.93	1.138
3	4.25	4.25	1.02	1.200
4	4.75	5.50	1.09	1.300

* Dimensions in inches. Threads conform to ASME B1.20.1.
† *L* = minimum length of perfect thread.
‡ EL = effective length of external thread.

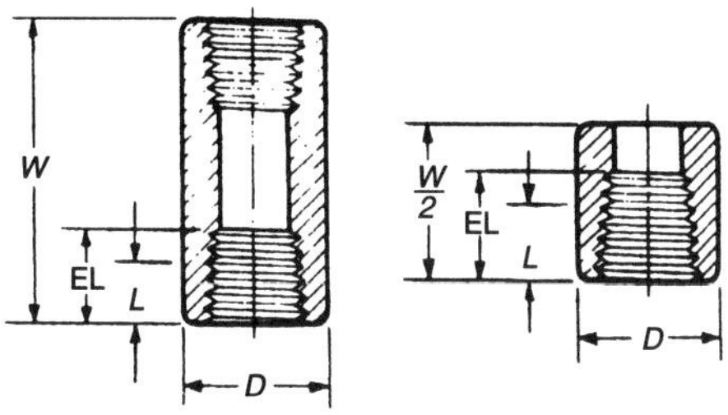

Coupling **Half-coupling**

TABLE 5.36M Class 3000 Forged Threaded Couplings and Half-Couplings (Metric Units)*

DN	Nom. pipe size	End to end, *W*	Outside diameter, *D*	Length of thread, min.	
				L†	EL‡
6	1/8	32	16	6.4	6.7
8	1/4	35	19	8.1	10.2
10	3/8	38	22	9.1	10.4
15	1/2	48	28	10.9	13.6
20	3/4	51	35	12.7	13.9
25	1	60	44	14.7	17.3
32	1 1/4	67	57	17.0	18.0
40	1 1/2	79	64	17.8	18.4
50	2	86	76	19.0	19.2
65	2 1/8	92	92	23.6	28.9
80	3	108	108	25.9	30.5
100	4	121	140	27.7	33.0

* Dimensions in millimeters. Threads must conform to ASME B1.20.1.
† *L* = minimum length of perfect thread.
‡ EL = effective length of external thread.

Dimensions of Class 6000 Threaded Couplings and Half-Couplings

See Tables 5.37 and 5.37M.

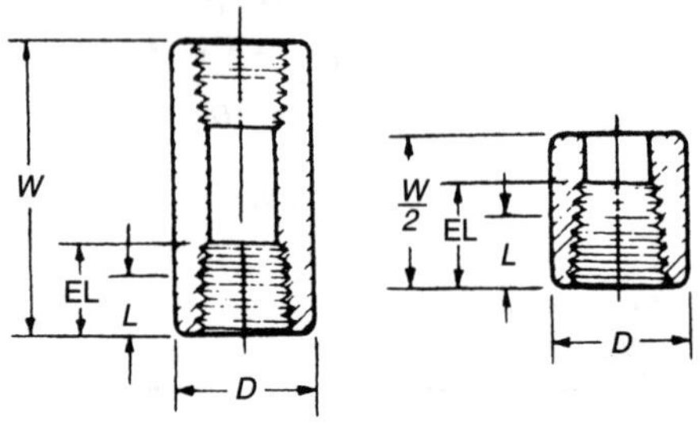

Coupling **Half-coupling**

TABLE 5.37 Class 6000 Forged Threaded Couplings and Half-Couplings (U.S. Customary Units)*

Nom. pipe size	End to end, *W*	Outside diameter, *D*	Length of thread, min.	
			L†	EL‡
1/8	1.25	0.88	0.25	0.2639
1/4	1.38	1.00	0.32	0.4018
3/8	1.50	1.25	0.36	0.4078
1/2	1.88	1.50	0.43	0.5337
3/4	2.00	1.75	0.50	0.5457
1	2.38	2.25	0.58	0.6828
1 1/4	2.62	2.50	0.67	0.7068
1 1/2	3.12	3.00	0.70	0.7235
2	3.38	3.62	0.75	0.7565
2 1/2	3.62	4.25	0.93	1.138
3	4.25	5.00	1.02	1.200
4	4.75	6.25	1.09	1.300

* Dimensions in inches. Threads must conform to ASME B1.20.1.
† L = minimum length of perfect thread.
‡ EL = effective length of external thread.

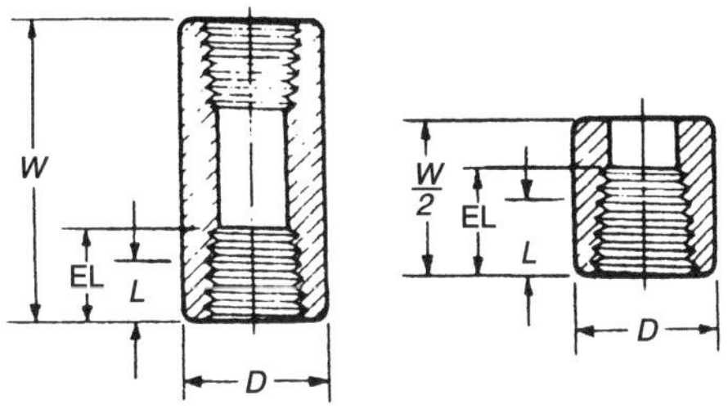

Coupling **Half-coupling**

TABLE 5.37M Class 6000 Forged Threaded Couplings and Half-Couplings (Metric Units)*

DN	Nom. pipe size	End to end, *W*	Outside diameter, *D*	Length of thread, min. *L*†	EL‡
6	1/8	32	22	6.4	6.7
8	1/4	35	25	8.1	10.2
10	3/8	38	32	9.1	10.4
15	1/2	48	38	10.9	13.6
20	3/4	51	44	12.7	13.9
25	1	60	57	14.7	17.3
32	1 1/4	67	64	17.0	18.0
40	1 1/2	79	76	17.8	18.4
50	2	86	92	19.0	19.2
65	2 1/2	92	108	23.6	28.9
80	3	108	127	25.9	30.5
100	4	121	159	27.7	33.0

* Dimensions in millimeters. Threads must conform to ASME B1.20.1.
† *L* = minimum length of perfect thread.
‡ EL = effective length of external thread.

Dimensions of Class 3000 Threaded Caps

See Tables 5.38 and 5.38M.

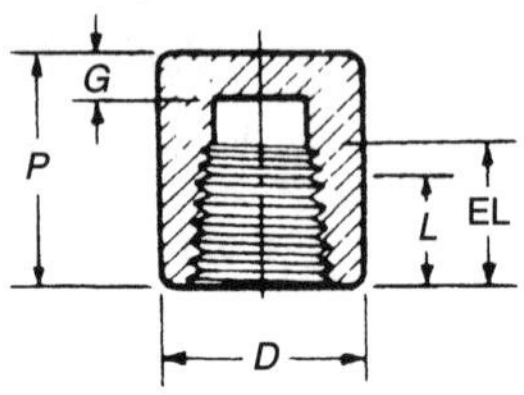

Cap

TABLE 5.38 Class 3000 Forged Threaded Caps (U.S. Customary Units)*

Nom. pipe size	End to end, P	Outside diameter, D	End wall thickness, G, min.	Length of thread, min.	
				L[†]	EL[‡]
1/8	0.75	0.62	0.19	0.25	0.2639
1/4	1.00	0.75	0.19	0.32	0.4018
3/8	1.00	0.88	0.19	0.36	0.4078
1/2	1.25	1.12	0.25	0.43	0.5337
3/4	1.44	1.38	0.25	0.50	0.5457
1	1.62	1.75	0.38	0.58	0.6828
1 1/4	1.75	2.25	0.38	0.67	0.7068
1 1/2	1.75	2.50	0.44	0.70	0.7235
2	1.88	3.00	0.50	0.75	0.7565
2 1/2	2.38	3.62	0.62	0.93	1.138
3	2.56	4.25	0.75	1.02	1.200
4	2.69	5.50	0.88	1.09	1.300

* Dimensions in inches. Threads must conform to ASME B1.20.1.
† L = minimum length of thread.
‡ EL = effective length of external thread.

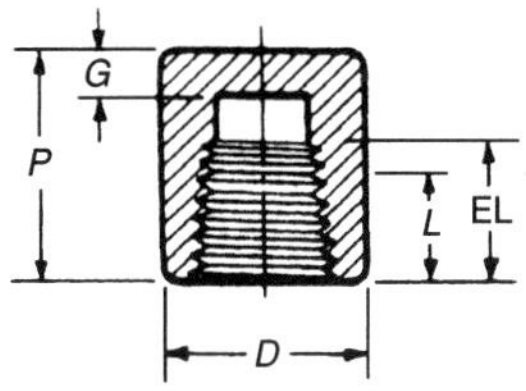

Cap

TABLE 5.38M Class 3000 Forged Threaded Caps (Metric Units)*

DN	Nom. pipe size	End to end, *P*	Outside diameter, *D*	End wall thickness, *G*, min.	Length of thread, min.	
					L†	EL‡
6	1/8	19	16	4.8	6.4	6.7
8	1/4	25	19	4.8	8.1	10.2
10	3/8	25	22	4.8	9.1	10.4
15	1/2	32	28	6.4	10.9	13.6
20	3/4	37	35	6.4	12.7	13.9
25	1	41	44	9.7	14.7	17.3
32	1 1/4	44	57	9.7	17.0	18.0
40	1 1/2	44	64	11.2	17.8	18.4
50	2	48	76	12.7	19.0	19.2
65	2 1/2	60	92	15.7	23.6	28.9
80	3	65	108	19.0	25.9	30.5
100	4	68	140	22.4	27.7	33.0

* Dimensions in millimeters. Threads must conform to ASME B1.20.1.
† *L* = minimum length of thread.
‡ EL = effective length of external thread.

Dimensions of Class 6000 Threaded Caps

See Tables 5.39 and 5.39M.

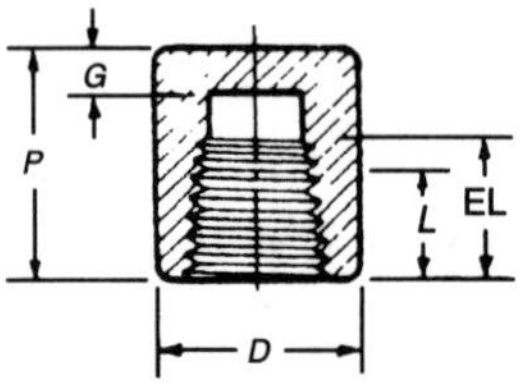

Cap

TABLE 5.39 Class 6000 Forged Threaded Caps (U.S. Customary Units)*

Nom. pipe size	End to end, *P*	Outside diameter, *D*	End wall thickness, *G*, min.	Length of thread, min.	
				L†	EL‡
1/4	1.06	1.00	0.25	0.32	0.4018
3/8	1.06	1.25	0.25	0.36	0.4078
1/2	1.31	1.50	0.31	0.43	0.5337
3/4	1.50	1.75	0.31	0.50	0.5457
1	1.69	2.25	0.44	0.58	0.6828
1 1/4	1.81	2.50	0.44	0.67	0.7068
1 1/2	1.88	3.00	0.50	0.70	0.7235
2	2.00	3.62	0.62	0.75	0.7565
2 1/2	2.50	4.25	0.75	0.93	1.138
3	2.69	5.00	0.88	1.02	1.200
4	2.94	6.25	1.12	1.09	1.300

* Dimensions in inches. Threads must conform to ASME B1.20.1.
† *L* = minimum length of thread.
‡ EL = effective length of external thread.

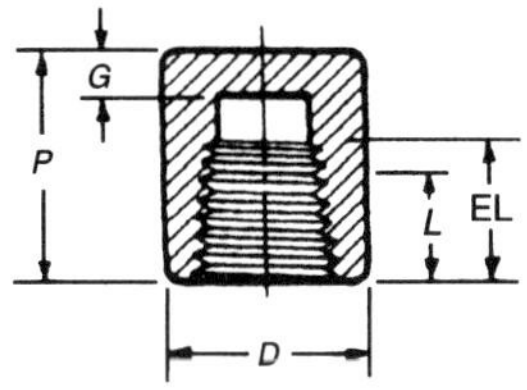

Cap

TABLE 5.39M Class 6000 Forged Threaded Caps (Metric Units)*

DN	Nom. pipe size	End to end, *P*	Outside diameter, *D*	End wall thickness, *G*, min.	Length of thread, min	
					L†	EL‡
8	1/4	27	25	6.4	8.1	10.2
10	3/8	27	32	6.4	9.1	10.4
15	1/2	33	38	7.9	10.9	13.6
20	3/4	38	44	7.9	12.7	13.9
25	1	43	57	11.2	14.7	17.3
32	1 1/4	46	64	11.2	17.0	18.0
40	1 1/2	48	76	12.7	17.8	18.4
50	2	51	92	15.7	19.0	19.2
65	2 1/2	64	108	19.0	23.6	28.9
80	3	68	127	22.4	25.9	30.5
100	4	75	159	28.4	27.7	33.0

* Dimensions in millimeters. Threads must conform to ASME B1.20.1.
† *L* = minimum length of thread.
‡ EL = effective length of external thread.

Dimensions of Threaded Plugs

See Tables 5.40 and 5.40M.

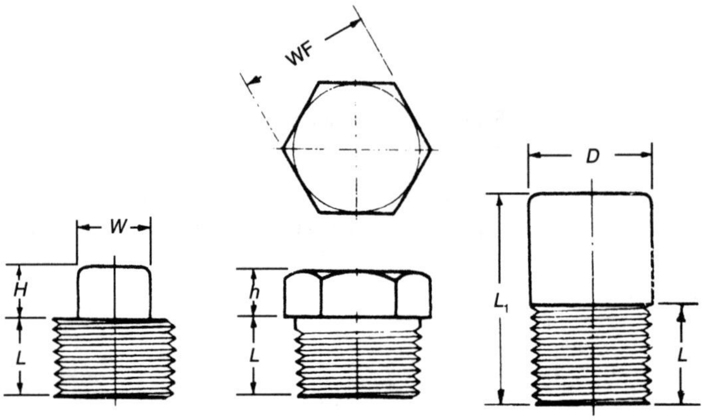

Square head plug **Hex head plug** **Round head plug**

TABLE 5.40 Forged Threaded Plugs (U.S. Customary Units)

		Square head		Round head		Hex plugs	
Nominal pipe size	Length (minimum), *L*	Height of square (minimum), *H*	Width flats (minimum), *W*	Nominal diameter of head, *D*	Length (minimum), L_1	Width flats (nominal), WF	Hex height (min.), *h*
1/8	0.38	0.25	0.28	0.41	1.38	0.44	0.25
1/4	0.44	0.25	0.38	0.53	1.62	0.62	0.25
3/8	0.50	0.31	0.44	0.69	1.62	0.69	0.31
1/2	0.56	0.38	0.56	0.84	1.75	0.88	0.31
3/4	0.62	0.44	0.62	1.06	1.75	1.06	0.38
1	0.75	0.50	0.81	1.31	2.00	1.38	0.38
1 1/4	0.81	0.56	0.94	1.69	2.00	1.75	0.56
1 1/2	0.81	0.62	1.12	1.91	2.00	2.00	0.62
2	0.88	0.69	1.31	2.38	2.50	2.50	0.69
2 1/2	1.06	0.75	1.50	2.88	2.75	3.00	0.75
3	1.12	0.81	1.69	3.50	2.75	3.50	0.81
4	1.25	1.00	2.50	4.50	3.00	4.62	1.00

Notes:
1. Plugs are not identified by class designation.
2. Plugs may be used for ratings up through Class 6000 designation.
3. Dimensions in inches.

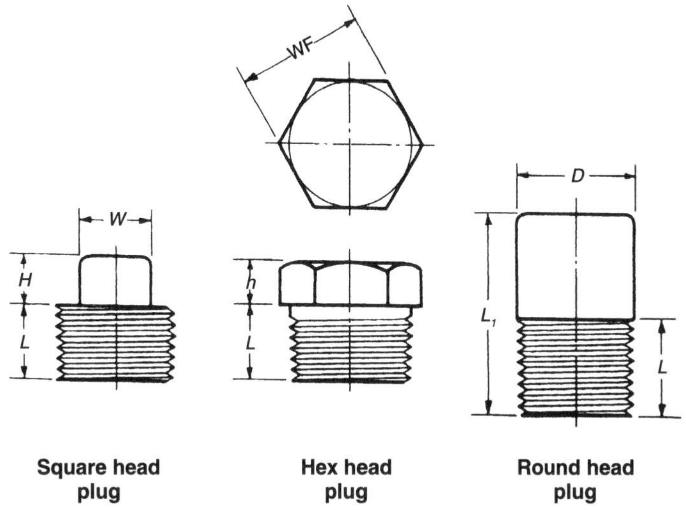

TABLE 5.40M Forged Threaded Plugs (Metric Units)

DN	Nominal pipe size	Length (minimum), *L*	Square head: Height of square (minimum), *H*	Square head: Width flats (minimum), *W*	Round head: Nominal diameter of head, *D*	Round head: Length (minimum), L_1	Hex plugs: Width flats (nominal), WF	Hex plugs: Hex height (min.), *h*
6	1/8	10	6	7	10	35	11	6
8	1/4	11	6	10	14	41	16	6
10	3/8	13	8	11	18	41	18	8
15	1/2	14	10	14	21	44	22	8
20	3/4	16	11	16	27	44	27	10
25	1	19	13	21	33	51	36	10
32	1 1/4	21	14	24	43	51	46	14
40	1 1/2	21	16	28	48	51	50	16
50	2	22	18	32	60	64	65	18
65	2 1/2	27	19	36	73	70	75	19
80	3	28	21	41	89	70	90	21
100	4	32	25	65	114	76	115	25

Notes:
1. Plugs are not identified by class designation.
2. Plugs may be used for ratings up through Class 6000 designation.
3. Dimensions in millimeters.

Dimensions of Threaded Bushings

See Tables 5.41 and 5.41M.

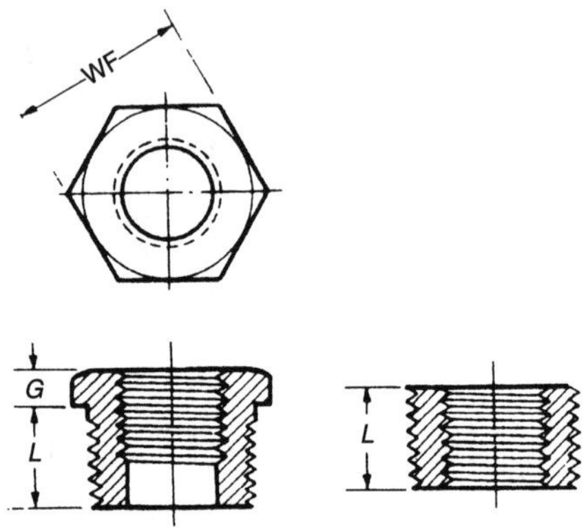

Hex head bushing **Flush bushing**

TABLE 5.41 Forged Bushings (U.S. Customary Units)

Nominal pipe size	Length (minimum), *L*	Width flats (nominal), WF	Hex height (min.), *G*
1/8	0.38	0.44	
1/4	0.44	0.62	0.12
3/8	0.50	0.69	0.16
1/2	0.56	0.88	0.19
3/4	0.62	1.06	0.22
1	0.75	1.38	0.25
1 1/4	0.81	1.75	0.28
1 1/2	0.81	2.00	0.31
2	0.88	2.50	0.34
2 1/2	1.06	3.00	0.38
3	1.12	3.50	0.41
4	1.25	4.62	0.50

Notes:
1. Bushings are not rated by class designation.
2. Bushings may be used for ratings up through Class 6000 designation.
3. Dimensions in inches.

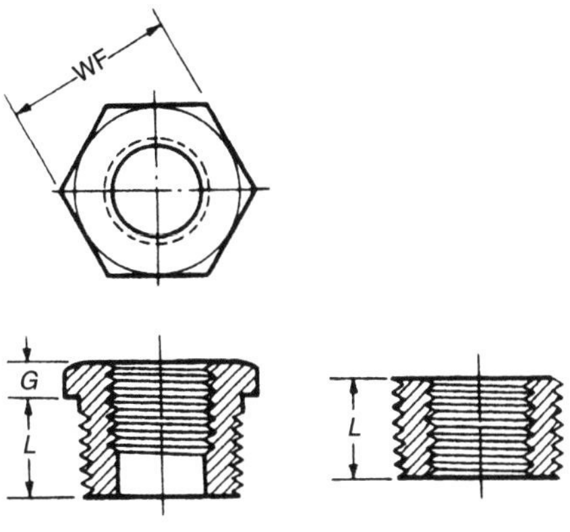

Hex head bushing

Flush bushing

TABLE 5.41M Forged Bushings (Metric Units)

DN	Nominal pipe size	Length (minimum), *L*	Width flats (nominal), WF	Hex height (min.), *G*
6	1/8	10	11	...
8	1/4	11	16	3
10	3/8	13	18	4
15	1/2	14	22	5
20	3/4	16	27	6
25	1	19	36	6
32	1 1/4	21	46	7
40	1 1/2	21	50	8
50	2	22	65	9
65	2 1/2	27	75	10
80	3	28	90	10
100	4	32	115	13

Notes:
1. Bushings are not rated by class designation.
2. Bushings may be used for ratings up through Class 6000 designation.
3. Dimensions in millimeters.

ASME B16.5 PIPE FLANGES

Material for ASME B16.5 Flanges and Flanged Fittings

ASME B16.5 pipe flanges may be manufactured from materials listed in Table 5.42. It is noted that only blind flanges may be made from plate, in addition to being cast or forged from the materials listed for forgings. Flanges and flanged fittings may be cast or forged.

Materials are grouped in 36 groups covering carbon steels (1.1 to 1.15), stainless steel (2.1 to 2.8), and high-alloy steels (3.1 to 3.17). It is noted that some group sequence numbers are missing because either they were merged with others or deleted.

Flange Sizes Covered by ASME B16.5

Pressure Class 150, 300, 400, 600, 900, and 1500 flanges are manufactured in sizes NPS 1/2 through NPS 24. Pressure Class 2500 flanges are manufactured in sizes NPS 1/2 through NPS 12.

TABLE 5.42 List of Material Specifications for ASME B16.5 Flanges

Material group	Nominal designation	Applicable ASTM specifications*		
		Forgings	Castings	Plates
1.1	C-Si C-Mn-Si C-Mn-Si-V	A105 A350 Gr. LF2 A350 Gr. LF6 Cl. 1	A216 Gr. WCB	A515 Gr. 70 A516 Gr. 70 A537 Cl. 1
1.2	C-Mn-Si C-Mn-Si-V 2½Ni 3½Ni	 A350 Gr. LF6 Cl. 2 A350 Gr. LF3	A216 Gr. WCC A352 Gr. LCC A352 Gr. LC2 A352 Gr. LC3	 A203 Gr. B A203 Gr. E
1.3	C-Si C-Mn-Si 2½Ni 3½Ni		A352 Gr. LCB	A515 Gr. 65 A516 Gr. 65 A203 Gr. A A203 Gr. D
1.4	C-Si C-Mn-Si	 A350 Gr. LF1 Cl. 1		A515 Gr. 60 A515 Gr. 60
1.5	C-½Mo	A182 Gr. F1	A217 Gr. WC1 A352 Gr. LC1	A204 Gr. A A204 Gr. B
1.7	C-½Mo ½Cr-½Mo Ni-½Cr-½Mo ¾Ni-¾Cr-1Mo	 A182 Gr. F2	 A217 Gr. WC4 A217 Gr. WC5	A204 Gr. C
1.9	1CR-½Mo 1¼Cr-½Mo 1¼Cr-½Mo-Si	A182 Gr. F12 Cl. 2 A182 Gr. F11 Cl. 2	 A217 Gr. WC6	 A387 Gr. 11 Cl. 2
1.10	2¼Cr-1Mo	A182 Gr. F22 Cl. 3	A217 Gr. WC9	A387 Gr. 22 Cl. 2
1.13	5Cr-½Mo	A182 Gr. F5 A182 Gr. F5a	 A217 Gr. C5	
1.14	9Cr-1Mo	A182 Gr. F9	A217 Gr. C12	
1.15	9Cr-1Mo-V	A1822 Gr. F91	A217 Gr. C12A	A387 Gr. 91 Cl. 2
2.1	18Cr-8Ni	A182 Gr. F304 A182 Gr. F304H	A351 Gr. CF3 A351 Gr. CF8	A240 Gr. 304 A240 Gr. 304H
2.2	16Cr-12Ni-2Mo 18Cr-13Ni-3Mo 19Cr-10Ni-3Mo	A182 Gr. F316 A182 Gr. F316H	A351 Gr. CF3M A351 Gr. CF8M A351 Gr. CG8M	A240 Gr. 316 A240 Gr. 316H A240 Gr. 317
2.3	18Cr-8Ni 16Cr-12Ni-2Mo	A182 Gr. F304L A182 Gr. F316L		A2240 Gr. 304L A240 Gr. 316L
2.4	18Cr-10Ni-Ti	A182 Gr. F321 A182 Gr. F321H		A240 Gr. 321 A240 Gr. 321H
2.5	18Cr-10Ni-Cb	A182 Gr. F347 A182 Gr. F347H A182 Gr. F348 A182 Gr. F348H	A351 Gr. CF8C	A240 Gr. 347 A240 Gr. 347H A240 Gr. 348 A240 Gr. 348H
2.6	25Cr-12Ni 23Cr-12Ni		A351 Gr. CH8 A351 Gr. CH20	 A240 Gr. 309S A240 Gr. 309H

TABLE 5.42 (Continued)

Material group	Nominal designation	Applicable ASTM specifications*		
		Forgings	Castings	Plates
2.7	25Cr-20Ni	A182 Gr. F310	A351 Gr. CK20	A240 Gr. 310S A240 Gr. 310H
2.8	20Cr-18Ni-6Mo 22Cr-5Ni-3Mo-N 25Cr-7Ni-4Mo-N 24Cr-10Ni-4Mo-V 25Cr-4Ni-2Mo-3Cu 25Cr-7Ni-3.5Mo-W-Cb 25Cr-7Ni-3.5Mo-N-Cu-W	A182 Gr. F44 A182 Gr. F51 A182 Gr. F53 A182 Gr. F55	A351 Gr. CK3MCuN A351 Gr. CE8MN A351 Gr. CD4MCu A351 Gr. CD3MWCuN	A240 Gr. S31254 A240 Gr. S31803 A240 Gr. S32750 A240 Gr. S32760
3.1	35Ni-35Fe-20Cr-Cb	B462 Gr. N08020		B463 Gr. N08020
3.2	99.0Ni	B160Gr. N02200		B162 Gr. N02200
3.3	99.0Ni-Low C	B160 Gr. N02201		B162 Gr. N02201
3.4	67Ni-30Cu 67Ni-30Cu-S	B564 Gr. N04400 B164 Gr. N04405		B127 Gr. N04400
3.5	72Ni-15Cr-8Fe	B564 Gr. N06600		B168 Gr. N06600
3.6	33Ni-42Fe-21Cr	B564 Gr. N08800		B409 Gr. N08800
3.7	65Ni-28Mo-2Fe	B335 Gr. N10665		B333 Gr. N10665
3.8	54Ni-16Mo-15Cr 60Ni-22Cr-9Mo-3.5Cb 62Ni-28Mo-5Fe 70Ni-16Mo-7Cr-5Fe 61Ni-16Mo-16Cr 42Ni-21.5Cr-3Mo-2.3Cu	B564 Gr. N10276 B564 Gr. N06625 B335 Gr. N10001 B573 Gr. N10003 B574 Gr. N06455 B564 Gr. N08825		B575 Gr. N10276 B443 Gr. N06625 B333 Gr. N10001 B434 Gr. N10003 B575 Gr. N06455 B424 Gr. N08825
3.9	47Ni-22Cr-9Mo-18Fe	B572 Gr. N06002		B435 Gr. N06002
3.10	25Ni-46Fe-21Cr-5Mo	B672 Gr. N08700		B599 Gr. N08700
3.11	44Fe-25Ni-21Cr-Mo	B649 Gr. N08904		B625 Gr. N08904
3.12	26Ni-43Fe-22Cr-5Mo 47Ni-22Cr-20Fe-7Mo	B621 Gr. N08320 B581 Gr. N06985		B620 Gr. N08320 B582 Gr. N06985
3.13	49Ni-25Cr-18Fe-6Mo Ni-Fe-Cr-Mo-Low Cu	B581 Gr. N06975 B564 Gr. N08031		B582 Gr. N06975 B625 Gr. N08031
3.14	47Ni-22Cr-19Fe-6Mo	B581 Gr. N06007		B582 Gr. N06007
3.15	33Ni-42Fe-21Cr	B564 Gr. N08810		B409 Gr. N08810
3.16	35Ni-19Cr-1¼Si	B511 Gr. N08330		B536 Gr. N08330
3.17	29Ni-20.5Cr-3.5Cu-2.5Mo		A351 Gr. CN7M	

* ASME Boiler and Pressure Vessel Code, Section II, materials that also meet the requirements of the listed ASTM specifications may also be used.

Pressure-Temperature Ratings of ASME B16.5 Flanges

ASME B16.5 lists the pressure-temperature ratings for flanges made from each group of materials listed in Table 5.42. We have selected only nine (9) groups of materials that are commonly used. Ratings are maximum allowable working gage pressure at the corresponding temperature shown in the pressure-temperature rating tables.

Pressure-Temperature Ratings of Class 150 Flanges per ASME B16.5

See Table 5.43.

Dimensions of Class 150 Flanges per ASME B16.5

See Table 5.44.

TABLE 5.43 Pressure-Temperature Ratings for ASME B16.5, Class 150 Flanges

	Material group								
Temp., °F	1.1	1.2	1.9	1.10	1.15	2.1	2.2	2.3	3.1
−20 to 100	285	290	290	290	290	275	275	230	290
200	260	260	260	260	260	230	235	195	260
300	230	230	230	230	230	205	215	175	230
400	200	200	200	200	200	190	195	160	200
500	170	170	170	170	170	170	170	145	170
600	140	140	140	140	140	140	140	140	140
650	125	125	125	125	125	125	125	125	125
700	110	110	110	110	110	110	110	110	110
750	95	95	95	95	95	95	95	95	95
800	80	80	80	80	80	80	80	80	80
850	65	65	65	65	65	65	65	65	
900	50	50	50	50	50	50	50		
950	35	35	35	35	35	35	35		
1000	20	20	20	20	20	20	20		

Notes:
1. For material groups refer to Table 5.42.
2. Ratings are maximum allowable working gage pressure (psig) at the temperature listed.

TABLE 5.44 Dimensions of Class 150 Flanges per ASME B16.5

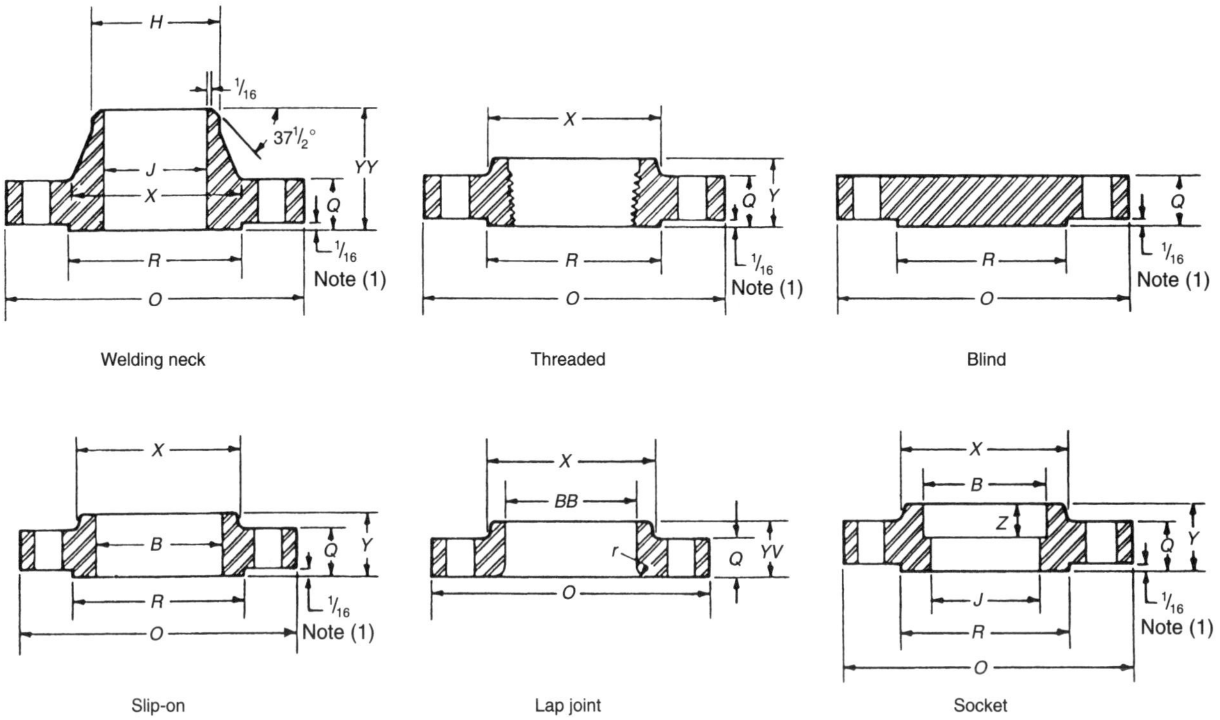

Nom. pipe size	O	Q (1)	R	X	No. and dia. of bolt holes (2)	Bolt circle dia.	YY (1)	H	J	Y (1)	B	r	YV	BB	Z
1/2	3.50	0.44	1.38	1.19	4–0.62	2.38	1.88	0.84	0.62	0.62	0.88	0.12	0.62	0.90	0.38
3/4	3.88	0.50	1.69	1.50	4–0.62	2.75	2.06	1.05	0.82	0.62	1.09	0.12	0.62	1.11	0.44
1	4.25	0.56	2.00	1.94	4–0.62	3.12	2.19	1.32	1.05	0.69	1.36	0.12	0.69	1.38	0.50
1 1/4	4.62	0.62	2.50	2.31	4–0.62	3.50	2.25	1.66	1.38	0.81	1.70	0.19	0.81	1.72	0.56
1 1/2	5.00	0.68	2.88	2.56	4–0.62	3.88	2.44	1.90	1.61	0.88	1.95	0.25	0.88	1.97	0.62
2	6.00	0.75	3.62	3.06	4–0.75	4.75	2.50	2.38	2.07	1.00	2.44	0.31	1.00	2.46	0.69
2 1/2	7.00	0.88	4.12	3.56	4–0.75	5.50	2.75	2.88	2.47	1.12	2.94	0.31	1.12	2.97	0.75
3	7.50	0.94	5.00	4.25	4–0.75	6.00	2.75	3.50	3.07	1.19	3.57	0.38	1.19	3.60	0.81
3 1/2	8.50	0.94	5.50	4.81	8–0.75	7.00	2.81	4.00	3.55	1.25	4.07	0.38	1.25	4.10	
4	9.00	0.94	6.19	5.31	8–0.75	7.50	3.00	4.50	4.03	1.31	4.57	0.44	1.31	4.60	
5	10.00	0.94	7.31	6.44	8–0.88	8.50	3.50	5.56	5.05	1.44	5.66	0.44	1.44	5.69	
6	11.00	1.00	8.50	7.56	8–0.88	9.50	3.50	6.63	6.07	1.56	6.72	0.50	1.56	6.75	
8	13.50	1.12	10.62	9.69	8–0.88	11.75	4.00	8.63	7.98	1.75	8.72	0.50	1.75	8.75	
10	16.00	1.19	12.75	12.00	12–1.00	14.25	4.00	10.75	10.02	1.94	10.88	0.50	1.94	10.92	
12	19.00	1.25	15.00	14.38	12–1.00	17.00	4.50	12.75	12.00	2.19	12.88	0.50	2.19	12.92	
14	21.00	1.38	16.25	15.75	12–1.12	18.75	5.00	14.00	To be specified by purchaser	2.25	14.14	0.50	3.12	14.18	
16	23.50	1.44	18.50	18.00	16–1.12	21.25	5.00	16.00		2.50	16.16	0.50	3.44	16.19	
18	25.00	1.56	21.00	19.88	16–1.25	22.75	5.50	18.00		2.69	18.18	0.50	3.81	18.20	
20	27.50	1.69	23.00	22.00	20–1.25	25.00	5.69	20.00		2.88	20.20	0.50	4.06	20.25	
24	32.00	1.88	27.25	26.12	20–1.38	29.50	6.00	24.00		3.25	24.25	0.50	4.38	24.25	

Notes:
(1) Raised face height, 1/16 in, is included in Q, Y, and YY dimensions.
(2) Bolt hole diameter is 1/8 in larger than the bolt diameter.
(3) Dimensions in inches.

Pressure-Temperature Ratings of Class 300 Flanges per ASME B16.5

See Table 5.45.

Dimensions of Class 300 Flanges per ASME B16.5

See Table 5.46.

TABLE 5.45 Pressure-Temperature Ratings for ASME 16.5, Class 300 Flanges

	Material group								
Temp., °F	1.1	1.2	1.9	1.10	1.15	2.1	2.2	2.3	3.1
–20 to 100	740	750	750	750	750	720	720	600	750
200	675	750	750	750	750	600	620	505	720
300	655	730	720	730	730	540	560	455	715
400	635	705	695	705	705	495	515	415	675
500	600	665	665	665	665	465	480	380	655
600	550	605	605	605	605	435	450	360	605
650	535	590	590	590	590	430	445	350	590
700	535	570	570	570	570	425	430	345	570
750	505	505	530	530	530	415	425	335	530
800	410	410	510	510	510	405	420	330	510
850	270	270	485	485	485	395	420	320	
900	170	170	450	450	450	390	415		
950	105	105	320	375	385	380	385		
1000	50	50	215	260	365	320	350		
1050			145	175	360	310	345		
1100			95	110	300	255	305		
1150			60	70	225	200	235		
1200			40	40	145	155	185		
1250						115	145		
1300						85	115		
1350						60	95		
1400						50	75		
1450						35	60		
1500						25	40		

Notes:
1. For material groups refer to Table 5.42.
2. Ratings are maximum allowable working gage pressure (psig) at the temperature listed.

TABLE 5.46 Dimensions of Class 300 Flanges per ASME B16.5*

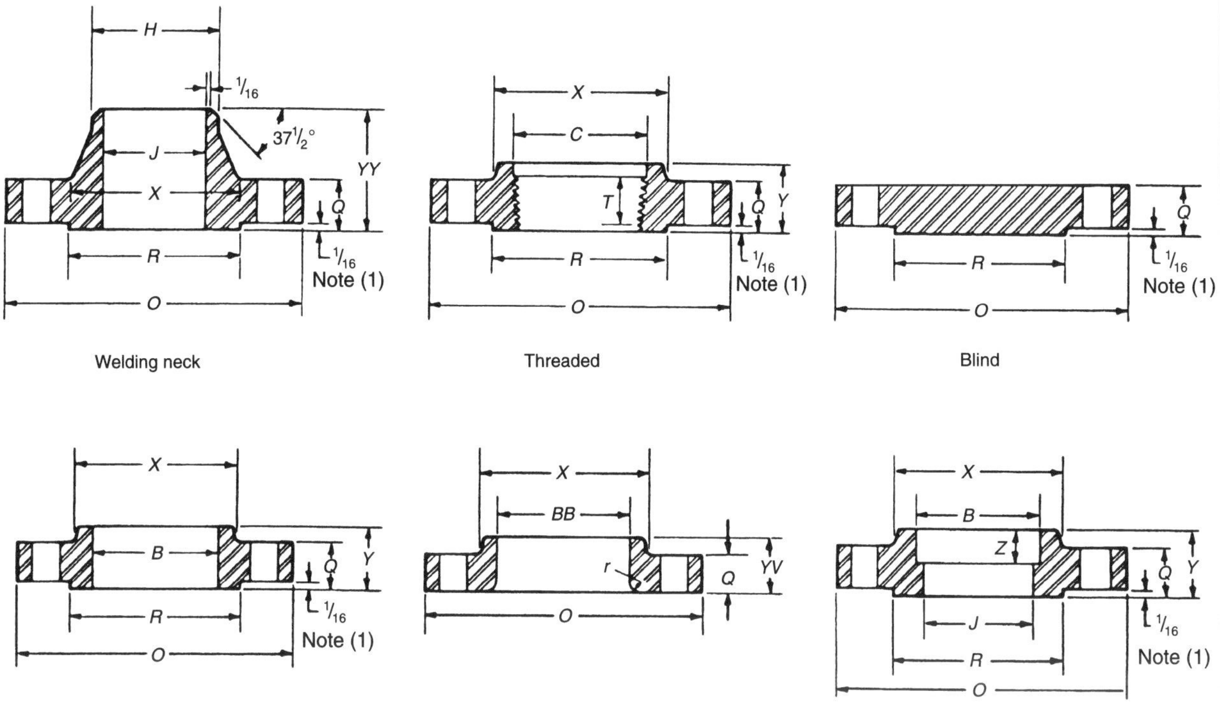

Nom. pipe size	O	Q (1)	R	X	No. and dia. of bolt holes (2)	Bolt circle dia.	YY (1)	H	J	Y (1)	B	r	YV	BB	Z	C	T
½	3.75	0.56	1.38	1.50	4–0.62	2.62	2.06	0.84	0.62	0.88	0.88	0.12	0.88	0.90	0.38	0.93	0.62
¾	4.62	0.62	1.69	1.88	4–0.75	3.25	2.25	1.05	0.82	1.00	1.09	0.12	1.00	1.11	0.44	1.14	0.62
1	4.88	0.69	2.00	2.12	4–0.75	3.50	2.44	1.32	1.05	1.06	1.36	0.12	1.06	1.38	0.50	1.41	0.69
1¼	5.25	0.75	2.50	2.50	4–0.75	3.88	2.56	1.66	1.38	1.06	1.70	0.19	1.06	1.72	0.56	1.75	0.81
1½	6.12	0.81	2.88	2.75	4–0.88	4.50	2.69	1.90	1.61	1.19	1.95	0.25	1.19	1.97	0.62	1.99	0.88
2	6.50	0.88	3.62	3.31	8–0.75	5.00	2.75	2.38	2.07	1.31	2.44	0.31	1.31	2.46	0.69	2.50	1.12
2½	7.50	1.00	4.12	3.94	8–0.88	5.88	3.00	2.88	2.47	1.50	2.94	0.31	1.50	2.97	0.75	3.00	1.25
3	8.25	1.12	5.00	4.62	8–0.88	6.62	3.12	3.50	3.07	1.69	3.57	0.38	1.69	3.60	0.81	3.63	1.25
3½	9.00	1.19	5.50	5.25	8–0.88	7.25	3.19	4.00	3.55	1.75	4.07	0.38	1.75	4.10		4.13	1.44
4	10.00	1.25	6.19	5.75	8–0.88	7.88	3.38	4.50	4.03	1.88	4.57	0.44	1.88	4.60		4.63	1.44
5	11.00	1.38	7.31	7.00	8–0.88	9.25	3.88	5.56	5.05	2.00	5.66	0.44	2.00	5.69		5.69	1.69
6	12.50	1.44	8.50	8.12	12–0.88	10.62	3.88	6.63	6.07	2.06	6.72	0.50	2.06	6.75		6.75	1.81
8	15.00	1.62	10.62	10.25	12–1.00	13.00	4.38	8.63	7.98	2.44	8.72	0.50	2.44	8.75		8.75	2.00
10	17.50	1.88	12.75	12.62	16–1.12	15.25	4.62	10.75	10.02	2.62	10.88	0.50	3.75	10.92		10.88	2.19
12	20.50	2.00	15.00	14.75	16–1.25	17.75	5.12	12.75	12.00	2.88	12.88	0.50	4.00	12.92		12.94	2.38
14	23.00	2.12	16.25	16.75	20–1.25	20.25	5.62	14.00	To be specified by purchaser	3.00	14.14	0.50	4.38	14.18		14.19	2.50
16	25.50	2.25	18.50	19.00	20–1.38	22.50	5.75	16.00		3.25	16.16	0.50	4.75	16.19		16.19	2.69
18	28.00	2.38	21.00	21.00	24–1.38	24.75	6.25	18.00		3.50	18.18	0.50	5.12	18.20		18.19	2.75
20	30.50	2.50	23.00	23.12	24–1.38	27.00	6.38	20.00		3.75	20.20	0.50	5.50	20.25		20.19	2.88
24	36.00	2.75	27.25	27.62	24–1.62	32.00	6.62	24.00		4.19	24.25	0.50	6.00	24.25		24.19	3.25

* Dimensions in inches.

Notes:

(1) Raised face height, 1/16 in, is included in Q, Y, and YY dimensions.

(2) Bolt hole diameter is ⅛ in larger than the bolt diameter.

Pressure-Temperature Ratings of Class 400 Flanges per ASME B16.5

See Table 5.47.

Dimensions of Class 400 Flanges per ASME B16.5

See Table 5.48.

TABLE 5.47 Pressure-Temperature Ratings for ASME B16.5, Class 400 Flanges

	Material group								
Temp., °F	1.1	1.2	1.9	1.10	1.15	2.1	2.2	2.3	3.1
−20 to 100	990	1000	1000	1000	1000	960	960	800	1000
200	900	1000	1000	1000	1000	800	825	675	960
300	875	970	965	970	970	720	745	605	950
400	845	940	925	940	940	660	685	550	900
500	800	885	885	885	885	620	635	510	875
600	730	805	805	805	805	580	600	480	805
650	715	785	785	785	785	575	590	470	785
700	710	755	755	755	755	565	580	460	755
750	670	670	710	710	710	555	570	450	710
800	550	550	675	675	675	540	565	440	675
850	355	355	650	650	650	530	555	430	
900	230	230	600	600	600	520	555		
950	140	140	425	505	515	510	515		
1000	70	70	290	345	485	430	465		
1050			190	235	480	410	460		
1100			130	145	400	345	405		
1150			80	90	295	265	315		
1200			50	55	190	205	245		
1250						150	195		
1300						115	155		
1350						80	130		
1400						65	100		
1450						45	80		
1500						35	55		

Notes:
1. For material groups refer to Table 5.42.
2. Ratings are maximum allowable working gage pressure (psig) at the temperature listed.

TABLE 5.48 Dimensions of Class 400 Flanges per ASME B16.5

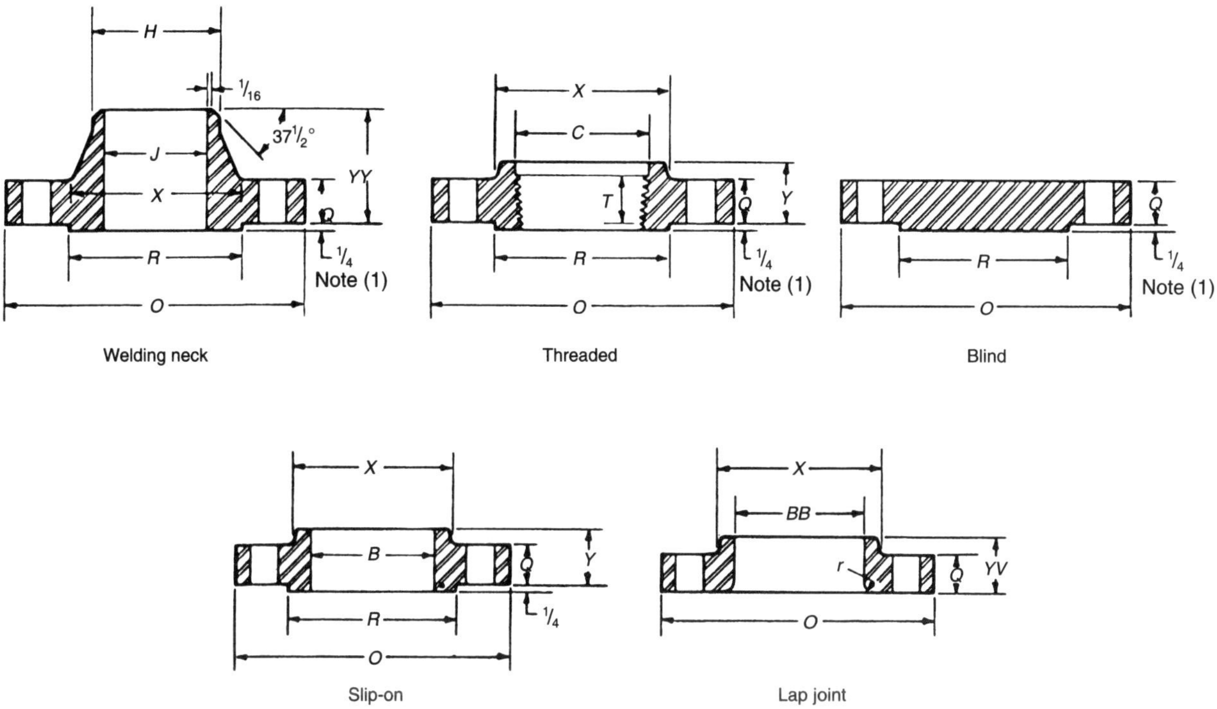

Nom. pipe size (1)	O	Q (2)	R	X	No. and dia. of bolt holes (3)	Bolt circle dia.	YY (2)	H	J	Y (2)	B	r	YV	BB	C	T
1/2	3.75	0.56	1.38	1.50	4–0.62	2.62	2.06	0.84	To be specified by purchaser	0.88	0.88	0.12	0.88	0.90	0.93	0.62
3/4	4.62	0.62	1.69	1.88	4–0.75	3.25	2.25	1.05		1.00	1.09	0.12	1.00	1.11	1.14	0.62
1	4.88	0.69	2.00	2.12	4–0.75	3.50	2.44	1.32		1.06	1.36	0.12	1.06	1.38	1.41	0.69
1 1/4	5.25	0.81	2.50	2.50	4–0.75	3.88	2.62	1.66		1.12	1.70	0.19	1.12	1.72	1.75	0.81
1 1/2	6.12	0.88	2.88	2.75	4–0.88	4.50	2.75	1.90		1.25	1.95	0.25	1.25	1.97	1.99	0.88
2	6.50	1.00	3.62	3.31	8–0.75	5.00	2.88	2.38		1.44	2.44	0.31	1.44	2.46	2.50	1.12
2 1/2	7.50	1.12	4.12	3.94	8–0.88	5.88	3.12	2.88		1.62	2.94	0.31	1.62	2.97	3.00	1.25
3	8.25	1.25	5.00	4.62	8–0.88	6.62	3.25	3.50		1.81	3.57	0.38	1.81	3.60	3.63	1.38
3 1/2	9.00	1.38	5.50	5.25	8–1.00	7.25	3.38	4.00		1.94	4.07	0.38	1.94	4.10	4.13	1.56
4	10.00	1.38	6.19	5.75	8–1.00	7.88	3.50	4.50		2.00	4.57	0.44	2.00	4.60	4.63	1.44
5	11.00	1.50	7.31	7.00	8–1.00	9.25	4.00	5.56		2.12	5.66	0.44	2.12	5.69	5.69	1.69
6	12.50	1.62	8.50	8.12	12–1.00	10.62	4.06	6.63		2.25	6.72	0.50	2.25	6.75	6.75	1.81
8	15.00	1.88	10.62	10.25	12–1.12	13.00	4.62	8.63		2.69	8.72	0.50	2.69	8.75	8.75	2.00
10	17.50	2.12	12.75	12.62	16–1.25	15.25	4.88	10.75		2.88	10.88	0.50	4.00	10.92	10.88	2.19
12	20.50	2.25	15.00	14.75	16–1.38	17.75	5.38	12.75		3.12	12.88	0.50	4.25	12.92	12.94	2.38
14	23.00	2.38	16.25	16.75	20–1.38	20.25	5.88	14.00		3.31	14.14	0.50	4.62	14.18	14.19	2.50
16	25.50	2.50	18.50	19.00	20–1.50	22.50	6.00	16.00		3.69	16.16	0.50	5.00	16.19	16.19	2.69
18	28.00	2.62	21.00	21.00	24–1.50	24.75	6.50	18.00		3.88	18.18	0.50	5.38	18.20	18.19	2.75
20	30.50	2.75	23.00	23.12	24–1.62	27.00	6.62	20.00		4.00	20.20	0.50	5.75	20.25	20.19	2.88
24	36.00	3.00	27.25	27.62	24–1.88	32.00	6.88	24.00		4.50	24.25	0.50	6.25	24.25	24.19	3.25

Notes:
(1) Dimensions of NPS 1/2 through NPS 3 1/2 Class 400 flanges are same as those of Class 600 flanges. All dimensions in inches.
(2) Raised face height, 1/4 in, is not included in *Q*, *Y*, and *YY* dimensions.
(3) Bolt hole diameter is 1/8 in larger than the bolt diameter.

Pressure-Temperature Ratings of Class 600 Flanges per ASME B16.5

See Table 5.49.

Dimensions of Class 600 Flanges per ASME B16.5

See Table 5.50.

TABLE 5.49 Pressure-Temperature Ratings for ASME B16.5, Class 600 Flanges

Temp., °F	Material group								
	1.1	1.2	1.9	1.10	1.15	2.1	2.2	2.3	3.1
–20 to 100	1480	1500	1500	1500	1500	1440	1440	1200	1500
200	1350	1500	1500	1500	1500	1200	1240	1015	1440
300	1315	1455	1445	1455	1455	1080	1120	910	1425
400	1270	1410	1385	1410	1410	995	1025	825	1345
500	1200	1330	1330	1330	1330	930	955	765	1310
600	1095	1210	1210	1210	1210	875	900	720	1210
650	1075	1175	1175	1175	1175	860	890	700	1175
700	1065	1135	1135	1135	1135	850	870	685	1135
750	1010	1010	1065	1065	1065	830	855	670	1065
800	825	825	1015	1015	1015	805	845	660	1015
850	535	535	975	975	975	790	835	645	
900	345	345	900	900	900	780	830		
950	205	205	640	755	775	765	775		
1000	105	105	430	520	725	640	700		
1050			290	350	720	615	685		
1100			190	220	605	515	610		
1150			125	135	445	400	475		
1200			75	80	290	310	370		
1250						225	295		
1300						170	235		
1350						125	190		
1400						90	150		
1450						70	115		
1500						55	85		

Notes:
1. For material groups refer to Table 5.42.
2. Ratings are maximum allowable working gage pressure (psig) at the temperature listed.

TABLE 5.50 Dimensions of Class 600 Flanges per ASME B16.5*

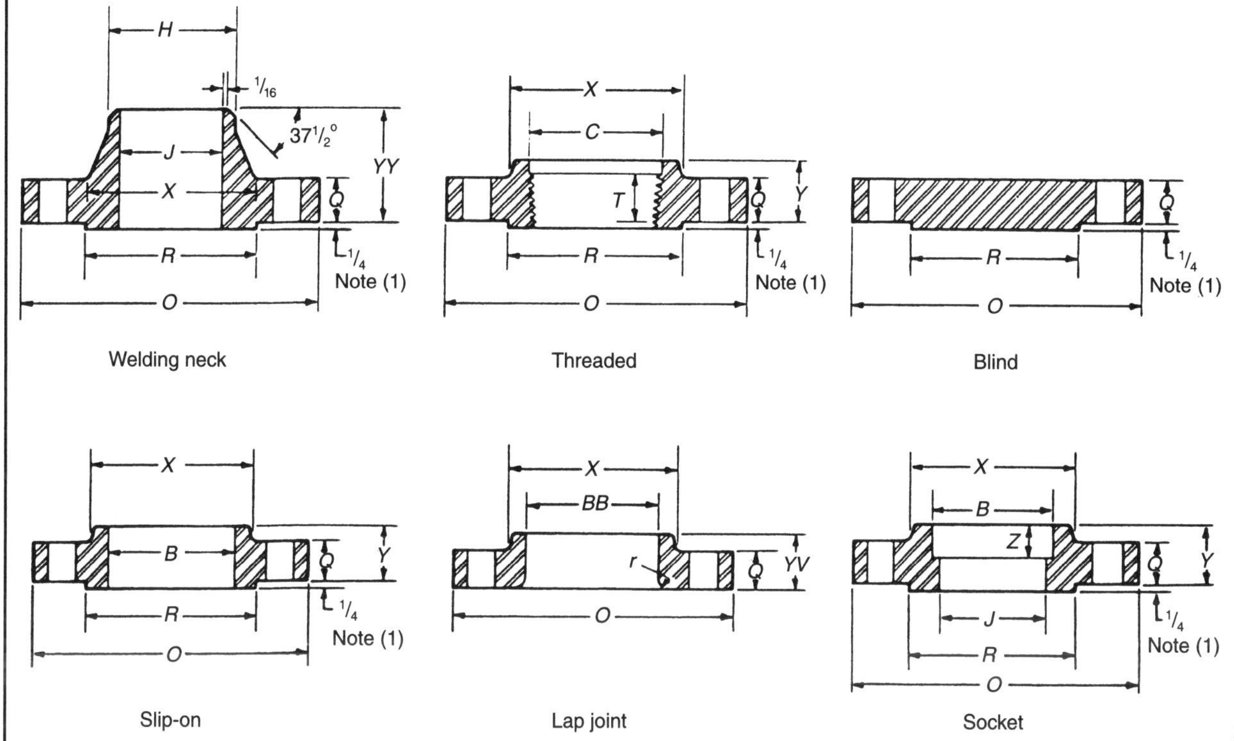

Nom. pipe size	O	Q (1)	R	X	No. and dia. of bolt holes (2)	Bolt circle dia.	YY (1)	H	J	Y	B	r	YV	BB	Z	C	T
½	3.75	0.56	1.38	1.50	4–0.62	2.62	2.06	0.84	To be specified by purchaser	0.88	0.88	0.12	0.88	0.90	0.38	0.93	0.62
¾	4.62	0.62	1.69	1.88	4–0.75	3.25	2.25	1.05		1.00	1.09	0.12	1.00	1.11	0.44	1.14	0.62
1	4.88	0.69	2.00	2.12	4–0.75	3.50	2.44	1.32		1.06	1.36	0.12	1.06	1.38	0.50	1.41	0.69
1¼	5.25	0.81	2.50	2.50	4–0.75	3.88	2.62	1.66		1.12	1.70	0.19	1.12	1.72	0.56	1.75	0.81
1½	6.12	0.88	2.88	2.75	4–0.88	4.50	2.75	1.90		1.25	1.95	0.25	1.25	1.97	0.62	1.99	0.88
2	6.50	1.00	3.62	3.31	8–0.75	5.00	2.88	2.38		1.44	2.44	0.31	1.44	2.46	0.69	2.50	1.12
2½	7.50	1.12	4.12	3.94	8–0.88	5.88	3.12	2.88		1.62	2.94	0.31	1.62	2.97	0.75	3.00	1.25
3	8.25	1.25	5.00	4.62	8–0.88	6.62	3.25	3.50		1.81	3.57	0.38	1.81	3.60	0.81	3.63	1.38
3½	9.00	1.38	5.50	5.25	8–1.00	7.25	3.38	4.00		1.94	4.07	0.38	1.94	4.10		4.13	1.56
4	10.75	1.50	6.19	6.00	8–1.00	8.50	4.00	4.50		2.12	4.57	0.44	2.12	4.60		4.63	1.62
5	13.00	1.75	7.31	7.44	8–1.12	10.50	4.50	5.56		2.38	5.66	0.44	2.38	5.69		5.69	1.88
6	14.00	1.88	8.50	8.75	12–1.12	11.50	4.62	6.63		2.62	6.72	0.50	2.62	6.75		6.75	2.00
8	16.50	2.19	10.62	10.75	12–1.25	13.75	5.25	8.63		3.00	8.72	0.50	3.00	8.75		8.75	2.25
10	20.00	2.50	12.75	13.50	16–1.38	17.00	6.00	10.75		3.38	10.88	0.50	4.38	10.92		10.88	2.56
12	22.00	2.62	15.00	15.75	20–1.38	19.25	6.12	12.75		3.62	12.88	0.50	4.62	12.92		12.94	2.75
14	23.75	2.75	16.25	17.00	20–1.50	20.75	6.50	14.00		3.69	14.14	0.50	5.00	14.18		14.19	2.88
16	27.00	3.00	18.50	19.50	20–1.62	23.75	7.00	16.00		4.19	16.16	0.50	5.50	16.19		18.19	3.06
18	29.25	3.25	21.00	21.50	20–1.75	25.75	7.25	18.00		4.62	18.18	0.50	6.00	18.20		18.10	3.12
20	32.00	3.50	23.00	24.00	24–1.75	28.50	7.50	20.00		5.00	20.20	0.50	6.50	20.25		20.19	3.25
24	37.00	4.00	27.25	28.25	24–2.00	33.00	8.00	24.00		5.50	24.25	0.50	7.25	24.25		24.19	3.62

* Dimensions in inches.

Notes:

(1) Raised face height, ¼ in, is not included in *Q*, *Y*, and *YY* dimensions.

(2) Bolt hole diameter is ⅛ in larger than the bolt diameter.

Pressure-Temperature Ratings of Class 900 Flanges per ASME B16.5

See Table 5.51.

Dimensions of Class 900 Flanges per ASME B16.5

See Table 5.52.

TABLE 5.51 Pressure-Temperature Ratings for ASME B16.5, Class 900 Flanges

	Material group								
Temp., °F	1.1	1.2	1.9	1.10	1.15	2.1	2.2	2.3	3.1
−20 to 100	2220	2250	2250	2250	2250	2160	2160	1800	2250
200	2025	2250	2250	2250	2250	1800	1860	1520	2160
300	1970	2185	2185	2185	2185	1620	1680	1360	2140
400	1900	2115	2080	2115	2115	1490	1540	1240	2020
500	1795	1995	1995	1995	1995	1395	1435	1145	1965
600	1640	1815	1815	1815	1815	1310	1355	1080	1815
650	1610	1765	1765	1765	1765	1290	1330	1050	1765
700	1600	1705	1705	1705	1705	1275	1305	1030	1705
750	1510	1510	1595	1595	1595	1245	1280	1010	1595
800	1235	1235	1525	1525	1525	1210	1265	985	1525
850	805	805	1460	1460	1460	1190	1255	965	
900	515	515	1350	1350	1350	1165	1245		
950	310	310	955	1130	1160	1145	1160		
1000	155	155	650	780	1090	965	1050		
1050			430	525	1080	925	1030		
1100			290	330	905	770	915		
1150			185	205	670	595	710		
1200			115	125	430	465	555		
1250						340	440		
1300						255	350		
1350						185	290		
1400						145	225		
1450						105	175		
1500						80	125		

Notes:
1. For material groups refer to Table 5.42.
2. Ratings are maximum allowable working gage pressure (psig) at the temperature listed.

TABLE 5.52 Dimensions of Class 900 Flanges per ASME B16.5

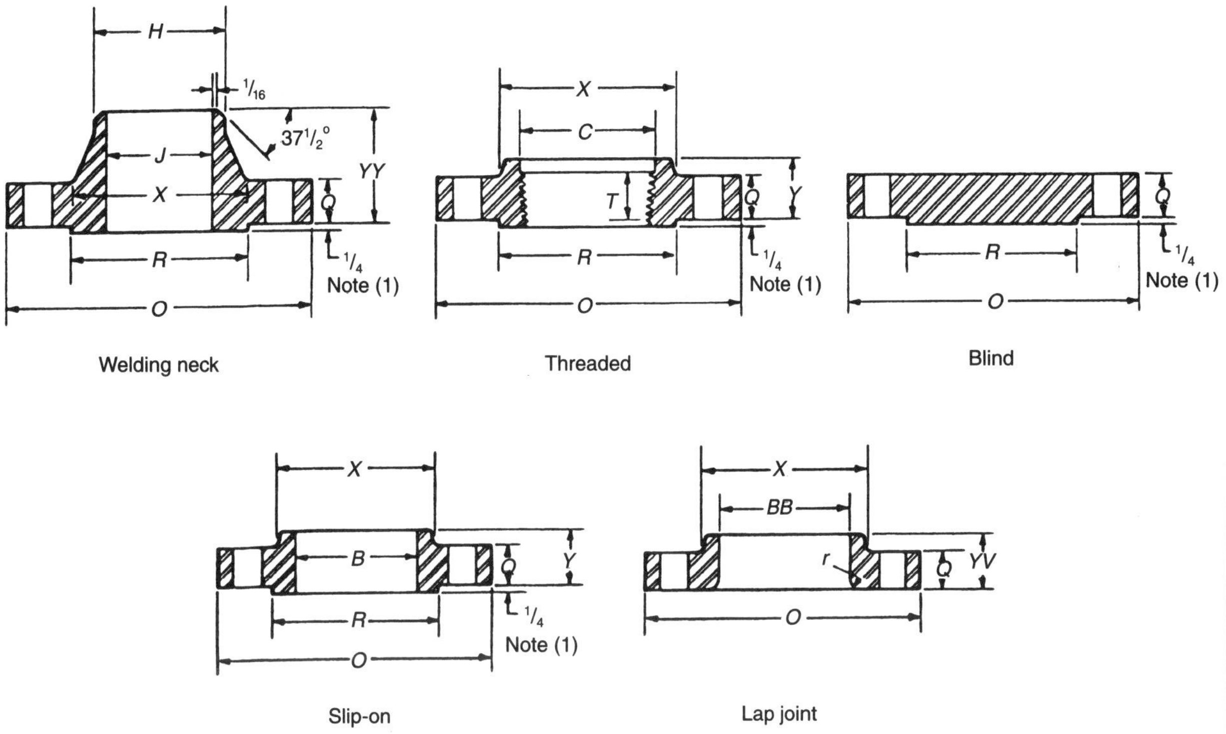

Nom. pipe size (1)	*O*	*Q* (2)	*R*	*X*	No. and dia. of bolt holes (3)	Bolt circle dia.	*YY* (2)	*H*	*J*	*Y* (2)	*B*	*r*	*YV*	*BB*	*C*	*T*
1/2	4.75	0.88	1.38	1.50	4–0.88	3.25	2.38	0.84	To be specified by purchaser	1.25	0.88	0.12	1.25	0.90	0.93	0.88
3/4	5.12	1.00	1.69	1.75	4–0.88	3.50	2.75	1.05		1.38	1.09	0.12	1.38	1.11	1.14	1.00
1	5.88	1.12	2.00	2.06	4–1.00	4.00	2.88	1.32		1.62	1.36	0.12	1.62	1.38	1.41	1.12
1 1/4	6.25	1.12	2.50	2.50	4–1.00	4.38	2.88	1.66		1.62	1.70	0.19	1.62	1.72	1.75	1.19
1 1/2	7.00	1.25	2.88	2.75	4–1.12	4.88	3.25	1.90		1.75	1.95	0.25	1.75	1.97	1.99	1.25
2	8.50	1.50	3.62	4.12	8–1.00	6.50	4.00	2.38		2.25	2.44	0.31	2.25	2.46	2.50	1.50
2 1/2	9.62	1.62	4.12	4.88	8–1.12	7.50	4.12	2.88		2.50	2.94	0.31	2.50	2.97	3.00	1.88
3	9.50	1.50	5.00	5.00	8–1.00	7.50	4.00	3.50		2.12	3.57	0.38	2.12	3.60	3.63	1.62
4	11.50	1.75	6.19	6.25	8–1.25	9.25	4.50	4.50		2.75	4.57	0.44	2.75	4.60	4.63	1.88
5	13.75	2.00	7.31	7.50	8–1.38	11.00	5.00	5.56		3.12	5.66	0.44	3.12	5.69	5.69	2.12
6	15.00	2.19	8.50	9.25	12–1.25	12.50	5.50	6.63		3.38	6.72	0.50	3.38	6.75	6.75	2.25
8	18.50	2.50	10.62	11.75	12–1.50	15.50	6.38	8.63		4.00	8.72	0.50	4.50	8.75	8.75	2.50
10	21.50	2.75	12.75	14.50	16–1.50	18.50	7.25	10.75		4.25	10.88	0.50	5.00	10.92	10.88	2.81
12	24.00	3.12	15.00	16.50	20–1.50	21.00	7.88	12.75		4.62	12.88	0.50	5.62	12.92	12.94	3.00
14	25.25	3.38	16.25	17.75	20–1.62	22.00	8.38	14.00		5.12	14.14	0.50	6.12	14.18	14.19	3.25
16	27.75	3.50	18.50	20.00	20–1.75	24.25	8.50	16.00		5.25	16.16	0.50	6.50	16.19	16.19	3.38
18	31.00	4.00	21.00	22.25	20–2.00	27.00	9.00	18.00		6.00	18.18	0.50	7.50	18.20	18.19	3.50
20	33.75	4.25	23.00	24.50	20–2.12	29.50	9.75	20.00		6.25	20.20	0.50	8.25	20.25	20.19	3.62
24	41.00	5.50	27.25	29.50	20–2.62	35.50	11.50	24.00		8.00	24.25	0.50	10.50	24.25	24.19	4.00

Notes:
(1) Dimensions of Class 900, NPS 1/2 through 2 1/2, flanges are same as those of Class 1500 flanges. All dimensions in inches.
(2) Raised face height, 1/4 in, is not included in *Q*, *Y*, and *YY* dimensions.
(3) Bolt hole diameter is 1/8 in larger than the bolt diameter.

Pressure-Temperature Ratings of Class 1500 Flanges per ASME B16.5

See Table 5.53.

Dimensions of Class 1500 Flanges per ASME B16.5

See Table 5.54.

TABLE 5.53 Pressure-Temperature Ratings for ASME B16.5, Class 1500 Flanges

	Material group								
Temp., °F	1.1	1.2	1.9	1.10	1.15	2.1	2.2	2.3	3.1
−20 to 100	3705	3750	3750	3750	3750	3600	3600	3000	3750
200	3375	3750	3750	3750	3750	3000	3095	2530	3600
300	3280	3640	3610	3640	3640	2700	2795	2270	3565
400	3170	3530	3465	3530	3530	2485	2570	2065	3365
500	2995	3325	3325	3325	3325	2330	2390	1910	3275
600	2735	3025	3025	3025	3025	2185	2255	1800	3025
650	2685	2940	2940	2940	2940	2150	2220	1750	2940
700	2665	2840	2840	2840	2840	2125	2170	1715	2840
750	2520	2520	2660	2660	2660	2075	2135	1680	2660
800	2060	2060	2540	2540	2540	2015	2110	1645	2540
850	1340	1340	2435	2435	2435	1980	2090	1610	
900	860	860	2245	2245	2245	1945	2075		
950	515	515	1595	1885	1930	1910	1930		
1000	260	260	1080	1305	1820	1605	1750		
1050			720	875	1800	1545	1720		
1100			480	550	1510	1285	1525		
1150			310	345	1115	995	1185		
1200			190	205	720	770	925		
1250						565	735		
1300						430	585		
1350						310	480		
1400						240	380		
1450						170	290		
1500						135	205		

Notes:
1. For material groups refer to Table 5.42.
2. Ratings are maximum allowable working gage pressure (psig) at the temperature listed.

TABLE 5.54 Dimensions of Class 1500 Flanges per ASME B16.5*

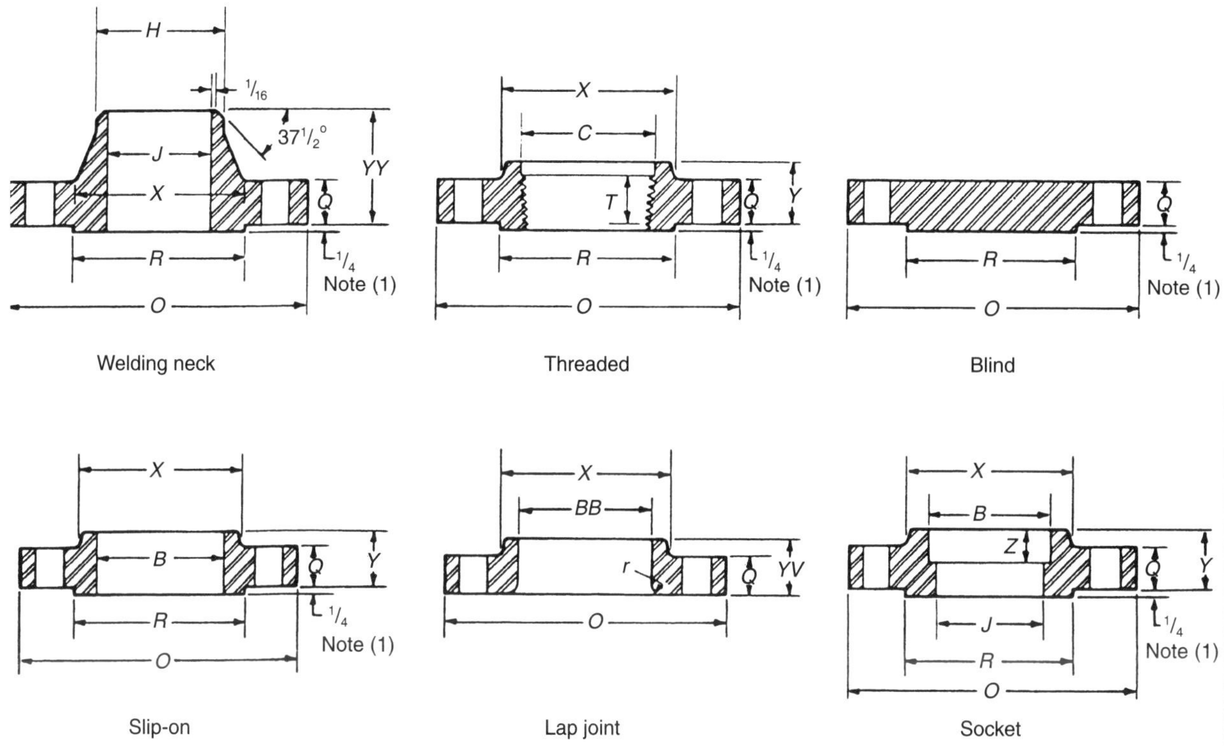

Nom. pipe size	O	Q	R	X	No. and dia. of bolt holes (2)	Bolt circle dia.	YY	H	J	Y	B	r	YV	BB	Z	C	T
1/2	4.75	0.88	1.38	1.50	4–0.88	3.25	2.38	0.84	To be specified by purchaser	1.25	0.88	0.12	1.25	0.90	0.38	0.93	0.88
3/4	5.12	1.00	1.69	1.75	4–0.88	3.50	2.75	1.05		1.38	1.09	0.12	1.38	1.11	0.44	1.14	1.00
1	5.88	1.12	2.00	2.06	4–1.00	4.00	2.88	1.32		1.62	1.36	0.12	1.62	1.38	0.50	1.41	1.12
1 1/4	6.25	1.12	2.50	2.50	4–1.00	4.38	2.88	1.66		1.62	1.70	0.19	1.62	1.72	0.56	1.75	1.19
1 1/2	7.00	1.25	2.88	2.75	4–1.12	4.88	3.25	1.90		1.75	1.95	0.25	1.75	1.97	0.62	1.99	1.25
2	8.50	1.50	3.62	4.12	8–1.00	6.50	4.00	2.38		2.25	2.44	0.31	2.25	2.46	0.69	2.50	1.50
2 1/2	9.62	1.62	4.12	4.88	8–1.12	7.50	4.12	2.88		2.50	2.94	0.31	2.50	2.97	0.75	3.00	1.88
3	10.50	1.88	5.00	5.25	8–1.25	8.00	4.62	3.50		2.88		0.38	2.88	3.60		3.63	2.00
4	12.25	2.12	6.19	6.38	8–1.38	9.50	4.88	4.50		3.56		0.44	3.56	4.60		4.63	2.25
5	14.75	2.88	7.31	7.75	8–1.62	11.50	6.12	5.56		4.12		0.44	4.12	5.69		5.69	2.50
6	15.50	3.25	8.50	9.00	12–1.50	12.50	6.75	6.63		4.69		0.50	4.69	6.75		6.75	2.75
8	19.00	3.62	10.62	11.50	12–1.75	15.50	8.38	8.63		5.62		0.50	5.62	8.75		8.75	3.00
10	23.00	4.25	12.75	14.50	12–2.00	19.00	10.00	10.75		6.25		0.50	7.00	10.92		10.88	3.31
12	26.50	4.88	15.00	17.75	16–2.12	22.50	11.12	12.75		7.12		0.50	8.62	12.92		12.94	3.62
14	29.50	5.25	16.25	19.50	16–2.38	25.00	11.75	14.00				0.50	9.50	14.18			
16	32.50	5.75	18.50	21.75	16–2.62	27.75	12.25	16.00				0.50	10.25	16.19			
18	36.00	6.38	21.00	23.50	16–2.88	30.50	12.88	18.00				0.50	10.88	18.20			
20	38.75	7.00	23.00	25.25	16–3.12	32.75	14.00	20.00				0.50	11.50	20.25			
24	46.00	8.00	27.25	30.00	16–3.62	39.00	16.00	24.00				0.50	13.00	24.25			

* Dimensions in inches.

Notes:

(1) Raised face height, 1/4 in, is not included in Q, Y, and YY dimensions.

(2) Bolt hole diameter is 1/8 in larger than the bolt diameter.

Pressure-Temperature Ratings of Class 2500 Flanges per ASME B16.5

See Table 5.55.

Dimensions of Class 2500 Flanges per ASME B16.5

See Table 5.56.

ASME B16.47 LARGE-DIAMETER STEEL FLANGES

Material for ASME B16.47 Large-Diameter Steel Flanges

ASME B16.47 flanges may be manufactured from materials listed in Table 5.57. It is noted that only blind flanges may be made from plate, in addition to being cast or forged from the materials listed for forgings.

Materials are grouped in 18 groups covering carbon steels (1.1 to 1.14) and stainless steels (2.1 to 2.8). It is noted that some group sequence numbers are missing because either they were merged with others or deleted.

TABLE 5.55 Pressure-Temperature Ratings for ASME B16.5, Class 2500 Flanges

	Material group								
Temp., °F	1.1	1.2	1.9	1.10	1.15	2.1	2.2	2.3	3.1
–20 to 100	6170	6250	6250	6250	6250	6000	6000	5000	6250
200	5625	6250	6250	6250	6250	5000	5160	4220	6000
300	5470	6070	6015	6070	6070	4500	4660	3780	5940
400	5280	5880	5775	5880	5880	4140	4280	3440	5610
500	4990	5540	5540	5540	5540	3880	3980	3180	5460
600	4560	5040	5040	5040	5040	3640	3760	3000	5040
650	4475	4905	4905	4905	4905	3580	3700	2920	4905
700	4440	4730	4730	4730	4730	3540	3620	2860	4730
750	4200	4200	4430	4430	4430	3460	3560	2800	4430
800	3430	3430	4230	4230	4230	3360	3520	2740	4230
850	2230	2230	4060	4060	4060	3300	3480	2680	
900	1430	1430	3745	3745	3745	3240	3460		
950	860	860	2655	3145	3220	3180	3220		
1000	430	430	1800	2170	3030	2675	2915		
1050			1200	1455	3000	2570	2865		
1100			800	915	2515	2145	2545		
1150			515	570	1855	1655	1970		
1200			315	345	1200	1285	1545		
1250						945	1230		
1300						715	970		
1350						515	800		
1400						400	630		
1450						285	485		
1500						230	345		

Notes:
1. For material groups refer to Table 5.42.
2. Ratings are maximum allowable working gage pressure (psig) at the temperature listed.

TABLE 5.56 Dimensions of Class 2500 Flanges per ASME B16.5*

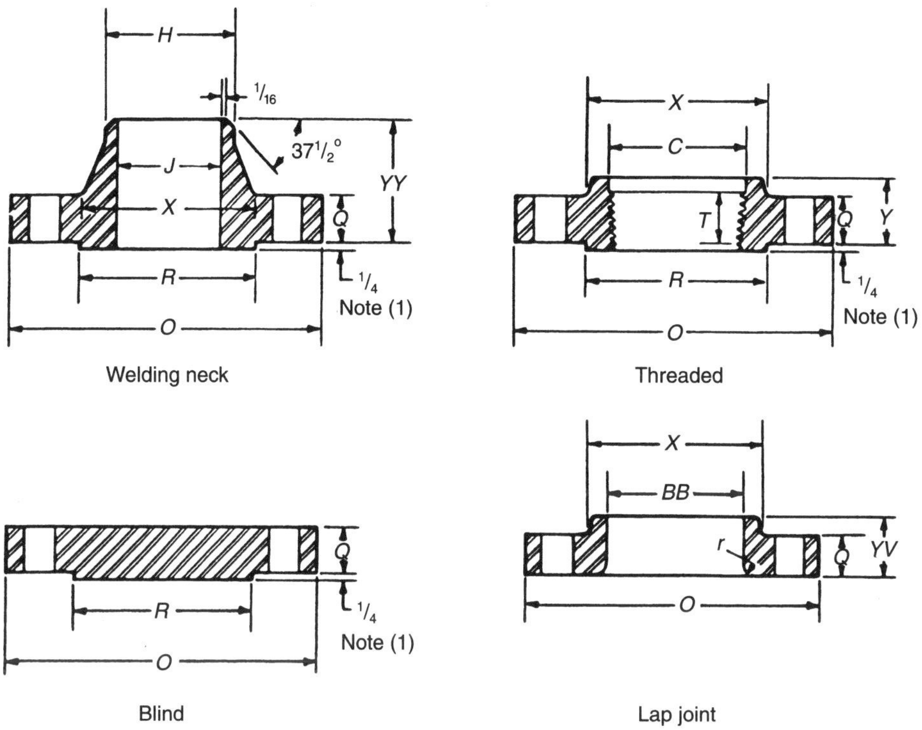

Nom. pipe size	O	Q (1)	R	X	No. and dia. of bolt holes	Bolt circle dia.	YY	H	J	Y*	r	YV	BB	C	T
½	5.25	1.19	1.38	1.69	4–0.88	3.50	2.88	0.84	To be specified by purchaser	1.56	0.12	1.56	0.90	0.93	1.12
¾	5.50	1.25	1.69	2.00	4–0.88	3.75	3.12	1.05		1.69	0.12	1.69	1.11	1.14	1.25
1	6.25	1.38	2.00	2.25	4–1.00	4.25	3.50	1.32		1.88	0.12	1.88	1.38	1.41	1.38
1¼	7.25	1.50	2.50	2.88	4–1.12	5.12	3.75	1.66		2.06	0.19	2.06	1.72	1.75	1.50
1½	8.00	1.75	2.88	3.12	4–1.25	5.75	4.38	1.90		2.38	0.25	2.38	1.97	1.99	1.75
2	9.25	2.00	3.62	3.75	8–1.12	6.75	5.00	2.38		2.75	0.31	2.75	2.46	2.50	2.00
2½	10.50	2.25	4.12	4.50	8–1.25	7.75	5.62	2.88		3.12	0.31	3.12	2.97	3.00	2.25
3	12.00	2.62	5.00	5.25	8–1.38	9.00	6.62	3.50		3.62	0.38	3.62	3.60	3.63	2.50
4	14.00	3.00	6.19	6.50	8–1.62	10.75	7.50	4.50		4.25	0.44	4.25	4.60	4.63	2.75
5	16.50	3.62	7.31	8.00	8–1.88	12.75	9.00	5.56		5.12	0.44	5.12	5.69	5.69	3.00
6	19.00	4.25	8.50	9.25	8–2.12	14.50	10.75	6.63		6.00	0.50	6.00	6.75	6.75	3.25
8	21.75	5.00	10.62	12.00	12–2.12	17.25	12.50	8.63		7.00	0.50	7.00	8.75	8.75	3.75
10	26.50	6.50	12.75	14.75	12–2.62	21.25	16.50	10.75		9.00	0.50	9.00	10.92	10.88	4.25
12	30.00	7.25	15.00	17.38	12–2.88	24.38	18.25	12.75		10.00	0.50	10.00	12.92	12.94	4.75

* Dimensions in inches.

Notes:

(1) Raised face height, ¼ in, is not included in *Q*, *Y*, and *YY* dimensions.

(2) Bolt hole diameter is ⅛ in larger than the bolt diameter.

TABLE 5.57 List of Material Specifications for ASME B16.47 Flanges

Material group	Nominal designation	Applicable ASTM specifications*		
		Forgings	Castings	Plates
1.1	C–Si C–Mn–Si	A105 A350 Gr. LF2	A216 Gr. WCB	A515 Gr. 70 A516 Gr. 70 A537 Cl. 1
1.2	C–Mn–Si 2½Ni 3½Ni	A350 Gr. LF3	A216 Gr. WCC A352 Gr. LCC A352 Gr. LC2 A352 Gr. LC3	A203 Gr. B A203 Gr. E
1.3	C–Si C–Mn–Si 2½Ni 3½Ni		A352 Gr. LCB	A515 Gr. 65 A516 Gr. 65 A203 Gr. A A203 Gr. D
1.4	C–Si C–Mn–Si	A350 Gr. LF1 Cl. 1		A515 Gr. 60 A516 Gr. 60
1.5	C–½Mo	A182 Gr. F1	A217 Gr. WC1 A352 Gr. LC1	A204 Gr. A A204 Gr. B
1.7	C–½Mo ½Cr–½Mo Ni–½Cr–½Mo ¾Ni–¾Cr–1Mo	A182 Gr. F2	A217 Gr. WC4 A217 Gr. WC5	A204 Gr. C
1.9	1Cr–½Mo 1¼Cr–½Mo 1¼Cr–½Mo–Si	A182 Gr. F12 Cl. 2 A182 Gr. F11 Cl. 2	A217 Gr. WC6	A387 Gr. 11 Cl. 2
1.10	2¼Cr–1Mo	A182 Gr. F22 Cl. 3	A217 Gr. WC9	A387 Gr. 22 Cl. 2
1.13	5Cr–½Mo ¾Ni–1Mo–¾Cr	A182 Gr. F5 A182 Gr. F5a	A217 Gr. C5	
1.14	9Cr–1Mo	A182 Gr. F9	A217 Gr. C12	
2.1	18Cr–8Ni	A182 Gr. F304 A182 Gr. F304H	A351 Gr. CF3 A351 Gr. CF8	A240 Gr. 304 A240 Gr. 304H
2.2	16Cr–12Ni–2Mo 18Cr–13Ni–3Mo 19Cr–10Ni–3Mo	A182 Gr. F316 A182 Gr. F316H	A351 Gr. CF3M A351 Gr. CF8M A351 Gr. CG8M	A240 Gr. 316 A240 Gr. 316H A240 Gr. 317
2.3	18Cr–8Ni 16Cr–12Ni–2Mo	A182 Gr. F304L A182 Gr. F316L		A240 Gr. 304L A240 Gr. 316L
2.4	18Cr–10Ni–Ti	A182 Gr. F321 A182 Gr. F321H		A240 Gr. 321 A240 Gr. 321H
2.5	18Cr–10Ni–Cb	A182 Gr. F347 A182 Gr. F347H A182 Gr. F348 A182 Gr. F348H	A351 Gr. CF8C	A240 Gr. 347 A240 Gr. 347H A240 Gr. 348 A240 Gr. 348H
2.6	25Cr–12Ni 23Cr–12Ni		A351 Gr. CH8 A351 Gr. CH20	A240 Gr. 309S A240 Gr. 309H
2.7	25Cr–20Ni	A182 Gr. F310	A351 Gr. CK20	A240 Gr. 310S A240 Gr. 310H
2.8	20Cr–18Ni–6Mo 22Cr–5Ni–3Mo–N 25Cr–7Ni–4Mo–N	A182 Gr. F44 A182 Gr. F51 A182 Gr. F53	A351 Gr. CK3MCuN	A240 Gr. S31254 A240 Gr. S31803 A240 Gr. S32750

* ASME Boiler and Pressure Vessel Code, Section II, materials that also meet the requirements of the listed ASTM specifications may also be used.

Pressure Classes of Flanges, Series, and Sizes Covered by ASME B16.47

Pressure Class 150, 300, 400, 600, and 900 Series A flanges are covered in ASME B16.47. Pressure Class 75, 150, 300, 400, 600, and 900 Series B flanges are also covered by ASME B16.47. Table 5.58 lists the pressure classes and sizes for Series A and B flanges.

Table 5.58 Pressure Classes and Sizes of ASME B16.47 Series A and B Flanges

	Flange Size (NPS)					
Flange series	Class 75	Class 150	Class 300	Class 400	Class 600	Class 900
Series A	26–60	26–60	26–60	26–60	26–60	26–48
Series B	—	26–60	26–60	26–36	26–36	26–36

Pressure-Temperature Ratings of ASME B16.47 Flanges

ASME B16.47 lists the pressure-temperature ratings for flanges made from each group of materials listed in Table 5.57. We have selected only nine groups of materials that are commonly used. Ratings are maximum allowable working gage pressures at the corresponding temperature shown in the pressure-temperature rating tables.

Pressure-Temperature Ratings of Class 75 Flanges per ASME B16.47

See Table 5.59.

TABLE 5.59 Pressure-Temperature Ratings for ASME B16.47 Class 75 Flanges

	Material group								
Temp., °F	1.1	1.2	1.9	1.10	1.15	2.1	2.2	2.3	2.4
−20 to 100	140	145	145	145	145	135	135	115	135
200	130	130	130	130	130	115	115	95	120
300	115	115	115	115	115	100	105	85	115
400	100	100	100	100	100	95	95	80	100
500	85	85	85	85	85	85	85	70	85
600	70	70	70	70	70	70	70	70	70
650	60	60	60	60	60	60	60	60	60

Notes:
1. For material groups refer to Table 5.57.
2. Ratings are the maximum allowable working gage pressure to the temperature listed.

Dimensions of Class 75, Series B Flanges per ASME B16.47

See Table 5.60. Class 75 flanges are only manufactured in Series B.

Pressure-Temperature Ratings of Class 150 Flanges per ASME B16.47

See Table 5.61.

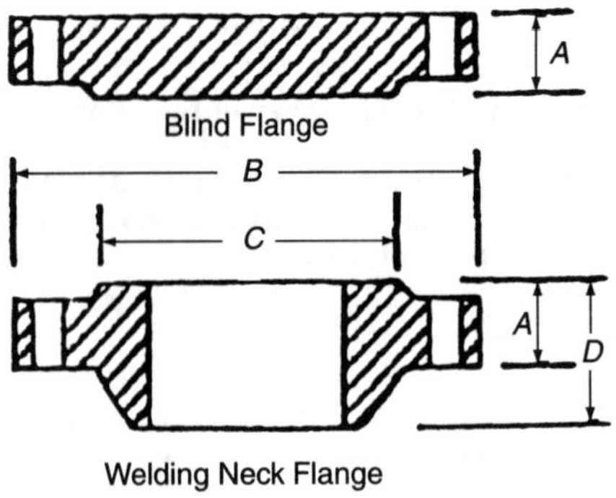

TABLE 5.60 Dimensions of ASME B16.47 Class 75 Series B Welding Neck and Blind Flanges*

Nominal pipe size (NPS)	Outer diameter of flange, *B*	Minimum thickness of flange†		Length through hub, *D*	Raised face diameter, *C*	Diameter of bolt circle	Number of bolts	Diameter of bolt
		Welding neck, *A*	Blind, *A*					
26	30.00	1.31	1.31	2.31	27.75	28.50	36	5/8
28	32.00	1.31	1.31	2.44	29.75	30.50	40	5/8
30	34.00	1.31	1.31	2.56	31.75	32.50	44	5/8
32	36.00	1.38	1.44	2.75	33.75	34.50	48	5/8
34	38.00	1.38	1.50	2.88	35.75	36.50	52	5/8
36	40.69	1.44	1.67	3.38	38.00	39.06	40	3/4
38	42.69	1.50	1.75	3.50	40.00	41.06	40	3/4
40	44.69	1.50	1.75	3.62	42.00	43.06	44	3/4
42	46.69	1.56	1.88	3.75	44.00	45.06	48	3/4
44	49.25	1.69	1.94	4.12	46.25	47.38	36	7/8
46	51.25	1.75	2.00	4.25	48.25	49.38	40	7/8
48	53.25	1.81	2.12	4.38	50.25	51.38	44	7/8
50	55.25	1.88	2.18	4.56	52.25	53.38	44	7/8
52	57.38	1.88	2.25	4.75	54.25	55.50	48	7/8
54	59.38	1.94	2.38	4.94	56.25	57.50	48	7/8
56	62.00	2.00	2.44	5.31	58.50	59.88	40	1
58	64.00	2.06	2.50	5.44	60.50	61.88	44	1
60	66.00	2.19	2.62	5.69	62.50	63.88	44	1

* Dimensions in inches.
† Raised face height of 0.06 in is included in minimum flange thickness *A*.

TABLE 5.61 Pressure-Temperature Ratings for ASME B16.47 Class 150 Flanges

Temp., °F	Material group								
	1.1	1.2	1.9	1.10	1.15	2.1	2.2	2.3	2.4
−20 to 100	285	290	290	290	290	275	275	230	275
200	260	260	260	260	260	230	235	195	245
300	230	230	230	230	230	205	215	175	230
400	200	200	200	200	200	190	195	160	200
500	170	170	170	170	170	170	170	145	170
600	140	140	140	140	140	140	140	140	140
650	125	125	125	125	125	125	125	125	125
700	110	110	110	110	110	110	110	110	110
750	95	95	95	95	95	95	95	95	95
800	80	80	80	80	80	80	80	80	80
850	65	65	65	65	65	65	65	65	65
900	50	50	50	50	50	50	50		50
950	35	35	35	35	35	35	35		35
1000	20	20	20	20	20	20	20		20

Notes:
1. For material groups refer to Table 5.57.
2. Ratings are the maximum allowable working gage pressure to the temperature listed.

Dimensions of Class 150, Series A Flanges per ASME B16.47

See Table 5.62.

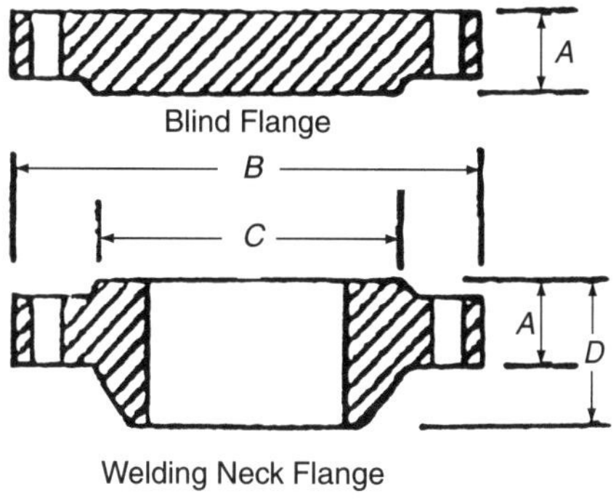

TABLE 5.62 Dimensions of ASME B16.47 Class 150 Series A Flanges*

Nominal pipe size (NPS)	Outer diameter of flange, *B*	Minimum thickness of flange†		Length through hub, *D*	Raised face diameter, *C*	Diameter of bolt circle	Number of bolts	Diameter of bolt
		Welding neck, *A*	Blind, *A*					
26	34.25	2.69	2.69	4.75	29.50	31.75	24	$1\frac{1}{4}$
28	36.50	2.81	2.81	4.94	31.50	34.00	28	$1\frac{1}{4}$
30	38.75	2.94	2.94	5.38	33.75	36.00	28	$1\frac{1}{4}$
32	41.75	3.19	3.19	5.69	36.00	38.50	28	$1\frac{1}{2}$
34	43.75	3.25	3.25	5.88	38.00	40.50	32	$1\frac{1}{2}$
36	46.00	3.56	3.56	6.19	40.25	42.75	32	$1\frac{1}{2}$
38	48.75	3.44	3.44	6.19	42.25	45.25	32	$1\frac{1}{2}$
40	50.75	3.56	3.56	6.44	44.25	47.25	36	$1\frac{1}{2}$
42	53.00	3.81	3.81	6.75	47.00	49.50	36	$1\frac{1}{2}$
44	55.25	4.00	4.00	7.00	49.00	51.75	40	$1\frac{1}{2}$
46	57.25	4.06	4.06	7.31	51.00	53.75	40	$1\frac{1}{2}$
48	59.50	4.25	4.25	7.56	53.50	56.00	44	$1\frac{1}{2}$
50	61.75	4.38	4.38	8.00	55.50	58.25	44	$1\frac{3}{4}$
52	64.00	4.56	4.56	8.25	57.50	60.50	44	$1\frac{3}{4}$
54	66.25	4.75	4.75	8.50	59.50	62.75	44	$1\frac{3}{4}$
56	68.75	4.88	4.88	9.00	62.00	65.00	48	$1\frac{3}{4}$
58	71.00	5.06	5.06	9.25	64.00	67.25	48	$1\frac{3}{4}$
60	73.00	5.19	5.19	9.44	66.00	69.25	52	$1\frac{3}{4}$

* Dimensions in inches.
† Raised face height of 0.06 in is included in minimum flange thickness *A*.

Dimensions of Class 150, Series B Flanges per ASME B16.47

See Table 5.63.

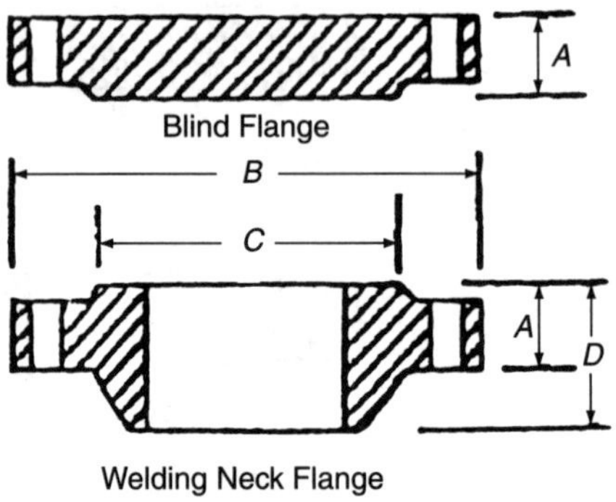

TABLE 5.63 Dimensions of ASME B16.47 Class 150 Series B Flanges*

Nominal pipe size (NPS)	Outer diameter of flange, *B*	Minimum thickness of flange†		Length through hub, *D*	Raised face diameter, *C*	Diameter of bolt circle	Number of bolts	Diameter of bolt
		Welding neck, *A*	Blind, *A*					
26	30.94	1.62	1.75	3.50	28.00	29.31	36	3/4
28	32.94	1.75	1.88	3.75	30.00	31.31	40	3/4
30	34.94	1.75	2.00	3.94	32.00	33.31	44	3/4
32	37.06	1.81	2.12	4.25	34.00	35.44	48	3/4
34	39.56	1.94	2.25	4.34	36.25	37.69	40	7/8
36	41.62	2.06	2.31	4.62	38.25	39.75	44	7/8
38	44.25	2.12	2.50	4.88	40.25	42.12	40	1
40	46.25	2.19	2.62	5.06	42.50	44.12	44	1
42	48.25	2.31	2.69	5.25	44.50	46.12	48	1
44	50.25	2.38	2.81	5.38	46.50	48.12	52	1
46	52.81	2.44	2.94	5.69	48.62	50.56	40	1 1/8
48	54.81	2.56	3.06	5.88	50.75	52.56	44	1 1/8
50	56.81	2.69	3.18	6.06	52.75	54.56	48	1 1/8
52	58.81	2.75	3.31	6.19	54.75	56.56	52	1 1/8
54	61.00	2.81	3.44	6.38	56.75	58.75	56	1 1/8
56	63.00	2.88	3.56	6.56	58.75	60.75	60	1 1/8
58	65.94	2.94	3.68	6.88	60.75	63.44	48	1 1/4
60	67.94	3.00	3.81	7.06	63.00	65.44	52	1 1/4

* Dimensions in inches.
† Raised face height of 0.06 in is included in minimum flange thickness *A*.

Pressure-Temperature Ratings of Class 300 Flanges per ASME B16.47

See Table 5.64.

TABLE 5.64 Pressure-Temperature Ratings for ASME B16.47 Class 300 Flanges

	Material group								
Temp., °F	1.1	1.2	1.9	1.10	1.15	2.1	2.2	2.3	2.4
–20 to 100	740	750	750	750	750	720	720	600	720
200	675	750	750	750	750	620	620	505	645
300	655	730	720	730	730	540	560	455	595
400	635	705	695	705	705	495	515	415	550
500	600	665	665	665	665	465	480	380	515
600	550	605	605	605	605	435	450	360	485
650	535	590	590	590	590	430	445	350	480
700	535	570	570	570	570	425	430	345	465
750	505	505	530	530	530	415	425	335	460
800	410	410	510	510	510	405	420	330	450
850	270	270	485	485	485	395	420	320	445
900	170	170	450	450	450	390	415		440
950	105	105	320	375	385	380	385		385
1000	50	50	215	260	365	320	350		355
1050			145	175	360	310	345		315
1100			95	110	300	255	305		270
1150			60	70	225	200	235		235
1200			40	40	145	155	185		185
1250						115	145		140
1300						85	115		110
1350						60	95		85
1400						50	75		65
1450						35	60		50
1500						25	40		40

Notes:
1. For material groups refer to Table 5.57.
2. Ratings are the maximum allowable working gage pressure to the temperature listed.

Dimensions of Class 300, Series A Flanges per ASME B16.47

See Table 5.65.

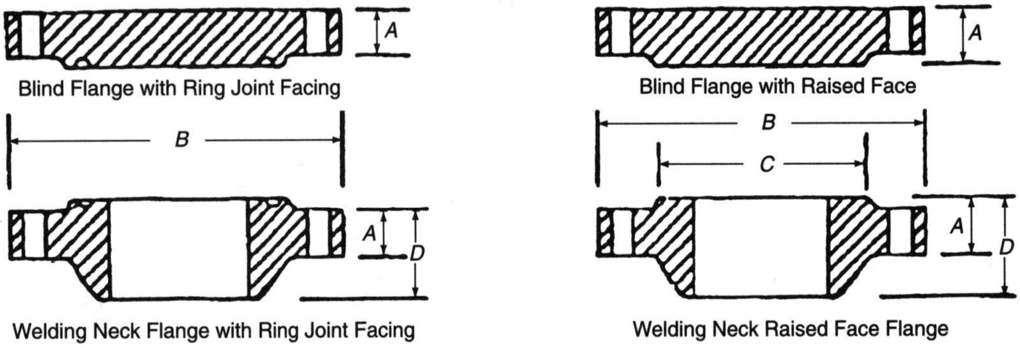

TABLE 5.65 Dimensions of ASME B16.47 Class 300 Series A Flanges*

Nominal pipe size (NPS)	Outer diameter of flange, *B*	Minimum thickness of flange† Welding neck, *A*	Minimum thickness of flange† Blind, *A*	Length through hub, *D*	Raised face diameter, *C*	Diameter of bolt circle	Number of bolts	Diameter of bolt
26	38.25	3.12	3.31	7.25	29.50	34.50	28	1 5/8
28	40.75	3.38	3.56	7.75	31.50	37.00	28	1 5/8
30	43.00	3.62	3.75	8.25	33.75	39.25	28	1 3/4
32	45.25	3.88	3.94	8.75	36.00	41.50	28	1 7/8
34	47.50	4.00	4.12	9.12	38.00	43.50	28	1 7/8
36	50.00	4.12	4.38	9.50	40.25	46.00	32	2
38	46.00	4.25	4.25	7.12	40.50	43.00	32	1 1/2
40	48.75	4.50	4.50	7.62	42.75	45.50	32	1 5/8
42	50.75	4.69	4.69	7.88	44.75	47.50	32	1 5/8
44	53.25	4.88	4.88	8.12	47.00	49.75	32	1 3/4
46	55.75	5.06	5.06	8.50	49.00	52.00	28	1 7/8
48	57.75	5.25	5.25	8.81	51.25	54.00	32	1 7/8
50	60.25	5.50	5.50	9.12	53.50	56.25	32	2
52	62.25	5.69	5.69	9.38	55.50	58.25	32	2
54	65.25	6.00	6.00	9.94	57.75	61.00	28	2 1/4
56	67.25	6.06	6.06	10.25	59.75	63.00	28	2 1/4
58	69.25	6.25	6.25	10.50	62.00	65.00	32	2 1/4
60	71.25	6.44	6.44	10.75	64.00	67.00	32	2 1/4

* Dimensions in inches.
† Raised face height of 0.06 in is included in minimum flange thickness *A*.

Dimensions of Class 300, Series B Flanges per ASME B16.47

See Table 5.66.

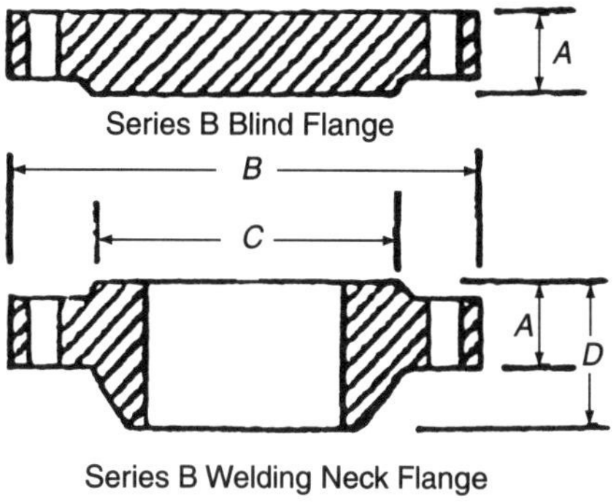

TABLE 5.66 Dimensions of ASME B16.47 Class 300 Series B Flanges*

Nominal pipe size (NPS)	Outer diameter of flange, *B*	Minimum thickness of flange†		Length through hub, *D*	Raised face diameter, *C*	Diameter of bolt circle	Number of bolts	Diameter of bolt
		Welding neck, *A*	Blind, *A*					
26	34.12	3.50	3.50	5.69	29.00	31.62	32	$1\frac{1}{4}$
28	36.25	3.50	3.50	5.88	31.00	33.75	36	$1\frac{1}{4}$
30	39.00	3.69	3.69	6.22	33.25	36.25	36	$1\frac{3}{8}$
32	41.50	4.06	4.06	6.62	35.50	38.50	32	$1\frac{1}{2}$
34	43.62	4.06	4.06	6.81	37.50	40.62	36	$1\frac{1}{2}$
36	46.12	4.06	4.06	7.12	39.75	42.88	32	$1\frac{5}{8}$
38	48.12	4.38	4.38	7.56	41.75	44.88	36	$1\frac{5}{8}$
40	50.12	4.56	4.56	7.81	43.88	46.88	40	$1\frac{5}{8}$
42	52.50	4.69	4.69	8.06	46.00	49.00	36	$1\frac{3}{4}$
44	54.50	5.00	5.00	8.44	48.00	51.00	40	$1\frac{3}{4}$
46	57.50	5.06	5.12	8.75	50.00	53.75	36	$1\frac{7}{8}$
48	59.50	5.06	5.31	8.81	52.25	55.75	40	$1\frac{7}{8}$
50	61.50	5.44	5.50	9.25	54.25	57.75	44	$1\frac{7}{8}$
52	63.50	5.62	5.68	9.56	56.25	59.75	48	$1\frac{7}{8}$
54	65.88	5.38	5.88	9.44	58.25	62.12	48	$1\frac{7}{8}$
56	69.50	6.06	6.18	10.56	60.50	65.00	36	$2\frac{1}{4}$
58	71.94	6.06	6.38	10.81	62.75	67.44	40	$2\frac{1}{4}$
60	73.94	5.94	6.56	10.69	65.00	69.44	40	$2\frac{1}{4}$

* Dimensions in inches.
† Raised face height of 0.06 in is included in minimum flange thickness *A*.

Pressure-Temperature Ratings of Class 400 Flanges per ASME B16.47

See Table 5.67.

TABLE 5.67 Pressure-Temperature Ratings for ASME B16.47 Class 400 Flanges

	Material group								
Temp., °F	1.1	1.2	1.9	1.10	1.15	2.1	2.2	2.3	2.4
–20 to 100	990	1000	1000	1000	1000	960	960	800	960
200	900	1000	1000	1000	1000	800	825	675	860
300	875	970	965	970	970	720	745	605	795
400	845	940	925	940	940	660	685	550	735
500	800	885	885	885	885	620	635	510	685
600	730	805	805	805	805	580	600	480	650
650	715	785	785	785	785	575	590	470	635
700	710	755	755	755	755	565	580	460	620
750	670	670	710	710	710	555	570	450	610
800	550	550	675	675	675	540	565	440	600
850	355	355	650	650	650	530	555	430	595
900	230	230	600	600	600	520	555		590
950	140	140	425	505	515	510	515		515
1000	70	70	290	345	485	430	465		475
1050			190	235	480	410	460		415
1100			130	145	400	345	405		360
1150			80	90	295	265	315		315
1200			50	55	190	205	245		245
1250						150	195		185
1300						115	155		145
1350						80	130		115
1400						65	100		85
1450						45	80		70
1500						35	55		50

Notes:
1. For material groups refer to Table 5.57.
2. Ratings are the maximum allowable working gage pressure to the temperature listed.

Dimensions of Class 400, Series A Flanges per ASME B16.47

See Table 5.68.

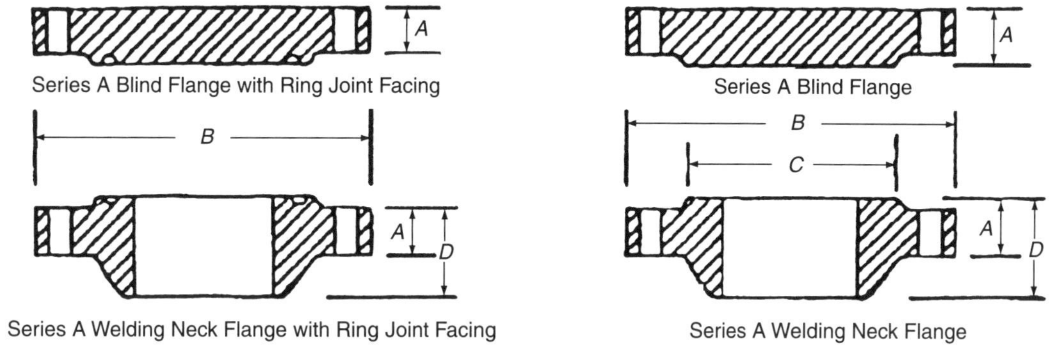

TABLE 5.68 Dimensions of ASME B16.47 Class 400 Series A Flanges*

Nominal pipe size (NPS)	Outer diameter of flange, *B*	Minimum thickness of flange†		Length through hub, *D*	Raised face diameter, *C*	Diameter of bolt circle	Number of bolts	Diameter of bolt
		Welding neck, *A*	Blind, *A*					
26	38.25	3.50	3.88	7.62	29.50	34.50	28	1¾
28	40.75	3.75	4.12	8.12	31.50	37.00	28	1⅞
30	43.00	4.00	4.38	8.62	33.75	39.25	28	2
32	45.25	4.25	4.56	9.12	36.00	41.50	28	2
34	47.50	4.38	4.81	9.50	38.00	43.50	28	2
36	50.00	4.50	5.06	9.88	40.25	46.00	32	2
38	47.50	4.88	4.88	8.12	40.75	44.00	32	1¾
40	50.00	5.12	5.12	8.50	43.00	46.25	32	1⅞
42	52.00	5.25	5.25	8.81	45.00	48.25	32	1⅞
44	54.50	5.50	5.50	9.18	47.25	50.50	32	2
46	56.75	5.75	5.75	9.62	49.50	52.75	36	2
48	59.50	6.00	6.00	10.12	51.50	55.25	28	2¼
50	61.75	6.19	6.25	10.56	53.62	57.50	32	2¼
52	63.75	6.38	6.44	10.88	55.62	59.50	32	2¼
54	67.00	6.69	6.75	11.38	57.88	62.25	28	2½
56	69.00	6.88	6.94	11.75	60.12	64.25	32	2½
58	71.00	7.00	7.12	12.06	62.12	66.25	32	2½
60	74.25	7.31	7.44	12.56	64.38	69.00	32	2¾

* Dimensions in inches.
† Flanges are furnished with 0.25-in raised face, which is not included in the minimum flange thickness *A*.

Dimensions of Class 400, Series B Flanges per ASME B16.47

See Table 5.69.

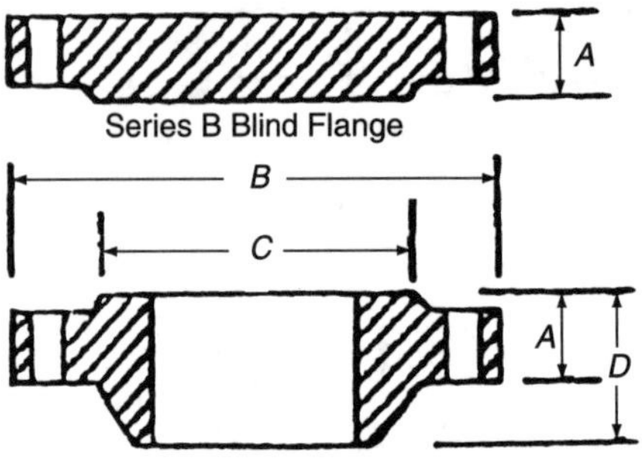

Series B Welding Neck Flange

TABLE 5.69 Dimensions of ASME B16.47 Class 400 Series B Flanges*

Nominal pipe size (NPS)	Outer diameter of flange, *B*	Minimum thickness of flange†		Length through hub, *D*	Raised face diameter, *C*	Diameter of bolt circle	Number of bolts	Diameter of bolt
		Welding neck, *A*	Blind, *A*					
26	33.50	3.50	3.50	5.88	28.00	30.75	28	13/8
28	36.00	3.75	3.75	6.25	30.00	33.00	24	11/2
30	38.25	4.00	4.00	6.69	32.25	35.25	28	11/2
32	40.75	4.25	4.25	7.06	34.38	37.50	28	15/8
34	42.75	4.38	4.38	7.38	36.50	39.50	32	15/8
36	45.50	4.69	4.69	7.88	38.62	42.00	28	13/4

* Dimensions in inches.
† Dimensions of NPS 38 and larger Series B flanges are the same as those of Series A flanges.

Pressure-Temperature Ratings of Class 600 Flanges per ASME B16.47

See Table 5.70.

TABLE 5.70 Pressure-Temperature Ratings for ASME B16.47 Class 600 Flanges

	Material group								
Temp., °F	1.1	1.2	1.9	1.10	1.15	2.1	2.2	2.3	2.4
–20 to 100	1480	1500	1500	1500	1500	1440	1440	1200	1440
200	1350	1500	1500	1500	1500	1200	1240	1015	1290
300	1315	1455	1445	1455	1455	1080	1120	910	1190
400	1270	1410	1385	1410	1410	995	1025	825	1105
500	1200	1330	1330	1330	1330	930	955	765	1030
600	1095	1210	1210	1210	1210	875	900	720	975
650	1075	1175	1175	1175	1175	860	890	700	955
700	1065	1135	1135	1135	1135	850	870	685	930
750	1010	1010	1065	1065	1065	830	855	670	915
800	825	825	1015	1015	1015	805	845	660	900
850	535	535	975	975	975	790	835	645	895
900	345	345	900	900	900	780	830		885
950	205	205	640	755	775	765	775		775
1000	105	105	430	520	725	640	700		715
1050			290	350	720	615	685		625
1100			190	220	605	515	610		545
1150			125	135	445	400	475		475
1200			75	80	290	310	370		370
1250						225	295		280
1300						170	235		220
1350						125	190		170
1400						90	150		130
1450						70	115		105
1500						55	85		75

Notes:
1. For material groups refer to Table 5.57.
2. Ratings are the maximum allowable working gage pressure to the temperature listed.

Dimensions of Class 600, Series A Flanges per ASME B16.47

See Table 5.71.

Series A Blind Flange with Ring Joint Facing

Series A Raised Face Blind Flange

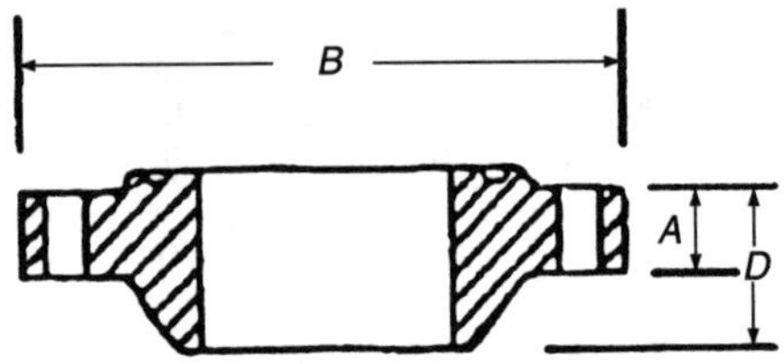

Series A Welding Neck Flange with Ring Joint Facing

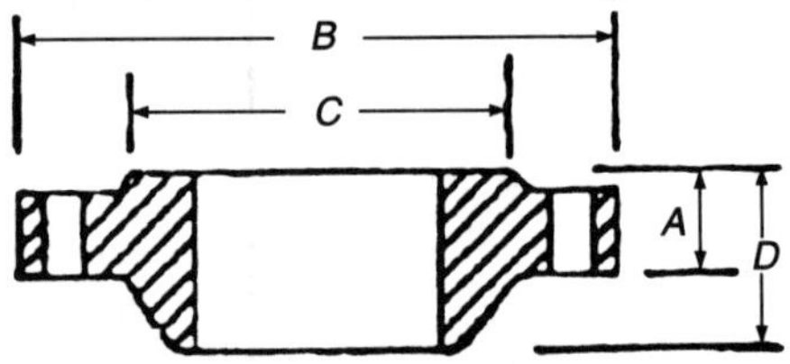

Series A Raised Face Welding Neck Flange

TABLE 5.71 Dimensions of ASME B16.47 Class 600 Series A Flanges*

Nominal pipe size (NPS)	Outer diameter of flange, *B*	Minimum thickness of flange† Welding neck, *A*	Minimum thickness of flange† Blind, *A*	Length through hub, *D*	Raised face diameter, *C*	Diameter of bolt circle	Number of bolts	Diameter of bolt
26	40.00	4.25	4.94	8.75	29.50	36.00	28	$1\frac{7}{8}$
28	42.25	4.38	5.19	9.25	31.50	38.00	28	2
30	44.50	4.50	5.50	9.75	33.75	40.25	28	2
32	47.00	4.62	5.81	10.25	36.00	42.50	28	$2\frac{1}{4}$
34	49.00	4.75	6.06	10.62	38.00	44.50	28	$2\frac{1}{4}$
36	51.75	4.88	6.38	11.12	40.25	47.00	28	$2\frac{1}{2}$
38	50.00	6.00	6.12	10.00	41.50	45.75	28	$2\frac{1}{4}$
40	52.00	6.25	6.38	10.38	43.75	47.75	32	$2\frac{1}{4}$
42	55.25	6.62	6.75	11.00	46.00	50.50	28	$2\frac{1}{2}$
44	57.25	6.81	7.00	11.38	48.25	52.50	32	$2\frac{1}{2}$
46	59.50	7.06	7.31	11.81	50.25	54.75	32	$2\frac{1}{2}$
48	62.75	7.44	7.69	12.44	52.50	57.50	32	$2\frac{3}{4}$
50	65.75	7.75	8.00	12.94	54.50	60.00	28	3
52	67.75	8.00	8.25	13.25	56.50	62.00	32	3
54	70.00	8.25	8.56	13.75	58.75	64.25	32	3
56	73.00	8.56	8.88	14.25	60.75	66.75	32	$3\frac{1}{4}$
58	75.00	8.75	9.12	14.56	63.00	68.75	32	$3\frac{1}{4}$
60	78.50	9.19	9.56	15.31	65.25	71.75	28	$3\frac{1}{2}$

* Dimensions in inches.
† Flanges are furnished with 0.25-in raised face, which is not included in the minimum flange thickness *A*.

Dimensions of Class 600, Series B Flanges per ASME B16.47

See Table 5.72.

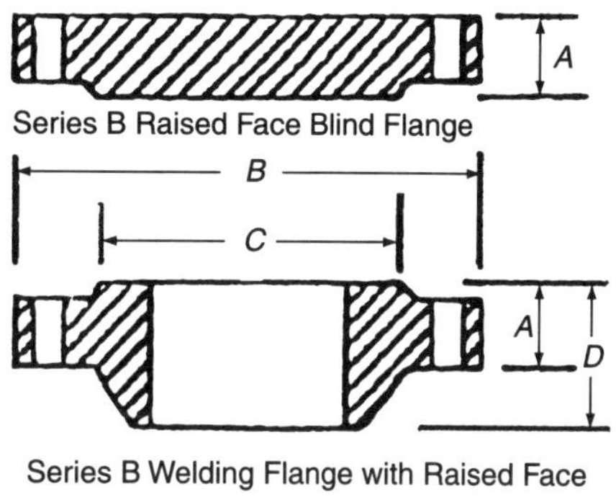

TABLE 5.72 Dimensions of ASME B16.47 Class 600 Series B Flanges*

Nominal pipe size (NPS)	Outer diameter of flange, *B*	Minimum thickness of flange†		Length through hub, *D*	Raised face diameter, *C*	Diameter of bolt circle	Number of bolts	Diameter of bolt
		Welding neck, *A*	Blind, *A*					
26	35.00	4.38	4.38	7.12	28.62	31.75	28	1⅝
28	37.50	4.56	4.56	7.50	30.88	34.00	28	1¾
30	40.25	4.94	5.00	8.06	33.12	36.50	28	1⅞
32	42.75	5.12	5.31	8.50	35.25	38.75	28	2
34	45.75	5.56	5.68	9.19	37.50	41.50	24	2¼
36	47.75	5.75	5.94	9.56	39.75	43.50	28	2¼

* Dimensions in inches.
† Dimensions of NPS 38 and larger Series B flanges are the same as Series A flanges.

Pressure-Temperature Ratings of Class 900 Flanges per ASME B16.47

See Table 5.73.

TABLE 5.73 Pressure-Temperature Ratings for ASME B16.47, Class 900 Flanges

Temp., °F	Material Group								
	1.1	1.2	1.9	1.10	1.15	2.1	2.2	2.3	2.4
–20 to 100	2220	2250	2250	2250	2250	2160	2160	1800	2160
200	2025	2250	2250	2250	2250	1800	1860	1520	1935
300	1970	2185	2165	2185	2185	1620	1680	1360	1785
400	1900	2115	2080	2115	2115	1490	1540	1240	1655
500	1795	1995	1995	1995	1995	1395	1435	1145	1545
600	1640	1815	1815	1815	1815	1310	1355	1080	1460
650	1610	1765	1765	1765	1765	1290	1330	1050	1435
700	1600	1705	1705	1705	1705	1275	1305	1030	1395
750	1510	1510	1595	1595	1595	1245	1280	1010	1375
800	1235	1235	1525	1525	1525	1210	1265	985	1355
850	805	805	1460	1460	1460	1190	1255	965	1340
900	515	515	1350	1350	1350	1165	1245		1325
950	310	310	955	1130	1160	1145	1160		1160
1000	155	155	650	780	1090	965	1050		1070
1050			430	525	1080	925	1030		940
1100			290	330	905	770	915		815
1150			185	205	670	595	710		710
1200			115	125	430	465	555		555
1250						340	440		420
1300						255	350		330
1350						185	290		255
1400						145	225		195
1450						105	175		155
1500						80	125		115

Notes:
1. For material groups refer to Table 5.57.
2. Ratings are the maximum allowable working gage pressure to the temperature listed.

Dimensions of Class 900, Series A Flanges per ASME B16.47

See Table 5.74.

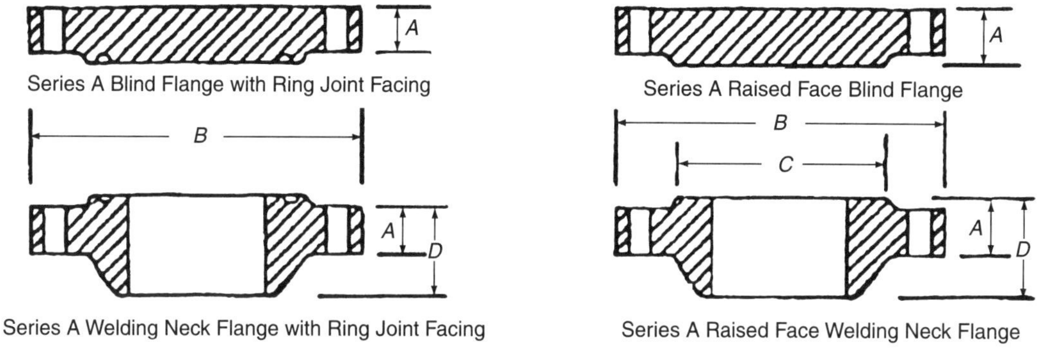

TABLE 5.74 Dimensions of ASME B16.47 Class 900 Series A Flanges*

Nominal pipe size (NPS)	Outer diameter of flange, *B*	Minimum thickness of flange†		Length through hub, *D*	Raised face diameter, *C*	Diameter of bolt circle	Number of bolts	Diameter of bolt
		Welding neck, *A*	Blind, *A*					
26	42.75	5.50	6.31	11.25	29.50	37.50	20	2¾
28	46.00	5.62	6.75	11.75	31.50	40.25	20	3
30	48.50	5.88	7.18	12.25	33.75	42.75	20	3
32	51.75	6.25	7.62	13.00	36.00	45.50	20	3¼
34	55.00	6.50	8.06	13.75	38.00	48.25	20	3½
36	57.50	6.75	8.44	14.25	40.25	50.75	20	3½
38	57.50	7.50	8.50	13.88	43.25	50.75	20	3½
40	59.50	7.75	8.81	14.31	45.75	52.75	24	3½
42	61.50	8.12	9.12	14.62	47.75	54.75	24	3½
44	64.88	8.44	9.56	15.38	50.00	57.62	24	3¾
46	68.25	8.88	10.06	16.18	52.50	60.50	24	4
48	70.25	9.19	10.38	16.50	54.50	62.50	24	4

* Dimensions in inches.
† Flanges are furnished with 0.25-in raised face, which is not included in the minimum flange thickness *A*.

Dimensions of Class 900, Series B Flanges per ASME B16.47

See Table 5.75.

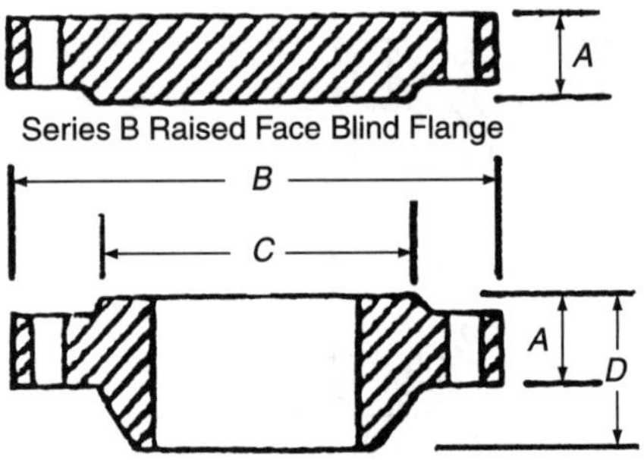

TABLE 5.75 Dimensions of ASME B16.47 Class 900 Series B Flanges*

Nominal pipe size (NPS)	Outer diameter of flange, *B*	Minimum thickness of flange[†]		Length through hub, *D*	Raised face diameter, *C*	Diameter of bolt circle	Number of bolts	Diameter of bolt
		Welding neck, *A*	Blind, *A*					
26	40.25	5.31	6.06	10.19	30.00	35.50	20	$2\frac{1}{2}$
28	43.50	5.81	6.56	10.88	32.25	38.25	20	$2\frac{3}{4}$
30	46.50	6.12	6.93	11.38	34.50	40.75	20	3
32	48.75	6.31	7.31	11.94	36.50	43.00	20	3
34	51.75	6.75	7.68	12.56	39.00	45.50	20	$3\frac{1}{4}$
36	53.00	6.81	7.94	12.81	40.50	47.25	24	3

* Dimensions in inches.
† Dimensions of NPS 38 and larger Series B flanges are the same as those of Series A flanges.

ASME B16.36 ORIFICE FLANGES

ASME B16.36 covers welding neck flanges in pressure Classes 300 through 2500 of orifice flanges. Slip-on and threaded flanges are manufactured in Class 300 only.

Materials for ASME B16.36 Orifice Flanges

Materials used for ASME B16.5 are also used for ASME B16.36 orifice flanges. Refer to Table 5.42.

Pressure-Temperature Ratings of ASME B16.36 Orifice Flanges

Pressure ratings of ASME B16.5 flanges apply to ASME B16.36 orifice flanges of the corresponding pressure class. Refer to:

- Table 5.45 for Class 300 flanges
- Table 5.47 for Class 400 flanges
- Table 5.49 for Class 600 flanges
- Table 5.51 for Class 900 flanges
- Table 5.53 for Class 1500 flanges
- Table 5.55 for Class 2500 flanges.

Dimensions of Class 300 Slip-on and Threaded End Raised Face Orifice Flanges

See Table 5.76.

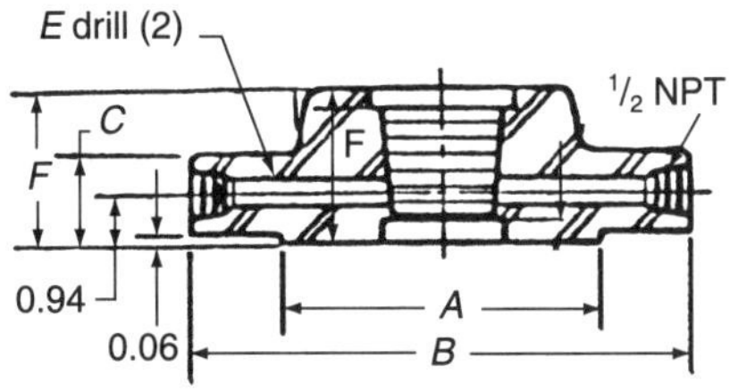

Raised Face Threaded Orifice Flange

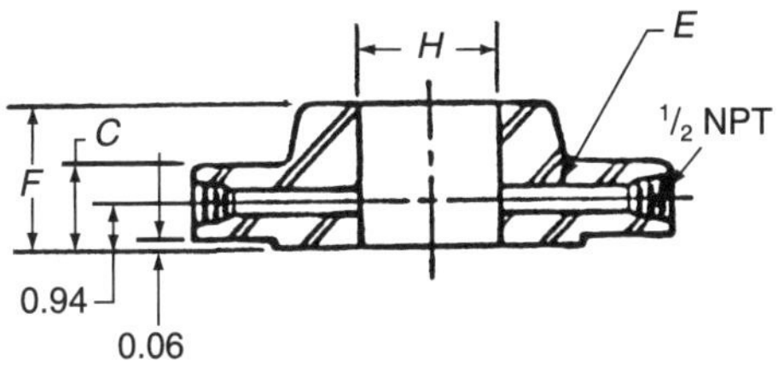

Raised Face Threaded Slip-on Orifice Flange

TABLE 5.76 Dimensions of ASME B16.36 Class 300 Slip-on and Threaded End, Raised Face Orifice Flanges*

Nominal pipe size	Outside diameter of raised face, *A*	Outside diameter of flange, *B*	Thickness of flange, min., *C*	Length through hub, *F*	Bore of slip-on flange, *H*	Diameter of pressure connection, *E*	Bolt circle	Number of holes	Diameter of bolts	Height of raised face
1	2.00	4.88	1.50	1.88	1.36	$\frac{1}{4}$	3.50	4	$\frac{5}{8}$	0.06
$1\frac{1}{2}$	2.88	6.12	1.50	1.88	1.95	$\frac{1}{4}$	4.50	4	$\frac{3}{4}$	0.06
2	3.62	6.50	1.50	1.94	2.44	$\frac{1}{4}$	5.00	8	$\frac{5}{8}$	0.06
$2\frac{1}{2}$	4.12	7.50	1.50	2.00	2.94	$\frac{1}{4}$	5.88	8	$\frac{3}{4}$	0.06
3	5.00	8.25	1.50	2.06	3.57	$\frac{3}{8}$	6.62	8	$\frac{3}{4}$	0.06
4	6.19	10.00	1.50	2.12	4.57	$\frac{1}{2}$	7.88	8	$\frac{3}{4}$	0.06
6	8.50	12.50	1.50	2.12	6.72	$\frac{1}{2}$	10.62	12	$\frac{3}{4}$	0.06
8	10.62	15.00	1.62	2.44	8.72	$\frac{1}{2}$	13.00	12	$\frac{7}{8}$	0.06
10	12.75	17.50	1.88	2.62	10.88	$\frac{1}{2}$	15.25	16	1	0.06
12	15.00	20.50	2.00	2.88	12.88	$\frac{1}{2}$	17.75	16	$1\frac{1}{8}$	0.06
14	16.25	23.00	2.12	3.00	14.14	$\frac{1}{2}$	20.25	20	$1\frac{1}{8}$	0.06
16	18.50	25.00	2.25	3.25	16.16	$\frac{1}{2}$	22.50	20	$1\frac{1}{4}$	0.06
18	21.00	28.00	2.38	3.50	18.18	$\frac{1}{2}$	24.75	24	$1\frac{1}{4}$	0.06
20	23.00	30.50	2.50	3.75	20.20	$\frac{1}{2}$	27.00	24	$1\frac{1}{4}$	0.06
24	27.25	36.00	2.75	4.19	24.25	$\frac{1}{2}$	32.00	24	$1\frac{1}{2}$	0.06

* Dimensions in inches.

Dimensions of Class 300 Welding Neck Raised Face Orifice Flanges

See Table 5.77.

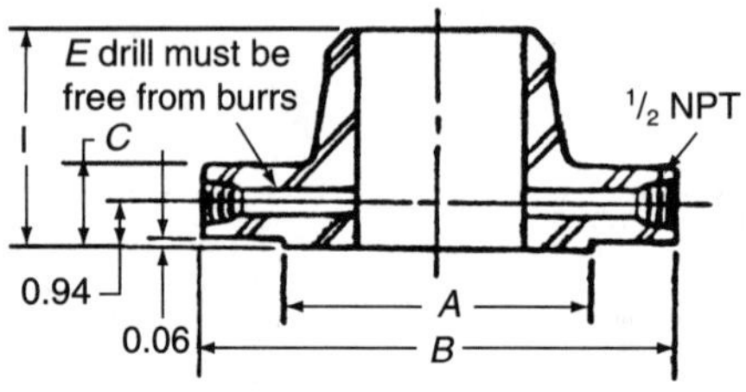

Welding Neck Raised Face Orifice Flange

TABLE 5.77 Dimensions of ASME B16.36 Class 300 Welding Neck Raised Face Orifice Flanges*

Nominal pipe size	Outside diameter of raised face, *A*	Outside diameter of flange, *B*	Thickness of flange, min., *C*	Length through hub of weld neck, *I*	Diameter of pressure connection, *E*	Bolt circle	Number of holes	Diameter of bolts	Height of raised face
1	2.00	4.88	1.50	3.25	1/4	3.50	4	5/8	0.06
1 1/2	2.88	6.12	1.50	3.38	1/4	4.50	4	3/4	0.06
2	3.62	6.50	1.50	3.38	1/4	5.00	8	5/8	0.06
2 1/2	4.12	7.50	1.50	3.50	1/4	5.88	8	3/4	0.06
3	5.00	8.25	1.50	3.50	3/8	6.62	8	3/4	0.06
4	6.19	10.00	1.50	3.62	1/2	7.88	8	3/4	0.06
6	8.50	12.50	1.50	3.94	1/2	10.62	12	3/4	0.06
8	10.62	15.00	1.62	4.38	1/2	13.00	12	7/8	0.06
10	12.75	17.50	1.88	4.62	1/2	15.25	16	1	0.06
12	15.00	20.50	2.00	5.12	1/2	17.75	16	1 1/8	0.06
14	16.25	23.00	2.12	5.62	1/2	20.25	20	1 1/8	0.06
16	18.50	25.00	2.25	5.75	1/2	22.50	20	1 1/4	0.06
18	21.00	28.00	2.38	6.25	1/2	24.75	24	1 1/4	0.06
20	23.00	30.50	2.50	6.38	1/2	27.00	24	1 1/4	0.06
24	27.25	36.00	2.75	6.62	1/2	32.00	24	1 1/2	0.06

* Dimensions in inches.

Dimensions of Class 400 Welding Neck Raised Face Orifice Flanges

See Table 5.78.

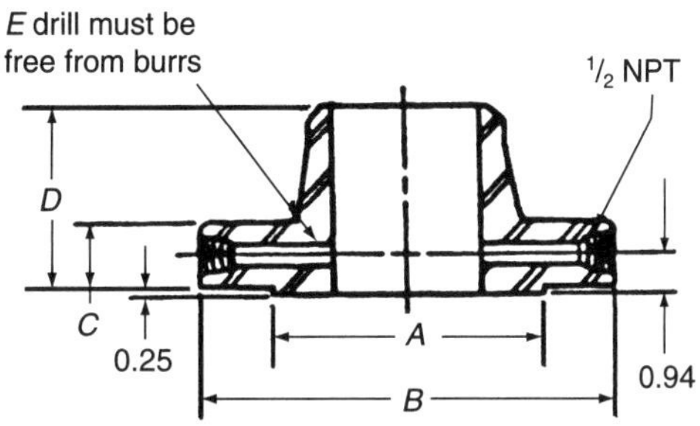

Welding Neck Raised Face Orifice Flange

TABLE 5.78 Dimensions of ASME B16.36 Class 400 Welding Neck Orifice Raised Face Flanges*

Nominal pipe size	Outside diameter of raised face, *A*	Outside diameter of flange, *B*	Thickness of flange, min., *C*	Length through hub, *D*	Diameter of pressure connection, *E*	Diameter of bolt circle	Number of holes	Diameter of bolts	Height of raised face
1 1½ 2 2½ 3	For nominal size and smaller, use Class 600								
4	6.19	10.00	1.50	3.50	½	7.88	8	⅞	0.25
6	8.50	12.50	1.62	4.06	½	10.62	12	⅞	0.25
8	10.62	15.00	1.88	4.62	½	13.00	12	1	0.25
10	12.75	17.50	2.12	4.88	½	15.25	16	1⅛	0.25
12	15.00	20.50	2.25	5.38	½	17.75	16	1¼	0.25
14	16.25	23.00	2.39	5.88	½	20.25	20	1¼	0.25
16	18.50	25.50	2.50	6.00	½	22.50	20	1⅜	0.25
18	21.00	28.00	2.62	6.50	½	24.75	24	1⅜	0.25
20	23.00	30.50	2.75	6.62	½	27.00	24	1½	0.25
24	27.25	36.00	3.00	6.88	½	32.00	24	1¾	0.25

* Dimensions in inches.

Dimensions of Class 600 Welding Neck Raised Face Orifice Flanges

See Table 5.79.

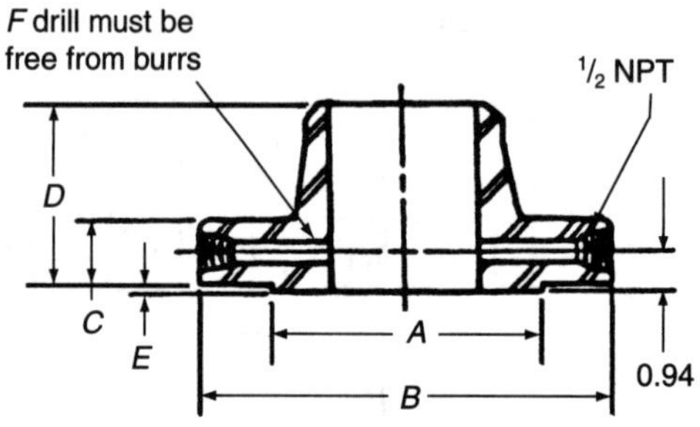

Welding Neck Raised Face Orifice Flange

TABLE 5.79 Dimensions of ASME B16.36 Class 600 Welding Neck Raised Face Orifice Flanges*

Nominal pipe size	Outside diameter of raised face, *A*	Outside diameter of flange, *B*	Thickness of flange, min., *C*	Length through hub, *D*	Height of raised face, *E*	Diameter of pressure connection, *F*	Diameter of bolt circle	Number of holes	Diameter of bolts
1	2.00	4.88	1.44	3.19	0.06	1/4	3.50	4	5/8
1 1/2	2.88	6.12	1.44	3.32	0.06	1/4	4.50	4	3/4
2	3.62	6.50	1.44	3.32	0.06	1/4	5.00	8	5/8
2 1/2	4.12	7.50	1.44	3.44	0.06	1/4	5.88	8	3/4
3	5.00	8.25	1.44	3.44	0.06	3/8	6.62	8	3/4
4	6.19	10.75	1.50	4.00	0.25	1/2	8.50	8	7/8
6	8.50	14.00	1.88	4.62	0.25	1/2	11.50	12	1
8	10.62	16.50	2.19	5.25	0.25	1/2	13.75	12	1 1/8
10	12.75	20.00	2.50	6.00	0.25	1/2	17.00	16	1 1/4
12	15.00	22.00	2.62	6.12	0.25	1/2	19.25	20	1 1/4
14	16.25	23.75	2.75	6.50	0.25	1/2	20.75	20	1 3/8
16	18.50	27.00	3.00	7.00	0.25	1/2	23.75	20	1 1/2
18	21.00	29.25	3.25	7.25	0.25	1/2	25.75	20	1 5/8
20	23.00	32.00	3.50	7.50	0.25	1/2	28.50	24	1 5/8
24	27.25	37.00	4.00	8.00	0.25	1/2	33.00	24	1 7/8

* Dimensions in inches.

Dimensions of Class 900 Welding Neck Raised Face Orifice Flanges

See Table 5.80.

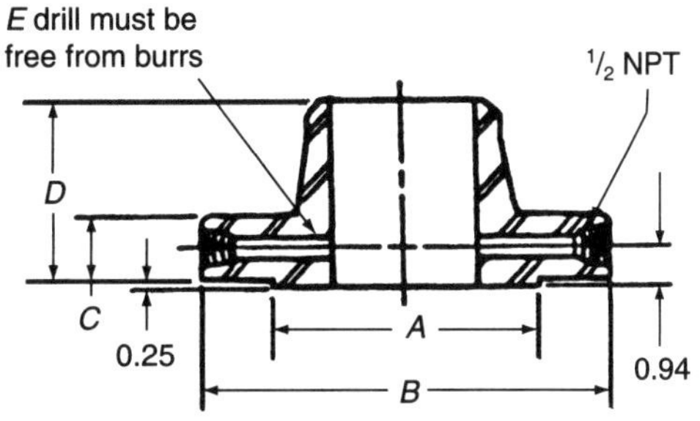

Welding Neck Raised Face Orifice Flange

TABLE 5.80 Dimensions of ASME B16.36 Class 900 Welding Neck Raised Face Orifice Flanges*

Nominal pipe size	Outside diameter of raised face, *A*	Outside diameter of flange, *B*	Thickness of flange, min., *C*	Length through hub, *D*	Diameter of pressure connection, *E*	Diameter of bolt circle	Number of holes	Diameter of bolts	Height of raised face
1 1½ 2 2½	For nominal size and smaller, use Class 1500								
3	5.00	9.50	1.50	4.00	⅜	7.50	8	⅞	0.25
4	6.19	11.50	1.75	4.50	½	9.25	8	1⅛	0.25
6	8.50	15.00	2.19	5.50	½	12.50	12	1⅛	0.25
8	10.62	18.50	2.50	6.38	½	15.50	12	1⅜	0.25
10	12.75	21.50	2.75	7.25	½	18.50	16	1⅜	0.25
12	15.00	24.00	3.12	7.88	½	21.00	20	1⅜	0.25
14	16.25	25.25	3.38	8.38	½	22.00	20	1½	0.25
16	18.50	27.75	3.50	8.50	½	24.25	20	1⅝	0.25
18	21.00	31.00	4.00	9.00	½	27.00	20	1⅞	0.25
20	23.00	33.75	4.25	9.75	½	29.50	20	2	0.25
24	27.25	41.00	5.50	11.50	½	35.50	20	2½	0.25

* Dimensions in inches.

Dimensions of Class 1500 Welding Neck Raised Face Orifice Flanges

See Table 5.81.

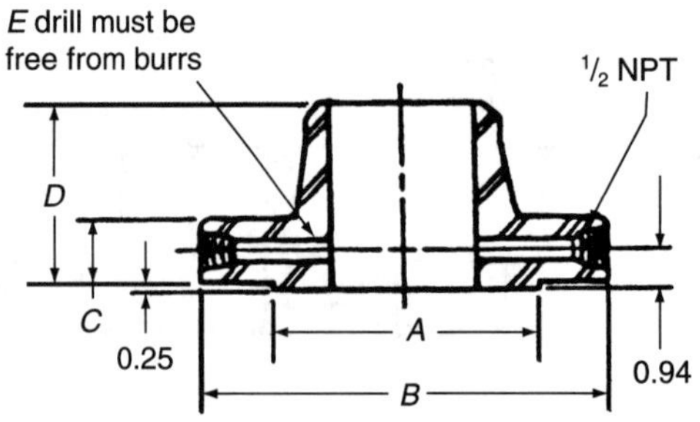

Welding Neck Raised Face Orifice Flange

TABLE 5.81 Dimensions of ASME B16.36 Class 1500 Welding Neck Raised Face Orifice Flanges*

Nominal pipe size	Outside diameter of raised face, *A*	Outside diameter of flange, *B*	Thickness of flange, min., *C*	Length through hub, *D*	Diameter of pressure connection, *E*	Diameter of bolt circle	Number of holes	Diameter of bolts	Height of raised face
1	2.00	5.88	1.50	3.25	¼	4.00	4	⅞	0.25
1½	2.88	7.00	1.50	3.50	¼	4.88	4	1	0.25
2	3.62	8.50	1.50	4.00	¼	6.50	8	⅞	0.25
2½	4.12	9.62	1.62	4.12	¼	7.50	8	1	0.25
3	5.00	10.50	1.88	4.62	⅜	8.00	8	1⅛	0.25
4	6.19	12.25	2.12	4.88	½	9.50	8	1¼	0.25
6	8.50	15.50	3.25	6.75	½	12.50	12	1⅜	0.25
8	10.62	19.00	3.62	8.38	½	15.50	12	1⅝	0.25
10	12.75	23.00	4.25	10.00	½	19.00	12	1⅞	0.25
12	15.00	26.50	4.88	11.12	½	22.50	16	2	0.25
14	16.25	29.50	5.25	11.75	½	25.00	16	2¼	0.25
16	18.50	32.50	5.75	12.25	½	27.75	16	2½	0.25
18	21.00	36.00	6.38	12.88	½	30.50	16	2¾	0.25
20	23.00	38.75	7.00	14.00	½	32.75	16	3	0.25
24	27.25	46.00	8.00	16.00	½	39.00	16	3½	0.25

* Dimensions in inches.

Dimensions of Class 2500 Welding Neck Raised Face Orifice Flanges

See Table 5.82.

TABLE 5.82 Dimensions of ASME B16.36 Class 2500 Welding Neck Raised Face Orifice Flanges*†

Nominal pipe size	Outside diameter of raised face, *A*	Outside diameter of flange, *B*	Thickness of flange, min., *C*	Length through hub, *D*	Diameter of pressure connection, *E*	Diameter of bolt circle	Number of holes	Diameter of bolts	Height of raised face
1	2.00	6.25	1.50	3.62	¼	4.25	4	⅞	0.25
1½	2.88	8.00	1.75	4.38	¼	5.75	4	1⅛	0.25
2	3.62	9.25	2.00	5.00	¼	6.75	8	1	0.25
2½	4.12	10.50	2.25	5.62	¼	7.75	8	1⅛	0.25
3	5.00	12.00	2.62	6.62	⅜	9.00	8	1¼	0.25
4	6.19	14.00	3.00	7.50	½	10.75	8	1½	0.25
6	8.50	19.00	4.25	10.75	½	14.50	8	2	0.25
8	10.62	21.75	5.00	12.50	½	17.25	12	2	0.25
10	12.75	26.50	6.50	16.50	½	21.25	12	2½	0.25
12	15.00	30.00	7.25	18.25	½	24.38	12	2¾	0.25

* Dimensions in inches.
† See diagram in Table 5.81.

Dimensions of Class 400 Welding Neck Ring Joint Orifice Flanges

See Table 5.83.

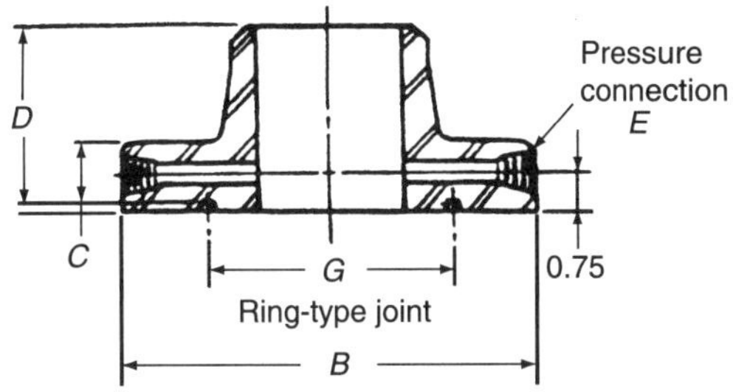

Welding Neck Orifice Flange with Ring Joint Facing

TABLE 5.83 Dimensions of ASME B16.36 Class 400 Welding Neck Ring Joint Orifice Flanges*

Nominal pipe size	Outside diameter of flange, *B*	Thickness of flange, min., *C*	Length through hub, *D*	Groove number	Pitch diameter, *G*	Diameter of pressure connection, *E*	Diameter of bolt circle	Number of holes	Diameter of bolts
1 1½ 2 2½ 3	For nominal pipe size 3 and smaller, use Class 600								
4	10.00	1.50	3.50	R37	5.875	½	7.88	8	⅞
6	12.50	1.62	4.06	R45	8.312	½	10.62	12	⅞
8	15.00	1.88	4.62	R49	10.625	½	13.00	12	1
10	17.50	2.12	4.88	R53	12.750	½	15.25	16	1⅛
12	20.50	2.25	5.38	R57	15.000	½	17.75	16	1¼
14	23.00	2.39	5.88	R61	16.500	½	20.25	20	1¼
16	25.50	2.50	6.00	R65	18.500	½	22.50	20	1⅜
18	28.00	2.62	6.50	R69	21.000	½	24.75	24	1⅜
20	30.50	2.75	6.62	R73	23.000	½	27.00	24	1½
24	36.00	3.00	6.88	R77	27.250	½	32.00	24	1¾

* Dimensions in inches.

Dimensions of Class 600 Welding Neck Ring Joint Orifice Flanges

See Table 5.84.

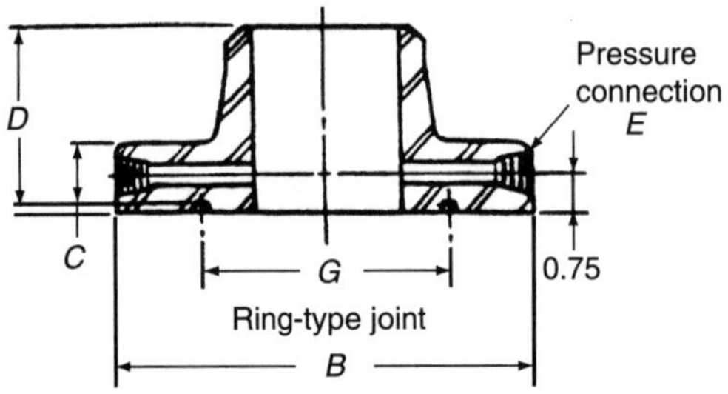

Welding Neck Orifice Flange with Ring Joint Facing

TABLE 5.84 Dimensions of ASME B16.36 Class 600 Welding Neck Ring Joint Orifice Flanges*

Nominal pipe size	Outside diameter of flange, *B*	Thickness of flange, min., *C*	Length through hub, *D*	Groove number	Pitch diameter, *G*	Diameter of pressure connection, *E*	Diameter of bolt circle	Number of holes	Diameter of bolts
1	4.88	1.44	3.19	R16	2.000	1/4	3.50	4	5/8
1 1/2	6.12	1.44	3.32	R20	2.688	1/4	4.50	4	3/4
2	6.50	1.44	3.32	R23	3.250	1/4	5.00	8	5/8
2 1/2	7.50	1.44	3.44	R26	4.000	1/4	5.88	8	3/4
3	8.25	1.44	3.44	R31	4.875	3/8	6.62	8	3/4
4	10.75	1.50	4.00	R37	5.875	1/2	8.50	8	7/8
6	14.00	1.88	4.62	R45	8.312	1/2	11.50	12	1
8	16.50	2.19	5.25	R49	10.625	1/2	13.75	12	1 1/8
10	20.00	2.50	6.00	R53	12.750	1/2	17.00	16	1 1/4
12	22.00	2.62	6.12	R57	15.000	1/2	19.25	20	1 1/4
14	23.75	2.75	6.50	R61	16.500	1/2	20.75	20	1 3/8
16	27.00	3.00	7.00	R65	18.500	1/2	23.75	20	1 1/2
18	29.25	3.25	7.25	R69	21.000	1/2	25.75	20	1 5/8
20	32.00	3.50	7.50	R73	23.000	1/2	28.50	24	1 5/8
24	37.00	4.00	8.00	R77	27.250	1/2	33.00	24	1 7/8

* Dimensions in inches.

Dimensions of Class 900 Welding Neck Ring Joint Orifice Flanges

See Table 5.85.

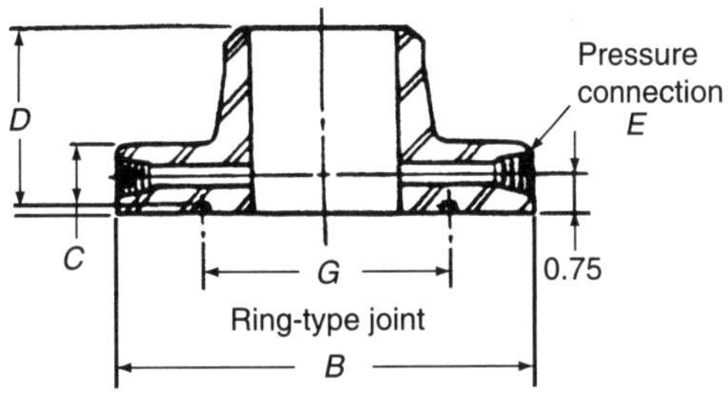

Welding Neck Orifice Flange with Ring Joint Facing

TABLE 5.85 Dimensions of ASME B16.36 Class 900 Welding Neck Ring Joint Orifice Flanges*

Nominal pipe size	Outside diameter of flange, *B*	Thickness of flange, min., *C*	Length through hub, *D*	Groove number	Pitch diameter, *G*	Diameter of pressure connection, *E*	Diameter of bolt circle	Number of holes	Diameter of bolts
1 1½ 2 2½	For nominal pipe size and smaller, use Class 1500								
3	9.50	1.50	4.00	R31	4.875	⅜	7.50	8	⅞
4	11.50	1.75	4.50	R37	5.875	½	9.25	8	1⅛
6	15.00	2.19	5.50	R45	8.312	½	12.50	12	1⅛
8	18.50	2.50	6.38	R49	10.625	½	15.50	12	1⅜
10	21.50	2.75	7.25	R53	12.750	½	18.50	16	1⅜
12	24.00	3.12	7.88	R57	15.000	½	21.00	20	1⅜
14	25.25	3.38	8.38	R62	16.500	½	22.00	20	1½
16	27.75	3.50	8.50	R66	18.500	½	24.25	20	1⅝
18	31.00	4.00	9.00	R70	21.000	½	27.00	20	1⅞
20	33.75	4.25	9.75	R74	23.000	½	29.50	20	2
24	41.00	5.50	11.50	R78	27.250	½	35.50	20	2½

* Dimensions in inches.

Dimensions of Class 1500 Welding Neck Ring Joint Orifice Flanges

See Table 5.86.

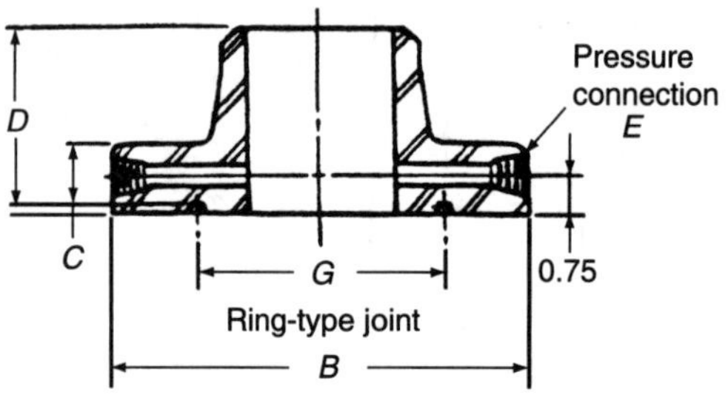

Welding Neck Orifice Flange with Ring Joint Facing

TABLE 5.86 Dimensions of ASME B16.36 Class 1500 Welding Neck Ring Joint Orifice Flanges*

Nominal pipe size	Outside diameter of flange, *B*	Thickness of flange, min., *C*	Length through hub, *D*	Groove number	Pitch diameter, *G*	Diameter of pressure connection, *E*	Diameter of bolt circle	Number of holes	Diameter of bolts
1	5.88	1.50	3.25	R16	2.000	1/4	4.00	4	7/8
1 1/2	7.00	1.50	3.50	R20	2.688	1/4	4.88	4	1
2	8.50	1.50	4.00	R24	3.750	1/4	6.50	8	7/8
2 1/2	9.62	1.62	4.12	R27	4.250	1/4	7.50	8	1
3	10.50	1.88	4.62	R35	5.375	3/8	8.00	8	1 1/8
4	12.25	2.12	4.88	R39	6.375	1/2	9.50	8	1 1/4
6	15.50	3.25	6.75	R46	8.312	1/2	12.50	12	1 3/8
8	19.00	3.62	8.38	R50	10.625	1/2	15.50	12	1 5/8
10	23.00	4.25	10.00	R54	12.750	1/2	19.00	12	1 7/8
12	26.50	4.88	11.12	R58	15.000	1/2	22.50	16	2
14	29.50	5.25	11.75	R63	16.500	1/2	25.00	16	2 1/4
16	32.50	5.75	12.25	R67	18.500	1/2	27.75	16	2 1/2
18	36.00	6.38	12.88	R71	21.000	1/2	30.50	16	2 3/4
20	38.75	7.00	14.00	R75	23.000	1/2	32.75	16	3
24	46.00	8.00	16.00	R79	27.250	1/2	39.00	16	3 1/2

* Dimensions in inches.

Dimensions of Class 2500 Welding Neck Ring Joint Orifice Flanges

See Table 5.87.

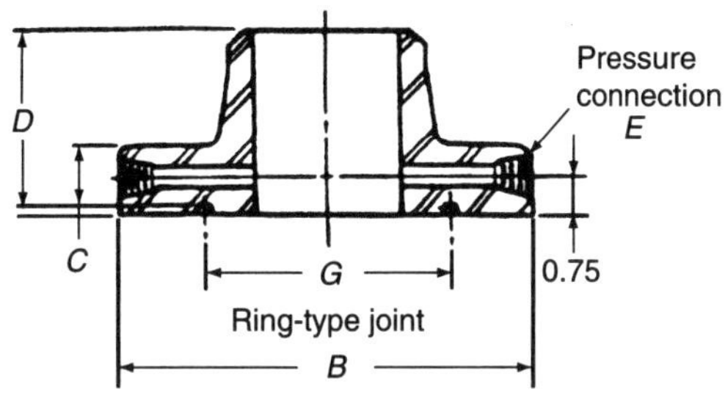

Welding Neck Orifice Flange with Ring Joint Facing

TABLE 5.87 Dimensions of ASME B16.36 Class 2500 Welding Neck Ring Joint Orifice Flanges*

Nominal pipe size	Outside diameter of flange, *B*	Thickness of flange, min., *C*	Length through hub, *D*	Groove number	Pitch diameter, *G*	Diameter of pressure connection, *E*	Diameter of bolt circle	Number of holes	Diameter of bolts
1	6.25	1.50	3.62	R18	2.375	1/4	4.25	4	7/8
1 1/2	8.00	1.75	4.38	R23	3.250	1/4	5.75	4	1 1/8
2	9.25	2.00	5.00	R26	4.000	1/4	6.75	8	1
2 1/2	10.50	2.25	5.62	R28	4.375	1/4	7.75	8	1 1/8
3	12.00	2.62	6.62	R32	5.000	3/8	9.00	8	1 1/4
4	14.00	3.00	7.50	R38	6.188	1/2	10.75	8	1 1/2
6	19.00	4.25	10.75	R47	9.000	1/2	14.50	8	2
8	21.75	5.00	12.50	R51	11.000	1/2	17.25	12	2
10	26.50	6.50	16.50	R55	13.500	1/2	21.25	12	2 1/2
12	30.00	7.25	18.25	R60	16.000	1/2	24.38	12	2 3/4

* Dimensions in inches.

Other Fittings and Flanges for Wrought Pipe

Refer to Chap. 8 for MSS fittings and flanges.

Piping Joints

Refer to Chap. 15.

Tables in Chapter 5

Tables in Chapter 5 (Continued)

CHAPTER 6

API PIPE, FLANGES, FITTINGS, AND JOINTS

See pp. 323–324 for a list of tables in Chap. 6.

API 5L LINE PIPE

Material Grades

API 5L line pipe is manufactured in material grades listed in Table 6.1. API 5L pipe specification covers seamless and welded steel pipe. Pipe is furnished with plain, threaded, and belled ends. It is also available as through-the-flow (TFL) pipe and pipe with ends prepared for special couplings.

TABLE 6.1 API 5L Line Pipe Material Grades

Grade	Symbol
A25, Class I	A25
A25, Class II	A25R
A	A
B	B
X42	X42
X46	X46
X52	X52
X56	X56
X60	X60
X65	X65
X70	X70
X80	X80

The two digits after X indicate the specified minimum yield strength. The symbols listed in Table 6.1 above are used for marking the pipe corresponding to the material grade.

Product Specification Level (PSL). API 5L pipe is manufactured in two product specification levels, PSL 1 and PSL 2. The product specification level defines the level of standard technical requirements. The standard technical requirements for PSL 2 pipe include mandatory requirements for carbon equivalent, notch toughness, maximum yield strength, and maximum tensile strength. Refer to App. J of the API 5L specification for a detailed listing of technical requirements for both PSL 1 and PSL2. PSL 1 pipe is manufactured in Grades A25 through X70. PSL 2 pipe is manufactured in Grades B through X80. Table 6.2 identifies manufacturing processes and the material grades of pipe classified as PSL 1. Table 6.3 lists the manufacturing processes and the material grades of pipe classified as PSL 2.

API 5L Standard Wall Threaded Line Pipe. Tables 6.4 and 6.4M provide dimensions and weights of API 5L, Grades A25, A and B, standard wall threaded line pipe in sizes 0.405 through 20. It is noted that Grade A25 pipe is manufactured in sizes 0.405 through $5\frac{9}{16}$ only. API 5L, Grades A and B, standard wall threaded line pipe are manufactured in sizes 0.405 through 20.

API 5L Heavy Wall Threaded Line Pipe. Tables 6.5 and 6.5M provide dimensions and weights of API 5L, Grades A25, A and B, heavy wall threaded line pipe in sizes 0.405 through $12\frac{3}{4}$. It is noted that Grade A25 pipe is manufactured in sizes 0.405 through $5^9/_{16}$ only. API 5L, Grades A and B, heavy wall threaded line pipe are manufactured in sizes 0.405 through $12^3/_4$.

API 5L Plain-End Line Pipe, Grades A25, A and B. Tables 6.6 and 6.6M provide dimensions and weights of API 5L, plain-end, Grades A25, A and B, line pipe.

API 5L Plain-End Line Pipe, Grades A25 through X80. Tables 6.7 and 6.7M provide dimensions and weights of API 5L line pipe, Grades A25 through X80, in sizes $2^3/_8$ through $5^9/_{16}$.

API 5L Plain-End Line Pipe, Grades A through X80. Tables 6.8 and 6.8M provide dimensions and weights of API 5L line pipe, Grades A through X80, in sizes $6^5/_8$ through 80. It is noted that API 5L, Grade A25 pipe is not produced in sizes $6^5/_8$ through 80.

TABLE 6.2 API 5L Pipe Grades and Manufacturing Process, for Product Specification Level (PSL) 1

Seamless or welded	Type of weld	Grade A25	Grades A and B	Grades X42 through X70	Notes
Seamless		X	X	X	1
Welded without 1 filler metal	Continuous weld	X			1, 2
Welded without filler metal	Electric welded	X	X	X	1
Welded without filler metal	Laser welded	X	X	X	1
Welded with filler metal	Longitudinal seam submerged arc welded (SMAW)		X	X	1
Welded with filler metal	Gas metal arc welded (GMAW)		X	X	1
Welded with filler metal	Combination GMAW		X	X	1
Welded with filler metal	Double-seam SMAW		X	X	1, 3
Welded with filler metal	Double-seam GMAW		X	X	1, 3
Welded with filler metal	Double-seam combination GMAW and SMAW		X	X	1, 3
Welded with filler metal	Helical seam SMAW		X	X	1, 4

Notes:

1. PSL 1 is limited to sizes from 0.405 through 80. PSL 1 pipe in Grades A25, A, B, and X42 through X70 is manufactured with plain, belled, and plain ends for special couplings. PSL 1 pipe with threaded ends is limited to Grades A25, A and B.
2. Grade A25 is limited to sizes $5^9/_{16}$ and smaller. Continuous-welded pipe is only manufactured in Grade A25; thus it is limited to sizes $5^9/_{16}$ and smaller.
3. Double-seam pipe is limited to sizes 36 and larger.
4. Helical seam pipe is limited to sizes $4^1/_2$ and larger. Also, helical seam pipe is not manufactured with threaded ends and couplings.

TABLE 6.3 API 5L Pipe Grades and Manufacturing Process for Product Specification Level (PSL) 2

Seamless or welded	Type of weld	Grades B through X80	Notes
Seamless		X	1
Welded without filler metal	Electric welded	X	1
Welded with filler metal	Longitudinal seam submerged arc welded (SMAW)	X	1
Welded with filler metal	Gas metal arc welded (GMAW)	X	1
Welded with filler metal	Combination GMAW	X	1
Welded with filler metal	Double-seam SMAW	X	1, 2
Welded with filler metal	Double-seam GMAW	X	1, 2
Welded with filler metal	Double-seam combination GMAW and SMAW	X	1, 2
Welded with filler metal	Helical seam SMAW	X	1, 3

Notes:

1. PSL 2 is limited to sizes from 4½ through 80 and manufactured with plain ends only.
2. Double-seam pipe is limited to sizes 36 and larger.
3. Helical seam pipe is limited to sizes 4½ and larger. Also, helical seam pipe is not manufactured with threaded ends and couplings.

TABLE 6.4 API 5L—Standard Wall Threaded Line Pipe Dimensions and Weights of Grades A25, A and B (U.S. Customary Units)

Size	Specified outside diameter D, in	Specified wall thickness t, in	Calculated weight: Plain-end weight per unit length w, lb/ft	Calculated weight: Threads and coupling e_w, lb	Calculated inside diameter d, in
0.405	0.405	0.068	0.24	0.20	0.269
0.540	0.540	0.088	0.43	0.20	0.364
0.675	0.675	0.091	0.57	0.20	0.493
0.840	0.840	0.109	0.85	0.20	0.622
1.050	1.050	0.113	1.13	0.20	0.824
1.315	1.315	0.133	1.68	0.20	1.049
1.660	1.660	0.140	2.27	0.60	1.380
1.900	1.900	0.145	2.72	0.40	1.610
$2\frac{3}{8}$	2.375	0.154	3.66	1.20	2.067
$2\frac{7}{8}$	2.875	0.203	5.80	1.80	2.469
$3\frac{1}{2}$	3.500	0.216	7.58	1.80	3.068
4	4.000	0.226	9.12	3.20	3.548
$4\frac{1}{2}$	4.500	0.237	10.80	4.40	4.026
$5\frac{9}{16}$	5.563	0.258	14.63	5.60	5.047
$6\frac{5}{8}$	6.625	0.280	18.99	7.20	6.065
$8\frac{5}{8}$	8.625	0.277	24.72	14.80	8.071
$8\frac{5}{8}$	8.625	0.322	28.58	14.00	7.981
$10\frac{3}{4}$	10.750	0.279	31.23	20.00	10.192
$10\frac{3}{4}$	10.750	0.307	34.27	19.20	10.136
$10\frac{3}{4}$	10.750	0.365	40.52	17.40	10.020
$12\frac{3}{4}$	12.750	0.330	43.81	32.60	12.090
$12\frac{3}{4}$	12.750	0.375	49.61	30.80	12.000
14	14.000	0.375	54.62	24.60	13.250
16	16.000	0.375	62.64	30.00	15.250
18	18.000	0.375	70.65	35.60	17.250
20	20.000	0.375	78.67	42.00	19.250

Notes:
1. Grade A25 pipe is manufactured in sizes 0.405 through $5\frac{9}{16}$ only.
2. Grades A and B pipe are manufactured in sizes 0.405 through 20.
3. Refers to Tables 6.9 through 6.13 for tolerances.

TABLE 6.4M API 5L—Standard Wall Threaded Line Pipe Dimensions and Weights of Grades A25, A and B (SI Units)

			Calculated weight		
Size	Specified outside diameter D, mm	Specified wall thickness t, mm	Plain-end weight per unit length w, kg/m	Threads and coupling e_w, kg	Calculated inside diameter d, mm
0.405	10.3	1.7	0.36	0.09	6.9
0.540	13.7	2.2	0.62	0.09	9.3
0.675	17.1	2.3	0.84	0.09	12.5
0.840	21.3	2.8	1.28	0.09	15.7
1.050	26.7	2.9	1.70	0.09	20.9
1.315	33.4	3.4	2.52	0.09	26.6
1.660	42.2	3.6	3.43	0.27	35.0
1.900	48.3	3.7	4.07	0.18	40.9
$2\frac{3}{8}$	60.3	3.9	5.42	0.54	52.5
$2\frac{7}{8}$	73.0	5.2	8.69	0.82	62.6
$3\frac{1}{2}$	88.9	5.5	11.31	0.82	77.9
4	101.6	5.7	13.48	1.45	90.2
$4\frac{1}{2}$	114.3	6.0	16.02	2.00	102.3
$5\frac{9}{16}$	141.3	6.6	21.92	2.54	128.1
$6\frac{5}{8}$	168.3	7.1	28.22	3.27	154.1
$8\frac{5}{8}$	219.1	7.0	36.61	6.72	205.1
$8\frac{5}{8}$	219.1	8.2	42.65	6.36	202.7
$10\frac{3}{4}$	273.1	7.1	46.57	9.08	258.9
$10\frac{3}{4}$	273.1	7.8	51.03	8.72	257.5
$10\frac{3}{4}$	273.1	9.3	60.50	7.90	254.5
$12\frac{3}{4}$	323.9	8.4	65.35	14.80	307.1
$12\frac{3}{4}$	323.9	9.5	73.65	13.98	304.9
14	355.6	9.5	81.08	11.17	336.6
16	406.4	9.5	92.98	13.62	387.4
18	457.0	9.5	104.84	16.16	438.0
20	508.0	9.5	116.78	19.07	489.0

Notes:
1. Grade A25 pipe is manufactured in sizes 0.405 through $5\frac{9}{16}$ only.
2. Grades A and B pipe are manufactured in sizes 0.405 through 20.
3. Refer to Tables 6.9 through 6.13 for tolerances.

TABLE 6.5 API 5L—Heavy Wall Threaded Line Pipe Dimensions and Weights for Grades A25, A and B (U.S. Customary Units)

Size	Specified outside diameter D, in	Specified wall thickness t, in	Calculated weight: Plain-end weight per unit length w, lb/ft	Calculated weight: Threads and coupling e_w, lb	Calculated inside diameter d, in
0.405	0.405	0.095	0.31	0.02	0.215
0.540	0.540	0.119	0.54	0.05	0.302
0.675	0.675	0.126	0.74	0.07	0.423
0.840	0.840	0.147	1.09	0.12	0.546
1.050	1.050	0.154	1.48	0.18	0.742
1.315	1.315	0.179	2.17	0.23	0.957
1.660	1.660	0.191	3.00	0.58	1.278
1.900	1.900	0.200	3.63	0.36	1.500
$2\frac{3}{8}$	2.375	0.218	5.03	1.09	1.939
$2\frac{7}{8}$	2.875	0.276	7.67	1.57	2.323
$3\frac{1}{2}$	3.500	0.300	10.26	1.76	2.900
4	4.000	0.318	12.52	3.00	3.364
$4\frac{1}{2}$	4.500	0.337	15.00	3.99	3.826
$5\frac{9}{16}$	5.563	0.375	20.80	4.94	4.813
$6\frac{5}{8}$	6.625	0.432	28.60	5.65	5.761
$8\frac{5}{8}$	8.625	0.500	43.43	11.47	7.625
$10\frac{3}{4}$	10.750	0.500	54.79	15.32	9.750
$12\frac{3}{4}$	12.750	0.500	65.48	28.74	11.750

Notes:
1. Grade A25 pipe is manufactured in sizes 0.405 through $5\frac{9}{16}$ only.
2. Grades A and B heavy-wall threaded line pipe are manufactured in sizes 0.405 through $12\frac{3}{4}$.
3. Refer to Tables 6.9 through 6.13 for tolerances.

TABLE 6.5M API 5L—Heavy Wall Threaded Line Pipe Dimensions and Weights for Grades A25, A and B (SI Units)

Size	Specified outside diameter D, mm	Specified wall thickness t, mm	Calculated weight: Plain-end weight per unit length w, kg/m	Calculated weight: Threads and coupling e_w, kg	Calculated inside diameter d, mm
0.405	10.3	2.4	0.47	0.01	5.5
0.540	13.7	3.0	0.79	0.02	7.7
0.675	17.1	3.2	1.10	0.03	10.7
0.840	21.3	3.7	1.61	0.05	13.9
1.050	26.7	3.9	2.19	0.08	18.9
1.315	33.4	4.5	3.21	0.10	24.4
1.660	42.2	4.9	4.51	0.56	32.4
1.900	48.3	5.1	5.43	0.16	38.1
$2\frac{3}{8}$	60.3	5.5	7.43	0.50	49.3
$2\frac{7}{8}$	73.0	7.0	11.39	0.71	59.0
$3\frac{1}{2}$	88.9	7.6	15.24	0.80	73.7
4	101.6	8.1	18.68	1.36	85.4
$4\frac{1}{2}$	114.3	8.6	22.42	1.81	97.1
$5\frac{9}{16}$	141.3	9.5	30.88	2.24	122.3
$6\frac{5}{8}$	168.3	11.0	42.67	2.56	146.3
$8\frac{5}{8}$	219.1	12.7	64.64	5.20	193.7
$10\frac{3}{4}$	273.1	12.7	81.55	6.95	247.7
$12\frac{3}{4}$	323.9	12.7	97.46	13.04	298.5

Notes:
1. Grade A25 pipe is manufactured in sizes 0.405 through $5\frac{9}{16}$ only.
2. Grades A and B heavy-wall threaded line pipe are manufactured in sizes 0.405 through $12\frac{3}{4}$.
3. Refer to Tables 6.9 through 6.13 for tolerances.

TABLE 6.6 API 5L—Plain-End Line Pipe Dimensions and Weights per Unit Length for Grades A25, A and B, in Sizes 0.405 through 1.900 (U.S. Customary Units)*

Size	Specified outside diameter D, in	Specified wall thickness t, in	Plain-end weight per unit length w, lb/ft	Calculated inside diameter d, in
0.405	0.405	0.068	0.24	0.269
0.405	0.405	0.095	0.31	0.215
0.540	0.540	0.088	0.43	0.364
0.540	0.540	0.119	0.54	0.302
0.675	0.675	0.091	0.57	0.493
0.675	0.675	0.126	0.74	0.423
0.840	0.840	0.109	0.85	0.622
0.840	0.840	0.147	1.09	0.546
0.840	0.840	0.294	1.72	0.252
1.050	1.050	0.113	1.13	0.824
1.050	1.050	0.154	1.48	0.742
1.050	1.050	0.308	2.44	0.434
1.315	1.315	0.133	1.68	1.049
1.315	1.315	0.179	2.17	0.957
1.315	1.315	0.358	3.66	0.599
1.660	1.660	0.140	2.27	1.380
1.660	1.660	0.191	3.00	1.278
1.660	1.660	0.382	5.22	0.896
1.900	1.900	0.145	2.72	1.610
1.900	1.900	0.200	3.63	1.500
1.900	1.900	0.400	6.41	1.100

* Refer to Tables 6.9 through 6.13 for tolerances.

TABLE 6.6M API 5L—Plain-End Line Pipe Dimensions, Weights per Unit Length for Grades A25, A and B, in Sizes 0.405 through 1.900 (Metric Units)*

Size	Specified outside diameter D, mm	Specified wall thickness t, mm	Plain-end weight per unit length, w_{pe}, kg/m	Calculated inside diameter d, mm
0.405	10.3	1.7	0.36	6.9
0.405	10.3	2.4	0.47	5.5
0.540	13.7	2.2	0.62	9.3
0.540	13.7	3.0	0.79	7.7
0.675	17.1	2.3	0.84	12.5
0.675	17.1	3.2	1.10	10.7
0.840	21.3	2.8	1.28	15.7
0.840	21.3	3.7	1.61	13.9
0.840	21.3	7.5	2.55	6.3
1.050	26.7	2.9	1.70	20.9
1.050	26.7	3.9	2.19	18.9
1.050	26.7	7.8	3.64	11.1
1.315	33.4	3.4	2.52	26.6
1.315	33.4	4.5	3.21	24.4
1.315	33.4	9.1	5.45	15.2
1.660	42.2	3.6	3.43	35.0
1.660	42.2	4.9	4.51	32.4
1.660	42.2	9.7	7.77	22.8
1.900	48.3	3.7	4.07	40.9
1.900	48.3	5.1	5.43	38.1
1.900	48.3	10.2	9.58	27.9

* Refer to Tables 6.9 through 6.13 for tolerances.

TABLE 6.7 API 5L—Plain-End Line Pipe Dimensions and Weights per Unit Length, Sizes $2\frac{3}{8}$ through $5\frac{9}{16}$ (U.S. Customary Units) (Note 1)

Size	Outside diameter D, in	Wall thickness t, in	Plain-end weight per unit length w, lb/ft	Inside diameter d, in	Notes
$2\frac{3}{8}$	2.375	0.083	2.03	2.209	2
$2\frac{3}{8}$	2.375	0.109	2.64	2.157	2
$2\frac{3}{8}$	2.375	0.125	3.01	2.125	2
$2\frac{3}{8}$	2.375	0.141	3.37	2.093	2
$2\frac{3}{8}$	2.375	0.154	3.66	2.067	
$2\frac{3}{8}$	2.375	0.172	4.05	2.031	
$2\frac{3}{8}$	2.375	0.188	4.40	1.999	
$2\frac{3}{8}$	2.375	0.218	5.03	1.939	
$2\frac{3}{8}$	2.375	0.250	5.68	1.875	
$2\frac{3}{8}$	2.375	0.281	6.29	1.813	
$2\frac{3}{8}$	2.375	0.436	9.04	1.503	
$2\frac{7}{8}$	2.875	0.083	2.48	2.709	2
$2\frac{7}{8}$	2.875	0.109	3.22	2.657	2
$2\frac{7}{8}$	2.875	0.125	3.67	2.625	2
$2\frac{7}{8}$	2.875	0.141	4.12	2.593	2
$2\frac{7}{8}$	2.875	0.156	4.53	2.563	
$2\frac{7}{8}$	2.875	0.172	4.97	2.531	
$2\frac{7}{8}$	2.875	0.188	5.40	2.499	
$2\frac{7}{8}$	2.875	0.203	5.80	2.469	
$2\frac{7}{8}$	2.875	0.216	6.14	2.443	
$2\frac{7}{8}$	2.875	0.250	7.02	2.375	
$2\frac{7}{8}$	2.875	0.276	7.67	2.323	
$2\frac{7}{8}$	2.875	0.552	13.71	1.771	
$3\frac{1}{2}$	3.500	0.083	3.03	3.334	2
$3\frac{1}{2}$	3.500	0.109	3.95	3.282	2
$3\frac{1}{2}$	3.500	0.125	4.51	3.250	2
$3\frac{1}{2}$	3.500	0.141	5.06	3.218	2
$3\frac{1}{2}$	3.500	0.156	5.58	3.188	2
$3\frac{1}{2}$	3.500	0.172	6.12	3.156	
$3\frac{1}{2}$	3.500	0.188	6.66	3.124	
$3\frac{1}{2}$	3.500	0.216	7.58	3.068	
$3\frac{1}{2}$	3.500	0.250	8.69	3.000	
$3\frac{1}{2}$	3.500	0.281	9.67	2.938	
$3\frac{1}{2}$	3.500	0.300	10.26	2.900	
$3\frac{1}{2}$	3.500	0.600	18.60	2.300	
4	4.000	0.083	3.48	3.834	2
4	4.000	0.109	4.53	3.782	2
4	4.000	0.125	5.18	3.750	2
4	4.000	0.141	5.82	3.718	2
4	4.000	0.156	6.41	3.688	2
4	4.000	0.172	7.04	3.656	
4	4.000	0.188	7.66	3.624	
4	4.000	0.226	9.12	3.548	
4	4.000	0.250	10.02	3.500	
4	4.000	0.281	11.17	3.438	
4	4.000	0.318	12.52	3.364	
$4\frac{1}{2}$	4.500	0.083	3.92	4.334	2
$4\frac{1}{2}$	4.500	0.125	5.85	4.250	2
$4\frac{1}{2}$	4.500	0.141	6.57	4.218	2
$4\frac{1}{2}$	4.500	0.156	7.24	4.188	2
$4\frac{1}{2}$	4.500	0.172	7.96	4.156	
$4\frac{1}{2}$	4.500	0.188	8.67	4.124	
$4\frac{1}{2}$	4.500	0.203	9.32	4.094	
$4\frac{1}{2}$	4.500	0.219	10.02	4.062	
$4\frac{1}{2}$	4.500	0.237	10.80	4.026	
$4\frac{1}{2}$	4.500	0.250	11.36	4.000	
$4\frac{1}{2}$	4.500	0.281	12.67	3.938	
$4\frac{1}{2}$	4.500	0.312	13.97	3.876	
$4\frac{1}{2}$	4.500	0.337	15.00	3.826	
$4\frac{1}{2}$	4.500	0.438	19.02	3.624	
$4\frac{1}{2}$	4.500	0.531	22.53	3.438	
$4\frac{1}{2}$	4.500	0.674	27.57	3.152	
$5\frac{9}{16}$	5.563	0.083	4.86	5.397	2
$5\frac{9}{16}$	5.563	0.125	7.27	5.313	2
$5\frac{9}{16}$	5.563	0.156	9.02	5.251	2
$5\frac{9}{16}$	5.563	0.188	10.80	5.187	
$5\frac{9}{16}$	5.563	0.219	12.51	5.125	
$5\frac{9}{16}$	5.563	0.258	14.63	5.047	
$5\frac{9}{16}$	5.563	0.281	15.87	5.001	
$5\frac{9}{16}$	5.563	0.312	17.51	4.939	
$5\frac{9}{16}$	5.563	0.344	19.19	4.875	
$5\frac{9}{16}$	5.563	0.375	20.80	4.813	
$5\frac{9}{16}$	5.563	0.500	27.06	4.563	
$5\frac{9}{16}$	5.563	0.625	32.99	4.313	
$5\frac{9}{16}$	5.563	0.750	38.59	4.063	

Notes:

1. This table covers pipe dimensions and weights per unit length of API 5L Line Pipe, Grades A25, A, B, X42, X46, X52, X56, X60, X65, X70, and X80 in sizes $2\frac{3}{8}$ through $5\frac{9}{16}$. Refer to Tables 6.9 through 6.13 for tolerances.

2. The pipe having this specified outside diameter and the specified wall thickness is termed *special plain-end pipe*. Other pipe is called *regular plain-end pipe*. Pipe that has a combination of specified outside diameter and specified wall thickness that is intermediate to the tabulated values is considered to be special plain-end pipe if the next lower tabulated value is for special plain-end pipe. Other intermediate combinations of outside diameter and wall thickness are considered to be regular plain-end pipe.

TABLE 6.7M API 5L Plain-End Pipe Dimensions and Weight per Unit Length, Sizes $2\frac{3}{8}$ through $5\frac{9}{16}$ (Metric Units) (Note 1)

Size	Outside diameter D, mm	Wall thickness t, mm	Plain-end weight per unit length w, kg/m	Inside diameter d, mm	Notes
$2\frac{3}{8}$	60.3	2.1	3.01	56.1	2
$2\frac{3}{8}$	60.3	2.8	3.97	54.7	2
$2\frac{3}{8}$	60.3	3.2	4.51	53.9	2
$2\frac{3}{8}$	60.3	3.6	5.03	53.1	2
$2\frac{3}{8}$	60.3	3.9	5.42	52.5	
$2\frac{3}{8}$	60.3	4.4	6.07	51.5	
$2\frac{3}{8}$	60.3	4.8	6.57	50.7	
$2\frac{3}{8}$	60.3	5.5	7.43	49.3	
$2\frac{3}{8}$	60.3	6.4	8.51	47.5	
$2\frac{3}{8}$	60.3	7.1	9.31	46.1	
$2\frac{3}{8}$	60.3	11.1	13.47	38.1	
$2\frac{7}{8}$	73.0	2.1	3.67	68.8	2
$2\frac{7}{8}$	73.0	2.8	4.85	67.4	2
$2\frac{7}{8}$	73.0	3.2	5.51	66.6	2
$2\frac{7}{8}$	73.0	3.6	6.16	65.8	2
$2\frac{7}{8}$	73.0	4.0	6.81	65.0	
$2\frac{7}{8}$	73.0	4.4	7.44	64.2	
$2\frac{7}{8}$	73.0	4.8	8.07	63.4	
$2\frac{7}{8}$	73.0	5.2	8.69	62.6	
$2\frac{7}{8}$	73.0	5.5	9.16	62.0	
$2\frac{7}{8}$	73.0	6.4	10.51	60.2	
$2\frac{7}{8}$	73.0	7.0	11.39	59.0	
$2\frac{7}{8}$	73.0	14.0	20.37	45.0	
$3\frac{1}{2}$	88.9	2.1	4.50	84.7	2
$3\frac{1}{2}$	88.9	2.8	5.95	83.3	2
$3\frac{1}{2}$	88.9	3.2	6.76	82.5	2
$3\frac{1}{2}$	88.9	3.6	7.57	81.7	2
$3\frac{1}{2}$	88.9	4.0	8.37	80.9	2
$3\frac{1}{2}$	88.9	4.4	9.17	80.1	
$3\frac{1}{2}$	88.9	4.8	9.95	79.3	
$3\frac{1}{2}$	88.9	5.5	11.31	77.9	
$3\frac{1}{2}$	88.9	6.4	13.02	76.1	
$3\frac{1}{2}$	88.9	7.1	14.32	74.7	
$3\frac{1}{2}$	88.9	7.6	15.24	73.7	
$3\frac{1}{2}$	88.9	15.2	27.63	58.5	
4	101.6	2.1	5.15	97.4	2
4	101.6	2.8	6.82	96.0	2
4	101.6	3.2	7.76	95.2	2
4	101.6	3.6	8.70	94.4	2
4	101.6	4.0	9.63	93.6	2
4	101.6	4.4	10.55	92.8	
4	101.6	4.8	11.46	92.0	
4	101.6	5.7	13.48	90.2	
4	101.6	6.4	15.02	88.8	
4	101.6	7.1	16.55	87.4	
4	101.6	8.1	18.68	85.4	
$4\frac{1}{2}$	114.3	2.1	5.81	110.1	2
$4\frac{1}{2}$	114.3	3.2	8.77	107.9	2
$4\frac{1}{2}$	114.3	3.6	9.83	107.1	2
$4\frac{1}{2}$	114.3	4.0	10.88	106.3	2
$4\frac{1}{2}$	114.3	4.4	11.92	105.5	
$4\frac{1}{2}$	114.3	4.8	12.96	104.7	
$4\frac{1}{2}$	114.3	5.2	13.99	103.9	
$4\frac{1}{2}$	114.3	5.6	15.01	103.1	
$4\frac{1}{2}$	114.3	6.0	16.02	102.3	
$4\frac{1}{2}$	114.3	6.4	17.03	101.5	
$4\frac{1}{2}$	114.3	7.1	18.77	100.1	
$4\frac{1}{2}$	114.3	7.9	20.73	98.5	
$4\frac{1}{2}$	114.3	8.6	22.42	97.1	
$4\frac{1}{2}$	114.3	11.1	28.25	92.1	
$4\frac{1}{2}$	114.3	13.5	33.56	87.3	
$4\frac{1}{2}$	114.3	17.1	40.99	80.1	
$5\frac{9}{16}$	141.3	2.1	7.21	137.1	2
$5\frac{9}{16}$	141.3	3.2	10.90	134.9	2
$5\frac{9}{16}$	141.3	4.0	13.54	133.3	2
$5\frac{9}{16}$	141.3	4.8	16.16	131.7	
$5\frac{9}{16}$	141.3	5.6	18.74	130.1	
$5\frac{9}{16}$	141.3	6.6	21.92	128.1	
$5\frac{9}{16}$	141.3	7.1	23.50	127.1	
$5\frac{9}{16}$	141.3	7.9	25.99	125.5	
$5\frac{9}{16}$	141.3	8.7	28.45	123.9	
$5\frac{9}{16}$	141.3	9.5	30.88	122.3	
$5\frac{9}{16}$	141.3	12.7	40.28	115.9	
$5\frac{9}{16}$	141.3	15.9	49.17	109.5	
$5\frac{9}{16}$	141.3	19.1	57.56	103.1	

Notes:

1. This table covers pipe dimensions and weights per unit length of API 5L line pipe. Grades A25, A, B, X42, X46, X52, X56, X60, X65, X70, and X80 in sizes $2\frac{3}{8}$ through $5\frac{9}{16}$. Refer to Tables 6.9 through 6.13 for tolerances.
2. The pipe having this specified outside diameter and the specified wall thickness is termed *special plain-end pipe*. Other pipe is called *regular plain-end pipe*. Pipe that has a combination of specified outside diameter and specified wall thickness that is intermediate to the tabulated values is considered to be special plain-end pipe if the next lower tabulated value is for special plain-end pipe. Other intermediate combinations of outside diameter and wall thickness are considered to be regular plain-end pipe.

TABLE 6.8 API 5L—Plain-End Line Pipe Dimensions and Weights per Unit Length, for Grades A through X80, Sizes $6^5/_8$ through 80 (U.S. Customary Units) (Note 1)

Size	Specified outside diameter *D*, in	Specified wall thickness *t*, in	Plain-end weight per unit length *w*, lb/ft	Calculated inside diameter *d*, in	Notes
$6^5/_8$	6.625	0.083	5.80	6.459	2
$6^5/_8$	6.625	0.109	7.59	6.407	2
$6^5/_8$	6.625	0.125	8.69	6.375	2
$6^5/_8$	6.625	0.141	9.77	6.343	2
$6^5/_8$	6.625	0.156	10.79	6.313	2
$6^5/_8$	6.625	0.172	11.87	6.281	
$6^5/_8$	6.625	0.188	12.94	6.249	
$6^5/_8$	6.625	0.203	13.94	6.219	
$6^5/_8$	6.625	0.219	15.00	6.187	
$6^5/_8$	6.625	0.250	17.04	6.125	
$6^5/_8$	6.625	0.280	18.99	6.065	
$6^5/_8$	6.625	0.312	21.06	6.001	
$6^5/_8$	6.625	0.344	23.10	5.937	
$6^5/_8$	6.625	0.375	25.05	5.875	
$6^5/_8$	6.625	0.432	28.60	5.761	
$6^5/_8$	6.625	0.500	32.74	5.625	
$6^5/_8$	6.625	0.562	36.43	5.501	
$6^5/_8$	6.625	0.625	40.09	5.375	
$6^5/_8$	6.625	0.719	45.39	5.187	
$6^5/_8$	6.625	0.750	47.10	5.125	
$6^5/_8$	6.625	0.864	53.21	4.897	
$6^5/_8$	6.625	0.875	53.78	4.875	
$8^5/_8$	8.625	0.125	11.36	8.375	2
$8^5/_8$	8.625	0.156	14.12	8.313	2
$8^5/_8$	8.625	0.188	16.96	8.249	
$8^5/_8$	8.625	0.203	18.28	8.219	
$8^5/_8$	8.625	0.219	19.68	8.187	
$8^5/_8$	8.625	0.250	22.38	8.125	
$8^5/_8$	8.625	0.277	24.72	8.071	
$8^5/_8$	8.625	0.312	27.73	8.001	
$8^5/_8$	8.625	0.322	28.58	7.981	
$8^5/_8$	8.625	0.344	30.45	7.937	
$8^5/_8$	8.625	0.375	33.07	7.875	
$8^5/_8$	8.625	0.438	38.33	7.749	
$8^5/_8$	8.625	0.500	43.43	7.625	
$8^5/_8$	8.625	0.562	48.44	7.501	
$8^5/_8$	8.625	0.625	53.45	7.375	
$8^5/_8$	8.625	0.719	60.77	7.187	
$8^5/_8$	8.625	0.750	63.14	7.125	
$8^5/_8$	8.625	0.812	67.82	7.001	
$8^5/_8$	8.625	0.875	72.49	6.875	
$8^5/_8$	8.625	1.000	81.51	6.625	
$10^3/_4$	10.750	0.156	17.67	10.438	2
$10^3/_4$	10.750	0.188	21.23	10.374	2
$10^3/_4$	10.750	0.203	22.89	10.344	2
$10^3/_4$	10.750	0.219	24.65	10.312	
$10^3/_4$	10.750	0.250	28.06	10.250	
$10^3/_4$	10.750	0.279	31.23	10.192	
$10^3/_4$	10.750	0.307	34.27	10.136	
$10^3/_4$	10.750	0.344	38.27	10.062	
$10^3/_4$	10.750	0.365	40.52	10.020	
$10^3/_4$	10.750	0.438	48.28	9.874	
$10^3/_4$	10.750	0.500	54.79	9.750	
$10^3/_4$	10.750	0.562	61.21	9.626	
$10^3/_4$	10.750	0.625	67.65	9.500	
$10^3/_4$	10.750	0.719	77.10	9.312	
$10^3/_4$	10.750	0.812	86.26	9.126	
$10^3/_4$	10.750	0.875	92.37	9.000	
$10^3/_4$	10.750	0.938	98.39	8.874	
$10^3/_4$	10.750	1.000	104.23	8.750	
$10^3/_4$	10.750	1.250	126.94	8.250	
$12^3/_4$	12.750	0.172	23.13	12.406	2
$12^3/_4$	12.750	0.188	25.25	12.374	2
$12^3/_4$	12.750	0.203	27.23	12.344	2
$12^3/_4$	12.750	0.219	29.34	12.312	2
$12^3/_4$	12.750	0.250	33.41	12.250	
$12^3/_4$	12.750	0.281	37.46	12.188	
$12^3/_4$	12.750	0.312	41.48	12.126	
$12^3/_4$	12.750	0.330	43.81	12.090	
$12^3/_4$	12.750	0.344	45.62	12.062	
$12^3/_4$	12.750	0.375	49.61	12.000	
$12^3/_4$	12.750	0.406	53.57	11.938	
$12^3/_4$	12.750	0.438	57.65	11.874	
$12^3/_4$	12.750	0.500	65.48	11.750	
$12^3/_4$	12.750	0.562	73.22	11.626	
$12^3/_4$	12.750	0.625	81.01	11.500	
$12^3/_4$	12.750	0.688	88.71	11.374	
$12^3/_4$	12.750	0.750	96.21	11.250	
$12^3/_4$	12.750	0.812	103.63	11.126	
$12^3/_4$	12.750	0.875	111.08	11.000	
$12^3/_4$	12.750	0.938	118.44	10.874	
$12^3/_4$	12.750	1.000	125.61	10.750	
$12^3/_4$	12.750	1.062	132.69	10.626	
$12^3/_4$	12.750	1.125	139.81	10.500	
$12^3/_4$	12.750	1.250	153.67	10.250	
14	14.000	0.188	27.76	13.624	2
14	14.000	0.203	29.94	13.594	2
14	14.000	0.210	30.96	13.580	2
14	14.000	0.219	32.26	13.562	2
14	14.000	0.250	36.75	13.500	2
14	14.000	0.281	41.21	13.438	2
14	14.000	0.312	45.65	13.376	
14	14.000	0.344	50.22	13.312	
14	14.000	0.375	54.62	13.250	
14	14.000	0.406	59.00	13.188	
14	14.000	0.438	63.50	13.124	
14	14.000	0.469	67.84	13.062	
14	14.000	0.500	72.16	13.000	
14	14.000	0.562	80.73	12.876	
14	14.000	0.625	89.36	12.750	
14	14.000	0.688	97.91	12.624	
14	14.000	0.750	106.23	12.500	
14	14.000	0.812	114.48	12.376	
14	14.000	0.875	122.77	12.250	
14	14.000	0.938	130.98	12.124	
14	14.000	1.000	138.97	12.000	
14	14.000	1.062	146.88	11.876	
14	14.000	1.125	154.84	11.750	
14	14.000	1.250	170.37	11.500	
16	16.000	0.188	31.78	15.624	2
16	16.000	0.203	34.28	15.594	2
16	16.000	0.219	36.95	15.562	2
16	16.000	0.250	42.09	15.500	2
16	16.000	0.281	47.22	15.438	2
16	16.000	0.312	52.32	15.376	
16	16.000	0.344	57.57	15.312	

TABLE 6.8 API 5L—Plain-End Line Pipe Dimensions and Weights per Unit Length, for Grades A through X80, Sizes $6^{5}/_{8}$ through 80 (U.S. Customary Units) (Note 1) (Continued)

Size	Specified outside diameter D, in	Specified wall thickness t, in	Plain-end weight per unit length w, lb/ft	Calculated inside diameter d, in	Notes
16	16.000	0.375	62.64	15.250	
16	16.000	0.406	67.68	15.188	
16	16.000	0.438	72.86	15.124	
16	16.000	0.469	77.87	15.062	
16	16.000	0.500	82.85	15.000	
16	16.000	0.562	92.75	14.876	
16	16.000	0.625	102.72	14.750	
16	16.000	0.688	112.62	14.624	
16	16.000	0.750	122.27	14.500	
16	16.000	0.812	131.84	14.376	
16	16.000	0.875	141.48	14.250	
16	16.000	0.938	151.03	14.124	
16	16.000	1.000	160.35	14.000	
16	16.000	1.062	169.59	13.876	
16	16.000	1.125	178.89	13.750	
16	16.000	1.188	188.11	13.624	
16	16.000	1.250	197.10	13.500	
18	18.000	0.188	35.80	17.624	2
18	18.000	0.219	41.63	17.562	2
18	18.000	0.250	47.44	17.500	2
18	18.000	0.281	53.23	17.438	2
18	18.000	0.312	58.99	17.376	
18	18.000	0.344	64.93	17.312	
18	18.000	0.375	70.65	17.250	
18	18.000	0.406	76.36	17.188	
18	18.000	0.438	82.23	17.124	
18	18.000	0.469	87.89	17.062	
18	18.000	0.500	93.54	17.000	
18	18.000	0.562	104.76	16.876	
18	18.000	0.625	116.09	16.750	
18	18.000	0.688	127.32	16.624	
18	18.000	0.750	138.30	16.500	
18	18.000	0.812	149.20	16.376	
18	18.000	0.875	160.18	16.250	
18	18.000	0.938	171.08	16.124	
18	18.000	1.000	181.73	16.000	
18	18.000	1.062	192.29	15.876	
18	18.000	1.125	202.94	15.750	
18	18.000	1.188	213.51	15.624	
18	18.000	1.250	223.82	15.500	
20	20.000	0.219	46.31	19.562	2
20	20.000	0.250	52.78	19.500	2
20	20.000	0.281	59.23	19.438	2
20	20.000	0.312	65.66	19.376	
20	20.000	0.344	72.28	19.312	
20	20.000	0.375	78.67	19.250	
20	20.000	0.406	85.04	19.188	
20	20.000	0.438	91.59	19.124	
20	20.000	0.469	97.92	19.062	
20	20.000	0.500	104.23	19.000	
20	20.000	0.562	116.78	18.876	
20	20.000	0.625	129.45	18.750	
20	20.000	0.688	142.03	18.624	
20	20.000	0.750	154.34	18.500	
20	20.000	0.812	166.56	18.376	
20	20.000	0.875	178.89	18.250	
20	20.000	0.938	191.14	18.124	
20	20.000	1.000	203.11	18.000	
20	20.000	1.062	215.00	17.876	
20	20.000	1.125	227.00	17.750	
20	20.000	1.188	238.91	17.624	
20	20.000	1.250	250.55	17.500	
20	20.000	1.312	262.10	17.376	
20	20.000	1.375	273.76	17.250	
22	22.000	0.219	50.99	21.562	2
22	22.000	0.250	58.13	21.500	2
22	22.000	0.281	65.24	21.438	2
22	22.000	0.312	72.34	21.376	
22	22.000	0.344	79.64	21.312	
22	22.000	0.375	86.69	21.250	
22	22.000	0.406	93.72	21.188	
22	22.000	0.438	100.96	21.124	
22	22.000	0.469	107.95	21.062	
22	22.000	0.500	114.92	21.000	
22	22.000	0.562	128.79	20.876	
22	22.000	0.625	142.81	20.750	
22	22.000	0.688	156.74	20.624	
22	22.000	0.750	170.37	20.500	
22	22.000	0.812	183.92	20.376	
22	22.000	0.875	197.60	20.250	
22	22.000	0.938	211.19	20.124	
22	22.000	1.000	224.49	20.000	
22	22.000	1.062	237.70	19.876	
22	22.000	1.125	251.05	19.750	
22	22.000	1.188	264.31	19.624	
22	22.000	1.250	277.27	19.500	
22	22.000	1.312	290.15	19.376	
22	22.000	1.375	303.16	19.250	
22	22.000	1.438	316.08	19.124	
22	22.000	1.500	328.72	19.000	
24	24.000	0.250	63.47	23.500	2
24	24.000	0.281	71.25	23.438	2
24	24.000	0.312	79.01	23.376	
24	24.000	0.344	86.99	23.312	
24	24.000	0.375	94.71	23.250	
24	24.000	0.406	102.40	23.188	
24	24.000	0.438	110.32	23.124	
24	24.000	0.469	117.98	23.062	
24	24.000	0.500	125.61	23.000	
24	24.000	0.562	140.81	22.876	
24	24.000	0.625	156.17	22.750	
24	24.000	0.688	171.45	22.624	
24	24.000	0.750	186.41	22.500	
24	24.000	0.812	201.28	22.376	
24	24.000	0.875	216.31	22.250	
24	24.000	0.938	231.25	22.124	
24	24.000	1.000	245.87	22.000	
24	24.000	1.062	260.41	21.876	
24	24.000	1.125	275.10	21.750	
24	24.000	1.188	289.71	21.624	
24	24.000	1.250	304.00	21.500	
24	24.000	1.312	318.21	21.376	
24	24.000	1.375	332.56	21.250	
24	24.000	1.438	346.83	21.124	
24	24.000	1.500	360.79	21.000	
24	24.000	1.562	374.66	20.876	
26	26.000	0.250	68.82	25.500	2
26	26.000	0.281	77.26	25.438	2
26	26.000	0.312	85.68	25.376	
26	26.000	0.344	94.35	25.312	

TABLE 6.8 API 5L—Plain-End Line Pipe Dimensions and Weights per Unit Length, for Grades A through X80, Sizes $6\frac{5}{8}$ through 80 (U.S. Customary Units) (Note 1) (Continued)

Size	Specified outside diameter D, in	Specified wall thickness t, in	Plain-end weight per unit length w, lb/ft	Calculated inside diameter d, in	Notes
26	26.000	0.375	102.72	25.250	
26	26.000	0.406	111.08	25.188	
26	26.000	0.438	119.69	25.124	
26	26.000	0.469	128.00	25.062	
26	26.000	0.500	136.30	25.000	
26	26.000	0.562	152.83	24.876	
26	26.000	0.625	169.54	24.750	
26	26.000	0.688	186.16	24.624	
26	26.000	0.750	202.44	24.500	
26	26.000	0.812	218.64	24.376	
26	26.000	0.875	235.01	24.250	
26	26.000	0.938	251.30	24.124	
26	26.000	1.000	267.25	24.000	
28	28.000	0.250	74.16	27.500	2
28	28.000	0.281	83.26	27.438	2
28	28.000	0.312	92.35	27.376	
28	28.000	0.344	101.70	27.312	
28	28.000	0.375	110.74	27.250	
28	28.000	0.406	119.76	27.188	
28	28.000	0.438	129.05	27.124	
28	28.000	0.469	138.03	27.062	
28	28.000	0.500	146.99	27.000	
28	28.000	0.562	164.84	26.876	
28	28.000	0.625	182.90	26.750	
28	28.000	0.688	200.87	26.624	
28	28.000	0.750	218.48	26.500	
28	28.000	0.812	236.00	26.376	
28	28.000	0.875	253.72	26.250	
28	28.000	0.938	271.36	26.124	
28	28.000	1.000	288.63	26.000	
30	30.000	0.250	79.51	29.500	2
30	30.000	0.281	89.27	29.438	2
30	30.000	0.312	99.02	29.376	
30	30.000	0.344	109.06	29.312	
30	30.000	0.375	118.76	29.250	
30	30.000	0.406	128.44	29.188	
30	30.000	0.438	138.42	29.124	
30	30.000	0.469	148.06	29.062	
30	30.000	0.500	157.68	29.000	
30	30.000	0.562	176.86	28.876	
30	30.000	0.625	196.26	28.750	
30	30.000	0.688	215.58	28.624	
30	30.000	0.750	234.51	28.500	
30	30.000	0.812	253.36	28.376	
30	30.000	0.875	272.43	28.250	
30	30.000	0.938	291.41	28.124	
30	30.000	1.000	310.01	28.000	
30	30.000	1.062	328.53	27.876	
30	30.000	1.125	347.26	27.750	
30	30.000	1.188	365.90	27.624	
30	30.000	1.250	384.17	27.500	
32	32.000	0.250	84.85	31.500	2
32	32.000	0.281	95.28	31.438	2
32	32.000	0.312	105.69	31.376	
32	32.000	0.344	116.41	31.312	
32	32.000	0.375	126.78	31.250	
32	32.000	0.406	137.12	31.188	
32	32.000	0.438	147.78	31.124	
32	32.000	0.469	158.08	31.062	
32	32.000	0.500	168.37	31.000	
32	32.000	0.562	188.87	30.876	
32	32.000	0.625	209.62	30.750	
32	32.000	0.688	230.29	30.624	
32	32.000	0.750	250.55	30.500	
32	32.000	0.812	270.72	30.376	
32	32.000	0.875	291.14	30.250	
32	32.000	0.938	311.47	30.124	
32	32.000	1.000	331.39	30.000	
32	32.000	1.062	351.23	29.876	
32	32.000	1.125	371.31	29.750	
32	32.000	1.188	391.30	29.624	
32	32.000	1.250	410.90	29.500	
34	34.000	0.250	90.20	33.500	2
34	34.000	0.281	101.29	33.438	2
34	34.000	0.312	112.36	33.376	
34	34.000	0.344	123.77	33.312	
34	34.000	0.375	134.79	33.250	
34	34.000	0.406	145.80	33.188	
34	34.000	0.438	157.14	33.124	
34	34.000	0.469	168.11	33.062	
34	34.000	0.500	179.06	33.000	
34	34.000	0.562	200.89	32.876	
34	34.000	0.625	222.99	32.750	
34	34.000	0.688	245.00	32.624	
34	34.000	0.750	266.58	32.500	
34	34.000	0.812	288.08	32.376	
34	34.000	0.875	309.84	32.250	
34	34.000	0.938	331.52	32.124	
34	34.000	1.000	352.77	32.000	
34	34.000	1.062	373.94	31.876	
34	34.000	1.125	395.36	31.750	
34	34.000	1.188	416.70	31.624	
34	34.000	1.250	437.62	31.500	
36	36.000	0.250	95.54	35.500	2
36	36.000	0.281	107.30	35.438	2
36	36.000	0.312	119.03	35.376	
36	36.000	0.344	131.12	35.312	
36	36.000	0.375	142.81	35.250	
36	36.000	0.406	154.48	35.188	
36	36.000	0.438	166.51	35.124	
36	36.000	0.469	178.14	35.062	
36	36.000	0.500	189.75	35.000	
36	36.000	0.562	212.90	34.876	
36	36.000	0.625	236.35	34.750	
36	36.000	0.688	259.71	34.624	
36	36.000	0.750	282.62	34.500	
36	36.000	0.812	305.44	34.376	
36	36.000	0.875	328.55	34.250	
36	36.000	0.938	351.57	34.124	
36	36.000	1.000	374.15	34.000	
36	36.000	1.062	396.64	33.876	
36	36.000	1.125	419.42	33.750	
36	36.000	1.188	442.10	33.624	
36	36.000	1.250	464.35	33.500	
38	38.000	0.312	125.70	37.376	
38	38.000	0.344	138.47	37.312	
38	38.000	0.375	150.83	37.250	
38	38.000	0.406	163.16	37.188	

TABLE 6.8 API 5L—Plain-End Line Pipe Dimensions and Weights per Unit Length, for Grades A through X80, Sizes $6\frac{5}{8}$ through 80 (U.S. Customary Units) (Note 1) (Continued)

Size	Specified outside diameter *D*, in	Specified wall thickness *t*, in	Plain-end weight per unit length *w*, lb/ft	Calculated inside diameter *d*, in	Notes
38	38.000	0.438	175.87	37.124	
38	38.000	0.469	188.17	37.062	
38	38.000	0.500	200.44	37.000	
38	38.000	0.562	224.92	36.876	
38	38.000	0.625	249.71	36.750	
38	38.000	0.688	274.42	36.624	
38	38.000	0.750	298.65	36.500	
38	38.000	0.812	322.80	36.376	
38	38.000	0.875	347.26	36.250	
38	38.000	0.938	371.63	36.124	
38	38.000	1.000	395.53	36.000	
38	38.000	1.062	419.35	35.876	
38	38.000	1.125	443.47	35.750	
38	38.000	1.188	467.50	35.624	
38	38.000	1.250	491.07	35.500	
40	40.000	0.312	132.37	39.376	
40	40.000	0.344	145.83	39.312	
40	40.000	0.375	158.85	39.250	
40	40.000	0.406	171.84	39.188	
40	40.000	0.438	185.24	39.124	
40	40.000	0.469	198.19	39.062	
40	40.000	0.500	211.13	39.000	
40	40.000	0.562	236.93	38.876	
40	40.000	0.625	263.07	38.750	
40	40.000	0.688	289.13	38.624	
40	40.000	0.750	314.69	38.500	
40	40.000	0.812	340.16	38.376	
40	40.000	0.875	365.97	38.250	
40	40.000	0.938	391.68	38.124	
40	40.000	1.000	416.91	38.000	
40	40.000	1.062	442.05	37.876	
40	40.000	1.125	467.52	37.750	
40	40.000	1.188	492.90	37.624	
40	40.000	1.250	517.80	37.500	
42	42.000	0.344	153.18	41.312	
42	42.000	0.375	166.86	41.250	
42	42.000	0.406	180.52	41.188	
42	42.000	0.438	194.60	41.124	
42	42.000	0.469	208.22	41.062	
42	42.000	0.500	221.82	41.000	
42	42.000	0.562	248.95	40.876	
42	42.000	0.625	276.44	40.750	
42	42.000	0.688	303.84	40.624	
42	42.000	0.750	330.72	40.500	
42	42.000	0.812	357.52	40.376	
42	42.000	0.875	384.67	40.250	
42	42.000	0.938	411.74	40.124	
42	42.000	1.000	438.29	40.000	
42	42.000	1.062	464.76	39.876	
42	42.000	1.125	491.57	39.750	
42	42.000	1.188	518.30	39.624	
42	42.000	1.250	544.52	39.500	
44	44.000	0.344	160.54	43.312	
44	44.000	0.375	174.88	43.250	
44	44.000	0.406	189.20	43.188	
44	44.000	0.438	203.97	43.124	
44	44.000	0.469	218.25	43.062	
44	44.000	0.500	232.51	43.000	
44	44.000	0.562	260.97	42.876	
44	44.000	0.625	289.80	42.750	
44	44.000	0.688	318.55	42.624	
44	44.000	0.750	346.76	42.500	
44	44.000	0.812	374.88	42.376	
44	44.000	0.875	403.38	42.250	
44	44.000	0.938	431.79	42.124	
44	44.000	1.000	459.67	42.000	
44	44.000	1.062	487.47	41.876	
44	44.000	1.125	515.63	41.750	
44	44.000	1.188	543.70	41.624	
44	44.000	1.250	571.25	41.500	
46	46.000	0.344	167.89	45.312	
46	46.000	0.375	182.90	45.250	
46	46.000	0.406	197.88	45.188	
46	46.000	0.438	213.33	45.124	
46	46.000	0.469	228.27	45.062	
46	46.000	0.500	243.20	45.000	
46	46.000	0.562	272.98	44.876	
46	46.000	0.625	303.16	44.750	
46	46.000	0.688	333.26	44.624	
46	46.000	0.750	362.79	44.500	
46	46.000	0.812	392.24	44.376	
46	46.000	0.875	422.09	44.250	
46	46.000	0.938	451.85	44.124	
46	46.000	1.000	481.05	44.000	
46	46.000	1.062	510.17	43.876	
46	46.000	1.125	539.68	43.750	
46	46.000	1.188	569.10	43.624	
46	46.000	1.250	597.97	43.500	
48	48.000	0.344	175.25	47.312	
48	48.000	0.375	190.92	47.250	
48	48.000	0.406	206.56	47.188	
48	48.000	0.438	222.70	47.124	
48	48.000	0.469	238.30	47.062	
48	48.000	0.500	253.89	47.000	
48	48.000	0.562	285.00	46.876	
48	48.000	0.625	316.52	46.750	
48	48.000	0.688	347.97	46.624	
48	48.000	0.750	378.83	46.500	
48	48.000	0.812	409.61	46.376	
48	48.000	0.875	440.80	46.250	
48	48.000	0.938	471.90	46.124	
48	48.000	1.000	502.43	46.000	
48	48.000	1.062	532.88	45.876	
48	48.000	1.125	563.73	45.750	
48	48.000	1.188	594.50	45.624	
48	48.000	1.250	624.70	45.500	
52	52.000	0.375	206.95	51.250	
52	52.000	0.406	223.93	51.188	
52	52.000	0.438	241.42	51.124	
52	52.000	0.469	258.36	51.062	
52	52.000	0.500	275.27	51.000	
52	52.000	0.562	309.03	50.876	
52	52.000	0.625	343.25	50.750	
52	52.000	0.688	377.39	50.624	
52	52.000	0.750	410.90	50.500	
52	52.000	0.812	444.33	50.376	
52	52.000	0.875	478.21	20.250	
52	52.000	0.938	512.01	50.124	
52	52.000	1.000	545.19	50.000	
52	52.000	1.062	578.29	49.876	

TABLE 6.8 API 5L—Plain-End Line Pipe Dimensions and Weights per Unit Length, for Grades A through X80, Sizes $6^5/_8$ through 80 (U.S. Customary Units) (Note 1) (Continued)

Size	Specified outside diameter *D*, in	Specified wall thickness *t*, in	Plain-end weight per unit length *w*, lb/ft	Calculated inside diameter *d*, in	Notes
52	52.000	1.125	611.84	49.750	
52	52.000	1.188	645.30	49.624	
52	52.000	1.250	678.15	49.500	
56	56.000	0.375	222.99	55.250	
56	56.000	0.406	241.29	55.188	
56	56.000	0.438	260.15	55.124	
56	56.000	0.469	278.41	55.062	
56	56.000	0.500	296.65	55.000	
56	56.000	0.562	333.06	54.876	
56	56.000	0.625	369.97	54.750	
56	56.000	0.688	406.80	54.624	
56	56.000	0.750	442.97	54.500	
56	56.000	0.812	479.05	54.376	
56	56.000	0.875	515.63	54.250	
56	56.000	0.938	552.12	54.124	
56	56.000	1.000	587.95	54.000	
56	56.000	1.062	623.70	53.876	
56	56.000	1.125	659.94	53.750	
56	56.000	1.188	696.10	53.624	
56	56.000	1.250	731.60	53.500	
60	60.000	0.375	239.02	59.250	
60	60.000	0.406	258.65	59.188	
60	60.000	0.438	278.88	59.124	
60	60.000	0.469	298.47	59.062	
60	60.000	0.500	318.03	59.000	
60	60.000	0.562	357.09	58.876	
60	60.000	0.625	396.70	58.750	
60	60.000	0.688	436.22	58.624	
60	60.000	0.750	475.04	58.500	
60	60.000	0.812	513.77	58.376	
60	60.000	0.875	553.04	58.250	
60	60.000	0.938	592.23	58.124	
60	60.000	1.000	630.71	58.000	
60	60.000	1.062	669.11	57.876	
60	60.000	1.125	708.05	57.750	
60	60.000	1.188	746.90	57.624	
60	60.000	1.250	785.05	57.500	
64	64.000	0.375	255.06	63.250	
64	64.000	0.406	276.01	63.188	
64	64.000	0.438	297.61	63.124	
64	64.000	0.469	318.52	63.062	
64	64.000	0.500	339.41	63.000	
64	64.000	0.562	381.12	62.876	
64	64.000	0.625	423.42	62.750	
64	64.000	0.688	465.64	62.624	
64	64.000	0.750	507.11	62.500	
64	64.000	0.812	548.49	62.376	
64	64.000	0.875	590.46	62.250	
64	64.000	0.938	632.34	62.124	
64	64.000	1.000	673.47	62.000	
64	64.000	1.062	714.52	61.876	
64	64.000	1.125	756.15	61.750	
64	64.000	1.188	797.69	61.624	
64	64.000	1.250	838.50	61.500	
68	68.000	0.469	338.57	67.062	
68	68.000	0.500	360.79	67.000	
68	68.000	0.562	405.15	66.876	
68	68.000	0.625	450.15	66.750	
68	68.000	0.688	495.06	66.624	
68	68.000	0.750	539.18	66.500	
68	68.000	0.812	583.21	66.376	
68	68.000	0.875	627.87	66.250	
68	68.000	0.938	672.45	66.124	
68	68.000	1.000	716.23	66.000	
68	68.000	1.062	759.93	65.876	
68	68.000	1.125	804.26	65.750	
68	68.000	1.188	848.49	65.624	
68	68.000	1.250	891.95	65.500	
72	72.000	0.500	382.17	71.000	
72	72.000	0.562	429.18	70.876	
72	72.000	0.625	476.87	70.750	
72	72.000	0.688	524.48	70.624	
72	72.000	0.750	571.25	70.500	
72	72.000	0.812	617.93	70.376	
72	72.000	0.875	665.29	70.250	
72	72.000	0.938	712.55	70.124	
72	72.000	1.000	758.99	70.000	
72	72.000	1.062	805.34	69.876	
72	72.000	1.125	852.36	69.750	
72	72.000	1.188	899.29	69.624	
72	72.000	1.250	945.40	69.500	
76	76.000	0.500	403.55	75.000	
76	76.000	0.562	453.21	74.876	
76	76.000	0.625	503.60	74.750	
76	76.000	0.688	553.90	74.624	
76	76.000	0.750	603.32	74.500	
76	76.000	0.812	652.65	74.376	
76	76.000	0.875	702.70	74.250	
76	76.000	0.938	752.66	74.124	
76	76.000	1.000	801.75	74.000	
76	76.000	1.062	850.75	73.876	
76	76.000	1.125	900.47	73.750	
76	76.000	1.188	950.09	73.624	
76	76.000	1.250	998.85	73.500	
80	80.000	0.562	477.25	78.876	
80	80.000	0.625	530.34	78.750	
80	80.000	0.688	583.32	78.624	
80	80.000	0.750	635.39	78.500	
80	80.000	0.812	687.37	78.376	
80	80.000	0.875	740.12	78.250	
80	80.000	0.938	792.77	78.124	
80	80.000	1.000	844.51	78.000	
80	80.000	1.062	896.17	77.876	
80	80.000	1.125	948.57	77.750	
80	80.000	1.188	1000.89	77.624	
80	80.000	1.250	1052.30	77.500	

Notes:

1. This table covers pipe dimensions and weights per unit length of API 5L line pipe, Grades A, B, X42, X46, X52, X56, X60, X65, X70, and X80 in sizes $6^3/_8$ through 80. Refer to Tables 6.9 through 6.13 for tolerances.
2. The pipe having this specified outside diameter and the specified wall thickness is termed *special plain-end pipe*. Other pipe is called *regular plain-end pipe*. Pipe that has a combination of specified outside diameter and specified wall thickness that is intermediate to the tabulated values is considered to be special plain-end pipe if the next lower tabulated value is for special plain-end pipe. Other intermediate combinations of outside diameter and wall thickness are considered to be regular plain-end pipe.

TABLE 6.8M API 5L Plain-End Pipe Dimensions and Weights per Unit Length for Grades A through X80 (SI Units) (Note 1)

Size	Specified outside diameter *D*, mm	Specified wall thickness *t*, mm	Plain-end weight per unit length *w*, kg/m	Calculated inside diameter *d*, mm	Notes
6 5/8	168.3	2.1	8.61	164.1	2
6 5/8	168.3	2.8	11.43	162.7	2
6 5/8	168.3	3.2	13.03	161.9	2
6 5/8	168.3	3.6	14.62	161.1	2
6 5/8	168.3	4.0	16.21	160.3	2
6 5/8	168.3	4.4	17.78	159.5	
6 5/8	168.3	4.8	19.35	158.7	
6 5/8	168.3	5.2	20.91	157.9	
6 5/8	168.3	5.6	22.47	157.1	
6 5/8	168.3	6.4	25.55	155.5	
6 5/8	168.3	7.1	28.22	154.1	
6 5/8	168.3	7.9	31.25	152.5	
6 5/8	168.3	8.7	34.24	150.9	
6 5/8	168.3	9.5	37.20	149.3	
6 5/8	168.3	11.0	42.67	146.3	
6 5/8	168.3	12.7	48.73	142.9	
6 5/8	168.3	14.3	54.31	139.7	
6 5/8	168.3	15.9	59.76	136.5	
6 5/8	168.3	18.3	67.69	131.7	
6 5/8	168.3	19.1	70.27	130.1	
6 5/8	168.3	21.9	79.06	124.5	
6 5/8	168.3	22.2	79.98	123.9	
8 5/8	219.1	3.2	17.04	212.7	2
8 5/8	219.1	4.0	21.22	211.1	2
8 5/8	219.1	4.8	25.37	209.5	
8 5/8	219.1	5.2	27.43	208.7	
8 5/8	219.1	5.6	29.48	207.9	
8 5/8	219.1	6.4	33.57	206.3	
8 5/8	219.1	7.0	36.61	205.1	
8 5/8	219.1	7.9	41.14	203.3	
8 5/8	219.1	8.2	42.65	202.7	
8 5/8	219.1	8.7	45.14	201.7	
8 5/8	219.1	9.5	49.10	200.1	
8 5/8	219.1	11.1	56.94	196.9	
8 5/8	219.1	12.7	64.64	193.7	
8 5/8	219.1	14.3	72.22	190.5	
8 5/8	219.1	15.9	79.67	187.3	
8 5/8	219.1	18.3	90.62	182.5	
8 5/8	219.1	19.1	94.20	180.9	
8 5/8	219.1	20.6	100.84	177.9	
8 5/8	219.1	22.2	107.79	174.7	
8 5/8	219.1	25.4	121.33	168.3	
10 3/4	273.1	4.0	26.54	265.1	2
10 3/4	273.1	4.8	31.76	263.5	2
10 3/4	273.1	5.2	34.35	262.7	2
10 3/4	273.1	5.6	36.94	261.9	
10 3/4	273.1	6.4	42.09	260.3	
10 3/4	273.1	7.1	46.57	258.9	
10 3/4	273.1	7.8	51.03	257.5	
10 3/4	273.1	8.7	56.72	255.7	
10 3/4	273.1	9.3	60.50	254.5	
10 3/4	273.1	11.1	71.72	250.9	
10 3/4	273.1	12.7	81.55	247.7	
10 3/4	273.1	14.3	91.26	244.5	
10 3/4	273.1	15.9	100.85	241.3	
10 3/4	273.1	18.3	114.99	236.5	
10 3/4	273.1	20.6	128.27	231.9	
10 3/4	273.1	22.2	137.36	228.7	
10 3/4	273.1	23.8	146.32	225.5	
10 3/4	273.1	25.4	155.15	222.3	
10 3/4	273.1	31.8	189.22	209.5	
12 3/4	323.9	4.4	34.67	315.1	2
12 3/4	323.9	4.8	37.77	314.3	2
12 3/4	323.9	5.2	40.87	313.5	2
12 3/4	323.9	5.6	43.96	312.7	2
12 3/4	323.9	6.4	50.11	311.1	
12 3/4	323.9	7.1	55.47	309.7	
12 3/4	323.9	7.9	61.56	308.1	
12 3/4	323.9	8.4	65.35	307.1	
12 3/4	323.9	8.7	67.62	306.5	
12 3/4	323.9	9.5	73.65	304.9	
12 3/4	323.9	10.3	79.65	303.3	
12 3/4	323.9	11.1	85.62	301.7	
12 3/4	323.9	12.7	97.46	298.5	
12 3/4	323.9	14.3	109.18	295.3	
12 3/4	323.9	15.9	120.76	292.1	
12 3/4	323.9	17.5	132.23	288.9	
12 3/4	323.9	19.1	143.56	285.7	
12 3/4	323.9	20.6	154.08	282.7	
12 3/4	323.9	22.2	165.17	279.5	
12 3/4	323.9	23.8	176.13	276.3	
12 3/4	323.9	25.4	186.97	273.1	
12 3/4	323.9	27.0	197.68	269.9	
12 3/4	323.9	28.6	208.27	266.7	
12 3/4	323.9	31.8	229.06	260.3	
14	355.6	4.8	41.52	346.0	2
14	355.6	5.2	44.93	345.2	2
14	355.6	5.3	45.78	345.0	2
14	355.6	5.6	48.33	344.4	2
14	355.6	6.4	55.11	342.8	2
14	355.6	7.1	61.02	341.4	2
14	355.6	7.9	67.74	339.8	
14	355.6	8.7	74.42	338.2	
14	355.6	9.5	81.08	336.6	
14	355.6	10.3	87.71	335.0	
14	355.6	11.1	94.30	333.4	
14	355.6	11.9	100.86	331.8	
14	355.6	12.7	107.39	330.2	
14	355.6	14.3	120.36	327.0	
14	355.6	15.9	133.19	323.8	
14	355.6	17.5	145.91	320.6	
14	355.6	19.1	158.49	317.4	
14	355.6	20.6	170.18	314.4	
14	355.6	22.2	182.52	311.2	
14	355.6	23.8	194.74	308.0	
14	355.6	25.4	206.83	304.8	
14	355.6	27.0	218.79	301.6	
14	355.6	28.6	230.63	298.4	
14	355.6	31.8	253.92	292.0	
16	406.4	4.8	47.54	396.8	2
16	406.4	5.2	51.45	396.0	2
16	406.4	5.6	55.35	395.2	2
16	406.4	6.4	63.13	393.6	2
16	406.4	7.1	69.91	392.2	2

TABLE 6.8M API 5L Plain-End Pipe Dimensions and Weights per Unit Length for Grades A through X80 (SI Units) (Note 1) (Continued)

Size	Specified outside diameter D, mm	Specified wall thickness t, mm	Plain-end weight per unit length w, kg/m	Calculated inside diameter d, mm	Notes
16	406.4	7.9	77.63	390.6	
16	406.4	8.7	85.32	389.0	
16	406.4	9.5	92.98	387.4	
16	406.4	10.3	100.61	385.8	
16	406.4	11.1	108.20	384.2	
16	406.4	11.9	115.77	382.6	
16	406.4	12.7	123.30	381.0	
16	406.4	14.3	138.27	377.8	
16	406.4	15.9	153.11	374.6	
16	406.4	17.5	167.83	371.4	
16	406.4	19.1	182.42	368.2	
16	406.4	20.6	195.98	365.2	
16	406.4	22.2	210.33	362.0	
16	406.4	23.8	224.55	358.8	
16	406.4	25.4	238.64	355.6	
16	406.4	27.0	252.61	352.4	
16	406.4	28.6	266.45	349.2	
16	406.4	30.2	280.17	346.0	
16	406.4	31.8	293.76	342.8	
18	457.0	4.8	53.53	447.4	2
18	457.0	5.6	62.34	445.8	2
18	457.0	6.4	71.12	444.2	2
18	457.0	7.1	78.77	442.8	2
18	457.0	7.9	87.49	441.2	
18	457.0	8.7	96.18	439.6	
18	457.0	9.5	104.84	438.0	
18	457.0	10.3	113.46	436.4	
18	457.0	11.1	122.05	434.8	
18	457.0	11.9	130.62	433.2	
18	457.0	12.7	139.15	431.6	
18	457.0	14.3	156.11	428.4	
18	457.0	15.9	172.95	425.2	
18	457.0	17.5	189.67	422.0	
18	457.0	19.1	206.25	418.8	
18	457.0	20.6	221.69	415.8	
18	457.0	22.2	238.03	412.6	
18	457.0	23.8	254.25	409.4	
18	457.0	25.4	270.34	406.2	
18	457.0	27.0	286.30	403.0	
18	457.0	28.6	302.14	399.8	
18	457.0	30.2	317.85	396.6	
18	457.0	31.8	333.44	393.4	
20	508.0	5.6	69.38	496.8	2
20	508.0	6.4	79.16	495.2	2
20	508.0	7.1	87.70	493.8	2
20	508.0	7.9	97.43	492.2	
20	508.0	8.7	107.12	490.6	
20	508.0	9.5	116.78	489.0	
20	508.0	10.3	126.41	487.4	
20	508.0	11.1	136.01	485.8	
20	508.0	11.9	145.58	484.2	
20	508.0	12.7	155.12	482.6	
20	508.0	14.3	174.10	479.4	
20	508.0	15.9	192.95	476.2	
20	508.0	17.9	216.34	472.2	
20	508.0	19.1	230.27	469.8	
20	508.0	20.6	247.60	466.8	
20	508.0	22.2	265.95	463.6	
20	508.0	23.8	284.18	460.4	
20	508.0	25.4	302.28	457.2	
20	508.0	27.0	320.26	454.0	
20	508.0	28.6	338.11	450.8	
20	508.0	30.2	355.83	447.6	
20	508.0	31.8	373.43	444.4	
20	508.0	33.3	389.81	441.4	
20	508.0	34.9	407.17	438.2	
22	559.0	5.6	76.42	547.8	2
22	559.0	6.4	87.21	546.2	2
22	559.0	7.1	96.63	544.8	2
22	559.0	7.9	107.36	543.2	
22	559.0	8.7	118.06	541.6	
22	559.0	9.5	128.73	540.0	
22	559.0	10.3	139.37	538.4	
22	559.0	11.1	149.97	536.8	
22	559.0	11.9	160.55	535.2	
22	559.0	12.7	171.09	533.6	
22	559.0	14.3	192.08	530.4	
22	559.0	15.9	212.95	527.2	
22	559.0	17.5	233.68	524.0	
22	559.0	19.1	254.30	520.8	
22	559.0	20.6	273.51	517.8	
22	559.0	22.2	293.87	514.6	
22	559.0	23.8	314.11	511.4	
22	559.0	25.4	334.23	508.2	
22	559.0	27.0	354.22	505.0	
22	559.0	28.6	374.08	501.8	
22	559.0	30.2	393.81	498.6	
22	559.0	31.8	413.42	495.4	
22	559.0	33.3	431.69	492.4	
22	559.0	34.9	451.06	489.2	
22	559.0	36.5	470.30	486.0	
22	559.0	38.1	489.41	482.8	
24	610.0	6.4	95.26	597.2	2
24	610.0	7.1	105.56	595.8	2
24	610.0	7.9	117.30	594.2	
24	610.0	8.7	129.00	592.6	
24	610.0	9.5	140.68	591.0	
24	610.0	10.3	152.32	589.4	
24	610.0	11.1	163.93	587.8	
24	610.0	11.9	175.51	586.2	
24	610.0	12.7	187.06	584.6	
24	610.0	14.3	210.07	581.4	
24	610.0	15.9	232.94	578.2	
24	610.0	17.5	255.69	575.0	
24	610.0	19.1	278.32	571.8	
24	610.0	20.6	299.41	568.8	
24	610.0	22.2	321.79	565.6	
24	610.0	23.8	344.05	562.4	
24	610.0	25.4	366.17	559.2	
24	610.0	27.0	388.17	556.0	
24	610.0	28.6	410.05	552.8	
24	610.0	30.2	431.80	549.6	
24	610.0	31.8	453.42	546.4	
24	610.0	33.3	473.57	543.4	

TABLE 6.8M API 5L Plain-End Pipe Dimensions and Weights per Unit Length for Grades A through X80 (SI Units) (Note 1) (Continued)

Size	Specified outside diameter *D*, mm	Specified wall thickness *t*, mm	Plain-end weight per unit length *w*, kg/m	Calculated inside diameter *d*, mm	Notes
24	610.0	34.9	494.95	540.2	
24	610.0	36.5	516.20	537.0	
24	610.0	38.1	537.33	533.8	
24	610.0	39.7	558.32	530.6	
26	660.0	6.4	103.15	647.2	2
26	660.0	7.1	114.31	645.8	2
26	660.0	7.9	127.04	644.2	
26	660.0	8.7	139.73	642.6	
26	660.0	9.5	152.39	641.0	
26	660.0	10.3	165.02	639.4	
26	660.0	11.1	177.62	637.8	
26	660.0	11.9	190.19	636.2	
26	660.0	12.7	202.72	634.6	
26	660.0	14.3	227.70	631.4	
26	660.0	15.9	252.55	628.2	
26	660.0	17.5	277.27	625.0	
26	660.0	19.1	301.87	621.8	
26	660.0	20.6	324.81	618.8	
26	660.0	22.2	349.16	615.6	
26	660.0	23.8	373.39	612.4	
26	660.0	25.4	397.49	609.2	
28	711.0	6.4	111.20	698.2	2
28	711.0	7.1	123.24	696.8	2
28	711.0	7.9	136.97	695.2	
28	711.0	8.7	150.67	693.6	
28	711.0	9.5	164.34	692.0	
28	711.0	10.3	177.98	690.4	
28	711.0	11.1	191.58	688.8	
28	711.0	11.9	205.15	687.2	
28	711.0	12.7	218.69	685.6	
28	711.0	14.3	245.68	682.4	
28	711.0	15.9	272.54	679.2	
28	711.0	17.5	299.28	676.0	
28	711.0	19.1	325.89	672.8	
28	711.0	20.6	350.72	669.8	
28	711.0	22.2	377.08	666.6	
28	711.0	23.8	403.32	663.4	
28	711.0	25.4	429.44	660.2	
30	762.0	6.4	119.25	749.2	2
30	762.0	7.1	132.17	747.8	2
30	762.0	7.9	146.91	746.2	
30	762.0	8.7	161.61	744.6	
30	762.0	9.5	176.29	743.0	
30	762.0	10.3	190.93	741.4	
30	762.0	11.1	205.54	739.8	
30	762.0	11.9	220.12	738.2	
30	762.0	12.7	234.67	736.6	
30	762.0	14.3	263.67	733.4	
30	762.0	15.9	292.54	730.2	
30	762.0	17.5	321.29	727.0	
30	762.0	19.1	349.91	723.8	
30	762.0	20.6	376.63	720.8	
30	762.0	22.2	405.00	717.6	
30	762.0	23.8	433.26	714.4	
30	762.0	25.4	461.38	711.2	
30	762.0	27.0	489.38	708.0	
30	762.0	28.6	517.25	704.8	
30	762.0	30.2	544.99	701.6	
30	762.0	31.8	572.61	698.4	
32	813.0	6.4	127.30	800.2	2
32	813.0	7.1	141.10	798.8	2
32	813.0	7.9	156.84	797.2	
32	813.0	8.7	172.56	795.6	
32	813.0	9.5	188.24	794.0	
32	813.0	10.3	203.88	792.4	
32	813.0	11.1	219.50	790.8	
32	813.0	11.9	235.09	789.2	
32	813.0	12.7	250.64	787.6	
32	813.0	14.3	281.65	784.4	
32	813.0	15.9	312.54	781.2	
32	813.0	17.5	343.30	778.0	
32	813.0	19.1	373.93	774.8	
32	813.0	20.6	402.54	771.8	
32	813.0	22.2	432.93	768.6	
32	813.0	23.8	463.19	765.4	
32	813.0	25.4	493.32	762.2	
32	813.0	27.0	523.33	759.0	
32	813.0	28.6	553.22	755.8	
32	813.0	30.2	582.98	752.6	
32	813.0	31.8	612.61	749.4	
34	864.0	6.4	135.35	851.2	2
34	864.0	7.1	150.03	849.8	2
34	864.0	7.9	166.78	848.2	
34	864.0	8.7	183.50	846.6	
34	864.0	9.5	200.18	845.0	
34	864.0	10.3	216.84	843.4	
34	864.0	11.1	233.46	841.8	
34	864.0	11.9	250.05	840.2	
34	864.0	12.7	266.61	838.6	
34	864.0	14.3	299.64	835.4	
34	864.0	15.9	332.53	832.2	
34	864.0	17.5	365.31	829.0	
34	864.0	19.1	397.95	825.8	
34	864.0	20.6	428.44	822.8	
34	864.0	22.2	460.85	819.6	
34	864.0	23.8	493.12	816.4	
34	864.0	25.4	525.27	813.2	
34	864.0	27.0	557.29	810.0	
34	864.0	28.6	589.19	806.8	
34	864.0	30.2	620.96	803.6	
34	864.0	31.8	652.60	800.4	
36	914.0	6.4	143.24	901.2	2
36	914.0	7.1	158.79	899.8	2
36	914.0	7.9	176.52	898.2	
36	914.0	8.7	194.22	896.6	
36	914.0	9.5	211.90	895.0	
36	914.0	10.3	229.54	893.4	
36	914.0	11.1	247.15	891.8	
36	914.0	11.9	264.72	890.2	
36	914.0	12.7	282.27	888.6	
36	914.0	14.3	317.27	885.4	
36	914.0	15.9	352.14	882.2	
36	914.0	17.5	386.88	879.0	
36	914.0	19.1	421.50	875.8	

TABLE 6.8M API 5L Plain-End Pipe Dimensions and Weights per Unit Length for Grades A through X80 (SI Units) (Note 1) (Continued)

Size	Specified outside diameter D, mm	Specified wall thickness t, mm	Plain-end weight per unit length w, kg/m	Calculated inside diameter d, mm	Notes
36	914.0	20.6	453.84	872.8	
36	914.0	22.2	488.22	869.6	
36	914.0	23.8	522.47	866.4	
36	914.0	25.4	556.59	863.2	
36	914.0	27.0	590.58	860.0	
36	914.0	28.6	624.45	856.8	
36	914.0	30.2	658.19	853.6	
36	914.0	31.8	691.81	850.4	
38	965.0	7.9	186.46	949.2	
38	965.0	8.7	205.17	947.6	
38	965.0	9.5	223.84	946.0	
38	965.0	10.3	242.49	944.4	
38	965.0	11.1	261.11	942.8	
38	965.0	11.9	279.69	941.2	
38	965.0	12.7	298.24	939.6	
38	965.0	14.3	335.25	936.4	
38	965.0	15.9	372.14	933.2	
38	965.0	17.5	408.89	930.0	
38	965.0	19.1	445.52	926.8	
38	965.0	20.6	479.75	923.8	
38	965.0	22.2	516.14	920.6	
38	965.0	23.8	552.40	917.4	
38	965.0	25.4	588.53	914.2	
38	965.0	27.0	624.54	911.0	
38	965.0	28.6	660.42	907.8	
38	965.0	30.2	696.18	904.6	
38	965.0	31.8	731.80	901.4	
40	1016.0	7.9	196.39	1000.2	
40	1016.0	8.7	216.11	998.6	
40	1016.0	9.5	235.79	997.0	
40	1016.0	10.3	255.45	995.4	
40	1016.0	11.1	275.07	993.8	
40	1016.0	11.9	294.66	992.2	
40	1016.0	12.7	314.22	990.6	
40	1016.0	14.3	353.24	987.4	
40	1016.0	15.9	392.13	984.2	
40	1016.0	17.5	430.90	981.0	
40	1016.0	19.1	469.55	977.8	
40	1016.0	20.6	505.66	974.8	
40	1016.0	22.2	544.06	971.6	
40	1016.0	23.8	582.33	968.4	
40	1016.0	25.4	620.48	965.2	
40	1016.0	27.0	658.50	962.0	
40	1016.0	28.6	696.39	958.8	
40	1016.0	30.2	734.16	955.6	
40	1016.0	31.8	771.80	952.4	
42	1067.0	8.7	227.05	1049.6	
42	1067.0	9.5	247.74	1048.0	
42	1067.0	10.3	268.40	1046.4	
42	1067.0	11.1	289.03	1044.8	
42	1067.0	11.9	309.62	1043.2	
42	1067.0	12.7	330.19	1041.6	
42	1067.0	14.3	371.22	1038.4	
42	1067.0	15.9	412.13	1035.2	
42	1067.0	17.5	452.91	1032.0	
42	1067.0	19.1	493.57	1028.8	
42	1067.0	20.6	531.57	1025.8	
42	1067.0	22.2	571.98	1022.6	
42	1067.0	23.8	612.26	1019.4	
42	1067.0	25.4	652.42	1016.2	
42	1067.0	27.0	692.45	1013.0	
42	1067.0	28.6	732.36	1009.8	
42	1067.0	30.2	772.14	1006.6	
42	1067.0	31.8	811.79	1003.4	
44	1118.0	8.7	237.99	1100.6	
44	1118.0	9.5	259.69	1099.0	
44	1118.0	10.3	281.35	1097.4	
44	1118.0	11.1	302.99	1095.8	
44	1118.0	11.9	324.59	1094.2	
44	1118.0	12.7	346.16	1092.6	
44	1118.0	14.3	389.21	1089.4	
44	1118.0	15.9	432.13	1086.2	
44	1118.0	17.5	474.92	1083.0	
44	1118.0	19.1	517.59	1079.8	
44	1118.0	20.6	557.47	1076.8	
44	1118.0	22.2	599.90	1073.6	
44	1118.0	23.8	642.19	1070.4	
44	1118.0	25.4	684.37	1067.2	
44	1118.0	27.0	726.41	1064.0	
44	1118.0	28.6	768.33	1060.8	
44	1118.0	30.2	810.12	1057.6	
44	1118.0	31.8	851.79	1054.4	
46	1168.0	8.7	248.72	1150.6	
46	1168.0	9.5	271.40	1149.0	
46	1168.0	10.3	294.05	1147.4	
46	1168.0	11.1	316.67	1145.8	
46	1168.0	11.9	339.26	1144.2	
46	1168.0	12.7	361.82	1142.6	
46	1168.0	14.3	406.84	1139.4	
46	1168.0	15.9	451.73	1136.2	
46	1168.0	17.5	496.50	1133.0	
46	1168.0	19.1	541.14	1129.8	
46	1168.0	20.6	582.87	1126.8	
46	1168.0	22.2	627.27	1123.6	
46	1168.0	23.8	671.54	1120.4	
46	1168.0	25.4	715.68	1117.2	
46	1168.0	27.0	759.70	1114.0	
46	1168.0	28.6	803.59	1110.8	
46	1168.0	30.2	847.36	1107.6	
46	1168.0	31.8	890.99	1104.4	
48	1219.0	8.7	259.66	1201.6	
48	1219.0	9.5	283.35	1200.0	
48	1219.0	10.3	307.01	1198.4	
48	1219.0	11.1	330.63	1196.8	
48	1219.0	11.9	354.23	1195.2	
48	1219.0	12.7	377.79	1193.6	
48	1219.0	14.3	424.82	1190.4	
48	1219.0	15.9	471.73	1187.2	
48	1219.0	17.5	518.51	1184.0	
48	1219.0	19.1	565.16	1180.8	
48	1219.0	20.6	608.78	1177.8	
48	1219.0	22.2	655.19	1174.6	
48	1219.0	23.8	701.47	1171.4	
48	1219.0	25.4	747.63	1168.2	

TABLE 6.8M API 5L Plain-End Pipe Dimensions and Weights per Unit Length for Grades A through X80 (SI Units) (Note 1) (Continued)

Size	Specified outside diameter *D*, mm	Specified wall thickness *t*, mm	Plain-end weight per unit length *w*, kg/m	Calculated inside diameter *d*, mm	Notes
48	1219.0	27.0	793.66	1165.0	
48	1219.0	28.6	839.56	1161.8	
48	1219.0	30.2	885.34	1158.6	
48	1219.0	31.8	930.99	1155.4	
52	1321.0	9.5	307.25	1302.0	
52	1321.0	10.3	332.92	1300.4	
52	1321.0	11.1	358.55	1298.8	
52	1321.0	11.9	384.16	1297.2	
52	1321.0	12.7	409.74	1295.6	
52	1321.0	14.3	460.79	1292.4	
52	1321.0	15.9	511.72	1289.2	
52	1321.0	17.5	562.53	1286.0	
52	1321.0	19.1	613.20	1282.8	
52	1321.0	20.6	660.60	1279.8	
52	1321.0	22.2	711.03	1276.6	
52	1321.0	23.8	761.34	1273.4	
52	1321.0	25.4	811.52	1270.2	
52	1321.0	27.0	861.57	1267.0	
52	1321.0	28.6	911.50	1263.8	
52	1321.0	30.2	961.30	1260.6	
52	1321.0	31.8	1010.98	1257.4	
56	1422.0	9.5	330.91	1403.0	
56	1422.0	10.3	358.57	1401.4	
56	1422.0	11.1	386.20	1399.8	
56	1422.0	11.9	413.80	1398.2	
56	1422.0	12.7	441.37	1396.6	
56	1422.0	14.3	496.41	1393.4	
56	1422.0	15.9	551.32	1390.2	
56	1422.0	17.5	606.11	1387.0	
56	1422.0	19.1	660.77	1383.8	
56	1422.0	20.6	711.91	1380.8	
56	1422.0	22.2	766.32	1377.6	
56	1422.0	23.8	820.61	1374.4	
56	1422.0	25.4	874.78	1371.2	
56	1422.0	27.0	928.82	1368.0	
56	1422.0	28.6	982.73	1364.8	
56	1422.0	30.2	1036.52	1361.6	
56	1422.0	31.8	1090.18	1358.4	
60	1524.0	9.5	354.80	1505.0	
60	1524.0	10.3	384.48	1503.4	
60	1524.0	11.1	414.12	1501.8	
60	1524.0	11.9	443.73	1500.2	
60	1524.0	12.7	473.31	1498.6	
60	1524.0	14.3	532.38	1495.4	
60	1524.0	15.9	591.32	1492.2	
60	1524.0	17.5	650.13	1489.0	
60	1524.0	19.1	708.82	1485.8	
60	1524.0	20.6	763.72	1482.8	
60	1524.0	22.2	822.16	1479.6	
60	1524.0	23.8	880.48	1476.4	
60	1524.0	25.4	938.67	1473.2	
60	1524.0	27.0	996.73	1470.0	
60	1524.0	28.6	1054.67	1466.8	
60	1524.0	30.2	1112.48	1463.6	
60	1524.0	31.8	1170.17	1460.4	
64	1626.0	9.5	378.70	1607.0	
64	1626.0	10.3	410.38	1605.4	
64	1626.0	11.1	442.04	1603.8	
64	1626.0	11.9	473.66	1602.2	
64	1626.0	12.7	505.26	1600.6	
64	1626.0	14.3	568.35	1597.4	
64	1626.0	15.9	631.31	1594.2	
64	1626.0	17.5	694.15	1591.0	
64	1626.0	19.1	756.86	1587.8	
64	1626.0	20.6	815.54	1584.8	
64	1626.0	22.2	878.00	1581.6	
64	1626.0	23.8	940.34	1578.4	
64	1626.0	25.4	1002.56	1575.2	
64	1626.0	27.0	1064.65	1572.0	
64	1626.0	28.6	1126.61	1568.8	
64	1626.0	30.2	1188.44	1565.6	
64	1626.0	31.8	1250.15	1562.4	
68	1727.0	11.9	503.30	1703.2	
68	1727.0	12.7	536.89	1701.6	
68	1727.0	14.3	603.96	1698.4	
68	1727.0	15.9	670.91	1695.2	
68	1727.0	17.5	737.73	1692.0	
68	1727.0	19.1	804.43	1688.8	
68	1727.0	20.6	866.84	1685.8	
68	1727.0	22.2	933.30	1682.6	
68	1727.0	23.8	999.62	1679.4	
68	1727.0	25.4	1065.82	1676.2	
68	1727.0	27.0	1131.89	1673.0	
68	1727.0	28.6	1197.84	1669.8	
68	1727.0	30.2	1263.66	1666.6	
68	1727.0	31.8	1329.36	1663.4	
72	1829.0	12.7	568.83	1803.6	
72	1829.0	14.3	639.93	1800.4	
72	1829.0	15.9	710.91	1797.2	
72	1829.0	17.5	781.75	1794.0	
72	1829.0	19.1	852.47	1790.8	
72	1829.0	20.6	918.66	1787.8	
72	1829.0	22.2	989.14	1784.6	
72	1829.0	23.8	1059.49	1781.4	
72	1829.0	25.4	1129.71	1778.2	
72	1829.0	27.0	1199.81	1775.0	
72	1829.0	28.6	1269.78	1771.8	
72	1829.0	30.2	1339.62	1768.6	
72	1829.0	31.8	1409.34	1765.4	
76	1930.0	12.7	600.46	1904.6	
76	1930.0	14.3	675.55	1901.4	
76	1930.0	15.9	750.51	1898.2	
76	1930.0	17.5	825.34	1895.0	
76	1930.0	19.1	900.05	1891.8	
76	1930.0	20.6	969.97	1888.8	
76	1930.0	22.2	1044.43	1885.6	
76	1930.0	23.8	1118.76	1882.4	
76	1930.0	25.4	1192.97	1879.2	
76	1930.0	27.0	1267.06	1876.0	
76	1930.0	28.6	1341.01	1872.8	
76	1930.0	30.2	1414.84	1869.6	
76	1930.0	31.8	1488.55	1866.4	
80	2032.0	14.3	711.52	2003.4	
80	2032.0	15.9	790.50	2000.2	

TABLE 6.8M API 5L Plain-End Pipe Dimensions and Weights per Unit Length for Grades A through X80 (SI Units) (Note 1) (Continued)

Size	Specified outside diameter D, mm	Specified wall thickness t, mm	Plain-end weight per unit length w, kg/m	Calculated inside diameter d, mm	Notes	Size	Specified outside diameter D, mm	Specified wall thickness t, mm	Plain-end weight per unit length w, kg/m	Calculated inside diameter d, mm	Notes
80	2032.0	17.5	869.36	1997.0		80	2032.0	25.4	1256.86	1981.2	
80	2032.0	19.1	948.09	1993.8		80	2032.0	27.0	1334.97	1978.0	
80	2032.0	20.6	1021.78	1990.8		80	2032.0	28.6	1412.95	1974.8	
80	2032.0	22.2	1100.27	1987.6		80	2032.0	30.2	1490.80	1971.6	
80	2032.0	23.8	1178.63	1984.4		80	2032.0	31.8	1568.53	1968.4	

Notes:

1. This table covers pipe dimensions and weights per unit length of API 5L line pipe, Grades A, B, X42, X46, X52, X56, X60, X65, X70, and X80 in sizes $6^3/_8$ through 80. Refer to Tables 6.9 through 6.13 for tolerances.

2. The pipe having this specified outside diameter and the specified wall thickness is termed *special plain-end pipe*. Other pipe is called *regular plain-end pipe*. Pipe that has a combination of specified outside diameter and specified wall thickness that is intermediate to the tabulated values is considered to be special plain-end pipe if the next lower tabulated value is for special plain-end pipe. Other intermediate combinations of outside diameter and wall thickness are considered to be regular plain-end pipe.

API 5L Line Pipe Tolerances.

Tolerances for Pipe Wall Thickness. Refer to Table 6.9.

Tolerances for Pipe Body Diameter. Refer to Tables 6.10 and 6.10M.

Tolerances for Diameter at Pipe Ends. Refer to Tables 6. 11 and 6.11M.

Tolerances on Length. Refer to Tables 6.12 and 6.12M.

Tolerances for Weight. Refer to Table 6.13.

API 5L Line Pipe Material Properties

Tensile and Yield Strength Requirements for PSL 1 Line Pipe. Refer to Tables 6.14 and 6.14M for tensile and yield strength requirements for Grades A25 through X80, PSL 1 line pipe.

Tensile and Yield Strength Requirements for PSL 2 Line Pipe. Refer to Tables 6.15 and 6.15M for tensile and yield strength requirements for Grades B through X80, PSL 2 line pipe.

TABLE 6.9 Tolerances for Wall Thickness*

Size	Type of pipe	Tolerance (percent of specified wall thickness) Grade B or lower	Grade X42 or higher
$\leq 2^7/_8$	All	+20.0, −12.5	+15.0, −12.5
$>2^7/_8$ and <20	All	+15.0, −12.5	+15.0, −12.5
≥20	Welded	+17.5, −12.5	+19.5, −8.0
≥20	Seamless	+15.0, −12.5	+17.5, −10.0

* When negative tolerances smaller than those listed are specified by the purchaser, the positive tolerance shall be increased to the applicable total tolerance range in percent less the wall thickness negative tolerance.

TABLE 6.10 Tolerances for Diameter of Pipe Body (U.S. Customary Units)

Size	Tolerance (with respect to specified outside diameter)*
<2³⁄₈	+0.016 in, −0.031 in
≥2³⁄₈ and ≤4¹⁄₂, continuous welded	±1.00%
≥2³⁄₈ and <20	±0.75%
≥20, seamless	±1.00%
≥20 and ≤36, welded	+0.75%, −0.25%
>36, welded	+¹⁄₄ in, −¹⁄₈ in

* In the case of pipe hydrostatically tested to pressures in excess of standard test pressures, other tolerances may be agreed upon between the manufacturer and the purchaser.

TABLE 6.10M Tolerances for Diameter of Pipe Body (Metric Units)

Size	Tolerance (with respect to specified outside diameter)*
<2³⁄₈	+0.41 mm, −0.79 mm
≥2³⁄₈> and ≤4¹⁄₂, continuous welded	±1.00%
≥2³⁄₈ and <20	±0.75%
≥20, seamless	±1.00%
≥20 and ≤36, welded	±0.75%, −0.25%
>36, welded	+6.35 mm, −3.20 mm

* In the case of pipe hydrostatically tested to pressures in excess of standard test pressures, other tolerances may be agreed upon between the manufacturer and the purchaser.

TABLE 6.11 Tolerance for Diameter at Pipe Ends (U.S. Customary Units)

				Out-of-roundness	
Size	Minus tolerance, in	Plus tolerance, in	End-to-end tolerance	Diameter, axis tolerance (percent of specified OD) (1)	Maximum differential between minimum and maximum diameters (applies only to pipe With $D/t \leq 75$), in
≤10³⁄₄	¹⁄₆₄	¹⁄₁₆	—	—	—
>10³⁄₄ and ≤20	¹⁄₃₂	³⁄₃₂	—	—	—
>20 and ≤42	¹⁄₃₂	³⁄₃₂	(2)	±1%	≤0.500
>42	¹⁄₃₂	³⁄₃₂	(2)	±1%	≤0.625

Notes:
(1) Out-of-roundness tolerances apply to maximum and minimum diameters as measured with a bar gage, caliper, or device measuring actual maximum and minimum diameters.
(2) The average diameter (as measured with a diameter tape) of one end of pipe shall not differ by more than ³⁄₃₂ in (2.38 mm) from that of the other end.

TABLE 6.11M Tolerance for Diameter at Pipe Ends (Metric Units)

				Out-of-roundness	
Size	Minus tolerance, mm	Plus tolerance, mm	End-to-end tolerance	Diameter, axis tolerance (percent of specified OD) (1)	Maximum differential between minimum and maximum diameters (applies only to pipe with $D/t \leq 75$), mm
≤10³⁄₄	0.40	1.59	—	—	—
>10³⁄₄ and ≤20	0.79	2.38	—	—	—
>20 and ≤42	0.79	2.38	(2)	±1%	12.7
>42	0.79	2.38	(2)	±1%	15.9

Notes:
(1) Out-of-roundness tolerances apply to maximum and minimum diameters as measured with a bar gage, caliper, or device measuring actual maximum and minimum diameters.
(2) The average diameter (as measured with a diameter tape) of one end of pipe shall not differ by more than ³⁄₃₂ in (2.38 mm) from that of the other end.

TABLE 6.12 Tolerances on Lengths (U.S. Customary Units)

Nominal length (1), ft	Minimum length, ft	Minimum average length for each order item (2), ft	Maximum length, ft
		Threaded-and-coupled pipe	
20	16.0	17.5	22.5
40	22.0	35.0	45.0
		Plain-end pipe	
20	9.0	17.5	22.5
40	14.0	35.0	45.0
50	17.5	43.8	55.0
60	21.0	52.5	65.0
80	28.0	70.0	85.0

Notes:

(1) Nominal lengths of 20 ft were formerly designated "single random lengths" and those of 40 ft (12 m) "double random lengths."

(2) By agreement between the purchaser and the manufacturer, these tolerances shall apply to each carload.

TABLE 6.12M Tolerances on Lengths (Metric Units)

Nominal length (1), m	Minimum length, m	Minimum average length for each order item (2), m	Maximum length, m
		Threaded-and-coupled pipe	
6	4.88	5.33	6.86
12	6.71	10.67	13.72
		Plain-end pipe	
6	2.74	5.33	6.86
12	4.27	10.67	13.72
15	5.33	13.35	16.76
18	6.40	16.00	19.81
24	8.53	21.34	25.91

Notes:

(1) Nominal lengths of 20 ft (6 m) were formerly designated "single random lengths" and those of 40 ft (12 m) "double random lengths."

(2) By agreement between the purchaser and the manufacturer, these tolerances shall apply to each carload.

TABLE 6.13 Tolerances for Weight

Quantity	Tolerance, %
Single lengths, special plain-end pipe or A25 pipe	+10, −5.0
Single lengths, other pipe	+10, −3.5
Carloads, Grade A25, 40,000 lb (18,144 kg) or more	−2.5
Carloads, other than Grade A25, 40,000 lb (18,144 kg) or more	−1.75
Carloads, all grades, less than 40,000 lb (18,144 kg)	−3.5
Order items, Grade A25, 40,000 lb (18,144 kg) or more	−3.5
Order items, other than Grade A25, 40,000 lb (18,144 kg) or more	−1.75
Order items, all grades, less than 40,000 lb (18,144 kg)	−3.5

Notes:

1. Weight tolerances apply to the calculated weights for threaded-and-coupled pipe and to the tabulated or calculated weights for plain-end pipe. Where negative wall thickness tolerances smaller than those listed in Table 6.9 are specified by the purchaser, the plus weight tolerance for single lengths shall be increased to 22.5 percent less the wall thickness negative tolerances.
2. For carloads composed of pipe from more than one order item, the carload tolerances are to be applied on an individual order item basis.
3. The tolerances for order items apply to the overall quantity of pipe shipped for the order item.

TABLE 6.14 Tensile and Yield Strengths of Grades A25 through X70, PSL 1 Pipe (U.S. Customary Units)

Grade	Yield strength, minimum, psi	Ultimate tensile strength, minimum, psi
A25	25,000	45,000
A	30,000	48,000
B	35,000	60,000
X42	42,000	60,000
X46	46,000	63,000
X52	52,000	66,000
X56	56,000	71,000
X60	60,000	75,000
X65	65,000	77,000
X70	70,000	82,000

TABLE 6.14M Tensile and Yield Strengths of Grades A25 through X70, PSL 1 Pipe (Metric Units)

Grade	Yield strength, minimum, MPa	Ultimate tensile strength, minimum, MPa
A25	172	310
A	207	331
B	241	414
X42	290	414
X46	317	434
X52	359	455
X56	386	490
X60	414	517
X65	448	531
X70	483	565

TABLE 6.15 Tensile and Yield Strengths of Grades B through X80, PSL 2 Pipe (U.S. Customary Units)

Grade	Yield strength, minimum, psi	Yield strength, maximum (1), psi	Ultimate tensile strength, minimum, psi	Ultimate tensile strength, maximum (2), psi
B	35,000	65,000 (3)	60,000	110,000
X42	42,000	72,000	60,000	110,000
X46	46,000	76,000	63,000	110,000
X52	52,000	77,000	66,000	110,000
X56	56,000	79,000	71,000	110,000
X60	60,000	82,000	75,000	110,000
X65	65,000	87,000	77,000	110,000
X70	70,000	90,000	82,000	110,000
X80	80,000	100,000 (4)	90,000	120,000

See notes below Table 6.15M.

TABLE 6.15M Tensile and Yield Strengths of Grades B through X80, PSL 2 Pipe (SI Units)

Grade	Yield strength, minimum, MPa	Yield strength, maximum (1), MPa	Ultimate tensile strength, minimum, MPa	Ultimate tensile strength, maximum (2), MPa
B	241	448 (3)	414	758
X42	290	496	414	758
X46	317	524	434	758
X52	359	531	455	758
X56	386	544	490	758
X60	414	565	517	758
X65	448	600	531	758
X70	483	621	565	758
X80	552	690 (4)	621	827

Notes:

(1) Maximum yield strength for an intermediate grade shall be the maximum for the next higher listed grade.

(2) All intermediate grades have a maximum ultimate tensile strength of 110,000 psi (758 MPa).

(3) Maximum yield strength for Grade B pipe in sizes subject to longitudinal testing is 72,000 psi (496 MPa).

(4) For wall thickness greater than 0.984 in (25.0 mm), the maximum yield strength shall be determined by mutual agreement between the purchaser and the manufacturer.

API 605 CARBON STEEL FLANGES

API 605 covers raised face (RF) welding neck and integral flanges. The integral flanges are either cast or forged as the integral ends of flanged valves, fittings, or nozzle necks. Table 6.16 lists the sizes and pressure classes of the API 605 welding neck flanges.

TABLE 6.16 API 605 Raised Face Welding Neck Flange Size and Classes*

NPS	Pressure Class 75	Pressure Class 150	Pressure Class 300	Pressure Class 400	Pressure Class 600	Pressure Class 900†
26	X	X	X	X	X	X
28	X	X	X	X	X	X
30	X	X	X	X	X	X
32	X	X	X	X	X	X
34	X	X	X	X	X	X
36	X	X	X	X	X	X
38	X	X	X	X	X	X
40	X	X	X	X	X	X
42	X	X	X	X	X	X
44	X	X	X	X	X	X
46	X	X	X	X	X	X
48	X	X	X	X	X	X
50	X	X	X	X	X	
52	X	X	X	X	X	
54	X	X	X	X	X	
56	X	X	X	X	X	
58	X	X	X	X	X	
60	X	X	X	X	X	

* Integral flanges are limited to Classes 75, 150, and 300.
† Class 900 welding neck flanges are manufactured in NPS 26 through NPS 48 only.

Materials

Welding neck flanges are forged from the following material specifications:

- ASTM A105, Carbon Steel Forgings for Piping Components.
- ASTM A694, Carbon and Alloy Steel Forgings for Pipe Flanges, Fittings, Valves, and Parts for High-Pressure Transmission Service.

The integral flanges are cast or forged from the following material specifications:

- ASTM A105, Carbon Steel Forgings for Piping Components
- ASTM A216, Grade WCB, Carbon Steel Castings Suitable for Fusion Welding for High-Temperature Service.

Pressure-Temperature Ratings

The pressure-temperature ratings for API 605 flanges are listed in two categories as follows.

Pressure-Temperature Ratings for Plant Piping and Pressure Vessel Flanges. Tables 6.17 and 6.17M list the pressure-temperature ratings for plant piping and pressure vessel flanges designed in accordance with the ASME Code or ASME B31.3.

TABLE 6.17 Pressure-Temperature Ratings for Plant Piping and Pressure Vessel Flanges (U.S. Customary Units)*

Metal temperature, °F	Maximum allowable nonshock working pressure, psig†					
	Class					
	75	150	300	400	600	900
100	140	285	740	990	1480	2220
200	130	260	675	900	1350	2025
300	115	230	655	875	1315	1970
400	100	200	635	845	1270	1900
500	85	170	600	800	1200	1795
600	70	140	550	730	1095	1640
650	60	125	535	715	1075	1610
700		110	535	710	1065	1600
750		95	505	670	1010	1510
800		80	410	550	825	1235
850		65	270	355	535	805
900		50	170	230	345	515
950		35	105	140	205	310
1000		20	50	70	105	155

* Pressure-temperature ratings are for flanges made from ASTM A105 and A216, Grade WCB.

† Pressure ratings are given in pounds per square inch gauge corresponding to the metal temperature listed.

TABLE 6.17M Pressure-Temperature Ratings for Plant Piping and Pressure Vessel Flanges (SI Units)*

Metal temperature, °C	Maximum allowable nonshock working pressure, barg†					
	Class					
	75	150	300	400	600	900
38	9.8	19.6	51.1	68.1	102.1	153.2
50	9.6	19.2	50.1	66.8	100.2	150.2
100	8.8	17.7	46.4	61.8	92.8	139.1
150	7.9	15.8	45.2	60.3	90.5	135.7
200	7.0	14.0	43.8	58.4	87.6	131.5
250	6.0	12.1	41.7	55.6	83.4	125.2
300	5.1	10.2	38.7	51.6	77.5	116.2
350	4.2	8.4	37.0	49.3	73.9	110.9
375		7.4	36.5	48.6	72.9	109.4
400		6.5	34.5	46.0	69.0	103.5
425		5.6	28.8	38.3	57.5	86.3
450		4.7	20.0	26.7	40.1	60.1
475		3.7	13.5	18.1	27.1	40.6
500		2.8	8.8	11.7	17.6	26.4
525		1.9	5.2	6.9	10.4	15.5
540		1.3	3.3	4.3	6.5	9.8

* Pressure-temperature ratings are for flanges made from ASTM A105 and A216, Grade WCB.

† Pressure ratings are given in bar gauge (barg) corresponding to the metal temperature listed.

Pressure-Temperature Ratings for Pipeline Flanges. Tables 6.18 and 6.18M provide pressure-temperature ratings for flanges meeting the requirements of the following codes:

- ASME B31.4, Liquid Transportation Systems for Hydrocarbons, Liquid Petroleum Gas, Anhydrous Ammonia, and Alcohol.
- ASME B31.8, Gas Transmission and Distribution Piping Systems.

TABLE 6.18 Pressure-Temperature Ratings for Pipeline Flanges Meeting the Requirements of ASME B31.4 and B31.8*

Metal temperature, °F	Maximum allowable nonshock working pressure, psig			
	Class			
	300	400	600	900
100	740	990	1480	2220
250	740	990	1480	2220
300	715	955	1430	2145
350	690	925	1380	2070
400	665	890	1330	2000
450	640	860	1285	1925

* Pressure ratings are in pounds per square inch gauge (psig) corresponding to the metal temperature.

TABLE 6.18M Pressure-Temperature Ratings for Pipeline Flanges Meeting the Requirements of ASME B31.4 and B31.8*

Metal temperature, °C	Maximum allowable nonshock working pressure, barg			
	Class			
	300	400	600	900
38	51.1	68.1	102.1	153.2
50	51.1	68.1	102.1	153.2
100	51.1	68.1	102.1	153.2
120	51.1	68.1	102.1	153.2
150	49.3	65.8	98.6	147.9
175	47.6	63.8	95.1	142.7
200	45.9	61.4	91.7	137.9
230	44.1	59.3	88.6	132.7

* Pressure ratings are in bar gauge (barg) corresponding to the metal temperature.

Dimensions of Welding Neck Flanges

Tables 6.19 through 6.30 and 6.19M through 6.30M provide dimensions of welding neck flanges and bolts rated Class 75, 150, 300, 400, 600, and 900 in U.S. Customary and SI units, respectively.

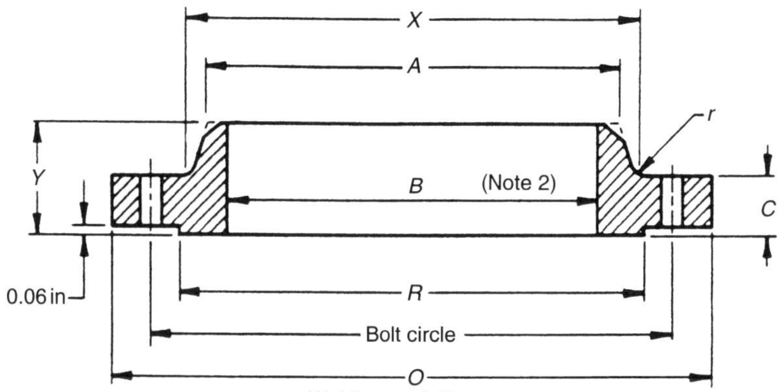

Welding neck flange

TABLE 6.19 Dimensions for Class 75 Welding Neck Flanges (U.S. Customary Units) (Note 1)

NPS	Flange outside diameter *O*	Flange thickness *C*	Raised face diameter *R*	Diameter at large end of hub, *X*	Diameter at small end of hub, *A*	Overall length *y*	Fillet radius *r*	Maximum inside diameter (2) ASME code *B*	ASME B31.3 *B*
26	30.00	1.31	27.75	26.62	26.06	2.31	0.31	25.50	25.50
28	32.00	1.31	29.75	28.62	28.06	2.44	0.31	27.50	27.50
30	34.00	1.31	31.75	30.62	30.06	2.56	0.31	29.50	29.50
32	36.00	1.38	33.75	32.62	32.06	2.75	0.31	31.50	31.50
34	38.00	1.38	35.75	34.62	34.06	2.88	0.31	33.50	33.50
36	40.69	1.44	38.00	36.81	36.06	3.38	0.38	35.50	35.50
38	42.69	1.50	40.00	38.81	38.06	3.50	0.38	37.50	37.50
40	44.69	1.50	42.00	40.81	40.06	3.62	0.38	39.50	39.50
42	46.69	1.56	44.00	42.81	42.06	3.75	0.38	41.50	41.50
44	49.25	1.69	46.25	44.88	44.06	4.12	0.38	43.50	43.50
46	51.25	1.75	48.25	46.88	46.06	4.25	0.38	45.50	45.50
48	53.25	1.81	50.25	48.88	48.06	4.38	0.38	47.50	47.50
50	55.25	1.88	52.25	50.94	50.06	4.56	0.38	49.50	49.50
52	57.38	1.88	54.25	52.94	52.06	4.75	0.38	51.50	51.50
54	59.38	1.94	56.25	55.00	54.06	4.94	0.38	53.50	53.50
56	62.00	2.00	58.50	57.12	56.06	5.31	0.44	55.50	55.50
58	64.00	2.06	60.50	59.12	58.06	5.44	0.44	57.50	57.50
60	66.00	2.19	62.50	61.12	60.06	5.69	0.44	59.50	59.50

***Notes*:**
1. Dimensions in inches. For tolerances refer to Table 6.31.
2. Maximum inside diameter is based on full pressure rating at 100°F and the gasket requirements of API 605. Purchaser must specify the required inside diameter based on the application and the governing code. Purchaser may specify a larger inside diameter than the maximum listed when the application pressure is lower or adjusted gasket dimensions are used.

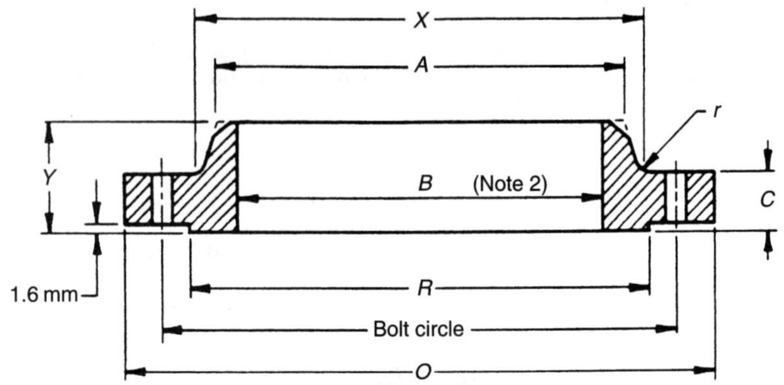

Welding neck flange

TABLE 6.19M Dimensions for Class 75 Welding Neck Flanges (SI Units) (Note 1)

NPS	Flange outside diameter *O*	Flange thickness *C*	Raised face diameter *R*	Diameter at large end of hub, *X*	Diameter at small end of hub, *A*	Overall length *y*	Fillet radius *r*	Maximum inside diameter (2) ASME code *B*	ASME B31.3 *B*
26	762	33.3	704.8	676.3	662.0	58.7	7.9	647.7	647.7
28	813	33.3	755.6	727.1	712.8	61.9	7.9	698.5	698.5
30	864	33.3	806.4	777.9	763.6	65.1	7.9	749.3	749.3
32	914	34.9	857.2	828.7	814.4	69.8	7.9	800.1	800.1
34	965	34.9	908.0	879.5	865.2	73.0	7.9	850.9	850.9
36	1033	36.5	965.2	935.0	916.0	85.7	9.5	901.7	901.7
38	1084	38.1	1016.0	985.8	966.8	88.9	9.5	952.5	952.5
40	1135	38.1	1066.8	1036.6	1017.6	92.1	9.5	1003.3	1003.3
42	1186	39.7	1117.6	1087.4	1068.4	95.2	9.5	1054.1	1054.1
44	1251	42.9	1174.8	1139.8	1119.2	104.8	9.5	1104.9	1104.9
46	1302	44.4	1225.6	1190.6	1170.0	108.0	9.5	1155.7	1155.7
48	1353	46.0	1276.4	1241.4	1220.8	111.1	9.5	1206.5	1206.5
50	1403	47.6	1327.2	1293.8	1271.6	115.9	9.5	1257.3	1257.3
52	1457	47.6	1378.0	1344.6	1322.4	120.7	9.5	1308.1	1308.1
54	1508	49.2	1428.8	1397.0	1373.2	125.4	9.5	1358.9	1358.9
56	1575	50.8	1485.9	1451.0	1424.0	134.9	11.1	1409.7	1409.7
58	1626	52.4	1536.7	1501.8	1474.8	138.1	11.1	1460.5	1460.5
60	1676	55.6	1587.5	1552.6	1525.6	144.5	11.1	1511.3	1511.3

Notes:

1. Dimensions in millimeters. For tolerances refer to Table 6.31M.
2. Maximum inside diameter is based on full pressure rating at 38°C and the gasket requirements of API 605. Purchaser must specify the required inside diameter based on the application and the governing code. Purchaser may specify a larger inside diameter than the maximum listed when the application pressure is lower or adjusted gasket dimensions are used.

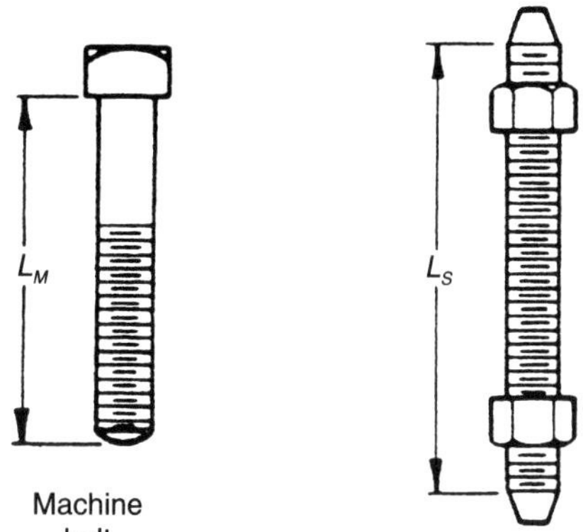

TABLE 6.20 Dimensions for Class 75 Welding Neck Flange Bolts (U.S. Customary Units)*

Bolt-circle diameter	Number of bolts	Bolt diameter	Bolt-hole diameter	Machine bolt length L_M	Stud bolt length L_S
28.50	36	5/8	0.75	4.00	4.50
30.50	40	5/8	0.75	4.00	4.50
32.50	44	5/8	0.75	4.00	4.50
34.50	48	5/8	0.75	4.25	4.75
36.50	52	5/8	0.75	4.25	4.75
39.06	40	3/4	0.88	4.50	5.00
41.06	40	3/4	0.88	4.50	5.25
43.06	44	3/4	0.88	4.50	5.25
45.06	48	3/4	0.88	4.75	5.25
47.38	36	7/8	1.00	5.25	5.75
49.38	40	7/8	1.00	5.25	6.00
51.38	44	7/8	1.00	5.50	6.00
53.38	44	7/8	1.00	5.50	6.25
55.50	48	7/8	1.00	5.75	6.25
57.50	48	7/8	1.00	5.75	6.25
59.88	40	1	1.12	6.00	6.75
61.88	44	1	1.12	6.00	6.75
63.88	44	1	1.12	6.25	7.00

* Dimensions in inches.

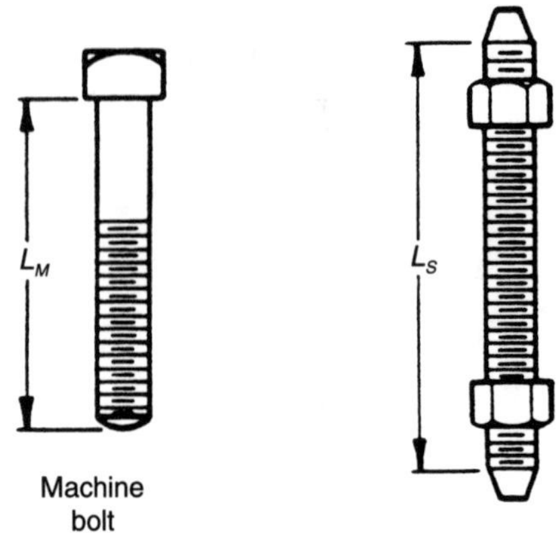

TABLE 6.20M Dimensions for Class 75 Welding Neck Flange Bolts (SI Units)*

Bolt-circle diameter	Number of bolts	Bolt diameter	Bolt-hole diameter	Machine bolt length L_M	Stud bolt length L_S
723.9	36	5/8	19.0	102	114
774.7	40	5/8	19.0	102	114
825.5	44	5/8	19.0	102	114
876.3	48	5/8	19.0	108	121
927.1	52	5/8	19.0	108	121
992.2	40	3/4	22.2	114	127
1043.0	40	3/4	22.2	114	133
1093.8	44	3/4	22.2	114	133
1144.6	48	3/4	22.2	121	133
1203.3	36	7/8	25.4	133	146
1254.1	40	7/8	25.4	133	152
1304.9	44	7/8	25.4	140	152
1355.7	44	7/8	25.4	140	159
1409.7	48	7/8	25.4	140	159
1460.5	48	7/8	25.4	146	159
1520.8	40	1	28.6	152	171
1571.6	44	1	28.6	152	171
1622.4	44	1	28.6	159	178

* Dimensions in millimeters.

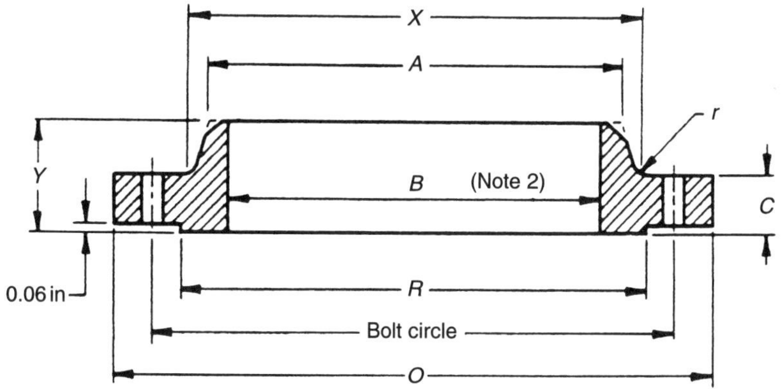

Welding neck flange

TABLE 6.21 Dimensions for Class 150 Welding Neck Flanges (U.S. Customary Units) (Note 1)

NPS	Flange outside diameter *O*	Flange thickness *C*	Raised face diameter *R*	Diameter at large end of hub, *X*	Diameter at small end of hub, *A*	Overall length *y*	Fillet radius *r*	Maximum inside diameter (2) ASME code *B*	Maximum inside diameter (2) ASME B31.3 *B*
26	30.94	1.62	28.00	26.94	26.06	3.50	0.38	25.44	25.50
28	32.94	1.75	30.00	28.94	28.06	3.75	0.38	27.44	27.50
30	34.94	1.75	32.00	31.00	30.06	3.94	0.38	29.44	29.50
32	37.06	1.81	34.00	33.06	32.06	4.25	0.38	31.44	31.50
34	39.56	1.94	36.25	35.12	34.06	4.34	0.38	33.38	33.50
36	41.62	2.06	38.25	37.19	36.06	4.62	0.38	35.38	35.50
38	44.25	2.12	40.25	39.25	38.12	4.88	0.38	37.38	37.50
40	46.25	2.19	42.50	41.31	40.12	5.06	0.38	39.31	39.50
42	48.25	2.31	44.50	43.38	42.12	5.25	0.44	41.31	41.50
44	50.25	2.38	46.50	45.38	44.12	5.38	0.44	43.25	43.50
46	52.81	2.44	48.62	47.44	46.12	5.69	0.44	45.19	45.50
48	54.81	2.56	50.75	49.50	48.12	5.88	0.44	47.19	47.44
50	56.81	2.69	52.75	51.50	50.12	6.06	0.44	49.12	49.44
52	58.81	2.75	54.75	53.56	52.12	6.19	0.44	51.12	51.44
54	61.00	2.81	56.75	55.62	54.12	6.38	0.44	53.06	53.31
56	63.00	2.88	58.75	57.69	56.12	6.56	0.56	55.06	55.38
58	65.94	2.94	60.75	59.69	58.12	6.88	0.56	57.06	57.38
60	67.94	3.00	63.00	61.81	60.12	7.06	0.56	59.00	59.25

Notes:
1. Dimensions in inches. For tolerances refer to Table 6.31.
2. Maximum inside diameter is based on full pressure rating at 100°F and the gasket requirements of API 605. Purchaser must specify the required inside diameter based on the application and the governing code. Purchaser may specify a larger inside diameter than the maximum listed when the application pressure is lower or adjusted gasket dimensions are used.

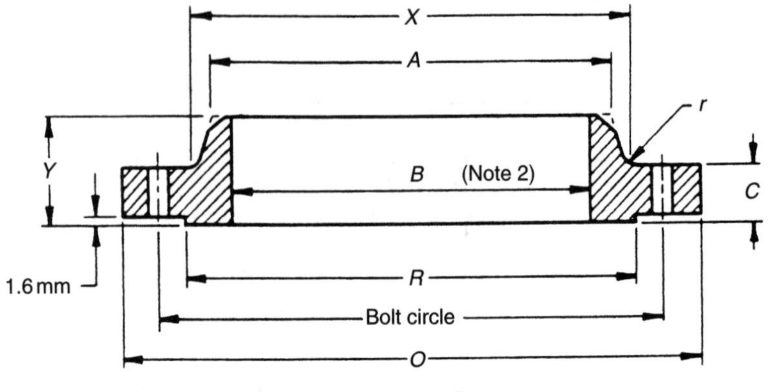

Welding neck flange

TABLE 6.21M Dimensions for Class 150 Welding Neck Flanges (SI Units) (Note 1)

NPS	Flange outside diameter *O*	Flange thickness *C*	Raised face diameter *R*	Diameter at large end of hub, *X*	Diameter at small end of hub, *A*	Overall length *y*	Fillet radius *r*	Maximum inside diameter (2) ASME code *B*	ASME B31.3 *B*
26	786	41.3	711.2	684.2	662.0	88.9	9.5	646.1	647.7
28	837	44.4	762.0	735.0	712.8	95.2	9.5	696.5	698.5
30	887	44.4	812.8	787.4	763.6	100.0	9.5	747.7	749.3
32	941	46.0	863.6	839.8	814.4	108.0	9.5	798.5	800.1
34	1005	49.2	920.8	892.2	865.2	110.3	9.5	847.7	850.9
36	1057	52.4	971.6	944.6	916.0	117.5	9.5	898.5	901.7
38	1124	54.0	1022.4	997.0	968.4	123.8	9.5	949.3	925.5
40	1175	55.6	1079.5	1049.3	1019.2	128.6	9.5	998.5	1003.3
42	1226	58.7	1130.3	1101.7	1070.0	133.4	11.1	1049.3	1054.1
44	1276	60.3	1181.1	1152.5	1120.8	136.5	11.1	1098.6	1104.9
46	1341	61.9	1235.1	1204.9	1171.6	144.5	11.1	1147.8	1155.7
48	1392	65.1	1289.0	1257.3	1222.4	149.2	11.1	1198.6	1204.9
50	1443	68.3	1339.8	1308.1	1273.2	154.0	11.1	1247.8	1255.7
52	1494	69.8	1390.6	1369.5	1324.0	157.2	11.1	1298.6	1306.5
54	1549	71.4	1441.4	1412.9	1374.8	161.9	11.1	1347.8	1354.1
56	1600	73.0	1492.3	1465.3	1425.6	166.7	14.3	1398.6	1406.5
58	1675	74.6	1543.0	1516.1	1476.4	174.6	14.3	1449.4	1457.3
60	1726	76.2	1600.2	1570.0	1527.2	179.4	14.3	1498.6	1505.0

Notes:

1. Dimensions in millimeters. For tolerances refer to Table 6.31M.
2. Maximum inside diameter is based on full pressure rating at 38°C and the gasket requirements of API 605. Purchaser must specify the required inside diameter based on the application and the governing code. Purchaser may specify a larger inside diameter than the maximum listed when the application pressure is lower or adjusted gasket dimensions are used.

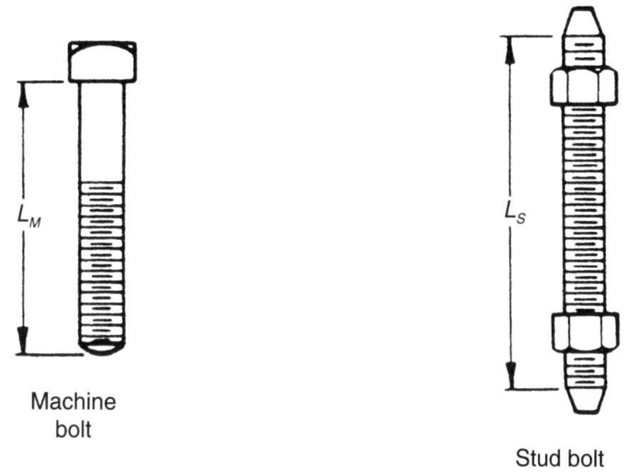

TABLE 6.22 Dimensions for Class 150 Welding Neck Flange Bolts (U.S. Customary Units)*

Bolt-circle diameter	Number of bolts	Bolt diameter	Bolt-hole diameter	Machine bolt length L_M	Stud bolt length L_S
29.31	36	3/4	0.88	4.75	5.50
31.31	40	3/4	0.88	5.00	5.75
33.31	44	3/4	0.88	5.00	5.75
35.44	48	3/4	0.88	5.25	5.75
37.69	40	7/8	1.00	5.75	6.25
39.75	44	7/8	1.00	6.00	6.50
42.12	40	1	1.12	6.25	7.00
44.12	44	1	1.12	6.50	7.00
46.12	48	1	1.12	6.50	7.25
48.12	52	1	1.12	6.75	7.50
50.56	40	1 1/8	1.25	7.00	8.00
52.56	44	1 1/8	1.25	7.25	8.00
54.56	48	1 1/8	1.25	7.50	8.25
56.56	52	1 1/8	1.25	7.75	8.50
58.75	56	1 1/8	1.25	7.75	8.50
60.75	60	1 1/8	1.25	8.00	8.75
63.44	48	1 1/4	1.38	8.25	9.00
65.44	52	1 1/4	1.38	8.25	9.25

* Dimensions in inches.

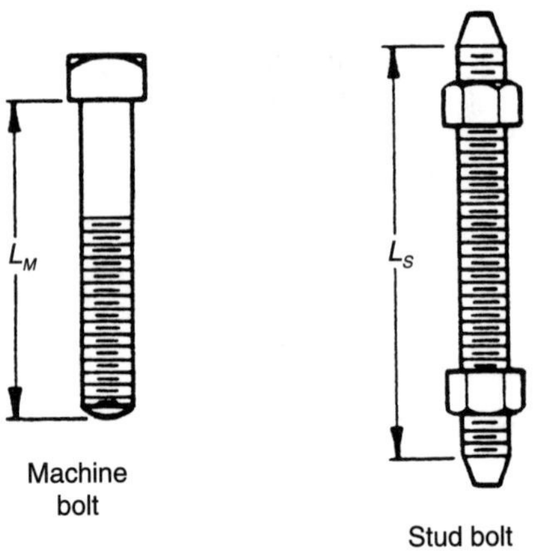

TABLE 6.22M Dimensions for Class 150 Welding Neck Flange Bolts (Metric Units)*

Bolt-circle diameter	Number of bolts	Bolt diameter	Bolt-hole diameter	Machine bolt length L_M	Stud bolt length L_S
744.5	36	3/4	22.2	121	140
795.3	40	3/4	22.2	127	146
846.1	44	3/4	22.2	127	146
900.1	48	3/4	22.2	133	146
957.3	40	7/8	25.4	146	159
1009.6	44	7/8	25.4	152	165
1070.0	40	1	28.6	159	178
1120.8	44	1	28.6	165	178
1171.6	48	1	28.6	165	184
1222.4	52	1	28.6	171	190
1284.3	40	1 1/8	31.8	178	203
1335.1	44	1 1/8	31.8	184	203
1385.9	48	1 1/8	31.8	190	210
1436.7	52	1 1/8	31.8	197	216
1492.2	56	1 1/8	31.8	197	216
1543.0	60	1 1/8	31.8	203	222
1611.3	48	1 1/4	34.9	210	229
1662.1	52	1 1/4	34.9	210	235

* Dimensions in millimeters.

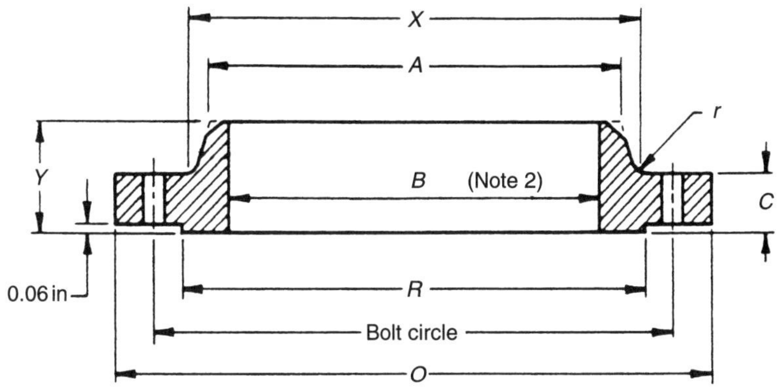

Welding neck flange

TABLE 6.23 Dimensions for Class 300 Welding Neck Flanges (U.S. Customary Units) (Note 1)

								Maximum inside diameter (2)		
NPS	Flange outside diameter *O*	Flange thickness *C*	Raised face diameter *R*	Diameter at large end of hub, *X*	Diameter at small end of hub, *A*	Overall length *y*	Fillet radius *r*	ASME code *B*	ASME B31.3 *B*	ASME B31.4/ B31.8 *B*
26	34.12	3.50	29.00	27.62	26.19	5.69	0.56	24.88	25.25	25.50
28	36.25	3.50	31.00	29.75	28.19	5.88	0.56	26.75	27.12	27.50
30	39.00	3.69	33.25	32.00	30.25	6.22	0.56	28.62	29.06	29.50
32	41.50	4.06	35.50	34.00	32.25	6.62	0.62	30.56	31.00	31.25
34	43.62	4.06	37.50	36.12	34.25	6.81	0.62	32.44	32.88	33.25
36	46.12	4.06	39.75	38.00	36.25	7.12	0.62	34.31	34.88	35.12
38	48.12	4.38	41.75	40.00	38.25	7.56	0.62	36.25	37.06	37.12
40	50.12	4.56	43.88	42.00	40.25	7.81	0.62	38.25	39.06	39.12
42	52.50	4.69	46.00	44.00	42.31	8.06	0.62	40.00	40.88	41.12
44	54.50	5.00	48.00	46.19	44.31	8.44	0.62	42.00	43.00	43.00
46	57.50	5.06	50.00	48.38	46.31	8.75	0.62	44.00	44.88	44.88
48	59.50	5.06	52.25	50.31	48.31	8.81	0.62	45.75	46.56	46.81
50	61.50	5.44	54.25	52.38	50.31	9.25	0.62	47.75	48.69	48.69
52	63.50	5.62	56.25	54.44	52.31	9.56	0.62	49.75	50.62	50.69
54	65.88	5.38	58.25	56.50	54.31	9.44	0.62	51.50	52.31	52.56
56	69.50	6.06	60.50	58.81	56.31	10.56	0.69	53.62	54.38	54.56
58	71.94	6.06	62.75	60.94	58.31	10.81	0.69	55.50	56.38	56.44
60	73.94	5.94	65.00	62.94	60.31	10.69	0.69	57.25	58.19	58.38

Notes:
1. Dimensions in inches. For tolerances refer to Table 6.31.
2. Maximum inside diameter is based on full pressure rating at 100°F and the gasket requirements of API 605. Purchaser must specify the required inside diameter based on the application and the governing code. Purchaser may specify a larger inside diameter than the maximum listed when the application pressure is lower or adjusted gasket dimensions are used.

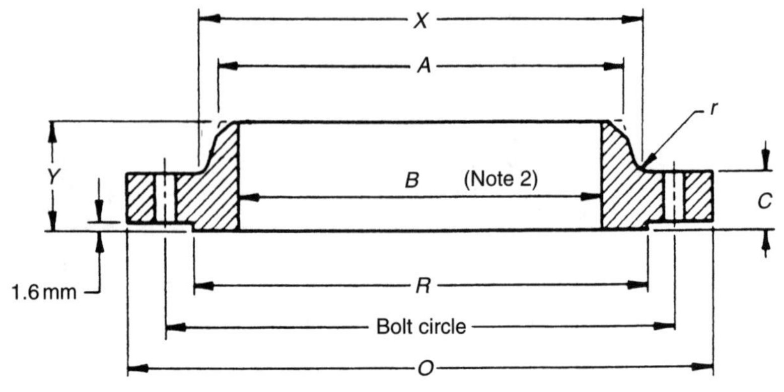

Welding neck flange

TABLE 6.23M Dimensions for Class 300 Welding Neck Flanges (SI Units) (Note 1)

NPS	Flange outside diameter *O*	Flange thickness *C*	Raised face diameter *R*	Diameter at large end of hub, *X*	Diameter at small end of hub, *A*	Overall length *y*	Fillet radius *r*	Maximum inside diameter (2) ASME code *B*	ASME B31.3 *B*	ASME B31.4/ B31.8 *B*
26	867	88.9	736.6	701.7	665.2	144.5	14.3	631.8	641.4	647.7
28	921	88.9	787.4	755.6	716.0	149.2	14.3	679.4	689.0	698.5
30	991	93.7	844.6	812.8	768.4	158.0	14.3	727.1	738.2	749.3
32	1054	103.2	901.7	863.6	819.2	168.3	15.9	776.3	787.4	793.7
34	1108	103.2	952.5	917.6	870.0	173.0	15.9	823.9	835.0	844.5
36	1172	103.2	1009.6	965.2	920.8	181.0	15.9	871.5	885.8	892.2
38	1222	111.1	1060.4	1016.0	971.6	192.1	15.9	920.8	941.4	943.0
40	1273	115.9	1114.4	1066.8	1022.4	198.4	15.9	971.6	992.2	993.8
42	1334	119.1	1168.4	1117.6	1074.7	204.8	15.9	1016.0	1038.2	1044.6
44	1384	127.0	1219.2	1173.2	1125.5	214.3	15.9	1066.8	1092.2	1092.2
46	1460	128.6	1270.0	1228.7	1176.3	222.2	15.9	1117.6	1139.8	1139.8
48	1511	128.6	1327.2	1277.9	1227.1	223.8	15.9	1162.0	1182.7	1189.0
50	1562	138.1	1378.0	1330.3	1277.9	235.0	15.9	1212.8	1236.7	1236.7
52	1613	142.9	1428.8	1382.7	1328.7	242.9	15.9	1263.6	1285.9	1287.5
54	1673	136.5	1479.6	1435.1	1379.5	239.7	15.9	1308.1	1328.7	1335.0
56	1765	154.0	1536.7	1493.8	1430.3	268.3	17.5	1362.1	1381.1	1385.8
58	1827	154.0	1593.8	1547.8	1481.1	274.6	17.5	1409.7	1431.9	1433.6
60	1878	150.8	1651.0	1598.6	1531.9	271.5	17.5	1454.2	1478.0	1482.8

Notes:

1. Dimensions in millimeters. For tolerances refer to Table 6.31M.
2. Maximum inside diameter is based on full pressure rating at 38°C and the gasket requirements of API 605. Purchaser must specify the required inside diameter based on the application and the governing code. Purchaser may specify a larger inside diameter than the maximum listed when the application pressure is lower or adjusted gasket dimensions are used.

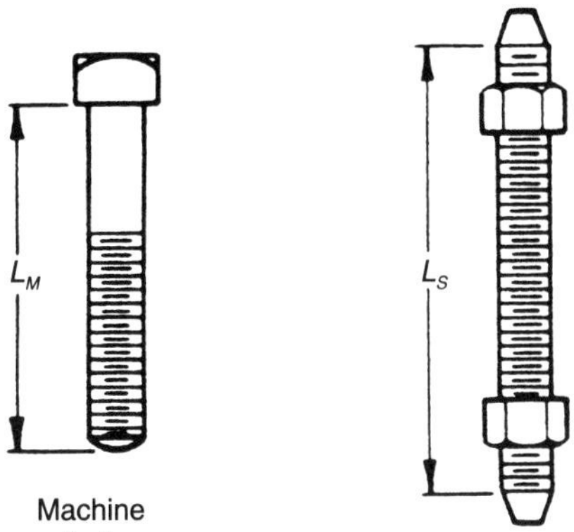

TABLE 6.24 Dimensions for Class 300 Welding Neck Flange Bolts (U.S. Customary Units)*

Bolt-circle diameter	Number of bolts	Bolt diameter	Bolt-hole diameter	Machine bolt length L_M	Stud bolt length L_S
31.62	32	1¼	1.38	9.25	10.25
33.75	36	1¼	1.38	9.25	10.25
36.25	36	1⅜	1.50	9.25	10.75
38.50	32	1½	1.62	10.75	11.75
40.62	36	1½	1.62	10.75	11.75
42.88	32	1⅝	1.75	10.75	12.00
44.88	36	1⅝	1.75	11.50	12.75
46.88	40	1⅝	1.75	11.75	13.00
49.00	36	1¾	1.88	12.25	13.50
51.00	40	1¾	1.88	12.75	14.25
53.75	36	1⅞	2.00	13.00	14.50
55.75	40	1⅞	2.00	13.00	14.50
57.75	44	1⅞	2.00	13.75	15.25
59.75	48	1⅞	2.00	14.25	15.75
62.12	48	1⅞	2.00	13.75	15.25
65.00	36	2¼	2.38	15.50	17.25
67.44	40	2¼	2.38	15.50	17.25
69.44	40	2¼	2.38	15.25	17.00

* Dimensions in inches.

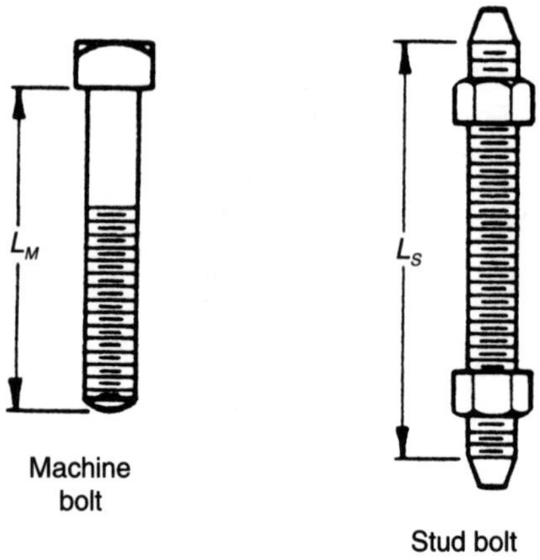

TABLE 6.24M Dimensions for Class 300 Welding Neck Flange Bolts (SI Units)*

Bolt-circle diameter	Number of bolts	Bolt diameter	Bolt-hole diameter	Machine bolt length L_M	Stud bolt length L_S
803.3	32	$1^1/_4$	34.9	235	260
857.2	36	$1^1/_4$	34.9	235	260
920.8	36	$1^3/_8$	38.1	248	273
977.9	32	$1^1/_2$	41.3	273	298
1031.9	36	$1^1/_2$	41.3	273	298
1089.0	32	$1^5/_8$	44.4	273	305
1139.8	36	$1^5/_8$	44.4	292	324
1190.6	40	$1^5/_8$	44.4	298	330
1244.6	36	$1^3/_4$	47.6	311	343
1295.4	40	$1^3/_4$	47.6	324	362
1365.2	36	$1^7/_8$	50.8	330	368
1416.0	40	$1^7/_8$	50.8	330	368
1466.8	44	$1^7/_8$	50.8	349	387
1517.6	48	$1^7/_8$	50.8	362	400
1578.0	48	$1^7/_8$	50.8	349	387
1651.0	36	$2^1/_4$	60.3	394	438
1712.9	40	$2^1/_4$	60.3	394	438
1763.7	40	$2^1/_4$	60.3	387	432

* Dimensions in millimeters.

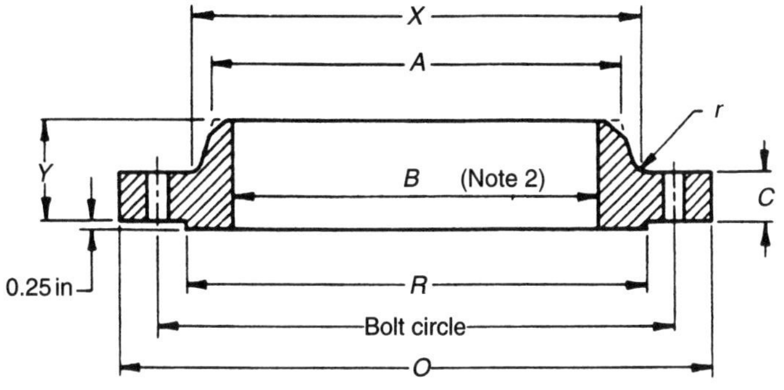

Welding neck flange

TABLE 6.25 Dimensions of Class 400 Welding Neck Flanges (U.S. Customary Units) (Note 1)

NPS	Flange outside diameter *O*	Flange thickness *C*	Raised face diameter *R*	Diameter at large end of hub, *X*	Diameter at small end of hub, *A*	Overall length *Y* (3)	Fillet radius *r*	Maximum inside diameter (2): ASME code *B*	ASME B31.3 *B*	ASME B31.4/B31.8 *B*
26	33.50	3.50	28.00	27.12	26.00	5.88	0.44	24.31	25.12	25.50
28	36.00	3.75	30.00	29.12	28.00	6.25	0.50	26.06	27.00	27.44
30	38.25	4.00	32.25	31.25	30.00	6.69	0.50	28.06	29.00	29.38
32	40.75	4.25	34.38	33.25	32.00	7.06	0.50	29.88	30.88	31.31
34	42.75	4.38	36.50	35.38	34.00	7.38	0.56	31.88	32.88	33.25
36	45.50	4.69	38.62	37.50	36.00	7.88	0.56	33.75	34.88	35.19
38	47.50	4.88	40.75	39.50	38.00	8.12	0.56	35.69	36.81	37.19
40	50.00	5.12	43.00	41.50	40.00	8.50	0.56	37.50	38.69	39.12
42	52.00	5.25	45.00	43.62	42.00	8.81	0.56	39.50	40.69	41.06
44	54.50	5.50	47.25	45.62	44.00	9.19	0.56	41.38	42.56	43.06
46	56.75	5.75	49.50	47.75	46.00	9.62	0.56	43.31	44.56	45.00
48	59.50	6.00	51.50	49.88	48.00	10.12	0.56	45.19	46.56	46.94
50	61.75	6.19	53.62	52.00	50.00	10.56	0.56	47.12	48.56	48.94
52	63.75	6.38	55.62	54.00	52.00	10.88	0.56	49.00	50.44	50.88
54	67.00	6.69	57.88	56.12	54.00	11.38	0.56	50.88	52.38	52.81
56	69.00	6.88	60.12	58.25	56.00	11.75	0.56	52.88	54.38	54.81
58	71.00	7.00	62.12	60.25	58.00	12.06	0.56	54.75	56.31	56.75
60	74.25	7.31	64.38	62.38	60.00	12.56	0.56	56.56	58.19	56.69

Notes:

1. Dimensions in inches. For tolerances refer to Table 6.31.
2. Maximum inside diameter is based on full pressure rating at 100°F and the gasket requirements of API 605. Purchaser must specify the required inside diameter based on the application and the governing code. Purchaser may specify a larger inside diameter than the maximum listed when the application pressure is lower or adjusted gasket dimensions are used.
3. Overall length does not include the height of the raised face for Class 400 through 900 flanges.

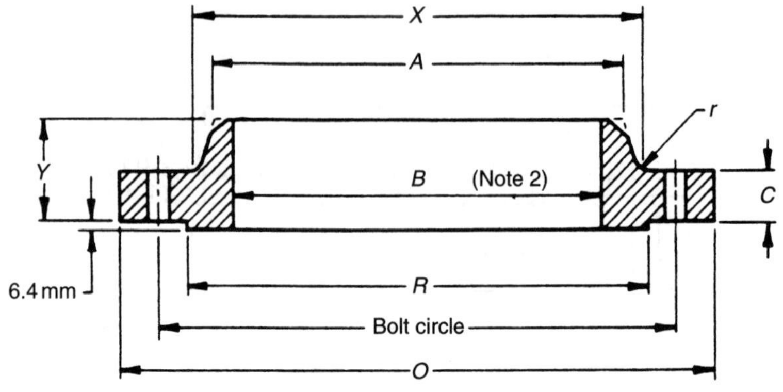

Welding neck flange

TABLE 6.25M Dimensions for Class 400 Welding Neck Flanges (SI Units) (Note 1)

NPS	Flange outside diameter *O*	Flange thickness *C*	Raised face diameter *R*	Diameter at large end of hub, *X*	Diameter at small end of hub, *A*	Overall length *Y* (3)	Fillet radius *r*	Maximum inside diameter (2) ASME code *B*	ASME B31.3 *B*	ASME B31.4/ B31.8 *B*
26	851	88.9	711.2	689.0	660.4	149.2	11.1	617.5	638.2	647.7
28	914	95.2	762.0	739.8	711.2	158.8	12.7	662.0	685.8	696.9
30	972	101.6	819.2	793.8	762.0	169.9	12.7	712.8	736.6	746.1
32	1035	108.0	873.1	844.6	812.8	179.4	12.7	758.8	784.2	795.3
34	1086	111.1	927.1	898.5	863.6	187.3	14.3	809.6	835.0	844.6
36	1156	119.1	981.1	952.5	914.4	200.0	14.3	857.2	885.8	893.8
38	1206	123.8	1035.0	1003.3	965.2	206.4	14.3	906.5	935.0	944.6
40	1270	130.2	1092.2	1054.1	1016.0	215.9	14.3	952.5	982.7	993.8
42	1321	133.4	1143.0	1108.1	1066.8	223.8	14.3	1003.3	1033.5	1043.0
44	1384	139.7	1200.2	1158.9	1117.6	233.4	14.3	1050.9	1081.1	1093.8
46	1441	146.0	1257.3	1212.8	1168.4	244.5	14.3	1100.1	1131.9	1143.0
48	1511	152.4	1308.1	1266.8	1219.2	257.2	14.3	1147.8	1182.7	1192.2
50	1568	157.2	1362.1	1320.8	1270.0	268.3	14.3	1197.0	1233.5	1243.0
52	1619	161.9	1412.9	1371.6	1320.8	276.2	14.3	1244.6	1281.1	1292.2
54	1702	169.7	1470.0	1425.6	1371.6	288.9	14.3	1292.2	1330.3	1341.4
56	1753	174.6	1527.2	1479.6	1422.4	298.4	14.3	1343.0	1381.1	1392.2
58	1803	177.8	1578.0	1530.4	1473.2	306.4	14.3	1390.6	1430.3	1441.4
60	1886	185.7	1635.1	1584.3	1524.0	319.1	14.3	1436.7	1478.0	1490.7

Notes:

1. Dimensions in millimeters. For tolerances refer to Table 6.31M.
2. Maximum inside diameter is based on full pressure rating at 38°C and the gasket requirements of API 605. Purchaser must specify the required inside diameter based on the application and the governing code. Purchaser may specify a larger inside diameter than the maximum listed when the application pressure is lower or adjusted gasket dimensions are used.
3. Overall length does not include the height of the raised face for Class 400 through 900 flanges.

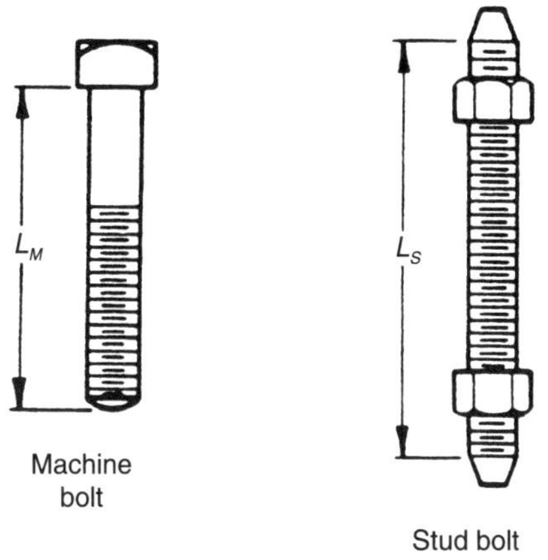

TABLE 6.26 Dimensions of Class 400 Welding Neck Flange Bolts (U.S. Customary Units)*

Bolt-circle diameter	Number of bolts	Bolt diameter	Bolt-hole diameter	Machine bolt length L_M	Stud bolt length L_S
30.75	28	1⅜	1.50	10.00	11.00
33.00	24	1½	1.62	10.50	11.75
35.25	28	1½	1.62	11.00	12.25
37.50	28	1⅝	1.75	11.75	13.00
39.50	32	1⅝	1.75	12.00	13.25
42.00	28	1¾	1.88	12.75	14.00
44.00	32	1¾	1.88	13.00	14.50
46.25	32	1⅞	2.00	13.75	15.25
48.25	32	1⅞	2.00	14.00	15.50
50.50	32	2	2.12	14.50	16.25
52.75	36	2	2.12	15.00	16.75
55.25	28	2¼	2.38	15.75	17.75
57.50	32	2¼	2.38	16.25	18.00
59.50	32	2¼	2.38	16.50	18.50
62.25	28	2½	2.62	17.50	19.75
64.25	32	2½	2.62	17.75	20.00
66.25	32	2½	2.62	18.00	20.25
69.00	32	2¾	2.88	19.00	21.50

* Dimensions in inches.

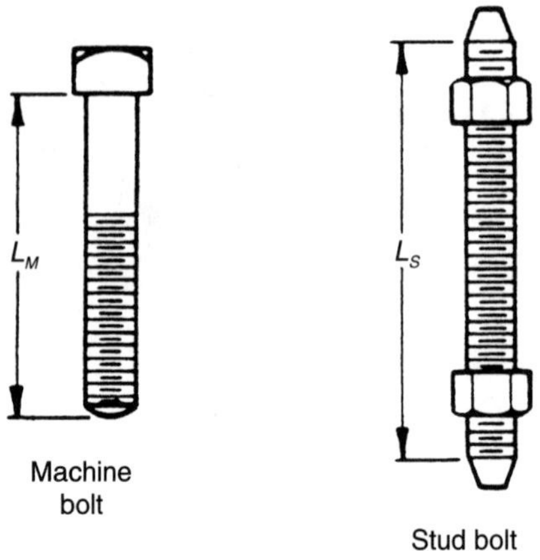

TABLE 6.26M Dimensions for Class 400 Welding Neck Flange Bolts (Metric Units)*

Bolt-circle diameter	Number of bolts	Bolt diameter	Bolt-hole diameter	Machine bolt length L_M	Stud bolt length L_S
781.0	28	$1\frac{3}{8}$	38.1	254	279
838.2	24	$1\frac{1}{2}$	41.3	267	298
895.4	28	$1\frac{1}{2}$	41.3	279	311
952.5	28	$1\frac{5}{8}$	44.4	298	330
1003.3	32	$1\frac{5}{8}$	44.4	305	337
1066.8	28	$1\frac{3}{4}$	47.6	324	356
1117.6	32	$1\frac{3}{4}$	47.6	330	368
1174.8	32	$1\frac{7}{8}$	50.8	349	387
1225.6	32	$1\frac{7}{8}$	50.8	356	394
1282.7	32	2	54.0	368	413
1339.8	36	2	54.0	381	425
1403.4	28	$2\frac{1}{4}$	60.3	400	451
1460.5	32	$2\frac{1}{4}$	60.3	413	457
1511.3	32	$2\frac{1}{4}$	60.3	419	470
1581.2	28	$2\frac{1}{2}$	66.7	444	502
1632.0	32	$2\frac{1}{2}$	66.7	451	508
1682.8	32	$2\frac{1}{2}$	66.7	457	514
1752.6	32	$2\frac{3}{4}$	73.0	483	546

* Dimensions in millimeters.

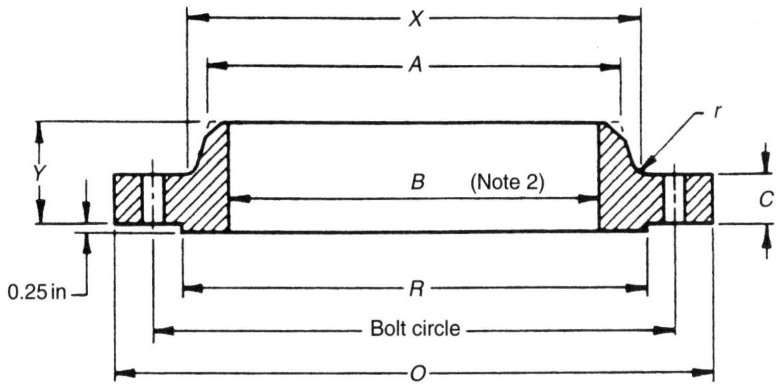

Welding neck flange

TABLE 6.27 Dimensions of Class 600 Welding Neck Flanges (U.S. Customary Units) (Note 1)

NPS	Flange outside diameter *O*	Flange thickness *C*	Raised face diameter *R*	Diameter at large end of hub, *X*	Diameter at small end of hub, *A*	Overall length *Y* (3)	Fillet radius *r*	Maximum inside diameter (2) ASME code *B*	ASME B31.3 *B*	ASME B31.4/ B31.8 *B*
26	35.00	4.38	28.62	27.50	26.00	7.12	0.50	23.56	24.88	25.12
28	37.50	4.56	30.88	29.62	28.00	7.50	0.50	25.44	26.81	27.06
30	40.25	4.94	33.12	31.75	30.00	8.06	0.50	27.31	28.81	29.00
32	42.75	5.12	35.25	33.88	32.00	8.50	0.50	29.19	30.69	30.94
34	45.75	5.56	37.50	36.00	34.00	9.19	0.56	31.00	32.62	32.88
36	47.75	5.75	39.75	38.12	36.00	9.56	0.56	33.00	34.62	34.81
38	50.00	6.00	41.50	40.25	38.00	10.00	0.56	34.81	36.62	36.75
40	52.00	6.25	43.75	42.25	40.00	10.38	0.56	36.69	38.38	38.69
42	55.25	6.62	46.00	44.38	42.00	11.00	0.56	38.50	40.25	40.62
44	57.25	6.81	48.25	46.50	44.00	11.38	0.56	40.44	42.25	42.56
46	59.50	7.06	50.25	48.62	46.00	11.81	0.56	42.31	44.19	44.50
48	62.75	7.44	52.50	50.75	48.00	12.44	0.56	44.12	46.12	46.44
50	65.75	7.75	54.50	52.88	50.00	12.94	0.56	46.00	48.06	48.38
52	67.75	8.00	56.50	54.88	52.00	13.25	0.56	47.88	49.88	50.31
54	70.00	8.25	58.75	57.00	54.00	13.75	0.56	49.69	51.81	52.25
56	73.00	8.56	60.75	59.12	56.00	14.25	0.62	51.50	53.69	54.19
58	75.00	8.75	63.00	61.12	58.00	14.56	0.62	53.31	55.56	56.12
60	78.50	9.19	65.25	63.38	60.00	15.31	0.69	55.25	57.56	58.06

Notes:
1. Dimensions in inches. For tolerances refer to Table 6.31.
2. Maximum inside diameter is based on full pressure rating at 100°F and the gasket requirements of API 605. Purchaser must specify the required inside diameter based on the application and the governing code. Purchaser may specify a larger inside diameter than the maximum listed when the application pressure is lower or adjusted gasket dimensions are used.
3. Overall length does not include the height of the raised face for Class 400 through 900 flanges.

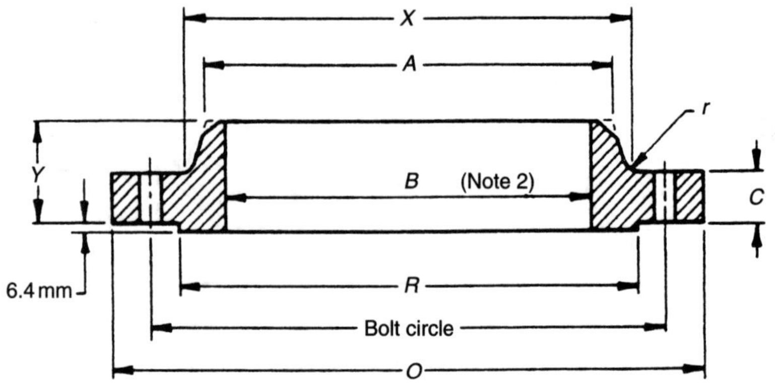

Welding neck flange

TABLE 6.27M Dimensions for Class 600 Welding Neck Flanges (Metric Units) (Note 1)

NPS	Flange outside diameter *O*	Flange thickness *C*	Raised face diameter *R*	Diameter at large end of hub, *X*	Diameter at small end of hub, *A*	Overall length *Y* (3)	Fillet radius *r*	Maximum inside diameter (2) ASME code *B*	ASME B31.3 *B*	ASME B31.4/ B31.8 *B*
26	889	111.1	727.1	698.5	660.4	181.0	12.7	598.5	631.8	638.2
28	952	115.9	784.2	752.5	711.2	190.5	12.7	646.1	681.0	687.4
30	1022	125.4	841.4	806.4	762.0	204.8	12.7	693.7	731.8	736.6
32	1086	130.2	895.4	860.4	812.8	215.9	12.7	741.4	779.5	785.8
34	1162	141.3	952.5	914.4	863.6	233.4	14.3	787.4	828.7	835.0
36	1213	146.0	1009.6	968.4	914.4	242.9	14.3	838.2	879.5	884.2
38	1270	152.4	1054.1	1022.4	965.2	254.0	14.3	884.2	930.3	933.4
40	1321	158.8	1111.2	1073.2	1016.0	263.5	14.3	931.9	974.7	982.7
42	1403	168.3	1168.4	1127.1	1066.8	279.4	14.3	977.9	1022.4	1031.9
44	1454	173.0	1225.6	1181.1	1117.6	288.9	14.3	1027.1	1073.2	1081.9
46	1511	179.4	1276.4	1235.1	1168.4	300.0	14.3	1074.7	1122.4	1130.3
48	1594	188.9	1333.5	1289.0	1219.2	315.9	14.3	1120.8	1171.6	1179.5
50	1670	196.8	1384.3	1343.0	1270.0	328.6	14.3	1168.4	1220.8	1228.7
52	1721	203.2	1435.1	1393.8	1320.8	336.6	14.3	1216.0	1266.8	1277.9
54	1778	209.6	1492.2	1447.8	1371.6	349.2	14.3	1262.1	1316.0	1327.2
56	1854	217.5	1543.0	1501.8	1422.4	362.0	15.9	1308.1	1363.7	1376.4
58	1905	222.2	1600.2	1552.6	1413.2	369.9	15.9	1354.1	1411.3	1425.6
60	1994	233.4	1657.4	1609.7	1524.0	388.9	17.5	1403.4	1462.1	1474.8

Notes:
1. Dimensions in millimeters. For tolerances refer to Table 6.31M.
2. Maximum inside diameter is based on full pressure rating at 38°C and the gasket requirements of API 605. Purchaser must specify the required inside diameter based on the application and the governing code. Purchaser may specify a larger inside diameter than the maximum listed when the application pressure is lower or adjusted gasket dimensions are used.
3. Overall length does not include the height of the raised face for Class 400 through 900 flanges.

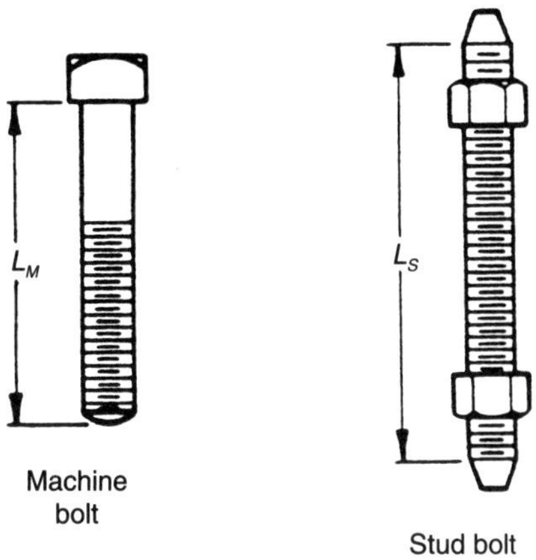

TABLE 6.28 Dimensions of Class 600 Welding Neck Flange Bolts (U.S. Customary Units)*

Bolt-circle diameter	Number of bolts	Bolt diameter	Bolt-hole diameter	Machine bolt length L_M	Stud bolt length L_S
31.75	28	$1^5/_8$	1.75	12.00	13.25
34.00	28	$1^3/_4$	1.88	12.50	13.75
36.50	28	$1^7/_8$	2.00	13.25	14.75
38.75	28	2	2.12	13.75	16.50
41.50	24	$2^1/_4$	2.38	15.00	16.75
43.50	28	$2^1/_4$	2.38	15.25	17.25
45.75	28	$2^1/_4$	2.38	15.75	17.75
47.75	32	$2^1/_4$	2.38	16.25	18.25
50.50	28	$2^1/_2$	2.62	17.25	19.50
52.50	32	$2^1/_2$	2.62	17.75	20.00
54.75	32	$2^1/_2$	2.62	18.25	20.50
57.50	32	$2^3/_4$	2.88	19.25	21.75
60.00	28	3	3.12	20.00	22.75
62.00	32	3	3.12	20.50	23.25
64.25	32	3	3.12	21.00	23.75
66.75	32	$3^1/_4$	3.38	22.00	25.00
68.75	32	$3^1/_4$	3.38	22.25	25.25
71.75	28	$3^1/_2$	3.62	23.50	26.75

* Dimensions in inches.

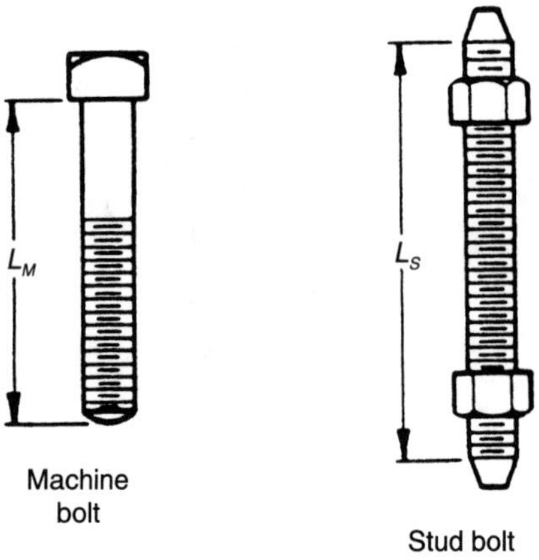

TABLE 6.28M Dimensions for Class 600 Welding Neck Flange Bolts (SI Units)*

Bolt-circle diameter	Number of bolts	Bolt diameter (inches)	Bolt-hole diameter	Machine bolt length L_M	Stud bolt length L_S
806.4	28	$1^5/_8$	44.4	305	337
863.6	28	$1^3/_4$	47.6	318	349
927.1	28	$1^7/_8$	50.8	337	375
984.2	28	2	54.0	349	394
1054.1	24	$2^1/_4$	60.3	381	425
1104.9	28	$2^1/_4$	60.3	387	438
1162.0	28	$2^1/_4$	60.3	400	451
1212.8	32	$2^1/_4$	60.3	413	464
1282.7	28	$2^1/_2$	66.7	438	495
1333.5	32	$2^1/_2$	66.7	451	508
1390.6	32	$2^1/_2$	66.7	464	521
1460.5	32	$2^3/_4$	73.0	489	552
1524.0	28	3	79.4	508	578
1574.8	32	3	79.4	521	591
1632.0	32	3	79.4	533	603
1691.4	32	$3^1/_4$	85.7	559	635
1746.2	32	$3^1/_4$	85.7	565	641
1822.4	28	$3^1/_2$	92.1	597	679

* Dimensions in millimeters.

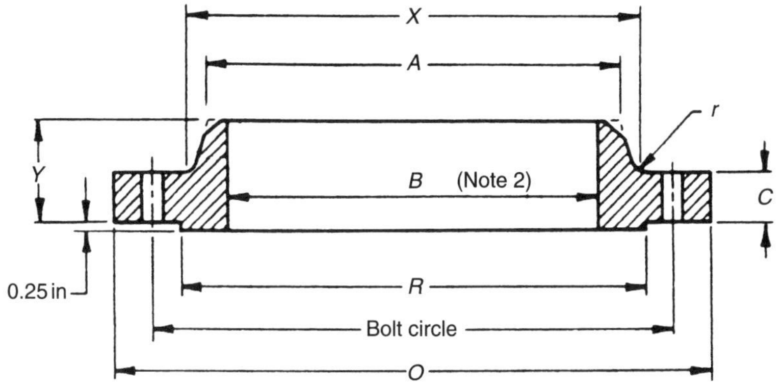

Welding neck flange

TABLE 6.29 Dimensions for Class 900 Welding Neck Flanges (U.S. Customary Units) (Note 1)

NPS	Flange outside diameter *O*	Flange thickness *C*	Raised face diameter *R*	Diameter at large end of hub, *X*	Diameter at small end of hub, *A*	Overall length *Y* (3)	Fillet Radius *r*	Maximum inside diameter (2) ASME B31.3 *B*	Maximum inside diameter (2) ASME B31.4/ B31.8 *B*
26	40.25	5.31	30.00	29.25	26.00	10.19	0.44	24.38	24.75
28	43.50	5.81	32.25	31.38	28.00	10.88	0.50	26.25	26.62
30	46.50	6.12	34.50	33.50	30.00	11.38	0.50	28.06	28.56
32	48.75	6.31	36.50	35.75	32.00	11.94	0.50	30.00	30.44
34	51.75	6.75	39.00	37.88	34.00	12.56	0.56	31.94	32.38
36	53.00	6.81	40.50	40.00	36.00	12.81	0.56	33.81	34.25
38	57.50	7.50	43.25	42.25	38.00	13.88	0.75	35.62	36.19
40	59.50	7.75	45.75	44.38	40.00	14.31	0.81	37.50	38.06
42	61.50	8.12	47.75	46.31	42.00	14.62	0.81	39.44	40.00
44	64.88	8.44	50.00	48.62	44.00	15.38	0.88	41.25	41.88
46	68.25	8.88	52.50	50.88	46.00	16.19	0.88	43.12	43.81
48	70.25	9.19	54.50	52.88	48.00	16.50	0.94	45.06	45.69

Notes:

1. Dimensions in inches. For tolerances refer to Table 6.31.
2. Maximum inside diameter is based on full pressure rating at 100°F and the gasket requirements of API 605. Purchaser must specify the required inside diameter based on the application and the governing code. Purchaser may specify a larger inside diameter than the maximum listed when the application pressure is lower or adjusted gasket dimensions are used.
3. Overall length does not include the height of the raised face for Class 400 through 900 flanges.

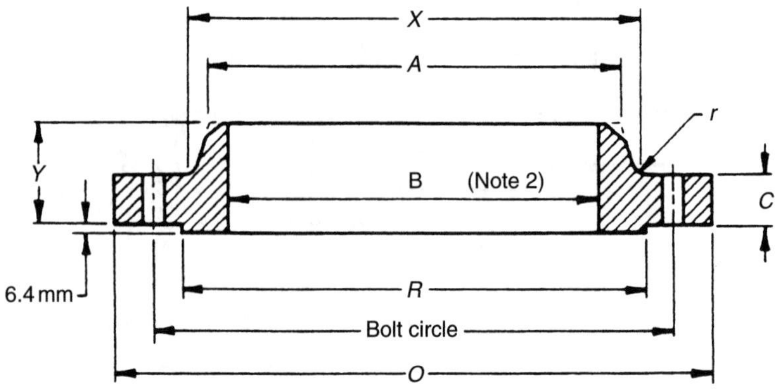

Welding neck flange

TABLE 6.29M Dimensions for Class 900 Welding Neck Flanges (SI Units) (Note 1)

NPS	Flange outside diameter *O*	Flange thickness *C*	Raised face diameter *R*	Diameter at large end of hub, *X*	Diameter at small end of hub, *A*	Overall length *Y* (3)	Fillet Radius *r*	Maximum inside diameter (2) ASME B31.3 *B*	Maximum inside diameter (2) ASME B31.4/ B31.8 *B*
26	1022	134.9	762.0	743.0	660.4	258.8	11.1	619.1	628.6
28	1105	147.6	819.2	796.9	711.2	276.2	12.7	666.8	676.3
30	1181	155.6	876.3	850.9	762.0	288.9	12.7	712.8	725.5
32	1238	160.3	927.1	908.0	812.8	303.2	12.7	762.0	773.1
34	1314	171.4	990.6	962.0	863.6	319.1	14.3	811.2	822.3
36	1346	173.0	1028.7	1016.0	914.4	325.4	14.3	858.8	870.0
38	1460	190.5	1099.0	1073.2	965.2	352.4	19.0	904.9	919.2
40	1511	196.8	1162.0	1127.1	1016.0	363.5	20.6	952.5	966.8
42	1562	206.4	1212.8	1176.3	1066.8	371.5	20.6	1001.7	1016.0
44	1648	214.3	1270.0	1235.1	1117.6	390.5	22.2	1047.8	1063.6
46	1734	225.4	1333.5	1292.2	1168.4	411.2	22.2	1095.4	1112.8
48	1784	233.4	1384.3	1343.0	1219.2	419.7	23.8	1144.6	1160.5

Notes:

1. Dimensions in millimeters. For tolerances refer to Table 6.31M.
2. Maximum inside diameter is based on full pressure rating at 38°C and the gasket requirements of API 605. Purchaser must specify the required inside diameter based on the application and the governing code. Purchaser may specify a larger inside diameter than the maximum listed when the application pressure is lower or adjusted gasket dimensions are used.
3. Overall length does not include the height of the raised face for Class 400 through 900 flanges.

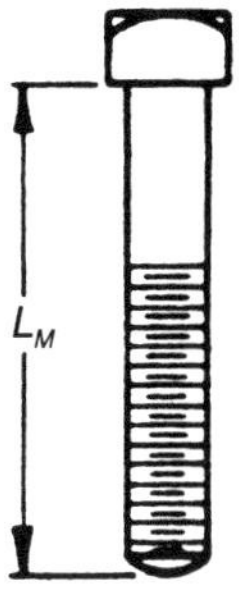

Machine bolt

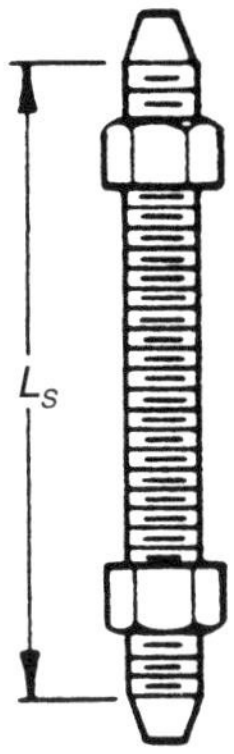

Stud bolt

TABLE 6.30 Dimensions for Class 900 Welding Neck Flange Bolts (U.S. Customary Units)*

Bolt-circle diameter	Number of bolts	Bolt diameter	Bolt-hole diameter	Machine bolt length L_M	Stud bolt length L_S
35.50	20	$2^1/_2$	2.62	14.75	16.75
38.25	20	$2^3/_4$	2.88	16.00	18.50
40.75	20	3	3.12	16.75	19.50
43.00	20	3	3.12	17.25	20.00
45.50	20	$3^1/_4$	3.38	18.25	21.25
47.25	24	3	3.12	18.25	21.00
50.75	20	$3^1/_2$	3.62	20.00	23.25
52.75	24	$3^1/_2$	3.62	20.50	23.75
54.75	24	$3^1/_2$	3.62	21.25	24.50
57.62	24	$3^3/_4$	3.88	22.25	25.75
60.50	24	4	4.12	23.25	27.00
62.50	24	4	4.12	24.00	27.75

* Dimensions in inches.

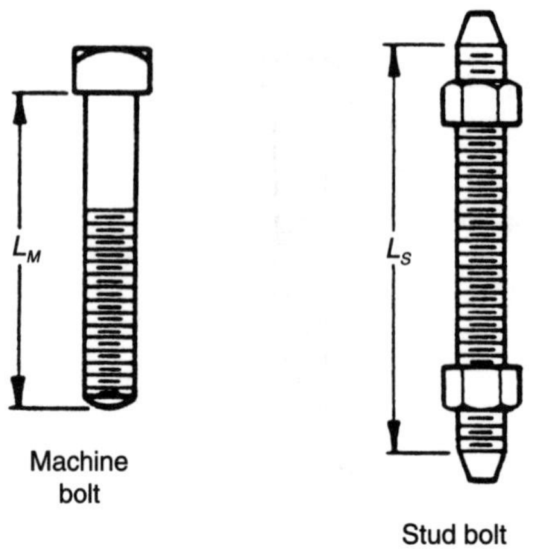

TABLE 6.30M Dimensions for Class 900 Welding Neck Flange Bolts (SI Units)*

Bolt-circle diameter	Number of bolts	Bolt diameter (inches)	Bolt-hole diameter	Machine bolt length L_M	Stud bolt length L_S
901.7	20	$2^1/_2$	66.7	375	425
971.6	20	$2^3/_4$	73.0	406	470
1035.0	20	3	79.4	425	495
1092.2	20	3	79.4	438	508
1155.7	20	$3^1/_4$	85.7	464	540
1200.2	24	3	79.4	464	533
1289.0	20	$3^1/_2$	92.1	508	591
1339.8	24	$3^1/_2$	92.1	521	603
1390.6	24	$3^1/_2$	92.1	540	622
1463.7	24	$3^3/_4$	98.4	565	654
1536.7	24	4	104.8	591	686
1587.5	24	4	104.8	610	705

* Dimensions in millimeters.

Tolerances

Tables 6.31 and 6.31M list the acceptable tolerances for flange dimensions in U.S. customary and SI units.

TABLE 6.31 Dimensional Tolerances for Flanges (U.S. Customary Units)*

Dimension	Tolerance, in
Outside diameter of raised face (*R*)	±0.03
Flange thickness (*C*)	+0.19, –0
Overall hub length (*Y*)	±0.12
Outside diameter of hub (*A*)	+0.16, –0.03
Diameter of bolt circle	±0.06
Center to center of adjacent bolt holes	±0.03
Inside diameter of welding neck flange (*B*)	+0.12, –0.06

* Bearing surfaces for bolting shall be parallel to the flange face within 1°.

TABLE 6.31M Dimensional Tolerances for Flanges (Metric Units)*

Dimension	Tolerance, mm
Outside diameter of raised face (*R*)	±0.8
Flange thickness (*C*)	+4.8, –0
Overall hub length (*Y*)	±3.2
Outside diameter of hub (*A*)	+4.0, –0.8
Diameter of bolt circle	±1.6
Center to center of adjacent bolt holes	±0.8
Inside diameter of welding neck flange (*B*)	+3.2, –1.6

* Bearing surfaces for bolting shall be parallel to the flange face within 1°.

Fittings for API Line Pipe

There are no API standards for fittings. Use fittings in accordance with the ASME B31.3, B31.4, and B31.8 codes and the fitting standards referenced therein.

Piping Joints

Use piping joints in accordance with the ASME B31.3, B31.4, and B31.8 sections of the ASME Pressure Piping Code. Refer to Chap. 15 for metallic piping joints and Sec. 5 for nonmetallic piping joints.

Tables in Chapter 6

Tables in Chapter 6 (Continued)

CHAPTER 7

AWWA PIPE, FITTINGS, FLANGES, AND JOINTS

See p. 344 for a list of tables in Chap. 7.

AWWA STEEL WATER PIPE

ANSI/AWWA C200, Steel Water Pipe—6 in (150 mm) and Larger, covers the following types of pipe that can be used for transmission and distribution of water:

- Butt-welded straight-seam pipe
- Spiral-seam welded pipe
- Seamless pipe

MATERIALS FOR ANSI/AWWA C200 PIPE

ASTM Pipe Material Specifications

Pipe can be made per the following ASTM pipe material specifications:

- ASTM A53, Type E or S
- ASTM A134, using the steels listed in Table 7.1
- ASTM A135, all grades listed in the specification
- ASTM A139/A139M, all grades listed in the specification

Even though the pipe can be made per the above listed ASTM material specifications, the pipe must meet the quality control requirements of ANSI/AWWA C200.

Alternative Materials

ANSI/AWWA C200 pipe may be fabricated from plates, sheets, or coils as listed in Table 7.1.

Other Pipe as ANSI/AWWA C200 Pipe

API 5L line pipe grades fully meet the requirements of ANSI/AWWA C200; as such, they may be used in place of C200 pipe.

Pipe Ends

ANSI/AWWA C200 pipe may be furnished with any of the following pipe ends, as specified by the purchaser:

- Plain, grooved, or banded ends for mechanically coupled field joints
- Ends for lap joints for field welding
- Plain ends
- Beveled ends for field butt welding

TABLE 7.1 Material Specifications for Steel Plates, Sheets, and Coils for AWWA C200 Fabricated Pipe

ASTM specification	Sheet or coil grade	Steel plate grade	Minimum yield strength, psi	Minimum yield strength, Mpa
A570/A570M	30		30,000	207
A570/A570M	33		33,000	228
A570/A570M	36		36,000	248
A570/A570M	40		40,000	276
A570/A570M	45		45,000	310
A570/A570M	50		50,000	345
A36/A36M	NA		36,000	248
A283/A283M		C	30,000	207
A283/A283M		D	33,000	228
A572/A572M		42	42,000	290
A572/A572M		50	50,000	345
A572/A572M		60	60,000	414

NA = not applicable.

- Ends fitted with butt straps for field welding
- Bell-and-spigot ends for rubber gasket joints
- Special ends, as agreed between the purchaser and the manufacturer.

Pipe Dimensions

Like any commercial pipe, the ANSI/AWWA C200 pipe is specified by nominal diameter or size and the nominal wall thickness.

Pipe manufactured to ASTM material specifications may be ordered by the sizes covered by those specifications. Refer to ASTM specifications and Table 5.1 and Table 5.1M of Chap. 5 of this book.

Pipe fabricated from plates, sheets or coils listed in Table 7.1 may be ordered by the nominal diameter and the nominal thickness required by the application design criteria and meeting the design requirements of AWWA M11.

Permissible Variations

Tolerances for pipe manufactured using ASTM pipe material specifications shall be those stated in the ASTM specification or ASTM 530.

Pipe Wall Thickness. The nominal wall thickness of the pipe may vary within the tolerances applicable to the plate, sheet, or coil from which the pipe is to be fabricated/manufactured.

Pipe Weight. Pipe weight may vary in accordance with the weight tolerances applicable to the plate, sheet, or coil from which pipe is to be manufactured.

Circumference. May vary by not more than plus or minus 1 percent, but not to exceed ¾ in (19 mm) from the nominal outside circumference based on the diameter specified.

Straightness. Straight pipe shall not deviate by more than ⅛ in (3.2 mm) from a 10-ft- (3-m-) long straight-edge held against the pipe.

Length. The specified length shall not vary by more than ±2 in (51 mm). This tolerance does not apply to the shorter lengths from which test coupons have been cut.

Random lengths shall have an average length of 29 ft (8.84 m), in cases where the minimum length is not less than 20 ft (6.10 m). In no case shall 5 percent of the random lengths be less than 25 ft (7.62 m).

By mutual agreement between the purchaser and the manufacturer pipe lengths may contain girth welds that have met the testing requirements of ANSI/AWWA C200.

Design of Pipe

AWWA Manual M11, *Steel Pipe—A Guide for Design and Installation*, is used for determining the wall thickness of the pipe. The pipe wall thickness must meet the loadings resulting from the internal pressure and the external loads.

AWWA STEEL PIPE FITTING DIMENSIONS

ANSI/AWWA C208 specifies dimensions for fabricated steel fittings for steel water pipe conforming to ANSI/AWWA C200 pipe for water service. It covers dimensions for the following fittings:

- Tees and crosses
- Reducing tees and crosses
- Laterals
- Wyes
- Reducers
- Two- to five-piece elbows (miters)

Design of Fittings

The design of fittings should be made in accordance with AWWA Manual M11—*Steel Water Pipe—A Guide for Design and Installation.*

The fittings conforming to the following ASME and ASTM standards are acceptable subject to the requirements of ANSI/AWWA C208.

- ASME B16.9, Factory-Made Wrought Buttwelding Fittings
- ASTM A234, Standard Specification for Piping Fittings of Wrought Carbon Steel and Alloy Steel for Moderate and Elevated Temperatures

Dimensions of Tees and Crosses

Table 7.2 lists the dimensions of straight tees and crosses in U.S. Customary units. Table 7.2M provides metric dimensions.

Dimensions of Reducing Tees

Depending on the diameter of the reducing branch, the center to end dimension B can be calculated by the following equation. See Fig. 7.1.

$$B = 0.5D_0 + f$$

where B = center to end dimension, in or mm
D_0 = diameter of reducing branch, in or mm
f = factor listed in Tables 7.2 and 7.2M

The A dimension shall be the same as given in Tables 7.2 and 7.2M.

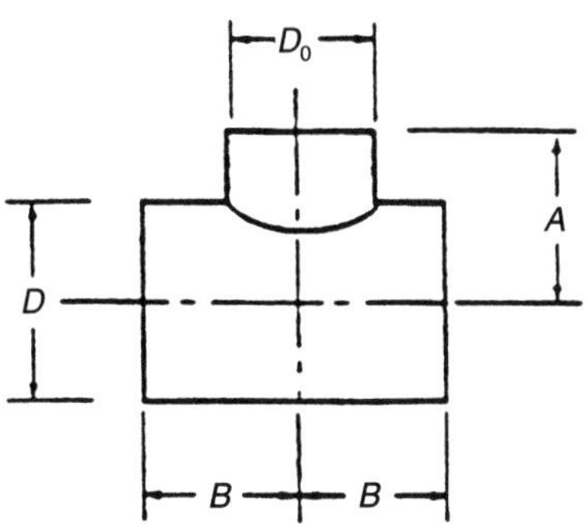

FIGURE 7.1 Reducing tee.

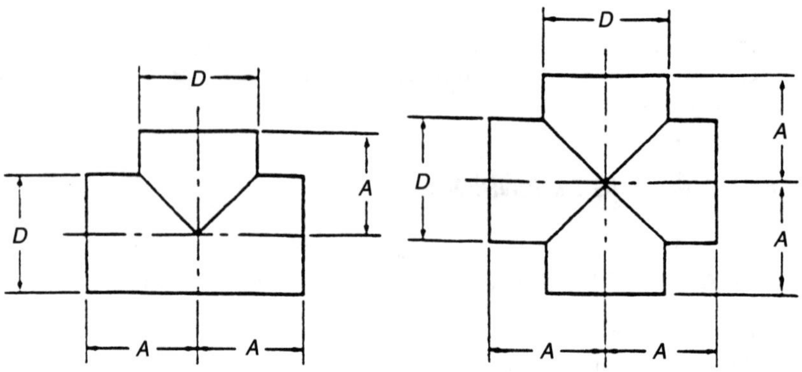

TABLE 7.2 Dimensions of Steel Water Pipe Tees and Crosses (U.S. Customary Units)*

Nominal diameter D	Factor f	Center to end dimension, $A = 0.5D + f$
6	5	8
8	5	9
10	6	11
12	6	12
14	7	14
16	7	15
18	7.5	16
20	8	18
22	9	20
24	10	22
30	10	25
36	11	29
42	12	33
48	12	36
54	14	41
60	14	45
66	15	48
72	15	51
78	16	55
84	17	59
90	17	62
96	18	66
102	19	70
108	20	74
114	21	78
120	22	82
126	23	86
132	24	90
138	25	94
144	26	98

* Dimensions in inches.

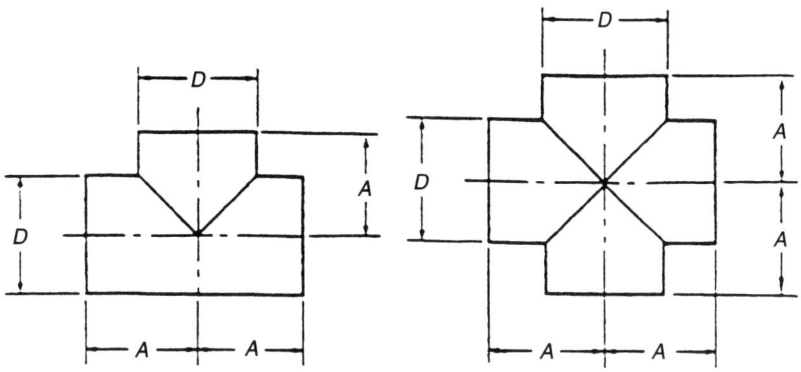

TABLE 7.2M Dimensions of Steel Water Pipe Tees and Crosses (Metric Units)*

Nominal diameter D	Factor f	Center to end dimension, $A = 0.5D + f$
150	5	203
200	5	229
250	6	279
300	6	305
350	7	356
400	7	381
450	7.5	406
500	8	457
550	9	508
600	10	559
750	10	635
900	11	737
1050	12	838
1200	12	914
1350	14	1041
1500	14	1143
1650	15	1219
1800	15	1295
1950	16	1397
2100	17	1499
2250	17	1575
2400	18	1676
2550	19	1778
2700	20	1880
2850	21	1981
3000	22	2083
3150	23	2184
3300	24	2286
3450	25	2388
3600	26	2489

* Dimensions in millimeters.

Dimensions of 90° Wyes

Dimensions of 90° wyes are listed in Tables 7.3 and 7.3M in U.S. Customary Units and metric units, respectively.

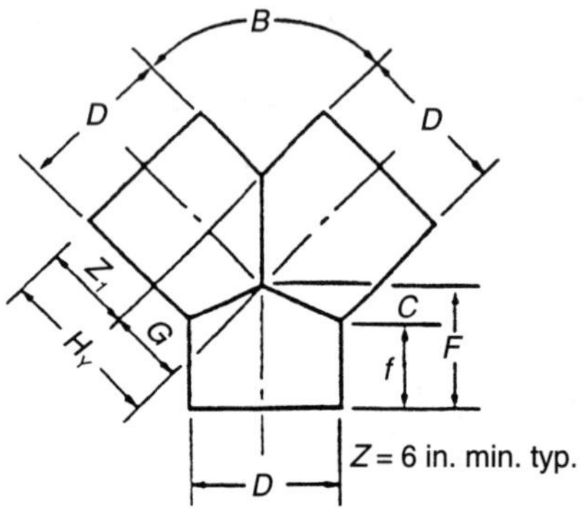

TABLE 7.3 Dimensions of Steel Water Pipe 90° Wye (U.S. Customary Units)*

Nominal diameter D	Factor f	Center to end dimension,† $F = C + f$
6	5	6
8	5	7
10	6	8
12	6	9
14	7	10
16	7	10
18	7.5	11
20	8	12
22	9	14
24	10	15
30	10	16
36	11	19
42	12	21
48	12	22
54	14	25
60	14	26
66	15	29
72	15	30
78	16	32
84	17	34
90	17	36
96	18	38
102	19	40
108	20	42
114	21	45
120	22	47
126	23	49
132	24	51
138	25	54
144	26	56

* Dimensions in inches.

† $C = [0.5D/\cos(\theta/2) - 0.5D]/\tan(\theta/2)$, where θ = angle of the fitting. For other angles between 90 and 30°, C may be calculated. Then F can be calculated.

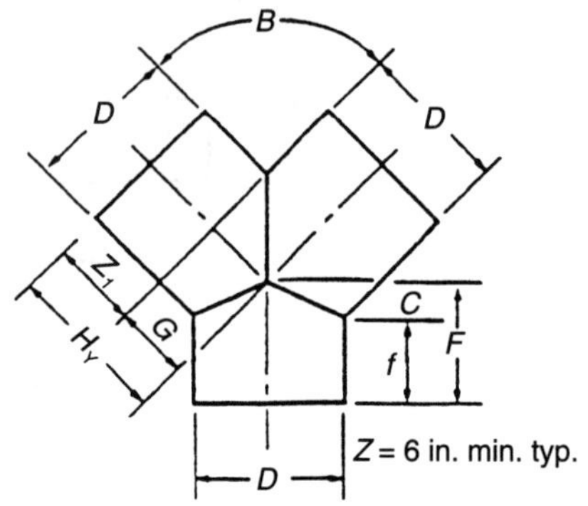

TABLE 7.3M Dimensions of Steel Water Pipe 90° Wye (Metric Units)*

Nominal diameter D	Factor f	Center to end dimension,† $F = C + f$
150	5	152
200	5	178
250	6	203
300	6	229
350	7	254
400	7	254
450	7.5	279
500	8	305
550	9	356
600	10	381
750	10	406
900	11	483
1050	12	533
1200	12	559
1350	14	635
1500	14	660
1650	15	737
1800	15	762
1950	16	813
2100	17	864
2250	17	914
2400	18	965
2550	19	1016
2700	20	1067
2850	21	1143
3000	22	1194
3150	23	1245
3300	24	1295
3450	25	1372
3600	26	1422

* Dimensions in millimeters.

† $C = [0.5D/\cos(\theta/2) - 0.5D]/\tan(\theta/2)$, where θ = angle of the fitting. For other angles between 90 and 30°, C may be calculated. Then F can be calculated.

Dimensions of Laterals

Tables 7.4 and 7.4M list the dimensions of laterals having a minimum theta (θ) angle of 30°. The dimensions listed are good for 30 to 70° laterals. The illustrations show equal and unequal diameter laterals.

For angles greater than 70°, use dimensions given for tees in Tables 7.2 and 7.2M. For angles less than 30°, use dimensions of a 30° lateral wye plus an elbow to fit.

Dimensions of Reducers

The length of an eccentric or a centric reducer can be calculated by the following equation, depending upon the diameters of the large and small ends. See Fig. 7.2.

$$L_r = 4(D_2 - D_1)$$

where L_r = length of reducer, in or mm
D_2 = diameter of large end, in or mm
D_1 = diameter of small end, in or mm

Dimensions of Elbows

The dimensions of fabricated elbows or miters would vary with the size of connecting pipe, angle of elbow, and the number of pieces to be joined together. Users should refer to ANSI/AWWA C208 and AWWA Manual M11.

AWWA STEEL PIPE FLANGES

ANSI/AWWA C207, AWWA Standard for Steel Pipe Flanges for Waterworks Service—Size 4in through 144in (100mm through 3, 600mm), covers two types of slip-on, ring and hub types, and blind flanges. See Fig. 7.3.

Ratings of Steel-Ring-Type Flanges

Tables 7.5 and 7.5M summarize the working pressure ratings of the different classes of steel-ring-type slip-on flanges depending on size, in U.S. Customary and metric units, respectively.

Ratings of Steel-Hub-Type Flanges

Tables 7.6 and 7.6M summarize the ratings of AWWA steel-hub-type slip-on flanges in U.S. Customary and metric units, respectively.

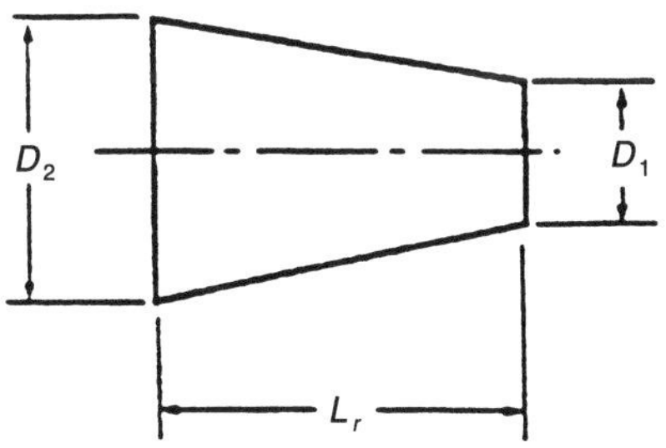

FIGURE 7.2 Reducer.

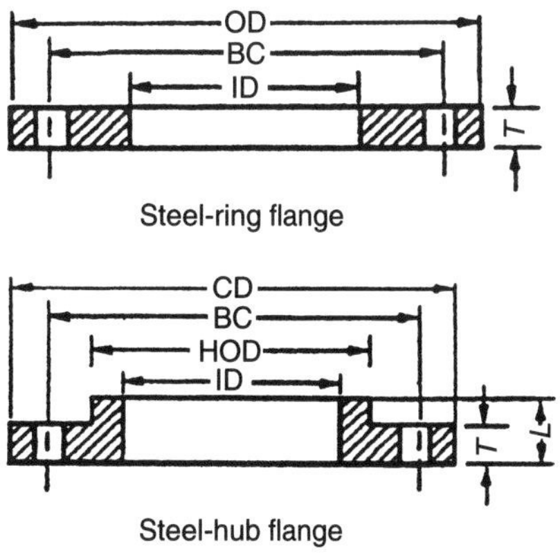

FIGURE 7.3 AWWAC 207.

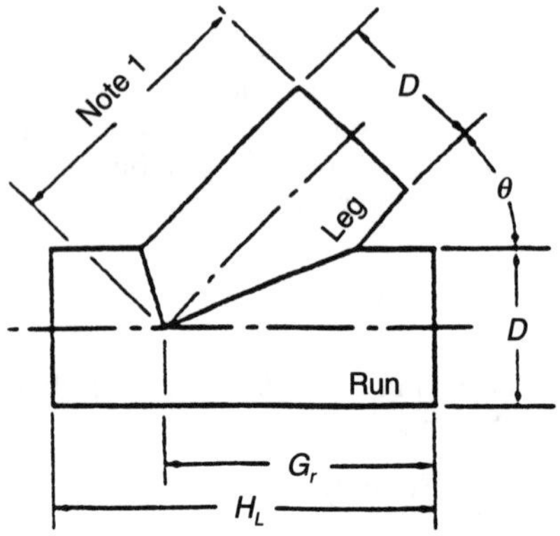

Equal diameter lateral

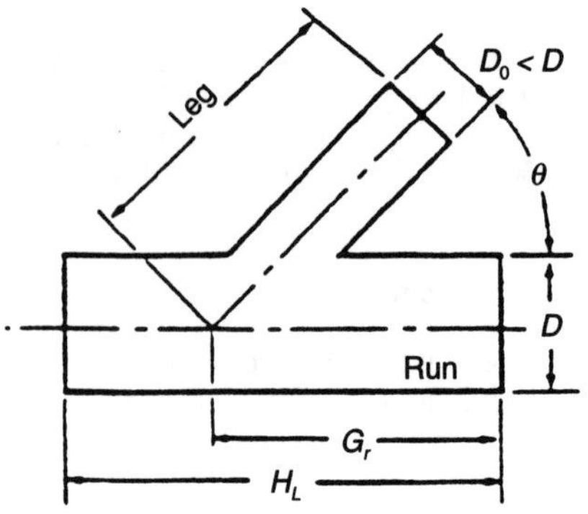

Unequal diameter lateral

TABLE 7.4 Dimensions of Steel Water Pipe 30° Laterals (U.S. Customary Units)*

Nominal diameter D	Factor f	G_r	H_L
6	5	23	29
8	5	27	33
10	6	32	40
12	6	36	44
14	7	41	49
16	7	44	53
18	7.5	49	59
20	8	54	64
22	9	60	71
24	10	65	78
30	10	76	90
36	11	90	105
42	12	103	120
48	12	114	132
54	14	129	150
60	14	140	162
66	15	154	177
72	15	165	189
78	16	178	204
84	17	191	219
90	17	202	231
96	18	216	246
102	19	229	261
108	20	242	276
114	21	255	291
120	22	268	306
126	23	282	321
132	24	295	336
138	25	308	351
144	26	321	366

* Dimension in inches. Dimensions in this table cover 30 to 70° angles. For angles greater than 70°, use dimensions given for tees in Tables 7.2 and 7.2M. For angles less than 30°, use 30° lateral wye plus an elbow to fit.

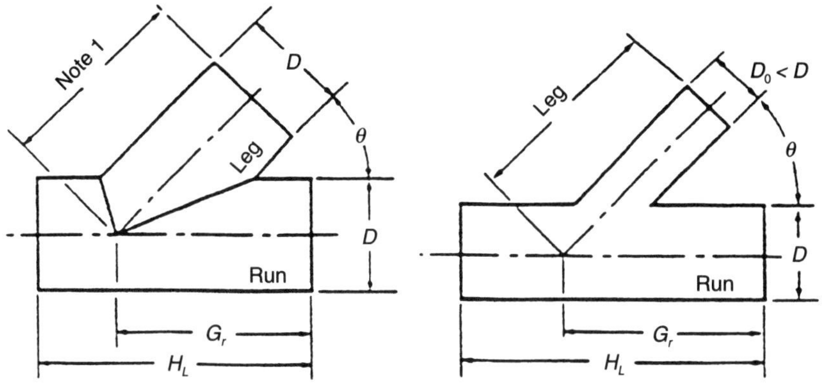

TABLE 7.4M Dimensions of Steel Water Pipe 30° Laterals (Metric Units)*

Nominal diameter D	Factor f	G_r	H_L
150	5	584	737
200	5	686	838
250	6	813	1016
300	6	914	1118
350	7	1041	1245
400	7	1118	1325
450	7.5	1245	1499
500	8	1372	1626
550	9	1524	1803
600	10	1651	1981
750	10	1930	2286
900	11	2286	2667
1050	12	2616	3048
1200	12	2896	3353
1350	14	3277	3810
1500	14	3566	4115
1650	15	3912	4496
1800	15	4191	4801
1950	16	4521	5182
2100	17	4851	5563
2250	17	5131	5867
2400	18	5486	6248
2550	19	5817	6629
2700	20	6147	7010
2850	21	6447	7391
3000	22	6807	7772
3150	23	7163	8153
3300	24	7493	8534
3450	25	7823	8915
3600	26	8153	9296

* Dimensions in millimeters. Dimensions in this table cover 30 to 70° angles. For angles greater than 70°, use dimensions given for tees in Tables 7.2 and 7.2M. For angles less than 30°, use 30° lateral wye plus an elbow to fit.

TABLE 7.5 AWWA Steel-Ring-Type Slip-on Flange Classes, Sizes, and Ratings (U.S. Customary Units)

Flange class	Flange size (nominal pipe size)	Working pressure rating, psi
B	4–24	86
B	26–144	86
D	4–12	175
D	14–144	150
E	4–144	275
F	4–48	300

TABLE 7.5M AWWA Steel-Ring-Type Slip-on Flange Classes, Sizes, and Ratings (Metric Units)

Flange class	Flange size (nominal pipe size)	Working pressure rating, kPa
B	100–600	593
B	650–3600	593
D	100–300	1207
D	350–3600	1034
E	100–3600	1896
F	100–1200	2068

TABLE 7.6 Steel-Hub-Type Slip-on Flange Classes, Sizes, and Ratings (U.S. Customary Units)

Flange class	Flange size (nominal pipe size)	Working pressure rating, psi
B	4–96	86
D	4–12	175
D	14–96	150
E	4–96	275

TABLE 7.6M Steel-Hub-Type Slip-on Flange Classes, Sizes, and Ratings (Metric Units)

Flange class	Flange size (nominal pipe size)	Working pressure rating, kPa
B	100–2400	593
D	100–300	1207
D	350–2400	1034
E	100–2400	1896

Ratings of Blind Flange

Tables 7.7 and 7.7M list the ratings of AWWA steel blind flanges in U.S. Customary and metric units, respectively.

TABLE 7.7 AWWA Blind Flange Classes, Sizes, and Ratings (U.S. Customary Units)

Flange class	Flange size (nominal pipe size)	Working pressure rating, psi
B	4–72	86
D	4–12	175
D	14–72	150
E	4–72	275
F	4–48	300

TABLE 7.7M AWWA Blind Flange Classes, Sizes, and Ratings (Metric Units)

Flange class	Flange size (nominal pipe size)	Working pressure rating, kPa
B	100–1800	593
D	100–300	1207
D	350–1800	1034
E	100–1800	1896
F	100–1200	2068

MATERIALS FOR FLANGES

AWWA C207 flanges may be made from seamless forgings meeting the requirements of the following ASTM material specifications:

- A105, Standard Specification for Forgings, Carbon Steel, for Piping Components
- A181, Standard Specification for Forgings, Carbon Steel, for General Purpose Piping

Flanges may also be made from a plate as a single piece, welded bar rings, or segmented and welded plates. The following ASTM material specifications for plate can be used:

- A36, Standard Specification for Structural Steel
- A516, Grades 60, 65, or 70, Standard Specification for Pressure Vessel Plates, Carbon Steel, for Moderate- and Lower-Temperature Service

Flange Bolting Materials

ASTM A307, Grade B bolts shall be used for Class B and D flanges. They shall not be used for Class E and F flanges. ASTM A193, Grade B7 bolts shall be used for Class E and F flanges.

Gaskets

Gaskets used shall be of red rubber (SBR) conforming to ASTM D1330, Grades I and II, having Shore A hardness of 80 ± 5 and suitable for water service at temperatures up to 200°F.

DIMENSIONS OF FLANGES

Dimensions of AWWA Steel-Ring Class B Flanges

Refer to Table 7.8 for dimensions of Class B steel-ring flanges.

Dimensions of AWWA Steel-Ring Class D Flanges

Refer to Table 7.9 for dimensions of Class D steel-ring flanges.

Dimensions of AWWA Steel-Ring Class E Flanges

Refer to Table 7.10 for dimensions of Class E steel-ring flanges.

Dimensions of AWWA Steel-Ring Class F Flanges

Refer to Table 7.11 for dimensions of Class F steel-ring flanges.

Dimensions of AWWA Steel-Hub-Type Class B and Class D Flanges

Refer to Table 7.12 for dimensions of Class B and Class D steel-hub-type flanges.

Dimensions of AWWA Steel-Hub-Type Class E Flanges

Refer to Table 7.13 for dimensions of Class E steel-hub-type flanges.

Thickness of AWWA Class B Blind Steel Flanges

Refer to Tables 7.14 and 7.14M for thickness of Class B blind steel flanges in U.S. Customary and metric units, respectively.

Thickness of AWWA Class D Blind Steel Flanges

Refer to Tables 7.15 and 7.15M for thickness of Class D blind steel flanges in U.S. Customary and metric units, respectively.

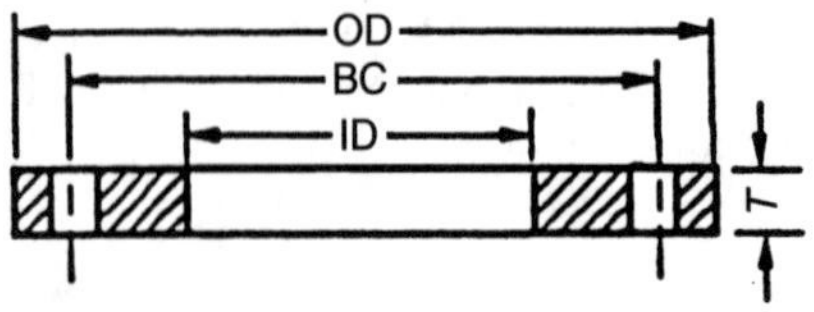

TABLE 7.8 AWWA Standard Class B Steel-Ring Flanges [Notes (1), (2)]

Flange size (nominal) pipe size	OD of flange (3)	ID of flange (4)	Thickness of flange, T	Number of bolts	Dia. of bolt circle, BC	Dia. of bolts (5)
4	9.00	4.57	0.625	8	7.50	0.625
5	10.00	5.66	0.625	8	8.50	0.750
6	11.00	6.72	0.688	8	9.50	0.750
8	13.50	8.72	0.688	8	11.75	0.750
10	16.00	10.88	0.688	12	14.25	0.875
12	19.00	12.88	0.688	12	17.00	0.875
14	21.00	14.19	0.688	12	18.75	1.000
16	23.50	16.19	0.688	16	21.25	1.000
18	25.00	18.19	0.688	16	22.75	1.125
20	27.50	20.19	0.688	20	25.00	1.125
22	29.50	22.19	0.750	20	27.25	1.250
24	32.00	24.19	0.750	20	29.50	1.250
26	34.25		0.812	24	31.75	1.250
28	36.50		0.875	28	34.00	1.250
30	38.75		0.875	28	36.00	1.250
32	41.75		0.938	28	38.50	1.500
34	43.75		0.938	32	40.50	1.500
36	46.00		1.000	32	42.75	1.500
38	48.75		1.000	32	45.25	1.500
40	50.75		1.000	36	47.25	1.500
42	53.00		1.125	36	49.50	1.500
44	55.25		1.125	40	51.75	1.500
46	57.25		1.125	40	53.75	1.500
48	59.50		1.250	44	56.00	1.500
50	61.75		1.250	44	58.25	1.750
52	64.00		1.250	44	60.50	1.750
54	66.25		1.375	44	62.75	1.750
60	73.00		1.500	52	69.25	1.750
66	80.00		1.625	52	76.00	1.750
72	86.50		1.750	60	82.50	1.750
78	93.00		2.000	64	89.00	2.000
84	99.75		2.000	64	95.50	2.000
90	106.50		2.250	68	102.00	2.250
96	113.25		2.250	68	108.50	2.250
102	120.00		2.500	72	114.50	2.500
108	126.75		2.500	72	120.75	2.500
114	133.50		2.750	76	126.75	2.750
120	140.25		2.750	76	132.75	2.750
126	147.00		3.000	80	139.25	3.000
132	153.75		3.000	80	145.75	3.000
138	160.50		3.250	84	152.00	3.250
144	167.25		3.250	84	158.25	3.250

Notes:
(1) Dimensions in inches.
(2) Class B flanges are rated for 86 psi at atmospheric temperature.
(3) Outside diameter and drilling dimensions match those of Class 125 cast iron flanges per ASME B16.1. NPS 24 and smaller flanges also match Class 150 flanges per ASME B16.5.
(4) Purchaser must specify ID of the flange for NPS 26 and larger flanges. In no case shall the ID of the flange exceed pipe outside diameter by more than 0.19 in.
(5) Bolt holes are drilled $^1/_8$ in larger than bolt nominal diameter.

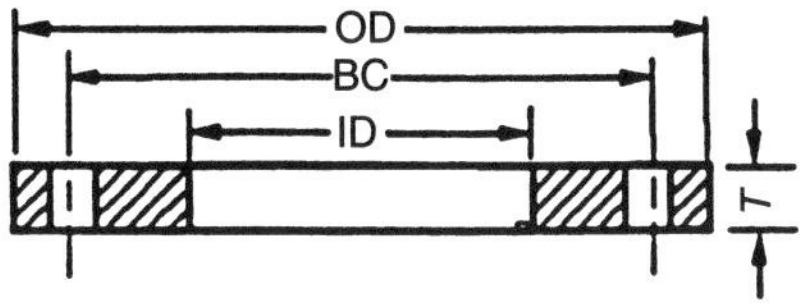

TABLE 7.9 AWWA Standard Class D Steel-Ring Flanges [Notes (1), (2)]

Flange size (Nominal pipe size)	OD of flange (3)	ID of flange (4)	Thickness of flange, T	Number of bolts	Dia. of bolt circle BC	Dia. of bolts (5)
4	9.00	4.57	0.625	8	7.50	0.625
5	10.00	5.66	0.625	8	8.50	0.750
6	11.00	6.72	0.688	8	9.50	0.750
8	13.50	8.72	0.688	8	11.75	0.750
10	16.00	10.88	0.688	12	14.25	0.875
12	19.00	12.88	0.812	12	17.00	0.875
14	21.00	14.19	0.938	12	18.75	1.000
16	23.50	16.19	1.000	16	21.25	1.000
18	25.00	18.19	1.062	16	22.75	1.125
20	27.50	20.09	1.125	20	25.00	1.125
22	29.50	22.19	1.188	20	27.25	1.250
24	32.00	24.19	1.250	20	29.50	1.250
26	34.25		1.312	24	31.75	1.250
28	36.50		1.312	28	34.00	1.250
30	38.75		1.375	28	36.00	1.250
32	41.75		1.500	28	38.50	1.500
34	43.75		1.500	32	40.50	1.500
36	46.00		1.625	32	42.75	1.500
38	48.75		1.625	32	45.25	1.500
40	50.75		1.625	36	47.25	1.500
42	53.00		1.750	36	49.50	1.500
44	55.25		1.750	40	51.75	1.500
46	57.25		1.750	40	53.75	1.500
48	59.50		1.875	44	56.00	1.500
50	61.75		2.000	44	58.25	1.750
52	64.00		2.000	44	60.50	1.750
54	66.25		2.125	44	62.75	1.750
60	73.00		2.250	52	69.25	1.750
66	80.00		2.500	52	76.00	1.750
72	86.50		2.625	60	82.50	1.750
78	93.00		2.750	64	89.00	2.000
84	99.75		2.875	64	95.50	2.000
90	106.50		3.000	68	102.00	2.250
96	113.25		3.250	68	108.50	2.250
102	120.00		3.250	72	114.50	2.500
108	126.75		3.375	72	120.75	2.500
114	133.50		3.500	76	126.75	2.750
120	140.25		3.500	76	132.75	2.750
126	147.00		3.750	80	139.25	3.000
132	153.75		3.875	80	145.75	3.000
138	160.50		4.000	84	152.00	3.250
144	167.25		4.125	84	158.25	3.250

Notes:

(1) Dimensions in inches.

(2) Class D flanges, NPS 4 through 12, are rated for 175 psi at atmospheric temperature. Class D flanges, NPS 14 and larger, are rated for 150 psi at atmospheric temperature.

(3) Outside diameter and drilling dimensions match those of Class 125 cast iron flanges per ASME B31.1. NPS and smaller flanges also match Class 150 flanges per ASME B16.5.

(4) Purchaser must specify ID of the flange for NPS 26 and larger sizes. In no case shall the ID of the flange exceed the outside diameter of the pipe by more than 0.19 in.

(5) Bolt holes are drilled $^1/_8$ in larger than nominal bolt diameter.

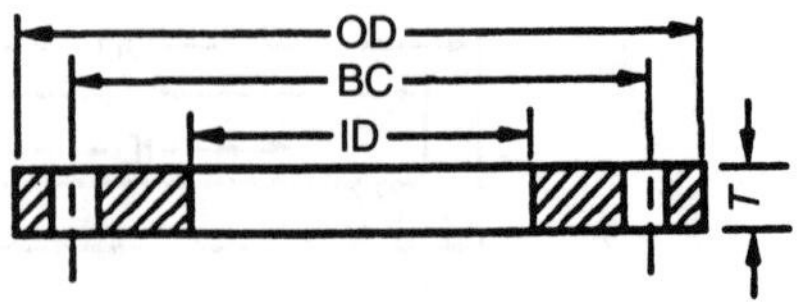

TABLE 7.10 AWWA Standard Class E Steel-Ring Flanges [Notes (1), (2)]

Nominal pipe size	OD of flange (3)	ID of flange (4)	Thickness of flange, *T*	Number of bolts	Dia. of bolt circle, BC	Dia. of bolts (5)
4	9.00	4.57	1.125	8	7.50	0.625
5	10.00	5.66	1.188	8	8.50	0.750
6	11.00	6.72	1.313	8	9.50	0.750
8	13.50	8.72	1.500	8	11.75	0.750
10	16.00	10.88	1.563	12	14.25	0.875
12	19.00	12.88	1.750	12	17.00	0.875
14	21.00	14.19	1.875	12	18.75	1.000
16	23.50	16.19	2.000	16	21.25	1.000
18	25.00	18.19	2.125	16	22.75	1.125
20	27.50	20.19	2.375	20	25.00	1.125
22	29.50	22.19	2.500	20	27.25	1.250
24	32.00	24.19	2.625	20	29.50	1.250
26	34.25		2.750	24	31.75	1.250
28	36.50		2.750	28	34.00	1.250
30	38.75		2.875	28	36.00	1.250
32	41.75		3.000	28	38.50	1.500
34	43.75		3.000	32	40.50	1.500
36	46.00		3.125	32	42.75	1.500
38	48.75		3.125	32	45.25	1.500
40	50.75		3.250	36	47.25	1.500
42	53.00		3.375	36	49.50	1.500
44	55.25		3.375	40	51.75	1.500
46	57.25		3.438	40	53.75	1.500
48	59.50		3.500	44	56.00	1.500
50	61.75		3.500	44	58.25	1.750
52	64.00		3.625	44	60.50	1.750
54	66.25		3.750	44	62.75	1.750
60	73.00		3.875	52	69.25	1.750
66	80.00		4.250	52	76.00	1.750
72	86.50		4.375	60	82.50	1.750
78	93.00		4.750	64	89.00	2.000
84	99.75		4.750	64	95.50	2.000
90	106.50		5.125	68	102.00	2.250
96	113.25		5.125	68	108.50	2.250
102	120.00		5.500	72	114.50	2.500
108	126.75		5.500	72	120.75	2.500
114	133.50		5.875	76	126.75	2.750
120	140.25		5.875	76	132.75	2.750
126	147.00		6.250	80	139.25	3.000
132	153.75		6.250	80	145.75	3.000
138	160.50		6.750	84	152.00	3.250
144	167.25		6.750	84	158.25	3.250

Notes:

(1) Dimensions in inches.

(2) Class E flanges are rated at 275 psi at atmospheric temperature.

(3) Outside diameter and drilling dimensions match Class 125 cast iron flanges per ASME B16.1. NPS 24 and smaller flanges also match Class 150 flanges per ASME B16.5.

(4) Purchaser must specify ID for NPS 26 and larger flanges. It is recommended that ID of the flange be $^3/_{16}$ in larger than outside diameter of the pipe.

(5) Bolt holes are drilled $^1/_8$ in larger than nominal bolt diameter.

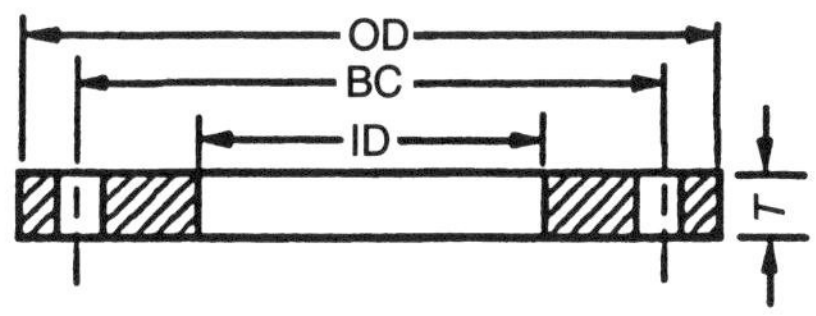

TABLE 7.11 AWWA Standard Class F Steel-Ring Flanges [Notes (1), (2)]

Nominal pipe size	OD of flange (3)	ID of flange	Thickness of flange, T	Number of bolts	Dia. of bolt circle, BC	Dia. of bolts (4)
4	10.00	4.57	1.13	8	7.88	0.750
5	11.00	5.66	1.21	8	9.25	0.750
6	12.50	6.73	1.31	12	10.62	0.750
8	15.00	8.73	1.31	12	13.00	0.875
10	17.50	10.88	1.50	16	15.25	1.000
12	20.50	12.88	1.63	16	17.75	1.125
14	23.00	14.19	1.94	20	20.25	1.125
16	25.50	16.19	2.14	20	22.50	1.250
18	28.00	18.19	2.25	24	24.75	1.250
20	30.50	20.19	2.33	24	27.00	1.250
22	33.00	22.19	2.50	24	29.25	1.250
24	36.00	24.19	2.69	24	32.00	1.500
26	38.25	26.25	3.00	28	34.50	1.750
28	40.75	28.25	3.13	28	37.00	1.750
30	43.00	30.25	3.15	28	39.25	1.750
32	45.25	32.25	3.25	28	41.50	1.750
34	47.50	34.25	3.38	28	43.50	1.750
36	50.00	36.25	3.46	32	46.00	2.000
38	52.25	38.25	3.50	32	48.00	2.000
40	54.25	40.25	3.63	36	50.25	2.000
42	57.00	42.25	3.81	36	52.75	2.000
44	59.25	44.25	4.00	36	55.00	2.000
46	61.50	46.25	4.13	40	57.25	2.000
48	65.00	48.25	4.50	40	60.75	2.000

Notes:
(1) Dimensions in inches.
(2) Class F steel-ring flanges are rated for 300 psi at atmospheric temperature.
(3) These flanges have the same outside diameter and drilling dimensions as ASME 16.1, Class 250 cast iron flanges.
(4) Bolt holes are drilled $^1/_8$ in larger than nominal bolt diameter.

Thickness of AWWA Class E Blind Steel Flanges

Refer to Tables 7.16 and 7.16M for thickness of Class E blind steel flanges in U.S. Customary and metric units, respectively.

Thickness of AWWA Class F Blind Steel Flanges

Refer to Tables 7.17 and 7.17M for thickness of Class F blind steel flanges in U.S. Customary and metric units, respectively.

TOLERANCES

Table 7.18 lists the tolerances for flange dimensions listed in Tables 7.8 through 7.17M.

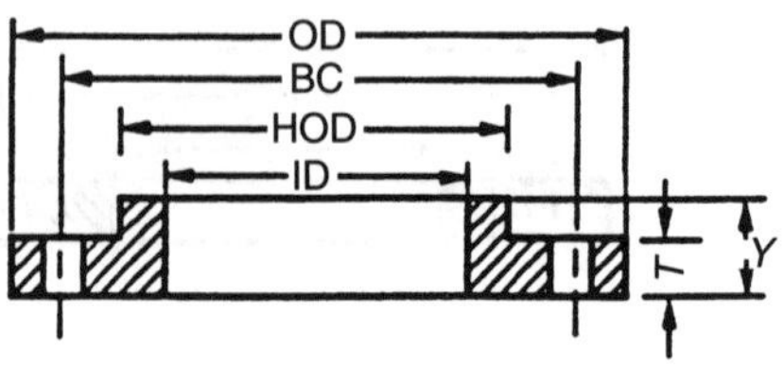

TABLE 7.12 AWWA Standard Class B and Class D Steel-Hub Flanges and Bolting Dimensions [Notes (1), (2)]

Nominal pipe size	OD of flange (3)	ID of flange	Thickness T	Flange height Y	Hub outside dia., HOD	Number of bolts	Dia. of bolt circle	Dia. of bolts (4)
4	9.00	4.57	0.500	0.875	5.312	8	7.50	0.625
5	10.00	5.66	0.562	1.250	6.312	8	8.50	0.750
6	11.00	6.72	0.562	1.250	7.562	8	9.50	0.750
8	13.50	8.72	0.562	1.250	9.688	8	11.75	0.750
10	16.00	10.88	0.688	1.250	12.000	12	14.25	0.875
12	19.00	12.88	0.688	1.250	14.375	12	17.00	0.875
14	21.00	14.19	0.750	1.250	15.750	12	18.75	1.000
16	23.50	16.19	0.750	1.250	18.000	16	21.25	1.000
18	25.00	18.19	0.750	1.250	19.875	16	22.75	1.125
20	27.50	20.19	0.750	1.250	22.000	20	25.00	1.125
22	29.50	22.19	1.000	1.750	24.250	20	27.25	1.250
24	32.00	24.19	1.000	1.750	26.125	20	29.50	1.250
26	34.25	26.19	1.000	1.750	28.500	24	31.75	1.250
28	36.50	28.19	1.000	1.750	30.500	28	34.00	1.250
30	38.75	30.19	1.000	1.750	32.500	28	36.00	1.250
32	41.75	32.19	1.125	1.750	34.750	28	38.50	1.500
34	43.75	34.19	1.125	1.750	36.750	32	40.50	1.500
36	46.00	36.19	1.125	1.750	38.750	32	42.75	1.500
38	48.75	38.19	1.125	1.750	40.750	32	45.25	1.500
40	50.75	40.19	1.125	1.750	43.000	36	47.25	1.500
42	53.00	42.19	1.250	1.750	45.000	36	49.50	1.500
44	55.25	44.19	1.250	2.250	47.000	40	51.75	1.500
46	57.25	46.19	1.250	2.250	49.000	40	53.75	1.500
48	59.50	48.19	1.375	2.500	51.000	44	56.00	1.500
50	61.75	50.19	1.375	2.500	53.000	44	58.25	1.750
52	64.00	52.19	1.375	2.500	55.000	44	60.50	1.750
54	66.25	54.19	1.375	2.500	57.000	44	62.75	1.750
60	73.00	60.19	1.500	2.750	63.000	52	69.25	1.750
66	80.00	66.19	1.500	2.750	69.000	52	76.00	1.750
72	86.50	72.19	1.500	2.750	75.000	60	82.50	1.750
78	93.00	78.19	1.750	3.000	81.250	64	89.00	2.000
84	99.75	84.19	1.750	3.000	87.500	64	95.50	2.000
90	106.50	90.19	2.000	3.250	93.750	68	102.00	2.250
96	113.25	96.19	2.000	3.250	100.000	68	108.50	2.250

Notes:

(1) Dimensions in inches.

(2) Class B flanges are rated for 86 psi at atmospheric temperature. Class D flanges, NPS 4 through 12, are rated for 175 psi; NPS 14 through 96 are rated for 150 psi at atmospheric temperature.

(3) Outside diameter and drilling dimensions match those of Class 125 cast iron flanges per ASME B16.1. NPS 24 and smaller flanges also match Class 150 flanges per ASME B16.5.

(4) Bolt holes are drilled $^1/_8$ in larger than nominal bolt diameter.

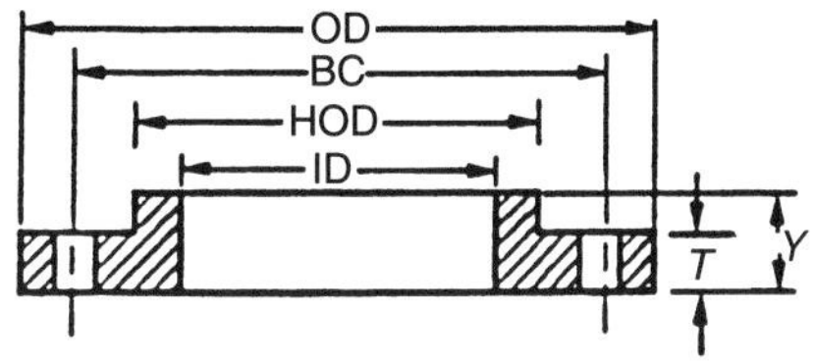

TABLE 7.13 AWWA Standard Class E Steel-Hub Flanges [Notes (1), (2)]

Flange size (nominal pipe size)	OD of flange (3)	ID of flange	Thickness *T*	Flange length, *Y*	Hub outside dia., HOD	Number of bolts	Dia. of bolt circle	Dia. of bolts (4)
4	9.00	4.57	0.938	1.312	5.312	8	7.50	0.625
5	10.00	5.66	0.938	1.438	6.438	8	8.50	0.750
6	11.00	6.72	1.000	1.562	7.562	8	9.50	0.750
8	13.50	8.72	1.125	1.750	9.688	8	11.75	0.750
10	16.00	10.88	1.188	1.938	12.000	12	14.25	0.875
12	19.00	12.88	1.250	2.188	14.375	12	17.00	0.875
14	21.00	14.19	1.375	2.250	15.750	12	18.75	1.000
16	23.50	16.19	1.438	2.500	18.000	16	21.25	1.000
18	25.00	18.19	1.562	2.688	19.875	16	22.75	1.125
20	27.50	20.19	1.688	2.875	22.000	20	25.00	1.125
22	29.50	22.19	1.812	3.125	24.000	20	27.25	1.250
24	32.00	24.19	1.875	3.250	26.125	20	29.50	1.250
26	34.25	26.19	2.000	3.375	28.500	24	31.75	1.250
28	36.50	28.19	2.062	3.438	30.750	28	34.00	1.250
30	38.75	30.19	2.125	3.500	32.750	28	36.00	1.250
32	41.75	32.19	2.250	3.625	35.000	28	38.50	1.500
34	43.75	34.19	2.312	3.688	37.000	32	40.50	1.500
36	46.00	36.19	2.375	3.750	39.250	32	42.75	1.500
38	48.75	38.19	2.375	3.750	41.750	32	45.25	1.500
40	50.75	40.19	2.500	3.875	43.750	36	47.25	1.500
42	53.00	42.19	2.625	4.000	46.000	36	49.50	1.500
44	55.25	44.19	2.625	4.000	48.000	40	51.75	1.500
46	57.25	46.19	2.688	4.062	50.000	40	53.75	1.500
48	59.50	48.19	2.750	4.125	52.250	44	56.00	1.500
50	61.75	50.19	2.750	4.125	54.250	44	58.25	1.750
52	64.00	52.19	2.875	4.250	56.500	44	60.50	1.750
54	66.25	54.19	3.000	4.375	58.750	44	62.75	1.750
60	73.00	60.19	3.125	4.500	65.250	52	69.25	1.750
66	80.00	66.19	3.375	4.875	71.500	52	76.00	1.750
72	86.50	72.19	3.500	5.000	78.500	60	82.50	1.750
78	93.00	78.19	3.875	5.375	84.500	64	89.00	2.000
84	99.75	84.19	3.875	5.375	90.500	64	95.50	2.000
90	106.50	90.19	4.250	5.750	96.750	68	102.00	2.250
96	113.25	96.19	4.250	5.750	102.750	68	108.50	2.250

***Notes*:**
(1) Dimensions in inches.
(2) Class E steel-hub flanges are rated for 275 psi at atmospheric temperature.
(3) These flanges have outside diameter and drilling dimensions as Class 125 Cast Iron flanges per ASME B16.1. NPS 24 and smaller flanges also match outside diameters and drilling dimensions of Class 150 flanges per ASME B16.5.
(4) Bolt holes are drilled $^1/_8$ in larger than nominal bolt diameter.

TABLE 7.14 AWWA Class B Blind Flange Thickness (U.S. Customary Units)

Flange size (nominal pipe size), in	Thickness, in
4	0.625
6	0.688
8	0.688
10	0.688
12	0.752
14	0.835
16	0.941
18	1.013
20	1.108
24	1.275
30	1.530
36	1.834
42	2.084
48	2.341
54	2.634
60	2.892
66	3.139
72	3.399

Notes:
1. AWWA blind flanges are designed per ASME Boiler and Pressure Vessel Code, Section VIII, Division 1.
2. AWWA Class D blind flanges, NPS 12 and smaller, are designed for 175 psi working pressure. NPS 14 and larger flanges are rated for 150 psi.

TABLE 7.14M AWWA Class B Blind Flange Thickness (Metric Units)

Flange size (nominal pipe size), mm	Thickness, mm
100	15.88
150	17.48
200	17.48
250	17.48
300	19.10
350	21.21
400	23.90
450	25.73
500	28.14
600	32.38
750	38.86
900	46.58
1050	52.93
1200	59.46
1350	66.90
1500	73.46
1650	79.73
1800	86.33

Notes:
1. AWWA blind flanges are designed per ASME Boiler and Pressure Vessel Code, Section VIII, Division 1.
2. AWWA Class D blind flanges, NPS 12 and smaller, are designed for 175 psi working pressure. NPS 14 and larger flanges are rated for 150 psi.

TABLE 7.15 AWWA Class D Blind Flange Thickness (U.S. Customary Units)

Flange size (nominal pipe size), in	Thickness, in
4	0.625
6	0.692
8	0.805
10	0.947
12	1.110
14	1.127
16	1.258
18	1.326
20	1.442
24	1.657
30	2.003
36	2.369
42	2.725
48	3.067
54	3.431
60	3.774
66	4.132
72	4.476

Notes:
1. AWWA blind flanges are designed per ASME Boiler and Pressure Vessel Code, Section VIII, Division 1.
2. AWWA Class D blind flanges, NPS 12 and smaller, are designed for 175 psi working pressure. NPS 14 and larger flanges are rated for 150 psi.

TABLE 7.15M AWWA Class D Blind Flange Thickness (Metric Units)

Flange size (nominal pipe size), mm	Thickness, mm
100	15.88
150	17.58
200	20.45
250	24.05
300	28.19
350	28.63
400	31.95
450	33.68
500	36.63
600	42.09
750	50.88
900	60.17
1050	69.21
1200	77.90
1350	87.15
1500	95.86
1650	104.95
1800	113.69

Notes:
1. AWWA blind flanges are designed per ASME Boiler and Pressure Vessel Code, Section VIII, Division 1.
2. AWWA Class D blind flanges, NPS 12 and smaller, are designed for 175 psi working pressure. NPS 14 and larger flanges are rated for 150 psi.

TABLE 7.16 AWWA Class E Blind Flange Thickness (U.S. Customary Units)

Flange size (nominal pipe size), in	Thickness, in
4	1.125
6	1.313
8	1.500
10	1.563
12	1.750
14	1.875
16	2.000
18	2.125
20	2.375
24	2.625
30	2.875
36	3.344
42	3.789
48	4.246
54	4.776
60	5.236
66	5.674
72	6.137

Notes:

1. AWWA blind flanges are designed per ASME Boiler and Pressure Vessel Code, Section VIII, Division 1.
2. AWWA Class D blind flanges, NPS 12 and smaller, are designed for 175 psi working pressure. NPS 14 and larger flanges are rated for 150 psi.

TABLE 7.16M AWWA Class E Blind Flange Thickness (Metric Units)

Flange size (nominal pipe size), mm	Thickness, mm
100	28.58
150	33.35
200	38.10
250	39.70
300	44.45
350	47.62
400	50.80
450	53.98
500	60.32
600	66.68
750	73.02
900	84.93
1050	96.24
1200	107.85
1350	121.31
1500	132.99
1650	144.12
1800	155.88

Notes:

1. AWWA blind flanges are designed per ASME Boiler and Pressure Vessel Code, Section VIII, Division 1.
2. AWWA Class D blind flanges, NPS 12 and smaller, are designed for 175 psi working pressure. NPS 14 and larger flanges are rated for 150 psi.

TABLE 7.17 AWWA Class F Blind Flange Thickness (U.S. Customary Units)

Flange size (nominal pipe size), in	Thickness, in
4	1.125
6	1.316
8	1.316
10	1.534
12	1.730
14	1.938
16	2.139
18	2.294
20	2.401
24	2.799
30	3.419
36	4.017
42	4.450
48	4.991

Notes:

1. AWWA blind flanges are designed per ASME Boiler and Pressure Vessel Code, Section VIII, Division 1.
2. AWWA Class D blind flanges, NPS 12 and smaller, are designed for 175 psi working pressure. NPS 14 and larger flanges are rated for 150 psi.

TABLE 7.17M AWWA Class F Blind Flange Thickness (Metric Units)

Flange size (nominal pipe size), mm	Thickness, mm
100	28.58
150	33.43
200	33.43
250	38.96
300	43.94
350	49.23
400	54.33
450	58.27
500	60.99
600	71.09
750	86.84
900	102.03
1050	113.03
1200	126.77

Notes:

1. AWWA blind flanges are designed per ASME Boiler and Pressure Vessel Code, Section VIII, Division 1.
2. AWWA Class D blind flanges, NPS 12 and smaller, are designed for 175 psi working pressure. NPS 14 and larger flanges are rated for 150 psi.

TABLE 7.18 Tolerances for AWWA C207 Flanges

Flange size (nominal pipe size), in	Inside diameter tolerance	Outside diameter tolerance	Thickness tolerance	Tolerance for hub length	Tolerance for bolt circle diameter	Tolerance for bolt-hole spacing
All sizes	+1/16 in (+1.6 mm) −0	+1/8 in (+3.2 mm) −1/8 in (−3.2 mm)			+1/16 in (+1.6 mm) −1/16 in (−1.6 mm)	+1/32 in (+0.79 mm) −1/32 in (−0.79 mm)
18 and smaller			+1/8 in (+3.2 mm) −0	+1/8 in (+3.2 mm) −1/32 in (−0.79 mm)		
20 and larger			+3/16 in (+4.8 mm) −0	+3/16 in (+4.8 mm) −1/16 in (−1.6 mm)		

PIPING JOINTS

Refer to Chap. 15.

Tables in Chapter 7

CHAPTER 8
MSS FITTINGS AND FLANGES

See p. 353 for a listing of tables in Chap. 8.

INTEGRALLY REINFORCED BRANCH FITTINGS

MSS SP-97 covers integrally reinforced forged fittings for making 90° branch connection to the run pipe. Branch connections made by using these fittings do not require reinforcements, such as saddles and pads. However, the fitting to be used must be suitable for the run pipe size, wall thickness, or schedule. Figure 8.1 shows butt welding, socket welding end, and threaded end fittings.

Figure 8.2 shows that the skirt or the crotch gap must not exceed $^1/_{16}$ in.

Table 8.1 provides the correlation between butt welding branch fittings and the run pipe.

Table 8.2 shows the correlation between the socket welding end and threaded end integrally reinforced fittings conforming to MSS SP-97.

It is noted that the manufacturers of these fittings may provide a wider range of these fittings that may be used with run pipes having schedules or wall thicknesses listed. The user should contact the manufacturer for availability of integrally reinforced fittings. Refer to Chap. 29 for fitting manufacturers.

Dimensions of Standard Integrally Reinforced Butt Welding Fittings for Standard-Weight Run Pipe

See Table 8.3.

Dimensions of Extra-Strong Integrally Reinforced Butt Welding Fittings for Extra-Strong Run Pipe

See Table 8.4.

Dimensions of Schedule 160 Integrally Reinforced Butt Welding Fittings for Schedule 160 Run Pipe

See Table 8.5.

Tolerances for Fittings Listed in Tables 8.3, 8.4, and 8.5.

See Table 8.6.

Dimensions of Class 3000 Integrally Reinforced Socket Welding Fittings

See Table 8.7.

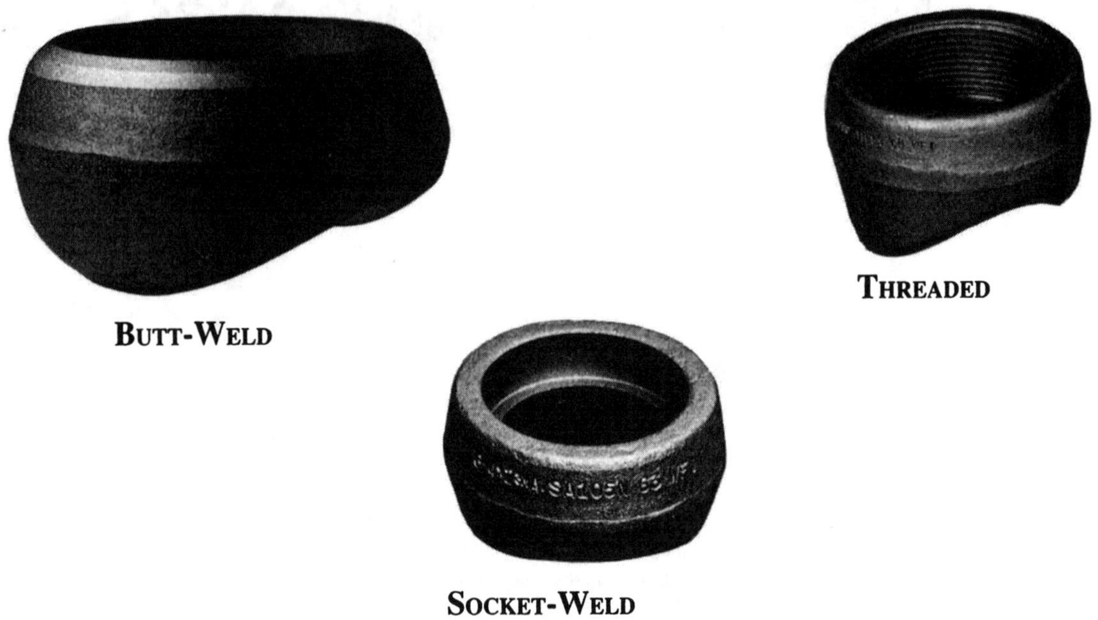

FIGURE 8.1 Integrally reinforced fittings. (*Courtesy WFI, Houston, Texas.*)

TABLE 8.1 Relationship between Integrally Reinforced Butt Welding Fittings and Run Pipe*

Run pipe wall thickness/ schedule	Branch size, NPS	Type of fitting	Fitting schedule
Standard	$\frac{1}{8}$–24	Butt welding	Standard
Extra strong	$\frac{1}{8}$–24	Butt welding	Extra strong
Schedule 160	$\frac{1}{2}$–6	Butt welding	Schedule 160

* Do not use run pipe thinner or thicker than shown for the corresponding fitting.

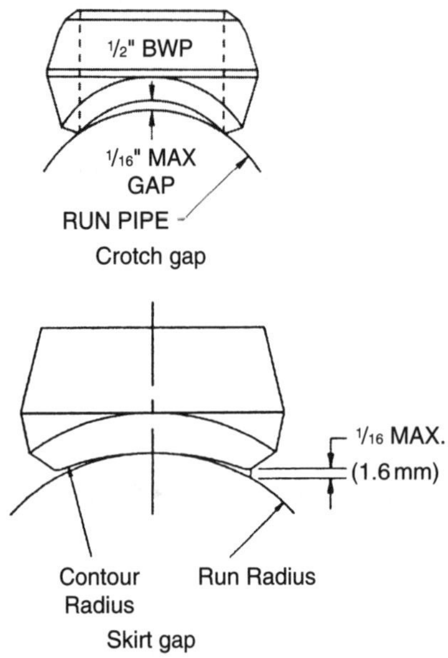

FIGURE 8.2 Crotch and skirt gap.

TABLE 8.2 Correlation between Socket Welding End and Threaded End Integrally Reinforced Fittings and Run Pipe*

Socket welding end fittings			Threaded end fittings		
Class	Branch Size, NPS	Run pipe schedule	Class	Branch size, NPS	Run pipe schedule
3000	1/8–4	Extra strong	3000	1/8–4	Extra strong
6000	1/2–2	160	6000	1/2–2	160

* Do not use run pipe thinner or thicker than shown for the corresponding fitting.

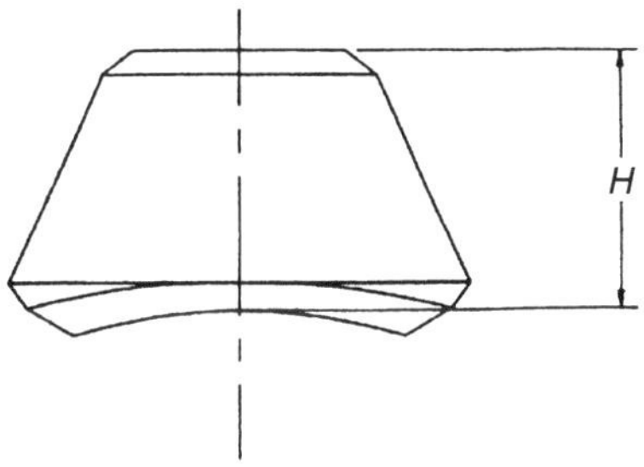

TABLE 8.3 Dimensions of Standard Butt Welding Integrally Reinforced Fittings*

Outlet NPS	H, in Reducing	H, in Full
1/8	0.62	
1/4	0.62	
3/8	0.75	
1/2	0.75	0.75
3/4	0.88	0.88
1	1.06	1.06
1 1/4	1.25	1.25
1 1/2	1.31	1.31
2	1.50	1.50
2 1/2	1.62	1.62
3	1.75	1.75
3 1/2	1.88	2.00
4	2.00	2.00
5	2.25	2.25
6	2.38	2.38
8	2.75	2.75
10	3.06	3.06
12	3.38	3.38
14	3.50	3.50
16	3.69	3.69
18	3.81	4.06
20	4.00	4.62
24	4.56	5.38

* For tolerances refer to Table 8.6.

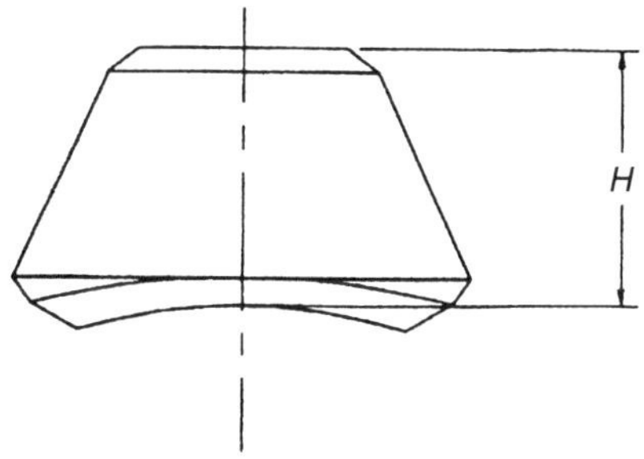

TABLE 8.4 Dimensions of Extra-Strong Butt Welding Integrally Reinforced Fittings*

Outlet NPS	H, in Reducing	H, in Full
1/8	0.62	
1/4	0.62	
3/8	0.75	
1/2	0.75	0.75
3/4	0.88	0.88
1	1.06	1.06
1 1/4	1.25	1.25
1 1/2	1.31	1.31
2	1.50	1.50
2 1/2	1.62	1.62
3	1.75	1.75
3 1/2	1.88	2.00
4	2.00	2.00
5	2.25	2.25
6	3.06	3.06
8	3.88	3.88
10	3.69	3.50
12	4.06	3.94
14	3.94	4.12
16	4.18	4.44
18	4.38	4.69
20	4.69	5.00
24	5.50	5.50

* For tolerances refer to Table 8.6.

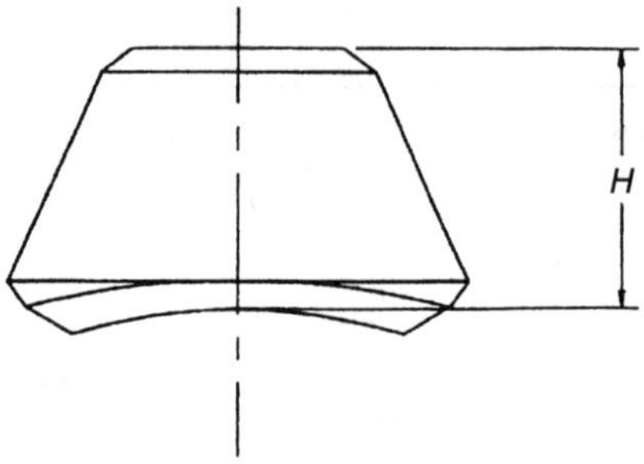

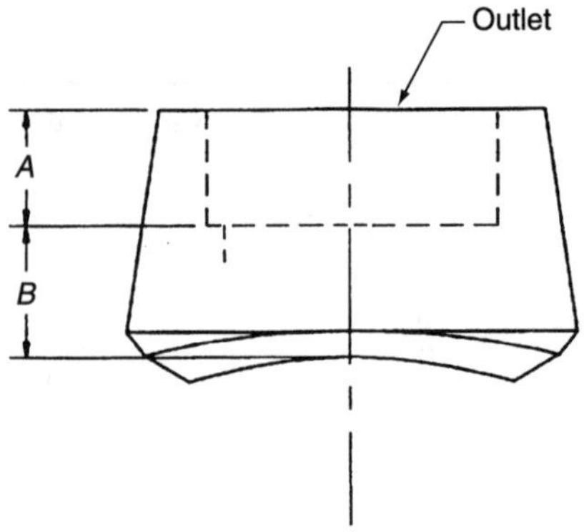

TABLE 8.5 Dimensions of Schedule 160 Butt Welding Integrally Reinforced Fittings*

Outlet NPS	*H*, in Reducing	*H*, in Full
½	1.12	1.12
¾	1.25	1.25
1	1.50	1.50
1¼	1.75	1.75
1½	2.00	2.00
2	2.18	2.18
2½	2.44	2.44
3	2.88	2.88
3½		
4	3.31	3.31
5	3.69	3.69
6	4.12	4.12

* For tolerances refer to Table 8.6.

TABLE 8.6 Tolerances for the Height *H* of Butt Welding Fittings

Outlet NPS	Tolerances for *H* dimension, in
⅛–¾	±0.03
1–4	±0.06
5–12	±0.12
14–24	±0.19

TABLE 8.7 Dimensions of Class 3000 Socket Welding Integrally Reinforced Fillings

Outlet NPS	*A*,* in	*B*,† in
⅛	0.38	0.41
¼	0.38	0.41
⅜	0.38	0.50
½	0.38	0.63
¾	0.50	0.63
1	0.50	0.88
1¼	0.50	0.88
1½	0.50	0.94
2	0.62	0.94
2½	0.62	1.00
3	0.62	1.19
4	0.75	1.19

* Socket depth *A* shall be no less than one required by ASME B16.11 for Class 3000 socket welding fittings.

† Dimension *B* listed is the maximum.

Dimensions of Class 6000 Integrally Reinforced Socket Welding Fittings

See Table 8.8.

Dimensions of Class 3000 Integrally Reinforced Threaded Fittings

See Table 8.9.

Dimensions of Class 6000 Integrally Reinforced Threaded Fittings

See Table 8.10.

MSS SP-83 STEEL PIPE UNIONS, SOCKET WELDING AND THREADED

Socket welding and threaded unions are made from carbon steel (A105) and stainless steels (A182, TP F304, F316, and F316L) in conformance with MSS SP-83.

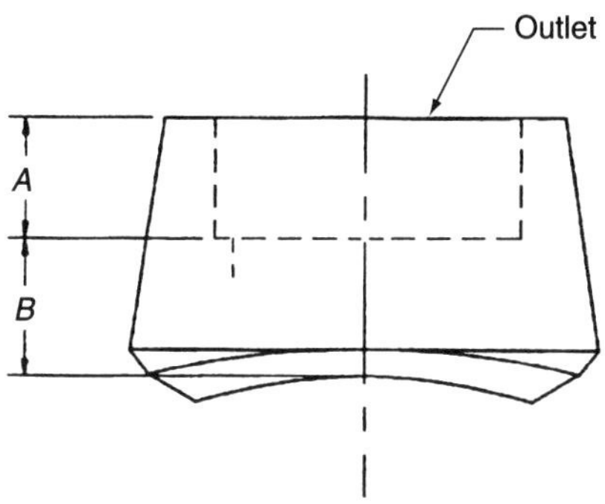

TABLE 8.8 Dimensions of Class 6000 Socket Welding Integrally Reinforced Fittings

Outlet NPS	A,* in	B,† in
1/2	0.38	0.94
3/4	0.50	1.00
1	0.50	1.13
1 1/4	0.50	1.19
1 1/2	0.50	1.25
2	0.62	1.44

* Socket depth A shall be no less than one required by ASME B16.11 for Class 6000 socket welding fittings.

† Dimension B listed is the maximum.

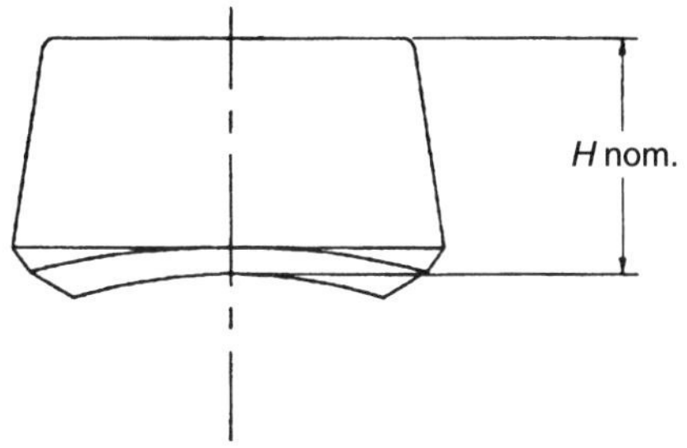

TABLE 8.9 Dimensions of Class 3000 Threaded Integrally Reinforced Fitting

Outlet NPS	H, in
1/8	0.75
1/4	0.75
3/8	0.81
1/2	1.00
3/4	1.06
1	1.31
1 1/4	1.31
1 1/2	1.38
2	1.50
2 1/2	1.81
3	2.00
4	2.25

Notes:
1. Threads shall be taper pipe threads per ASME B1.20.1.
2. Tolerances: NPS 1/8–3/4, ±0.03 in
NPS 1–4, ±0.06 in

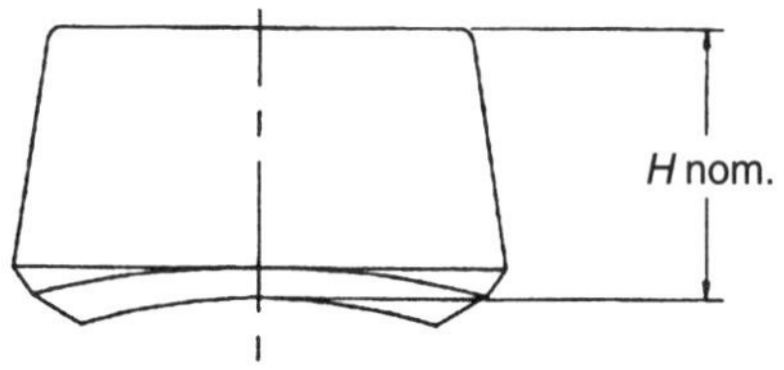

TABLE 8.10 Dimension of Class 6000 Threaded Integrally Reinforced Fittings

Outlet NPS	H, in
1/2	1.25
3/4	1.44
1	1.56
1 1/4	1.62
1 1/2	1.69
2	2.06

Notes:
1. Taper pipe threads per ASME B1.20.1.
2. Tolerances: NPS 1/2–3/4, ±0.03 in
NPS 1–2, ±0.06 in

Pressure-Temperature Ratings of Steel Pipe Unions

See Table 8.11.

TABLE 8.11 Pressure-Temperature Ratings of Class 3000 Steel Unions

Service Temp., °F	Pressure rating,* psig			
	ASTM A105 carbon steel	ASTM A182, F316	ASTM A182, F304L, F316L	ASTM A182, F304
100	3000	2915	2430	2915
200	2735	2510	2050	2430
300	2655	2265	1835	2140
400	2565	2080	1670	1905
500	2425	1935	1545	1770
600	2220	1830	1460	1680
650	2180	1800	1420	1650
700	2155	1750	1390	1630
750		1710	1360	1610
800		1680	1330	1595
850		1645	1300	1575
900		1595		1555
950		1565		1515
1000		1470		1300

* Ratings are nonshock working pressures corresponding to the temperature listed.

Dimensions of Class 3000 Socket Welding Steel Unions

See Table 8.12.

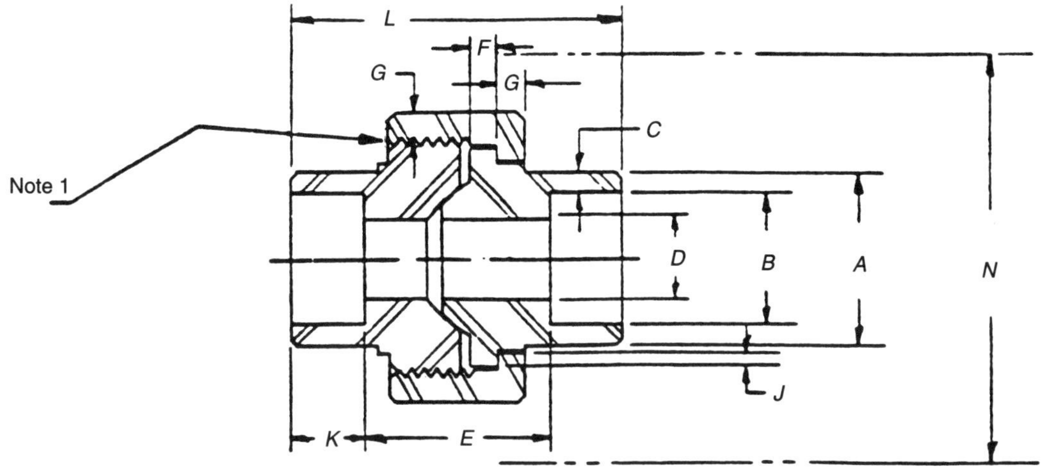

TABLE 8.12 Dimensions of Class 3000 Socket Welding Steel Unions (Inches)

Nom. pipe size	Pipe end, min., *A*	Socket bore dia., *B*	Socket wall, min., *C*	Water way bore, *D*	Laying length, *E*	Male flange, min., *F*	Nut, min., *G*	Threads per inch	Bearing, min., *J*	Depth of socket, min., *K*	Length assem., nominal, *L*	Clear assem. nut, *N*
1/8	0.86	0.440 0.420	0.125	0.299 0.239	0.88 0.75	0.125	0.125	16	0.049	0.38	1.63	2.0
1/4	0.86	0.575 0.555	0.130	0.394 0.334	0.88 0.75	0.125	0.125	16	0.049	0.38	1.63	2.0
3/8	1.02	0.710 0.690	0.138	0.523 0.463	1.06 0.81	0.135	0.135	14	0.054	0.38	1.81	2.2
1/2	1.23	0.875 0.855	0.161	0.652 0.592	1.06 0.81	0.145	0.145	14	0.059	0.38	1.93	2.3
3/4	1.46	1.085 1.065	0.168	0.854 0.794	1.25 1.00	0.160	0.160	11	0.066	0.50	2.24	2.6
1	1.79	1.350 1.330	0.196	1.079 1.019	1.35 1.03	0.180	0.175	11	0.073	0.50	2.44	3.1
1 1/4	2.16	1.695 1.675	0.208	1.410 1.350	1.60 1.28	0.210	0.205	10	0.084	0.50	2.80	3.7
1 1/2	2.42	1.935 1.915	0.218	1.640 1.580	1.66 1.34	0.230	0.220	10	0.091	0.50	3.01	4.4
2	2.96	2.426 2.406	0.238	2.097 2.037	1.79 1.47	0.260	0.250	10	0.106	0.62	3.39	5.2
2 1/2	3.61	2.931 2.906	0.302	1.529 2.409	2.43 2.05	0.295	0.280	8	0.121	0.62	4.03	5.9
3	4.30	3.560 3.535	0.327	3.128 3.008	2.51 2.11	0.325	0.315	8	0.139	0.62	4.29	6.9

Notes:
1. Minimum 4 threads full engagement is required.
2. Minimum wall thickness other than socket wall must be equal to or greater than the nominal wall thickness of Schedule 80 pipe.

Dimensions of Class 3000 Threaded Steel Unions

See Table 8.13.

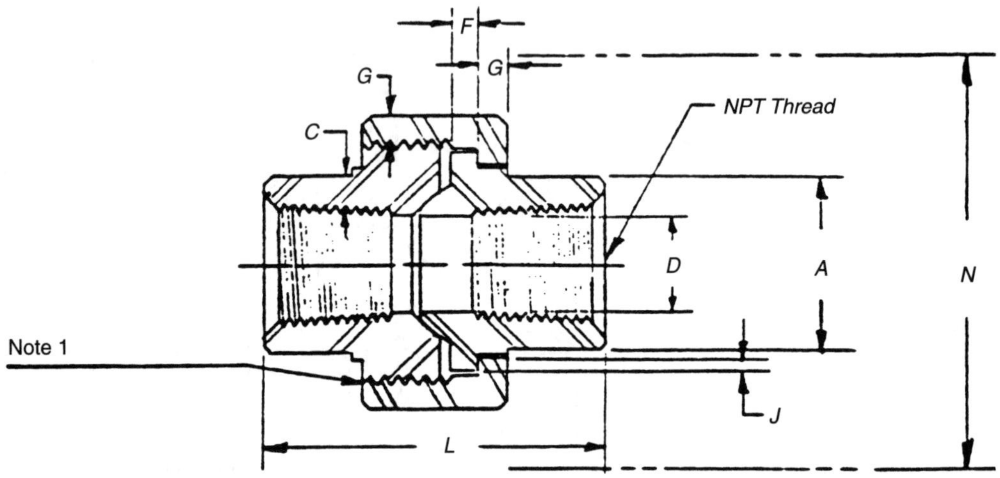

TABLE 8.13 Dimensions of Class 3000 Threaded Steel Unions (Inches)

Nom. pipe size	Pipe end, min., *A*	Wall, min., *C*	Water way bore, *D*	Male flange min., *F*	Nut, min., *G*	Threads per inch	Bearing, min., *J*	Length assem., nominal, *L*	Clear assem. nut, *N*
1/8	0.58	0.095	0.332 0.253	0.125	0.125	16	0.049	1.63	2.0
1/4	0.75	0.119	0.438 0.372	0.125	0.125	16	0.049	1.63	2.0
3/8	0.90	0.126	0.562 0.532	0.135	0.135	14	0.054	1.81	2.2
1/2	1.09	0.147	0.703 0.672	0.145	0.145	14	0.059	1.93	2.3
3/4	1.32	0.154	0.906 0.842	0.160	0.160	11	0.066	2.24	2.6
1	1.63	0.179	1.141 1.092	0.180	0.175	11	0.073	2.44	3.1
1 1/4	1.99	0.191	1.484 1.392	0.210	0.205	10	0.084	2.80	3.7
1 1/2	2.25	0.200	1.714 1.622	0.230	0.220	10	0.091	3.01	4.4
2	2.76	0.218	2.188 2.052	0.260	0.250	10	0.106	3.39	5.2
2 1/2	3.36	0.276	2.609 2.532	0.295	0.280	8	0.121	4.03	5.9
3	4.03	0.300	3.250 3.042	0.325	0.315	8	0.139	4.29	6.9

Notes:
1. Minimum 4 full threads engagement is required.
2. Minimum wall thickness at the root of the pipe thread must be equal to or exceed the nominal wall thickness of Schedule 80 pipe.

OTHER MSS FITTINGS

There are several other MSS standard practices that cover pipe fittings other than those listed above. Please refer to the relevant MSS Standard Practice (SP).

MSS SP-44 FLANGES

Refer to ASME B16.47, Series A flanges. For flanges other than ASME B16.47, Series A flanges covered by MSS SP-44, refer to SP-44.

Tables in Chapter 8

CHAPTER 9
CAST IRON PIPE, FITTINGS, FLANGES AND JOINTS

See pp. 393–394 for a complete list of tables in Chap. 9.

CAST IRON PRESSURE AND GRAVITY PIPE

Cast iron pressure pipe is no longer manufactured in the United States and other countries; however, it is possible that it might be manufactured in some countries. Ductile iron pressure pipe has replaced cast iron pressure pipe. Therefore, readers should refer to Chap. 10 for ductile iron pipe, fittings, and flanges.

Dimensions of cast iron pipe and fittings for gravity applications are covered by this chapter following the cast iron pressure flanges and flanged fittings conforming to ASME B16.1, Cast Iron Pipe Flanges and Flanged Fittings.

CAST IRON PIPE FLANGES AND FLANGED FITTINGS

ASTM Material for Cast Iron Flanges and Flanged Fittings

The applicable document is ASTM A126, Specification for Gray Iron Castings for Valves, Flanges and Flanged Pipe Fittings. Classes A and B iron can be used for manufacturing Class 25, 125, and 250 flanges and flanged fittings as shown in Table 9.1.

TABLE 9.1 Iron Class A or B for Flanges and Flanged Fittings

Rating class	Size (NPS)	ASTM A126, Class A or B iron
25	All	A
125	1 to 12	A or B
125	14 and above	B
250	1 to 12	A or B
250	14 and above	B

Presssure-Temperature Ratings of Cast Iron Flanges and Flanged Fittings

See Table 9.2 to 9.6.

TABLE 9.2 Pressure-Temperature Ratings of Class 25, ASTM A126, Class A Flanges*

	Pressure, psig	
Temperature, °F	NPS 4–36	NPS 42–96
–20 to 150	45	25
200	40	25
225	35	25
250	30	25

* The pressure rating is the maximum non-shock pressure corresponding to the temperature listed.

TABLE 9.3 Pressure-Temperature Ratings of Class 125, ASTM A126, Class A Flanges*

Temperature, °F	Pressure, psig, NPS 1–12
−20 to 150	175
200	165
225	155
250	150
275	145
300	140
325	130
353†	125

* The pressure rating is the maximum nonshock pressure corresponding to the temperature listed.

† 353°F is the temperature of 125 psig saturated steam.

TABLE 9.4 Pressure-Temperature Ratings of Class 125, ASTM A126, Class B Flanges*

Temperature, °F	Pressure, psig		
	NPS 1–12	NPS 14–24	NPS 30–48
−20 to 150	200	150	150
200	190	135	115
225	180	130	100
250	175	125	85
275	170	120	65
300	165	110	50
325	155	105	
353†	150	100	
375	145		
406‡	140		
425	130		
450	125		

* The pressure rating is the maximum nonshock pressure corresponding to the temperature listed.

† 353°F is the temperature of 125 psig saturated steam.

‡ 406°F is the temperature of 250 psig saturated steam.

TABLE 9.5 Pressure-Temperature Ratings of Class 250, ASTM A126, Class A Flanges*

Temperature, °F	Pressure, psig, NPS 1–12
−20 to 150	400
200	370
225	355
250	340
275	325
300	310
325	295
353†	280
375	265
406‡	250

* The pressure rating is the maximum nonshock pressure corresponding to the temperature listed.

† 353°F is the temperature of 125 psig saturated steam.

‡ 406°F is the temperature of 250 psig saturated steam.

TABLE 9.6 Pressure-Temperature Ratings of Class 250, ASTM A126, Class B Flanges*

Temperature, °F	Pressure, psig NPS 1–12	NPS 14–24	NPS 30–48
–20 to 150	500	300	300
200	460	280	250
225	440	270	225
250	415	260	200
275	395	250	175
300	375	240	150
325	355	230	125
353†	335	220	100
375	315	210	
406‡	290	200	

* The pressure rating is the maximum nonshock pressure corresponding to the temperature listed.
† 353°F is the temperature of 125 psig saturated steam.
‡ 406°F is the temperature of 250 psig saturated steam.

Dimensions of Class 25 Cast Iron Flanges and Flanged Fittings

See Tables 9.7, 9.8, and 9.9 and Figs. 9.1 and 9.2.

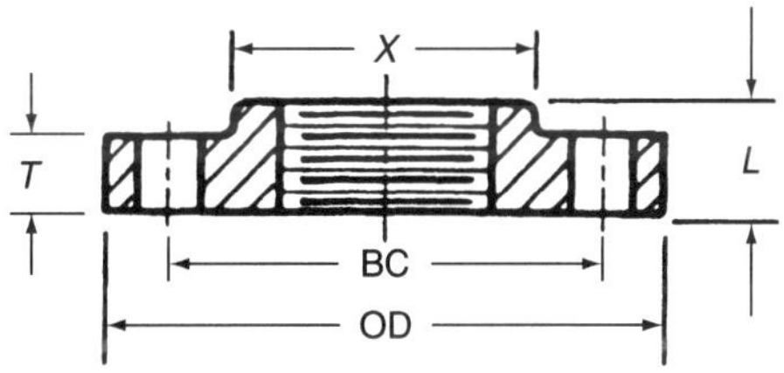

TABLE 9.7 Dimensions of Class 25 Cast Iron Flanges*

Nominal pipe size	Flange diameter OD	Flange thickness, min., T	Hub diameter, min., X	Hub length, min., L	Diameter of bolt circle BC	Number of bolts	Diameter of bolts	Length of bolts
4	9.00	0.75	5.31	1.31	7.50	8	0.62	2.50
5	10.00	0.75	6.44	1.44	8.50	8	0.62	2.50
6	11.00	0.75	7.56	1.56	9.50	8	0.62	2.50
8	13.50	0.75	9.69	1.75	11.75	8	0.62	2.50
10	16.00	0.88	11.94	1.94	14.25	12	0.62	3.00
12	19.00	1.00	14.06	2.19	17.00	12	0.62	3.00
14	21.00	1.12	15.37	2.25	18.75	12	0.75	3.50
16	23.50	1.12	17.50	2.50	21.25	16	0.75	3.50
18	25.00	1.25	19.62	2.69	22.75	16	0.75	4.00
20	27.50	1.25	21.75	2.88	25.00	20	0.75	4.00
24	32.00	1.38	26.00	3.25	29.50	20	0.75	4.00
30	38.75	1.50	...	...	36.00	28	0.88	4.50
36	46.00	1.62	...	...	42.75	32	0.88	5.00
42	53.00	1.75	...	...	49.50	36	1.00	5.50
48	59.50	2.00	...	...	56.00	44	1.00	6.00
54	66.25	2.25	...	...	62.75	44	1.00	6.50
60	73.00	2.25	...	...	69.25	52	1.12	6.50
72	86.50	2.50	...	...	82.50	60	1.12	7.00
84	99.75	2.75	...	...	95.50	64	1.25	7.50
96	113.25	3.00	...	...	108.50	68	1.25	8.00

* Dimensions in inches.

TABLE 9.8 Dimensions of Class 25 Cast Iron Straight Elbows, Tees, and Crosses (Refer to Fig. 9.1)*

Nominal pipe size	Diameter of flange	Thickness of flange, min.	Wall thickness of body	Inside diameter of fitting	Center-to-face elbow, tee, and cross, *A*	Center-to-face long-radius elbow, *B*	Center-to-face 45° elbow, *C*
4	9.00	0.75	0.42	4	6.5	9.00	4.0
5	10.00	0.75	0.44	5	7.5	10.25	4.5
6	11.00	0.75	0.44	6	8.0	11.50	5.0
8	13.50	0.75	0.46	8	9.0	14.00	5.5
10	16.00	0.88	0.50	10	11.0	16.50	6.5
12	19.00	1.00	0.54	12	12.0	19.00	7.5
14	21.00	1.12	0.57	14	14.0	21.50	7.5
16	23.50	1.12	0.60	16	15.0	24.00	8.0
18	25.00	1.25	0.64	18	16.5	26.50	8.5
20	27.50	1.25	0.67	20	18.0	29.00	9.5
24	32.00	1.38	0.76	24	22.0	34.00	11.0
30	38.75	1.50	0.88	30	25.0	41.50	15.0
36	46.00	1.62	0.99	36	28.0	49.00	18.0
42	53.00	1.75	1.10	42	31.0	56.50	21.0
48	59.50	2.00	1.26	48	34.0	64.00	24.0
54	66.25	2.25	1.35	54	39.0	71.50	27.0
60	73.00	2.25	1.39	60	44.0	79.00	30.0
72	86.50	2.50	1.62	72	53.0	94.00	36.0

* Dimensions in inches.

TABLE 9.9 Dimensions of Class 25 Reducing Tees and Crosses (Short-Body Pattern) and Elbows (Refer to Fig. 9.2)*

Nominal pipe size	Center-to-face elbow, *A*	NPS size of outlet and smaller†	Center-to-face, run, *H*†	Center-to-face, outlet, *J*†
4†	6.5			
5†	7.5			
6†	8.0			
8†	9.0			
10†	11.0			
12†	12.0			
14†	14.0			
16†	15.0			
18	16.5	12.0	13.0	15.5
20	18.0	14.0	14.0	17.0
24	22.0	16.0	15.0	19.0
30	25.0	20.0	18.0	23.0
36	28.0	24.0	20.0	26.0
42	31.0	24.0	23.0	30.0
48	34.0	30.0	26.0	34.0
54	39.0	36.0	29.0	37.0
60	44.0	40.0	33.0	41.0
72	53.0	48.0	40.0	48.0

* Dimensions in inches.

† NPS 16 and smaller reducing filtings have the same center-to-face dimensions as straight-size fittings. NPS 16 and smaller tees, side-outlet tees, and crosses reducing on the outlet have the same dimensions center to face and face to face as straight-size filtings corresponding to the larger opening. See Table 9.8.

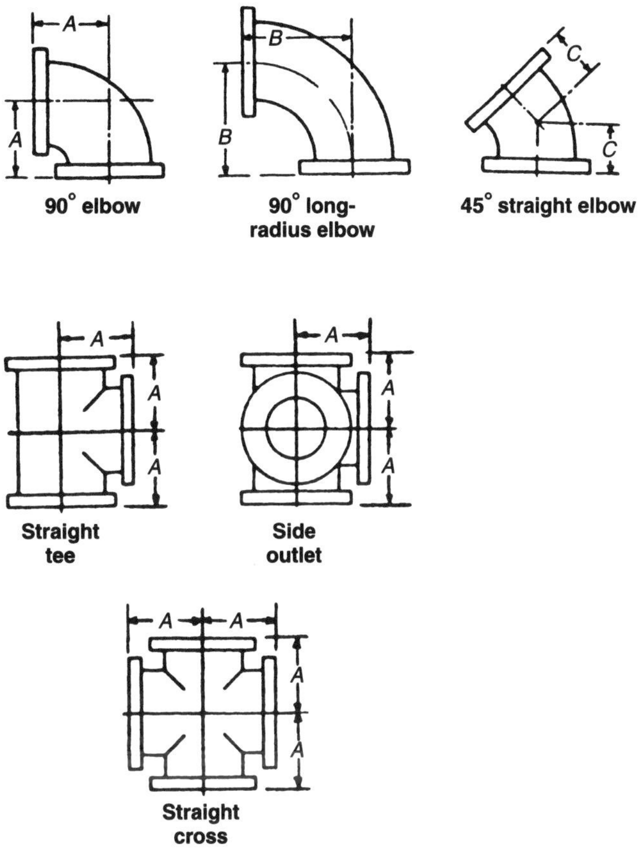

FIGURE 9.1 Class 25 cast iron straight elbows, tees, and crosses. (*Courtesy of ASME.*)

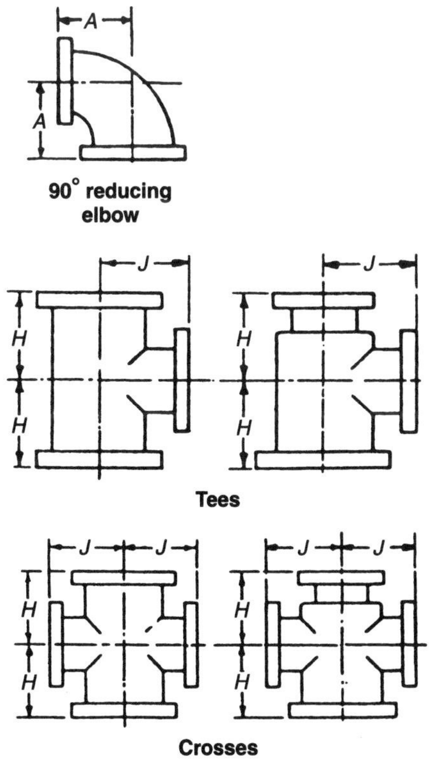

FIGURE 9.2 Class 25 cast iron reducing elbows, tees, and crosses. (*Courtesy of ASME.*)

Dimensions of Class 125 Cast Iron Flanges and Flanged Fittings

See Tables 9.10 to 9.14 and Figs. 9.3 to 9.7.

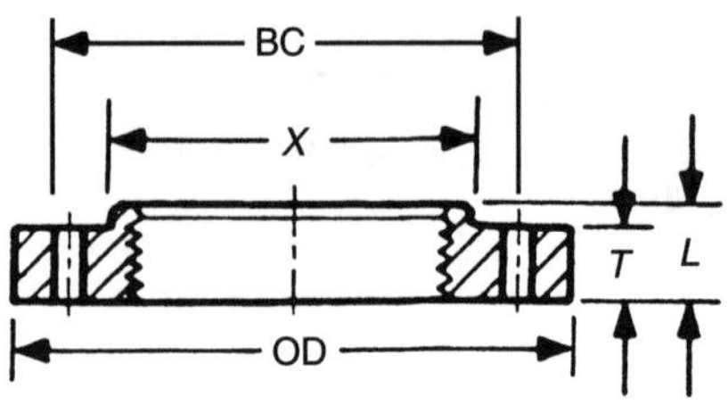

FIGURE 9.3 Class 125 cast iron threaded flange.

TABLE 9.10 Dimensions of Class 125 Cast Iron Flanges (Refer to Fig. 9.3)*

Nominal pipe size	Diameter of flange, OD	Thickness of flange, min., *T*	Diameter of hub, min., *X*	Length of hub and threads, min., *L*	Diameter of bolt circle, BC	Number of bolts	Diameter of bolts	Length of bolts	Length of bolt stud with two nuts
1	4.25	0.44	1.94	0.69	3.12	4	0.50	2.00	. . .
1¼	4.62	0.50	2.31	0.81	3.50	4	0.50	2.00	. . .
1½	5.00	0.56	2.56	0.88	3.88	4	0.50	2.00	. . .
2	6.00	0.62	3.06	1.00	4.75	4	0.62	2.50	. . .
2½	7.00	0.69	3.56	1.12	5.50	4	0.62	2.50	. . .
3	7.50	0.75	4.25	1.19	6.00	4	0.62	2.50	. . .
3½	8.50	0.81	4.81	1.25	7.00	8	0.62	3.00	. . .
4	9.00	0.94	5.31	1.31	7.50	8	0.62	3.00	. . .
5	10.00	0.94	6.44	1.44	8.50	8	0.75	3.00	. . .
6	11.00	1.00	7.56	1.56	9.50	8	0.75	3.50	. . .
8	13.50	1.12	9.69	1.75	11.75	8	0.75	3.50	. . .
10	16.00	1.19	11.94	1.94	14.25	12	0.88	4.00	. . .
12	19.00	1.25	14.06	2.19	17.00	12	0.88	4.00	. . .
14	21.00	1.38	15.38	2.25	18.75	12	1.00	4.50	. . .
16	23.50	1.44	17.50	2.50	21.25	16	1.00	4.50	. . .
18	25.00	1.56	19.62	2.69	22.75	16	1.12	5.00	. . .
20	27.50	1.69	21.75	2.88	25.00	20	1.12	5.00	. . .
24	32.00	1.88	26.00	3.25	29.50	20	1.25	5.50	. . .
30	38.75	2.12	. . .	. . .	36.00	28	1.25	6.50	. . .
36	46.00	2.38	. . .	. . .	42.75	32	1.50	7.00	9.00
42	53.00	2.62	. . .	. . .	49.50	36	1.50	7.50	9.50
48	59.50	2.75	. . .	. . .	56.00	44	1.50	8.00	9.50
54†	66.25	3.00	. . .	. . .	62.75	44	1.75	8.50	10.50
60†	73.00	3.12	. . .	. . .	69.25	52	1.75	9.00	11.00
72†	86.50	3.50	. . .	. . .	82.50	60	1.75	9.50	11.50
84†	99.75	3.88	. . .	. . .	95.50	64	2.00	10.50	13.00
96†	113.25	4.25	. . .	. . .	108.50	68	2.25	11.50	14.00

* Dimensions in inches.

† Pressure-temperature ratings for Class 125 cast iron flanges, NPS 54 and larger, and not covered by ASME B16.1.

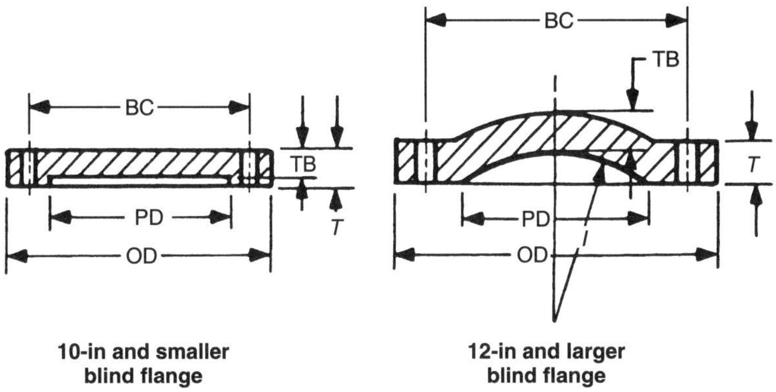

FIGURE 9.4 Class 125 cast iron blind flanges.

TABLE 9.11 Dimensions of Class 125 Cast Iron Blind Flanges (Refer to Fig. 9.4)*

Nominal pipe size	Diameter of flange, OD	Thickness of flange, min., *T*	Diameter of port, PD	Wall thickness, TB	Diameter of bolt circle, BC	Number of bolts	Diameter of bolts	Length of bolts	Length of bolt stud with two nuts
1	4.25	0.44	1.00	0.38	3.12	4	0.50	2.00	. . .
$1\frac{1}{4}$	4.62	0.50	1.25	0.44	3.50	4	0.50	2.00	. . .
$1\frac{1}{2}$	5.00	0.56	1.50	0.50	3.88	4	0.50	2.00	. . .
2	6.00	0.62	2.00	0.56	4.75	4	0.62	2.50	. . .
$2\frac{1}{2}$	7.00	0.69	2.50	0.63	5.50	4	0.62	2.50	. . .
3	7.50	0.75	3.00	0.69	6.00	4	0.62	2.50	. . .
$3\frac{1}{2}$	8.50	0.81	3.50	0.75	7.00	8	0.62	3.00	. . .
4	9.00	0.94	4.00	0.88	7.50	8	0.62	3.00	. . .
5	10.00	0.94	5.00	0.88	8.50	8	0.75	3.00	. . .
6	11.00	1.00	6.00	0.94	9.50	8	0.75	3.50	. . .
8	13.50	1.12	8.00	1.06	11.75	8	0.75	3.50	. . .
10	16.00	1.19	10.00	1.12	14.25	12	0.88	4.00	. . .
12†	19.00	1.25	12.00	0.81	17.00	12	0.88	4.00	. . .
14†	21.00	1.38	14.00	0.88	18.75	12	1.00	4.50	. . .
16†	23.50	1.44	16.00	1.00	21.25	16	1.00	4.50	. . .
18†	25.00	1.56	18.00	1.06	22.75	16	1.12	5.00	. . .
20†	27.50	1.69	20.00	1.12	25.00	20	1.12	5.00	. . .
24†	32.00	1.88	24.00	1.25	29.50	20	1.25	5.50	. . .
30†	38.75	2.12	30.00	1.44	36.00	28	1.25	6.50	. . .
36†	46.00	2.38	36.00	1.62	42.75	32	1.50	7.00	9.00
42†	53.00	2.62	42.00	1.81	49.50	36	1.50	7.50	9.50
48†	59.50	2.75	48.00	2.00	56.00	44	1.50	8.00	9.50

* Dimensions in inches.
† All NPS 12 and larger blind flanges must be dished with inside radius equal to the port diameter (PD).

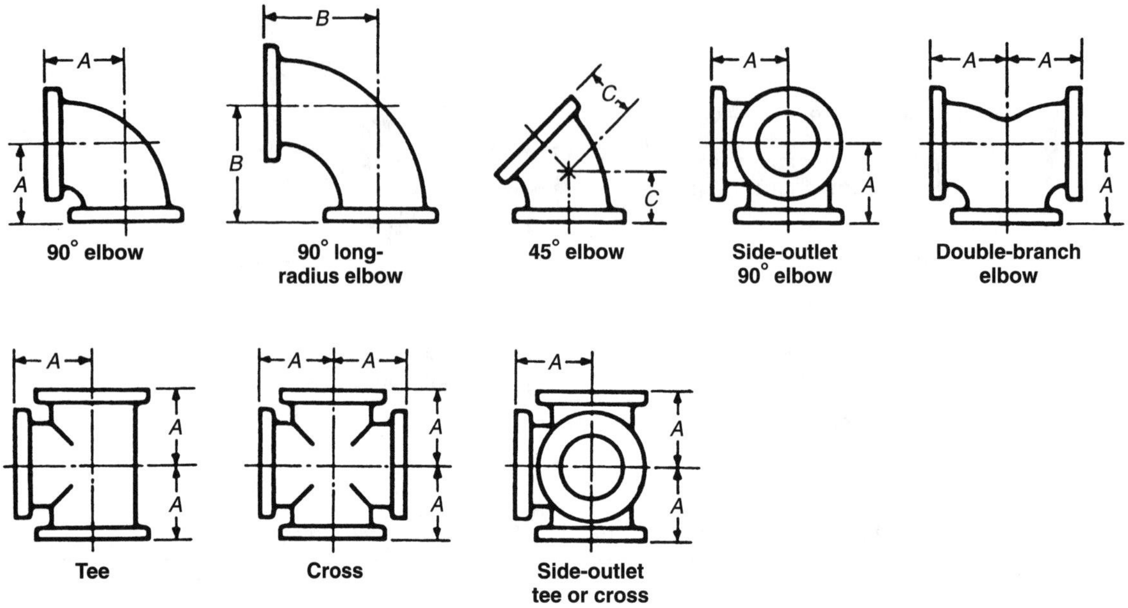

FIGURE 9.5 Class 125 cast iron straight elbows, tees, and crosses.

TABLE 9.12 Dimensions of Class 125 Cast Iron Straight Elbows, Tees, and Crosses (Refer to Fig. 9.5)

Nominal pipe size	Diameter of flange	Thickness of flange, min.	Inside diameter of fittings	Wall thickness	Center-to-face 90° elbow, tees, crosses, and double-branch elbow, *A*	Center-to-face 90° long-radius elbow, *B*	Center-to-face 45° elbow, *C*
1	4.25	0.44	1.00	0.31	3.50	5.00	1.75
$1\frac{1}{4}$	4.62	0.50	1.25	0.31	3.75	5.50	2.00
$1\frac{1}{2}$	5.00	0.56	1.50	0.31	4.00	6.00	2.25
2	6.00	0.62	2.00	0.31	4.50	6.50	2.50
$2\frac{1}{2}$	7.00	0.69	2.50	0.31	5.00	7.00	3.00
3	7.50	0.75	3.00	0.38	5.50	7.75	3.00
$3\frac{1}{2}$	8.50	0.81	3.50	0.44	6.00	8.50	3.50
4	9.00	0.94	4.00	0.50	6.50	9.00	4.00
5	10.00	0.94	5.00	0.50	7.50	10.25	4.50
6	11.00	1.00	6.00	0.56	8.00	11.50	5.00
8	13.50	1.12	8.00	0.62	9.00	14.00	5.50
10	16.00	1.19	10.00	0.75	11.00	16.50	6.50
12	19.00	1.25	12.00	0.81	12.00	19.00	7.50
14	21.00	1.38	14.00	0.88	14.00	21.50	7.50
16	23.50	1.44	16.00	1.00	15.00	24.00	8.00
18	25.00	1.56	18.00	1.06	16.50	26.50	8.50
20	27.50	1.69	20.00	1.12	18.00	29.00	9.50
24	32.00	1.88	24.00	1.25	22.00	34.00	11.00
30	38.75	2.12	30.00	1.44	25.00	41.50	15.00
36	46.00	2.38	36.00	1.62	28.00	49.00	18.00
42	53.00	2.62	42.00	1.81	31.00	56.50	21.00
48	59.50	2.75	48.00	2.00	34.00	64.00	24.00

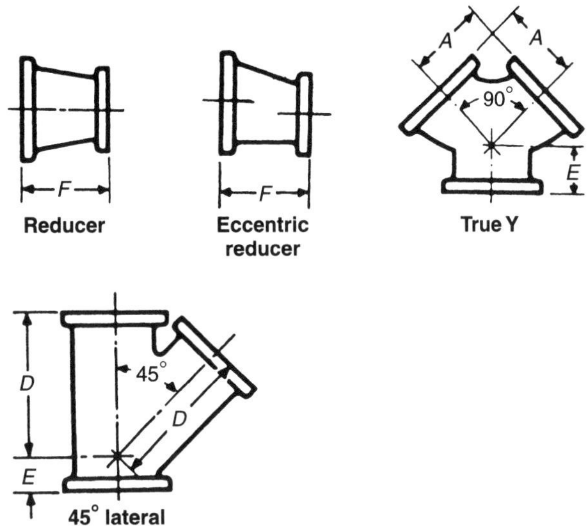

FIGURE 9.6 Class 125 cast iron flanged laterals, reducers, and true Y's. *(Courtesy of ASME.)*

TABLE 9.13 Dimensions of Class 125 Cast Iron Flanged Lateral, Reducers, and True Y's (Refer to Fig. 9.6)*

Nominal pipe size	Diameter of flange	Thickness of flange, min.	Inside diameter of fittings	Wall thickness	Center-to-face true Y, *A*	Center-to-face lateral, *D*	Short center-to-face true Y and lateral, *E*	Face-to-face reducer, *F*
1	4.25	0.44	1.00	0.31	3.50	5.75	1.75	. . .
1¼	4.62	0.50	1.25	0.31	3.75	6.25	1.75	. . .
1½	5.00	0.56	1.50	0.31	4.00	7.00	2.00	. . .
2	6.00	0.62	2.00	0.31	4.50	8.00	2.50	5.0
2½	7.00	0.69	2.50	0.31	5.00	9.50	2.50	5.5
3	7.50	0.75	3.00	0.38	5.50	10.00	3.00	6.0
3½	8.50	0.81	3.50	0.44	6.00	11.50	3.00	6.5
4	9.00	0.94	4.00	0.50	6.50	12.00	3.00	7.0
5	10.00	0.94	5.00	0.50	7.50	13.50	3.50	8.0
6	11.00	1.00	6.00	0.56	8.00	14.50	3.50	9.0
8	13.50	1.12	8.00	0.62	9.00	17.50	4.50	11.0
10	16.00	1.19	10.00	0.75	11.00	20.50	5.00	12.0
12	19.00	1.25	12.00	0.81	12.00	24.50	5.50	14.0
14	21.00	1.38	14.00	0.88	14.00	27.00	6.00	16.0
16	23.50	1.44	16.00	1.00	15.00	30.00	6.50	18.0
18	25.00	1.56	18.00	1.06	16.50	32.00	7.00	19.0
20	27.50	1.69	20.00	1.12	18.00	35.00	8.00	20.0
24	32.00	1.88	24.00	1.25	22.00	40.50	9.00	24.0
30	38.75	2.12	30.00	1.44	25.00	49.00	10.00	30.0
36	46.00	2.38	36.00	1.62	28.00	. . .	. . .	36.0
42	53.00	2.62	42.00	1.81	31.00	. . .	. . .	42.0
48	59.50	2.75	48.00	2.00	34.00	. . .	. . .	48.0

* Dimensions in inches.

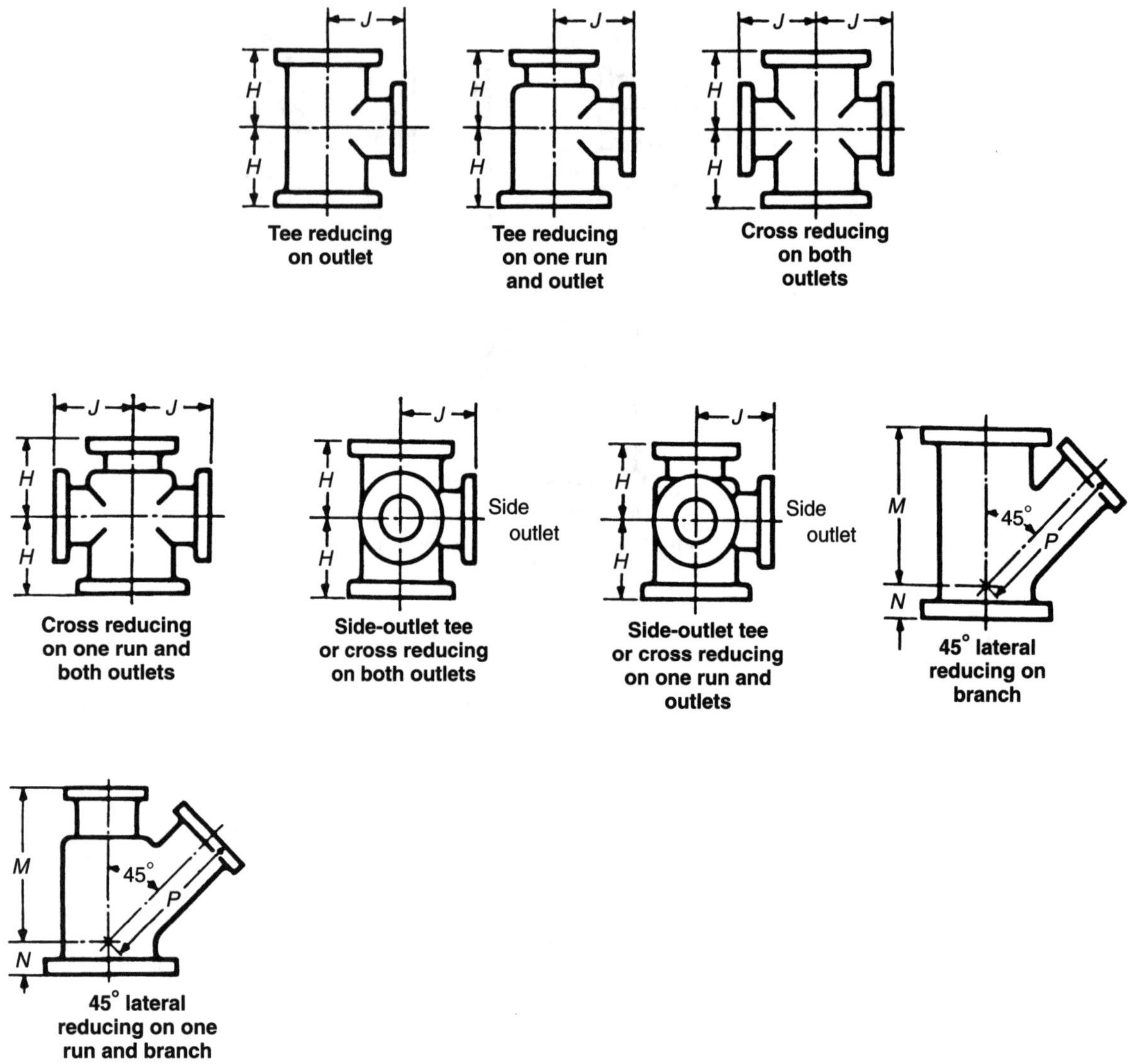

FIGURE 9.7 Class 125 cast iron short-body-pattern reducing tees, crosses, and laterals. (*Courtesy of ASME.*)

TABLE 9.14 Dimensions of Class 125 Cast Iron Short-Body-Pattern Reducing Tees, Crosses, and Laterals (Refer to Fig. 9.7)*

					Tees and crosses†			Laterals†			
Nominal pipe size	Diameter of flange	Thickness of flange, min.	Inside diameter of fittings	Wall thickness	NPS size of outlet and smaller	Center-to-face run, *H*	Center-to-face outlet or side outlet, *J*	NPS size of branch and smaller	Center-to-face run, *M*	Center-to-face run, *N*	Center-to-face branch, *P*
1	4.25	0.44	1.00	0.31							
1¼	4.62	0.50	1.25	0.31							
1½	5.00	0.56	1.50	0.31							
2	6.00	0.62	2.00	0.31							
2½	7.00	0.69	2.50	0.31							
3	7.50	0.75	3.00	0.38							
3½	8.50	0.81	3.50	0.44							
4	9.00	0.94	4.00	0.50							
5	10.00	0.94	5.00	0.50							
6	11.00	1.00	6.00	0.56							
8	13.50	1.12	8.00	0.62							
10	16.00	1.19	10.00	0.75							
12	19.00	1.25	12.00	0.81							
14	21.00	1.38	14.00	0.88							
16	23.50	1.44	16.00	1.00							
18	25.00	1.56	18.00	1.06	12.0	13.0	15.5	8	25.0	1.0	27.5
20	27.50	1.69	20.00	1.12	14.0	14.0	17.0	10	27.0	1.0	29.5
24	32.00	1.88	24.00	1.25	16.0	15.0	19.0	12	31.5	0.5	34.5
30	38.75	2.12	30.00	1.44	20.0	18.0	23.0	14	39.0	0.0	42.0
36	46.00	2.38	36.00	1.62	24.0	20.0	26.0				
42	53.00	2.62	42.00	1.81	24.0	23.0	30.0				
48	59.50	2.75	48.00	2.00	30.0	26.0	34.0				

* Dimensions in inches.

† Center-to-face dimensions of NPS 16 and smaller reducing tees, crosses, and laterals are the same as those of straight-size fittings corresponding to the largest opening.

Dimensions of Class 250 Cast Iron Flanges and Flanged Fittings

See Tables 9.15 to 9.19 and Figs. 9.8 to 9.12.

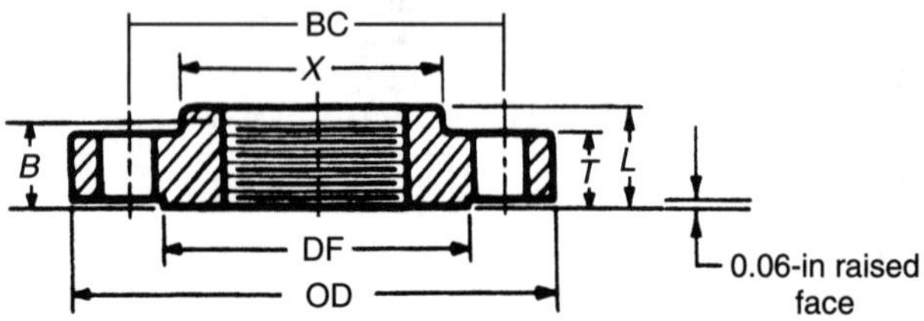

FIGURE 9.8 Class 250 cast iron flange.

TABLE 9.15 Dimensions of Class 250 Cast Iron Flanges (Refer to Fig. 9.8)*

Nominal pipe size	Diameter of flange, OD	Thickness of flange, min., *T*	Diameter of raised face, DF	Diameter of hub, min., *X*	Length through hub, min., *L*	Length of threads, min., B	Diameter of bolt circle, BC	Number of bolts	Diameter of bolts	Length of bolts	Length of stud bolts
1	4.88	0.69	2.69	2.06	0.88	0.68	3.50	4	0.62	2.50	. . .
$1\frac{1}{4}$	5.25	0.75	3.06	2.50	1.00	0.81	3.88	4	0.62	2.50	. . .
$1\frac{1}{2}$	6.12	0.81	3.56	2.75	1.12	0.87	4.50	4	0.75	3.00	. . .
2	6.50	0.88	4.19	3.31	1.25	1.00	5.00	8	0.62	3.00	. . .
$2\frac{1}{2}$	7.50	1.00	4.94	3.94	1.43	1.13	5.88	8	0.75	3.50	. . .
3	8.25	1.12	5.69	4.62	1.56	1.19	6.62	8	0.75	3.50	. . .
$3\frac{1}{2}$	9.00	1.19	6.31	5.25	1.62	1.25	7.25	8	0.75	3.50	. . .
4	10.00	1.25	6.94	5.75	1.75	1.31	7.88	8	0.75	4.00	. . .
5	11.00	1.38	8.31	7.00	1.88	1.44	9.25	8	0.75	4.00	. . .
6	12.50	1.44	9.69	8.12	1.94	1.56	10.62	12	0.75	4.00	. . .
8	15.00	1.62	11.94	10.25	2.19	1.75	13.00	12	0.88	4.50	. . .
10	17.50	1.88	14.06	12.62	2.38	1.94	15.25	16	1.00	5.50	. . .
12	20.50	2.00	16.44	14.75	2.56	2.19	17.75	16	1.12	5.50	. . .
14	23.00	2.12	18.94	16.25	2.69	2.25	20.25	20	1.12	6.00	. . .
16	25.50	2.25	21.06	18.38	2.88	2.50	22.50	20	1.25	6.50	. . .
18	28.00	2.38	23.31	. . .	. . .	. . .	24.75	24	1.25	6.50	. . .
20	30.50	2.50	25.56	. . .	. . .	. . .	27.00	24	1.25	7.00	. . .
24	36.00	2.75	30.31	. . .	. . .	. . .	32.00	24	1.50	7.50	9.50
30	43.00	3.00	37.19	. . .	. . .	. . .	39.25	28	1.75	8.50	10.50
36†	50.00	3.38	43.69	. . .	. . .	. . .	46.00	32	2.00	9.50	12.00
42†	57.00	3.69	50.44	. . .	. . .	. . .	52.75	36	2.00	10.00	12.50
48†	65.00	4.00	58.44	. . .	. . .	. . .	60.75	40	2.00	11.00	13.00

* Dimensions in inches.
† NPS 36 and larger flanges are not commonly used or commercially produced. Dimensions are given for convenience.

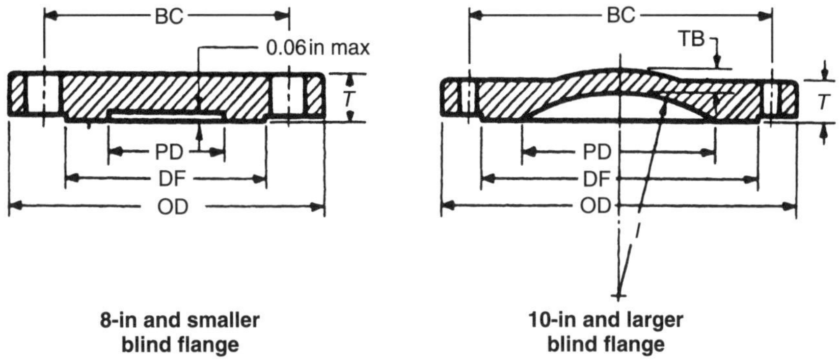

FIGURE 9.9 Class 250 cast iron blind flanges.

TABLE 9.16 Dimensions of Class 250 Cast Iron Blind Flanges (Refer to Fig. 9.9)*

Nominal pipe size	Diameter of flange, OD	Thickness of flange, min., *T*	Diameter of raised face, DF	Diameter of port, PD	Wall thickness, TB	Diameter of bolt circle, BC	Number of bolts	Diameter of bolts	Length of bolts	Length of stud bolts
1	4.88	0.69	2.69	1.00	. . .	3.50	4	0.62	2.50	. . .
1¼	5.25	0.75	3.06	1.25	. . .	3.88	4	0.62	2.50	. . .
1½	6.12	0.81	3.56	1.50	. . .	4.50	4	0.75	3.00	. . .
2	6.50	0.88	4.19	2.00	. . .	5.00	8	0.62	3.00	. . .
2½	7.50	1.00	4.94	2.50	. . .	5.88	8	0.75	3.50	. . .
3	8.25	1.12	5.69	3.00	. . .	6.62	8	0.75	3.50	. . .
3½	9.00	1.19	6.31	3.50	. . .	7.25	8	0.75	3.50	. . .
4	10.00	1.25	6.94	4.00	. . .	7.88	8	0.75	4.00	. . .
5	11.00	1.38	8.31	5.00	. . .	9.25	8	0.75	4.00	. . .
6	12.50	1.44	9.69	6.00	. . .	10.62	12	0.75	4.00	. . .
8	15.00	1.62	11.94	8.00	. . .	13.00	12	0.88	4.50	. . .
10†	17.50	1.88	14.06	10.00	0.94	15.25	16	1.00	5.50	. . .
12†	20.50	2.00	16.44	12.00	1.00	17.75	16	1.12	5.50	. . .
14†	23.00	2.12	18.94	13.25	1.12	20.25	20	1.12	6.00	. . .
16†	25.50	2.25	21.06	15.25	1.25	22.50	20	1.25	6.50	. . .
18†	28.00	2.38	23.31	17.00	1.38	24.75	24	1.25	6.50	. . .
20†	30.50	2.50	25.56	19.00	1.50	27.00	24	1.25	7.00	. . .
24†	36.00	2.75	30.31	23.00	1.62	32.00	24	1.50	7.50	9.50
30†	43.00	3.00	37.19	29.00	2.00	39.25	28	1.75	8.50	10.50

* Dimensions in inches.
† NPS 10 and larger blind flanges must be dished with inside radius equal to port diameter.

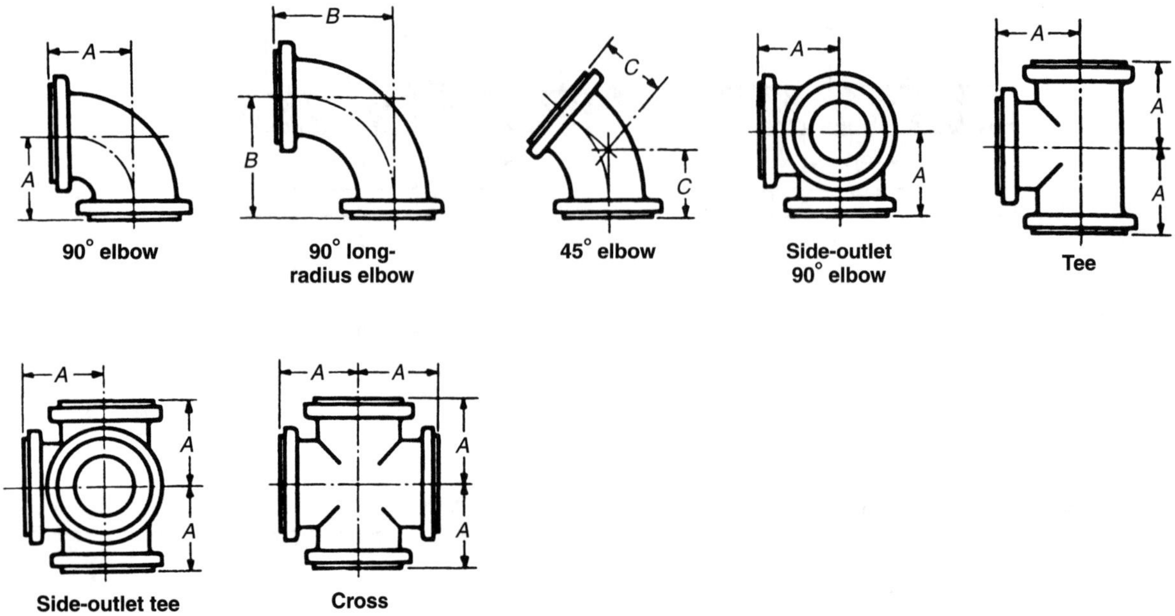

FIGURE 9.10 Class 250 cast iron straight elbows, tees, and crosses.

TABLE 9.17 Dimensions of Class 250 Cast Iron Straight Elbows, Tees, and Crosses (Refer to Fig. 9.10)*

Nominal pipe size	Diameter of flange	Thickness of flange, min., *T*	Diameter of raised face	Inside diameter of fittings	Body wall thickness	Center-to-face 90° elbows, tees, crosses, *A*	Center-to-face 90° long-radius elbows, *B*	Center-to-face 45° elbows, *C*
1	4.88	0.69	2.69	1.00	0.44	4.00	5.00	2.00
1¼	5.25	0.75	3.06	1.25	0.44	4.25	5.50	2.50
1½	6.12	0.81	3.56	1.50	0.44	4.50	6.00	2.75
2	6.50	0.88	4.19	2.00	0.44	5.00	6.50	3.00
2½	7.50	1.00	4.94	2.50	0.50	5.50	7.00	3.50
3	8.25	1.12	5.69	3.00	0.56	6.00	7.75	3.50
3½	9.00	1.19	6.31	3.50	0.56	6.50	8.50	4.00
4	10.00	1.25	6.94	4.00	0.62	7.00	9.00	4.50
5	11.00	1.38	8.31	5.00	0.69	8.00	10.25	5.00
6	12.50	1.44	9.69	6.00	0.75	8.50	11.50	5.50
8	15.00	1.62	11.94	8.00	0.81	10.00	14.00	6.00
10	17.50	1.88	14.06	10.00	0.94	11.50	16.50	7.00
12	20.50	2.00	16.44	12.00	1.00	13.00	19.00	8.00
14	23.00	2.12	18.94	13.25	1.12	15.00	21.50	8.50
16	25.50	2.25	21.06	15.25	1.25	16.50	24.00	9.50
18	28.00	2.38	23.31	17.00	1.38	18.00	26.50	10.00
20	30.50	2.50	25.56	19.00	1.50	19.50	29.00	10.50
24	36.00	2.75	30.31	23.00	1.62	22.50	34.00	12.00
30	43.00	3.00	37.19	29.00	2.00	27.50	41.50	15.00

* Dimensions in inches.

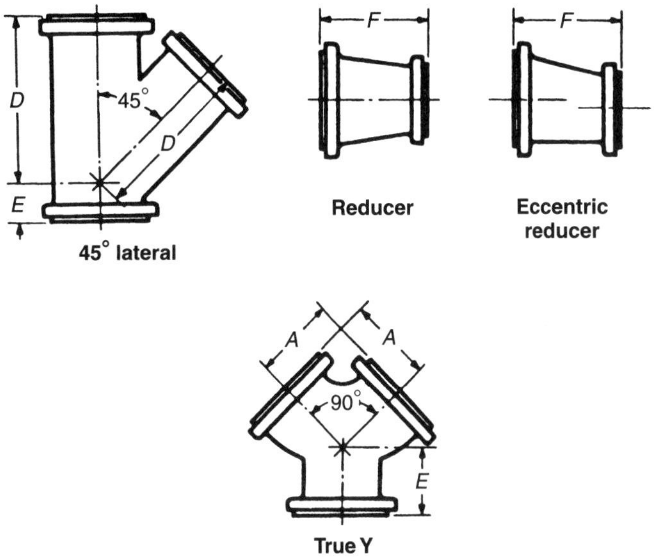

FIGURE 9.11 Class 250 cast iron flanged laterals, reducers, and true Y's.

TABLE 9.18 Dimensions of Class 250 Cast Iron Laterals, Reducers, and True Y's (Refer to Fig. 9.11)*

Nominal pipe size	Diameter of flange	Thickness of flange, min., *T*	Diameter of raised face	Inside diameter of fittings	Body wall thickness	True Y, *A*	Center-to-face lateral, *D*	Short center-to-face true Y and lateral, *E*	Face-to-face reducer, *F*
1	4.88	0.69	2.69	1.00	0.44	4.00	6.50	2.00	...
$1\frac{1}{4}$	5.25	0.75	3.06	1.25	0.44	4.25	7.25	2.25	...
$1\frac{1}{2}$	6.12	0.81	3.56	1.50	0.44	4.50	8.50	2.50	...
2	6.50	0.88	4.19	2.00	0.44	5.00	9.00	2.50	5.0
$2\frac{1}{2}$	7.50	1.00	4.94	2.50	0.50	5.50	10.50	2.50	5.5
3	8.25	1.12	5.69	3.00	0.56	6.00	11.00	3.00	6.0
$3\frac{1}{2}$	9.00	1.19	6.31	3.50	0.56	6.50	12.50	3.00	6.5
4	10.00	1.25	6.94	4.00	0.62	7.00	13.50	3.00	7.0
5	11.00	1.38	8.31	5.00	0.69	8.00	15.00	3.50	8.0
6	12.50	1.44	9.69	6.00	0.75	8.50	17.50	4.00	9.0
8	15.00	1.62	11.94	8.00	0.81	10.00	20.50	5.00	11.0
10	17.50	1.88	14.06	10.00	0.94	11.50	24.00	5.50	12.0
12	20.50	2.00	16.44	12.00	1.00	13.00	27.50	6.00	14.0
14	23.00	2.12	18.94	13.25	1.12	15.00	31.00	6.50	16.0
16	25.50	2.25	21.06	15.25	1.25	16.50	34.50	7.50	18.0
18	28.00	2.38	23.31	17.00	1.38	18.00	37.50	8.00	19.0
20	30.50	2.50	25.56	19.00	1.50	19.50	40.50	8.50	20.0
24	36.00	2.75	30.31	23.00	1.62	22.50	47.50	10.00	24.0
30	43.00	3.00	37.19	29.00	2.00	27.50	...	...	30.0

* Dimensions in inches.

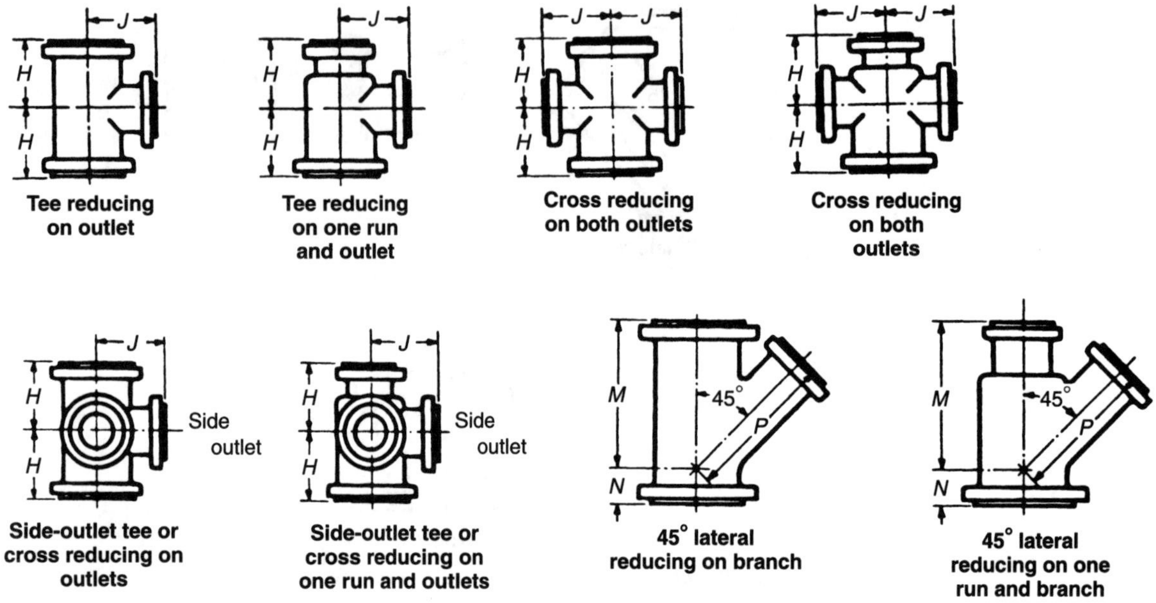

FIGURE 9.12 Class 250 cast iron short-body-pattern reducing tees, crosses, and laterals.

TABLE 9.19 Dimensions of Class 250 Cast Iron Short-Body-Pattern Reducing Tees, Crosses, and Laterals (Refer to Fig. 9.12)*

						Tees and crosses†			Laterals†			
Nominal pipe size	Diameter of flange	Thickness of flange, min.	Diameter of raised face	Inside diameter of fittings	Body wall thickness	NPS size of outlet and smaller	Center-to-face run, *H*	Center-to-face outlet or side outlet, *J*	NPS size of branch and smaller	Center-to-face run, *M*	Center-to-face run, *N*	Center-to-face branch, *P*
1	4.88	0.69	2.69	1.00	0.44							
1¼	5.25	0.75	3.06	1.25	0.44							
1½	6.12	0.81	3.56	1.50	0.44							
2	6.50	0.88	4.19	2.00	0.44							
2½	7.50	1.00	4.94	2.50	0.50							
3	8.25	1.12	5.69	3.00	0.56							
3½	9.00	1.19	6.31	3.50	0.56							
4	10.00	1.25	6.94	4.00	0.62							
5	11.00	1.38	8.31	5.00	0.69							
6	12.50	1.44	9.69	6.00	0.75							
8	15.00	1.62	11.94	8.00	0.81							
10	17.50	1.88	14.06	10.00	0.94							
12	20.50	2.00	16.44	12.00	1.00							
14	23.00	2.12	18.94	13.25	1.12							
16	25.50	2.25	21.06	15.25	1.25							
18	28.00	2.38	23.31	17.00	1.38	12	14.0	17.0	8	31	3	32.5
20	30.50	2.50	25.56	19.00	1.50	14	15.5	18.5	10	34	3	36
24	36.00	2.75	30.31	23.00	1.62	16	17.0	21.5	12	41	3	43
30	43.00	3.00	37.19	29.00	2.00	20	20.5	25.5				

* Dimensions in inches.
† Center-to-face dimensions of NPS 16 and smaller reducing tees, crosses, and laterals are the same as those of straight-size fittings corresponding to the largest opening.

Minimum Thread Length for Drain Tappings in Cast Iron Flanges and Flanged Fittings

See Table 9.20.

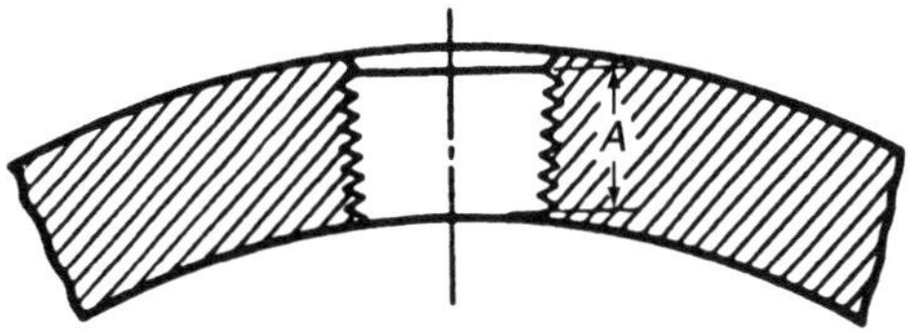

Drain tappings

TABLE 9.20 Minimum Thread Length

Size of tapping, in	Length of thread, A, in
$\frac{3}{8}$	0.41
$\frac{1}{2}$	0.53
$\frac{3}{4}$	0.55
1	0.68
$1\frac{1}{4}$	0.71
$1\frac{1}{2}$	0.72
2	0.76

ASTM A74 CAST IRON SOIL PIPE FOR GRAVITY APPLICATION

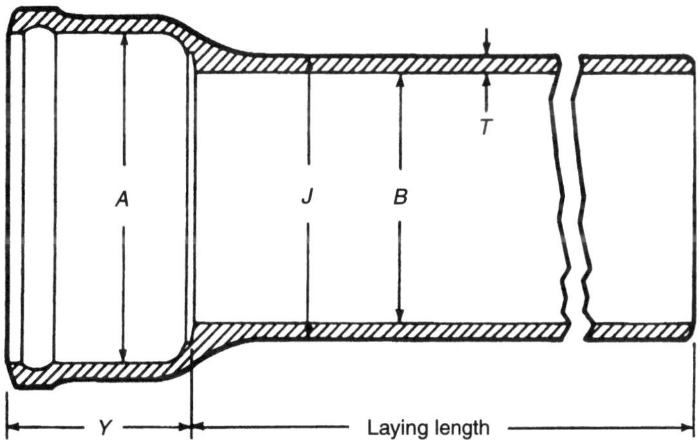

FIGURE 9.13 Single-hub cast iron soil pipe. Laying length can be 3.5, 5.0, or 10.0 ft, as specified in ASTM A74.

ASTM A74 Cast Iron Soil Pipe and fittings for gravity applications are manufactured in extra-heavy (XH) and service (SV)weights. See Tables 9.21 through 9.24.

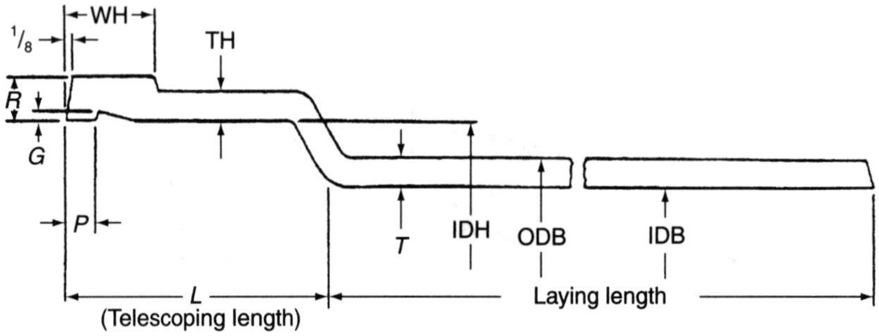

TABLE 9.21 Extra-Heavy Cast Iron Pipe Dimensions*

Size (nominal inside diameter)	Inside diameter of hub, IDH	Outside diameter of barrel, ODB	Telescoping length, L	Inside diameter of barrel, IDB	Thickness of barrel, T Nom.	Thickness of barrel, T Min.
2	3.06	2.38	2.50	2.00	0.19	0.16
3	4.19	3.50	2.75	3.00	0.25	0.22
4	5.19	4.50	3.00	4.00	0.25	0.22
5	6.19	5.50	3.00	5.00	0.25	0.22
6	7.19	6.50	3.00	6.00	0.25	0.22
8	9.50	8.62	3.50	8.00	0.31	0.25
10	11.62	10.75	3.50	10.00	0.37	0.31
12	13.75	12.75	4.25	12.00	0.37	0.31
15	16.95	15.88	4.25	15.00	0.44	0.38

* Dimensions in inches.

TABLE 9.22 Hub Dimensions of Extra-Heavy Cast Iron Pipe*

Size	Thickness of Hub: Hub body, TH (min.)	Thickness of Hub: Over bead, R (min.)	Width of hub bead, WH	Distance from lead groove to end, pipe and fittings, P	Depth of lead groove: G (min.)	Depth of lead groove: G (max.)
2	0.18	0.37	0.75	0.22	0.10	0.19
3	0.25	0.43	0.81	0.22	0.10	0.19
4	0.25	0.43	0.88	0.22	0.10	0.19
5	0.25	0.43	0.88	0.22	0.10	0.19
6	0.25	0.43	0.88	0.22	0.10	0.19
8	0.34	0.59	1.19	0.38	0.15	0.22
10	0.40	0.65	1.19	0.38	0.15	0.22
12	0.40	0.65	1.44	0.47	0.15	0.22
15	0.46	0.71	1.44	0.47	0.15	0.22

* Dimensions in inches.

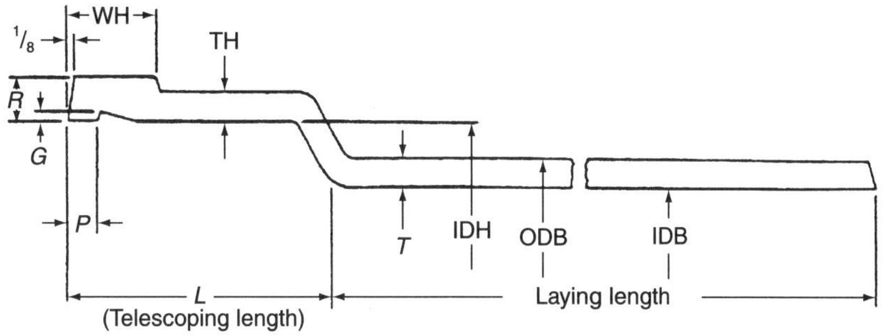

TABLE 9.23 Dimensions of Service Cast Iron Soil Pipe*

Size (nominal inside diameter)	Inside diameter of hub, IDH	Outside diameter of barrel, ODB	Telescoping length, *L*	Inside diameter of barrel, IDB	Thickness of barrel, *T*	
					Nom.	Min.
2	2.94	2.30	2.50	1.96	0.17	0.14
3	3.94	3.30	2.75	2.96	0.17	0.14
4	4.94	4.30	3.00	3.94	0.18	0.15
5	5.94	5.30	3.00	4.94	0.18	0.15
6	6.94	6.30	3.00	5.94	0.18	0.15
8	9.25	8.38	3.50	7.94	0.23	0.17
10	11.38	10.50	3.50	9.94	0.28	0.22
12	13.50	12.50	4.25	11.94	0.28	0.22
15	16.95	15.88	4.25	15.16	0.36	0.30

* Dimensions in inches.

TABLE 9.24 Hub Dimensions of Service Cast Iron Soil Pipe*

Size	Thickness of hub		Width of hub bead, *F*	Distance from lead groove to end, pipe and fittings, *P*	Depth of lead groove	
	Hub body, *S* (min.)	Over bead, *R* (min.)			*G* (min.)	*G* (max.)
2	0.13	0.34	0.75	0.22	0.10	0.19
3	0.16	0.37	0.81	0.22	0.10	0.19
4	0.16	0.37	0.88	0.22	0.10	0.19
5	0.16	0.37	0.88	0.22	0.10	0.19
6	0.18	0.37	0.88	0.22	0.10	0.19
8	0.19	0.44	1.19	0.38	0.15	0.22
10	0.27	0.53	1.19	0.38	0.15	0.22
12	0.27	0.53	1.44	0.47	0.15	0.22
15	0.30	0.58	1.44	0.47	0.15	0.22

* Dimensions in inches.

CAST IRON SOIL PIPE FITTINGS

See Tables 9.25 through 9.53.

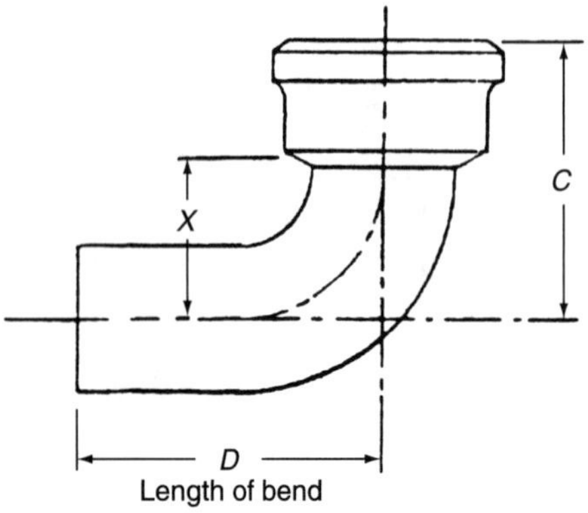

TABLE 9.25 Dimensions of One-Quarter Bends (Extra-Heavy and Service Weights)

	Dimensions, in		
Size, in	*C*	*D*	*X*
2	5 3/4	6	3 1/4
3	6 3/4	7	4
4	7 1/2	8	4 1/2
5	8	8 1/2	5
6	8 1/2	9	5 1/2
8	10 1/6	11 1/2	6 5/6
10	11 1/6	12 1/2	7 5/6
12	13	15	8 3/4
15	14 1/2	16 1/2	10 1/4

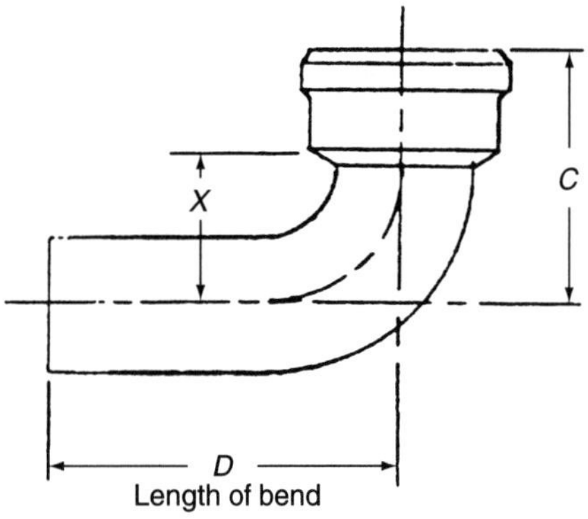

TABLE 9.26 Dimensions of Long One-Quarter Bends (Extra-Heavy and Service Weights)

	Dimensions, in		
Size, in	*C*	*D*	*X*
2 × 12	5 3/4	12	3 1/4
2 × 18	5 3/4	18	3 1/4
2 × 24	5 3/4	24	3 1/4
3 × 12	6 3/4	12	4
3 × 18	6 3/4	18	4
3 × 24	6 3/4	24	4
4 × 12	7 1/2	12	4 1/2
4 × 18	7 1/2	18	4 1/2
4 × 24	7 1/2	24	4 1/2

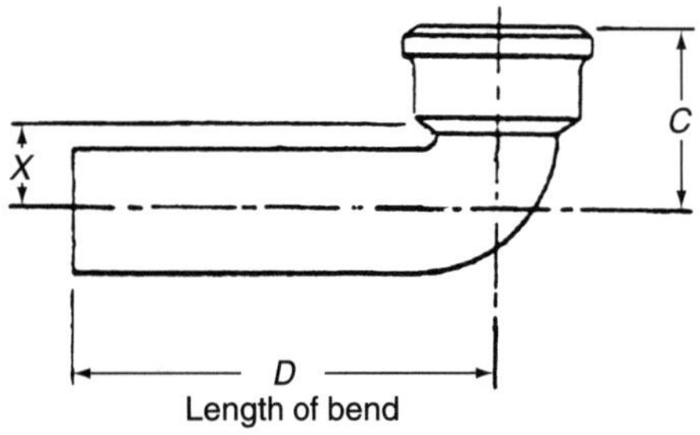

TABLE 9.27 Dimensions of Long Low-Hub One-Quarter Bends (Service Weight)

	Dimensions, in		
Size, in	*C*	*D*	*X*
4 × 12	5 3/4	12	2 3/4
4 × 14	5 3/4	14	2 3/4
4 × 16	5 3/4	16	2 3/4
4 × 18	5 3/4	18	2 3/4

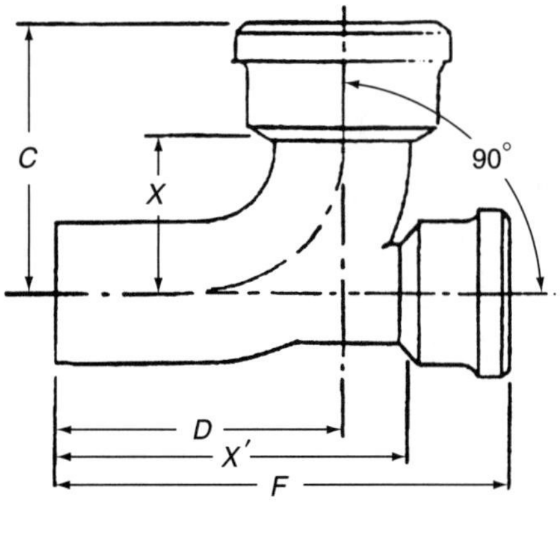

TABLE 9.28 Dimensions of One-Quarter Bends with Low-Heel Inlet (Extra-Heavy and Service Weights)

	Dimensions, in				
Size, in	*C*	*D*	*F*	*X*	*X′*
3 × 2	6 3/4	7	11 1/2	4	9
4 × 2	7 1/2	8	13	4 1/2	10 1/2
4 × 3	7 1/2	8	13 1/4	4 1/2	10 1/2

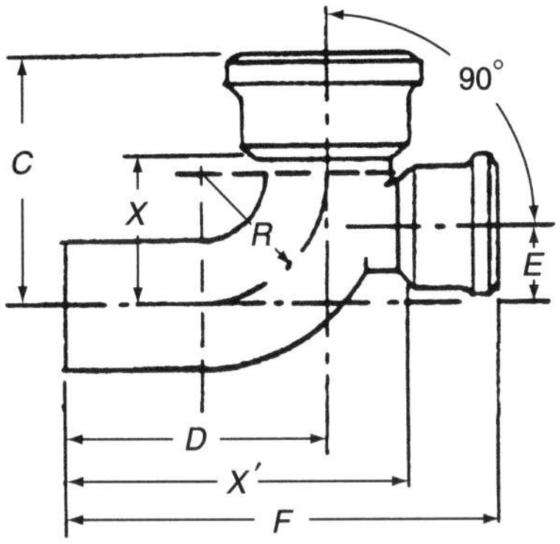

TABLE 9.29 Dimensions of High-Heel One-Quarter Bends (Extra-Heavy and Service Weights)

Size, in	Dimensions, in: C	D	F	X	X′
3 × 2	6 3/4	7	11 1/2	4	9
4 × 2	7 1/2	8	13	4 1/2	10 1/2
4 × 3	7 1/2	8	13 1/4	4 1/2	10 1/2

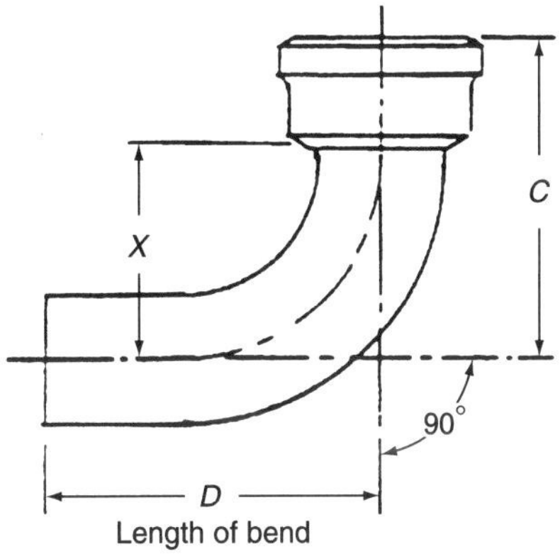

TABLE 9.30 Dimensions of Short- and Long-Sweep Bends (Extra-Heavy and Service Weights)*

	Short-sweep bend dimensions, in			Long-sweep bend dimensions, in		
Size, in	C	D	X	C	D	X
2	7 3/4	8	5 1/4	10 3/4	11	8 1/4
3	8 3/4	9	6	11 3/4	12	9
4	9 1/2	10	6 1/2	12 1/2	13	9 1/2
5	10	10 1/2	7	13	13 1/2	10
6	10 1/2	11	7 1/2	13 1/2	14	10 1/2
8	12 1/5	13 1/2	8 5/6	15 1/5	16 1/2	11 5/6
10	13 1/5	14 1/2	9 5/6	16 1/5	17 1/2	12 5/6
12	15	17	10 3/4	18	20	13 3/4
15	16 1/2	18 1/2	12 1/4	19 1/2	21 1/2	15 1/4

* Size 15 is made in service weight only.

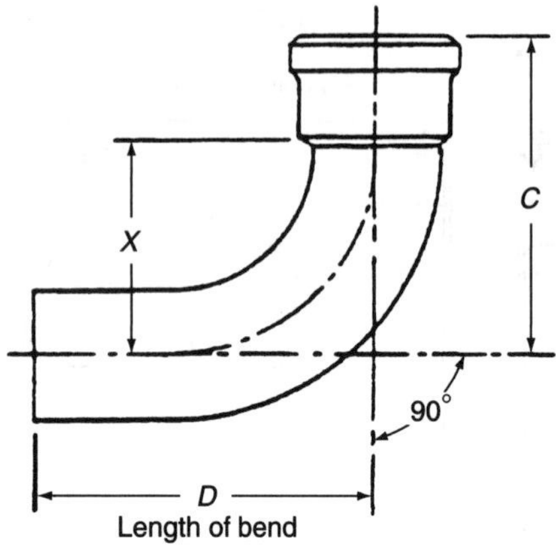

TABLE 9.31 Dimensions of Reducing Long-Sweep Bends (Extra-Heavy and Service Weights)

	Dimensions, in		
Size, in	*C*	*D*	*X*
3 × 2	$11\frac{1}{2}$	12	9
4 × 3	$12\frac{1}{4}$	13	$9\frac{1}{2}$

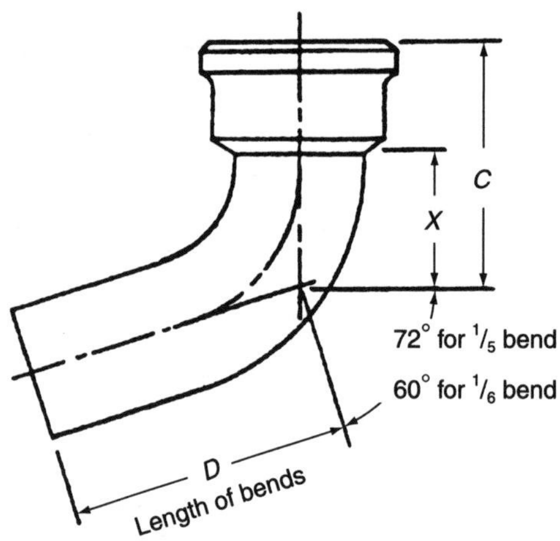

TABLE 9.32 Dimensions of One-Fifth and One-Sixth Bends (Extra-Heavy and Service Weights)*

	One-fifth bend dimensions, in			One-sixth bend dimensions, in		
Size, in	*C*	*D*	*X*	*C*	*D*	*X*
2	$4\frac{15}{16}$	$5\frac{3}{16}$	$2\frac{7}{16}$	$4\frac{1}{2}$	$4\frac{3}{4}$	2
3	$5\frac{13}{16}$	$6\frac{1}{16}$	$3\frac{1}{16}$	$5\frac{1}{4}$	$5\frac{1}{2}$	$2\frac{1}{2}$
4	$6\frac{7}{16}$	$6\frac{15}{16}$	$3\frac{7}{16}$	$5\frac{13}{16}$	$6\frac{5}{14}$	$2\frac{13}{16}$
5	$6\frac{3}{4}$	$7\frac{1}{4}$	$3\frac{3}{4}$	$6\frac{1}{8}$	$5\frac{3}{8}$	$3\frac{1}{8}$
6	$7\frac{1}{8}$	$7\frac{5}{8}$	$4\frac{1}{8}$	$6\frac{3}{8}$	$6\frac{7}{8}$	$3\frac{3}{8}$
8				$7\frac{5}{8}$	9	$4\frac{1}{8}$
10				$8\frac{3}{16}$	$9\frac{9}{14}$	$4\frac{11}{16}$
12				$9\frac{5}{8}$	$11\frac{5}{8}$	$5\frac{3}{8}$
15				$10\frac{1}{2}$	$12\frac{1}{2}$	$6\frac{1}{4}$

* Size 10 and smaller made in both extra-heavy and service weights. Sizes 12 and 15 made only in extra-heavy weight.

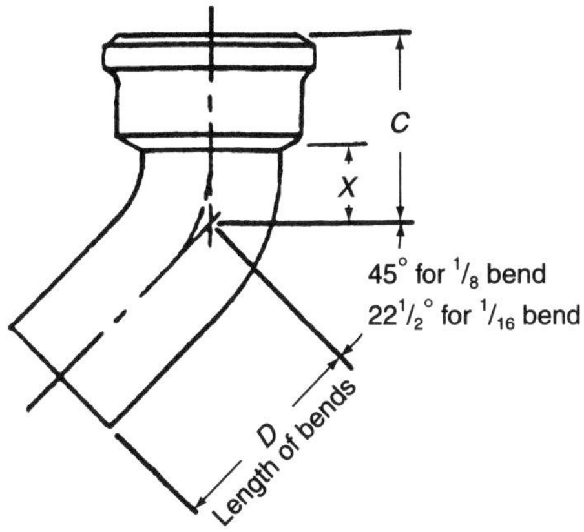

TABLE 9.33 Dimensions of One-Eighth, Long One-Eighth, and One-Sixteenth Bends*

Size, in	Dimensions, in C	D	X
One-eighth bends			
2	4	4 1/4	1 1/2
3	4 1/16	4 15/16	1 15/16
4	5 3/16	5 11/16	2 3/16
5	5 3/8	5 7/8	2 3/8
6	5 5/16	6 1/16	2 9/16
8	6 5/8	8	3 1/8
10	7	8 3/8	3 1/2
12	8 5/16	10 5/16	4 1/16
15	8 15/16	10 15/16	4 11/16
Long one-eighth bends			
2 × 12	4	12	1 1/2
2 × 18	4	18	1 1/2
2 × 24	4	24	1 1/2
3 × 12	4 11/16	12	1 15/16
3 × 18	4 11/16	18	1 15/16
3 × 24	4 11/16	24	1 15/16
4 × 12	5 3/16	12	2 3/16
4 × 18	5 3/16	18	2 3/16
4 × 24	5 3/16	24	2 3/16
One-sixteenth bends			
2	3 3/8	3 5/8	7/8
3	3 15/16	4 3/16	1 3/16
4	4 5/16	4 13/16	1 5/16
5	4 3/8	4 7/8	1 3/8
6	4 1/2	5	1 1/2
8	5 5/16	6 11/16	1 13/16
10	5 1/2	6 7/8	2
12	6 5/8	8 5/8	2 3/8
15	6 7/8	8 7/8	2 5/8

* All sizes listed in this table are made in both extra-heavy and service weights.

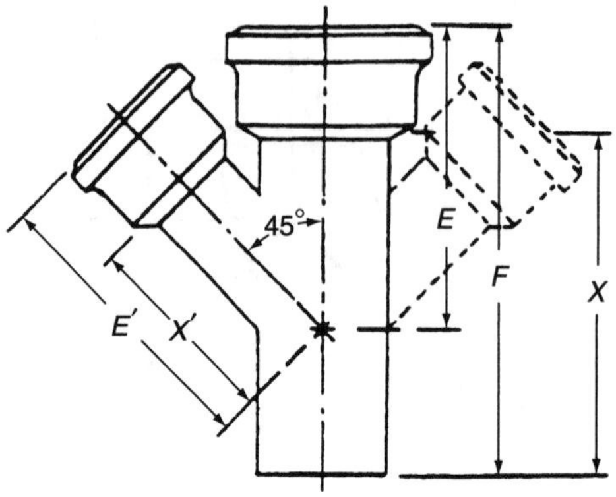

TABLE 9.34 Dimensions of Y Branches, Single and Double

Size, in	E	E'	F	X	X'
Single dimensions, in*					
2	$6\frac{1}{2}$	$6\frac{1}{2}$	$10\frac{1}{2}$	8	4
3	$8\frac{1}{4}$	$8\frac{1}{4}$	$13\frac{1}{4}$	$10\frac{1}{2}$	$5\frac{1}{2}$
4	$9\frac{3}{4}$	$9\frac{3}{4}$	15	12	$6\frac{3}{4}$
5	11	11	$16\frac{1}{2}$	$13\frac{1}{2}$	8
6	$12\frac{1}{4}$	$12\frac{1}{4}$	18	15	$9\frac{1}{4}$
8	$15\frac{5}{16}$	$15\frac{5}{16}$	23	$19\frac{1}{2}$	$11\frac{13}{16}$
10	18	18	26	$22\frac{1}{2}$	$14\frac{1}{2}$
12	$21\frac{1}{8}$	$21\frac{1}{8}$	$31\frac{1}{4}$	27	$16\frac{7}{8}$
15	25	25	$35\frac{3}{4}$	$31\frac{1}{2}$	$20\frac{3}{4}$
3 × 2	$7\frac{9}{16}$	$7\frac{1}{2}$	$11\frac{3}{4}$	9	5
4 × 2	$8\frac{3}{8}$	$8\frac{1}{4}$	12	9	$5\frac{3}{4}$
4 × 3	$9\frac{1}{16}$	9	$13\frac{1}{2}$	$10\frac{1}{2}$	$6\frac{1}{4}$
5 × 2	$8\frac{7}{8}$	9	12	9	$6\frac{1}{2}$
5 × 3	$9\frac{5}{8}$	$9\frac{3}{4}$	$13\frac{1}{2}$	$10\frac{1}{2}$	7
5 × 4	$10\frac{5}{16}$	$10\frac{1}{2}$	15	12	$7\frac{1}{2}$
6 × 2	$9\frac{7}{16}$	$9\frac{3}{4}$	12	9	$7\frac{1}{4}$
6 × 3	$10\frac{1}{8}$	$10\frac{1}{2}$	$13\frac{1}{2}$	$10\frac{1}{2}$	$7\frac{3}{4}$
6 × 4	$10\frac{13}{16}$	$11\frac{1}{4}$	15	12	$8\frac{1}{4}$
6 × 5	$11\frac{9}{16}$	$11\frac{3}{4}$	$16\frac{1}{2}$	$13\frac{1}{2}$	$8\frac{3}{4}$
8 × 2	$10\frac{7}{8}$	11	14	$10\frac{1}{2}$	$8\frac{1}{2}$
8 × 3	$11\frac{9}{16}$	$11\frac{3}{4}$	$15\frac{1}{2}$	12	9
8 × 4	$12\frac{1}{4}$	$12\frac{1}{2}$	17	$13\frac{1}{2}$	$9\frac{1}{2}$
8 × 5	13	13	$18\frac{1}{2}$	15	10
8 × 6	$13\frac{11}{16}$	$13\frac{1}{2}$	20	$16\frac{1}{2}$	$10\frac{1}{2}$
10 × 4	$13\frac{7}{16}$	$14\frac{1}{8}$	17	$13\frac{1}{2}$	$11\frac{1}{8}$
10 × 5	$14\frac{3}{16}$	$14\frac{5}{8}$	$18\frac{1}{2}$	15	$11\frac{5}{8}$
10 × 6	$14\frac{7}{8}$	$15\frac{1}{8}$	20	$16\frac{1}{2}$	$12\frac{1}{8}$
10 × 8	$16\frac{1}{2}$	$16\frac{15}{16}$	23	$19\frac{1}{2}$	$13\frac{7}{16}$
12 × 4	$15\frac{1}{8}$	$15\frac{7}{16}$	$19\frac{1}{4}$	15	$12\frac{7}{16}$
12 × 5	$15\frac{7}{8}$	$15\frac{15}{16}$	$20\frac{3}{4}$	$16\frac{1}{2}$	$12\frac{15}{16}$
12 × 6	$16\frac{5}{16}$	$16\frac{7}{16}$	$22\frac{1}{4}$	18	$13\frac{7}{16}$
12 × 8	$18\frac{3}{16}$	$18\frac{1}{4}$	$25\frac{1}{4}$	21	$14\frac{3}{4}$
12 × 10	$19\frac{11}{16}$	$19\frac{5}{16}$	$28\frac{1}{4}$	24	$15\frac{15}{16}$
15 × 6	$18\frac{1}{4}$	$18\frac{3}{4}$	$22\frac{1}{4}$	18	$15\frac{3}{4}$
15 × 8	$19\frac{7}{8}$	$20\frac{5}{16}$	$25\frac{1}{4}$	21	$17\frac{1}{16}$
15 × 10	$21\frac{3}{8}$	$21\frac{5}{8}$	$28\frac{1}{4}$	24	$18\frac{1}{8}$
15 × 12	$22\frac{13}{16}$	$23\frac{7}{16}$	$31\frac{1}{4}$	27	$19\frac{3}{16}$

Size, in	E	E'	F	X	X'
Double dimensions, in†					
2	$6\frac{1}{2}$	$6\frac{1}{2}$	$10\frac{1}{2}$	8	4
3	$8\frac{1}{4}$	$8\frac{1}{4}$	$13\frac{1}{4}$	$10\frac{1}{2}$	$5\frac{1}{2}$
4	$9\frac{3}{4}$	$9\frac{3}{4}$	15	12	$9\frac{1}{4}$
5	11	11	$16\frac{1}{2}$	$13\frac{1}{2}$	8
6	$12\frac{1}{4}$	$12\frac{1}{4}$	18	15	$9\frac{1}{4}$
8	$15\frac{5}{16}$	$15\frac{5}{16}$	23	$19\frac{1}{2}$	$11\frac{13}{16}$
10	18	18	26	$22\frac{1}{2}$	$14\frac{1}{2}$
12	$21\frac{1}{8}$	$21\frac{1}{8}$	$31\frac{1}{4}$	27	$16\frac{7}{8}$
15	25	25	$35\frac{3}{4}$	$31\frac{1}{2}$	$20\frac{3}{4}$
3 × 2	$7\frac{5}{16}$	$7\frac{1}{2}$	$11\frac{3}{4}$	9	5
4 × 2	$8\frac{3}{8}$	$8\frac{1}{4}$	12	9	$5\frac{3}{4}$
4 × 3	$9\frac{1}{16}$	9	$13\frac{1}{2}$	$10\frac{1}{2}$	$6\frac{1}{4}$
5 × 2	$8\frac{7}{8}$	9	12	9	$6\frac{1}{2}$
5 × 3	$9\frac{5}{8}$	$9\frac{3}{4}$	$13\frac{1}{2}$	$10\frac{1}{2}$	7
5 × 4	$10\frac{5}{16}$	$10\frac{1}{2}$	15	12	$7\frac{1}{2}$
6 × 2	$9\frac{7}{16}$	$9\frac{3}{4}$	12	9	$7\frac{1}{4}$
6 × 3	$10\frac{1}{8}$	$10\frac{1}{2}$	$13\frac{1}{2}$	$10\frac{1}{2}$	7
6 × 4	$10\frac{13}{16}$	$11\frac{1}{4}$	15	12	$8\frac{1}{4}$
6 × 5	$11\frac{5}{16}$	$11\frac{3}{4}$	$16\frac{1}{2}$	$13\frac{1}{2}$	$8\frac{3}{4}$
8 × 4	$12\frac{1}{4}$	$12\frac{1}{2}$	17	$13\frac{1}{2}$	$9\frac{1}{2}$
8 × 5	13	13	$18\frac{1}{2}$	15	10
8 × 6	$13\frac{11}{16}$	$13\frac{1}{2}$	20	$16\frac{1}{2}$	$10\frac{1}{2}$
10 × 4	$13\frac{7}{16}$	$14\frac{1}{8}$	17	$13\frac{1}{2}$	$11\frac{1}{8}$
10 × 6	$14\frac{7}{8}$	$15\frac{1}{8}$	20	$16\frac{1}{2}$	$12\frac{1}{8}$
10 × 8	$16\frac{1}{2}$	$16\frac{15}{16}$	23	$19\frac{1}{2}$	$13\frac{7}{16}$
12 × 6	$16\frac{9}{16}$	$16\frac{7}{16}$	$22\frac{1}{4}$	18	$13\frac{7}{16}$
12 × 8	$18\frac{3}{16}$	$18\frac{1}{4}$	$25\frac{1}{4}$	21	$14\frac{3}{4}$

* All sizes made in both extra-heavy and service weights.

† The 12 × 6 double Y branch is made only in service weight. All other sizes are made in both extra-heavy and service weights.

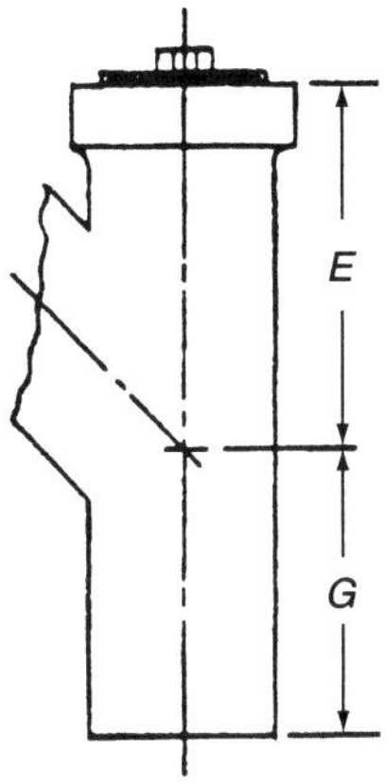

TABLE 9.35 Dimensions of Y-Branch Cleanout with Screw Plug on Main*,†

	Dimensions, in			
Size, in	*E*	*E′*	*G*	*X′*
2	$5\frac{1}{8}$	$6\frac{1}{2}$	4	4
3	$6\frac{5}{8}$	$8\frac{1}{4}$	5	$5\frac{1}{2}$
4	$7\frac{7}{8}$	$9\frac{3}{4}$	$5\frac{1}{4}$	$6\frac{3}{4}$

* Made in both extra-heavy and service weights.
† Refer to drawing on Table 9.34 for *E′*, *X′*, *E*, *G*, and *X*.

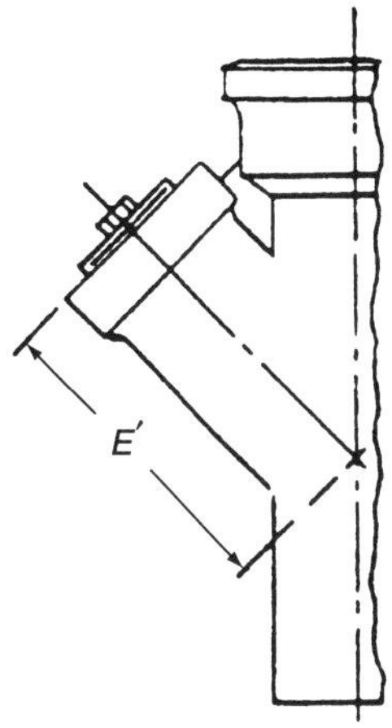

TABLE 9.36 Dimensions of Y-Branch Cleanout with Screw Plug on Branch*,†

	Dimensions, in			
Size, in	*E*	*E′*	*G*	*X*
2	$6\frac{1}{8}$	$5\frac{1}{4}$	4	8
3	$8\frac{1}{4}$	$6\frac{5}{8}$	5	$10\frac{1}{2}$
4	$9\frac{3}{4}$	$7\frac{7}{8}$	$5\frac{1}{4}$	12

* Made in both extra-heavy and service weights.
† Refer to drawing on Table 9.34 for *E′*, *X′*, *E*, *G*, and *X*.

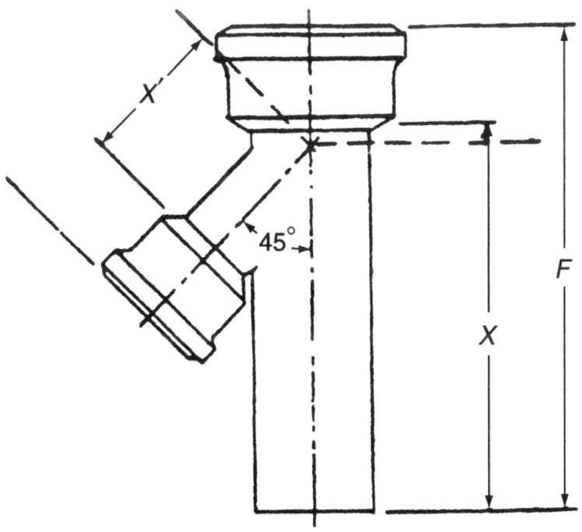

TABLE 9.37 Dimensions of Y Branches, Inverted, Single*

	Dimensions, in		
Size, in	*F*	*X*	*X′*
2	12	$9\frac{1}{2}$	$3\frac{3}{8}$
3	$15\frac{1}{4}$	$12\frac{1}{2}$	$4\frac{5}{8}$
4	17	14	$5\frac{7}{8}$
3 × 2	$13\frac{3}{4}$	11	$4\frac{1}{8}$
4 × 2	14	11	$4\frac{7}{8}$
4 × 3	$15\frac{1}{2}$	$12\frac{1}{2}$	$5\frac{3}{8}$

* Made in both extra-heavy and service weights.

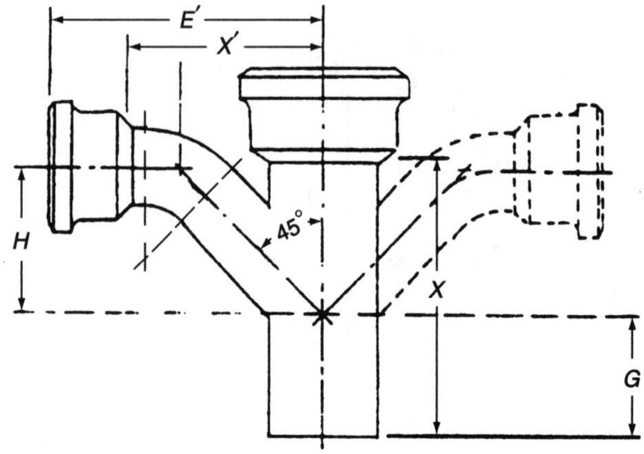

TABLE 9.38 Dimensions of Combination Y and One-Eighth Bend, Single and Double*

Size, in	Dimensions, in				
	E′	*G*	*H*	*X*	*X′*
			Single		
2	$7\frac{3}{8}$	4	$3\frac{3}{8}$	8	$4\frac{7}{8}$
3	$9\frac{3}{4}$	5	$5\frac{1}{16}$	$10\frac{1}{2}$	7
4	12	$5\frac{1}{4}$	$6\frac{13}{16}$	12	9
5	14	$5\frac{1}{2}$	$8\frac{5}{8}$	$13\frac{1}{2}$	11
6	$15\frac{7}{8}$	$5\frac{3}{4}$	$10\frac{5}{16}$	15	$12\frac{7}{8}$
8	$20\frac{1}{2}$	$7\frac{11}{16}$	$13\frac{7}{8}$	$19\frac{1}{2}$	17
3 × 2	$8\frac{1}{4}$	$4\frac{3}{16}$	4	9	$5\frac{3}{4}$
4 × 2	$8\frac{3}{4}$	$3\frac{11}{16}$	$4\frac{1}{2}$	9	$6\frac{1}{4}$
4 × 3	$10\frac{1}{4}$	$4\frac{1}{2}$	$5\frac{5}{16}$	$10\frac{1}{2}$	$7\frac{1}{2}$
5 × 2	$9\frac{1}{4}$	$3\frac{3}{8}$	5	9	$6\frac{3}{4}$
5 × 3	$10\frac{3}{4}$	4	$6\frac{1}{16}$	$10\frac{1}{2}$	8
5 × 4	$12\frac{1}{2}$	$4\frac{1}{4}$	$7\frac{5}{16}$	12	$9\frac{1}{2}$
6 × 2	$9\frac{3}{4}$	$2\frac{11}{16}$	$5\frac{1}{2}$	9	$7\frac{1}{4}$
6 × 3	$11\frac{1}{4}$	$3\frac{1}{8}$	$6\frac{9}{16}$	$10\frac{1}{4}$	$8\frac{1}{2}$
6 × 4	13	$4\frac{1}{4}$	$7\frac{13}{16}$	12	10
6 × 5	$14\frac{1}{2}$	$5\frac{1}{16}$	$9\frac{1}{8}$	$13\frac{1}{2}$	$11\frac{1}{2}$
8 × 2	$10\frac{3}{4}$	$3\frac{1}{8}$	$6\frac{1}{2}$	$10\frac{1}{2}$	$8\frac{1}{4}$
8 × 4	14	$4\frac{3}{4}$	$8\frac{13}{16}$	$13\frac{1}{2}$	11
8 × 5	$15\frac{1}{2}$	$5\frac{1}{2}$	$10\frac{1}{8}$	15	$12\frac{1}{2}$
8 × 6	$16\frac{7}{8}$	$6\frac{5}{16}$	$11\frac{5}{16}$	$16\frac{1}{2}$	$13\frac{7}{8}$
			Double		
2	$7\frac{3}{8}$	4	$3\frac{3}{8}$	8	$4\frac{7}{8}$
3	$9\frac{3}{4}$	5	$5\frac{1}{16}$	$10\frac{1}{2}$	7
4	12	$5\frac{1}{4}$	$6\frac{13}{16}$	12	9
5	14	$5\frac{1}{2}$	$8\frac{5}{8}$	$13\frac{1}{2}$	11
6	$15\frac{7}{8}$	$5\frac{3}{4}$	$10\frac{5}{16}$	15	$12\frac{7}{8}$
3 × 2	$8\frac{1}{4}$	$4\frac{3}{16}$	4	9	$5\frac{3}{4}$
4 × 2	$8\frac{3}{4}$	$3\frac{11}{16}$	$4\frac{1}{2}$	9	$6\frac{1}{4}$
4 × 3	$10\frac{1}{4}$	$4\frac{1}{2}$	$5\frac{9}{16}$	$10\frac{1}{2}$	$7\frac{1}{2}$
5 × 4	$12\frac{1}{2}$	$4\frac{3}{4}$	$7\frac{5}{16}$	12	$9\frac{1}{2}$
6 × 4	13	$4\frac{1}{4}$	$7\frac{13}{16}$	12	10

* Made in both extra-heavy and service weights.

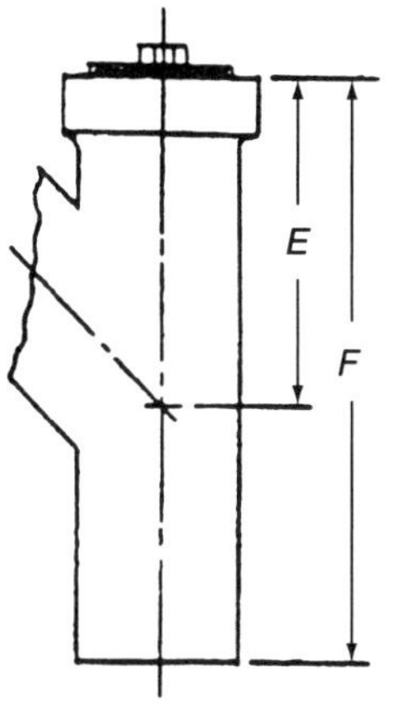

TABLE 9.39 Dimensions of Combination Y and One-Eighth-Bend Cleanout with Screw Plug on Main*

Size, in	Dimensions, in	
	E	*F*
2	$5\frac{1}{4}$	$9\frac{1}{4}$
3	$6\frac{5}{8}$	$11\frac{5}{8}$
4	$7\frac{1}{8}$	$13\frac{1}{8}$
5	$9\frac{1}{8}$	
6	$10\frac{3}{8}$	

* Made in both extra-heavy and service weights.

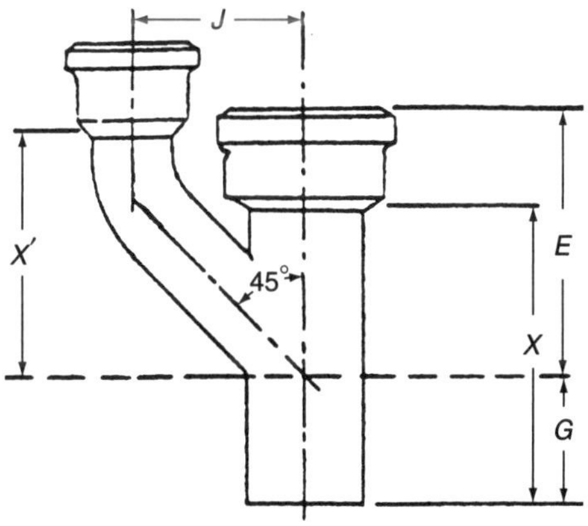

TABLE 9.40 Dimensions of Y Branches, Upright, Single*

Size, in	Dimensions, in				
	E	*G*	*J*	*X*	*X′*
2	$6\frac{1}{2}$	4	$4\frac{1}{2}$	8	6
3	$8\frac{1}{4}$	5	$5\frac{1}{2}$	$10\frac{1}{2}$	$7\frac{7}{16}$
4	$9\frac{3}{4}$	$5\frac{1}{4}$	$6\frac{1}{2}$	12	$8\frac{11}{16}$
3 × 2	$7\frac{9}{16}$	$4\frac{3}{16}$	5	9	$6\frac{3}{4}$
4 × 2	$8\frac{5}{16}$	$3\frac{11}{16}$	$5\frac{1}{2}$	9	$7\frac{1}{4}$
4 × 3	9	$4\frac{1}{2}$	6	$10\frac{1}{2}$	$6\frac{15}{16}$

* Made in both extra-heavy and service weights.

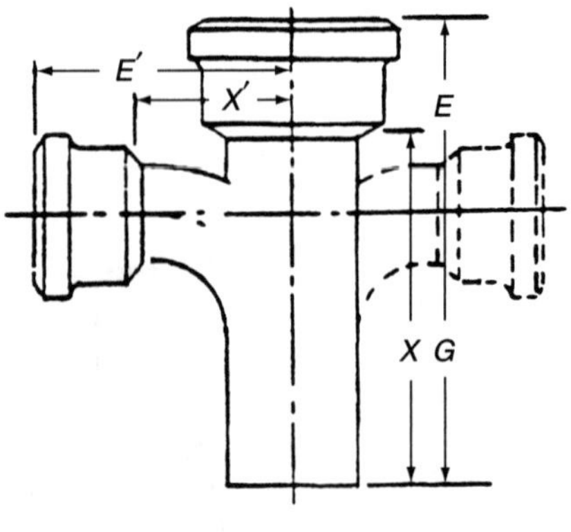

TABLE 9.41 Dimensions of Sanitary T Branches, Single and Double*

Size, in	Dimensions, in				
	E	*E′*	*G*	*X*	*X′*
Single					
2	4¼	5¼	6¼	8	2¾
3	5¼	6¾	7½	10	4
4	6	7½	8	11	4½
5	6½	8	8½	12	5
6	7	8½	9	13	5½
8	8¾	10⅛	11¾	17	6⅝
10	9¾	11⅛	12¾	19	7⅝
12	11¾	13	15	22½	8¾
15	13¼	14½	16½	25½	10¼
3 × 2	4¾	6½	7	9	4
4 × 2	5	7	7	9	4½
4 × 3	5½	7¼	7½	10	4½
5 × 2	5	7½	7	9	5
5 × 3	5½	7¾	7½	10	5
5 × 4	6	8	8	11	5
6 × 2	5	8	7	9	5½
6 × 3	5½	8¼	7½	10	5½
6 × 4	6	8½	8	11	5½
6 × 5	6½	8½	8½	12	5½
8 × 4	6¾	9½	9¾	13	6½
8 × 5	7¼	9½	10¼	14	6½
8 × 6	7¾	9½	10¾	15	6½
10 × 4	6¾	10½	9¾	13	7½
10 × 6	7¾	10½	10¾	15	7½
10 × 8	8¾	11⅛	11¾	17	7½
12 × 6	8¾	11½	12	16½	8½
12 × 8	9¾	12⅛	13	18½	8½
12 × 10	10¾	12⅛	14	20½	8½
15 × 6	8¾	13	12	16½	10
15 × 8	9¾	13⅝	13	18½	10⅛
15 × 10	10¾	13⅝	14	20½	10⅛
15 × 12	11¾	14½	15	22½	10¼
Double					
2	4¼	5¼	6¼	8	2¾
3	5¼	6¾	7½	10	4
4	6	7½	8	11	4½
5	6½	8	8½	12	5
6	7	8½	9	13	5½
8	8¾	10⅛	11¾	17	6⅝
3 × 2	4¾	6½	7	9	4
4 × 2	5	7	7	9	4½
4 × 3	5½	7¼	7½	10	4½
5 × 3	5½	7¾	7½	10	5
5 × 4	6	8	8	11	5
6 × 4	6	8½	8	11	5½
8 × 4	6¾	9½	9¾	13	6½
8 × 6	7¾	9½	10¾	15	6½

* Made in both extra-heavy and service weights.

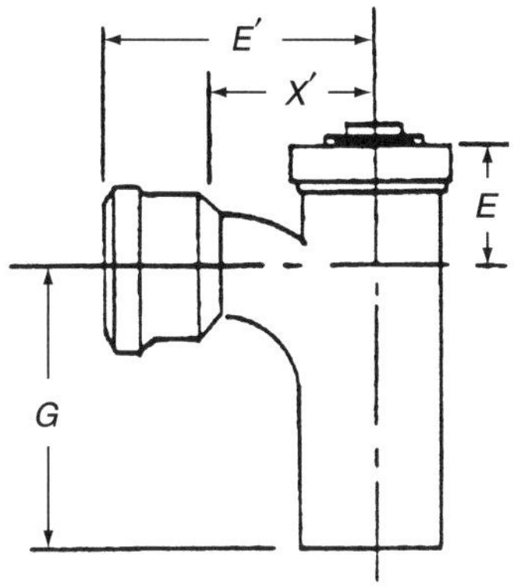

TABLE 9.42 Dimensions of Sanitary T Branches, Cleanout Plug on Main*

	Dimensions, in			
Size, in	*E*	*E′*	*G*	*X′*
2	3	$5\frac{1}{4}$	$6\frac{1}{4}$	$2\frac{3}{4}$
3	$3\frac{5}{8}$	$6\frac{3}{4}$	$7\frac{1}{8}$	4
4	$4\frac{1}{8}$	$7\frac{1}{2}$	8	$4\frac{1}{2}$

* Made in both extra-heavy and service weights.

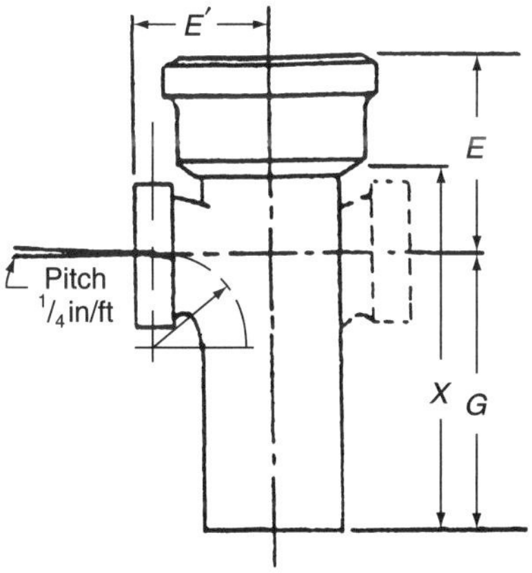

TABLE 9.43 Dimensions of Sanitary T Branches, Tapped, Single and Double*

	Dimensions, in			
Size, in	*E*	*E′*	*G*	*X*
	Single			
2 × 2	$4\frac{1}{2}$	$3\frac{1}{16}$	$6\frac{1}{4}$	8
3 × 2	$4\frac{3}{4}$	$3\frac{9}{16}$	7	9
4 × 2	5	$4\frac{1}{16}$	7	9
5 × 2	5	$4\frac{9}{16}$	7	9
6 × 2	5	$5\frac{1}{16}$	7	9
2	. . .	$2\frac{13}{16}$	. . .	. . .
3	. . .	$3\frac{5}{16}$	. . .	. . .
4	. . .	$3\frac{13}{16}$	. . .	. . .
5*	. . .	$4\frac{5}{16}$	. . .	. . .

	Dimensions, in			
Size, in	*E*	*E′*	*G*	*X*
	Double			
2 × 2	$4\frac{1}{2}$	$3\frac{1}{16}$	$6\frac{1}{4}$	8
3 × 2	$4\frac{3}{4}$	$3\frac{9}{16}$	7	9
4 × 2	5	$4\frac{1}{16}$	7	9
2	. . .	$2\frac{13}{16}$	. . .	. . .
3	. . .	$3\frac{5}{16}$	. . .	. . .
4	. . .	$3\frac{13}{16}$	. . .	. . .

* Made in both extra-heavy and service weights, except for size 5.

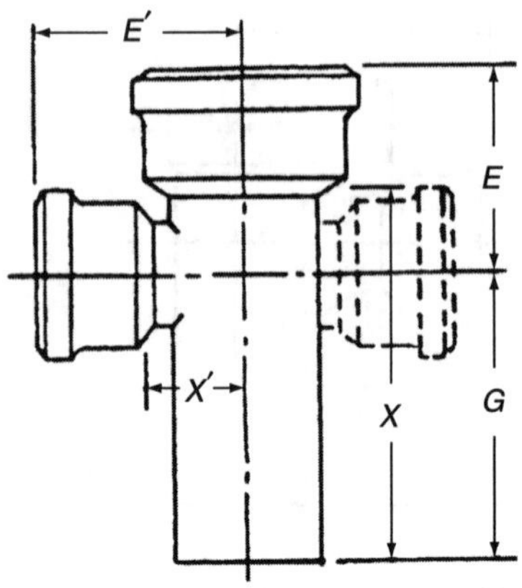

TABLE 9.44 Dimensions of T Branches, Single and Double*

Size, in	Dimensions, in				
	E	E′	G	X	X′
			Single		
2	4¼	4¼	6¼	8	1¾
3	5¼	5¼	7½	10	2⅛
4	6	6	8	11	3
5	6½	6½	8½	12	3⅛
6	7	7	9	13	4
3 × 2	4¾	5	7	9	2½
4 × 2	5	5½	7	9	3
4 × 3	5½	5¾	7½	10	3
5 × 2	5	6	7	9	3½
5 × 3	5½	6¼	7½	10	3½
5 × 4	6	6¼	8	11	3½
6 × 2*	5	6½	7	9	4
6 × 3	5½	6¾	7½	10	4
6 × 4	6	7	8	11	4
6 × 5	6½	7	8½	12	4
			Double		
2	4¼	4¼	6¼	8	1¾
3	5¼	5¼	7½	10	2½
4	6	6	8	11	3
3 × 2	4¾	5	7	9	2½
4 × 2	5	5½	7	9	3
4 × 3	5½	5¾	7½	10	3

* Size 6 × 2 made in service weight only. All other sizes are made in both extra-heavy and service weights.

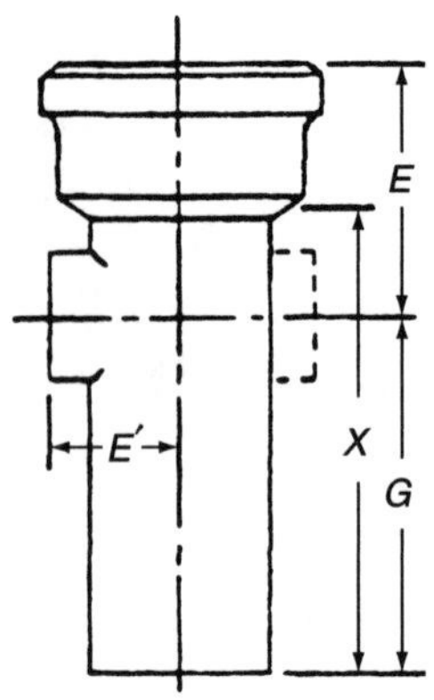

TABLE 9.45 Dimensions of Tapped T Branches, Single and Double*

Size, in	Dimensions, in			
	E	E′	G	X
		Single		
2	4¼	2	6¼	8
3	4¾	2½	7	9
4	5	3	7	9
5	5	3½	7	9
6*	5	4	7	9
4 × 3	6	3	8	11
5 × 4	6½	3½	8½	12
6 × 4	7	4	9	13
		Double		
2*	4¼	2	6¼	8
3	4¾	2½	7	9
4	5	3	7	9

* Sizes 6 single and 2 double made in service weight only. All other sizes are made in both extra-heavy and service weights.

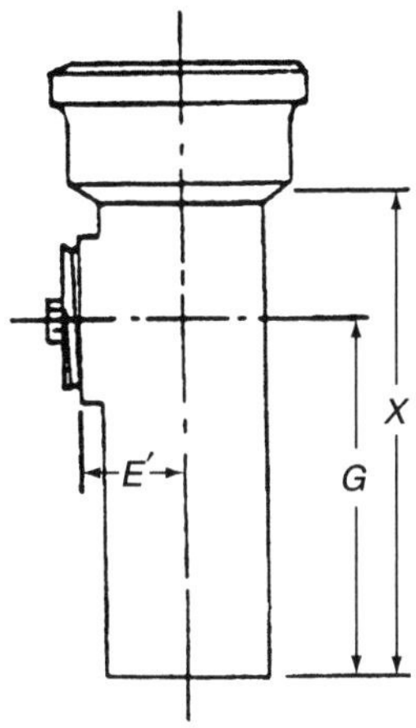

TABLE 9.46 Dimensions of T Branch Cleanout with Screw Plug*

Size, in	Dimensions, in		
	E′	*G*	*X*
2 × 1½	2	6¼	8
3 × 2½	2½	7½	10
4 × 3½	3	8	11
5 × 4	3½	8½	12
6 × 4	4	9	13

* Made in both extra-heavy and service weights.

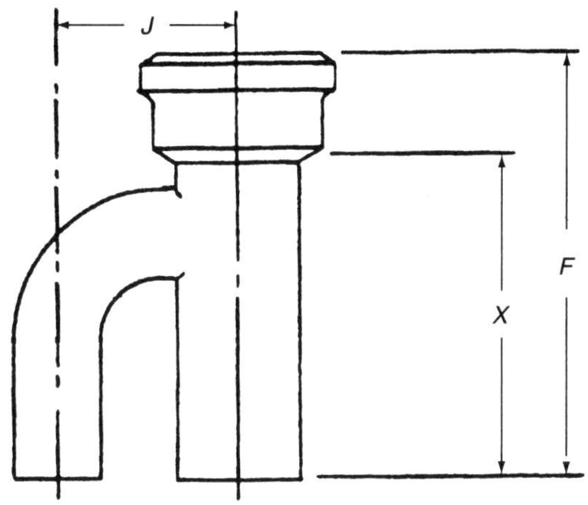

TABLE 9.47 Dimensions of Vent Branches, Single*

Size, in	Dimensions, in		
	F	*J*	*X*
2	10½	4½	8
3	12¾	5½	10
4	14	6½	11
3 × 2*	11¾	5	9
4 × 2	12	5½	9
4 × 3	13	6	10

* Size 3 × 2 made in service weight only. All other sizes are made in both extra-heavy and service weights.

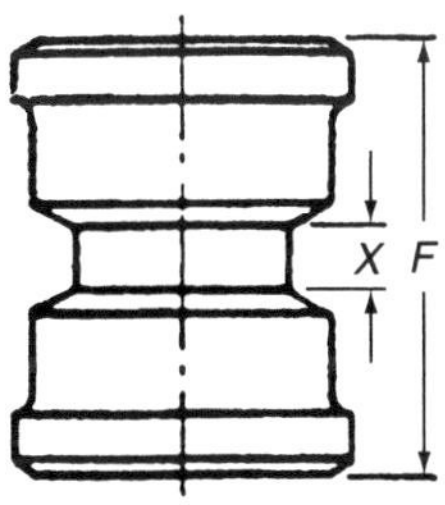

TABLE 9.48 Dimensions of Double Hub and Long Double Hub*

Size, in	*F*, in	*X*, in
	Double hub	
2	6	1
3	6½	1
4	7	1
5	7	1
6	7	1
8	8¼	1¼
10	8¼	1¼
12	10	1½
15	10	1½
	Long double hub	
2 × 30	30	25
3 × 30	30	24½
4 × 30	30	24

* Made in both extra-heavy and service weights.

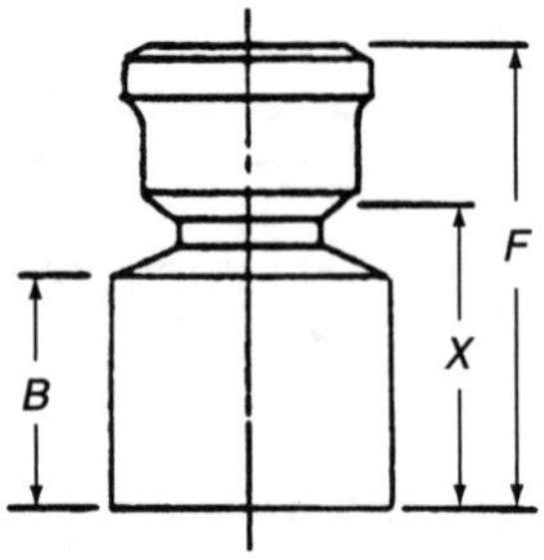

TABLE 9.49 Dimensions of Reducers*

Size, in	Dimensions, in		
	B	F	X
3 × 2	3¾	7¼	4¾
4 × 2	4	7½	5
4 × 3	4	7¾	5
5 × 2	4	7½	5
5 × 3	4	7¾	5
5 × 4	4	8	5
6 × 2	4	7½	5
6 × 3	4	7¾	5
6 × 4	4	8	5
6 × 5	4	8	5
8 × 2	4½	8½	6
8 × 3	4½	8¾	6
8 × 4	4½	9	6
8 × 5	4½	9	6
8 × 6	4½	9	6
10 × 4	4½	9	6
10 × 5	4½	9	6
10 × 6	4½	9	6
10 × 8	4½	9½	6
12 × 4	5¼	9½	6½
12 × 5	5¼	9½	6½
12 × 6	5¼	9½	6½
12 × 8	5¼	10	6½
12 × 10	5¼	10	6½
15 × 6	5¼	9½	6½
15 × 8	5¼	10	6½
15 × 10	5¼	10	6½
15 × 12	5¼	10¾	6½

* Made in both extra-heavy and service weights.

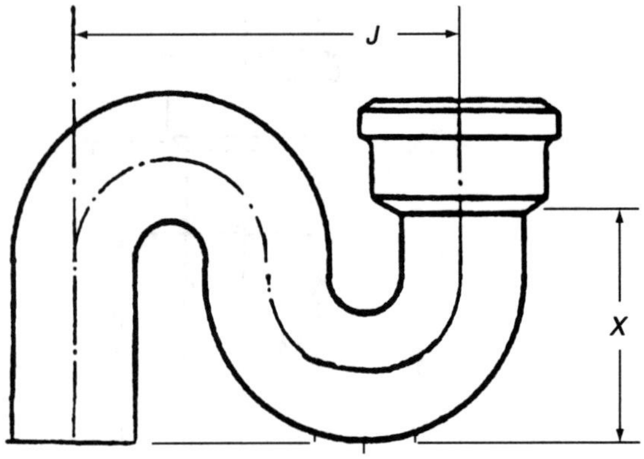

TABLE 9.50 Dimensions of S Traps*

Size, in, trap × vent	Dimensions, in	
	J	X
2 × 2	8	4
4 × 4	12	7½

* Made in both extra-heavy and service weights.

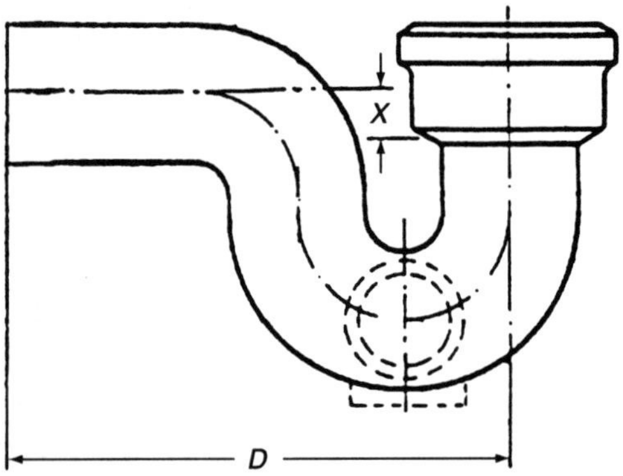

TABLE 9.51 Dimensions of P Traps with Cleanout*

Size, in	Dimensions, in	
	D	X
	9½	1½
	12	1¼
	14	1

* Made in both extra-heavy and service weights.

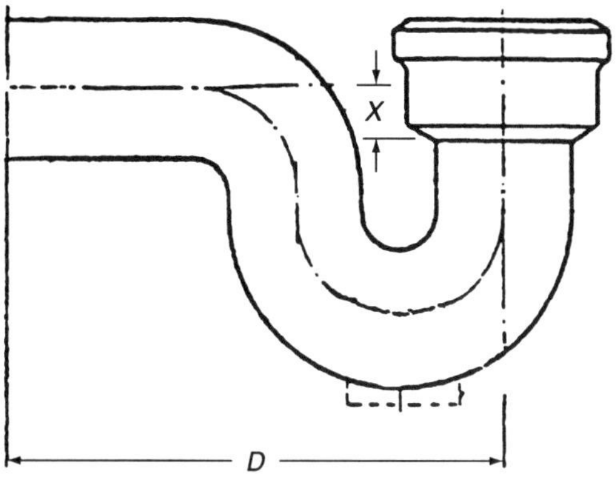

TABLE 9.52 Dimensions of Plain P Trap*

Size, in, trap × vent	Dimensions, in	
	J	*X*
2	$9\frac{1}{2}$	$1\frac{1}{2}$
3	12	$1\frac{1}{4}$
4	14	1
5	$15\frac{1}{2}$	$\frac{1}{2}$
6	17	. . .

* Made in both extra-heavy and service weights.

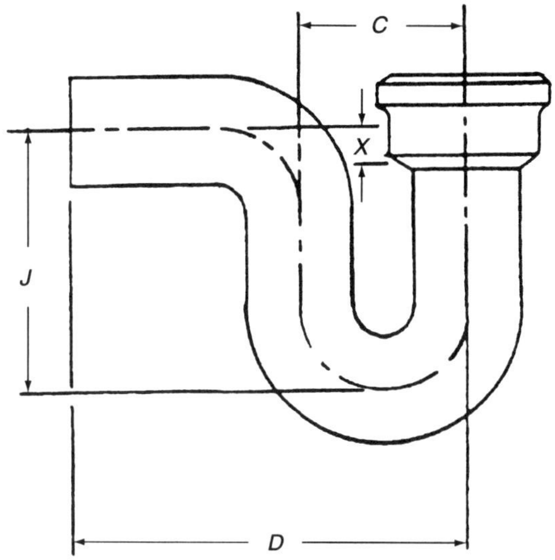

TABLE 9.53 Dimensions of Deep Seal P Trap*

Size, in	Dimensions, in			
	C	*D*	*J*	*X*
2	4	$9\frac{1}{2}$	6	$1\frac{1}{2}$
3	5	12	7	$1\frac{1}{4}$
4	6	14	8	$1\frac{1}{4}$

* Made in both extra-heavy and service weights.

ASME B16.12 CAST IRON THREADED DRAINAGE FITTINGS

ASME B16.12 cast iron threaded drainage fittings are intended for use in gravity drainage systems subjected to gravity head of waste liquids at temperatures from ambient to 212°F.

Materials for ASME B16.12 Fittings

The applicable document is ASTM A126, Specification for Gray Iron Castings for Valves, Flanges, and Pipe Fittings. Manufacturers may use ASTM A197, Specification for Cupola Malleable Iron, for these fittings at their option.

Threading of Fittings

Threads must comply with ASME B1.20.1, American National Standard Taper Pipe Threads.

Dimensions of ASME B16.12 Cast Iron Soil Fittings

See Tables 9.54 through 9.64.

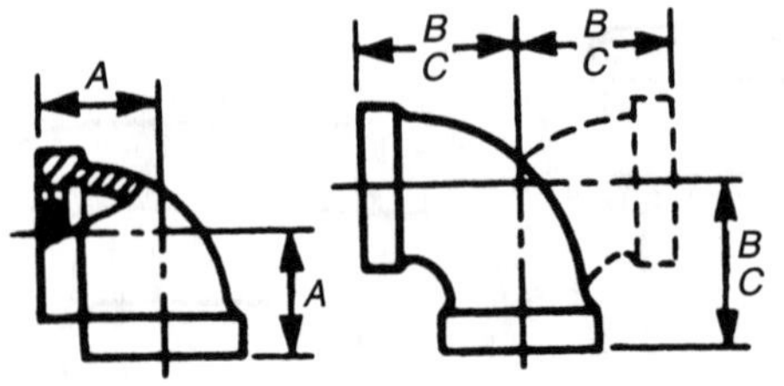

90° short elbow

90° long elbow, 90° extra-long elbow, and three-way elbow

TABLE 9.54 Center-to-End Dimensions of 90° Elbows*

Nominal pipe size	Short,† A	Long‡ and 3-way elbow, B	Extra long, C
$1\frac{1}{4}$	1.75	2.25	3.00
$1\frac{1}{2}$	1.94	2.50	3.50
2	2.25	3.06	4.00
$2\frac{1}{2}$	2.69	3.69	4.50
3	3.06	4.25	5.25
4	3.81	5.19	6.25
5	4.50	6.13	. . .
6	5.13	7.13	. . .
8	6.56	9.00	. . .

* Dimensions in inches.
† Same dimensions as adopted for Class 125 Cast Iron Threaded Fittings (ASME B16.4).
‡ Three-way elbows have the same dimensions as 90° long-radius elbows.

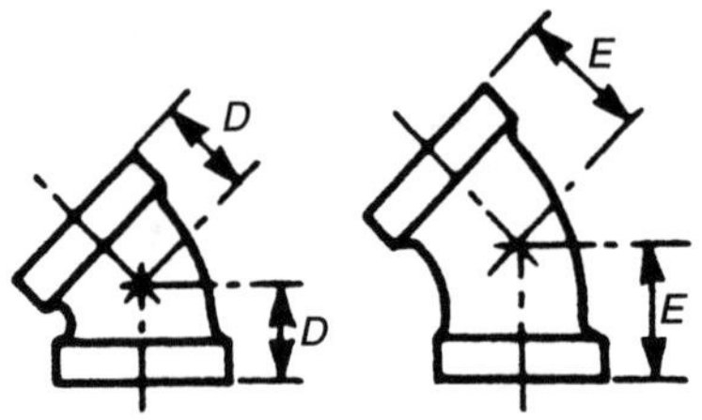

45° short elbow

45° long elbow

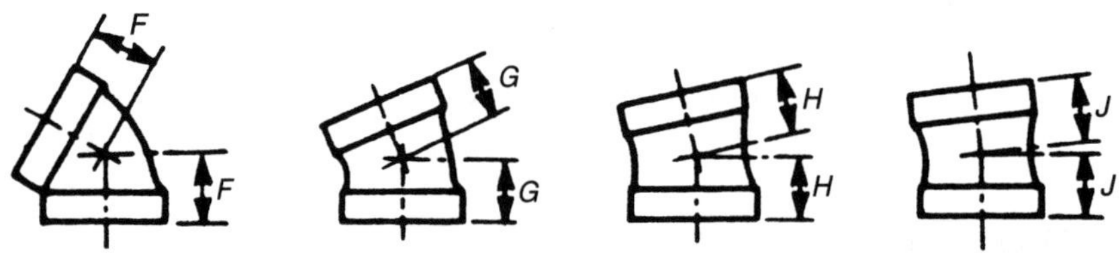

60° elbow

$22\frac{1}{2}$° elbow

$11\frac{1}{4}$° elbow

$5\frac{5}{8}$° elbow

TABLE 9.55 Center-to-End Dimensions of Elbows at Various Angles*

	45° elbow					
Nominal pipe size	Short, D	Long, E	60° elbow, F	$22\frac{1}{2}$° elbow, G	$11\frac{1}{4}$° elbow, H	$5\frac{5}{8}$° elbow, J
$1\frac{1}{4}$	1.31	1.75	1.56	1.13	1.06	1.19
$1\frac{1}{2}$	1.44	1.88	1.75	1.25	1.25	1.31
2	1.69	2.25	2.06	1.44	1.38	1.50
$2\frac{1}{2}$	1.94	2.63	2.50	1.75	1.63	1.63
3	2.19	2.94	2.88	2.00	1.81	1.75
4	2.63	3.50	3.38	2.31	2.00	1.88
5	3.06	4.13	3.88	2.63	2.25	2.00
6	3.44	4.88	4.25	2.94	2.38	2.25
8	4.25	. . .	. . .	. . .	. . .	. . .

* Dimensions in inches.

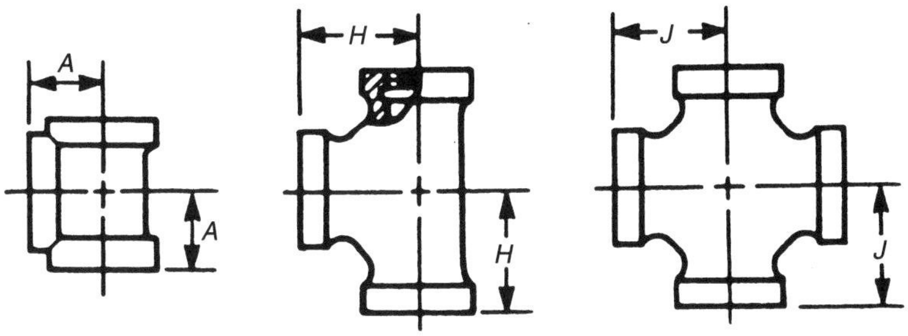

Tee Basin tee Basin cross

TABLE 9.56 Dimensions of Tees and Crosses*

Nominal pipe size	Center to end of tee,† *A*	Center to end of basin tee, *H*	Center to end of basin cross, *J*
$1\frac{1}{4}$	1.75	2.31	2.31
$1\frac{1}{2}$	1.94	2.69	2.69
2	2.25	3.50	3.50
$2\frac{1}{2}$	2.70	4.25	
3	3.08		
4	3.79		
5	4.50		
6	5.13		
8	6.56		

* Dimensions in inches.
† Same dimensions as adopted for Class 125 Cast Iron Threaded Fittings (ASME B16.4).

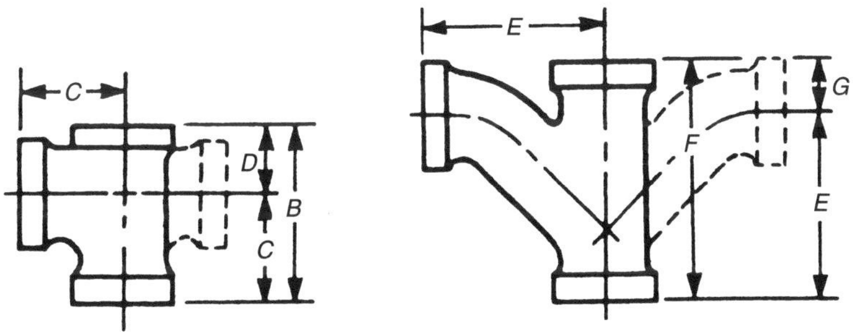

90° short Y branch 90° long Y branch

TABLE 9.57 Dimensions of Y Branches*

	90° short Y branch†			90° long Y branch†		
Nominal pipe size	End to end, *B*	Center to end, *C*	Center to end, *D*	Center to end, *E*	End to end, *F*	Center to end, *G*
$1\frac{1}{4}$	3.75	2.25	1.50	3.63	4.75	1.13
$1\frac{1}{2}$	4.25	2.50	1.75	4.13	5.38	1.25
2	5.19	3.06	2.13	5.25	7.00	1.75
$2\frac{1}{2}$	6.31	3.69	2.63	6.25	8.25	2.00
3	7.25	4.25	3.00	7.50	9.88	2.38
4	8.75	5.19	3.56	9.88	13.00	3.13
5	10.31	6.13	4.19	12.25	15.75	3.50
6	11.94	7.13	4.81	14.63	18.75	4.13

* Dimensions in inches.
† Double Y branches have the same dimensions as single Y branches.

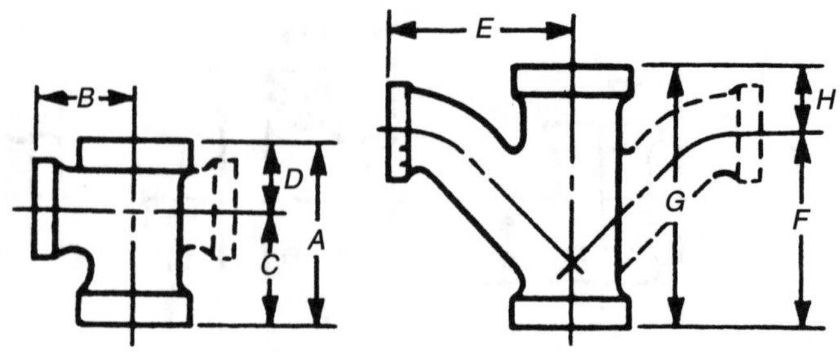

90° short Y branch reducing

90° long Y branch reducing

TABLE 9.58 Dimensions of Reducing Size Y Branches*

	90° short Y branch				90° long Y branch			
Nominal pipe size	End to end, *A*	Center to end, *B*	Center to end, *C*	Center to end, *D*	End to end, *G*	Center to end, *E*	Center to end, *F*	Center to end, *H*
$1\frac{1}{2} \times 1\frac{1}{4}$	3.88	2.50	2.38	1.50	5.13	3.88	3.94	1.19
$2 \times 1\frac{1}{2}$	4.63	2.94	2.69	1.94	5.57	4.38	4.44	1.31
$2\frac{1}{2} \times 1\frac{1}{2}$	4.88	3.06	2.81	2.06	5.75	4.56	4.44	1.31
$2\frac{1}{2} \times 2$	5.50	3.31	3.25	2.25	7.38	5.75	5.75	1.63
$3 \times 1\frac{1}{2}$	5.06	3.31	2.94	2.13	5.94	5.00	4.63	1.31
3×2	5.69	3.63	3.31	2.38	7.56	6.06	5.94	1.63
$4 \times 1\frac{1}{2}$	5.25	3.81	3.00	2.25	6.06	5.44	4.75	1.31
4×2	5.81	4.06	3.38	2.44	7.69	6.63	6.06	1.63
$4 \times 2\frac{1}{2}$	6.63	4.44	3.81	2.81	8.63	7.06	6.63	2.00
4×3	7.38	4.75	4.31	3.06	10.00	8.06	7.63	2.38
$5 \times 1\frac{1}{2}$	5.56	4.38	3.19	2.38	6.31	6.00	4.94	1.38
5×2	6.13	4.63	3.56	2.56	7.75	7.13	6.13	1.63
5×3	7.75	5.25	4.50	3.25	10.25	8.63	7.88	2.38
5×4	9.13	5.69	5.38	3.75	13.00	10.44	10.13	2.88
6×2	6.25	5.13	3.63	2.63	7.94	7.75	6.31	1.63
6×3	7.88	5.75	4.56	3.31	10.38	9.13	8.06	2.31
6×4	9.25	6.19	5.44	3.81	13.06	11.00	10.19	2.88
6×5	10.56	6.63	6.25	4.31	16.06	12.81	12.56	3.50

* Dimensions in inches. Double Y branches have the same dimensions as single Y branches.

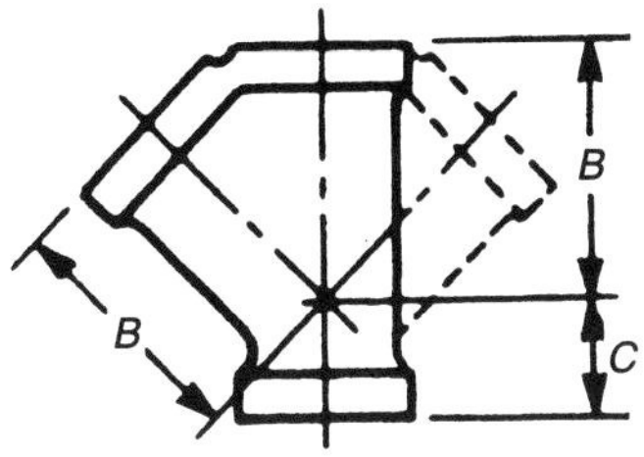

TABLE 9.59 Dimensions of 45° Y Branches*

Nominal pipe size	Center to end, *B*	Center to end, *C*
$1\frac{1}{4}$	3.25	1.75
$1\frac{1}{2}$	3.63	1.88
2	4.38	2.13
$2\frac{1}{2}$	5.38	2.50
3	6.19	2.81
4	7.69	3.19
5	9.19	3.75
6	10.75	4.13
8	13.56	5.25

* Dimensions in inches. Double Y branches have the same dimensions as single Y branches.

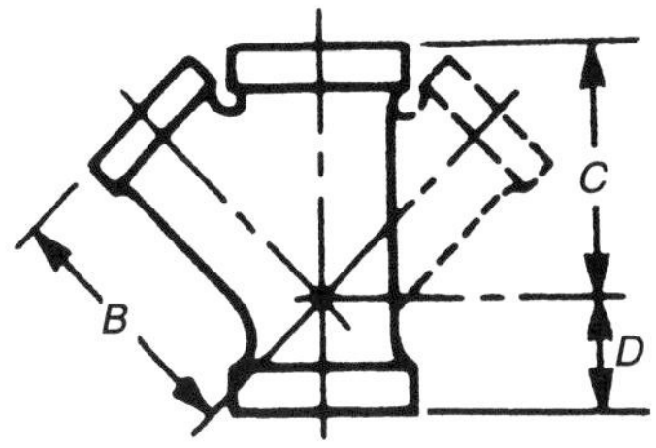

TABLE 9.60 Dimensions of (Reducing-Size) 45° Y Branches*

Nominal pipe size	Center to end, *B*	Center to end, *C*	Center to end, *D*
$1\frac{1}{2} \times 1\frac{1}{4}$	3.44	3.50	1.75
$2 \times 1\frac{1}{2}$	4.13	4.06	1.81
$2\frac{1}{2} \times 1\frac{1}{2}$	4.63	4.56	1.88
$2\frac{1}{2} \times 2$	5.13	4.94	2.13
$3 \times 1\frac{1}{2}$	5.06	4.94	1.69
3×2	5.38	5.31	2.06
$3 \times 2\frac{1}{2}$	5.81	5.69	2.31
$4 \times 1\frac{1}{2}$	6.06	5.75	1.44
4×2	6.31	6.00	1.69
4×3	7.19	6.88	2.38
5×2	7.19	6.88	1.38
5×3	7.88	7.63	2.19
5×4	8.50	8.44	2.94
6×2	8.06	7.56	0.88
6×3	8.75	8.31	1.69
6×4	9.38	9.06	2.81
6×5	10.00	9.81	3.19
8×4	11.00	10.44	1.00
8×6	12.38	11.88	3.06

* Dimensions in inches. Double Y branches have the same dimensions as single Y branches.

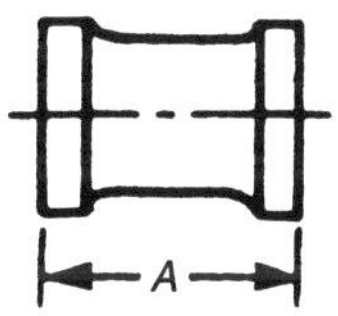

TABLE 9.61 Dimensions of Couplings

Nominal pipe size, in	End to end, *A*, in
$1\frac{1}{4}$	3.00
$1\frac{1}{2}$	3.38
2	3.75
$2\frac{1}{2}$	4.00
3	4.25
4	4.50
5	4.75
6	5.00
8	5.62

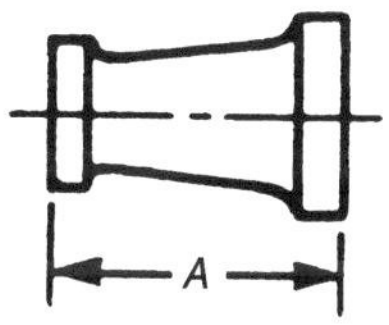

TABLE 9.62 Dimensions of Increasers

Nominal pipe size, in	End to end, *A*, in
$1\frac{1}{2} \times 2$	9.00
$2 \times 2\frac{1}{2}$	9.00
2×3	9.00
$2\frac{1}{2} \times 3$	9.00
2×4	9.00
3×4	9.00
2×5	9.00
3×5	9.00
4×5	9.00
4×6	9.00
5×6	9.00
4×8	9.00
5×8	9.00
6×8	9.00

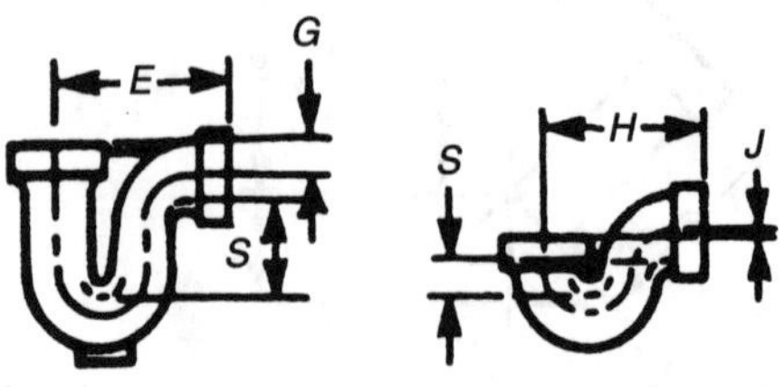

P trap **Bath P trap**

TABLE 9.63 Dimensions of P Traps and Bath P Traps*

Nominal pipe size	P traps *E*	P traps *G*	Bath P traps *H*	Bath P traps *J*	Water seal, min., *S*	Cleanout, nominal pipe size
1¼	3.94	0.75			2.00	¾
1½	4.38	0.88	4.50	0.25	2.00	1
2	5.31	0.88	5.31	0	2.00	1
3	7.13	1.19			2.50	1¼
4	9.25	1.50			2.50	2

* Dimensions in inches.

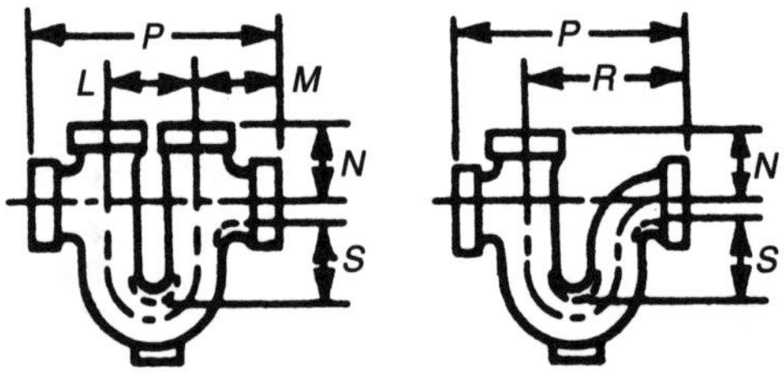

Running trap, double vent **Running trap, single vent**

TABLE 9.64 Dimensions of P Traps, Bath P Traps, and Running Traps*

Nominal pipe size	Running traps *P*	Running traps *L*	Running traps *M*	Running traps *N*	Running traps *R*	Water seal, min., *S*	Cleanout, nominal pipe size
1¼	7.75	3.13	2.31	2.00	5.44	2.00	¾
1½	8.38	3.38	2.50	2.25	5.88	2.00	1
2	10.13	4.00	3.06	2.56	7.06	2.00	1
3	13.75	5.38	4.19	3.50	9.56	2.50	1¼
4	17.25	7.00	5.13	4.19	12.13	2.50	2

* Dimensions in inches.

Tolerances for ASME B16.12 Cast Iron Soil Fittings

TABLE 9.65 Tolerances for ASME B16.12 Cast Iron Threaded Drainage Fittings

Nominal pipe size (NPS)	Tolerance for outside diameter and width of band, in	Tolerance for inside diameter of fitting, in (plus and minus)	Tolerance for center-to-end dimension, in (plus and minus)	Tolerance for end-to-end dimension, in (plus and minus)
$1\frac{1}{4}$	0.035	0.03	0.07	0.14
$1\frac{1}{2}$	0.04	0.03	0.08	0.16
2	0.04	0.03	0.08	0.16
$2\frac{1}{2}$	0.05	0.03	0.10	0.20
3	0.05	0.03	0.10	0.20
4	0.06	0.03	0.12	0.24
5	0.06	0.06	0.12	0.24
6	0.07	0.06	0.14	0.28
8	0.09	0.06	0.16	0.32

Tables in Chapter 9

Tables in Chapter 9 (Continued)

CHAPTER 10
DUCTILE IRON PIPE, FITTINGS, FLANGES, AND JOINTS

See p. 412 for a complete list of tables in Chap. 10.

REFERENCES, STANDARDS, AND SPECIFICATIONS

ANSI/AWWA C110/A21.10:	American National Standard for Gray-Iron Fittings, 3 in. through 48 in. (75 mm through 1200 mm), for Water and Other Liquids
ANSI/AWWA C111/A21.11:	AWWA Standard for Rubber-Gasket Joints for Ductile-Iron Pressure Pipe and Fittings
ANSI/AWWA C150/A21.50:	American National Standard for the Thickness Design of Ductile-Iron Pipe
ANSI/AWWA C153/A21.53:	American National Standard for Ductile-Iron Compact Fittings, 3 in. through 24 in. (76 mm through 610 mm) and 54 in. through 64 in. (1400 mm through 1600 mm), for Water Service
ASME B16.42:	Ductile-Iron Pipe Flanges and Flanged Fittings

DUCTILE IRON PIPES

This section covers the following tables:

- Table 10.1 Standard Laying Conditions for Ductile Iron Pipes. The maximum depths of cover are provided for various types of laying conditions in the subsequent tables.
- Table 10.2 Standard Pressure Classes and Nominal Wall Thickness for Ductile Iron Pipe. Thicknesses are provided for pressure classes 150, 200, 250, 300, and 350.
- Table 10.3 Rated Pressures and Maximum Depths of Cover for Minimum Pressure Classes of Ductile Iron Pipe. Maximum depths of cover are provided for various types of laying conditions.
- Table 10.4 Rated Pressures and Maximum Depths of Cover for Standard Pressure Classes of Ductile Iron Pipe. Maximum depths of cover are provided for various types of laying conditions.
- Table 10.5 Nominal Wall Thickness for Special Thickness Classes for Ductile Iron Pipe. Wall thicknesses are provided for special thickness classes 50, 51, 52, 53, 54, 55, and 56. These thickness classes were earlier considered as standard thicknesses.

TABLE 10.1 Standard Laying Conditions for Ductile Iron Pipe (Conforms to AWWA C150)

Laying condition	Description
 Type 1 (Note 1)	Flat-bottom trench. Loose backfill (note 2).
 Type 2	Flat-bottom trench. Backfill lightly consolidated to centerline of pipe (note 2).
 Type 3	Pipe bedded in 4-in minimum loose soil. Backfill lightly consolidated to top of pipe (note 3).
 Type 4	Pipe bedded in sand, gravel, or crushed stone to depth of $^1/_8$ pipe diameter, 4 in minimum. Backfill compacted to top of pipe (approximately 80% Standard Proctor, AASHTO T-99).
 Type 5	Pipe bedded to its centerline in compacted granular material, 4 in minimum under pipe. Compacted granular or select material (note 3) to top of pipe (approximately 90% Standard Proctor, AASHTO T-99).

Notes:
1. For 14-in and larger pipe, consideration should be given to the use of laying conditions other than type 1.
2. "Flat bottom" is defined as undisturbed earth.
3. "Loose soil" or "select material" is defined as native soil excavated from the trench, free of rocks, foreign material, and frozen earth.

TABLE 10.2 Standard Pressure Classes and Nominal Wall Thickness for Ductile Iron Pipe (Conforms to AWWA C150)

Pipe size, in	Outside diameter, in	Pressure class, psi—nominal thickness in inches				
		150	200	250	300	350
3	3.96					0.25 (note 1)
4	4.80					0.25 (note 1)
6	6.90					0.25 (note 1)
8	9.05					0.25 (note 1)
10	11.10					0.26
12	13.20					0.28
14	15.30			0.28	0.30	0.31
16	17.40			0.30	0.32	0.34
18	19.50			0.31	0.34	0.36
20	21.60			0.33	0.36	0.38
24	25.80		0.33	0.37	0.40	0.43
30	32.00	0.34	0.38	0.42	0.45	0.49
36	38.30	0.38	0.42	0.47	0.51	0.56
42	44.50	0.41	0.47	0.52	0.57	0.63
48	50.80	0.46	0.52	0.58	0.64	0.70
54	57.56	0.51	0.58	0.65	0.72	0.79
60	61.61	0.54	0.61	0.68	0.76	0.83
64	65.67	0.56	0.64	0.72	0.80	0.87

Notes:

(1) Calculated thicknesses for these sizes and pressure ratings are less than those shown above. Presently, these are the lowest nominal thicknesses available in these sizes.

(2) Pressure classses are defined as the rated water working pressure of the pipe in psi. The thicknesses shown are adequate for the rated water working pressure plus a surge allowance of 100 psi. Calculations result in net thickness and are based on a minimum yield strength in tension of 42,000 psi and 2.0 safety factor times the sum of working pressure and 100 psi surge allowance.

TABLE 10.3 Rated Pressures and Maximum Depths of Cover for Minimum Pressure Classes of Ductile Iron Pipe (Conforms to AWWA C150)

Pipe size, in	Pressure class, psi	Thickness, in	Laying conditions—maximum depth of cover in feet (1)				
			Type 1	Type 2	Type 3	Type 4	Type 5
3	350	0.25	78	88	99	100 (2)	100 (2)
4	350	0.25	53	61	69	85	100 (2)
6	350	0.25	26	31	37	47	65
8	350	0.25	16	20	25	34	50
10	350	0.26	11 (3)	15	19	28	45
12	350	0.28	10 (3)	15	19	28	44
14	250	0.28	(4)	11 (3)	15	23	36
16	250	0.30	(4)	11 (3)	15	24	34
18	250	0.31	(4)	10 (3)	14	22	31
20	250	0.33	(4)	10	14	22	30
24	200	0.33	(4)	8 (3)	12	17	25
30	150	0.34	(4)		9	14	22
36	150	0.38	(4)		9	14	21
42	150	0.41	(4)		9	13	20
48	150	0.46	(4)		9	13	20
54	150	0.51	(4)		9	13	20
60	150	0.54	(4)	5 (3)	9	13	20
64	150	0.56	(4)	5 (3)	9	13	20

Notes:

(1) An allowance for single H-20 truck with 1.5 impact factor is included for all sizes and all depths of cover.

(2) Calculated maximum depth of cover exceeds 100 ft.

(3) Minimum allowable depth of cover is 3 ft.

(4) Laying condition type 1 is limited to 12-in and smaller pipes. For 14-in and larger pipes, laying condition type 1 should not be used.

TABLE 10.4 Rated Pressures and Maximum Depths of Cover for Standard Pressure Classes of Ductile Iron Pipe (Conforms to AWWA C150)

Pipe size, in	Pressure class, psi (1)	Thickness, in	Laying conditions—maximum depth of cover in feet (2)				
			Type 1	Type 2	Type 3	Type 4	Type 5
3	350	0.25	78	88	99	100 (3)	100 (3)
4	350	0.25	53	61	69	85	100 (3)
6	350	0.25	26	31	37	47	65
8	350	0.25	16	20	25	34	50
10	350	0.26	11 (4)	15	19	28	45
12	350	0.28	10 (4)	15	19	28	44
14	250	0.28	(5)	11 (4)	15	23	36
	300	0.30	(5)	13	17	26	42
	350	0.31	(5)	14	19	27	44
16	250	0.30	(5)	11 (4)	15	24	34
	300	0.32	(5)	13	17	26	39
	350	0.34	(5)	15	20	28	44
18	250	0.31	(5)	10 (4)	14	22	31
	300	0.34	(5)	13	17	26	36
	350	0.36	(5)	15	19	28	41
20	250	0.33	(5)	10	14	22	30
	300	0.36	(5)	13	17	26	35
	350	0.38	(5)	15	19	28	38
24	200	0.33	(5)	8 (4)	12	17	25
	250	0.37	(5)	11	15	20	29
	300	0.40	(5)	13	17	24	32
	350	0.43	(5)	15	19	28	37
30	150	0.34	(5)		9	14	22
	200	0.38	(5)	8 (4)	12	16	24
	250	0.42	(5)	11	15	19	27
	300	0.45	(5)	12	16	21	29
	350	0.49	(5)	15	19	25	33
36	150	0.38	(5)		9	14	21
	200	0.42	(5)	8 (4)	12	15	23
	250	0.47	(5)	10	14	18	25
	300	0.51	(5)	12	16	20	28
	350	0.56	(5)	15	19	24	32
42	150	0.41	(5)		9	13	20
	200	0.47	(5)	8	12	15	22
	250	0.52	(5)	10	14	17	25
	300	0.57	(5)	12	16	20	27
	350	0.63	(5)	15	19	23	32(5)
48	150	0.46	(5)		9	13	20
	200	0.52	(5)	8	11	15	22
	250	0.58	(5)	10	13	17	24
	300	0.64	(5)	12	15	19	27
	350	0.70	(5)	15	18	22	30
54	150	0.51	(5)		9	13	20
	200	0.58	(5)	8	11	14	22
	250	0.65	(5)	10	13	16	24
	300	0.72	(5)	13	15	19	27
	350	0.79	(5)	15	18	22	30
60	150	0.54	(5)	5 (4)	9	13	20
	200	0.61	(5)	8	11	14	22
	250	0.68	(5)	10	13	16	24
	300	0.76	(5)	13	15	19	26
	350	0.83	(5)	15	18	22	30

TABLE 10.4 Rated Pressures and Maximum Depths of Cover for Standard Pressure Classes of Ductile Iron Pipe (Conforms to AWWA C150) (Continued)

Pipe size, in	Pressure class, psi (1)	Thickness, in	Laying conditions—maximum depth of cover in feet (2)				
			Type 1	Type 2	Type 3	Type 4	Type 5
64	150	0.56	(5)	5 (4)	9	13	20
	200	0.64	(5)	8	11	14	21
	250	0.72	(5)	10	13	16	24
	300	0.80	(5)	12	15	19	26
	350	0.87	(5)	18	17	21	29

Notes:
(1) Ductile iron pipe is adequate for the rated working pressure indicated for each nominal size plus a surge allowance of 100 psi. Calculations are based on 2.0 safety factor times the sum of working pressure and 100 psi surge allowance. Ductile iron pipe for working pressure higher than 350 psi is available.
(2) An allowance for single H-20 truck with 1.5 impact factor is included for all sizes and all depths of cover.
(3) Calculated maximum depth of cover exceeds 100 ft.
(4) Minimum allowable depth of cover is 3 ft.
(5) For pipe 14 in and larger, consideration should be given to the use of laying conditions other than Type 1.

TABLE 10.5 Nominal Wall Thickness for Special Thickness Classes for Ductile Iron Pipe (Conforms to AWWA C150)*

Pipe size, in.	Outside diameter, in.	Special thickness classes—wall thickness in inches						
		50	51	52	53	54	55	56
3	3.96		0.25	0.28	0.31	0.34	0.37	0.40
4	4.80		0.26	0.29	0.32	0.35	0.38	0.41
6	6.90	0.25	0.28	0.31	0.34	0.37	0.40	0.43
8	9.05	0.27	0.30	0.33	0.36	0.39	0.42	0.45
10	11.10	0.29	0.32	0.35	0.38	0.41	0.44	0.47
12	13.20	0.31	0.34	0.37	0.40	0.43	0.46	0.49
14	15.30	0.33	0.36	0.39	0.42	0.45	0.48	0.51
16	17.40	0.34	0.37	0.40	0.43	0.46	0.49	0.52
18	19.50	0.35	0.38	0.41	0.44	0.47	0.50	0.53
20	21.60	0.36	0.39	0.42	0.45	0.48	0.51	0.54
24	25.80	0.38	0.41	0.44	0.47	0.50	0.53	0.56
30	32.00	0.39	0.43	0.47	0.51	0.55	0.59	0.63
36	38.30	0.43	0.48	0.53	0.58	0.63	0.68	0.73
42	44.50	0.47	0.53	0.59	0.65	0.71	0.77	0.83
48	50.80	0.51	0.58	0.65	0.72	0.79	0.86	0.93
54	57.56	0.57	0.65	0.73	0.81	0.89	0.97	1.05

* These are special thickness classes as shown in AWWA C150 and C151. They were previously designated standard thickness classes. Special classes are most appropriately used for some threaded, grooved, or ball and socket pipes or for extraordinary design conditions, and they are generally less available than standard pressure class pipe.

DUCTILE IRON FLANGES

This section covers the following tables:

- Table 10.6 Pressure-Temperature Ratings for Ductile Iron Pipe Flanges. The ratings are provided for Class 150 and Class 300.
- Table 10.7 Dimensions of Class 150 Flanges for Ductile Iron Pipe.
- Table 10.8 Dimensions of Class 300 Flanges for Ductile Iron Pipe.
- Table 10.9 Dimensions of 30- to 64-in Flanges for Ductile Iron Pipe. These flanges are used in water service of 250 psi working pressure.
- Table 10.10 Base Flange Details for Ductile Iron Fittings.

TABLE 10.6 Pressure-Temperature Ratings for Ductile Iron Pipe Flanges (Conforms to ASME B16.42)

	Working pressure, psi gage	
Temperature, °F	Class 150	Class 300
−20 to 100	250	640
200	235	600
300	215	565
400	200	525
500	170	495
600	140	465
650	125	450

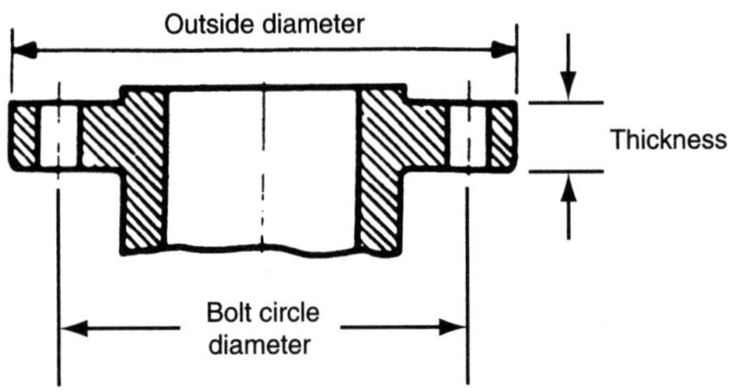

TABLE 10.7 Dimensions of Class 150 Flanges for Ductile Iron Pipe (Conforms to ASME B16.42)

Size, in	Outside diameter, in	Bolt circle diameter, in	Thickness, in	Bolt hole diameter, in	Number of bolts	Bolt size, dia. × length, in
1	4.25	3.12	0.56	0.62	4	$^1/_2 \times 2.75$
$1^1/_4$	4.62	3.50	0.62	0.62	4	$^1/_2 \times 2.75$
$1^1/_2$	5.00	3.88	0.69	0.62	4	$^1/_2 \times 3.00$
2	6.00	4.75	0.75	0.75	4	$^5/_8 \times 3.25$
$2^1/_2$	7.00	5.50	0.88	0.75	4	$^5/_8 \times 3.50$
3	7.50	6.00	0.94	0.75	4	$^5/_8 \times 3.75$
$3^1/_2$	8.50	7.00	0.94	0.75	8	$^5/_8 \times 3.75$
4	9.00	7.50	0.94	0.75	8	$^5/_8 \times 3.75$
5	10.00	8.50	0.94	0.88	8	$^3/_4 \times 4.00$
6	11.00	9.50	1.00	0.88	8	$^3/_4 \times 4.00$
8	13.50	11.75	0.12	0.88	8	$^3/_4 \times 4.25$
10	16.00	14.25	1.19	1.00	12	$^7/_8 \times 4.75$
12	19.00	17.00	1.25	1.00	12	$^7/_8 \times 4.75$
14	21.00	18.75	1.38	1.12	12	1×5.25
16	23.50	21.25	1.44	1.12	16	1×5.50
18	25.00	22.75	1.56	1.25	16	$1^1/_8 \times 6.00$
20	27.50	25.00	1.69	1.25	20	$1^1/_8 \times 6.25$
24	32.00	29.50	1.88	1.38	20	$1^1/_4 \times 7.00$

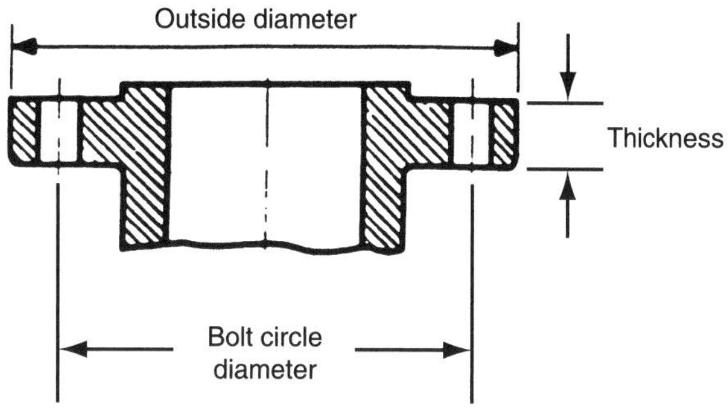

TABLE 10.8 Dimensions of Class 300 Flanges for Ductile Iron Pipe (Conforms to ASME B16.42)

Size, in	Outside diameter, in	Bolt circle diameter, in	Thickness, in	Bolt hole diameter, in	Number of bolts	Bolt size, dia. × length, in
1	4.88	3.50	0.69	0.75	4	$^5/_8 \times 3.00$
$1^1/_4$	5.25	3.88	0.75	0.75	4	$^5/_8 \times 3.25$
$1^1/_2$	6.12	4.50	0.81	0.88	4	$^3/_4 \times 3.50$
2	6.50	5.00	0.88	0.75	8	$^5/_8 \times 3.50$
$2^1/_2$	7.50	5.88	1.00	0.88	8	$^3/_4 \times 4.00$
3	8.25	6.62	1.12	0.88	8	$^3/_4 \times 4.25$
$3^1/_2$	9.00	7.25	1.19	0.88	8	$^3/_4 \times 4.25$
4	10.00	7.88	1.25	0.88	8	$^3/_4 \times 4.50$
5	11.00	9.25	1.38	0.88	8	$^3/_4 \times 4.75$
6	12.50	10.62	1.44	0.88	12	$^3/_4 \times 4.75$
8	15.00	13.00	1.62	1.00	12	$^7/_8 \times 5.50$
10	17.50	15.25	1.88	1.12	16	1×6.25
12	20.50	17.75	2.00	1.25	16	$1^1/_8 \times 6.75$
14	23.00	20.25	2.12	1.25	20	$1^1/_8 \times 7.00$
16	25.50	22.50	2.25	1.38	20	$1^1/_4 \times 7.50$
18	28.00	24.75	2.38	1.38	24	$1^1/_4 \times 7.75$
20	30.50	27.00	2.50	1.38	24	$1^1/_4 \times 8.00$
24	36.00	32.00	2.75	1.62	24	$1^1/_2 \times 9.00$

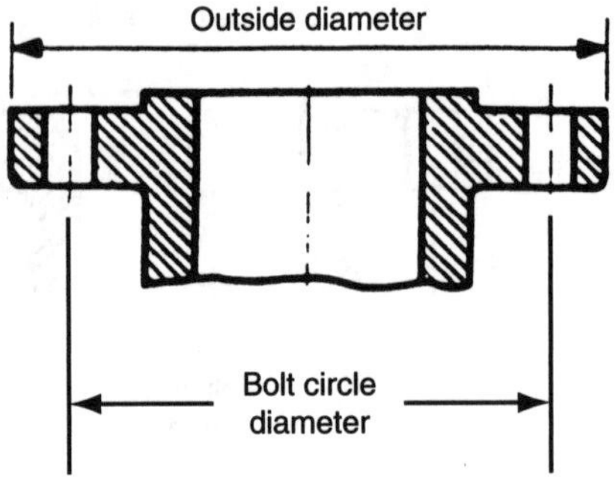

TABLE 10.9 Dimensions of 30- to 64-in Flanges for Ductile Iron Pipe (For water service of 250psi working pressure. Conforms to AWWA C153)

Size, in	Outside diameter, in	Bolt circle diameter, in	Thickness, in	Bolt hole diameter, in	Number of bolts	Bolt size, dia. × length, in
30	38.75	36.00	2.12	$1^3/_8$	28	$1^1/_4 \times 6^1/_2$
36	46.00	42.75	2.38	$1^5/_8$	32	$1^1/_2 \times 7$
42	53.00	49.50	2.62	$1^5/_8$	36	$1^1/_2 \times 7^1/_2$
48	59.50	56.00	2.75	$1^5/_8$	44	$1^1/_2 \times 8$
54	66.25	62.75	3.00	2	44	$1^3/_4 \times 8^1/_2$
60	73.00	69.25	3.12	2	52	$1^3/_4 \times 9$
64	80.00	76.00	3.38	2	52	$1^3/_4 \times 9$

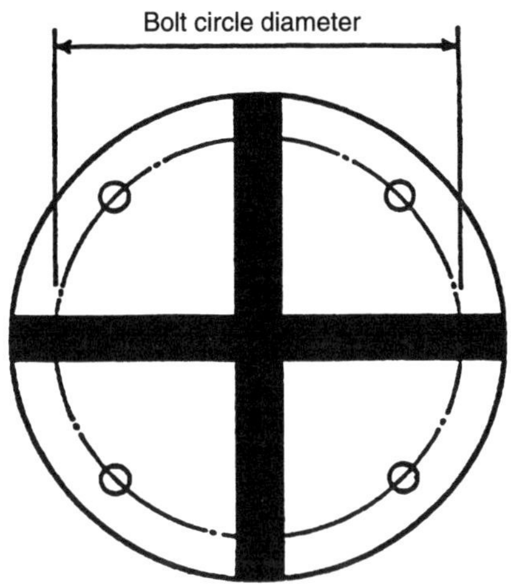

TABLE 10.10 Base Flange Details for Ductile Iron Fittings (Conforms to AWWA C110)*

Fitting size, in	Dimensions in inches: Bolt circle diameter	Bolt hole diameter	Number of bolts
3	3.88	5/8	4
4	4.75	3/4	4
6	5.50	3/4	4
8	7.50	3/4	4
10	7.50	3/4	4
12	9.50	7/8	4
14	9.50	7/8	4
16	9.50	7/8	4
18	11.75	7/8	4
20	11.75	7/8	4
24	11.75	7/8	4
30	14.25	1	4
36	17.00	1	4
42	21.25	1 1/8	4
48	22.75	1 1/4	4

* Contact the manufacturer for availability of sizes.

DUCTILE IRON FITTINGS

This section covers the following tables:

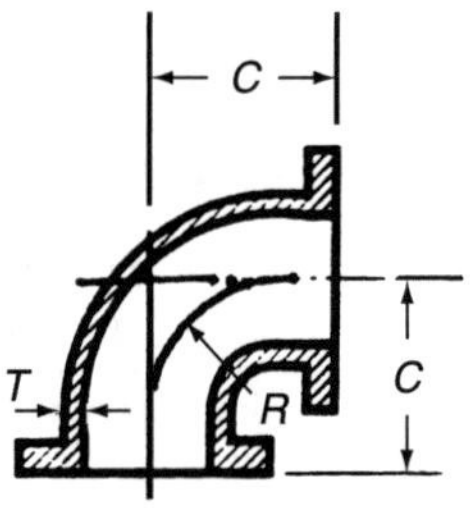

Short radius

TABLE 10.11 Dimensions for Ductile Iron Flanged 90° Bends (Conforms to AWWA C110 and C153)*

Size, in	Pressure rating, psi	*T*, in	*C*, in	*R*, in
3	250	0.48	5.5	4.0
4	250	0.52	6.5	4.5
6	250	0.55	8.0	6.0
8	250	0.60	9.0	7.0
10	250	0.68	11.0	9.0
12	250	0.75	12.0	10.0
14	250	0.66	14.0	11.5
16	250	0.70	15.0	12.5
18	250	0.75	16.5	14.0
20	250	0.80	18.0	15.5
24	250	0.89	22.0	18.5
30	250	1.03	25.0	21.5
36	250	1.15	28.0	24.5
42	250	1.28	31.0	27.5
48	250	1.42	34.0	30.5
54	150	0.90	39.0	34.3
60	150	0.94	43.0	36.5
64	150	0.99	48.0	38.8

* Contact the manufacturer for the availability of sizes.

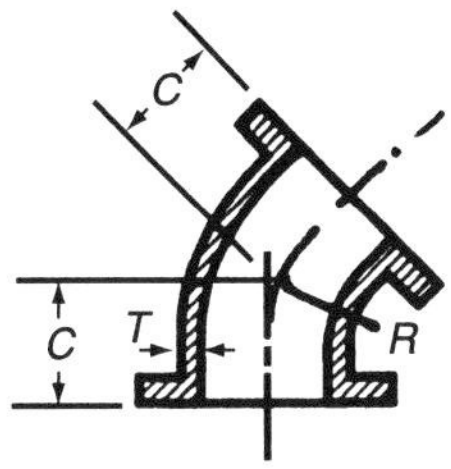

45° bend

TABLE 10.12 Dimensions for Ductile Iron Flanged 45° Bends (Conforms to AWWA C110 and C153)*

Size, in	Pressure rating, psi	*T*, in	*C*, in	*R*, in
3	250	0.48	3.0	3.62
4	250	0.52	4.0	4.81
6	250	0.55	5.0	7.25
8	250	0.60	5.5	8.44
10	250	0.68	6.5	10.88
12	250	0.75	7.5	13.25
14	250	0.66	7.5	12.06
16	250	0.70	8.0	13.25
18	250	0.75	8.5	14.50
20	250	0.80	9.5	16.88
24	250	0.89	11.0	18.12
30	250	1.03	15.0	27.75
36	250	1.15	18.0	35.00
42	250	1.28	21.0	42.25
48	250	1.42	24.0	49.50
54	150	0.90	20.5	42.32
60	150	0.94	23.5	44.08
64	150	0.99	25.0	45.85

* Contact the manufacturer for the availability of sizes.

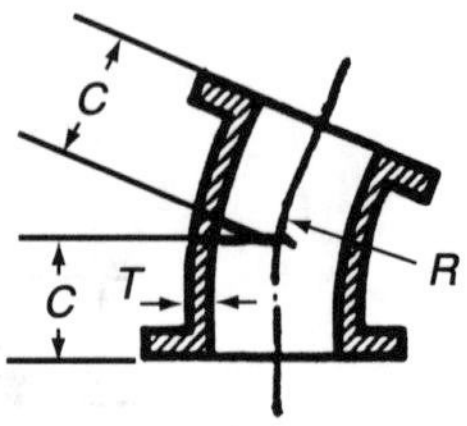

$22^1/_2°$ bend

TABLE 10.13 Dimensions for Ductile Iron Flanged 22½° Bends (Conforms to AWWA C110 and C153)*

Size, in	Pressure rating, psi	*T*, in	*C*, in	*R*, in
3	250	0.48	3.0	7.56
4	250	0.52	4.0	10.06
6	250	0.55	5.0	15.06
8	250	0.60	5.5	17.62
10	250	0.68	6.5	22.62
12	250	0.75	7.5	27.62
14	250	0.66	7.5	25.12
16	250	0.70	8.0	27.62
18	250	0.75	8.5	30.19
20	250	0.80	9.5	35.19
24	250	0.89	11.0	37.69
30	250	1.03	15.0	57.81
36	250	1.15	18.0	72.88
42	250	1.28	21.0	88.00
48	250	1.42	24.0	103.06
54	150	0.90	14.0	37.65
60	150	0.94	14.0	38.36
64	150	0.99	13.5	39.06

* Contact the manufacturer for the availability of sizes.

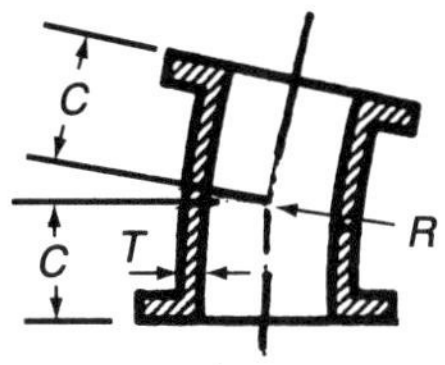

$11^1/_4°$ bend

TABLE 10.14 Dimensions for Ductile Iron Flanged $11^1/_4°$ Bends (Conforms to AWWA C110 and C153)*

Size, in	Pressure rating, psi	*T*, in	*C*, in	*R*, in
3	250	0.48	3.0	15.25
4	250	0.52	4.0	20.31
6	250	0.55	5.0	30.50
8	250	0.60	5.5	35.50
10	250	0.68	6.5	45.69
12	250	0.75	7.5	55.81
14	250	0.66	7.5	50.75
16	250	0.70	8.0	55.81
18	250	0.75	8.5	60.94
20	250	0.80	9.5	71.06
24	250	0.89	11.0	76.12
30	250	1.03	15.0	116.75
36	250	1.15	18.0	147.25
42	250	1.28	21.0	177.69
48	250	1.42	24.0	208.12
54	150	0.90	12.0	38.07
60	150	0.94	10.5	40.61
64	150	0.99	9.5	38.07

* Contact the manufacturer for the availability of sizes.

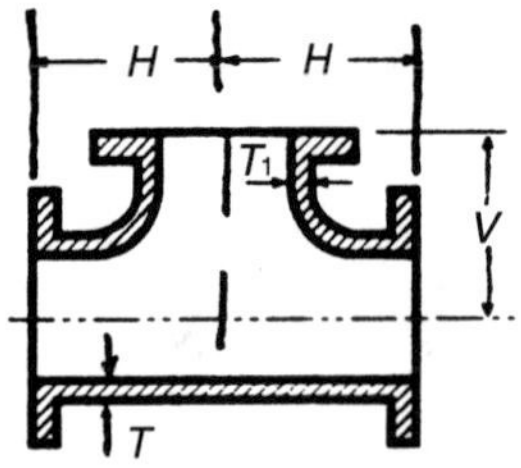

Tee

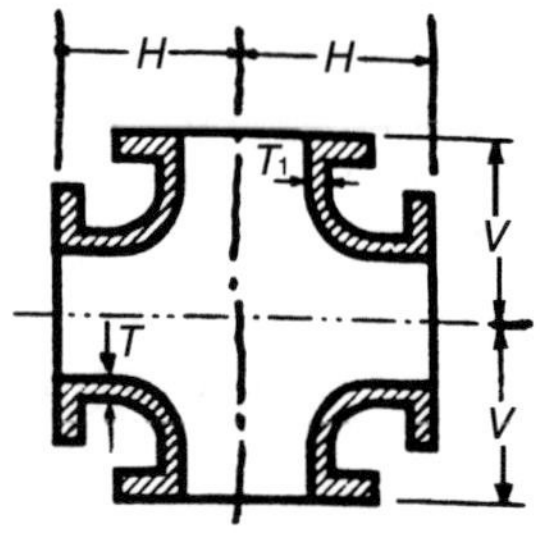

Cross

TABLE 10.15 Dimensions for Ductile Iron Flanged Tees and Crosses (Conforms to AWWA C110 and C153)*

Size, in		Pressure rating,	Dimensions in inches			
Run	Branch	psi	T	T_1	H	V
3	3	250	0.48	0.48	5.5	5.5
4	3	250	0.52	0.48	6.5	6.5
4	4	250	0.52	0.52	6.5	6.5
6	3	250	0.55	0.48	8.0	8.0
6	4	250	0.55	0.52	8.0	8.0
6	6	250	0.55	0.55	8.0	8.0
8	4	250	0.60	0.52	9.0	9.0
8	6	250	0.60	0.55	9.0	9.0
8	8	250	0.60	0.60	9.0	9.0
10	4	250	0.68	0.52	11.0	11.0
10	6	250	0.68	0.55	11.0	11.0
10	8	250	0.68	0.60	11.0	11.0
10	10	250	0.80	0.80	11.0	11.0
12	4	250	0.75	0.52	12.0	12.0
12	6	250	0.75	0.55	12.0	12.0
12	8	250	0.75	0.60	12.0	12.0
12	10	250	0.87	0.80	12.0	12.0
12	12	250	0.87	0.87	12.0	12.0
14	6	250	0.66	0.55	14.0	14.0
14	8	250	0.66	0.60	14.0	14.0
14	10	250	0.66	0.68	14.0	14.0
14	12	250	0.66	0.75	14.0	14.0
14	14	250	0.66	0.66	14.0	14.0
16	6	250	0.70	0.55	15.0	15.0
16	8	250	0.70	0.60	15.0	15.0
16	10	250	0.70	0.68	15.0	15.0
16	12	250	0.70	0.75	15.0	15.0
16	14	250	0.70	0.66	15.0	15.0
16	16	250	0.70	0.70	15.0	15.0
18	6	250	0.75	0.55	13.0	15.5
18	8	250	0.75	0.60	13.0	15.5
18	10	250	0.75	0.68	13.0	15.5
18	12	250	0.75	0.75	13.0	15.5
18	14	250	0.75	0.66	16.5	16.5
18	16	250	0.75	0.70	16.5	16.5
18	18	250	0.75	0.75	16.5	16.5
20	6	250	0.80	0.55	14.0	17.0
20	8	250	0.80	0.60	14.0	17.0
20	10	250	0.80	0.68	14.0	17.0
20	12	250	0.80	0.75	14.0	17.0
20	14	250	0.80	0.66	14.0	17.0
20	16	250	0.80	0.70	18.0	18.0
20	18	250	0.80	0.75	18.0	18.0
20	20	250	0.80	0.80	18.0	18.0
24	6	250	0.89	0.55	15.0	19.0
24	8	250	0.89	0.60	15.0	19.0
24	10	250	0.89	0.68	15.0	19.0
24	12	250	0.89	0.75	15.0	19.0
24	14	250	0.89	0.66	15.0	19.0
24	16	250	0.89	0.70	15.0	19.0
24	18	250	0.89	0.75	22.0	22.0
24	20	250	0.89	0.80	22.0	22.0
24	24	250	0.89	0.89	22.0	22.0
30	12	250	1.03	0.75	18.0	23.0
30	14	250	1.03	0.66	18.0	23.0
30	16	250	1.03	0.70	18.0	23.0
30	18	250	1.03	0.75	18.0	23.0
30	20	250	1.03	0.80	18.0	23.0
30	24	250	1.03	0.89	25.0	25.0
30	30	250	1.03	1.03	25.0	25.0
36	12	250	1.15	0.75	20.0	26.0
36	14	250	1.15	0.66	20.0	26.0
36	16	250	1.15	0.70	20.0	26.0
36	18	250	1.15	0.75	20.0	26.0
36	20	250	1.15	0.80	20.0	26.0
36	24	250	1.15	0.89	20.0	26.0
36	30	250	1.15	1.03	28.0	28.0
36	36	250	1.15	1.15	28.0	28.0
42	12	250	1.28	0.75	23.0	30.0
42	14	250	1.28	0.66	23.0	30.0
42	16	250	1.28	0.70	23.0	30.0
42	18	250	1.28	0.75	23.0	30.0
42	20	250	1.28	0.80	23.0	30.0
42	24	250	1.28	0.89	23.0	30.0
42	30	250	1.28	1.03	31.0	31.0
42	36	150	1.28	1.15	31.0	31.0
42	36	250	1.78	1.58	31.0	31.0
42	42	150	1.28	1.28	31.0	31.0
42	42	250	1.78	1.78	31.0	31.0
48	12	250	1.42	0.75	26.0	34.0
48	14	250	1.42	0.66	26.0	34.0
48	16	250	1.42	0.70	26.0	34.0
48	18	250	1.42	0.75	26.0	34.0
48	20	250	1.42	0.80	26.0	34.0
48	24	250	1.42	0.89	26.0	34.0
48	30	250	1.42	1.03	26.0	34.0
48	36	250	1.42	1.15	34.0	34.0
48	42	150	1.42	1.28	34.0	34.0
48	42	250	1.96	1.78	34.0	34.0
48	48	150	1.42	1.42	34.0	34.0
48	48	250	1.96	1.96	34.0	34.0
54	36	150	1.05	1.15	31.0	37.0
54	42	150	1.05	1.28	39.0	39.0
54	48	150	1.05	1.42	39.0	39.0
54	54	150	1.05	1.05	39.0	39.0
60	36	150	1.10	1.15	33.0	42.0
60	42	150	1.10	1.28	43.0	43.0
60	48	150	1.10	1.42	43.0	43.0
60	54	150	1.10	1.05	43.0	43.0
60	60	150	1.10	1.10	43.0	43.0
64	36	150	1.16	1.15	32.0	44.0
64	42	150	1.16	1.28	48.0	45.0
64	48	150	1.16	1.42	48.0	45.0
64	54	150	1.16	1.05	48.0	46.0
64	60	150	1.16	1.10	48.0	46.0
64	64	150	1.16	1.16	48.0	48.0

* Contact the manufacturer for the availability of sizes.

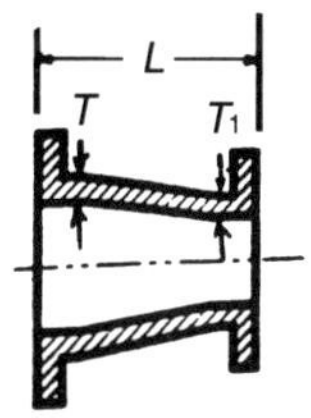

Concentric reducer

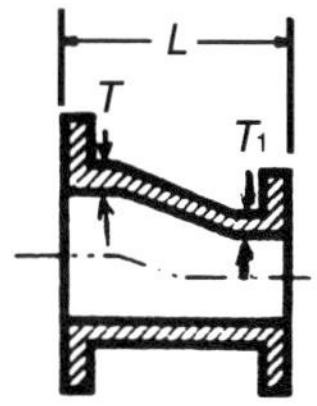

Eccentric reducer

TABLE 10.16 Dimensions for Ductile Iron Flanged Reducers (Conforms to AWWA C110 and C153)*

Size, in		Pressure rating, psi	Dimensions in inches		
Large end	Small end		T	T_1	L
4	3	250	0.52	0.48	7
6	3	250	0.55	0.48	9
6	4	250	0.55	0.52	9
8	4	250	0.60	0.52	11
8	6	250	0.60	0.55	11
10	4	250	0.68	0.52	12
10	6	250	0.68	0.55	12
10	8	250	0.68	0.60	12
12	4	250	0.75	0.52	14
12	6	250	0.75	0.55	14
12	8	250	0.75	0.60	14
12	10	250	0.75	0.68	14
14	6	250	0.66	0.55	16
14	8	250	0.66	0.60	16
14	10	250	0.66	0.68	16
14	12	250	0.66	0.75	16
16	6	250	0.70	0.55	18
16	8	250	0.70	0.60	18
16	10	250	0.70	0.68	18
16	12	250	0.70	0.75	18
16	14	250	0.70	0.66	18
18	8	250	0.75	0.60	19
18	10	250	0.75	0.68	19
18	12	250	0.75	0.75	19
18	14	250	0.75	0.66	19
18	16	250	0.75	0.70	19
20	10	250	0.80	0.68	20
20	12	250	0.80	0.75	20
20	14	250	0.80	0.66	20
20	16	250	0.80	0.70	20
20	18	250	0.80	0.75	20
24	10	250	0.89	0.68	24
24	12	250	0.89	0.75	24
24	14	250	0.89	0.66	24
24	16	250	0.89	0.70	24
24	18	250	0.89	0.75	24
24	20	250	0.89	0.80	24
30	18	250	1.03	0.75	30
30	20	250	1.03	0.80	30
30	24	250	1.03	0.89	30
36	20	250	1.15	0.80	36
36	24	250	1.15	0.89	36
36	30	250	1.15	1.03	36
42	24	250	1.28	0.89	42
42	30	250	1.28	1.03	42
42	36	250	1.28	1.15	42
48	30	250	1.42	1.03	48
48	36	250	1.42	1.15	48
48	42	250	1.42	1.28	48
54	30	150	0.90	1.00	32
54	36	150	0.90	1.15	28
54	42	150	0.90	1.25	25
54	48	150	0.90	1.40	18
60	36	150	0.94	1.15	32
60	42	150	0.94	1.25	26
60	48	150	0.94	1.40	20
60	54	150	0.94	0.90	15
64	36	150	0.99	1.15	35
64	42	150	0.99	1.25	30
64	48	150	0.99	1.40	25
64	54	150	0.99	0.90	18
64	60	150	0.99	0.94	17

* Contact the manufacturer for the availability sizes.

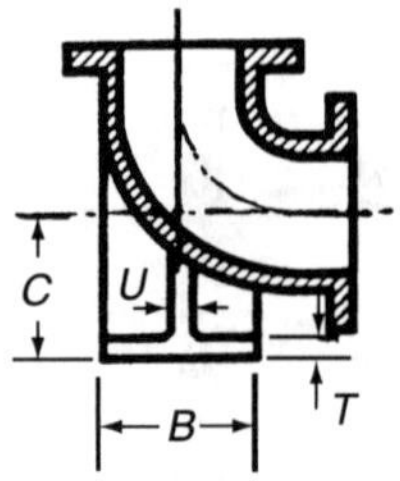

TABLE 10.17 Dimensions for Ductile Iron Flanged Base 90° Bends (Conforms to AWWA C110)*

Size, in	Pressure rating, psi	Dimensions in inches			
		C	*B*, dia.	*T*	*U*
3	250	4.88	5.00	0.56	0.50
4	250	5.50	6.00	0.62	0.50
6	250	7.00	7.00	0.69	0.62
8	250	8.38	9.00	0.94	0.88
10	250	9.75	9.00	0.94	0.88
12	250	11.25	11.00	1.00	1.00
14	250	12.50	11.00	1.00	1.00
16	250	13.75	11.00	1.00	1.00
18	250	15.00	13.50	1.12	1.12
20	250	16.00	13.50	1.12	1.12
24	250	18.50	13.50	1.12	1.12
30	150	23.00	16.00	1.19	1.15
30	250	23.00	16.00	1.19	1.15
36	250	26.00	19.00	1.25	1.15
42	250	30.00	23.50	1.44	1.28
48	250	34.00	25.00	1.56	1.42

* Contact the manufacturer for the availability of sizes.

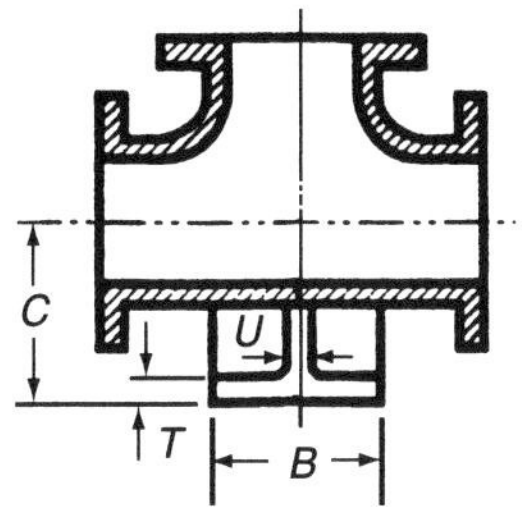

Base tee

TABLE 10.18 Dimensions for Ductile Iron Flanged Base Tees (Conforms to AWWA C110 and C153)*

Size, in	Pressure rating, psi	Dimensions in inches			
		C	*B*, dia.	*T*	*U*
3	250	4.88	5.00	0.56	0.50
4	250	5.50	6.00	0.62	0.50
6	250	7.00	7.00	0.69	0.62
8	250	8.38	9.00	0.94	0.88
10	250	9.75	9.00	0.94	0.88
12	250	11.25	11.00	1.00	1.00
14	250	12.50	11.00	1.00	1.00
16	250	13.75	11.00	1.00	1.00
18	250	15.00	13.50	1.12	1.12
20	250	16.00	13.50	1.12	1.12
24	250	18.50	13.50	1.12	1.12
30	250	23.00	16.00	1.19	1.15
36	250	26.00	19.00	1.25	1.15
42	150	30.00	23.50	1.44	1.28
42	250	30.00	23.50	1.44	1.28
48	150	34.00	25.00	1.56	1.42
48	250	34.00	25.00	1.56	1.42

* Contact the manufacturer for the availability of sizes.

DUCTILE IRON JOINTS

For information regarding ductile iron joints, refer to Chap. 15.

Tables in Chapter 10

CHAPTER 11
MALLEABLE IRON AND GRAY IRON FITTINGS

See p. 438 for a list of tables in Chap. 11.

MALLEABLE IRON THREADED FITTINGS

ASME Standard for Malleable Iron Threaded Fittings

ASME B16.3, Malleable Iron Threaded Fittings, covers the following Class 150 and Class 300 malleable iron threaded fittings:

- 90 and 45° elbows
- Tees and crosses
- Y branches
- Couplings
- Caps
- Return bends

Pressure-Temperature Ratings of Class 150 Malleable Iron Threaded Fittings

See Table 11.1.

TABLE 11.1 Pressure-Temperature Ratings of Class 150 Fittings

Temperature, °F	Pressure, psig
–20 to 150	300
200	265
250	225
300	185
350*	150*

* This rating can be used for temperatures up to 366°F, which is the temperature of 150 psig saturated steam.

Pressure-Temperature Ratings of Class 300 Malleable Iron Threaded Fittings

See Table 11.2.

Materials for Malleable Iron Threaded Fittings

- ASTM A197, Specification for Cupola Malleable Iron.
- Class 150 couplings and caps, NPS ⅛, ¼, and ⅜, may be made from steel rod or bar having a minimum yield strength of 30 ksi.

TABLE 11.2 Pressure-Temperature Ratings of Class 300 Threaded Fittings

	Pressure, psig		
Temperature, °F	NPS ¼–1	NPS 1¼–2	NPS 2½–3
–20 to 150	2000	1500	1000
200	1785	1350	910
250	1575	1200	825
300	1360	1050	735
350	1150	900	650
400	935	750	560
450	725	600	475
500	510	450	385
550	300	300	300

Galvanization or Zinc Coating of Malleable Iron Threaded Fittings

- 0.0034-in minimum hot-dipped zinc coating per ASTM A153, only when specified.
- 0.001-in electrodeposited zinc coating in accordance with ASTM B633, Type I, Service Condition, only when specified.

Threads

- Taper threads in accordance with ASME B1.20.1, Pipe Threads, General Purpose (inch).
- Class 150 wrought couplings and caps, NPS ⅛, ¼, and ⅜, may have straight threads as an alternative to taper threads per ASME B1.20.1.

Dimensions of ASME B16.3 Class 150 and 300 Malleable Iron Fittings

See Tables 11.3 through 11.20.

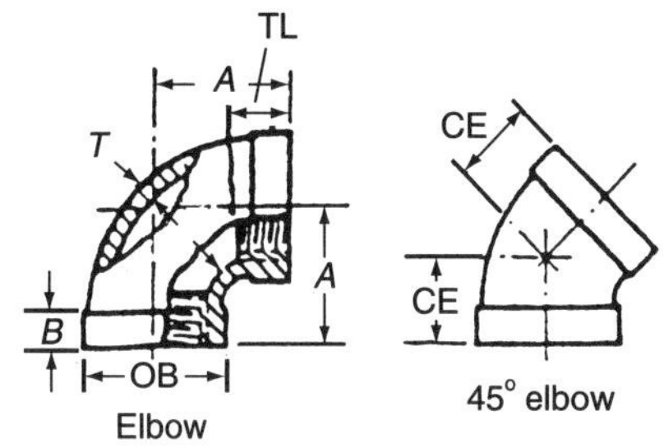

TABLE 11.3 Dimensions of Class 150 Malleable Iron 90 and 45° Elbows*

Nominal pipe size	Center-to-end, 90° elbows, *A*	Center-to-end, 45° elbows, CE	Length of thread, min., TL	Width of band, min., *B*	Minimum metal thickness, *T*	Minimum outside diameter of band, OB
1/8	0.69		0.25	0.20	0.09	0.69
1/4	0.81	0.73	0.32	0.21	0.09	0.84
3/8	0.95	0.80	0.36	0.23	0.10	1.01
1/2	1.12	0.88	0.43	0.25	0.10	1.20
3/4	1.31	0.98	0.50	0.27	0.12	1.46
1	1.50	1.12	0.58	0.30	0.13	1.77
1 1/4	1.75	1.29	0.67	0.34	0.14	2.15
1 1/2	1.94	1.43	0.70	0.37	0.15	2.43
2	2.25	1.68	0.75	0.42	0.17	2.96
2 1/2	2.70	1.95	0.92	0.48	0.21	3.59
3	3.08	2.17	0.98	0.55	0.23	4.28
3 1/2	3.42	2.39	1.03	0.60	0.25	4.84
4	3.79	2.61	1.08	0.66	0.26	5.40
5	4.50	3.05	1.18	0.78	0.30	6.58
6	5.13	3.46	1.28	0.90	0.34	7.77

* Dimensions in inches.

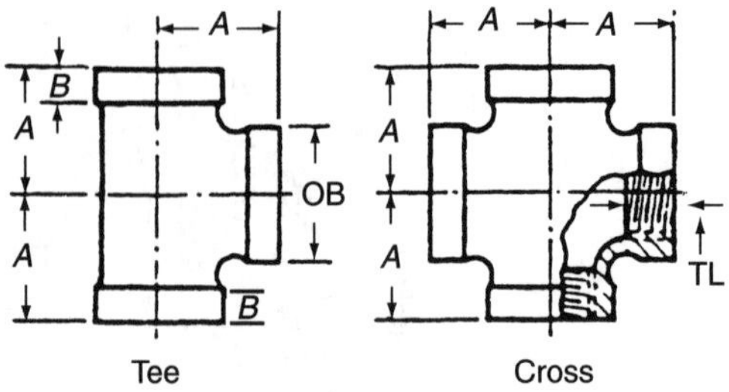

TABLE 11.4 Dimensions of Class 150 Malleable Iron Straight Tees and Crosses*

Nominal pipe size	Center-to-end, tees and crosses, *A*	Minimum length of thread, TL	Minimum width of band, *B*	Minimum metal thickness, *T*	Outside diameter of band, OB
1/8	0.69	0.25	0.20	0.09	0.69
1/4	0.81	0.32	0.21	0.09	0.84
3/8	0.95	0.36	0.23	0.10	1.01
1/2	1.12	0.43	0.25	0.10	1.20
3/4	1.31	0.50	0.27	0.12	1.46
1	1.50	0.58	0.30	0.13	1.77
1 1/4	1.75	0.67	0.34	0.14	2.15
1 1/2	1.94	0.70	0.37	0.15	2.43
2	2.25	0.75	0.42	0.17	2.96
2 1/2	2.70	0.92	0.48	0.21	3.59
3	3.08	0.98	0.55	0.23	4.28
3 1/2	3.42	1.03	0.60	0.25	4.84
4	3.79	1.08	0.66	0.26	5.40
5	4.50	1.18	0.78	0.30	6.58
6	5.13	1.28	0.90	0.34	7.77

* Dimensions in inches.

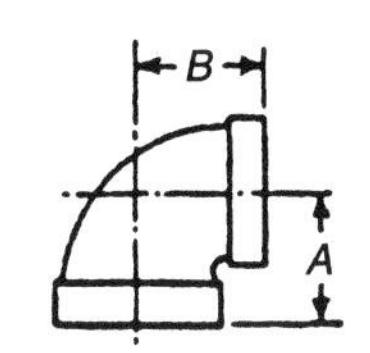

90° reducing elbow

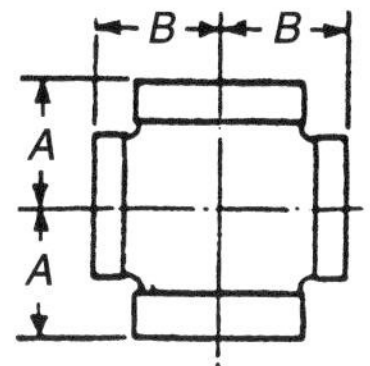

Reducing cross

TABLE 11.5 Dimensions of Class 150 Malleable Iron, 90° Reducing Elbows*

Nominal pipe size	Center to end A	Center to end B
¼ × ⅛	0.74	0.76
⅜ × ¼	0.88	0.90
⅜ × ⅜	0.81	0.85
½ × ⅜	1.04	1.03
½ × ¼	0.97	0.98
¾ × ½	1.20	1.22
¾ × ⅜	1.12	1.13
¾ × ¼	1.05	1.08
1 × ¾	1.37	1.45
1 × ½	1.26	1.36
1 × ⅜	1.18	1.27
1¼ × 1	1.58	1.67
1¼ × ¾	1.45	1.62
1¼ × ½	1.34	1.53
1½ × 1¼	1.82	1.88
1½ × 1	1.65	1.80
1½ × ¾	1.52	1.75
2 × 1½	2.02	2.16
2 × 1¼	1.90	2.10
2 × 1	1.73	2.02
2 × ¾	1.60	1.97
2½ × 2	2.39	2.60
2½ × 1½	2.16	2.51
3 × 2½	2.83	2.99
3 × 2	2.52	2.89
4 × 3	3.30	3.60

* Dimensions in inches. For dimensions of band width (*B*), thread length (TL), outside band diameter (OB), and metal thickness (*T*), refer to Table 11.4 for the corresponding size.

TABLE 11.6 Dimensions of Class 150 Malleable Iron Reducing Cross*

Nominal pipe size	Center to end A	Center to end B
¾ × ¾ × ½ × ½	1.20	1.22
1 × 1 × ¾ × ¾	1.37	1.45
1 × 1 × ½ × ½	1.26	1.36
1¼ × 1¼ × 1 × 1	1.58	1.67
1¼ × 1¼ × ¾ × ¾	1.45	1.62
1½ × 1½ × 1¼ × 1¼	1.82	1.88
1½ × 1½ × 1 × 1	1.65	1.80
1½ × 1½ × ¾ × ¾	1.52	1.75
2 × 2 × 1½ × 1½	2.02	2.16
2 × 2 × 1¼ × 1¼	1.90	2.10
2 × 2 × 1 × 1	1.73	2.02
2 × 2 × ¾ × ¾	1.60	1.97
2½ × 2½ × 2 × 2	2.39	2.60
3 × 3 × 2 × 2	2.52	2.89

* Dimensions in inches. For width of band (*B*), outside diameter of band (OB), thread length (TL), and metal thickness (*T*), refer to Table 11.4 for the corresponding size.

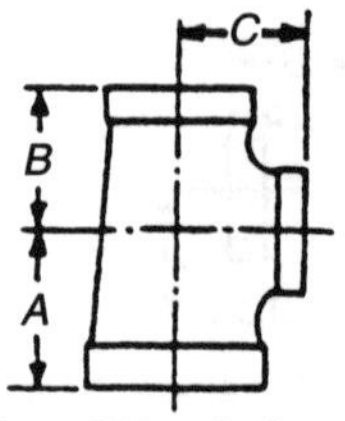

Class 150 reducing tee

TABLE 11.7 Dimensions of Class 150 Malleable Iron Reducing Tees*

Nominal pipe size	Center to end			Nominal pipe size	Center to end		
	A	*B*	*C*		*A*	*B*	*C*
1/4 × 1/4 × 1/8	0.74	0.74	0.76	1 1/4 × 3/4 × 3/4	1.45	1.31	1.62
1/8 × 1/8 × 1/4	0.76	0.76	0.74	1 1/4 × 1/2 × 1 1/4	1.75	1.53	1.75
3/8 × 3/8 × 1/4	0.88	0.88	0.90	1 1/4 × 1/2 × 1	1.58	1.36	1.67
3/8 × 3/8 × 1/8	0.81	0.81	0.85	1 × 1 × 1 1/4	1.67	1.67	1.58
3/8 × 1/4 × 3/8	0.95	0.90	0.95	3/4 × 3/4 × 1 1/4	1.62	1.62	1.45
3/8 × 1/4 × 1/4	0.88	0.81	0.90	1 1/2 × 1 1/2 × 1 1/4	1.82	1.82	1.88
1/4 × 1/4 × 3/8	0.90	0.90	0.88	1 1/2 × 1 1/2 × 1	1.65	1.65	1.80
1/2 × 1/2 × 3/8	1.04	1.04	1.03	1 1/2 × 1 1/2 × 3/4	1.52	1.52	1.75
1/2 × 1/2 × 1/4	0.97	0.97	0.98	1 1/2 × 1 1/2 × 1/2	1.41	1.41	1.66
1/2 × 3/8 × 1/2	1.12	1.03	1.12	1 1/2 × 1 1/4 × 1 1/2	1.94	1.88	1.94
1/2 × 3/8 × 3/8	1.04	0.95	1.03	1 1/2 × 1 1/4 × 1 1/4	1.82	1.75	1.88
1/2 × 1/4 × 1/2	1.12	0.98	1.12	1 1/2 × 1 1/4 × 1	1.65	1.58	1.80
3/8 × 3/8 × 1/2	1.03	1.03	1.04	1 1/2 × 1 1/4 × 3/4	1.52	1.45	1.75
3/4 × 3/4 × 1/2	1.20	1.20	1.22	1 1/2 × 1 1/4 × 1/2	1.41	1.34	1.66
3/4 × 3/4 × 3/8	1.12	1.12	1.13	1 1/2 × 1 × 1 1/2	1.94	1.80	1.94
3/4 × 3/4 × 1/4	1.05	1.05	1.08	1 1/2 × 1 × 1 1/4	1.82	1.67	1.88
3/4 × 1/2 × 3/4	1.31	1.22	1.31	1 1/2 × 1 × 1	1.65	1.50	1.80
3/4 × 1/2 × 1/2	1.20	1.12	1.22	1 1/2 × 3/4 × 1 1/2	1.94	1.75	1.94
3/4 × 1/2 × 3/8	1.12	1.04	1.13	1 1/2 × 1/2 × 1 1/2	1.94	1.66	1.94
3/4 × 3/8 × 3/4	1.31	1.13	1.31	1 1/4 × 1 1/4 × 1 1/2	1.88	1.88	1.82
3/4 × 3/8 × 3/8	1.12	0.95	1.13	1 × 1 × 1 1/2	1.80	1.80	1.65
3/4 × 1/4 × 3/4	1.31	1.08	1.31	2 × 2 × 1 1/2	2.02	2.02	2.16
1/2 × 1/2 × 3/4	1.22	1.22	1.20	2 × 2 × 1 1/4	1.90	1.90	2.10
1 × 1 × 3/4	1.37	1.37	1.45	2 × 2 × 1	1.73	1.73	2.02
1 × 1 × 1/2	1.26	1.26	1.36	2 × 2 × 3/4	1.60	1.60	1.97
1 × 1 × 3/8	1.18	1.18	1.27	2 × 2 × 1/2	1.49	1.49	1.88
1 × 1 × 1/4	1.11	1.11	1.22	2 × 1 1/2 × 2	2.25	2.16	2.25
1 × 3/4 × 1	1.50	1.45	1.50	2 × 1 1/2 × 1 1/2	2.02	1.94	2.16
1 × 3/4 × 3/4	1.37	1.31	1.45	2 × 1 1/2 × 1 1/4	1.90	1.82	2.10
1 × 3/4 × 1/2	1.26	1.20	1.36	2 × 1 1/2 × 1	1.73	1.65	2.02
1 × 1/2 × 1	1.50	1.36	1.50	2 × 1 1/4 × 2	2.25	2.10	2.25
1 × 1/2 × 3/4	1.37	1.22	1.45	2 × 1 1/4 × 1 1/2	2.02	1.88	2.16
1 × 1/2 × 1/2	1.26	1.12	1.36	2 × 1 1/4 × 1 1/4	1.90	1.75	2.10
3/4 × 3/4 × 1	1.45	1.45	1.37	2 × 1 × 2	2.25	2.02	2.25
1/2 × 1/2 × 1	1.36	1.36	1.26	2 × 1 × 1 1/2	2.02	1.80	2.16
1 1/4 × 1 1/4 × 1	1.58	1.58	1.67	2 × 3/4 × 2	2.25	1.97	2.25
1 1/4 × 1 1/4 × 3/4	1.45	1.45	1.62	2 × 1/2 × 2	2.25	1.88	2.25
1 1/4 × 1 1/4 × 1/2	1.34	1.34	1.53	1 1/2 × 1 1/2 × 2	2.16	2.16	2.02
1 1/4 × 1 1/4 × 3/8	1.26	1.26	1.44	1 1/4 × 1 1/4 × 2	2.10	2.10	1.90
1 1/4 × 1 × 1 1/4	1.75	1.67	1.75	1 × 1 × 2	2.02	2.02	1.73
1 1/4 × 1 × 1	1.58	1.50	1.67	2 1/2 × 2 1/2 × 2	2.39	2.39	2.60
1 1/4 × 1 × 3/4	1.45	1.37	1.62	2 1/2 × 2 1/2 × 1 1/2	2.16	2.16	2.51
1 1/4 × 1 × 1/2	1.34	1.26	1.53	2 1/2 × 2 1/2 × 1 1/4	2.04	2.04	2.45
1 1/4 × 3/4 × 1 1/4	1.75	1.62	1.75	2 1/2 × 2 1/2 × 1	1.87	1.87	2.37
1 1/4 × 3/4 × 1	1.58	1.45	1.67	2 1/2 × 2 1/2 × 3/4	1.74	1.74	2.32

TABLE 11.7 Dimensions of Class 150 Malleable Iron Reducing Tees* (Continued)

Nominal pipe size	Center to end A	B	C	Nominal pipe size	Center to end A	B	C
$2\frac{1}{2} \times 2 \times 2\frac{1}{2}$	2.70	2.60	2.70	$3 \times 2 \times 2$	2.52	2.25	2.89
$2\frac{1}{2} \times 2 \times 2$	2.39	2.25	2.60	$2\frac{1}{2} \times 2\frac{1}{2} \times 3$	2.99	2.99	2.83
$2\frac{1}{2} \times 1\frac{1}{2} \times 2\frac{1}{2}$	2.70	2.51	2.70	$3\frac{1}{2} \times 3\frac{1}{2} \times 2\frac{1}{2}$	2.93	2.93	3.24
$2\frac{1}{2} \times 1\frac{1}{2} \times 2$	2.39	2.16	2.60	$4 \times 4 \times 3$	3.30	3.30	3.60
$2 \times 2 \times 2\frac{1}{2}$	2.60	2.60	2.39	$4 \times 4 \times 2\frac{1}{2}$	3.05	3.05	3.51
$3 \times 3 \times 2\frac{1}{2}$	2.83	2.83	2.99	$4 \times 4 \times 2$	2.74	2.74	3.41
$3 \times 3 \times 2$	2.52	2.52	2.89	$4 \times 4 \times 1\frac{1}{2}$	2.51	2.51	3.32
$3 \times 3 \times 1\frac{1}{2}$	2.29	2.29	2.80	$4 \times 3 \times 4$	3.79	3.60	3.79
$3 \times 3 \times 1\frac{1}{4}$	2.17	2.17	2.74	$3 \times 3 \times 4$	3.60	3.60	3.30
$3 \times 3 \times 1$	2.00	2.00	2.66	$5 \times 5 \times 3$	3.51	3.51	4.22
$3 \times 3 \times \frac{3}{4}$	1.87	1.87	2.61	$6 \times 6 \times 4$	4.13	4.13	4.94
$3 \times 2\frac{1}{2} \times 3$	3.08	2.99	3.08	$6 \times 6 \times 3$	3.64	3.64	4.75
$3 \times 2\frac{1}{2} \times 2\frac{1}{2}$	2.83	2.70	2.99	$6 \times 6 \times 2\frac{1}{2}$	3.39	3.39	4.66
$3 \times 2\frac{1}{2} \times 2$	2.52	2.39	2.89	$6 \times 6 \times 2$	3.08	3.08	4.56
$3 \times 2 \times 3$	3.08	2.89	3.08				

* Dimensions in inches. For thread length, metal thickness, width, and outsider diameter of the band, refer to Table 11.4 for corresponding size.

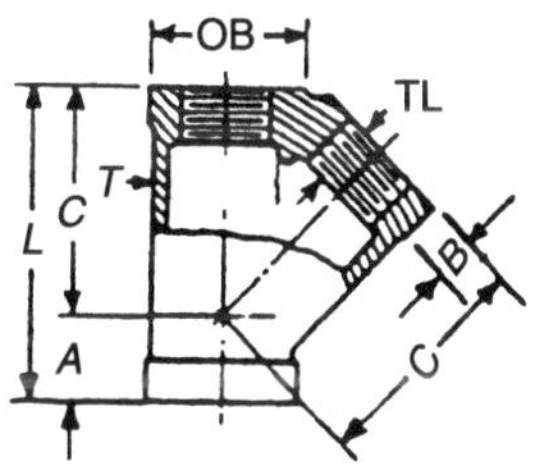

Class 150 straight Y branch

TABLE 11.8 Dimensions of Class 150 Straight Y Branch*

Nominal pipe size	Minimum length of thread, TL	Minimum width of band, *B*	Metal thickness, *T*	Minimum outside diameter of band, OB	Center-to-end inlet, *A*	Center-to-end outlet, *C*	End to end, *L*
$\frac{3}{8}$	0.36	0.23	0.10	1.01	0.50	1.43	1.93
$\frac{1}{2}$	0.43	0.25	0.10	1.20	0.61	1.71	2.32
$\frac{3}{4}$	0.50	0.27	0.12	1.46	0.72	2.05	2.77
1	0.58	0.30	0.13	1.77	0.85	2.43	3.28
$1\frac{1}{4}$	0.67	0.34	0.14	2.15	1.02	2.92	3.94
$1\frac{1}{2}$	0.70	0.37	0.15	2.43	1.10	3.28	4.38
2	0.75	0.42	0.17	2.96	1.24	3.93	5.17
$2\frac{1}{2}$	0.92	0.48	0.21	3.59	1.52	4.73	6.25
3	0.98	0.55	0.23	4.28	1.71	5.55	7.26
4	1.08	0.66	0.26	5.40	2.01	6.97	8.98

* Dimensions in inches.

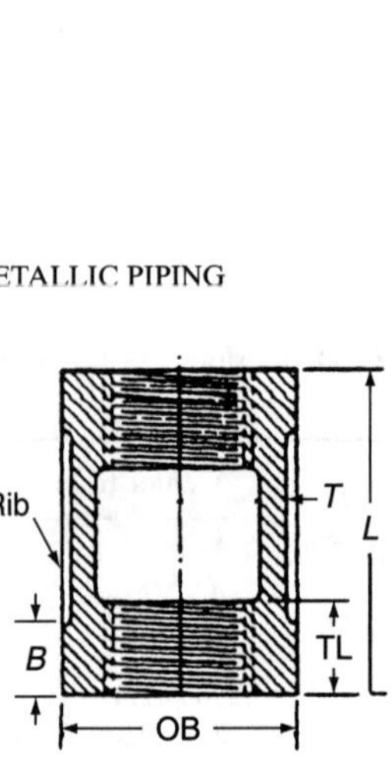

Class 150 malleable iron straight coupling (cast)

TABLE 11.9 Dimensions of Class 150 Malleable Iron Straight Couplings*

Nominal pipe size	Minimum length of thread, TL	Minimum width of band, *B*	Metal thickness, *T*	Minimum outside diameter of band, OB	Length of straight coupling, *L*
1/8	0.25	0.20	0.09	0.69	0.96
1/4	0.32	0.21	0.09	0.84	1.06
3/8	0.36	0.23	0.10	1.01	1.16
1/2	0.43	0.25	0.10	1.20	1.34
3/4	0.50	0.27	0.12	1.46	1.52
1	0.58	0.30	0.13	1.77	1.67
1 1/4	0.67	0.34	0.14	2.15	1.93
1 1/2	0.70	0.37	0.15	2.43	2.15
2	0.75	0.42	0.17	2.96	2.53
2 1/2	0.92	0.48	0.21	3.59	2.88
3	0.98	0.55	0.23	4.28	3.18
4	1.08	0.66	0.26	5.40	3.69

* Dimensions in inches.

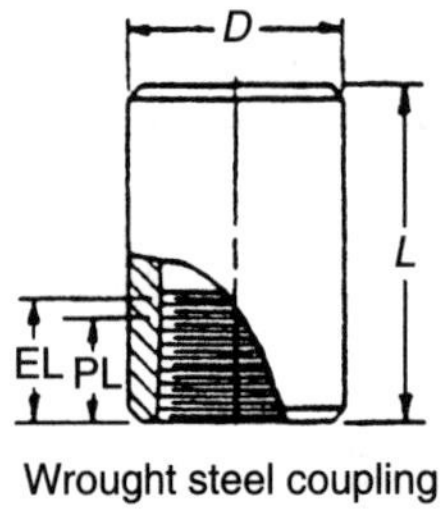

Wrought steel coupling

TABLE 11.10 Dimensions of Class 150 Wrought Steel Straight Couplings—NPS 1/8 through NPS 3/8*

Nominal pipe size	Minimum length of thread		Length, *L*	Outside diameter, *D*
	PL†	EL‡		
1/8	0.25	0.2639	0.81	0.56
1/4	0.32	0.4018	1.19	0.72
3/8	0.36	0.4078	1.19	0.87

* Dimensions in inches.
† PL is the minimum length of perfect thread.
‡ EL is the length of perfect thread (PL) plus threads with fully formed roots and flat crests.

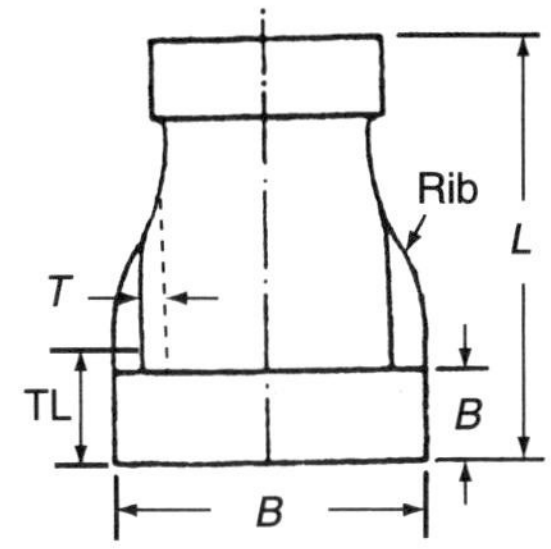

Class 150 malleable iron reducing coupling

TABLE 11.11 Dimensions of Class 150 Reducing Couplings*

Nominal pipe size	Minimum length of thread, TL	Minimum width of band, *B*	Metal thickness, *T*	Minimum outside diameter of band, OB	Length of reducing concentric coupling, *L*
1/8	0.25	0.20	0.09	0.69	
1/4	0.32	0.21	0.09	0.84	1.00
3/8	0.36	0.23	0.10	1.01	1.13
1/2	0.43	0.25	0.10	1.20	1.25
3/4	0.50	0.27	0.12	1.46	1.44
1	0.58	0.30	0.13	1.77	1.69
1 1/4	0.67	0.34	0.14	2.15	2.06
1 1/2	0.70	0.37	0.15	2.43	2.31
2	0.75	0.42	0.17	2.96	2.81
2 1/2	0.92	0.48	0.21	3.59	3.25
3	0.98	0.55	0.23	4.28	3.69
4	1.08	0.66	0.26	5.40	4.38

* Dimensions in inches.

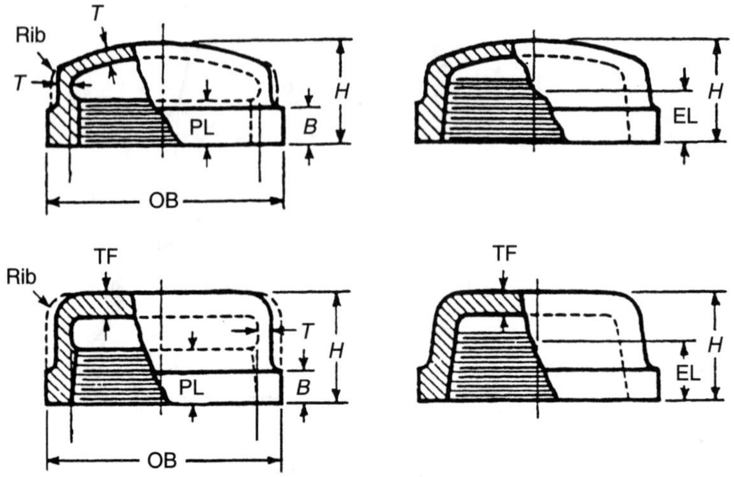

TABLE 11.12 Dimensions of Class 150 Caps*

Nominal pipe size	Minimum length of thread PL	Minimum length of thread EL	Minimum width of band, *B*	Metal thickness of side, *T*	Minimum outside diameter of band, OB	Minimum height, *H*	Thickness of flat-top caps, TF
1/8	0.25	0.2639	0.20	0.09	0.69†	0.53	
1/4	0.32	0.4018	0.21	0.09	0.84†	0.63	
3/8	0.36	0.4078	0.23	0.10	1.01†	0.74	
1/2	0.43	0.5337	0.25	0.10	1.20	0.87	
3/4	0.50	0.5457	0.27	0.12	1.46	0.97	0.13
1	0.58	0.6828	0.30	0.13	1.77	1.16	0.15
1 1/4	0.67	0.7068	0.34	0.14	2.15	1.28	0.17
1 1/2	0.70	0.7235	0.37	0.15	2.43	1.33	0.19
2	0.75	0.7565	0.42	0.17	2.46	1.45	0.22
2 1/2	0.92	1.1375	0.48	0.21	3.59	1.70	0.25
3	0.98	1.2000	0.55	0.23	4.28	1.80	0.29
3 1/2	1.03	1.2500	0.60	0.25	4.84	1.90	0.30
4	1.08	1.3000	0.66	0.26	5.40	2.08	0.36
5	1.18	1.4063	0.78	0.30	6.58	2.32	
6	1.28	1.5125	0.90	0.34	7.77	2.55	

* Dimensions in inches.

† NPS 1/8, 1/4, and 3/8 caps made from steel rod with a minimum yield strength of 30 ksi shall have the following diameters:
0.50 in for NPS 1/8
0.69 in for NPS 1/4
0.84 in for NPS 3/8

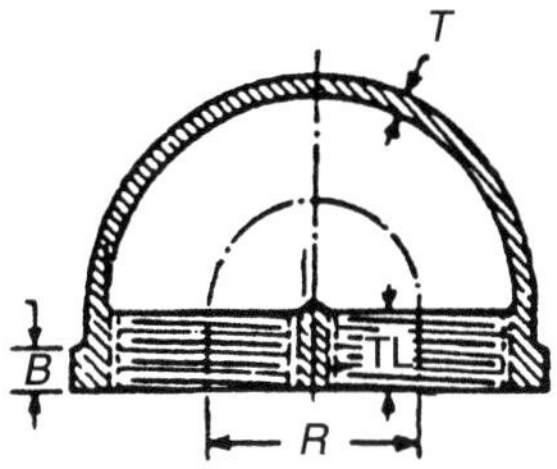

Closed

TABLE 11.13 Dimensions of Class 150 Closed-Pattern Return Bends*

Nominal pipe size	Minimum length of thread, TL	Minimum width of band, *B*	Metal thickness, *T*	Minimum outside diameter of band, OB	Center to center, *R*
½	0.43	0.25	0.12	1.20	1.00
¾	0.50	0.27	0.13	1.46	1.25
1	0.58	0.30	0.15	1.77	1.50
1¼	0.67	0.34	0.16	2.15	1.75
1½	0.70	0.37	0.18	2.43	2.19
2	0.75	0.42	0.20	2.96	2.62
2½	0.92	0.48	0.24	3.59	
3	0.98	0.55	0.27	4.28	
4	1.08	0.66	0.31	5.40	

* Dimensions in inches. Closed-pattern return bends may be furnished without band.

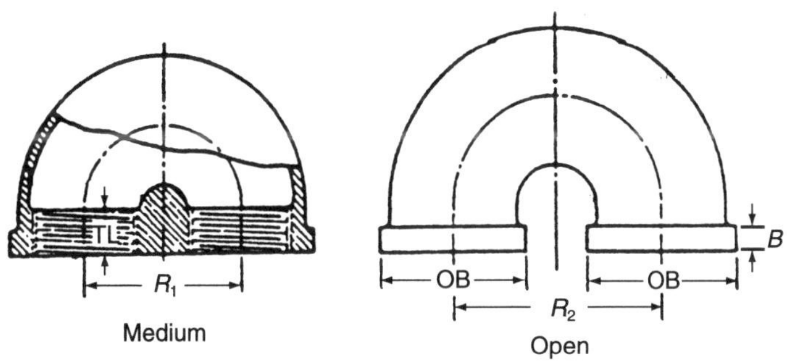

Class 150 medium and open-pattern return bends

TABLE 11.14 Dimensions of Class 150 Medium and Open-Pattern Return Bends*

Medium center-to-center, R_1	Open center-to-center, R_2
1.25	1.50
1.50	2.00
1.87	2.50
2.25	3.00
2.50	3.50
3.00	4.00
	4.50
	5.00
	6.00

* Dimensions in inches.

Note: For bandwidth *B*, outside diameter of band OB, minimum length of thread TL, and metal thickness *T*, refer to Table 11.13.

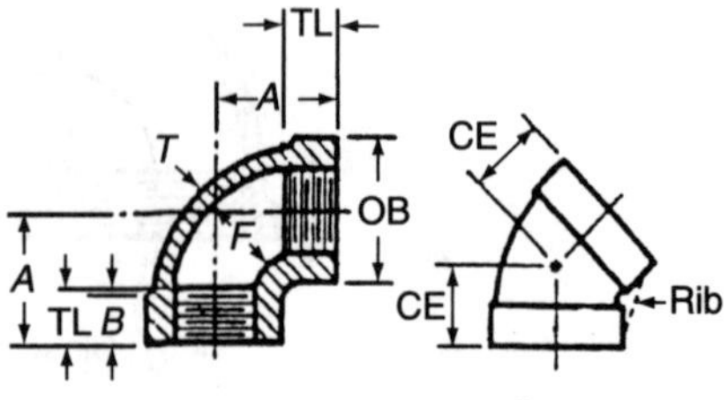

TABLE 11.15 Dimensions of Class 300 Straight 90 and 45° Elbows*

Nominal pipe size	Center to end, 90° elbows, *A*	Center to end, 45° elbows, CE	Minimum length of thread, TL	Minimum width of band, *B*	Minimum metal thickness, *T*	Minimum outside diameter of band, OB
1/4	0.94	0.81	0.43	0.38	0.14	0.93
3/8	1.06	0.88	0.47	0.44	0.15	1.12
1/2	1.25	1.00	0.57	0.50	0.16	1.34
3/4	1.44	1.13	0.64	0.56	0.18	1.63
1	1.63	1.31	0.75	0.62	0.20	1.95
1 1/4	1.94	1.50	0.84	0.69	0.22	2.39
1 1/2	2.13	1.69	0.87	0.75	0.24	2.68
2	2.50	2.00	1.00	0.84	0.26	3.28
2 1/2	2.94	2.25	1.17	0.94	0.31	3.86
3	3.38	2.50	1.23	1.00	0.35	4.62

* Dimensions in inches.

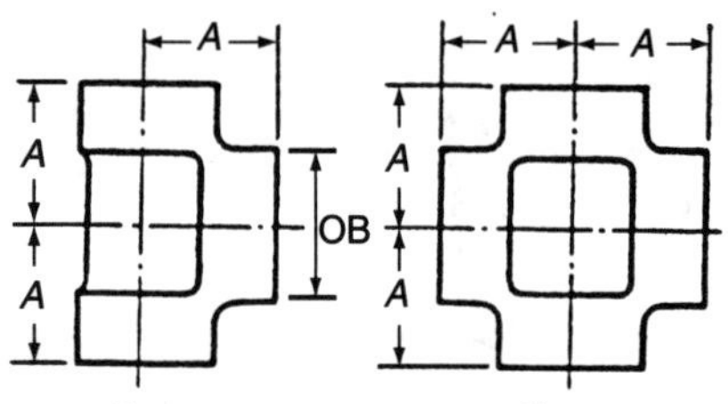

TABLE 11.16 Dimensions of Class 300 Straight Tees and Crosses*

Nominal pipe size	Center to end, *A*	Minimum length of thread, TL	Minimum width of band, *B*	Minimum metal thickness, *T*	Minimum outside diameter of band, OB
1/4	0.94	0.43	0.38	0.14	0.93
3/8	1.06	0.47	0.44	0.15	1.12
1/2	1.25	0.57	0.50	0.16	1.34
3/4	1.44	0.64	0.56	0.18	1.63
1	1.63	0.75	0.62	0.20	1.95
1 1/4	1.94	0.84	0.69	0.22	2.39
1 1/2	2.13	0.87	0.75	0.24	2.68
2	2.50	1.00	0.84	0.26	3.28
2 1/2	2.94	1.17	0.94	0.31	3.86
3	3.38	1.23	1.00	0.35	4.62

* Dimensions in inches.
Note: For pictorial depiction of dimensions TL, *B*, and *T*, refer to Table 11.15.

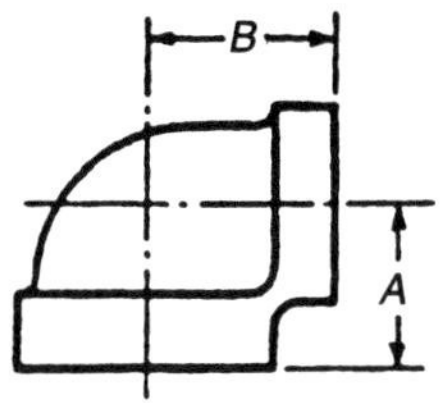

TABLE 11.17 Dimensions of Class 300 Reducing 90° Elbows*

Nominal pipe size	Center to end A	Center to end B
$\frac{1}{2} \times \frac{3}{8}$	1.19	1.19
$\frac{3}{4} \times \frac{1}{2}$	1.31	1.38
$1 \times \frac{3}{4}$	1.50	1.56
$1\frac{1}{4} \times 1$	1.75	1.81
$1\frac{1}{2} \times 1\frac{1}{4}$	2.00	2.06
$2 \times 1\frac{1}{2}$	2.25	2.38
$2\frac{1}{2} \times 2$	2.69	2.75
$3 \times 2\frac{1}{2}$	3.06	3.31

* Dimensions in inches.

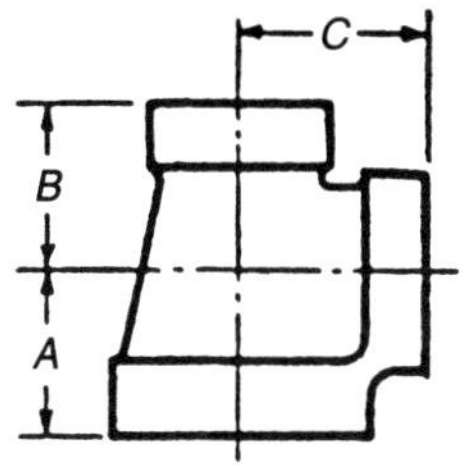

TABLE 11.18 Dimensions of Class 300 Malleable Iron Reducing Tees*

Nominal pipe size	Center to end A	Center to end B	Center to end C
$\frac{1}{2} \times \frac{1}{2} \times \frac{3}{8}$	1.19	1.19	1.19
$\frac{1}{2} \times \frac{3}{8} \times \frac{1}{2}$	1.25	1.19	1.25
$\frac{3}{4} \times \frac{3}{4} \times \frac{1}{2}$	1.31	1.31	1.38
$\frac{3}{4} \times \frac{1}{2} \times \frac{3}{4}$	1.44	1.38	1.44
$1 \times 1 \times \frac{3}{4}$	1.50	1.50	1.56
$1 \times 1 \times \frac{1}{2}$	1.44	1.44	1.50
$1 \times 1 \times \frac{3}{8}$	1.31	1.31	1.44
$1 \times \frac{3}{4} \times 1$	1.63	1.56	1.63
$1\frac{1}{4} \times 1\frac{1}{4} \times 1$	1.75	1.75	1.81
$1\frac{1}{4} \times 1\frac{1}{4} \times \frac{3}{4}$	1.63	1.63	1.75
$1\frac{1}{4} \times 1\frac{1}{4} \times \frac{1}{2}$	1.50	1.50	1.69
$1\frac{1}{4} \times 1 \times 1\frac{1}{4}$	1.94	1.81	1.94
$1\frac{1}{2} \times 1\frac{1}{2} \times 1\frac{1}{4}$	2.00	2.00	2.06
$1\frac{1}{2} \times 1\frac{1}{2} \times 1$	1.81	1.81	2.00
$1\frac{1}{2} \times 1\frac{1}{2} \times \frac{3}{4}$	1.69	1.69	1.88
$1\frac{1}{2} \times 1\frac{1}{2} \times \frac{1}{2}$	1.63	1.63	1.81
$1\frac{1}{2} \times 1\frac{1}{4} \times 1\frac{1}{2}$	2.13	2.06	2.13
$2 \times 2 \times 1\frac{1}{2}$	2.25	2.25	2.38
$2 \times 2 \times 1\frac{1}{4}$	2.13	2.13	2.31
$2 \times 2 \times 1$	2.00	2.00	2.25
$2 \times 2 \times \frac{3}{4}$	1.81	1.81	2.13
$2 \times 2 \times \frac{1}{2}$	1.75	1.75	2.06
$2 \times 1\frac{1}{2} \times 2$	2.50	2.38	2.50
$2\frac{1}{2} \times 2\frac{1}{2} \times 2$	2.69	2.69	2.75
$2\frac{1}{2} \times 2\frac{1}{2} \times 1\frac{1}{2}$	2.44	2.44	2.63
$2\frac{1}{2} \times 2 \times 2\frac{1}{2}$	2.94	2.75	2.94
$3 \times 3 \times 2\frac{1}{2}$	3.06	3.06	3.31
$3 \times 3 \times 2$	2.81	2.81	3.13
$3 \times 2\frac{1}{2} \times 3$	3.38	3.31	3.38

* Dimensions in inches.

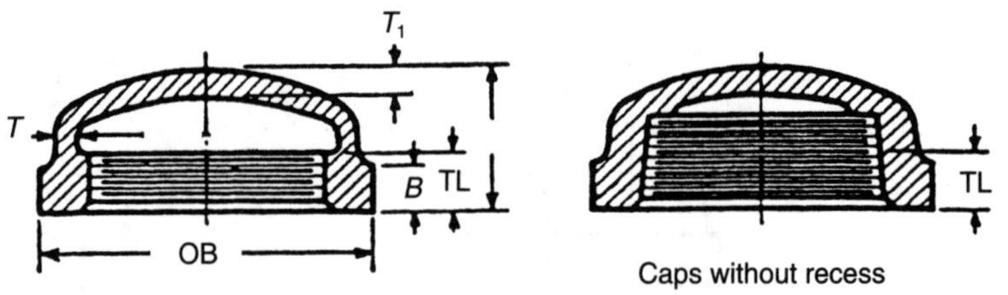

Caps with recess

Caps without recess

TABLE 11.19 Dimensions of Class 300 Caps*

Nominal pipe size	Minimum length of thread, TL	Minimum width of band, *B*	Metal thickness† *T*	Metal thickness† Recommended T_1	Minimum outside diameter of band, OB	Minimum height, *H*
1/4	0.43	0.38	0.14	0.18	0.93	0.78
3/8	0.47	0.44	0.15	0.19	1.12	0.83
1/2	0.57	0.50	0.16	0.20	1.34	0.98
3/4	0.64	0.56	0.18	0.23	1.63	1.08
1	0.75	0.62	0.20	0.25	1.95	1.26
1 1/4	0.84	0.69	0.22	0.28	2.39	1.38
1 1/2	0.87	0.75	0.24	0.30	2.68	1.43
2	1.00	0.84	0.26	0.33	3.28	1.68
2 1/2	1.17	0.94	0.31	0.39	3.86	2.06
3	1.23	1.00	0.35	0.44	4.62	2.17

* Dimensions in inches.
† In no case shall the thickness T_1 be less than *T*. Recommended thickness T_1 may be equal to *T*.

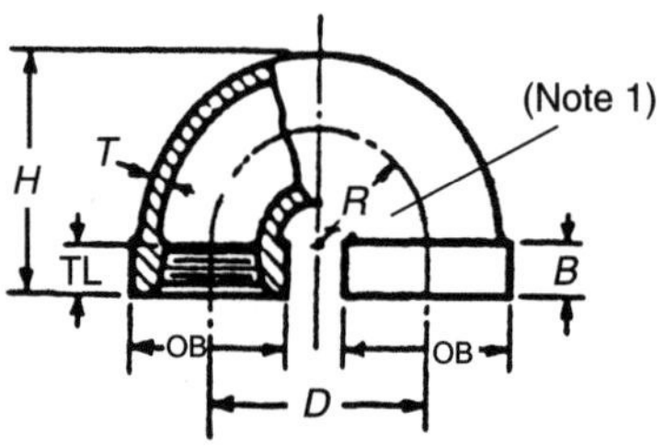

TABLE 11.20 Dimensions of Class 300 Return Bends (Note 2)

Nominal pipe size	Center to center, *D* Closed pattern	Center to center, *D* Medium pattern	Center to center, *D* Open pattern	Minimum length of thread, TL	Minimum width of band, *B*	Metal thickness, *T* (Note 3)	Minimum outside diameter of band, OB
1	1.75	2.50	3.00	0.75	0.62	0.20	1.95
1 1/4	2.25	2.50	3.00	0.84	0.69	0.22	2.39
1 1/2	3.00	3.50	6.00	0.87	0.75	0.24	2.68
2	4.00	6.00	8.00	1.00	0.84	0.26	3.28

Notes:
1. *R* is recommended to be one-half of dimension *D*.
2. Dimensions in inches.
3. For 180° return bend, it is recommended that the thickness *T* be increased by 10 percent.

Tolerances for Center-to-End Dimensions of Malleable Iron Fittings

Tolerances given in Table 11.21 apply to fittings other than return bends.

Tolerances for End-to-End Dimensions and Lengths of Reducers and Couplings

Tolerances for end-to-end dimensions and lengths of reducers and couplings shall be twice the tolerances listed in Table 11.21 corresponding to the size.

GRAY IRON THREADED FITTINGS

ASME Standard for Gray Iron Threaded Fittings

ASME B16.4, Gray Iron Threaded Fittings, covers the following Class 125 and Class 250 gray iron threaded fittings:

- 90 and 45° elbows
- Tees and crosses
- Y branches
- Couplings
- Caps
- Return bends

Pressure-Temperature Ratings of Class 125 Gray Iron Threaded Fittings

See Table 11.22.

Pressure-Temperature Ratings of Class 250 Gray Iron Threaded Fittings

See Table 11.23.

Materials for Gray Iron Threaded Fittings

- ASTM A126, Specification for Gray Iron Castings for Valves, Flanges, and Pipe Fittings, Class A, B, or C.

TABLE 11.21 Tolerances for Center-to-End Dimensions*

Nominal pipe size (NPS) of fitting, in.	Tolerance, in. (plus or minus)
1/8	0.03
1/4	0.04
3/8	0.05
1/2, 3/4	0.06
1, 1 1/4	0.07
1 1/2, 2	0.08
2 1/2, 3, 3 1/2	0.10
4, 5	0.12
6	0.14

* The largest opening in a reducing fitting governs the tolerance.

TABLE 11.22 Pressure-Temperature Ratings of Class 125 Gray Iron Threaded Fittings

Temperature, °F	Pressure, psig
–20 to 150	175
200	165
250	150
300	140
350*	125*

* This rating can be used for temperatures up to 353°F, which is the temperature of 125 psig saturated steam.

TABLE 11.23 Pressure-Temperature Ratings of Class 250 Gray Iron Threaded Fittings

Temperature, °F	Pressure, psig
–20 to 150	400
200	370
250	340
300	310
350*	300
400	250*

* This pressure rating can be used up to 406°F, which is the temperature of 250 psig saturated steam.

Galvanization or Zinc Coating of Gray Iron Threaded Fittings

- 0.0034-in minimum hot-dipped zinc coating per ASTM A153, only when specified.
- 0.001-in electrodeposited zinc coating in accordance with ASTM B633, Type I, Service Condition, only when specified.

Threads

- Taper threads in accordance with ASME B1.20.1, Pipe Threads, General Purpose (inch).

Dimensions of ASME B16.4 Class 125 and 250 Gray Iron Threaded Fittings

See Tables 11.24 through 11.32.

Tolerances for Center-to-End Dimensions of Gray Iron Fittings

Tolerances given in Table 11.33 apply to fittings other than return bends.

Tolerances for End-to-End Dimensions and Lengths of Gray Iron Reducers and Couplings

Tolerances for end-to-end dimensions and lengths of reducers and couplings shall be twice the tolerances listed in Table 11.33 corresponding to the size.

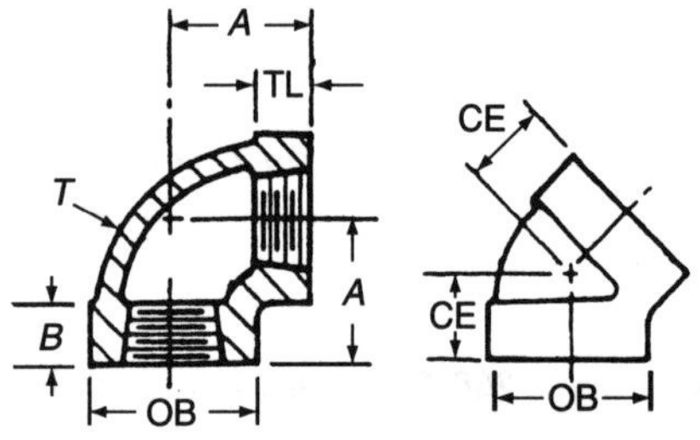

TABLE 11.24 Dimensions of Class 125 Gray Iron Straight 90 and 45° Elbows*

Nominal pipe size	Center to end, 90° elbows, *A*	Center to end, 45° elbows, CE	Minimum length of thread, TL	Minimum width of band, *B*	Metal thickness, *T*	Outside diameter of band, OB
1/4	0.81	0.73	0.32	0.38	0.11	0.93
3/8	0.95	0.80	0.36	0.44	0.12	1.12
1/2	1.12	0.88	0.43	0.50	0.13	1.34
3/4	1.31	0.98	0.50	0.56	0.15	1.63
1	1.50	1.12	0.58	0.62	0.17	1.95
1 1/4	1.75	1.29	0.67	0.69	0.18	2.39
1 1/2	1.94	1.43	0.70	0.75	0.20	2.68
2	2.25	1.68	0.75	0.84	0.22	3.28
2 1/2	2.70	1.95	0.92	0.94	0.24	3.86
3	3.08	2.17	0.98	1.00	0.26	4.62
3 1/2	3.42	2.39	1.03	1.06	0.28	5.20
4	3.79	2.61	1.08	1.12	0.31	5.79
5	4.50	3.05	1.18	1.18	0.38	7.05
6	5.13	3.46	1.28	1.28	0.43	8.28
8	6.56	4.28	1.47	1.47	0.55	10.63
10	8.08	5.16	1.68	1.68	0.69	13.12
12	9.50	5.97	1.88	1.88	0.80	15.47

* Dimensions in inches.

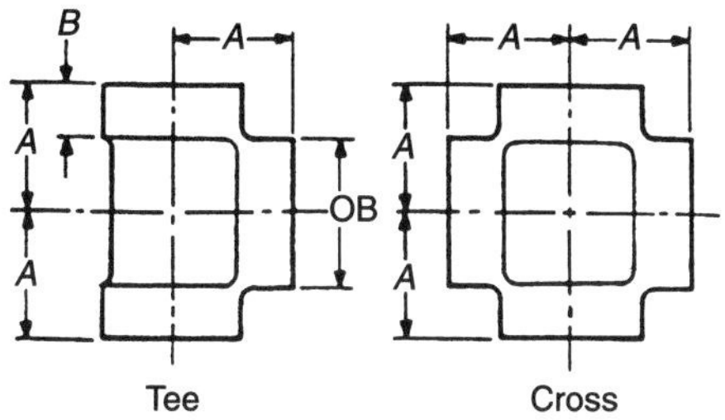

TABLE 11.25 Dimensions of Class 125 Gray Iron Straight Tees and Crosses*

Nominal pipe size	Center to end, *A*	Minimum length of thread	Minimum width of band, *B*	Metal thickness, *T*	Minimum outside diameter of band, OB
1/4	0.81	0.32	0.38	0.11	0.93
3/8	0.95	0.36	0.44	0.12	1.12
1/2	1.12	0.43	0.50	0.13	1.34
3/4	1.31	0.50	0.56	0.15	1.63
1	1.50	0.58	0.62	0.17	1.95
1 1/4	1.75	0.67	0.69	0.18	2.39
1 1/2	1.94	0.70	0.75	0.20	2.68
2	2.25	0.75	0.84	0.22	3.28
2 1/2	2.70	0.92	0.94	0.24	3.86
3	3.08	0.98	1.00	0.26	4.62
3 1/2	3.42	1.03	1.06	0.28	5.20
4	3.79	1.08	1.12	0.31	5.79
5	4.50	1.18	1.18	0.38	7.05
6	5.13	1.28	1.28	0.43	8.28
8	6.56	1.47	1.47	0.55	10.63
10	8.08	1.68	1.68	0.69	13.12
12	9.50	1.88	1.88	0.80	15.47

* Dimensions in inches.

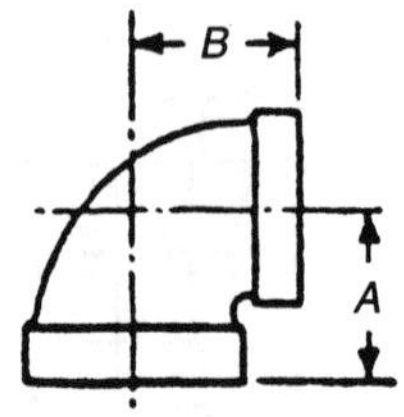

Reducing elbow

TABLE 11.26 Dimensions of Class 125 Gray Iron Reducing Elbows*

Nominal pipe size	Center to end A	Center to end B	Nominal pipe size	Center to end A	Center to end B
$\frac{1}{2} \times \frac{3}{8}$	1.04	1.03	$2\frac{1}{2} \times 1\frac{1}{4}$	2.04	2.45
$\frac{3}{4} \times \frac{1}{2}$	1.20	1.22	$2\frac{1}{2} \times 1$	1.87	2.37
$1 \times \frac{3}{4}$	1.37	1.45	$3 \times 2\frac{1}{2}$	2.83	2.99
$1 \times \frac{1}{2}$	1.26	1.36	3×2	2.52	2.89
$1\frac{1}{4} \times 1$	1.58	1.67	$3 \times 1\frac{1}{2}$	2.29	2.80
$1\frac{1}{4} \times \frac{3}{4}$	1.45	1.62	$3 \times 1\frac{1}{4}$	2.17	2.74
$1\frac{1}{4} \times \frac{1}{2}$	1.34	1.53	$3\frac{1}{2} \times 3$	3.18	3.33
$1\frac{1}{2} \times 1\frac{1}{4}$	1.82	1.88	$4 \times 3\frac{1}{2}$	3.54	3.69
$1\frac{1}{2} \times 1$	1.65	1.80	4×3	3.30	3.60
$1\frac{1}{2} \times \frac{3}{4}$	1.52	1.75	$4 \times 2\frac{1}{2}$	3.05	3.51
$1\frac{1}{2} \times \frac{1}{2}$	1.41	1.66	4×2	2.74	3.41
$2 \times 1\frac{1}{2}$	2.02	2.16	5×4	4.00	4.41
$2 \times 1\frac{1}{4}$	1.90	2.10	5×3	3.51	4.22
2×1	1.73	2.02	$5 \times 2\frac{1}{2}$	3.26	4.13
$2 \times \frac{3}{4}$	1.60	1.97	6×5	4.63	5.03
$2 \times \frac{1}{2}$	1.49	1.88	6×4	4.13	4.94
$2\frac{1}{2} \times 2$	2.39	2.60	6×3	3.64	4.75
$2\frac{1}{2} \times 1\frac{1}{2}$	2.16	2.51	8×6	5.56	6.37

* Dimensions in inches.

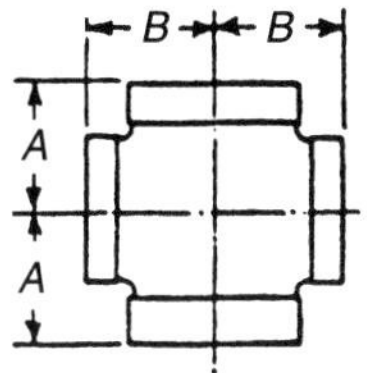

Reducing cross

TABLE 11.27 Dimensions of Class 125 Gray Iron Reducing Crosses*

Nominal pipe size	Center to end A	Center to end B	Nominal pipe size	Center to end A	Center to end B
$\frac{3}{4} \times \frac{3}{4} \times \frac{1}{2} \times \frac{1}{2}$	1.20	1.22	$3 \times 3 \times 1 \times 1$	2.00	2.66
$1 \times 1 \times \frac{3}{4} \times \frac{3}{4}$	1.37	1.45	$3\frac{1}{2} \times 3\frac{1}{2} \times 2\frac{1}{2} \times 2\frac{1}{2}$	2.93	3.24
$1\frac{1}{4} \times 1\frac{1}{4} \times 1 \times 1$	1.58	1.67	$3\frac{1}{2} \times 3\frac{1}{2} \times 2 \times 2$	2.62	3.14
$1\frac{1}{4} \times 1\frac{1}{4} \times \frac{3}{4} \times \frac{3}{4}$	1.45	1.62	$3\frac{1}{2} \times 3\frac{1}{2} \times 1\frac{1}{2} \times 1\frac{1}{2}$	2.39	3.05
$1\frac{1}{2} \times 1\frac{1}{2} \times 1\frac{1}{4} \times 1\frac{1}{4}$	1.82	1.88	$4 \times 4 \times 3 \times 3$	3.30	3.60
$1\frac{1}{2} \times 1\frac{1}{2} \times 1 \times 1$	1.65	1.80	$4 \times 4 \times 2\frac{1}{2} \times 2\frac{1}{2}$	3.05	3.51
$1\frac{1}{2} \times 1\frac{1}{2} \times \frac{3}{4} \times \frac{3}{4}$	1.52	1.75	$4 \times 4 \times 2 \times 2$	2.74	3.41
$2 \times 2 \times 1\frac{1}{2} \times 1\frac{1}{2}$	2.02	2.16	$4 \times 4 \times 1\frac{1}{2} \times 1\frac{1}{2}$	2.51	3.32
$2 \times 2 \times 1\frac{1}{4} \times 1\frac{1}{4}$	1.90	2.10	$5 \times 5 \times 4 \times 4$	4.00	4.41
$2 \times 2 \times 1 \times 1$	1.73	2.02	$5 \times 5 \times 3 \times 3$	3.51	4.22
$2 \times 2 \times \frac{3}{4} \times \frac{3}{4}$	1.60	1.97	$5 \times 5 \times 2 \times 2$	2.95	4.03
$2\frac{1}{2} \times 2\frac{1}{2} \times 2 \times 2$	2.39	2.60	$6 \times 6 \times 4 \times 4$	4.13	4.94
$2\frac{1}{2} \times 2\frac{1}{2} \times 1\frac{1}{2} \times 1\frac{1}{2}$	2.16	2.51	$6 \times 6 \times 3 \times 3$	3.64	4.75
$2\frac{1}{2} \times 2\frac{1}{2} \times 1\frac{1}{4} \times 1\frac{1}{4}$	2.04	2.45	$6 \times 6 \times 2\frac{1}{2} \times 2\frac{1}{2}$	3.39	4.66
$2\frac{1}{2} \times 2\frac{1}{2} \times 1 \times 1$	1.87	2.37	$6 \times 6 \times 2 \times 2$	3.08	4.56
$3 \times 3 \times 2 \times 2$	2.57	2.89	$8 \times 8 \times 6 \times 6$	5.56	6.37
$3 \times 3 \times 1\frac{1}{2} \times 1\frac{1}{2}$	2.29	2.80	$8 \times 8 \times 4 \times 4$	4.50	6.17
$3 \times 3 \times 1\frac{1}{4} \times 1\frac{1}{4}$	2.17	2.74			

* Dimensions in inches.

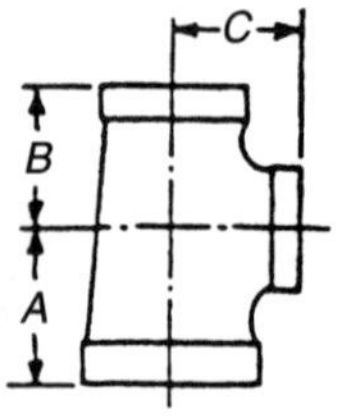

Class 12 gray iron reducing tee

TABLE 11.28 Dimensions of Class 125 Gray Iron Reducing Tees*

Nominal pipe size	Center to end A	Center to end B	Center to end C
1/2 × 1/2 × 3/8	1.04	1.04	1.03
1/2 × 1/2 × 1/4	0.97	0.97	0.98
3/4 × 3/4 × 1/2	1.20	1.20	1.22
3/4 × 3/4 × 3/8	1.12	1.12	1.13
3/4 × 3/4 × 1/4	1.05	1.05	1.08
3/4 × 1/2 × 3/4	1.31	1.22	1.31
3/4 × 1/2 × 1/2	1.20	1.12	1.22
1/2 × 1/2 × 3/4	1.22	1.22	1.20
1 × 1 × 3/4	1.37	1.37	1.45
1 × 1 × 1/2	1.26	1.26	1.36
1 × 1 × 3/8	1.18	1.18	1.27
1 × 3/4 × 1	1.50	1.45	1.50
1 × 3/4 × 3/4	1.37	1.31	1.45
1 × 3/4 × 1/2	1.26	1.20	1.36
1 × 1/2 × 1	1.50	1.36	1.50
1 × 1/2 × 3/4	1.37	1.22	1.45
1 × 1/2 × 1/2	1.26	1.12	1.36
1 × 3/8 × 1	1.50	1.27	1.50
3/4 × 3/4 × 1	1.45	1.45	1.37
1 1/4 × 1 1/4 × 1	1.58	1.58	1.67
1 1/4 × 1 1/4 × 3/4	1.45	1.45	1.62
1 1/4 × 1 1/4 × 1/2	1.34	1.34	1.53
1 1/4 × 1 × 1 1/4	1.75	1.67	1.75
1 1/4 × 1 × 1	1.58	1.50	1.67
1 1/4 × 1 × 3/4	1.45	1.37	1.62
1 1/4 × 1 × 1/2	1.34	1.26	1.53
1 1/4 × 3/4 × 1 1/4	1.75	1.62	1.75
1 1/4 × 3/4 × 1	1.58	1.45	1.67
1 1/4 × 3/4 × 3/4	1.45	1.31	1.62
1 1/4 × 1/2 × 1 1/4	1.75	1.53	1.75
1 1/4 × 1/2 × 1	1.58	1.36	1.67
1 × 1 × 1 1/4	1.67	1.67	1.58
1 1/2 × 1 1/2 × 1 1/4	1.82	1.82	1.88
1 1/2 × 1 1/2 × 1	1.65	1.65	1.80
1 1/2 × 1 1/2 × 3/4	1.52	1.52	1.75
1 1/2 × 1 1/2 × 1/2	1.41	1.41	1.66
1 1/2 × 1 1/4 × 1 1/2	1.94	1.88	1.94
1 1/2 × 1 1/4 × 1 1/4	1.82	1.75	1.88
1 1/2 × 1 1/4 × 1	1.65	1.58	1.80
1 1/2 × 1 1/4 × 3/4	1.52	1.45	1.75
1 1/2 × 1 1/4 × 1/2	1.41	1.34	1.66
1 1/2 × 1 × 1 1/2	1.94	1.80	1.94
1 1/2 × 1 × 1 1/4	1.82	1.67	1.88
1 1/2 × 1 × 1	1.65	1.50	1.80
1 1/2 × 3/4 × 1 1/2	1.94	1.75	1.94
1 1/2 × 3/4 × 1 1/4	1.82	1.62	1.88
1 1/2 × 1/2 × 1 1/2	1.94	1.66	1.94
1 1/4 × 1 1/4 × 1 1/2	1.88	1.88	1.82
1 1/4 × 1 × 1 1/2	1.88	1.80	1.82
1 × 1 × 1 1/2	1.80	1.80	1.65
2 × 2 × 1 1/2	2.02	2.02	2.16
2 × 2 × 1 1/4	1.90	1.90	2.10
2 × 2 × 1	1.73	1.73	2.02
2 × 2 × 3/4	1.60	1.60	1.97
2 × 2 × 1/2	1.49	1.49	1.88
2 × 1 1/2 × 2	2.25	2.16	2.25
2 × 1 1/2 × 1 1/2	2.20	1.94	2.16
2 × 1 1/2 × 1 1/4	1.90	1.82	2.10
2 × 1 1/2 × 1	1.73	1.65	2.02
2 × 1 1/2 × 3/4	1.60	1.52	1.97
2 × 1 1/2 × 1/2	1.49	1.41	1.88
2 × 1 1/4 × 2	2.25	2.10	2.25
2 × 1 1/4 × 1 1/2	1.02	1.88	2.16
2 × 1 1/4 × 1 1/4	1.90	1.75	2.10
2 × 1 1/4 × 1	1.73	1.58	2.02
2 × 1 × 2	2.25	2.02	2.25
2 × 1 × 1 1/2	2.02	1.80	2.16
2 × 1 × 1 1/4	1.90	1.67	2.10
2 × 3/4 × 2	2.25	1.97	2.25
2 × 1/2 × 2	2.25	1.88	2.25
1 1/2 × 1 × 2	2.16	2.02	2.02
1 1/4 × 1 1/4 × 2	2.10	2.10	1.90
1 1/2 × 1 1/2 × 2	2.16	2.16	2.02
1 1/2 × 1 1/4 × 2	2.16	2.10	2.02
2 1/2 × 2 1/2 × 2	2.39	2.39	2.60
2 1/2 × 2 1/2 × 1 1/2	2.16	2.16	2.51
2 1/2 × 2 1/2 × 1 1/4	2.04	2.04	2.45
2 1/2 × 2 1/2 × 1	1.87	1.87	2.37
2 1/2 × 2 1/2 × 3/4	1.74	1.74	2.32
2 1/2 × 2 1/2 × 1/2	1.63	1.63	2.23
2 1/2 × 2 × 2 1/2	2.70	2.60	2.70
2 1/2 × 2 × 2	2.39	2.25	2.60
2 1/2 × 2 × 1 1/2	2.16	2.02	2.51
2 1/2 × 2 × 1 1/4	2.04	1.90	2.45
2 1/2 × 2 × 1	1.87	1.73	2.37
2 1/2 × 2 × 3/4	1.74	1.60	2.32
2 1/2 × 2 × 1/2	1.63	1.49	2.23
2 1/2 × 1 1/2 × 2 1/2	2.70	2.51	2.70
2 1/2 × 1 1/2 × 2	2.39	2.16	2.60
2 1/2 × 1 1/2 × 1 1/2	2.16	1.94	2.51
2 1/2 × 1 1/4 × 2 1/2	2.70	2.45	2.70
2 1/2 × 1 1/4 × 2	2.39	2.10	2.60
2 1/2 × 1 × 2 1/2	2.70	2.37	2.70
2 1/2 × 1 × 2	2.39	2.02	2.60
2 1/2 × 3/4 × 2 1/2	2.70	2.32	2.70

TABLE 11.28 Dimensions of Class 125 Gray Iron Reducing Tees* (Continued)

Nominal pipe size	Center to end A	Center to end B	Center to end C
2½ × ½ × 2½	2.70	2.23	2.70
2 × 2 × 2½	2.60	2.60	2.39
2 × 1½ × 2½	2.60	2.51	2.39
2 × 1¼ × 2½	2.60	2.45	2.39
1½ × 1½ × 2½	2.51	2.51	2.16
3 × 3 × 2½	2.83	2.83	2.99
3 × 3 × 2	2.52	2.52	2.89
3 × 3 × 1½	2.29	2.29	2.80
3 × 3 × 1¼	2.17	2.17	2.74
3 × 3 × 1	2.00	2.00	2.66
3 × 3 × ¾	1.87	1.87	2.61
3 × 3 × ½	1.76	1.76	2.52
3 × 2½ × 3	3.08	2.99	3.08
3 × 2½ × 2½	2.83	2.70	2.99
3 × 2½ × 2	2.52	2.39	2.89
3 × 2½ × 1½	2.29	2.16	2.80
3 × 2½ × 1¼	2.17	2.04	2.74
3 × 2½ × 1	2.00	1.87	2.66
3 × 2 × 3	3.08	2.89	3.08
3 × 2 × 2½	2.83	2.60	2.99
3 × 2 × 2	2.52	2.25	2.89
3 × 2 × 1½	2.29	2.02	2.80
3 × 1½ × 3	3.08	2.80	3.08
3 × 1¼ × 3	3.08	2.74	3.08
3 × 1 × 3	3.08	2.66	3.08
3 × ¾ × 3	3.08	2.61	3.08
2½ × 2½ × 3	2.99	2.99	2.83
2½ × 2 × 3	2.99	2.89	2.83
2 × 2 × 3	2.89	2.89	2.52
3½ × 3½ × 3	3.18	3.18	3.33
3½ × 3½ × 2½	2.93	2.93	3.24
3½ × 3½ × 2	2.62	2.62	3.14
3½ × 3½ × 1½	2.39	2.39	3.05
3½ × 3½ × 1¼	2.27	2.27	2.99
3½ × 3½ × 1	2.10	2.10	2.91
3½ × 3 × 3	3.18	3.08	3.33
3½ × 3 × 2½	2.93	2.83	3.24
3½ × 3 × 2	2.62	2.52	3.14
3½ × 3 × 1½	2.39	2.29	3.05
3½ × 2½ × 3½	3.42	3.24	3.42
3½ × 2½ × 3	3.18	2.99	3.33
3½ × 2½ × 2½	2.93	2.70	3.24
3½ × 2 × 3½	3.42	3.14	3.42
3½ × 1½ × 3½	3.42	3.05	3.42
3½ × 1¼ × 3½	3.42	2.99	3.42
3½ × 1 × 3½	3.42	2.91	3.42
3 × 3 × 3½	3.33	3.33	3.18
4 × 4 × 3½	3.54	3.54	3.69
4 × 4 × 3	3.30	3.30	3.60
4 × 4 × 2½	3.05	3.05	3.51
4 × 4 × 2	2.74	2.74	3.41
4 × 4 × 1½	2.51	2.51	3.32
4 × 4 × 1¼	2.39	2.39	3.26
4 × 4 × 1	2.22	2.22	3.18
4 × 4 × ¾	2.09	2.09	3.13
4 × 3½ × 4	3.79	3.69	3.79
4 × 3½ × 3½	3.54	3.42	3.69
4 × 3½ × 3	3.30	3.18	3.60
4 × 3½ × 2½	3.05	2.93	3.51
4 × 3½ × 2	2.74	2.62	3.41
4 × 3½ × 1½	2.51	2.39	3.32
4 × 3½ × 1¼	2.39	2.27	3.26
4 × 3 × 4	3.79	3.60	3.79
4 × 3 × 3	3.30	3.08	3.60
4 × 3 × 2½	3.05	2.83	3.51
4 × 3 × 2	2.74	2.52	3.41
4 × 2½ × 4	3.79	3.51	3.79
4 × 2½ × 3	3.30	2.99	3.60
4 × 2½ × 2½	3.05	2.70	3.51
4 × 2 × 4	3.79	3.41	3.79
4 × 2 × 3	3.30	2.89	3.60
4 × 2 × 2	2.74	2.25	3.41
4 × 1½ × 4	3.79	3.32	3.79
4 × 1¼ × 4	3.79	3.26	3.79
4 × 1 × 4	3.79	3.18	3.79
3½ × 3½ × 4	3.69	3.69	3.54
3 × 3 × 4	3.60	3.60	3.30
2½ × 2½ × 4	3.51	3.51	3.05
2 × 2 × 4	3.41	3.41	2.74
5 × 5 × 4	4.00	4.00	4.41
5 × 5 × 3½	3.75	3.75	4.31
5 × 5 × 3	3.51	3.51	4.22
5 × 5 × 2½	3.26	3.26	4.13
5 × 5 × 2	2.95	2.95	4.03
5 × 5 × 1½	2.72	2.72	3.94
5 × 5 × 1¼	2.60	2.60	3.88
5 × 5 × 1	2.43	2.43	3.80
5 × 4 × 5	4.50	4.41	4.50
5 × 4 × 4	4.00	3.79	4.41
5 × 4 × 3½	3.75	3.54	4.31
5 × 4 × 3	3.51	3.30	4.22
5 × 4 × 2½	3.26	3.05	4.13
5 × 4 × 2	2.95	2.74	4.03
5 × 4 × 1½	2.72	2.51	3.94
5 × 3 × 5	4.50	4.22	4.50
5 × 3 × 4	4.00	3.60	4.41
5 × 3 × 3	3.51	3.08	4.22
5 × 2½ × 5	4.50	4.13	4.50
5 × 2 × 5	4.50	4.03	4.50
4 × 4 × 5	4.41	4.41	4.00
6 × 6 × 5	4.63	4.63	5.03
6 × 6 × 4	4.13	4.13	4.94
6 × 6 × 3	3.64	3.64	4.75
6 × 6 × 2½	3.39	3.39	4.66
6 × 6 × 2	3.08	3.08	4.56
6 × 6 × 1½	2.85	2.85	4.47
6 × 6 × 1¼	2.73	2.73	4.41
6 × 6 × 1	2.56	2.56	4.33
6 × 5 × 6	5.13	5.03	5.13
6 × 5 × 5	4.63	4.50	5.03
6 × 5 × 4	4.13	4.00	4.94
6 × 5 × 3	3.64	3.51	4.75
6 × 5 × 2½	3.39	3.26	4.66
6 × 5 × 2	3.08	2.95	4.56

TABLE 11.28 Dimensions of Class 125 Gray Iron Reducing Tees* (Continued)

Nominal pipe size	Center to end A	Center to end B	Center to end C	Nominal pipe size	Center to end A	Center to end B	Center to end C
6 × 4 × 6	5.13	4.94	5.13	8 × 8 × 6	5.56	5.56	6.37
6 × 4 × 5	4.63	4.41	5.03	8 × 8 × 5	5.03	5.03	6.27
6 × 4 × 4	4.13	3.79	4.94	8 × 8 × 4	4.50	4.50	6.17
6 × 3 × 6	5.13	4.75	5.13	8 × 8 × 4	4.00	4.00	6.07
6 × 3 × 3	3.64	3.08	4.75	8 × 8 × 2½	3.69	3.69	6.01
6 × 2½ × 6	5.13	4.66	5.13	8 × 8 × 2	3.44	3.44	5.84
6 × 2 × 6	5.13	4.56	5.13	8 × 6 × 8	6.56	6.37	6.56
5 × 5 × 6	5.03	5.03	4.63	8 × 6 × 6	5.56	5.13	6.37
5 × 3 × 6	5.03	4.75	4.63	8 × 4 × 8	6.56	6.17	6.56
4 × 4 × 6	4.94	4.94	4.13	6 × 6 × 8	6.37	6.37	5.56

* Dimensions in inches.

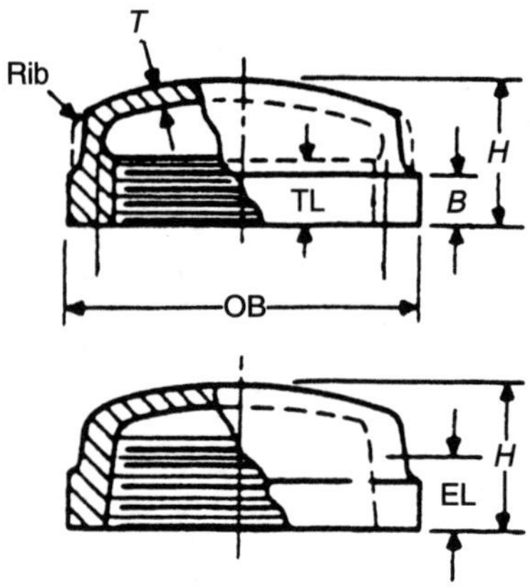

TABLE 11.29 Dimensions of Class 125 Gray Iron Caps*

Nominal pipe size	Minimum length of thread TL	Minimum length of thread EL	Width of band, *B*	Metal thickness, *T*	Outside diameter of band, OB	Height, *H*
½	0.43	...	0.50	0.13	1.34	...
¾	0.50	...	0.56	0.15	1.63	...
1	0.58	...	0.62	0.17	1.95	...
1¼	0.67	...	0.69	0.18	2.39	...
1½	0.70	...	0.75	0.20	2.68	...
2	0.75	...	0.84	0.22	3.28	...
2½	0.92	1.14	0.94	0.24	3.86	1.81
3	0.98	1.20	1.00	0.26	4.62	1.91
3½	1.03	1.25	1.06	0.28	5.20	2.03
4	1.08	1.30	1.12	0.31	5.79	2.22
5	1.18	1.41	1.18	0.38	7.05	2.38
6	1.28	1.51	1.28	0.43	8.28	2.63
8	1.47	1.71	1.47	0.55	10.63	2.88
10	1.68	1.92	1.68	0.69	13.12	3.50
12	1.88	2.12	1.88	0.80	15.47	3.88

* Dimensions in inches.

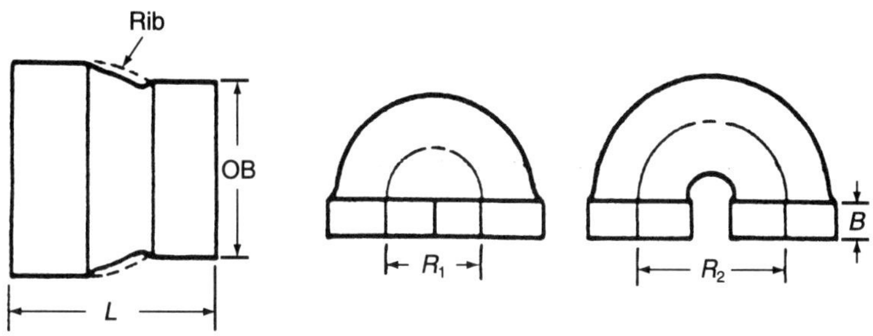

TABLE 11.30 Dimensions of Class 125 Gray Iron Reducing Couplings and Closed- and Open-Pattern Return Bends*

Nominal pipe size	Minimum width of band, B	Metal thickness	Minimum outside diameter of band, OB	Length of reducing concentric couplings, $L^{\dagger}$	Center to center: Closed R_1	Center to center: Open R_2
½	0.50	0.13	1.34	1.38	1.25	1.75
¾	0.56	0.15	1.63	1.50	1.50	1.88
1	0.62	0.17	1.95	1.70	1.75	2.50
1¼	0.69	0.18	2.39	2.13	2.25	3.00
1½	0.75	0.20	2.68	2.25	2.50	3.50
2	0.84	0.22	3.28	2.32	3.25	4.50
2½	0.94	0.24	3.86	2.63	3.75	5.50
3	1.00	0.26	4.62	2.88	4.50	6.50
3½	1.06	0.28	5.20	3.13	...	...
4	1.12	0.31	5.79	3.38	6.00	7.50
5	1.18	0.38	7.05	3.57	...	...
6	1.28	0.43	8.28	3.81	...	...
8	1.47	0.55	10.63	5.25	...	...
10	1.68	0.69	13.12	...	...	...
12	1.88	0.80	15.47	...	...	...

* Dimensions in inches.

† Length L of concentric reducing coupling is the length listed corresponding to the largest opening.

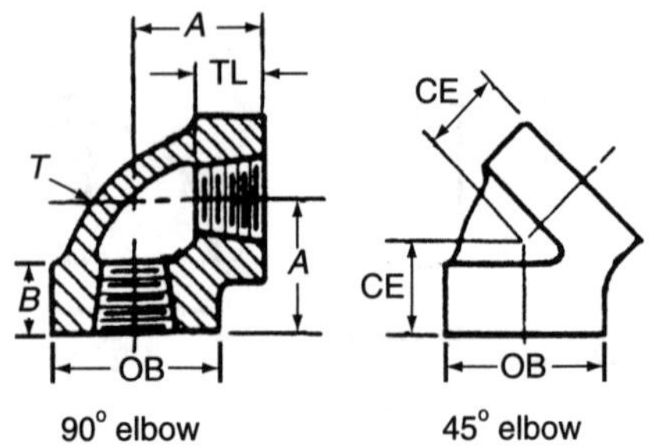

TABLE 11.31 Dimensions of Class 250 Gray Iron Straight 90 and 45° Elbows*

Nominal pipe size	Center to end, 90° elbow, *A*	Center to end, 45° elbow, CE	Minimum length of thread, TL	Minimum width of band, *B*	Metal thickness	Minimum outside diameter of band, OB
1/4	0.94	0.81	0.43	0.49	0.18	1.17
3/8	1.06	0.88	0.47	0.55	0.18	1.36
1/2	1.25	1.00	0.57	0.60	0.20	1.59
3/4	1.44	1.13	0.64	0.68	0.23	1.88
1	1.63	1.31	0.75	0.76	0.28	2.24
1 1/4	1.94	1.50	0.84	0.88	0.33	2.73
1 1/2	2.13	1.69	0.87	0.97	0.35	3.07
2	2.50	2.00	1.00	1.12	0.39	3.74
2 1/2	2.94	2.25	1.17	1.30	0.43	4.60
3	3.38	2.50	1.23	1.40	0.48	5.36
3 1/2	3.75	2.63	1.28	1.49	0.52	5.98
4	4.13	2.81	1.33	1.57	0.56	6.61
5	4.88	3.19	1.43	1.74	0.66	7.92
6	5.63	3.50	1.53	1.91	0.74	9.24
8	7.00	4.31	1.72	2.24	0.90	11.73
10	8.63	5.19	1.93	2.58	1.08	14.37
12	10.00	6.00	2.13	2.91	1.24	16.84

* Dimensions in inches.

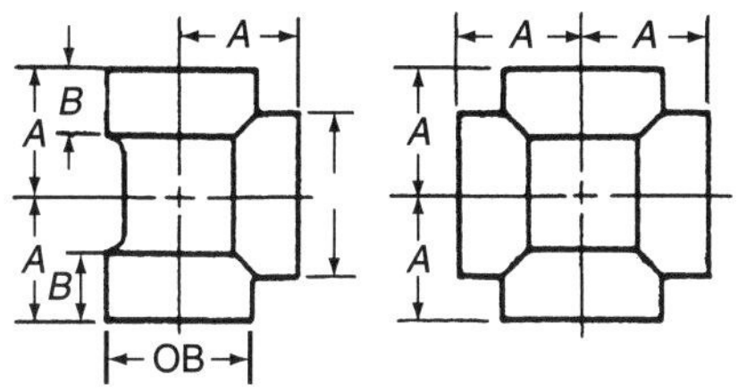

Tee Cross

TABLE 11.32 Dimensions of Class 250 Gray Iron Straight Tees and Crosses*

Nominal pipe size	Center to end, *A*	Minimum length of thread, TL	Minimum width of band, *B*	Metal thickness, *T*	Outside diameter of band, OB
1/4	0.94	0.43	0.49	0.18	1.17
3/8	1.06	0.47	0.55	0.18	1.36
1/2	1.25	0.57	0.60	0.20	1.59
3/4	1.44	0.64	0.68	0.23	1.88
1	1.63	0.75	0.76	0.28	2.24
1 1/4	1.94	0.84	0.88	0.33	2.73
1 1/2	2.13	0.87	0.97	0.35	3.07
2	2.50	1.00	1.12	0.39	3.74
2 1/2	2.94	1.17	1.30	0.43	4.60
3	3.38	1.23	1.40	0.48	5.36
3 1/2	3.75	1.28	1.49	0.52	5.98
4	4.13	1.33	1.57	0.56	6.61
5	4.88	1.43	1.74	0.66	7.92
6	5.63	1.53	1.91	0.74	9.24
8	7.00	1.72	2.24	0.90	11.73
10	8.63	1.93	2.58	1.08	14.37
12	10.00	2.13	2.91	1.24	16.84

* Dimensions in inches.
Note: For pictorial view of dimensions TL, *T*, and OB, refer to Table 11.31.

TABLE 11.33 Tolerances for Center-to-End Dimensions of Gray Iron Fittings*

Nominal pipe size (NPS) of fitting, in	Tolerance, in (plus or minus)
1/4	0.04
3/8	0.05
1/2, 3/4	0.06
1, 1 1/4	0.07
1 1/2, 2	0.18
2 1/2, 3, 3 1/2	0.10
4, 5	0.12
6	0.14
8	0.16
10	0.19
12	0.21

* The largest opening in a reducing fitting governs the tolerance.

Tables in Chapter 11

CHAPTER 12
COPPER AND COPPER ALLOY PIPE, FITTINGS, FLANGES, AND JOINTS

See pp. 484–485 for a list of tables in Chap. 12.

COPPER AND COPPER ALLOY PIPES

This section covers the following pipe specifications:

- ASTM B42 Seamless Copper Pipe
- ASTM B43 Seamless Red Brass Pipe
- ASTM B315 Seamless Copper Alloy Pipe
- ASTM B466 Seamless Copper-Nickel Pipe
- ASTM B467 Welded Copper-Nickel Pipe

Seamless Copper Pipe (ASTM B42)

Scope. This specification covers seamless copper pipe in all nominal or standard pipe sizes, both regular and extra-strong, suitable for use in plumbing, boiler feed lines, and for similar purposes. The pipe may be produced from one of the following coppers:

- C10200 Oxygen-free without residential deoxidants
- C12000 Phosphorized, low residual phosphorus
- C12200 Phosphorized, high residual phosphorus

Chemical Composition. The pipe shall conform to the chemical requirements in accordance with Table 12.1.

TABLE 12.1 Chemical Requirements for Seamless Copper Pipes (ASTM B42)

Copper UNS number	Copper (incl. silver), min., %	Phosphorus, %
C10200	99.95	
C12000	99.90	0.004 to 0.012
C12200	99.9	0.015 to 0.40

Temper. All pipe shall normally be furnished in one of the following tempers:

- O61 (annealed) All sizes
- H55 (light drawn) 2 in to 12 in NPS, inclusive
- H80 (hard drawn) ⅛ in to 2 in NPS, inclusive

Pipe Dimensions. The standard dimensions, wall thickness, and tolerances for regular pipes shall be in accordance with Table 12.2 and those for extra-strong pipes shall be in accordance with Table 12.3.

TABLE 12.2 Standard Dimensions, Wall Thickness, and Tolerances for Regular Seamless Copper Pipes (ASTM B42)

Nominal size (NPS)	Outside diameter, in*	Thickness, in	Tolerance, in	
			Outside diameter	Thickness†
1/8	0.405	0.062	−0.004	0.004
1/4	0.540	0.082	−0.004	0.005
3/8	0.675	0.090	−0.005	0.005
1/2	0.840	0.107	−0.005	0.006
3/4	1.050	0.114	−0.006	0.006
1	1.315	0.126	−0.006	0.007
1 1/4	1.660	0.146	−0.006	0.008
1 1/2	1.900	0.150	−0.006	0.008
2	2.375	0.156	−0.008	0.009
2 1/2	2.875	0.187	−0.008	0.010
3	3.500	0.219	−0.010	0.012
3 1/2	4.000	0.250	−0.010	0.013
4	4.500	0.250	−0.012	0.014
5	5.562	0.250	−0.014	0.014
6	6.625	0.250	−0.016	0.014
8	8.625	0.312	−0.020	0.022
10	10.750	0.365	−0.022	0.030
12	12.750	0.375	−0.024	0.030

* The outside diameter is the average of maximum and minimum outside diameters as determined at any one cross section of the pipe.
† All tolerances are plus and minus except as otherwise indicated.

TABLE 12.3 Standard Dimensions, Wall Thickness, and Tolerances for Extra-Strong Seamless Copper Pipes (ASTM B42)

Nominal size (NPS)	Outside diameter, in*	Thickness, in	Tolerance, in	
			Diameter	Thickness†
1/8	0.405	0.100	−0.004	0.006
1/4	0.540	0.123	−0.004	0.007
3/8	0.675	0.127	−0.005	0.007
1/2	0.840	0.149	−0.005	0.008
3/4	1.050	0.157	−0.006	0.009
1	1.315	0.182	−0.006	0.010
1 1/4	1.660	0.194	−0.006	0.010
1 1/2	1.900	0.203	−0.006	0.011
2	2.375	0.221	−0.008	0.012
2 1/2	2.875	0.280	−0.008	0.015
3	3.500	0.304	−0.010	0.016
3 1/2	4.000	0.321	−0.010	0.017
4	4.500	0.341	−0.012	0.018
5	5.562	0.375	−0.014	0.019
6	6.625	0.437	−0.016	0.027
8	8.625	0.500	−0.020	0.035
10	10.750	0.500	−0.022	0.040

* The outside diameter is the average of maximum and minimum outside diameters as determined at any one cross section of the pipe.
† All tolerances are plus and minus except as otherwise indicated.

Seamless Red Brass Pipe (ASTM B43)

Scope. This specification covers seamless red brass pipe in all nominal pipe sizes, both regular and extra-strong. In the annealed temper (O61), the pipe is suitable for use in plumbing, boiler feed lines, and for similar purposes. The pipe is produced from copper alloy C23000.

Temper. All pipe shall normally be furnished in O61 (annealed) condition.

Pipe Dimensions. The standard dimensions, wall thickness, and tolerances for regular pipes shall be in accordance with Table 12.4 and those for extra-strong pipes shall be in accordance with Table 12.5.

TABLE 12.4 Standard Dimensions, Wall Thickness, and Tolerances for Regular Seamless Red Brass Pipes (ASTM B43)

Nominal size (NPS)	Outside diameter, in*	Thickness, in	Tolerance, in	
			Outside diameter	Thickness†
1/8	0.405	0.062	−0.004	0.004
1/4	0.540	0.082	−0.004	0.005
3/8	0.675	0.090	−0.005	0.005
1/2	0.840	0.107	−0.005	0.006
3/4	1.050	0.114	−0.006	0.006
1	1.315	0.126	−0.006	0.007
1 1/4	1.660	0.146	−0.006	0.008
1 1/2	1.900	0.150	−0.006	0.008
2	2.375	0.156	−0.008	0.009
2 1/2	2.875	0.187	−0.008	0.010
3	3.500	0.219	−0.010	0.012
3 1/2	4.000	0.250	−0.010	0.013
4	4.500	0.250	−0.012	0.014
5	5.562	0.250	−0.014	0.014
6	6.625	0.250	−0.016	0.014
8	8.625	0.312	−0.020	0.022
10	10.750	0.365	−0.022	0.030
12	12.750	0.375	−0.024	0.030

* The outside diameter is the average of maximum and minimum outside diameters as determined at any one cross section of the pipe.

† All tolerances are plus and minus except as otherwise indicated.

TABLE 12.5 Standard Dimensions, Wall Thickness, and Tolerances for Extra-Strong Seamless Red Brass Pipes (ASTM B43)

Nominal size (NPS)	Outside diameter, in*	Thickness, in	Tolerance, in	
			Diameter	Thickness†
1/8	0.405	0.100	−0.004	0.006
1/4	0.540	0.123	−0.004	0.007
3/8	0.675	0.127	−0.005	0.007
1/2	0.840	0.149	−0.005	0.008
3/4	1.050	0.157	−0.006	0.009
1	1.315	0.182	−0.006	0.010
1 1/4	1.660	0.194	−0.006	0.010
1 1/2	1.900	0.203	−0.006	0.011
2	2.375	0.221	−0.008	0.012
2 1/2	2.875	0.280	−0.008	0.015
3	3.500	0.304	−0.010	0.016
3 1/2	4.000	0.321	−0.010	0.017
4	4.500	0.341	−0.012	0.018
5	5.562	0.375	−0.014	0.019
6	6.625	0.437	−0.016	0.027
8	8.625	0.500	−0.020	0.035
10	10.750	0.500	−0.022	0.040

* The outside diameter is the average of maximum and minimum outside diameters as determined at any one cross section of the pipe.

† All tolerances are plus and minus except as otherwise indicated.

Seamless Copper Alloy Pipe (ASTM B315)

Scope. This specification covers seamless copper alloy pipe in nominal or standard pipe sizes, both regular and extra-strong, suitable for general engineering purposes. The pipe is produced from copper alloy C65500.

Temper. All pipe shall normally be furnished in the following conditions:

- O30 (extruded and annealed)
- O61 (cold-worked and annealed)

Pipe supplied in cold-worked and annealed condition (O61) shall show complete recrystallization.

Pipe Dimensions and Variations. The standard dimensions, wall thickness, and tolerances for regular pipes shall be in accordance with Table 12.6 and those for extra-strong pipes shall be in accordance with Table 12.7.

The standard method of specifying wall thickness shall be in decimal fractions of an inch. Tolerances on a given pipe may be specified with respect to any two, but not all three, of the following: outside diameter, inside diameter, or wall thickness.

TABLE 12.6 Standard Dimensions, Wall Thickness, and Tolerances for Regular Seamless Copper Alloy Pipes (ASTM B315)

Nominal size (NPS)	Outside diameter, in			Thickness, in		
	Nominal	Minimum	Maximum	Nominal	Minimum	Maximum
1/8	0.405	0.374	0.421	0.068	0.065	0.083
1/4	0.540	0.509	0.556	0.088	0.084	0.102
3/8	0.675	0.644	0.691	0.091	0.086	0.103
1/2	0.840	0.809	0.856	0.109	0.104	0.122
3/4	1.050	1.019	1.066	0.113	0.107	0.124
1	1.315	1.284	1.331	0.133	0.126	0.145
1 1/4	1.660	1.629	1.676	0.140	0.133	0.151
1 1/2	1.900	1.869	1.916	0.145	0.138	0.156
2	2.375	2.351	2.399	0.154	0.146	0.164
2 1/2	2.875	2.846	2.904	0.203	0.193	0.217
3	3.500	3.465	3.535	0.216	0.205	0.230
3 1/2	4.000	3.960	4.040	0.226	0.215	0.240
4	4.500	4.455	4.545	0.237	0.225	0.252
5	5.562	5.506	5.618	0.258	0.245	0.275
6	6.625	6.559	6.691	0.280	0.266	0.298
8	8.625	8.539	8.711	0.322	0.299	0.349
10	10.750	10.643	10.858	0.365	0.336	0.400
12	12.750	12.623	12.878	0.375	0.345	0.410

TABLE 12.7 Standard Dimensions, Wall Thickness, and Tolerances for Extra-Strong Seamless Copper Alloy Pipes (ASTM B315)

Nominal size (NPS)	Outside diameter, in			Thickness, in		
	Nominal	Minimum	Maximum	Nominal	Minimum	Maximum
1/8	0.405	0.374	0.421	0.095	0.090	0.123
1/4	0.540	0.509	0.556	0.119	0.107	0.144
3/8	0.675	0.644	0.691	0.126	0.120	0.146
1/2	0.840	0.809	0.856	0.147	0.140	0.166
3/4	1.050	1.019	1.066	0.154	0.146	0.171
1	1.315	1.284	1.331	0.179	0.170	0.196
1 1/4	1.660	1.629	1.676	0.191	0.181	0.207
1 1/2	1.900	1.869	1.916	0.200	0.190	0.216
2	2.375	2.351	2.399	0.218	0.207	0.233
2 1/2	2.875	2.846	2.904	0.276	0.262	0.295
3	3.500	3.465	3.535	0.300	0.285	0.321
3 1/2	4.000	3.960	4.040	0.318	0.302	0.340
4	4.500	4.455	4.545	0.337	0.320	0.360
5	5.562	5.506	5.618	0.375	0.356	0.400
6	6.625	6.559	6.691	0.432	0.410	0.461
8	8.625	8.539	8.711	0.500	0.465	0.544
10	10.750	10.643	10.858	0.500	0.460	0.548

Seamless Copper-Nickel Pipe (ASTM B466)

Scope. This specification covers seamless copper-nickel alloy pipe suitable for general engineering purposes. The following three alloys are covered:

- C70600 Nominal composition—90% copper, 10% nickel
- C71000 Nominal composition—80% copper, 20% nickel
- C71500 Nominal composition—70% copper, 30% nickel

Temper. All pipe shall normally be furnished in the annealed temper only.

Pipe Dimensions and Variations. The wall thickness tolerances shall be in accordance with Table 12.8. The diameter tolerances shall be in accordance with Table 12.9. The preferred diameter and wall thickness for this pipe specification are indicated in Table 12.10.

Welded Copper-Nickel Pipe (ASTM B467)

Scope. This specification covers welded copper-nickel alloy pipe suitable for general engineering purposes. The following alloys are covered:

C70600 Nominal composition—90% copper, 10% nickel
C71500 Nominal composition—70% copper, 30% nickel

Temper. All pipe shall be supplied in any one of the following tempers as specified:

- As welded from annealed sheet, strip, or plate (WM50)
- As welded from cold-worked sheet, strip, or plate (WM00, WM01, etc.)
- Welded and annealed (WO50)
- Welded and cold drawn in either light drawn (Alloy C70600 only) or hard drawn, stress relieved (WR00, WR04)
- Fully finished as annealed (WO61) light drawn (Alloy C70600 only) or hard drawn, stress relieved (WH00, WH04)

TABLE 12.8 Wall Thickness Tolerances for Seamless Copper-Nickel Pipes (ASTM B466)*

Wall thickness, in	Outside diameter, in						
	$^1/_{32}$ to $^1/_8$, incl.	Over $^1/_8$ to $^5/_8$, incl.	Over $^5/_8$ to 1, incl.	Over 1 to 2, incl.	Over 2 to 4, incl.	Over 4 to 7, incl.	Over 7 to 10, incl.
Up to 0.017	0.0025	0.0015	0.002	0.0025	. . .	. . .	. . .
Over 0.017 to 0.024, incl.	0.004	0.0025	0.0025	0.003	. . .	. . .	. . .
Over 0.024 to 0.034, incl.	0.004	0.003	0.003	0.004	0.005	. . .	. . .
Over 0.034 to 0.057, incl.	0.004	0.004	0.0045	0.0045	0.0065	0.009	. . .
Over 0.057 to 0.082, incl.	. . .	0.0045	0.005	0.005	0.0075	0.010	0.013
Over 0.082 to 0.119, incl.	. . .	0.005	0.0065	0.0065	0.009	0.011	0.014
Over 0.119 to 0.164, incl.	. . .	0.007	0.007	0.0075	0.010	0.013	0.015
Over 0.164 to 0.219, incl.	. . .	. . .	0.009	0.010	0.012	0.015	0.018
Over 0.219 to 0.283, incl.	. . .	. . .	0.012	0.013	0.015	0.018	0.020
Over 0.283 to 0.379, incl.	. . .	. . .	. . .	0.015	0.018	0.020	0.023
Over 0.379†	. . .	. . .	. . .	6	6	8	8

* All tolerances are plus and minus; if tolerances all plus or all minus are desired, double the values given.
† The values given for wall thickness over 0.379 in are percent of specified wall thickness expressed to the nearest 0.001 in.

TABLE 12.9 Average Diameter Tolerances for Seamless Copper-Nickel Pipes (ASTM B466)*

Specified diameter, in*	Tolerance applies to	Tolerances, plus and minus, in†
Up to 1/8, incl.	Inside diameter	0.003
Up to 1/8, incl.	Outside diameter	0.0025
Over 1/8 to 5/8, incl.	Inside or outside	0.0025
Over 5/8 to 1, incl.	Inside or outside	0.003
Over 1 to 2, incl.	Inside or outside	0.004
Over 2 to 3, incl.	Inside or outside	0.005
Over 3 to 4, incl.	Inside or outside	0.006
Over 4 to 5, incl.	Inside or outside	0.008
Over 5 to 6, incl.	Inside or outside	0.009
Over 6 to 8, incl.	Inside or outside	0.010
Over 8 to 10, incl.	Inside or outside	0.013

* The average outside or inside diameter is the average of the maximum and minimum outside diameters, or of the maximum and minimum inside diameters, whichever is applicable, as determined at one cross section of the pipe.

† If tolerances all plus or minus are desired, double the values given.

TABLE 12.10 Preferred Diameter and Wall Thickness for Seamless Copper-Nickel Pipes (ASTM B466)

Nominal size	Outside diameter, in	Wall thickness, in						
		Specials					Regular	Extra-strong
1/8	0.405	0.058	...	...	...	...	0.062	0.100
1/4	0.540	0.065	0.072	...	...	...	0.082	0.123
3/8	0.675	0.065	0.072	0.095	0.148	...	0.090	0.127
1/2	0.840	0.065	0.072	0.120	0.203	...	0.107	0.149
3/4	1.050	0.065	0.083	0.148	0.238	...	0.114	0.157
1	1.315	0.065	0.095	0.203	0.340	...	0.126	0.182
1 1/4	1.660	0.072	0.095	0.120	0.220	0.380	0.146	0.194
1 1/2	1.900	0.072	0.109	0.134	0.250	0.425	0.150	0.203
2	2.375	0.083	0.120	0.165	0.340	0.520	0.156	0.221
2 1/2	2.875	0.083	0.134	0.203	0.380	...	0.187	0.280
3	3.500	0.095	0.165	0.250	0.458	...	0.219	0.304
3 1/2	4.000	0.095	0.180	0.284	...	...	0.250	0.321
4	4.500	0.109	0.203	0.340	...	...	0.250	0.341
5	5.562	0.125	0.220	0.425	...	...	0.250	0.375
6	6.625	0.134	0.259	0.457	...	...	0.250	0.437

Pipe Dimensions and Variations. The wall thickness tolerances shall be in accordance with Table 12.11, with some exceptions. The tolerances in Table 12.11 do not apply to that portion of the "as welded" wall that contains the weld flash or bead. The portion of the "as-welded" wall that contains the weld zone shall be increased by 100 percent.

The outside diameter tolerances shall be in accordance with Table 12.12 with an exception. These outside diameter tolerances do not apply to the "as-welded" pipe when measured across that portion that contains the weld zone.

The suggested diameter and wall thickness for this pipe specification are indicated in Table 12.13.

TABLE 12.11 Wall Thickness Tolerances for Welded Copper-Nickel Pipes (ASTM B467)*

	Outside diameter, in					
Wall thickness, in	Up to 2½, incl.	Over 2½ to 4½, incl.	Over 4½ to 6½, incl.	Over 6½ to 9, incl.	Over 9 to 11½, incl.	Over 11½
To 0.017, incl.	0.0013	...	...	...	...	...
Over 0.017 to 0.021, incl.	0.0015	...	...	...	...	...
Over 0.021 to 0.026, incl.	0.002	...	...	...	...	...
Over 0.026 to 0.037, incl.	0.0025	0.003	...	...	...	...
Over 0.037 to 0.050, incl.	0.003	0.0035	0.0035	...	...	...
Over 0.050 to 0.073, incl.	0.0035	0.004	0.004	0.007	...	...
Over 0.073 to 0.130, incl.	0.004	0.0045	0.0045	0.008	...	...
Over 0.130 to 0.205, incl.	0.0045	0.005	0.005	0.010	0.012	0.014
Over 0.205 to 0.300, incl.	0.005	0.006	0.006	0.012	0.014	0.018
Over 0.300	0.006	0.007	0.007	0.019	0.017	0.023

* The tolerances are plus and minus; if tolerances all plus or all minus are desired, double the values given.

TABLE 12.12 Average Outside Diameter Tolerances for Welded Copper-Nickel Pipes (ASTM B467)

Specified diameter, in*	Tolerances, in†
Over 2 to 3, incl.	0.005
Over 3 to 4, incl.	0.006
Over 4 to 5, incl.	0.008
Over 5 to 6, incl.	0.009
Over 6 to 8, incl.	0.010
Over 8 to 10, incl.	0.013
Over 10 to 12, incl.	0.015
Over 12	0.5%

* The average outside diameter of a pipe is the average of the maximum and minimum outside diameters, as determined at any one cross section.
† If tolerances all plus or minus are desired, double the values given.

TABLE 12.13 Suggested Diameter and Wall Thickness for Welded Copper-Nickel Pipes (ASTM B467)

		Wall thickness, in		
Nominal size	Outside diameter, in	A	B	C
2.5	2.875	...	0.083	0.134
3	3.500	...	0.095	0.165
3.5	4.000	...	0.095	0.180
4	4.500	...	0.109	0.203
4.5	5.000	...	0.120	0.203
5	5.563	...	0.125	0.220
6	6.625	...	0.134	0.259
7	7.625	...	0.134	0.284
8	8.625	...	0.148	0.340
9	9.625	...	0.187	0.340
10	10.750	0.134	0.187	0.380
12	12.750	0.156	0.250	0.454
14	14.0	0.165	...	...
16	16.0	0.165	...	...
18	18.0	0.180	...	...
20	20.0	0.180	...	...
24	24.0	0.180	...	...
30	30.0	0.250	...	...

COPPER AND COPPER ALLOY TUBES

This section covers the following tube specifications:

ASTM B75	Seamless Copper Tube
ASTM B88	Seamless Copper Water Tube
ASTM B111	Seamless Copper and Copper Alloy Condenser Tube and Ferrule Stock
ASTM B135	Seamless Brass Tube
ASTM B315	Seamless Copper Alloy Tube
ASTM B359	Seamless Copper and Copper Alloy Condenser and Heat Exchanger Tubes with Integral fins

ASTM B395 Seamless U-Bend Copper and Copper Alloy Condenser and Heat Exchanger Tubes with Integral fins

ASTM B466 Seamless Copper-Nickel Tube

ASTM B543 Welded Copper and Copper Alloy Heat Exchanger Tube

Seamless Copper Tube (ASTM B75)

Scope. This specification establishes the requirements for seamless, round, rectangular, and square copper tube suitable for general engineering applications. The tubes may be produced from any one of the following coppers:

C10200 Oxygen-free without residential deoxidants

C12000 Phosphorized, low residual phosphorus

C12200 Phosphorized, high residual phosphorus

Temper. Rectangular, including square, tube is normally supplied only in H58 temper; however, tube in H55 or H80 temper may be supplied as agreed upon between the manufacturer and the purchaser.

Permissible Variations. Tolerances for diameter or distance between parallel surfaces are specified in Table 12.14 for round tubes and Table 12.15 for rectangular including square tubes.

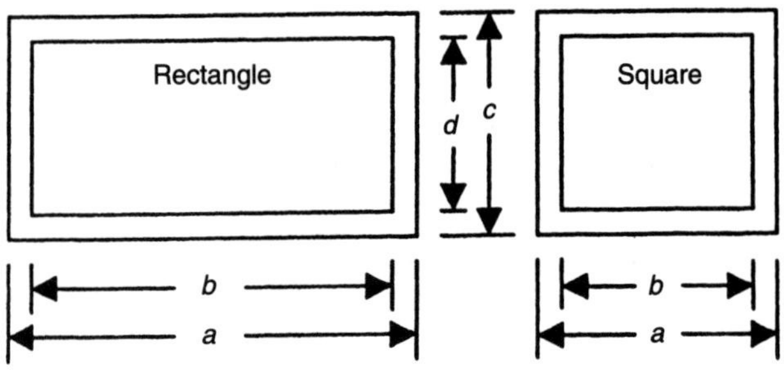

TABLE 12.14 Average Diameter Tolerances (Round Tubes Only) for Seamless Copper Tube (ASTM B75)*

Specified diameter, in	Tolerances, plus and minus, in
Up to 1/8, incl.	0.002
Over 1/8–5/8, incl.	0.002
Over 5/8–1, incl.	0.0025
Over 1–2, incl.	0.003
Over 2–3, incl.	0.004
Over 3–4, incl.	0.005
Over 4–5, incl.	0.006
Over 5–6, incl.	0.007
Over 6–8, incl.	0.008
Over 8–10, incl.	0.010

* Tolerances are applicable to inside or outside diameter.

TABLE 12.15 Average Tolerances on Distance Between Parallel Surfaces for Rectangular, Including Square, Tube for Seamless Brass Tube (ASTM B75)

Dimension *a* or *b*, in (1), (2)	Tolerances, in (3)
Up to 1/8, incl.	0.003
Over 1/8–5/8, incl.	0.004
Over 5/8–1, incl.	0.005
Over 1–2, incl.	0.006
Over 2–3, incl.	0.007
Over 3–4, incl.	0.008
Over 4–5, incl.	0.009
Over 5–6, incl.	0.010
Over 6–8, incl.	0.011
Over 8–10, incl.	0.012

Notes:

(1) Nominal dimension *a* determines tolerance applicable to both *a* and *c*.

(2) Nominal dimension *b* determines tolerance applicable to both *b* and *d*.

(3) Tolerances are plus and minus; if tolerances all plus or all minus are desired, double the values given.

Tolerances for wall thickness are specified in Table 12.16 for round tubes and Table 12.17 for rectangular, including square, tubes.

TABLE 12.16 Wall Thickness Tolerances (Round Tubes Only) for Seamless Copper Tube (ASTM B75) (1)

	Outside diameter, in (2)						
Wall thickness, in	1/32 to 1/8, incl.	Over 1/8 to 5/8, incl.	Over 5/8 to 1, incl.	Over 1 to 2, incl.	Over 2 to 4, incl.	Over 4 to 7, incl.	Over 7 to 10, incl.
Up to 0.017, incl.	0.002	0.001	0.0015	0.002	...	...	...
Over 0.017–0.024, incl.	0.003	0.002	0.002	0.0025	...	...	...
Over 0.024–0.034, incl.	0.003	0.0025	0.0025	0.003	0.004	...	...
Over 0.034–0.057, incl.	0.003	0.003	0.0035	0.0035	0.005	0.007	...
Over 0.057–0.082, incl.	...	0.0035	0.004	0.004	0.006	0.008	0.010
Over 0.082–0.119, incl.	...	0.004	0.005	0.005	0.007	0.009	0.011
Over 0.119–0.164, incl.	...	0.005	0.006	0.006	0.008	0.010	0.012
Over 0.164–0.219, incl.	...	0.007	0.009	0.009	0.011	0.012	0.014
Over 0.219–0.283, incl.	...	...	0.011	0.012	0.014	0.015	0.016
Over 0.283–0.379, incl.	...	...	0.014	6% (3)	6% (3)	7% (3)	7% (3)
Over 0.379	...	...	...	6% (3)	6% (3)	7% (3)	7% (3)

Notes:
(1) Tolerances are plus and minus; if tolerances all plus or all minus are desired, double the values given.
(2) When round tube is ordered by outside and inside diameters, the maximum plus and minus deviation of the wall thickness from the nominal at any point shall not exceed the values given in the table by more than 50 percent.
(3) Percent of the specified wall expressed to the nearest 0.001 in.

TABLE 12.17 Wall Thickness Tolerances on Distance between Parallel Surfaces for Rectangular, Including Square, Tube for Seamless Copper Tube (ASTM B75)*

	Outside diameter, in†						
Wall thickness, in	1/32 to 1/8, incl.	Over 1/8 to 5/8, incl.	Over 5/8 to 1, incl.	Over 1 to 2, incl.	Over 2 to 4, incl.	Over 4 to 7, incl.	Over 7 to 10, incl.
Up to 0.017, incl.	0.002	0.002	0.0025	0.003	...	...	...
Over 0.017–0.024, incl.	0.003	0.0025	0.003	0.0035	...	...	...
Over 0.024–0.034, incl.	0.0035	0.0035	0.0035	0.004	0.006	...	...
Over 0.034–0.057, incl.	0.004	0.004	0.0045	0.005	0.007	0.009	...
Over 0.057–0.082, incl.	...	0.005	0.006	0.007	0.008	0.010	0.012
Over 0.082–0.119, incl.	...	0.007	0.008	0.009	0.010	0.012	0.014
Over 0.119–0.164, incl.	...	0.009	0.010	0.011	0.012	0.014	0.016
Over 0.164–0.219, incl.	...	0.011	0.012	0.013	0.015	0.017	0.019
Over 0.219–0.283, incl.	...	...	0.015	0.016	0.018	0.020	0.022

* The above tolerances are plus and minus; if tolerances all plus or all minus are desired, double the values given.
† In case of rectangular tube, the major dimension determines the thickness tolerance applicable to all walls.

Recommended Applications. Tube in the H55 temper is recommended when a tube of some stiffness, yet capable of being bent when necessary, is required. Tube in H58 temper is recommended for general application where there is no specific need for high strength or bending qualities. Tube in H80 temper is recommended for applications where there is a need for a tube as strong as technically feasible for the size indicated.

Seamless Copper Water Tube (ASTM B88)

Scope. This specification covers copper UNS No. C12200 seamless copper water tube suitable for general plumbing, and similar applications for the conveyance of fluids, and commonly used with solder-, flared-, or compression-type fittings.

Temper. Seamless copper water tube shall be furnished in the tempers designated below:

- Annealed (O)
- Drawn (H)

Standard Dimensions, Wall Thickness, and Diameter Tolerances. Standard dimensions shall be in accordance with Table 12.18. Diameter tolerances shall be in accordance with Table 12.19. Wall thickness tolerances shall be in accordance with Table 12.20.

TABLE 12.18 Dimensions for Seamless Copper Water Tubes (ASTM B88)

Standard size	Outside diameter, in	Wall thickness, in Type K	Type L	Type M
1/4	0.375	0.035	0.030	. . .
3/8	0.500	0.049	0.035	0.025
1/2	0.625	0.049	0.040	0.028
5/8	0.750	0.049	0.042	. . .
3/4	0.875	0.065	0.045	0.032
1	1.125	0.065	0.050	0.035
1 1/4	1.375	0.065	0.055	0.042
1 1/2	1.625	0.072	0.060	0.049
2	2.125	0.083	0.070	0.058
2 1/2	2.625	0.095	0.080	0.065
3	3.125	0.109	0.090	0.072
3 1/2	3.625	0.120	0.100	0.083
4	4.125	0.134	0.110	0.095
5	5.125	0.160	0.125	0.109
6	6.125	0.192	0.140	0.122
8	8.125	0.271	0.200	0.170
10	10.125	0.338	0.250	0.212
12	12.125	0.405	0.280	0.254

TABLE 12.19 Diameter Tolerances for Seamless Copper Water Tubes (ASTM B88)

Standard size	Outside diameter, in	Average outside diameter tolerance, in* Annealed	Drawn
1/4	0.375	0.002	0.001
3/8	0.500	0.0025	0.001
1/2	0.625	0.0025	0.001
5/8	0.750	0.0025	0.001
3/4	0.875	0.003	0.001
1	1.125	0.0035	0.0015
1 1/4	1.375	0.004	0.0015
1 1/2	1.625	0.0045	0.002
2	2.125	0.005	0.002
2 1/2	2.625	0.005	0.002
3	3.125	0.005	0.002
3 1/2	3.625	0.005	0.002
4	4.125	0.005	0.002
5	5.125	0.005	0.002
6	6.125	0.005	0.002
8	8.125	0.006	+0.002 −0.004
10	10.125	0.008	+0.002 −0.006
12	12.125	0.008	+0.002 −0.006

* All tolerances are plus and minus except as otherwise indicated.

TABLE 12.20 Wall Thickness Tolerances for Seamless Copper Water Tubes (ASTM B88)*

Standard size	Wall thickness tolerances, in*					
	Type K		Type L		Type M	
	Wall thick.	Tolerance	Wall thick.	Tolerance	Wall thick.	Tolerance
1/4	0.035	0.0035	0.030	0.003	...	...
3/8	0.049	0.005	0.035	0.004	0.025	0.002
1/2	0.049	0.005	0.040	0.004	0.028	0.003
5/8	0.049	0.005	0.042	0.004	...	...
3/4	0.065	0.006	0.045	0.004	0.032	0.003
1	0.065	0.006	0.050	0.005	0.035	0.004
1 1/4	0.065	0.006	0.055	0.006	0.042	0.004
1 1/2	0.072	0.007	0.060	0.006	0.049	0.005
2	0.083	0.008	0.070	0.007	0.058	0.006
2 1/2	0.095	0.010	0.080	0.008	0.065	0.006
3	0.109	0.011	0.090	0.009	0.072	0.007
3 1/2	0.120	0.012	0.100	0.010	0.083	0.008
4	0.134	0.013	0.110	0.011	0.095	0.010
5	0.160	0.016	0.125	0.012	0.109	0.011
6	0.192	0.019	0.140	0.014	0.122	0.012
8	0.271	0.027	0.200	0.020	0.170	0.017
10	0.338	0.034	0.250	0.025	0.212	0.021
12	0.405	0.040	0.280	0.028	0.254	0.025

* All tolerances are plus and minus except as otherwise indicated.

Seamless Copper and Copper Alloy Condenser Tubes and Ferrule Stock (ASTM B111)

Scope. This specification covers seamless tube and ferrule stock of copper and various copper alloys up to and including 3 1/8 in. in diameter, for use in surface condensers, evaporators, and heat exchangers. The tubes may be produced from any one of the following coppers:

C10200 Oxygen-free without residential deoxidants
C12000 Phosphorized, low residual phosphorus
C12200 Phosphorized, high residual phosphorus
C14200 Phosphorized, arsenical
C19200 Phosphorized, 1% iron
C23000 Red Brass
C28000 Muntz metal
C44300 Admiralty metals B, C, and D
C44400
C44500
C60800 Aluminum bronze
C68700 Aluminum brass, B
C70400 95-5 Copper-nickel
C70600 90-10 Copper-nickel
C71000 80-20 Copper-nickel
C71500 70-30 Copper-nickel
C71640 Copper-nickel-iron-manganese
C72200

Temper. Tubes of copper alloy C23000, C28000, C44300, C44400, C44500, C60800, C61300, C61400, C68700, and C71000 shall be furnished in the annealed (O) temper unless otherwise specified on the purchase order.

Tubes of copper alloy C71500 and C71640 shall be supplied in one of the following tempers as specified: annealed (O) or drawn, stress relieved (HR50).

Tubes of copper alloy C10100, C10200, C10300, C10800, C12000, C12200, and C14200 shall be supplied in any one of the following tempers, one of which shall be specified: light drawn (H55), hard drawn (H80), or hard drawn, end annealed.

Tubes of copper alloy C19200 shall be supplied in any one of the following tempers, one of which shall be specified: annealed (O), light drawn (H55), hard drawn (H80), or hard drawn, end annealed.

Tubes of copper alloy C70400, C70600, and C72200 may be supplied in either light-drawn (H55) or annealed (O) temper.

Tubes for ferrule stock shall be annealed sufficiently to be fully recrystallized.

Permissible Variations. The outside diameter of the tube shall not vary from that specified by more than the amounts shown in Table 12.21 as measured by go and no-go ring gages. The wall thickness tolerances shall be in accordance with Table 12.22.

TABLE 12.21 Diameter Tolerances for Seamless Copper and Copper Alloy Condenser Tubes and Ferrule Stock (ASTM B111)

	Wall thickness, in				
	0.020* 0.022 0.025 0.028	0.032	0.035	0.042	0.049 and over
Outside diameter, in	Diameter tolerances, plus and minus, in				
Up to 0.500, incl.	0.003	0.0025	0.0025	0.0025	0.0025
Over 0.500–0.740, incl.	0.0040	0.004	0.004	0.0035	0.003
Over 0.740–1.000, incl.	0.0060	0.006	0.005	0.0045	0.004
Over 1.000–1.250, incl.	...	0.009	0.008	0.006	0.0045
Over 1.250–1.375, incl.	...	...	...	0.008	0.005
Over 1.375–2.000, incl.	...	...	...	...	0.006

* Tolerances in this column are applicable to light-drawn and drawn tempers only. Tolerances for annealed tempers shall be as agreed upon between the manufacturer and the purchaser.

TABLE 12.22 Wall Thickness Tolerances, Plus and Minus, for Seamless Copper and Copper Alloy Condenser Tubes and Ferrule Stock (ASTM B111)

	Outside diameter, in		
Wall thickness, in	Over $^1/_8$ to $^5/_8$, incl.	Over $^5/_8$ to 1, incl.	Over 1 to 2, incl.
0.020 to 0.032, incl.	0.003	0.003	...
0.032 to 0.035, incl.	0.003	0.003	0.004
0.035 to 0.058, incl.	0.004	0.0045	0.0045
0.058 to 0.083, incl.	0.0045	0.005	0.005
0.083 to 0.120, incl.	0.005	0.0065	0.0065
0.120 to 0.134, incl.	0.007	0.007	0.0075

Notes:

1. For tubes ordered to minimum wall thickness, no tube wall at its thinnest point shall be less than the specified wall thickness. The maximum plus deviation from the specified wall at any point shall not exceed twice the values shown above.
2. For tubes ordered to nominal wall thickness, the maximum plus and minus deviation from the nominal wall at any point shall not exceed the values shown above.

Seamless Brass Tube (ASTM B135)

Scope. This specification covers seamless round and rectangular, including square, copper alloy tube in straight lengths. The tube is produced from copper alloy C23000.

Temper. The tempers of the annealed tube shall be designated as light anneal (O50) and soft anneal (O60).

Permissible Variations. The diameter tolerances for round tubes shall be in accordance with Table 12.23. Tolerances on distance between parallel surfaces for rectangular, including square, tube shall be in accordance with Table 12.24.

Wall thickness tolerances for round tubes only shall be in accordance with Table 12.25. Wall thickness tolerances for rectangular, including square, tube shall be in accordance with Table 12.26.

TABLE 12.23 Average Diameter Tolerances (Round Tubes Only) for Seamless Brass Tube (ASTM B135)

Specified diameter, in	Tolerances, plus and minus, in*
Up to 1/8, incl.	0.002
Over 1/8–5/8, incl.	0.002
Over 1/8–1, incl.	0.0025
Over 1–2, incl.	0.003
Over 2–3, incl.	0.004
Over 3–4, incl.	0.005
Over 4–5, incl.	0.006
Over 5–6, incl.	0.007
Over 6–8, incl.	0.008
Over 8–10, incl.	0.010

* Tolerances are applicable to inside or outside diameter.

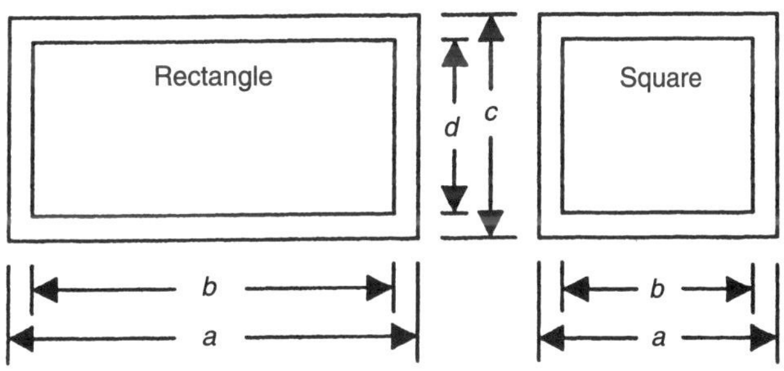

TABLE 12.24 Average Tolerances on Distance between Parallel Surfaces for Rectangular, Including Square, Tube for Seamless Brass Tube (ASTM B135)

Dimension *a* or *b* (see sketches), in (1), (2)	Tolerances, in (3)
Up to 1/8, incl.	0.003
Over 1/8–5/8, incl.	0.004
Over 5/8–1, incl.	0.005
Over 1–2, incl.	0.006
Over 2–3, incl.	0.007
Over 3–4, incl.	0.008
Over 4–5, incl.	0.009
Over 5–6, incl.	0.010
Over 6–8, incl.	0.011
Over 8–10, incl.	0.012

Notes:

(1) Nominal dimension *a* determines tolerance applicable to both *a* and *c*.

(2) Nominal dimension *b* determines tolerance applicable to both *b* and *d*.

(3) Tolerances are plus and minus; if tolerances all plus or all minus are desired, double the values given.

TABLE 12.25 Wall Thickness Tolerances (Round Tubes Only) for Seamless Brass Tube (ASTM B135)

	Outside diameter, in (1), (2)						
Wall thickness, in	1/32 to 1/8, incl.	Over 1/8 to 5/8, incl.	Over 5/8 to 1, incl.	Over 1 to 2, incl.	Over 2 to 4, incl.	Over 4 to 7, incl.	Over 7 to 10, incl.
Up to 0.017, incl.	0.002	0.001	0.0015	0.002	...	...	...
Over 0.017–0.024, incl.	0.003	0.002	0.002	0.0025	...	...	...
Over 0.024–0.034, incl.	0.003	0.0025	0.0025	0.003	0.004	...	...
Over 0.034–0.057, incl.	0.003	0.003	0.0035	0.0035	0.005	0.007	...
Over 0.057–0.082, incl.	...	0.0035	0.004	0.004	0.006	0.008	0.010
Over 0.082–0.119, incl.	...	0.004	0.005	0.005	0.007	0.009	0.011
Over 0.119–0.164, incl.	...	0.005	0.006	0.006	0.008	0.010	0.012
Over 0.164–0.219, incl.	...	0.007	0.009	0.009	0.011	0.012	0.014
Over 0.219–0.283, incl.	...	...	0.011	0.012	0.014	0.015	0.016
Over 0.283–0.379, incl.	...	...	0.014	6% (3)	6% (3)	7% (3)	7% (3)
Over 0.379	...	...	...	6% (3)	6% (3)	7% (3)	7% (3)

Notes:
(1) When round tube is ordered by outside and inside diameters, the maximum plus and minus deviation of the wall thickness from the nominal at any point shall not exceed the values given in the table by more than 50 percent.
(2) Tolerances are plus and minus; if tolerances all plus or all minus are desired, double the values given.
(3) Percent of the specified wall expressed to the nearest 0.001 in.

TABLE 12.26 Wall Thickness Tolerances on Distance between Parallel Surfaces for Rectangular, Including Square, Tube for Seamless Brass Tube (ASTM B135)*

	Outside diameter, in†						
Wall thickness, in	1/32 to 1/8, incl.	Over 1/8 to 5/8, incl.	Over 5/8 to 1, incl.	Over 1 to 2, incl.	Over 2 to 4, incl.	Over 4 to 7, incl.	Over 7 to 10, incl.
Up to 0.017, incl.	0.002	0.002	0.0025	0.003	...	...	...
Over 0.017–0.024, incl.	0.003	0.0025	0.003	0.0035	...	...	...
Over 0.024–0.034, incl.	0.0035	0.0035	0.0035	0.004	0.006	...	...
Over 0.034–0.057, incl.	0.004	0.004	0.0045	0.005	0.007	0.009	...
Over 0.057–0.082, incl.	...	0.005	0.006	0.007	0.008	0.010	0.012
Over 0.082–0.119, incl.	...	0.007	0.008	0.009	0.010	0.012	0.014
Over 0.119–0.164, incl.	...	0.009	0.010	0.011	0.012	0.014	0.016
Over 0.164–0.219, incl.	...	0.011	0.012	0.013	0.015	0.017	0.019
Over 0.219–0.283, incl.	...	...	0.015	0.016	0.018	0.020	0.022

* Tolerances are plus and minus; if tolerances all plus or all minus are desired, double the values given.
† In case of rectangular tube, the major dimension determines the thickness tolerance applicable to all walls.

Seamless Copper Alloy Tube (ASTM B315)

Scope. This specification covers seamless copper alloy tubes in straight lengths for general engineering purposes. The tube is produced from copper alloy C65500.

Temper. All tubes shall normally be furnished in following conditions:

- O30 (extruded and annealed)
- O61 (cold-worked and annealed)

Tube supplied in cold-worked and annealed condition (O61) shall show complete recrystallization.

Permissible Variations. Diameter tolerances for tube shall be in accordance with Table 12.27. Wall thickness tolerances for tube shall be in accordance with Table 12.28.

The standard method of specifying wall thickness shall be in decimal fractions of an inch. Tolerances on a given tube may be specified with respect to any two, but not all three, of the following: outside diameter, inside diameter, or wall thickness.

TABLE 12.27 Average Diameter Tolerances for Seamless Copper Alloy Tubes (ASTM B315)

Specified diameter, in	Tolerance, in (plus and minus) (1)
Up to 1/8, incl.	0.003 (2)
Up to 1/8, incl.	0.0025 (3)
Over 1/8 to 5/8, incl.	0.0025
Over 5/8 to 1, incl.	0.003
Over 1 to 2, incl.	0.004
Over 2 to 3, incl.	0.005
Over 3 to 4, incl.	0.006
Over 4 to 5, incl.	0.008
Over 5 to 6, incl.	0.009
Over 6 to 8, incl.	0.010
Over 8 to 10, incl.	0.013

Notes:
(1) Tolerance applies to inside or outside diameters, except as noted.
(2) On inside diameter.
(3) On outside diameter.

TABLE 12.28 Wall Thickness Tolerance for Seamless Copper Alloy Tube (ASTM B315) (1)

	Outside diameter, in (2)						
Wall thickness, in	1/32 to 1/8, incl.	Over 1/8 to 5/8, incl.	Over 5/8 to 1, incl.	Over 1 to 2, incl.	Over 2 to 4, incl.	Over 4 to 7, incl.	Over 7 to 10, incl.
Up to 0.017, incl.	0.0025	0.0015	0.002	0.0025	...	...	...
Over 0.017–0.024, incl.	0.004	0.0025	0.0025	0.003	...	...	...
Over 0.024–0.034, incl.	0.004	0.003	0.003	0.004	0.005	...	...
Over 0.034–0.057, incl.	0.004	0.004	0.0045	0.0045	0.0065	0.009	...
Over 0.057–0.082, incl.	...	0.0045	0.005	0.005	0.0075	0.010	0.013
Over 0.082–0.119, incl.	...	0.005	0.0065	0.0065	0.009	0.011	0.014
Over 0.119–0.164, incl.	...	0.007	0.007	0.0075	0.010	0.013	0.015
Over 0.164–0.219, incl.	...	...	0.009	0.010	0.012	0.015	0.018
Over 0.219–0.283, incl.	...	...	0.012	0.013	0.015	0.018	0.020
Over 0.283–0.379, incl.	...	...	...	6% (3)	6% (3)	8% (3)	8% (3)
Over 0.379	...	...	...	6% (3)	6% (3)	8% (3)	8% (3)

Notes:
(1) Tolerances are plus and minus; if tolerances all plus or all minus are desired, double the values given.
(2) When round tube is ordered by outside and inside diameters, the maximum plus and minus deviation of the wall thickness from the nominal at any point shall not exceed the values given in the table by more than 50 percent.
(3) Percent of the specified wall expressed to the nearest 0.001 in.

Seamless Copper and Copper Alloy Condenser and Heat Exchanger Tubes with Integral Fins (ASTM B359)

Scope. This specification covers seamless copper and copper alloy tubes on which the external or internal surface, or both, has been modified by a cold-forming process to produce an integral enhanced surface for improved heat transfer. The tubes are used in surface condensers, evaporators, and heat exchangers and are normally made from the following copper and copper alloys:

C10200 Oxygen-free without residential deoxidants
C12000 Phosphorized, low residual phosphorus
C12200 Phosphorized, high residual phosphorus
C14200 Phosphorized, arsenical

C19200 Phosphorized, 1% iron
C23000 Red brass
C44300 Admiralty metals B
C44400 Admiralty metals C
C44500 Admiralty metals D
C60800 Aluminum bronze
C68700 Aluminum brass, B
C70400 95-5 Copper-nickel
C70600 90-10 Copper-nickel
C71000 80-20 Copper-nickel
C71500 70-30 Copper-nickel

General Requirements. Tubes covered by this specification are normally furnished with unenhanced ends, but may be furnished with enhanced ends or stripped ends from which the outside diameter enhancement has been removed by machining.

Temper. The tube after enhancing shall be supplied in either of the following tempers, one of which must be specified on the purchase order: annealed temper or as-fabricated temper.

Tubes of copper alloy C23000, C44300, C44400, C44500, C60800, and C68700 furnished in the as-finned temper must be stress-relief annealed after enhancing.

Permissible Variations. The outside diameter of the unfinned sections shall not vary from that specified by more than the amounts shown in Table 12.29 as measured by go and no-go ring gages. The diameters over the fins shall at no point exceed the diameter of the plain sections involved, as determined by a go ring gage unless otherwise specified.

No tube at its thinnest point beneath the fins or in the plain section shall be less than the minimum thickness specified.

TABLE 12.29 Diameter Tolerances for Seamless Copper and Copper Alloy Condenser and Heat Exchanger Tubes with Integral Fins (ASTM B359)

Specified diameter, in	Tolerance, in (plus and minus)
0.500 and under	0.002
Over 0.500–0.740, incl.	0.0025
Over 0.740–1.000, incl.	0.003

Seamless U-Bend Copper and Copper Alloy Condenser and Heat Exchanger Tubes (ASTM B395)

Scope. This specification covers U-bend seamless condenser, evaporator, and heat exchanger tubes of copper and copper alloys up to 2 in, inclusive, in diameter. The tubes for this application are normally made from the following copper and copper alloys:

C10200 Oxygen-free without residential deoxidants
C12000 Phosphorized, low residual phosphorus
C12200 Phosphorized, high residual phosphorus
C14200 Phosphorized, arsenical
C19200 Phosphorized, 1% iron
C23000 Red brass
C44300 Admiralty metals B

C44400 Admiralty metals C
C44500 Admiralty metals D
C60800 Aluminum bronze
C68700 Aluminum brass, B
C70600 90-10 Copper-nickel
C71000 80-20 Copper-nickel
C71500 70-30 Copper-nickel

Temper. Prior to U bending, tubes of copper alloy C23000, C44300, C44400, C44500, C60800, C68700, C70600, C71000, and C71500 shall be in the annealed (O61) temper unless otherwise specified on the purchase order.

Prior to U bending, tubes of copper alloys C10200, C12000, C12200, and C14200 shall normally be in light-drawn temper (H55). Tubes of copper alloy C70600 shall, if specified, be made in light-drawn temper (H55).

Prior to U bending, tubes of copper alloy C19200 shall normally be in the annealed (O61) or light-drawn temper (H55) as specified.

Prior to U bending, tubes of copper alloy C71500 shall, if specified, be made in the drawn, stress-relieved temper (HR58).

The U-bend portions furnished in copper alloy C23000, C44300, C44400, C44500, C60800, and C68700 shall be relief annealed (HR) after bending. If specified, the U-bend portions of tubes furnished in the copper alloys C70600, C71000, and C71500 shall be relief annealed (HR) after bending.

Permissible Variations. The outside diameter of the straight leg portion of the tube shall not vary from that specified by more than the amounts shown in Table 12.30 as measured by go and no-go ring gages.

Neither the major nor the minor outside diameter of the tube at any one cross section included within the points of tangency of the bend shall deviate from the nominal diameter prior to bending by more than 10 percent.

For tubes ordered to minimum wall, prior to U bending the wall thickness of the single-gage tubes at the thinnest point shall not be less than the thickness specified. The maximum plus deviation from the specified wall at any point shall not exceed twice the value shown in Table 12.31.

TABLE 12.30 Diameter Tolerances for Seamless U-Bend Copper and Copper Alloy Condenser and Heat Exchanger Tubes (ASTM B395)

	Wall thickness, in			
	0.032	0.035	0.042	0.049 and over
Outside diameter, in	Diameter tolerances, plus and minus, in			
Up to 0.500, incl.	0.0025	0.0025	0.0025	0.0025
Over 0.500 to 0.740, incl.	0.004	0.004	0.0035	0.003
Over 0.740 to 1.000, incl.	0.006	0.005	0.0045	0.004
Over 1.000 to 1.250, incl.	0.009	0.008	0.006	0.0045
Over 1.250 to 1.375, incl.	...	...	0.008	0.005
Over 1.375 to 2.000, incl.	...	...	...	0.006

TABLE 12.31 Wall Thickness Tolerances for Seamless U-Bend Copper and Copper Alloy Condenser and Heat Exchanger Tubes (ASTM B395)

	Outside diameter, in		
	Over $\frac{1}{8}$ to $\frac{5}{8}$, incl.	Over $\frac{5}{8}$ to 1, incl.	Over 1 to 2, incl.
Wall thickness, in	Wall thickness tolerances, plus and minus, in		
0.032 to 0.035, incl.	0.003	0.003	0.004
0.035 to 0.058, incl.	0.004	0.0045	0.0045
0.058 to 0.083, incl.	0.0045	0.005	0.005
0.083 to 0.120, incl.	0.005	0.0065	0.0065
0.120 to 0.134, incl.	0.007	0.007	0.0075

The wall thickness of the tube at the apex of the U-bend section shall not be less than the value determined by the following equation:

$$t_f = t(2R)/(2R + D)$$

where t_f = thickness after bending
t = specified thickness of minimum wall or specified nominal wall minus the permissible wall thickness tolerance, in
R = centerline bend radius, in
D = nominal outside diameter of the tube, in

Seamless Copper-Nickel Tube (ASTM B466)

Scope. This specification covers seamless copper-nickel alloy tube in straight lengths suitable for general engineering purposes. The following three alloys are covered:

C70600 Nominal composition—90% copper, 10% nickel
C71000 Nominal composition—80% copper, 20% nickel
C71500 Nominal composition—70% copper, 30% nickel

Temper. All pipe shall normally be furnished in the annealed temper only.

Permissible Variations. The diameter tolerances shall be in accordance with Table 12.32. The wall thickness tolerances shall be in accordance with Table 12.33.

TABLE 12.32 Average Diameter Tolerances for Seamless Copper-Nickel Tubes (ASTM B466)*

Specified diameter, in	Tolerance applies to	Tolerances, plus and minus, in†
Up to 1/8, incl.	Inside diameter	0.003
Up to 1/8, incl.	Outside diameter	0.0025
Over 1/8 to 5/8, incl.	Inside or outside	0.0025
Over 5/8 to 1, incl.	Inside or outside	0.003
Over 1 to 2, incl.	Inside or outside	0.004
Over 2 to 3, incl.	Inside or outside	0.005
Over 3 to 4, incl.	Inside or outside	0.006
Over 4 to 5, incl.	Inside or outside	0.008
Over 5 to 6, incl.	Inside or outside	0.009
Over 6 to 8, incl.	Inside or outside	0.010
Over 8 to 10, incl.	Inside or outside	0.013

* If tolerances all plus or minus are desired, double the values given.
† The average outside or inside diameters is the average of the maximum and minimum outside diameters, or of the maximum and minimum inside diameters, whichever is applicable, as determined at one cross section of the pipe.

TABLE 12.33 Wall Thickness Tolerances for Seamless Copper-Nickel Tubes (ASTM B466) (1)

	Outside diameter, in (2)						
Wall thickness, in	$^1/_{32}$ to $^1/_8$, incl.	Over $^1/_8$ to $^5/_8$, incl.	Over $^5/_8$ to 1, incl.	Over 1 to 2, incl.	Over 2 to 4, incl.	Over 4 to 7, incl.	Over 7 to 10, incl.
Up to 0.017	0.0025	0.0015	0.002	0.0025	...	...	...
Over 0.017 to 0.024, incl.	0.004	0.0025	0.0025	0.003	...	...	...
Over 0.024 to 0.034, incl.	0.004	0.003	0.003	0.004	0.005	...	...
Over 0.034 to 0.057, incl.	0.004	0.004	0.0045	0.0045	0.0065	0.009	...
Over 0.057 to 0.082, incl.	...	0.0045	0.005	0.005	0.0075	0.010	0.013
Over 0.082 to 0.119, incl.	...	0.005	0.0065	0.0065	0.009	0.011	0.014
Over 0.119 to 0.164, incl.	...	0.007	0.007	0.0075	0.010	0.013	0.015
Over 0.164 to 0.219, incl.	...	...	0.009	0.010	0.012	0.015	0.018
Over 0.219 to 0.283, incl.	...	...	0.012	0.013	0.015	0.018	0.020
Over 0.283 to 0.379, incl.	...	...	...	0.015	0.018	0.020	0.023
Over 0.379	...	...	...	6 (3)	6 (3)	8 (3)	8 (3)

Notes:
(1) All tolerances are plus and minus; if tolerances all plus or all minus are desired, double the values given.
(2) When tube is ordered by outside and inside diameters, the maximum plus and minus deviation of the wall thickness from the nominal at any point shall not exceed the values given in this table by more than 50 percent.
(3) Percent of specified wall thickness expressed to the nearest 0.001 in.

Welded Copper and Copper Alloy Heat Exchanger Tube (ASTM B543)

Scope. This specification covers welded tube of copper and various copper alloys up to $3^1/_8$ in, inclusive, in diameter, for use in surface condensers, evaporators, heat exchangers, and general applications. The tubes may be produced from any one of the following coppers:

C12200 Phosphorized, high residual phosphorus
C19400 Copper-iron alloy
C23000 Red brass
C28000 Muntz metal
C44300 Admiralty metals B
C44400 Admiralty metals C
C44500 Admiralty metals D
C68700 Aluminum brass, B
C70400 95-5 Copper-nickel
C70600 90-10 Copper-nickel
C71500 70-30 Copper-nickel
C71640 Copper-nickel-iron-manganese

Temper. Tube tempers shall be designated as follows:

Annealed tempers: Welded and annealed; fully finished—annealed

Light-cold-worked tempers: As welded from annealed strip; as welded, light cold worked; fully finished—light drawn

Other tempers shall be produced to the mechanical properties as agreed upon between the manufacturer or supplier and the purchaser.

Tubes of copper alloy C23000, C44300, C44400, C44500, and C68700 shall be furnished in the annealed temper or the stress-relieved condition as specified in the purchase order unless otherwise agreed upon between the purchaser and the manufacturer or supplier.

Tubes of copper alloy C12200, C19400, C70400, C70600, C71500, and C71640 are normally supplied in the temper specified in the purchase order without stress relief treatment.

Permissible Variations. The outside diameter of the tube shall not vary from that specified by more than the amounts shown in Table 12.34 as measured by go and no-go ring gages.

TABLE 12.34 Diameter Tolerances for Welded Copper and Copper Alloy Heat Exchanger Tube (ASTM B543)

	Wall thickness, in				
	0.020* 0.022 0.025 0.028	0.032	0.035	0.042	0.049 and over
Outside diameter, in	Diameter tolerances, plus and minus, in				
Up to 0.500, incl.	0.003	0.0025	0.0025	0.0025	0.0025
Over 0.500–0.740, incl.	0.004	0.004	0.004	0.0035	0.003
Over 0.740–1.000, incl.	0.006	0.006	0.005	0.0045	0.004
Over 1.000–1.250, incl.	...	0.009	0.008	0.006	0.0045
Over 1.250–1.375, incl.	...	...	...	0.008	0.005
Over 1.375–2.000, incl.	...	...	...	...	0.006
Over 2.000–3.125, incl.	...	...	...	...	0.0065

* Thin wall thicknesses are supplied only in light-cold-worked tubes.

TABLE 12.35 Wall Thickness Tolerances for Welded Copper and Copper Alloy Heat Exchanger Tube (ASTM B543)

	Outside diameter, in			
Wall thickness, in	Over 1/8 to 5/8, incl.	Over 5/8 to 1, incl.	Over 1 to 2, incl.	Over 2 to 3.125, incl.
0.020 to 0.032, incl.	0.003	0.003	...	...
0.032 to 0.035, incl.	0.003	0.003	0.004	...
0.035 to 0.058, incl.	0.004	0.0045	0.0045	0.005
0.058 to 0.083, incl.	0.0045	0.005	0.005	0.0055
0.083 to 0.120, incl.	0.005	0.0065	0.0065	0.0065
0.120 to 0.135, incl.	0.007	0.007	0.0075	0.008

For tubes ordered to minimum wall thickness, no tube wall at its thinnest point shall be less than the specified wall thickness. The maximum plus deviation from the specified wall at any point shall not exceed twice the values shown in Table 12.35.

For tubes ordered to nominal wall thickness, the maximum plus and minus deviation from the nominal wall at any point shall not exceed the values shown in Table 12.35.

COPPER AND COPPER ALLOY FLANGES AND FITTINGS

This section covers the following tube specifications:

ANSI B16.24 Cast Copper Alloy Pipe Flanges and Flanged Fittings

ANSI B16.22 Wrought Copper and Copper Alloy Solder Joint Pressure Fittings

ANSI B16.15 Cast Bronze Threaded Fittings

ANSI B16.18 Cast Copper Alloy Solder Joint Pressure Fittings

Cast Copper Alloy Pipe Flanges and Flanged Fittings (ANSI B16.24)

Scope. This specification establishes the requirements for cast copper and copper alloy flanges and flanged fittings and covers the chemical requirements, dimensions, and pressure-temperature ratings. The flanges and flanged fittings may be produced from any one of the following coppers:

- C83600
- C92200
- C95200

Materials. Flanges shall be in accordance with one of the material requirements listed below:

- Alloy C83600 Castings shall meet the requirements of ASTM B62.
- Alloy C92200 Castings shall meet the requirements of ASTM B61.
- Alloy C95200 Castings shall meet the requirements of ASTM B148.

Flanged fittings shall be in accordance with one of the material requirements listed below:

- Alloy C83600 Castings shall meet the requirements of ASTM B62.
- Alloy C92200 Castings shall meet the requirements of ASTM B61.

Flange Dimensions. Dimensions of Class 150 and Class 300 threaded companion and blind flanges for alloys C83600 and C92200 are shown as follows. The flange dimensions for alloy C95200 shall be in accordance with the dimensional requirements of ANSI/ASME B16.5.

Table 12.36 Dimensions of Class 150 threaded companion flanges for alloys C83600 and C92200
Table 12.37 Dimensions of Class 150 blind flanges for alloys C83600 and C92200
Table 12.38 Dimensions of Class 300 threaded companion flanges for alloys C83600 and C92200
Table 12.39 Dimensions of Class 300 blind flanges for alloys C83600 and C92200

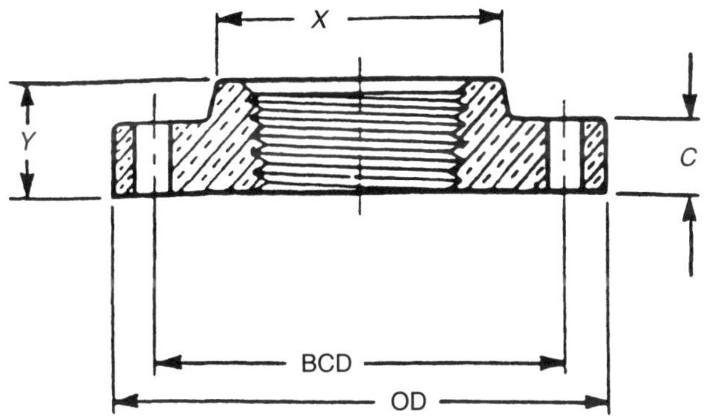

TABLE 12.36 Dimensions of Class 150 Threaded Companion Flanges for Alloys C83600 and C92200*

Pipe size (NPS)	Diameter of flange, OD	Thickness of flange, min., C	Bolt circle diameter, BCD	Number of bolts	Diameter of bolt†	Diameter of hub, min., X	Length overall, min., Y
$^1/_2$	3.50	0.31	2.38	4	$^1/_2$	1.19	0.59
$^3/_4$	3.88	0.34	2.75	4	$^1/_2$	1.25	0.62
1	4.25	0.38	3.12	4	$^1/_2$	1.94	0.69
$1^1/_4$	4.62	0.41	3.50	4	$^1/_2$	2.31	0.81
$1^1/_2$	5.00	0.44	3.88	4	$^1/_2$	2.56	0.88
2	6.00	0.50	4.75	4	$^5/_8$	3.06	1.00
$2^1/_2$	7.00	0.56	5.50	4	$^5/_8$	3.56	1.12
3	7.50	0.62	6.00	4	$^5/_8$	4.25	1.19
$3^1/_2$	8.50	0.69	7.00	8	$^5/_8$	4.81	1.25
4	9.00	0.69	7.50	8	$^5/_8$	5.31	1.31
5	10.00	0.75	8.50	8	$^3/_4$	6.44	1.44
6	11.00	0.81	9.50	8	$^3/_4$	7.56	1.56
8	13.50	0.94	11.75	8	$^3/_4$	9.69	1.75
10	16.00	1.00	14.25	12	$^7/_8$	12.00	1.94
12	19.00	1.06	17.00	12	$^7/_8$	14.38	2.19

* Dimensions are in inches.
† Diameter of bolt hole is $^1/_8$ in larger than the diameter of the bolt.

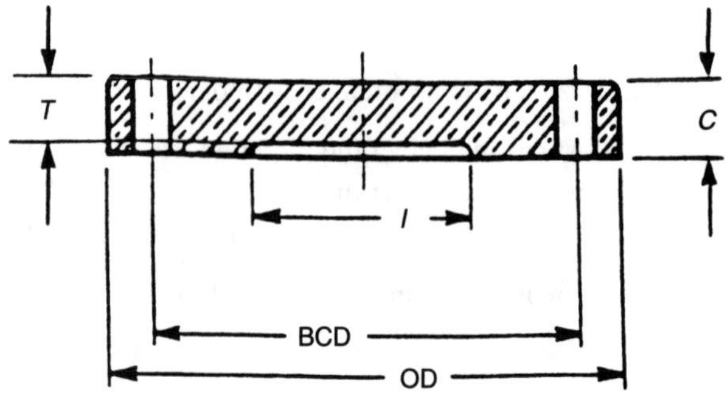

TABLE 12.37 Dimensions of Class 150 Blind Flanges for Alloys C83600 and C92200*

Pipe size (NPS)	Diameter of flange, OD	Thickness of flange, *C*	Bolt circle diameter, BCD	Number of bolts	Diameter of bolt†	Diameter of counterbore, max., *I*	Thickness at recess, min., *T*
$^{1}/_{2}$	3.50	0.31	2.38	4	$^{1}/_{2}$	0.50	0.25
$^{3}/_{4}$	3.88	0.34	2.75	4	$^{1}/_{2}$	0.75	0.28
1	4.25	0.38	3.12	4	$^{1}/_{2}$	1.00	0.31
$1^{1}/_{4}$	4.62	0.41	3.50	4	$^{1}/_{2}$	1.25	0.34
$1^{1}/_{2}$	5.00	0.44	3.88	4	$^{1}/_{2}$	1.50	0.38
2	6.00	0.50	4.75	4	$^{5}/_{8}$	2.00	0.44
$2^{1}/_{2}$	7.00	0.56	5.50	4	$^{5}/_{8}$	2.50	0.50
3	7.50	0.62	6.00	4	$^{5}/_{8}$	3.00	0.56
$3^{1}/_{2}$	8.50	0.69	7.00	8	$^{5}/_{8}$	3.50	0.62
4	9.00	0.69	7.50	8	$^{5}/_{8}$	4.00	0.62
5	10.00	0.75	8.50	8	$^{3}/_{4}$	5.00	0.69
6	11.00	0.81	9.50	8	$^{3}/_{4}$	6.00	0.75
8	13.50	0.94	11.75	8	$^{3}/_{4}$	8.00	0.88
10	16.00	1.00	14.25	12	$^{7}/_{8}$	10.00	0.94
12	19.00	1.06	17.00	12	$^{7}/_{8}$	12.00	1.00

* Dimensions are in inches.
† Diameter of bolt hole is $^{1}/_{8}$ in larger than the diameter of the bolt.

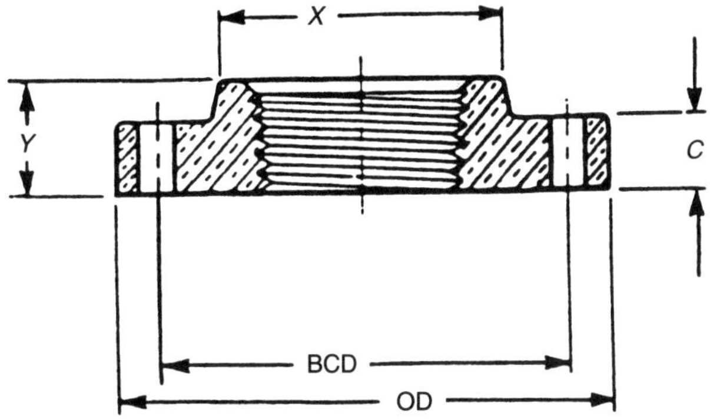

TABLE 12.38 Dimensions of Class 300 Threaded Companion Flanges for Alloys C83600 and C92200*

Pipe size (NPS)	Diameter of flange, OD	Thickness of flange, min., C	Bolt circle diameter, BCD	Number of bolts	Diameter of bolt†	Diameter of hub, min., X	Length overall, min., Y
1/2	3.75	0.50	2.62	4	1/2	1.19	0.59
3/4	4.62	0.53	3.25	4	5/8	1.50	0.62
1	4.88	0.59	3.50	4	5/8	1.94	0.69
1 1/4	5.25	0.62	3.88	4	5/8	2.31	0.81
1 1/2	6.12	0.69	4.50	4	3/4	2.56	0.88
2	6.50	0.75	5.00	8	5/8	3.06	1.00
2 1/2	7.50	0.81	5.88	8	3/4	3.56	1.12
3	8.25	0.91	6.62	8	3/4	4.25	1.19
3 1/2	9.00	0.97	7.25	8	3/4	4.81	1.25
4	10.00	1.06	7.88	8	3/4	5.31	1.31
5	11.00	1.12	9.25	8	3/4	6.44	1.44
6	12.50	1.19	10.62	12	3/4	7.56	1.56
8	15.00	1.38	13.00	12	7/8	9.69	1.75

* Dimensions are in inches.
† Diameter of bolt hole is 1/8 in larger than the diameter of the bolt.

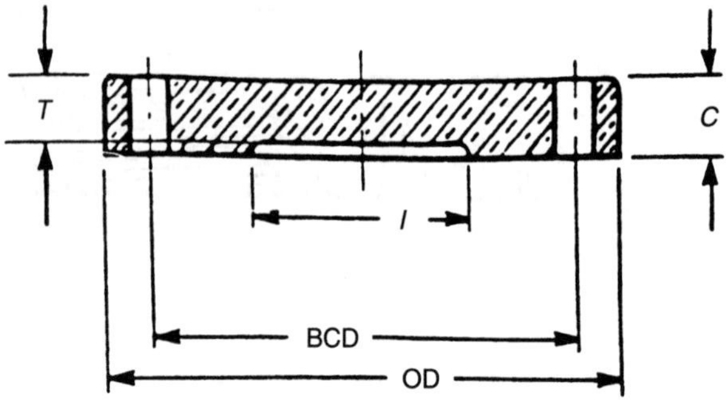

TABLE 12.39 Dimensions of Class 300 Blind Flanges for Alloys C83600 and C92200*

Pipe size (NPS)	Diameter of flange, OD	Thickness of flange, C	Bolt circle diameter, BCD	Number of bolts	Diameter of bolt†	Diameter of counterbore, max., I	Thickness at recess, min., T
$^1/_2$	3.75	0.50	2.62	4	$^1/_2$	0.50	0.44
$^3/_4$	4.62	0.53	3.25	4	$^5/_8$	0.75	0.47
1	4.88	0.59	3.50	4	$^5/_8$	1.00	0.53
$1^1/_4$	5.25	0.62	3.88	4	$^5/_8$	1.25	0.56
$1^1/_2$	6.12	0.69	4.50	4	$^3/_4$	1.50	0.62
2	6.50	0.75	5.00	8	$^5/_8$	2.00	0.69
$2^1/_2$	7.50	0.81	5.88	8	$^3/_4$	2.50	0.75
3	8.25	0.91	6.62	8	$^3/_4$	3.00	0.84
$3^1/_2$	9.00	0.97	7.25	8	$^3/_4$	3.50	0.91
4	10.00	1.06	7.88	8	$^3/_4$	4.00	1.00
5	11.00	1.12	9.25	8	$^3/_4$	5.00	1.06
6	12.50	1.19	10.62	12	$^3/_4$	6.00	1.12
8	15.00	1.38	13.00	12	$^7/_8$	8.00	1.31

* Dimensions are in inches.

† Diameter of bolt hole is $^1/_8$ in larger than the diameter of the bolt.

Fitting Dimensions. Dimensions of fittings shall be in accordance with the following tables:

Table 12.40 Dimensions of Class 150 elbows and reducers
Table 12.41 Dimensions of Class 300 elbows and reducers
Table 12.42 Dimensions of Class 150 tees and crosses
Table 12.43 Dimensions of Class 300 tees and crosses
Table 12.44 Dimensions of Class 150 laterals and true Y's
Table 12.45 Dimensions of Class 300 laterals and true Y's

90° elbow **90° long-radius elbow** **45° elbow** **90° reducing elbow** **Reducer** **Eccentric reducer**

TABLE 12.40 Dimensions of Class 150 Elbows and Reducers*

Nominal pipe size	Center to face, *A*	Face to face, *A* + *A*	Center to face, *B*	Center to face, *C*	Face to face, *G*	Wall thickness, *t*
1/2	3.00	6.00	...	1.62	...	0.09
3/4	3.25	6.50	...	1.75	...	0.11
1	3.50	7.00	5.00	1.75	...	0.12
1 1/4	3.75	7.50	5.50	2.00	...	0.14
1 1/2	4.00	8.00	6.00	2.25	...	0.16
2	4.50	9.00	6.50	2.50	5.00	0.19
2 1/2	5.00	10.00	7.00	3.00	5.50	0.20
3	5.50	11.00	7.75	3.00	6.00	0.22
3 1/2	6.00	12.00	8.50	3.50	6.50	0.25
4	6.50	13.00	9.00	4.00	7.00	0.27
5	7.50	15.00	10.25	4.50	8.00	0.30
6	8.00	16.00	11.50	5.00	9.00	0.33
8	9.00	18.00	14.00	5.50	11.00	0.41
10	11.00	22.00	16.50	6.50	12.00	0.48
12	12.00	24.00	19.00	7.50	14.00	0.56

* Dimensions in inches.

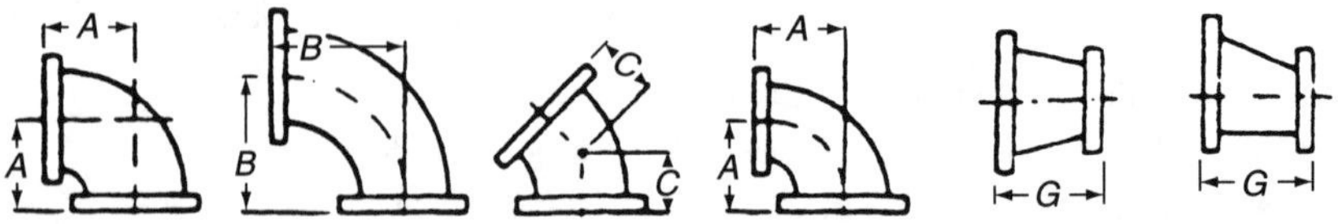

90° elbow | 90° long-radius elbow | 45° elbow | 90° reducing elbow | Reducer | Eccentric reducer

TABLE 12.41 Dimensions of Class 300 Elbows and Reducers*

Nominal pipe size	Center to face, A	Face to face, $A + A$	Center to face, B	Center to face, C	Face to face, G	Wall thickness, t
1/2	3.00	6.00	. . .	1.75	. . .	0.12
3/4	3.50	7.00	. . .	2.25	. . .	0.16
1	4.00	8.00	5.00	2.25	. . .	0.17
1 1/4	4.25	8.50	5.50	2.50	. . .	0.19
1 1/2	4.50	9.00	6.00	2.75	. . .	0.20
2	5.00	10.00	6.50	3.00	5.00	0.25
2 1/2	5.50	11.00	7.00	3.50	5.50	0.28
3	6.00	12.00	7.75	3.50	6.00	0.33
3 1/2	6.50	13.00	8.50	4.00	6.50	0.36
4	7.00	14.00	9.00	4.50	7.00	0.41
5	8.00	16.00	10.25	5.00	8.00	0.48
6	8.50	17.00	11.50	5.50	9.00	0.56
8	10.00	20.00	14.00	6.00	11.00	0.72

* Dimensions in inches.

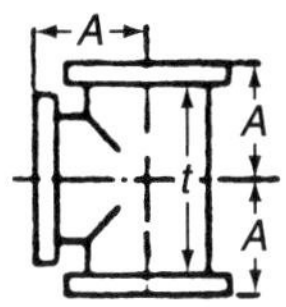

Tee

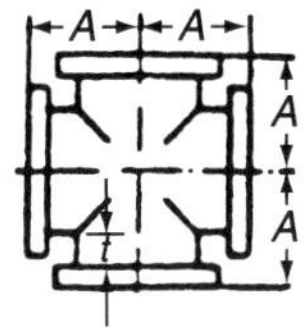

Cross

TABLE 12.42 Dimensions of Class 150 Tees and Crosses*

Nominal pipe size	Center to face, A	Face to face, $A + A$	Wall thickness, t
$^1/_2$	3.00	6.00	0.09
$^3/_4$	3.25	6.50	0.11
1	3.50	7.00	0.12
$1^1/_4$	3.75	7.50	0.14
$1^1/_2$	4.00	8.00	0.16
2	4.50	9.00	0.19
$2^1/_2$	5.00	10.00	0.20
3	5.50	11.00	0.22
$3^1/_2$	6.00	12.00	0.25
4	6.50	13.00	0.27
5	7.50	15.00	0.30
6	8.00	16.00	0.33
8	9.00	18.00	0.41
10	11.00	22.00	0.48
12	12.00	24.00	0.56

* Dimensions in inches.

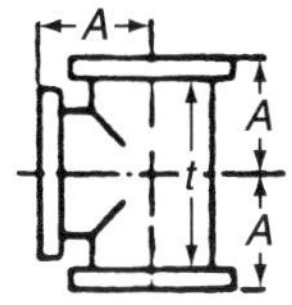

Tee

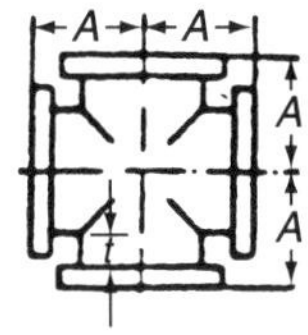

Cross

TABLE 12.43 Dimensions of Class 300 Tees and Crosses*

Nominal pipe size	Center to face, A	Face to face, $A + A$	Wall thickness, t
$^1/_2$	3.00	6.00	0.12
$^3/_4$	3.50	7.00	0.16
1	4.00	8.00	0.17
$1^1/_4$	4.25	8.50	0.19
$1^1/_2$	4.50	9.00	0.20
2	5.00	10.00	0.25
$2^1/_2$	5.50	11.00	0.28
3	6.00	12.00	0.33
$3^1/_2$	6.50	13.00	0.36
4	7.00	14.00	0.41
5	8.00	16.00	0.48
6	8.50	17.00	0.56
8	10.00	20.00	0.72

* Dimensions in inches.

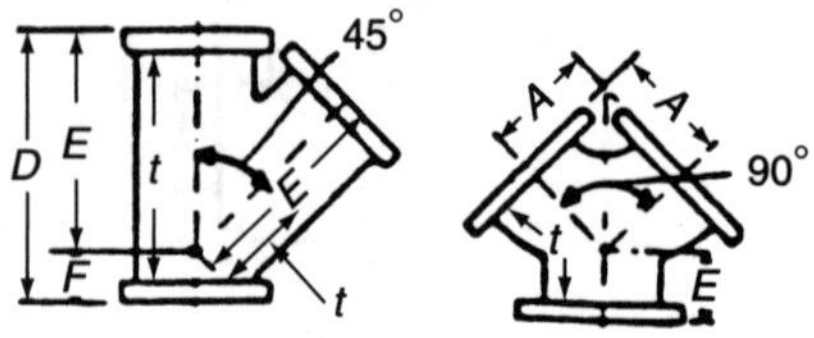

Lateral True Y

TABLE 12.44 Dimensions of Class 150 Laterals and True Y's (Straight Sizes)*

Nominal pipe size	Center to face, *A*	Face to face, *D*	Center to face, *E*	Center to face, *F*	Wall thickness, *t*
$^1/_2$	3.00	...	...	...	0.09
$^3/_4$	3.25	...	...	...	0.11
1	3.50	7.50	5.75	1.75	0.12
$1^1/_4$	3.75	8.00	6.25	1.75	0.14
$1^1/_2$	4.00	9.00	7.00	2.00	0.16
2	4.50	10.50	8.00	2.50	0.19
$2^1/_2$	5.00	12.00	9.50	2.50	0.20
3	5.50	13.00	10.00	3.00	0.22
$3^1/_2$	6.00	14.50	11.50	3.00	0.25
4	6.50	15.00	12.00	3.00	0.27
5	7.50	17.00	13.50	3.50	0.30
6	8.00	18.00	14.50	3.50	0.33
8	9.00	22.00	17.50	4.50	0.41
10	11.00	25.50	20.50	5.00	0.48
12	12.00	30.00	24.50	5.50	0.56

* Dimensions in inches.

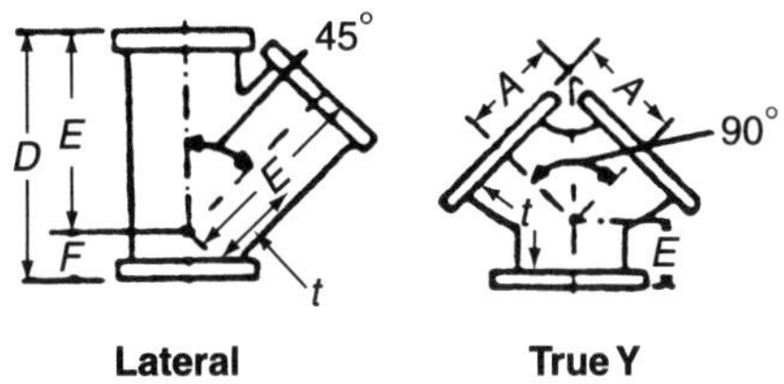

TABLE 12.45 Dimensions of Class 300 Laterals and True Y's (Straight Sizes)*

Nominal pipe size	Center to face, *A*	Face to face, *D*	Center to face, *E*	Center to face, *F*	Wall thickness, *t*
1/2	3.00	...	...	...	0.12
3/4	3.50	...	...	...	0.16
1	4.00	8.50	6.50	2.00	0.17
1 1/4	4.25	9.50	7.25	2.25	0.19
1 1/2	4.50	11.00	8.50	2.50	0.20
2	5.00	11.50	9.00	2.50	0.25
2 1/2	5.50	13.00	10.50	2.50	0.28
3	6.00	14.00	11.00	3.00	0.33
3 1/2	6.50	15.50	12.50	3.00	0.36
4	7.00	16.50	13.50	3.00	0.41
5	8.00	18.50	15.00	3.50	0.48
6	8.50	21.50	17.50	4.00	0.56
8	10.00	25.50	20.50	5.00	0.72

* Dimensions in inches.

Pressure-Temperature Ratings. The pressure-temperature ratings for alloys C83600 and C92200 are shown in Table 12.46 and those for alloy C95200 are shown in Table 12.47.

TABLE 12.46 Pressure-Temperature Ratings for Alloys C83600 and C92200

	Working pressure, psig			
	Class 150		Class 300	
Service temperature, °F	ASTM B62, C83600	ASTM B61, C92200	ASTM B62, C83600	ASTM B61, C92200
–20 to 150	225	225	500	500
175	220	220	480	490
200	210	215	465	475
225	205	210	445	465
250	195	205	425	450
275	190	200	410	440
300	180	195	390	425
350	165	180	350	400
400		170		375
406	150			
450	135	160	280	350
500		150		325
550		140		300
Test pressure	350	350	750	750

TABLE 12.47 Pressure-Temperature Ratings for ASTM B148, Alloy C95200

Service temperature, °F	Working pressure, psig						
	Class 150	Class 300	Class 400	Class 600	Class 900	Class 1500	Class 2500
−20 to 100	195	515	685	1030	1545	2575	4290
150	165	430	570	855	1285	2140	3570
200	155	400	535	800	1205	2005	3340
250	145	385	510	770	1150	1920	3200
300	140	370	495	740	1110	1850	3085
350	140	365	490	735	1100	1835	3060
400	140	365	485	725	1090	1820	3030
450	140	360	480	725	1085	1805	3010
500	140	360	480	720	1080	1800	3000
Test pressure	300	775	1050	1550	2325	3875	6450

Wrought Copper and Copper Alloy Solder Joint Pressure Fittings (ANSI B16.22)

Scope. This specification establishes the requirements for copper and wrought copper alloy solder joint seamless fittings, designed for use with copper water tube. The fittings shall be made from any one of the following coppers:

- C10200
- C12000
- C12200

Abbreviations. The following symbols are used to designate the types of fitting end:

C = solder joint fitting end made to receive copper tube diameter (female)

FTG = solder joint fitting end made to copper tube diameter (male)

F = internal ANSI standard taper pipe thread (female) NPTI

M = external ANSI standard taper pipe thread (male) NPTE

Fitting Dimensions. Fittings are designated by the size of the openings in the sequence illustrated in Fig. 12.1. The dimensions of the fittings shown in Table 12.48 correspond to standard water tube size as shown in ASTM B88, Specification for Seamless Copper Water Tube.

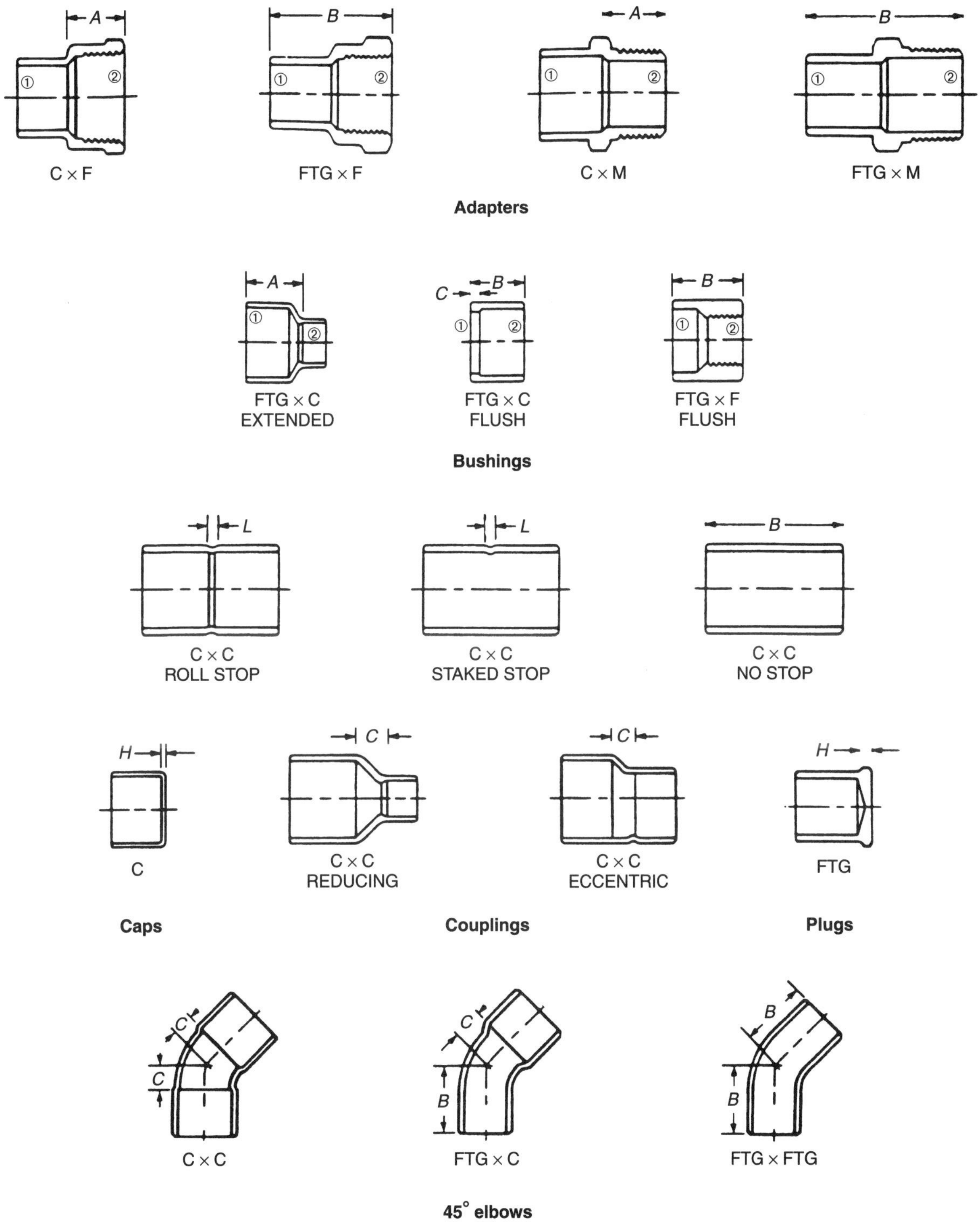

FIGURE 12.1 Method of designating laying lengths of fittings and openings of reducing fittings. Fittings are designated by size in the order: 1 × 2 × 3. (*Courtesy of ASME.*)

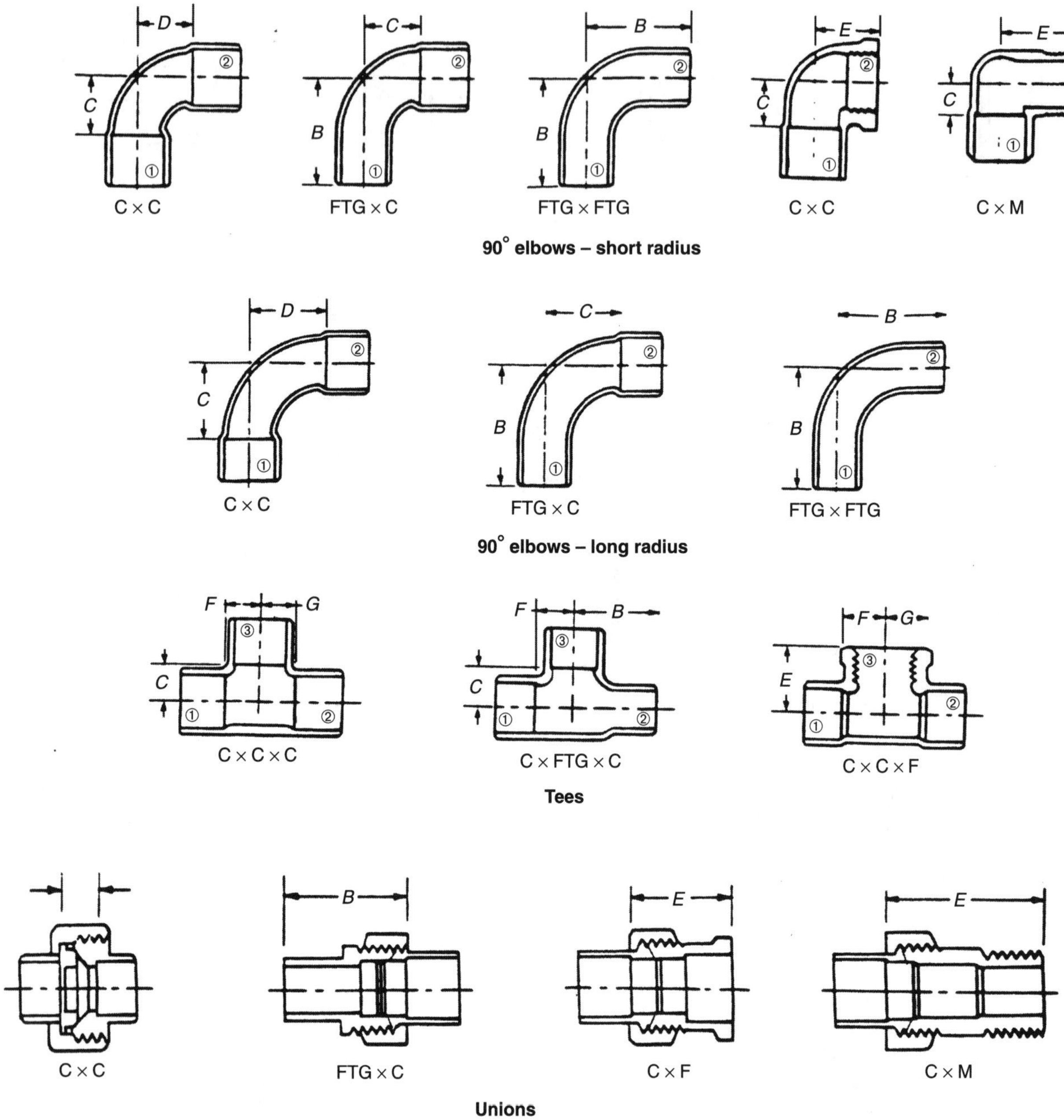

FIGURE 12.1 Method of designating laying lengths of fittings and openings of reducing fittings. Fittings are designated by size in the order: 1 × 2 × 3. (*Courtesy of ASME.*) (*Continued*)

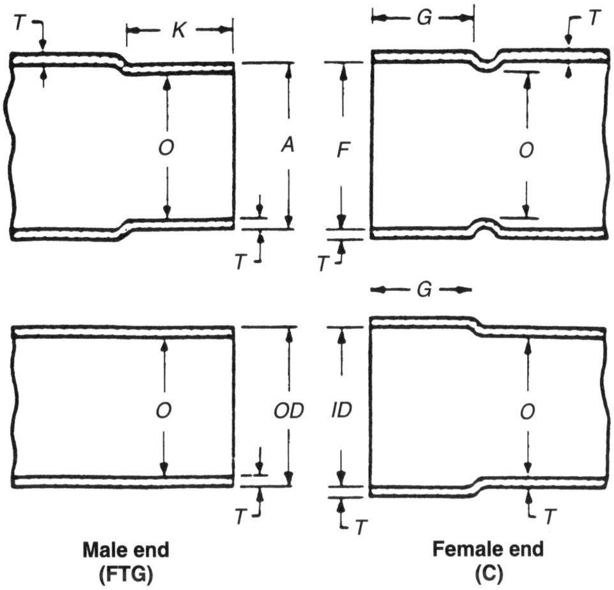

TABLE 12.48 Dimensions of Solder Joint Ends*

Standard tube size	Male end: Outside diameter, *A* Min.	Male end: Outside diameter, *A* Max.	Male end: Length, *K*, min.	Female end: Inside diameter, *F* Min.	Female end: Inside diameter, *F* Max.	Female end: Depth, *G*, min.	Wall thickness, *T*, min.	Inside diameter of fitting, *O*, min.
1/8	0.248	0.251	0.31	0.252	0.256	0.25	0.019	0.18
1/4	0.373	0.376	0.38	0.377	0.381	0.31	0.023	0.30
3/8	0.497	0.501	0.44	0.502	0.506	0.38	0.026	0.39
1/2	0.622	0.626	0.56	0.627	0.631	0.50	0.029	0.52
5/8	0.747	0.751	0.69	0.752	0.756	0.62	0.031	0.63
3/4	0.872	0.876	0.81	0.877	0.881	0.75	0.033	0.74
1	1.122	1.127	0.97	1.128	1.132	0.91	0.040	0.98
1 1/4	1.372	1.377	1.03	1.378	1.382	0.97	0.044	1.23
1 1/2	1.621	1.627	1.16	1.628	1.633	1.09	0.051	1.47
2	2.121	2.127	1.41	2.128	2.133	1.34	0.059	1.94
2 1/2	2.621	2.627	1.53	2.628	2.633	1.47	0.067	2.42
3	3.121	3.127	1.72	3.128	3.133	1.66	0.075	2.89
3 1/2	3.621	3.627	1.97	3.628	3.633	1.91	0.086	3.37
4	4.121	4.127	2.22	4.128	4.133	2.16	0.096	3.84
5	5.121	5.127	2.72	5.128	5.133	2.66	0.111	4.70
6	6.121	6.127	3.22	6.128	6.133	3.09	0.124	5.72
8	8.119	8.127	4.09	8.128	8.133	3.97	0.173	7.55

* Dimensions in inches.

Pressure-Temperature Ratings. The pressure-temperature ratings for solder joints made to the dimensions shown in Table 12.45 are shown in Table 12.49.

Cast Bronze Threaded Fittings, Classes 125 and 250 (ANSI B16.15)

Scope. This specification pertains primarily to cast Class 125 and Class 250 bronze threaded pipe fittings. Certain requirements also pertain to wrought and cast bushings, couplings, and plugs. The fittings shall be made from any one of the following coppers:

C83600	ASTM B62
C83800 or C84400	ASTM B584
C36000	ASTM B16
C32000 or C31400	ASTM B140

Fitting Dimensions. Refer to the following tables for the fitting dimensions:

Table 12.50	Dimensions of 90° Elbows, Tees, and Crosses, and 45° Elbows—Class 125
Table 12.51	Dimensions of Reducers and Return Bends—Class 125
Table 12.52	Dimensions of Caps—Class 125
Table 12.53	Dimensions of 90° Elbows, Tees, and Crosses and 45° Elbows—Class 250
Table 12.54	Dimensions of Square Head and Square Socket Plugs—Class 250

The size of the pipe fittings scheduled in these tables is identified by the corresponding nominal pipe size.

Pressure-Temperature Ratings. The pressure-temperature ratings for these pipe fittings are shown in Table 12.55.

TABLE 12.49 Rated Internal Working Pressure for Copper Fittings, psi*

Standard tube size	Temperature, °F						
	–20 to 100	150	200	250	300	350	400
1/4	912	775	729	729	714	608	456
3/8	779	662	623	623	610	519	389
1/2	722	613	577	577	565	481	361
5/8	631	537	505	505	495	421	316
3/4	582	495	466	466	456	388	291
1	494	420	395	395	387	330	247
1 1/4	439	373	351	351	344	293	219
1 1/2	408	347	327	327	320	272	204
2	364	309	291	291	285	242	182
2 1/2	336	285	269	269	263	224	168
3	317	270	254	254	248	211	159
3 1/2	304	258	243	243	238	202	152
4	293	249	235	235	230	196	147
5	269	229	215	215	211	179	135
6	251	213	201	201	196	167	125
8	270	230	216	216	212	180	135

* The fitting pressure rating applies to the largest opening of the fitting.

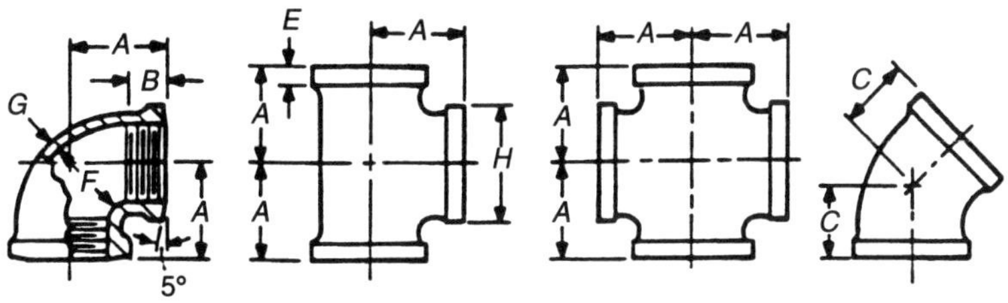

90° elbow **Tee** **Cross** **45° elbow**

TABLE 12.50 Dimensions of Class 125, 90° Elbows, Tees, and Crosses and 45° Elbows*

Nominal pipe size	Center to end, elbows, tees, and crosses, *A*	Length of thread, *B*, min.	Center to end, 45° elbows, *C*	Diameter, wrought coupling, *D*	Band length, *E*, min.	Inside diameter of cast fitting, *F*		Metal thickness, *G*	Band diameter, *H*, min.
						Min.	Max.		
1/8	0.54	0.25	0.42	0.56	0.14	0.41	0.44	0.08	0.67
1/4	0.71	0.32	0.56	0.69	0.16	0.54	0.58	0.08	0.81
3/8	0.82	0.36	0.63	0.84	0.17	0.68	0.72	0.09	1.00
1/2	1.01	0.43	0.78	1.06	0.19	0.84	0.90	0.09	1.17
3/4	1.18	0.50	0.89	1.31	0.23	1.05	1.11	0.10	1.42
1	1.43	0.58	1.06	...	0.27	1.32	1.39	0.11	1.72
1 1/4	1.69	0.67	1.22	...	0.31	1.66	1.73	0.12	2.10
1 1/2	1.84	0.70	1.30	...	0.34	1.90	1.97	0.13	2.38
2	2.12	0.75	1.45	...	0.41	2.38	2.45	0.15	2.92
2 1/2	2.70	0.92	1.95	...	0.48	2.88	2.98	0.17	3.49
3	3.08	0.98	2.17	...	0.55	3.50	3.60	0.19	4.20
4	3.79	1.08	2.61	...	0.66	4.50	4.60	0.22	5.31

* Dimensions in inches.

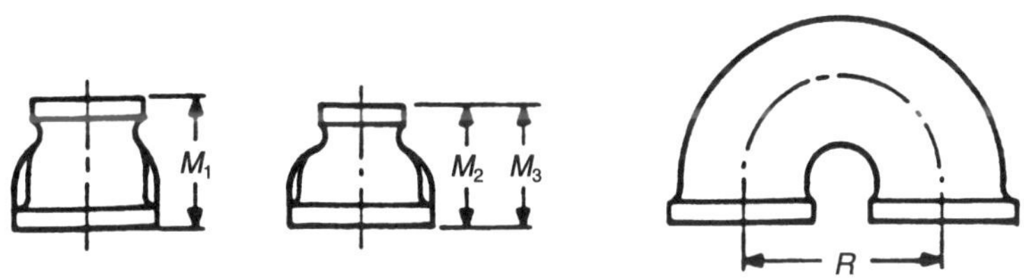

Reducer (1 size) **Reducer (2 and 3 sizes)** **Return bend**

TABLE 12.51 Dimensions of Class 125 Reducers and Return Bends (Straight Sizes)*

Nominal pipe size	Reducers, end-to-end reducing			Return bends, center to center, *R*
	1 size M_1	2 size M_2	3 size M_3	
1/4	0.88	...	...	...
3/8	1.01	0.92	...	...
1/2	1.17	1.13	...	1.50
3/4	1.36	1.24	1.24	2.00
1	1.56	1.49	...	2.50
1 1/4	1.77	1.65	...	3.00
1 1/2	1.89	1.80	1.80	3.50
2	2.06	2.03	2.03	4.00
2 1/2	3.25	...	...	...
3	3.69	3.69	...	...
4	4.38	...	...	...

* Dimensions in inches.

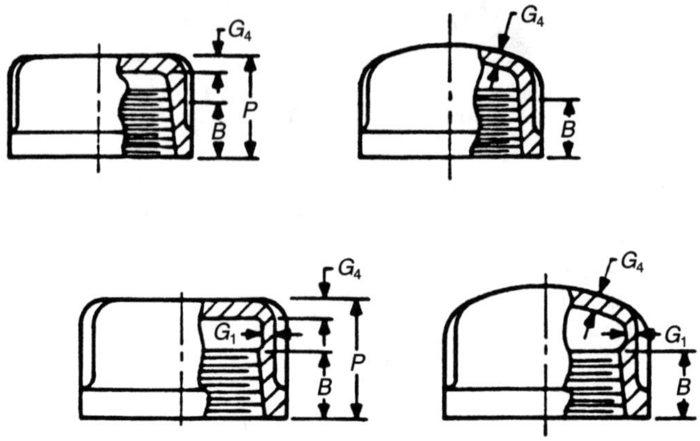

TABLE 12.52 Dimensions of Class 125 Caps*

Nominal pipe size	Length of thread		Metal thickness		Height of cap, P, min., cast
	B, min.	L_2, min.	G_1	G_4	
1/8	0.25	0.2639	0.08	0.09	0.49
1/4	0.32	0.4018	0.08	0.10	0.59
3/8	0.36	0.4078	0.09	0.11	0.64
1/2	0.43	0.5337	0.09	0.12	0.76
3/4	0.50	0.5457	0.10	0.13	0.84
1	0.58	0.6828	0.11	0.15	0.99
1 1/4	0.67	0.7068	0.12	0.17	1.10
1 1/2	0.70	0.7235	0.13	0.19	1.15
2	0.75	0.7565	0.15	0.22	1.32
2 1/2	0.92	1.1375	0.17	0.25	1.70
3	0.98	1.2000	0.19	0.29	1.80
4	1.08	1.3000	0.22	0.36	2.08

* Dimensions in inches.

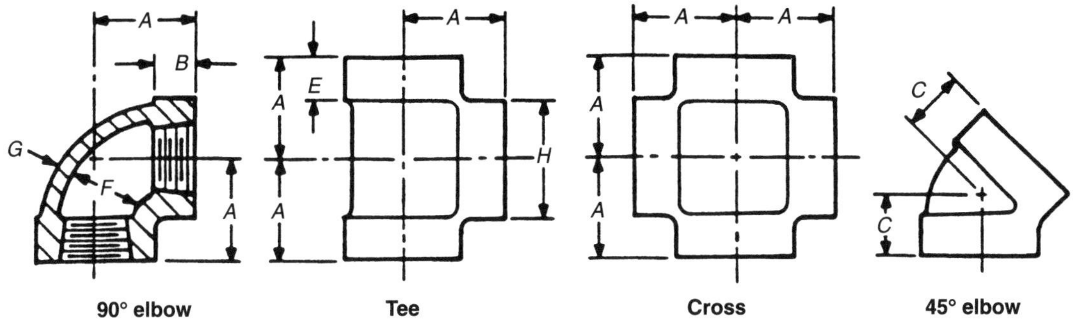

TABLE 12.53 Dimensions of Class 250 90° Elbows, Tees, and Crosses and 45° Elbows (Straight Sizes)*

Nominal pipe size	Center to end, elbows, tees, and crosses, A	Length of thread, B, min.	Center to end, 45° elbows, C	Width of band, E, min.	Inside diameter of fitting, F		Metal thickness, G	Outside diameter of band, H, min.
					Min.	Max.		
1/4	0.81	0.32	0.73	0.38	0.54	0.58	0.11	0.93
3/8	0.95	0.36	0.80	0.44	0.68	0.72	0.12	1.12
1/2	1.12	0.43	0.88	0.50	0.84	0.90	0.13	1.34
3/4	1.31	0.50	0.98	0.56	1.05	1.11	0.16	1.63
1	1.50	0.58	1.12	0.62	1.32	1.38	0.17	1.95
1 1/4	1.75	0.67	1.29	0.69	1.66	1.73	0.19	2.39
1 1/2	1.94	0.70	1.43	0.75	1.90	1.97	0.20	2.68
2	2.25	0.75	1.68	0.84	2.38	2.45	0.22	3.28
2 1/2	2.70	0.92	1.95	0.94	2.88	2.98	0.24	3.86
3	3.08	0.98	2.17	1.00	3.50	3.60	0.26	4.62
4	3.79	1.08	2.61	1.12	4.50	4.60	0.31	5.79

* Dimensions in inches.

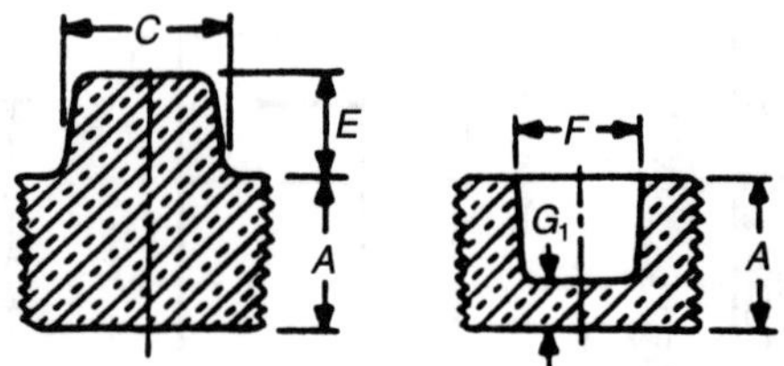

Solid pattern **Countersunk pattern**

TABLE 12.54 Dimensions of Class 250 Square Head and Square Socket Plugs*

Nominal pipe size	Thread length, A, min.	Nominal width across flats, C	Height of plug square, E, min.	Metal thickness, G_1	Nominal size of square socket, F
1/8	0.27	9/32	0.24	. . .	. . .
1/4	0.41	3/8	0.28	. . .	. . .
3/8	0.41	7/16	0.31	. . .	. . .
1/2	0.54	9/16	0.38	0.09	3/8
3/4	0.55	5/8	0.44	0.10	1/2
1	0.69	13/16	0.50	0.11	1/2
1 1/4	0.71	15/16	0.56	0.12	3/4
1 1/2	0.73	1 1/8	0.62	0.13	3/4
2	0.76	1 5/16	0.68	0.15	7/8
2 1/2	1.07	1 1/2	0.74	0.17	1 1/8
3	1.13	1 11/16	0.80	0.19	1 3/8
4	1.22	2 1/4	0.92	0.22	2

* Dimensions in inches.

TABLE 12.55 Pressure-Temperature Ratings for Cast Bronze Threaded Fittings (ANSI B16.15)

Temperature, °F	Class 125, psi	Class 250, psi
–20 to 150	200	400
200	190	385
250	180	365
300	165	335
350	150	300
400	125	250

Cast Copper Alloy Solder Joint Pressure Fittings (ANSI B16.18)

Scope. This specification is for cast copper alloy solder joint pressure fittings designed for use with copper water tube. The fittings shall be made from any one of the following coppers:

- C83600 ASTM B62
- C83800 or C84400 ASTM B584

Fitting Dimensions. Refer to the following tables for the fitting dimensions:

Table 12.56 Dimensions of Solder Joint Ends
Table 12.57 Dimensions of Elbows, Tees, and 45° Elbows
Table 12.58 Dimensions of Couplings
Table 12.59 Dimensions of Caps
Table 12.60 Dimensions of Solder Joint Elbows and Tees with Pipe Thread Ends
Table 12.61 Dimensions of Solder Joint Adapters and Fitting Adapters with Pipe Thread Ends
Table 12.62 Dimensions of Return Bends

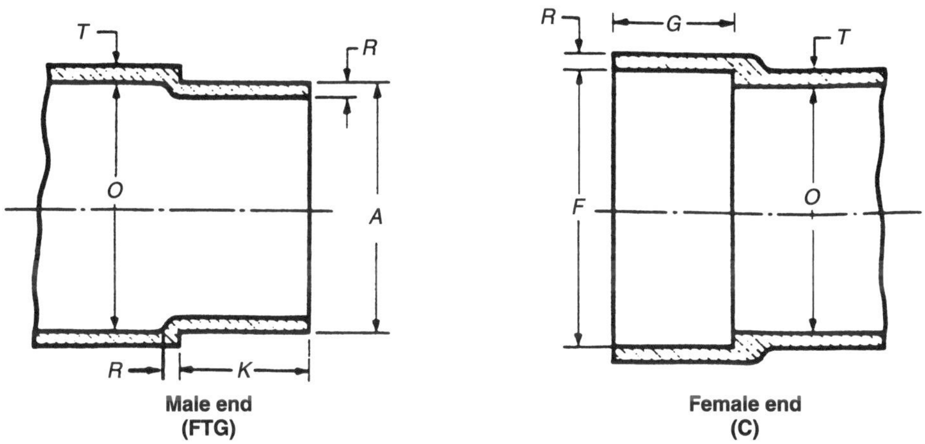

TABLE 12.56 Dimensions of Solder Joint Ends*

Standard water tube size	Male end: Outside diameter, *A*, Min.	Male end: Outside diameter, *A*, Max.	Male end: Length, *K*, min.	Female end: Inside diameter, *F*, Min.	Female end: Inside diameter, *F*, Max.	Female end: Depth, *G*, min.	Metal thickness: Body, *T*	Metal thickness: Joint, *R*	Inside diameter of fitting, *O*, min.
1/4	0.373	0.376	0.38	0.377	0.381	0.31	0.08	0.05	0.31
3/8	0.497	0.501	0.44	0.502	0.506	0.38	0.09	0.05	0.43
1/2	0.622	0.626	0.56	0.627	0.631	0.50	0.09	0.05	0.54
3/4	0.872	0.876	0.81	0.877	0.881	0.75	0.10	0.06	0.78
1	1.122	1.127	0.97	1.128	1.132	0.91	0.11	0.07	1.02
1 1/4	1.372	1.377	1.03	1.378	1.382	0.97	0.12	0.07	1.26
1 1/2	1.621	1.627	1.16	1.628	1.633	1.09	0.13	0.08	1.50
2	2.121	2.127	1.41	2.128	2.133	1.34	0.15	0.09	1.98
2 1/2	2.621	2.627	1.53	2.628	2.633	1.47	0.17	0.10	2.46
3	3.121	3.127	1.72	3.128	3.133	1.66	0.19	0.11	2.94
3 1/2	3.621	3.627	1.97	3.628	3.633	1.91	0.20	0.12	3.42
4	4.121	4.127	2.22	4.128	4.133	2.16	0.22	0.13	3.90
5	5.121	5.127	2.72	5.128	5.133	2.66	0.28	0.17	4.87
6	6.121	6.127	3.22	6.128	6.133	3.09	0.34	0.20	5.84
8	8.119	8.127	4.09	8.128	8.133	3.97	0.38	0.31	7.72
10	10.119	10.127	4.12	10.128	10.133	4.00	0.48	0.48	9.62
12	12.119	12.127	4.62	12.128	12.133	4.50	0.56	0.56	11.56

* Dimensions in inches.

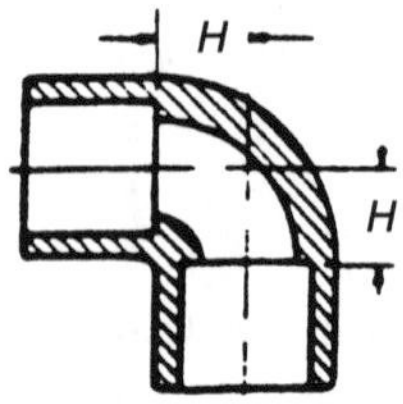

90° elbow
C × C

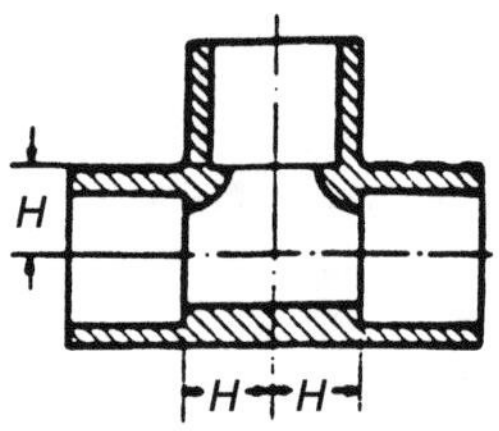

Tee
C × C × C

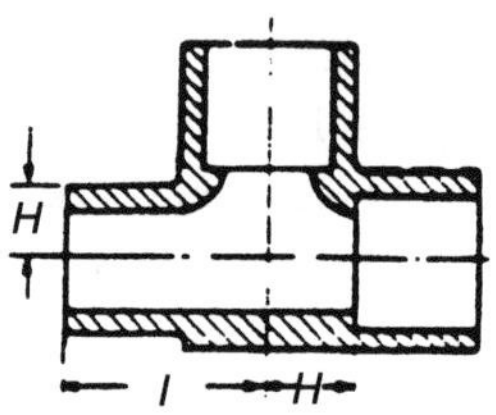

Fitting tee
FTG × C × C

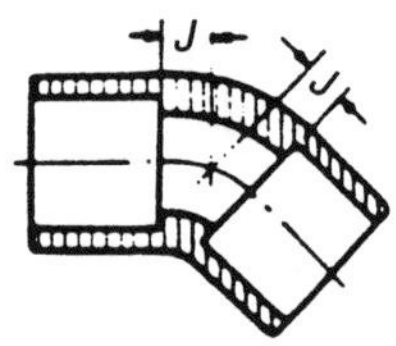

45° elbow
C × C

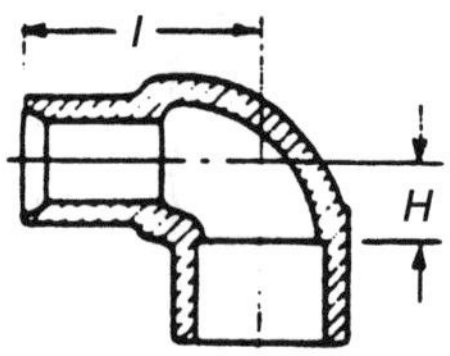

90° elbow
FTG × C

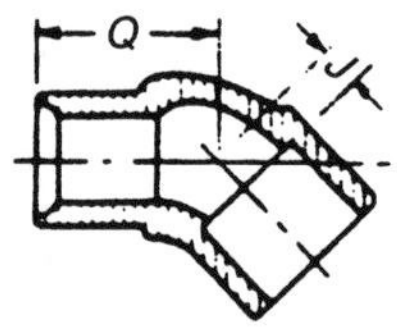

45° elbow
FTG × C

TABLE 12.57 Dimensions of Elbows, Tees, and 45° Elbows*

Standard water tube size	Laying length, tee and 90° elbow, *H*	Center to end, 90° elbow and tee, *I*	Laying length, 45° elbow, *J*	Center to end, 45° elbow, *Q*
1/4	0.25	0.75	...	...
3/8	0.31	0.88	0.19	0.75
1/2	0.44	1.12	0.19	0.88
3/4	0.56	1.50	0.25	1.19
1	0.75	1.84	0.31	1.31
1 1/4	0.88	2.03	0.44	1.56
1 1/2	1.00	2.28	0.50	1.75
2	1.25	2.78	0.56	2.12
2 1/2	1.50	3.16	0.62	...
3	1.75	3.59	0.75	...
3 1/2	2.00	...	0.88	...
4	2.25	4.59	0.94	...
5	3.12	...	1.44	...
6	3.62	...	1.62	...
8	4.88	...	2.12	...

* Dimensions in inches.

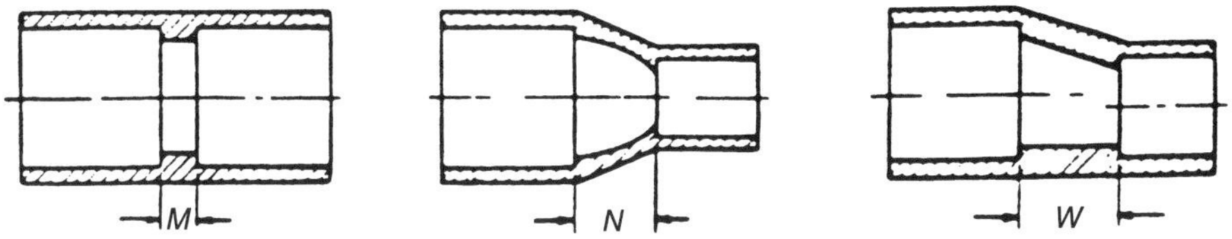

Coupling C × C **Reducing coupling** C × C **Eccentric coupling** C × C

TABLE 12.58 Dimensions of Couplings*

Straight		Reducing		Eccentric	
Standard water tube size	Laying length, *M*	Standard water tube size	Laying length, *N*	Standard water tube size	Laying length, *W*
1/4	0.06	3/4 × 1/2	0.31	3/4 × 1/2	0.62
3/8	0.06	1 × 3/4	0.38	1 × 3/4	0.69
1/2	0.12	1 1/4 × 1	0.38	1 1/4 × 1	0.75
3/4	0.12	1 1/4 × 3/4	0.38	1 1/4 × 3/4	0.75
1	0.12	1 1/2 × 1 1/4	0.38	1 1/2 × 1 1/4	0.69
1 1/4	0.12	1 1/2 × 1	0.38	1 1/2 × 1	0.69
1 1/2	0.12	1 1/2 × 3/4	0.44	2 × 1 1/2	1.12
2	0.19	2 × 1 1/2	0.50	2 × 1 1/4	0.94
2 1/2	0.19	2 × 1 1/4	0.50	2 1/2 × 2	1.19
3	0.19	2 × 1	0.50	3 × 2 1/2	1.25
3 1/2	0.25	2 × 3/4	0.50	3 × 2	1.31
4	0.25	2 1/2 × 2	0.56	4 × 3	2.00
5	0.25	2 1/2 × 1 1/2	0.56	...	...
6	0.25	2 1/2 × 1 1/4	0.62	...	...
8	0.62	2 1/2 × 1	0.62	...	...
...	...	3 × 2 1/2	0.62	...	...
...	...	3 × 2	0.62	...	...
...	...	4 × 3	0.69	...	...
...	...	4 × 2 1/2	1.12	...	...
...	...	4 × 2	1.19	...	...
...	...	6 × 4	1.31	...	...
...	...	8 × 6	1.38	...	...

* Dimensions in inches.

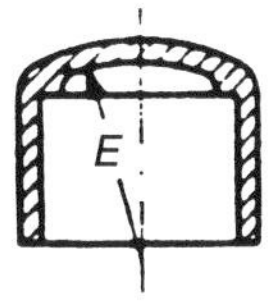

Cap C

TABLE 12.59 Dimensions of Caps*

Standard water tube size	Cap radius, *E*
1/4	0.38
3/8	0.50
1/2	0.62
3/4	0.88
1	1.12
1 1/4	1.38
1 1/2	1.62
2	2.12
2 1/2	2.62
3	3.12
3 1/2	3.62
4	4.12
5	5.12
6	6.12

* Dimensions in inches.

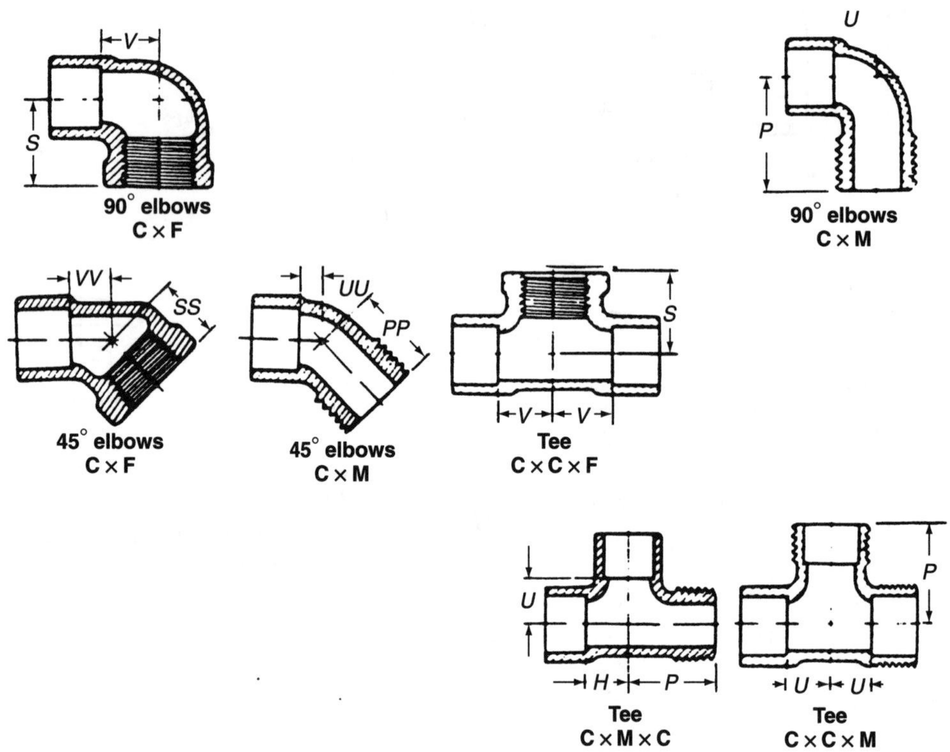

TABLE 12.60 Dimensions of Solder Joint Elbows and Tees with Pipe Thread Ends (Straight Sizes)*

Standard water tube and pipe thread size	Internal threads					External threads				
	Center to end		Laying length			Center to end		Laying length		
	Elbow and tee, *S*	45° elbow, *SS*	Elbow and tee, *V*	Tee, *H*	45° elbow, VV	Elbow and tee, *P*	45° elbow, *PP*	Elbow and tee, *U*	Tee, *H*	45° elbow, *UU*
1/4	0.56	...	0.38	...	...	0.94	...	0.25	...	...
3/8	0.69	0.69	0.44	0.31	0.19	1.06	0.81	0.31	0.31	0.19
1/2	0.88	0.94	0.56	0.44	0.19	1.31	1.00	0.44	0.44	0.19
3/4	1.00	1.00	0.69	0.56	0.25	1.50	1.19	0.56	0.56	0.25
1	1.25	1.19	0.88	0.75	0.31	1.81	1.31	0.75	0.75	0.31
1 1/4	1.50	...	1.00	0.88	...	2.00	...	0.88	...	...
1 1/2	1.62	...	1.12	1.00	...	2.19	...	1.00	...	...
2	1.94	...	1.38	1.25	...	2.62	...	1.25	...	...
2 1/2	2.50	...	1.62	...	...	...	...	...	...	...
3	2.81	...	1.94	...	...	...	...	...	...	...
4	3.44	...	2.44	...	...	...	...	...	...	...
6	4.88	...	3.88	...	...	...	...	...	...	...

* Dimensions in inches.

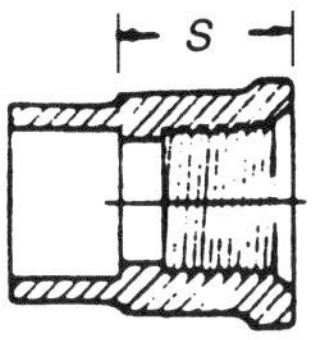

Adapter
C × F

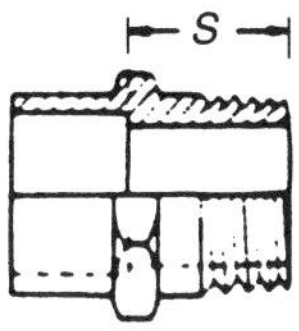

Adapter
C × M

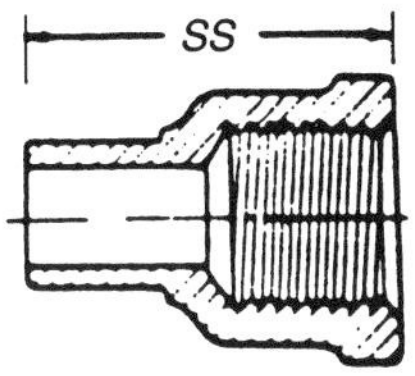

Fitting adapter
FTG × F

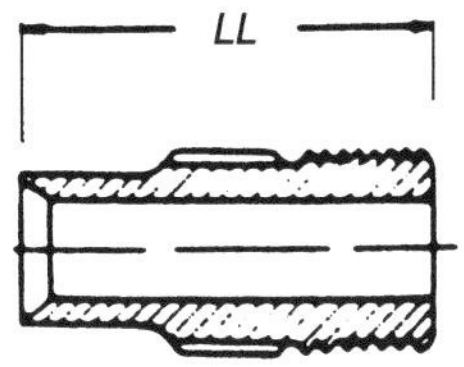

Fitting adapter
FTG × M

TABLE 12.61 Dimensions of Solder Joint Adapters and Fitting Adapters with Pipe Thread Ends*

Standard water tube size		Shoulder to end,	End to end	
Solder joint	Pipe thread	*S*	*SS*	*LL*
1/4	1/4	0.62	...	1.00
3/8	3/8	0.62	1.06	1.12
1/2	1/2	0.75	1.38	1.38
3/4	3/4	0.88	1.66	1.72
1	1	1.00	1.97	2.09
1 1/4	1 1/4	1.06	2.03	2.22
1 1/2	1 1/2	1.06	2.22	2.41
2	2	1.12	2.53	2.75
2 1/2	2 1/2	1.38	3.09	3.12
3	3	1.50	3.22	3.41
4	4	1.69	3.81	4.12
6	6	2.00	5.34	5.75
8	8	2.25	...	...

* Dimensions in inches.

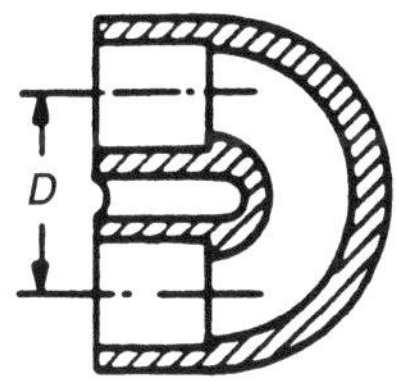

Return bend
C × C

TABLE 12.62 Dimensions of Return Bends (Straight Sizes)

Standard water tube size	Center to center, *D*, in
1/2	1.00
3/4	1.31
1	1.88
1 1/4	2.00
1 1/2	2.50
2	3.00
3	4.00
4	5.00

The size of the fittings scheduled in these tables is designated by the size of the openings in the sequence illustrated in Fig. 12.2.

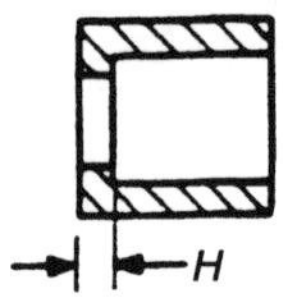

Flush bushing
FTG × C

TABLE 12.63 Dimensions of Flush Bushings

Standard water tube size	Laying length, H, in
¼ × ⅛	0.06
⅜ × ¼	0.06
½ × ⅜	0.09
½ × ¼	0.25
⅝ × ½	0.06
¾ × ⅝	0.19
¾ × ½	0.09
¾ × ⅜	0.44
1 × ¾	0.12
1 × ½	0.47
1¼ × 1	0.09
1½ × 1¼	0.16
2 × 1½	0.12

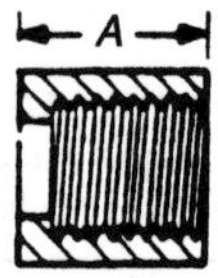

Flush bushing
FTG × F

TABLE 12.64 Dimensions of Flush Bushings

Standard water tube size	Laying length, A, in
½ × ⅛	0.56
¾ × ⅜	0.81
1 × ½	0.97
1¼ × ¾	1.03
1½ × 1	1.16
2 × 1½	1.41

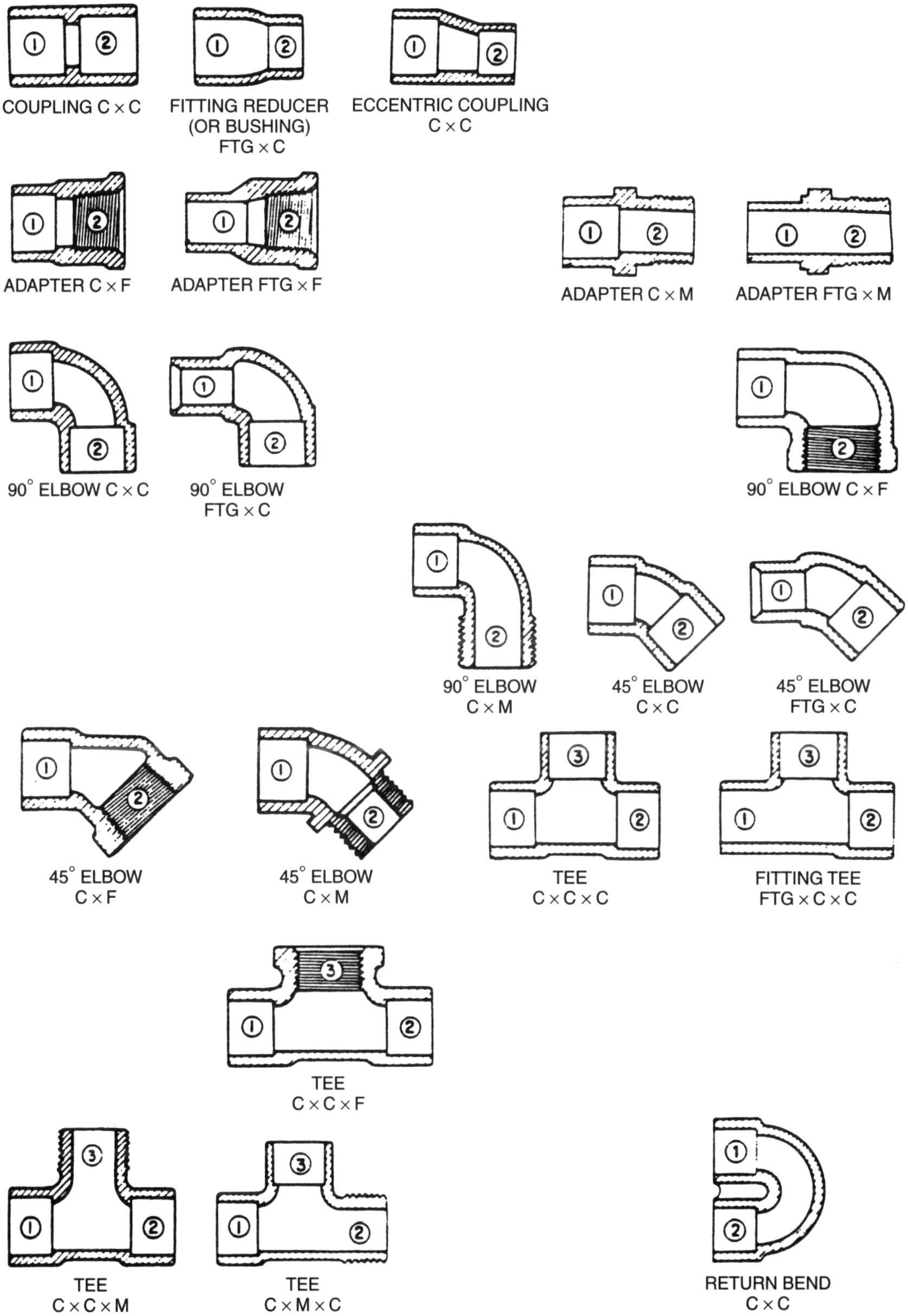

FIGURE 12.2 Method of designating openings of reducing fittings (ANSI B16.18). Fittings are designated by size in the order shown—i.e., 1 × 2 × 3.

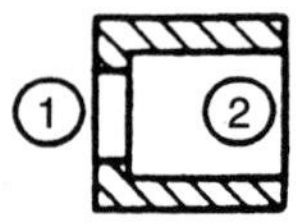

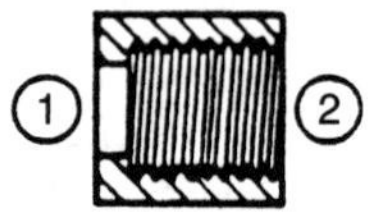

FIGURE 12.2 Method of designating openings of reducing fittings (ANSI B16.18). Fittings are designated by size in the order shown—i.e., 1 × 2 × 3. (Continued)

COPPER AND COPPER ALLOY JOINTS

Refer to Chap. 15.

Tables in Chapter 12

Tables in Chapter 12 (Continued)

CHAPTER 13
ALUMINUM AND ALUMINUM ALLOY PIPE AND TUBE, FITTINGS, FLANGES, AND JOINTS

See p. 496 for a list of tables in Chap. 13.

ALUMINUM AND ALUMINUM ALLOY PIPE

Seamless and welded pipe is manufactured in accordance with the following ASTM specification:

ASTM B241   Aluminum and Aluminum-Alloy Seamless Pipe and Seamless Extruded Tube.

Table 13.1 lists the alloys, tempers, pipe sizes, and tensile properties of the seamless pipe manufactured per ASTM B241M.

ASTM B241 Aluminum Alloy Pipe Dimensions and Weight

The standard sizes of seamless pipe are according to ANSI H35.2/H35.2M. Table 13.2 lists the nominal pipe size, outside diameter, pipe wall thickness, and weight per foot of pipe length corresponding to the pipe schedule.

Tolerance for Outside Diameter of ASTM B241 Pipe

Table 13.3 provides tolerances for the outside diameter of ASTM B241 pipe listed in Table 13.2.

Permissible Variation for Pipe Wall Thickness of ASTM B241 Pipe

See Table 13.4.

Permissible Variations in Weight of ASTM B241 Pipe

See Table 13.5.

Permissible Variation in Length of ASTM B241 Plain-End Pipe

The following variations in length are allowed:

- 0.250 in for lengths up to 20 ft.
- 0.50 in for lengths over 20 through 40 ft.

TABLE 13.1 ASTM B241 Pipe Sizes, Aluminum Alloys, Tempers, and Tensile Properties

Aluminum alloy	Temper	Pipe size	Minimum tensile strength, ksi	Minimum tensile strength, MPa	Minimum yield strength, ksi	Minimum yield strength, MPa
3003	H18	Under 1	27.0	185	24.0	165
3003	H112	1 and over	14.0	95	5.0	35
6061	T6	Under 1	42.0*	290*	35.0	240
6061	T6	1 and over	38.0	260	35.0	240
6063	T6	All	30.0	205	25.0	170

* Applies to extruded and drawn pipe. For extruded-only pipe, the tensile strength is minimum 38.0 ksi (260 MPa).

PIPING COMPONENTS OTHER THAN PIPE

Piping components other than pipe may be manufactured from the following ASTM material specifications, as permitted by the governing code or standard of jurisdiction:

- B26 Aluminum Alloy Sand Castings
- B108 Aluminum Alloy Mold Castings
- B209 Aluminum and Aluminum Alloy Sheet and Plate
- B210 Aluminum and Aluminum Alloy Drawn Seamless Tubes
- B211 Aluminum Alloy Bar, Rod, and Wire
- B221 Aluminum and Aluminum Alloy Extruded Bars, Rods, and Shapes
- B234 Aluminum and Aluminum Alloy Drawn Seamless Tubes for Condensers and Heat Exchangers
- B247 Aluminum Alloy Die and Hand Forgings

ASTM B241 ALUMINUM AND ALUMINUM ALLOY SEAMLESS EXTRUDED TUBE

Diameter and Tolerances for ASTM B241 Seamless Extruded Tube

See Table 13.6.

Wall Thickness and Tolerances for ASTM B241 Seamless Extruded Tube

See Table 13.7.

Length and Tolerances for ASTM B241 Seamless Extruded Tube

See Table 13.8.

TABLE 13.2 Diameters, Wall Thickness, and Weights of ASTM B241 Aluminum Alloy Pipe

Nominal pipe size (NPS)	Schedule number	Nominal outside diameter, in	Nominal wall thickness, in	Nominal inside diameter, in	Weight, lb/ft*
1/8	40	0.405	0.068	0.269	0.085
	80	0.405	0.095	0.215	0.109
1/4	40	0.540	0.088	0.364	0.147
	80	0.540	0.119	0.302	0.185
3/8	40	0.675	0.091	0.493	0.196
	80	0.675	0.126	0.423	0.256
1/2	5	0.840	0.065	0.710	0.186
	10	0.840	0.083	0.674	0.232
	40	0.840	0.109	0.622	0.294
	80	0.840	0.147	0.546	0.376
	160	0.840	0.188	0.464	0.453
3/4	5	1.050	0.065	0.920	0.237
	10	1.050	0.083	0.884	0.297
	40	1.050	0.113	0.824	0.391
	80	1.050	0.154	0.742	0.510
	160	1.050	0.219	0.612	0.672
1	5	1.315	0.065	1.185	0.300
	10	1.315	0.109	1.097	0.486
	40	1.315	0.133	1.049	0.581
	80	1.315	0.179	0.957	0.751
	160	1.315	0.250	0.815	0.984
1 1/4	5	1.660	0.065	1.530	0.383
	10	1.660	0.109	1.442	0.625
	40	1.660	0.140	1.380	0.786
	80	1.660	0.191	1.278	1.037
	160	1.660	0.250	1.160	1.302
1 1/2	5	1.900	0.065	1.770	0.441
	10	1.900	0.109	1.682	0.721
	40	1.900	0.145	1.610	0.940
	80	1.900	0.200	1.500	1.256
	160	1.900	0.281	1.338	1.681
2	5	2.375	0.065	2.245	0.555
	10	2.375	0.109	2.157	0.913
	40	2.375	0.154	2.067	1.264
	80	2.375	0.218	1.939	1.737
	160	2.375	0.344	1.687	2.581
2 1/2	5	2.875	0.083	2.709	0.856
	10	2.875	0.120	2.635	1.221
	40	2.875	0.203	2.469	2.004
	80	2.875	0.276	2.323	2.650
	160	2.875	0.375	2.125	3.464
3	5	3.500	0.083	3.334	1.048
	10	3.500	0.120	3.260	1.498
	40	3.500	0.216	3.068	2.621
	80	3.500	0.300	2.900	3.547
	160	3.500	0.438	2.624	4.955
3 1/2	5	4.000	0.083	3.834	1.201
	10	4.000	0.120	3.760	1.720
	40	4.000	0.226	3.548	3.151
	80	4.000	0.318	3.364	4.326

TABLE 13.2 Diameters, Wall Thickness, and Weights of ASTM B241 Aluminum Alloy Pipe (Continued)

Nominal pipe size (NPS)	Schedule number	Nominal outside diameter, in	Nominal wall thickness, in	Nominal inside diameter, in	Weight, lb/ft*
4	5	4.500	0.083	4.334	1.354
	10	4.500	0.120	4.260	1.942
	40	4.500	0.237	4.026	3.733
	80	4.500	0.337	3.826	5.183
	120	4.500	0.438	3.624	6.573
	160	4.500	0.531	3.438	7.786
5	5	5.563	0.109	5.345	2.196
	10	5.563	0.134	5.295	2.688
	40	5.563	0.258	5.047	5.057
	80	5.563	0.375	4.813	7.188
	120	5.563	0.500	4.563	9.353
	160	5.563	0.625	4.313	11.400
6	5	6.625	0.109	6.407	2.264
	10	6.625	0.134	6.357	3.213
	40	6.625	0.280	6.065	6.264
	80	6.625	0.432	5.761	9.884
	120	6.625	0.562	5.501	12.590
	160	6.625	0.719	5.187	15.690
8	5	8.625	0.109	8.407	3.429
	10	8.625	0.148	8.329	4.635
	20	8.625	0.250	8.125	7.735
	30	8.625	0.277	8.071	8.543
	40	8.625	0.322	7.981	9.878
	60	8.625	0.405	7.813	12.330
	80	8.625	0.500	7.625	15.010
	100	8.625	0.594	7.437	17.620
	120	8.625	0.719	7.187	21.000
	140	8.625	0.812	7.001	23.440
	160	8.625	0.906	6.813	25.840
10	5	10.750	0.134	10.482	5.256
	10	10.750	0.165	10.420	6.453
	20	10.750	0.250	10.250	9.698
	30	10.750	0.307	10.136	11.840
	40	10.750	0.365	10.020	14.000
	60	10.750	0.500	9.750	18.930
	80	10.750	0.594	9.562	22.290
	100	10.750	0.719	9.312	26.850
12	5	12.750	0.156	12.438	7.258
	10	12.750	0.180	12.390	8.359
	20	12.750	0.250	12.250	11.550
	30	12.750	0.330	12.090	15.140
	40	12.750	0.406	11.938	18.520
	60	12.750	0.562	11.626	25.310
	80	12.750	0.688	11.374	30.660

* Based on nominal dimensions, plain ends, and a density of 0.098 lb/in^3, the density of 6061 and 6063 alloys. For alloy 3003 multiply by 1.01.

TABLE 13.3 Tolerances for Nominal Outside Diameter of ASTM B241 Pipe

Nominal pipe size (NPS)	Tolerance for nominal outside diameter, in or % of nominal outside diameter			
	Schedules 5 and 10		Schedules 20 and greater	
Under 2	+0.015	–0.031	+0.015	–0.031
2–4	+0.031	–0.031	+1%	–1%
5–7	+0.062	–0.031	+1%	–1%
8–12	+0.093	–0.031	+1%	–1%

TABLE 13.4 Tolerances for Pipe Wall Thickness of ASTM B241 Pipe

Nominal pipe size (NPS)	Pipe schedule	Pipe wall thickness tolerance, % of nominal wall thickness	
		Plus	Minus
All sizes	5 and 10	+12.5%	–12.5% (1)
	20 and greater	(2)	–12.5%

Notes:
(1) In no case shall the minimum tolerance for pipe wall thickness be less than ±0.012 in.
(2) Maximum wall thickness shall be controlled by weight tolerance.

TABLE 13.5 Permissible Variations in Weight of ASTM B241 Pipe

Schedule number	Allowable variation from nominal weight
5 and 10	(1)
20 and greater	+8% (2)

Notes:
(1) Weight tolerance does not apply to Schedule 5 and 10 pipe. The outside diameter, pipe wall thickness, and length tolerances govern Schedule 5 and 10 pipe.
(2) Minimum weight is controlled by the outside diameter and pipe wall thickness tolerances.

TABLE 13.6 ASTM B241 Extruded Round Tube Diameter and Tolerances
(Except for T3510, T4510, T6510, T73510, and T8510 tempers)

Specified diameter, in	Allowable deviation of mean diameter from specified diameter		Allowable deviation of diameter at any point from specified diameter	
	Alloys 5083, 5086, and 5454	Other alloys	Alloys 5083, 5086, and 5454	Other alloys
0.500–0.999	0.015	0.010	0.030	0.020
1.000–1.999	0.018	0.012	0.038	0.025
2.000–3.999	0.023	0.015	0.045	0.030
4.000–5.999	0.038	0.025	0.075	0.050
6.000–7.999	0.053	0.035	0.113	0.075
8.000–9.999	0.068	0.045	0.150	0.100
10.000–11.999	0.083	0.055	0.188	0.125
12.000–13.999	0.098	0.065	0.225	0.150
14.000–15.999	0.113	0.075	0.263	0.175
16.000–17.999	0.128	0.085	0.300	0.200

TABLE 13.7 ASTM B241 Extruded Round Tube Wall Thickness and Tolerances

	Tolerance, in (plus and minus) (allowable deviation of mean wall thickness from specified wall thickness)								Allowable deviation of wall thickness at any point from mean wall thickness (eccentricity)
	Outside diameter, in								
	Under 1.250		1.250–2.999		3.000–4.999		5.000 and over		
Specified wall thickness, in	Alloys 5083, 5086, 5454	Other alloys	Alloys 5083, 5086, 5454	Other alloys	Alloys 5083, 5086, 5454	Other alloys	Alloys 5083, 5086, 5454	Other alloys	All alloys
Under 0.047	0.009	0.006							
0.047–0.061	0.011	0.007	0.012	0.008	0.012	0.008	0.015	0.010	
0.062–0.077	0.012	0.008	0.012	0.008	0.014	0.009	0.018	0.012	±10% of mean wall thickness
0.078–0.124	0.014	0.009	0.014	0.009	0.015	0.010	0.023	0.015	
0.125–0.249	0.014	0.009	0.014	0.009	0.020	0.013	0.030	0.020	
0.250–0.374	0.017	0.011	0.017	0.011	0.024	0.016	0.038	0.025	
0.375–0.499			0.023	0.015	0.032	0.028	0.053	0.035	
0.500–0.749			0.030	0.020	0.042	0.028	0.068	0.045	max. ± 0.060, min. ± 0.010
0.750–0.999					0.053	0.035	0.083	0.055	
1.000–1.499					0.068	0.045	0.098	0.065	
1.500–2.000							0.113	0.075	
2.001–2.499							0.128	0.085	
2.500–2.999							0.143	0.095	±0.120
3.000–4.000							0.158	0.105	
							0.173	0.115	

TABLE 13.8 ASTM B241 Extruded Round Tube Length and Tolerances

	Tolerance, in (plus except as noted allowable deviation from specified length)							
	Straight				Coiled			
	Specified Length—ft.							
Specified outside diameter, in	Up through 12	Over 12 through 30	Over 30 through 50	Over 50	Up through 100	Over 100 to 250	250 to 500	500 and over
0.500–1.249	1/8	1/4	3/8	1	+5%, −0%	±10%	±15%	±20%
1.250–2.999	1/8	1/4	3/8	1				
3.000–7.999	3/16	5/16	7/16	1				
8.000 and over	1/4	3/8	1/2	1				

ASTM B234 DRAWN SEAMLESS TUBES FOR CONDENSERS AND HEAT EXCHANGERS

Aluminum and aluminum alloy drawn seamless tubes for condensers and heat exchangers are manufactured in accordance with the following ASTM specification:

B234 Aluminum and Aluminum Alloy Drawn Seamless Tubes for Condensers and Heat Exchangers

ASTM B234 Drawn Seamless Tube Dimensions

See Tables 13.9 to 13.12.

TABLE 13.9 Outside Diameter and Tolerance for Non-Heat-Treatable Tube per ASTM B234

Outside diameter of tube, in	Tolerance, in, plus and minus
Up through 0.500	0.0020
0.501–0.749	0.0025
0.750–1.000	0.0030
1.001–1.250	0.0035
1.251–2.000	0.0040

TABLE 13.10 Outside Diameter and Tolerance for Heat-Treatable Tube per ASTM B234

Outside diameter of tube, in	Tolerance, in, plus and minus
Up through 0.500	0.003
0.501–1.000	0.004
1.001–1.500	0.005
1.501–2.000	0.006

TABLE 13.11 Wall Thickness and Tolerances for Drawn Seamless Tube per ASTM B234

Outside diameter, in	Wall thickness tolerance, in	Tolerance for mean wall thickness of tube, in
Up through 0.035	0.006	0.004
0.036–0.049	0.009	0.006
0.050–0.058	0.010	0.007
0.059–0.065	0.012	0.008
0.066–0.072	0.013	0.009
0.073–0.083	0.015	0.010
0.084–0.095	0.017	0.011
0.096–0.109	0.020	0.013
0.110–0.120	0.022	0.014
0.121–0.134	0.024	0.016
0.135–0.145	0.026	0.017
0.146–0.154	0.028	0.018
0.155–0.179	0.032	0.021
0.180–0.200	0.036	0.024

TABLE 13.12 Length Tolerances for Drawn Seamless Tube per ASTM B234

Specified length, ft	Tolerance, in (plus)
Up through 15	3/32
Over 15–20	1/8
Over 20–30	5/16
Over 30–36	1/4

ASTM B210 ALUMINUM AND ALUMINUM ALLOY DRAWN SEAMLESS TUBE

ASTM B210 specification covers drawn seamless tubes for general pressure applications. Coiled tubes are available only in round tubes with a wall thickness not exceeding 0.083 in. They are made from non-heat-treatable alloys only.

Diameter and Tolerances for ASTM B210 Drawn Seamless Tube

See Table 13.13.

Wall Thickness and Tolerances for ASTM B210 Drawn Seamless Tube

See Table 13.14.

TABLE 13.13 Diameter and Tolerances for ASTM B210 Drawn Seamless Round Tube

		Allowable deviation of diameter at any point from specified diameter, in		
Specified diameter, in	Allowable deviation of mean diameter from specified diameter, in	Nonannealed and non-heat-treated tube	Heat-treated tube	Annealed tube
Up through 0.500	0.003	0.003	0.006	0.018
0.501–1.000	0.004	0.004	0.008	0.024
1.001–2.000	0.005	0.005	0.010	0.030
2.001–3.000	0.006	0.006	0.012	0.036
3.001–4.000	0.008	0.008	0.016	0.048
5.001–5.000	0.010	0.010	0.020	0.060
6.001–8.000	0.015	0.015	0.030	0.090
8.001–10.000	0.020	0.020	0.040	0.120
10.001–12.000	0.025	0.025	0.050	0.150

TABLE 13.14 Wall Thickness and Tolerances for ASTM B210 Drawn Seamless Round Tube

	Tolerance, in (plus and minus)	
Specified thickness, in	Allowable deviation of mean wall thickness* from specified wall thickness (round, non-heat-treatable alloys)	Allowable deviation of wall thickness at any point from specified wall thickness (eccentricity)
0.010–0.035	0.002	0.002
0.036–0.049	0.003	0.003
0.050–0.083	0.004	0.004
0.084–0.120	0.005	0.006
0.121–0.203	0.006	0.008
0.204–0.300	0.008	0.012
0.301–0.375	0.015	0.020
0.376–0.500	0.020	0.030

* Mean wall thickness is equal to one-half of the maximum and minimum wall thickness measured at the opposite ends of a tube diameter.

ALUMINUM ALLOY FLANGES

There is no ASME or ANSI standard published for aluminum alloy flanges. Appendix L, Aluminum Alloy Pipe Flanges, to ASME B31.3, Process Piping Code, is commonly used for aluminum alloy flanges.

Types of Flanges

The following types are covered by Appendix L to ASME B31.3:

- Welding Neck
- Slip-on
- Socket Welding
- Lapped
- Blind

Pressure Classes and Pressure-Temperature Ratings

Aluminum alloy flanges are manufactured in:

- Class 150 (PN 20)
- Class 300 (PN 50)
- Class 600 (PN 110)

Pressure-Temperature Ratings of Class 150 Flanges

Table 13.15 provides the pressure-temperature ratings of Class 150 (PN 20) flanges in U.S. Customary and metric units.

Pressure-Temperature Ratings of Class 300 Flanges

Table 13.16 provides the pressure-temperature ratings of Class 300 (PN 50) flanges in U.S. Customary and metric units.

TABLE 13.15 Pressure-Temperature Ratings of Class 150 (PN 20) Flanges*

ASTM B247 alloy and temper	Pressure at 100°F (38°C)		Pressure at 150°F (66°C)		Pressure at 200°F (93°C)		Pressure at 250°F (121°C)	
	psig	kPa	psig	kPa	psig	kPa	psig	kPa
3003-H112	40	275	40	275	35	240	35	240
6061-T6†	275	1895	270	1895	265	1825	260	1795
6061-T6‡	180	1240	180	1240	175	1205	170	1170

* These flanges can be used to the minimum temperature of –452°F. The ratings at 100°F (38°C) also apply to temperatures below 100°F (38°C) to –452°F.
† These ratings apply to welding neck, lapped, and blind flanges.
‡ These ratings apply to slip-on and socket welding flanges.

TABLE 13.16 Pressure-Temperature Ratings of Class 300 (PN 50) Flanges*

ASTM B247 alloy and temper	Pressure at 100°F (38°C)		Pressure at 150°F (66°C)		Pressure at 200°F (93°C)		Pressure at 250°F (121°C)	
	psig	kPa	psig	kPa	psig	kPa	psig	kPa
3003-H112	105	725	100	690	95	655	95	655
6061-T6†	720	4965	710	4895	700	4825	675	4651
6061-T6‡	480	3310	470	3240	465	3205	450	3105

* These flanges can be used to the minimum temperature of –452°F. The ratings at 100°F (38°C) also apply to temperatures below 100°F (38°C) to –452°F.
† These ratings apply to welding neck, lapped, and blind flanges.
‡ These ratings apply to slip-on and socket welding flanges.

Pressure-Temperature Ratings of Class 600 (PN 110) Flanges

Table 13.17 provides the pressure-temperature ratings of Class 600 (PN 110) flanges in U.S. Customary and metric units.

Dimensions of Aluminum Alloy Flanges

The aluminum alloy flanges shall have dimensions conforming to ASME Class 150, Class 300, and Class 600 flanges according ASME B16.5. For dimensions of ASME B16.5 flanges, refer to Chap. 6.

Materials for Aluminum Alloy Flanges

Flanges are forged in accordance with ASTM B247, and they are manufactured in the alloys and tempers in Table 13.18.

TABLE 13.17 Pressure-Temperature Ratings of Class 600 (PN 110) Flanges*

ASTM B247 alloy and temper	Pressure at 100°F (38°C)		Pressure at 150°F (66°C)		Pressure at 200°F (93°C)		Pressure at 250°F (121°C)	
	psig	kPa	psig	kPa	psig	kPa	psig	kPa
3003-H112	205	1415	200	1380	195	1345	185	1275
6061-T6†	1440	9930	1420	9790	1400	9655	1355	9345
6061-T6‡	960	6620	945	6515	930	6410	900	6205

* These flanges can be used to the minimum temperature of –452°F. The ratings at 100°F (38°C) also apply to temperatures below 100°F (38°C) to –452°F.
† These ratings apply to welding neck, lapped, and blind flanges.
‡ These ratings apply to slip-on and socket welding flanges.

TABLE 13.18 ASTM B247 Aluminum Alloys and Tempers for Flanges

Aluminum alloy	Temper
3003	H112
6061	T6

ALUMINUM ALLOY PIPING JOINTS

Refer to Chap. 15.

Tables in Chapter 13

CHAPTER 14

TITANIUM AND TITANIUM ALLOY PIPE AND TUBE, ZIRCONIUM AND ZIRCONIUM ALLOY PIPE AND TUBE

See p. 501 for a list of tables in Chap. 14.

TITANIUM AND TITANIUM ALLOY PIPE

Seamless and welded pipe is manufactured in accordance with the following ASTM specification:

ASTM B337 Seamless and Welded Titanium and Titanium Alloy Pipe

Titanium and titanium alloy pipe is commercially available in Grades 1, 2, 3, 7, 9, and 12. Grade 9 pipe may be furnished in either the annealed or stress-relief-annealed condition, while all other grades of titanium and titanium alloy pipe are furnished in annealed condition.

Pipe Dimensions

Table 14.1 lists the nominal pipe size (NPS), outside diameters, and the nominal wall thickness for various pipe sizes and Schedules 5S, 10S, 40S, and 80S. These dimensions are in accordance with the ASME B36.19, Stainless Steel Pipe.

TABLE 14.1 Dimensions of Welded and Seamless Pipe*

Nominal pipe size	Outside diameter, in	Nominal wall thickness, in			
		Schedule 5S	Schedule 10S	Schedule 40S	Schedule 80S
1/8	0.405	. . .	0.049	0.068	0.095
1/4	0.540	. . .	0.065	0.088	0.119
3/8	0.675	. . .	0.065	0.091	0.126
1/2	0.840	0.065	0.083	0.109	0.147
3/4	1.050	0.065	0.083	0.113	0.154
1	1.315	0.065	0.109	0.133	0.179
1 1/4	1.660	0.065	0.109	0.140	0.191
1 1/2	1.900	0.065	0.109	0.145	0.200
2	2.375	0.065	0.109	0.154	0.218
2 1/2	2.875	0.083	0.120	0.203	0.276
3	3.500	0.083	0.120	0.216	0.300
3 1/2	4.000	0.083	0.120	0.226	0.318
4	4.500	0.083	0.120	0.237	0.337
5	5.563	0.109	0.134	0.258	0.375
6	6.625	0.109	0.134	0.280	0.432
8	8.625	0.109	0.148	0.322	0.500
10	10.750	0.134	0.165	0.365	0.500
12	12.750	0.156	0.180	0.375	0.500

* Dimensions listed in this table are in accordance with ASME B36.19, Stainless Steel Pipe. The dimensions in this table apply to pipes manufactured per ASTM B337 and B658 (titanium and zirconium pipe).

Permissible Variations for Pipe Outside Diameter

Tables 14.2 and 14.2M list the acceptable variations for the outside diameter of pipe manufactured according to ASTM B337.

Permissible Variation for Pipe Wall Thickness

The wall thickness of pipe at any point shall not vary more than ±12.5 percent of the nominal wall thickness specified.

Permissible Variations in Length

Pipe length shall not be under the specified length, and not more than $\frac{1}{4}$ in (6.4 mm) over the specified length.

Piping Components Other Than Pipe

Piping components other than pipe may be manufactured from the following ASTM material specifications, as permitted by the governing code or standard of jurisdiction:

- B265 Titanium and Titanium Alloy Strip, Sheet, and Plate
- B348 Titanium and Titanium Alloy Bars and Billets
- B367 Titanium and Titanium Alloy Castings
- B381 Titanium and Titanium Alloy Forgings

TABLE 14.2 Permissible Variations in Pipe Outside Diameter (U.S. Customary Units)*

Nominal pipe size (NPS)	Permissible variations in outside diameter, in	
	Over	Under
$\frac{1}{8}$ to $1\frac{1}{2}$, incl.	$\frac{1}{64}$	$\frac{1}{32}$
Over $1\frac{1}{2}$ to 4, incl.	$\frac{1}{32}$	$\frac{1}{32}$
Over 4 to 8, incl.	$\frac{1}{16}$	$\frac{1}{32}$
Over 8 to 12, incl.	$\frac{3}{32}$	$\frac{1}{32}$

* Dimensions apply to ASTM B337 and B658 pipe.

TABLE 14.2M Permissible Variations in Pipe Outside Diameter (Metric Units)*

Nominal pipe size (NPS)	Permissible variations in outside diameter, mm	
	Over	Under
$\frac{1}{8}$ to $1\frac{1}{2}$, incl.	0.397	0.794
Over $1\frac{1}{2}$ to 4, incl.	0.794	0.794
Over 4 to 8, incl.	1.588	0.794
Over 8 to 12, incl.	2.382	0.794

* Dimensions apply to pipes manufactured according to ASTM B337 and B658.

TITANIUM AND TITANIUM ALLOY TUBE

Titanium and titanium alloy seamless and welded tubes for condensers and heat exchangers are manufactured in accordance with the following ASTM specification:

- B338 Seamless and Welded Titanium and Titanium Alloy Tubes for Condensers and Heat Exchangers

Tube Dimensions

Tubes are specified by outside diameter and the wall thickness, as required. Therefore, no table is included for tube dimensions in this chapter.

Titanium and Titanium Alloy tubes according to ASTM B338 are furnished in several grades. These include Grades 1, 2, 3, 7, 9, and 11 through 18. All grades of tube shall be furnished in annealed condition with the exception that Grades 9 and 18 can be furnished either in annealed or cold-worked and stress-relieved conditions, as specified.

Permissible Variations for Tube Dimensions

Tolerance for Outside Diameter. Refer to Tables 14.3 and 14.3M for tubes complying with ASTM B338.

Tube Wall Thickness Tolerance. Tolerances for tube wall thickness are listed in Table 14.4.

Tolerance for Straightness. Tubes, manufactured according to ASTM B338, shall be free of bends or kinks, and the maximum uniform bow shall not exceed the limits shown in Tables 14.5 and 14.5M.

TABLE 14.3 Outside Diameter Tolerance for B338 and B523 Tubes (U.S. Customary Units)

Outside diameter, in	Outside diameter tolerance, in
Under 1, excl.	±0.004
1 to 1½, excl.	±0.005
1½ to 2, excl.	±0.006
2 to 2½, excl.	±0.007
2½ to 3½, excl.*	±0.010

* For sizes equal to or greater than 3½, tolerances shall be as agreed upon by the purchaser and the manufacturer.

TABLE 14.3M Outside Diameter Tolerance for B338 and B523 Tubes (Metric Units)

Outside diameter, mm	Outside diameter tolerance, mm
Under 25.4, excl.	±0.102
25.4 to 38.1, excl.	±0.127
38.1 to 50.8, excl.	±0.152
50.8 to 63.5, excl.	±0.175
63.5 to 88.9, excl.*	±0.254

* For sizes equal to or greater than 88.9, tolerances shall be as agreed between the purchaser and the manufacturer.

TABLE 14.4 Tolerance for Wall Thickness of B338 and B523 Tubes*

Outside diameter, in	Permissible variations in wall thickness t, %
Under 1, excl.	10
1 to 1½, excl.	10
1½ to 2, excl.	10
2 to 2½, excl.	10
2½ to 3½, excl.	10

* When tubes are ordered by minwall, tolerances are all plus and shall be double the values shown.

TABLE 14.5 Straightness Tolerance for B338 and B523 Tubes (U.S. Customary Units)

Length, ft	Maximum curvature, depth of arc
Over 3 to 6, incl.	⅛ in
Over 6 to 8, incl.	3/16 in
Over 8 to 10, incl.	¼ in
Over 10	¼ in/any 10 ft

TABLE 14.5M Straightness Tolerance for B338 and B523 Tubes (Metric Units)

Length, mm	Maximum curvature, depth of arc
Over 0.91 to 1.83, incl.	3.2 mm
Over 1.83 to 2.44, incl.	4.8 mm
Over 2.44 to 3.05, incl.	6.4 mm
Over 3.05	2.1 mm/m

ZIRCONIUM AND ZIRCONIUM ALLOY PIPE

Zirconium and zirconium alloy seamless and welded pipe is manufactured in accordance with the following ASTM specification:

- ASTM B658 Seamless and Welded Zirconium and Zirconium Alloy Pipe

Titanium and titanium alloy pipe is commercially available in following grades:

- Grade R60702, unalloyed zirconium
- Grade R60704, zirconium-tin alloy
- Grade R60705, zirconium-niobium alloy

Pipe shall be furnished in either the annealed or stress-relieved condition.

Pipe Dimensions

Table 14.1 applies to the pipe dimensions of zirconium and zirconium alloy pipe also. Table 14.1 lists the nominal pipe size (NPS), outside diameters, and the nominal wall thickness for various pipe sizes and schedules 5S, 10S, 40S, and 80S. These dimensions are in accordance with the ASME B36.19, Stainless Steel Pipe.

Permissible Variations for Pipe Outside Diameter

Tables 14.2 and 14.2M list the acceptable variations for the outside diameter of zirconium and zirconium alloy pipe also.

Permissible Variation for Pipe Wall Thickness

The wall thickness of pipe at any point shall not vary more than ±12.5 percent of the nominal wall thickness specified.

Permissible Variations in Length

Pipe length shall not be under the specified length, and not more than $\frac{1}{4}$ in (6.4 mm) over the specified length.

Piping Components Other Than Pipe

Piping components other than pipe may be manufactured from the following ASTM material specifications, as permitted by the governing code or standard of jurisdiction:

- B551 Zirconium and Zirconium Alloy Strip, Sheet, and Plate
- B550 Zirconium and Zirconium Alloy Bars and Wire
- B493 Zirconium and Zirconium Alloy Forgings

ZIRCONIUM AND ZIRCONIUM ALLOY TUBE

Zirconium and zirconium alloy seamless and welded tubes are manufactured in accordance with the following ASTM specification:

- B523 Seamless and Welded Zirconium and Zirconium Alloy Tubes

These tubes are manufactured in the following grades:

- Grade R60702, unalloyed zirconium
- Grade R60704, zirconium-tin alloy
- Grade R60705, zirconium-niobium alloy

Tube Dimensions

Tubes are specified by outside diameter and the wall thickness, as required. As such, no table is included for tube dimensions in this chapter.

Permissible Variations for Tube Dimensions

Tolerance for Outside Diameter. Tolerances listed in Tables 14.3 and 14.3M also apply to zirconium and zirconium alloy tubes manufactured according to ASTM B523.

Tube Wall Thickness Tolerance. Tolerances for tube wall thickness listed in Table 14.4 also apply to zirconium and zirconium alloy tubes manufactured according to ASTM B523.

Tolerance for Straightness. Tube manufactured in accordance with ASTM B523 shall be free of bends or kinks, and the maximum uniform bow shall not exceed the limits shown In Tables 14.5 and 14.5M.

Tables in Chapter 14

S • E • C • T • I • O • N • 4

PIPING JOINTS AND FABRICATION OF PIPING

CHAPTER 15
PIPING JOINTS

STEEL AND ALLOY STEEL PIPING JOINTS

Butt Weld Joint

Butt welds are the most common type of joints for 2.5-in and larger piping systems; however, they are used for smaller piping joints also. Butt weld joints are used to join pipe to pipe, pipe to flange, pipe to valve, and pipe to fitting. Figure 15.1 shows a butt weld joint.

Socket Weld Joint

The socket weld joint is fitted up by slipping the plain end of a pipe into a socket. It is used to join pipe to pipe, pipe to flanges, pipe to valves, and pipe to socket fittings. Figure 15.2 shows a socket weld joint.

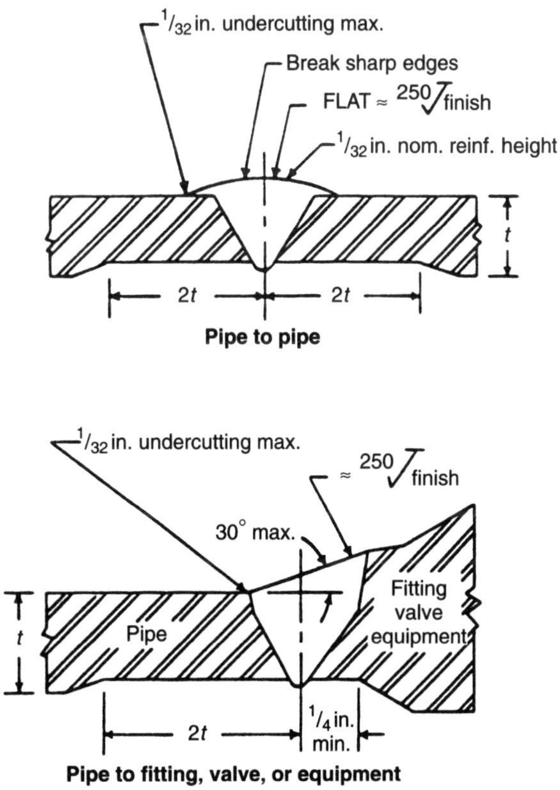

FIGURE 15.1 Butt weld joint.

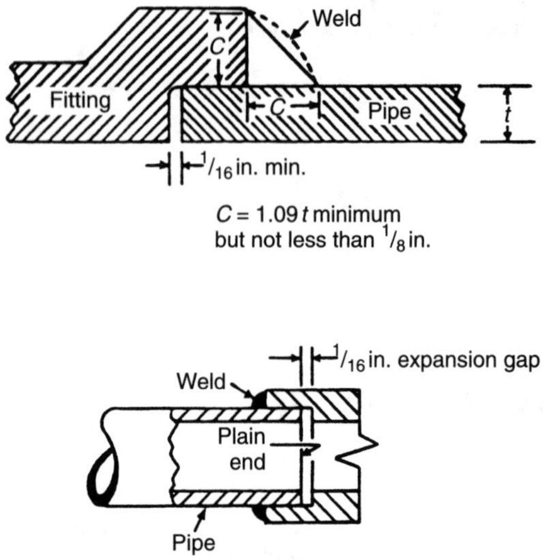

FIGURE 15.2 Socket weld joint.

Threaded Joint

Threaded joints are normally used in low-pressure, small-bore, nonflammable service; however, they may be used in other applications involving higher pressures and temperatures. Figure 15.3 shows a threaded joint.

Flanged Joint

Flanged joints are made by bolting together two flanges with a gasket between them. Figure 15.4 shows a flanged joint.

Compression Sleeve Coupling

This type of joint is used with plain-end pipe. It is widely known under the trade names of the manufacturers, such as Dresser coupling and Dayton coupling. Figure 15.5 shows a compression sleeve coupling.

Grooved Segmented-Ring Coupling

This type of joint is used for pipe that has grooves near the ends that enable the coupling to grip the pipe in order to prevent disengagement of the pipe. Figure 15.5 shows a grooved segmented-ring coupling.

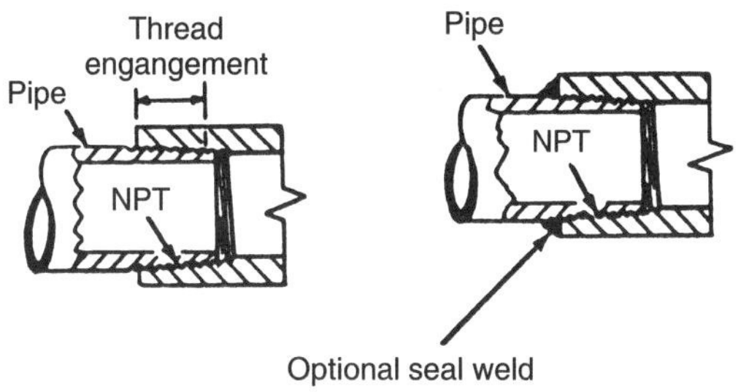

FIGURE 15.3 Threaded joint.

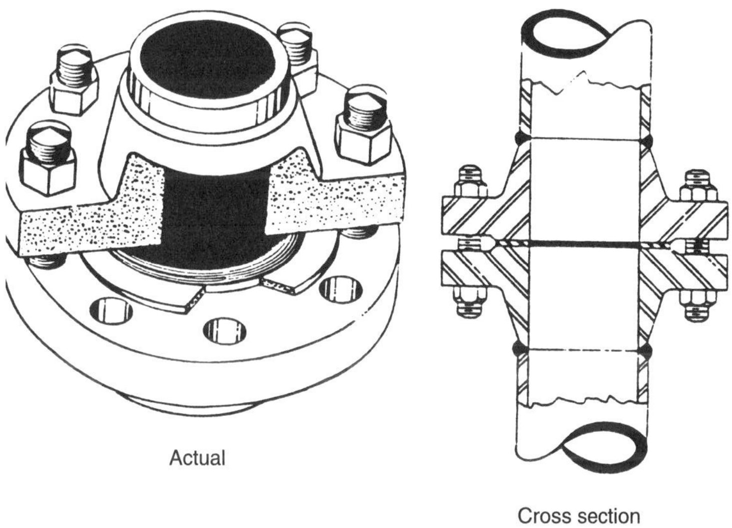

FIGURE 15.4 Flanged joint.

CAST IRON PIPING JOINTS

Bell-and-Spigot Joint

Figure 15.6 shows a bell-and-spigot joint. This type of joint is used for underground piping. The joint may be made up with lead and oakum, sulfur compounds, or cement.

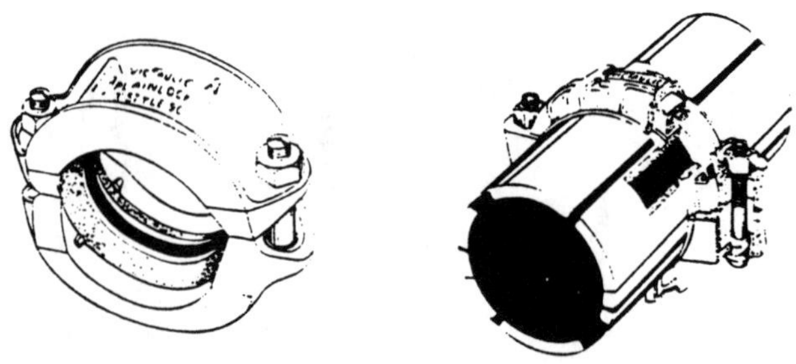

Compression sleeve Grooved end

Victaulic couplings

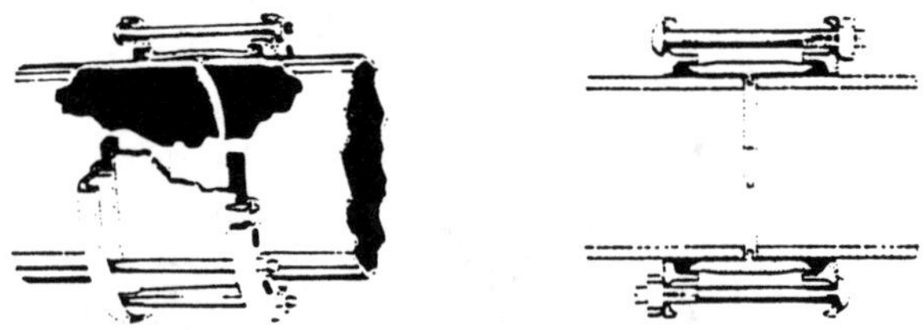

Actual Cross section

Dresser compression sleeve coupling

FIGURE 15.5 Compression sleeve coupling and grooved segmented-ring coupling.

Mechanical (Gland-Type) Joint

Figure 15.7 shows a mechanical (gland-type) joint. The lead and oakum of the conventional bell-and-spigot joint are supplanted by a stuffing box in which a rubber or composition packing ring, with or without a metal or canvas tip or canvas backing, is compressed by a ductile cast iron follower ring drawn up with bolts.

Tyton Joint

Figure 15.8 shows a Tyton joint. It is designed to contain an elongated grooved gasket. The inside contour of the socket bell provides a seat for the circular rubber in a modified bulb-shaped gasket. An internal ridge in the socket fits into the groove of the gasket. A slight taper on the plain end of the pipe facilitates assembly.

Mechanical Lock-Type Joint

This joint is similar to the gland-type mechanical joint except that in the locked joint the spigot end of the pipe is grooved or has a recess to grip the gasket. This type of joint is used to resist end pull where the joints may tend to come apart owing to sag or lateral thrust in the pipeline.

Mechanical Push-on-Type Joint

In this joint, a round rubber gasket is placed over the spigot end and is pulled into the bell by mechanical means, thus pulling the ring into place in the bottom of the bell. Outside the rubber gasket, braided jute is wedged behind a projecting ridge in the bell.

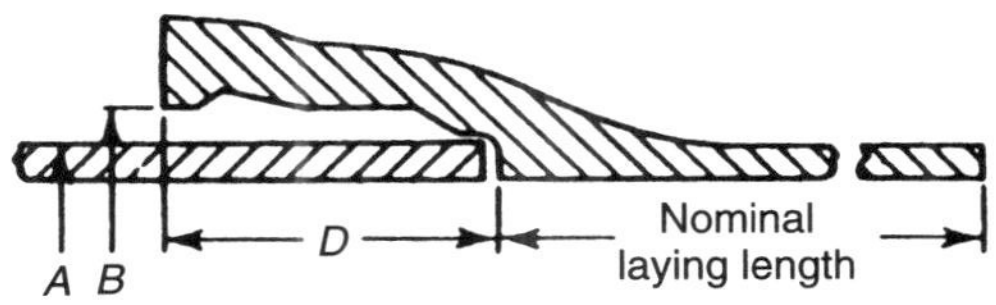

Nominal pipe size, in	Class	Thickness designation	Thickness of pipe, in	OD of pipe, *A*, in	Diameter of socket, *B*, in	Depth of socket, *D*, in	Weight (approx), lb: Barrel per ft	Weight (approx), lb: Bell	Weight (approx), lb: 18-ft laying length	Joint compound, lb per 2½-in depth	Jute, lb per joint	Lead, lb per 2½-in depth
3	Through 350	22	0.32	3.96	4.76	3.30	11.4	11	215			
4	Through 350	22	0.35	4.80	5.60	3.30	15.3	14	290	2.00	0.21	8.00
6	Through 350	22	0.38	6.90	7.70	3.88	24.3	25	460	3.00	0.31	11.25
8	Through 350	22	0.41	9.05	9.85	4.38	34.7	41	665	4.00	0.44	14.50
10	Through 250	22	0.44	11.10	11.90	4.38	46.0	54	880	5.00	0.53	17.50
	300	23	0.48	11.10	11.90	4.38	50.0	54	955			
	350	24	0.52	11.10	11.90	4.38	53.9	54	1025			
12	Through 200	22	0.48	13.20	14.00	4.38	59.8	66	1140	6.00	0.61	20.50
	250, 300	23	0.52	13.20	14.00	4.38	64.6	66	1230			
	350	24	0.56	13.20	14.00	4.38	69.4	66	1315			

FIGURE 15.6 Bell-and-spigot joint.

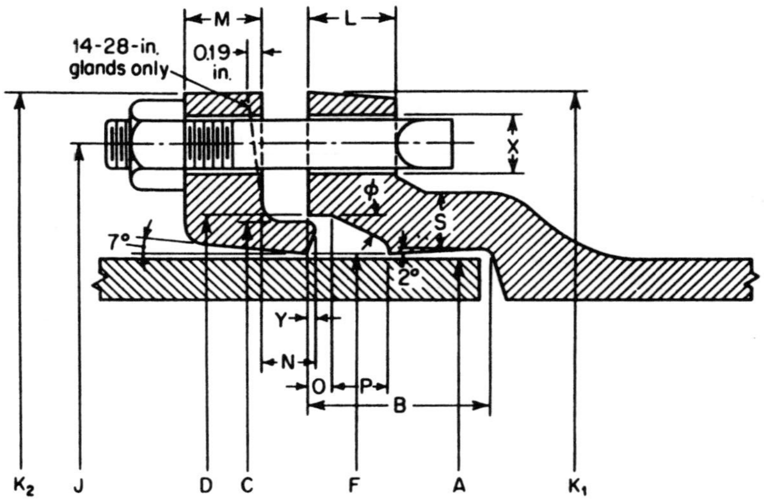

Nominal pipe size	A, Plain end	B	C	D	F	φ deg	X	J	K₁ Centrifugal pipe	K₁ Pit cast pipe and fittings	K_2	L	M	N	O	P	S Centrifugal pipe	S Pit cast pipe and fittings	Y	Bolts No.	Bolts Size	Bolts Length
2	±0.05 2.50	2.50	±0.05 3.39	±0.05 3.50	±0.05 2.61	28°	+0.06 −0.0 ¾	±0.05 4.75	−0.05 6.00	−0.10 6.25	−0.10 6.25	−0.05 0.75	−0.05 0.62	0.50	0.31	0.63	−0.05 0.37	−0.07 0.44	0.08	2	⅝	2½
2½	±0.05 2.75	2.50	±0.05 3.64	±0.05 3.75	±0.05 2.86	28°	+0.06 −0.0 ¾	±0.05 5.00	−0.05 6.25	−0.10 6.50	−0.10 6.50	−0.05 0.75	−0.03 0.62	0.50	0.31	0.63	−0.05 0.37	−0.07 0.44	0.08	2	⅝	2½
3	±0.06 3.96	2.50	±0.04 4.84	+0.06 −0.04 4.94	+0.07 −0.03 4.06	28°	+0.06 −0.0 ¾	±0.06 6.19	−0.06 7.62	−0.12 7.69	−0.12 7.69	−0.06 0.94	−0.06 0.62	0.75	0.31	0.63	−0.05 0.47	−0.10 0.52	0.12	4	⅝	3
4	±0.06 4.80	2.50	±0.04 5.92	+0.06 −0.04 6.02	+0.07 −0.03 4.90	28°	+0.06 −0.0 ⅞	±0.06 7.50	−0.06 9.06	−0.12 9.12	−0.12 9.12	−0.06 1.00	−0.06 0.75	0.75	0.31	0.75	−0.05 0.55	−0.10 0.65	0.12	4	¾	3½
6	±0.06 6.90	2.50	±0.04 8.02	+0.06 −0.04 8.12	+0.07 −0.03 7.00	28°	+0.06 −0.0 ⅞	±0.06 9.50	−0.06 11.06	−0.12 11.12	−0.12 11.12	−0.06 1.06	−0.06 0.88	0.75	0.31	0.75	−0.05 0.60	−0.10 0.70	0.12	6	¾	3½
8	±0.06 9.05	2.50	±0.04 10.17	+0.06 −0.04 10.27	+0.07 −0.03 9.15	28°	+0.06 −0.0 ⅞	±0.06 11.75	−0.06 13.31	−0.12 13.37	−0.12 13.37	−0.08 1.12	−0.08 1.00	0.75	0.31	0.75	−0.05 0.66	−0.12 0.75	0.12	6	¾	4

FIGURE 15.7 Mechanical (gland-type) joint.

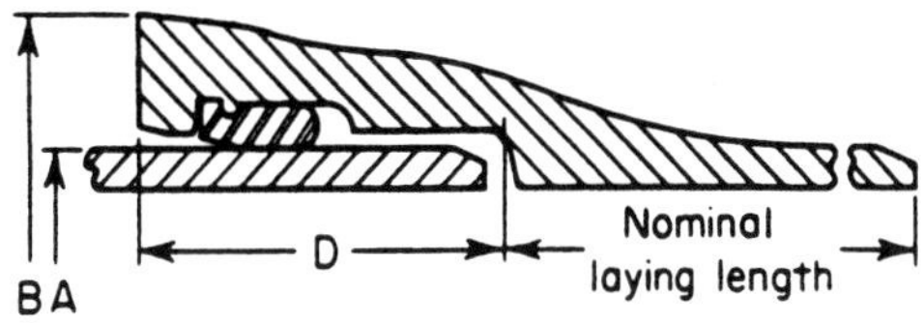

Nominal pipe size	Class	Thickness designation	Thickness of pipe	OD of pipe, *A*	OD of bell, *B*	Depth of socket, *D*	Weight (approx), lb: Barrel per ft	Bell	18 ft length*
3	Through 350	22	0.32	3.96	6.08	3.00	11.4	11	215
4	Through 350	22	0.35	4.80	7.22	3.15	15.3	14	290
6	Through 350	22	0.38	6.90	9.47	3.38	24.3	25	460
8	Through 350	22	0.41	9.05	12.00	3.69	34.7	41	665
10	Through 350	22	0.44	11.10	14.20	3.75	46.0	54	880
	300	23	0.48				50.0		955
	350	24	0.52				53.9		1025
12	Through 200	22	0.48	13.20	16.35	3.75	59.8	66	1140
	250, 300	23	0.52				64.6		1230
	350	24	0.56				69.4		1315
14	50	21	0.48	15.30	19.15	5.00	69.7	78	1335
	100, 150	22	0.51				73.9		1410
	200	23	0.55				79.5		1510
	250, 300	24	0.59				85.1		1610
	350	25	0.64				92.0		1735
16	Through 150	22	0.54	17.40	21.36	5.00	89.2	96	1700
	200	23	0.58				95.6		1815
	250	24	0.63				103.6		1960
	300, 350	25	0.68				111.4		2100
18	50	21	0.54	19.50	23.56	5.00	100.4	114	1920
	100, 150	22	0.58				107.6		2050
	200	23	0.63				116.5		2210
	250	24	0.68				125.4		2370
	300	25	0.73				134.3		2530
	350	26	0.79				144.9		2720
20	50	21	0.57	21.60	25.80	5.00	117.5	133	2250
	100, 150	22	0.62				127.5		2430
	200	23	0.67				137.5		2610
	250	24	0.72				147.4		2785
	300	25	0.78				159.2		3000
	350	26	0.84				170.9		3210
24	50	21	0.63	25.80	30.32	5.00	155.4	179	2975
	100	22	0.68				167.4		3190
	150	23	0.73				179.4		3410
	200, 250	24	0.79				193.7		3665
	300	25	0.85				207.9		3920
	350	26	0.92				224.4		4220

* Includes weight of bell.

FIGURE 15.8 Standard dimensions of Tyton joint.

Mechanical Screw-Gland-Type Joint

This joint makes use of a coarse-threaded screw gland drawn up by means of a spanner wrench to compress a standard rubber or composition packing gasket. A lead ring, inserted in the bell ahead of the gasket, seals off the contents of the line from the gasket.

Ball-and-Socket Joint

Figure 15.9 shows a ball-and-socket joint.

Universal Pipe Joint

Figure 15.10 shows a universal pipe joint. The joint is pulled up snugly with two bolts, after which the nuts are backed off slightly, thus enabling the lock washers to relax enough to avoid overstressing the socket or lugs.

Compression Sleeve Coupling

This type of joint is used with plain-end pipe. It is widely known under the trade names of the manufacturers, such as Dresser coupling and Dayton coupling. Figure 15.11 shows a compression sleeve coupling.

Grooved Segmented-Ring Coupling

This type of joint is used for pipe that has grooves near the ends that enable the coupling to grip the pipe in order to prevent disengagement of the pipe. Figure 15.12 shows a grooved segmented-ring coupling.

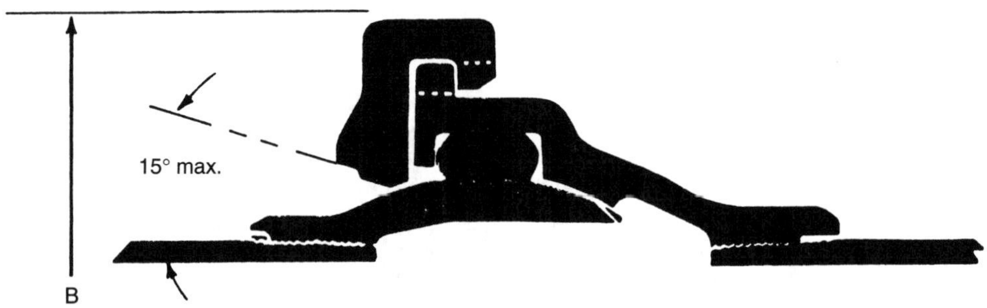

FIGURE 15.9 Ball-and-socket joint.

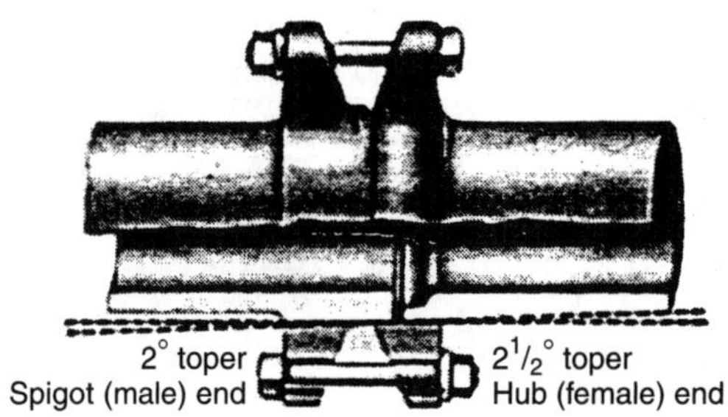

FIGURE 15.10 Universal pipe joint.

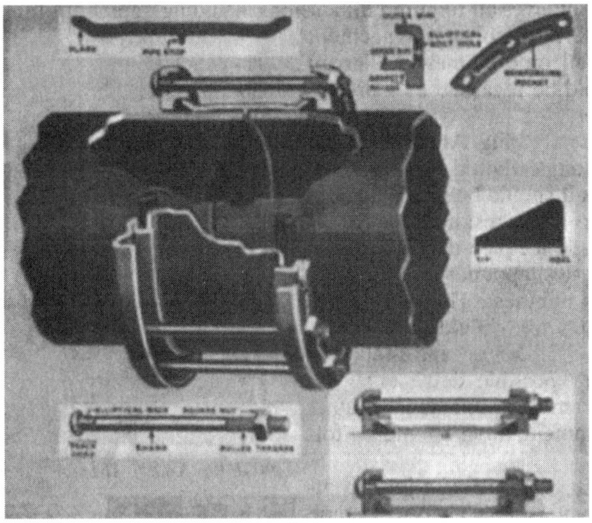

FIGURE 15.11 Compression sleeve (Dresser) coupling for plain-end cast iron or steel pipe.

FIGURE 15.12 Victualic coupling for grooved-end cast iron or steel pipe.

Flanged Joint

Figure 15.4 shows a flanged joint. Cast iron pipe is made with both integrally cast flanges and with threaded companion flanges for screwing onto the pipe. In the latter case, the outside diameter of the pipe conforms to iron pipe size dimensions to allow for the threads provided.

DUCTILE IRON PIPING JOINTS

Bell-and-Spigot Joint

Figure 15.6 shows a bell-and-spigot joint. This type of joint is used for underground piping. The joint may be made up with lead and oakum, sulfur compounds, or cement.

Mechanical (Gland-Type) Joint

Figure 15.7 shows a mechanical (gland-type) joint. The lead and oakum of the conventional bell-and-spigot joint are supplanted by a stuffing box in which a rubber or composition packing ring, with or without a metal or canvas tip or canvas backing, is compressed by a ductile cast iron follower ring drawn up with bolts.

Tyton Joint

Figure 15.8 shows a Tyton joint. It is designed to contain an elongated grooved gasket. The inside contour of the socket bell provides a seat for the circular rubber in a modified bulb-shaped gasket. An internal ridge in the socket fits into the groove of the gasket. A slight taper on the plain end of the pipe facilitates assembly.

Mechanical Lock-Type Joint

This joint is similar to the gland-type mechanical joint except that in the locked joint the spigot end of the pipe is grooved or has a recess to grip the gasket. This type of joint is used to resist end pull where the joints may tend to come apart owing to sag or lateral thrust in the pipeline.

Mechanical Push-on-Type Joint

In this joint, a round rubber gasket is placed over the spigot end and is pulled into the bell by mechanical means, thus pulling the ring into place in the bottom of the bell. Outside the rubber gasket, braided jute is wedged behind a projecting ridge in the bell.

Ball-and-Socket Joint

Figure 15.9 shows a ball-and-socket joint.

Grooved Segmented-Ring Coupling

This type of joint is used for pipe that has grooves near the ends that enable the coupling to grip the pipe in order to prevent disengagement of the pipe. Figure 15.12 shows a grooved segmented-ring coupling.

Flanged Joint

Figure 15.4 shows a flanged joint.

ALUMINUM PIPING JOINTS

Aluminum piping joints include butt weld joints, socket weld joints, threaded joints, and flanged joints. Refer to "Steel and Alloy Steel Piping Joints" for a description of these joints.

TITANIUM PIPING JOINTS

Titanium piping joints include butt weld joints, socket weld joints, threaded joints, and flanged joints. Refer to "Steel and Alloy Steel Piping Joints" for a description of these joints.

AWWA WROUGHT PIPING JOINTS

AWWA piping joints can be steel piping joints, cast iron piping joints, ductile iron piping joints, thermoplastic piping joints, and fiberglass piping joints. Refer to individual sections for detailed joint description.

COPPER AND COPPER ALLOY PIPE JOINTS

Brazed Joint

Figure 15.13 shows a brazed joint. The joint is achieved by a nonferrous filler metal having a melting point above 800°F but lower than that of the base metals joined. The filler metal is distributed between the closely fitted surfaces of the joint by capillary action.

Soldered Joint

Figure 15.13 shows a soldered joint. The joint is made by heating to a suitable temperature and by using a nonferrous alloy fusible at temperatures below 800°F and having a melting point below that of the base metals being joined.

Threaded Joint

Figure 15.3 shows a threaded joint.

THERMOPLASTIC PIPING JOINTS

Hot Gas Welded Joint

Figure 15.14 shows a hot gas welded joint.

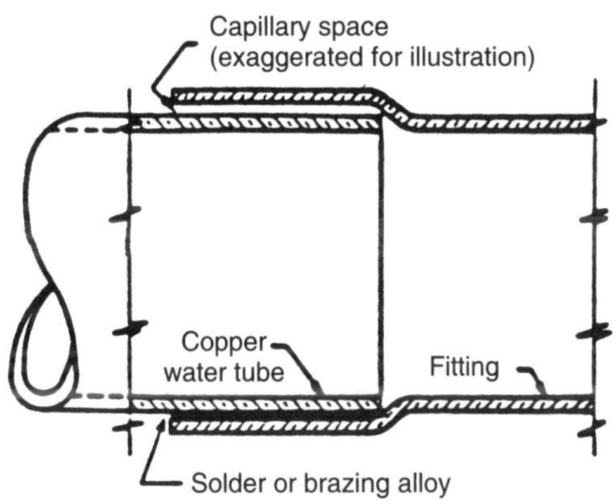

FIGURE 15.13 Soldered and brazed joint.

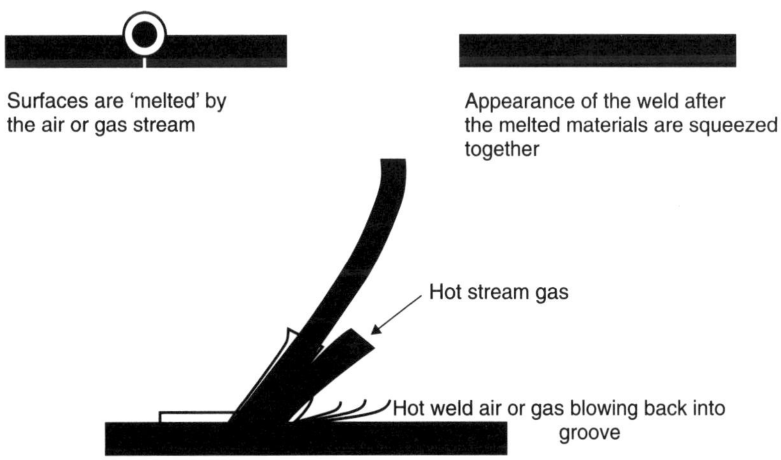

FIGURE 15.14 Hot gas welding.

Solvent-Cemented Joint

Figure 15.15 shows a solvent-cemented joint (socket joint). A continuous bond is produced by the application of cement to both surfaces to be joined and assembly of these surfaces.

Heat Fusion Joint

Figure 15.16 shows heat fusion joints (socket joint and butt joint). A continuous homogenous bond is produced by uniform heating of both surfaces to be joined and assembly of these surfaces.

Adhesive Joint

Figure 15.17 shows an adhesive joint. A continuous bond is produced by the application of adhesives to the surfaces to be joined and assembly of these surfaces.

Butt-and-Wrap Joint

Figure 15.18 shows butt-and-wrap joint types. The joint is made by the application of plies of reinforcement saturated with catalyzed resin to the surfaces to be joined.

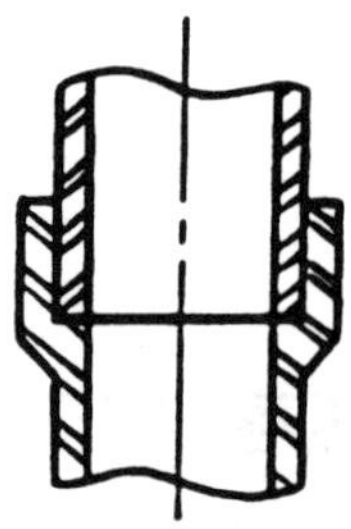

FIGURE 15.15 Thermoplastic solvent-cemented socket joint.

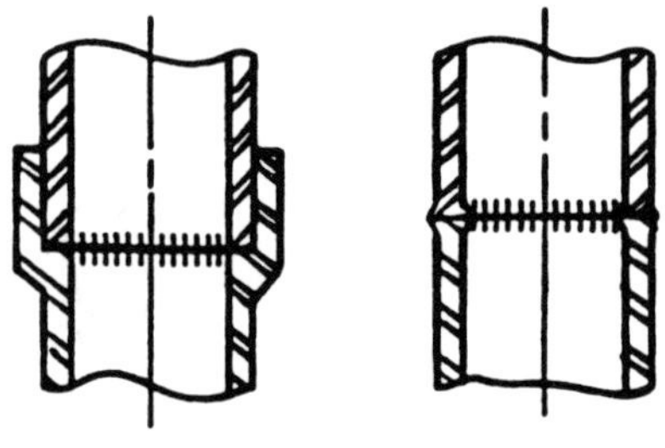

FIGURE 15.16 Thermoplastic heat fusion joints.

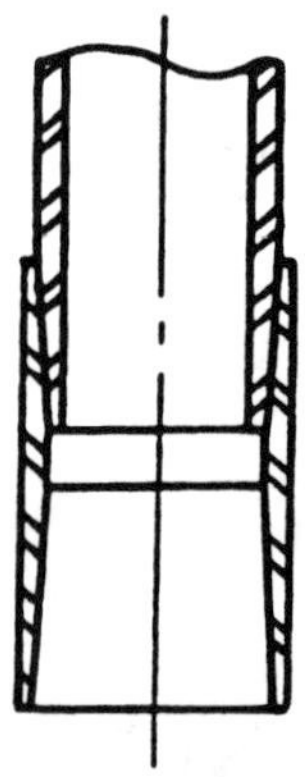

FIGURE 15.17 Fully tapered thermosetting adhesive joint.

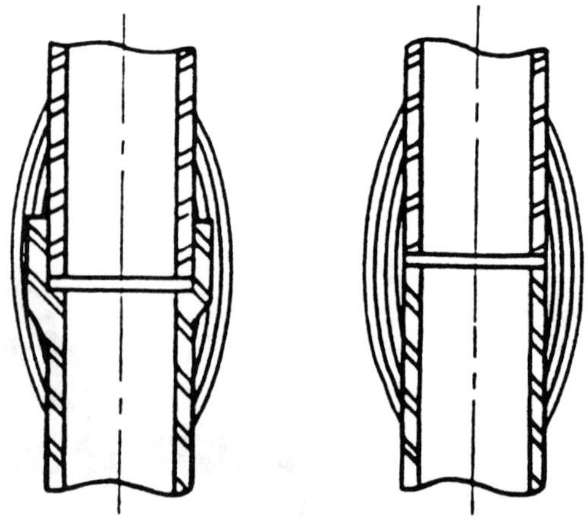

FIGURE 15.18 Thermosetting wrapped joints.

Electrofusion Joint

Figure 15.19 shows different types of electrofusion joints. The joints are made by applying electric current, producing heat and thereby melting and joining the components.

FIBERGLASS PIPING JOINTS

Bell-and-Spigot Joint

Figure 15.20 shows a bell-and-spigot joint. The joints are adhesively bonded, usually with a two-part adhesive. The joint includes a machined taper for an interference fit which gives a very high strength.

Straight-Socket Adhesive Joint

Figure 15.21 shows a straight-socket adhesive joint. This type of joint is adhesively bonded.

Flanged Joint

Figure 15.22 shows a flanged joint. The flanges are available with industry-standard bolt patterns such as ANSI, DIN, ISO, and JSI bolt patterns.

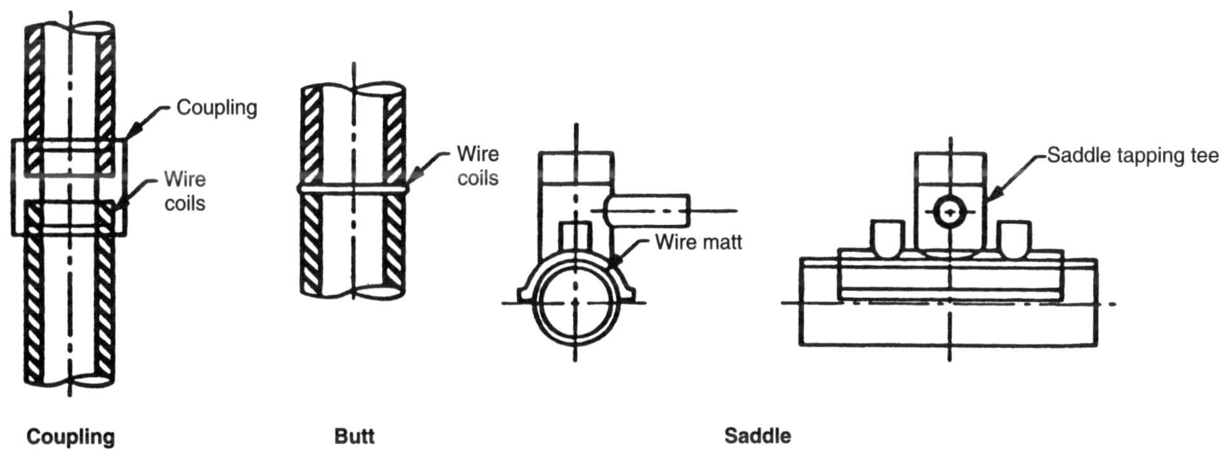

FIGURE 15.19 Thermoplastic electrofusion joints.

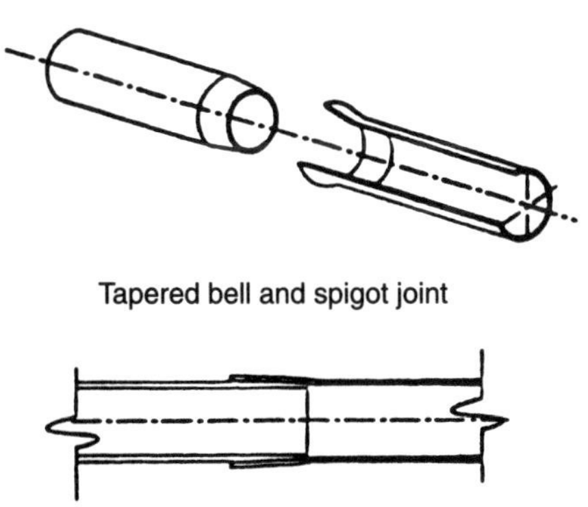

FIGURE 15.20 Bell-and-spigot joint.

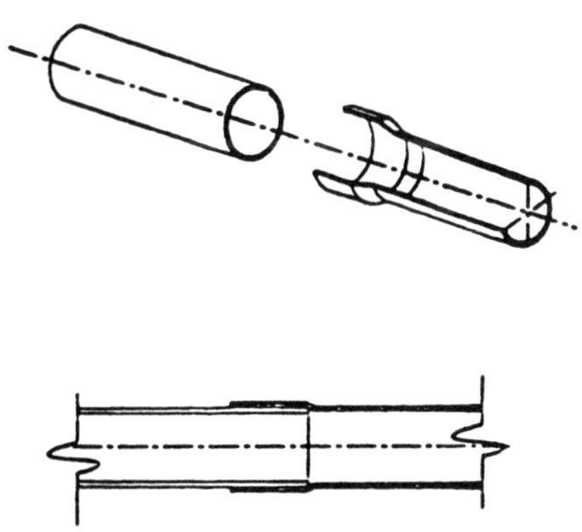

FIGURE 15.21 Straight-socket adhesive joint.

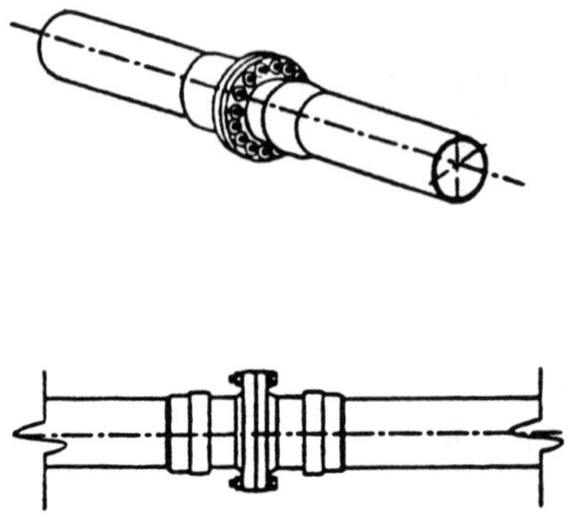

FIGURE 15.22 Flanged joint.

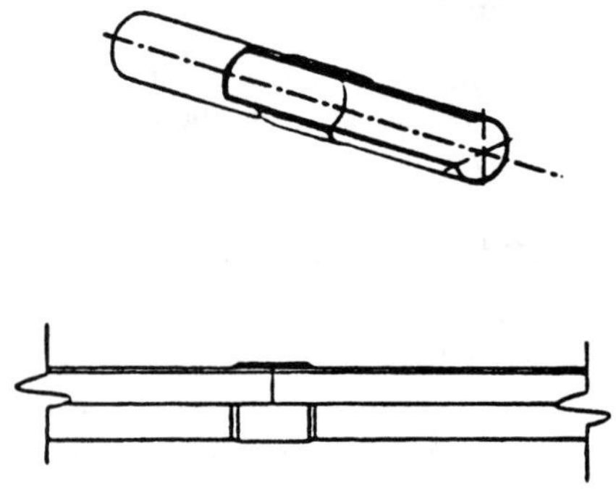

FIGURE 15.23 Butt-and-wrap joint.

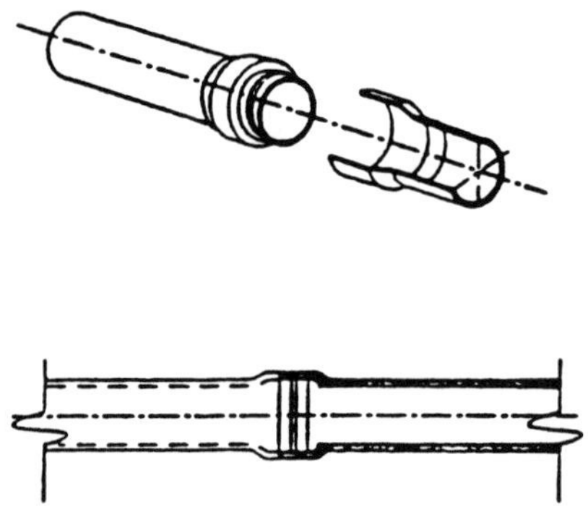

FIGURE 15.24 O-ring bell-and-spigot mechanical joint.

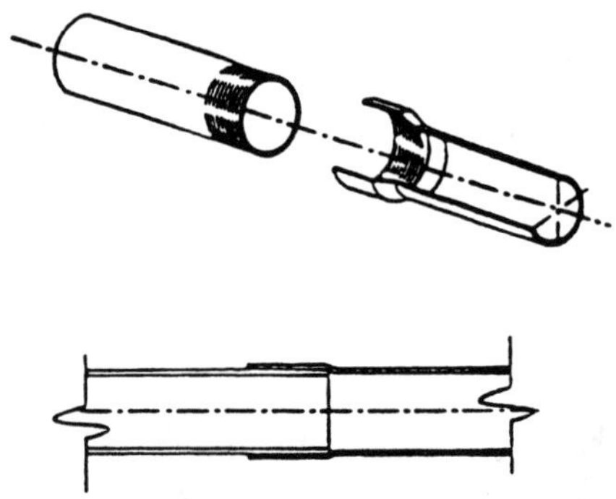

FIGURE 15.25 Threaded joint.

Butt-and-Wrap Joint

Figure 15.23 shows a butt-and-wrap joint. It is used to join two pieces of plain-end pipe or a pipe and fitting. The joint is made by first placing a corrosion barrier. After the corrosion barrier is formed, the mechanical strength of the system is ensured by using saturated reinforcement, which is laid up out of mat and woven roving.

O-Ring Bell-and-Spigot Joint

Figure 15.24 shows an O-ring bell-and-spigot joint. It is available in NPS 14 and larger sizes of filament-wound pipe and all sizes of polymer concrete pipe.

Threaded Joint

Figure 15.25 shows a threaded joint. It is available in API eight round and ten round EU (external upset) and NPT (nominal pipe taper) threads.

Mechanical Joint

Figure 15.24 shows a mechanical joint. Mechanical joints using O-rings for sealing purposes are available in NPS 2 through NPS 36. One type of mechanical joint uses a retainer pin made of either metal cable or thermoplastic rod to retain the axial loads. Also available are mechanical joints that use coarse threads to retain the axial loads.

CONCRETE AND REINFORCED AND PRESTRESSED CONCRETE PRESSURE PIPE JOINTS

Bell-and-Spigot Joint

Figure 15.26 shows a bell-and-spigot joint. This joint configuration consists of a steel bell ring and a steel spigot ring that compresses a rubber O-ring when assembled.

Snap Ring Joint

Figure 15.27 shows a snap ring joint. This joint incorporates a split ring, which is preassembled in the manufacturing plant so that it is recessed into a groove in the bell ring. After the spigot is pushed home into the bell, the installer tightens a single bolt, which draws the split ring down around the shank of the spigot ring. This effectively locks the joint together.

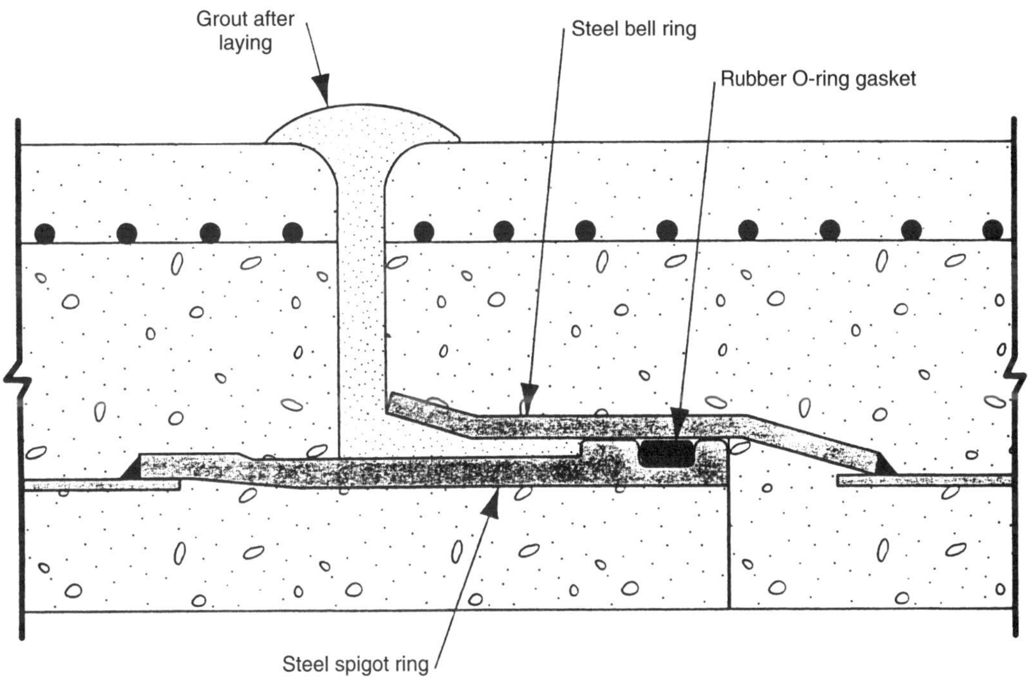

FIGURE 15.26 Bell-and-spigot joint.

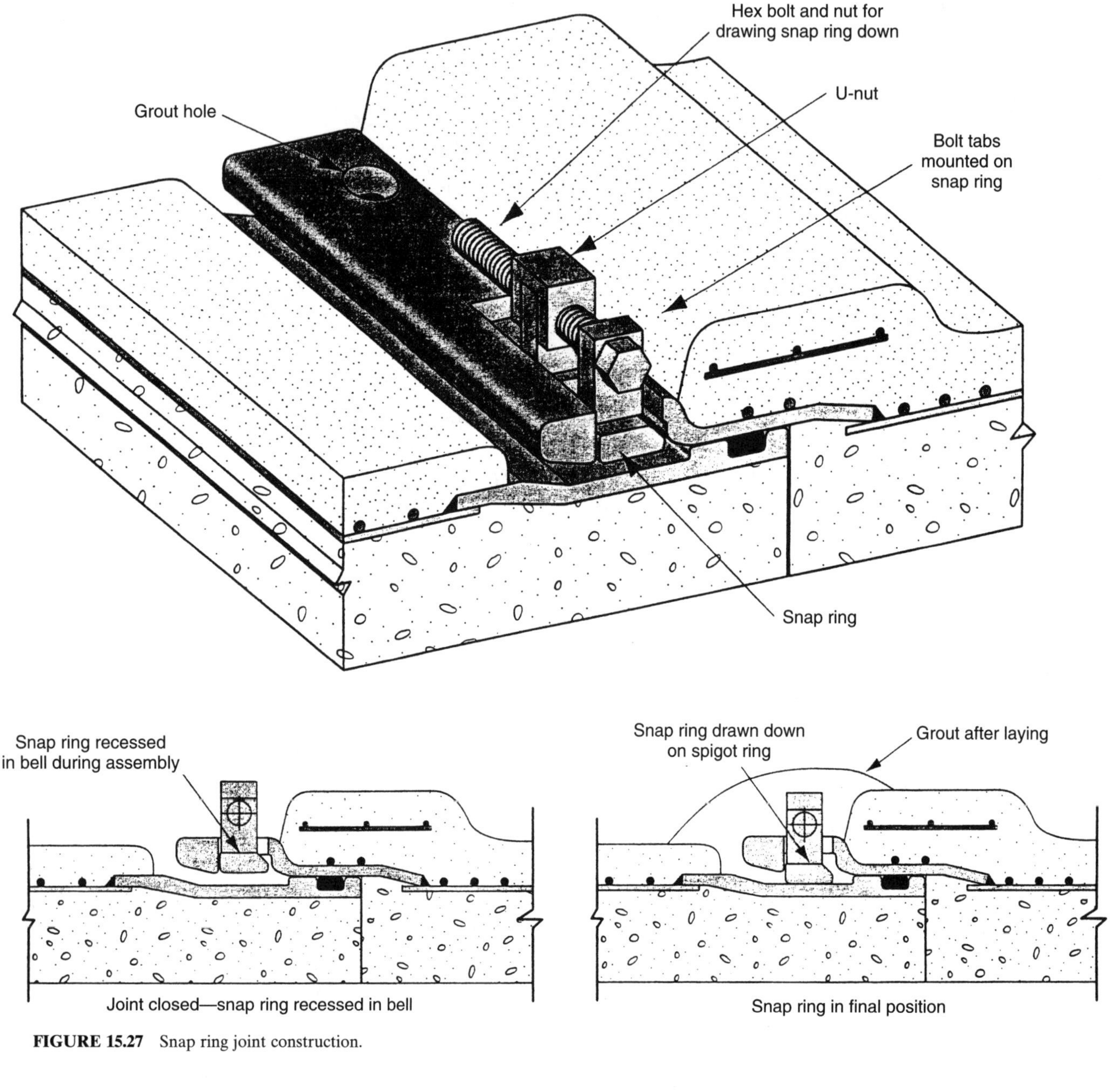

FIGURE 15.27 Snap ring joint construction.

Testable Joint

Figure 15.28 shows a testable joint. This type of joint is normally used when pipe is installed in a manner that necessitates immediate verification of the gasket seal. Testable joints are available for pipe with diameters of 54 in and larger.

Subaqueous Joint

Figure 15.29*a* and *b* shows a subaqueous joint for lined cylinder pipe and embedded cylinder pipe. This joint incorporates external lug and drawbolt assemblies. The joint engaging assemblies consist of steel anchor sockets solidly cast into the walls at the ends of the pipe at the springline. Lugs are then attached to the sockets to receive a drawbolt which is used by divers to draw the bell and spigot together.

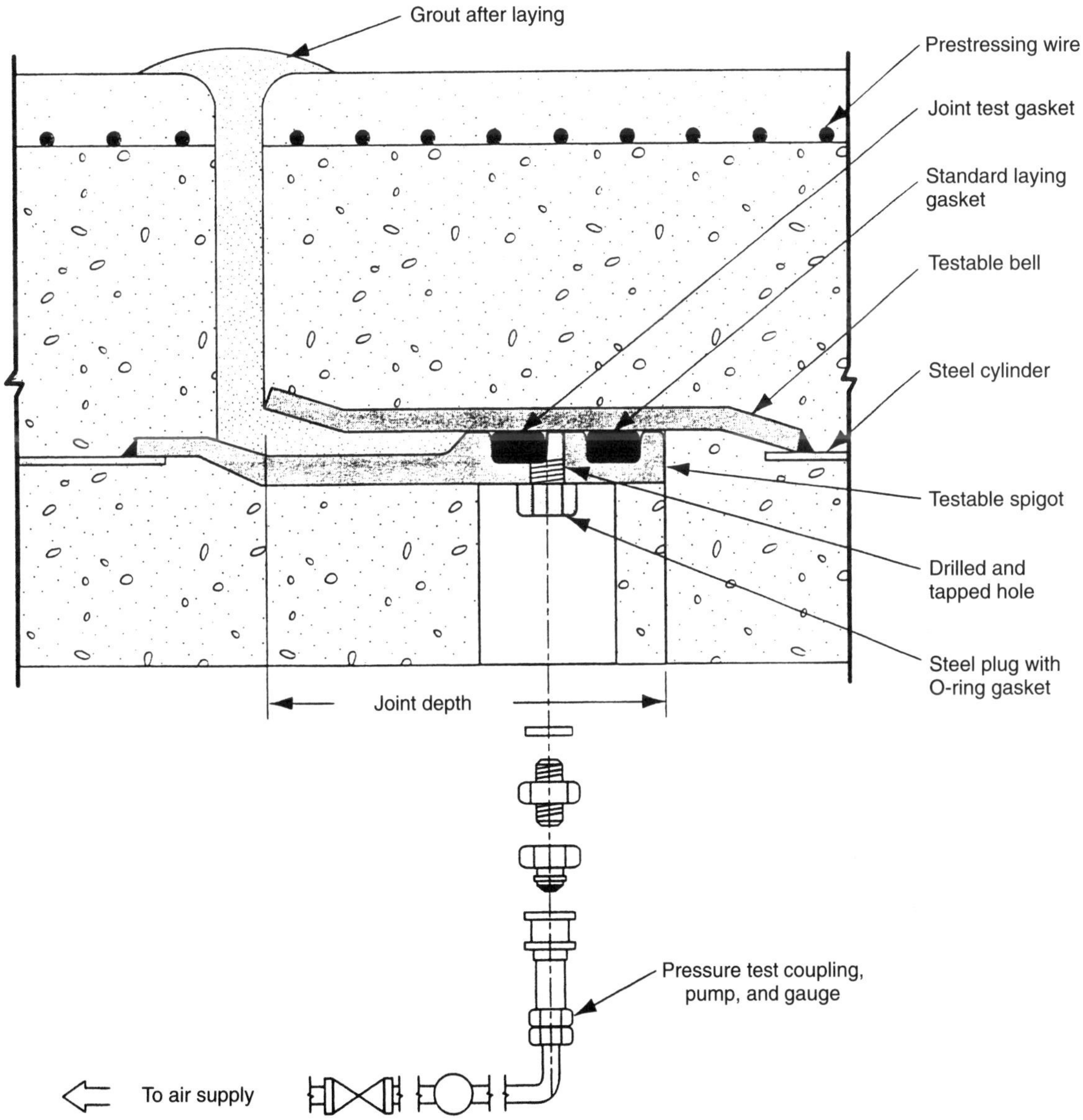

FIGURE 15.28 Testable joint construction.

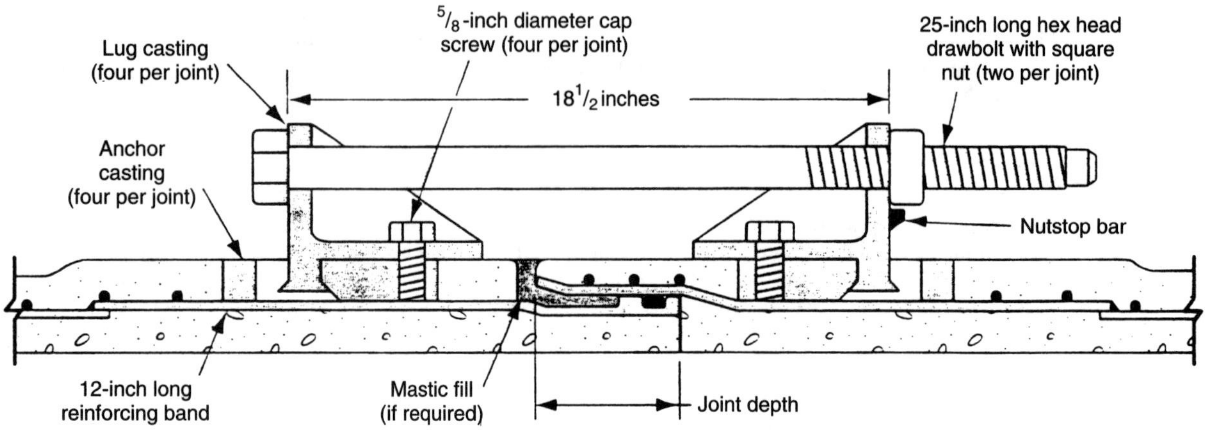

(*a*)

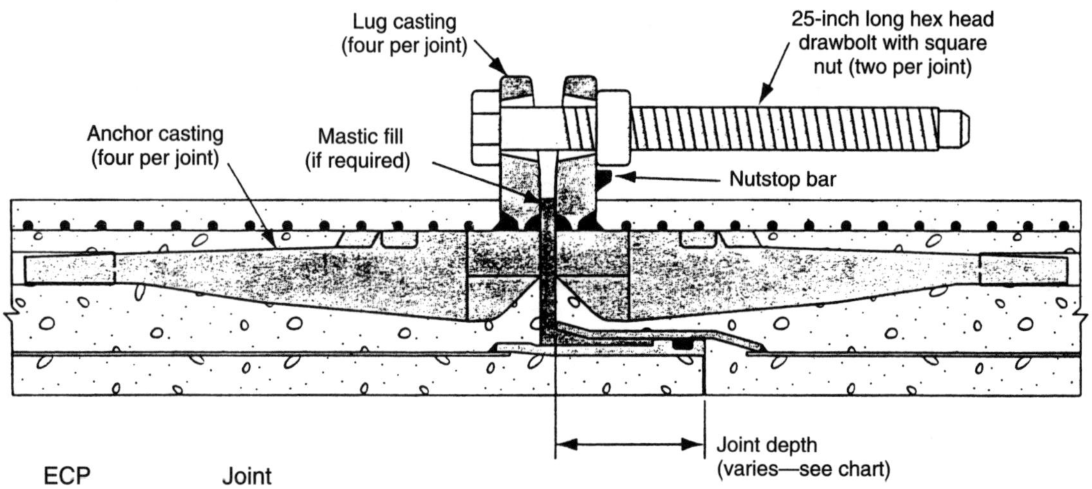

ECP diameter (in)	Joint depth (in)
54	5⅛
60	5¼
66	5⅜
72	5½
78	5⅝
84	5¾
90–96	5⅞
102–144	6

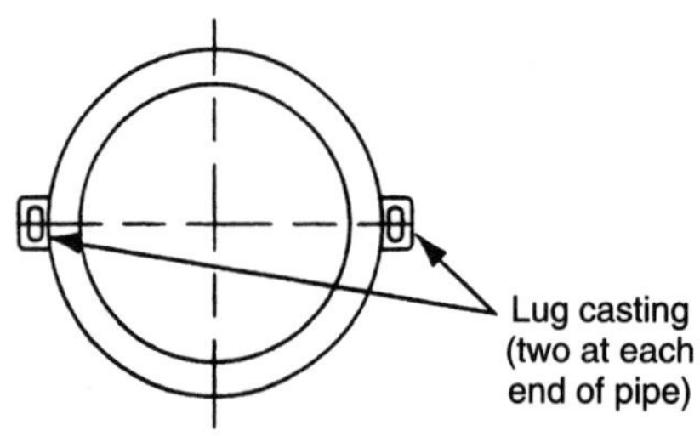

(*b*)

FIGURE 15.29 Subaqueous joints: (*a*) lined cylinder pipe; (*b*) embedded cylinder pipe.

Bonded Joint

Figure 15.30 shows a bonded joint. This joint allows for monitoring of environmental conditions. It is designed to provide electrical continuity across the joint for the future monitoring of electrical activity on the pipeline or for the application of cathodic protection.

Harnessed Clamp Joint

Figure 15.31 shows a harnessed clamp joint. This joint provides restraint for diameters 66 in and larger. This joint is restrained by a two-part harness clamp.

Ball Joint

Figure 15.32 shows a ball joint. This joint is used mainly for river crossings. The ball joint combines longitudinal thrust restraint with a high degree of rotation of up to 15 degrees in any direction.

Deep Joint

Figure 15.33 shows a deep joint. This joint is constructed in the same fashion as the bell-and-spigot joint, except the joint depths are greater to provide additional deflection capability.

Extra-Deep Joint

Figure 15.34 shows an extra-deep joint. This joint is used on lined cylinder pipe sections where additional joint flexibility is required.

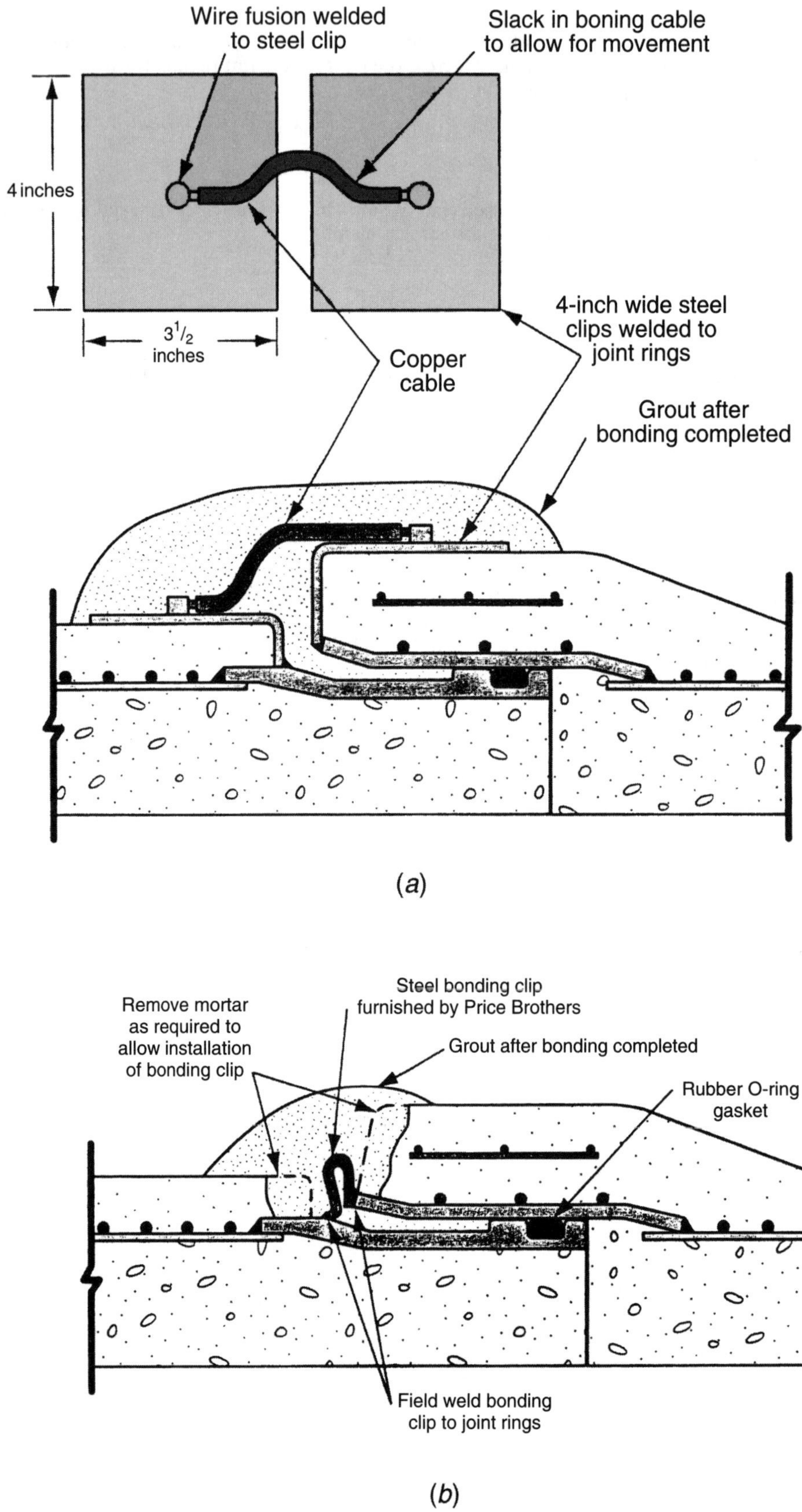

FIGURE 15.30 Bonded joint construction—lined cylinder pipe: (*a*) exothermic fusion welded copper cable method; (*b*) field-welded U-clip method.

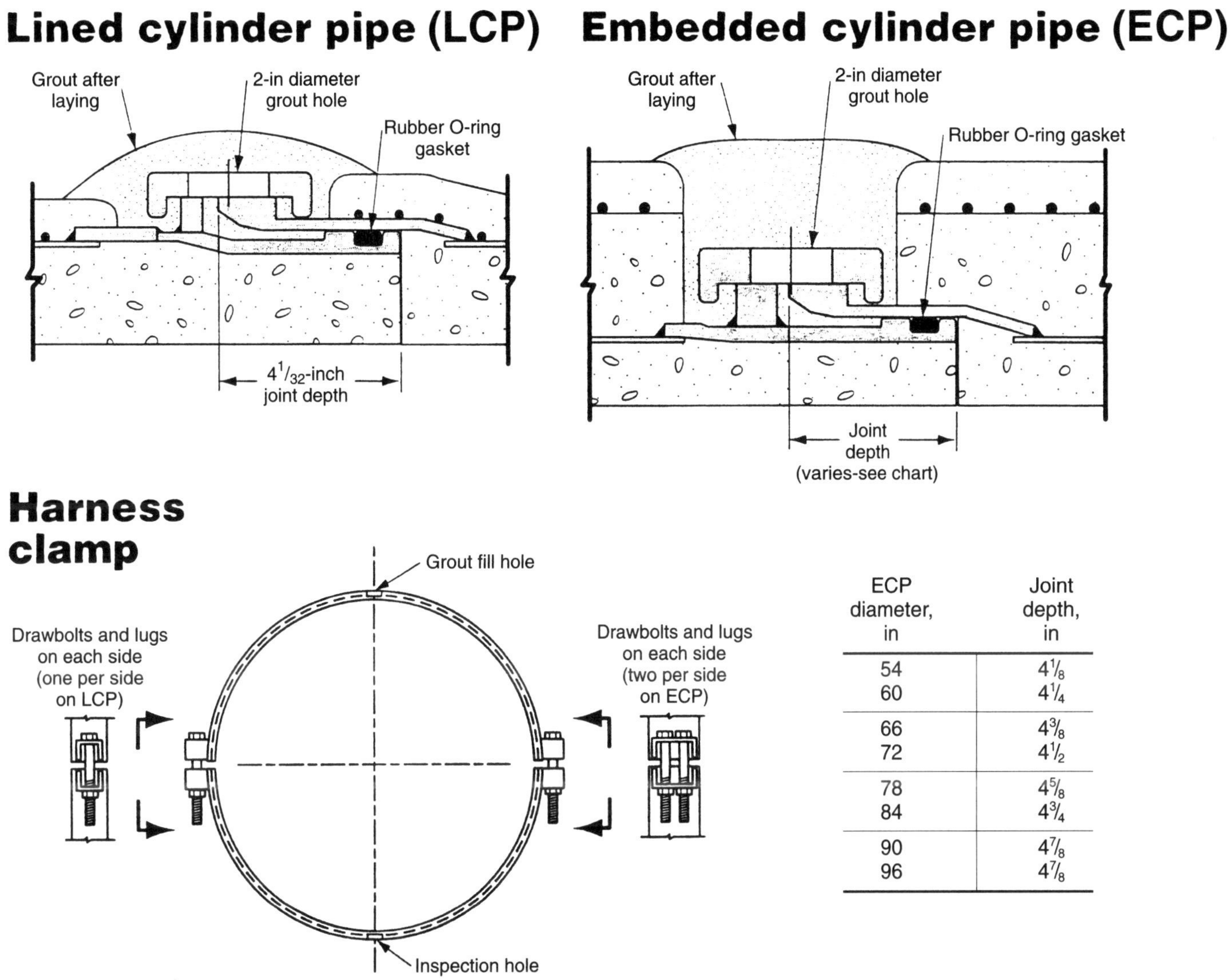

ECP diameter, in	Joint depth, in
54	$4\frac{1}{8}$
60	$4\frac{1}{4}$
66	$4\frac{3}{8}$
72	$4\frac{1}{2}$
78	$4\frac{5}{8}$
84	$4\frac{3}{4}$
90	$4\frac{7}{8}$
96	$4\frac{7}{8}$

FIGURE 15.31 Harnessed clamp joint construction.

Pipe diameter, in	Rotation, degrees
16–42	15
48	12½

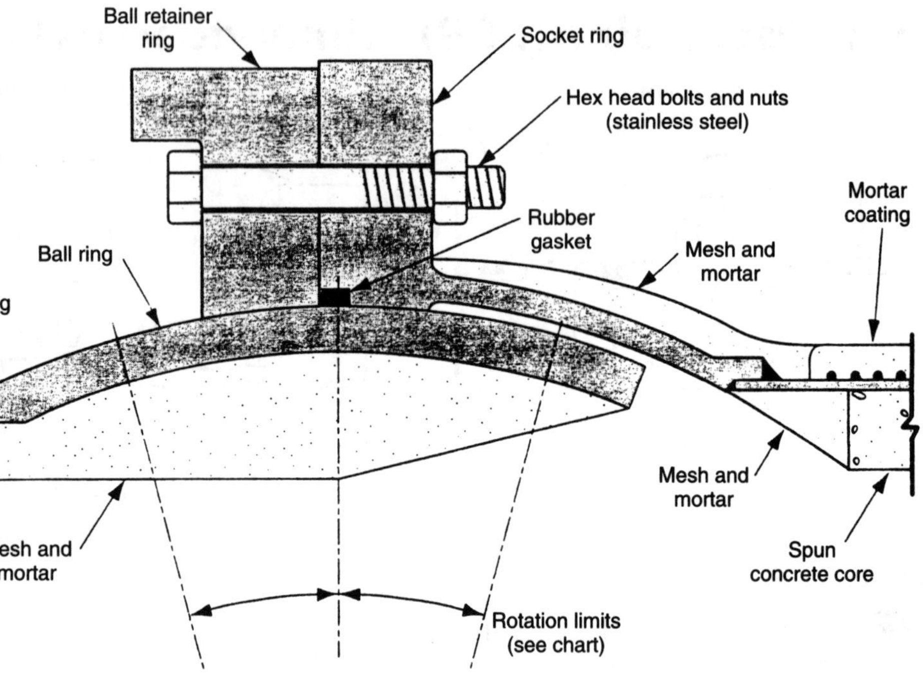

FIGURE 15.32 Ball joint construction.

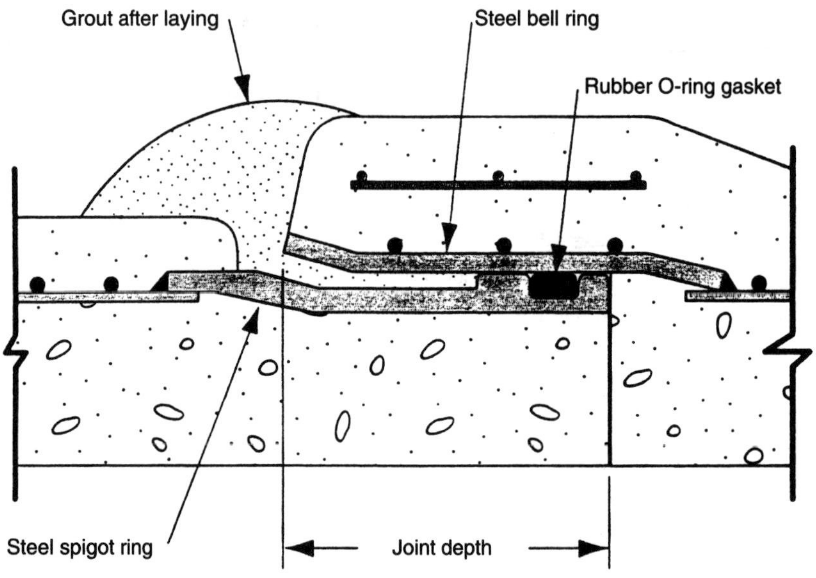

FIGURE 15.33 Deep joint construction.

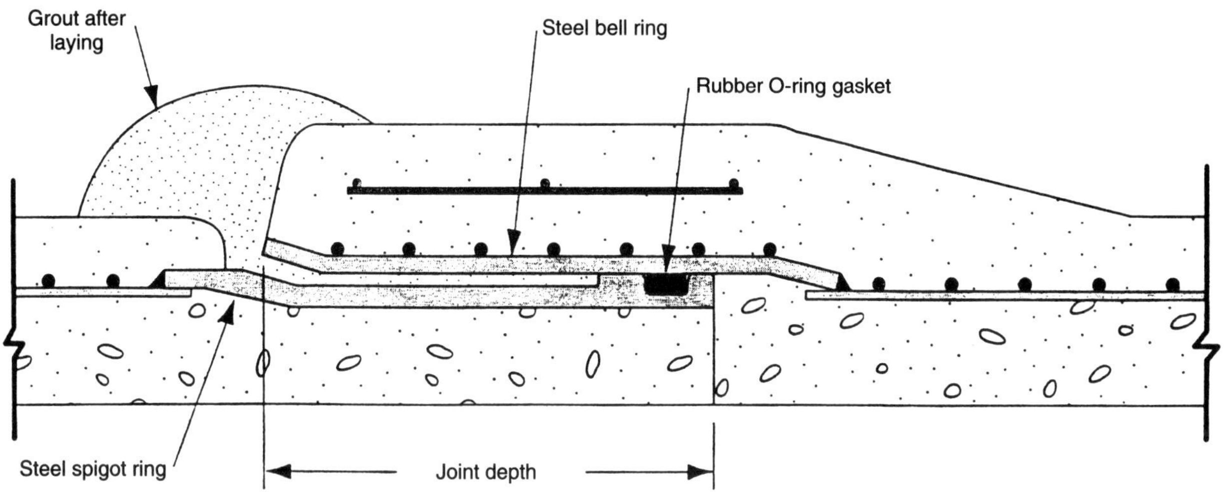

FIGURE 15.34 Extra-deep joint construction.

PIPING JOINTS FOR VITRIFIED CLAY PIPING

Figure 15.35 illustrates an O-ring joint. An O-ring gasket is fitted in the groove on the spigot end, and the bell is lubricated prior to slipping the spigot with O-ring installed into the bell.

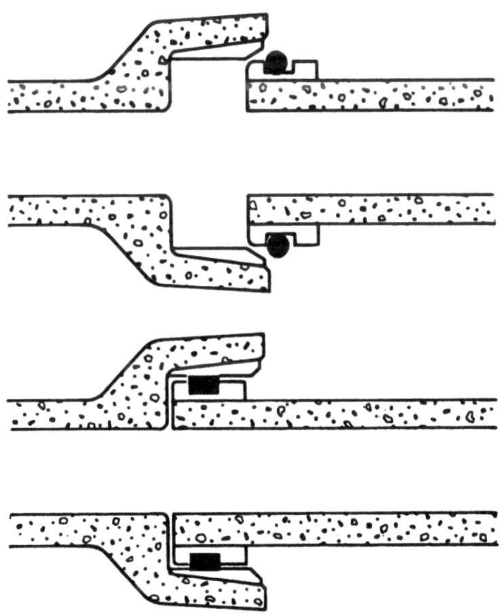

FIGURE 15.35 O-ring compression joint for vitrified clay pipe.

CHAPTER 16
FABRICATION OF PIPING

See p. 539 for a list of tables in Chap. 16.

REFERENCE CODES AND STANDARDS FOR FABRICATION OF PIPING

Piping codes and standards contain requirements for materials, design, fabrication, examination, examination, testing, inspection, and stamping of piping. The following codes and standards are of importance to piping:

- ASME Boiler and Pressure Vessel Code
- ASME B31 Pressure Piping Code
- ASME B16 standards
- API piping and valve standards
- ASTM standards and specifications
- AWWA standards, specifications, and manuals
- AWS handbooks, manuals, guides, etc.
- Ductile Iron Pipe Research Association specifications
- MSS standard practices
- Piping Fabrication Institute standards and bulletins
- Steel Structures Painting Council standard specifications

FABRICATION OF PIPING

The following key activities are involved in piping fabrication:

- Welding
- Brazing
- Soldering
- Bending
- Forming
- Threading
- Grooving
- Cutting
- Assembly
- Bolting
- Installation
- Erection

WELDING

ASME Section IX classifies base metals in various P numbers, which are based on the composition of metal, weldability, and mechanical properties. Each P number is further subdivided into group numbers depending upon fracture toughness properties. Refer to Table 16.1 for P numbers and group numbers.

Table 16.1 ASME P Numbers and Group Numbers for Some Typical Piping Materials

Nominal compositon	P no.	Group no.
Carbon steel—65 ksi and under	1	1
—65 ksi to 75 ksi	1	2
C-$\frac{1}{2}$Mo and $\frac{1}{2}$Cr-$\frac{1}{2}$Mo—65 ksi and under	3	1
—70 to 75 ksi	3	2
1Cr-$\frac{1}{2}$Mo and 1$\frac{1}{4}$Cr-$\frac{1}{2}$Mo-Si	4	1
2$\frac{1}{4}$Cr-1Mo	5A	1
5Cr-$\frac{1}{2}$Mo and 9Cr-1Mo	5B	1
9Cr-1Mo-V	5B	2
Type 304 and 316 stainless	8	1
Type 309 and 310 stainless	8	2
3$\frac{1}{2}$ Ni Steel	9B	1
Al and Al alloys	21 through 25	
Cu and Cu alloys	31 through 35	
Ni and Ni alloys	41 through 47	

Source: Selected from ASME Boiler and Pressure Vessel Code, Sec. IX, 1995.

Weld End Preparation

Butt welding ends are required to be prepared in accordance with ASME B16.25 and the code of jurisdiction. Figure 16.1 shows the maximum envelope for welding end transitions. Figure 16.2 illustrates the butt weld end preparation details for a weld end with a wall thickness equal to or less than 0.875 in (22 mm), with or without counterboring. Figure 16.3 illustrates the butt weld end preparation details for a weld end with a wall thickness over 0.875 in (22 mm), with or without counterboring.

Counterboring of Inside Diameter of Connecting Items to Be Butt Welded

To assure alignment of components to be joined by butt welding within the acceptable limits, it is common to machine or counterbore the inside diameters to a specified dimension, referred to as the C dimension. This also helps in minimizing or eliminating false indications when radiographic or volumetric examinations of the welds are conducted.

The C dimension is determined by the following equations:

$$C = A - 0.041 - 1.75t \text{ in}$$

$$C = A - 1.04 - 1.75t \text{ mm}$$

where A = pipe outside diameter and t = nominal pipe wall thickness.

Table 16.2 lists the values of pipe outside diameter A, pipe inside diameter B, and the C dimension for commercially available seamless or seam-welded wrought pipe, subjected to manufacturing pipe wall thickness tolerances of no more than 12.5 percent below the nominal wall thickness.

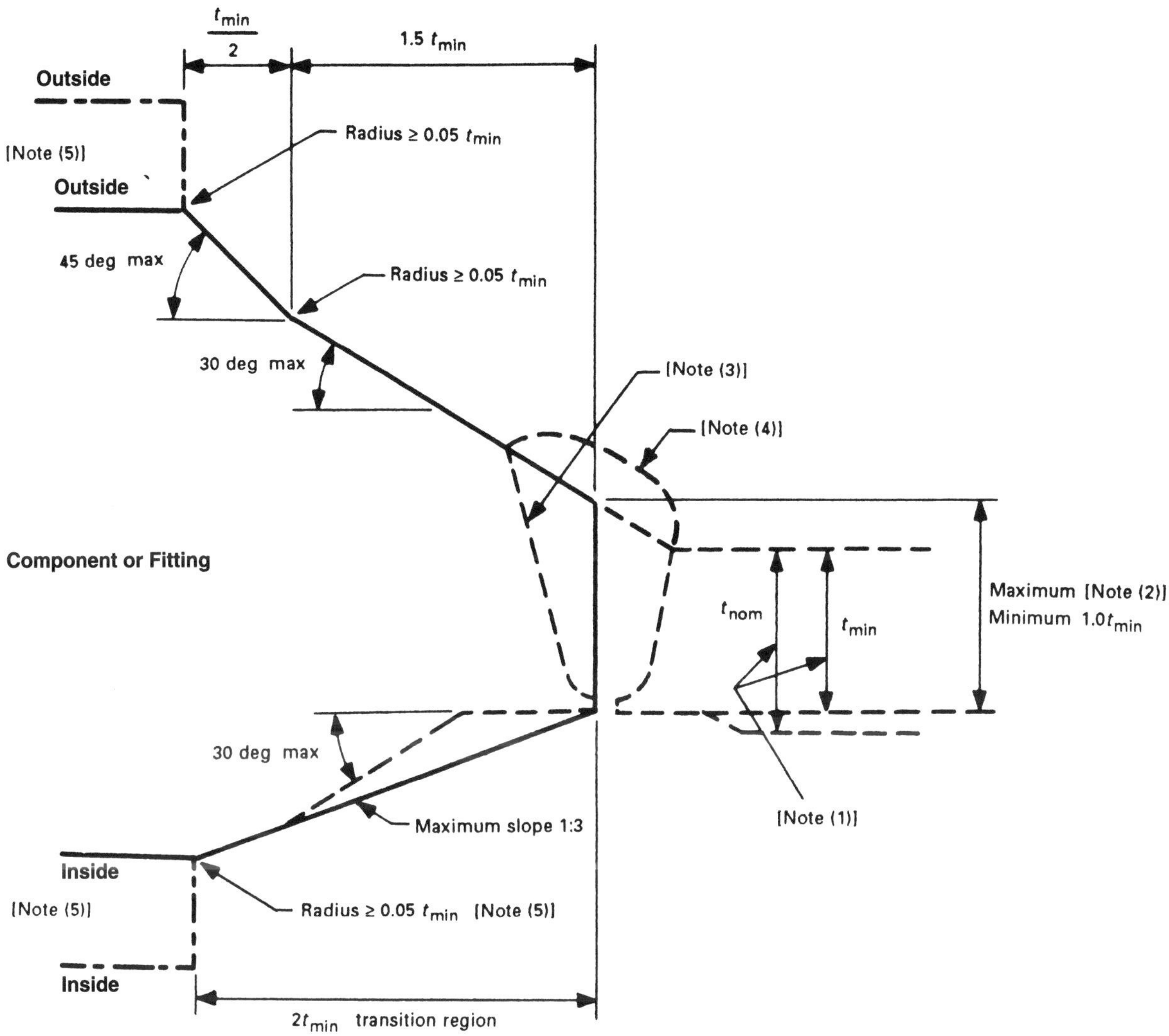

NOTES:

(1) The value of t_{min} is whichever of the following is applicable:
  (a) the minimum ordered wall thickness of the pipe;
  (b) 0.875 times the nominal wall thickness of pipe ordered to a pipe schedule wall thickness that has an under-tolerance of 12.5%;
  (c) the minimum ordered wall thickness of the cylindrical welding end of a component or fitting (or the thinner of the two) when the joint is between two components.

(2) The maximum thickness at the end of the component is:
  (a) the greater of t_{min} + 4 mm (0.16 in.) or 1.15t_{min} when ordered on a minimum wall basis;
  (b) the greater of t_{min} + 4 mm (0.16 in.) or 1.10t_{nom} when ordered on a nominal wall basis.

(3) Weld bevel shown is for illustration only.

(4) The weld reinforcement permitted by applicable code may lie outside the maximum envelope.

(5) Where transitions using maximum slope do not intersect inside or outside surface, as shown by phantom outlines, maximum slopes shown or alternate radii shall be used.

FIGURE 16.1 Maximum envelope for welding end transitions. (*Courtesy ASME International; adopted from ASME B16.25, Butt Welding Ends.*)

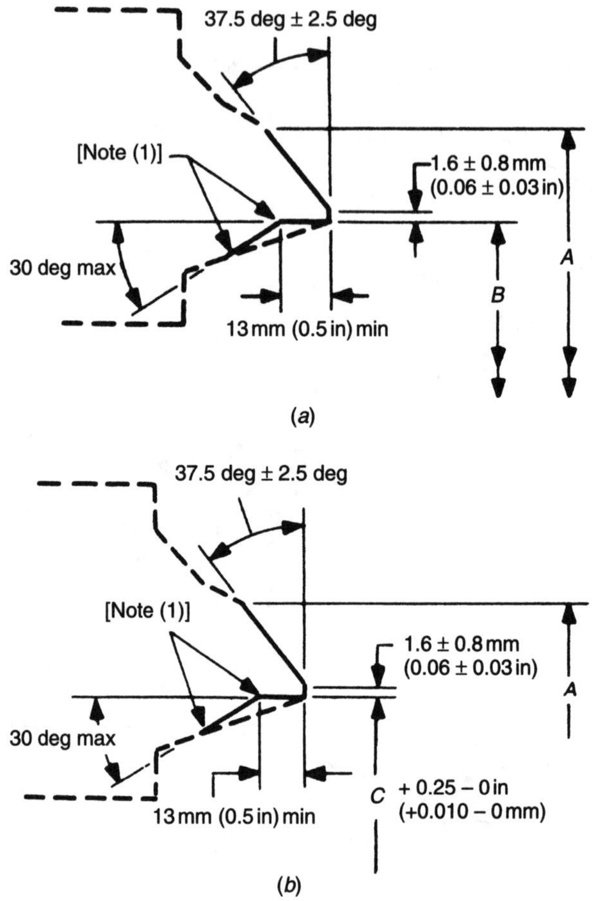

FIGURE 16.2 Butt welding ends for wall thickness equal to or less than 0.875 in (22 mm). Intersections shall be slightly rounded. (*a*) Weld end detail without counterbore. (*b*) Weld end detail with counterbore.

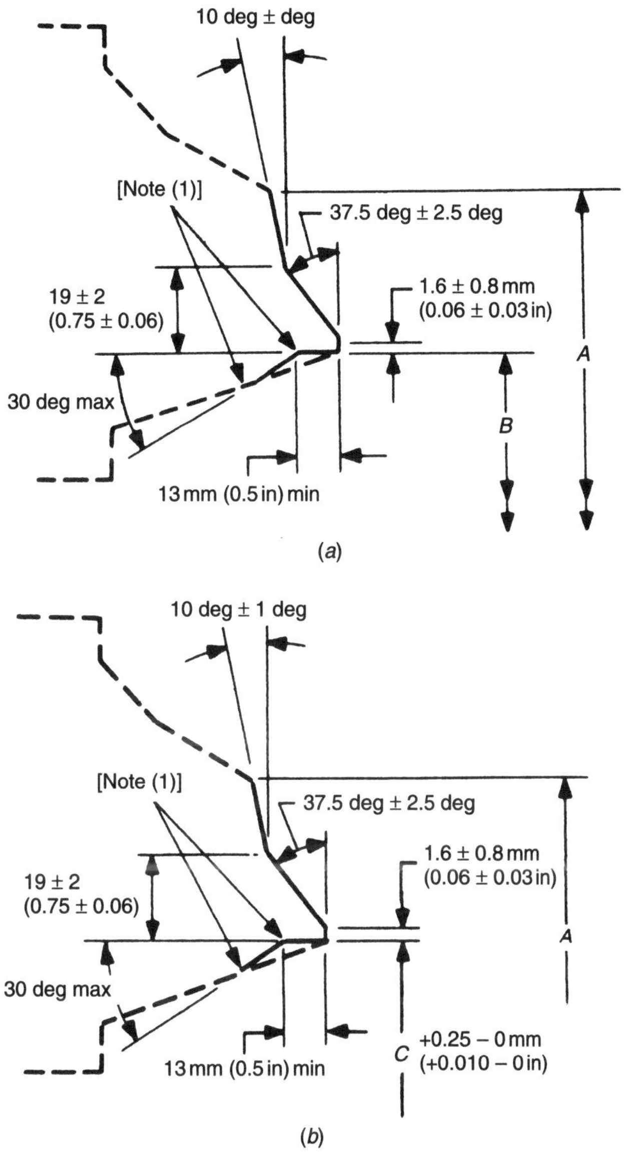

FIGURE 16.3 Butt welding ends for wall thickness over 0.875 in (22 mm). Intersections shall be slightly rounded. (*a*) Weld end detail without counterbore. (*b*) Weld end detail with counterbore.

TABLE 16.2 Dimensions of Welding Ends*

Nominal pipe size	Schedule number	Nominal wall thickness	Typical outside diameter of wrought or fabricated components, *A*	Nominal inside diameter, *B*	Diameter counterbore, *C*
$2\frac{1}{2}$	40	0.203	2.875	2.469	2.479
$2\frac{1}{2}$	80	0.276	2.875	2.323	2.351
$2\frac{1}{2}$	160	0.375	2.875	2.125	2.178
$2\frac{1}{2}$	XXS	0.552	2.875	1.771	1.868
3	40	0.216	3.500	3.068	3.081
3	80	0.300	3.500	2.900	2.934
3	160	0.438	3.500	2.624	2.692
3	XXS	0.600	3.500	2.300	2.409
$3\frac{1}{2}$	40	0.226	4.000	3.548	3.564
$3\frac{1}{2}$	80	0.318	4.000	3.364	3.402
4	40	0.237	4.500	4.026	4.044
4	80	0.337	4.500	3.826	3.869
4	120	0.438	4.500	3.624	3.692
4	160	0.531	4.500	3.438	3.530
4	XXS	0.674	4.500	3.152	3.279
5	40	0.258	5.563	5.047	5.070
5	80	0.375	5.563	4.813	4.866
5	120	0.500	5.563	4.563	4.647
5	160	0.625	5.563	4.313	4.428
5	XXS	0.750	5.563	4.063	4.209
6	40	0.280	6.625	6.065	6.094
6	80	0.432	6.625	5.761	5.828
6	120	0.562	6.625	5.501	5.600
6	160	0.719	6.625	5.187	5.326
6	XXS	0.864	6.625	4.897	5.072
8	40	0.322	8.625	7.981	8.020
8	60	0.406	8.625	7.813	7.873
8	80	0.500	8.625	7.625	7.709
8	100	0.594	8.625	7.437	7.544
8	120	0.719	8.625	7.187	7.326
8	140	0.812	8.625	7.001	7.163
8	XXS	0.875	8.625	6.875	7.053
8	160	0.906	8.625	6.813	6.998
10	40	0.365	10.750	10.020	10.070
10	60	0.500	10.750	9.750	9.834
10	80	0.594	10.750	9.562	9.670
10	100	0.719	10.750	9.312	9.451
10	120	0.844	10.750	9.062	9.232
10	140	1.000	10.750	8.750	8.959
10	160	1.125	10.750	8.500	8.740
12	STD	0.375	12.750	12.000	12.053
12	40	0.406	12.750	11.938	11.999
12	XS	0.500	12.750	11.750	11.834
12	60	0.562	12.750	11.626	11.725
12	80	0.688	12.750	11.374	11.505
12	100	0.844	12.750	11.062	11.232
12	120	1.000	12.750	10.750	10.959
12	140	1.125	12.750	10.500	10.740
12	160	1.312	12.750	10.126	10.413
14	STD	0.375	14.000	13.250	13.303
14	40	0.438	14.000	13.124	13.192
14	XS	0.500	14.000	13.000	13.084
14	60	0.594	14.000	12.812	12.920
14	80	0.750	14.000	12.500	12.646
14	100	0.938	14.000	12.124	12.318
14	120	1.094	14.000	11.812	12.044
14	140	1.250	14.000	11.500	11.771
14	160	1.406	14.000	11.188	11.498
16	STD	0.375	16.000	15.250	15.303
16	40	0.500	16.000	15.000	15.084
16	60	0.656	16.000	14.688	14.811
16	80	0.844	16.000	14.312	14.482

TABLE 16.2 Dimensions of Welding Ends* (Continued)

Nominal pipe size	Schedule number	Nominal wall thickness	Typical outside diameter of wrought or fabricated components, *A*	Nominal inside diameter, *B*	Diameter counterbore, *C*
16	100	1.031	16.000	13.938	14.155
16	120	1.219	16.000	13.562	13.826
16	140	1.438	16.000	13.124	13.442
16	160	1.594	16.000	12.812	13.170
18	STD	0.375	18.000	17.250	17.303
18	XS	0.500	18.000	17.000	17.084
18	40	0.562	18.000	16.876	16.975
18	60	0.750	18.000	16.500	16.646
18	80	0.938	18.000	16.124	16.318
18	100	1.156	18.000	15.688	15.936
18	120	1.375	18.000	15.250	15.553
18	140	1.562	18.000	14.876	15.225
18	160	1.781	18.000	14.438	14.842
20	STD	0.375	20.000	19.250	19.303
20	XS	0.500	20.000	19.000	19.084
20	40	0.594	20.000	18.812	18.920
20	60	0.812	20.000	18.376	18.538
20	80	1.031	20.000	17.938	18.155
20	100	1.281	20.000	17.438	17.717
20	120	1.500	20.000	17.000	17.334
20	140	1.750	20.000	16.500	16.896
20	160	1.969	20.000	16.062	16.513
22	STD	0.375	22.000	21.250	21.303
22	XS	0.500	22.000	21.000	21.084
22	60	0.875	22.000	20.250	20.428
22	80	1.125	22.000	19.750	19.990
22	100	1.375	22.000	19.250	19.553
22	120	1.625	22.000	18.750	19.115
22	140	1.875	22.000	18.250	18.678
22	160	2.125	22.000	17.750	18.240
24	STD	0.375	24.000	23.250	23.303
24	XS	0.500	24.000	23.000	23.084
24	30	0.562	24.000	22.876	22.975
24	40	0.688	24.000	22.624	22.755
24	60	0.969	24.000	22.062	22.263
24	80	1.219	24.000	21.562	21.826
24	100	1.531	24.000	20.938	21.280
24	120	1.812	24.000	20.376	20.788
24	140	2.062	24.000	19.876	20.350
24	160	2.344	24.000	19.312	19.857
26	10	0.312	26.000	25.376	25.413
26	20	0.500	26.000	25.000	25.084
28	10	0.312	28.000	27.376	27.413
28	20	0.500	28.000	27.000	27.084
28	30	0.625	28.000	26.750	26.865
30	10	0.312	30.000	29.376	29.413
30	20	0.500	30.000	29.000	29.084
30	30	0.625	30.000	28.750	28.865
32	10	0.312	32.000	31.376	31.413
32	20	0.500	32.000	31.000	31.084
32	30	0.625	32.000	30.750	30.865
32	40	0.688	32.000	30.624	30.755
34	10	0.312	34.000	33.376	33.413
34	20	0.500	34.000	33.000	33.084
34	30	0.625	34.000	32.750	32.865
34	40	0.688	34.000	32.624	32.755
36	10	0.312	36.000	35.376	35.413
36	20	0.500	36.000	35.000	35.084
36	30	0.625	36.000	34.750	34.865
36	40	0.750	36.000	34.500	34.646

* Dimensions in inches.

Preheat Temperatures for Welding

Preheating of base metals prior to welding slows the cooling rate of the weld joint and results in more ductile metallurgical structure in the weld metal and the heat-affected zone (HAZ). Also, it helps to reduce shrinkage, distortion, and possible cracking caused by residual stresses. Table 16.3 lists the typical preheat temperatures for some of the P-number materials.

Table 16.3 Typical Preheat Requirements

P no.	Temp., °F	Composition/thickness limits
1	175	For *both* a max. specified carbon content >0.30% *and* thickness >1 in.
	50	For all others.
3	175	For *either* a min. specified tensile strength >60 ksi, *or* thickness >½ in.
	50	For all others.
4	250	For *either* a min. specified tensile strength >60 ksi, *or* thickness >½ in.
	50	For all others.
5A and 5B	400	For *either* a min. specified tensile strength >60 ksi, *or both* a min. specified Cr content >6.0% *and* thickness >½ in.
	300	For all others.
6	400	For all materials.
7	50	For all materials.
8	50	For all materials.
9A	250	For all materials.
9B	300	For all materials.
10I	300	With a max. interpass temperature of 450°F.

Source: ASME B31.1, 1995.

Postweld Heat Treatment (PWHT)

Table 16.4 provides postweld heat treatment temperature range, holding time, and cooling rate for various materials.

BENDING OF PIPE

Depending on the method of bending and the radius of bend, additional pipe wall thickness or bending allowance must be provided to overcome thinning of pipe on the extrados. Table 16.5 lists recommended pipe wall thickness prior to bending for various methods of bending used in the industry.

Table 16.4 Some Typical Time and Temperature Cycles for Heat Treatment

P no.	Heating rate	Holding temperature range,* °F				Minimum holding time at temperature	Cooling program
		SR or T	N	A	CST		
P-1	Above 800°F heat at a rate of 400°F/h divided by the thickness in inches but not faster than 400°F or less than 100°F	1100–1250	1600–1700	1500–1600	NA	1 h/in of thickness but not less than 30 min or more than 2 h plus 15 min for each additional inch over 2 in	SR or T—Cool at 400°F/h divided by the thickness in inches but not faster than 400°F/h; need not be lower than 100°F/h down to 800°F N—Remove from furnace at normalizing temperature and cool in still air to 800°F; temper as necessary A—Furnace cool to 800°F at a rate of 400°F/h divided by the thickness in inches but not faster than 400°F/h; need not be slower than 100°F/h
P-3	Same as P-1	1100–1250	1600–1700	1500–1600	NA		
P-4	Same as P-1	1300–1375	1725–1775	1625–1675	NA	1 h/in of thickness but not less than 30 min or more than 5 h plus 15 min for each additional inch over 5 in	
P-5A and P-5B Gr.-1	Same as P-1	1300–1400	1725–1775	1625–1675	NA		
P-8	Same as P-1	Not required	N/A	N/A	1900–2000	1 h/in of thickness but not less than 30 min or more than 2 h plus 15 min for each additional inch over 2 in	CST—Remove from furnace at holding temperature and quench in water to 300°F within 2 min

* SR = stress relief, T = temper, N = normalize, A = anneal, CST = carbide solution treatment.
Source: Pullman Power Products Corporation.

Table 16.5 Recommended Minimum Pipe Wall Thickness Prior to Bending for Various Bending Methods

Furnace bending		Induction and incremental bending		Rotary draw bending		Ram and roll bending	
Bend radius	Minimum thickness prior to bending	Bend radius	Minimum thickness prior to bending	Bend radius	Minimum thickness prior to bending	Bend radius	Minimum thickness prior to bending
$6D$	$1.06t_m$	$6D_n$	$1.06t_m$	$6D_n$	$1.09t_m$	$6D_n$	$1.08t_m$
$5D$	$1.08t_m$	$5D_n$	$1.08t_m$	$5D_n$	$1.14t_m$	$5D_n$	$1.10t_m$
$4D$	$1.14t_m$	$4D_n$	$1.10t_m$	$4D_n$	$1.20t_m$	$4D_n$	$1.13t_m$
$3D$	$1.25t_m$	$3D_n$	$1.14t_m$	$3D_n$	$1.28t_m$	$3D_n$	$1.17t_m$
		$2D_n$	$1.22t_m$				
		$1.5D_n$	$1.30t_m$				

D = diameter, D_n = nominal outside diameter, t_m = minimum thickness.

Hot Bending of Piping

Hot bending of pipe is performed at a temperature above the lower critical temperature of the pipe material minus 100°F (56°C). Table 16.6 lists lower critical temperatures of some piping materials. Similarly, Table 16.7 lists the temperature range for hot bending of copper and copper alloy piping. Table 16. 8 lists the temperature range for hot bending of nickel and nickel alloy piping. Figure 16.4 shows hot bending of pipe by induction bending or incremental bending method.

Cold Bending of Piping

Cold bending of piping is performed at a temperature below the lower critical temperature minus 100°F (56°C). Refer to Tables 16.6 through 16.8.

FABRICATING TOLERANCES

Fabricating tolerances for piping are covered by PFI Standard ES-3. Figure 16.5 illustrates fabricating tolerance for a piping subassembly. Linear tolerances for dimension A are listed in Table 16.9.

Table 16.6 Approximate Lower Critical Temperatures

Material	Approximate lower critical temperature, °F (°C)
Carbon steel	1340 (725)
Carbon molybdenum steel	1350 (730)
$1\frac{1}{4}$Cr-$\frac{1}{2}$Mo	1430 (775)
$2\frac{1}{4}$Cr-1Mo, 3Cr-1Mo	1480 (805)
5Cr-$\frac{1}{2}$Mo	1505 (820)
9Cr-$\frac{1}{2}$Mo	1490 (810)

Notes:
1. Hot bending is performed at a temperature above lower critical temperature minus 100°F (56°C).
2. Cold bending is performed at a temperature below lower critical temperature minus 100°F (56°C).

Source: From ASME B31.1, 1998.

Table 16.7 Temperature Range for Hot Bending of Copper and Copper Alloy Pipe

Material	Spec no.	Alloy	Temperature range, °F
Copper deoxidized	SB-42, SB-75	C10200	1400–1600
Red brass	SB-43, SB-135	C23000	1450–1650
Copper silicon A	SB-315	C65500	1300–1600
70-30 CuNi	SB-466, SB-467	C71500	1700–2000
80-20 CuNi	SB-466	C71000	1600–1900
90-10 CuNi	SB-466, SB-467	C70600	1400–1800

Source: Adapted from ASME Boiler and Pressure Vessel Code, 1995, Sec. VIII, Div. 1, Table NF-4.

Table 16.8 Temperature Range for Hot Bending of Nickel and Nickel Alloy Pipe and Tube

Material	Spec. no.	Alloy	Temperature range, °F
Nickel	SB-161	N02200	1200–2300
Low carbon nickel	SB-161	N02201	1200–2300
Nickel-copper	SB-165	N04400	1700–2150
Ni-Cr-Fe	SB-167, SB-517	N06600, N06690	1850–2300
Ni-Fe-Cr	SB-407, SB-514	N08800, N08810	1850–2200

Source: Adapted from ASME Boiler and Pressure Vessel Code, 1995, Sec. VIII, Div. 1, Table NF-4.

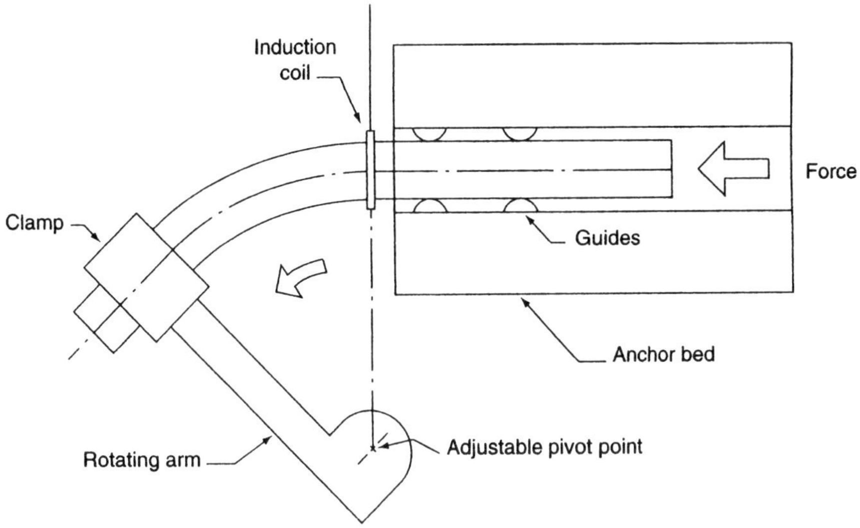

FIGURE 16.4 Induction bending.

Table 16.9 Fabricating Tolerances per PFI ES-3

Nominal pipe size (NPS)	Linear tolerance on *A*	
	Inch	Millimeter
10 and under	±$^1/_8$	±3.0
12 through 24	±$^3/_{16}$	±5.0
Over 24 through 36	±$^1/_4$	±6.0
Over 36	±$^1/_4$ plus*	±6.0 plus*

* Add ±$^1/_{16}$ in (2.0 mm) for each 12-in diameter over 36 in.

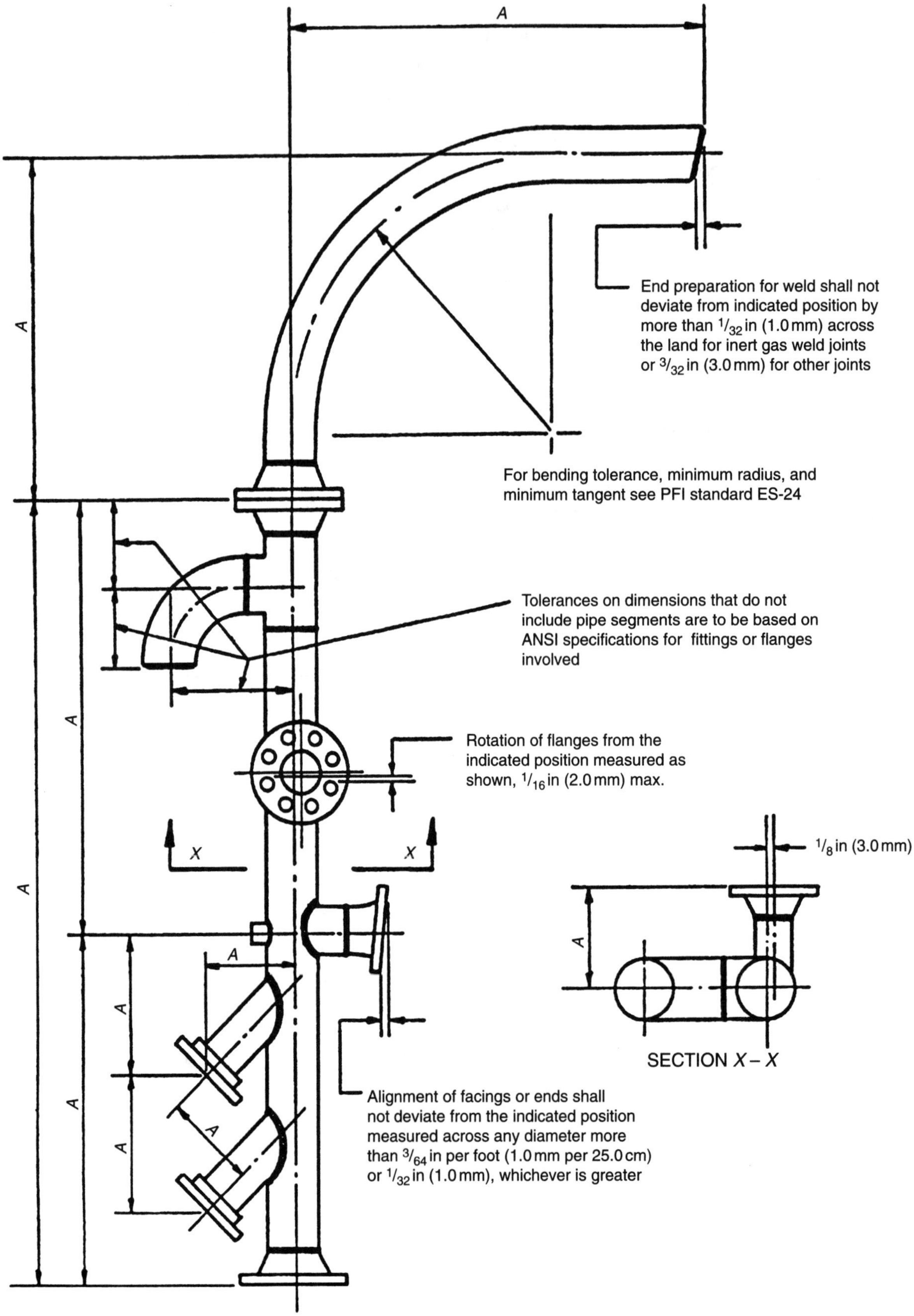

FIGURE 16.5 Appplication of pipe fabrication tolerances.

Tables in Chapter 16

S · E · C · T · I · O · N · 5

NONMETALLIC PIPING

CHAPTER 17
PVC PLASTIC PIPES, FITTINGS, AND JOINTS

See p. 556 for a list of tables in Chap. 17.

Poly vinyl chloride (PVC) plastic pipes are made of thermoplastic compounds prepared by combining PVC resin with modifiers to attain desirable properties and with stabilizers, lubricants, and pigments for processing, property control, and coloring. This chapter provides various data for PVC plastic pipes, fittings, and joints.

The length changes for PVC pipe due to expansion at various temperatures shall be in accordance with Table 17.1.

The pressure-class ratings for PVC material are generally provided at service temperatures of 73.4°F and lower. The pressure-class ratings must be appropriately reduced for use at temperatures greater than 73.4°F in accordance with Table 17.2.

Table 17.1 Thermal Expansion Table for PVC Plastic Pipe and Fittings

	Temperature change, °F						
	40	50	60	70	80	90	100
Length, ft	Length change, in						
20	0.28	0.35	0.42	0.49	0.58	0.63	0.70
40	0.56	0.70	0.84	0.97	1.11	1.25	1.39
60	0.84	1.04	1.25	1.46	1.67	1.88	2.09
80	1.13	1.39	1.67	1.95	2.23	2.51	2.78
100	1.39	1.74	2.09	2.44	2.78	3.13	3.48

Table 17.2 Temperature Design Factors

Maximum service temperature, °F	Temperature design factor for PVC
80	0.88
90	0.75
100	0.62
120	0.40
130	0.30
140	0.22

PVC PLASTIC PIPES

This section covers the following pipe standard specifications:

- ANSI/AWWA C900 PVC Pressure Pipe, 4 in. through 12 in., for Water Distribution
- ASTM D1785 PVC Plastic Pipe, Schedules 40, 80 and 120
- ASTM D2241 PVC Pressure-Rated Pipe (SDR Series)
- ASTM D2665 PVC Plastic Drain, Waste, and Vent Pipe and Fittings
- ASTM F891 PVC Plastic Pipe with a Cellular Core

PVC Pressure Pipe, 4 in through 12 in, for Water Distribution (ANSI/AWWA C900)

This standard pertains to 4-in through 12-in PVC pressure pipe providing a hydrostatic design basis (HDB) of 4000 psi and with cast-iron (CI)-pipe-equivalent outside diameter (OD) dimensions and with wall thicknesses equivalent to a dimension ratio (DR) series 14, 18, or 25. For pipe covered by this standard, the DR is determined by dividing the average outside diameter of the pipe by its minimum wall thickness and rounding off the quotient, when necessary, to the nearest whole number.

Table 17.3 Dimensions and Pressure Classes for Dimension Ratio (DR) for PVC Pipe with Cast-Iron-Pipe-Equivalent ODs

Nominal size, in	Pressure class at 73.4°F, psi	DR	Average OD, in	Minimum wall thickness, in
4	100	25	4.800	0.192
4	150	18	4.800	0.267
4	200	14	4.800	0.343
6	100	25	6.900	0.276
6	150	18	6.900	0.383
6	200	14	6.900	0.493
8	100	25	9.050	0.362
8	150	18	9.050	0.503
8	200	14	9.050	0.646
10	100	25	11.100	0.444
10	150	18	11.100	0.617
10	200	14	11.100	0.793
12	100	25	13.200	0.528
12	150	18	13.200	0.733
12	200	14	13.200	0.943

Table 17.4 Outside Diameters and Tolerances for PVC Plastic Pipe Schedules 40, 80, and 120*

Nominal pipe size	Outside diameter	Average tolerance, plus and minus
1/8	0.405	0.004
1/4	0.540	0.004
3/8	0.675	0.004
1/2	0.840	0.004
3/4	1.050	0.004
1	1.315	0.005
1 1/4	1.660	0.005
1 1/2	1.900	0.006
2	2.375	0.006
2 1/2	2.875	0.007
3	3.500	0.008
3 1/2	4.000	0.008
4	4.500	0.009
5	5.563	0.010
6	6.625	0.011
8	8.625	0.015
10	10.750	0.015
12	12.750	0.015
14	14.000	0.015
16	16.000	0.019
18	18.000	0.019
20	20.000	0.023
24	24.000	0.031

* Dimensions in inches.

The requirements for dimensions and pressure classes for DR for PVC pipes with CI-pipe-equivalent ODs shall be in accordance with Table 17.3.

PVC Plastic Pipe, Schedules 40, 80, and 120 (ASTM D1785)

This specification covers PVC pipe made in schedule 40, 80, and 120 sizes and pressure rated for water. PVC pipe made to this specification is often belled for use as line pipe.

The outside diameters and tolerances shall be in accordance with Table 17.4. The minimum wall thickness and tolerances shall be in accordance with Table 17.5. The water pressure ratings shall be in accordance with Table 17.6.

PVC Pressure-Rated Pipe—SDR Series (ASTM D2241)

This specification covers PVC pipe made in standard pipe dimension ratios and pressure rated for water.

The average outside diameter and tolerance for IPS pipe shall be in accordance with Table 17.7. The wall thickness and tolerances for pipe with IPS outside diameters shall be in accordance with Table 17.8. The water pressure ratings shall be in accordance with Table 17.9.

PVC Plastic Drain, Waste, and Vent Pipe and Fittings (ASTM D2665)

This specification covers requirements and test methods for materials, dimensions and tolerances, pipe stiffness, crush resistance, impact resistance, hydrostatic burst resistance, and solvent cement for PVC plastic drain, waste, and vent pipe and fittings.

The dimension and tolerances for outside diameter and thickness shall be in accordance with Table 17.10. The dimensions for fitting sockets shall be in accordance with Table 17.11.

TABLE 17.5 Wall Thicknesses and Tolerances for PVC Plastic Pipe, Schedules 40, 80, and 120

Nominal pipe size, in	Wall thickness, in					
	Schedule 40		Schedule 80		Schedule 120	
	Minimum	Tolerance	Minimum	Tolerance	Minimum	Tolerance
1/8	0.068	+0.020	0.095	+0.020	...	...
1/4	0.088	+0.020	0.119	+0.020	...	...
3/8	0.091	+0.020	0.126	+0.020	...	...
1/2	0.109	+0.020	0.147	+0.020	0.170	+0.020
3/4	0.113	+0.020	0.154	+0.020	0.170	+0.020
1	0.133	+0.020	0.179	+0.021	0.200	+0.024
1 1/4	0.140	+0.020	0.191	+0.023	0.215	+0.026
1 1/2	0.145	+0.020	0.200	+0.024	0.225	+0.027
2	0.154	+0.020	0.218	+0.026	0.250	+0.030
2 1/2	0.203	+0.024	0.276	+0.033	0.300	+0.036
3	0.216	+0.026	0.300	+0.036	0.350	+0.042
3 1/2	0.226	+0.027	0.318	+0.038	0.350	+0.042
4	0.237	+0.028	0.337	+0.040	0.437	+0.052
5	0.258	+0.031	0.375	+0.045	0.500	+0.060
6	0.280	+0.034	0.432	+0.052	0.562	+0.067
8	0.322	+0.039	0.500	+0.060	0.718	+0.086
10	0.365	+0.044	0.593	+0.071	0.843	+0.101
12	0.406	+0.049	0.687	+0.082	1.000	+0.120
14	0.437	+0.053	0.750	+0.090	...	...
16	0.500	+0.060	0.843	+0.101	...	...
18	0.562	+0.067	0.937	+0.112	...	...
20	0.593	+0.071	1.031	+0.124	...	...
24	0.687	+0.082	1.218	+0.146	...	...

* Dimensions in inches.

TABLE 17.6 Water Pressure Ratings at 73°F for Schedule 40 PVC Plastic Pipe

Nominal pipe size, in	Pressure ratings,* psi			
	PVC1120 PVC1220 PVC2120	PVC2116	PVC2112	PVC2110
1/8	810	650	500	400
1/4	780	620	490	390
3/8	620	500	390	310
1/2	600	480	370	300
3/4	480	390	300	240
1	450	360	280	220
1 1/4	370	290	230	180
1 1/2	330	260	210	170
2	280	220	170	140
2 1/2	300	240	190	150
3	260	210	160	130
3 1/2	240	190	150	120
4	220	180	140	110
5	190	160	120	100
6	180	140	110	90
8	160	120	100	80
10	140	110	90	70
12	130	110	80	70
14	130	100	80	60
16	130	100	80	60
18	130	100	80	60
20	120	100	80	60
24	120	90	70	60

* These pressure ratings apply only to unthreaded pipe. Schedule 40 pipe is not recommended to be threaded.

TABLE 17.7 IPS PVC Pipe—Outside Diameters and Tolerances*

Nominal pipe size	Average outside diameter	Tolerance, plus and minus
1/8	0.405	0.004
1/4	0.540	0.004
3/8	0.675	0.004
1/2	0.840	0.004
3/4	1.050	0.004
1	1.315	0.005
1 1/4	1.660	0.005
1 1/2	1.900	0.006
2	2.375	0.006
2 1/2	2.875	0.007
3	3.500	0.008
3 1/2	4.000	0.008
4	4.500	0.009
5	5.563	0.010
6	6.625	0.011
8	8.625	0.015
10	10.750	0.015
12	12.750	0.015
14	14.000	0.015
16	16.000	0.019
18	18.000	0.019
20	20.000	0.023
24	24.000	0.031
30	30.000	0.041
36	36.000	0.050

* Dimensions in inches.

TABLE 17.8 Wall Thicknesses and Tolerances for PVC Plastic Pipe with IPS Outside Diameters

Nominal pipe size, in	Wall thickness, in													
	SDR64		SDR41		SDR32.5		SDR26		SDR21		SDR17		SDR13.5	
	Minimum	Tolerance	Minimum	Tolerance	Minimum	Tolerance	Minimum	Tolerance	Minimum	Tolerance	Minimum	Tolerance	Minimum	Tolerance
1/8	...	...	...	...	...	...	...	...	...	...	...	...	0.060	+0.020
1/4	...	...	...	...	...	...	...	...	...	...	...	...	0.060	+0.020
3/8	...	...	...	...	...	...	...	...	...	...	...	...	0.060	+0.020
1/2	...	...	...	...	...	...	...	...	...	...	...	...	0.062	+0.020
3/4	...	...	...	...	...	...	...	...	0.060	+0.020	0.062	+0.020	0.078	+0.020
1	...	...	...	...	...	...	0.060	+0.020	0.063	+0.020	0.077	+0.020	0.097	+0.020
1 1/4	...	...	...	...	0.060	+0.020	0.064	+0.020	0.079	+0.020	0.098	+0.020	0.123	+0.020
1 1/2	...	...	...	...	0.060	+0.020	0.073	+0.020	0.090	+0.020	0.112	+0.020	0.141	+0.020
2	...	...	...	...	0.073	+0.020	0.091	+0.020	0.113	+0.020	0.140	+0.020	0.176	+0.020
2 1/2	...	...	...	...	0.088	+0.020	0.110	+0.020	0.137	+0.020	0.169	+0.020	0.213	+0.026
3	...	...	0.085	+0.020	0.108	+0.020	0.135	+0.020	0.167	+0.020	0.206	+0.025	0.259	+0.031
3 1/2	...	...	0.098	+0.020	0.123	+0.020	0.154	+0.020	0.190	+0.023	0.235	+0.028	0.296	+0.036
4	0.070	+0.020	0.110	+0.020	0.138	+0.020	0.173	+0.020	0.214	+0.026	0.265	+0.032	0.333	+0.040
5	0.087	+0.020	0.136	+0.020	0.171	+0.021	0.214	+0.027	0.265	+0.032	0.327	+0.039	0.412	+0.049
6	0.104	+0.020	0.162	+0.020	0.204	+0.024	0.255	+0.031	0.316	+0.038	0.390	+0.047	0.491	+0.059
8	0.135	+0.020	0.210	+0.025	0.265	+0.032	0.332	+0.040	0.410	+0.049	0.508	+0.061	...	...
10	0.168	+0.020	0.262	+0.031	0.331	+0.040	0.413	+0.050	0.511	+0.061	0.632	+0.076	...	...
12	0.199	+0.024	0.311	+0.037	0.392	+0.047	0.490	+0.059	0.606	+0.073	0.750	+0.090	...	...
14	...	...	0.341	+0.048	0.430	+0.052	0.538	+0.064	0.666	+0.080	0.823	+0.099	...	...
16	...	...	0.390	+0.055	0.492	+0.059	0.615	+0.074	0.762	+0.091	0.941	+0.113	...	...
18	...	...	0.439	+0.061	0.554	+0.066	0.692	+0.083	0.857	+0.103	1.059	+0.127	...	...
20	...	...	0.488	+0.068	0.615	+0.074	0.769	+0.092	0.952	+0.114	1.176	+0.141	...	...
24	...	...	0.585	+0.082	0.738	+0.088	0.923	+0.111	1.143	+0.137	1.412	+0.169	...	...
30	...	...	0.732	+0.102	0.923	+0.111	1.154	+0.138	1.428	+0.171	1.765	+0.212	...	...
36	...	...	0.878	+0.123	1.108	+0.133	1.385	+0.166	1.714	+0.205	2.118	+0.254	...	...

TABLE 17.9 Water Pressure Ratings (PR) at 73°F for Nonthreaded PVC Plastic Pipe

SDR	PVC 1120, PVC 1220, PVC 2120, psi
11	400
13.5	315
17	250
21	200

TABLE 17.10 Dimensions and Tolerances for Outside Diameters and Thicknesses of PVC Plastic Drain, Waste, and Vent Pipe*

Nominal pipe size	Outside diameter		Wall thickness	
	Average	Tolerance on average	Minimum	Tolerance
$1^1/_4$	1.660	±0.005	0.140	+0.020 −0.000
$1^1/_2$	1.900	±0.006	0.145	+0.020 −0.000
2	2.375	±0.006	0.154	+0.020 −0.000
3	3.500	±0.008	0.216	+0.026 −0.000
4	4.500	±0.009	0.237	+0.028 −0.000
6	6.625	±0.011	0.280	+0.034 −0.000
8	8.625	±0.015	0.322	+0.039 −0.000
10	10.750	±0.015	0.365	+0.044 −0.000
12	12.750	±0.015	0.406	+0.049 −0.000

* Dimensions in inches.

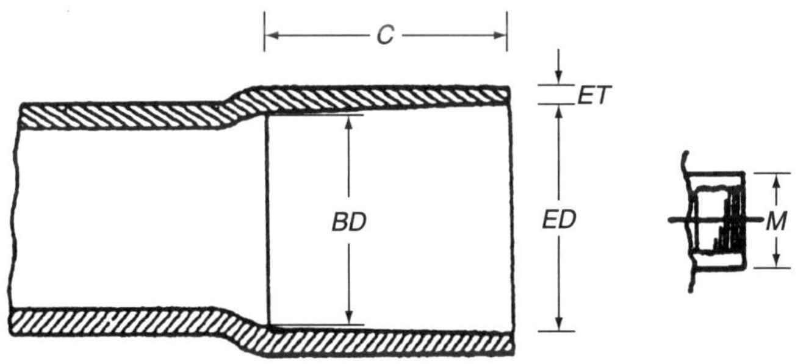

TABLE 17.11 Dimensions and Tolerances for Fitting Sockets for PVC Plastic Drain, Waste, and Vent Pipe Fittings*

Nominal pipe size	*ED*, socket entrance diameter	*BD*, socket bottom diameter	*C*, socket depth, min.	*ET*, wall thickness, min.	Internal threads	
					Outside diameter of hub, *M*, min.	Thread length, min.
$1^1/_4$	1.675	1.655	0.687	0.156	1.871	0.687
$1^1/_2$	1.915	1.895	0.687	0.156	2.127	0.687
2	2.390	2.370	0.750	0.156	2.634	0.750
3	3.520	3.495	1.500	0.219	3.841	1.187
4	4.520	4.495	1.750	0.250	4.907	1.281
6	6.647	6.614	3.000	0.281	7.203	1.500
8	8.655	8.610	4.000	0.328	9.320	1.687
10	10.780	10.735	5.000	0.365		
12	12.780	12.735	6.000	0.406		

* Dimensions in inches.

Coextruded PVC Plastic Pipe with a Cellular Core (ASTM F891)

This specification covers coextruded PVC pipe with a cellular core and concentric inner and outer solid layers, produced by using a multilayer coextrusion die for nonpressure use in three series: an IPS schedule 40 series; a PS series with an iron pipe size outside diameter with varying wall thickness as required for pipe stiffness of 25, 50, and 100; and a sewer and drain series.

The outside diameter and tolerance shall be in accordance with Table 17.12. The minimum wall thickness, pipe stiffness, and impact strength for IPS schedule 40 series pipe shall be in accordance with Table 17.13. The minimum wall thickness for PS series pipe or sewer and drain series pipe for minimum pipe stiffness shall be in accordance with Table 17.14.

TABLE 17.12 Outside Diameter and Tolerances*

Nominal pipe size, in	Outside diameter: IPS schedule 40 series, PS series	Outside diameter: Sewer and drain series	Tolerance on average outside diameter, plus and minus
$1^1/_4$	1.660	...	0.005
$1^1/_2$	1.900	...	0.006
2	2.375	...	0.006
3	3.500	3.250	0.008
$3^1/_2$	4.000	...	0.008
4	4.500	4.215	0.009
5	5.563	...	0.010
6	6.625	6.275	0.011
8	8.625	8.400	0.015
10	10.750	10.500	0.015
12	12.750	12.500	0.015
15	...	15.300	0.023
18	...	18.700	0.028

* Dimensions in inches.

TABLE 17.13 Minimum Wall Thickness, Pipe Stiffness, and Impact Strength for IPS Schedule 40 Series

Nominal pipe size, in	Wall thickness, in*	Pipe stiffness, lbf/in^2	Impact strength, ft·lbf
$1^1/_4$	0.140	600	40
$1^1/_2$	0.145	600	50
2	0.154	300	80
3	0.216	300	100
$3^1/_2$	0.226	250	100
4	0.237	200	100
5	0.258	120	100
6	0.280	120	120
8	0.322	100	120
10	0.365	60	120
12	0.406	50	120

* The maximum wall thickness shall not be greater than 1.25 times the minimum wall thickness.

TABLE 17.14 Minimum Wall Thickness for PS Series or Sewer and Drain Series for Minimum Pipe Stiffness

Nominal pipe size, in	Wall thickness, in*						
	PS series			Sewer and drain series			
	PS 25	PS 50	PS 100	PS 12.5	PS 25	PS 50	PS 100
2	...	...	0.100	...	...	...	...
3	...	0.103	0.130	...	...	0.100	0.120
$3^1/_2$	0.100	0.118	0.148	...	...	...	...
4	0.106	0.133	0.167	...	0.100	0.124	0.156
5	0.131	0.164	0.206	...	...	...	...
6	0.156	0.195	0.245	0.118	0.148	0.185	0.232
8	...	...	...	...	0.198	0.248	0.310
10	...	...	...	...	...	0.310	0.388
12	...	...	...	...	...	0.369	0.462
15	...	...	...	...	...	0.452	0.565
18	...	...	...	...	...	0.548	0.691

* The maximum wall thickness shall not be greater than 1.25 times the minimum wall thickness.

PVC PLASTIC FITTINGS

This section covers the following fitting specifications:

- ASTM D2464 Threaded PVC Plastic Pipe Fittings, Schedule 80
- ASTM D2466 PVC Plastic Pipe Fittings, Schedule 40
- ASTM D2467 PVC Plastic Pipe Fittings, Schedule 80
- ASTM D2665 PVC Plastic Drain, Waste, and Vent Pipe and Fittings

Threaded PVC Plastic Pipe Fittings, Schedule 80 (ASTM D2464)

This specification covers PVC threaded schedule 80 pipe fittings. For dimensions for 90° elbows, refer to Table 17.15. For dimensions for 45° elbows, refer to Table 17.16. For dimensions for tees and crosses, refer to Table 17.17. For dimensions for couplings, refer to Table 17.18. For dimensions for caps, refer to Table 17.19.

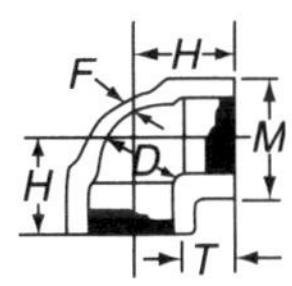

TABLE 17.15 Dimensions of 90° Elbows*

Nominal pipe size	Center to thread end, H, min.	Length of thread, T, min.	Inside diameter of fitting, D, min.	Nominal wall thickness, F, min.	Outside diameter of band, M, min.
$^1/_8$	0.688	0.38	0.171	0.108	0.645
$^1/_4$	0.812	0.50	0.258	0.135	0.840
$^3/_8$	0.938	0.50	0.379	0.144	1.000
$^1/_2$	1.125	0.64	0.502	0.198	1.280
$^3/_4$	1.250	0.65	0.698	0.207	1.500
1	1.500	0.81	0.911	0.225	1.810
$1^1/_4$	1.750	0.85	1.227	0.261	2.200
$1^1/_2$	1.938	0.85	1.446	0.270	2.500
2	2.250	0.90	1.881	0.297	3.000
$2^1/_2$	2.688	1.21	2.250	0.315	3.580
3	3.063	1.30	2.820	0.405	4.300
4	3.625	1.38	3.737	0.450	5.430
6	5.125	1.50	5.646	0.504	7.625

* Dimensions in inches.

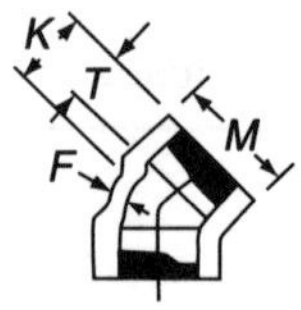

TABLE 17.16 Dimensions of 45° Elbows*

Nominal pipe size	Length of thread, T, min.	Center to thread end, K, min.	Nominal wall thickness, F, min.	Outside diameter of band, M, min.
1/8	0.38	0.625	0.108	0.645
1/4	0.50	0.688	0.135	0.840
3/8	0.50	0.750	0.144	1.000
1/2	0.64	0.750	0.198	1.280
3/4	0.65	1.000	0.207	1.500
1	0.81	1.125	0.225	1.810
1 1/4	0.85	1.313	0.261	2.200
1 1/2	0.85	1.438	0.270	2.500
2	0.90	1.625	0.297	3.000
2 1/2	1.21	1.938	0.315	3.580
3	1.30	2.125	0.405	4.300
4	1.38	2.625	0.450	5.430
6	1.50	3.250	0.504	7.625

* Dimensions in inches.

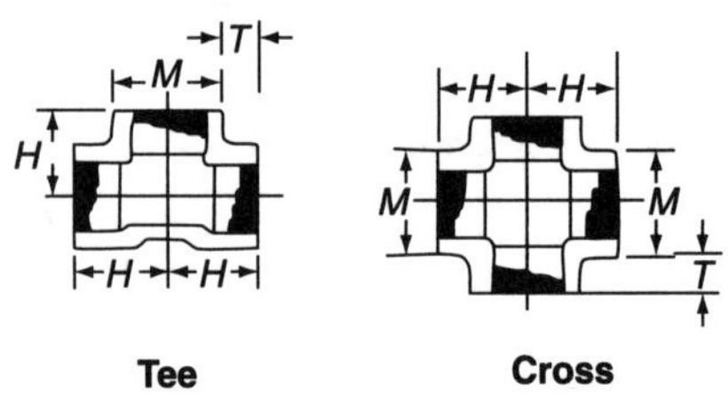

TABLE 17.17 Dimensions of Tees and Crosses*

Nominal pipe size	Center to thread end, H, min.	Length of thread, T, min.	Outside diameter of band, M, min.
1/8	0.688	0.38	0.645
1/4	0.812	0.50	0.840
3/8	0.938	0.50	1.000
1/2	1.125	0.64	1.280
3/4	1.250	0.65	1.500
1	1.500	0.81	1.810
1 1/4	1.750	0.85	2.200
1 1/2	1.938	0.85	2.500
2	2.250	0.90	3.000
2 1/2	2.688	1.21	3.580
3	3.063	1.30	4.300
4	3.625	1.38	5.430
6	5.125	1.50	7.625

* Dimensions in inches.

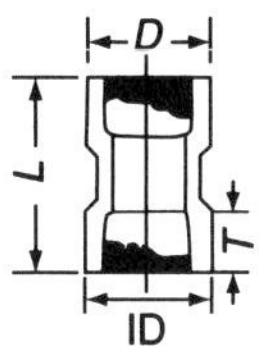

TABLE 17.18 Dimensions of Couplings*

Nominal pipe size	Length of thread, T, min.	Inside diameter of fitting, ID, min.	Thread end to thread end of coupling, L, min.
1/8	0.38	0.171	0.813
1/4	0.50	0.258	1.063
3/8	0.50	0.379	1.063
1/2	0.64	0.502	1.344
3/4	0.65	0.698	1.500
1	0.81	0.911	1.688
1 1/4	0.85	1.227	1.750
1 1/2	0.85	1.446	2.000
2	0.90	1.881	2.063
2 1/2	1.21	2.250	2.625
3	1.30	2.820	2.750
4	1.38	3.737	3.000
6	1.50	5.646	3.250

* Dimensions in inches.

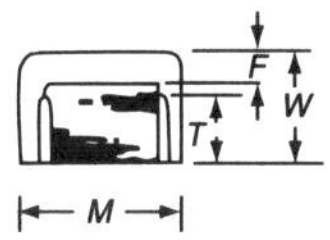

TABLE 17.19 Dimensions of Caps*

Nominal pipe size	Length of female thread, T, min.	Nominal wall thickness, F, min.	Cap height, W, min.	Outside diameter of band, M, min.
1/8	0.38	0.108	0.625	0.645
1/4	0.50	0.135	0.688	0.840
3/8	0.50	0.144	0.688	1.000
1/2	0.64	0.198	0.875	1.280
3/4	0.65	0.207	1.000	1.450
1	0.81	0.225	1.188	1.810
1 1/4	0.85	0.261	1.250	2.200
1 1/2	0.85	0.270	1.250	2.450
2	0.90	0.297	1.375	3.000
2 1/2	1.21	0.369	1.625	3.560
3	1.30	0.405	1.750	4.250
4	1.38	0.450	2.000	5.350
6	1.50	0.504	2.125	7.625

* Dimensions in inches.

PVC Plastic Pipe Fittings, Schedule 40 (ASTM D2466)

This specification covers PVC schedule 40 pipe fittings. The tapered socket diameter, wall thickness, and dimensions shall be in accordance with Table 17.20. The dimensions for couplings, tees, and elbows shall be in accordance with Table 17.21.

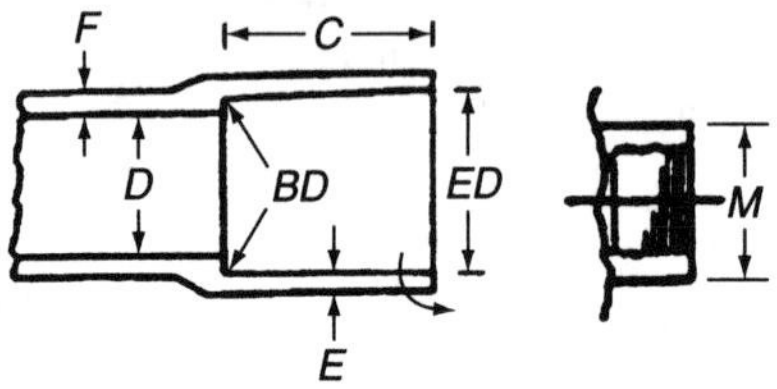

TABLE 17.20 Tapered Sockets for PVC Pipe Fittings, Schedule 40*

Nominal pipe size	*ED*, socket entrance diameter	*BD*, socket bottom diameter	*C*, socket length, min.	*D*, inside diameter, min.	Wall thickness, min.		Min. outside dimater of hub, *M*
					E	*F*	
1/8	0.417	0.401	0.500	0.225	0.088	0.085	0.526
1/4	0.552	0.536	0.500	0.320	0.088	0.110	0.672
3/8	0.687	0.671	0.594	0.449	0.091	0.114	0.821
1/2	0.848	0.836	0.688	0.578	0.109	0.136	0.996
3/4	1.058	1.048	0.719	0.740	0.113	0.141	1.221
1	1.325	1.310	0.875	0.990	0.133	0.166	1.504
1 1/4	1.670	1.655	0.938	1.335	0.140	0.175	1.871
1 1/2	1.912	1.894	1.094	1.564	0.145	0.181	2.127
2	2.387	2.369	1.158	2.021	0.154	0.193	2.634
2 1/2	2.889	2.868	1.750	2.414	0.203	0.254	3.170
3	3.516	3.492	1.875	3.008	0.216	0.270	3.841
3 1/2	4.016	3.992	2.000	3.488	0.226	0.283	4.374
4	4.518	4.491	2.000	3.961	0.237	0.296	4.907
5	5.583	5.553	3.000	4.975	0.258	0.323	6.039
6	6.647	6.614	3.000	5.988	0.280	0.350	7.203
8	8.655	8.610	4.000	7.888	0.322	0.403	9.320
10	10.780	10.735	5.000	9.917	0.365	0.456	11.614
12	12.780	12.735	6.000	11.825	0.406	0.506	13.786

* Dimensions in inches.

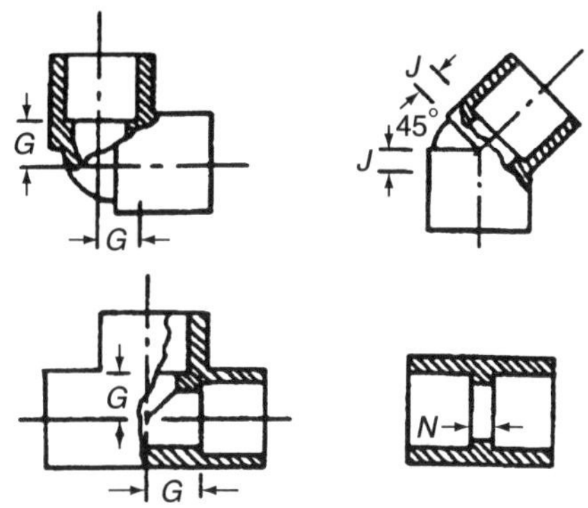

TABLE 17.21 Minimum Dimension from Center to End of Sockets (Laying Length) for Couplings, Tees, 90° and 45° Elbows, PVC Socket-Type Pipe Fittings, Schedule 40*

Nominal pipe size	G, min.	J, min.	N, min.
1/8	1/4	1/8	1/16
1/4	5/16	3/32	1/16
3/8	3/8	5/16	3/32
1/2	1/2	1/4	3/32
3/4	9/16	5/16	3/32
1	11/16	5/16	3/32
1 1/4	7/8	3/8	3/32
1 1/2	1	7/16	3/32
2	1 1/4	3/8	3/32
2 1/2	1 1/2	11/16	3/16
3	1 13/16	3/4	3/16
3 1/2	2 1/8	1	3/16
4	2 5/16	1	3/16
5	3	1 3/8	3/16
6	3 1/2	1 3/4	1/4
8	4 1/2	2	1/4
10	5 13/16	2 3/16	1/4
12	6 13/16	3 1/16	1/4

* Dimensions in inches.

PVC Plastic Pipe Fittings, Schedule 80 (ASTM D2467)

This specification covers PVC schedule 80pipe fittings. The tapered socket diameter, wall thickness, and dimensions shall be in accordance with Table 17.22. The dimensions for couplings, tees, crosses, and elbows shall be in accordance with Table 17.23. The minimum dimension of sockets for couplings, tees, and elbows shall be in accordance with Table 17.24. The dimension of caps shall be in accordance with Table 17.25.

PVC Plastic Drain, Waste, and Vent Pipe and Fittings (ASTM D2665)

Refer to the section of the same title under "PVC Plastic Pipes" above.

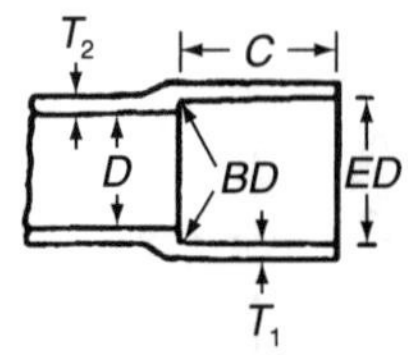

TABLE 17.22 Tapered Sockets for PVC Pipe Fittings, Schedule 80*

Nominal pipe size	*ED*, socket entrance diameter	*BD*, socket bottom diameter	*C*, socket length, min.	*D*, inside diameter, min.	Wall thickness, min.	
					T_1	T_2
1/8	0.417	0.401	0.500	0.171	0.095	0.118
1/4	0.552	0.536	0.625	0.258	0.119	0.149
3/8	0.687	0.671	0.750	0.379	0.126	0.160
1/2	0.848	0.836	0.875	0.502	0.147	0.185
3/4	1.058	1.046	1.000	0.698	0.154	0.195
1	1.325	1.310	1.125	0.911	0.179	0.225
1 1/4	1.670	1.655	1.250	1.227	0.191	0.239
1 1/2	1.912	1.894	1.375	1.446	0.200	0.250
2	2.387	2.369	1.500	1.881	0.218	0.275
2 1/2	2.889	2.868	1.750	2.250	0.276	0.345
3	3.516	3.492	1.875	2.820	0.300	0.375
3 1/2	4.016	3.992	2.125	3.280	0.318	0.400
4	4.518	4.491	2.250	3.737	0.337	0.420
5	5.583	5.553	2.625	4.703	0.375	0.470
6	6.647	6.614	3.000	5.646	0.432	0.540
8	8.655	8.610	4.000	7.490	0.500	0.625
10	10.780	10.735	5.000	9.407	0.593	0.741
12	12.780	12.735	6.000	11.197	0.687	0.859

* Dimensions in inches.

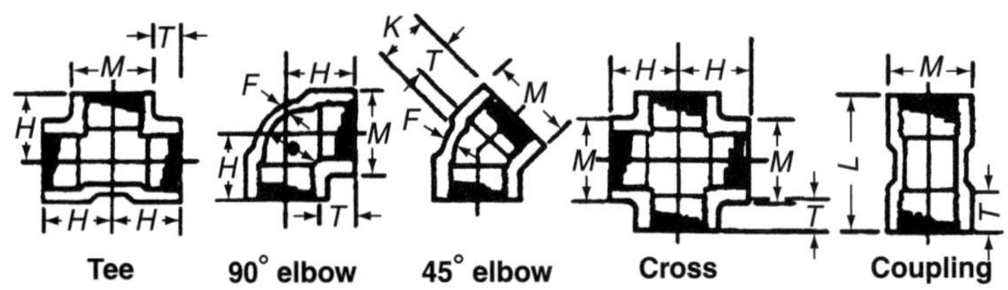

TABLE 17.23 Dimensions of 90° Elbows, Tees, Crosses, 45° Elbows, and Couplings (Straight Sizes)*

Nominal pipe size	Center to thread end, 90° elbows, tees, crosses, *H*, min.	Length of thread, *T*, min.	Center to thread end, 45° elbow, *K*, min.	Inside diameter of fitting, *D*, min.	Nominal wall thickness, *F*, min.	Outside diameter of band, *M*, min.	Thread end to thread end of coupling, *L*, min.
1/8	0.688	0.38	0.625	0.215	0.118	0.645	0.813
1/4	0.812	0.50	0.688	0.302	0.149	0.840	1.063
3/8	0.938	0.50	0.750	0.423	0.160	1.000	1.063
1/2	1.125	0.64	0.750	0.550	0.185	1.280	1.344
3/4	1.250	0.65	1.000	0.750	0.195	1.500	1.500
1	1.500	0.81	1.125	0.960	0.225	1.810	1.688
1 1/4	1.750	0.85	1.313	1.280	0.261	2.200	1.750
1 1/2	1.938	0.85	1.438	1.500	0.270	2.500	2.000
2	2.250	0.90	1.625	1.940	0.297	3.000	2.083
2 1/2	2.688	1.21	1.938	2.320	0.345	3.580	2.625
3	3.063	1.30	2.125	2.900	0.405	4.300	2.750
4	3.625	1.38	2.625	3.830	0.450	5.430	3.000
6	5.125	1.50	3.250	5.761	0.540	7.625	3.250

* Dimensions in inches.

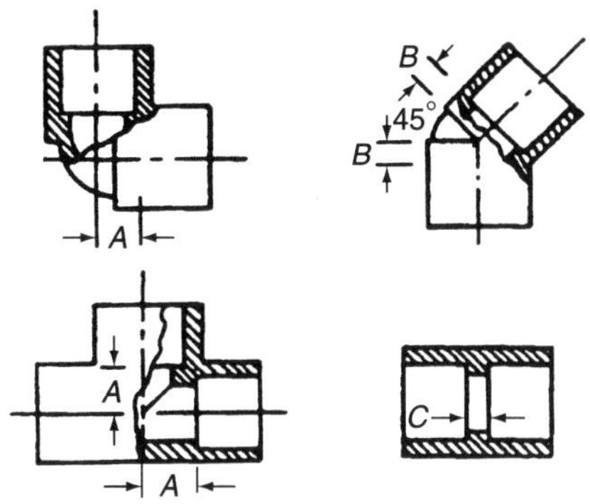

TABLE 17.24 Minimum Dimensions from Center to End of Sockets (Laying Length) for Couplings, Tees, 90° and 45° Elbows, PVC Socket-Type Pipe Fittings, Schedule 80*

Nominal pipe size	*A*, min.	*B*, min.	*C*, min.
1/8	1/4	1/8	1/16
1/4	5/16	5/32	1/16
3/8	3/8	3/16	3/32
1/2	1/2	1/4	3/32
3/4	9/16	5/16	3/32
1	11/16	5/16	3/32
1 1/4	7/8	3/8	3/32
1 1/2	1	7/16	3/32
2	1 1/4	5/8	3/32
2 1/2	1 1/2	11/16	3/16
3	1 13/16	3/4	3/16
3 1/2	2 1/8	1	3/16
4	2 5/16	1	3/16
5	3	1 3/8	3/16
6	3 1/2	1 3/4	1/4
8	4 1/12	2	1/4
10	5 11/16	2 1/4	1/4
12	6 7/8	2 1/2	3/8

* Dimensions in inches.

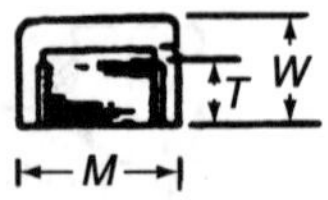

TABLE 17.25 Dimensions of Caps*

Nominal pipe size	Length of female thread, *T*, min.	Cap height, *W*, min.	Outside diameter of band, *M*, min.
1/8	0.38	0.625	0.645
1/4	0.50	0.688	0.840
3/8	0.50	0.688	1.000
1/2	0.64	0.875	1.280
3/4	0.65	1.000	1.450
1	0.81	1.188	1.810
1 1/4	0.85	1.250	2.200
1 1/2	0.85	1.250	2.450
2	0.90	1.375	3.000
2 1/2	1.21	1.625	3.560
3	1.30	1.750	4.250
4	1.38	2.000	5.350
6	1.50	2.125	7.625

* Dimensions in inches.

PVC PLASTIC JOINTS

Refer to Chap. 15.

Tables in Chapter 17

CHAPTER 18
CPVC PIPE, TUBE, FITTINGS, FLANGES, AND JOINTS

See pp. 580–581 for a list of tables in Chap. 18.

STANDARDS AND SPECIFICATIONS FOR CPVC PIPE, TUBE, AND FITTINGS

Table 18.1 lists the standards and specifications for CPVC pipe, tube, and fittings.

ASTM D2846 CPVC Pipe, Tube, and Fittings

ASTM D2846 covers CPVC 4120, SDR 11 pipe, tube, and fittings for hot and cold water service at temperatures up to and including 180°F (82°C). Only SDR 11 pipe, tube, and fittings are manufactured according to ASTM D2846.

CPVC 4120 pipe, tube, and fittings meet the material requirements for CPVC 23447-B per specification ASTM D1784. Previously, this material was designated as CPVC Type IV, Grade 1.

Pressure Ratings of SDR 11 Pipe, Tube, and Fittings per ASTM D2846

See Tables 18.2 and 18.2M.

TABLE 18.1 Principal CPVC Pipe, Tube, and Fitting Standards and Specifications

Piping material and product standard	Subject product, or abbreviated title of standard	Nominal size range, in	Principal applications
ASTM D2846	CPVC hot and cold water piping	3/8–2	Hot and cold water
ASTM F437	CPVC fittings, threaded, Schedule 80	1/4–6	Hot and cold water, industrial
ASTM F438	CPVC fittings, socket, Schedule 40	1/4–6	Hot and cold water, industrial
ASTM F439	CPVC fittings, socket, Schedule 80	1/4–6	Hot and cold water, industrial
ASTM F441	CPVC pipe, Schedules 40 and 80	1/4–12	Hot and cold water, industrial
ASTM F442	CPVC pipe, dimension ratio series	1/4–12	Hot and cold water, industrial
CSA B137.6	CPVC pipe, tubing, and fittings for hot and cold water distributing systems		Hot and cold water, plumbing

TABLE 18.2 Hydrostatic Design Stresses and Pressure Ratings for CPVC 4120, SDR 11, Hot-Water Distribution System (U.S. Customary Units)

Rated temperature, °F	Hydrostatic design stress, psi	Pressure rating for water, psi
73.4	2000	400
180	500	100

TABLE 18.2M Hydrostatic Design Stresses and Pressure Ratings for CPVC 4120, SDR 11, Hot-Water Distribution System (Metric Units)

Rated temperature, °C	Hydrostatic design stress, MPa	Pressure rating for water, MPa
23	13.8	2.8
82	3.5	0.7

Dimensions of ASTM D2846 Pipe and Tubing

See Tables 18.3 and 18.3M.

Dimensions of CPVC Fittings per ASTM D2846 for SDR 11 Pipe and Tubing

Center-to-End or Laying Length of Fitting Sockets. See Tables 18.4 and 18.4M.

Dimensions of Tapered Socket of CPVC Fittings for SDR 11 Pipe and Tubing per ASTM D2846. See Tables 18.5 and 18.5M.

ASTM F441 CPVC PLASTIC SCHEDULES 40 AND 80 PIPE

ASTM F441 covers pressure-rated CPVC 4120, Schedules 40 and 80 pipe in sizes ranging from NPS 1/4 through 12.

Pressure Rating of ASTM F441 CPVC Schedule 40 Pipe

See Tables 18.6 and 18.6M.

TABLE 18.3 Outside Diameters, Wall Thicknesses, and Tolerances for CPVC 41, SDR 11, Plastic Pipe and Tubing (U.S. Customary Units)

Nominal tube or pipe size	Outside diameter, in			Wall thickness, in (SDR 11)	
	Average	Tolerance on average	Maximum out-of-round	Minimum	Tolerance
1/4 tube	0.375	±0.003	±0.003	0.055	+0.020
3/8 tube	0.500	±0.003	±0.003	0.068	+0.020
1/2 tube	0.625	±0.003	±0.004	0.068	+0.020
3/4 tube	0.875	±0.003	±0.005	0.080	+0.020
1 tube	1.125	±0.003	±0.006	0.102	+0.020
1 1/4 tube	1.375	±0.003	±0.007	0.125	+0.020
1 1/2 tube	1.625	±0.004	±0.008	0.148	+0.020
2 tube	2.125	±0.004	±0.010	0.193	+0.023
1 1/2 pipe	1.900	+0.006, −0.002	±0.008	0.173	+0.021
2 pipe	2.375	+0.006, −0.002	±0.010	0.216	+0.026

TABLE 18.3M Outside Diameters, Wall Thicknesses, and Tolerances for CPVC 41, SDR 11, Plastic Pipe and Tubing (Metric Units)

Nominal tube or pipe size	Outside diameter, mm			Wall thickness, mm (SDR 11)	
	Average	Tolerance on average	Maximum out-of-round	Minimum	Tolerance
6 tube	9.5	±0.08	±0.08	1.40	+0.51
10 tube	12.7	±0.08	±0.08	1.73	+0.51
15 tube	15.9	±0.08	±0.10	1.73	+0.51
20 tube	22.2	±0.08	±0.13	2.03	+0.51
25 tube	28.6	±0.08	±0.15	2.59	+0.51
32 tube	34.9	±0.08	±0.18	3.18	+0.51
40 tube	41.3	±0.10	±0.20	3.76	+0.51
50 tube	54	±0.10	±0.25	4.90	+0.58
40 pipe	48.6	+0.15, −0.05	±0.20	4.39	+0.53
50 pipe	60.7	+0.15, −0.05	±0.25	5.49	+0.66

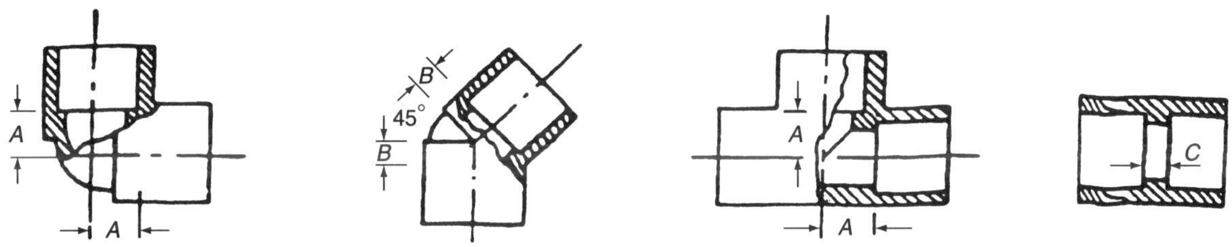

TABLE 18.4 Minimum Dimensions from Center to End of Socket (Laying Length) for CPVC 41, SDR 11, Plastic Pipe and Tubing Fittings (U.S. Customary Units)

Nominal tube or pipe size	*A*, min., in	*B*, min., in	*C*, min., in
$^{3}/_{8}$ tube	0.359	0.174	0.102
$^{1}/_{2}$ tube	0.382	0.183	0.102
$^{3}/_{4}$ tube	0.507	0.235	0.102
1 tube	0.633	0.287	0.102
$1^{1}/_{4}$ tube	0.758	0.339	0.102
$1^{1}/_{2}$ tube	0.884	0.391	0.102
2 tube	1.134	0.495	0.102
$1^{1}/_{2}$ pipe	1.022	0.448	0.102
2 pipe	1.260	0.547	0.102

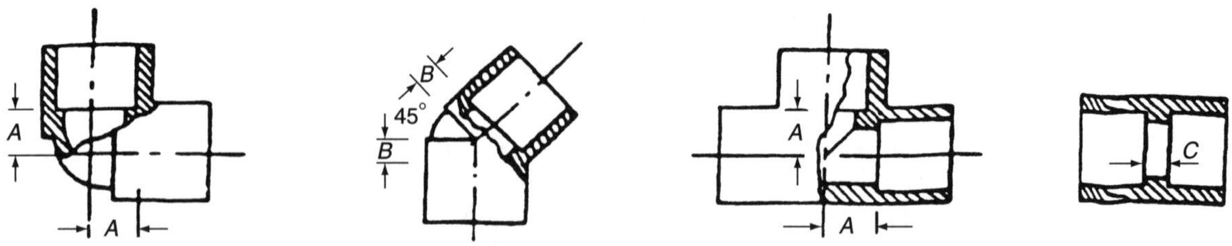

TABLE 18.4M Minimum Dimensions from Center to End of Socket (Laying Length) for CPVC 41, SDR 11, Plastic Pipe and Tubing Fittings (Metric Units)

Nominal tube or pipe size	*A*, min., mm	*B*, min., mm	*C*, min., mm
10 tube	9.12	4.42	2.59
15 tube	9.70	4.65	2.59
20 tube	12.88	5.97	2.59
25 tube	16.08	7.29	2.59
32 tube	19.25	8.61	2.59
40 tube	22.45	9.93	2.59
50 tube	28.83	12.57	2.59
40 pipe	25.96	11.38	2.59
50 pipe	32.00	13.89	2.59

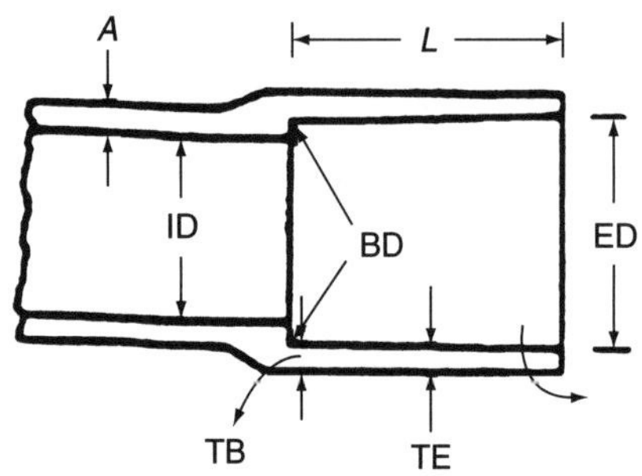

TABLE 18.5 Tapered Socket Dimensions for CPVC 41, SDR 11, Plastic Pipe and Tube Fittings (U.S. Customary Units)

	ED, socket entrance diameter, in			BD, socket bottom diameter, in					Min. wall thickness, in		
Nominal tube or pipe size	Average	Tolerance on average	Max. out-of-round	Average	Tolerance on average	Max. out-of-round	*L*, socket length, min.	ID, inside diameter, min.	TE, socket entrance	TB, socket bottom	*A*
3/8 tube	0.508	±0.003	±0.003	0.494	±0.003	±0.003	0.500	0.364	0.068	0.102	0.128
1/2 tube	0.633	±0.003	±0.004	0.619	±0.003	±0.004	0.500	0.489	0.068	0.102	0.128
3/4 tube	0.884	±0.003	±0.005	0.870	±0.003	±0.005	0.700	0.716	0.080	0.102	0.128
1/2 tube	1.135	±0.003	±0.006	1.121	±0.003	±0.006	0.900	0.921	0.102	0.102	0.128
1 1/4 tube	1.386	±0.003	±0.007	1.372	±0.003	±0.007	1.100	1.125	0.125	0.125	0.156
1 1/2 tube	1.640	±0.004	±0.008	1.622	±0.004	±0.008	1.300	1.329	0.148	0.148	0.185
2 tube	2.141	±0.004	±0.010	2.123	±0.004	±0.010	1.700	1.739	0.193	0.193	0.241
1 1/2 pipe	1.918	±0.004	±0.008	1.900	±0.004	±0.008	1.375	1.494	0.173	0.173	0.216
2 pipe	2.393	±0.004	±0.010	2.375	±0.004	±0.010	1.500	1.933	0.216	0.216	0.270

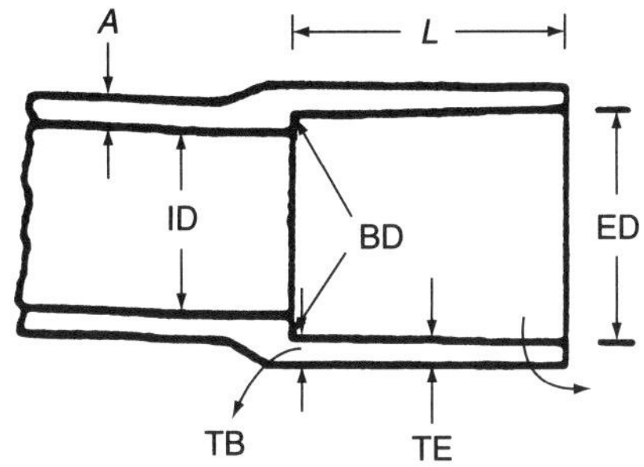

TABLE 18.5M Tapered Socket Dimensions for CPVC 41, SDR 11, Plastic Pipe and Tube Fittings (Metric Units)

Nominal tube or pipe size	ED, socket entrance diameter, mm: Average	Tolerance on average	Max. out-of-round	BD, socket bottom diameter, mm: Average	Tolerance on average	Max. out-of-round	*L*, socket length, min.	ID, inside diameter, min.	Min. wall thickness, mm: TE, socket entrance	TB, socket bottom	*A*
10 tube	12.90	±0.08	±0.08	12.55	±0.08	±0.08	12.70	9.25	1.73	2.59	3.25
15 tube	16.08	±0.08	±0.10	15.72	±0.08	±0.10	12.70	12.42	1.73	2.59	3.25
20 tube	22.45	±0.08	±0.13	22.10	±0.08	±0.13	17.78	18.16	2.03	2.59	3.25
25 tube	28.83	±0.08	±0.15	28.47	±0.08	±0.15	22.86	23.39	2.59	2.59	3.25
32 tube	35.20	±0.08	±0.18	34.85	±0.08	±0.18	27.94	28.58	3.18	3.18	3.96
40 tube	41.66	±0.10	±0.20	41.20	±0.10	±0.20	33.02	33.76	3.76	3.76	4.70
50 tube	54.38	±0.10	±0.25	53.92	±0.10	±0.25	43.18	44.17	4.90	4.90	6.12
40 pipe	48.72	±0.10	±0.20	48.26	±0.10	±0.20	34.92	37.95	4.39	4.39	5.49
50 pipe	60.78	±0.10	±0.25	60.33	±0.10	±0.25	38.10	49.10	5.49	5.49	6.86

TABLE 18.6 Pressure Rating of Schedule 40 CPVC 4120 Plastic Pipe for Water at 73°F (U.S. Customary Units)

Nominal pipe size	Pressure rating, psi
1/4	780
3/8	620
1/2	600
3/4	480
1	450
1 1/4	370
1 1/2	330
2	280
2 1/2	300
3	260
3 1/2	240
4	220
5	190
6	180
8	160
10	140
12	130

Notes:
1. Ratings listed in the table are for unthreaded Schedule 40 pipe. It is recommended that threaded joints should not be used with Schedule 40 pipe.
2. To establish pressure rating at a temperature higher than 73°F for water service, use derating factor values from Table 18.9.

TABLE 18.6M Pressure Rating of Schedule 40 CPVC 4120 Plastic Pipe for Water at 23°C (Metric Units)

Nominal pipe size	Pressure rating, kPa
8	5380
10	4270
15	4140
20	3310
25	3100
32	2550
40	3280
50	1930
65	2070
80	1790
90	1650
100	1520
125	1310
150	1240
200	1100
250	970
300	900

Notes:
1. The ratings listed above are for unthreaded pipe. Threaded joints are not recommended to be used with Schedule 40 CPVC pipe.
2. To establish pressure rating at a temperature higher than 23°C for water service, use derating factor values given in Table 18.9M.

Pressure Rating of ASTM F441 CPVC Schedule 80 Pipe

See Tables 18.7 and 18.7M and Tables 18.8 and 18.8M.

Derating Factors for Pressure Rating of ASTM F441 CPVC Pipe

See Tables 18.9 and 18.9M.

TABLE 18.7 Pressure Ratings of CPVC 4120 Schedule 80 Pipe for Water at 73°F (U.S. Customary Units)*

Nominal pipe size	Unthreaded, psi	Threaded, psi
1/4	1130	570
3/8	920	460
1/2	850	420
3/4	690	340
1	630	320
1 1/4	520	260
1 1/2	470	240
2	400	200
2 1/2	420	210
3	370	190
3 1/2	350	170
4	320	160
5	290	140
6	280	140
8	250	120
10	230	120
12	230	110

* To establish pressure rating at a temperature higher than 73°F for water service, use derating factor values from Table 18.9.

TABLE 18.7M Pressure Ratings of CPVC 4120 Schedule 80 Pipe for Water at 23°C (Metric Units)*

Nominal pipe size	Unthreaded, kPa	Threaded, kPa
8	7790	3930
10	6340	3170
15	5860	2900
20	4760	2340
25	4340	2210
32	3590	1790
40	3240	1650
50	2760	1380
65	2900	1450
80	2550	1310
90	2410	1170
100	2210	1100
125	2000	970
150	1930	970
200	1720	830
250	1590	830
300	1590	760

* To establish pressure rating at a temperature higher than 23°C, use derating factor values from Table 18.9M.

TABLE 18.8 Pressure Ratings of CPVC 4120 Schedule 40 and 80 Pipe for Water at 180°F (U.S. Customary Units)

Nominal pipe size	Schedule 40, psi	Schedule 80 unthreaded, psi	Schedule 80 threaded, psi
1/4	195	280	140
3/8	155	230	115
1/2	150	210	105
3/4	120	170	85
1	110	155	80
1 1/4	90	130	65
1 1/2	80	115	60
2	70	100	50
2 1/2	75	105	50
3	65	90	45
3 1/2	60	85	40
4	55	80	40
5	45	70	35
6	45	70	35
8	40	60	30
10	35	55	30
12	30	55	25

TABLE 18.8M Pressure Ratings of CPVC 4120 Schedule 40 and 80 Pipe for Water at 82°C (Metric Units)

Nominal pipe size	Schedule 40, kPa	Schedule 80 unthreaded, kPa	Schedule 80 threaded, kPa
8	1340	1930	970
10	1070	1590	790
15	1030	1450	720
20	830	1170	590
25	760	1070	550
32	620	900	450
40	550	790	410
50	480	690	340
65	520	720	340
80	450	620	310
90	410	590	280
100	380	550	280
125	310	480	240
150	310	480	240
200	280	410	210
250	240	380	210
300	210	380	170

TABLE 18.9 Derating Factors for CPVC Pipe for Water Service at Elevated Temperatures (U.S. Customary Units)*

Temperature, F°	Derating factor
73 to 80	1.00
90	0.91
100	0.82
120	0.65
140	0.50
160	0.40
180	0.25
200	0.20

* To determine pressure rating at an elevated temperature, multiply 73°F pressure rating from Tables 18.6 and 18.7 by appropriate factor.

TABLE 18.9M Derating Factors for CPVC Pipe for Water Service at Elevated Temperatures (Metric Units)*

Temperature, C°	Derating factor
22.5 to 26.7	1.00
32.2	0.91
37.8	0.82
48.9	0.65
60.0	0.50
71.1	0.40
82.2	0.25
93.3	0.20

* To determine pressure rating at an elevated temperature for water service, multiply 23°C pressure rating from Tables 18.6M and 18.7M by appropriate factor.

Outside Diameter and Tolerances for ASTM F441 CPVC Pipe

See Tables 18.10 and 18.10M.

TABLE 18.10 Outside Diameters and Tolerances for CPVC Schedules 40 and 80 Pipe (U.S. Customary Units)

Nominal pipe size, in	Average outside diameter, in	Tolerance, in: For average outside diameter	Tolerance, in: Maximum out-of-roundness (maximum minus minimum diameter): Schedule 40 sizes $3^1/_2$ in and over; Schedule 80 sizes 8 in and over	Tolerance, in: Maximum out-of-roundness (maximum minus minimum diameter): Schedule 40 sizes 3 in and less; Schedule 80 sizes 6 in and less
$^1/_4$	0.540	±0.004		0.016
$^3/_8$	0.675	±0.004		0.016
$^1/_2$	0.840	±0.004		0.016
$^3/_4$	1.050	±0.004		0.020
1	1.315	±0.005		0.020
$1^1/_4$	1.660	±0.005		0.024
$1^1/_2$	1.900	±0.006		0.024
2	2.375	±0.006		0.024
$2^1/_2$	2.875	±0.007		0.030
3	3.500	±0.008		0.030
$3^1/_2$	4.000	±0.008	0.100	0.030
4	4.500	±0.009	0.100	0.030
5	5.563	±0.010	0.100	0.060
6	6.625	±0.011	0.100	0.070
8	8.625	±0.015	0.150	
10	10.750	±0.015	0.150	
12	12.750	±0.015	0.150	

TABLE 18.10M Outside Diameters and Tolerances for CPVC Schedules 40 and 80 Pipe (Metric Units)

Nominal pipe size, mm	Average outside diameter, mm	Tolerance, mm: For average outside diameter	Tolerance, mm: Maximum out-of-roundness (maximum minus minimum diameter): Schedule 40 sizes 90 and over; Schedule 80 sizes 200 and over	Tolerance, mm: Maximum out-of-roundness (maximum minus minimum diameter): Schedule 40 sizes 90 and less; Schedule 80 sizes 150 and less
8	13.7	±0.10		0.41
10	17.1	±0.10		0.41
15	21.3	±0.10		0.41
20	26.7	±0.10		0.51
25	33.4	±0.13		0.51
32	42.2	±0.13		0.61
40	48.3	±0.15		0.61
50	60.3	±0.15		0.61
65	73.0	±0.18		0.76
80	88.9	±0.20		0.76
90	101.6	±0.20	2.54	0.76
100	114.3	±0.23	2.54	0.76
125	141.3	±0.25	2.54	1.52
150	168.3	±0.28	2.54	1.78
200	219.1	±0.38	3.81	
250	273.1	±0.38	3.81	
300	323.9	±0.38	3.81	

Pipe Wall Thickness and Tolerances for ASTM F441 CPVC Pipe

See Tables 18.11 and 18.11M.

TABLE 18.11 Wall Thickness and Tolerance for CPVC Schedule 40 and 80 Pipe (U.S. Customary Units)

	Wall thickness, in			
	Schedule 40		Schedule 80	
Nominal pipe size, in	Min.	Tolerance	Min.	Tolerance
1/4	0.088	+0.020	0.119	+0.020
3/8	0.091	+0.020	0.126	+0.020
1/2	0.109	+0.020	0.147	+0.020
3/4	0.113	+0.020	0.154	+0.020
1	0.133	+0.020	0.179	+0.021
1 1/4	0.140	+0.020	0.191	+0.023
1 1/2	0.145	+0.020	0.200	+0.024
2	0.154	+0.020	0.218	+0.026
2 1/2	0.203	+0.024	0.276	+0.033
3	0.216	+0.026	0.300	+0.036
3 1/2	0.226	+0.027	0.318	+0.038
4	0.237	+0.028	0.337	+0.040
5	0.258	+0.031	0.375	+0.045
6	0.280	+0.034	0.432	+0.052
8	0.322	+0.039	0.500	+0.060
10	0.365	+0.044	0.593	+0.071
12	0.406	+0.049	0.687	+0.082

TABLE 18.11M Wall Thickness and Tolerance for CPVC Schedule 40 and 80 Pipe (Metric Units)

	Wall thickness, mm			
	Schedule 40		Schedule 80	
Nominal pipe size, mm	Min.	Tolerance	Min.	Tolerance
8	2.24	0.51	3.02	0.51
10	2.31	0.51	3.20	0.51
15	2.77	0.51	3.73	0.51
20	2.87	0.51	3.91	0.51
25	3.38	0.51	4.55	0.53
32	3.56	0.51	4.85	0.58
40	3.68	0.51	5.08	0.61
50	3.91	0.51	5.54	0.66
65	5.16	0.61	7.01	0.84
80	5.49	0.66	7.62	0.91
90	5.74	0.68	8.06	0.96
100	6.02	0.71	8.56	1.02
125	6.55	0.79	9.52	1.14
150	7.11	0.86	10.97	1.32
200	8.18	0.99	12.70	1.52
250	9.27	1.12	15.06	1.80
300	10.31	1.24	17.45	2.08

ASTM F442 CPVC PLASTIC SDR (11, 13.5, 17, 21, 26, 32.5) PIPE

ASTM F442 covers pressure-rated CPVC 4120 SDR (11, 13.5, 17, 21, 26, 32.5) pipe in sizes ranging from NPS 1/4 through 12.

Pressure Rating of ASTM F442 CPVC SDR Pipe

See Tables 18.12 and 18.12M.

Derating Factors for Pressure Rating of ASTM F442 CPVC Pipe

See Tables 18.13 and 18.13M.

Outside Diameter and Tolerances for ASTM F442 CPVC Pipe

See Tables 18.14 and 18.14M.

TABLE 18.12 Pressure Ratings (PR) of ASTM F442, Nonthreaded CPVC 4120 Pipe for Water Service at 73°F*

SDR	Pressure rating, psi
11	400
13.5	315
17	250
21	200
26	160
32.5	125

* To establish pressure rating at a temperature higher than 73°F for water service, use derating factors from Table 18.13.

TABLE 18.12M Pressure Ratings (PR) of ASTM F442, Nonthreaded CPVC 4120 Pipe for Water Service at 23°C (Metric Units)*

SDR	Pressure rating, kPa
11	2760
13.5	2170
17	1720
21	1380
26	1100
32.5	860

* To establish pressure rating at a temperature higher than 23°C for water service, use derating factor values from Table 18.13M.

TABLE 18.13 Derating Factors for ASTM F442, CPVC Pipe for Water Service*

Temperature, °F	Derating factor
73 to 80	1.0
90	0.91
100	0.82
120	0.65
140	0.50

* To determine pressure rating at an elevated temperature, multiply 73°F pressure rating from Table 18.12 by appropriate factor.

TABLE 18.13M Derating Factors for ASTM F442, CPVC Pipe for Water Service*

Temperature, °C	Derating factor
23 to 26.7	1.0
32.2	0.91
37.8	0.82
48.9	0.65
60	0.50

* To determine pressure rating at an elevated temperature, multiply 23°C pressure rating from Table 18.12M by appropriate derating factor.

TABLE 18.14 Outside Diameters and Tolerances for ASTM F442, CPVC Pipe (U.S. Customary Units)

Nominal pipe size, in	Average outside diameter, in	Tolerance, in		
		For average outside diameter	Maximum out-of-roundness (maximum minus minimum diameter)	
			SDR32.5, SDR26, SDR21	SDR17, SDR13.5, SDR11
1/4	0.540	±0.004	0.030	0.016
3/8	0.675	±0.004	0.030	0.016
1/2	0.840	±0.004	0.030	0.016
3/4	1.050	±0.004	0.030	0.020
1	1.315	±0.005	0.030	0.020
1 1/4	1.660	±0.005	0.030	0.024
1 1/2	1.900	±0.005	0.060	0.024
2	2.375	±0.006	0.060	0.024
2 1/2	2.875	±0.007	0.060	0.030
3	3.500	±0.008	0.060	0.030
3 1/2	4.000	±0.008	0.100	0.030
4	4.500	±0.009	0.100	0.030
5	5.563	±0.010	0.100	0.060
6	6.625	±0.011	0.100	0.070
8	8.625	±0.015	0.150	0.090
10	10.750	±0.015	0.150	0.100
12	12.750	±0.015	0.150	0.120

TABLE 18.14M Outside Diameters and Tolerances for ASTM F442, CPVC Pipe (Metric Units)

Nominal pipe size, mm	Average outside diameter, mm	Tolerance, mm		
		For average outside diameter	Maximum out-of-roundness (maximum minus minimum diameter)	
			SDR32.5, SDR26, SDR21	SDR17, SDR13.5, SDR11
8	13.7	±0.10	0.76	0.41
10	17.1	±0.10	0.76	0.41
15	21.3	±0.10	0.76	0.41
20	26.7	±0.10	0.76	0.51
25	33.4	±0.13	0.76	0.51
32	42.2	±0.13	0.76	0.61
40	48.2	±0.15	1.52	0.61
50	60.3	±0.15	1.52	0.61
65	73.0	±0.18	1.52	0.76
80	88.9	±0.20	1.52	0.76
90	101.6	±0.20	2.54	0.76
100	114.3	±0.23	2.54	0.76
125	141.3	±0.25	2.54	1.52
150	168.3	±0.28	2.54	1.78
200	219.1	±0.38	3.81	2.29
250	273.1	±0.38	3.81	2.54
300	323.9	±0.38	3.81	3.05

Pipe Wall Thickness and Tolerances for ASTM F442 CPVC Pipe

See Tables 18.15 and 18.15M.

TABLE 18.15 Wall Thickness and Tolerance for ASTM F442, CPVC Pipe (U.S. Customary Units)

Nominal pipe size, in	Wall thickness, in											
	SDR32.5		SDR26		SDR21		SDR17		SDR13.5		SDR11	
	Min.	Tolerance	Min.	Tolerance	Min.	Tolerance	Min.	Tolerance	Min.	Tolerance	Min.	Tolerance
1/4									0.060	+0.020	0.060	+0.020
3/8									0.060	+0.020	0.061	+0.020
1/2									0.062	+0.020	0.076	+0.020
3/4					0.060	+0.020	0.062	+0.020	0.078	+0.020	0.095	+0.020
1			0.060	+0.020	0.063	+0.020	0.077	+0.020	0.097	+0.020	0.119	+0.020
1 1/4			0.064	+0.020	0.079	+0.020	0.098	+0.020	0.123	+0.020	0.151	+0.020
1 1/2			0.073	+0.020	0.090	+0.020	0.112	+0.020	0.141	+0.020	0.173	+0.021
2			0.091	+0.020	0.113	+0.020	0.140	+0.020	0.176	+0.021	0.216	+0.026
2 1/2			0.110	+0.020	0.137	+0.020	0.169	+0.020	0.213	+0.026	0.261	+0.031
3	0.108	+0.020	0.135	+0.020	0.167	+0.020	0.206	+0.025	0.259	+0.031	0.318	+0.039
3 1/2	0.123	+0.020	0.154	+0.020	0.190	+0.023	0.235	+0.028	0.296	+0.036	0.363	+0.044
4	0.138	+0.020	0.173	+0.021	0.214	+0.026	0.265	+0.032	0.333	+0.040	0.409	+0.049
5	0.171	+0.021	0.214	+0.027	0.265	+0.032	0.327	+0.039	0.412	+0.049	0.506	+0.061
6	0.204	+0.024	0.255	+0.031	0.316	+0.038	0.390	+0.047	0.491	+0.059	0.602	+0.073
8	0.265	+0.032	0.332	+0.040	0.410	+0.049	0.508	+0.061	0.639	+0.077	0.785	+0.095
10	0.331	+0.040	0.413	+0.050	0.511	+0.061	0.632	+0.076	0.797	+0.096	0.978	+0.118
12	0.392	+0.047	0.490	+0.059	0.606	+0.073	0.750	+0.090	0.945	+0.114	1.160	+0.140

TABLE 18.15M Wall Thickness and Tolerance for ASTM F442, CPVC Pipe (Metric Units)

Nominal pipe size, mm	Wall thickness, mm											
	SDR32.5		SDR26		SDR21		SDR17		SDR13.5		SDR11	
	Min.	Tolerance	Min.	Tolerance	Min.	Tolerance	Min.	Tolerance	Min.	Tolerance	Min.	Tolerance
8									1.52	0.51	1.52	0.51
10									1.52	0.51	1.55	0.51
15									1.57	0.51	1.93	0.51
20					1.52	0.51	1.57	0.51	1.98	0.51	2.41	0.51
25			1.52	0.51	1.60	0.51	1.96	0.51	2.46	0.51	3.02	0.51
32			1.63	0.51	2.01	0.51	2.49	0.51	3.12	0.51	3.84	0.51
40			1.85	0.51	2.29	0.51	2.84	0.51	3.58	0.51	4.39	0.53
50			2.31	0.51	2.87	0.51	3.56	0.51	4.47	0.53	5.49	0.66
65			2.79	0.51	3.48	0.51	4.29	0.51	5.41	0.66	6.63	0.79
80	2.74	0.51	3.43	0.51	4.24	0.51	5.23	0.64	6.58	0.79	8.08	0.99
90	3.12	0.51	3.91	0.51	4.83	0.58	5.97	0.71	7.52	0.91	9.22	1.12
100	3.50	0.51	4.39	0.53	5.44	0.66	6.73	0.81	8.46	1.02	10.39	1.24
125	4.34	0.53	5.44	0.69	6.73	0.81	8.30	0.99	10.46	1.24	12.85	1.55
150	5.18	0.61	6.48	0.79	8.03	0.96	9.91	1.19	12.47	1.50	15.29	1.85
200	6.73	0.81	8.43	1.02	10.41	1.24	12.90	1.55	16.23	1.95	19.94	2.41
250	8.41	1.02	10.49	1.27	12.98	1.55	16.05	1.93	20.24	2.44	24.84	2.99
300	9.96	1.19	12.45	1.50	15.39	1.85	19.05	2.29	24.00	2.89	29.46	3.56

ASTM F438 SOCKET-TYPE CPVC PIPE FITTINGS, SCHEDULE 40

ASTM F438 covers socket-type CPVC Schedule 40 pipe fittings.

Minimum Center-to-End Dimension (Laying Length) of Sockets of ASTM F438 Couplings, Tees, and Elbows

See Tables 18.16 and 18.16M.

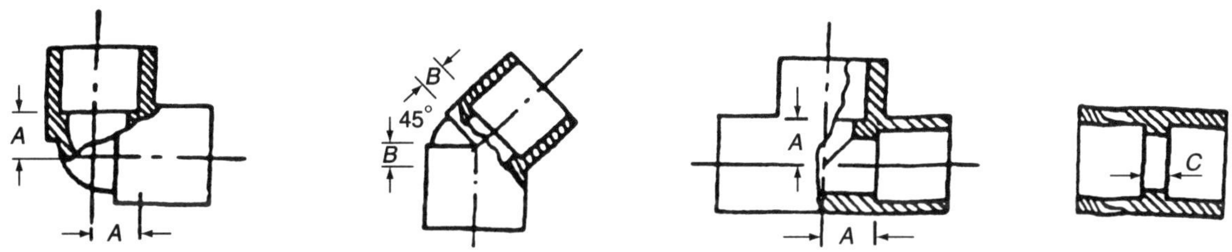

TABLE 18.16 Minimum Center-to-End (Laying Length) Dimension of Sockets for Couplings, Tees, 90° and 45° Elbows, CPVC Socket-Type Pipe Fittings, Schedule 40, per ASTM F438 (U.S. Customary Units)*

Nominal pipe size, in	A, min., in	B, min., in	C, min., in
1/4	5/16	5/32	1/16
3/8	3/8	3/16	3/32
1/2	1/2	1/4	3/32
3/4	9/16	5/16	3/32
1	11/16	5/16	3/32
1 1/4	7/8	3/8	3/32
1 1/2	1	7/16	3/32
2	1 1/4	5/8	3/32
2 1/2	1 1/2	11/16	3/16
3	1 13/16	3/4	3/16
3 1/2	2 1/8	1	3/16
4	2 5/16	1	3/16
5	3	1 3/8	3/16
6	3 1/2	1 3/4	1/4

* Refer to Table 18.18 for other socket dimensions.

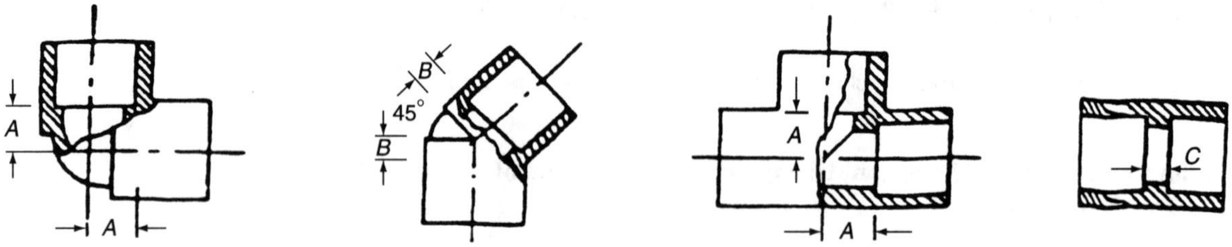

TABLE 18.16M Minimum Center-to-End (Laying Length) Dimension of Sockets for Couplings, Tees, 90° and 45° Elbows, CPVC Socket-Type Pipe Fittings, Schedule 40, per ASTM F438 (Metric Units)*

Nominal pipe size, mm	*A*, min., mm	*B*, min., mm	*C*, min., mm
8	7.94	3.97	1.59
10	9.53	4.76	2.38
15	12.70	6.35	2.38
20	14.29	7.94	2.38
25	17.46	7.94	2.38
32	22.22	9.53	2.38
40	25.40	11.11	2.38
50	31.75	15.88	2.38
65	38.10	17.46	4.76
80	46.04	19.05	4.76
90	53.98	25.40	4.76
100	58.74	25.40	4.76
125	76.20	34.92	4.76
150	88.20	44.45	6.35

* Refer to Table 18.18M for other socket dimensions.

Dimensions and Tolerances of Tapered Sockets of ASTM F438 Socket-Type Fittings

See Tables 18.17 and 18.17M.

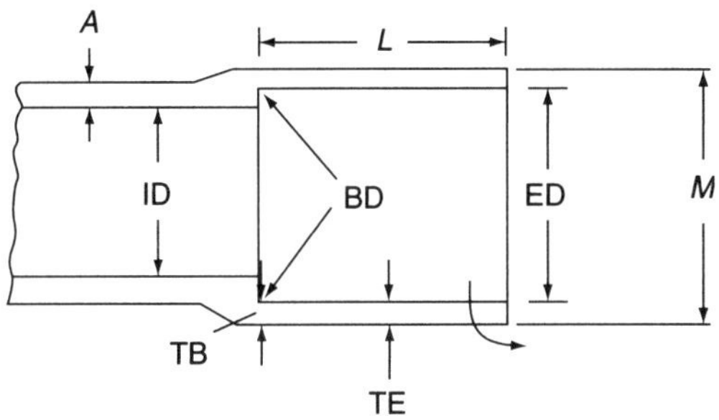

TABLE 18.17 Dimensions of Tapered Sockets for CPVC Pipe Fittings, Schedule 40, per ASTM F438 (U.S. Customary Units)

Nominal pipe size, in	ED, socket entrance diameter, in			BD, socket bottom diameter, in			L, socket length, min., in	ID, inside diameter, min., in	Minimum wall thickness, in		Minimum outside diameter of hub, M, in
	Diameter	Tolerance on diameter	Maximum out-of-round	Diameter	Tolerance on diameter	Maximum out-of-round			TE	TB	
1/4	0.552	±0.004	0.016	0.536	±0.004	0.016	0.500	0.320	0.088	0.110	0.672
3/8	0.687	±0.004	0.016	0.671	±0.004	0.016	0.594	0.449	0.091	0.114	0.821
1/2	0.848	±0.004	0.016	0.836	±0.004	0.016	0.688	0.578	0.109	0.136	0.998
3/4	1.058	±0.004	0.020	1.046	±0.004	0.020	0.719	0.740	0.113	0.141	1.221
1	1.325	±0.005	0.020	1.310	±0.005	0.020	0.875	0.990	0.133	0.166	1.504
1 1/4	1.670	±0.005	0.024	1.655	±0.005	0.024	0.938	1.335	0.140	0.175	1.871
1 1/2	1.912	±0.006	0.024	1.894	±0.006	0.024	1.094	1.564	0.145	0.181	2.127
2	2.387	±0.006	0.024	2.369	±0.006	0.024	1.156	2.021	0.154	0.193	2.634
2 1/2	2.889	±0.007	0.030	2.868	±0.007	0.030	1.750	2.414	0.203	0.254	3.170
3	3.516	±0.008	0.030	3.492	±0.008	0.030	1.875	3.008	0.216	0.270	3.841
3 1/2	4.016	±0.008	0.030	3.992	±0.008	0.030	2.000	3.486	0.226	0.283	4.374
4	4.518	±0.009	0.030	4.491	±0.009	0.030	2.000	3.961	0.237	0.296	4.907
5	5.583	±0.010	0.060	5.553	±0.010	0.060	3.000	4.975	0.258	0.323	6.039
6	6.647	±0.011	0.060	6.614	±0.011	0.060	3.000	5.986	0.280	0.350	7.203

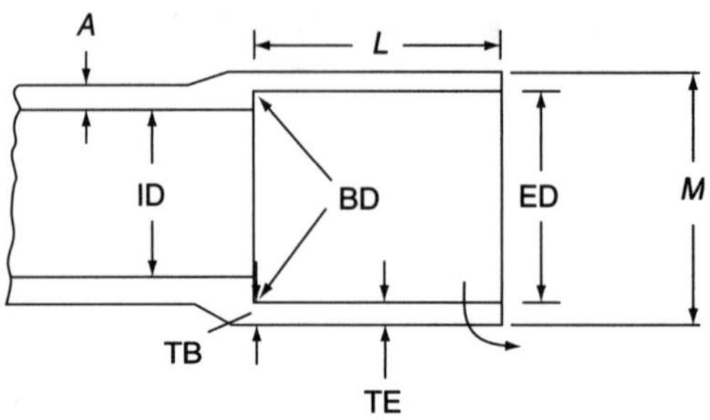

TABLE 18.17M Dimensions of Tapered Sockets for CPVC Pipe Fittings, Schedule 40, per ASTM F438 (Metric Units)

Nominal pipe size, mm	ED, socket entrance diameter, mm			BD, socket bottom diameter, mm			L, min. socket length, mm	ID, minimum inside diameter, mm	Minimum wall thickness, mm		Minimum outside diameter of hub, M, mm
	Diameter	Tolerance on diameter	Maximum out-of-round	Diameter	Tolerance on diameter	Maximum out-of-round			TE	TB	
8	14.02	0.10	0.41	13.60	0.10	0.41	12.70	8.13	2.23	2.79	0.672
10	17.45	0.10	0.41	17.04	0.10	0.41	15.08	11.40	2.31	2.90	0.821
15	21.54	0.10	0.41	21.23	0.10	0.41	17.48	14.68	2.76	3.45	0.998
20	26.87	0.10	0.51	26.57	0.10	0.51	18.26	18.80	2.87	3.58	1.221
25	33.66	0.13	0.51	33.27	0.13	0.51	22.23	25.15	3.38	4.22	1.504
32	42.42	0.13	0.61	42.04	0.13	0.61	23.83	33.91	3.56	4.45	1.871
40	48.56	0.15	0.61	48.10	0.15	0.61	27.79	39.73	3.68	4.60	2.127
50	60.63	0.15	0.61	60.17	0.15	0.61	29.36	51.33	3.91	4.90	2.634
65	73.38	0.18	0.76	72.85	0.18	0.76	44.45	61.32	5.16	6.45	3.170
80	89.31	0.20	0.76	88.70	0.20	0.76	47.63	76.40	5.49	6.86	3.841
90	102.01	0.20	0.76	101.40	0.20	0.76	50.80	88.54	5.74	7.19	4.374
100	114.76	0.23	0.76	114.07	0.23	0.76	50.80	100.61	6.02	7.52	4.907
125	141.81	0.25	1.52	141.05	0.25	1.52	76.20	126.37	6.55	8.20	6.039
150	168.83	0.28	1.52	168.00	0.28	1.52	76.20	152.04	7.11	8.89	7.203

ASTM F437 SCHEDULE 80 THREADED CPVC PIPE FITTINGS

ASTM F437 covers NPS 1/4 through 6 Schedule 80 CPVC threaded pipe fittings.

Dimensions of ASTM F437 CPVC Threaded Elbows, Tees, and Couplings

See Tables 18.18 and 18.18M.

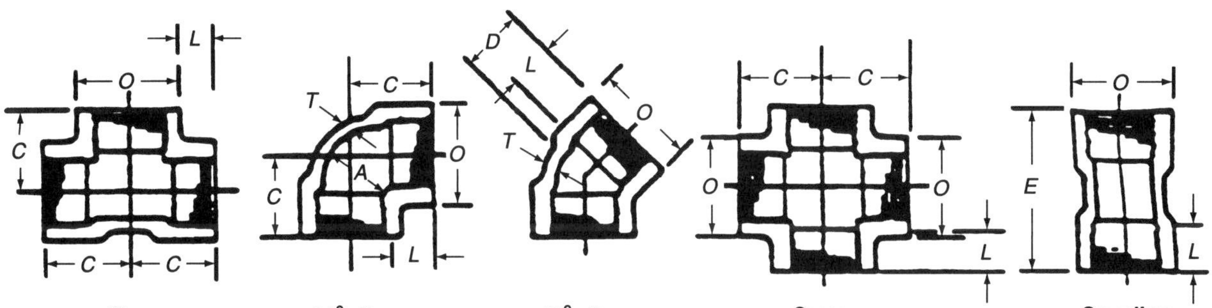

TABLE 18.18 Dimensions of Straight Sizes, Schedule 80, Threaded 90° Ells, Tees, Crosses, 45° Elbows, and Couplings per ASTM F437 (U.S. Customary Units)*

Nominal pipe size	Center to thread end, 90° elbows, tees, crosses, *C*, min.	Length of thread, *L*, min.	Center to thread end, 45° elbow, *D*, min.	Inside of diameter of fitting, *A*, min.	Nominal wall thickness, *T*, min.	Outside diameter of hub, *O*, min.	Thread end to thread end of coupling, *E*, min.
1/4	0.812	0.50	0.688	0.258	0.135	0.840	1.063
3/8	0.938	0.50	0.750	0.379	0.144	1.000	1.063
1/2	1.125	0.64	0.750	0.502	0.198	1.280	1.344
3/4	1.250	0.65	1.000	0.698	0.207	1.500	1.500
1	1.500	0.81	1.125	0.911	0.225	1.810	1.688
1 1/4	1.750	0.85	1.313	1.277	0.261	2.200	1.750
1 1/2	1.938	0.85	1.438	1.446	0.270	2.500	2.000
2	2.250	0.90	1.625	1.881	0.297	3.000	2.063
2 1/2	2.688	1.21	1.938	2.250	0.315	3.560	2.625
3	3.063	1.30	2.125	2.820	0.406	4.300	2.750
4	3.625	1.38	2.625	3.737	0.450	5.430	3.000
6	5.125	1.50	3.250	5.646	0.504	7.625	3.250

* All dimensions are in inches.

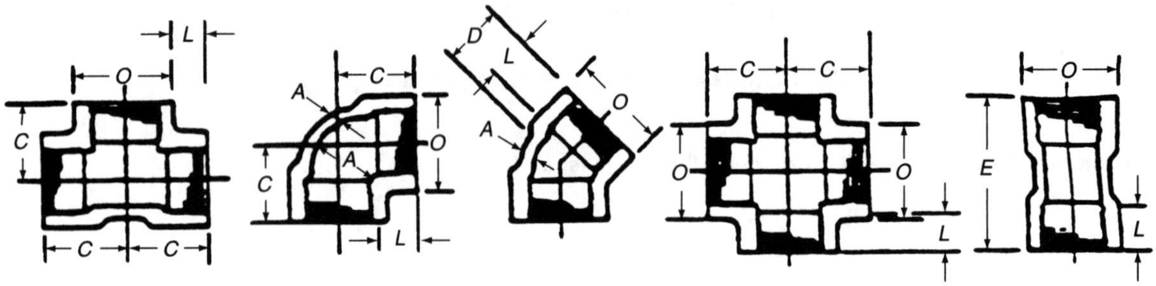

TABLE 18.18M Dimensions of Straight Sizes, Schedule 80, Threaded 90° Ells, Tees, Crosses, 45° Elbows, and Couplings per ASTM F437 (Metric Units)*

Nominal pipe size	Center to thread end, 90° elbows, tees, crosses, *C*, min.	Length of thread, *L*, min.	Center to thread end, 45° elbow, *D*, min.	Inside of diameter of fitting, *A*, min.	Nominal wall thickness, *T*, min.	Outside diameter of hub, *O*, min.	Thread end to thread end of coupling, *E*, min.
8	20.62	12.70	17.48	6.55	3.43	21.34	27.00
10	23.83	12.70	19.05	9.63	3.66	25.40	27.00
15	28.58	16.26	19.05	12.75	5.03	32.51	34.14
20	31.75	16.51	25.40	17.73	5.25	38.10	38.10
25	28.10	20.51	28.58	23.14	5.72	45.97	42.88
32	44.45	21.59	23.35	31.17	6.63	55.88	44.45
40	49.23	21.59	36.83	36.73	6.85	63.50	50.80
50	57.15	22.86	41.28	47.78	7.54	76.20	52.40
65	68.28	30.73	49.23	57.15	8.00	90.42	66.68
80	77.80	33.02	53.98	71.63	10.29	109.22	69.85
100	92.08	35.05	66.68	94.92	11.43	137.92	76.20
150	130.18	38.10	82.55	143.41	12.80	193.68	82.55

* All dimensions are in millimeters.

ASTM F439 SCHEDULE 80 SOCKET-TYPE PIPE FITTINGS

ASTM F439 covers socket-type, CPVC, Schedule 80 pipe fittings.

Minimum Center-to-End Dimensions (Laying Length) of Sockets of ASTM F439 Couplings, Tees, and Elbows

See Tables 18.19 and 18.19M.

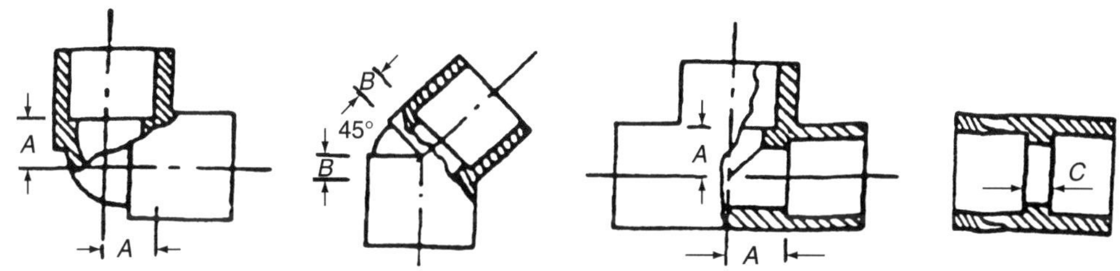

TABLE 18.19 Minimum Dimensions from Center to End (Laying Length) of Sockets for Couplings, Tees, 90° and 45° Elbows, CPVC Socket-Type Pipe Fittings, Schedule 80, per ASTM F439 (U.S. Customary Units)*

Nominal pipe size	A, min.	B, min.	C, min.
1/4	0.31	0.16	0.06
3/8	0.38	0.19	0.09
1/2	0.50	0.25	0.09
3/4	0.56	0.31	0.09
1	0.69	0.31	0.09
1 1/4	0.88	0.38	0.09
1 1/2	1.00	0.44	0.09
2	1.25	0.63	0.09
2 1/2	1.50	0.69	0.19
3	1.81	0.75	0.19
3 1/2	2.13	1.00	0.19
4	2.31	1.00	0.19
5	3.00	1.38	0.19
6	3.50	1.75	0.25
8	4.50	2.00	0.25

* All dimensions are in inches. Refer to Table 18.20 for other dimensions of sockets.

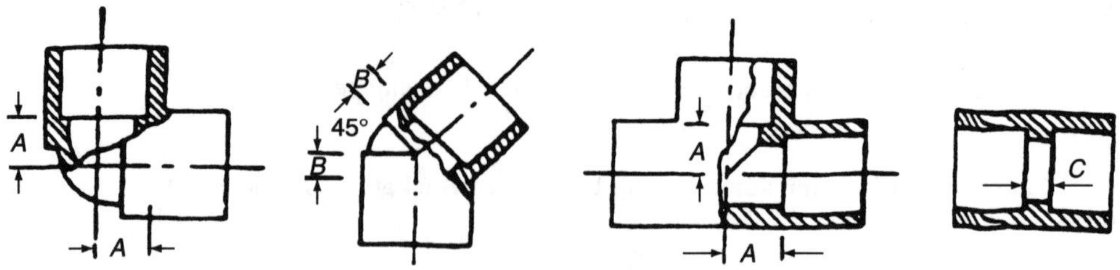

TABLE 18.19M Minimum Dimensions from Center to End (Laying Length) of Sockets for Couplings, Tees, 90° and 45° Elbows, CPVC Socket-Type Pipe Fittings, Schedule 80, per ASTM F439 (Metric Units)*

Nominal pipe size	*A*, min.	*B*, min.	*C*, min.
8	7.94	3.97	1.59
10	9.53	4.76	2.38
15	12.70	6.35	2.38
20	14.29	7.94	2.38
25	17.46	7.94	2.38
32	22.22	9.53	2.38
40	25.40	11.11	2.38
50	31.75	15.88	2.38
65	38.10	17.46	4.76
80	46.04	19.05	4.76
90	53.96	25.40	4.76
100	58.74	25.40	4.76
125	76.20	34.92	4.76
150	88.90	44.45	6.35
200	114.30	50.80	6.35

* All dimensions are in millimeters. Refer to Table 18.20M for other dimensions of sockets.

Dimensions and Tolerances of Tapered Sockets of ASTM F439 Schedule 80 Socket-Type Fittings

See Tables 18.20 and 18.20M.

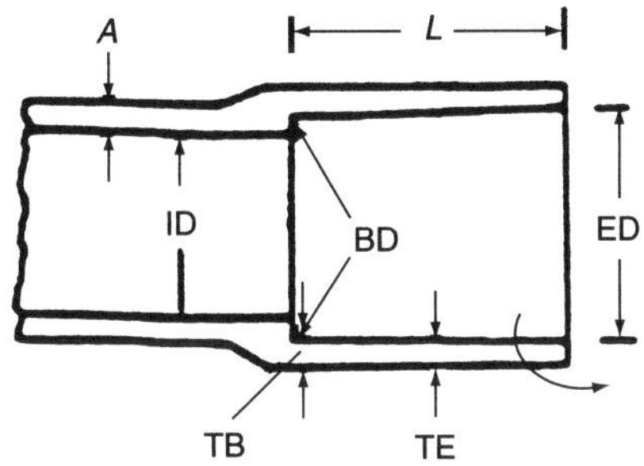

TABLE 18.20 Dimensions of Tapered Sockets for CPVC Pipe Fittings, Schedule 80, per ASTM F439 (U.S. Customary Units)*

Nominal pipe size	ED, socket entrance diameter: Diameter	ED: Tolerance on nominal diameter	ED: Out-of-round, max.-min.	BD, socket bottom diameter: Diameter	BD: Tolerance on nominal diameter	BD: Out-of-round, max.-min.	*L*, min. socket length	ID, min. inside diameter	Minimum wall thickness: TE	Minimum wall thickness: *A*
1/4	0.552	±0.004	0.016	0.536	±0.004	0.016	0.625	0.258	0.119	0.149
3/8	0.687	±0.004	0.016	0.671	±0.004	0.016	0.750	0.379	0.126	0.160
1/2	0.848	±0.004	0.016	0.836	±0.004	0.016	0.875	0.502	0.147	0.185
3/4	1.058	±0.004	0.020	1.046	±0.004	0.020	1.000	0.698	0.154	0.195
1	1.325	±0.005	0.020	1.310	±0.005	0.020	1.125	0.911	0.179	0.225
1 1/4	1.670	±0.005	0.024	1.655	±0.005	0.024	1.250	1.227	0.191	0.240
1 1/2	1.912	±0.006	0.024	1.894	±0.006	0.024	1.375	1.446	0.200	0.250
2	2.387	±0.006	0.024	2.369	±0.006	0.024	1.500	1.881	0.218	0.275
2 1/2	2.889	±0.007	0.030	2.868	±0.007	0.030	1.750	2.250	0.276	0.345
3	3.518	±0.008	0.030	3.492	±0.008	0.030	1.875	2.820	0.300	0.375
3 1/2	4.016	±0.008	0.030	3.992	±0.008	0.030	2.125	3.280	0.318	0.400
4	4.518	±0.009	0.030	4.491	±0.009	0.030	2.250	3.737	0.337	0.420
5	5.583	±0.010	0.060	5.553	±0.010	0.060	2.625	4.703	0.375	0.470
6	6.647	±0.011	0.060	6.614	±0.011	0.060	3.000	5.646	0.432	0.540
8	8.655	±0.015	0.090	8.610	±0.015	0.090	4.000	7.610	0.500	0.625

* Dimensions in inches.

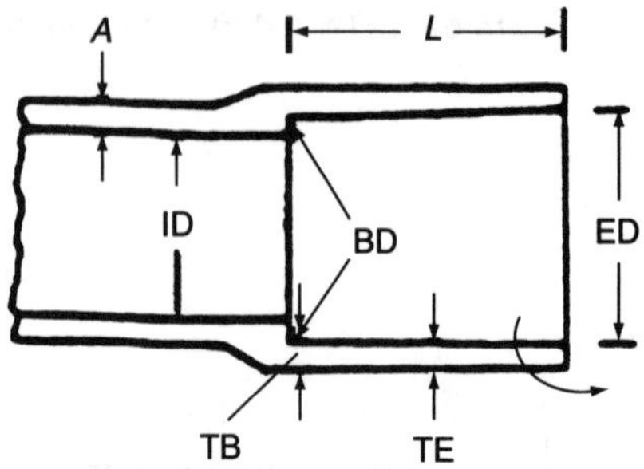

TABLE 18.20M Dimensions of Tapered Sockets for CPVC Pipe Fittings, Schedule 80, per ASTM F439 (Metric Units)*

	ED, socket entrance diameter			BD, socket bottom diameter					Minimum wall thickness	
Nominal pipe size	Diameter	Tolerance on nominal diameter	Out-of-round, max.-min.	Diameter	Tolerance on nominal diameter	Out-of-round, max.-min.	*L*, min. socket length	ID, min. inside diameter	TE	*A*
8	14.02	0.10	0.41	13.61	0.10	0.41	15.88	6.55	3.02	3.78
10	17.45	0.10	0.41	17.04	0.10	0.41	19.05	9.63	3.20	4.06
15	21.54	0.10	0.41	21.23	0.10	0.41	22.22	12.75	3.73	4.70
20	26.87	0.10	0.51	26.57	0.10	0.51	25.40	17.73	3.91	4.95
25	33.66	0.13	0.51	33.27	0.13	0.51	28.58	23.14	4.55	5.72
32	42.42	0.13	0.61	42.04	0.13	0.61	31.75	31.17	4.85	6.10
40	48.56	0.15	0.61	48.11	0.15	0.61	34.93	36.73	5.08	6.35
50	60.63	0.15	0.61	60.17	0.15	0.61	38.10	47.78	5.54	6.99
65	73.38	0.18	0.76	72.85	0.18	0.76	44.45	57.15	7.01	8.76
80	89.31	0.20	0.76	88.70	0.20	0.76	47.63	71.63	7.62	9.53
90	102.01	0.20	0.76	101.40	0.20	0.76	53.98	83.31	8.08	10.16
100	114.76	0.23	0.76	114.07	0.23	0.76	57.15	94.92	8.56	10.67
125	141.81	0.25	1.52	141.05	0.25	1.52	66.68	119.46	9.53	11.94
150	168.83	0.28	1.52	168.00	0.28	1.52	76.20	143.41	10.97	13.72
200	219.84	0.38	2.29		0.38	2.29	101.60	193.29	12.70	15.88

* Dimensions in millimeters.

Dimensions of Threaded Couplings, Tees, Crosses, and Elbows per ASTM F439

See Tables 18.21 and 18.21M.

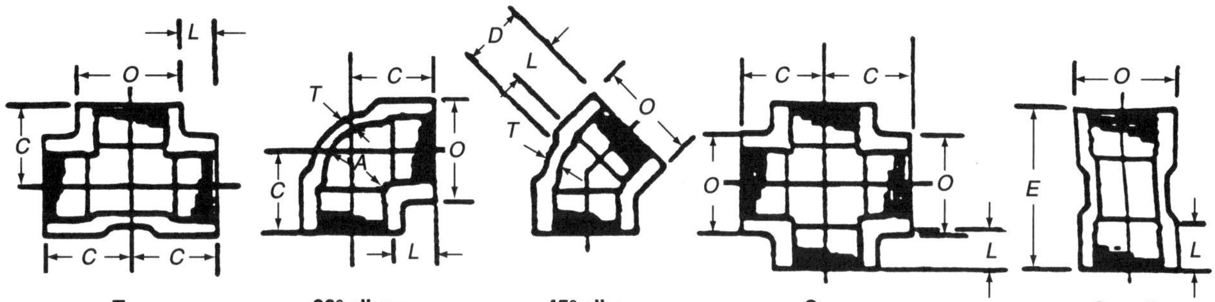

TABLE 18.21 Dimensions of Threaded 90° Ells, Tees, Crosses, 45° Elbows, and Couplings (Straight Sizes) per ASTM F439 (U.S. Customary Units)*

Nominal pipe size	Center to thread end, 90° elbows, tees, crosses, *C*, min.	Length of thread, *L*, min.	Center to thread end, 45° elbow, *D*, min.	Inside diameter of fitting, *A*, min.	Nominal wall thickness, *T*, min.	Outside diameter of hub, *O*, min.	Thread end to thread end of coupling, *E*, min.
1/4	0.812	0.50	0.688	0.258	0.135	0.840	1.063
3/8	0.938	0.50	0.750	0.379	0.144	1.000	1.063
1/2	1.125	0.64	0.750	0.502	0.198	1.280	1.344
3/4	1.250	0.65	1.000	0.698	0.207	1.500	1.500
1	1.500	0.81	1.125	0.911	0.225	1.810	1.688
1 1/4	1.750	0.85	1.313	1.227	0.261	2.200	1.750
1 1/2	1.938	0.85	1.438	1.446	0.270	2.500	2.000
2	2.250	0.90	1.625	1.881	0.297	3.000	2.063
2 1/2	2.688	1.21	1.938	2.250	0.315	3.560	2.625
3	3.063	1.30	2.125	2.820	0.405	4.300	2.750
4	3.625	1.38	2.625	3.737	0.450	5.430	3.000
6	5.125	1.50	3.250	5.646	0.504	7.625	3.250

* All dimensions are in inches.

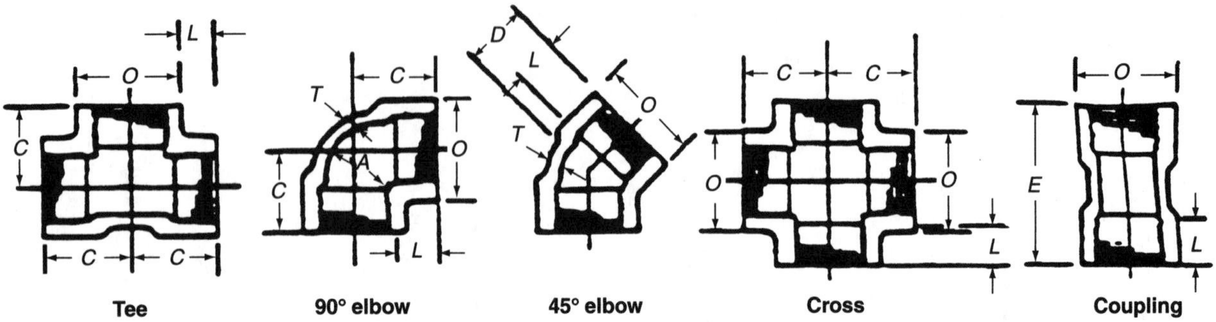

TABLE 18.21M Dimensions of Threaded 90° Ells, Tees, Crosses, 45° Elbows, and Couplings (Straight Sizes) per ASTM F439 (Metric Units)*

Nominal pipe size	Center to thread end, 90° elbows, tees, crosses, *C*, min.	Length of thread, *L*, min.	Center to thread end, 45° elbow, *D*, min.	Inside diameter of fitting, *A*, min.	Nominal wall thickness, *T*, min.	Outside diameter of hub, *O*, min.	Thread end to thread end of coupling, *E*, min.
8	20.62	12.70	17.48	6.55	3.43	21.34	27.00
10	23.83	12.70	19.05	9.63	3.66	25.40	27.00
15	28.58	16.26	19.05	12.75	5.03	32.51	34.14
20	31.75	16.51	25.40	17.73	5.25	38.10	38.10
25	28.10	20.51	28.58	23.14	5.72	45.97	42.88
32	44.45	21.59	23.35	31.17	6.63	55.88	44.45
40	49.23	21.59	36.83	36.73	6.85	63.50	50.80
50	57.15	22.86	41.28	47.78	7.54	76.20	52.40
65	68.28	30.73	49.23	57.15	8.00	90.42	66.68
80	77.80	33.02	53.98	71.63	10.29	109.22	69.85
100	92.08	35.05	66.68	94.92	11.43	137.92	76.20
150	130.18	38.10	82.55	143.41	12.80	193.68	82.55

* All dimensions are in millimeters.

Tables in Chapter 18

Tables in Chapter 18 (Continued)

CHAPTER 19
POLYETHYLENE PIPE, FITTINGS, FLANGES, AND JOINTS

See pp. 613–614 for a list of tables in Chap. 19.

STANDARDS AND SPECIFICATIONS FOR PE PIPE, TUBE, AND FITTINGS

Table 19.1 lists the standards, specifications, and applications for PE pipe, tube, and fittings.

STANDARDS AND SPECIFICATIONS FOR PEX PIPE, TUBE, AND FITTINGS

Table 19.2 lists the standards, specifications and applications for PEX pipe, tube, and fittings.

ASTM D2104, PE PIPE, SCHEDULE 40, ID BASED

ASTM D2104 covers polyethylene pipe made in schedule 40 size for use with insert fittings and pressure rated for water.

Dimensions of ASTM D2104 Pipe

See Tables 19.3 and 19.4.

Pressure Ratings of PE Plastic Schedule 40 Pipes per ASTM D2104

See Table 19.5.

ASTM D2239, PE PIPE, DIMENSION RATIO SERIES, ID BASED

ASTM D2239 covers polyethylene pipe made in standard thermoplastic pipe dimension ratios and pressure rated for water.

Dimensions of ASTM D2239 Pipe

See Tables 19.6, 19.7*a*, 19.7*b*, and 19.8.

TABLE 19.1 Principal PE Piping Standards

Piping material and product standard	Subject product or abbreviated title of standard	Nominal size range, in	Principal applications
ASTM D2513	Thermoplastic gas pressure pipe and fittings	¼–24	Natural gas and LPG distribution
CSA B181.3	Polyolefin drainage systems		Industrial
AASHTO M252	Corrugated PE drainage tubing	3–15	Subsurface drainage
AASHTO M294	Corrugated PE pipe	12–24	Subsurface drainage
API 15LE	PE line pipe	½–12	Oil and gas production; cold water
ASTM D2104	PE pipe, schedule 40, ID based	½–6	Cold water; industrial
ASTM D2239	PE pipe, dimension ratio series, ID based	½–6	Cold water; industrial
ASTM D2447	PE Pipe, schedules 40 and 80	½–12	Cold water; industrial
ASTM D2609	Plastic insert fittings for PE pipe	½–6	Cold water
ASTM D2683	PE fittings, socket fusion type	½–4	Cold water; natural gas; industrial
ASTM D2737	PE tubing	½–2	Cold water
ASTM D3035	PE pipe, dimension ratio series	½–6	Cold water; industrial
ASTM D3261	PE fittings, butt fusion type	½–48	Cold water; natural gas; industrial
ASTM F405	PE corrugated tubing and fittings	3–6	Drainage; leaching fields
ASTM F667	PE corrugated tubing, larger diameter	8–24	Drainage; leaching fields
ASTM F714	PE pipe, large diameter	3–48	Cold water; sewer and drain; industrial
ASTM F771	PE pipe for irrigation	½–6	Irrigation
ASTM F810	PE pipe for drainage and waste disposal	3–6	Drainage; leaching fields
ASTM F892	PE corrugated pipe with smooth interior	4	Sewer; drain; conduit
ASTM F894	PE profile wall pipe, large diameter	18–120	Sewer; drain; industrial
ASTM F1055	PE electrofusion fittings	½–12	Water; natural gas; industrial
ASTM F1533	Deformed PE liner for pipeline rehabilitation	3–18	Rehabilitation of existing pressure and nonpressure pipelines
ASTM F1759	PE manholes for subsurface applications		Sewer and drain; industrial
AWWA C901	PE pipe and tubing for water service	½–3	Cold water service
AWWA C906	PE pipe for water distribution	4–63	Water distribution and transmission
CSA B137.1	PE pipe, tubing, and fittings for cold water	½–6	Cold water; industrial
CSA B137.4	PE piping for gas service	½–8	Natural gas distribution
CSA B137.4.1	Electrofusion-type PE fittings for gas service	½–8	Natural gas distribution
CSA B182.6	Profile PE pipe and fittings	4–48	Drainage; leaching fields
ASTM F1282	Polyethylene/aluminum/polyethylene composite pressure piping	¼–1	
CSA B137.9	Polyethylene/aluminum/polyethylene composite pressure piping	¼–1	Cold water service; industrial

TABLE 19.2 Principal PEX Piping Standards

Piping material and product standard	Subject product or abbreviated title of standard	Nominal size range, in	Principal applications
ASTM F876	Cross-linked PE tubing	¼–2	Hydronic heating
ASTM F877	Cross-linked PE tubing	¼–2	Hot and cold water distributing
ASTM F1807	Metal insert fittings utilizing a copper crimp ring for cross-linked PE tubing	⅜–1	Hot and cold water
CSA B137.5	Cross-linked PE tubing systems	¼–1	Hot and cold water
ASTM F1281	Cross-linked PE/aluminum/cross-linked PE composite pressure piping		Hot and cold water; hydronic heating
CSA B137.10	Cross-linked PE/aluminum/cross-linked PE composite pressure piping	¼–1	Hot and cold water; hydronic heating

TABLE 19.3 Inside Diameters and Tolerances for ASTM D2104 PE Plastic Pipe, Schedule 40*

Nominal pipe size	Average inside diameter	Tolerances
1/2	0.622	+0.010, −0.010
3/4	0.824	+0.010, −0.015
1	1.049	+0.010, −0.020
1 1/4	1.380	+0.010, −0.020
1 1/2	1.610	+0.015, −0.020
2	2.067	+0.015, −0.020
2 1/2	2.469	+0.015, −0.025
3	3.068	+0.015, −0.030
4	4.026	+0.015, −0.035
6	6.065	+0.020, −0.035

* Dimensions in inches.

TABLE 19.4 Wall Thicknesses and Tolerances for ASTM D2104 PE Plastic Pipe, Schedule 40*

Nominal pipe size	Wall thickness	
	Minimum	Tolerance
1/2	0.109	+0.020
3/4	0.113	+0.020
1	0.133	+0.020
1	0.140	+0.020
1 1/2	0.145	+0.020
2	0.154	+0.020
2 1/2	0.203	+0.024
3	0.216	+0.026
4	0.237	+0.028
6	0.280	+0.034

* Dimensions in inches.

TABLE 19.5 Water Pressure Ratings at 23°C (73.4°F) for ASTM D2104 Schedule 40 PE Plastic Pipe*

Nominal pipe size, in	Pressure ratings, psi		
	PE2306 PE2406 PE3306 PE3406	PE2305	PE1404
1/2	190	150	120
3/4	150	120	100
1	140	110	90
1 1/4	120	90	70
1 1/2	100	80	70
2	90	70	60
2 1/2	100	80	60
3	80	70	50
4	70	60	NR†
6	60	NR†	NR†

* These pressure ratings are for unthreaded pipe.
† NR = not rated.

TABLE 19.6 Inside Diameters and Tolerances for SIDR-PR ASTM D2239 PE Plastic Pipe*

Nominal pipe size	Average inside diameter	Tolerance
1/2	0.622	+0.010, −0.010
3/4	0.824	+0.010, −0.015
1	1.049	+0.010, −0.020
1 1/4	1.380	+0.010, −0.020
1 1/2	1.610	+0.015, −0.020
2	2.067	+0.015, −0.020
2 1/2	2.469	+0.015, −0.025
3	3.068	+0.015, −0.030
4	4.026	+0.015, −0.035
6	6.065	+0.020, −0.035

* Dimensions in inches.

TABLE 19.7*a* Wall Thicknesses and Tolerances for SIDR-PR ASTM D2239 PE Plastic Pipe*

Nominal pipe size	Wall thickness					
	SIDR 19		SIDR 15		SIDR 11.5	
	Minimum	Tolerance	Minimum	Tolerance	Minimum	Tolerance
1/2	0.060	+0.020	0.060	+0.020	0.060	+0.020
3/4	0.060	+0.020	0.060	+0.020	0.072	+0.020
1	0.060	+0.020	0.070	+0.020	0.091	+0.020
1 1/4	0.073	+0.020	0.092	+0.020	0.120	+0.020
1 1/2	0.085	+0.020	0.107	+0.020	0.140	+0.020
2	0.109	+0.020	0.138	+0.020	0.180	+0.022
2 1/2	0.130	+0.020	0.165	+0.020	0.215	+0.025
3	0.161	+0.020	0.205	+0.020	0.267	+0.032
4	0.212	+0.025	0.268	+0.032	0.350	+0.042
6	0.319	+0.038	0.404	+0.048	0.527	+0.063

* Dimensions in inches.

TABLE 19.7*b* Wall Thicknesses and Tolerances for SIDR-PR ASTM D2239 PE Plastic Pipe*

Nominal pipe size	Wall thickness					
	SIDR 9		SIDR 7		SIDR 5.3	
	Minimum	Tolerance	Minimum	Tolerance	Minimum	Tolerance
½	0.069	+0.020	0.089	+0.020	0.117	+0.020
¾	0.092	+0.020	0.118	+0.020	0.155	+0.020
1	0.117	+0.020	0.150	+0.020	0.198	+0.024
1¼	0.153	+0.020	0.197	+0.024	0.260	+0.031
1½	0.179	+0.020	0.230	+0.028	0.304	+0.036
2	0.230	+0.028	0.295	+0.035	0.390	+0.047

* Dimensions in inches.

TABLE 19.8 Water Pressure Ratings (PR) of ASTM D2239 Pipe at 23°C (73°F) or SIDR-PR PE Plastic Pipe

Standard dimension ratio	PE pipe materials				
	PE3408	PE3306 PE3406	PE2306 PE2406	PE2305	PE1404
	Pressure rating, psi				
5.3	250	200	200	160	125
7	200	160	160	125	100
9	160	125	125	100	80
11.5	125	100	100	80	
15	100	80	80		
19	80				
	Pressure rating, psi				
250		5.3			
200		7	5.3		
160		9	7	5.3	
125		11.5	9	7	5.3
100		15	11.5	9	7
80		19	15	11.5	9

ASTM D2447, PE PIPE, SCHEDULES 40 AND 80

ASTM D2447 covers polyethylene (PE) pipe that is outside-diameter controlled, made in Schedule 40 and 80 sizes for use with socket-type fittings and butt fittings joined by heat fusion, and pressure rated for water.

Dimensions of ASTM D2247 Pipe

See Tables 19.9, 19.10*a*, and 19.10*b*.

TABLE 19.9 Outside Diameters and Tolerances for ASTM D2447 PE Plastic Pipe, Schedules 40 and 80*

Nominal pipe size	Average outside diameter	Tolerance (plus and minus): For average diameter	Tolerance (plus and minus): For maximum and minimum diameter (out-of-roundess)
1/2	0.840	0.004	0.015
3/4	1.050	0.004	0.020
1	1.315	0.005	0.025
1 1/4	1.660	0.005	0.025
1 1/2	1.900	0.006	0.030
2	2.375	0.006	0.035
2 1/2	2.875	0.007	0.035
3	3.500	0.008	0.040
3 1/2	4.000	0.008	0.050
4	4.500	0.009	0.050
5	5.563	0.010	0.050
6	6.625	0.011	0.050
8	8.625	0.015	0.075
10	10.750	0.015	0.075
12	12.750	0.015	0.075

* Dimensions in inches.

TABLE 19.10*a* Wall Thicknesses and Tolerances for ASTM D2447 PE Plastic Pipe, Schedule 40*

Nominal pipe size	Wall thickness: Minimum	Wall thickness: Tolerance (plus)
1/2	0.109	0.020
3/4	0.113	0.020
1	0.133	0.020
1 1/4	0.140	0.020
1 1/2	0.145	0.020
2	0.154	0.020
2 1/2	0.203	0.024
3	0.216	0.026
3 1/2	0.226	0.027
4	0.237	0.028
5	0.258	0.031
6	0.280	0.034
8	0.322	0.039
10	0.365	0.044
12	0.406	0.049

* Dimensions in inches.

TABLE 19.10*b* Wall Thicknesses and Tolerances for ASTM D2447 PE Plastic Pipe, Schedule 80*

Nominal pipe size	Wall thickness: Minimum	Wall thickness: Tolerance (plus)
1/2	0.147	0.020
3/4	0.154	0.020
1	0.179	0.021
1 1/4	0.191	0.023
1 1/2	0.200	0.024
2	0.218	0.026
2 1/2	0.276	0.033
3	0.300	0.036
3 1/2	0.318	0.038
4	0.337	0.040
5	0.375	0.045
6	0.432	0.052

* Dimensions in inches.

Pressure Ratings of PE Plastic Schedule 40 and 80 Pipes per ASTM D2447

See Table 19.11.

TABLE 19.11 Water Pressure Ratings at 73.4°F for ASTM D2447 PE Plastic Pipe, Schedules 40 and 80*

Nominal pipe size, in	Pressure rating, psi					
	PE2306 PE2406 PE3306 PE3406		PE2305		PE1404	
	Schedule 40	Schedule 80	Schedule 40	Schedule 80	Schedule 40	Schedule 80
1/2	188	267	149	212	119	170
3/4	152	217	120	172	96	137
1	142	199	113	158	90	126
1 1/4	116	164	92	130	74	104
1 1/2	104	148	83	118	66	94
2	87	127	69	101	55	81
2 1/2	96	134	76	106	61	85
3	83	118	66	94	53	75
3 1/2	75	109	60	86	50	69
4	70	102	55	81	NR†	65
5	61	91	50	72	NR	58
6	55	88	NR	70	NR	56
8	50	...	NR	...	NR	...
10	NR	...	NR	...	NR	...
12	NR	...	NR	...	NR	...

* These pressure ratings are for unthreaded pipe.
† NR = not rated.

ASTM D3035, PE PIPE, DIMENSION RATIO SERIES

ASTM D3035 covers polyethylene pipe made in dimension ratios (DRs) based on outside diameter and pressure rated for water. All pipes produced under this specification may be used for the transport of water, industrial process liquids, effluents, slurries, municipal sewage, etc.

Dimensions of ASTM D3035 Pipe

See Tables 19.12 and 19.13*a* and 19.13*b*.

TABLE 19.12 Outside Diameters and Tolerances for ASTM D3035 DR-PR PE Plastic Pipe*

Nominal pipe size	Outside diameter	Tolerance (plus and minus)
1/2	0.840	0.004
3/4	1.050	0.004
1	1.315	0.005
1 1/4	1.660	0.005
1 1/2	1.900	0.006
2	2.375	0.006
3	3.500	0.008
4	4.500	0.009
6	6.625	0.011
8	8.625	0.013
10	10.750	0.015
12	12.750	0.017
14	14.000	0.063
16	16.000	0.072
18	18.000	0.081
20	20.000	0.090
22	22.000	0.099
24	24.000	0.108

* Dimensions in inches.

TABLE 19.13*a* Wall Thicknesses and Tolerances for ASTM D3035 DR-PR PE Plastic Pipe*

Nominal pipe size, IPS	DR 32.5		DR 26		DR 21		DR 17		DR 15.5	
	Mininum	Tolerance	Minimum	Tolerance	Minimum	Tolerance	Minimum	Tolerance	Minimum	Tolerance
1/2	0.062	0.020	0.062	0.020	0.062	0.020	0.062	0.020	0.062	0.020
3/4	0.062	0.020	0.062	0.020	0.062	0.020	0.062	0.020	0.068	0.020
1	0.062	0.020	0.062	0.020	0.063	0.020	0.077	0.020	0.084	0.020
1 1/4	0.062	0.020	0.064	0.020	0.079	0.020	0.098	0.020	0.107	0.020
1 1/2	0.062	0.020	0.073	0.020	0.090	0.020	0.112	0.020	0.123	0.020
2	0.073	0.020	0.091	0.020	0.113	0.020	0.140	0.020	0.153	0.020
3	0.108	0.020	0.135	0.020	0.167	0.020	0.206	0.025	0.226	0.027
4	0.138	0.020	0.173	0.021	0.214	0.026	0.265	0.032	0.290	0.035
5	0.171	0.021	0.214	0.026	0.265	0.032	0.327	0.039	0.359	0.043
6	0.204	0.024	0.255	0.031	0.315	0.038	0.390	0.047	0.427	0.051
8	0.265	0.032	0.332	0.040	0.411	0.049	0.507	0.061	0.556	0.067
10	0.331	0.040	0.413	0.050	0.512	0.061	0.632	0.076	0.694	0.083
12	0.392	0.047	0.490	0.059	0.607	0.073	0.750	0.090	0.823	0.099
14	0.431	0.052	0.538	0.065	0.667	0.080	0.824	0.099	0.903	0.108
16	0.492	0.059	0.615	0.074	0.762	0.091	0.941	0.113	1.032	0.124
18	0.554	0.068	0.692	0.083	0.857	0.103	1.059	0.127	1.161	0.139
20	0.615	0.074	0.769	0.092	0.952	0.114	1.176	0.141	1.290	0.155
22	0.677	0.081	0.846	0.102	1.048	0.126	1.294	0.155	1.419	0.170
24	0.738	0.089	0.923	0.111	1.143	0.137	1.412	0.169	1.548	0.188

* Dimensions in inches.

TABLE 19.13*b* Wall Thicknesses and Tolerances for ASTM D3035 DR-PR PE Plastic Pipe*

Nominal pipe size, IPS	DR 13.5		DR 11		DR 9.3		DR 9		DR 7	
	Mininum	Tolerance	Minimum	Tolerance	Minimum	Tolerance	Minimum	Tolerance	Minimum	Tolerance
1/2	0.062	0.020	0.076	0.020	0.090	0.020	0.093	0.020	0.120	0.020
3/4	0.078	0.020	0.095	0.020	0.113	0.020	0.117	0.020	0.150	0.020
1	0.097	0.020	0.120	0.020	0.141	0.020	0.146	0.020	0.188	0.023
1 1/4	0.123	0.020	0.151	0.020	0.178	0.021	0.184	0.022	0.237	0.028
1 1/2	0.141	0.020	0.173	0.021	0.204	0.024	0.211	0.025	0.271	0.033
2	0.176	0.021	0.216	0.026	0.255	0.031	0.264	0.032	0.339	0.041
3	0.259	0.031	0.318	0.038	0.376	0.045	0.389	0.047	0.500	0.060
4	0.333	0.040	0.409	0.049	0.484	0.058	0.500	0.060	0.643	0.077
5	0.412	0.049	0.506	0.061	0.598	0.072	0.618	0.074	0.795	0.095
6	0.491	0.059	0.602	0.072	0.712	0.085	0.736	0.088	0.946	0.114
8	0.639	0.077	0.784	0.094	0.927	0.111	0.958	0.115	1.232	0.147
10	0.796	0.096	0.977	0.117	1.156	0.139	1.194	0.143	1.536	0.184
12	0.944	0.113	1.159	0.139	1.371	0.165	1.417	0.170	1.821	0.219
14	1.037	0.124	1.273	0.153	1.505	0.181	1.556	0.187	2.000	0.240
16	1.185	0.142	1.455	0.175	1.720	0.206	1.778	0.213	2.286	0.274
18	1.333	0.160	1.636	0.196	1.935	0.232	2.000	0.240	2.571	0.309
20	1.481	0.178	1.818	0.218	2.151	0.258	2.222	0.267	2.857	0.343
22	1.630	0.196	2.000	0.240	2.366	0.284	2.444	0.293	3.143	0.377
24	1.778	0.213	2.182	0.262	2.581	0.310	2.667	0.320	3.429	0.411

* Dimensions in inches.

Pressure Ratings of DR-PR PE Plastic Pipe per ASTM D3035

See Table 19.14.

ASTM F771, PE PIPE FOR IRRIGATION

ASTM F771 covers polyethylene thermoplastic pipelines used to convey, at rated pressures of 80 to 200 psi, water that is to be used for irrigation purposes.

Dimensions of ASTM F771 Pipe

See Tables 19.15, 19.16, 19.17, and 19.18.

TABLE 19.14 Thermoplastic Pipe Dimension Ratios (DR) and Water Pressure Ratings (PR) at 73°F for DR-PR PE Plastic Pipe

	PE pipe material			
Dimension ratio	PE 3408	PE 2306 PE 2406 PE 3306 PE 3406	PE 2305	PE 1404
	Pressure rating, psi			
7	267	210	167	133
9	200	158	125	100
9.3	193	152	120	96
11	160	126	100	80
13.5	128	100	80	64
15.5	110	87	69	55
17	100	79	63	50
21	80	63	50	40
26	64	50	40	32
32.5	51	40	32	25

TABLE 19.15 Outside Diameters and Tolerances for ASTM F771 SDR-PR PE Plastic Pipe*

Nominal pipe size	Outside diameter	Tolerance (plus and minus)
1/2	0.840	0.004
3/4	1.050	0.004
1	1.315	0.005
1 1/4	1.660	0.005
1 1/2	1.900	0.006
2	2.375	0.006
3	3.500	0.008
4	4.500	0.009
6	6.625	0.011

* Dimensions in inches.

TABLE 19.16 Inside Diameters and Tolerances for ASTM F771 SIDR-PR PE Plastic Pipe*

Nominal pipe size	Average inside diameter	Tolerance
1/2	0.622	+0.010, –0.010
3/4	0.824	+0.010, –0.015
1	1.049	+0.010, –0.020
1 1/4	1.380	+0.010, –0.020
1 1/2	1.610	+0.015, –0.020
2	2.067	+0.015, –0.020
2 1/2	2.469	+0.015, –0.025
3	3.068	+0.015, –0.030
4	4.026	+0.015, –0.035
6	6.065	+0.020, –0.035

* Dimensions in inches.

Water Pressure Ratings for PE Plastic Pipe per ASTM F771

See Table 19.19.

TABLE 19.17 Wall Thicknesses and Tolerances for ASTM F771 SDR-PR PE Plastic Pipe with Controlled Outside Diameters*

Nominal pipe size	SDR 21 Minimum	SDR 21 Tolerance (plus)	SDR 17 Minimum	SDR 17 Tolerance (plus)	SDR 13.5 Minimum	SDR 13.5 Tolerance (plus)	SDR 11 Minimum	SDR 11 Tolerance (plus)
1/2	0.062	0.020	0.062	0.020	0.062	0.020	0.076	0.020
3/4	0.062	0.020	0.062	0.020	0.078	0.020	0.095	0.021
1	0.062	0.020	0.077	0.020	0.097	0.020	0.119	0.026
1 1/4	0.079	0.020	0.098	0.020	0.123	0.020	0.151	0.026
1 1/2	0.090	0.020	0.112	0.020	0.141	0.020	0.173	0.026
2	0.113	0.020	0.140	0.021	0.176	0.020	0.216	0.026
3	0.167	0.021	0.206	0.025	0.259	0.031	0.318	0.038
4	0.214	0.026	0.264	0.032	0.333	0.040	0.409	0.049
6	0.315	0.038	0.390	0.047	0.491	0.059	0.602	0.068

* Dimensions in inches.

TABLE 19.18 Wall Thicknesses and Tolerances for ASTM F771 SIDR-PR PE Plastic Pipe with Controlled Inside Diameter Wall Thickness*

Nominal pipe size	SIDR 9 Minimum	SIDR 9 Tolerance (plus)	SIDR 7 Minimum	SIDR 7 Tolerance (plus)	SIDR 5.3 Minimum	SIDR 5.3 Tolerance (plus)	SIDR 19 Minimum	SIDR 19 Tolerance (plus)	SIDR 15 Minimum	SIDR 15 Tolerance (plus)	SIDR 11.5 Minimum	SIDR 11.5 Tolerance (plus)
1/2	0.069	0.020	0.089	0.020	0.117	0.020	0.060	0.020	0.060	0.020	0.060	0.020
3/4	0.092	0.020	0.118	0.020	0.155	0.020	0.060	0.020	0.060	0.020	0.072	0.020
1	0.117	0.020	0.150	0.020	0.198	0.024	0.060	0.020	0.070	0.020	0.091	0.020
1 1/4	0.153	0.020	0.197	0.024	0.260	0.031	0.073	0.020	0.092	0.020	0.120	0.020
1 1/2	0.179	0.020	0.230	0.028	0.304	0.036	0.085	0.020	0.107	0.020	0.140	0.020
2	0.230	0.028	0.295	0.035	0.390	0.047	0.109	0.020	0.138	0.020	0.180	0.022
2 1/2							0.130	0.020	0.165	0.020	0.215	0.025
3							0.162	0.020	0.205	0.020	0.267	0.032
4							0.212	0.025	0.268	0.032	0.350	0.042
6							0.319	0.038	0.404	0.048	0.527	0.063

* Dimensions in inches.

TABLE 19.19 Water Pressure Ratings for ASTM F771 at 73°F for PE Plastic Pipe

	PE pipe materials	
	PE 3408	PE 2306 PE 2406 PE 3306 PE 3406
Standard	Pressure rating, psi	
. . .	. . .	200
. . .	200	160
. . .	160	125
11	. . .	125
	125	100
13.5	125	100
	100	80
17	100	80
	80	. . .
21	80	. . .

	PE 3408		PE 2306 PE 2406 PE 3306 PE 3406	
Pressure rating, psi	Standard dimension ratio			
200	. . .	(7)	. . .	(5.3)
160	. . .	(9)	. . .	(7)
125	13.5	(11.5)	11	(9)
100	17	(15)	13.5	(11.5)
80	21	(19)	17	(15)

ASTM F810, PE PIPE FOR DRAINAGE AND WASTE DISPOSAL

ASTM F810 covers requirements and test methods for workmanship, flattening dimensions, tolerances, stiffness, perforation, environmental stress cracking, bonding, and impact resistance for smooth-wall perforated and nonperforated polyethylene (PE) pipe, including coextruded pipe.

Dimensions of ASTM F810 Pipe

See Table 19.20.

TABLE 19.20 Diameters, Tolerance, and Minimum Wall Thickness for PE Smooth-Wall Pipe*

Nominal pipe size	Average outside diameter	Average inside diameter, min.	Minimum wall thickness
3	3.250 ± 0.008	2.875	0.090
4	4.215 ± 0.009	3.875	0.095
6	6.275 ± 0.011	5.875	0.125

* Dimensions in inches.

ASTM F714, PE PIPE, LARGE DIAMETER

ASTM F714 covers polyethylene (PE) pipe made in dimensions based on outside diameters of 90 mm and larger. Three standard outside-diameter sizing systems are detailed: (1) the ISO metric system, (2) the IPS system, and (3) the DIPS system.

The piping is intended for new construction and insertion renewal of old piping systems used for the transport of the water, municipal sewage, industrial process liquids, effluent, slurries, etc., in both pressure and nonpressure systems. All pipes produced under this specification are pressure rated. The specification also includes criteria for choice of raw materials and test methods for evaluation of raw material, together with performance requirements and test methods for determining conformance with the requirements.

Pressure Ratings of PE Plastic Pipe per ASTM F714

See Table 19.21.

TABLE 19.21 Pressure Rating for ASTM F714 PE Pipe

HDB		DR 41		DR 32.5		DR 26		DR 21		DR 17		DR 15.5		DR 11		DR 9.3		DR 9		DR 7.3	
MPa	psi	kPa	psi	kPa	psi	kPa	psi	kPa	psi	kPa	psi	kPa	psi	kPa	psi	kPa	psi	kPa	psi	kPa	psi
8.6	1250	215	31	275	40	345	50	430	63	540	78	595	86	860	125	1035	150	1075	156	1365	198
10	1450	250	36	315	46	400	58	500	72	625	91	690	100	1000	145	1205	175	1250	181	1585	230
11	1600	275	40	350	50	440	64	550	80	690	100	760	110	1100	160	1325	192	1380	200	1750	254

Dimensions of ASTM F714 Pipe

See Tables 19.22 to 19.27.

TABLE 19.22 Outside Diameters and Tolerances for ASTM F714 PE Pipe (Metric Units)

Nominal pipe size, mm	Outside diameter, D_o, mm	
	Min.	Max.
90	90	90.8
110	110	111.0
160	160	161.4
200	200	201.8
250	250	252.3
280	280	282.5
315	315	317.8
355	355	358.2
400	400	403.6
450	450	454.1
500	500	504.5
560	560	565.0
630	630	635.7
710	710	716.4
800	800	807.2
900	900	908.1
1000	1000	1009.0
1200	1200	1210.8
1400	1400	1412.6
1600	1600	1614.4

TABLE 19.23 Outside Diameters and Tolerances for ASTM F714 PE Pipe (ASME B36.10)

Nominal pipe size, in	Actual outside diameter, in	
	Average	Tolerance (plus and minus)
3	3.500	0.016
4	4.500	0.020
5	5.375	0.025
5	5.563	0.025
6	6.625	0.030
7	7.125	0.034
8	8.625	0.039
10	10.750	0.048
12	12.750	0.057
13	13.375	0.060
14	14.000	0.063
16	16.000	0.072
18	18.000	0.081
20	20.000	0.090
21.5	21.500	0.097
22	22.000	0.099
24	24.000	0.108
26	26.000	0.117
28	28.000	0.126
30	30.000	0.135
32	32.000	0.144
34	34.000	0.153
36	36.000	0.162
42	42.000	0.189
48	48.000	0.216

TABLE 19.24 Outside Diameters and Tolerances for ASTM F714 PE Pipe, DIPS Sizing System

Nominal DIPS size, in	Actual outside diameter, in	
	Average	Tolerance (plus and minus)
3	3.96	0.016
4	4.80	0.022
6	6.90	0.031
8	9.05	0.041
10	11.10	0.050
12	13.20	0.059
14	15.30	0.069
16	17.40	0.078
18	19.50	0.088
20	21.60	0.097
24	25.80	0.116
30	32.00	0.144
36	38.30	0.172
42	44.50	0.200
48	50.80	0.229

TABLE 19.25 Minimum Wall Thickness of ASTM F714 PE Pipe, ISO Sizing System*

Nominal pipe size	DR 41	32.5	26	21	17	11
90	...	...	3.5	4.3	5.3	8.2
110	...	3.4	4.2	5.2	6.5	10.0
160	...	4.9	6.2	7.6	9.4	14.5
200	...	6.2	7.7	9.5	11.8	18.2
250	...	7.7	9.6	11.9	14.7	22.7
280	...	8.6	10.8	13.3	16.5	25.5
315	...	9.7	12.1	15.0	18.5	28.6
355	...	10.9	13.7	16.9	20.9	32.3
400	...	12.3	15.4	19.0	23.5	36.4
450	...	13.8	17.3	21.4	26.5	...
500	...	15.4	19.2	23.8	29.4	...
560	...	17.2	21.5	26.7	32.9	...
630	...	19.4	24.2	30.0	37.1	...
710	...	21.8	27.3	33.8	41.8	...
800	...	24.6	30.8	38.1	47.1	...
900	...	27.7	34.6	42.9	...	...
1000	24.4	30.8	38.5	47.6	...	...
1200	29.3	36.9	46.2	...	...	...
1400	34.1	43.1	...	...	...	...
1600	39.0	49.2	...	...	...	...

* Dimensions in millimeters.

TABLE 19.26 Minimum Wall Thickness of ASTM F714 PE Pipe (U.S. Customary Units)*

Nominal IPS pipe size	Actual pipe size	Dimension ratio 32.5	26	21	17	15.5	13.5	11	9.3	9	8.3	7.3
3	3.500	0.108	0.135	0.167	0.206	0.226	0.259	0.318	0.376	0.389	0.422	0.479
4	4.500	0.138	0.173	0.214	0.265	0.290	0.333	0.409	0.484	0.500	0.542	0.616
5	5.375	0.165	0.207	0.256	0.316	0.347	0.398	0.489	0.578	0.597	0.648	0.736
5	5.563	0.171	0.214	0.265	0.327	0.359	0.412	0.506	0.598	0.618	0.670	0.762
6	6.625	0.204	0.255	0.315	0.390	0.427	0.491	0.602	0.712	0.736	0.798	0.908
7	7.125	0.219	0.274	0.340	0.420	0.460	0.528	0.648	0.766	0.792	0.858	0.976
8	8.625	0.265	0.332	0.411	0.507	0.556	0.639	0.784	0.927	0.958	1.039	1.182
10	10.750	0.331	0.413	0.512	0.632	0.694	0.796	0.977	1.156	1.194	1.295	1.473
12	12.750	0.392	0.490	0.607	0.750	0.823	0.944	1.159	1.371	1.417	1.536	1.747
13	13.375	0.412	0.514	0.637	0.787	0.863	0.991	1.216	1.438	1.486	1.611	1.832
14	14.000	0.431	0.538	0.667	0.824	0.903	1.037	1.273	1.505	1.556	1.687	1.918
16	16.000	0.492	0.615	0.762	0.941	1.032	1.185	1.455	1.720	1.778	1.928	2.192
18	18.000	0.554	0.692	0.857	1.059	1.161	1.333	1.636	1.935	2.000	2.169	2.466
20	20.000	0.615	0.769	0.952	1.176	1.290	1.481	1.818	2.151	2.222	2.409	...
21.5	21.500	0.662	0.827	1.024	1.265	1.387	1.593	...	...	...	...	...
22	22.000	0.677	0.846	1.048	1.294	1.419	1.630	2.000	2.366	2.444	...	...
24	24.000	0.738	0.923	1.143	1.412	1.548	1.778	2.162	2.581	2.667	...	...
26	26.000	0.800	1.000	1.238	1.529	1.677	1.926	2.364	2.796	...	...	...
28	28.000	0.862	1.077	1.333	1.647	1.806	2.074	2.545	3.011	...	...	...
30	30.000	0.923	1.154	1.429	1.765	1.935	2.222	2.727	3.226	...	...	...
32	32.000	0.985	1.231	1.524	1.882	2.065	2.370	2.909	...	...	...	...
34	34.000	1.046	1.308	1.619	2.000	2.194	2.519	3.091	...	...	...	...
36	36.000	1.108	1.385	1.714	2.118	2.323	2.667	3.273	...	...	...	...
42	42.000	1.292	1.615	2.000	2.471	2.710	...	...	...	...	...	...
48	48.000	1.477	1.846	2.286	2.824	3.097	...	...	...	...	...	...

* Dimensions in inches.

TABLE 19.27 Minimum Wall Thickness of F714 PE Pipe, DIPS Sizing System*

Nominal DIPS pipe size	Actual OD pipe size	Dimension ratio					
		32.5	26	21	17	13.5	11
3	3.96	0.122	0.153	0.189	0.233	0.294	0.360
4	4.80	0.148	0.189	0.229	0.283	0.356	0.437
6	6.90	0.213	0.266	0.329	0.406	0.512	0.628
8	9.05	0.279	0.348	0.431	0.533	0.670	0.823
10	11.10	0.342	0.427	0.529	0.653	0.823	1.009
12	13.20	0.407	0.508	0.629	0.777	0.978	1.200
14	15.30	0.471	0.589	0.729	0.900	1.134	1.391
16	17.40	0.536	0.670	0.829	1.024	1.289	1.582
18	19.50	0.600	0.750	0.929	1.147	1.445	1.773
20	21.60	0.665	0.831	1.029	1.271	1.600	1.964
24	25.80	0.794	0.993	1.229	1.518	1.912	2.346
30	32.00	0.985	1.231	1.524	1.883	2.371	2.909
36	38.30	1.179	1.473	1.824	2.253	2.837	3.482
42	44.50	1.370	1.712	2.119	2.618	3.297	4.046
48	50.80	1.563	1.954	2.419	2.989	3.763	4.619

* Dimensions in inches.

ASTM D2737, PE TUBING

ASTM D2737 covers polyethylene (PE) tubing pressure rated for water. PE plastic tubing is often used with fittings that require flaring the tubing.

Dimensions of ASTM D2737 Tubing

See Tables 19.28 and 19.29.

TABLE 19.28 Outside Diameter and Tolerance for ASTM D2737 PE Plastic Tubing*

Nominal tubing size	Outside diameter	Tolerance: For average (plus and minus)	Tolerance: For max. and min. (out of roundness) (plus and minus)
1/2	0.625	0.004	0.015
5/8	0.750	0.004	0.015
3/4	0.875	0.004	0.015
1	1.125	0.005	0.015
1 1/4	1.375	0.005	0.015
1 1/2	1.625	0.006	0.015
2	2.125	0.006	0.015

* Dimensions in inches.

TABLE 19.29 Wall Thickness and Tolerances for PE Plastic Tubing

Nominal tubing size, in	Wall thickness, in					
	PE 2305 SDR 7.3		PE 2306 PE 2406 PE 3306 PE 3406 PE 3408 SDR 9		PE 3408 SDR 11	
	Minimum	Tolerance	Minimum	Tolerance	Minimum	Tolerance
$^1/_2$	0.086	+0.010	0.069	+0.010	0.062	+0.010
$^5/_8$	0.103	+0.010	0.083	+0.010	0.068	+0.010
$^3/_4$	0.120	+0.012	0.097	+0.010	0.080	+0.010
1	0.154	+0.015	0.125	+0.012	0.102	+0.010
$1^1/_4$	0.188	+0.019	0.153	+0.015	0.125	+0.012
$1^1/_2$	0.233	+0.022	0.181	+0.018	0.148	+0.015
2	0.291	+0.029	0.236	+0.024	0.193	+0.019

Maximum Internal Pressures per ASTM D2737

See Table 19.30.

TABLE 19.30 Maximum Internal Pressures for ASTM D2737 PE Pipe

Material	SDR	Internal pressure, psi	
		73.4°F	100°F
PE 2305	7.3	1.10	0.86
PE 2306, PE 2406, PE 3306, PE 3406	9	1.10	0.86
PE 3408	9	1.38	1.10
PE 3408	11	1.10	0.86

ASTM D2513, THERMOPLASTIC GAS PRESSURE PIPE AND FITTINGS

ASTM D2513 covers requirements and test methods for material dimensions and tolerances, hydrostatic burst strength, and chemical and impact resistance of plastic pipe, tubing, and fittings for use in fuel gas mains and services for direct burial and reliner applications. This specification does not cover threaded pipes.

Dimensions of ASTM D2513 Pipes and Fittings

See Tables 19.31 to 19.35.

TABLE 19.31 ASTM D2513 Tubing Diameters and Tolerances*

Nominal tubing size (CTS)	Outside diameter	Tolerance (plus and minus)
1/4	0.375	0.004
3/8	0.500	0.004
1/2	0.625	0.004
1/2	0.625	0.004
1/2	0.625	0.004
3/4	0.875	0.004
3/4	0.875	0.004
3/4	0.875	0.004
1	1.125	0.005
1	1.125	0.005
1	1.125	0.005
1	1.125	0.005
1	1.125	0.005
1 1/4	1.375	0.005
1 1/4	1.375	0.005
1 1/4	1.375	0.005
1 3/4	1.875	0.006

* Dimensions in inches.

TABLE 19.32 ASTM D2513 Tubing Wall Thicknesses and Tolerances*

Nominal tubing size (CTS)	Minimum wall thickness	Wall thickness tolerance (plus)
1/4	0.062	0.006
3/8	0.062	0.006
1/2	0.062	0.006
1/2	0.090	0.009
1/2	0.104	0.010
3/4	0.062	0.006
3/4	0.077	0.008
3/4	0.090	0.009
1	0.062	0.007
1	0.090	0.011
1	0.099	0.012
1	0.101	0.012
1	0.121	0.015
1 1/4	0.062	0.007
1 1/4	0.090	0.011
1 1/4	0.121	0.015
1 3/4	0.062	0.007

* Dimensions in inches.

TABLE 19.33 Diameter Ratios, Wall Thicknesses, and Tolerances for ASTM D2513 Plastic Pipe*

Nominal pipe size (IPS)	Diameter ratio	Wall thickness		Nominal pipe size (IPS)	Diameter ratio	Wall thickness	
		Minimum	Tolerance			Minimum	Tolerance
1/2		0.062	+0.007		13.5	0.296	+0.036
	11.0	0.076	+0.009		11	0.363	+0.044
	9.33	0.090	+0.011	4	21	0.214	+0.026
3/4		0.090	+0.011		19	0.237	+0.028
	11.0	0.095	+0.011		17	0.264	+0.032
	Sched. 40	0.113	+0.014		13.5	0.333	+0.040
1		0.090	+0.011		11.5	0.391	+0.047
	13.5	0.097	+0.012		11.0	0.409	+0.049
	11.0	0.119	+0.014		9.33	0.482	+0.058
	9.9	0.133	+0.016	5	21.6	0.258	+0.031
	9.33	0.140	+0.017		21	0.265	+0.032
1 1/4		0.090	+0.011		17	0.328	+0.039
	17.0	0.098	+0.012		13.5	0.413	+0.050
	13.5	0.123	+0.015		11	0.506	+0.061
	Sched. 40	0.140	+0.017	6	32.5	0.204	+0.024
	11.0	0.151	+0.018		26	0.255	+0.031
	10.0	0.166	+0.020		23.7	0.280	+0.034
	9.33	0.178	+0.021		21	0.316	+0.038
	6.0	0.277	+0.033		17	0.390	+0.047
1 1/2		0.090	+0.011		13.5	0.491	+0.059
	17	0.112	+0.013		11.5	0.576	+0.069
	13.5	0.141	+0.017		11.0	0.602	+0.072
	Sched. 40	0.145	+0.017	8	32.5	0.265	+0.032
	11	0.173	+0.021		26	0.332	+0.040
2	21	0.113	+0.014		21	0.410	+0.049
	17	0.140	+0.017		17	0.508	+0.061
	Sched. 40	0.154	+0.018		13.5	0.639	+0.077
	13.5	0.176	+0.021		11.5	0.750	+0.090
	11	0.216	+0.026		11	0.785	+0.094
	9.33	0.255	+0.031	10	32.5	0.331	+0.040
2 1/2	21	0.137	+0.016		26	0.413	+0.050
	17	0.169	+0.020		21	0.511	+0.061
	13.5	0.213	+0.026		17	0.633	+0.076
	11	0.261	+0.031		13.5	0.797	+0.096
3	21	0.167	+0.020		11.5	0.935	+0.112
	17	0.206	+0.025		11	0.978	+0.117
	Sched. 40	0.216	+0.026	12	32.5	0.392	+0.047
	13.5	0.259	+0.031		26	0.490	+0.059
	11.5	0.304	+0.036		21	0.608	+0.073
	11	0.318	+0.038		17	0.750	+0.090
	9.33	0.375	+0.045		13.5	0.945	+0.113
3 1/2	21	0.190	+0.023		11.5	1.109	+0.133
	17	0.236	+0.028		11	1.160	+0.139

* Dimensions in inches.

TABLE 19.34 Outside Diameters and Tolerances, 14-in and Larger Pipe per ASTM D2513

Nominal pipe size	Actual outside diameter, in	
	Average	Tolerance (plus and minus)
14	14.000	0.063
16	16.000	0.072
18	18.000	0.081
20	20.000	0.090
22	22.000	0.099
24	24.000	0.108

TABLE 19.35 Minimum Wall Thickness and Tolerance, 14-in and Larger Pipe, per ASTM D2513*

Nominal pipe size	SDR 32.5	SDR 26	SDR 21	SDR 17	SDR 15.5	SDR 13.5	SDR 11	SDR 9	SDR 7.3
14	0.431	0.538	0.667	0.824	0.903	1.037	1.273	1.556	1.918
	+0.052	+0.065	+0.080	+0.099	+0.108	+0.124	+0.153	+0.187	+0.230
16	0.492	0.615	0.762	0.941	1.032	1.185	1.455	1.778	. . .
	+0.059	+0.074	+0.091	+0.113	+0.124	+0.142	+0.175	+0.213	
18	0.554	0.692	0.857	1.059	1.161	1.333	1.636	2.000	. . .
	+0.066	+0.083	+0.103	+0.127	+0.139	+0.160	+0.196	+0.240	
20	0.615	0.769	0.952	1.176	1.290	1.481	1.818	. . .	. . .
	+0.074	+0.092	+0.114	+0.141	+0.155	+0.178	+0.218		
22	0.677	0.846	1.048	1.294	1.419	1.630	2.000	. . .	. . .
	+0.081	+0.102	+0.126	+0.155	+0.170	+0.196	+0.240		
24	0.738	0.923	1.143	1.412	1.548	1.778	2.182	. . .	. . .
	+0.089	+0.111	+1.137	+0.169	+0.186	+0.213	+0.262		

* Dimensions in inches.

ASTM D2683, PE FITTINGS, SOCKET FUSION TYPE

ASTM D2683 covers polyethylene socket-type fittings for use with outside-diameter-controlled polyethylene pipe as specified by specifications D3035 and D2513.

Dimensions of ASTM D2683 Fittings

See Tables 19.36, 19.37, and 19.38.

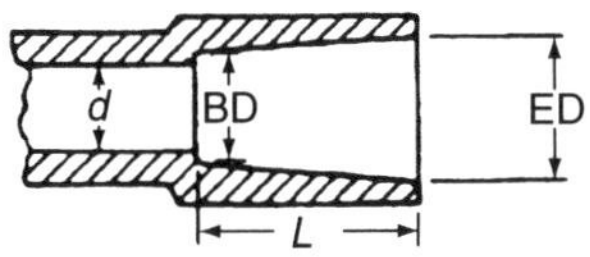

TABLE 19.36 ASTM D2683 Fittings for Outside-Diameter-Controlled Polyethylene Pipe*

Nominal pipe size	Socket entrance diameter, ED: Average diameter	Tolerance on diameter	Maximum out-of-round	Socket bottom diameter, BD: Average diameter	Tolerance on diameter	Maximum out-of-round	Socket depth, L, min.	Inside diameter, d: Min.	Max.
1/2	0.810	+0.005, −0.010	±0.008	0.804	+0.005, −0.026	±0.008	0.625	0.648	0.804
3/4	1.020	±0.008	±0.010	1.012	+0.008, −0.012	±0.010	0.625	0.814	1.020
1	1.275	±0.008	±0.010	1.267	+0.008, −0.012	±0.010	0.687	1.020	1.275
1 1/4	1.620	±0.008	±0.012	1.612	+0.008, −0.016	±0.012	0.875	1.301	1.620
1 1/2	1.860	±0.010	±0.012	1.849	+0.010, −0.020	±0.012	0.875	1.496	1.859
2	2.335	±0.010	±0.012	2.324	+0.010, −0.020	±0.012	0.875	1.885	2.334
3	3.455	±0.015	±0.015	3.439	+0.015, −0.025	±0.015	1.000	2.780	3.454
4	4.450	±0.015	±0.015	4.434	+0.015, −0.035	±0.015	1.125	3.575	4.449

* Dimensions in inches.

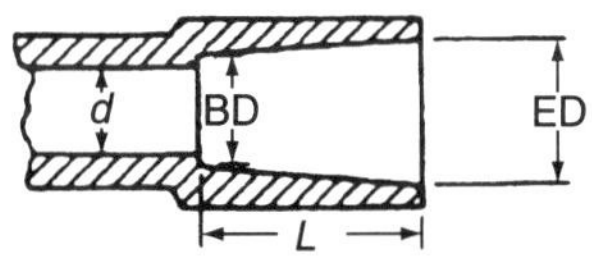

TABLE 19.37 ASTM D2683 Fittings for Outside-Diameter-Controlled Polyethylene Tubing*

Nominal tubing size	Socket entrance diameter, ED: Average diameter	Tolerance on diameter	Maximum out-of-round	Socket bottom diameter, BD: Average diameter	Tolerance on diameter	Maximum out-of-round
1/2	0.585	+0.010, −0.007	±0.010	0.575	+0.010, −0.007	±0.010
3/4	0.835	±0.010	±0.010	0.825	±0.010	±0.010
1	1.075	±0.008	±0.008	1.070	+0.008, −0.012	±0.008
1 1/4	1.330	±0.008	±0.010	1.322	±0.008	±0.010

* Dimensions in inches.

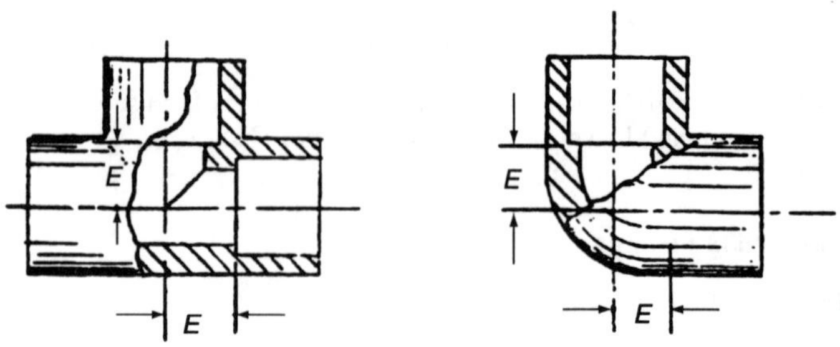

TABLE 19.38 Minimum Dimensions from Center to End of Sockets (Laying Length) for Tees and for SDR 11.0 Pipe per ASTM D2683

Nominal pipe size	E, min., in
$^1/_2$	$^5/_8$
$^3/_4$	$^3/_4$
1	$^7/_8$
$1^1/_4$	1
$1^1/_2$	$1^1/_4$
2	$1^1/_2$
3	$2^3/_4$
4	4

ASTM D3261, PE FITTINGS, BUTT FUSION TYPE

ASTM D3261 covers polyethylene butt fusion fittings for use with polyethylene pipe (IPS and ISO) and tubing (CTS).

Dimensions of ASTM D3261 Pipe

See Tables 19.39 to 19.42.

TABLE 19.39 Outside Diameters and Tolerances for Fit for Use with Polyethylene Pipe per ASTM D3261*

Nominal pipe size	Average outside diameter at area of fusion	
	Min.	Max.
90	90.0	90.8
110	110.0	111.0
160	160.0	161.4
200	200.0	201.8
250	250.0	252.3
280	280.0	282.5
315	315.0	317.8
355	355.0	358.2
400	400.0	403.6
450	450.0	454.1
500	500.0	504.5
560	560.0	565.0
630	630.0	635.7
710	710.0	716.4
800	800.0	807.2
900	900.0	908.1
1000	1000.0	1009.0
1200	1200.0	1210.8
1400	1400.0	1412.6
1600	1600.0	1614.4

* Dimensions in millimeters.

TABLE 19.40 Diameter, Wall Thickness, and Tolerances for Fittings for Use with Plastic Tubing per ASTM D3261*

Tubing type, in	Nominal tubing size	Diameter at area of fusion				Minimum wall thickness
		Outside		Inside		
		Average	Tolerance (plus and minus)	Average	Tolerance (plus and minus)	
0.082	½ CTS	0.625	0.010	0.495	0.004	0.062
	¾ CTS	0.875	0.010	0.745	. . .	. . .
0.090	½ CTS	0.625	0.010	0.437	0.004	0.090
	¾ CTS	0.875	0.010	0.687	0.004	0.090
	1 CTS	1.125	0.013	0.937	0.005	0.090
	1¼ CTS	1.375	0.013	1.187	0.005	0.090
DR 11	¾ CTS	0.875	0.010	0.715	0.004	0.077
	1 CTS	1.125	0.013	0.915	0.005	0.101
	1¼ CTS	1.375	0.013	1.125	0.005	0.121
DR 9.3	½ CTS	0.625	0.010	0.483	0.004	0.067
	¾ CTS	0.875	0.010	0.679	0.004	0.094
	1 CTS	1.125	0.013	0.873	0.005	0.121
	1¼ CTS	1.375	0.013	1.069	0.005	0.148

* Dimensions in inches.

TABLE 19.41 IPS Sizing System Wall Thickness and Tolerance at the Area of Fusion for Fittings for Use with Polyethylene Pipe per ASTM D3261*

	Minimum wall thickness									
Nominal pipe size	Sched. 40	Sched. 80	SDR 21	SDR 17	SDR 13.5	DR 10	DR 11.5	SDR 11	DR 9.3	SDR 9
$^1/_2$	0.109	0.147	. . .	. . .	. . .	. . .	. . .	0.076	0.090	. . .
$^3/_4$	0.113	0.154	. . .	. . .	. . .	. . .	. . .	0.095	0.113	0.117
1	0.133	0.179	. . .	. . .	. . .	. . .	. . .	0.119	0.142	0.146
$1^1/_4$	0.140	0.191	. . .	. . .	. . .	0.166	. . .	0.151	0.179	0.184
$1^1/_2$	0.145	0.200	. . .	. . .	. . .	. . .	. . .	0.173	0.204	0.211
2	0.154	0.218	. . .	. . .	. . .	. . .	. . .	0.216	0.256	0.264
3	0.216	0.300	. . .	. . .	0.259	. . .	0.305	0.318	0.377	0.389
4	0.237	0.337	. . .	0.264	0.333	. . .	0.392	0.409	0.484	0.500
6	0.280	0.432	0.316	0.390	0.491	. . .	0.576	0.603	0.713	0.736
8	0.322	. . .	0.410	0.508	0.639	. . .	0.750	0.785	0.928	0.958
10	0.365	. . .	0.511	0.633	0.797	. . .	0.935	0.978	1.156	1.194
12	0.406	. . .	0.608	0.750	0.945	. . .	1.109	1.160	1.371	1.417
14	. . .	. . .	0.667	0.824	. . .	. . .	. . .	1.273	1.505	1.556
16	. . .	. . .	0.762	0.941	. . .	. . .	. . .	1.455	1.720	1.778
18	. . .	. . .	0.857	1.059	. . .	. . .	. . .	1.636	1.935	2.000
20	. . .	. . .	0.952	1.176	. . .	. . .	. . .	1.818	2.151	2.222
21.5	. . .	. . .	1.024	1.265	. . .	. . .	. . .	. . .	. . .	. . .
22	. . .	. . .	1.048	1.294	. . .	. . .	. . .	2.000	2.366	2.444
24	. . .	. . .	1.143	1.412	. . .	. . .	. . .	2.182	2.581	. . .
28	. . .	. . .	1.333	1.647	. . .	. . .	. . .	2.545	. . .	. . .
32	. . .	. . .	1.524	1.882	. . .	. . .	. . .	2.909	. . .	. . .
36	. . .	. . .	1.714	2.118	. . .	. . .	. . .	. . .	. . .	. . .
42	. . .	. . .	2.000	2.471	. . .	. . .	. . .	. . .	. . .	. . .
48	. . .	. . .	2.286	. . .	. . .	. . .	. . .	. . .	. . .	. . .

* Dimensions in inches.

TABLE 19.42 ISO Sizing System Wall Thickness and Tolerance at the Area of Fusion for Fittings for Use with Polyethylene Pipe per ASTM D3261*

	Minimal wall thickness					
Nominal pipe size	DR 41	DR 32.5	DR 26	DR 21	DR 17	DR 11
90	. . .	. . .	3.5	4.3	5.3	8.2
110	. . .	3.4	4.2	5.2	6.5	10.0
160	. . .	4.9	6.2	7.6	9.4	14.5
200	. . .	6.2	7.7	9.5	11.8	18.2
250	. . .	7.7	9.6	11.9	14.7	22.7
280	. . .	8.6	10.8	13.3	16.5	25.5
315	. . .	9.7	12.1	15.0	18.5	28.6
355	. . .	10.9	13.7	16.9	20.9	32.3
400	. . .	12.3	15.4	19.0	23.5	36.4
450	. . .	13.8	17.3	21.4	26.5	. . .
500	. . .	15.4	19.2	23.8	29.4	. . .
560	. . .	17.2	21.5	26.7	32.9	. . .
630	. . .	19.4	24.2	30.0	37.1	. . .
710	. . .	21.8	27.3	33.8	41.8	. . .
800	. . .	24.6	30.8	38.1	47.1	. . .
900	. . .	27.7	34.6	42.9	. . .	. . .
1000	24.4	30.8	38.5	47.6	. . .	. . .
1200	29.3	36.9	46.2	. . .	. . .	. . .
1400	34.1	43.1	. . .	. . .	. . .	. . .
1600	39.0	49.2	. . .	. . .	. . .	. . .

* Dimensions in millimeters.

ASTM F1055, PE ELECTROFUSION FITTINGS

ASTM F1055 covers electrofusion polyethylene fittings for use with outside-diameter-controlled polyethylene pipe, covered by specifications D2447, D2513, D2737, D3035, and F714.

AWWA C901, PE PIPE AND TUBING FOR WATER SERVICES

AWWA C901 covers polyethylene pressure pipe and tubing for use primarily as service lines in the construction of underground water distribution systems.

Materials

See Table 19.43.

TABLE 19.43 Materials for PE Pipe or Tubing per AWWA C901

Standard PE code	Call classification per ASTM D3350
PE 2406	PE 213323
PE 3406	PE 324433
PE 3408	PE 334434

Dimensions of AWWA C901 Pipe and Tubing

See Tables 19.44 to 19.49.

TABLE 19.44 Inside Diameter, Wall Thickness, and Tolerance for ID-Based PE Pipe per AWWA C901*

	Inside diameter		Wall thickness									
			PC 80 IDR 15		PC 100 IDR 11.5		PC 125 IDR 9		PC 160 IDR 7		PC 200 IDR 5.3	
Nominal pipe size	Average	Tolerance	Min.	Tolerance (plus)	Min.	Tolerance (plus)	Min.	Tolerance (plus)	Min.	Tolerance (plus)	Min.	Tolerance (plus)
$^1/_2$	0.622	–0.010, +0.010					0.069	0.020	0.089	0.020	0.117	0.020
$^3/_4$	0.824	–0.015, +0.010					0.092	0.020	0.118	0.020	0.155	0.020
1	1.049	–0.020, +0.010					0.117	0.020	0.150	0.020	0.198	0.024
$1^1/_4$	1.380	–0.020, +0.010					0.153	0.020	0.197	0.024	0.260	0.031
$1^1/_2$	1.610	–0.020, +0.015					0.179	0.020	0.230	0.028	0.304	0.036
2	2.067	–0.020, +0.015	0.138	0.020	0.180	0.020	0.230	0.028	0.295	0.035	0.390	0.047
$2^1/_2$	2.469	–0.025, +0.015	0.165	0.020	0.215	0.025	0.272	0.033	0.353	0.042	0.466	0.056
3	3.068	–0.030, +0.015	0.205	0.025	0.267	0.032	0.341	0.041	0.438	0.053	0.579	0.069

* Dimensions in inches.

Note:

PC = pressure class.

PC 80 = pressure class rated for 80 psi for water at 73.4°F. Other pressure classes are designated similarly.

IDR = inside-diameter-based dimension ratio.

TABLE 19.45 Inside Diameter, Wall Thickness, and Tolerance for Inside-Diameter-Based PE Pipe of PE 3408 per AWWA C901*

Nominal pipe size	Inside diameter		Wall thickness									
			PC 80 IDR 19		PC 100 IDR 15		PC 125 IDR 11.5		PC 160 IDR 9		PC 200 IDR 7	
	Average	Tolerance	Min.	Tolerance (plus)	Min.	Tolerance (plus)	Min.	Tolerance (plus)	Min.	Tolerance (plus)	Min.	Tolerance (plus)
1/2	0.622	−0.010, +0.010							0.069	0.020	0.089	0.020
3/4	0.824	−0.015, +0.010					0.072	0.020	0.092	0.020	0.118	0.020
1	1.049	−0.020, +0.010					0.091	0.020	0.117	0.020	0.150	0.020
1 1/4	1.380	−0.020, +0.010					0.120	0.020	0.153	0.020	0.197	0.024
1 1/2	1.610	−0.020, +0.015					0.140	0.020	0.179	0.020	0.230	0.028
2	2.067	−0.020, +0.015	0.109	0.020	0.138	0.020	0.180	0.022	0.230	0.028	0.295	0.035
2 1/2	2.469	−0.025, +0.015	0.130	0.020	0.165	0.020	0.215	0.025	0.272	0.033	0.353	0.042
3	3.068	−0.030, +0.015	0.161	0.020	0.205	0.020	0.267	0.032	0.341	0.041	0.438	0.053

* Dimensions in inches.
Note:
PC = pressure class.
PC 80 = pressure class rated for 80 psi for water at 73.4°F. Other pressure classes are designated similarly.
IDR = inside-diameter-based dimension ratio.

TABLE 19.46 Outside Diameter, Wall Thickness, and Tolerance of PE 2406 and PE 3406 Pipe (Outside Diameter Based) per AWWA C901*

Nominal pipe size	Outside diameter		Wall thickness							
			PC 80 DR 17		PC 100 DR 13.5		PC 125 DR 11		PC 160 DR 9	
	Average	Tolerance	Min.	Tolerance (plus)	Min.	Tolerance (plus)	Min.	Tolerance (plus)	Min.	Tolerance (plus)
1/2	0.840	±0.004					0.076	0.020	0.093	0.020
3/4	1.050	±0.004					0.095	0.020	0.117	0.020
1	1.315	±0.005					0.119	0.020	0.146	0.020
1 1/4	1.660	±0.005					0.151	0.020	0.184	0.022
1 1/2	1.900	±0.006					0.173	0.021	0.211	0.025
2	2.375	±0.006	0.140	0.020	0.176	0.021	0.216	0.026	0.264	0.032
3	3.500	± 0.008	0.206	0.025	0.259	0.031	0.318	0.038	0.389	0.047

* Dimensions in inches.
Note:
PC = pressure class.
DR = dimension ratio based on outside diameter.

TABLE 19.47 Outside Diameter, Wall Thickness, and Tolerance of Outside-Diameter-Based PE 3408 Pipe per AWWA C901*

			Wall thickness									
			PC 80 DR 21		PC 100 DR 17		PC 125 DR 13.5		PC 160 DR 11		PC 200 DR 9	
	Inside diameter											
Nominal pipe size	Average	Tolerance	Min.	Tolerance (plus)	Min.	Tolerance (plus)	Min.	Tolerance (plus)	Min.	Tolerance (plus)	Min.	Tolerance (plus)
$^1/_2$	0.840	±0.004							0.076	0.020	0.093	0.020
$^3/_4$	1.050	±0.004					0.078	0.020	0.095	0.061	0.117	0.020
1	1.315	±0.005					0.097	0.020	0.119	0.020	0.146	0.020
$1^1/_4$	1.660	±0.005					0.123	0.020	0.151	0.020	0.184	0.022
$1^1/_2$	1.900	±0.006					0.141	0.020	0.173	0.021	0.211	0.025
2	2.375	±0.006	0.113	0.020	0.140	0.020	0.176	0.021	0.216		0.264	0.032
3	3.500	±0.008	0.167	0.020	0.206	0.025	0.259	0.031	0.318	0.038	0.389	0.047

* Dimensions in inches.
Note:
PC = pressure class.
DR = dimension ratio based on outside diameter.

TABLE 19.48 Dimensions of PE Tubing (PE 2406 and PE 3406) per AWWA C901*

	Outside diameter		Wall thickness, PC 160 DR 9	
Nominal tubing size	Average	Tolerance (plus and minus)	Minimum	Tolerance (plus)
$^1/_2$	0.625	0.004	0.069	0.010
$^3/_4$	0.875	0.004	0.097	0.010
1	1.125	0.005	0.125	0.012
$1^1/_4$	1.375	0.005	0.153	0.015
$1^1/_2$	1.625	0.006	0.181	0.018
2	2.125	0.006	0.236	0.024

* Dimensions in inches.
Note:
PC = pressure class.
DR = dimension ration based on outside diameter.

TABLE 19.49 Dimensions of Outside-Diameter-Based PE 3408 Tubing per AWWA C901*

			Wall thickness			
	Outside diameter		PC 160 DR 11		PC 200 DR 9	
Nominal tubing size	Average	Tolerance (plus and minus)	Min.	Tolerance (plus)	Min.	Tolerance (plus)
$^1/_2$	0.625	0.004			0.069	0.010
$^3/_4$	0.875	0.004	0.080	0.010	0.097	0.010
1	1.125	0.005	0.102	0.010	0.125	0.012
$1^1/_4$	1.375	0.005	0.125	0.012	0.153	0.015
$1^1/_2$	1.625	0.006	0.148	0.015	0.181	0.018
2	2.125	0.006	0.193	0.019	0.236	0.024

* Dimensions in inches.
Note:
PC = pressure class.
DR = dimension ratio based on outside diameter.

AWWA C906, PE PIPE FOR WATER DISTRIBUTION

AWWA C906 covers polyethylene pressure pipe made from materials conforming to standard PE code designations PE 2406, PE 3406, and PE 3408. The pipe is primarily intended for use in transporting potable water in either buried or aboveground installations.

AWWA C906 PE Pipe and Fitting Materials

See Table 19.50.

TABLE 19.50 AWWA C906 PE Pipe and Fitting Materials

Standard PE code designation	PE material, minimum call classification, ASTM D3350	Hydrostatic design basis at 73.4°F, ASTM D2837
PE 2406	PE 213333C	1250 psi
PE 3406	PE 324433C	1250 psi
PE 3408	PE 334434C	1600 psi

Dimensions of AWWA C906 Pipe and Fittings

See Tables 19.51 to 19.54.

TABLE 19.51 Outside Diameter and Tolerance of AWWA C906 Pipe per IPS Sizing System (ASME B36.10)*

Nominal pipe size	Outside diameter	
	Average	Tolerance (plus and minus)
4	4.500	0.020
5	5.563	0.025
6	6.625	0.030
7	7.125	0.034
8	8.625	0.039
10	10.750	0.048
12	12.750	0.057
13	13.375	0.060
14	14.000	0.063
16	16.000	0.072
18	18.000	0.081
20	20.000	0.090
21.5	21.500	0.097
22	22.000	0.099
24	24.000	0.108
28	28.000	0.126
32	32.000	0.144
36	36.000	0.162
42	42.000	0.189
48	48.000	0.216
54	54.000	0.243
63	63.000	0.284

* Dimensions in inches.

TABLE 19.52 Outside Diameter and Tolerance of AWWA C906 Pipe per ISO Sizing System

Nominal pipe size, mm	Outside diameter, mm		Equivalent outside diameter, in	
	Min.	Max.	Min.	Max.
110	110	111.0	4.331	4.370
160	160	161.4	6.299	6.354
200	200	201.8	7.874	7.945
250	250	252.3	9.843	9.933
280	280	282.5	11.024	11.122
315	315	317.8	12.402	12.512
355	355	358.2	13.976	14.102
400	400	403.6	15.748	15.890
450	450	454.1	17.717	17.878
500	500	504.5	19.685	19.862
560	560	565.0	22.047	22.244
630	630	636.7	24.803	25.028
710	710	716.4	27.953	28.205
800	800	807.2	31.496	31.800
900	900	908.1	35.433	35.752
1000	1000	1009.0	39.370	39.724
1200	1200	1210.8	47.244	47.669
1400	1400	1412.5	55.118	55.614
1600	1600	1614.4	62.992	63.559

TABLE 19.53 Minimum Wall Thickness of AWWA C906 PE Pipe per IPS Sizing System*

Nominal pipe size (OD)	Dimension ratio†									
	32.5	26.0	21.0	17.0	15.5	13.5	11.0	9.3	9.0	7.3
4	0.138	0.173	0.214	0.236	0.290	0.333	0.409	0.482	0.500	0.616
5	0.171	0.214	0.265	0.327	0.359	0.412	0.506	0.598	0.518	0.762
6	0.204	0.255	0.316	0.390	0.427	0.491	0.602	0.710	0.736	0.908
7†	0.220	0.274	0.340	0.420	0.460	0.528	0.648	0.766	0.792	0.978
8	0.264	0.332	0.411	0.507	0.556	0.639	0.784	0.927	0.958	1.182
10	0.331	0.413	0.512	0.632	0.694	0.796	0.977	1.158	1.194	1.473
12	0.392	0.490	0.607	0.750	0.823	0.944	1.159	1.371	1.417	1.747
13†	0.412	0.515	0.638	0.788	0.863	0.991	1.216	1.438	1.486	1.832
14	0.431	0.538	0.667	0.824	0.903	1.037	1.273	1.505	1.556	1.918
16	0.492	0.615	0.782	0.941	1.032	1.185	1.455	1.720	1.778	2.192
18	0.554	0.692	0.857	1.059	1.161	1.323	1.636	1.935	2.000	2.466
20	0.615	0.769	0.952	1.176	1.290	1.481	1.818	2.151	2.222	2.740
21.5†	0.662	0.827	1.024	1.265	1.387	1.593	1.955			
22	0.677	0.846	1.048	1.294	1.419	1.630	2.000	2.366	2.444	
24	0.738	0.923	1.143	1.412	1.548	1.778	2.182	2.581	2.667	
28	0.862	1.077	1.333	1.647	1.806	2.074	2.545			
32	0.985	1.231	1.524	1.882	2.065	2.370	2.909			
36	1.108	1.385	1.714	2.118	2.323	2.657				
42	1.292	1.615	2.000	2.471	2.710					
48	1.477	1.846	2.286	2.834						
54	1.662	2.077	2.571							
63	1.938	2.428	3.000							

* Dimensions in inches.
† Dimension ratios (DRs) are from standard dimension ratio (SDR) series.

TABLE 19.54 Minimum Wall Thickness of AWWA C906 PE Pipe Based on ISO Sizing System*

Nominal pipe size (OD)	Dimension ratio									
	32.5	25.0	21.0	17.0	15.5	13.5	11.0	9.3	9.0	7.8
110	3.4	4.2	5.2	6.5	7.1	8.1	10.0	11.8	12.2	15.1
160	4.9	6.2	7.6	9.4	10.3	11.9	14.5	17.2	17.8	21.9
200	6.2	7.7	9.5	11.8	12.9	14.8	18.2	21.5	22.2	27.4
250	7.7	9.6	11.9	14.7	16.1	18.5	22.7	26.9	27.8	34.2
280	8.6	10.8	13.3	16.5	18.1	20.7	25.5	30.1	31.1	38.4
215	9.7	12.1	15.0	18.5	20.3	23.3	28.6	33.9	35.0	43.2
255	10.9	13.7	16.9	20.9	22.9	26.8	32.3	38.2	39.4	48.6
400	12.3	15.4	19.0	23.5	25.8	29.6	36.4	43.0	44.4	54.8
450	13.8	17.8	21.4	26.5	29.0	33.3	40.9	48.4	50.0	61.6
500	15.4	19.2	23.3	29.4	32.3	37.0	45.5	53.8	55.6	68.5
530	17.2	21.5	26.7	32.9	36.1	41.5	50.9	60.2	62.2	
630	19.4	24.2	30.0	37.1	40.6	46.7	57.3	67.7	70.0	
710	21.2	27.3	33.8	41.8	45.8	52.5	64.5			
800	24.6	30.8	38.1	47.1	51.6	59.3	72.7			
900	27.7	34.6	42.9	52.9	58.1	66.7				
1000	30.5	38.5	47.6	58.8	64.5	74.1				
1200	36.9	46.2	56.1	70.6						
1400	43.1	53.8	66.7							
1800	49.2	61.5	76.2							

* Dimensions in millimeters.

PRESSURE CLASS

See Table 19.55.

TABLE 19.55 Pressure Class

Dimension ratio	Pressure, psig	
	PE 2406, PE 3406*	PE 3408†
32.5	40	51
26	50	64
21	63	80
17	78	100
15.5	86	110
13.5	100	128
11	125	160
9.3	151	193
9	156	200
7.3	198	254

* Hydrostatic design basis (HDB) and hydrostatic design stress (HDS) for PE 2406 and 3406 are 1250 psi and 625 psi, respectively.

† HDB and HDS for PE 3408 are 1600 psi and 800 psi, respectively.

Tables in Chapter 19

Tables in Chapter 19 (Continued)

CHAPTER 20
ACRYLONITRILE-BUTADIENE-STYRENE (ABS) PIPE, FITTINGS, AND JOINTS

See p. 623 for a list of tables in Chap. 20.

ABS PIPING AND APPLICATIONS

ABS piping is intended for use in the following pressure and nonpressure liquid applications:

- Drain, waste, and vent
- Sewer and drain
- Industrial cold water applications

ABS conduit is used as electrical duct.

REFERENCE STANDARDS AND SPECIFICATIONS FOR ABS PIPE, FITTINGS, CONDUIT, AND JOINTS

Refer to Table 20.1.

TABLE 20.1 Principal ABS Piping Standards

Piping material and product standard	Subject product or abbreviated title of standard	Nominal size range, in	Principal applications
ASTM D3311	DWV fitting patterns	$1^1/_4$–8	Drain, waste, and vent
ASTM D2680	ABS and PVC sewer pipe of composite wall construction	4–15	Sewer and drain
ASTM F409	Accessible and replaceable tube and fittings	$1^1/_4$–$1^1/_2$	Drain, waste, and vent
ASTM F480	Thermoplastic water well casing	2–16	Water well casing
ASTM F1499	Coextruded composite DWV pipe	$1^1/_4$–8	Drain, waste, and vent
CSA B181.5	Coextruded ABS/PVC DWV pipe		Drain, waste, and vent
ASTM F1488	Coextruded composite pipe	2–12	Drain, waste, and vent; sewer and drain; electrical and communications conduit
ASTM D1527	ABS pipe, Schedules 40 and 80	$^1/_8$–12	Cold water; industrial
ASTM D2282	ABS pipe, dimension ratio series	$^1/_8$–12	Cold water; industrial
ASTM D2468	ABS socket fittings, Schedule 40	$^1/_8$–8	Cold water; industrial
ASTM D2661	ABS DWV pipe and fittings	$1^1/_4$–6	Drain, waste, and vent
ASTM D2750	ABS utility conduit and fittings	1–6	Electrical duct
ASTM D2751	ABS sewer pipe and fittings	3–12	Sewer and drain
ASTM F 628	ABS foam core DWV	$1^1/_4$–6	Drain, waste, and vent
CSA B181.1	ABS DWV pipe and fittings		Drain, waste, and vent
ASTM D2235	Solvent cements for ABS plastic pipe and fittings		

ASTM D1527 ABS PLASTIC PIPE

ABS pipe conforming to ASTM D1527 is intended for distribution of pressurized liquids only. The liquids must be chemically compatible with the piping materials. It is cautioned that ABS piping is not safe for use to convey pressurized air or compressed gases.

ASTM D1527 Pipe Materials and Hydrostatic Design Stresses

See Table 20.2.

Pressure Ratings of ASTM D1527 ABS Schedule 40 Pipes

See Table 20.3.

Pressure Ratings of ASTM D1527 ABS Schedule 80 Unthreaded Pipe

See Table 20.4.

Pressure Ratings of ASTM D1527 ABS Schedule 80 Threaded Pipe

See Table 20.5.

TABLE 20.2 ASTM D1527 Material Designations

Type and grade of material	ABS designation	Hydrostatic design stress, psi
Type 1, Grade 1	ABS 1208	800
Type 1, Grade 2	ABS 1210	1000
Type 1, Grade 3	ABS 2112	1250
Type 2, Grade 1	ABS 1316	1600

TABLE 20.3 Pressure Ratings of ASTM D1527 Schedule 40 ABS Pipe*

	Pressure ratings, psi			
Nominal pipe size, in	ABS 1208	ABS 1210	ABS 1316	ABS 2112
1/8	320	400	650	500
1/4	310	390	620	490
3/8	250	310	500	390
1/2	240	300	480	370
3/4	190	240	390	300
1	180	220	360	280
1 1/4	150	180	290	230
1 1/2	130	170	260	210
2	110	140	220	170
2 1/2	120	150	240	190
3	100	130	210	160
3 1/2	90	120	190	150
4	90	110	180	140
5	80	100	160	120
6	70	90	140	110
8	60	80	120	100
10	60	70	110	90
12	50	70	110	80

* Pressure ratings are the maximum water pressures at 73°F (23°C) the pipe is capable of withstanding continuously. These ratings are for unthreaded pipe. Schedule 40 ABS pipe is not recommended to be threaded.

TABLE 20.4 Pressure Ratings of ASTM D1527 Schedule 80 Unthreaded ABS Pipe*

	Pressure ratings, psi			
Nominal pipe size, in	ABS 1208	ABS 1210	ABS 1316	ABS 2112
1/8	...	...	980	...
1/4	...	...	900	...
3/8	...	...	730	...
1/2	340	420	680	530
3/4	280	340	550	430
1	250	320	500	390
1 1/4	210	260	420	330
1 1/2	190	240	380	290
2	160	200	320	250
2 1/2	170	210	340	270
3	150	190	300	230
3 1/2	140	170	280	220
4	130	160	260	200
5	120	140	230	180
6	110	140	220	170
8	100	120	200	150
10	90	120	190	150
12	90	110	180	140

* Pressure ratings are the maximum water pressures at 73°F (23°C) the pipe is capable of withstanding continuously. Pressure ratings for threaded schedule 80 ABS pipe are listed in Table 20.5.

TABLE 20.5 Pressure Ratings of ASTM D1527 Schedule 80 Threaded ABS Pipe

Nominal pipe size, in	Pressure ratings, psi ABS 1208	ABS 1210	ABS 1316	ABS 2112
1/8	...	...	490	...
1/4	...	...	450	...
3/8	...	...	370	...
1/2	170	210	340	260
3/4	140	170	280	210
1	130	160	250	200
1 1/4	100	130	210	160
1 1/2	90	120	190	150
2	80	100	160	130
2 1/2	80	110	170	130
3	70	90	150	120
3 1/2	70	90	140	110
4	60	80	130	100
5	60	70	120	90
6	60	70	110	90
8	50	60	100	80
10	50	60	90	70
12	50	60	90	70

TABLE 20.6 Outside Diameter and Tolerance for ASTM D1527 ABS Pipe*

Nominal pipe size	Average outside diameter†	For average diameter (plus and minus)	For max. and min. diameter (out-of-roundness): Schedule 40 sizes 3 1/2 in and over, Schedule 80 sizes 8 in and over (plus and minus)	For max. and min. diameter (out-of-roundness): Schedule 40 sizes 3 in and less, Schedule 80 sizes 6 in and less (plus and minus)
1/8	0.405	0.004	0.015	0.008
1/4	0.540	0.004	0.015	0.008
3/8	0.675	0.004	0.015	0.008
1/2	0.840	0.004	0.015	0.008
3/4	1.050	0.004	0.015	0.010
1	1.315	0.005	0.015	0.010
1 1/4	1.660	0.005	0.015	0.012
1 1/2	1.900	0.006	0.030	0.012
2	2.375	0.006	0.030	0.012
2 1/2	2.875	0.007	0.030	0.015
3	3.500	0.008	0.030	0.015
3 1/2	4.000	0.008	0.050	0.015
4	4.500	0.009	0.050	0.015
5	5.563	0.010	0.050	0.030
6	6.625	0.011	0.050	0.035
8	8.625	0.015	0.075	0.045
10	10.750	0.015	0.075	0.050
12	12.750	0.015	0.075	0.060

* Dimensions in inches.
† Average outside diameter is the mean of the maximum and minimum diameters at any cross section.

ABS PIPE FOR POTABLE WATER SERVICE

ABS pipe for potable water service must be evaluated, tested, and certified to one of the following:

- ANSI/NSF Standard 61, Drinking Water System Components—Health Effects
- ANSI/NSF Standard 14, Plastic Piping Components and Related Materials

DIMENSIONS AND TOLERANCES FOR ASTM D1527 ABS PIPE

Outside Diameter and Tolerances

See Table 20.6.

Pipe Wall Thickness and Tolerances

See Table 20.7.

TABLE 20.7 Pipe Wall Thickness and Tolerances for ASTM D1527 ABS Pipe

	Schedule 40		Schedule 80	
Nominal pipe size	Min.	Tolerance (plus)	Min.	Tolerance (plus)
1/8	0.068	0.020	0.095	0.020
1/4	0.088	0.020	0.119	0.020
3/8	0.091	0.020	0.126	0.020
1/2	0.109	0.020	0.147	0.020
3/4	0.113	0.020	0.154	0.020
1	0.133	0.020	0.179	0.021
1 1/4	0.140	0.020	0.191	0.023
1 1/2	0.145	0.020	0.200	0.024
2	0.154	0.020	0.218	0.026
2 1/2	0.203	0.024	0.276	0.033
3	0.216	0.026	0.300	0.036
3 1/2	0.226	0.027	0.318	0.038
4	0.237	0.028	0.337	0.040
5	0.258	0.031	0.375	0.045
6	0.280	0.034	0.432	0.052
8	0.322	0.039	0.500	0.060
10	0.365	0.044	0.593	0.071
12	0.406	0.049	0.687	0.082

Notes:
1. Dimensions in inches.
2. Pipe wall thickness at any point shall not be less than the minimum wall thickness listed.
3. The maximum wall thickness at any point shall be the minimum plus the plus tolerance listed above.

ASTM D2282 ABS STANDARD DIMENSION RATIO (SDR) PRESSURE-RATED (PR) PIPE

ASTM D2282 covers standard dimension ratio (SDR) pipe in sizes NPS 1/8 through 12. The SDR 13.5, 17, 21, and 26 pipes are made from the grades of material listed in Table 20.8.

ASTM D2282 Pipe Materials and Hydrostatic Design Stresses

See Table 20.8.

TABLE 20.8 ASTM D2282 Material Designations, Hydrostatic Design Stress, and Standard Dimension Ratios

Type and grade of material	ABS designation	Hydrostatic design stress, psi	Standard dimension ratio (SDR)
Type 1, Grade 1	ABS 1208	800	13.5, 17, 21
Type 1, Grade 2	ABS 1210	1000	13.5, 17, 21, 26
Type 1, Grade 3	ABS 2112	1250	13.5, 17, 21, 26
Type 2, Grade 1	ABS 1316	1600	13.5, 17, 21, 26

Pressure Ratings of Different SDR ABS Pipe Per ASTM D2282

Each SDR pipe is rated for the pressures listed in Table 20.9. The pressures listed are the maximum working pressures for water at 73°F (23°C). These pressure ratings are not to be used for air or gases. The ABS pipe is intended to be used for pressurized liquids only, which are chemically compatible with the piping material. Contact the manufacturer for assistance.

TABLE 20.9 Pressure Ratings of ASTM D2282 ABS SDR Series Pipe for Water at 73°F (23°C)

	Pressure rating, psi			
SDR	ABS 1316	ABS 2112	ABS 1210	ABS 1208
13.5	250	200	160	125
17	200	160	125	100
21	160	125	100	80
26	125	100	80	

Dimensions of ASTM D2282 ABS Pipe

See Tables 20.10 and 20.11.

TABLE 20.10 Outside Diameter and Tolerances for ASTM D2282 SDR Series Pipe*

		Tolerance		
			For max. and min. (out-of-roundness)‡	
Nominal pipe size	Outside diameter†	Average (plus and minus)	SDR 26 SDR 21 (plus and minus)	SDR 17 SDR 13.5 (plus and minus)
1/8	0.405	0.004	0.015	0.008
1/4	0.540	0.004	0.015	0.008
3/8	0.675	0.004	0.015	0.008
1/2	0.840	0.004	0.015	0.008
3/4	1.050	0.004	0.015	0.010
1	1.315	0.005	0.015	0.010
1 1/4	1.660	0.005	0.015	0.012
1 1/2	1.900	0.006	0.030	0.012
2	2.375	0.006	0.030	0.012
2 1/2	2.875	0.007	0.030	0.015
3	3.500	0.008	0.030	0.015
3 1/2	4.000	0.008	0.030	0.015
4	4.500	0.009	0.050	0.015
5	5.563	0.010	0.050	0.030
6	6.625	0.011	0.050	0.035
8	8.625	0.015	0.075	0.045
10	10.750	0.015	0.075	0.050
12	12.750	0.015	0.075	0.060

* Dimensions in inches.

† The average outside diameter is the arithmetic mean of the maximum and minimum diameters at any cross section.

‡ Out-of-roundeness tolerance applies on pipe prior to shipment.

TABLE 20.11 Pipe Wall Thickness and Tolerances for ASTM D2282 ABS SDR Series Pipe

	Wall thickness, in*							
	SDR 26		SDR 21		SDR 17		SDR 13.5	
Nominal pipe size, in	Min.	Tolerance (plus)	Min.	Tolerance (plus)	Min.	Tolerance (plus)	Min.	Tolerance (plus)
$^1/_8$	...	...	...	...	...	...	0.060	0.020
$^1/_4$	...	...	...	...	...	...	0.060	0.020
$^3/_8$	...	...	...	...	...	...	0.060	0.020
$^1/_2$	...	...	...	...	0.060	0.020	0.062	0.020
$^3/_4$	...	...	0.060	0.020	0.062	0.020	0.078	0.020
1	0.060	0.020	0.063	0.020	0.077	0.020	0.097	0.020
$1^1/_4$	0.064	0.020	0.079	0.020	0.098	0.020	0.123	0.020
$1^1/_2$	0.073	0.020	0.090	0.020	0.112	0.020	0.141	0.020
2	0.091	0.020	0.113	0.020	0.140	0.020	0.176	0.021
$2^1/_2$	0.110	0.020	0.137	0.020	0.169	0.020	0.231	0.026
3	0.135	0.020	0.167	0.020	0.206	0.025	0.259	0.031
$3^1/_2$	0.154	0.020	0.190	0.023	0.235	0.028	0.296	0.036
4	0.173	0.021	0.214	0.026	0.265	0.032	0.333	0.040
5	0.214	0.027	0.265	0.032	0.327	0.039	0.412	0.049
6	0.255	0.031	0.316	0.038	0.390	0.047	0.491	0.059
8	0.332	0.040	0.410	0.049	...	...	...	...
10	0.413	0.050	0.511	0.061	...	...	...	...
12	0.490	0.059	0.606	0.073	...	...	...	...

* Pipe wall thickness shall not be less than the minimum listed for the SDR and the nominal pipe size. The maximum pipe wall thickness is equal to the minimum plus the plus tolerance.

ASTM D2468 ABS PIPE FITTINGS

ASTM D2468 covers molded in-line pipe fittings for ABS pipe in sizes NPS $^1/_8$ through NPS 8. The types of the fittings covered include couplings, unions, bushings, caps, and nipples. Fittings in sizes larger than NPS 8 are usually fabricated as mitered fittings, and they are not covered by ASTM D2468. For fabricated fittings, contact the manufacturer.

DIMENSIONS OF SOCKET-TYPE COUPLINGS, TEES, AND ELBOWS FOR SCHEDULE 40 ABS PIPE

See Table 20.12.

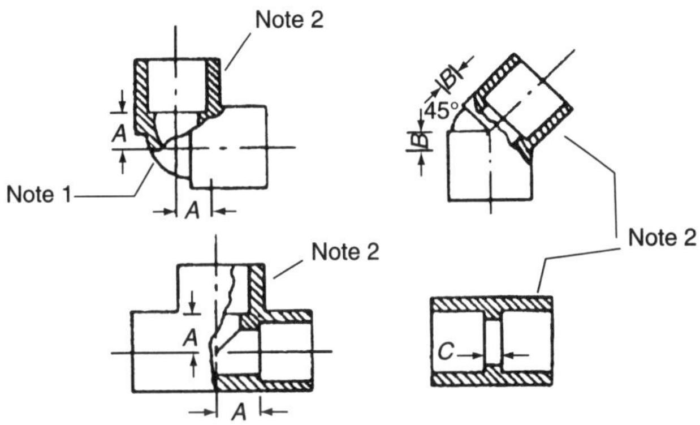

TABLE 20.12 Center to End of Socket Dimensions of Socket-Type Elbows, Tees, and Couplings for Schedule 40 ABS Pipe

Nominal pipe size	*A*, min.	*B*, min.	*C*, min.
1/8	1/4	1/8	1/16
1/4	5/16	5/32	1/16
3/8	3/8	3/16	3/32
1/2	1/2	1/4	3/32
3/4	9/16	5/16	3/32
1	11/16	5/16	3/32
1 1/4	7/8	3/8	3/32
1 1/2	1	7/16	3/32
2	1 1/4	5/8	3/32
2 1/2	1 1/2	11/16	3/16
3	1 13/16	3/4	3/16
3 1/2	2 1/8	1	3/16
4	2 5/16	1	3/16
5	3	1 3/8	3/16
6	3 1/2	1 3/4	1/4
8	4 1/2	2	1/4

Notes:

1. The minimum wall thickness of fittings shall be 125 percent of the minimum wall thickness of the corresponding size of Schedule 40 ABS pipe.
2. The socket wall thickness shall be at least equal to the minimum wall thickness of the corresponding size of Schedule 40 pipe.
3. In case of threaded fittings, the minimum wall thickness of the threaded portion shall be at least equal to the thickness of material under the thread root of the threaded Schedule 80 pipe of the corresponding size.
4. Refer to Table 20.13 for other dimensions.
5. Dimensions in inches.

DIMENSIONS OF TAPERED SOCKETS OF ABS PIPE FITTINGS

See Table 20.13.

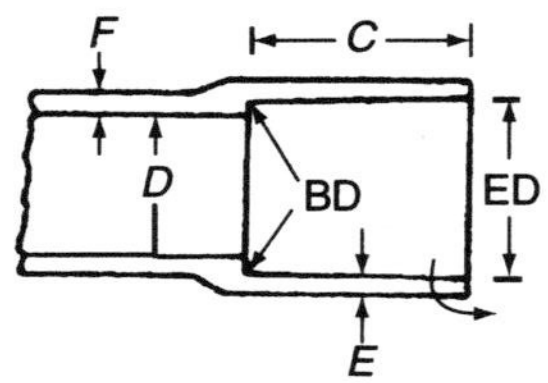

TABLE 20.13 Dimensions and Tolerances for Tapered Sockets of ABS Pipe Fittings*

Nominal pipe size	ED socket entrance diameter: Diameter	ED: Tolerance on diameter (plus and minus)	ED: Maximum out-of-round (plus and minus)	BD socket bottom diameter: Diameter	BD: Tolerance on diameter (plus and minus)	BD: Maximum out-of-round (plus and minus)	C, socket length, min.	D, inside diameter, min.	Wall thickness, min.: $E^{\dagger}$	$F^{\dagger}$	$G^{\ddagger}$
1/8	0.417	0.004	0.008	0.401	0.004	0.008	0.500	0.265	0.068	0.085	0.056
1/4	0.552	0.004	0.008	0.536	0.004	0.008	0.500	0.360	0.088	0.110	0.063
3/8	0.687	0.004	0.008	0.671	0.004	0.008	0.594	0.489	0.091	0.114	0.068
1/2	0.848	0.004	0.008	0.836	0.004	0.008	0.688	0.618	0.109	0.136	0.072
3/4	1.058	0.004	0.010	1.046	0.004	0.010	0.719	0.820	0.113	0.141	0.079
1	1.325	0.005	0.010	1.310	0.005	0.010	0.875	1.044	0.133	0.166	0.087
1 1/4	1.670	0.005	0.012	1.655	0.005	0.012	0.938	1.375	0.140	0.175	0.098
1 1/2	1.912	0.006	0.012	1.894	0.006	0.012	1.094	1.604	0.145	0.181	0.106
2	2.387	0.006	0.012	2.369	0.006	0.012	1.156	2.061	0.154	0.193	0.123
2 1/2	2.889	0.007	0.015	2.868	0.007	0.015	1.750	2.462	0.203	0.254	0.139
3	3.516	0.008	0.015	3.492	0.008	0.015	1.875	3.060	0.216	0.270	0.160
3 1/2	4.016	0.008	0.015	3.992	0.008	0.015	2.000	3.540	0.226	0.283	0.183
4	4.518	0.009	0.015	4.491	0.009	0.015	2.000	4.017	0.237	0.296	0.194
5	5.583	0.010	0.030	5.553	0.010	0.030	3.000	5.037	0.258	0.323	0.228
6	6.647	0.011	0.030	6.614	0.011	0.030	3.000	6.054	0.280	0.350	0.281
8	8.655	0.015	0.045	8.610	0.015	0.045	4.000	7.966	0.322	0.403	0.341

* Dimensions in inches.
† Refer to notes for Table 20.12.
‡ For threaded fittings the minimum wall thickness of the threaded portion shall be at least equal to the thickness of material under the thread root of threaded Schedule 80 pipe of the same size.

ABS PIPE FITTINGS FOR POTABLE WATER SERVICE

ABS pipe fittings for potable water service must be evaluated, tested, and certified to one of the following:

- ANSI/NSF Standard 61, Drinking Water System Components—Health Effects
- ANSI/NSF Standard 14, Plastic Piping Components and Related Materials

ABS PIPING JOINTS

Primarily two types of piping joints are used to join pipe and fittings. These include solvent cement and threaded joints. Threaded joints are not used with Schedule 40 pipe. Refer to Chap. 15 for joints for nonmetallic piping.

Tables in Chapter 20

CHAPTER 21
FIBERGLASS PIPE, FITTINGS, FLANGES, AND JOINTS

See p. 639 for a list of tables in Chap. 21.

FIBERGLASS PIPE

This chapter provides dimensions of fiberglass pipe conforming to the following material specifications for fiberglass piping:

- ASTM D2996, Standard Specification for Filament-Wound "Fiberglass" (Glass-Fiber-Reinforced Thermosetting-Resin) Pipe
- ANSI/AWWA C950, AWWA Standard for Fiberglass Pressure Pipe
- ASTM D3517, Standard Specification for "Fiberglass" (Glass-Fiber-Reinforced Thermosetting-Resin) Pressure Pipe

ASTM D2996 FIBERGLASS PIPE

ASTM D2996 fiberglass pipe is classified by type, grade, class, and hydrostatic design basis (HDB).

Types. Type 1 is known as the filament-wound pipe. Type 1 is the only type covered by ASTM D2996.

Grades. 1, 2, and 3, as defined below.

Grade 1—Glass-fiber-reinforced epoxy resin pipe
Grade 2—Glass-fiber-reinforced polyester resin pipe
Grade 3—Glass-fiber-reinforced furan resin pipe.

Classes

Class A—No liner
Class B—Polyester resin liner (nonreinforced)
Class C—Epoxy resin liner (nonreinforced)
Class E—Polyester resin liner (reinforced)
Class F—Epoxy resin liner (reinforced)
Class H—Thermoplastic resin liner (to be specified by the purchaser)
Class I—Furan resin liner (reinforced)

Hydrostatic Design Basis (HDB)

The hydrostatic design basis is equal to the circumferential stress due to internal hydrostatic pressure that will fail the pipe when extrapolated to 150×10^6 pressure cycles (Procedure A) or to 100,000 hours under continuously applied pressure (Procedure B). Procedure A is called the cyclic test method and Procedure B is called the static test method.

Internal and External Pressure Rating of ASTM D2996 Pipe

Refer to Table 21.1.

TABLE 21.1 Fiberglass Process Piping (Internal and External Pressure Ratings for Pipe Manufactured According to ASTM D2996 or ASTM D2997)

NPS (DN)	Units	Internal pressure rating ranges*				External pressure rating ranges*	
		Vinyl ester		Epoxy		Vinyl ester	Epoxy
		Pipe	Fittings and joints	Pipe	Fittings and joints	Pipe	Pipe
1 in (25)	psi MPa	735–950 5.07–6.55	50–950 0.34–6.55	450–950 3.1–6.55	150–950 1.03–6.55	1975–6400 13.6–44.10	360–2125 2.48–14.70
2 in (50)	psi MPa	200–850 1.38–5.86	50–850 0.34–5.86	450–1250 3.1–8.62	150–1250 1.03–8.62	330–2700 2.28–18.60	55–1170 0.38–8.07
3 in (80)	psi MPa	150–660 1.03–4.55	50–660 0.34–4.55	300–830 2.07–5.72	150–830 1.03–5.72	97–800 0.67–5.52	25–335 0.17–2.31
4 in (100)	psi MPa	125–510 0.86–3.52	50–510 0.34–3.52	175–785 1.21–5.41	150–785 1.03–5.41	45–340 0.31–2.34	16–225 0.11–1.55
6 in (150)	psi MPa	125–420 0.86–2.9	50–420 0.34–2.90	125–525 0.86–3.62	100–525 0.69–3.62	37–100 0.26–0.69	10–62 0.07–0.43
8 in (200)	psi MPa	100–360 0.69–2.48	50–360 0.34–2.48	125–450 0.86–3.10	75–450 0.52–3.10	21–56 0.14–0.39	10–45 0.07–0.31
10 in (250)	psi MPa	100–335 0.69–2.31	50–335 0.34–2.31	125–415 0.86–2.86	50–415 0.34–2.86	13–33 0.09–0.23	10–35 0.07–0.24
12 in (300)	psi MPa	100–280 0.69–1.93	45–280 0.31–1.93	125–350 0.86–2.41	50–350 0.34–2.41	10–23 0.07–0.16	7–23 0.05–0.16

* Ratings at ambient temperatures.

ASTM D2996 Outside Diameter (OD) Series Pipe

The outside diameter (OD) series pipe is manufactured in two different series:

- Outside diameter (OD) series pipe with steel pipe equivalent iron pipe size (IPS)
- Outside diameter (OD) series pipe with cast iron pipe equivalent

Dimensions of ASTM D2996 OD Series Pipe

Dimensions of ASTM D2996 OD Series Pipe with Steel Pipe Equivalent (IPS). Tables 21.2 and 21.2M provide outside diameter dimensions of ASTM D2996 OD series pipe with steel pipe equivalent (IPS).

Tolerances for Outside Diameter of OD Series Pipe with Steel Pipe Equivalent (IPS). Table 21.3 provides tolerances for outside diameter of the pipe in U.S. Customary and metric units.

Dimensions of ASTM D2996 OD Series Pipe with Cast Iron Pipe Equivalent. Tables 21.4 and 21.4M provide dimensions of OD series pipe with cast iron pipe equivalent in U.S. Customary and metric units, respectively.

Tolerances for Outside Diameter of OD Series Pipe with Cast Iron Pipe Equivalent. Table 21.5 provides tolerances for outside diameter of the OD series pipe with cast iron pipe equivalent in U.S. Customary and metric units.

TABLE 21.2 Dimensions for Outside Diameter (OD) Series Pipe with Steel Pipe Equivalent (Iron Pipe Size)

Nominal pipe size, in	OD, in
1	1.315
$1\frac{1}{2}$	1.900
2	2.375
$2\frac{1}{2}$	2.875
3	3.500
4	4.500
6	6.625
8	8.625
10	10.750
12	12.750
14	14.000
16	16.000

TABLE 21.2M Dimensions for Outside Diameter (OD) Series Pipe with Steel Pipe Equivalent (Iron Pipe Size)

Nominal pipe size, in	OD, mm
1	33.40
$1\frac{1}{2}$	48.26
2	60.32
$2\frac{1}{2}$	73.02
3	88.90
4	114.30
6	168.28
8	219.08
10	273.05
12	323.85
14	355.60
16	406.40

TABLE 21.3 Tolerances for Outside Diameter of OD Series Pipe

Nominal pipe size (NPS), in	OD tolerance, in		OD tolerance, mm	
	Plus	Minus	Plus	Minus
1	0.060	0.016	1.52	0.41
$1\frac{1}{2}$ to 4	0.060	0.018	1.52	0.46
6	0.066	0.028	1.68	0.64
8	0.066	0.040	2.18	1.02
10	0.108	0.048	2.74	1.22
12	0.128	0.056	3.25	1.42
14	0.145	0.064	3.68	1.63
16	0.165	0.074	4.19	1.88

TABLE 21.4 Dimensions for Outside Diameter (OD) Series Pipe with Cast Iron Pipe Equivalent

Nominal pipe size, in	OD, in
2	2.50
3	3.96
4	4.80
6	6.90
8	9.05
10	11.10
12	13.20
14	15.30
16	17.40

TABLE 21.4M Dimensions for Outside Diameter (OD) Series Pipe with Cast Iron Pipe Equivalent

Nominal pipe size, in	OD, mm
2	63.50
3	100.58
4	121.92
6	175.26
8	229.87
10	281.94
12	335.28
14	388.62
16	441.96

TABLE 21.5 Tolerances for OD Series Pipe with Cast Iron Pipe Equivalent

Nominal pipe size (NPS), in	OD tolerance, in		OD tolerance, mm	
	Plus	Minus	Plus	Minus
2	0.050	0.050	1.27	1.27
3, 4, 6, 8, 10, 12	0.060	0.060	1.52	1.52
14, 16	0.050	0.080	1.27	2.03

ASTM D2996 Inside Diameter (ID) Series Pipe with Iron Pipe Size (IPS) Equivalent

Table 21.6 provides dimensions of ID series pipe conforming to ASTM D2996.

Tolerances of Inside Diameter of ID Series Pipe with Iron Pipe Size Equivalent. Table 21.7 lists the applicable tolerances for the dimensions of ID series pipe with iron pipe size equivalent shown in Table 21.6.

ASTM D2996 Inside Diameter (ID) Series Pipe

Refer to Table 21.8 for dimensions of ID series pipe per ASTM D2996.

Tolerances for ID Series Pipe (Table 21.8) per ASTM D2996. Table 21.9 lists the applicable tolerances.

TABLE 21.6 Dimensions for Inside Diameter (ID) Series Pipe with Iron Pipe Size Equivalent

Nominal pipe size, in	ID, in	ID, mm
2	2.25	57.15
3	3.34	84.84
4	4.37	111.00
6	6.43	163.32
8	8.39	213.11
10	10.43	264.92
12	12.38	314.45
14	13.60	345.44
16	15.40	391.16

TABLE 21.7 Tolerances for ID Series Pipe with Iron Pipe Size Equivalent

Nominal pipe size (NPS), in	Tolerance for ID, in		Tolerance for ID, mm	
	Plus	Minus	Plus	Minus
2	0.05	0.05	1.27	1.27
3, 4, 6, 8, 10, 12	0.06	0.06	1.52	1.52
14, 16	0.05	0.08	1.27	2.03

TABLE 21.8 Dimensions for Inside Diameter (ID) Series Pipe

Nominal pipe size, in	ID, in	ID, mm
1	1.000	25.4
$1\frac{1}{2}$	1.500	38.1
2	2.000	50.8
$2\frac{1}{2}$	2.500	63.5
3	3.000	76.2
4	4.000	101.6
6	6.000	152.4
8	8.000	203.2
10	10.000	254.0
12	12.000	304.8
14	14.000	355.6
15	15.000	381.0
16	16.000	406.4

TABLE 21.9 Tolerances for ID Series Pipe

Nominal pipe size (NPS), in	Tolerance for inside diameter, in		Tolerance for inside diameter, mm	
	Plus	Minus	Plus	Minus
1, 1½, 2, 2½	0.06	0.06	1.52	1.52
3, 4	0.12	0.12	3.05	3.05
6, 8, 10, 14, 15, 16	0.25	0.25	6.35	6.35

Tolerance for Pipe Wall Thickness of OD and ID Series Pipe Per ASTM D2996

The minimum pipe wall thickness at any point shall not be less than 87.5 percent of the nominal wall thickness listed in the manufacturer's publications.

Liner Thickness of ASTM D2996 Pipe

The average liner thickness for Class B, C, E, F, H, and I pipe shall be determined in accordance with ASTM D3567.

ANSI/AWWA C950 FIBERGLASS PRESSURE PIPE

AWWA C950 covers fiberglass pressure pipe in sizes 1 in through 144 in (25 mm through 3600 mm). The pipe is available in the pressure classes, types, grades, and liners indicated in Table 21.10. This pipe is intended for use in both aboveground and underground water systems.

TABLE 21.10 AWWA C950 Fiberglass Pressure Pipe Pressure and Stiffness Classes, Types, Grades, and Liners*

Pressure class, psig (kPa)†	Type	Grade	Liner	Stiffness class, psi (kPa)
50 (345)	Type I	1, Glass-fiber-reinforced epoxy (RTRP epoxy)	A, No liner	9 (62)
100 (689)	Type II	2, Glass-fiber-reinforced polyester (RTRP polyester)	B, Thermoplastic liner	18 (124)
150 (1034)		3, Glass-fiber-reinforced epoxy mortar (RPMP epoxy)	C, Reinforced thermoset liner	36 (248)
200 (1379)		4, Glass-fiber-reinforced polyester mortar (RPMP polyester)	D, Nonreinforced thermoset liner	72 (496)
250 (1724)				

* When ordering pipe, the purchaser must specify the pipe size, type, pressure class, grade, liner, and stiffness class.

† This refers to the maximum sustained working pressure. The sustained working pressure should be adjusted in accordance with the manufacturer's recommended temperature-derating factors for service at temperatures other than 73.4°F (23°C).

Dimensions of AWWA C950 Outside Diameter (OD) Series Pipe

AWWA C950 OD Series Pipe with Steel Pipe Equivalent. Tables 21.11 and 21.11M provide the nominal pipe size, outside diameter, and applicable tolerance for the outside diameter in U.S. Customary and metric units, respectively.

AWWA C950 OD Series Pipe with Cast Iron Pipe Equivalent. Tables 21.12 and 21.12M provide the nominal pipe size, outside diameter, and applicable tolerance for the outside diameter in U.S. Customary and metric units, respectively.

TABLE 21.11 AWWA C950 Fiberglass Pressure Pipe Dimensions for Outside Diameter (OD) Series Pipe with Steel Pipe Equivalent (IPS) ODs (U.S. Customary Units)

Nominal pipe size, in	Outside diameter, in	Tolerance, in
1	1.315	+0.060, −0.016
1½	1.900	+0.060, −0.018
2	2.375	+0.060, −0.018
2½	2.875	+0.060, −0.018
3	3.500	+0.060, −0.018
4	4.500	+0.060, −0.018
6	6.625	+0.066, −0.028
8	8.625	+0.086, −0.040
10	10.750	+0.108, −0.048
12	12.750	+0.128, −0.056
14	14.000	+0.140, −0.062
16	16.000	+0.160, −0.070

TABLE 21.11M AWWA C950 Fiberglass Pressure Pipe Dimensions for Outside Diameter (OD) Series Pipe with Steel Pipe Equivalent (IPS) ODs (Metric Units)

Nominal pipe size, mm	Outside diameter, mm	Tolerance, mm
25	33.40	+1.52, −0.41
40	48.26	+1.52, −0.46
50	60.33	+1.52, −0.46
65	73.03	+1.52, −0.46
80	88.90	+1.52, −0.46
100	144.30	+1.52, −0.46
150	168.28	+1.68, −0.71
200	219.08	+2.18, −1.02
250	273.05	+2.74, −1.22
300	323.85	+3.25, −1.42
350	355.60	+3.56, −1.57
400	406.40	+4.06, −1.78

TABLE 21.12 AWWA C950 Fiberglass Pressure Pipe Dimensions for Outside Diameter (OD) Series Pipe with Cast Iron Pipe Equivalent Outside Diameters. (U.S. Customary Units)

Nominal pipe size, in	Outside diameter, in	Tolerance, in
2	2.50	+0.05, −0.05
3	3.96	+0.06, −0.06
4	4.80	+0.06, −0.06
6	6.90	+0.06, −0.06
8	9.05	+0.06, −0.06
10	11.10	+0.06, −0.06
12	13.20	+0.06, −0.06
14	15.30	+0.05, −0.08
16	17.40	+0.05, −0.08
18	19.50	+0.05, −0.08
20	21.60	+0.05, −0.08
24	25.80	+0.05, −0.08
30	32.00	+0.08, −0.06
36	38.30	+0.08, −0.06
42	44.50	+0.08, −0.06
48	50.80	+0.08, −0.06
54	57.56	+0.08, −0.06
60	61.61	+0.08, −0.06

TABLE 21.12M AWWA C950 Fiberglass Pressure Pipe Dimensions for Outside Diameter (OD) Series Pipe with Cast Iron Pipe Equivalent Outside Diameters (Metric Units)

Nominal pipe size, mm	Outside diameter, mm	Tolerance, mm
50	63.5	+1.3, −1.3
80	100.6	+1.5, −1.5
100	121.9	+1.5, −1.5
150	175.3	+1.5, −1.5
200	229.9	+1.5, −1.5
250	281.9	+1.5, −1.5
300	335.3	+1.5, −1.5
350	388.6	+1.3, −2.0
400	442.0	+1.3, −2.0
450	495.3	+1.3, −2.0
500	548.6	+1.3, −2.0
600	655.3	+1.3, −2.0
800	812.8	+2.0, −1.5
900	972.8	+2.0, −1.5
1100	1130.3	+2.0, −1.5
1200	1290.3	+2.0, −1.5
1350	1462.0	+2.0, −1.5
1500	1564.9	+2.0, −1.5

AWWA C950 OD Series Pipe with Ductile Iron Pipe Equivalent. Table 21.13 lists metric dimensions of OD series pipe with ductile iron pipe equivalent conforming to ISO 2531.

AWWA C950 OD Series Pipe with Metric Dimensions per International Practice. Refer to Table 21.14.

TABLE 21.13 Metric Dimensions for Outside Diameter (OD) Series Pipe with Ductile Iron Pipe Equivalent*

Nominal pipe size, mm	Outside diameter, mm	Outside diameter tolerance, mm
80	98	+1.0, –2.7
100	118	+1.0, –2.8
150	170	+1.0, –2.9
200	222	+1.0, –3.0
250	274	+1.0, –3.1
300	326	+1.0, –3.3
350	378	+1.0, –3.4
400	429	+1.0, –3.5
500	532	+1.0, –3.8
600	635	+1.0, –4.0
700	738	+1.0, –4.3
800	842	+1.0, –4.5
900	945	+1.0, –4.8
1000	1048	+1.0, –5.0
1200	1255	+1.0, –5.5
1400	1462	+1.0, –6.0
1600	1668	+1.0, –7.4
1800	1875	+1.0, –8.2
2000	2082	+1.0, –9.0

* Dimensions per ISO 2531.

TABLE 21.14 Metric Dimensions for Outside Diameter (OD) Series Pipe*

Nominal pipe size, mm	Outside diameter, mm	Outside diameter tolerance, mm
25	30	+1.5, –2.0
40	45	+1.6, –2.0
50	55	+1.6, –2.0
80	86	+1.7, –2.0
100	106	+1.8, –2.0
150	157	+1.9, –2.0
200	208	+2.0, –2.0
250	259	+2.1, –2.0
300	310	+2.3, –2.0
400	412	+2.5, –2.0
500	514	+2.8, –2.0
600	616	+3.0, –2.0
700	718	+3.3, –2.0
800	820	+3.5, –2.0
900	922	+3.8, –2.0
1000	1024	+4.0, –2.0
1200	1228	+4.5, –2.0
1400	1432	+5.0, –2.0
1600	1636	+5.5, –2.0
1800	1840	+6.0, –2.0
2000	2044	+6.5, –2.0
2200†	2248	+7.0, –2.0
2400	2452	+7.5, –2.0
2600†	2656	+8.0, –2.0
2800	2860	+8.5, –2.0
3000†	3064	+9.0, –2.0
3200	3268	+9.5, –2.0
3400†	3472	+10.0, –2.0
3600	3676	+10.5, –2.0

* OD series commonly used in informational practice.

† Nonpreferred diameters.

Dimensions of AWWA C950 Inside Diameter (ID) Series Pipe

Tables 21.15 and 21.15M provide the dimensions of inside diameter (ID) series pipe in U.S. Customary and metric units, respectively.

Tolerances for Pipe Wall Thickness of AWWA C950 Pipe

The minimum pipe wall thickness at any point shall not be less than 87.5 percent of the nominal wall thickness listed in the manufacturer's publications.

TABLE 21.15 AWWA C950 Fiberglass Pressure Pipe, Dimensions for Inside Diameter Series Pipe (U.S. Customary Units)

Nominal pipe size, in	Inside diameter, in	Tolerance, in (plus and minus)
1	1.00	0.06
$1^1/_2$	1.50	0.06
2	2.00	0.06
$2^1/_2$	2.50	0.06
3	3.00	0.12
4	4.00	0.12
6	6.00	0.25
8	8.00	0.25
10	10.00	0.25
12	12.00	0.25
14	14.00	0.25
15	15.00	0.25
16	16.00	0.25
18	18.00	0.25
20	20.00	0.25
21	21.00	0.25
24	24.00	0.25
27	27.00	0.27
30	30.00	0.30
33	33.00	0.33
36	36.00	0.36
39	39.00	0.39
42	42.00	0.42
45	45.00	0.45
48	48.00	0.48
51	51.00	0.51
54	54.00	0.54
60	60.00	0.60
66	66.00	0.66
72	72.00	0.72
78	78.00	0.78
84	84.00	0.84
90	90.00	0.90
96	96.00	0.96
102	102.00	1.00
108	108.00	1.00
114	114.00	1.00
120	120.00	1.00
132	132.00	1.00
144	144.00	1.00

TABLE 21.15M Metric Dimensions for Inside Diameter (ID) Series Pipe*

Nominal pipe size, mm	Inside diameter range, mm		Inside diameter tolerance, mm (plus and minus)
	Min.	Max.	
25	23.5	26.5	1.5
40	38	42	1.5
50	48	52	1.5
80	78	82	1.5
100	97	103	1.5
150	147	153	1.5
200	196	204	1.5
250	246	255	1.5
300	296	306	1.8
350	346	357	2.1
400	396	408	2.4
450	446	459	2.7
500	496	510	3.0
600	595	612	3.6
700	695	714	4.2
800	795	816	4.2
900	895	918	4.2
1000	995	1020	4.2
1200	1195	1220	5.0
1400	1395	1420	5.0
1600	1595	1620	5.0
1800	1795	1820	5.0
2000	1995	2020	5.0
2200†	2195	2220	5.0
2400	2395	2420	6.0
2600†	2595	2620	6.0
2800	2795	2820	6.0
3000†	2995	3020	6.0
3200	3195	3220	6.0
3400†	3395	3420	7.0
3600	3595	3620	7.0

* Dimensions are per ISO 7370, Glass Fibre Reinforced Thermosetting Plastics (GRP) Pipe and Fittings—Nominal Diameters, Specified Diameters and Standard Lengths.

† These are classified as nonpreferred diameters.

Hydrostatic Test Pressure for AWWA C950 Pipe

Table 21.16 lists the hydrostatic test pressures for various pressure classes of pipe. It is noted that the test pressures are twice the pressure class rating.

TABLE 21.16 Hydrostatic Test Pressures

Pressure class, psig (kPa)	Hydrostatic pressure, psig (kPa)
50 (345)	100 (689)
100 (689)	200 (1379)
150 (1034)	300 (2068)
200 (1379)	400 (2758)
250 (1724)	500 (3447)

Design, Hydraulics, and Installation of AWWA C950 Pipe

Refer to AWWA Manual M45, Fiberglass Pipe Design.

ASTM D3517, STANDARD SPECIFICATION FOR "FIBERGLASS" (GLASS-FIBER-REINFORCED THERMOSETTING-RESIN) PIPE

ASTM D3517 covers machine-made fiberglass pipe in sizes from 8 in (200 mm) through 144 in (3700 mm). This pipe is intended for conveyance of water at a pressure of 250 psig or less. The pipe conforming to this standard is intended for buried applications; however, it can also be used for slip lining and rehabilitation of existing pipelines. The application of this pipe is not limited to the services listed. ASTM D3517 fiberglass pressure pipe is available in the pressure classes, types, grades, stiffness classes, and types of liners shown in Table 21.17.

TABLE 21.17 ASTM D3517 Fiberglass Pressure Pipe Pressure and Stiffness Classes, Types, Grades, and Liners*

Pressure class†	Type	Grade	Liner	Pipe stiffness, psi (kPa)
C50	1, Glass-fiber-reinforced thermosetting polyester resin mortar (RPMP polyester)	1, Polyester resin surface layer-reinforced	1, Reinforced thermoset liner	A, 9 (62)
C75	2, Glass-fiber-reinforced thermosetting polyester resin (RTRP polyester)	2, Polyester resin surface layer-nonreinforced	2, Nonreinforced thermoset liner	B, 18 (124)
C100	3, Glass-fiber-reinforced thermosetting epoxy resin mortar (RPMP epoxy)	3, Polyester resin and sand surface layer-nonreinforced	3, Thermoplastic liner	C, 36 (248)
C125			4, No liner	
C150	4, Glass-fiber-reinforced thermosetting epoxy resin (RTRP epoxy)	4, Epoxy resin surface layer-reinforced		
		5, Epoxy resin surface layer-nonreinforced		
C175		6, No surface layer		D, 72 (496)
C200				
C225				
C250				

* When ordering pipe, the purchaser must specify the pipe by designating the type, liner, grade, class, and the pipe stiffness. For example, ASTM D3517-Type-Liner-Grade-Class-Pipe stiffness (ASTM D3517-1-1-4-C150-C).

† Arabic digits following the letter C designate the operating pressure of the class. The operating pressure must be adjusted in accordance with the manufacturer's recommended temperature-derating factors for service at temperatures other than 73.4°F (23°C).

Dimensions of ASTM D3517 Fiberglass Pipe

Dimensions of ASTM D3517 Inside-Diameter-Contolled Pipe. See Table 21.18.

Dimensions of ASTM D3517 Inside-Diameter-Controlled Pipe (SI Units). Table 21.19 provides nominal inside diameter and the minimum and maximum inside diameter of ID-controlled pipe per ISO standards.

TABLE 21.18 Nominal Inside Diameter (ID) and Tolerances (U.S. Customary Units)

Nominal inside diameter, in	Tolerance for ID, in	
	Plus	Minus
8	0.25	0.25
10	0.25	0.25
12	0.25	0.25
14	0.25	0.25
15	0.25	0.25
16	0.25	0.25
18	0.25	0.25
20	0.25	0.25
21	0.25	0.25
24	0.25	0.25
27	0.27	0.27
30	0.30	0.30
33	0.33	0.33
36	0.36	0.36
39	0.39	0.39
42	0.42	0.42
45	0.45	0.45
48	0.48	0.48
51	0.54	0.54
54	0.60	0.60
60	0.66	0.66
66	0.72	0.72
72	0.78	0.78
78	0.84	0.84
84	0.90	0.90
90	0.96	0.96
96	1.00	1.00
102	0.96	0.96
108	1.00	1.00
114	1.00	1.00
120	1.00	1.00
132	1.00	1.00
144	1.00	1.00

TABLE 21.19 Dimensions of ASTM D3517 Inside-Diameter-Controlled Pipe

Nominal diameter*	Inside diameter, mm	
	Min.	Max.
200	196	204
250	246	255
300	296	306
400	396	408
500	496	510
600	595	612
700	695	714
800	795	816
900	895	918
1000	995	1020
1200	1195	1220
1400	1395	1420
1600	1595	1620
1800	1795	1820
2000	1995	2020
2200†	2195	2220
2400	2395	2420
2600†	2595	2620
2800	2795	2820
3000†	2995	3020
3200	3195	3220
3400†	3395	3420
3600	3595	3620
3800†	3795	3820
4000	3995	4020

* Dimensions are per International Organization for Standardization (ISO).
† These are nonpreferred diameters.

Tolerances for Inside Diameter of ID-Controlled Pipe (SI Units). Table 21.20 lists the applicable tolerances for the specified inside diameter of ID-controlled pipe.

Hydrostatic Test Pressure for ASTM D3517 Fiberglass Pipe

Table 21.21 lists the hydrostatic test pressures for different classes of ASTM D3517 fiberglass pipe.

FIBERGLASS PIPE FITTINGS

The following types of fiberglass fittings are used in fiberglass pipe:

1. Hand-layup or contact molding fittings
2. Press-molded fittings
3. Fliament-wound fittings
4. Mitered fittings

Hand-Layup or Contact Pressure Molding Fittings

These fittings are manufactured in sizes NPS 2 through NPS 42. The dimensions of these fittings conform to National Bureau of Standards (NBS) Voluntary Product Standard PS 15–69. Even though this standard was issued in November 1969 and has not been renewed thereafter, it is still used by the fiberglass pipe and fitting manufacturers. The user is cautioned to check with the manufacturer to verify the dimensions of fittings.

Hand-layup fittings are typically used for critical or severe applications. They are made up to a 10:1 safety factor. They are usually provided with plain ends and joined with pipe by butt-and-wrap joints.

Table 21.22 provides dimensions of fiberglass pipe fittings shown in Fig. 21.1, including hand-layup plain-end fittings.

TABLE 21.20 Tolerances for Specified Inside Diameter (SI Units)

Nominal metric diameter, mm	Plus tolerance, mm*	Minus tolerance, mm*
200, 250	1.5	1.5
300	1.8	1.8
400	2.4	2.4
500	3.0	3.0
600	3.6	3.6
700, 800, 900	4.2	4.2
1000, 1200, 1400, 1600, 1800, 2000	5.0	5.0
2200, 2400, 2600, 2800, 3000	6.0	6.0
3200, 3400, 3600, 3800, 4000	7.0	7.0

* Tolerances apply to the specified inside diameter. (See Table 21.19 for minimum and maximum ID.)

TABLE 21.21 Hydrostatic Test Pressures

Class	Hydrostatic test pressure	
	psig	kPa
C50	100	689
C75	150	1034
C100	200	1379
C125	250	1723
C150	300	2068
C175	350	2412
C200	400	2757
C225	450	3102
C250	500	3445

TABLE 21.22 Dimension of Hand-Layup Fiberglass Fittings (Figure 21.1)

Dimensions, in							
D	*A*	*B*	*C*	*E*	*F*	*G*	*H*
2	6	10	6	4	16	$1\frac{5}{8}$	6
3	7	12	6	6	18	$2\frac{1}{2}$	6
4	8	14	6	6	20	$2\frac{1}{2}$	6
6	10	16	8	9	24	$3\frac{3}{4}$	8
8	12	20	10	12	30	5	8
10	14	24	10	15	34	$6\frac{1}{4}$	10
12	16	26	12	18	38	$7\frac{1}{2}$	10
14	18	30	12	21	42	$8\frac{3}{4}$	12
16	20	32	14	24	46	10	12
18	21	36	14	27	50	$11\frac{1}{4}$	12
20	22	38	16	30	54	$12\frac{1}{2}$	12
24	24	42	18	36	60	15	12
30	30	52	20	45	72	$18\frac{5}{8}$	15
36	33	62	22	54	84	$22\frac{1}{2}$	15
42	36	72	24	63	96	26	15

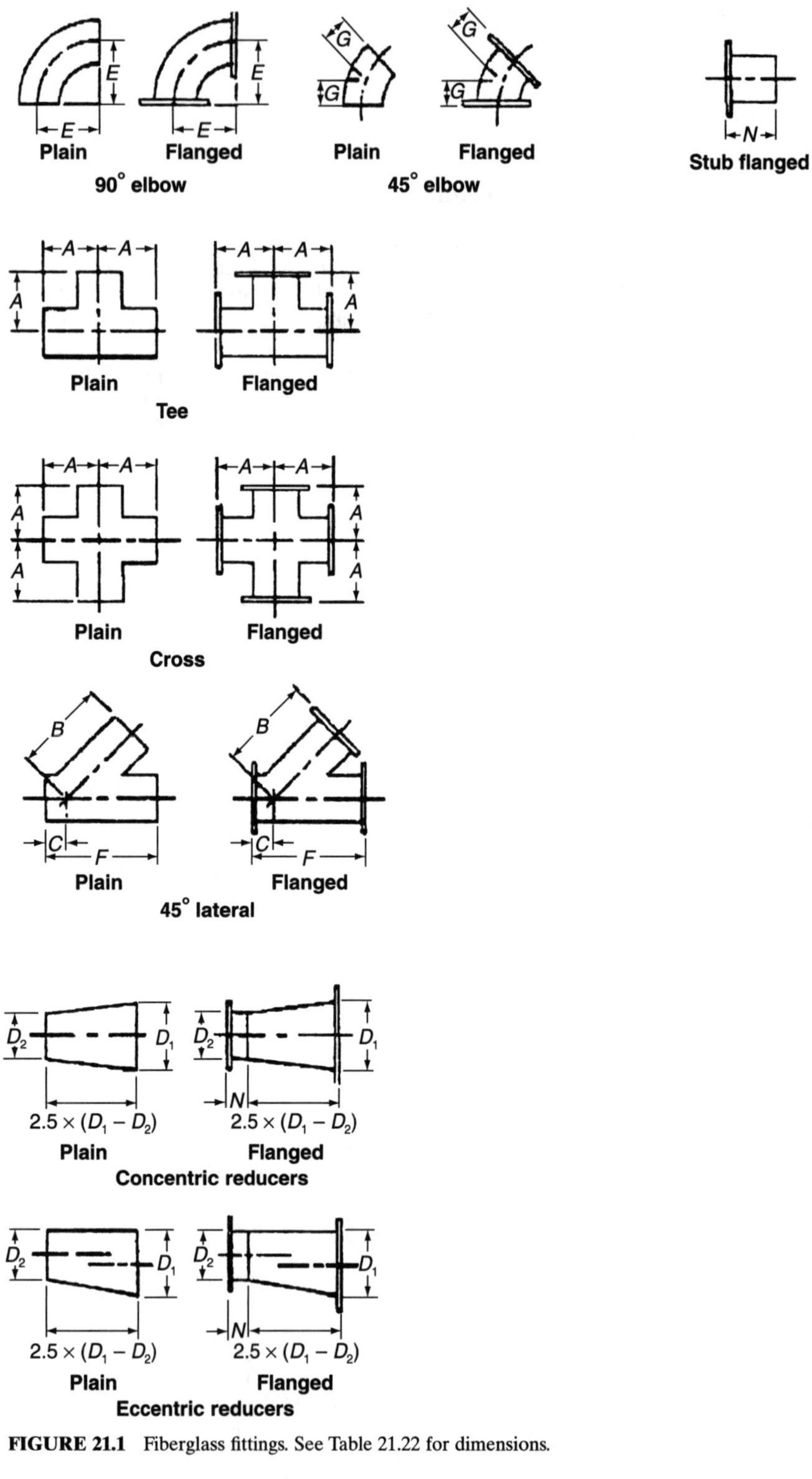

FIGURE 21.1 Fiberglass fittings. See Table 21.22 for dimensions.

Press-Molded Fittings

Press-molded fittings are available in sizes ranging from NPS 1 through 14. They are suitable for bell-and-spigot, straight-socket, and flanged joints. They can be very corrosion resistant, but typically have heavy walls to achieve desired strength.

Dimensions of press-molded fittings are per the manufacturer's standard. Refer to the manufacturer's literature or contact the manufacturer to get the fitting dimensions.

Filament-Wound Fittings

These fittings are usually made in sizes NPS 1 through NPS 24. They are available with the complete range of joining methods used in fiberglass piping.

Flanged-end dimensions conform to ASME B16.5, Class 150 flanges. Refer to Chap. 5. For complete fitting dimensions refer to the manufacturer's literature.

Mitered Fittings

These are made from pipe sections, which are joined together by butt-and-wrap joints. They are available in sizes ranging from NPS 1 through 120 and they are rated for full pipe pressure. The mitered fittings are provided with plain ends suitable for butt-and-wrap joints. Dimensions of mitered fittings vary with the size and the covered angle of the miter. Contact the manufacturer for dimensions.

Marine Pipe and Fittings

Marine pipe and fittings are manufactured in accordance with ASTM F1173.

FIBERGLASS PIPE FLANGES

Flanges are available in all pipe sizes conforming to dimensional requirements and bolting patterns of ASME, DIN, ISO, and JSI standards.

NPS 1 through NPS 24 flanges conforming to ASME B16.5, Class 150 flange dimensions are commonly used in the industry. Flanges larger than NPS 24 are manufactured in compliance with the dimensional requirements of ASME B16.1, Cast Iron Flanges and Flanged Fittings.

The dimensions and bolting pattern of fiberglass pipe flanges should be the same as those of the mating flanges on the equipment or nozzle.

FIBERGLASS PIPING JOINTS

Refer to Chap. 15.

TYPICAL APPLICATIONS OF FIBERGLASS PIPE

Refer to Table 21.23 for typical applications of fiberglass pipe by industry.

TABLE 21.23 Typical Fiberglass Pipe Applications by Industry

Applications	Industry								
	Chemical process	Food processing	Marine and offshore platforms	Mining and metal refining	Petrochemical and petroleum	Pharmaceutical	Power plants	Pulp and paper	Wastewater treatment
Aeration lines									x
Brine slurry lines	x								
Chemical feed lines	x	x		x	x		x	x	x
Column piping			x						
Condensate return	x	x	x		x	x	x	x	
Conduit			x				x	x	
Cooling water lines	x	x			x	x	x		
Disposal well systems	x		x	x	x				x
Downhole tubing and casing			x						x
Effluent drain lines	x	x	x	x	x	x	x	x	x
Fire mains			x	x	x		x	x	
Guttering and downspouts	x	x					x	x	
Oily water			x	x	x				
Scrubber headers	x				x		x		
Seawater lines			x		x		x		
Slurry lines	x						x		
Vent lines	x	x	x	x	x	x		x	x
Water lines	x	x	x		x		x	x	
Waste treatment	x	x	x	x		x	x	x	x
Buried gasoline piping*					x				

* At gasoline service stations.

Tables in Chapter 21

CHAPTER 22

PRESTRESSED CONCRETE CYLINDER AND REINFORCED CONCRETE PRESSURE PIPE, FITTINGS, AND JOINTS

See p. 658 for a list of tables in Chap. 22.

PRESTRESSED CONCRETE CYLINDER PRESSURE PIPE (PCCP)

Prestressed concrete pressure pipe is also referred to as *prestressed concrete cylinder pipe* (PCCP). Two types of PCCP are manufactured:

- *Lined cylinder pipe (LCP).* Refer to Fig. 22.1. The entire concrete core is placed inside the cylinder. The cured concrete core and the steel cylinder are then helically wrapped with prestressing wire, which is subsequently coated with cement mortar.
- *Embedded cylinder pipe (ECP).* Refer to Fig. 22.2. The concrete core is placed both outside and inside the steel cylinder by a vertical casting operation. The cured concrete core and the steel cylinder are then helically wrapped with prestressing wire, which is subsequently coated with cement mortar, like the LCP.

Figures 22.3 and 22.4 show configurations of a typical spigot ring and bell ring. The spigot ring is the protruding end of a PCCP joint, which contains a shaped groove to retain the O-ring rubber gasket, and the bell ring is the receiving portion of a PCCP joint.

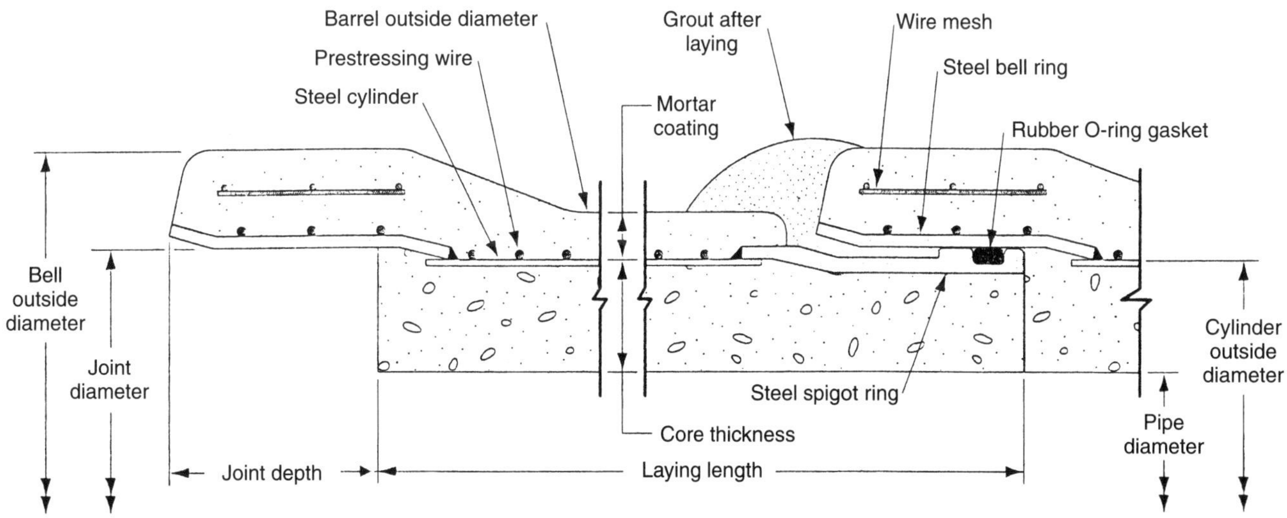

FIGURE 22.1 Line cylinder pipe construction. (*Courtesy of Price Brothers.*)

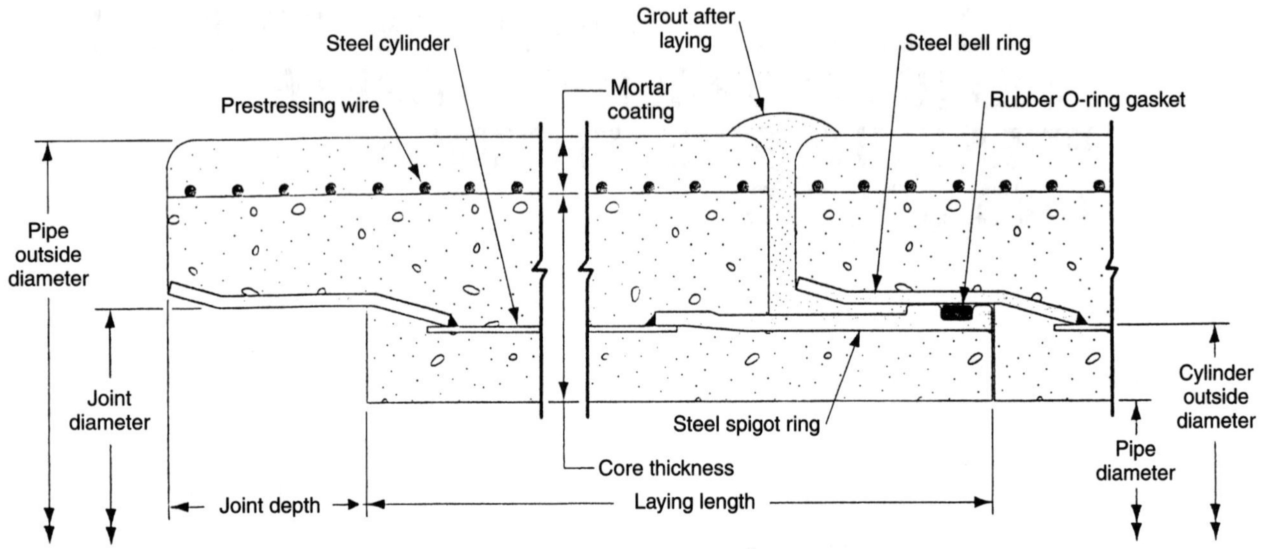

FIGURE 22.2 Construction of embedded cylinder pipe (ECP). (*Courtesy of Price Brothers.*)

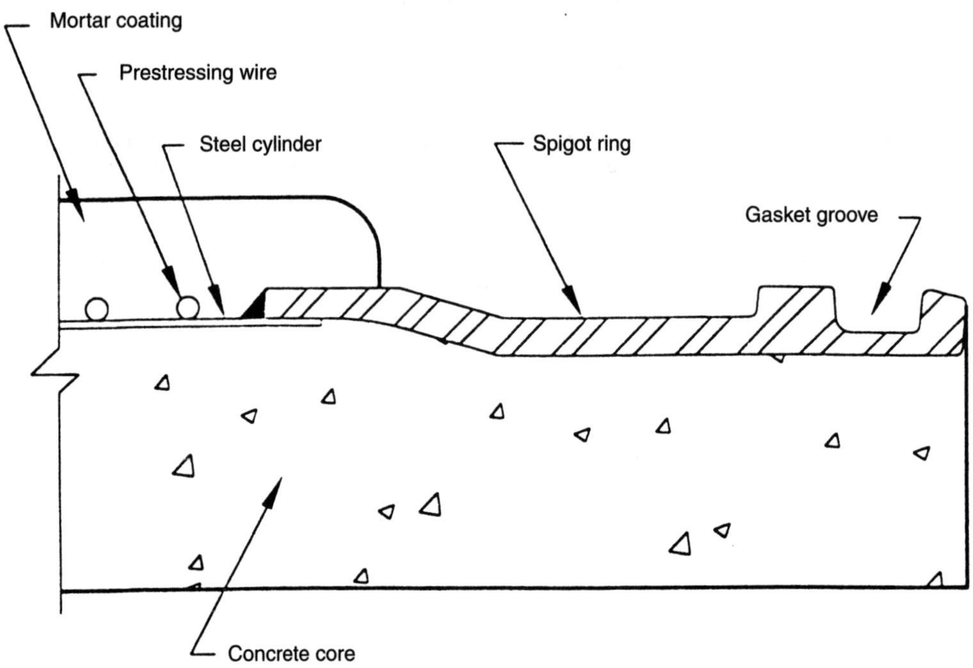

FIGURE 22.3 Spigot ring.

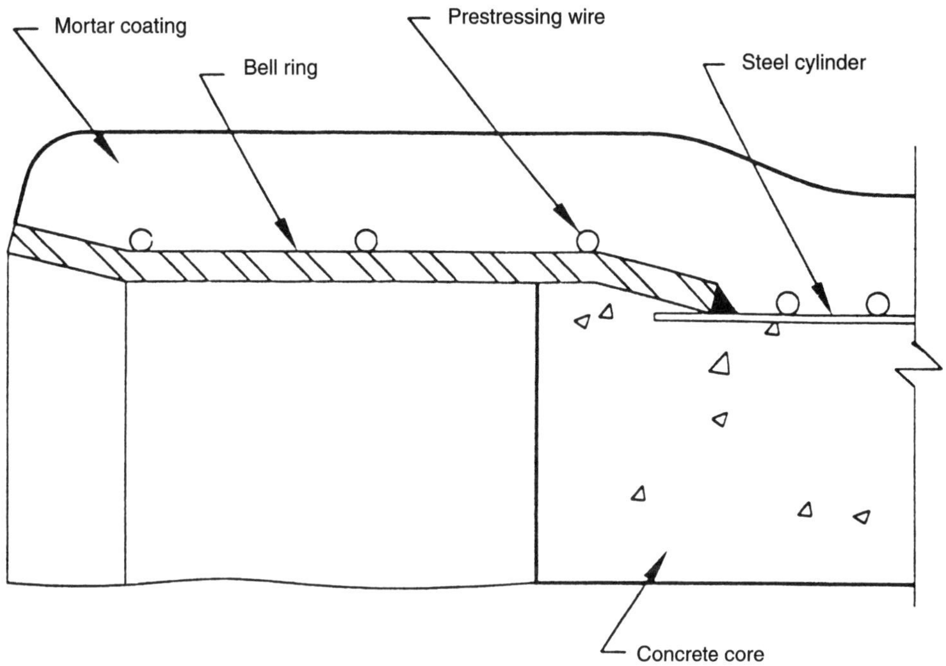

FIGURE 22.4 Bell ring.

Reference Standards and Specifications for PCCP

See Table 22.1.

Pipe Diameter

The inside diameter (waterway) of the pipe is termed the *pipe diameter*.

Pipe Diameter Tolerance for PCCP

See Table 22.2.

PCCP Sizes and Laying Lengths

See Tables 22.3, 22.4, and 22.5.

Out-of-Roundness and Squareness Tolerances for PCCP

See Table 22.6.

TABLE 22.1 Standards/Specifications for PCCP

Number	Title
ANSI/AWWA C301	Prestressed Concrete Pressure Pipe, Steel-Cylinder Type for Water and Other Liquids
ANSI/AWWA C304	Design of Prestressed Concrete Cylinder Pipe
AWWA Manual of Water Supply Practices M9	Concrete Pressure Pipe Provides information for manufacturing and use of PCCP

TABLE 22.2 Tolerance for Pipe Diameter

Pipe diameter	Tolerance
36 in and smaller	$^1/_4$ in
42 in and 48 in	$^3/_8$ in
54 in to 78 in	$^1/_2$ in
84 in and larger	$^3/_4$ in

TABLE 22.3 PCCP Pipe Sizes and Laying Lengths

Type of PCCP	Normally manufactured sizes, NPS (DN)*	Laying Length, ft (m)	Remarks
LCP	NPS 16–NPS 48 (DN 400–1200)	20 (6)	See Table 22.4 for dimensions
ECP	NPS 54–NPS 144 (DN 1400–DN 3700)	20 (6)[†] 16 (4.9 m)[‡]	See Table 22.5 for dimensions

* Contact the manufacturer for sizes other than those listed in this table.
[†] For pipe sizes up to and including NPS 114 (DN 2900).
[‡] For pipe sizes greater than NPS 114 (DN 2900).

TABLE 22.4 Dimensions of Line Cylinder Pipe (LCP)*

Pipe diameter	Core thickness	Joint depth	Joint diameter	Cylinder outside diameter	Bell outside diameter	Barrel outside diameter	Standard laying length, ft
16	1	$4^1/_2$	$18^1/_2$	18	$22^1/_2$	20	19.93
18	$1^1/_8$	$4^1/_2$	$20^3/_4$	$20^1/_4$	$24^3/_4$	$22^1/_4$	19.93
20	$1^1/_4$	$4^1/_2$	23	$22^1/_2$	27	$24^1/_2$	19.93
24	$1^1/_2$	$4^1/_2$	$27^1/_2$	27	$31^1/_2$	29	19.93
30	$1^7/_8$	$4^1/_2$	$34^1/_4$	$33^3/_4$	$38^1/_4$	$35^3/_4$	19.93
36	$2^1/_4$	$4^1/_2$	41	$40^1/_2$	45	$42^1/_2$	19.94
42	$2^5/_8$	$4^1/_2$	$47^1/_4$	$47^1/_4$	$51^1/_4$	$49^1/_4$	19.97
48	3	$4^1/_2$	54	54	58	56	19.98

* Dimensions in inches except for standard laying length. These dimensions of LCP are for Price Brother's pipe conforming to ANSI/AWWA C301. Use dimensions of pipe per the manufacturer.

TABLE 22.5 Dimensions of Embedded Cylinder Pipe (ECP)

Pipe diameter	Core thickness	Joint depth	Joint diameter	Cylinder outside diameter	Pipe outside diameter	Standard laying length, ft
36	$2\frac{1}{4}$	$4\frac{1}{2}$	39	$38\frac{1}{2}$	$42\frac{1}{2}$	19.94
42	$3\frac{1}{2}$	$4\frac{1}{2}$	45	$44\frac{1}{2}$	51	19.97
48	4	$4\frac{1}{2}$	$51\frac{1}{4}$	$50\frac{3}{4}$	58	19.98
54	4	$4\frac{1}{8}$	$57\frac{3}{4}$	$56\frac{7}{8}$	64	20.03
60	$4\frac{1}{2}$	$4\frac{1}{4}$	$63\frac{7}{8}$	63	71	20.03
66	5	$4\frac{3}{8}$	$70\frac{1}{8}$	$69\frac{1}{4}$	78	20.04
72	$5\frac{1}{2}$	$4\frac{1}{2}$	$76\frac{3}{8}$	$75\frac{1}{2}$	85	20.04
78	6	$4\frac{5}{8}$	$82\frac{1}{2}$	$81\frac{5}{8}$	92	20.04
84	$6\frac{1}{2}$	$4\frac{3}{4}$	$88\frac{3}{4}$	$87\frac{7}{8}$	$99\frac{1}{8}$	20.04
90	$6\frac{1}{2}$	$4\frac{7}{8}$	$94\frac{7}{8}$	94	$105\frac{1}{8}$	20.05
96	$6\frac{1}{2}$	$4\frac{7}{8}$	$101\frac{1}{8}$	$100\frac{1}{4}$	$111\frac{1}{8}$	20.05
102	$6\frac{1}{2}$	6	$106\frac{7}{8}$	$106\frac{1}{4}$	$117\frac{1}{8}$	20.05
108	$6\frac{3}{4}$	6	$113\frac{1}{8}$	$112\frac{1}{2}$	$123\frac{5}{8}$	20.05
114	$7\frac{1}{8}$	6	$120\frac{5}{8}$	120	$130\frac{3}{8}$	20.05
120	$7\frac{1}{2}$	6	$126\frac{5}{8}$	126	$137\frac{1}{8}$	16.06
126	$8\frac{1}{2}$	6	$132\frac{5}{8}$	132	$145\frac{1}{8}$	16.06
132	$8\frac{1}{4}$	6	$138\frac{5}{8}$	138	$150\frac{5}{8}$	16.06
138	$8\frac{5}{8}$	6	$144\frac{5}{8}$	144	$157\frac{3}{8}$	16.06
144	9	6	$150\frac{5}{8}$	150	$164\frac{1}{8}$	16.06

Notes:
1. Dimensions in inches except for standard laying length.
2. Some pipe may be supplied with a core thickness other than shown on this chart. AWWA C301 standard allows a 16:1 ratio between pipe internal diameter and core thickness.
3. Contact the pipe manufacturer for exact dimensions.

Source: Price Brothers.

TABLE 22.6 Out-of-Roundness and Squareness Tolerance

Pipe diameter	Squareness	Out-of-roundness
Up to and including 36 in	$\frac{1}{4}$ in	At any transverse section, difference between the maximum and minimum diameters must be less than 1 percent of the average of these diameters
>36 in to and including 120 in	$\frac{3}{8}$ in	
>120 in	$\frac{1}{2}$ in	

PCCP Core and Coating Tolerances

See Table 22.7.

Maximum Design Pressure for PCCP Lined Cylinder Pipe (LCP)

See Table 22.8.

TABLE 22.7 Minimum Core Thickness and Tolerances for PCCP

Pipe size	Minimum core thickness	Tolerance for core thickness	Minimum thickness of coating
36 in and smaller	$^1/_{16}$ of the design pipe diameter*	$^1/_8$ in	$^3/_4$ in over the wire
42 in and 48 in		$^3/_{16}$ in	
54 in to 72 in		$^1/_4$ in	
>72 in		$^3/_8$ in	

* This is for normal applications. A normal application involves a combination of design pressure of 40 psi and 6 ft of earth cover for Type 3 trench bedding. Type 3 trench bedding is defined in Chap. 10. The design pressure may be adjusted higher or lower depending on whether the depth of earth cover is less than or greater than 6 ft.

TABLE 22.8 Maximum Design Pressure for Lined Cylinder Pipe (LCP)

Pipe diameter, in	Maximum design pressure, psi*
16 to 20	250
24 to 36	200
42	175
48	150
54 and 60	100

* Higher design pressures are permissible with thicker cores. The pressures listed in the table above are for pipes with concrete cores meeting the following strength requirements:

- 3000 psi minimum compressive strength of concrete test cylinder at the time of wrapping
- 4500 psi compressive strength of concrete test cylinder within 28 days
- 4000 psi minimum compressive strength of centrifugal test cylinder at the time of wrapping
- 6000 psi compressive strength of centrifugal test cylinder within 28 days

PCCP Pipe Fittings and Flange Adapters

Construction and Dimensions of Typical PCCP Elbows. See Fig. 22.5 and Table 22.9.

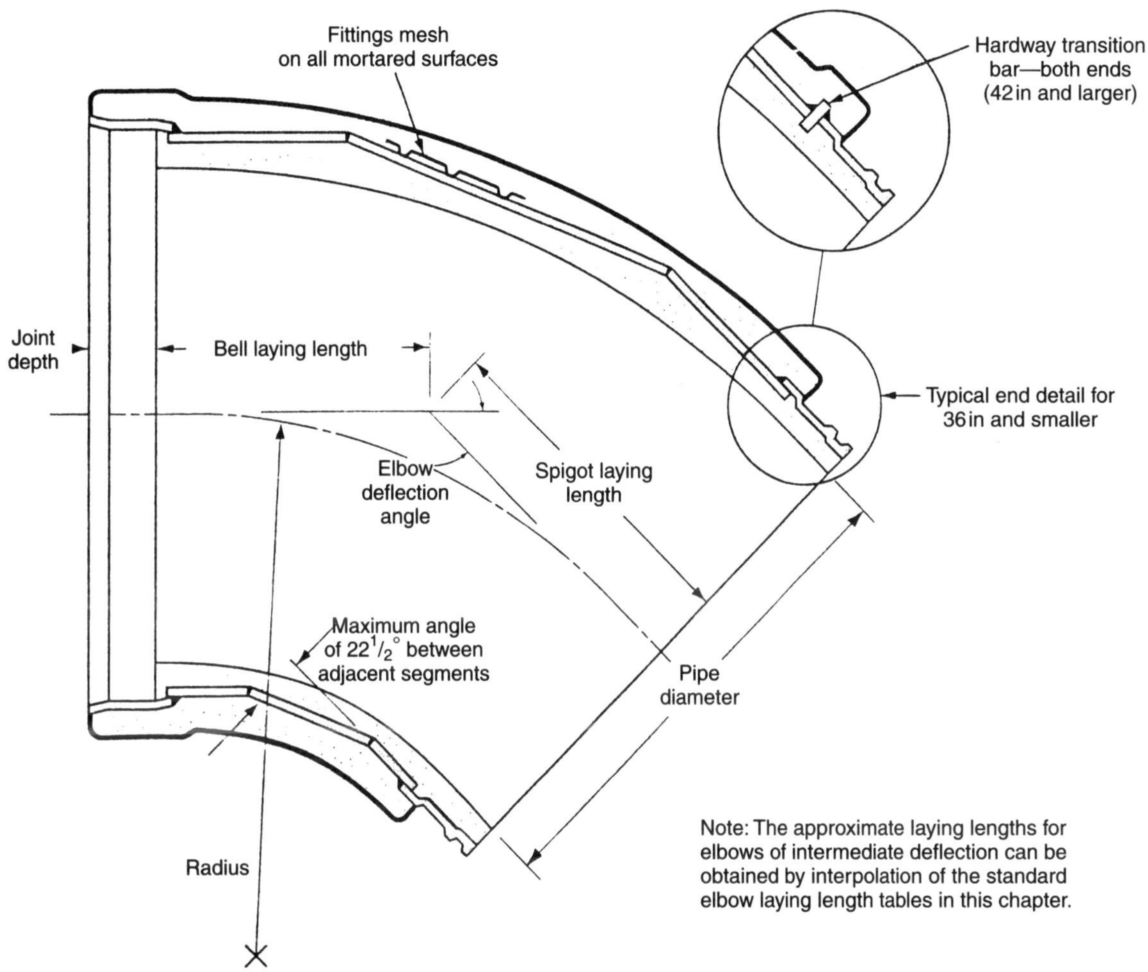

FIGURE 22.5 PCCP elbow construction. (*Courtesy of Price Brothers.*)

TABLE 22.9 Elbow Laying Lengths (LL)

Elbow angle		$11\frac{1}{4}°$		15°		$22\frac{1}{2}°$		30°		45°		60°		75°		90°	
Pipe diameter, in	Elbow radius, in	Spigot LL, ft	Bell LL, ft	Spigot LL, ft	Bell LL, ft	Spigot LL, ft	Bell LL, ft	Spigot LL, ft	Bell LL, ft	Spigot LL, ft	Bell LL, ft	Spigot LL, ft	Bell LL, ft	Spigot LL, ft	Bell LL, ft	Spigot LL, ft	Bell LL, ft
16	24	0.66	0.26	0.72	0.33	0.86	0.46	0.99	0.60	1.29	0.89	1.61	1.22	1.99	1.60	2.46	2.06
18	26	0.67	0.28	0.74	0.35	0.89	0.49	1.04	0.64	1.36	0.96	1.71	1.31	2.12	1.73	2.62	2.23
20	28	0.69	0.29	0.77	0.37	0.92	0.53	1.08	0.69	1.42	1.03	1.81	1.41	2.25	1.85	2.79	2.40
24	30	0.70	0.31	0.79	0.39	0.96	0.56	1.13	0.73	1.49	1.10	1.90	1.51	2.38	1.98	2.96	2.56
30	34	0.74	0.34	0.83	0.44	1.02	0.63	1.22	0.82	1.63	1.24	2.09	1.70	2.63	2.24	3.29	2.90
36	38	0.77	0.37	0.88	0.48	1.09	0.69	1.31	0.91	1.77	1.37	2.29	1.89	2.89	2.49	3.62	3.23
42	42	0.88	0.54	0.99	0.66	1.23	0.89	1.47	1.14	1.98	1.65	2.55	2.22	3.22	2.88	4.03	3.70
48	48	0.93	0.59	1.06	0.72	1.33	0.99	1.60	1.27	2.19	1.85	2.84	2.51	3.60	3.27	4.53	4.20
54	54	1.08	0.63	1.23	0.78	1.53	1.08	1.84	1.39	2.50	2.05	3.23	2.79	4.09	3.64	5.14	4.69
60	60	1.14	0.69	1.30	0.86	1.64	1.19	1.99	1.54	2.72	2.27	3.53	3.08	4.48	4.03	5.65	5.20
66	66	1.21	0.74	1.39	0.92	1.76	1.29	2.14	1.67	2.94	2.48	3.84	3.37	4.89	4.42	6.17	5.70
72	72	1.26	0.78	1.46	0.98	1.86	1.38	2.27	1.80	3.15	2.67	4.13	3.65	5.27	4.79	6.67	6.19
78	78	1.31	0.82	1.52	1.03	1.96	1.47	2.41	1.92	3.36	2.87	4.42	3.93	5.65	5.16	7.17	6.68
84	84	1.36	0.86	1.59	1.09	2.06	1.56	2.54	2.04	3.57	3.07	4.71	4.21	6.04	5.54	7.67	7.17
90	90	1.43	0.92	1.67	1.16	2.18	1.67	2.70	2.19	3.79	3.28	5.02	4.51	6.44	5.93	8.19	7.68
96	96	1.48	0.97	1.74	1.23	2.28	1.77	2.83	2.32	4.00	3.49	5.31	4.80	6.83	6.32	8.69	8.18

Notes:
1. Contact the manufacturer for exact dimensions.
2. Elbows larger than 96 in are also manufactured.

Construction and Dimensions of Typical PCCP Tee Fitting. See Fig. 22.6 and Table 22.10.

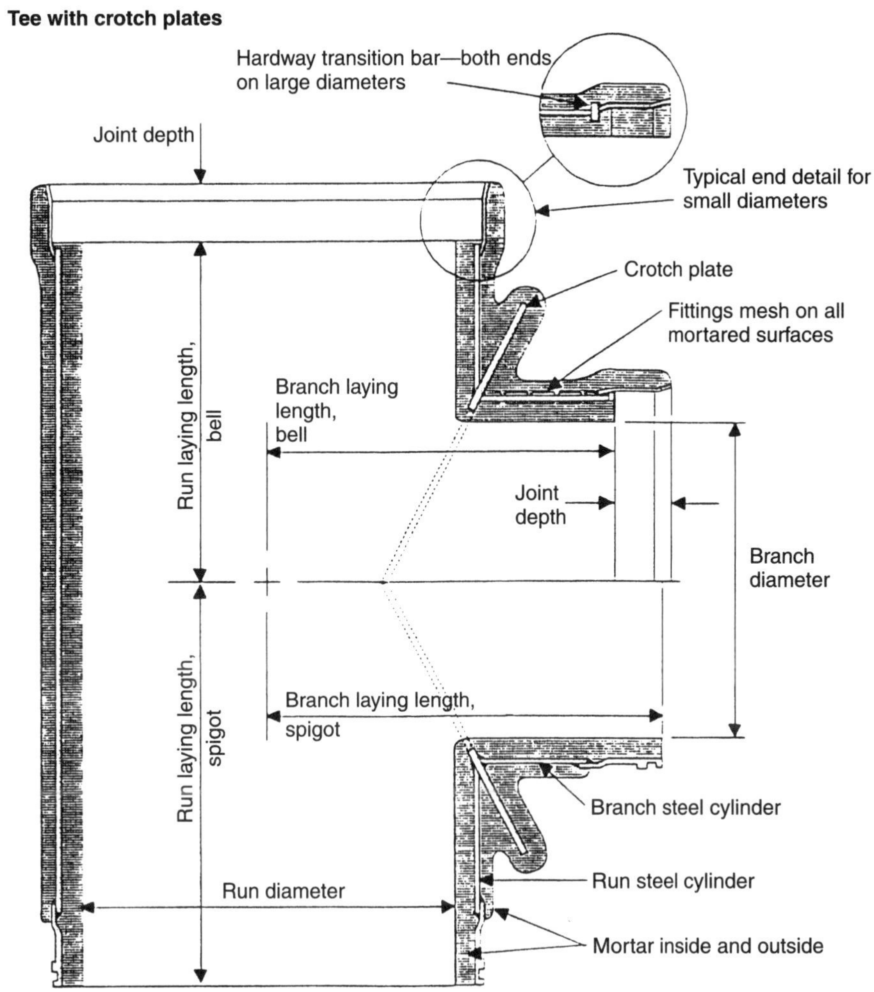

FIGURE 22.6 Typical tee fitting.

TABLE 22.10 Typical Tee Laying Lengths*

Diameter, in		Run laying length, ft				Branch laying length, ft			
Run	Branch	Spigot	Bell	Flange	Mechanical joint bell	Spigot	Bell	Flange	Mechanical joint bell
16	16	2.46	2.08	3.04	2.69	2.21	1.83	2.76	2.44
18	18	2.46	2.08	3.04	2.69	2.30	1.93	2.85	2.53
	16	2.46	2.08	3.04	2.69	2.30	1.93	2.90	2.53
20	20	2.46	2.08	3.05	2.70	2.40	2.02	2.95	2.64
	18	2.46	2.08	3.05	2.70	2.40	2.02	2.99	2.63
	16	2.46	2.08	3.05	2.70	2.40	2.02	2.99	2.63
24	24	2.96	2.58	3.56	3.20	2.58	2.21	3.19	2.83
	20	2.46	2.08	3.06	2.70	2.58	2.21	3.18	2.83
	18	2.46	2.08	3.06	2.70	2.58	2.21	3.18	2.82
30	30	2.96	2.58	3.57	3.18	2.86	2.49	3.48	3.09
	24	2.96	2.58	3.57	3.18	2.86	2.49	3.47	3.11
	20	2.46	2.08	3.07	2.68	2.86	2.49	3.46	3.11
36	36	3.46	3.08	4.09	3.70	3.15	2.77	3.78	3.39
	30	2.96	2.58	3.59	3.20	3.15	2.77	3.76	3.37
	24	2.96	2.58	3.59	3.20	3.15	2.77	3.75	3.39
42	42	3.53	3.16	4.10	3.70	3.50	3.13	4.07	3.67
	36	3.53	3.16	4.10	3.70	3.43	3.05	4.06	3.67
	30	3.03	2.66	3.60	3.20	3.43	3.05	4.04	3.65
48	48	4.03	3.66	4.63	4.20	3.78	3.41	4.38	3.95
	42	3.53	3.16	4.13	3.70	3.78	3.41	4.35	3.95
	36	3.53	3.16	4.13	3.70	3.71	3.33	4.34	3.95
54	54	4.14	3.69	4.88		4.01	3.56	4.47	
	48	4.14	3.69	4.88		3.90	3.53	4.49	4.07
	42	3.64	3.19	4.38		3.90	3.53	4.47	4.07
	36	3.64	3.19	4.38		3.83	3.45	4.46	4.07
60	60	4.65	4.19	5.40		4.27	3.90	4.98	
	54	4.15	3.69	4.90		4.26	3.81	4.73	
	48	4.15	3.69	4.90		4.16	3.78	4.75	4.33
	42	3.65	3.19	4.40		4.16	3.78	4.73	4.33
66	66	4.67	4.29	5.44		4.55	4.17	5.33	
	60	4.67	4.28	5.44		4.53	4.16	5.29	
	54	4.17	3.78	4.94		4.52	4.07	5.27	
	48	4.17	3.78	4.94		4.42	4.01	5.02	4.59
72	72	5.17	4.77	5.95		4.81	4.42	5.60	
	66	4.67	4.27	5.45		4.81	4.43	5.59	
	60	4.67	4.27	5.45		4.79	4.42	5.55	
	54	4.17	3.77	4.95		4.78	4.33	5.53	
	48	4.17	3.77	4.95		4.68	4.27	5.28	4.85
78	78	5.17	4.76	5.96		5.07	4.66	5.86	
	72	5.17	4.76	5.96		5.07	4.67	5.85	
	66	4.67	4.26	5.46		5.07	4.68	5.84	
	60	4.67	4.26	5.46		5.05	4.65	5.80	
	54	4.17	3.76	4.96		5.04	4.59	5.78	
84	84	5.67	5.25	6.46		5.33	4.91	6.12	
	78	5.17	4.75	5.96		5.33	4.92	6.12	
	72	5.17	4.75	5.96		5.33	4.93	6.11	
	66	4.67	4.25	5.46		5.33	4.94	6.10	
	60	4.67	4.25	5.46		5.31	4.93	6.06	
90	90	5.69	5.26	6.50		5.60	5.18	6.42	
	84	5.69	5.26	6.50		5.58	5.17	6.38	
	78	5.19	4.76	6.00		5.58	5.18	6.38	
	72	5.19	4.76	6.00		5.58	5.19	6.37	
	66	4.69	4.26	5.50		5.58	5.20	6.36	
96	96	6.18	5.76	7.00		5.86	5.43	6.68	
	90	5.69	5.26	6.50		5.86	5.44	6.68	
	84	5.69	5.26	6.50		5.84	5.43	6.64	
	78	5.19	4.76	6.00		5.84	5.44	6.64	
	72	5.19	4.76	6.00		5.84	5.45	6.63	
	66	4.69	4.26	5.50		5.84	5.46	6.62	

* These dimensions are of tees manufactured by Price Brothers. Contact other manufacturers for their exact dimensions.

Construction and Dimensions of Typical PCCP Reducers. See Fig. 22.7.

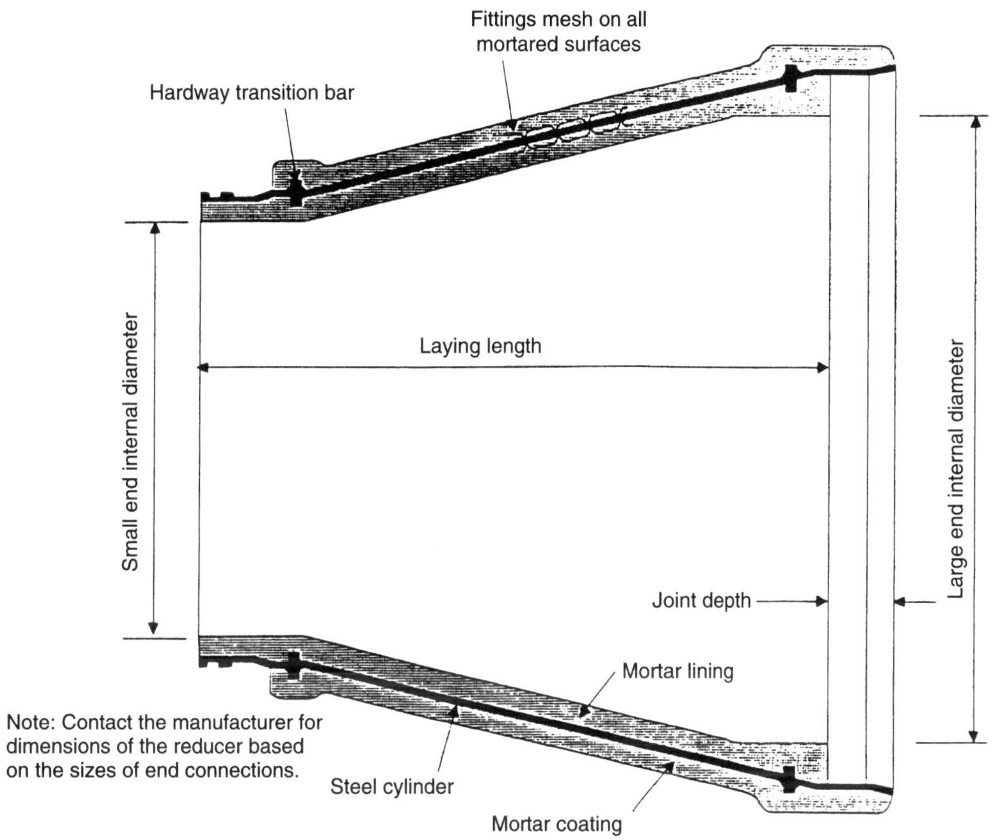

FIGURE 22.7 Typical concentric reducer fitting.

Construction and Dimensions of Typical 45° Wye Fittings. See Fig. 22.8.

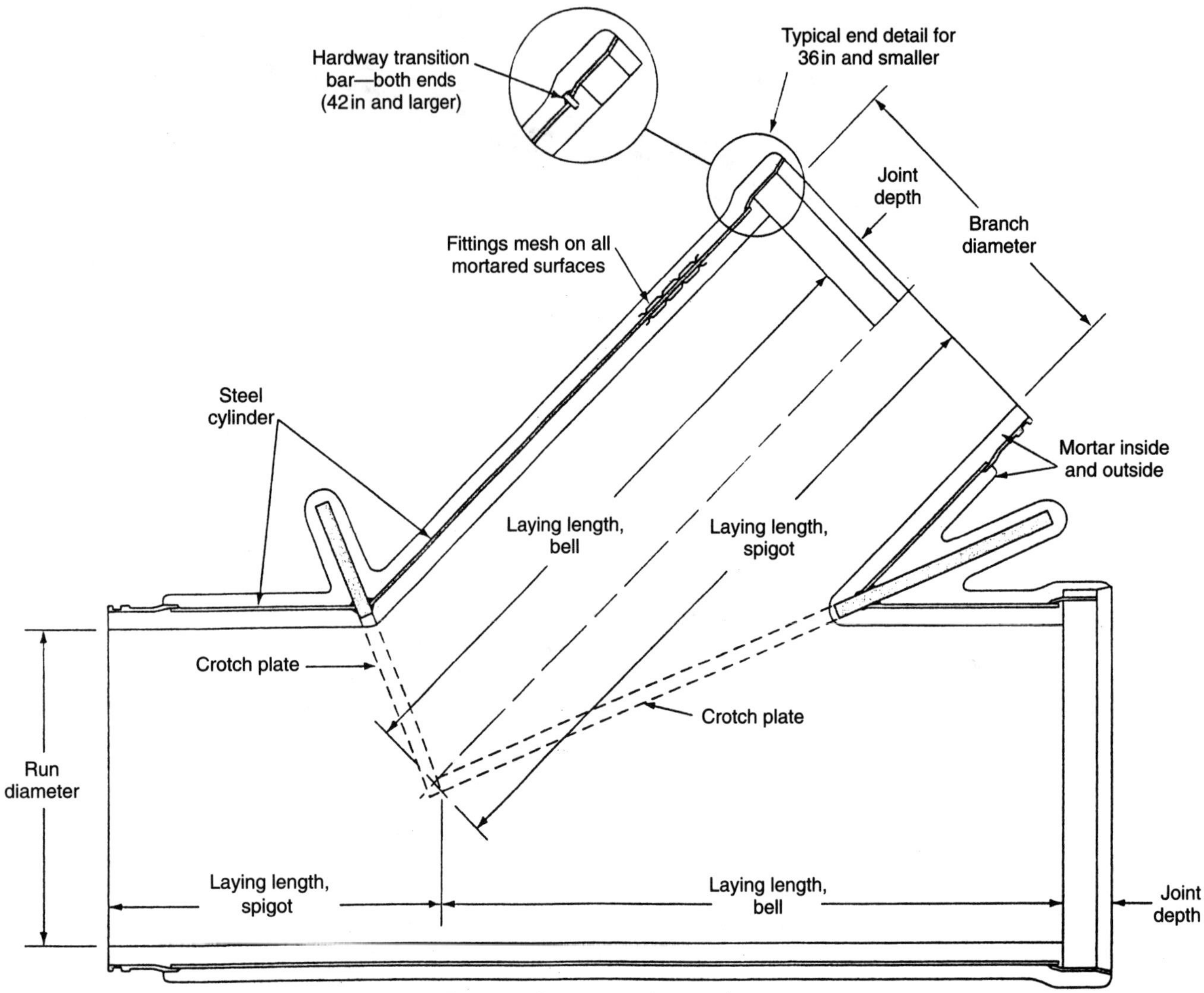

FIGURE 22.8 Typical 45° wye. (*Courtesy of Price Brothers.*)

Typical Flange Adapters for PCCP Joints. See Fig. 22.9.

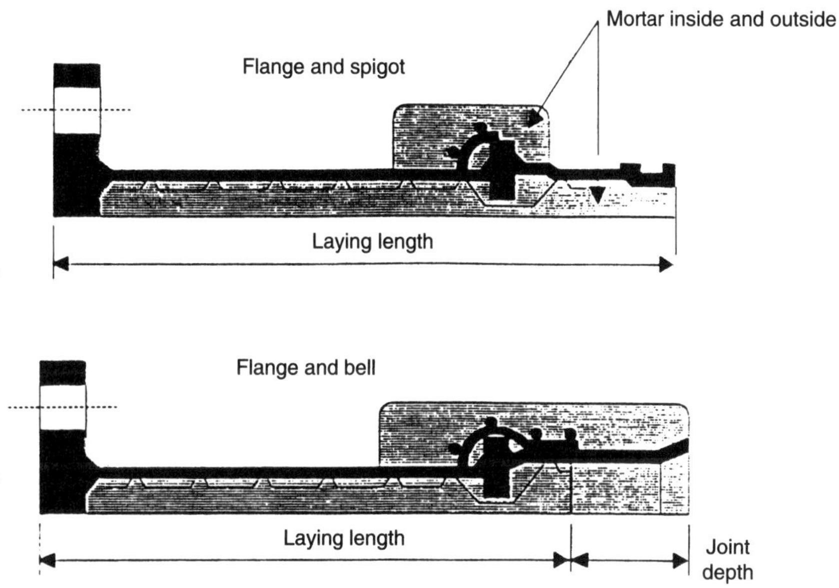

FIGURE 22.9 Typical flange adapters.

REINFORCED CONCRETE PRESSURE PIPE (RCPP)

Reinforced concrete pressure pipe is of two types:

1. Noncylinder type
2. Steel cylinder type

Reference Standards for Concrete Pressure Pipe

See Table 22.11.

TABLE 22.11 Reference Standards for Reinforced Concrete Pressure Pipe (RCPP)

ANSI/AWWA standard	Title
C 302	Reinforced Concrete Pressure Pipe, Noncylinder Type, for Water and Other Liquids
C 303	Reinforced Concrete Pressure Pipe, Steel Cylinder Type, Pretensioned, for Water and Other Liquids
AWWA Manual M9	Concrete Pressure Pipe

AWWA C302 REINFORCED CONCRETE PRESSURE PIPE, NONCYLINDER PIPE

AWWA C302 Pipe Sizes

Sizes range from 12 to 144 in. Pipes larger than 144 in may be manufactured in accordance with AWWA C302.

Limits of Operating Pressure for AWWA C302 Pipe

Operating pressure is not to exceed 55 psi. Operating plus surge pressure is not to exceed 65 psi.

AWWA C302 Pipe Diameter Tolerance

See Table 22.12.

Maximum Pipe Laying Lengths for AWWA C302 Pipe

See Table 22.13.

TABLE 22.12 Tolerance for Pipe Diameter

Pipe diameter	Tolerance
36 in and smaller	$^1/_4$ in
42 in and 48 in	$^3/_8$ in
54 in to 78 in	$^1/_2$ in
84 in and larger	$^3/_4$ in

TABLE 22.13 Maximum Laying Lengths

Pipe internal diameter, in	Laying length, ft
12–18 inclusive	8*
20–24 inclusive	16
27–30 inclusive	18
33–36 inclusive	20
39 and larger	24

* Maximum laying length of 12 ft may be used for pipe with wall thickness of 2 in or more.

Minimum Wall Thicknesses and Tolerances for AWWA C302 Pipe

See Table 22.14.

TABLE 22.14 Pipe Wall Thicknesses and Tolerances

Pipe inside diameter, in	Minimum wall thickness, in*	Wall thickness tolerance, in
12	2	1/8
15	2	1/8
16	2 1/8	1/8
18	2 1/4	1/8
20	2 3/8	1/8
24	2 1/2	1/8
27	2 5/8	1/8
30	2 3/4	1/8
36	3	1/8
42	3 1/2	3/16
48	4	3/16
54	4 1/2	1/4
60	5	1/4
66	5 1/2	1/4
72	6	1/4
78	6 1/2	3/8
84	7	3/8
90	7 1/2	3/8
96	8	3/8
Larger than 96†		3/8

* Minimum wall thickness listed in the table above may not be adequate for some conditions of internal design pressure.

† Dimensions and other details for pipe larger than 96 in are to be approved by the purchaser.

Fittings for AWWA C302 Noncylinder Pipe

Fittings are made from steel sheet or plate welded to form the fitting. The fitting is lined and coated with concrete and cement-mortar lining. Wire fabric reinforcement is applied to the interior and exterior surfaces except where concrete or cement-mortar linings are applied by a centrifugal process.

Minimum Lining and Coating Thickness for Fittings

See Table 22.15.

TABLE 22.15 Minimum Lining and Coating Thickness for AWWA C302 Pipe Fittings

Fitting size, in	Minimum lining thickness, in*	Minimum coating thickness, in
12 to 16 inclusive	1/2	1.0
Larger than 16	3/4	1.0

* Lining thickness at adapter ends or outlets may be less than the minimum listed, but in no case shall it be less than 3/8 in.

AWWA C303 RCPP—STEEL CYLINDER TYPE, PRETENSIONED PIPE

AWWA C303 Pipe Sizes

Sizes range from 10 to 54 in. Pipes larger than 54 in may be manufactured in accordance with AWWA C302.

Limits of Operating Pressure for AWWA C303 Pipe

Design pressures up to and including 400 psi are allowed. The design approach of AWWA C303 provides for an additional capacity of surge pressure of at least 50 percent.

AWWA C303 Pipe Diameter Tolerance

See Table 22.16.

Maximum Pipe Laying Lengths for AWWA C303 Pipe

See Table 22.17.

TABLE 22.16 Tolerance for AWWA C303 Pipe Diameter

Pipe diameter	Tolerance
16 in and smaller	$^1/_4$ in
18 in and larger	$^3/_8$ in

TABLE 22.17 Maximum Laying Lengths

Pipe internal diameter, in	Laying length, ft*
18 and smaller	36
20–54	32–40

* Maximum laying length of 18 in and smaller pipe shall not exceed 36 ft.

Thicknesses and Tolerances for Concrete or Cement-Mortar Linings for AWWA C303 Pipe

See Table 22.18.

TABLE 22.18 Thicknesses and Tolerances for Concrete or Cement-Mortar Linings

Pipe diameter range, in	Lining thickness, in	Tolerances, in (minus)
10–16	0.50	0.125
18–54	0.75	0.1875

Minimum Cylinder Thickness for AWWA C303 Pipe

See Table 22.19.

Circumferential Tolerances for Steel Cylinders of AWWA C303 Pipe

See Table 22.20.

Fittings for AWWA C303 Steel Cylinder Pipe

Fittings include bends, tees, wyes, closures, etc. Fittings are made from steel sheet or plate welded to form the fitting. The fitting is lined and coated with concrete and cement-mortar lining. Wire fabric reinforcement is applied to the interior and exterior surfaces.

Minimum Thickness of Sheet or Plate for Fittings

See Table 22.21.

CONCRETE PRESSURE PIPING JOINTS

Refer to Chap. 15.

TABLE 22.19 Minimum Cylinder Thickness for AWWA C303 Pipe

Pipe diameter range, in	Minimum cylinder thickness	
	Gauge	Inch
10–21	16	0.060
24–33	14	0.075
36–39	13	0.090
42–48	12	0.105
51–54	10	0.135

TABLE 22.20 Circumferential Tolerances for Steel Cylinders

Pipe diameter range, in	Tolerance, in (plus and minus)
10–16	$^3/_{16}$
18–54	$^1/_4$

TABLE 22.21 Minimum Thickness of Plate or Sheet for AWWA C303 Pipe Fittings

Fitting size range, in	Minimum sheet or plate thickness	
	Gauge	Inch
10–21	10	0.135
24–36	7	0.179
39–48		0.250
51–54		0.3125

Tables in Chapter 22

CHAPTER 23
GLASS PIPE, FITTINGS, AND JOINTS

See p. 679 for a list of tables in Chap. 23.

GLASS PIPE AND FITTINGS

Two different types of glass pipe and fittings are available for pressure and gravity applications, process glass and borosilicate glass, covered by the documents listed below.

Reference Standards and Specifications

ASTM specification number	Title
C599	Standard Specification for Conical Process Glass Pipe
C600	Standard Test Method of Thermal Shock on Glass Pipe
C601	Standard Test Method for Pressure Test on Glass Pipe
C1053	Standard Specification for Borosilicate Glass Pipe and Fittings for Drain, Waste, and Vent (DWV) Applications

ASTM C599 PROCESS GLASS PIPE AND FITTINGS

ASTM C599 pipe is available in the sizes listed in Table 23.1.

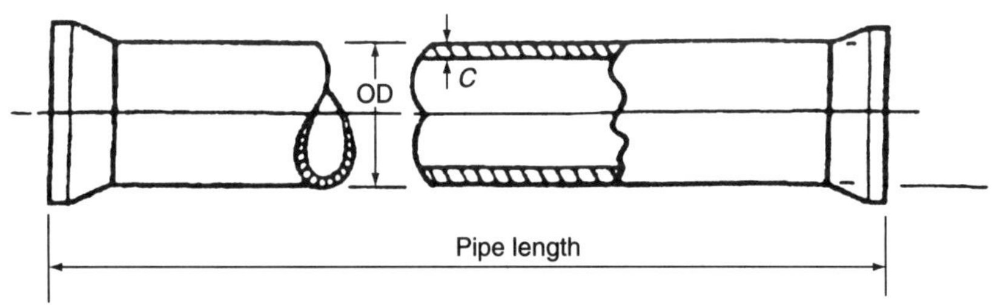

TABLE 23.1 ASTM C599 Process Glass Pipe Dimensions and Tolerances*

Nominal pipe size, in	Outside diameter (OD)		OD tolerance (plus and minus)		Pipe wall thickness, C		Pipe wall thickness, C, tolerance (plus and minus)	
	in	mm	in	mm	in	mm	in	mm
1	$1\frac{5}{16}$	33.3	$\frac{3}{64}$	1.2	$\frac{5}{32}$	4.0	$\frac{1}{32}$	0.8
$1\frac{1}{2}$	$1\frac{27}{32}$	46.8	$\frac{1}{16}$	1.6	$\frac{11}{64}$	4.4	$\frac{1}{32}$	0.8
2	$2\frac{11}{32}$	59.5	$\frac{5}{64}$	2.0	$\frac{11}{64}$	4.4	$\frac{1}{32}$	0.8
3	$3\frac{13}{32}$	86.5	$\frac{3}{32}$	2.4	$\frac{13}{64}$	5.2	$\frac{1}{32}$	0.8
4	$4\frac{17}{32}$	115.1	$\frac{7}{64}$	2.8	$\frac{17}{64}$	6.7	$\frac{3}{64}$	1.2
6	$6\frac{5}{8}$	168.3	$\frac{5}{32}$	4.0	$\frac{21}{64}$	8.3	$\frac{3}{64}$	1.2

* Dimensions in millimeters are obtained by multiplying inch dimension by 25.4. The inch dimensions are to be regarded as standard. Dimensions in millimeters are for information only.

Pressure Rating of ASTM C599 Process Glass Pipe and Fittings

See Table 23.2.

Temperature Capabilities and Limitations of Process Glass Pipe

See Table 23.3.

Dimensions and Tolerances for ASTM C599 Pipe Flange

See Table 23.4.

Dimensions of ASTM C599 Process Glass Pipe Fittings

See Tables 23.5 through 23.16.

TABLE 23.2 Recommended Maximum Working Pressure

Nominal pipe size, in	Maximum recommended working pressure	
	psi	kPa
1 to 2	50	345
3	40	276
4	35	241
6	20	138

TABLE 23.3 Temperature Rating and Limitations of ASTM C599 Process Glass Pipe

Nominal pipe size, in	Minimum rated operating temperature,* °F (°C)	Maximum rated operating temperature, °F (°C)	Maximum temperature differential for thermal endurance,† °F (°C)
1 to 2	−50 (−46)	450 (232)	200 (93)
3	−50 (−46)	450 (232)	140 (78)
4	−50 (−46)	450 (232)	122 (68)
6	−50 (−46)	450 (232)	180 (82)

* The minimum rated operating temperature is based upon the conditions that the temperature of the flow medium is above the freezing point and that aluminum alloy flanges conforming to alloys listed in ASTM specifications B26 and B85 are used.

† Maximum temperature differential refers to an instantaneous temperature change. It applies to both the internal and external pipe surfaces.

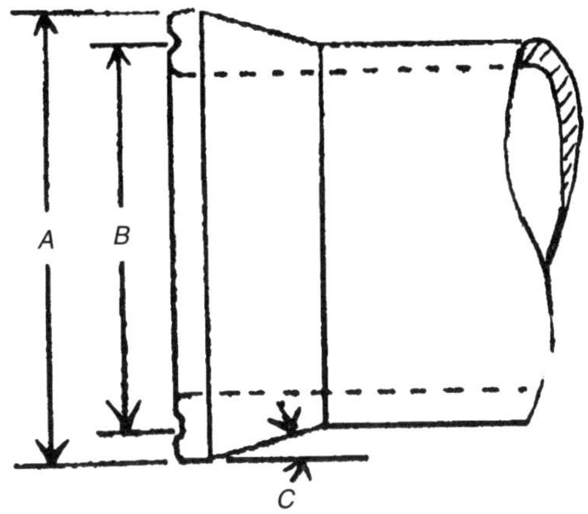

TABLE 23.4 Dimensions and Tolerances of Process Pipe Flanges

NPS, in	Outside diameter, *A* in.	Tolerance for *A*, in (plus and minus)	Groove diameter, *B*, in	Tolerance for *B*, in (plus and minus)	Slope *C*, degrees	Tolerance for *C*, degrees (plus and minus)
1	$1\frac{9}{16}$	$\frac{1}{32}$	$1\frac{9}{32}$	$\frac{1}{32}$	12	2
$1\frac{1}{2}$	$2\frac{1}{8}$	$\frac{1}{32}$	$1\frac{13}{16}$	$\frac{1}{32}$	12	2
2	$2\frac{5}{8}$	$\frac{3}{64}$	$2\frac{5}{16}$	$\frac{1}{32}$	12	2
3	$3\frac{25}{32}$	$\frac{3}{64}$	$3\frac{25}{64}$	$\frac{1}{32}$	12	2
4	$5\frac{11}{32}$	$\frac{3}{64}$	$4\frac{27}{64}$	$\frac{1}{32}$	21	2
6	$7\frac{31}{32}$	$\frac{1}{16}$	$6\frac{27}{64}$	$\frac{1}{32}$	21	2

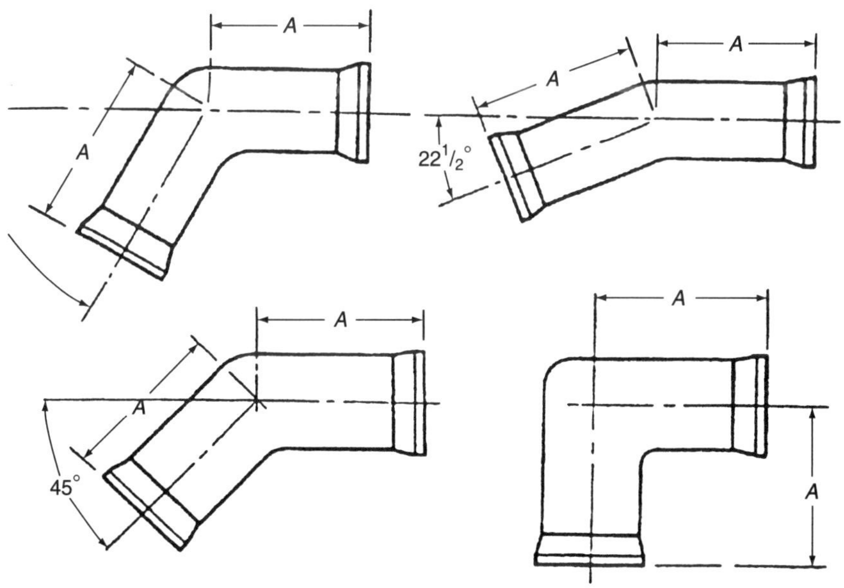

TABLE 23.5 Dimensions and Tolerances of Glass Straight Elbows*

Nominal size, in	*A*, in
1	$2\frac{3}{4}$
$1\frac{1}{2}$	$3\frac{1}{2}$
2	4
3	5
4	7
6	9

* Tolerances for *A* are $\pm\frac{1}{8}$ in. Multiply inch dimension by 25.4 to get millimeters.

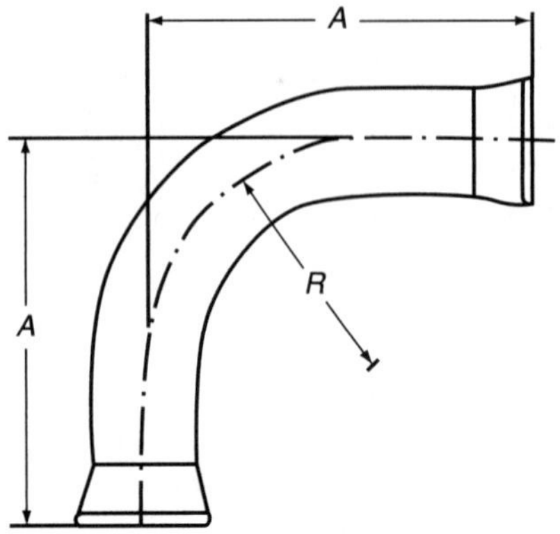

TABLE 23.6 Dimensions and Tolerances for Sweep Elbows

Nominal size, in	A, in (Note 2)	R, in (Note 3)
1	$4\frac{3}{4}$	$2\frac{3}{4}$
$1\frac{1}{2}$	5	$3\frac{1}{2}$
2	6	$3\frac{1}{2}$
3	7	$4\frac{1}{2}$
4	10	8

Notes:
1. Multiply by 25.4 to get dimensions and tolerances in millimeters.
2. Tolerance for A is $\pm\frac{1}{8}$ in.
3. Tolerance for R is $\pm\frac{1}{2}$ in.

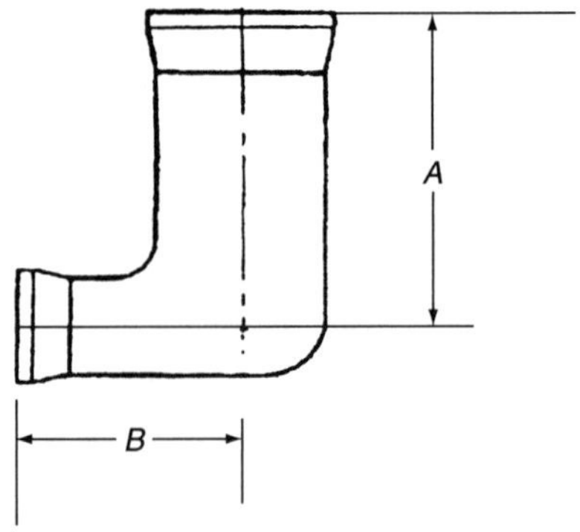

TABLE 23.7 Dimensions and Tolerances of Reducing Glass Elbows*

Nominal size, in	A, in	B, in
$1\frac{1}{2}$ by 1	$3\frac{1}{2}$	3
2 by 1	4	3
2 by $1\frac{1}{2}$	4	$3\frac{1}{2}$
3 by 1	5	$3\frac{1}{2}$
3 by $1\frac{1}{2}$	5	4
3 by 2	5	$4\frac{1}{2}$
4 by 1	7	4
4 by $1\frac{1}{2}$	7	$4\frac{1}{2}$
4 by 2	7	5
4 by 3	7	$5\frac{1}{2}$
6 by 1	9	5
6 by $1\frac{1}{2}$	9	$5\frac{1}{2}$
6 by 2	9	6
6 by 3	9	$6\frac{1}{2}$
6 by 4	9	8

* Multiply by 25.4 to get dimensions and tolerances in millimeters. Tolerance for A and B dimension is $\pm\frac{1}{8}$ in.

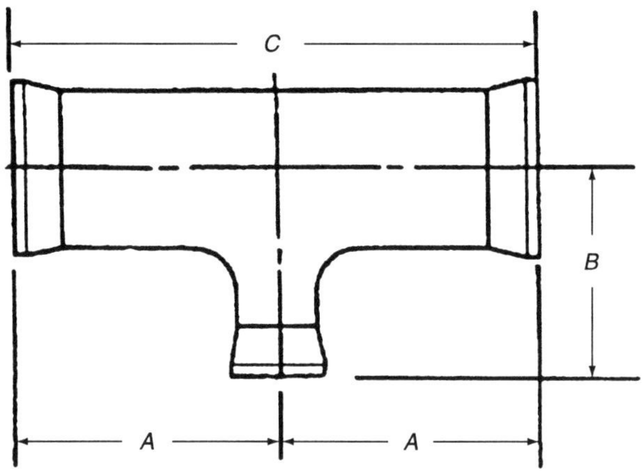

TABLE 23.8 Dimensions and Tolerances for Reducing Glass Tees*

Nominal size, in	*A*, in	*B*, in	*C*, in
1 by 1	2¾	2¾	5½
1½ by 1	3½	3	7
1½ by 1½	3½	3½	7
2 by 1	4	3	8
2 by 1½	4	3½	8
2 by 2	4	4	8
3 by 1	5	3½	10
3 by 1½	5	4	10
3 by 2	5	4½	10
3 by 3	5	5	10
4 by 1	7	4	14
4 by 1½	7	4½	14
4 by 2	7	5	14
4 by 3	7	5½	14
4 by 4	7	7	14
6 by 1	9	5	18
6 by 1½	9	5½	18
6 by 2	9	6	18
6 by 3	9	6½	18
6 by 4	9	8	18
6 by 6	9	9	18

* Multiply by 25.4 to get dimensions and tolerances in millimeters. Tolerance for *A*, *B*, and *C* dimensions is ±⅛ in.

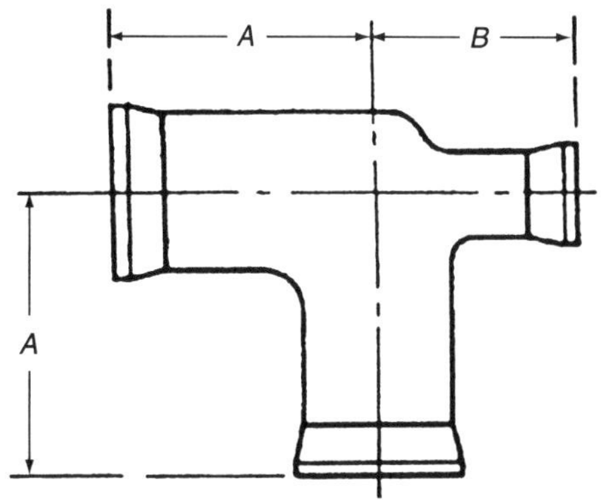

TABLE 23.9 Dimensions and Tolerances of Reducing Glass Tees*

Nominal size, in	*A*, in	*B*, in
1½ by 1 by 1½	3½	3
2 by 1 by 2	4	3
2 by 1½ by 2	4	3½
3 by 1 by 3	5	3½
3 by 1½ by 3	5	4
3 by 2 by 3	5	4½
4 by 1 by 4	7	4
4 by 1½ by 4	7	4½
4 by 2 by 4	7	5
4 by 3 by 4	7	5½
6 by 1 by 6	9	5
6 by 1½ by 6	9	5½
6 by 2 by 6	9	6
6 by 3 by 6	9	6½
6 by 4 by 6	9	8

* Multiply inch dimensions and tolerances by 25.4 to obtain millimeters. Tolerance for *A* and *B* dimensions is ±⅛ in.

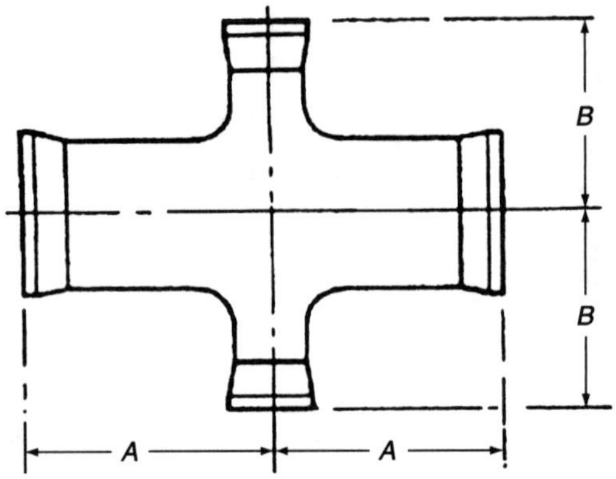

TABLE 23.10 Dimensions and Tolerances for Glass Crosses*

Nominal size, in	A, in	B, in
1 by 1	$2\frac{2}{3}$	$2\frac{3}{4}$
$1\frac{1}{2}$ by 1	$3\frac{1}{2}$	3
$1\frac{1}{2}$ by $1\frac{1}{2}$	$3\frac{1}{2}$	$3\frac{1}{2}$
2 by 1	4	3
2 by $1\frac{1}{2}$	4	$3\frac{1}{2}$
2 by 2	4	4
3 by 1	5	$3\frac{1}{2}$
3 by $\frac{1}{2}$	5	4
3 by 2	5	$4\frac{1}{2}$
3 by 3	5	5
4 by 1	7	4
4 by $1\frac{1}{2}$	7	$4\frac{1}{2}$
4 by 2	7	5
4 by 3	7	$5\frac{1}{2}$
4 by 4	7	7
6 by 1	9	5
6 by $1\frac{1}{2}$	9	$5\frac{1}{2}$
6 by 2	9	6
6 by 3	9	$6\frac{1}{2}$
6 by 4	9	8
6 by 6	9	9

* Tolerance for both A and B dimensions is $\pm\frac{1}{8}$ in. Multiply inch dimensions by 25.4 to obtain millimeters.

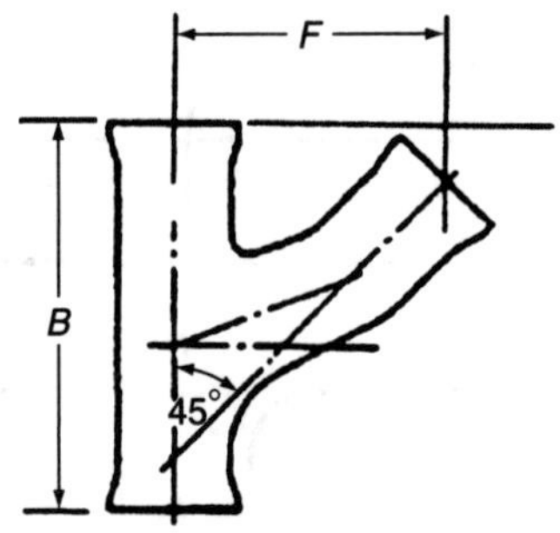

TABLE 23.11 Dimensions and Tolerances of Laterals*

Nominal size, in	B, in	F, in
1	$5\frac{1}{2}$	$3\frac{7}{8}$
$1\frac{1}{2}$	7	$5\frac{1}{16}$
2	8	$5\frac{5}{8}$
3	10	$7\frac{1}{16}$
4	14	$9\frac{3}{16}$
6	18	$11\frac{27}{32}$

* Tolerance for B and F dimensions is $\pm\frac{1}{8}$ in. Multiply by 25.4 to obtain millimeters.

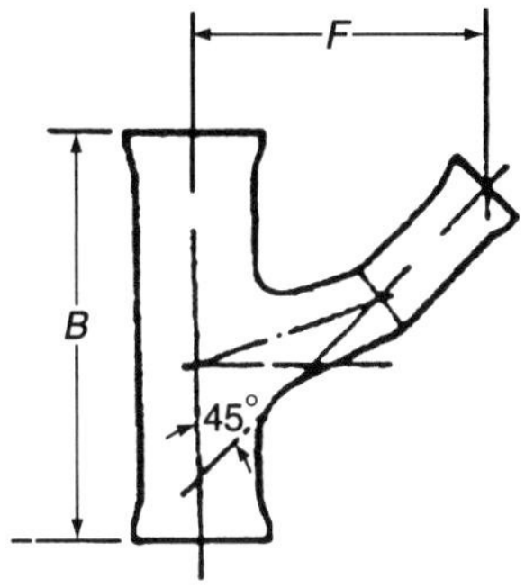

TABLE 23.12 Dimensions and Tolerances for Reducing Laterals*

Nominal size, in	*B*, in	*F*, in
2 by 1	8	$5\frac{5}{8}$
2 by $1\frac{1}{2}$	8	$5\frac{5}{8}$
3 by $1\frac{1}{2}$	10	$7\frac{1}{16}$
3 by 2	10	$7\frac{1}{16}$
4 by $1\frac{1}{2}$	14	$9\frac{3}{16}$
4 by 2	14	$9\frac{3}{16}$
4 by 3	14	$9\frac{3}{16}$
6 by 3	18	$11\frac{27}{32}$

* Tolerance for *B* and *F* dimensions is $\pm\frac{1}{8}$ in. Multiply by 25.4 to get millimeters.

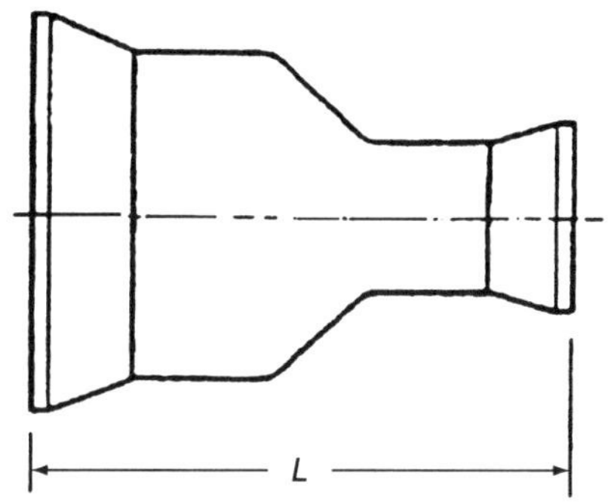

TABLE 23.13 Dimensions and Tolerances for Concentric Reducers*

Nominal size, in	*L*, in
$1\frac{1}{2}$ by 1	4
2 by 1	4
2 by $1\frac{1}{2}$	4
3 by 1	5
3 by $1\frac{1}{2}$	5
3 by 2	5
4 by 1	7
4 by $1\frac{1}{2}$	7
4 by 2	7
4 by 3	7
6 by 1	9
6 by $1\frac{1}{2}$	9
6 by 2	9
6 by 3	9
6 by 4	9

* Tolerance for *L* dimension is $\pm\frac{1}{8}$ in. Multiply by 25.4 by get millimeters.

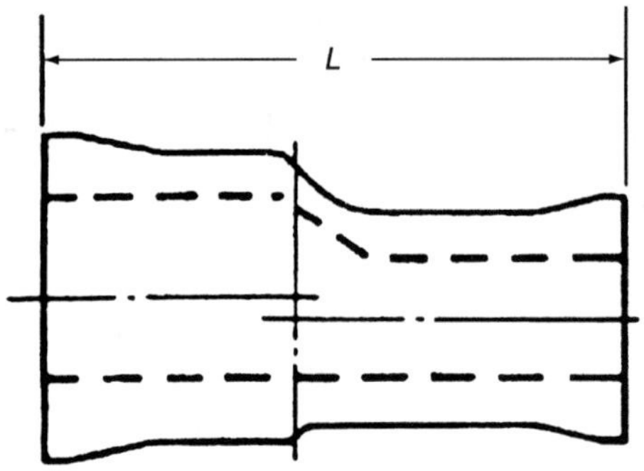

TABLE 23.14 Dimensions and Tolerances for Eccentric Reducers*

Nominal size, in	L, in
$1\frac{1}{2}$ by 1	4
2 by 1	4
2 by $1\frac{1}{2}$	4
3 by 1	5
3 by $1\frac{1}{2}$	5
3 by 2	5
4 by 1	7
4 by $1\frac{1}{2}$	7
4 by 2	7
4 by 3	7
6 by $1\frac{1}{2}$	9
6 by 2	9
6 by 3	9
6 by 4	9

* Tolerance for L dimension is $\pm\frac{1}{8}$ in. Multiply by 25.4 to get millimeters.

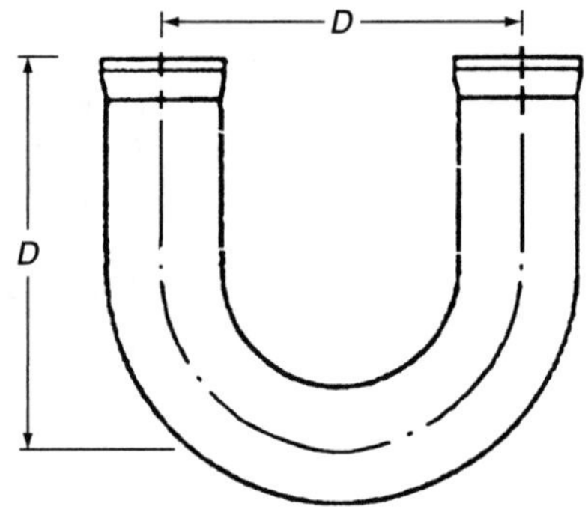

TABLE 23.15 Dimensions and Tolerances for U Bends*

Nominal size, in	D, in
1	$5\frac{1}{2}$
$1\frac{1}{2}$	7
2	7
3	9
4	12

* Tolerance for D dimension is $\pm\frac{1}{8}$ in. Multiply by 25.4 to obtain millimeters.

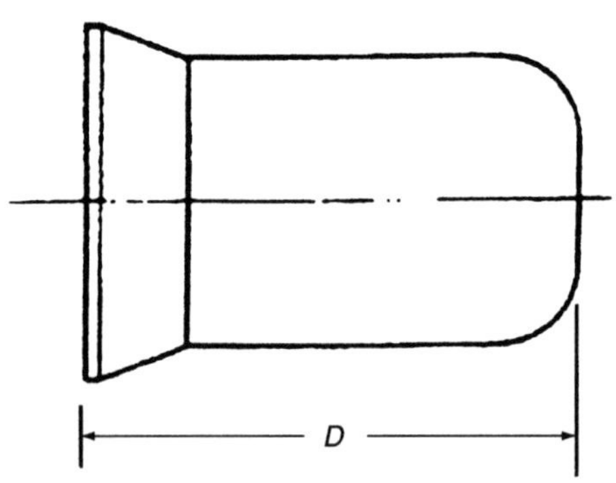

TABLE 23.16 Dimensions and Tolerances for Caps*

Nominal size, in	D, max., in
1	$2\frac{13}{16}$
$1\frac{1}{2}$	$3\frac{9}{16}$
2	4
3	$4\frac{13}{16}$
4	$6\frac{1}{16}$
6	$7\frac{1}{8}$

* Multiply inch dimension by 25.4 to obtain millimeters.

BOROSILICATE GLASS PIPE AND FITTINGS FOR DRAIN, WASTE, AND VENT (DWV) APPLICATIONS

Pipe and fittings conforming to ASTM C1053 specification are suitable for corrosion-resistant gravity flow and vacuum service applications.

Temperature Capabilities and Limitations of DWV Glass Pipe

See Table 23.17.

TABLE 23.17 Temperature Rating and Limitations of ASTM C1053 DWV Glass Pipe

Nominal pipe size, in	Minimum rated operating temperature,* °F (°C)	Maximum rated operating temperature, °F (°C)	Maximum temperature differential for thermal endurance, °F (°C)
1 to 3	–50 (–46)	212 (100)	200 (93)
4	–50 (–46)	212 (100)	175 (78)
6	–50 (–46)	212 (100)	160 (71)

* Maximum temperature differential refers to an instantaneous temperature change. It applies to both the internal and external pipe surfaces.

Dimensions of DWV Glass Pipe

See Table 23.18.

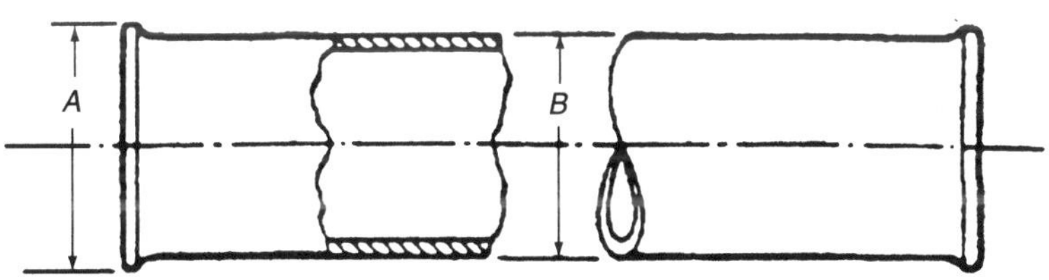

For aboveground applications

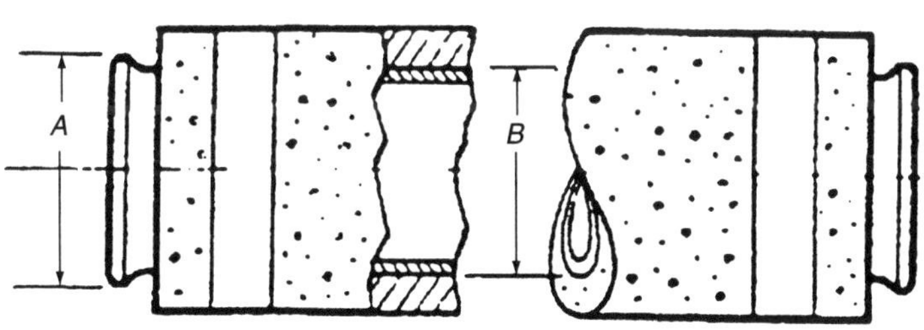

For underground applications covered in expanded polystyrene (EPS)

TABLE 23.18 Dimensions of DWV Glass Pipe for Above- and Underground Applications

Size, in	Flange outside diameter, A, in	Pipe outside diameter, B, in	Pipe wall thickness, in	Nominal OD over EPS,* in	Pipe length, ft
$1^1/_2$	2.06	1.84	0.17	$2^{13}/_{16}$	5 and 10
2	2.58	2.34	0.17	$3^5/_{16}$	5 and 10
3	3.69	3.41	0.20	5	5 and 10
4	4.84	4.53	0.27	$6^1/_4$	5 and 10
6	7.12	6.56	0.33	$8^9/_{16}$	5 and 10

* Underground pipe is covered in expanded polystyrene (EPS). For underground installations, the pipe is available in 5-ft lengths only.

Dimensions of DWV Glass Pipe Fittings

See Tables 23.19 through 23.37. For additional information on fittings, please contact the fitting manufacturer and/or the supplier.

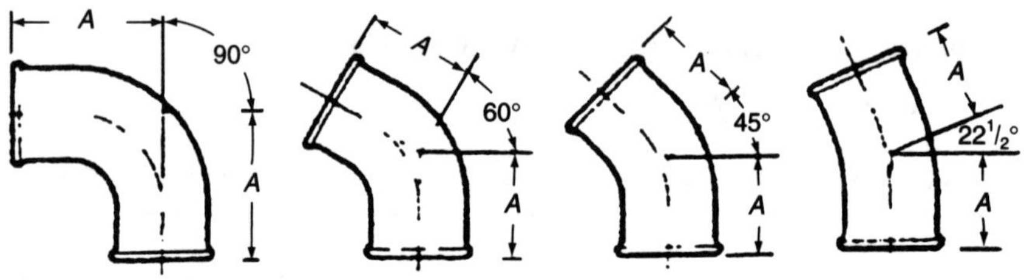

¼ bend (90°) ⅙ bend (60°) ⅛ bend (45°) 1/16 bend (22½°)

TABLE 23.19 Dimensions of DWV Glass Pipe Elbows

	A dimension, in			
Size, in	90°	60°	45°	22½°
1½	3	2½	2	
2	3¼	2¾	2¼	2¼
3	5	3½	2¾	2¾
4	7	4½	3¼	3¼
6			7	

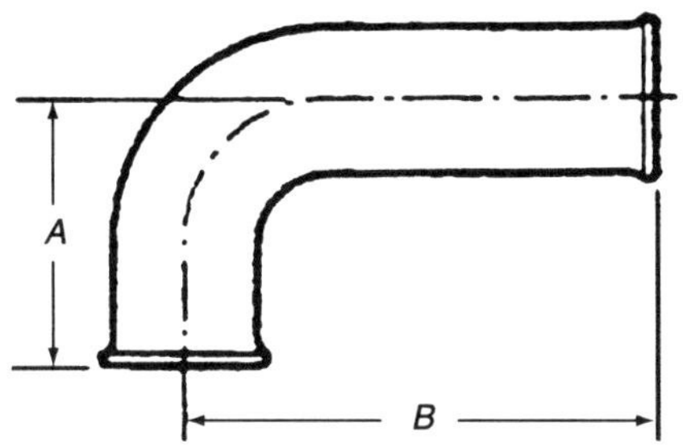

TABLE 23.20 Dimensions of 90° Long Elbows

Size, in	A, in	B, in
1½	3⅜	6
2	3⅝	7½

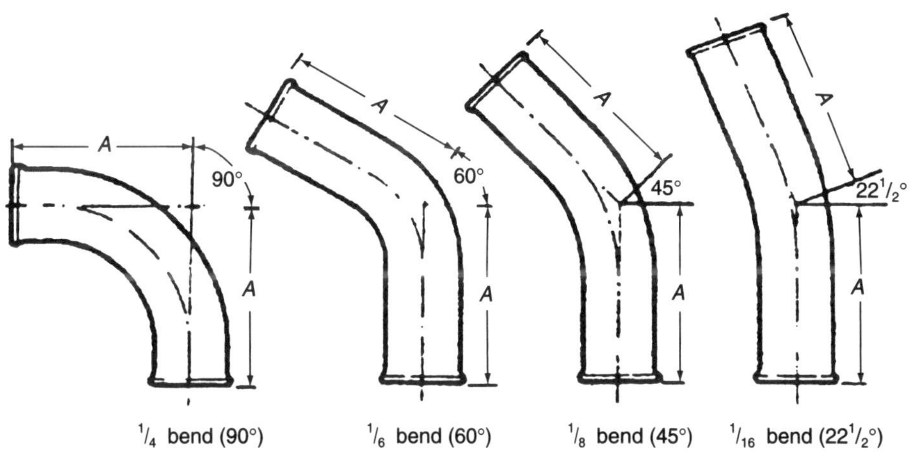

TABLE 23.21 Dimensions of DWV Glass Sweep Elbows

	A dimension, in			
Size, in	90°	60°	45°	22½°
1½	4½	4½	4½	4½
2	5	5	5	5
3	6½	6½	6½	6½
4	9	9	9	9
6	12	12	12	12

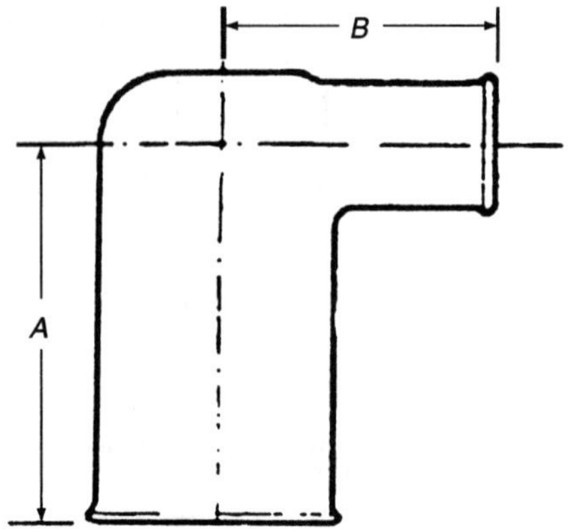

TABLE 23.22 Dimensions of DWV Glass Quarter-Bend Reducers

Size, in	*A*, in	*B*, in
2 by 1½	4	3½
3 by 1½	5	4
3 by 2	5	4½
4 by 1½	7	4½
4 by 2	7	5
4 by 3	7	5½
6 by 1½	9	5½
6 by 2	9	6
6 by 3	9	6½
6 by 4	9	8

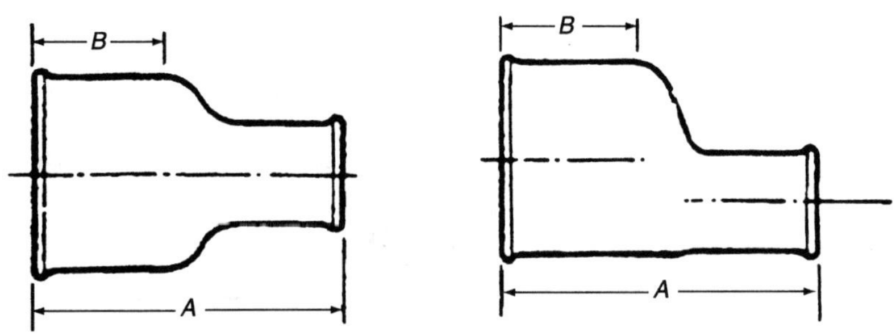

TABLE 23.23 Dimensions of DWV Glass Concentric and Eccentric Reducers

	Dimension, in	
Size, in	*A*	*B*, min.
2 by 1½	4	1¾
3 by 1½	5	2¼
3 by 2	5	2¼
4 by 1½	7	3
4 by 2	7	3
4 by 3	7	3
6 by 1½	9	4
6 by 2	9	4
6 by 3	9	4
6 by 4	9	4

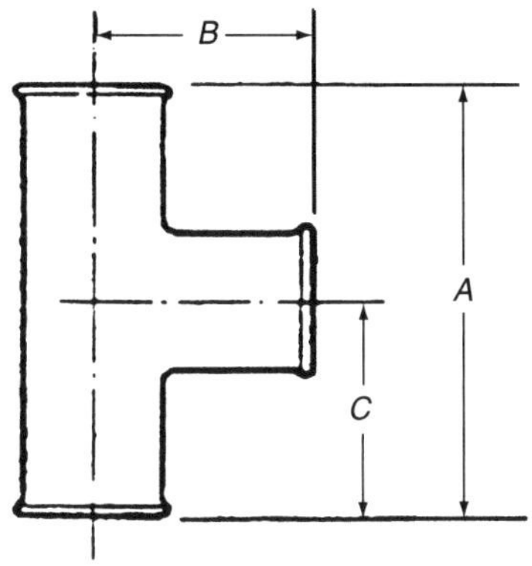

TABLE 23.24 Dimensions of DWV Glass Straight Tees

Size, in	Dimension, in		
	A	*B*	*C*
$1\frac{1}{2}$ by $1\frac{1}{2}$	6	3	3
2 by $1\frac{1}{2}$	8	$3\frac{1}{4}$	4
2 by 2	8	4	4
3 by $1\frac{1}{2}$	12	$4\frac{1}{2}$	6
3 by 2	12	$4\frac{1}{2}$	6
3 by 3	12	6	6
4 by $1\frac{1}{2}$	14	5	7
4 by 2	14	5	7
4 by 3	14	$6\frac{1}{2}$	7
4 by 4	14	8	7
6 by 3	20	$7\frac{3}{4}$	10
6 by 4	20	9	10
6 by 6	20	10	10

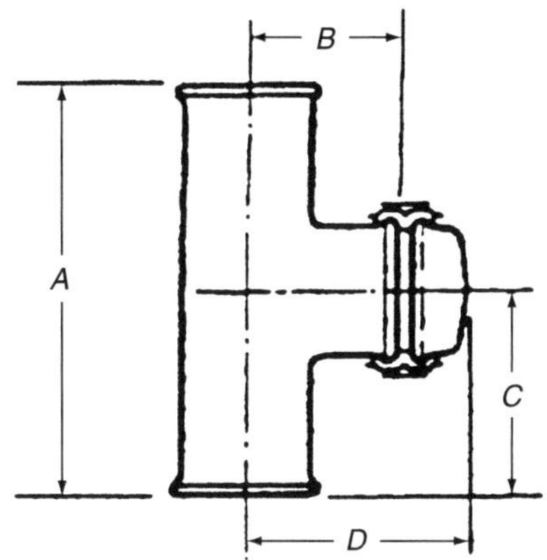

TABLE 23.25 Dimensions of Test Tee with Cleanout*

Size	*A*	*B*	*C*	*D*
$1\frac{1}{2} \times 1\frac{1}{2}$	6	$2\frac{3}{16}$	3	$3\frac{1}{16}$
2×2	8	$2\frac{11}{16}$	4	$3\frac{11}{16}$
3×3	12	$3\frac{3}{16}$	6	$4\frac{11}{16}$
4×4	14	$4\frac{5}{8}$	7	$5\frac{7}{8}$

* Test tee and cleanout come as complete assembly including cap and coupling. Dimensions in inches.

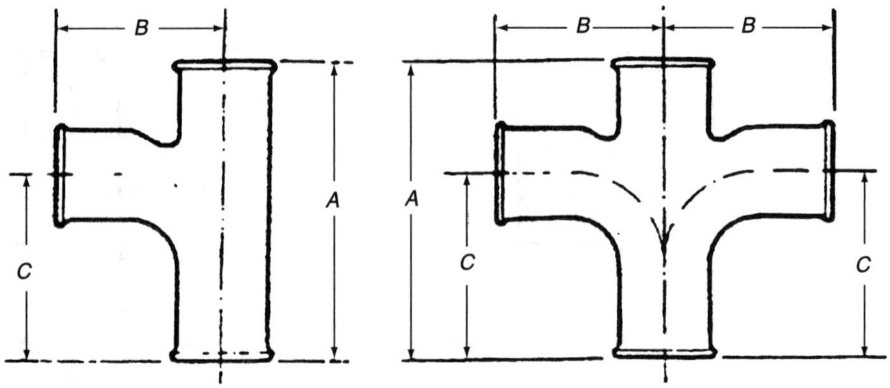

TABLE 23.26 Dimensions of Sanitary Tee for DWV Glass Pipe*

Size	*A*	*B*	*C*
$1\frac{1}{2} \times 1\frac{1}{2}$	6	$3\frac{1}{2}$	$3\frac{3}{4}$
$2 \times 1\frac{1}{2}$	8	$3\frac{3}{4}$	5
2×2	8	$4\frac{1}{2}$	5
$3 \times 1\frac{1}{2}$	12	$4\frac{1}{4}$	$7\frac{7}{16}$
3×2	12	5	$7\frac{7}{16}$
3×3	12	$6\frac{3}{8}$	$7\frac{7}{16}$
$4 \times 1\frac{1}{2}$	14	$4\frac{7}{4}$	$8\frac{3}{4}$
4×2	14	$5\frac{5}{4}$	$8\frac{3}{4}$
4×3	14	7	$8\frac{3}{4}$
4×4	14	$8\frac{1}{4}$	$8\frac{3}{4}$
6×2	20	$6\frac{3}{4}$	$12\frac{3}{8}$
6×3	20	8	$12\frac{3}{8}$
6×4	20	$9\frac{5}{16}$	$12\frac{3}{8}$
6×6	20	12	$12\frac{3}{8}$

* Dimensions in inches.

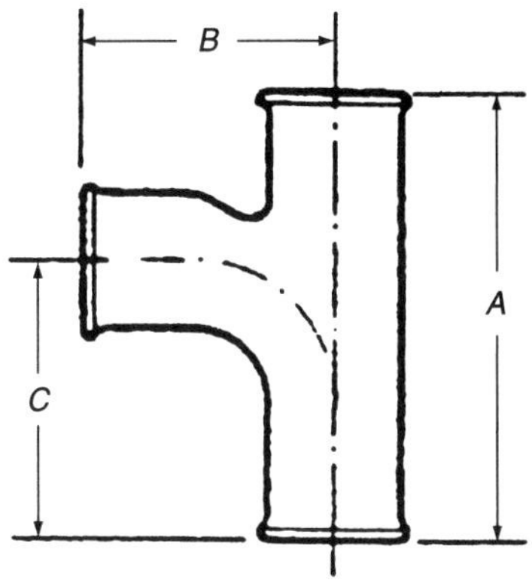

TABLE 23.27 Dimensions of Compact Single Sanitary Tee for Glass Pipe*

Size	*A*	*B*	*C*
$1\frac{1}{2} \times 1\frac{1}{2}$	$4\frac{1}{2}$	$2\frac{1}{3}$	$2\frac{1}{4}$
$2 \times 1\frac{1}{2}$	$4\frac{3}{4}$	$2\frac{3}{8}$	$2\frac{3}{8}$
2×2	$5\frac{3}{8}$	$2\frac{3}{4}$	$2\frac{3}{4}$
3×2	9	$4\frac{1}{16}$	$5\frac{1}{16}$
3×3	9	$4\frac{1}{16}$	$5\frac{1}{16}$
$4 \times 1\frac{1}{2}$	10	$4\frac{5}{8}$	$6\frac{3}{8}$
4×2	10	$4\frac{7}{8}$	$6\frac{3}{8}$
4×3	10	$5\frac{3}{8}$	$6\frac{1}{8}$
4×4	10	$6\frac{1}{4}$	$6\frac{1}{8}$
6×2	16	$5\frac{7}{8}$	$9\frac{1}{2}$
6×3	16	$6\frac{3}{4}$	$9\frac{1}{2}$
6×4	16	$7\frac{7}{8}$	$9\frac{1}{2}$
6×6	16	$9\frac{7}{8}$	$9\frac{1}{4}$

* Dimensions in inches.

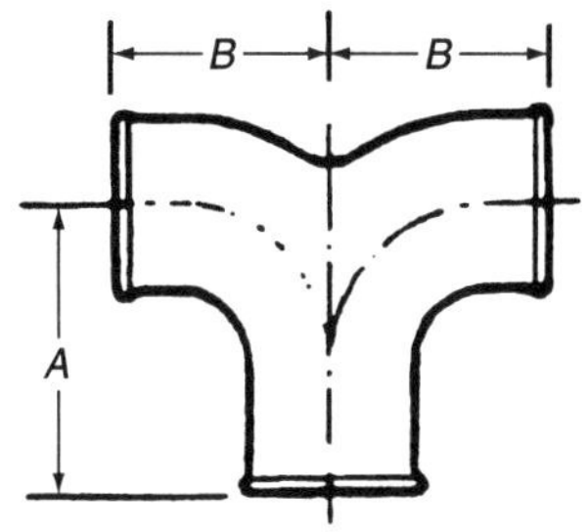

TABLE 23.28 Dimensions of Double Quarter Bend for DWV Glass Pipe*

Size	*A*	*B*
$1\frac{1}{2}$	3	$2\frac{1}{8}$
$2 \times 2 \times 1\frac{1}{2}$	$3\frac{1}{4}$	$2\frac{1}{4}$
2	$3\frac{1}{2}$	$2\frac{1}{4}$
3	5	$4\frac{1}{2}$
4	$6\frac{1}{2}$	$5\frac{3}{4}$
6	$8\frac{1}{2}$	7

* Dimensions in inches.

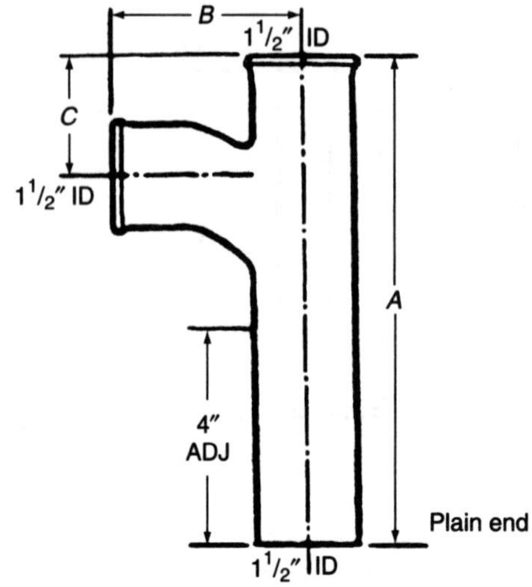

Adustable sanitary tee

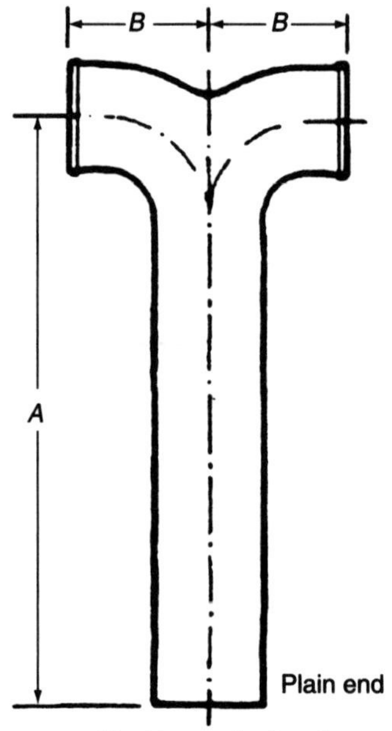

Double quarter bend

TABLE 23.29 Dimensions of Adjustable Sanitary Tee (Plain-End Outlet) and Double Quarter Bend (Plain-End Outlet)*

Size	*A*	*B*	*C*
Adustable sanitary tee:			
$1\frac{1}{2} \times 1\frac{1}{2} \times 1\frac{1}{2}$	$8\frac{3}{4}$	$3\frac{1}{2}$	$3\frac{1}{2}$
Double quarter bend:			
$1\frac{1}{2}$	10	$2\frac{1}{8}$	
$2 \times 2 \times 1\frac{1}{2}$	10	$2\frac{1}{4}$	
2	10	$2\frac{1}{4}$	
3			
4			
6			

* Plain end can be field cut to adjust for fit-up.

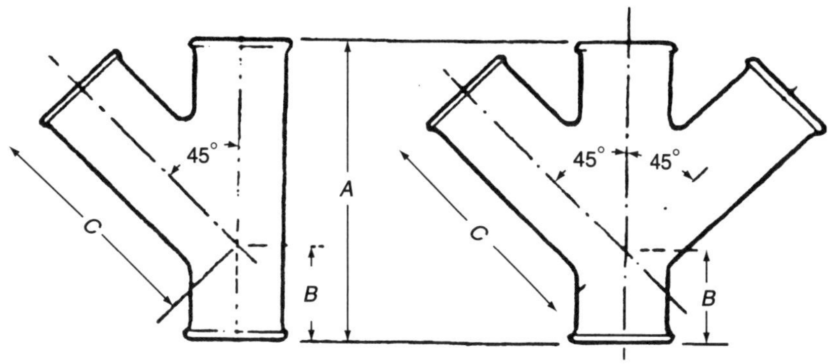

TABLE 23.30 Dimensions of Drainline Wye for DWV Glass Pipe*

Size	*A*	*B*	*C*
$1\frac{1}{2} \times 1\frac{1}{2}$	6	$1\frac{7}{8}$	$4\frac{1}{2}$
$2 \times 1\frac{1}{2}$	8	$2\frac{1}{2}$	$4\frac{3}{4}$
2×2	8	$2\frac{1}{2}$	6
$3 \times 1\frac{1}{2}$	12	$3\frac{3}{4}$	$5\frac{1}{2}$
3×2	12	$3\frac{3}{4}$	$6\frac{3}{4}$
3×3	12	$3\frac{3}{4}$	8
$4 \times 1\frac{1}{2}$	14	$4\frac{1}{2}$	$6\frac{3}{8}$
4×2	14	$4\frac{1}{2}$	$7\frac{1}{2}$
4×3	14	$4\frac{1}{2}$	$8\frac{3}{4}$
4×4	14	$4\frac{1}{2}$	10
6×2	20	$5\frac{3}{4}$	9
6×3	20	$5\frac{3}{4}$	$10\frac{3}{8}$
6×4	20	$5\frac{3}{4}$	$11\frac{1}{2}$
6×6	20	$5\frac{3}{4}$	14

* Dimensions in inches.

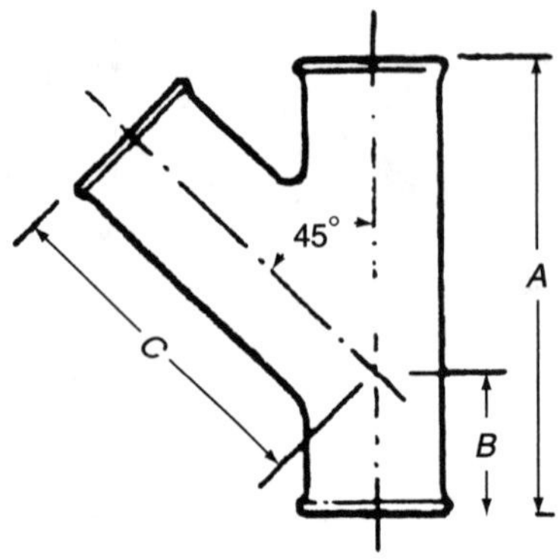

TABLE 23.31 Dimensions of Compact Single Drainline Y for DWV Glass Pipe*

Size	*A*	*B*	*C*
$1\frac{1}{2} \times 1\frac{1}{2}$	$5\frac{3}{8}$	$1\frac{1}{8}$	4
2×2	$5\frac{7}{8}$	$1\frac{1}{4}$	$4\frac{5}{8}$
3×2	9	$2\frac{3}{16}$	$6\frac{5}{16}$
3×3	9	$2\frac{3}{16}$	$7\frac{1}{16}$
$4 \times 1\frac{1}{2}$	10	$3\frac{1}{8}$	$6\frac{1}{4}$
4×2	10	$2\frac{7}{8}$	$7\frac{3}{4}$
4×3	10	2	$8\frac{1}{8}$
4×4	10	2	$8\frac{3}{8}$
6×2	16	$4\frac{7}{8}$	$9\frac{7}{8}$
6×3	16	$5\frac{1}{8}$	$10\frac{3}{8}$
6×4	16	$4\frac{1}{2}$	$11\frac{5}{8}$
6×6	16	$4\frac{5}{8}$	$12\frac{5}{8}$

* Dimensions in inches.

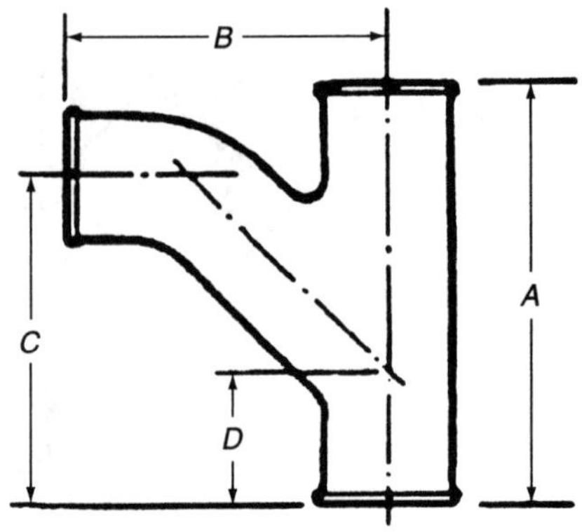

TABLE 23.32 Dimensions of Compact Combination Y and One-Eighth Bend*

Size	*A*	*B*	*C*	*D*
$2 \times 1\frac{1}{2}$	$5\frac{1}{8}$	$3\frac{3}{4}$	$3\frac{3}{4}$	$1\frac{5}{16}$
2×2	6	$4\frac{3}{8}$	$4\frac{5}{8}$	$1\frac{1}{2}$
3×2	9	$6\frac{1}{16}$	$6\frac{3}{16}$	$2\frac{3}{4}$
3×3	9	$6\frac{1}{16}$	$6\frac{11}{16}$	$2\frac{1}{4}$
$4 \times 1\frac{1}{2}$	10	$5\frac{3}{8}$	$6\frac{3}{8}$	$2\frac{7}{8}$
4×2	10	$5\frac{3}{8}$	$6\frac{3}{8}$	$2\frac{7}{8}$
4×3	10	$6\frac{5}{8}$	$7\frac{5}{8}$	$3\frac{1}{8}$
4×4	10	$6\frac{7}{8}$	$7\frac{3}{8}$	$2\frac{1}{16}$
6×2	16	$7\frac{5}{8}$	$9\frac{5}{8}$	$4\frac{5}{8}$
6×3	16	$8\frac{1}{8}$	$9\frac{5}{8}$	$4\frac{3}{4}$
6×4	16	$9\frac{1}{8}$	$10\frac{3}{8}$	$4\frac{5}{8}$
6×6	16	$11\frac{1}{4}$	$10\frac{1}{8}$	$3\frac{1}{2}$

* Dimensions in inches.

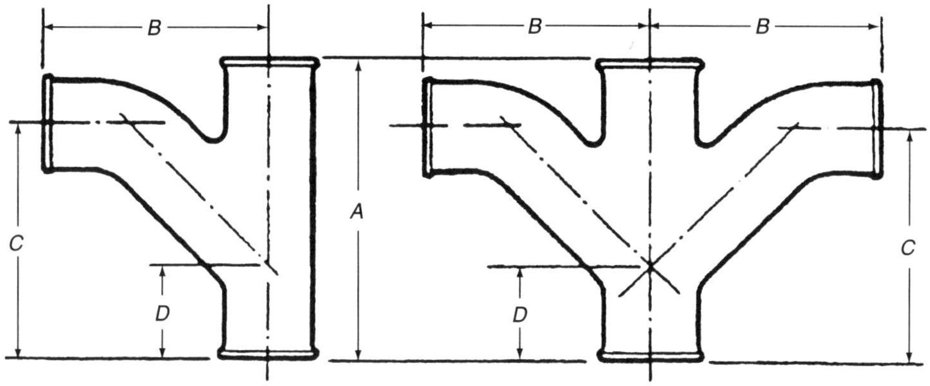

TABLE 23.33 Dimensions of Combination Y and One-Eighth Bend for DWV Glass Pipe*

Size	*A*	*B*	*C*	*D*
1½ × 1½	6	4½	4⅝	1⅞
2 × 1½	8	4¾	5½	2½
2 × 2	8	6	6¼	2½
3 × 1½	12	5⅜	7¼	3¾
3 × 2	12	6½	8	3¾
3 × 3	12	8½	9	3¾
4 × 1½	14	6	8½	4½
4 × 2	14	7	9¼	4½
4 × 3	14	9	10¼	4½
4 × 4	14	11	11	4½
6 × 2	20	8¼	11⅝	5¾
6 × 3	20	10	12½	5¾
6 × 4	20	12	13½	5¾
6 × 6	20	15	14½	5¾

* Dimensions in inches.

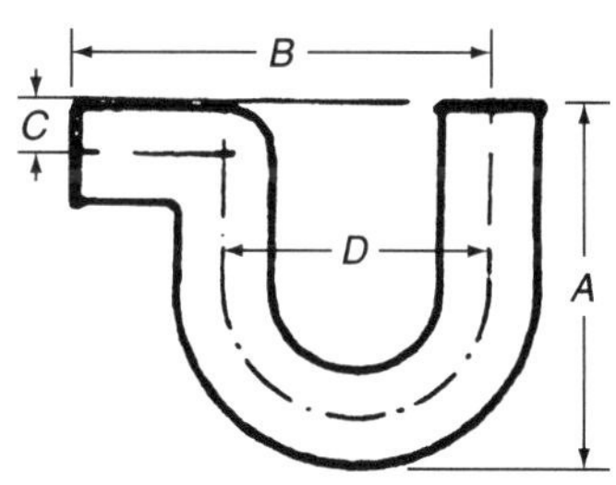

TABLE 23.34 Dimensions of P-Trap Style for DWV Glass Pipe*

Size, inlet by outlet	*A*	*B*	*C*	*D*
1½ × 1½	7	8	1	5
2 × 1½	7½	8	1½	5
2 × 2	8³⁄₁₆	8¾	1½	5½
3 × 3	10¼	10½	2	6½
4 × 4	12¼	12½	2½	7½
6 × 6	18½	31	3⅜	24

* Dimensions in inches.

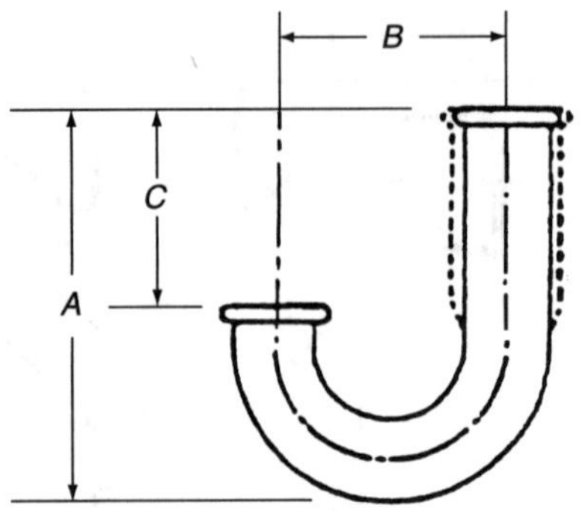

TABLE 23.35 Dimensions of J Bend for DWV Glass Pipe*

Size	*A*	*B*	*C*
$1\frac{1}{2} \times 1\frac{1}{2}$	$7\frac{13}{16}$	5	$3\frac{1}{2}$
$2 \times 1\frac{1}{2}$	$8\frac{11}{16}$	5	$4\frac{1}{2}$
2×2	$8\frac{7}{8}$	$5\frac{1}{2}$	$4\frac{1}{8}$

* Dimensions in inches. J bend can be used in combination with quarter bend and return bend to make a variety of custom swivel traps.

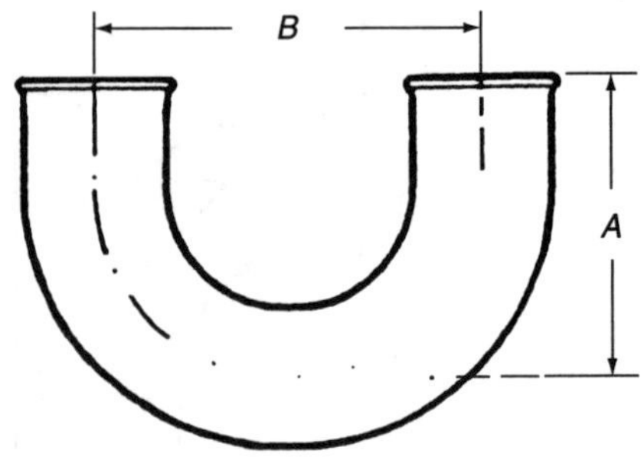

TABLE 23.36 Dimensions of U Bend for DWV Glass Pipe*

Size	*A*	*B*
$1\frac{1}{2}$	4	5
2	$4\frac{1}{2}$	$5\frac{1}{2}$
3	$5\frac{1}{2}$	$6\frac{1}{2}$
4	$6\frac{1}{2}$	$7\frac{1}{2}$

* Dimensions in inches.

TABLE 23.37 Dimensions of Cleanout Plug for DWV Glass Pipe*

Size	*A*
$1\frac{1}{2}$	1
2	1
3	$1\frac{1}{8}$
4	$1\frac{1}{4}$
6	$1\frac{1}{2}$

* Dimensions in inches.

DWV Glass Piping Joints

Refer to Chap. 15.

Tables in Chapter 23

CHAPTER 24

VITRIFIED CLAY PIPE, FITTINGS, AND JOINTS

See pp. 692–693 for a list of tables in Chap. 24.

VITRIFIED CLAY PIPE AND FITTINGS

Vitrified clay piping is used for carrying sewage, industrial waste, and storm water. It is available in standard-strength and extra-strength designs. It is noted that some manufacturers produce only extra-strength vitrified clay pipe because of economic considerations related to manufacturing and inventory. Perforated vitrified clay pipe is used for underdrainage, filter fields, leaching fields, and other such subdrainage installations.

Vitrified clay pipe and fittings are manufactured in accordance with ASTM C700, Standard Specification for Vitrified Clay Pipe, Extra Strength, Standard Strength, and Perforated.

Reference Standards and Specifications

See Table 24.1.

TABLE 24.1 Standards and Specification for Vitrified Clay Pipe

Standard/specification number	Title
ASTM C301	Standard Test Methods for Vitrified Clay Pipe
ASTM C425	Standard Specification for Compression Joints for Vitrified Clay Pipe and Fittings
ASTM C700	Standard Specification for Vitrified Clay Pipe, Extra Strength, Standard Strength, and Perforated
ASTM C1091	Standard Test Method for Hydrostatic Infiltration and Exfiltration Testing of Vitrified Clay Pipelines
ASTM C12	Practice of Installing Vitrified Clay Pipe Lines
ASTM C896	Terminology Relating to Clay Products
ASTM C828	Test Method for Low-Pressure Air Test of Vitrified Clay Pipe Lines

Minimum Bearing Strength of Extra-Strength and Standard-Strength Pipe

Table 24.2 lists the minimum bearing strength required for extra-strength and standard-strength vitrified clay pipe. Table 24.3 lists the minimum bearing strength required for perforated vitrified clay pipe complying with ASTM C700. Table 24.4 lists permissible variation of inside diameter.

Wall Thickness of Pipe and Fittings

The wall thickness of vitrified clay pipe and fittings may vary from one manufacturer to another depending upon the type and properties of the clay used. ASTM C700 does not specify wall thickness requirements for pipe or fittings; however, both the pipe and fittings must meet technical and other requirements of ASTM C700.

Standard Lengths and Tolerances for ASTM C700 Vitrified Clay Pipe

See Table 24.5.

TABLE 24.2 Bearing Strength Requirements of ASTM C700 for Vitrified Clay Pipe

Nominal size, in	Nominal size, mm	Minimum bearing strength, lbf/linear ft	
		Extra-strength pipe	Standard-strength pipe
3	75	2000	
4	100	2000	1200
6	150	2000	1200
8	205	2200	1400
10	255	2400	1600
12	305	2600	1800
15	380	2900	2000
18	455	3300	2200
21	535	3850	2400
24	610	4400	2600
27	685	4700	2800
30	760	5000	3300
33	890	5500	3600
36	915	6000	4000
39	990	6600	
42	1065	7000	

TABLE 24.3 Bearing Strength Requirements of ASTM C700 for Perforated Vitrified Clay Pipe

Nominal size, in	Nominal size, mm	Minimum bearing strength, lbf/linear ft	
		Perforated extra-strength pipe	Perforated standard-strength pipe
3	75		
4	100	1250	1000
6	150	1600	1000
8	205	1600	1000
10	255	1600	1100
12	305	1800	1200
15	380	2200	1400
18	455	2640	1700
21	535	3100	2000
24	610	3520	2400

TABLE 24.4 Permissible Variation for Average Inside Diameter of Vitrified Clay Pipe per ASTM C700

Nominal size, in	Nominal size, mm	Maximum minus variation from nominal size* in	mm
3	75	3/16	5
4	100	3/16	5
6	150	1/4	6
8	205	5/16	8
10	255	3/8	10
12	305	7/16	11
15	380	9/16	14
18	455	11/16	17
21	535	13/16	21
24	610	15/16	24
27	685	1 1/16	30
30	760	1 3/16	30
33	890	1 5/16	33
36	915	1 7/16	37
39	990	1 7/16	37
42	1065	1 7/16	37

* The inside diameter must not vary from a true circle by more than 3 percent of its nominal diameter. The average inside diameter shall be determined by averaging two readings taken at any two 90° opposing measurements.

TABLE 24.5 Standard Lengths and Tolerances for ASTM C700 Vitrified Clay Pipe

Nominal pipe size (NPS), in	Nominal size (DN), mm	Standard length: Length,* ft	Standard length: Tolerance,† minus, in/ft
3	75	4	1/4
4	100	4	1/4
6	150	5	1/4
8	205	5	1/4
10	255	5	1/4
12	305	6	1/4
15	380	7	1/4
18	455	7	1/4
21	535	7	3/8
24	610	7	3/8
27	685		3/8
30	760		3/8
33	840		3/8
36	915		3/8
39	990		3/8
42	1065		3/8

* Contact the manufacturer for availability of pipe in lengths other than those listed in this table.
† Manufacturers may comply with tighter tolerances.

Dimensions of Pipe and Fittings

Tables 24.6 through 24.29 provide dimensions for extra-strength vitrified clay pipe and fittings conforming to ASTM C700. These dimensions may differ from one manufacturer to another. Variations in dimensions may be due to type of clay used, capabilities of manufacturing equipment, tolerance controls, etc. Therefore, the user should contact the manufacturer for exact dimensions of the pipe and fittings to be used. It is important that the pipe and fittings comply with the requirements of ASTM C700.

TABLE 24.6 Dimensions of Extra-Strength Vitrified Clay Pipe Conforming to ASTM C700

Pipe size, in	Average bell OD, in	Average spigot OD, in
4	7.20	4.95
6	10.70	7.50
8	13.18	9.89
10	15.51	12.10
12	18.30	14.60
15	22.63	18.20
18	27.29	21.93
21	31.59	25.91
24	36.63	29.25

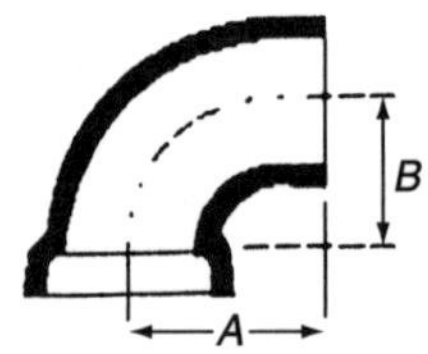

TABLE 24.7 Dimensions of 90° Short-Radius Elbow for ASTM C700 Extra-Strength Vitrified Clay Pipe

Pipe size, in	Dimensions (approx.), in	
	A	*B*
4	$6\frac{3}{4}$	7
6	$8\frac{1}{2}$	10

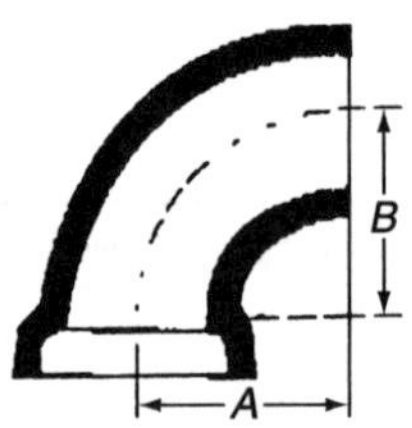

TABLE 24.8 Dimensions of 90° Medium-Radius Elbow for ASTM C700 Extra-Strength Vitrified Clay Pipe

Pipe size, in	Dimensions (approx.), in	
	A	*B*
4	$12\frac{1}{2}$	$12\frac{1}{4}$
6	$13\frac{1}{4}$	13

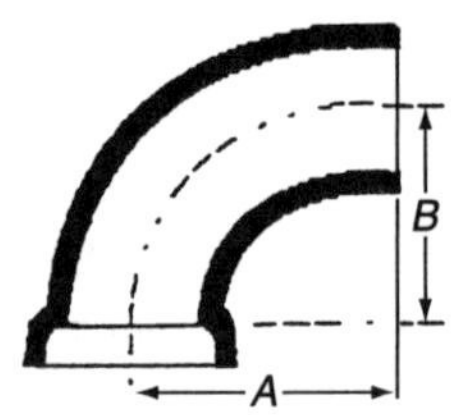

TABLE 24.9 Dimensions of 90° Long-Radius Elbow for ASTM C700 Extra-Strength Vitrified Clay Pipe

Pipe size, in	Dimensions (approx.), in	
	A	*B*
4	$16\frac{1}{2}$	$14\frac{3}{4}$
6	15	13

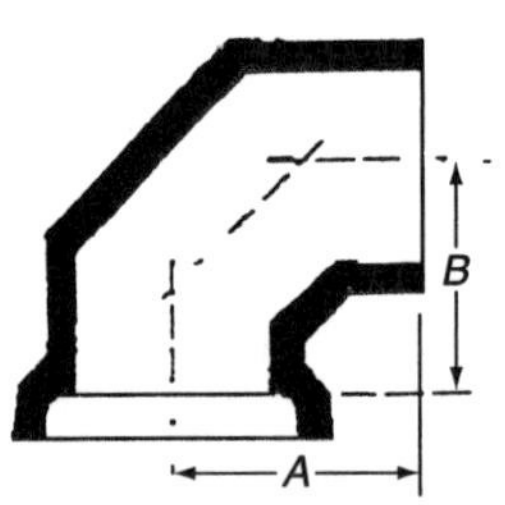

TABLE 24.10 Dimensions of 90° Cut Elbow for ASTM C700 Extra-Strength Vitrified Clay Pipe

Pipe size, in	Dimensions (approx.), in	
	A	*B*
8	9	$10\frac{1}{2}$
10	13	12
12	$14\frac{1}{4}$	$13\frac{3}{4}$
15	$16\frac{1}{4}$	$15\frac{1}{2}$
18	17	$17\frac{1}{4}$

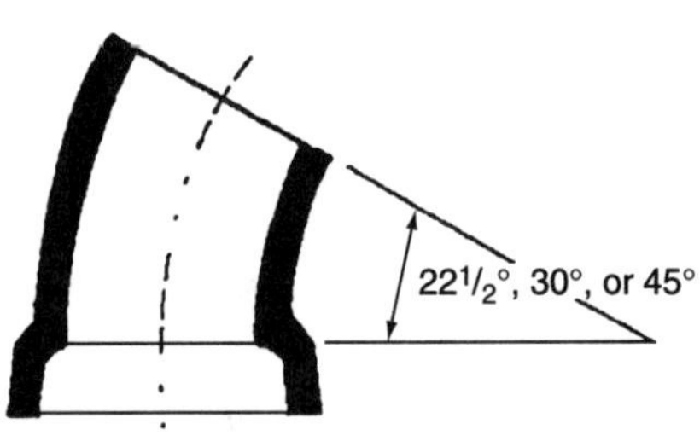

TABLE 24.11 Dimensions of $22\frac{1}{2}$°, 30°, and 45° Short-Radius Curves for ASTM C700 Extra-Strength Vitrified Clay Pipe

Pipe size, in	Centerline length, in
4	12
6	12
8	17

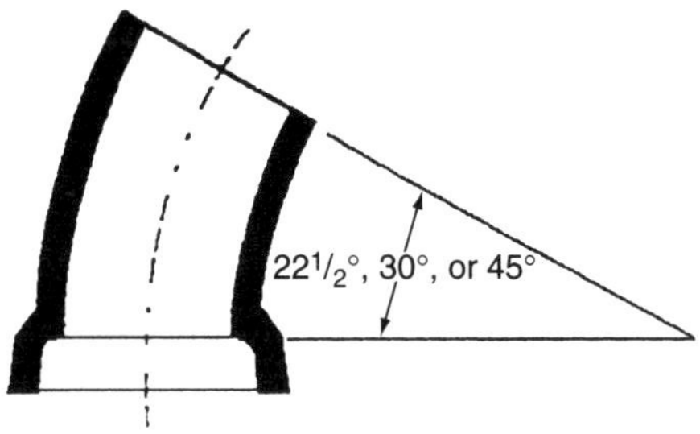

TABLE 24.12 Dimensions of 22½°, 30°, and 45° Medium-Radius Curves for ASTM C700 Extra-Strength Vitrified Clay Pipe

Pipe size, in	Centerline length, in
4	18
6	18

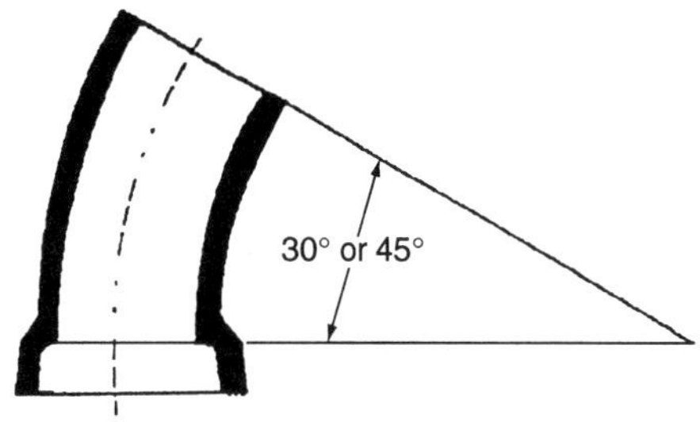

TABLE 24.13 Dimensions of 30° and 45° Long-Radius Curves for ASTM C700 Extra-Strength Vitrified Clay Pipe

Pipe size, in	Centerline length, in
4	24
6	24
8	24

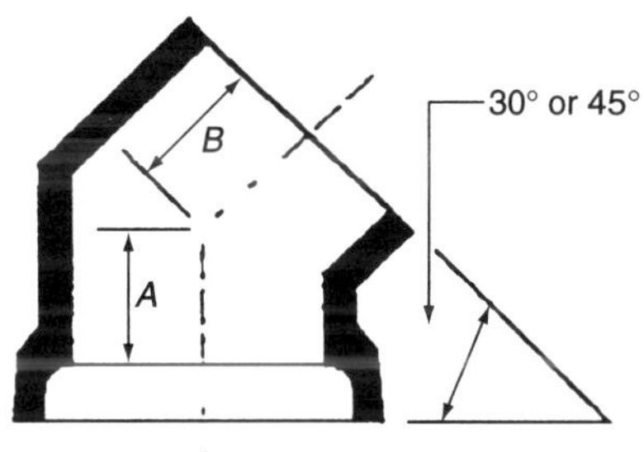

TABLE 24.14 Dimensions of 30° and 45° Cut Curves for ASTM C700 Extra-Strength Vitrified Clay Pipe

Pipe size, in	Angle, degrees	Dimensions (approx.), in	
		A	*B*
10	22½°	9½	3⅞
10	30°	8¾	4¾
10	45°	7¼	5⅝
12	30°	9¼	5⅜
12	45°	8½	6¾
15	30°	12¾	5
15	45°	10¼	6¾
18	30°	13	6
18	45°	10½	9
21	30°	16⅝	7½
21	45°	11⅝	12
24	30°	22½	11¼
24	45°	19¾	12

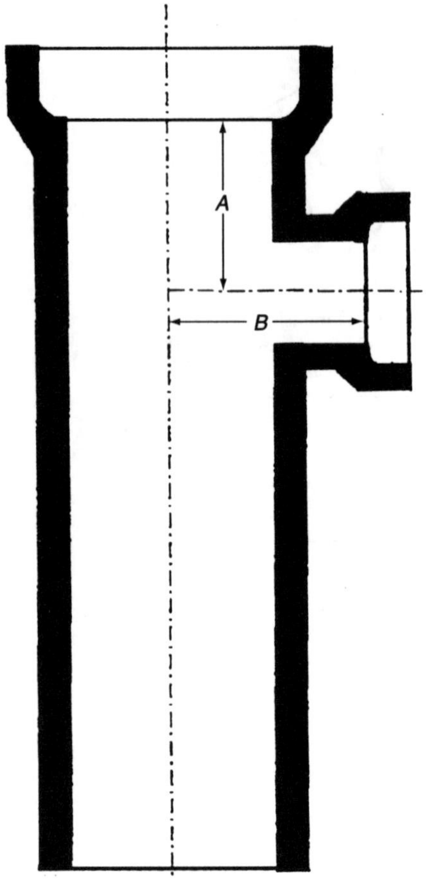

TABLE 24.15 Dimensions of T Branches for ASTM C700 Extra-Strength Vitrified Clay Pipe*

Barrel diameter	Spur diameter	Laying length	Dimensions (approx.) A	B
4	4	1	5	4¼
4	4	2	5	4¼
6	4	2	5¾	5¾
6	6	2	6¼	6
8	4	2	5¼	7
8	6	2	6½	7
8	8	2	7	8
10	4	2	6	8¾
10	6	2	7½	8½
10	8	2	9	9
10	10	2	9¾	9
12	4	2	7½	10
12	6	2	8	9½
12	8	2	9½	9½
12	10	2	11	10½
12	12	2	11¼	10¾
15	6	3	8¾	11¼
15	8	3	10¼	12
15	10	3	13	14
15	12	3	12½	12
15	15	3	15½	14½
18	6	3	9	13¾
18	8	3	11½	13¾
18	10	3	11¾	14¾
18	12	3	12½	14
18	15	3	16¼	16
18	18	3	16¼	15½
21	6	3	10½	16¼
21	8	3	12	16½
21	10	3	13	16¼
21	12	3	14½	15½
21	15	3	16½	17
24	6	3	11½	16¾
24	8	3	11½	17¼
24	10	3	11¾	17
24	12	3	12	18½
24	15	3	16	18½
24	18	3	18	18½

* Dimensions in inches.

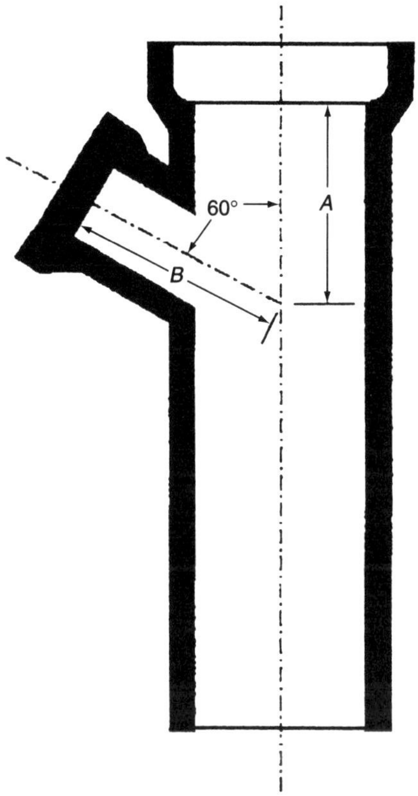

TABLE 24.16 Dimensions of Y Branches for ASTM C700 Extra-Strength Vitrified Clay Pipe*

Barrel diameter	Spur diameter	Laying length	Dimensions (approx.) *A*	*B*
4	4	1	8¼	6¾
4	4	2	8¼	6¾
6	4	2	9¼	8
6	6	2	12¼	10½
8	4	2	9	9¼
8	6	2	12	10¼
8	8	2	14¾	12¾
10	4	2	11¼	12
10	6	2	12¾	11¼
10	8	2	16¼	14¾
10	10	2	18¼	16
12	6	2	13¼	14¼
12	4	2	12	13
12	8	2	17	16¼
12	10	2	19¼	17½
12	12	2	18¾	19¼
15	6	3	14½	13½
15	8	3	17	16½
15	10	3	19½	18
15	12	3	21	20
15	15	3	28¼	26¼
18	6	3	16	17¼
18	8	3	17¼	18
18	10	3	20½	21½
18	12	3	22½	20¼
18	15	3	24	24¾
18	18	3	32	29¼
21	6	3	19¼	22½
21	8	3	18	21
21	10	3	23¾	22
21	12	3	23	19½
21	15	3	26	29½
24	6	3	17½	19¼
24	8	3	18½	21½
24	10	3	22¾	26¼
24	12	3	21½	24¾
24	15	3	26	28
24	18	3	27¼	30

* Dimensions in inches.

TABLE 24.17 Nominal Lengths of Wyes and Tees for ASTM C700 Extra-Strength Vitrified Clay Pipe

Nominal pipe size (NPS), in	Nominal size (DN), mm	Nominal length of wyes and tees, ft
4	100	2
6	150	2
8	200	2
10	250	2
12	300	2
15	375	3
18	450	3
21	525	3
24	600	3

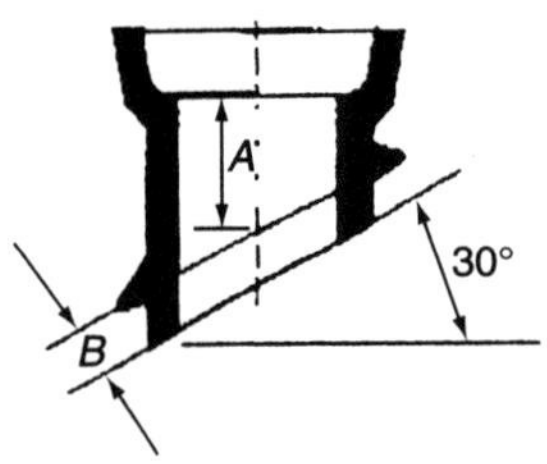

TABLE 24.18 Dimensions of Y Saddle (with Flange) for ASTM C700 Extra-Strength Vitrified Clay Pipe

Pipe size, in	Dimensions (approx.), in	
	A	*B*
4	$5\frac{1}{4}$	$1\frac{1}{2}$
6	6	$1\frac{1}{2}$
8	$5\frac{1}{2}$	$1\frac{1}{2}$

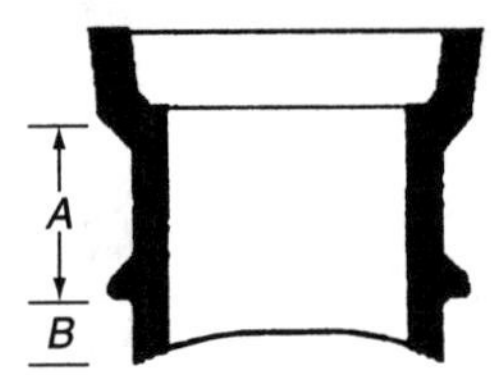

TABLE 24.19 Dimensions of T Saddle (with Flange) for ASTM C700 Extra-Strength Vitrified Clay Pipe

Pipe size, in	Dimensions (approx.), in	
	A	*B*
4	$4\frac{1}{2}$	$1\frac{1}{2}$
6	6	$1\frac{1}{2}$
8	8	$1\frac{1}{2}$

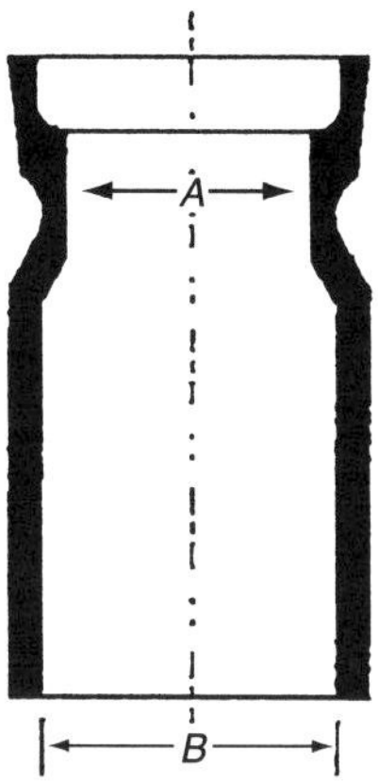

TABLE 24.20 Dimensions of Increasers for ASTM C700 Extra-Strength Vitrified Clay Pipe

Pipe size, in		Nominal laying length, in
A	*B*	
4	5	12
4	6	3
4	8	11
4	10	12
5	6	5
6	8	4
6	10	12
6	12	12
8	10	12
8	12	12
8	15	12
10	12	12
10	15	13
12	15	14
12	18	14
15	18	15

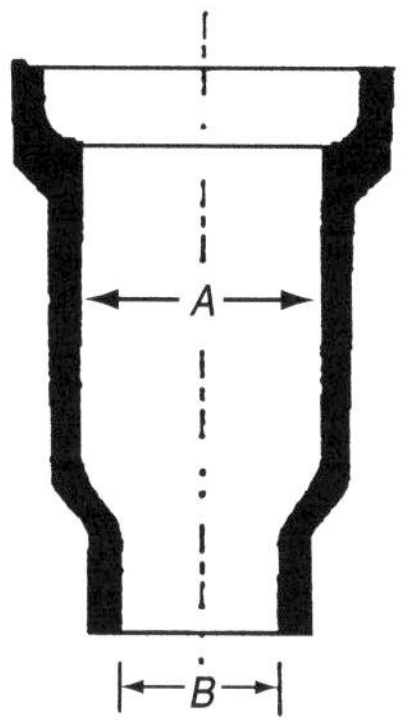

TABLE 24.21 Dimensions of Reducers for ASTM C700 Extra-Strength Vitrified Clay Pipe

Pipe size, in		Nominal laying length, in
A	*B*	
6	4	12
6	5	10
8	4	12
8	5	12
8	6	12
10	6	12
10	8	12
12	8	12
12	10	12
15	10	12
15	12	12

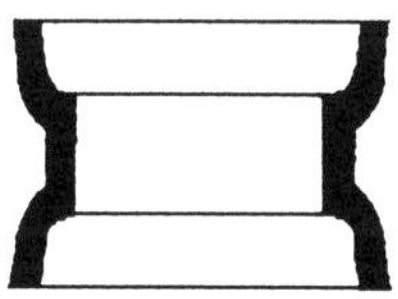

TABLE 24.22 Dimensions of Double Hubs for ASTM C700 Extra-Strength Vitrified Clay Pipe

Diameter, in	Laying length, in
4	3
6	4
8	5
10	5
12	7

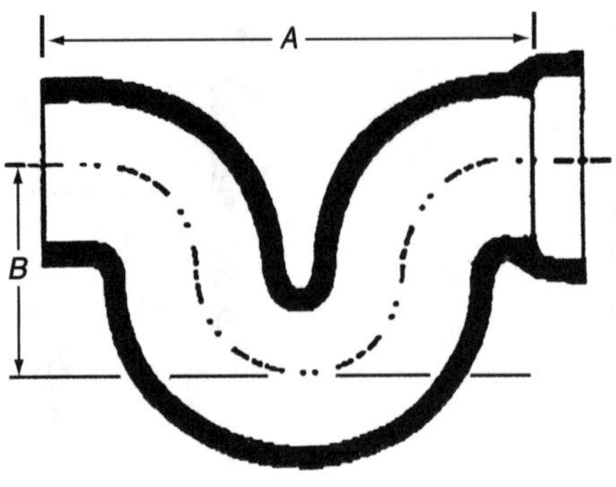

TABLE 24.23 Dimensions of Running Traps for ASTM C700 Extra-Strength Vitrified Clay Pipe

Size, in	Dimensions (approx.), in *A*	*B*
4	19	8
6	$21^3/_4$	9

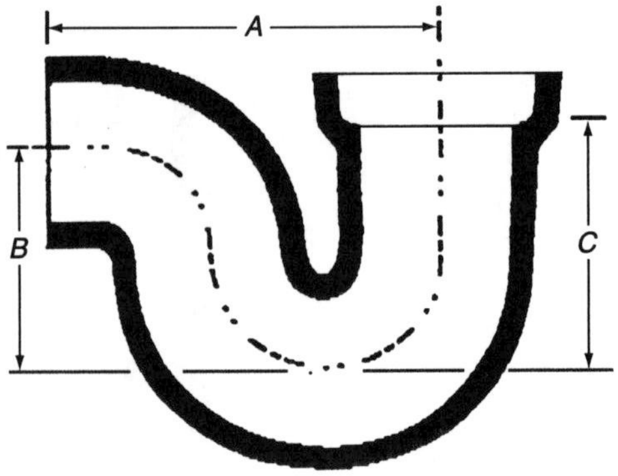

TABLE 24.24 Dimensions of P Traps for ASTM C700 Extra-Strength Vitrified Clay Pipe

Size, in	Dimensions (approx.), in *A*	*B*	*C*
4	15	$7^1/_4$	8
6	17	9	10

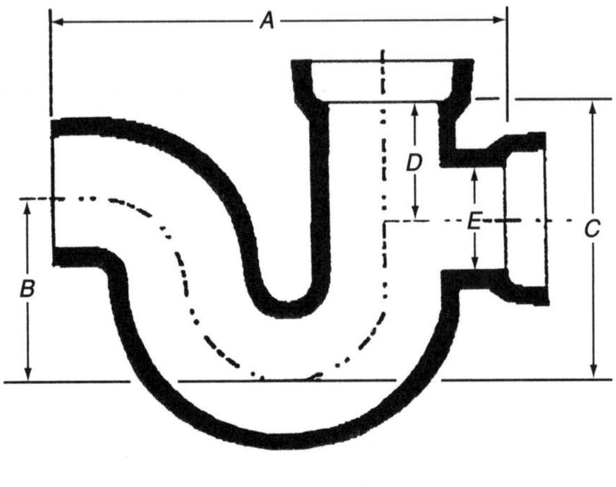

TABLE 24.25 Dimensions of Shoulder Hand Hole Traps (SHH) for ASTM C700 Extra-Strength Vitrified Clay

Pipe size, in	Dimensions (approx.), in *A*	*B*	*C*	*D*	*E*
4 4 4	18	7	$11^1/_4$	4	4
6 6 6	$23^3/_4$	11	$15^1/_4$	$6^1/_4$	6

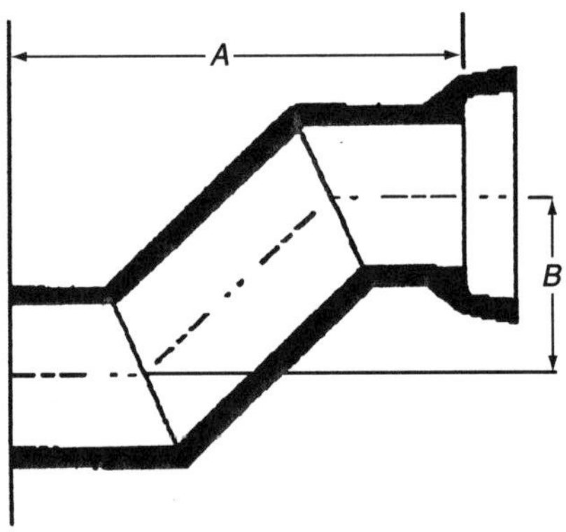

TABLE 24.26 Dimensions of Half Traps for ASTM C700 Extra-Strength Vitrified Clay Pipe

Pipe size, in	*A*, laying length, in	*B*, in
6	21	$8^1/_4$
8	24	$10^1/_4$
10	25	$11^1/_4$
12	27	$11^1/_2$

TABLE 24.27 Dimensions of Stoppers for ASTM C700 Extra-Strength Vitrified Clay Pipe

	O-ring		Plain clay		Concrete	
Size	Outer diameter	Thickness	Outer diameter	Thickness	Outer diameter	Thickness
4	5 1/2	1 3/8	5	7/8	NA	NA
5	6 3/4	1 3/8	6	1	NA	NA
6	8 1/4	1 3/8	8 3/4	1 1/8	NA	NA
8	10 3/4	1 5/8	10	1 1/8	NA	NA
10	13 1/4	1 3/4	12 1/2	1 1/4	13 1/4	2
12	15 5/8	2	15	1 1/4	15 1/4	2
15	19 1/4	2 1/8	18	1 1/2	18 3/4	2 1/8
18	23 1/8	2 3/8	20 5/8	1 1/2	21	2 1/8
21	27	3	NA	NA	26	2 1/8
24	30 7/8	3	NA	NA	29	2 1/8

Notes:
1. Dimensions in inches.
2. O-ring stoppers are furnished with handles.
3. Size 10- and 12-in concrete stoppers are provided with lifting holes; 15-in concrete stoppers are furnished with handles.
4. Plain clay stoppers are not provided with handles.
5. Contact the manufacturer for specific design details.
6. NA = not applicable.

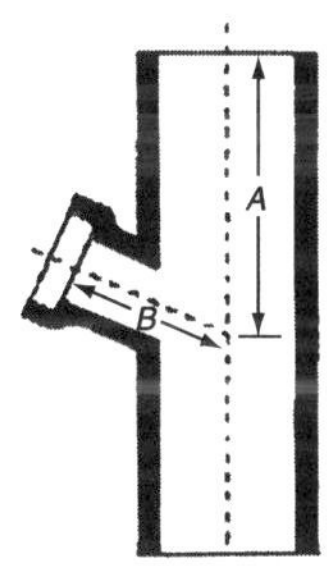

TABLE 24.28 Dimensions of Repair Wyes for Vitrified Clay Pipe

			Dimensions (approx.), in	
Barrel diameter, in	Spur diameter, in	Laying length, ft	*A*	*B*
6	6	2	14	10 1/2
8	6	2	14	10 1/4
8	8	2	14	12 3/4
10	6	2	14	11 1/4
10	8	2	14	14 3/4
12	6	2	14	14 1/4

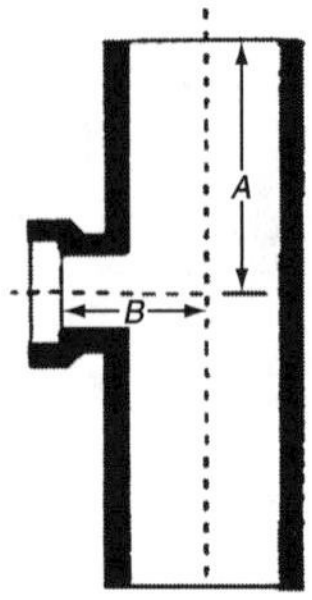

TABLE 24.29 Dimensions of Repair Tees for Vitrified Clay Pipe*

Barrel diameter, in	Spur diameter, in	Laying length, ft	Dimensions (approx.), in A	B
6	6	2	12	6
8	6	2	12	7
8	8	2	12	8
10	6	2	12	$8^1/_2$
10	8	2	12	9
12	6	2	12	$9^1/_2$

* Plain ends are joined with existing pipe by couplings. Branch end connections may either be plain ends suitable for coupling joints or bell types for O-ring joints.

JOINTS FOR VITRIFIED CLAY PIPE AND FITTINGS

Refer to Chap. 15.

Tables in Chapter 24

Tables in Chapter 24 (Continued)

S · E · C · T · I · O · N · 6

LINED PIPING

CHAPTER 25
LINED PIPING

See p. 719 for a list of tables in Chap. 25.

TYPES OF LINING FOR PIPING

The common types of lining are:

1. Cement mortar and concrete
2. Epoxy
3. Rubber
4. Plastic

CEMENT MORTAR AND CONCRETE LINING FOR PIPE AND FITTINGS

Reference Standards for Cement Mortar and Concrete Lining

Refer to Table 25.1 for AWWA standards that are commonly used for cement mortar and concrete lining of steel, cast iron, and ductile iron pipe and fittings. Refer to Table 25.2 for AWWA manuals for cement mortar and concrete lining.

TABLE 25.1 AWWA Reference Standards for Cement Mortar and Concrete Linings for Pipe

Number	Title	Remarks
American National Standards Institute (ANSI)/American Water Works Association (AWWA) C104/A21.4	Standard for Cement-Mortar Lining for Ductile-Iron Pipe and Fittings for Water	For factory-applied mortar linings in ductile iron pipe
ANSI/AWWA C205	Standard for Cement-Mortar Protective Lining and Coating for Steel Water Pipe—NPS 4 (DN 100) and Larger—Shop Applied	For factory-applied mortar linings in steel pipe
ANSI/AWWA C300	Standard for Reinforced Concrete Pressure Pipe, Steel-Cylinder Type	For concrete and mortar linings in concrete pressure pipe
ANSI/AWWA C301	Standard for Prestressed Concrete Pressure Pipe, Steel-Cylinder Type	For concrete and mortar linings in concrete pressure pipe
ANSI/AWWA C302	Standard for Reinforced Concrete Pressure Pipe, Non-Cylinder Type	For concrete and mortar linings in concrete pressure pipe
ANSI/AWWA C303	Standard for Concrete Pressure Pipe, Bar-Wrapped, Steel-Cylinder Type	For mortar linings in concrete pressure pipe
ANSI/AWWA C602	Standard for Cement-Mortar Lining of Water Pipelines in Place—NPS 4 (DN 100) and Larger	For field-applied mortar linings in steel pipe, cast iron pipe, and ductile iron pipe

Suitability of Cement Mortar and Concrete–Lined Piping

Table 25.3 lists the fluids for which cement mortar and concrete–lined piping is suitable.

Lining Application and Curing Methods

Table 25.4 lists the application and curing methods for cement mortar and concrete lining of pipe and fittings.

Cement Mortar and Concrete Lining of Ductile Iron Pipe and Fittings

Table 25.5 summarizes the mortar lining application methods, curing methods, and mix proportions used in the manufacturing plant and in the field for ductile iron pipe and fittings.

Shop-Applied Lining Thickness for Ductile Iron Pipe and Fittings. Refer to Table 25.6 for shop-applied lining thickness for ductile iron pipe and fittings.

Field-Applied Lining Thickness for Ductile Iron Pipe and Fittings. Refer to Table 25.7 for field-applied lining thickness for ductile iron pipe and fittings.

TABLE 25.2 AWWA Reference Manuals for Cement Mortar and Concrete Linings for Pipe

Title	Remarks
American Water Works Association Manual of Water Supply Practices, "M9—Concrete Pressure Pipe"	Covers the four types of concrete pressure pipe and fittings*
American Water Works Association Manual of Water Supply Practices, "M11—Steel Pipe-A Guide for Design and Installation"	Covers steel pipe and fittings
American Water Works Association Manual of Water Supply Practices, "M41—Ductile Iron Pipe and Fittings"	Covers ductile iron pipe and fittings

* The four types of concrete pressure pipe are
- ANSI/AWWA C300
- ANSI/AWWA C301
- ANSI/AWWA C302
- ANSI/AWWA C303

TABLE 25.3 Transmitted Fluids for Which Cement Mortar and Concrete Linings Are Suitable

Transmitted fluid
Raw freshwater
Potable water
Raw sewage
Treated sewage
Seawater
Power plant cooling water (both seawater and freshwater)

TABLE 25.4 Concrete Core and Mortar Linng Application and Curing Methods for Concrete Pressure Pipe

		ANSI/AWWA C300	ANSI/AWWA C301	ANSI/AWWA C302	ANSI/AWWA C303
Pipe	Application	Vertical casting or radial compaction; concrete mix temperature must be between 40°F (4°C) and 90°F (32°C)	Centrifugal casting, vertical casting, or radial compaction; concrete mix temperature must be between 40°F (4°C) and 90°F (32°C)	Vertical casting or radial compaction; concrete mix temperature must be between 40°F (4°C) and 90°F (32°C)	Centrifugal casting; concrete mix temperature must be over 40°F (4°C)
	Curing	Accelerated method with steam for 12h minimum, or with water for 12h minimum	Accelerated method with steam for 12h minimum, or with water for 12h minimum	Accelerated method with steam for 12h minimum, or with water for 12h minimum	Accelerated method with steam for 6h minimum, or with water for 24h minimum
Fittings	Application	Hand placed or pneumatic placement	Hand placed or pneumatic placement	Hand placed or pneumatic placement	Centrifugal casting, hand placed, or pneumatic placement
	Curing	Accelerated method with steam for 12h minimum, with water for 12h minimum, or curing compound per ASTM C309	Accelerated method with steam for 12h minimum, with water for 24h minimum, or curing compound per ASTM C309	Accelerated method with steam for 12h minimum, with water for 12h minimum, or curing compound per ASTM C309	Accelerated method with steam for 6h minimum, with water for 24h minimum, or curing compound per ASTM C309

TABLE 25.5 Cement Mortar Lining Application, Curing, and Mix Proportions for Ductile Iron Pipe

		Applied at manufacturing plant	Applied in the field
Pipe	Application	Centrifugal casting or pneumatic placement	Mechanical placement and troweling
	Curing	Asphaltic seal coat, steam, or water	Water
	Mix proportions	1 part portland cement, not more than 2 parts fine aggregate, and sufficient water to achieve required workability	1 part portland cement to 1 to $1^1/_2$ parts fine aggregate, and sufficient water to achieve required workability
Fittings	Application	Hand placed or pneumatic placement	Hand placed if not by mechanical placement and troweling
	Curing	Asphaltic seal coat, steam, or water	Water
	Mix proportions	1 part portland cement, not more than 2 parts fine aggregate, and sufficient water to achieve required workability	1 part portland cement to 1 to $1^1/_2$ parts fine aggregate, and sufficient water to achieve required workability

TABLE 25.6 Factory-Applied Mortar Lining Thickness for Ductile Iron Pipe and Fittings*

Pipe size range	Mortar lining thickness
NPS 3–NPS 12 (DN 75–DN 300)	0.0625 in (1.6 mm)
NPS 14–NPS 24 (DN 350–DN 600)	0.0938 in (2.4 mm)
NPS 30–NPS 64 (DN 750–DN 1600)	0.125 in (3.2 mm)

* Lining with twice the thickness listed in this table can be applied for extra protection if specified by the user.

TABLE 25.7 Field-Applied Mortar Lining Thickness for Ductile Iron Pipe and Fittings

Pipe size range	Mortar lining thickness for old and new ductile iron
4–10 in (100–250 mm)	0.1875 in (4.8 mm)
11–23 in (280–580 mm)	0.25 in (6.4 mm)
24–36 in (600–900 mm)	0.3125 in (8 mm)
>36 in (900 mm)	0.3125 in (8 mm)

Cement Mortar and Concrete Lining of Steel Pipe and Fittings

Table 25.8 summarizes the mortar lining application methods, curing methods, and mix proportions used in the manufacturing plant and in the field for steel pipe and fittings.

Factory-Applied Lining Thickness for Steel Pipe and Fittings. Refer to Table 25.9 for shop-applied lining thickness for steel pipe and fittings.

Field-Applied Lining Thickness for Steel Pipe and Fittings. Refer to Table 25.10 for field-applied lining thickness for steel pipe and fittings.

TABLE 25.8 Cement Mortar Lining Application, Curing, and Mix Proportions for Steel Pipe and Fittings

		Applied at manufacturing plant	Applied in the field
Pipe	Application	Centrifugal casting	Mechanical placement and troweling
	Curing	Steam or water	Water
	Mix proportions	1 part portland cement, not more than 3 parts fine aggregate, and sufficient water to achieve required workability	1 part portland cement to 1 to $1^1/_2$ parts fine aggregate, and sufficient water to achieve required workability
Fittings	Application	Hand placed or pneumatic placement if not by centrifugal casting	Hand placed if not by mechanical placement and troweling
	Curing	Steam or water	Water
	Mix proportions	1 part portland cement, not more than 3 parts fine aggregate, and sufficient water to achieve required workability	1 part porland cement to 1 to $1^1/_2$ parts fine aggregate, and sufficient water to achieve required workability

TABLE 25.9 Factory-Applied Mortar Lining Thickness for Steel Pipe and Fittings

Pipe size range	Mortar lining thickness
4–10 in (100–250 mm)	0.25 in (6 mm)
11–23 in (280–580 mm)	0.3125 in (8 mm)
24–36 in (600–900 mm)	0.375 in (10 mm)
>36 in (900 mm)	0.50 in (13 mm)

TABLE 25.10 Field-Applied Mortar Lining Thickness for Steel Pipe and Fittings

Pipe size range	Mortar lining thickness for old steel	Mortar lining thickness for new steel
4–10 in (100–250 mm)	0.3125 in (8 mm)	0.25 in (6 mm)
11–23 in (280–580 mm)	0.375 in (10 mm)	0.3125 in (8 mm)
24–36 in (600–900 mm)	0.4375 in (11 mm)	0.375 in (10 mm)
>36 in (900 mm)	0.5625 in (14 mm)	0.50 in (13 mm)

Joint Protection

Fig. 25.1 illustrates a steel pipe lap-welded slip joint. The lining contractor must hand place lining to uncoated surfaces after welding.

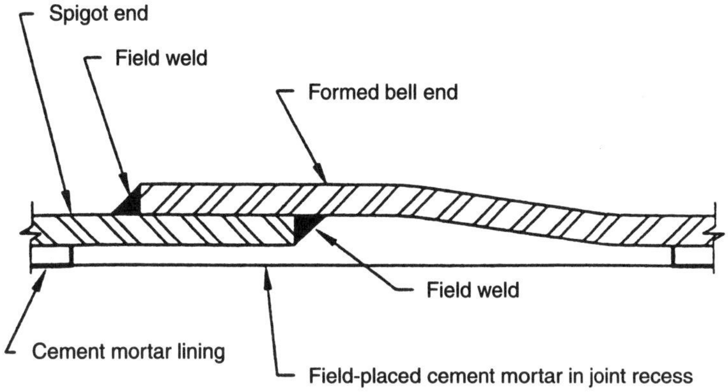

FIGURE 25.1 Mortaring of interior joint recess for field-welded lap joint steel pipe.

EPOXY LINING FOR PIPE AND FITTINGS

Reference Codes, Standards, Specifications, and Regulations

Table 25.11 lists various codes, standards, specifications, and regulations that relate to epoxy lining and coating of piping systems.

Types of Internal Organic Linings

See Table 25.12.

Types of External Pipe Coating Systems

See Table 25.13.

Selection of Internal Lining

See Tables 25.14 and 25.15.

TABLE 25.11 Reference Codes, Standards, Specifications, and Regulations

American Society of Mechanical Engineers
B31.1: Power Piping
B31.11: Slurry Transportation Piping Systems, Special Notice—1990
B31.3: Process Piping
B31.4: Liquid Transportation Systems for Hydrocarbons, Liquid Petroleum Gas, Anhydrous Ammonia, and Alcohol
B31.8: Gas Transmission and Distribution Piping Systems
American Society for Testing and Materials (ASTM)
G8-96: Test Method for Cathodic Disbonding of Pipeline Coatings
G12-83: Method for Nondestructive Measurement of Film Thickness of Pipeline Coatings on Steel
G14-88: Test Method for Impact Resistance of Pipeline Coatings (Falling Weight Test)
G17-88: Test Method for Penetration Resistance of Pipeline Coatings (Blunt Rod)
G20-88: Test Method for Chemical Resistance of Pipeline Coatings
G80-88: Test Method for Specific Cathodic Disbonding of Pipeline Coatings
G95-87: Test Method for Cathodic Disbondment Test of Pipeline Coatings
American Petroleum Institute
RP 5L1 Recommended Practice for Railroad Transportation of Line Pipe
RP 5L2-87 Recommended Practice for Internal Coating of Line Pipe for Non-Corrosive Gas Transmission Service; Third Edition
RP 5LW Recommended Practice for Transportation of Line Pipe on Barges and Marine Vessels
RP 5L7-88 Recommended Practices for Unprimed Internal Fusion Bonded Epoxy Coating of Line Pipe; Second Edition
Manufacturers Standardization Society of the Valve and Fittings Industry
SP-98-97 Protective Epoxy Coatings for the Interior of Valves and Hydrants
ES-29-79 Abrasive Blast Cleaning of Ferritic Piping Materials (R 1984)
NACE International
RP0394-94 Standard Recommended Practice: Application, Performance, and Quality Control of Plant-Applied, Fusion-Bonded Epoxy External Pipe Coating
NACE Standard Recommended Practice, RP0188, Discontinuity (Holiday) Testing of Protective Coatings
RP0490-95 NACE Standard Recommended Practice, Holiday Detection of Fusion-Bonded Epoxy External Pipeline Coatings of 250 to 760 microns (10 to 30 mils)
Steel Structures Painting Council
SSPC-Vis 1 Pictorial Surface Preparation Standard for Painting Steel Surfaces
NACE/SSPC
NACE No. 1/SSPC-SP 5 White Metal Blast Cleaning
NACE No. 2/SSPC-SP 10 Near White Metal Blast Cleaning
AWWA Standards for Water Systems Piping
C213-96 Standard for Fusion Bonded Epoxy Coating for the Interior and Exterior of Steel Water Pipelines
C550-90 Protective Interior Coatings for Valves and Hydrants
National Standards of Canada
CAN/CSA-Z245.20-M92, External Fusion Bond Epoxy Coating for Steel Pipe
CAN/CSA-Z245.21-M92, External Polyethylene Coating for Pipe
ECISS: European Committee of Iron and Steel Standardization* Technical Committee 29: Steel Tubes and Fittings Subcommittee 4: Coatings
EC 029016 External Three-layer, Polyethylene-based Coating Systems
EC 029063 External Three-layer, Polypropylene-based Coating Systems
EC 029062 External Thermosetting Fusion-bonded Epoxy Coating Systems

* Standards are under development.

TABLE 25.12 Types of Organic Internal Linings

Type	Advantages	Disadvantages	Upper temperature	Single coat	Type
Phenolic	• Low water-vapor permeability • Good carbon dioxide, solvent and abrasion resistance	• Low impact resistance • Low flexibility • Moderate caustic resistance	392°F (200°C)	N	TS L
Urethane	• Good abrasion • Flexible • Smooth, resists paraffin buildup		220°F (105°C)		TS L
Nylon	• Good abrasion • Flexible • Good saltwater resistance • Resists installation handling damage	• Low hydrogen sulfide and acid resistance	220°F (105°C)	Y	TP P
Epoxy-phenolic	• Improved flexibility and caustic resistance compared to phenolic • Resists depressurization	• Increased water-vapor permeability • Reduced acid resistance compared to phenolic	250°F (120°C)	N	TS L
Epoxy-novolac—FBE	• Improved flexibility and caustic resistance compared to phenolic • Resists depressurization	• Increased water-vapor permeability • Reduced acid resistance compared to phenolic	300°F (150°C)	Y	TS P
Epoxy—FBE	• Flexible • Good saltwater resistance • Resists installation handling damage	• Increased water-vapor permeability • Reduced acid resistance compared to phenolic	250°F (120°C)	Y	TS

Notes: *Single coat*: Y = yes, the coating can be applied in a single pass. For example, FBE coatings are frequently applied at a coating thickness of 25 mils (625 μm). N = No, materials like phenolics are applied in several layers of 25 to 50 μm (1 to 2 mils).

Type: TS = thermoset coating system (there is a chemical reaction that causes the coating to cure and take on final properties), TP = thermoplastic, L = liquid, P = powder.

TABLE 25.13 Types of External Pipe Coating Systems

Material	Advantages	Disadvantages	Temperature ±2°F (5°C)
Polyethylene tapes	Simple application in field.	Cathodic shielding—SCC.† Very poor resistance to soil stress.	95°F (35°C)
Extruded polyethylene	Good resistance to handling and installation damage. Self-healing adhesive.	Limited to NPS 24 (DN 600) or smaller pipe. Limited at elevated temperatures under soil stress.	105°F (40°C)
Wax/vinyl tapes	Simple technology. Low moisture permeation.	Very sensitive to temperature. Prone to cathodic disbondment.	85°F (30°C)
Asphalt enamels	Simple technology.	Poor resistance to moisture and soil stress. Very sensitive to temperature.	95°F (35°C)
Coal-tar enamels	Good corrosion resistance.	Low soil stress resistance. Properties sensitive to temperature.	120°F (50°C)
Fusion bond epoxy	Good corrosion resistance. Does not shield CP* systems. No evidence of SCC.†	Moisture absorption and permeation. Susceptible to cathodic disbondment.	175°F (80°C)
Two-layer FBE	Good corrosion resistance. Good resistance to handing and installation damage.	Moisture absorption and permeation.	230°F (110°C)
Three-layer polyethylene	Good corrosion resistance. Good resistance to handing and installation damage.	Cathodic shielding concern.	175°F (80°C)
Three-layer polypropylene	Good corrosion resistance. Good resistance to handing and installation damage.	Cathodic shielding concern.	250°F + (120°C+)

* CP: cathodic protection.
† SCC: stress-corrosion cracking.

TABLE 25.14 Autoclave Testing (Internal Lining Selection Depends on Environment)

Temperature	200°F (93°C)	200°F (93°C)	275°F (135°C)
Duration, hours	16	24	24
Liquid phase	5% NaCl Brine	15% HCl	34% Brine 33% Kerosene 33% Toluene
Gas phase	5% CO_2 95% Methane	Air	25% CO_2 75% Methane
Discharge	Cool for 4 hours then rapidly release pressure	Force-cool to ambient, release pressure over 5-minute period	Release pressure over 5-minute period at test temperautre
Results			
Lining A	Use	Don't use	Don't use
Lining B	Use	Use	Don't use
Lining C	Use	Use	Use
Lining D	Use	Use	Use

Lining A: Bisphenol-A type epoxy with NSF listing for potable drinking water.
Lining B: Modified bisphenol-A type epoxy for harsh environment.
Lining C: Novolac epoxy.
Lining D: Novolac epoxy designed for girth-weld field application. Requires no primer or post cure.

TABLE 25.15 Application and Handling Characteristics of Different Types of Lining (Internal Lining Selection Depends on the Environment, Required Handling Characteristics, and Application Characteristics of the Material)

Lining	Elongation, %	Primer required	Postcure	Wire line*	NSF[†]	Test[‡]
Lining A	6	No	No	+	Yes	Water
Lining E	4	No	Yes	++	Yes	Hot water
Lining E	4	Yes	Yes	++	No	3% CO_2
Lining B	6	No	Yes	++	No	3% CO_2
Lining B	6	Yes	Yes	++	No	100% CO_2
Lining F	1.5	No	No	+++++	No	100% CO_2

Lining A: Bisphenol-A type epoxy with NSF listing for potable drinking water
Lining E: Modified bisphenol-A type epoxy with NSF listing for potable drinking water
Lining B: Modified bisphenol-A type epoxy for harsh environment
Lining F: Novolac epoxy designed for wire-line damage resistance
* Wire line: For downhole tubing, equipment is moved via a cable, or wire line. Resistance to damage by the wire line is based on a 1-to-5 scale with +++++ being the best rating.
[†] NSF: National Sanitation Foundation, U.S. certification agency for potable drinking water systems.
[‡] Testing: An increase in either the temperature or level of CO_2 creates a more hostile environment for internal linings.

Welded Internal Coupling Systems

Couplings are available to isolate the welded area of the pipe from the corrosive effects of long-term fluid transmission. An attachment system acts as a welding guide for joining the two sections of pipe and becomes an integral part of the weld. The coupling allows pigging for cleaning and inspection of the pipeline.

The coupling (see Fig. 25.2) is inserted into the end of an internally lined pipe to the weld guide. The second internally lined pipe is slipped over the protruding half to the weld nubs. The center portion of the outer coupling circumference has a machined recess holding heat-resistant material and a steel band with the attached weld nubs. The heat-resistant material acts as an insulator to protect the lining on the inside of the coupling from weld-induced heat damage. O-rings isolate the weld from corrosive attack. The welding process unifies the two pipes and the coupling.

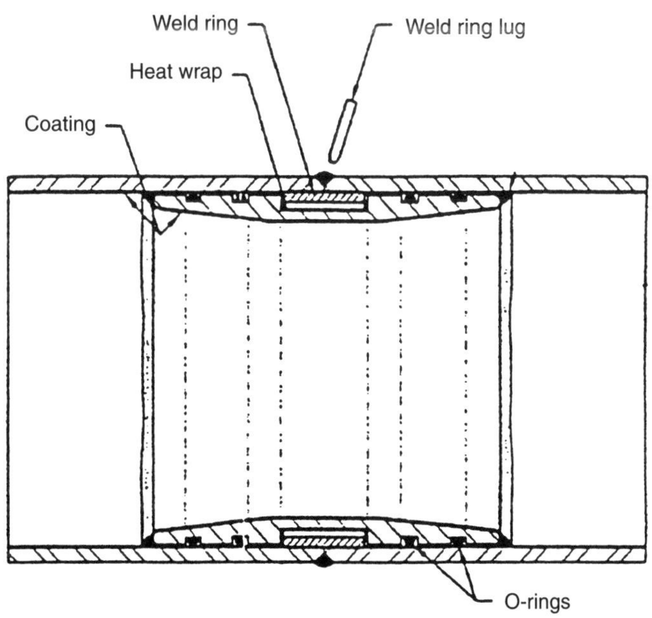

FIGURE 25.2 Welded internal coupling system. The coupling is internally coated with FBE and externally coated to include the O-rings.

Weldless Coupling System

This system consists of a coupling that slips over the outside of the coated pipeline (see Fig. 25.3). The coupling is machined with serrated grooves and is smaller than the outside of the line pipe to provide a compression fitting.

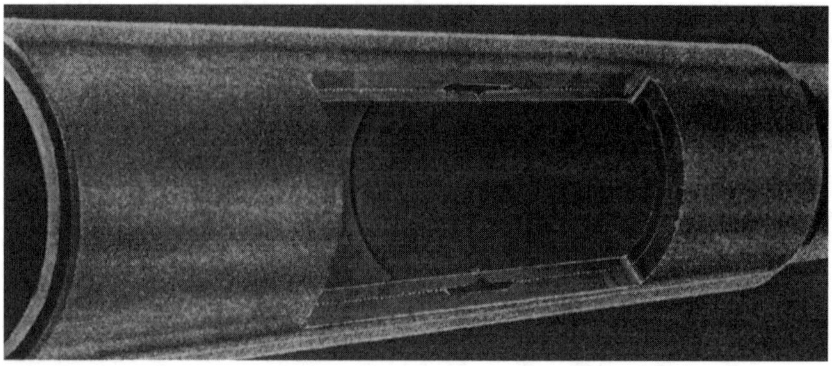

FIGURE 25.3 Weldless coupling system for internal and/or external FBE coating.

A two-part epoxy is applied to the inside of the coupling and to the outside of the lined pipe. It acts as a lubricant during the compression-fit installation and as a seal at the pipe ends, where it forms a bead to provide a continously lined internal pipe. Normally, the coupling is installed in one end of each pipe in the coating-plant yard. In the field, a hydraulic ram is used to join the two pieces of pipe together inside the coupling.

The system is available for pipe in the range of NPS 2 to 12 (ND 50 to 300). It has been used in highly corrosive applications including carbon dioxide injection, brine-water disposal, and produced fluids of hydrogen sulfide and carbon dioxide containing oil, saltwater, and natural gas.

RUBBER LINING OF PIPE AND FITTINGS

Elastomeric Lining Material Properties and Limitations

See Table 25.16.

Chemical Resistance of Various Linings

See Table 25.17.

Rubber Lining Thickness Selection

Lining thickness may vary from $^1/_8$ to $^1/_4$ in (3 to 6 mm). In some very abrasive conditions, up to 1 in (25 mm) has been applied. Experience has proved that not only the rate of abrasion must be considered but temperature also. For temperatures greater than 140°F (60°C), $^1/_4$-in (6-mm)-thick lining will provide better service. Table 25.18 shows the recommended maximum lining thicknesses for different pipe sizes.

The most commonly used and economical lining thickness is $^1/_4$ in (6 mm). It offers long life for a wide range of applications.

Pipe Size, Length, and Other Physical Constraints for Rubber Lining

Pipe Size and Typical Lengths for Straight Pipe. See Table 25.19.

Maximum Straight Length of Pipe. See Table 25.20.

Maximum Length of Pipe with a 45° Elbow. See Table 25.21.

TABLE 25.16 Hardness and Temperature Limitations of Elstomeric Lining Materials

Lining material	Hardness (Shore A)	Temperature range
Natural rubber	30 to 100	Up to 190°F (88°C)
Butyl	50 to 70	Up to 260°F (127°C)
EPDM	50 to 70	Up to 200°F (93°C)
Nitrile rubber	50 to 90	Up to 200°F (93°C)
Urethane	50 to 90	Up to 120°F (49°C)
Neoprene	50 to 70	Up to 200°F (93°C)
Hypalon	50 to 60	Up to 200°F (93°C)

TABLE 25.17 Chemical Resistance of Various Linings

Corrosion medium	Max. temp., °F (°C)	Concentration, %	Natural rubber		Butyl	EPDM	Neoprene	Nitrile	Hypalon	Urethanes
			Soft	Hard						
Abrasion			R				R			R
Acetic acid	140 (60)	100		R	R					
Ammonium sulfate	200 (93)	Saturated		R	R		R			
Chlorine gas wet	175 (79)	Saturated		R						
Copper cyanide	160 (71)	Saturated	R	R						
Ferric chloride	175 (79)	Saturated	R	R						
Hydrochloric	125 (52)	37	R							
Hydrofluoric	90 (32)	50			R					
Kerosene	90 (32)	Saturated						R		
Nickel chloride	190 (88)	Saturated		R	R					
Nitric acid	150 (66)	10			R				R	
Phosphoric acid	140 (60)	85			R					
Sodium chlorite	190 (88)	Saturated	R							
Sodium hypochlorite	150 (66)	15			R					
Sodium hydroxide	194 (90)	50			R	R				
Sodium sulfate	190 (88)	Saturated	R	R	R					
Sulfuric acid	180 (82)	20			R					
Sulfuric acid	150 (66)	50							R	
Water—fresh	175 (79)			R		R				
Water—salt	175 (79)						R			
Water—mine tailings	120 (49)			R			R			R

R = recommended for the corrosion medium shown. Additional service conditions, such as exposure to sunlight, must also be considered before a lining is selected.

TABLE 25.18 Recommended Maximum Lining Thicknesses

Pipe size NPS (DN)	Lining thickness, in (mm)				
	1/4 (6.4)	3/8 (9.53)	1/2 (12.7)	3/4 (19.1)	1 (25.4)
2 (50)	X				
3 (80)	X				
4 (100)	X				
6 (150)	X	X			
8 (200)	X	X	X		
10 (250)	X	X	X		
12 (300)	X	X	X	X	
14 (350)	X	X	X	X	
16 (400)	X	X	X	X	
18 (450)	X	X	X	X	
24 and larger (600 and larger)	X	X	X	X	X

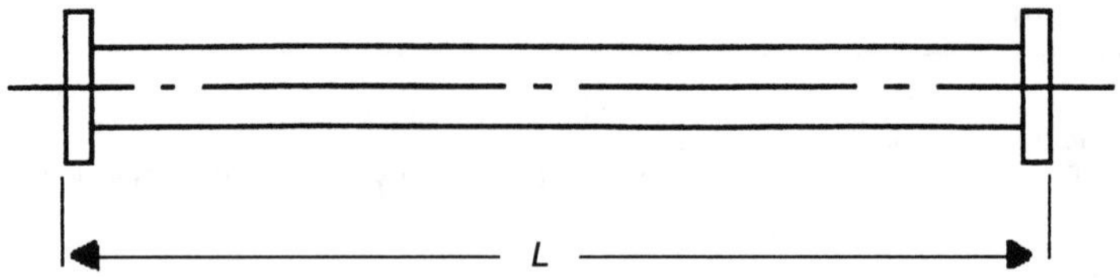

TABLE 25.19 Typical Lengths for Straight Pipe

National pipe size NPS (DN):	2 (50)	3 (80)	4 (100)	6 (150)	8 (200)	10–42 (250–1050)
Maximum recommended length, ft (m)	6 (2)	10 (3)	20 (6)	30 (9)	40 (12)	40 (12)

TABLE 25.20 Maximum Straight Lengths

Pipe size		Maximum straight length, ft (m)
NPS	DN	
2–3	50–80	14 (4.3)
4–8	100–200	24 (7.3)
8–42	200–1050	40 (12.20)

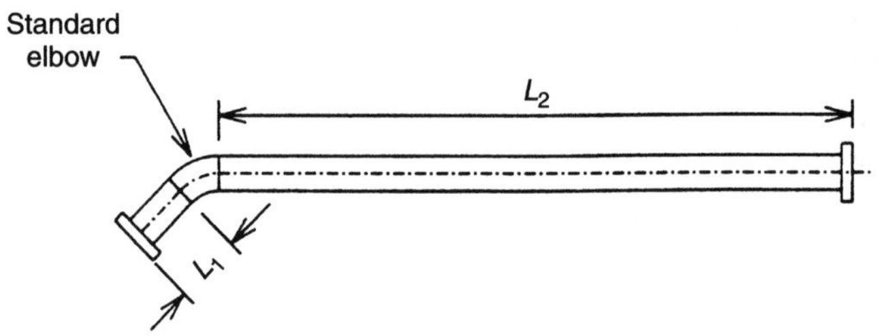

TABLE 25.21 Pipe with Standard 45° Elbow Limitations

Pipe size, NPS (DN)	L_1, max., in (mm)	L_2, max., ft (mm)
2 (50)	2 (51)	6 (1829)
3 (80)	4 (102)	6 (1829)
4 (100)	5 (127)	8 (2438)
6 (150)	8 (203)	10 (3048)
8 (200)	12 (305)	14 (4267)
10 (250)	12 (305)	14 (4267)
12 (300)	18 (457)	14 (4267)
14 (350)	24 (610)	14 (4267)
16 (400)	24 (610)	14 (4267)
18 (450)	24 (610)	20 (6096)
20 (500)	24 (610)	20 (6096)
24 (600)	60 (1524)	30 (9144)

Maximum Length of Pipe with a 90° Elbow. See Table 25.22.

Maximum Length of Pipe with a Branch Connection. See Table 25.23.

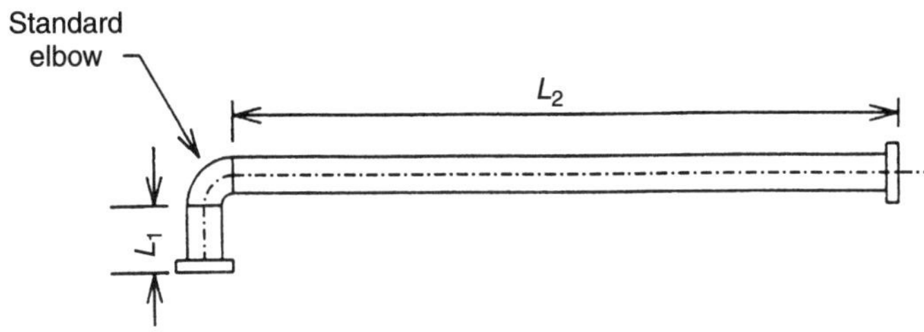

TABLE 25.22 Pipe with Standard 90° Elbow Limitations

Pipe size, NPS (DN)	L_1, max., in (mm)	L_2, max., ft (mm)
2 (50)	2 (51)	4 (1219)
3 (80)	6 (152)	4 (1219)
4 (100)	6 (152)	4 (1219)
6 (150)	6 (152)	6 (1829)
8 (200)	18 (457)	10 (3048)
10 (250)	18 (457)	10 (3048)
12 (300)	24 (610)	10 (3048)
14 (350)	30 (762)	10 (3048)
16 (400)	36 (914)	10 (3048)
18 (450)	36 (914)	10 (3048)
20 (500)	36 (914)	12 (3658)
24 (600)	48 (1219)	30 (9144)

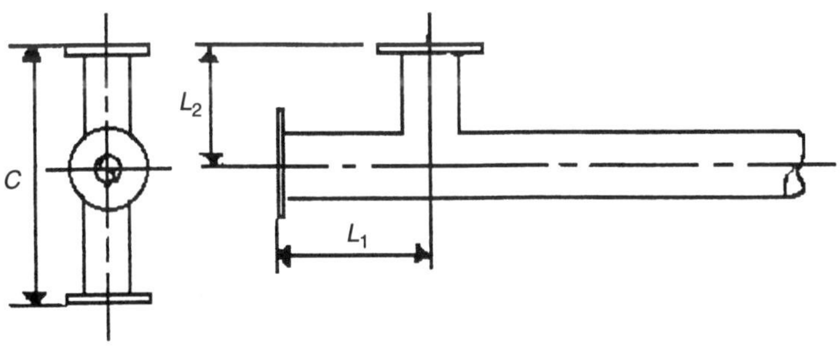

TABLE 25.23 Header Branch Limitations, Branches Near End of Fabricated Piece

HDR size (DN)	Max. L_1, in (mm)	Max. L_2, in (mm)
2 (50)	4 (102)	6 (152)
3 (80)	6 (152)	6 (152)
4 (100)	9 (229)	9 (229)
6 (150)	12 (305)	18 (457)
8 (200)	24 (610)	18 (457)
10 (250)	24 (610)	32 (813)
12 (300)	24 (610)	34 (864)
14 (350)	24 (610)	36 (914)
16 (400)	24 (610)	40 (1016)
18 (450)	36 (914)	48 (1219)
20 (500)	120 (3048)	48 (1219)
24 (600)	240 (6096)	48 (1219)

If branch connections are required in opposite directions off main run, dimension C shall not exceed 3 ft.

Rubber Lining of Pipe, Flanges, and Nozzles

Rubber Lining of Pipe. See Fig. 25.4.

Rubber Lining of Grooved Pipe. See Fig. 25.5.

Flange Face Rubber Lining. See Figs. 25.6 and 25.7.

Rubber Lining of 90° Branch Connection Nozzle. See Fig. 25.8.

Rubber Lining of Lateral Branch Connection. See Fig. 25.9.

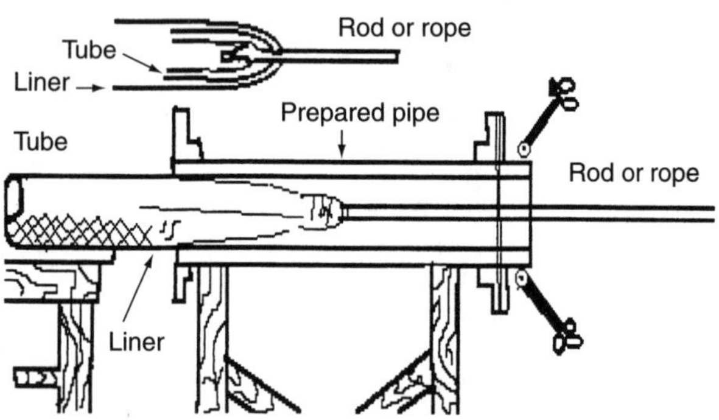

FIGURE 25.4 Rubber lining of pipe (pulling in tube).

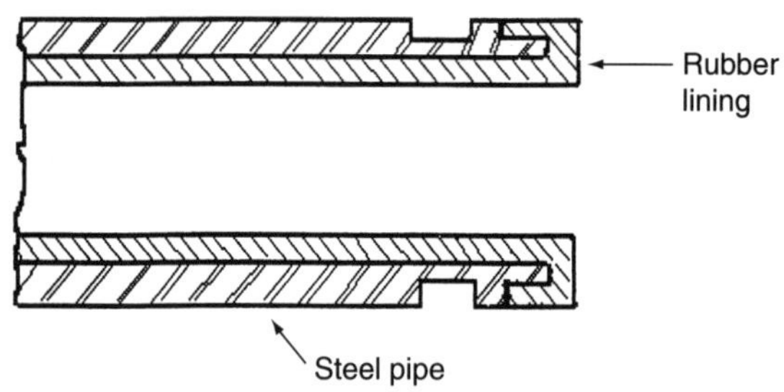

FIGURE 25.5 Rubber lining of grooved pipe.

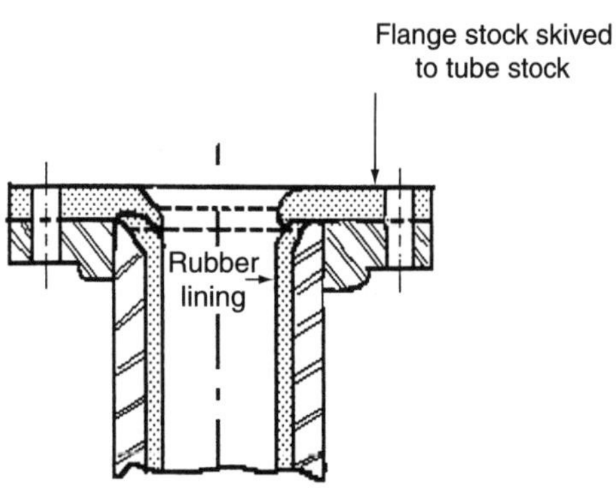

FIGURE 25.6 Flange face rubber lining.

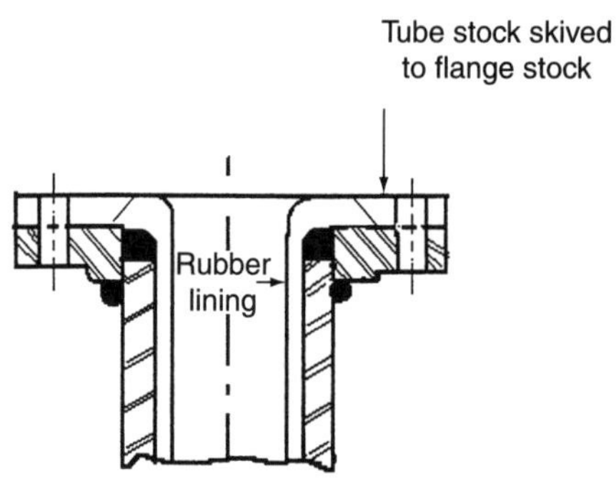

FIGURE 25.7 Flange face rubber lining.

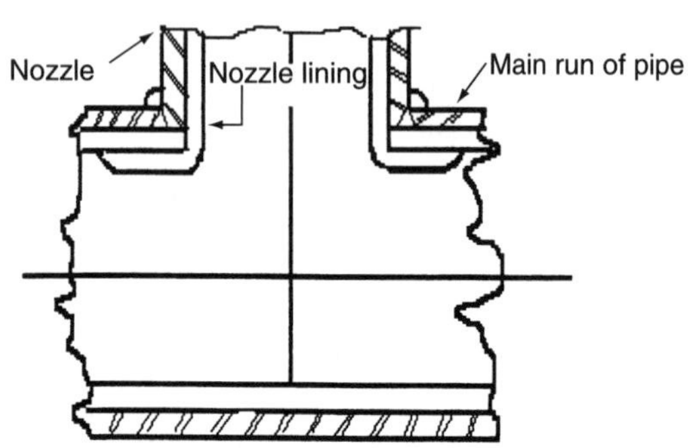

FIGURE 25.8 Rubber lining of nozzle or branch connection.

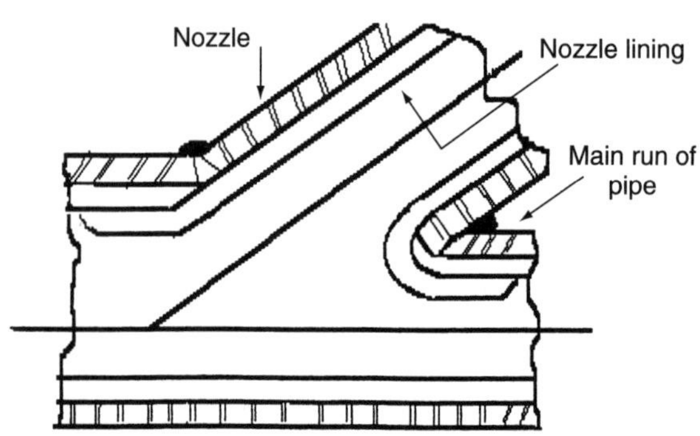

FIGURE 25.9 Rubber lining of lateral nozzle or branch connection.

PLASTIC LINING OF PIPE AND FITTINGS

Applicable Industry Standards

ASTM F423, Standard Specification for Polytetrafluoroethylene (PTFE) Plastic Lined Ferrous Metal Pipe, Fittings, and Flanges

ASTM F491, Standard Specification for Poly (Vinylidene Fluoride) (PVDF) Plastic-Lined Ferrous Metal Pipe and Fittings

ASTM F492, Standard Specification for Propylene and Polypropylene (PP) Plastic-Lined Ferrous Metal Pipe and Fittings

ASTM F546, Standard Specification for Perfluoro (Ethylene-Propylene) Copolymer (FEP) Plastic-Lined Ferrous Metal Pipe and Fittings

ASTM F599, Standard Specification for Poly (Vinylidene Chloride) (PVDC) Plastic-Lined Ferrous Metal Pipe and Fittings

ASTM F781, Standard Specification for Perfluoro (Alkoxyalkane) Copolymer (PFA) Plastic-Lined Ferrous Metal Pipe and Fittings

ASTM F1545, lists requirements for materials of construction, qualification, testing, and inspection requirements for plastic-lined piping.

Plastic Lining Material Specifications and Color Codes

See Table 25.24.

Temperature Limitations of Plastic Lining Materials

See Table 25.25.

Mechanical Properties of Plastic Lining Materials

See Table 25.26.

TABLE 25.24 Polymer Standard Specifications*

Lining material—resin type	Standard resin specification	Allowable resin classification	Standard liner color	Maximum filler material (by weight)
Polypropylene	ASTM D4101	Type I and II	Orange	<30% glass fiber
Poly (vinylidene chloride) (PVDC)	ASTM D729		Gray	<20% glass fiber
Poly (vinylidene fluoride) (PVDF)	ASTM D3222		Black	
Poly (vinylidene fluoride) copolymer (PVDF)	ASTM D5575		Black	
Polytetrafluoroethylene (PTFE)	ASTM D1457 ASTM D4894 ASTM D4895		White	
Perfluoro (ethylene-propylene) copolymer (FEP)	ASTM D2116	Type III	Green	
Perfluoro (alkoxyalkane) copolymer (PFA)	ASTM D3307	Type II	Natural	
Ethylyene tetrafluoroethylene copolymer (ETFE)	ASTM D3159	Type I	Natural	

* A maximum of 1% by weight of additives or colorants, or both, is permissible. Colorants, if used, shall be identified in the manufacturer's specification.

Clean reworked resins may be used, provided all mechanical property requirements are maintained. Only virgin PTFE resin may be used.

TABLE 25.25 Temperature Specifications/Limitations

Material	Temperature range, °F (°C)
PTFE	–20 to 500 (–29 to 260)
PFA	–20 to 500 (–29 to 260)
FEP	–20 to 300 (–29 to 149)
ETFE	–20 to 300 (–29 to 149)
PVDF*	–20 to 275 (–29 to 135)
PP	0 to 225 (–18 to 107)
PVDC	0 to 175 (–18 to 79)

* Homopolymer and copolymer.

TABLE 25.26 Polymer Mechanical Properties

Lining material—resin type	Minimum tensile strength at break, psi (MPa)	Minimum elongation at break, %
Polypropylene (PP) Type I	4000 (27.6)*	10*
Polypropylene (PP) Type II	3000 (20.7)*	10*
Polypropylene (PP) 30% glass filled	2500 (17.3)*	2*
Poly (vinylidene chloride) (PVDC)	1500 (10.3)*	2*
Poly (vinylidene fluoride) (PVDF)	4500 (31.0)	10
Poly (vinylidene fluoride) copolymer (PVDF)	4000 (27.6)	300
Polytetrafluoroethylene (PTFE)	3000 (20.7)	250
Perfluoro (ethylene-propylene) copolymer (FEP)	3000 (20.7)	250
Perfluoro (alkoxyalkane) copolymer (PFA)	3800 (26.2)	300
Ethylyene tetrafluoroethylene copolymer (ETFE)	6500 (44.8)	275

* Minimum tensile strength and elongation at yield.

Chemical Resistance of Plastic Lining Materials

See Table 25.27.

Mechanical Properties of Ferrous Metal Pipe, Fittings, and Flanges

ASTM F1545 requires that ferrous metal pipe, fittings, and flanges be lined with plastic to comply with the ASTM specifications listed in Table 25.28 corresponding to the plastic lining specified.

PTFE Liner Thickness versus Pipe Size

See Table 25.29.

Examples of Plastic-Lined Pipe, Fittings, and Joint Configurations

See Figs. 25.10, 25.11, and 25.12.

Available Sizes of Plastic-Lined Piping

NPS 1 through 12 pipe sizes are readily available. For sizes NPS 14 and larger, contact the manufacturer of plastic-lined piping.

Flow Velocity Limitations

Flow velocities should be maintained less than 12 ft/s (3.7 m/s). Gas velocities in plastic-lined piping should be kept less than 35 ft/s (10.7 m/s).

Total Installed Cost Ratios of Lined and Unlined Piping Systems

See Table 25.30.

TABLE 25.27 Chemical Resistance of Liner Material

Liners made from PTFE and PFA are resistant to the following chemicals to 450°F (232°C). Liners made from FEP are resistant to the following chemicals to 300°F (149°C).

Chemical	Liner material maximum-use temperature, °F			
	ETFE	PVDF	PP	PVDC
Acetic acid (glacial)	230	NR	70	125
Acetone (10%)	150	75	120	75
Ammonia (dry gas)	300	NR	140	
Ammonia aqua (30%)	230	175	150	
Amyl acetate	250	125	NR	125
Benzene	212	170	NR	75
Bromine liquid	150	150	NR	NR
Chlorine liquid	212	200	NR	NR
Chlorine gas	212	175	NR	75
Chlorine dioxide (15%)	250	150	NR	125
Chlorosulfonic acid	75	NR	NR	NR
Cyclohexane	300	275	NR	125
Diethylamine	230	70	120	NR
Ethyl acrylate	212	70	NR	
Formaldehyde (37%)	230	120	140	125
Formic acid	275	250	140	150
Hydrochloric acid (10%)	300	275	185	175
Hydrochloric acid (20%)	300	275	175	175
Hydrochloric acid (36%)	300	275	150	
Hydrofluoric acid (35%)	275	250	200	175
Hydrofluoric acid (100%)	230	212	70	NR
Hydrogen peroxide (30%)	250	212	70	125
Hydrogen peroxide (90%)	150	75	70	125
Methyl ethyl ketone	230	NR	70	NR
Methylene chloride	212	70	70	NR
Nitric acid (10%)	212	225	150	150
Nitric acid (50%)	150	120	70	125
Nitric acid (90%)	NR	NR	NR	NR
Phenol	212	120	140	75
Phosgene (gas or liquid)	212	120	NR	
Phosphoric acid	275	250	225	125
Propyl alcohol	212	120	140	150
Sodium hydroxide (10%)	230	NR	225	150
Sodium hydroxide (50%)	230	NR	225	75
Sodium hypochlorite	300	125	150	125
Sulfuric acid (30%)	300	230	200	75
Sulfuric acid (50%)	300	230	125	NR
Sulfuric acid (93%)	300	200	NR	NR
Sulfuric acid (98%)	200	120	NR	NR
Sulfuric acid—fuming	120	NR	NR	NR
Toluene	250	170	NR	75
Trisodium phosphate	275	275	150	150

To convert from °F to °C, subtract 32 and multiply by 5/9.
NR = not recommended.

TABLE 25.28 Approved Ferrous-Metal Flange and Fitting Material Standards (ASTM)

Lining material	Pipe specifications	Flange specifications	Fitting specifications
PVDF, PTFE, FEP, ETFE, and PFA	A53	A105	A105
	A106	A181	A181
	A135	A182	A182
	A312	A216	A216
	A513	A395	A234
	A587	A536	A351
			A395
			A403
			A536 (60-40-18)
PP and PVDC	A53	A105	A48
	A106	A181	A105
	A135	A182	A126
	A312	A216	A181
	A513	A395	A182
	A587	A536	A216
			A234
			A278
			A351
			A395
			A403
			A536

TABLE 25.29 Published PTFE-Liner Thickness versus Pipe Size

Pipe size, NPS (DN)	PTFE-liner thickness range, in (mm)
1 (25)	0.125 (3.2)–0.138 (3.5)
1.5 (40)	0.125 (3.2)–0.150 (3.8)
2 (50)	0.125 (3.2)–0.160 (4.1)
3 (80)	0.125 (3.2)–0.185 (4.7)
4 (100)	0.125 (3.2)–0.187 (4.7)
6 (150)	0.145 (3.7)–0.275 (7.0)
8 (200)	0.150 (3.8)–0.315 (8.0)
10 (250)	0.190 (4.8)–0.325 (8.2)
12 (300)	0.200 (5.1)–0.425 (10.8)

FIGURE 25.10 Lined pipe flanged end. (*Courtesy of Crane Resistoflex.*)

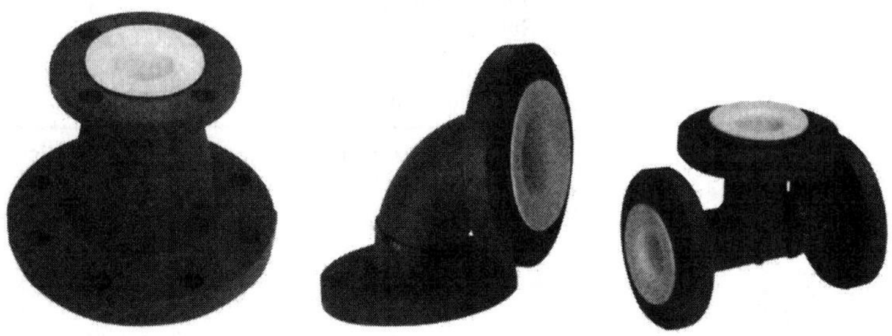

FIGURE 25.11 Concentric reducer, elbow, and tee. (*Courtesy of Crane Resistoflex.*)

Connections of plastic-lined pipe to other types of piping

Plastic-lined pipe to flanged plastic pipe

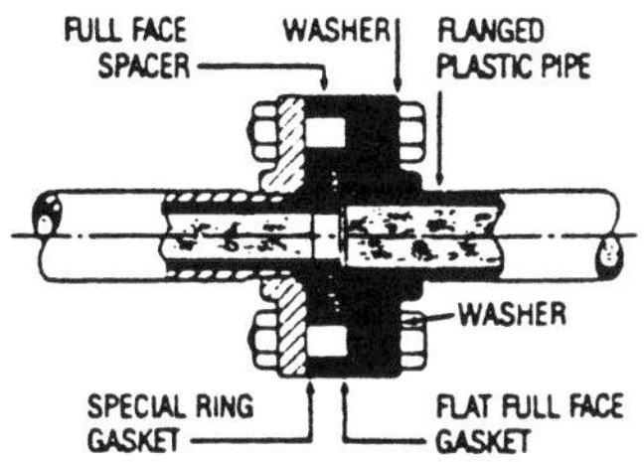

Note: Extreme care must be used to avoid bending of plastic flange. A special nng gasket is required around the O.D. of the flared plastic face of the plastic-lined pipe.

Plastic-lined pipe to flanged metal pipe, valves, fittings, and pumps

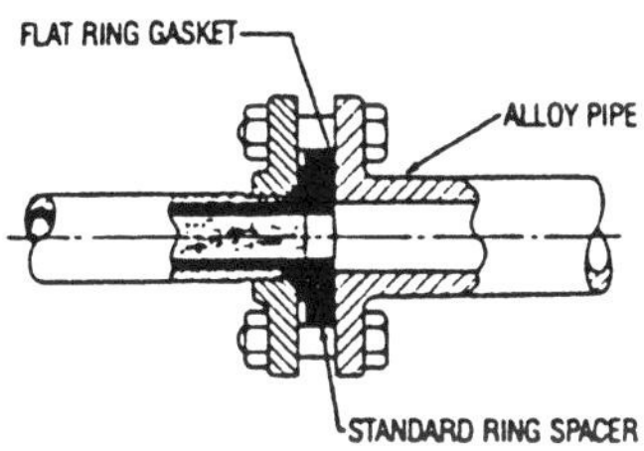

NOTE: No flat gasket required if flange face is smooth and non-fragile. However. most alloy piping uses raised face flanges, and therefore, a gasket should be used.

Plastic-lined pipe to Schedule 10 alloy pipe with lap joint flange

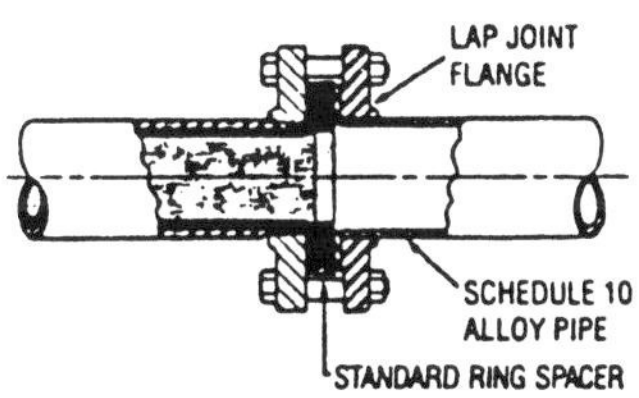

FIGURE 25.12 Recommended installation of plastic-lined pipe to carbon-steel, alloy, and non metallic piping systems. (*Courtesy of Crane Resistoflex.*)

Pipe-to-pipe connections between different joint types

Connection of reinforced flared face to gasketed plastic-lined pipe

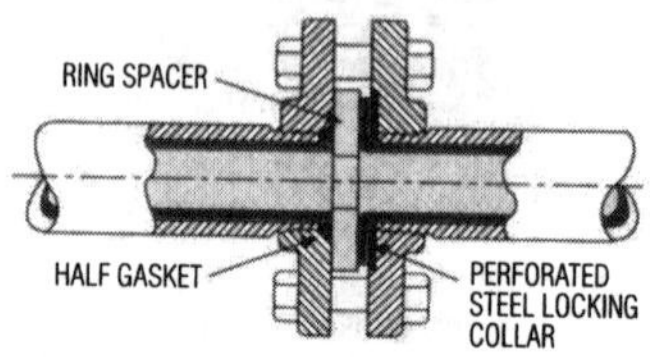

Providing low-angle bend using tapered face spacer

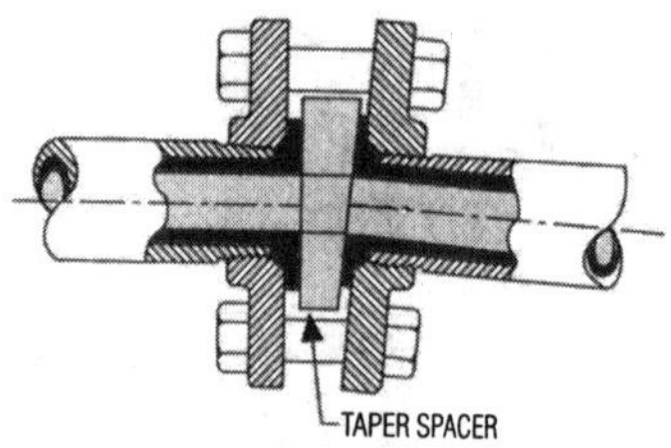

NOTE: Avoid axial forces on this joint. Such forces will tend to push the spacer outward. Consider bent pipe for an alternative to tapered spacers.

Pipe-to-pipe reducing connections

With standard reducing spacer

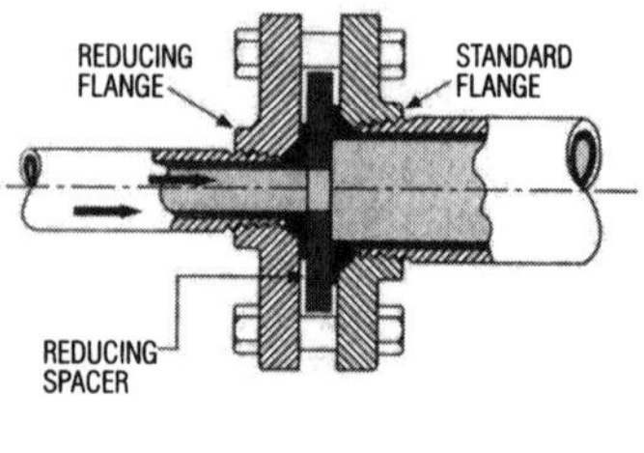

With taper reducing spacer

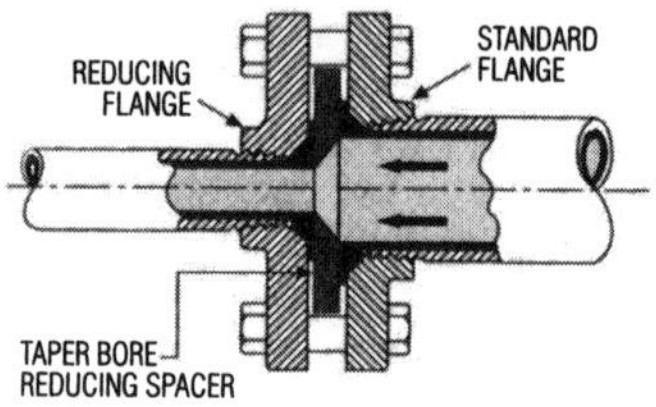

With reducing filler flange

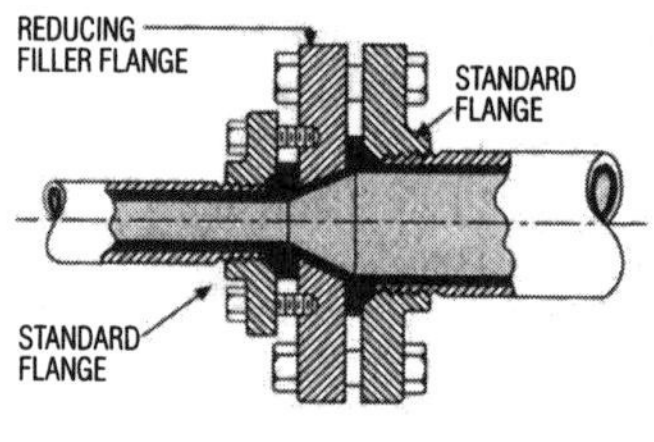

FIGURE 25.12 (Continued)

Plastic-lined pipe to porcelain, glass or carbon pipe[1]

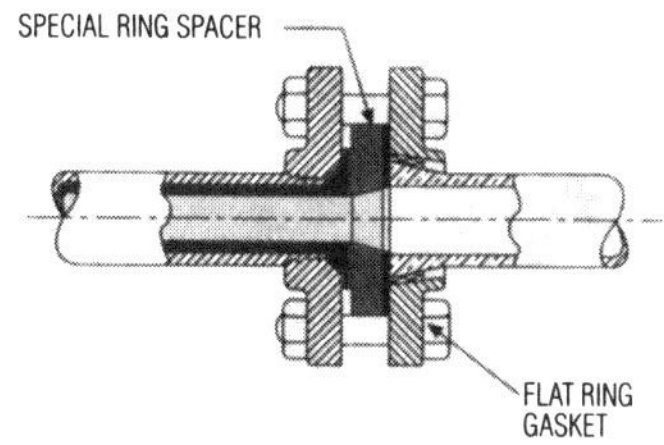

[1]In order to ensure that the special ring spacer supplied has the correct I.D., the I.D. of the porcelain, glass or carbon pipe should be specified on the order or inquiry

Plastic-lined pipe to glass fiber-reinforced plastic (FRP) flanged fitting or pipe

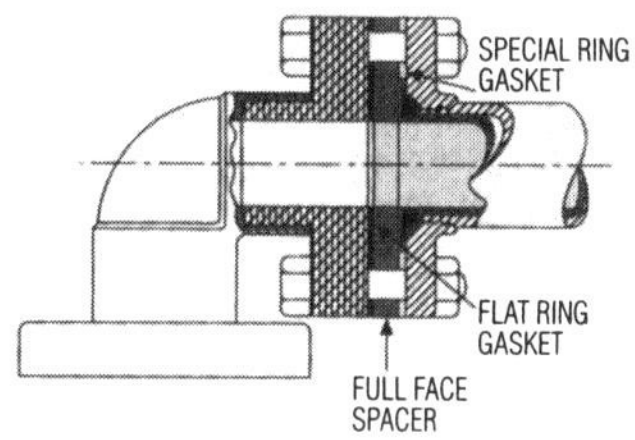

NOTE: Extreme care must be used to avoid bending of FRP flange. Special spacer may be required. A special ring gasket is required around the O.D. of the flared plastic face of the plastic-lined pipe

Blanking one outlet of fitting

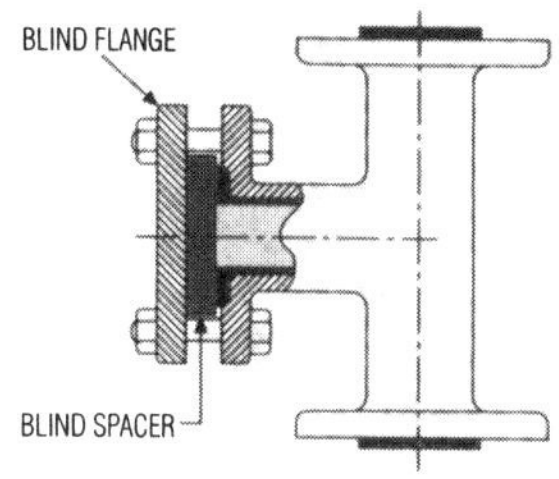

Valve connections

Plastic-lined pipe to butterfly valve

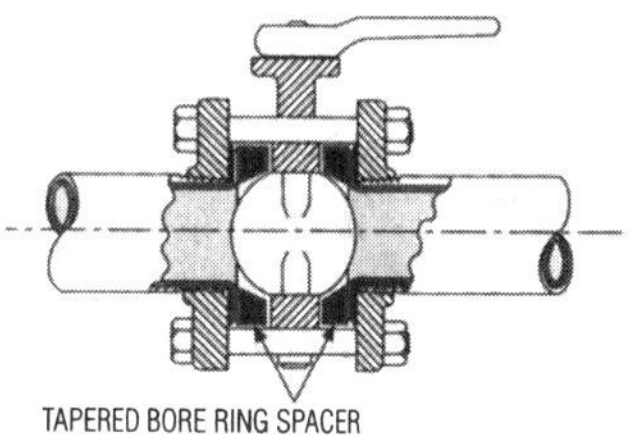

FIGURE 25.12 (Continued)

TABLE 25.30 Piping Material Cost Ratios

Installed cost ratio (carbon steel as 1.00) of plastic-lined piping and other selected piping materials. Cost ratios based on a shop-fabricated complex system NPS 2 (DN 50) and 400 ft (102 m).	
PVC (Sch. 80)	0.56
CPVC (Sch. 80)	0.63
Carbon steel (Sch. 40)	1.00
304L S.S. (Sch. 10)	1.13
Rubber-lined steel (Sch. 40)	1.16
316L S.S. (Sch. 10)	1.20
304L S.S. (Sch. 40)	1.31
316L S.S. (Sch. 40)	1.45
FRP/vinyl ester	1.78
FRP/epoxy	1.86
FRP/polyester	1.86
Polypropylene lined steel (Sch. 40)	1.90
Saran lined steel (Sch. 40)	1.91
PVDF-lined steel (Sch. 40)	2.47
Alloy 20 (Sch. 10)	2.60
Monel (Sch. 10)	2.61
Glass-lined steel (Sch. 40)	2.69
PTFE-lined FRP	3.20
PVDF (Sch. 80)	2.71
PTFE-lined steel (Sch. 40)	2.94
Titanium (Sch. 10)	2.99
FEP-lined steel (Sch. 40)	2.99
Monel (Sch. 40)	3.24
Alloy 20 (Sch. 40)	3.32
Nickel (Sch. 10)	3.34
Hastelloy C-276 (Sch. 10)	3.52
PTFE-lined 304L SS (Sch. 10)	4.12
Nickel (Sch. 40)	4.27
Titanium (Sch. 10)	4.46
Hastelloy C-276 (Sch. 40)	4.46
Hastelloy B (Sch. 40)	5.71
Zirconium (Sch. 10)	5.95
Zirconium (Sch. 40)	7.04

Tables in Chapter 25

S · E · C · T · I · O · N · 7

PIPE SUPPORTS

CHAPTER 26
PIPE SUPPORTS AND HANGERS FOR METALLIC PIPING

See p. 736 for a list of tables in Chap. 26.

The design of pipe supports and hangers involves many parameters that must be closely evaluated prior to choosing the proper supporting components and design. Consideration must be given to the pipe material and its contents, insulation thickness if applicable, deadweight spans and loads, support spans and loads to satisfy seismic or wind or transient conditions if applicable, operating temperature conditions, and sufficient flexibility within the piping system to avoid overstressing the pipe or equipment nozzle connections.

In all cases the pipe support engineer or designer must follow the specific ordering instructions and limitations stated in the supplier's catalog.

SUPPORT COMPONENTS

To assist in selecting the proper support, brief descriptions of the various components and their functions follow.

Typical Rod Support. See Fig. 26.1. This assembly provides a rigid means of supporting the pipe in the vertical downward direction only.

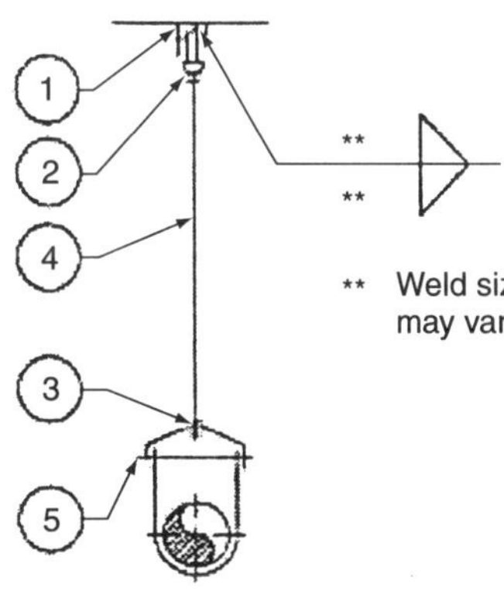

Elevation view

Item	Quantity	Description
1.	1	Welded beam attachment with bolt and nut
2.	1	Weldless eyenut
3.	3	Hex nut
4.	1	Continuous-thread rod
5.	1	Adjustable clevis for insulated line

FIGURE 26.1 Typical rod support.

Various Structural and Pipe Attachments. See Figs. 26.2 and 26.3.

- An example illustrating various structural and pipe attachments that can be selected.
- Other structural attachments not shown are concrete inserts, brackets, ceiling flanges, horizontal travelers, etc. The pipe support engineer or designer should consult with a specific pipe support hardware supplier for component availability.
- Other pipe attachments not shown are pipe rings, socket clamps, pipe stanchion, pipe roll, pipe alignment guides, pipe slides, etc. The pipe support engineer or designer should consult with a specific pipe support hardware supplier for component availability.
- Proper selection of structural and pipe attachments depends upon component load ratings, piping material and operating temperatures, insulation thickness, parts compatibility and functionality of the assembly, rod swing from various operating conditions, installation requirements such as welded versus nonwelded attachments, and material and installation costs.

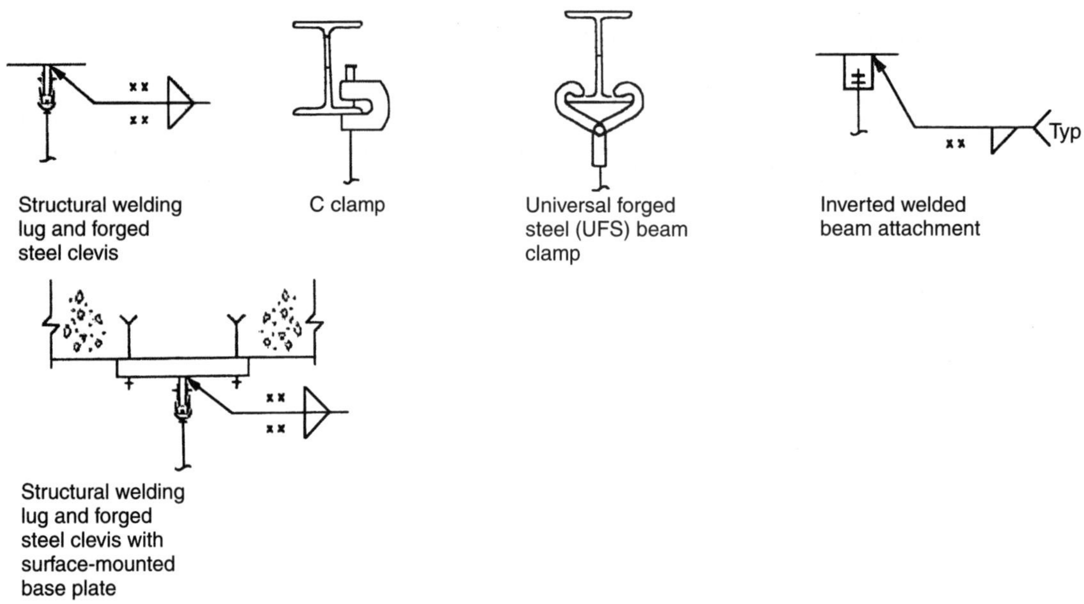

FIGURE 26.2 (*a*) Various structural attachments (elevation view).

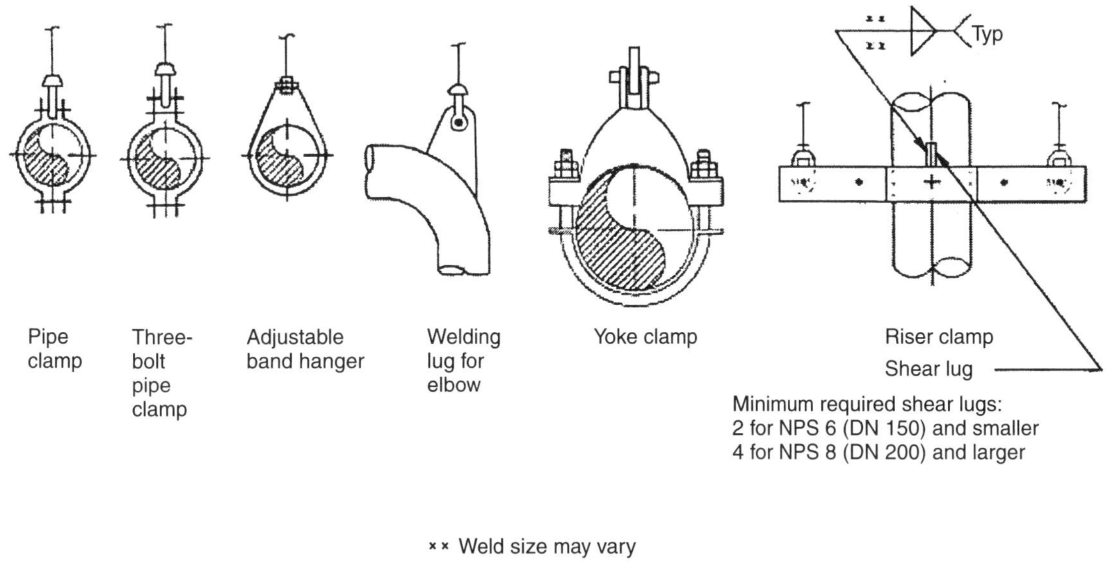

FIGURE 26.2 (*b*) Various pipe attachments (elevation view).

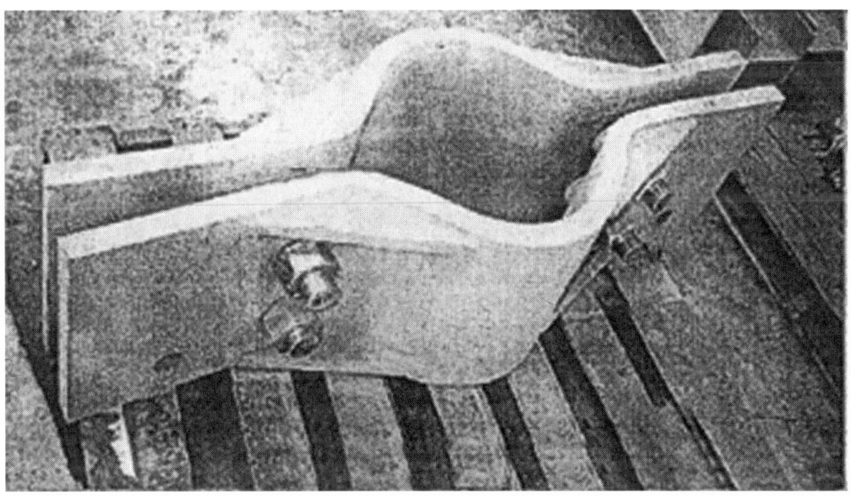

FIGURE 26.3 Specially designed riser clamp. (*Courtesy Piping Technology & Products Inc.*)

SPACING OF PIPING SUPPORTS

Steel Pipe

See Table 26.1. When welded shear lugs are utilized, the recommendations in Table 26.2 shall be used.

Ductile Iron Pipe

A maximum support spacing of 20 ft (6.1 m) is allowed. A minimum of one hanger is placed as close as practical to the joint behind the bell, and at changes of directions and branch connections. For NPS 6 and under (DN 150 and under) used on ANSI B31 projects and subjected to loading other than weight of pipe and

TABLE 26.1 Suggested Piping Support Spacing

NPS	(DN)	Suggested maximum span, ft (m) Water service	Suggested maximum span, ft (m) Steam, gas, or air service
1	(25)	7 (2.13)	9 (2.74)
2	(50)	10 (3.05)	13 (3.96)
3	(80)	12 (3.66)	15 (4.57)
4	(100)	14 (4.27)	17 (5.18)
6	(150)	17 (5.18)	21 (6.40)
8	(200)	19 (5.79)	24 (7.32)
12	(300)	23 (7.01)	30 (9.14)
16	(400)	27 (8.23)	35 (10.7)
20	(500)	30 (9.14)	39 (11.9)
24	(600)	32 (9.75)	42 (12.8)

Reference: ASME B31.1 Power Piping Code, 1998.

TABLE 26.2 Suggested Support Spacing with Welded Shear Lugs

NPS	DN	Minimum number of lugs	Number of lugs designed to support load
6 and under	150 and under	2	2
8 and larger	200 and under	4	2

Reference: Pipe Fabrication Institute, ES-26, 1993.

contents, and span is limited to the maximum spacing for water service steel pipe. (*Reference*: MSS SP-69, 1996.)

Cast Iron Pipe

A maximum support spacing of 10 ft (3.0 m) is allowed. A minimum of one hanger is placed as close as practical to the joint on the barrel, and at changes of directions and branch connections. (*Reference*: MSS SP-69, 1996.)

Copper Pipe (Types L and K)

See Table 26.3. The table does not apply where there are concentrated loads between supports or where temperatures exceed 200°F (93°C).

Fire Protection Pipe

See Table 26.4. Pipe supports for dead weight applications must be designed for loads that are 5 times the weight of the water-filled pipe plus 250 pounds (1110 newtons). All pipe support components must be ferrous. No welded attachments are permitted.

Design Loads

See Table 26.5.

Variable Spring Support. See Figs. 26.4 and 26.5.

- This assembly is used to support the gravity load of the piping while allowing vertical thermal movement.
- This type of support permits a variance in loads between the hot and cold condition of the piping system. This variance is called variability and is calculated as follows:

$$\text{Variability} = (\text{thermal movement} \times \text{spring constant})/\text{hot load}$$

TABLE 26.3 Suggested Piping Support Spacing

NPS (DN)	Suggested maximum span, ft (m)	
	Water service	Vapor service
1/4 (5)	5 (1.52)	5 (1.52)
3/8 (10)	5 (1.52)	6 (1.83)
1/2 (15)	5 (1.52)	6 (1.83)
3/4 (20)	5 (1.52)	7 (2.13)
1 (25)	6 (1.83)	8 (2.44)
1 1/4 (30)	7 (2.13)	9 (2.74)
1 1/2 (40)	8 (2.44)	10 (3.05)
2 (50)	8 (2.44)	11 (3.35)
2 1/2 (65)	9 (2.74)	13 (3.96)
3 (80)	10 (3.05)	14 (4.27)
3 1/2 (95)	11 (3.35)	15 (4.57)
4 (100)	12 (3.66)	16 (4.88)
5 (125)	13 (3.96)	18 (5.49)
6 (150)	14 (4.27)	20 (6.10)
8 (200)	16 (4.88)	23 (7.01)
10 (250)	18 (5.49)	25 (7.62)
12 (300)	19 (5.79)	28 (8.53)

Reference: MSS SP-69, 1996.

TABLE 26.4 Maximum Distance between Dead Weight Supports, ft (m)

NPS	DN	Steel pipe, except threaded light wall	Threaded light wall steel pipe	Copper tube	CPVC	Polybutylene (IPS)	Polybutylene (CTS)
3/4	20	NA	NA	8.00 (2.44)	5.50 (1.68)	NA	2.92 (0.89)
1	25	12.00 (3.66)	12.00 (3.66)	8.00 (2.44)	6.00 (1.83)	3.75 (1.14)	3.33 (1.02)
1 1/4	30	12.00 (3.66)	12.00 (3.66)	10.00 (3.05)	6.50 (1.98)	4.58 (1.40)	3.92 (1.20)
1 1/2	40	15.00 (4.57)	12.00 (3.66)	10.00 (3.05)	7.00 (2.13)	5.00 (1.52)	4.42 (1.35)
2	50	15.00 (4.57)	12.00 (3.66)	12.00 (3.66)	8.00 (2.44)	5.92 (1.80)	5.42 (1.65)
2 1/2	65	15.00 (4.57)	12.00 (3.66)	12.00 (3.66)	9.00 (2.74)	NA	NA
3	80	15.00 (4.57)	12.00 (3.66)	12.00 (3.66)	10.00 (3.05)	NA	NA
3 1/2	950	15.00 (4.57)	NA	15.00 (4.57)	NA	NA	NA
4	100	15.00 (4.57)	NA	15.00 (4.57)	NA	NA	NA
5	125	15.00 (4.57)	NA	15.00 (4.57)	NA	NA	NA
6	150	15.00 (4.57)	NA	15.00 (4.57)	NA	NA	NA
8	200	15.00 (4.57)	NA	15.00 (4.57)	NA	NA	NA

Note: NA = not applicable.
IPS = iron pipe size.
CTS = copper tube size.
CPVC = Chlorinated Polyvinyl Chloride.
Reference: NFPA 13, Standard for the Installation of Sprinkler Systems, 1999.

TABLE 26.5 Minimum Design Loads for Pipe Supports for steel Pipe

NPS	DN	Vertical design load: Pounds	Vertical design load: Newtons
3/8	10	150	670
1/2	15	150	670
3/4	20	150	670
1	25	150	670
1 1/2	40	150	670
2	50	150	670
2 1/2	65	170	760
3	80	210	935
4	100	300	1,335
6	150	530	2,360
8	200	760	3,380
10	250	1120	4,980
12	300	1480	6,585
14	350	1710	7,610
16	400	2030	9,030
18	450	2580	11,500
20	500	2920	13,000
24	600	4180	18,600

Source: MSS SP-58, 1993.

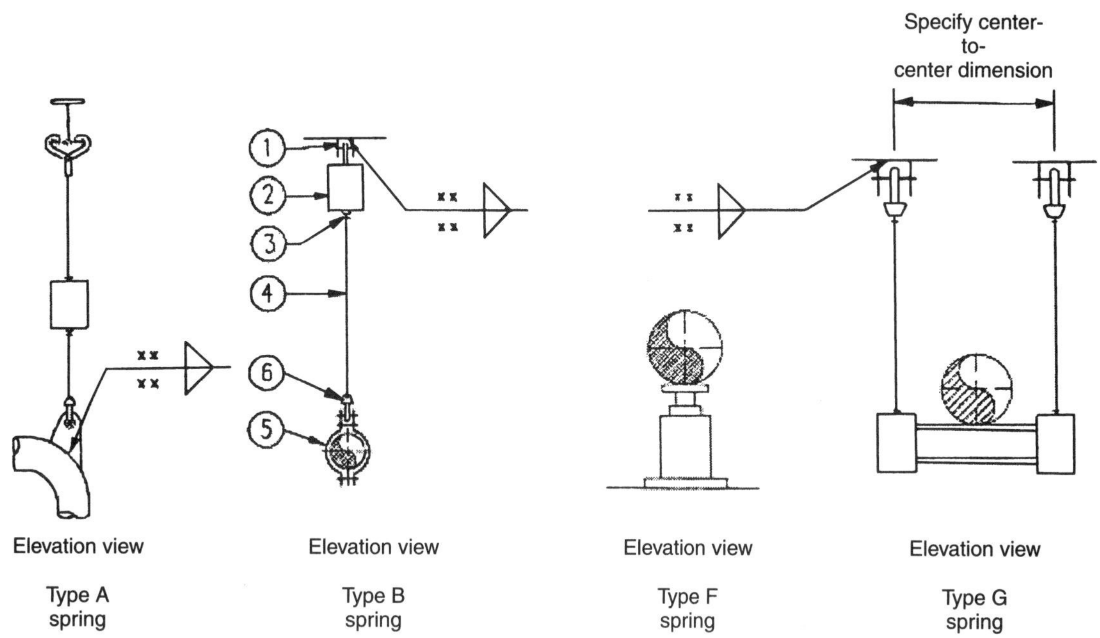

xx Weld size may vary

Item	Quantity	Description
1.	1	Welded beam attachment
2.	1	Variable spring, type B
3.	2	Hex nut
4.	1	Continuous-thread rod
5.	1	Heavy-duty double-bolt pipe clamp
6.	1	Weldless eyenut

FIGURE 26.4 Variable spring supports, elevation views.

FIGURE 26.5 Type B variable spring support assembly. Travel stops are shown banded around the spring at the cold position setting. (*Courtesy Piping Technology & Products Inc.*)

The variability shall not exceed 25 percent. It may be necessary to have a much lower variability when spring supports are adjacent to equipment nozzles.

- Structural and pipe attachments are similar to those shown in Fig. 26.2.
- This support may also be used with trapeze supports such as those shown in Fig. 26.7.
- Other variable spring support types are available that are not shown. Consult with a specific pipe support hardware supplier for additional spring types.

Saddle Support. See Fig. 26.6.

- This component is used to prevent crushing of insulation and to prevent heat loss at the support point.
- It can rest on existing steel or be used with a box frame to restrain the pipe. It can also be used in conjunction with a variable spring support Type F or G (Fig. 26.3) or a trapeze support (Fig. 26.7).
- Clearance must be provided on top and both sides between the saddle and box frame in order to prevent binding of the support.

Trapeze Support. See Fig. 26.7.

- This type of support is used where a single rod support cannot be used because of an interference with another commodity directly above.
- This support can also be used with a saddle, as shown in Fig. 26.6.
- This support can provide an increased pin-to-pin distance for rod swing angle, whereas a single rod will not.

Constant Support. See Fig. 26.8.

- This assembly is used to support the gravity load of the piping while allowing vertical thermal movement.

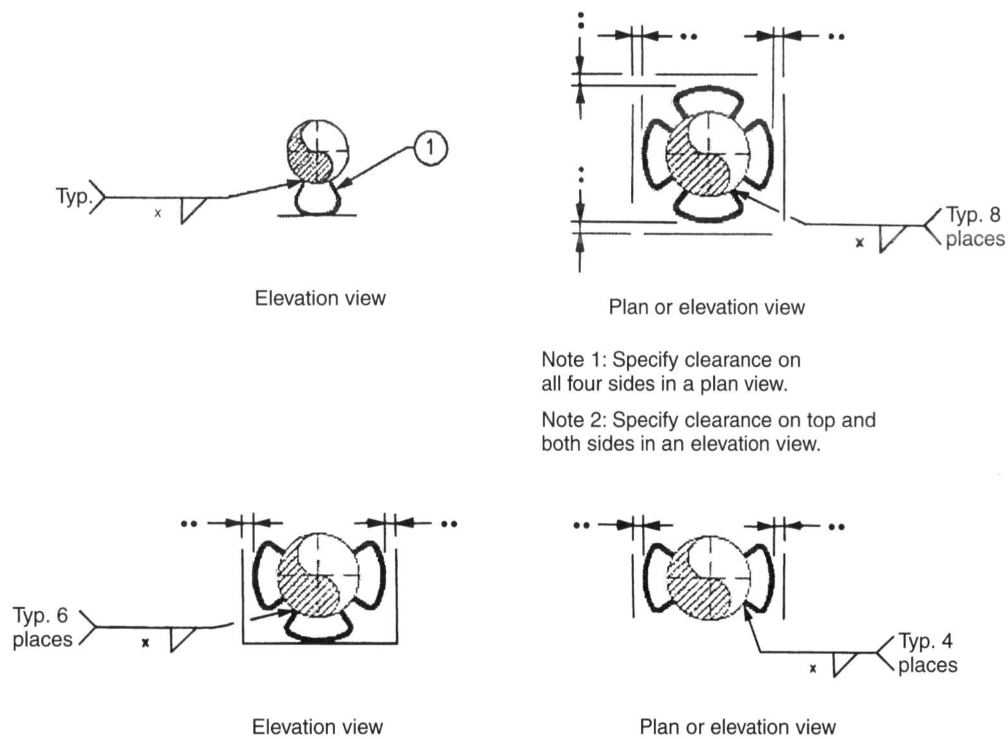

× Weld size may vary.

•• A predetermined clearance must be specified

Item	Quantity	Description
1.	1, 2, 3, or 4, as required	Pipe covering protection saddle

FIGURE 26.6 Saddle supports.

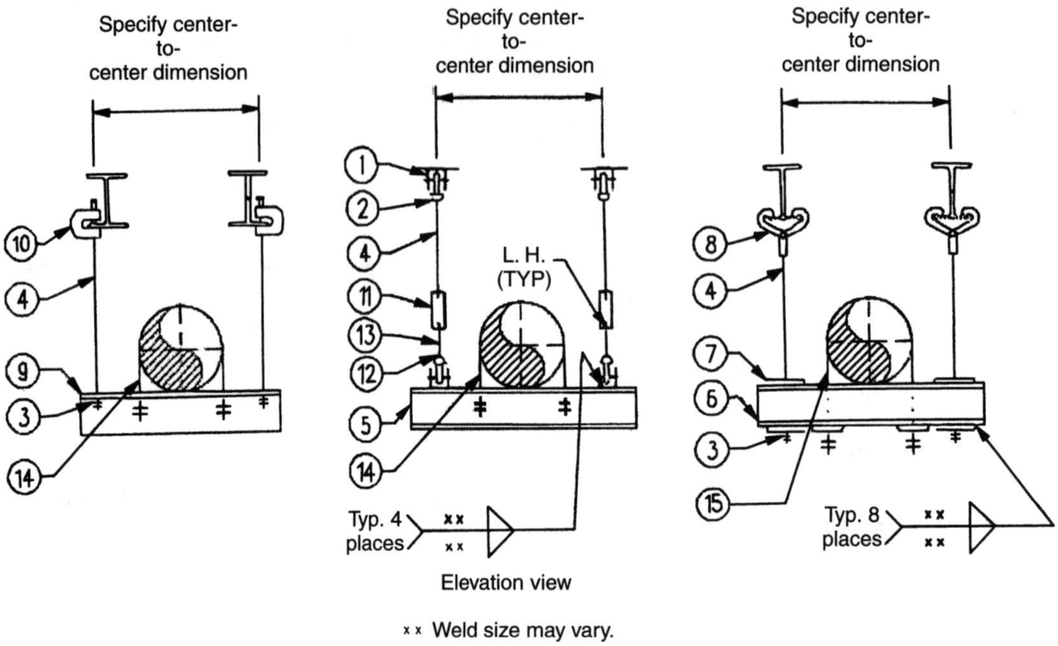

Item	Quantity	Description
1.	4	Welded beam attachment with bolt and nut
2.	4	Weldless eyenut
3.	4	Hex nut
4.	2	Continuous-thread rod
5.	1	Wide-flange beam
6.	2	Channel iron, back to back
7.	6	Steel washer plate
8.	2	Universal forged steel beam clamp
9.	1	Angle iron
10.	2	C clamp
11.	2	Turnbuckle
12.	6	Hex nut
13.	2	Right- and left-hand threaded rod
14.	1	Standard U bolt, as required
15.	1	Special U bolt, as required

FIGURE 26.7 Trapeze support.

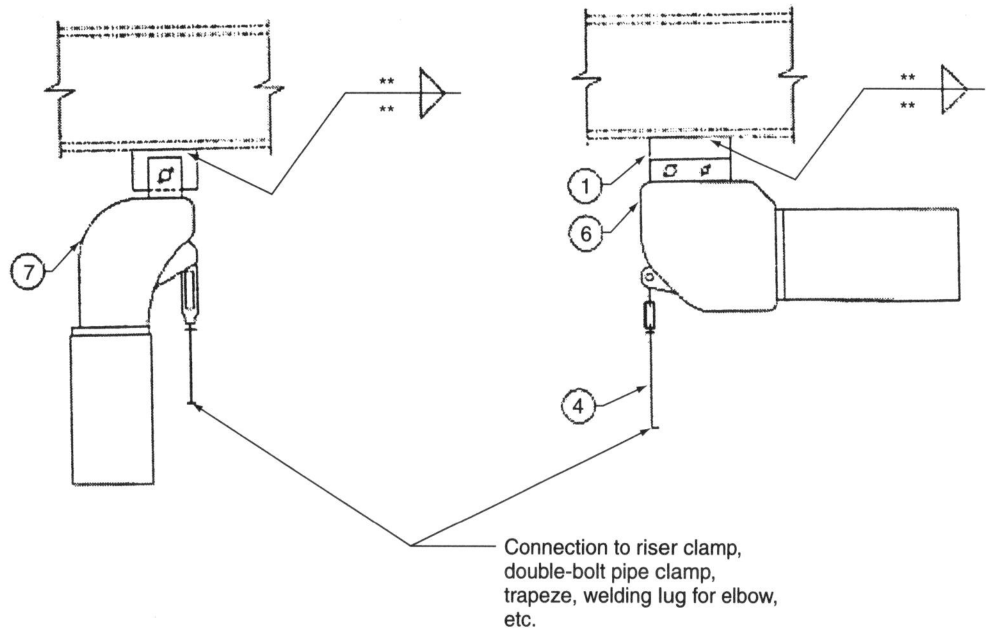

Elevation view

** Weld size may vary.

Item	Quantity	Description
1.	1	Two-hole welding beam lug
4.	1	Continuous-thread rod
6.	1	Horizontal constant support, Type C
7.	1	Vertical constant support, Type C

FIGURE 26.8 Constant support.

- This type of support allows no variance in loads between the hot and cold condition of the piping system.
- Structural and pipe attachments are similar to those shown in Fig. 26.2.
- This support may also be used with trapeze supports such as those shown in Fig. 26.7.
- There are many configurations of constant supports, other than those shown, to satisfy different physical conditions in the plant. Consult with a specific pipe support hardware supplier for additional types of constant supports.

Snubber Support. Mechanical snubbers are illustrated in Figs. 26.9 and 26.10.

- This device allows unrestrained thermal movement along the axis of the snubber while providing a means of rigid restraint during seismic-, wind-, or flow-induced transients in both the tension and compression directions of the snubber.
- For the double-snubber application, the pipe support engineer or designer must calculate the resultant load in each snubber if the snubbers are not oriented in the same directions as the applied loads from the piping system.
- Mechanical and hydraulic snubbers (Fig. 26.11) are available. Both types can be ordered with a rear bracket at one end and a pipe clamp at the other end, or with rear brackets at both ends for special applications such as welding to a pipe stanchion or valve operator.

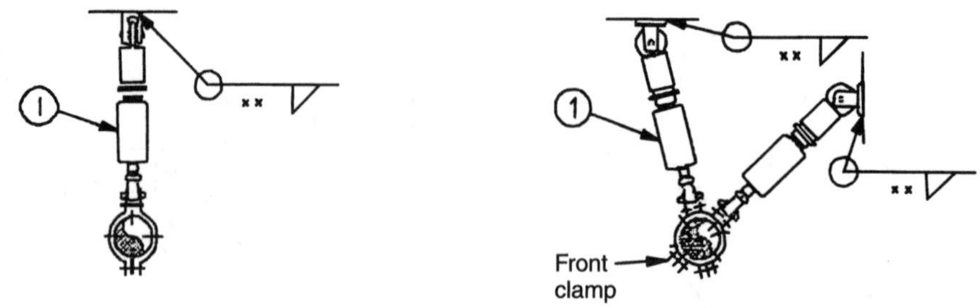

Plan or elevation view
Single snubber application

Plan or elevation view
Double snubber application

x x Weld size may vary.

Item	Quantity	Description
1.	1 (2 required for double-snubber application)	Mechanical snubber

Note: Snubbers of different lengths and/or sizes must be ordered separately.

FIGURE 26.9 Diagrams of mechanical snubbers.

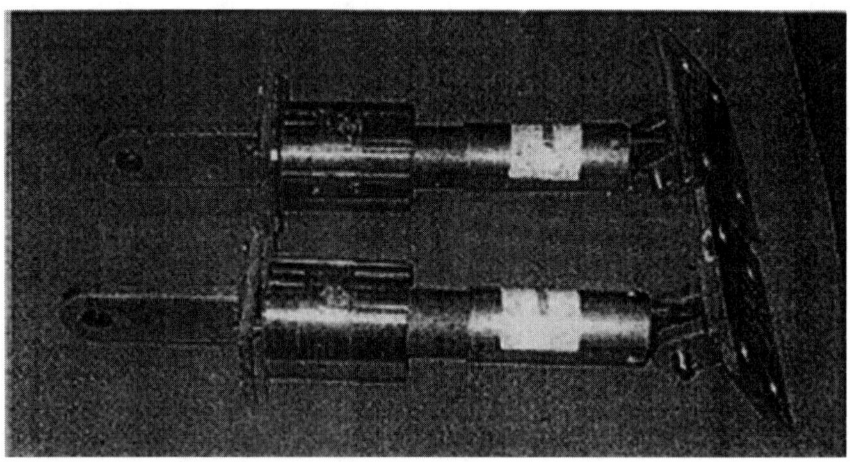

FIGURE 26.10 Photograph of mechanical snubbers. (*Courtesy Piping Technology & Products Inc.*)

Sway Strut Support. See Figs. 26.12, 26.13, and 26.14.

- This device provides a rigid means of restraining the pipe in both the tension and compression directions of the sway strut.
- For the double-strut application, the pipe support engineer or designer must calculate the resultant load in each strut if the struts are not oriented in the same directions as the applied loads from the piping system.
- Sway struts can be ordered with a rear bracket at one end and a pipe clamp at the other end, or with rear brackets at both ends for special applications such as welding to a pipe stanchion or valve operator.

FIGURE 26.11 Hydraulic snubber.

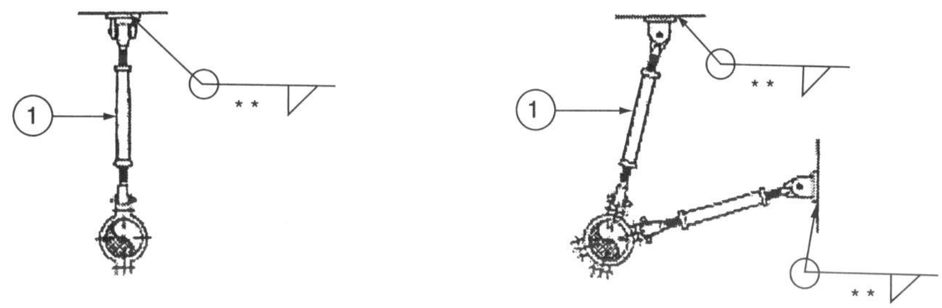

Plan or elevation view
Single-strut application

Plan or elevation view
Double-strut application

* * Weld size may vary.

Item	Quantity	Description
1.	1 (2 required for double-strut application)	Double strut assembly

Note: Struts of different lengths and /or sizes must be ordered separately.

FIGURE 26.12 Diagram of sway strut supports.

FIGURE 26.13 Photograph of Sway struts. (*Courtesy Piping Technology & Products Inc.*)

FIGURE 26.14 Specially designed sway strut clamp. (*Courtesy Piping Technology and Products Inc.*)

Tables in Chapter 26

CHAPTER 27

PIPE SUPPORTS AND HANGERS FOR NONMETALLIC PIPING

See p. 744 for a list of tables in Chap. 27.

THERMOPLASTIC PIPING MATERIAL

Important support considerations for thermoplastic piping are as follows:

- The piping must be properly supported to avoid excessive stresses and sagging.
- Valves and other heavy concentrated loads must be supported as close to the concentrated load as possible.
- Piping must be supported in a manner that avoids mechanical damage such as crushing, scratching, and denting.
- Support configurations must allow for sufficient piping flexibility in order to avoid excessive bending and axial stresses and fatigue effects induced by repetitive expansion and contraction. Piping with terminal points consisting of elastomeric gasket or compression joints must be adequately restrained to resist pullout forces or minimize pipe translation and rotations.
- Pipe supports used for steel pipe may be used for thermoplastic pipe provided there is protection between the pipe and the steel support. It is suggested that a plastic-coated pipe clevis or a neoprene wrap be used to cushion the pipe from the steel support. A plastic sleeve may be bonded to the pipe at the support location as an alternative method of protection.
- Vertical piping must be supported at intervals to reduce loads on the lower fittings. Use of riser clamps or double-bolt pipe clamps located just below a coupling or other fitting will avoid excessive compression on the pipe.
- When anchors are required to prevent translation and rotation of the piping at a particular location, they should be placed as close to elbows and tees as possible. This will prevent excessive bending and axial stresses at the fittings.
- If guides are required to prevent lateral movements but allow axial movement, then they must be located and configured to prevent overstressing the pipe.
- Suggested maximum pipe support spacing for various thermoplastic piping materials is given in Tables 27.1 to 27.4.

THERMOSETTING FIBERGLASS PIPING MATERIAL

Important support considerations for thermosetting fiberglass piping are as follows:

- The piping must be supported to avoid point loads. Point loads can cause significant damage, which will affect the life of the system.
- Piping must be supported in a manner that avoids mechanical damage such as crushing, scratching, and denting. (*List continued on p. 740.*)

TABLE 27.1 Polyvinyl Chloride (PVC) Pipe: Suggested Piping Support Spacing

NPS	DN	Pipe schedule	Pipe temperature, °F (°C) 60 (16)	100 (38)	140 (60)
			Suggested maximum span, ft (m)		
½	15	40	4.50 (1.37)	4.00 (1.22)	2.50 (0.76)
¾	20	40	5.00 (1.52)	4.00 (1.22)	2.50 (0.76)
1	25	40	5.50 (1.68)	4.50 (1.37)	2.50 (0.76)
1¼	30	40	5.50 (1.68)	5.00 (1.52)	3.00 (0.91)
1½	40	40	6.00 (1.83)	5.00 (1.52)	3.00 (0.91)
2	50	40	6.00 (1.83)	5.00 (1.52)	3.00 (0.91)
3	80	40	7.00 (2.13)	6.00 (1.83)	3.50 (1.07)
4	100	40	7.50 (2.29)	6.50 (1.98)	4.00 (1.22)
6	150	40	8.50 (2.59)	7.50 (2.29)	4.50 (1.37)
8	200	40	9.00 (2.74)	8.00 (2.44)	4.50 (1.37)
½	15	80	5.00 (1.52)	4.50 (1.37)	2.50 (0.76)
¾	20	80	5.50 (1.68)	4.50 (1.37)	2.50 (0.76)
1	25	80	6.00 (1.83)	5.00 (1.52)	3.00 (0.91)
1¼	30	80	NA	NA	NA
1½	40	80	6.50 (1.98)	5.50 (1.68)	3.50 (1.07)
2	50	80	7.00 (2.13)	6.00 (1.83)	3.50 (1.07)
3	80	80	8.00 (2.44)	7.00 (2.13)	4.00 (1.22)
4	100	80	9.00 (2.74)	7.50 (2.29)	4.50 (1.37)
6	150	80	10.00 (3.05)	9.00 (2.74)	5.00 (1.52)
8	200	80	11.00 (3.35)	9.50 (2.90)	5.50 (1.68)

Notes: 1. The table assumes continuous spans, uninsulated pipe, and no concentrated loads.
2. The table is for fluids with specific gravity up to 1.35.
3. Add individual supports for concentrated loads.
4. Check local building codes, if applicable, for spans that differ from above.
5. Coefficient of linear expansion is 30×10^6 in/in-°F.
6. NA = not applicable.

Source: *Piping Handbook*, 7th ed., McGraw-Hill, New York.

TABLE 27.2 Chlorinated Polyvinyl Chloride (CPVC) Pipe: Suggested Piping Support Spacing

NPS	DN	Pipe schedule	Pipe temperature, °F (°C) 60 (16)	100 (38)	140 (60)	180 (82)
			Suggested maximum span, ft (m)			
½	15	40	5.00 (1.52)	4.50 (1.37)	4.00 (1.22)	2.50 (0.76)
¾	20	40	5.50 (1.68)	5.00 (1.52)	4.00 (1.22)	2.50 (0.76)
1	25	40	6.00 (1.83)	5.50 (1.68)	4.50 (1.37)	2.50 (0.76)
1¼	30	40	6.00 (1.83)	5.50 (1.68)	5.00 (1.52)	3.00 (0.91)
1½	40	40	6.50 (1.98)	6.00 (1.83)	5.00 (1.52)	3.00 (0.91)
2	50	40	6.50 (1.98)	6.00 (1.83)	5.00 (1.52)	3.00 (0.91)
3	80	40	8.00 (2.44)	7.00 (2.13)	6.00 (1.83)	3.50 (1.07)
4	100	40	8.50 (2.59)	7.50 (2.29)	6.50 (1.98)	4.00 (1.22)
6	150	40	9.50 (2.90)	8.50 (2.59)	7.50 (2.29)	4.50 (1.37)
½	15	80	5.50 (1.68)	5.00 (1.52)	4.50 (1.37)	2.50 (0.76)
¾	20	80	6.00 (1.83)	5.50 (1.68)	4.50 (1.37)	2.50 (0.76)
1	25	80	6.50 (1.98)	6.00 (1.83)	5.00 (1.52)	3.00 (0.91)
1¼	30	80	NA	NA	NA	NA
1½	40	80	7.00 (2.13)	6.50 (1.98)	5.50 (1.68)	3.50 (1.07)
2	50	80	7.50 (2.29)	7.00 (2.13)	6.00 (1.83)	3.50 (1.07)
3	80	80	9.00 (2.74)	8.00 (2.44)	7.00 (2.13)	4.00 (1.22)
4	100	80	10.00 (3.05)	9.00 (2.74)	7.50 (2.29)	4.50 (1.37)
6	150	80	11.00 (3.35)	10.00 (3.05)	9.00 (2.74)	5.00 (1.52)

Notes: 1. The table assumes continuous spans, uninsulated pipe, and no concentrated loads.
2. The table is for fluids with specific gravity up to 1.35.
3. Add individual supports for concentrated loads.
4. Check local building codes, if applicable, for spans that differ from above.
5. Coefficient of linear expansion is 30×10^6 in/in-°F.
6. NA = not applicable.

Source: *Piping Handbook*, 7th ed., McGraw-Hill, New York.

TABLE 27.3 Polyvinylidene Fluoride (PVDF) Pipe: Suggested Piping Support Spacing

			Pipe temperature, °F (°C)			
			80 (27)	100 (38)	140 (60)	160 (71)
NPS	DN	Pipe schedule	Suggested maximum span, ft (m)			
½	15	40	3.75 (1.14)	3.50 (1.07)	2.00 (0.61)	*
¾	20	40	4.00 (1.22)	3.75 (1.14)	2.50 (0.76)	*
1	25	40	4.50 (1.37)	4.00 (1.22)	2.50 (0.76)	*
1¼	30	40	NA	NA	NA	NA
1½	40	40	4.50 (1.37)	4.50 (1.37)	2.50 (0.76)	*
2	50	40	4.50 (1.37)	4.50 (1.37)	2.25 (0.69)	*
½	15	80	4.50 (1.37)	4.50 (1.37)	2.50 (0.76)	*
¾	20	80	4.50 (1.37)	4.50 (1.37)	3.00 (0.91)	*
1	25	80	5.00 (1.52)	4.75 (1.45)	3.00 (0.91)	*
1¼	30	80	NA	NA	NA	NA
1½	40	80	5.50 (1.68)	5.00 (1.52)	3.00 (0.91)	*
2	50	80	5.50 (1.68)	5.25 (1.60)	3.00 (0.91)	*

* Continuous support recommended.

Notes: 1. The table assumes continuous spans, uninsulated pipe, and no concentrated loads.
2. The table is for fluids with specific gravity up to 1.35.
3. Add individual supports for concentrated loads.
4. Check local building codes, if applicable, for spans that differ from above.
5. Coefficient of linear expansion is 30×10^6 in/in-°F.
6. NA = not applicable.

Source: *Piping Handbook*, 7th ed., McGraw-Hill, New York.

TABLE 27.4 Polypropylene (PP) Pipe: Suggested Piping Support Spacing

			Pipe temperature, °F (°C)			
			60 (16)	100 (38)	140 (60)	180 (82)
NPS	DN	Pipe schedule	Suggested maximum span, ft (m)			
½	15	40	1.75 (0.53)	1.75 (0.53)	1.50 (0.46)	1.25 (0.38)
¾	20	40	2.00 (0.61)	2.00 (0.61)	1.75 (0.53)	1.75 (0.53)
1	25	40	2.00 (0.61)	2.00 (0.61)	2.00 (0.61)	1.75 (0.53)
1¼	30	40	2.50 (0.76)	2.25 (0.69)	2.00 (0.61)	2.00 (0.61)
1½	40	40	2.50 (0.76)	2.50 (0.76)	2.25 (0.69)	2.00 (0.61)
2	50	40	3.00 (0.91)	2.25 (0.69)	2.50 (0.76)	2.25 (0.69)
3	80	40	3.50 (1.07)	2.25 (0.69)	3.00 (0.91)	2.75 (0.84)
4	100	40	4.00 (1.22)	3.25 (0.99)	3.50 (1.07)	3.00 (0.91)
½	15	80	2.00 (0.61)	2.00 (0.61)	2.00 (0.61)	1.50 (0.46)
¾	20	80	2.50 (0.76)	2.50 (0.76)	2.25 (0.69)	2.00 (0.61)
1	25	80	2.50 (0.76)	2.50 (0.76)	2.25 (0.69)	2.00 (0.61)
1¼	30	80	3.00 (0.91)	2.75 (0.84)	2.50 (0.76)	2.50 (0.76)
1½	40	80	3.00 (0.91)	3.00 (0.91)	2.75 (0.84)	2.50 (0.76)
2	50	80	3.50 (1.07)	3.25 (0.99)	3.00 (0.91)	2.75 (0.84)
3	80	80	4.00 (1.22)	4.00 (1.22)	3.50 (1.07)	3.50 (1.07)
4	100	80	4.50 (1.37)	4.50 (1.37)	4.00 (1.22)	3.50 (1.07)

Notes: 1. The table assumes continuous spans, uninsulated pipe, and no concentrated loads.
2. The table is for fluids with specific gravity up to 1.35.
3. Add individual supports for concentrated loads.
4. Check local building codes, if applicable, for spans that differ from above.

Source: *Piping Handbook*, 7th ed., McGraw-Hill, New York.

- Pipe supports used for steel pipe may be used for thermoplastic pipe provided there is protection between the pipe and the steel support. It is suggested that a plastic-coated pipe clevis or a neoprene wrap be used to cushion the pipe from the steel support. A fiberglass or metal sleeve or cradle may be bonded to the pipe at the support location as an alternative method of protection.
- Vertical piping must be supported at intervals to reduce loads on the lower fittings. Use of riser clamps or double-bolt pipe clamps located just below a coupling or fiberglass pipe buildup is recommended. This will avoid excessive compression on the pipe.
- Refer to Figs. 27.1 through 27.4 for typical details used for supporting fiberglass pipe. See Table 27.5 for maximum support spacing and Table 27.6 for minimum support width.

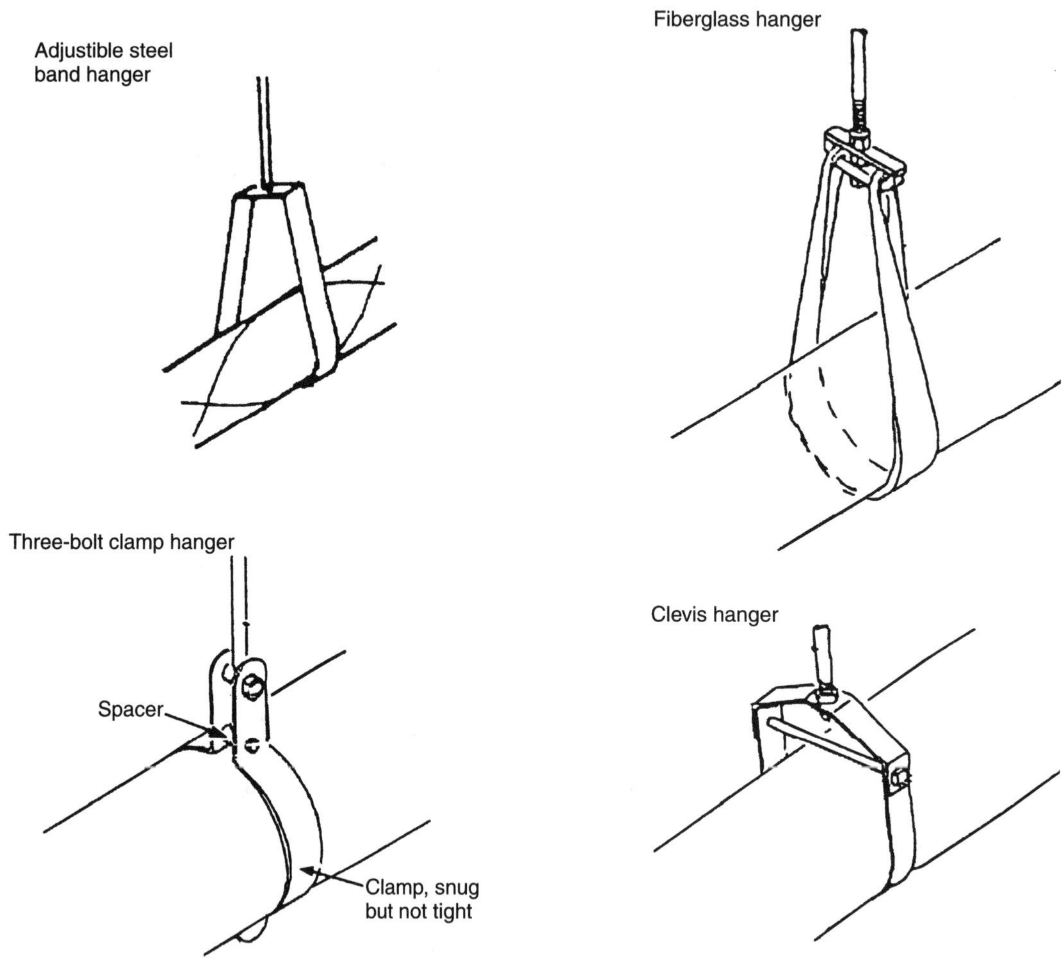

FIGURE 27.1 Typical hanger details. Pipe can move laterally and axially. (*Illustrations courtesy of Smith Fiberglass Products Inc., Ameron Fiberglass Pipe Systems, and Fibercast Co.*)

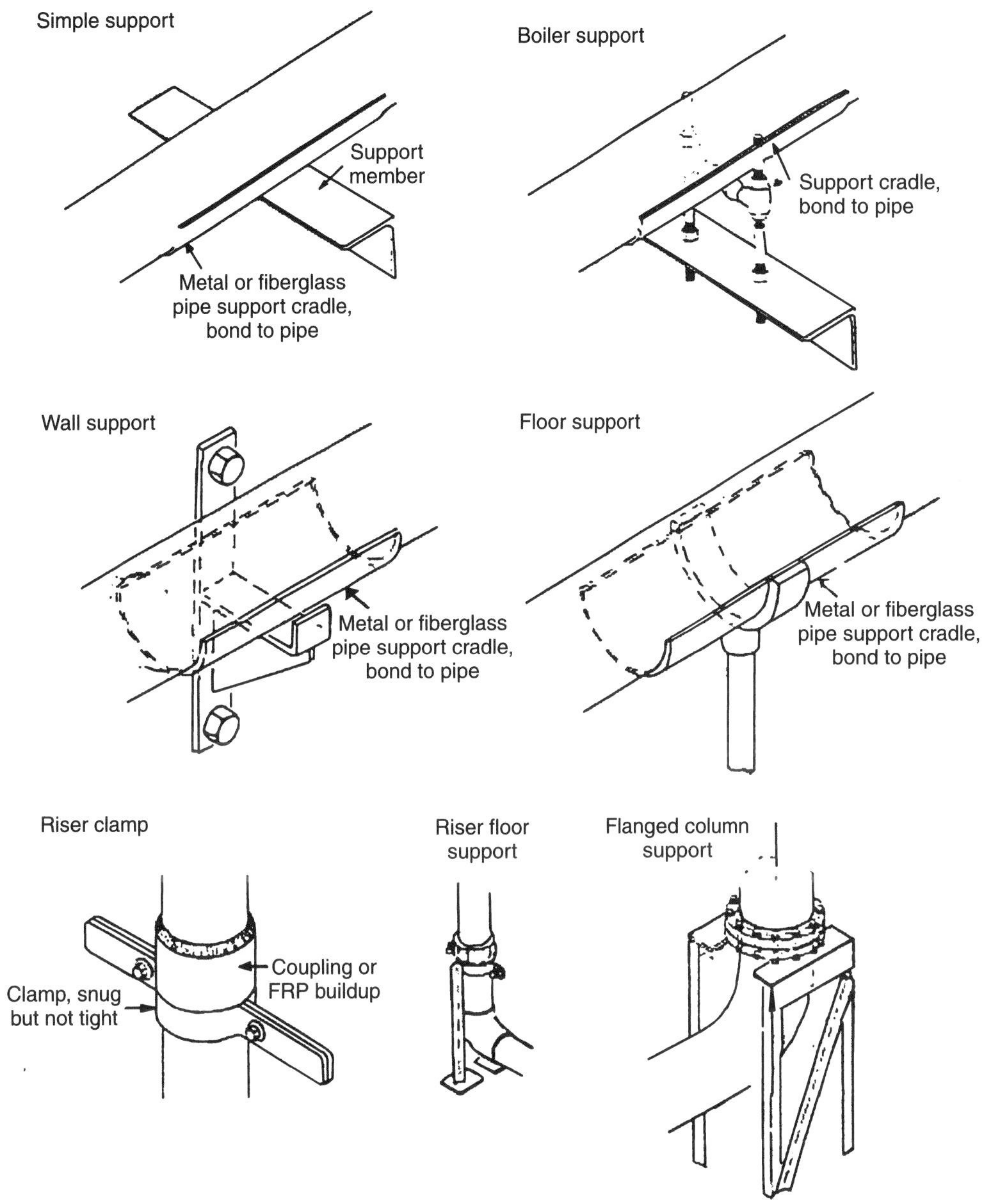

FIGURE 27.2 Typical support details. Pipe can move laterally and axially. (*Illustrations courtesy of Smith Fiberglass Products Inc., Ameron Fiberglass Pipe Systems, and Fibercast Co.*)

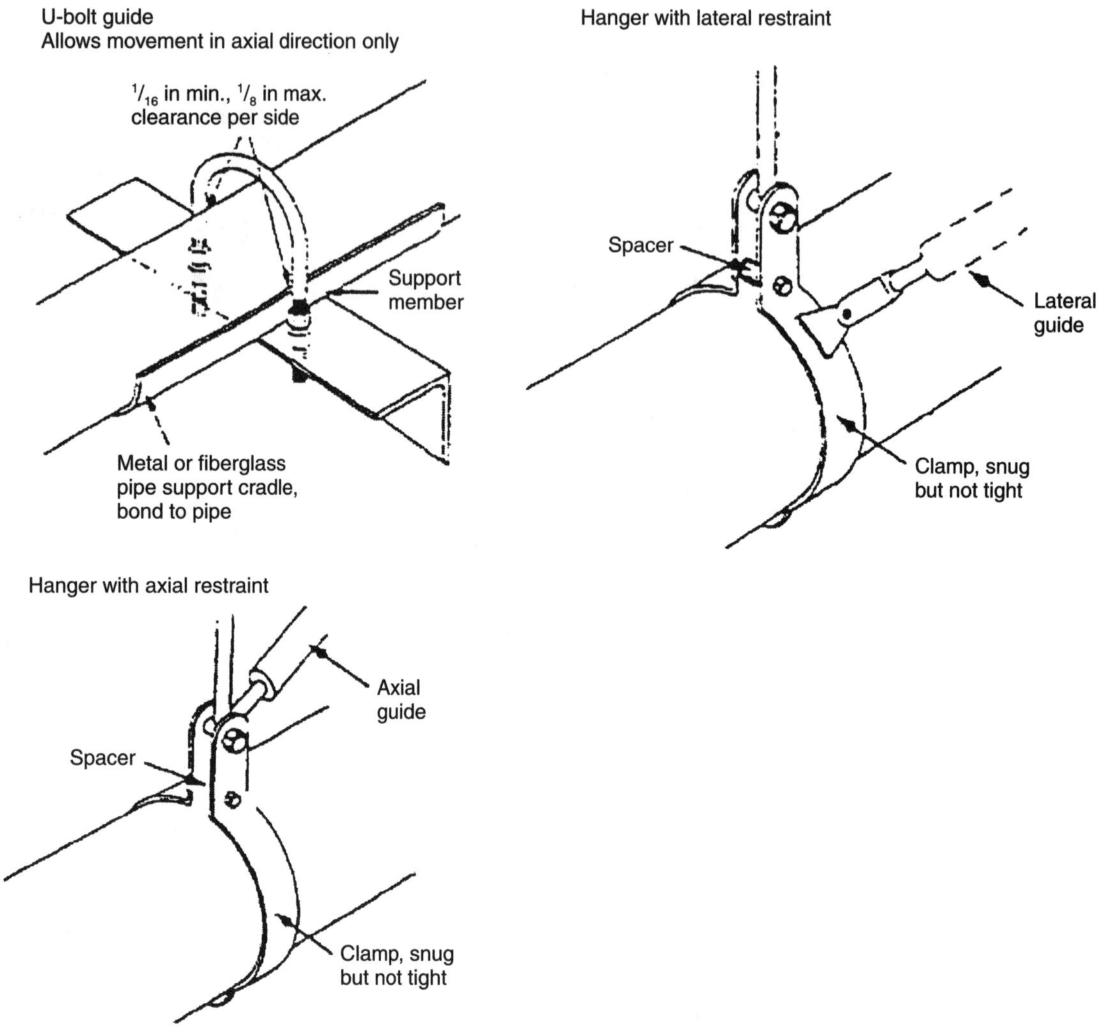

FIGURE 27.3 Typical guide details. Pipe is restrained axially or laterally. (*Courtesy of Fibercast Co.*)

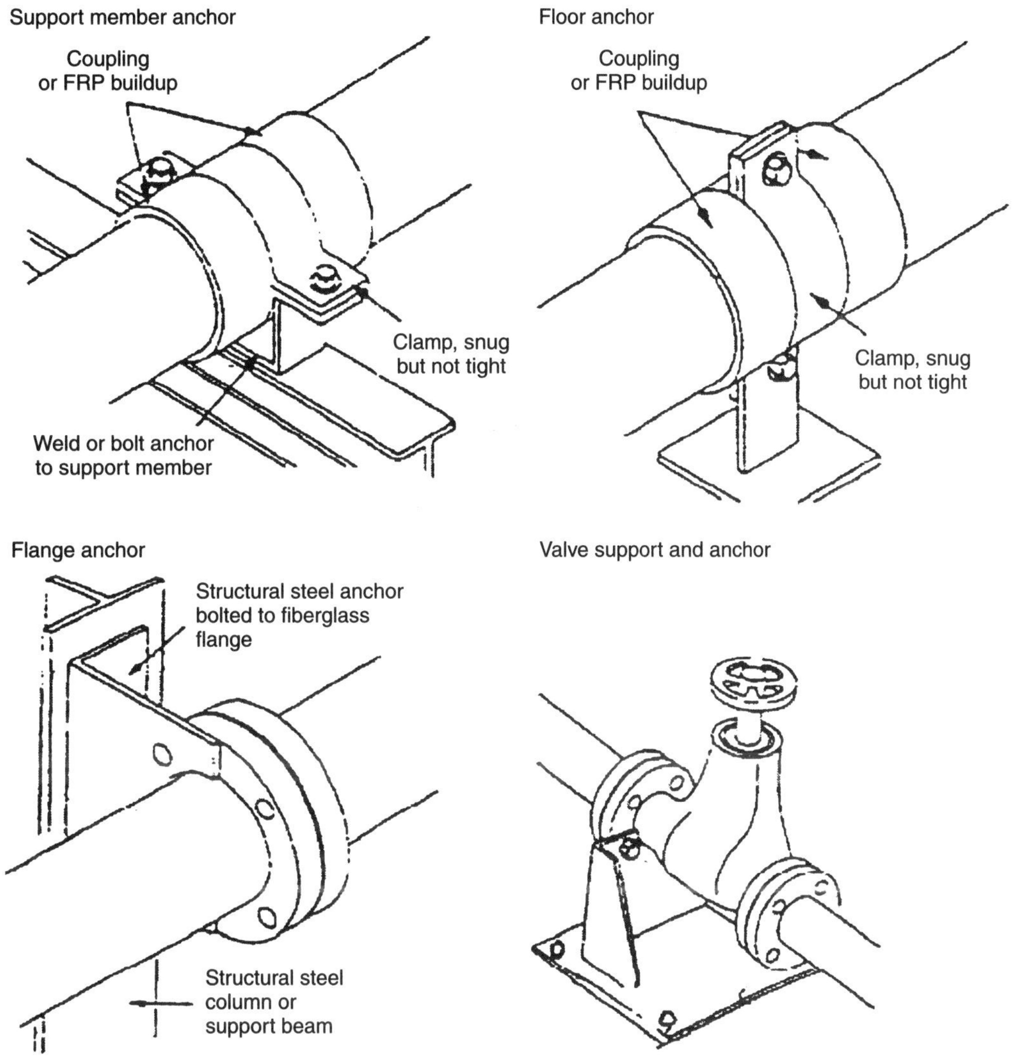

FIGURE 27.4 Typical anchor details. Pipe is restrained in all directions. (*Courtesy of Fibercast Co.*)

TABLE 27.5 Fiberglass Pipe: Suggested Piping Support Spacing*

NPS	DN	Wall thickness, in (mm)	Pipe Temperature, °F (°C) 100 (38)	150 (66)	200 (93)	250 (121)
			Suggested maximum span, ft (m)			
2	50	0.160 (4)	11.8 (3.6)	11.2 (3.4)	10.4 (3.2)	9.4 (2.9)
3	80	0.160 (4)	13.6 (4.1)	12.8 (3.9)	11.9 (3.6)	10.8 (3.3)
4	100	0.206 (5)	15.4 (4.7)	14.6 (4.5)	13.6 (4.1)	12.3 (3.7)
6	150	0.206 (5)	17.2 (5.2)	16.4 (5.0)	15.1 (4.6)	13.8 (4.2)
8	200	0.229 (6)	19.2 (5.9)	18.1 (5.5)	16.9 (5.2)	15.2 (4.6)
10	250	0.229 (6)	20.3 (6.2)	19.2 (5.9)	17.9 (5.5)	16.1 (4.9)
12	300	0.229 (6)	21.3 (6.5)	20.1 (6.1)	18.7 (5.7)	16.9 (5.2)
14	350	0.250 (6)	22.3 (6.8)	21.2 (6.5)	19.6 (6.0)	17.7 (5.4)
16	400	0.286 (7)	23.3 (7.1)	22.3 (6.8)	20.5 (6.2)	18.6 (5.7)

* Span recommendations are intended for normal horizontal piping support arrangements (a compromise between continuous spans and simple spans), but include no provision for weight from fittings, valves, flanges, etc., or thrusts from branches or turns.

Source: Ameron catalog for Bondstrand product data (FP215C, 12/93, series 4000).

TABLE 27.6 Minimum Support Width*

NPS	DN	Width, in (mm)
2	50	0.875 (22)
3	80	1.250 (32)
4	100	1.250 (32)
6	150	1.500 (38)
8	200	1.750 (45)
10	250	1.750 (45)
12	300	2.000 (51)
14	350	2.000 (51)

* Minimum support width is based on fluid specific gravity of 1.25 and suggested maximum spans for filament wound fiberglass pipe at 75°F (24°C), and 120° circumferential contact around the pipe.

Source: *Piping Handbook*, 7th ed., McGraw-Hill, New York.

Tables in Chapter 27

S · E · C · T · I · O · N · 8

GASKETS

CHAPTER 28
TYPES OF GASKETS: SELECTION, APPLICATION, AND INSTALLATION

See pp. 787–788 for a list of tables in Chap. 28.

THE GASKET

A gasket is a compressible material, or a combination of materials that, when clamped between two stationary members, prevents the passage of the medium across those members. The gasket material selected must be capable of sealing mating surfaces, resistant to the medium being sealed, and able to withstand the application temperatures and pressures.

How Does It Work?

A seal is effected by the action of force upon the gasket surface. This force, which compresses the gasket, causes it to flow into the flange macro- and microimperfections. The combination of contact stress, generated by the applied force between the gasket and the flange, and the densification of the gasket material prevents the escape of the confined fluid from the assembly.

Flange Imperfections

On seating, the gasket must be capable of overcoming the macro- and microimperfections. Macrodefects are imperfections such as flange distortions, nonparallelism, scoring, and troughs, while superficial imperfections such as minor scratches and minor scores are considered microimperfections. See Fig. 28.1.

Forces on the Gasket

In order to ensure the maintenance of the seal throughout the life expectancy of the assembly, sufficient stresses must remain on the gasket surface to prevent leakage. The residual bolt load on the gasket should at all times be greater than the hydrostatic end force acting against it. The hydrostatic end force is the force produced by the internal pressure that acts to separate the flanges. See Fig. 28.2.

Gasket Standards

Refer to Table 28.1 for commonly used gasket standards.

TYPES OF GASKETS

Usually nonmetallic composite sheet materials are used for low-pressure and low-temperature applications. They are suitable for a wide range of general and corrosive chemical services. Nonmetallic gaskets (Fig. 28.3) are manufactured with nonasbestos material or compressed asbestos fibers (CAF). Check applicable rules, regulations, and laws when using compressed asbestos fiber gaskets. Full face gaskets are used with flat-face flanges. Flat ring gaskets are used with raised-face flanges.

Types of nonmetallic gaskets are elastomers, nonasbestos fiber sheets, PTFE gaskets, biaxially orientated reinforced PTFE, graphite sheets, and insulating gaskets (PTFE stands for polytetrafluorethylene).

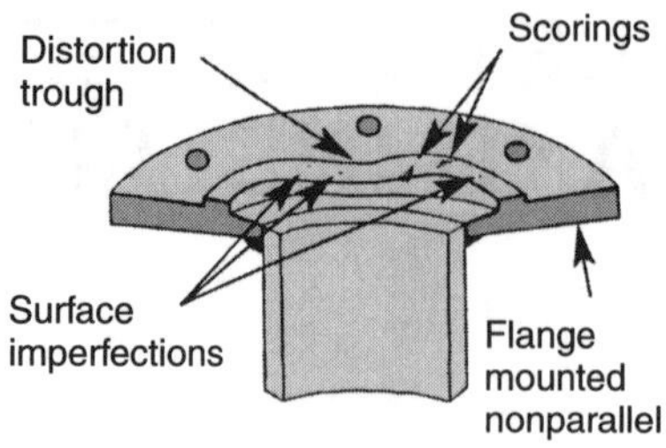

FIGURE 28.1 Overcoming flange imperfections.

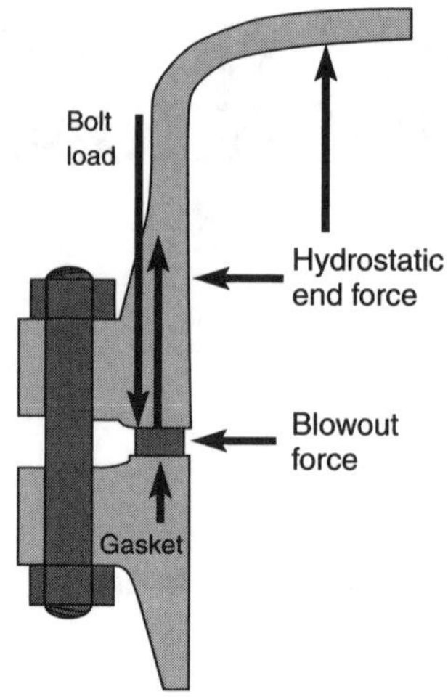

FIGURE 28.2 Internal pressure is exerted against both the flange and the gasket.

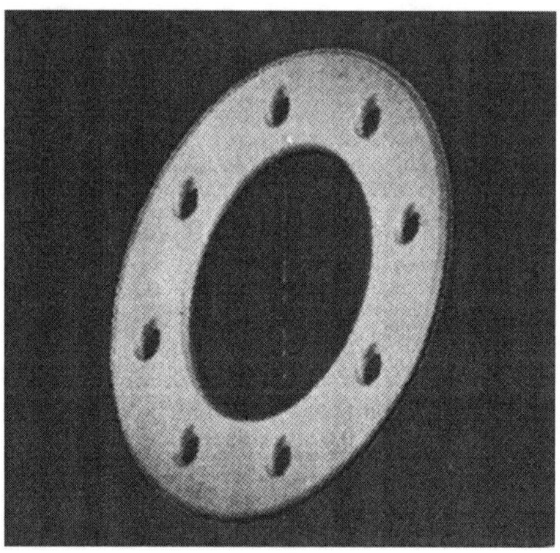

FIGURE 28.3 Nonmetallic gasket.

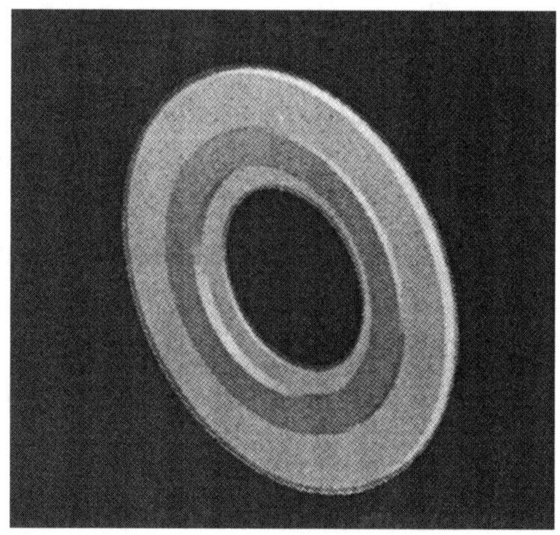

FIGURE 28.4 Composite gasket.

A composite gasket (Fig. 28.4) consists of both metallic and nonmetallic materials. The metal provides the strength and the resilience of the gasket. These gaskets are suitable for low and high pressure and temperature applications and are available in a wide range of materials.

Types of composite gaskets are spiral-wound gaskets, metal-jacketed gaskets, metal-reinforced gaskets (MRGs), and metal serrated core gaskets with soft facings.

TABLE 28.1 Referenced Gasket Standards

Designation	Title
ASME B16.20	Metallic Gasket for Pipe Flanges—Ring Joint, Spiral Wound and Jacketed
ASME B16.21	Non-Metallic Flat Gaskets for Pipe Flanges
BS 4865 Part 1	Flat Ring Gaskets to Suit BS 4505 and DIN Flanges
BS 3381	Spiral Wound Gaskets to Suit BS 1560 Flanges
API 6A	Specification for Wellhead and Christmas Tree Equipment

Spiral-wound gaskets are the most common gaskets used on raised-face flanges. They are used in all pressure classes from Class 150 to Class 2500. The part of the gasket that creates the seal between the flange faces is the spiral-wound section. It is manufactured by winding a preformed metal strip and a soft filler material around a metal mandrel. The inside and outside diameters are reinforced by several additional metal windings without filler.

Metal gaskets (Fig. 28.5) are fabricated from one or a combination of metallic materials in various shapes and sizes. They are suitable for use in high-pressure/temperature applications. Except for weld rings, high loads are required to seat metallic gaskets by the deformation or coining of the material into the flange surfaces.

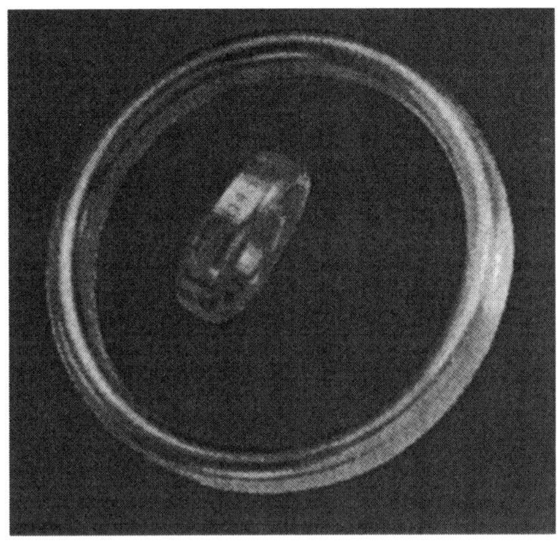

FIGURE 28.5 Metal gasket.

Types of metal gaskets are ring-type joints, lens rings, weld gaskets, and solid metal gaskets.

FUNCTION OF A GASKET

The function of a gasket is to conform to the irregularities of the flange faces to effect a seal, preventing the inside fluid from leaking out. The leak performance of the gasket is dependent on the stress on the gasket during operation. Each type of gasket has its own inherent leaktightness capabilities. The higher the gasket stress, the higher the leaktightness capability.

The ideal gasket is composed of a body with good load-bearing and recovery characteristics, with soft conformable surface layers. Gaskets have a combination of elastic and plastic characteristics. Ideal gaskets should have the following properties:

1. *Compressibility.* Sufficient compressibility to suit the style and surface finish of the flange, ensuring that all the imperfections will be filled by the gasket material.
2. *Resilience.* High resilience enables the gasket to move with dynamic loads of the flange to maintain its seating stress.
3. *No change in thickness.* No continued deformation under varying load cycles of temperature and pressure or under constant load at elevated temperature (creep and relaxation).

Unfortunately, most gaskets available on the market are not ideal gaskets. Most gaskets usually exhibit one or two of the above properties. For critical applications, designers are always on the lookout for gaskets that have the optimum properties.

The most difficult, often critical, property required of a gasket is the ability to resist creep during operation. In high-temperature services, the flanges will heat up at a faster rate than the bolts, and under steady-state conditions will continue to be hotter than the bolts as a result of thermal gradient. This results in a higher thermal expansion of the flanges with respect to the bolts, increasing the bolt load and concurrently the gasket stress. The gasket will then deform under the higher applied load during this cycle. Many gaskets will deform permanently and will not rebound when the load cycle is removed because of varying conditions. The permanent set or plastic deformation that occurred during operation will cause loss of bolt load and consequently loss of gasket stress. As gasket stress decreases, leak rate increases.

SELECTION AND INSTALLATION OF GASKETS

Criteria for Selection of a Gasket

1. Mechanical properties of a gasket
2. Temperature of the media to be contained
3. Pressure of the media to be contained
4. Corrosive nature of the application

Application of Types of Gaskets

See Table 28.2 and consult with the gasket manufacturer. Consider the past experience feedback.

TABLE 28.2 Application of Types of Gaskets

Gasket type	Pressure class			
	Low class 150–300	Medium class 600–900	High class 1500–2500	Maximum temperature of materials, °F
Nonmetallic				
CAF	×			650–1000
Nonasbestos fiber	×			550
PTFE	×			390–550
Graphite	×			750
Semimetallic				
Spiral-wound gaskets		×	×	750+*
Metal-jacketed		×	×	750+*
Metal-reinforced graphite	×	×	×	750+*
Serrated metal core with soft facings	×	×	×	750+*
Metallic				
Ring joint gaskets			×	650+*
Lens ring			×	650+*
Machined ring			×	650+*

× Applicable.
* Depends on material.

Recommended Flange Surface Finish

See Table 28.3 for recommended flange surface finish for various types of gaskets.

TABLE 28.3 Recommended Flange Surface Finish for Various Gasket Types

Gasket type	Flange surface finish, μin R_a*	Flange surface finish, μm R_a
Soft cut sheet:		
Material <1.5 mm thick	125–250	3.2–6.3
Material = 1.5 mm thick	125–250	3.2–6.3
Spiral-wound	125–250	3.2–6.3
Metal-jacketed	100 max.	2.5 max.
Serrated metal with soft facings	125–250	3.2–6.3
Metal-reinforced graphite	125–250	3.2–6.3
Solid metal	63 max.	1.6 max.

* R_a = roughness average.

Bolt Material Selection

See Table 28.4 for bolting selection based on bolt material stresses and temperature.

Torque Calculations

Calculate bolt load for the required tightness class. Consult the gasket manufacturer.

Tightening Methods and Assembly Efficiencies

See Table 28.5 to select the tightening method.

TABLE 28.4 Material Specifications for Bolts and Nuts and Recommended Bolting Temperature Range

Material specifications	Alloy types	Mechanical properties: Tensile strength N/mm^2	Tensile strength psi	Yield strength, 0.2% proof stress min. N/mm^2	Yield strength psi	Recommended bolting temperature range,* °C Min.	Max.	Recommended corresponding nut grades
B7, L7 BS 1506-621A	1% CrMo steel	860	123,000	730	103,000	−100	400	2H, 4, 7, or 8 (L4 and/or 8 with L7 bolts)
B16 BS 1506-661	1% CrMo and V	860	123,000	730	103,000	0	520	4, 7, or 8
B8, L8 BS 1506-801B	Austenitic CrNi 18/8 steel	540	77,000	210	30,000	−250	575	8, 8F
B8, CX BS 1506-821T	Stabilized austenitic CrNi 18/8 steel, cold worked after solution treatment	860	123,000	700	99,000	−250	575	8 CX
B17B	Precipitation hardened austenitic NiCr steel	900	128,000	590	84,000	−250	650	17B
B80A BS 3076 NA20	Precipitation hardened NiCr TiA1 alloy	1000	143,000	620	88,000	−250	750	80A

* Temperature of bolting refers to actual metal temperature.

TABLE 28.5 Tightening Methods and Assembly Efficiencies

Method to control bolt preload	Tightening method	Stud-to-stud load variation from the mean, %	Assembly efficiency e
No torque/stretch control	Power impact, lever, or hammer wrench	>±50	0.75
Torque control	Calibrated torque wrench or hydraulic wrench	±30 to ±50	0.85
Tensioner load control	Multiple-stud tensioner	±10 to ±50	0.95
Direct measurement	Ultrasonic extensometer, calipers, strain gages	±10 or less	1.00

Chemical Compatibility Selection

See Fig. 28.6.

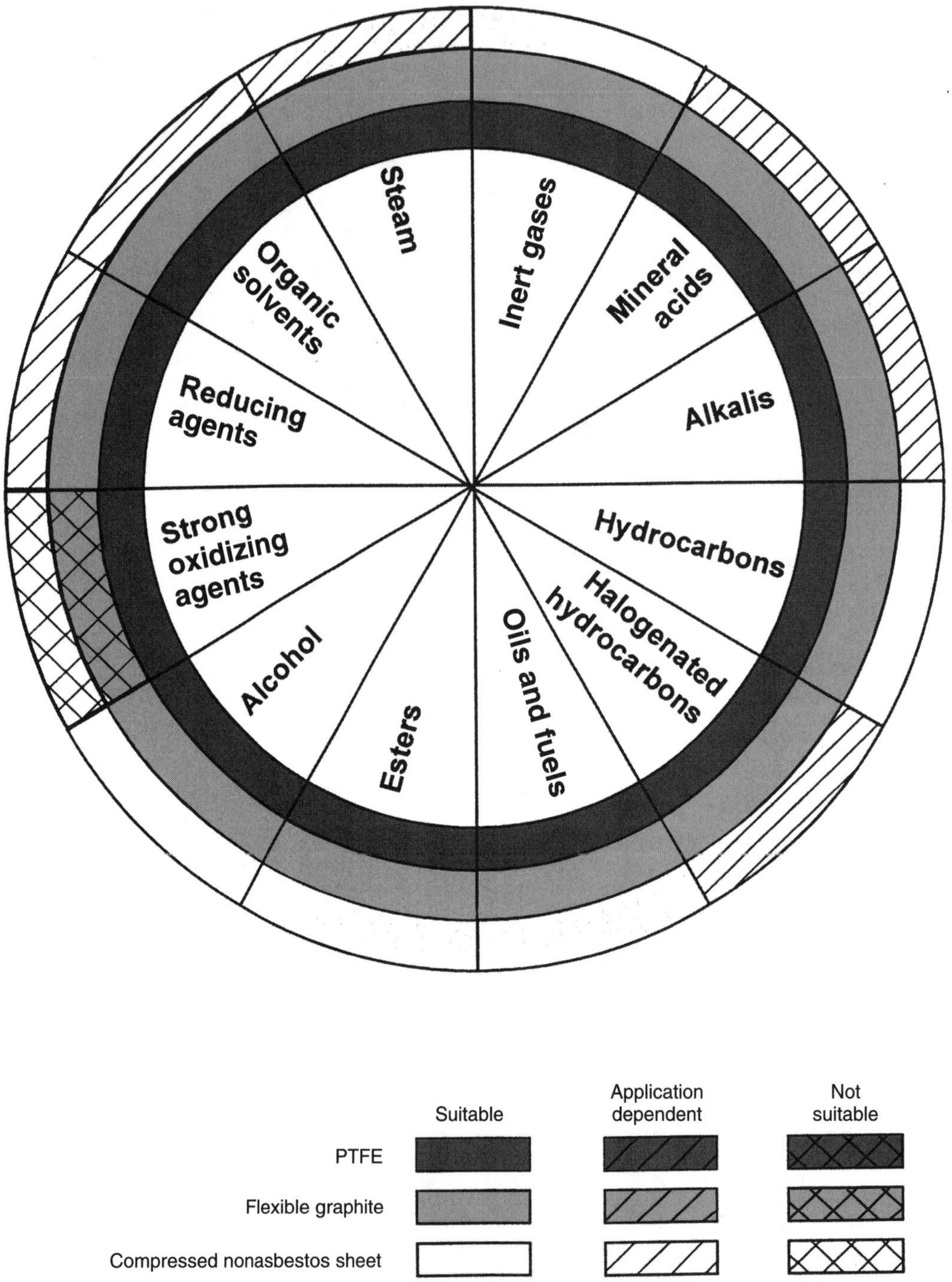

FIGURE 28.6 Chemical compatibility selector. (Guidelines only; refer to chemical application/compatibility data for specific application.)

Gasket Installation

Check and verify the quality of the gasket. *Never install a previously used gasket!*

Flange Surfaces

The condition of flange surfaces, as well as the proper flange material selection, plays an important part in achieving a leak-free joint assembly. Assure that flanges have the acceptable:

- Surface finish
- Flatness
- Parallelism
- Waviness
- Allowable imperfections

To optimize proper and even compression of the gasket, it is recommended that parallelism be 0.008 in and that flatness and waviness be kept at better than 0.008 in.

Under no circumstances should flange sealing surfaces be machined in a manner such that tool marks would extend radially across the sealing surface. Such tool marks are practically impossible to seal regardless of the type of gasket used.

Fasteners

It is important that proper studs, bolts, and nuts are selected to assure joint integrity. Improper selection of these components may compromise the entire joint assembly. The following list is to be considered when selecting fasteners:

- Type
- Grade
- Class
- Proper material
- Appropriate coating or plating
- Correct stud/bolt length

Assembly

In an effort to achieve a high degree of success in attaining a leak-free joint, several steps are required. It is imperative that a regimented bolt-up procedure be applied. As a minimum, the following is offered:

- Install a new gasket on the gasket seating surface and bring the mating flange in contact with the gasket.
- Do not apply any compounds on the gasket or gasket seating surfaces.
- Install all bolts, making sure that they are free of any foreign matter, and well lubricated. Lubricate nut bearing surfaces as well. (Lubrication will not be required for PTFE-coated fasteners.)
- Run-up all nuts finger tight.
- Develop the required bolt stress or torque incrementally in four steps in a crisscross pattern. The initial prestress should be no more than 30 percent of the final required bolt stress. After following this sequence, a final tightening pass should be performed bolt to bolt to ensure that all bolts have been evenly stressed.

Bolt-up Sequence

Stage 1. Torque bolts up to approximately 30 percent of the final torque value following the diametrically opposed sequence illustrated in Fig. 28.7.

Stage 2. Repeat stage 1, increasing the torque value to approximately 60 percent of the final torque value.

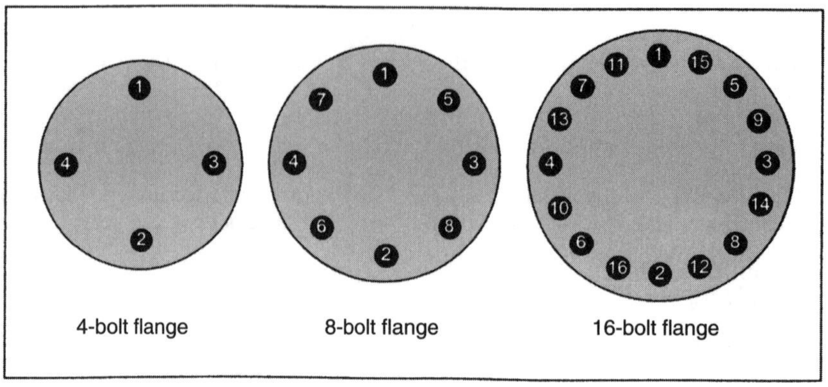

FIGURE 28.7 Bolt-tightening sequences.

Stage 3. Repeat stage 2, increasing the torque value to the final required torque value.

Stage 4. A final tightening should be performed by following an adjacent bolt-to-bolt sequence to ensure that all bolts have been evenly stressed.

SHEET GASKETS

Table 28.6 lists applications of nonmetallic elastomeric gaskets. Table 28.7 lists applications of compressed nonasbestos sheet gaskets. Table 28.8 presents properties and applications of biaxially oriented filled PTFE sheet gaskets.

Flexible Graphite Sheet Gaskets

Flexible graphite provides excellent sealing characteristics, exhibiting outstanding chemical resistance, thermal stability, and physical properties. This material is manufactured from graphite flakes thermally exfoliated and compressed into sheets. The material in the homogeneous form is 95 to 98 percent carbon with no binder, fibers, lubricating agents, or additives present.

Flexible graphite's ability to flow into flange imperfections, self-lubricating structure, exceptional chemical stability, and temperature resistance make it an excellent choice of sealing material. The material is used in a variety of other sealing products as well. Table 28.9 lists properties of flexible graphite products.

Temperature Resistance. Flexible graphite may be used for applications at cryogenic to elevated temperatures. To improve oxidation and corrosion resistance, flexible graphite can be supplied with an inorganic passivating inhibitor.

Chemical Stability. The high-purity grade of flexible graphite can be supplied with an inorganic passivating inhibitor. The high-purity grade of flexible graphite is chemically inert in most media.

Nuclear Grade. A 99.9 percent purity grade is available for nuclear applications. Nuclear-grade graphite is low in chloride, fluoride, and other elements of concern in the nuclear industry.

Sheet Types

Homogenous Foil. This type is manufactured without the benefit of any substrates or reinforcements and comes in various thicknesses ranging from 0.13 mm (0.005 in) to 1.5 mm (0.060 in).

Laminate Foils. Laminated sheets are available in thicknesses from 1.5 mm (0.060 in) to 6.35 mm (0.25 in).

TABLE 28.6 Nonmetallic Elastomeric Gaskets

Elastomer	Temperature range, °F (°C)	Suitable for	Not recommended for	Notes
Natural rubber (NR)	–60 to 250 (–50 to 120)	Inorganic salts Ammonia Mild acids Alkalies	Oils Solvents	Not recommended for applications where exposure to ozone, oxygen, or sunlight is prominent. Exhibits exceptional elongation, excellent tear strength, good resistance and recovery properties.
Neoprene (CR)	–60 to 230 (–50 to 110)	Moderate acids Alkalies Salt solutions Petroleum Solvents Commercial oils and fuels	Strong acids Fats Greases Most hydrocarbons	Offers excellent oil resistance, low permeability to gases, and resistance to ozone, weathering, and abrasion.
Nitrile (NBR)	–60 to 250 (–50 to 120)	Oils and solvents Aromatic and aliphatic hydrocarbons Petroleum oils and gasoline Animal fats and lacquer solvents	Strong oxidizing agents Chlorinated hydrocarbons Ketones and esters	Nitrile is a synthetic rubber. It offers improved chemical resistance and temperature capabilities over neoprene.
Butyl (IIR), also known as isobutylene and isoprene	–40 to 300 (–40 to 150)	Mild acids Alkalies Esters	Commercial oils Commercial fuels	Offers excellent resistance to wear, tear, and general abrasion. Offers good resistance to ozone and gas permeation.
Viton (FDM), fluorinated hydrocarbon	–40 to 392 (–40 to 200)	Acids Alphatic hydrocarbons Oils Gasoline Corrosive fluids Amines	Amines Esters Ketones Steam Low-temperature service	Offers excellent resistance to temperature, weathering, and gas permeability.
Ethylene (EPDM), propylene diene monomer	–60 to 300 (–50 to 150)	Ozone Heat Steam Strong acids Alkalis	Oils Solvents Aromatic hydrocarbons	Offers good resistance to ozone, heat, steam, strong acids.
Styrene butadiene (SBR)	–60 to 250 (–50 to 120)	Weak organic acids Moderate chemicals	Ozone Strong acids Most hydrocarbons	Offers excellent abrasion resistance.
Silicone range of rubbers	–95 to 480 (–70 to 250)	Ozone Heat Low temperature	Steam Hydrocarbons	Offers excellent high- and low-temperature properties. Unaffected by sunlight and ozone.

TABLE 28.7 Compressed Nonasbestos Sheet Gasket Application Guide*

Material	Composition	Applications	Relative cost (1 = lowest)
SF 1600	Glass/celluloses/natural rubber	Noncritical service Hydrotest Economical sheet Max. temp. 230 to 350°F (110 to 180°C)	1
AF 2100	Glass/NBR	Steam, water, gases, oils, mild acids, alkalies General chemicals As a covering on high-temperature transfer rollers of annealing furnaces at steel mills Max. temp. 350 to 890°F (180 to 475°C)	3
AF 2150	Glass/NBR/wire gauze	Same as AF 2100, fluctuating pressure, vibrations	3
SF 2400	Aramid/NBR	Excellent high-performance, general-purpose sheet for steam, water, gases, oils, mild solvents, and alkalies	2
SF 2420	Aramic/SBR	Same as SF 2400 except SBR binder. Ideal for the paper making industry. Max. temp. 350 to 750°F (180 to 400°C)	4
SF 2440	Aramid/chloroprene	Same as SF 2400 except chloroprene binder. Refrigerants and where self-extinguishing properties are required. Max. temp. 350 to 750°F (180 to 400°C)	5
SF 3300	Aramid/glass/NBR	Top-grade sheet for general industrial applications Max. temp. 350 to 750°F (180 to 400°C)	6
SF 3500	Aramid/glass/NBR	More aramid fiber than SF 3300 for increased strength, ideal for use in split casing pumps Max. temp. 750°F (400°C) at 1/64 in thickness	7
SF 5000	Carbon/aramid/NBR	Especially suitable for sealing caustic liquors. Max. temp. 350 to 825°F (180 to 440°C)	8

* These are for Flexitallic compressed nonasbestos sheet gaskets. There are other compressed nonasbestos sheet gaskets. Check with gasket manufacturers for comparable products.

TABLE 28.8 Biaxially Oriented Filled PTFE (Teflon®) Sheet Material*

Teflon is the registered trademark of the DuPont Company for its fluorocarbon resins.

Sigma grade†	500	511	522	533	577
Composition	PTFE and glass microspheres	PTFE and silica	PTFE and barium sulfate, with virgin PTFE facings	PTFE and barium	Sigma 511 core with virgin PTFE facings
Color	Blue	Fawn	White	Off-white	White with fawn core
Tensile strength: Across grain, psi With grain, psi	 1740 1250	 1940 2260	 2175 2275	 2230 1220	 1220 1250
Thickness increase: Oil no. 3, 300°F Fuel B, 70°F	 1% 1%	 2% 1%	 1% 1%	 1% 1%	 1% 1%
Weight Increase: Oil no. 3, 300°F Fuel B, 70°F	 3% 4%	 2% 3%	 12% 4%	 1% 2%	 12% 4%
Compressibility	42%	10%	33.6%	11%	33.6%
Recovery	40%	44%	23%	46%	23%
Creep relaxation	21%	23.9%	42%	16.8%	42%
Sealability Fuel A at 10 psi, (1000 psi gasket stress), mL/h N_2 permeability, mL/min	 0.12 0.01	 0.02 0.42	 0.42 0.01	 0.01 0.66	 0.66 0.01
Maximum pressure	42–1232 psi				
Maximum temperature	00°F				
Gasket constants G_b a G_s	 4 psi 0.804 0.115 psi	 209 psi 0.356 0.00498 psi	 472 psi 0.25 0.037 psi	 115 psi 0.382 0.000065 psi	 NA NA NA
Applications	Acids and caustics at moderate concentrations, hydrocarbons, solvents, hydrogen peroxide, low-bolt-load glass-lined flanges in place of PFTE envelope gaskets	Strong acids, sulfuric acid, solvents, hydrocarbons, steam, chlorine, general service	Hydrofluoric acid, warped or glass-lined flanges in place of PTFE envelope gaskets	Strong caustics, moderate acids, chlorine, hydrocarbons, food and drug, hydrofluoric acid, aqueous hydrogen fluoride, aluminum fluoride	Glass-lined and lightly loaded flanges
Misapplications‡	Hydrogen fluoride, hydrofluoric acid	Hydrogen fluoride	Anhydrous HF	Anhydrous HF	Hydrogen fluoride
Relative cost	2	1 (lowest)	5 (highest)	3	4

* This table lists Flexitallic biaxially orientated filled PTFE sheet gaskets. There are others; contact manufacturers for comparable biaxially oriented filled PTFE sheet gaskets.

† All sigma grades comply with FDA requirements.

‡ All sigma grades are incompatible with fluorine gas, molten alkali metals, e.g., molton sodium, potasium, lithium.

TABLE 28.9 Typical Flexible Graphite Properties

Melting point	3650°C (6602°F)
Tensile strength	5.5 MN/m^2 (800 psi)
Ash content	2%
Leachable chloride ion content	50 ppm (certificate on request)
Leachable fluoride ion content	50 ppm (certificate on request)
Temperature range	
Inert or reducing atmospheres	Cryogenic to 3000°C (5400°F)
Oxidizing atmospheres	Cryogenic to 500°C (930°F)
Gas permeability	<10–5 cm^3/s
Specific electrical resistance	
Parallel to surface	10 Ω-mm^2/m
Perpendicular to surface	600 Ω-mm^2/m
Thermal conductivity	
Parallel to surface	200 W/m-K
Perpendicular to surface	7 W/m-K

TABLE 28.10 Reinforced Graphite Sheets

Type	Temperature range	Suitable for	Notes
316 stainless steel core [0.05 mm (0.002 in) thick core]	Cryogenic to 700°C (1292°F) in an inert or reducing atmosphere Cryogenic to 50°C (932°F) in an oxidizing atmosphere*	Water, steam, acids, alkalies, organic compounds, heat transfer fluids, salt solutions	Recommended for Class 150 and 300 applications
316 stainless steel tanged core [0.1 mm (0.004 in) thick core]	Cryogenic to 700°C (1292°F) in an inert or reducing atmosphere Cryogenic to 500°C (932°F) in an oxidizing atmosphere*	Water, steam, acids, alkalies, organic compounds, heat transfer fluids, salt solutions	Recommended for Class 150 and 300 applications Not recommended for use on glass-lined, bronze, and aluminum flanges
Wire reinforced (316 stainless steel woven wire mesh)	Cryogenic to 700°C (1292°F) in an inert atmosphere Cryogenic to 500°C (932°F) in an oxidizing atmosphere*	Water, steam, acids, alkalies, organic compounds, heat transfer fluids, salt solutions	Recommended for Class 150 applications only

* Because of concerns of oxidation of flexible graphite sheets at elevated temperatures consult with the manufacturer for temperature limits.

Reinforced Laminates. These gasket materials are manufactured of graphite foil laminated on metallic cores giving it better handleability and mechanical strength. The metallic core is generally 316 stainless steel. Material thicknesses available are 0.40 mm (0.016 in), 0.75 mm (0.030 in), and 1.5 mm (0.060 in). Table 28.10 lists properties and applications.

Sheet Gasket Dimensions

Tables 28.11 to 28.21 give dimensions of sheet gaskets for various types of pipe and fittings.

TABLE 28.11 Flat Ring and Full Face Nonmetallic Gasket Dimensions for ASME B16.1 Class 25 Cast Iron Pipe Flanges and Flanged Fittings (U.S. Customary Units)

			Full face gasket			
Nominal pipe size (NPS)	Gasket ID	Flat ring OD	OD	Number of holes	Hole diameter	Bolt circle diameter
4	4.50	6.88	9.00	8	0.75	7.50
5	5.56	7.88	10.00	8	0.75	8.50
6	6.62	8.88	11.00	8	0.75	9.50
8	8.62	11.12	13.50	8	0.75	11.75
10	10.75	13.62	16.00	12	0.75	14.25
12	12.75	16.38	19.00	12	0.75	17.00
14	14.00	18.00	21.00	12	0.88	18.75
16	16.00	20.50	23.50	16	0.88	21.25
18	18.00	22.00	25.00	16	0.88	22.75
20	20.00	24.25	27.50	20	0.88	25.00
24	24.00	28.75	32.00	20	0.88	29.50
30	30.00	35.12	38.75	28	1.00	36.00
36	36.00	41.88	46.00	32	1.00	42.75
42	42.00	48.50	53.00	36	1.12	49.50
48	48.00	55.00	59.50	44	1.12	56.00
54	54.00	61.75	66.25	44	1.12	62.75
60	60.00	68.12	73.00	52	1.25	69.25
72	72.00	81.38	86.50	60	1.25	82.50
84	84.00	94.25	99.75	64	1.38	95.50
96	96.00	107.25	113.25	68	1.38	108.50

* Dimensions in inches. For tolerances refer to Table 28.21.

TABLE 28.12 Flat Ring and Full Face Nonmetallic Gasket Dimensions for ASME B16.1 Class 125 Cast Iron Pipe Flanges and Flanged Fittings (U.S. Customary Units)*

			Full face gasket			
Nominal pipe size (NPS)	Gasket ID	Flat ring OD	OD	Number of holes	Hole diameter	Bolt circle diameter
1	1.31	2.62	4.25	4	0.62	3.12
$1\frac{1}{4}$	1.66	3.00	4.62	4	0.62	3.50
$1\frac{1}{2}$	1.91	3.38	5.00	4	0.62	3.88
2	2.38	4.12	6.00	4	0.75	4.75
$2\frac{1}{2}$	2.88	4.88	7.00	4	0.75	5.50
3	3.50	5.38	7.50	4	0.75	6.00
$3\frac{1}{2}$	4.00	6.38	8.50	8	0.75	7.00
4	4.50	6.88	9.00	8	0.75	7.50
5	5.56	7.75	10.00	8	0.88	8.50
6	6.62	8.75	11.00	8	0.88	9.50
8	8.62	11.00	13.50	8	0.88	11.50
10	10.75	13.88	16.00	12	1.00	14.25
12	12.75	16.12	19.00	12	1.00	17.00
14	14.00	17.75	21.00	12	1.12	18.75
16	16.00	20.25	23.50	16	1.12	21.25
18	18.00	21.62	25.00	16	1.25	21.75
20	20.00	23.88	27.50	20	1.25	25.00
24	24.00	28.25	32.00	20	1.38	29.50
30	30.00	34.75	38.75	28	1.38	38.00
36	36.00	41.25	46.00	32	1.62	42.75
42	42.00	48.00	53.00	36	1.62	49.50
48	48.00	54.50	59.50	44	1.62	56.00

* Dimensions in inches. For tolerances refer to Table 28.21.

TABLE 28.13 Flat Ring and Full Face Nonmetallic Gasket Dimensions for ASME B16.1 Class 250 Cast Iron Pipe Flanges and Flanged Fittings (U.S. Customary Units)*

Nominal pipe size (NPS)	Gasket ID	Flat ring OD
1	1.31	2.88
1¼	1.66	3.25
1½	1.91	3.75
2	2.38	4.38
2½	2.88	5.12
3	3.50	5.88
3½	4.00	6.50
4	4.50	7.12
5	5.56	8.50
6	6.62	9.88
8	8.62	12.12
10	10.75	14.25
12	12.75	16.62
14	14.00	19.12
16	16.00	21.25
18	18.00	23.50
20	20.00	25.75
24	24.00	30.50
30	30.00	37.50
36	36.00	44.00

* Dimensions in inches. For tolerances refer to Table 28.21.

TABLE 28.14 Flat Ring and Full Face Nonmetallic Gasket Dimensions for ASME B16.1 Class 150 Cast Iron Pipe Flanges and Flanged Fittings (U.S. Customary Units)*

			Full face gasket			
Nominal pipe size (NPS)	Gasket ID	Flat ring OD	OD	Number of holes	Hole diameter	Bolt circle diameter
½	0.84	1.88	3.50	4	0.62	2.38
¾	1.06	2.25	3.88	4	0.62	2.75
1	1.31	2.62	4.25	4	0.62	3.12
1¼	1.66	3.00	4.63	4	0.62	3.50
1½	1.91	3.38	5.00	4	0.62	3.88
2	2.38	4.12	6.00	4	0.75	4.75
2½	2.88	4.88	7.00	4	0.75	5.50
3	3.50	5.38	7.50	4	0.75	6.00
3½	4.00	6.38	8.50	8	0.75	7.00
4	4.50	6.88	9.00	8	0.75	7.50
5	5.56	7.75	10.00	8	0.88	8.50
6	6.62	8.75	11.00	8	0.88	9.50
8	8.62	11.00	13.50	8	0.88	11.75
10	10.75	13.38	16.00	12	1.00	14.25
12	12.75	16.13	19.00	12	1.00	17.00
14	14.00	17.75	21.00	12	1.12	18.75
16	16.00	20.25	23.50	16	1.12	21.25
18	18.00	21.62	25.00	16	1.25	22.75
20	20.00	23.88	27.50	20	1.25	25.00
24	24.00	28.25	32.00	20	1.38	29.50

* Dimensions in inches. For tolerances refer to Table 28.21.

TABLE 28.15 Flat Ring Nonmetallic Gasket Dimensions for ASME B16.5 Classes 300, 400, 600, and 900 Pipe Flanges and Flanged Fittings (U.S. Customary Units)*

		Gasket OD			
Nominal pipe size (NPS)	Gasket ID	Class 300	Class 400	Class 600	Class 900
1/2	0.84	2.12	2.12	2.12	2.12
3/4	1.06	2.62	2.62	2.62	2.75
1	1.31	2.88	2.88	2.88	3.12
1 1/4	1.66	3.25	3.25	3.25	3.50
1 1/2	1.91	3.75	3.75	3.75	3.88
2	2.38	4.38	4.38	4.38	5.62
2 1/2	2.88	5.12	5.12	5.12	6.50
3	3.50	5.88	5.88	5.88	6.62
3 1/2	4.00	6.50	6.38	6.38	
4	4.50	7.12	7.00	7.62	8.12
5	5.56	8.50	8.38	9.50	9.75
6	6.62	9.88	9.75	10.50	11.38
8	8.62	12.12	12.00	12.62	14.12
10	10.75	14.25	14.12	15.75	17.12
12	12.75	16.62	16.50	18.00	19.62
14	14.00	19.12	19.00	19.38	20.50
16	16.00	21.25	21.12	22.25	22.62
18	18.00	23.50	23.38	24.12	25.12
20	20.00	25.75	25.50	26.88	27.50
24	24.00	30.50	30.25	31.12	33.00

* Dimensions in inches. For tolerances refer to Table 28.21.

TABLE 28.16 Flat Face Nonmetallic Gasket Dimensions for ASME B16.24 Class 125 Cast Copper Alloy Pipe Flanges and Flanged Fittings, Class 150 (U.S. Customary Units)*

Nominal pipe size (NPS)	Gasket ID	OD	Number of holes	Hole diameter	Bolt circle diameter
1/2	0.84	3.50	4	0.62	2.38
3/4	1.06	3.88	4	0.62	2.75
1	1.31	4.25	4	0.62	3.12
1 1/4	1.66	4.62	4	0.62	3.50
1 1/2	1.91	5.00	4	0.62	3.88
2	2.38	6.00	4	0.75	4.75
2 1/2	2.88	7.00	4	0.75	5.50
3	3.50	7.50	4	0.75	6.00
3 1/2	4.00	8.50	8	0.75	7.00
4	4.50	9.00	8	0.75	7.50
5	5.56	10.00	8	0.88	8.50
6	6.62	11.00	8	0.88	9.50
8	8.62	13.50	8	0.88	11.75
10	10.75	16.00	12	1.00	14.25
12	12.75	19.00	12	1.00	17.00

* Dimensions in inches. For tolerances refer to Table 28.21.

TABLE 28.17 Flat Face Nonmetallic Gasket Dimensions for ASME B16.24 Class 125 Cast Copper Alloy Pipe Flanges and Flanged Fittings, Class 300 (U.S. Customary Units)*

Nominal pipe size (NPS)	Gasket ID	OD	Number of holes	Hole diameter	Bolt circle diameter
$^1/_2$	0.84	3.75	4	0.62	2.62
$^3/_4$	1.06	4.62	4	0.75	3.25
1	1.31	4.88	4	0.75	3.50
$1^1/_4$	1.66	5.25	4	0.75	3.88
$1^1/_2$	1.91	6.12	4	0.88	4.50
2	2.38	6.50	8	0.75	5.00
$2^1/_2$	2.88	7.50	8	0.88	5.88
3	3.50	8.25	8	0.88	6.62
$3^1/_2$	4.00	9.00	8	0.88	7.25
4	4.50	10.00	8	0.88	7.88
5	5.56	11.00	8	0.88	9.25
6	6.62	12.50	12	0.88	10.63
8	8.62	15.00	12	1.00	13.00

* Dimensions in inches. For tolerances refer to Table 28.21.

TABLE 28.18 Flat Ring Nonmetallic Gasket Dimensions for ASME B16.47 Series A (MSS SP 44) Large-Diameter Steel Flanges, Classes 150, 300, 400, and 600 (U.S. Customary Units)*

		Gasket OD			
Nominal pipe size (NPS)	Gasket ID	Class 150	Class 300	Class 400	Class 600
26	26.00	30.50	32.88	32.75	34.12
28	28.00	32.75	35.38	35.12	36.00
30	30.00	34.75	37.50	37.25	38.25
32	32.00	37.00	39.62	39.50	40.25
34	34.00	39.00	41.62	41.50	42.25
36	36.00	41.25	44.00	44.00	44.50
38	38.00	43.75	41.50	42.26	43.50
40	40.00	45.75	43.88	44.58	45.50
42	42.00	48.00	45.88	46.38	48.00
44	44.00	50.25	48.00	48.50	50.00
46	46.00	52.25	50.12	50.75	52.25
48	48.00	54.50	52.12	53.00	54.75
50	50.00	56.50	54.25	55.25	57.00
52	52.00	58.75	56.25	57.25	59.00
54	54.00	61.00	58.75	59.75	61.25
56	56.00	63.25	60.75	61.75	63.50
58	58.00	65.50	62.75	63.75	65.50
60	60.00	67.50	64.75	66.25	67.75

* Dimensions in inches. For tolerances refer to Table 28.21.

TABLE 28.19 Flat Ring Nonmetallic Gasket Dimensions for ASME B16.47 Series B (API 605) Large-Diameter Steel Flanges, Classes 75, 150, 300, 400, and 600 (US Customary Units)*

Nominal pipe size (NPS)	Gasket ID	Gasket OD				
		Class 75	Class 150	Class 300	Class 400	Class 600
26	26.00	27.88	28.56	30.38	29.38	30.12
28	28.00	29.88	30.56	32.50	31.50	32.25
30	30.00	31.88	32.56	34.88	33.75	34.62
32	32.00	33.88	34.69	37.00	35.88	36.75
34	34.00	35.88	36.81	39.12	37.88	39.25
36	36.00	38.31	38.88	41.25	40.25	41.25
38	38.00	40.31	41.12	43.25		
40	40.00	42.31	43.12	45.25		
42	42.00	44.31	45.12	47.25		
44	44.00	46.50	47.12	49.25		
46	46.00	48.50	49.44	51.88		
48	48.00	50.50	51.44	53.88		
50	50.00	52.50	53.44	55.88		
52	52.00	54.62	55.44	57.88		
54	54.00	56.62	57.62	61.25		
56	56.00	58.88	59.62	62.75		
58	58.00	60.88	62.19	65.19		
60	60.00	62.88	64.19	67.12		

* Dimensions in inches. For tolerances refer to Table 28.21.

TABLE 28.20 Full Face Nonmetallic Gasket Dimensions for MSS SP-51 Class 150 LW Corrosion-Resistant Cast Flanges and Flanged Fittings (U.S. Customary Units)*

Nominal pipe size (NPS)	Gasket ID	OD	Number of holes	Hole diameter	Bolt circle diameter
1/4	0.56	2.50	4	0.44	1.69
	0.69	2.50	4	0.44	1.69
1/2	0.84	3.50	4	0.62	2.38
3/4	1.06	3.88	4	0.62	2.75
1	1.31	4.25	4	0.62	3.12
1 1/4	1.66	4.62	4	0.62	3.50
1 1/2	1.91	5.00	4	0.62	3.88
2	2.38	6.00	4	0.75	4.75
2 1/2	2.88	7.00	4	0.75	5.50
3	3.50	7.50	4	0.75	6.00
4	4.50	9.00	4	0.75	7.50
5	5.56	10.00	4	0.88	8.50
6	6.62	11.00	4	0.88	9.50
8	8.62	13.60	8	0.88	11.75
10	10.75	16.00	12	1.00	14.25
12	12.75	19.00	12	1.00	17.00
14	14.00	21.00	12	1.12	18.75
16	16.00	23.50	16	1.12	21.25
18	18.00	25.00	16	1.25	22.75
20	20.00	27.50	20	1.25	25.00
24	24.00	32.00	20	1.38	29.50

* Dimensions in inches. Dimensions also apply to MSS SP 42. For tolerances refer to Table 28.21.

TABLE 28.21 Tolerances for Flat Ring and Full Face Nonmetallic Gaskets per ASME B16.21 (U.S. Customary Units)*

Nominal pipe size	Gasket inside diameter	Gasket outside diameter	Bolt circle diameter	Bolt center-to-center distance
12 and smaller	±0.06	+0.00, −0.06	±0.06	±0.03
14 and larger	±0.12	+0.00, −0.12	±0.06	±0.03

* Dimensions in inches.

METALLIC GASKETS

Ring-Type Gaskets

The ring-type joint was initially developed for use in the petroleum industry, where high-pressure/temperature applications necessitated the need for a high-integrity seal. They are mainly used in the oil field on drilling and completion equipment. Ring-type joints are commonly used on valves and pipework assemblies, along with some high-integrity pressure vessels.

Style R. The style R ring gasket is manufactured in accordance with API 6A and ASME B16.20 to suit API 6B and ASME B16.5 flanges. Style R rings are manufactured in both oval and octagonal configurations (Fig. 28.8). Both styles are interchangeable on the modern flat-bottom groove; however, only the oval style can be used in the old-type round-bottom groove. Style R rings are designed to seal pressures up to 6250 psi in accordance with ASME B16.5 pressure ratings and up to 5000 psi in accordance with API 6A pressure ratings. (See Figs. 28.8*a* and 28.8*b*.)

Styles RX and SRX. The style RX ring gasket is manufactured in accordance with API 6A and ASME B16.20 to suit API 6B and ASME B16.5 flanges. The style RX ring is a pressure-energized adaptation of the style R ring gasket. The style RX is designed to fit the modern flat-bottom groove and is interchangeable with the standard style R ring. However, since the style RX is significantly taller than the style R, larger flange makeup distances will result. Style RX rings are designed to seal pressures up to 6250 psi in accordance with ASME B16.20 pressure ratings, and up to 5000 psi in accordance with API 6A pressure ratings. The design uses a pressure-energized effect, which improves the efficiency of the seal as the internal pressure system increases. Selected sizes incorporate a pressure passage hole to allow for pressure equalization of each side of the sealing face. Style SRX rings are manufactured in accordance with API 17D for use on subsea wellhead and Christmas tree equipment. (See Fig. 28.8*c*.)

Styles BX and SBX. The style BX ring gasket is manufactured in accordance with API 6A. Style BX ring design exhibits a pressure-energized effect which improves the efficiency of the seal as the internal pressure of the system increases. All BX rings incorporate a pressure passage hole to allow for pressure equalization each side of the sealing faces. On assembly, metal to metal contact of the flange faces is achieved. The style BX is not interchangeable with any other style, and is only suited for API 6BX flanges. Style BX rings are designed to seal up to 20,000 psi in accordance with API 6A pressure ratings. Style SBX rings are manufactured in accordance with API 17D for use on subsea wellhead and Christmas tree equipment. (See Fig. 28.8*d*.)

Ring-Type Joints

How They Work. Under axial compressive load, ring-type gaskets plastically deform and flow into the irregularities of the flange grooves. Since the load-bearing area of the ring gasket is relatively small, very high surface stresses result between the sealing faces of the ring and the groove. These stresses are further increased on the pressure-energized types, which allows very high internal pressures to be sealed.

Since ring-type gaskets are solid metal, their recovery characteristics are poor. The seal is maintained by the action of the axial load upon the gasket.

Surface Finish Requirements. With all metal-to-metal–type seals, it is imperative that the gasket and groove sealing surfaces are free from indentations, score marks, tool chatter marks, and other imperfections. The surface finish of the gaskets and groove sealing faces is also critical and should not exceed the following:

Styles R and RX	63 CLA μin maximum R_a 1.6 μm
Style BX	32 CLA μin maximum R_a 0.8 μm

Reuse. Ring gaskets are designed to have a limited amount of positive interference, which ensures that the ring gasket seats correctly into the groove on compression. Their reuse is not recommended for two reasons:

- The initial seating of the gasket will be impaired.
- When the gasket is plastically deformed, work hardening of the metal surface occurs. This may result in permanent damage to the groove if the gasket is reused.

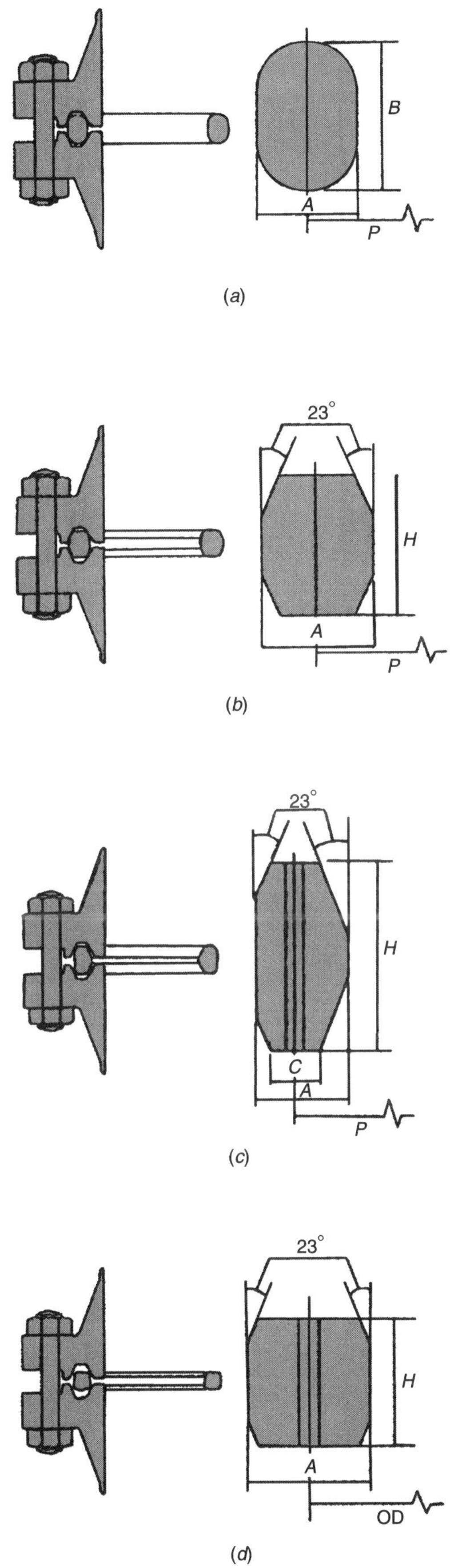

FIGURE 28.8 Types of ring gasket. (*a*) Oval, style R; (*b*) octagonal, style R; (*c*) style RX; (*d*) style BX.

Hardness of Materials. On compression of the flange assembly, it is imperative that the ring gasket be significantly softer than the flange groove so that the gasket plastically deforms and not the groove. The use of harder materials can result in flange groove damage. For this reason, ring gaskets are supplied with the maximum hardness values given in Table 28.22.

TABLE 28.22 Hardness Values

Material	Werkstoff no.*	Maximum hardness, Brinell†	Maximum hardness, Rockwell B‡	Identification
Soft iron		90	56	D
Low-carbon steel		120	68	S
4–6% Chrome $^1/_2$ Moly		130	72	F5
Type 304 stainless	1.4301	160	83	S304
Type 316 stainless	1.4401	160	83	S316
Type 347 stainless	1.4550	160	83	S347
Type 410 stainless	1.4406	170	86	S410

* Material designation per DIN standards.
† Measured with 3000-kg load, except soft iron which is measured with 500 kg.
‡ Measured with 100-kg load and $^1/_{16}$-in-diameter ball.

Protective Coating. In accordance with API specifications, soft iron and low-carbon steel ring gaskets are protected with electroplated zinc or cadmium to a maximum thickness of 0.0005 in. Other types of coatings may also be used.

Dimensions. Figure 28.9 defines ring-type joint dimensions and Tables 28.23 to 28.30 provide their specifications for various types of flange.

Lens Ring Gaskets

In certain applications, the specification of a high-integrity metallic seal has usually lead to the selection of the lens ring concept, rather than the more generally recognized ring-type joint solution. The lens ring is covered solely by the DIN 2696 specification. However, ASME B16.5 and other flange types can be modified to accept the lens ring.

The lens ring provides a metallic gasket design incorporating spherical seating faces designed to suit specifically mating flange recesses, providing the user with a high-integrity, high-pressure/temperature metal-to-metal seal.

As with all metallic gaskets, the lens ring material should be specified softer than the flange material, thus ensuring that an applied compressive load leads to the elastic/plastic deformation of the lens ring and not the flange sealing face. The distribution of high compressive loads leads to the spread of the gasket facings, ensuring overstressing of the gasket is prevented.

In accordance with DIN 2696, general materials are limited to a range of specified carbon steels and stainless steel grades, although alternative grades are available upon request.

Figure 28.10 specifies lens ring dimensions and Table 28.31 provides values for the dimensions.

COMPOSITE GASKETS

Spiral-Wound Gaskets

A requirement of any gasket is the ability to recover under variable loads. The effects of pressure and temperature fluctuations, the temperature difference across the flange face, flange rotation, bolt stress relaxation, and creep demand a gasket with adequate flexibility and recovery to maintain a seal under variable working conditions.

A spiral-wound gasket is manufactured by spirally winding a preformed metal strip and a filler on the outer periphery of winding mandrels. The winding mandrel outside diameter forms the inner diameter of the gasket and the superposed metal and nonmetallic windings are continually wound until the desired outer diameter is attained (Fig. 28.11). Normal practice is to reinforce the inner and outer diameters with several plies of metal, with no soft fillers being introduced.

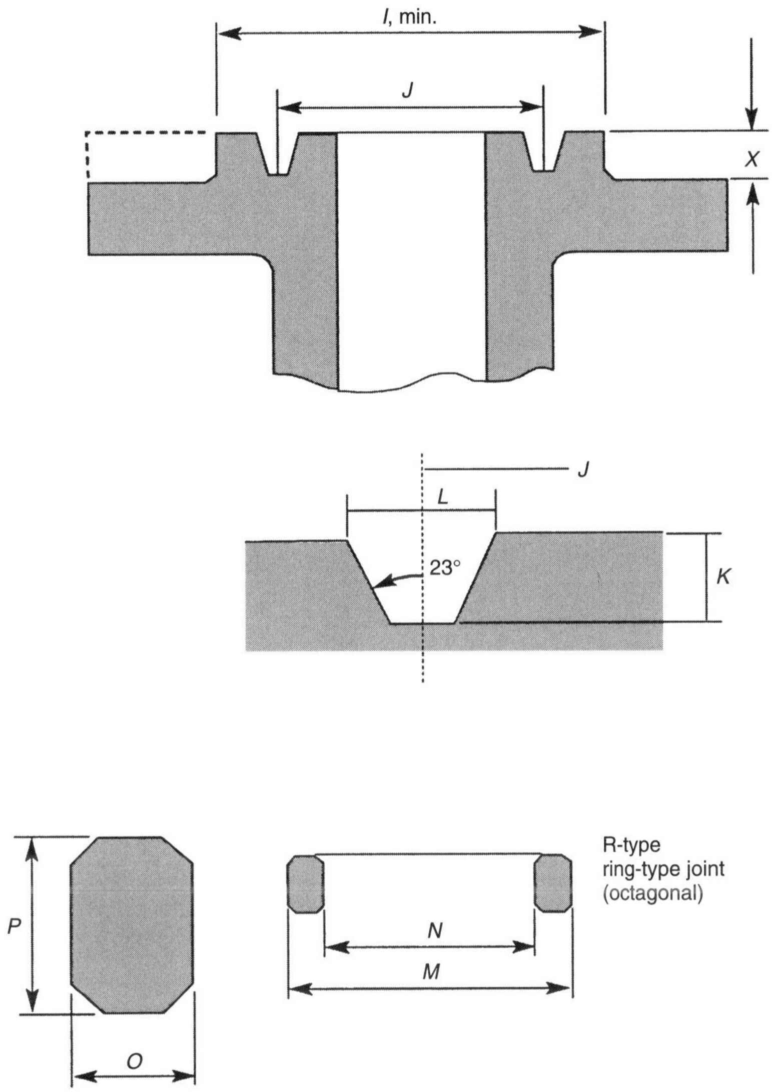

FIGURE 28.9 Ring joint facing and RTJ gasket dimensions.

TABLE 28.23 Ring Joint Facing and RTJ Gasket Dimensions for ASME B16.5 Class 150 Flanges*

Nominal pipe size (NPS)	1/2	3/4	1	1 1/2	2	3	4	6	8	10	12	14	16	18	20	24
Diameter of raised section, *I*	2	2 1/2	2 3/4	3 9/16	4 1/4	5 3/4	6 7/8	9 1/2	11 7/8	14	16 1/4	18	20	22 5/8	25	19 1/2
Groove pitch diameter *J*	1 11/32	1 11/16	2	2 11/16	3 1/4	4 7/8	5 7/8	8 5/16	10 5/8	12 3/4	15	16 1/2	18 1/2	21	23	27 1/4
Depth of groove, *K*	7/32	1/4	1/4	1/4	5/16	5/16	5/16	5/16	5/16	5/16	5/16	5/16	5/16	5/16	3/8	5/16
Width *L*	9/32	11/32	11/32	11/32	15/32	15/32	15/32	15/32	15/32	15/32	15/32	15/32	15/32	15/32	17/32	21/32
Outside diameter *M*	1 19/32	2	2 5/16	3	3 11/16	5 5/16	6 5/16	8 3/4	11 1/16	13 3/16	15 7/16	16 15/16	18 15/16	21 7/16	23 1/2	5/16
Inside diameter *N*	1 3/32	1 3/8	1 11/16	2 3/8	2 13/16	4 7/16	5 7/16	7 1/8	10 3/16	12 5/16	14 9/16	16 1/16	18 1/16	20 9/16	22 1/2	27 7/8
Width *O*	1/4	5/16	5/16	5/16	7/16	7/16	7/16	7/16	7/16	7/16	7/16	7/16	7/16	7/16	1/2	26 5/8
Thickness *P*	3/8	1/2	1/2	1/2	5/8	5/8	5/8	5/8	5/8	5/8	5/8	5/8	5/8	5/8	11/16	11/16
R number	11	13	16	20	23	31	37	45	49	53	57	61	65	69	73	77

* All dimensions in inches. Ring dimensions are per ASME B16.20. Refer to Fig. 28.9.

TABLE 28.24 Ring Joint Facing and RTJ Gasket Dimensions for ASME B16.5 Class 300 Flanges*

Nominal pipe size (NPS)	1/2	3/4	1	1 1/2	2	3	4	6	8	10	12	14	16	18	20	24
Diameter of raised section, *I*	2	2 1/2	2 3/4	3 9/16	4 1/4	5 3/4	6 7/8	9 1/2	11 7/8	14	16 1/4	18	20	22 5/8	25	19 1/2
Groove pitch diameter *J*	1 11/32	1 11/16	2	2 11/16	3 1/4	4 7/8	5 7/8	8 5/16	10 5/8	12 3/4	15	16 1/2	18 1/2	21	23	27 1/4
Depth of groove, *K*	7/32	1/4	1/4	1/4	5/16	5/16	5/16	5/16	5/16	5/16	5/16	5/16	5/16	5/16	3/8	5/16
Width *L*	9/32	11/32	11/32	11/32	15/32	15/32	15/32	15/32	15/32	15/32	15/32	15/32	15/32	15/32	17/32	21/32
Outside diameter *M*	1 19/32	2	2 5/16	3	3 11/16	5 5/16	6 5/16	8 3/4	11 1/16	13 3/16	15 7/16	16 15/16	18 15/16	21 7/16	23 1/2	5/16
Inside diameter *N*	1 3/32	1 3/8	1 11/16	2 3/8	2 13/16	4 7/16	5 7/16	7 1/8	10 3/16	12 5/16	14 9/16	16 1/16	18 1/16	20 9/16	22 1/2	27 7/8
Width *O*	1/4	5/16	5/16	5/16	7/16	7/16	7/16	7/16	7/16	7/16	7/16	7/16	7/16	7/16	1/2	26 5/8
Thickness *P*	3/8	1/2	1/2	1/2	5/8	5/8	5/8	5/8	5/8	5/8	5/8	5/8	5/8	5/8	11/16	11/16
R number	11	13	16	20	23	31	37	45	49	53	57	61	65	69	73	77

* All dimensions in inches. Ring dimensions are per ASME B16.20. Refer to Fig. 28.9.

TABLE 28.25 Ring Joint Facing and RTJ Gasket Dimensions for ASME B16.5 Class 600 Flanges*

Nominal pipe size (NPS)	1/2	3/4	1	1 1/2	2	3	4	6	8	10	12	14	16	18	20	24
Diameter of raised section, *I*	2	2 1/4	2 3/4	3 9/16	4 1/4	5 3/4	6 7/8	9 1/2	11 7/8	14	16 1/4	18	20	22 5/8	25	29 1/2
Groove pitch diameter *J*	1 11/32	1 11/16	2	2 11/16	3 1/4	4 7/8	5 7/8	8 5/16	10 5/8	12 3/4	15	16 1/2	18 1/2	21	23	27 1/4
Depth of groove, *K*	7/32	1/4	1/4	1/4	5/16	5/16	5/16	5/16	5/16	5/16	5/16	5/16	5/16	5/16	3/8	5/16
Width *L*	9/32	11/32	11/32	11/32	15/32	15/32	15/32	15/32	15/32	15/32	15/32	15/32	15/32	15/32	17/32	21/32
Outside diameter *M*	1 19/32	2	2 5/16	3	3 11/16	5 5/16	6 5/16	8 3/4	11 1/16	13 3/16	15 7/16	16 13/16	18 15/16	21 7/16	23 1/2	27 7/8
Inside diameter *N*	1 3/32	1 3/8	1 11/16	2 3/8	2 13/16	4 7/16	5 7/16	7 1/8	10 3/16	12 5/16	14 9/16	16 1/16	18 1/16	20 9/16	22 1/2	26 5/8
Width *O*	1/4	5/16	5/16	5/16	7/16	7/16	7/16	7/16	7/16	7/16	7/16	7/16	7/16	7/16	1/2	5/8
Thickness *P*	3/8	1/2	1/2	1/2	5/8	5/8	5/8	5/8	5/8	5/8	5/8	5/8	5/8	5/8	11/16	11/16
R number	11	13	16	20	23	31	37	45	49	53	57	61	65	69	73	77

* All dimensions in inches. Ring dimensions are per ASME B16.20. Refer to Fig. 28.9.

TABLE 28.26 Ring Joint Facing and RTJ Gasket Dimensions for ASME B16.5 Class 900 Flanges*

<table>
<tr><th>Nominal pipe size (NPS)</th><th>1/2</th><th>3/4</th><th>1</th><th>1 1/2</th><th>2</th><th>3</th><th>4</th><th>6</th><th>8</th><th>10</th><th>12</th><th>14</th><th>16</th><th>18</th><th>20</th><th>24</th></tr>
<tr><td>Diameter of raised section, I</td><td colspan="5" rowspan="8">Use Class 1500 dimensions in these sizes</td><td>6 1/8</td><td>7 1/8</td><td>9 1/2</td><td>12 1/8</td><td>14 1/4</td><td>16 1/2</td><td>18 3/8</td><td>20 5/8</td><td>23 3/8</td><td>25 1/2</td><td>30 3/8</td></tr>
<tr><td>Groove pitch diameter J</td><td>4 7/8</td><td>5 7/8</td><td>8 5/16</td><td>10 5/8</td><td>12 3/4</td><td>15</td><td>16 1/2</td><td>18 1/2</td><td>21</td><td>23</td><td>27 1/4</td></tr>
<tr><td>Depth of groove, K</td><td>5/16</td><td>5/16</td><td>5/16</td><td>5/16</td><td>5/16</td><td>5/16</td><td>7/16</td><td>7/16</td><td>1/2</td><td>1/2</td><td>5/8</td></tr>
<tr><td>Width L</td><td>15/32</td><td>15/32</td><td>15/32</td><td>15/32</td><td>15/32</td><td>15/32</td><td>21/32</td><td>21/32</td><td>25/32</td><td>25/32</td><td>1 1/16</td></tr>
<tr><td>Outside diameter M</td><td>5 5/16</td><td>6 5/16</td><td>8 3/4</td><td>11 1/16</td><td>13 3/16</td><td>15 7/16</td><td>17 1/8</td><td>19 1/8</td><td>21 3/4</td><td>23 3/4</td><td>28 1/4</td></tr>
<tr><td>Inside diameter N</td><td>4 7/16</td><td>5 7/16</td><td>7 7/8</td><td>10 3/16</td><td>12 5/16</td><td>14 9/16</td><td>15 7/8</td><td>17 7/8</td><td>20 1/8</td><td>22 1/4</td><td>26 1/4</td></tr>
<tr><td>Width O</td><td>7/16</td><td>7/16</td><td>7/16</td><td>7/16</td><td>7/16</td><td>7/16</td><td>5/8</td><td>5/8</td><td>3/4</td><td>3/4</td><td>1</td></tr>
<tr><td>Thickness P</td><td>5/8</td><td>5/8</td><td>5/8</td><td>5/8</td><td>5/8</td><td>5/8</td><td>13/16</td><td>13/16</td><td>15/16</td><td>15/16</td><td>1 1/4</td></tr>
<tr><td>R number</td><td colspan="5"></td><td>31</td><td>37</td><td>45</td><td>49</td><td>53</td><td>57</td><td>62</td><td>66</td><td>70</td><td>74</td><td>78</td></tr>
</table>

* All dimensions in inches. Ring dimensions are per ASME B16.20. Refer to Fig. 28.9.

TABLE 28.27 Ring Joint Facing and RTJ Gasket Dimensions for ASME B16.5 Class 1500 Flanges*

Nominal pipe size (NPS)	1/2	3/4	1	1 1/2	2	3	4	6	8	10	12	14	16	18	20	24
Diameter of raised section, *I*	2 3/8	2 5/8	2 13/16	3 5/8	4 7/8	6 3/8	7 5/8	9 3/4	12 1/2	14 3/8	17 1/4	19 1/4	21 1/2	24 1/8	26 1/2	31 1/4
Groove pitch diameter *J*	1 9/16	1 3/4	2	2 11/16	3 3/4	5 3/8	6 3/8	8 5/16	10 5/8	12 3/4	15	16 1/2	18 1/2	21	23	27 1/4
Depth of groove, *K*	1/4	1/4	1/4	1/4	5/16	5/16	5/16	3/8	5/16	7/16	9/16	5/8	11/16	11/16	11/16	13/16
Width *L*	11/32	11/32	11/32	11/32	15/32	15/32	15/32	17/32	21/32	21/32	29/32	1 1/16	1 3/16	1 3/16	1 5/16	1 7/16
Outside diameter *M*	1 7/8	2 1/16	2 5/16	3	4 3/16	5 13/16	6 13/16	8 13/16	11 1/4	13 3/8	15 7/8	17 1/2	19 5/8	22 1/8	24 1/4	28 5/8
Inside diameter *N*	1 1/4	1 7/16	1 11/16	2 3/8	3 5/16	4 15/16	5 15/16	7 15/16	10	12 1/8	14 1/8	15 1/2	17 3/8	19 7/8	21 3/4	25 7/8
Width *O*	5/16	5/16	5/16	5/16	7/16	7/16	7/16	1/2	5/8	5/8	7/8	1	1 1/8	1 1/8	1 1/4	1 3/8
Thickness *P*	1/2	1/2	1/2	1/2	5/8	5/8	5/8	11/16	13/16	13/16	11/16	1 1/4	1 3/8	1 3/8	1 1/2	1 5/8
R number	12	14	16	20	24	35	39	46	50	54	58	63	67	71	75	79

* All dimensions in inches. Ring dimensions are per ASME B16.20. Refer to Fig. 28.9.

TABLE 28.28 Ring Joint Facing and RTJ Gasket Dimensions for ASME B16.5 Class 2500 Flanges*

Nominal pipe size (NPS)	1/2	3/4	1	1 1/2	2	3	4	6	8	10	12
Diameter of raised section, *I*	2 9/16	2 7/8	3 1/4	4 1/2	5 1/4	6 3/8	8	11	13 3/8	16 3/4	19 1/2
Groove pitch diameter *J*	1 11/16	2	2 3/8	3 1/4	4	5	6 3/16	9	11	13 1/2	16
Depth of groove, *K*	1/4	1/4	1/4	5/16	5/16	3/8	7/16	1/2	9/16	11/16	11/16
Width *L*	11/32	11/32	11/32	15/32	15/32	17/32	21/32	29/32	29/32	1 3/16	1 5/16
Outside diameter *M*	2	2 5/16	2 5/16	3 11/16	4 7/16	5 1/2	6 13/16	9 3/4	11 7/8	14 5/8	17 1/4
Inside diameter *N*	1 3/8	1 11/16	1 11/16	2 13/16	3 9/16	4 1/2	5 9/16	8 1/4	10 1/8	12 3/8	14 3/4
Width *O*	5/16	5/16	5/16	7/16	7/16	1/2	5/8	3/4	7/8	1 1/8	1 1/4
Thickness *P*	1/2	1/2	1/2	5/8	5/8	11/16	13/16	13/16	1 1/16	1 3/8	1 1/2
R number	13	16	18	23	26	32	38	47	51	55	60

* All dimensions in inches. Ring dimensions are per ASME B16.20. Refer to Fig. 28.9.

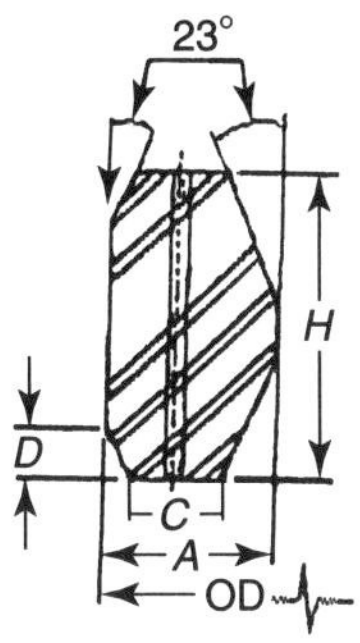

TABLE 28.29 Type RX Ring Gasket Dimensions*

Ring number	Outside diameter of ring, OD	Width of ring, A	Width of flat, C	Height of outside bevel, D	Height of ring, H	Radius in ring, R_1	Hole size, E
RX-20	3.000	0.344	0.182	0.125	0.750	0.06	
RX-23	3.672	0.469	0.254	0.167	1.000	0.06	
RX-24	4.172	0.469	0.254	0.167	1.000	0.06	
RX-25	4.313	0.344	0.182	0.125	0.750	0.06	
RX-26	4.406	0.469	0.254	0.167	1.000	0.06	
RX-27	4.656	0.469	0.254	0.167	1.000	0.06	
RX-31	5.297	0.469	0.254	0.167	1.000	0.06	
RX-35	5.797	0.469	0.254	0.167	1.000	0.06	
RX-37	6.297	0.469	0.254	0.167	1.000	0.06	
RX-39	6.797	0.469	0.254	0.167	1.000	0.06	
RX-41	7.547	0.469	0.254	0.167	1.000	0.06	
RX-45	8.734	0.469	0.254	0.167	1.000	0.06	
RX-46	8.750	0.531	0.263	0.188	1.125	0.06	
RX-47	9.656	0.781	0.407	0.271	1.625	0.09	
RX-49	11.047	0.469	0.254	0.167	1.000	0.06	
RX-50	11.156	0.656	0.335	0.208	1.250	0.06	
RX-53	13.172	0.469	0.254	0.167	1.000	0.06	
RX-54	13.281	0.656	0.335	0.208	1.250	0.06	
RX-57	15.422	0.469	0.254	0.167	1.000	0.06	
RX-63	17.391	1.063	0.582	0.333	2.000	0.06	
RX-65	18.922	0.469	0.254	0.167	1.000	0.06	
RX-66	18.031	0.656	0.335	0.208	1.250	0.06	
RX-69	21.422	0.469	0.254	0.167	1.000	0.09	
RX-70	21.656	0.781	0.407	0.271	1.625	0.09	
RX-73	23.469	0.531	0.263	0.208	1.250	0.06	
RX-74	23.656	0.781	0.407	0.271	1.625	0.09	
RX-82	2.672	0.469	0.254	0.167	1.000	0.06	0.06
RX-84	2.922	0.469	0.254	0.167	1.000	0.06	0.06
RX-85	3.547	0.531	0.263	0.167	1.000	0.06	0.06
RX-86	4.078	0.594	0.335	0.188	1.125	0.06	0.09
RX-87	4.453	0.594	0.335	0.188	1.125	0.06	0.09
RX-88	5.484	0.688	0.407	0.208	1.250	0.06	0.12
RX-89	5.109	0.719	0.407	0.208	1.250	0.06	0.12
RX-90	6.875	0.781	0.479	0.292	1.750	0.09	0.12
RX-91	11.297	1.188	0.780	0.297	1.781	0.09	0.12
RX-99	9.672	0.469	0.254	0.167	1.000	0.06	
RX-201	2.026	0.226	0.126	0.057	0.445	0.02	
RX-205	2.453	0.219	0.120	0.072	0.437	0.02	
RX-210	3.844	0.375	0.213	0.125	0.750	0.03	
RX-215	5.547	0.469	0.210	0.167	1.000	0.06	

* Dimensions in inches.

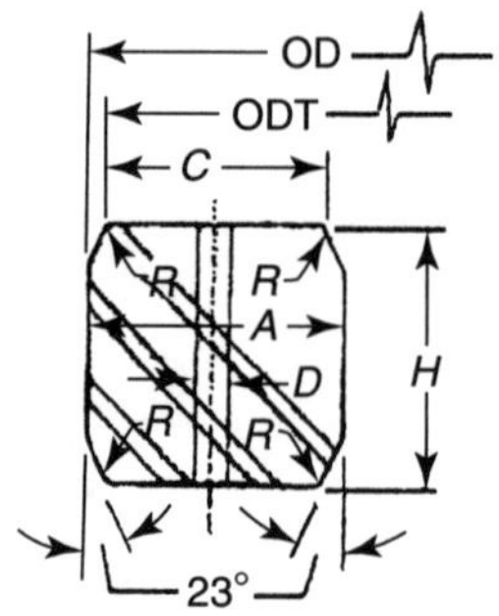

TABLE 28.30 Type BX Ring Gasket Dimensions*

Ring number	Nominal size	Outside diameter of ring, OD	Height of ring, *H*	Width of ring, *A*	Diameter of flat, ODT	Width of flat, *C*	Hole size, *D*
BX-150	1 11/16	2.842	0.366	0.366	2.790	0.314	0.06
BX-151	1 13/16	3.008	0.379	0.379	2.954	0.325	0.06
BX-152	2 1/16	3.334	0.403	0.403	3.277	0.346	0.06
BX-153	2 9/16	3.974	0.448	0.448	3.910	0.385	0.06
BX-154	3 1/16	4.600	0.488	0.488	4.531	0.419	0.06
BX-155	4 1/16	5.825	0.560	0.560	5.746	0.481	0.06
BX-156	7 1/16	9.367	0.733	0.733	9.263	0.629	0.12
BX-157	9	11.593	0.826	0.826	11.476	0.709	0.12
BX-158	11	13.860	0.911	0.911	13.731	0.782	0.12
BX-159	13 5/8	16.800	1.012	1.012	16.657	0.869	0.12
BX-160	13 5/8	15.850	0.938	0.541	15.717	0.408	0.12
BX-161	16 5/8	19.347	1.105	0.638	19.191	0.482	0.12
BX-162	16 5/8	18.720	0.560	0.560	18.641	0.481	0.06
BX-163	18 3/4	21.896	1.185	0.684	21.728	0.516	0.12
BX-164	18 3/4	22.463	1.185	0.968	22.295	0.800	0.12
BX-165	21 1/4	24.595	1.261	0.728	24.417	0.550	0.12
BX-166	21 1/4	25.198	1.261	1.029	25.020	0.851	0.12
BX-167	26 3/4	29.869	1.412	0.516	29.696	0.316	0.06
BX-168	26 3/4	30.128	0.142	0.632	29.928	0.432	0.06
BX-169	5 1/8	6.831	0.624	0.509	6.743	0.421	0.06
BX-170	6 5/8	8.584	0.560	0.560	8.505	0.481	0.06
BX-171	8 9/16	10.529	0.560	0.560	10.450	0.481	0.06
BX-172	11 5/32	13.113	0.560	0.560	13.034	0.481	0.06
BX-303	30	33.573	1.494	0.668	33.361	0.457	0.06

* Dimensions in inches.

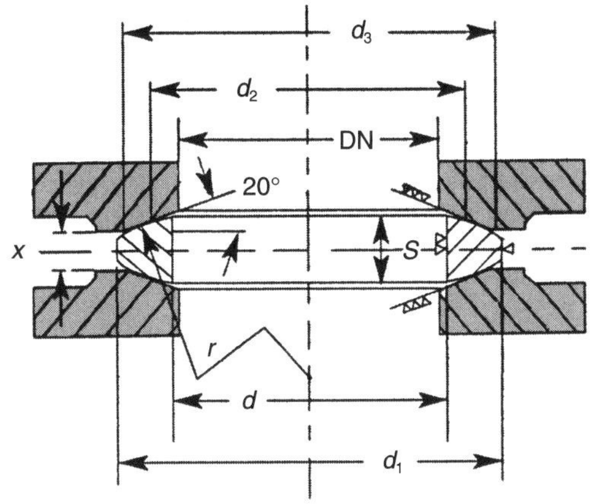

FIGURE 28.10 Lens ring geometry.

TABLE 28.31 Lens Ring Dimensions*

NPS size, DN	*d* Min.	*d* Max.	d_1	*S* for *d* max.	d_2, middle contact diameter	*r*	d_3	*x*
Nominal pressure PN64–400								
10	10	14	21	7	17.1	25	18	5.7
15	14	18	28	8.5	22	32	27	6
25	20	29	43	11	34	50	39	6
40	34	43	62	14	48	70	55	8
50	46	55	78	16	60	88	68	9
65	62	70	102	20	76.6	112	85	13
80	72	82	116	22	88.2	129	97	13
100	94	108	143	26	116	170	127	15
125	116	135	180	29	149	218	157	22
150	139	158	210	33	171	250	183	26
Nominal pressure PN64 and 100								
(175)†	176	183	243	31	202.5	296	218	28
200	198	206	276	35	225	329	243	27
250	246	257	332	37	277.7	406	298	25
300	295	305	385	40	323.5	473	345	26
350	330	348	425	41	368	538	394	23
400	385	395	475	42	417.2	610	445	24
Nominal pressure PN160–400								
(175)†	162	177	243	37	202.5	296	218	21
200	183	200	276	40	225	329	243	25
250	230	246	332	46	277.7	406	298	25
300	278	285	385	50	323.5	473	345	30

* Dimensions in millimeters.
† Avoid nominal pipe sizes in brackets.

This engineered product is made to be compatible with the flange closure in which it is to be used. For example, a closure designed for vacuum service may require a gasket of exactly the same dimensions as a closure designed for 1500-psi service. The closure designed for vacuum service would have relatively light bolting, indicating the necessity for a soft gasket, while the 1500-psi application would have heavy bolting, requiring a relatively dense gasket.

The service conditions under which a spiral-wound gasket is expected to hold its seal dictate the density of the gasket. Gaskets that have identical inside and outside diameters can be either hard or soft, as shown in Fig. 28.12. The available compressive force is the basis for calculating the density of the gasket structure to support specific loads.

Spiral-Wound Gasket Styles

Basic Spiral-Wound Gasket. In the basic construction, inner and outer diameters are reinforced with several plies without filler to give greater stability and better compression characteristics. Suitable for tongue and groove or male and female or groove to flat face flange assemblies. See Fig. 28.13.

With Solid Inner Metal Ring. A solid inner metal ring acts as a compression stop and fills the annular space between flange bore and the inside diameter. This style is designed to prevent accumulation of solids, reduce turbulent flow of process fluids, and minimize erosion of flange faces. Suitable for male and female pipe flanges. See Fig. 28.14.

With Solid External Metal Ring. This style utilizes an external ring that accurately centers the gasket on the flange face, provides additional radial strength to prevent gasket blowout, and acts as a compression stop. It is a general-purpose gasket suitable for use with flat face and raised face flanges. See Fig. 28.15.

With Inner and Outer Metal Rings. This style is suitable for use with flat face and raised face flanges and is specified for high-pressure/temperature service or where corrosive or toxic media are present. See Fig. 28.16. ASME B16.20, which covers spiral-wound gaskets, requires the use of solid metal inner rings in:

- Pressure Class 900, nominal pipe sizes NPS 24 (DN 600) and larger
- Pressure Class 1500, nominal pipe sizes NPS 12 (DN 300) and larger
- Pressure Class 2500, nominal pipe sizes NPS 4 (DN100) and larger
- All PTFE-filled gaskets

It is also recommended that inner rings be used if the user's experience has shown inward buckling of the gasket.

Specifications

Table 28.32 provides color codes and defines abbreviations for spiral-wound gaskets. Tables 28.33 to 28.42 give dimensions of spiral-wound gaskets.

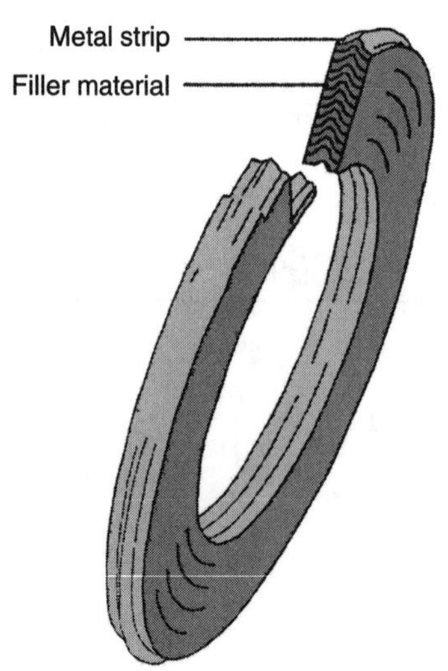

FIGURE 28.11 Structure of a spiral-wound gasket.

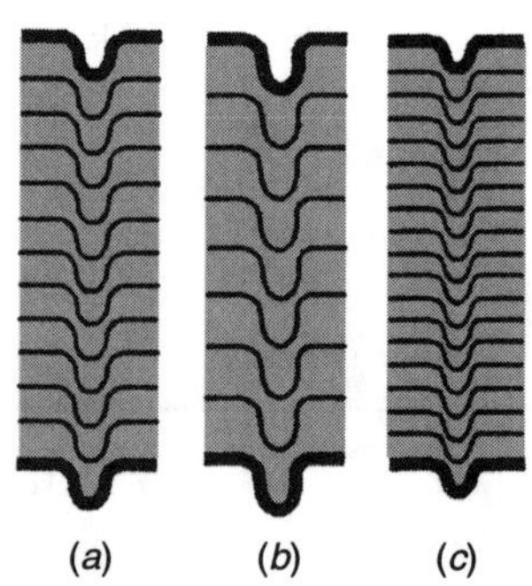

FIGURE 28.12 Spiral-wound gasket cross sections: (*a*) medium density, (*b*) low density, (*c*) high density.

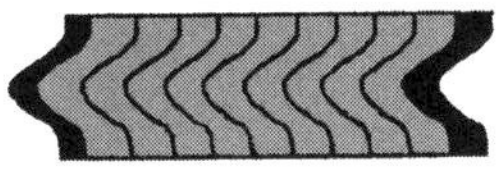

FIGURE 28.13 Basic construction.

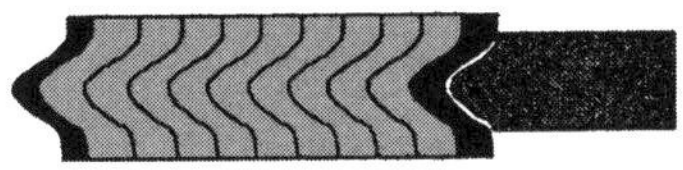

FIGURE 28.14 With solid inner metal ring.

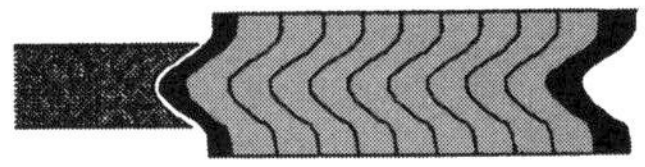

FIGURE 28.15 With solid external metal ring.

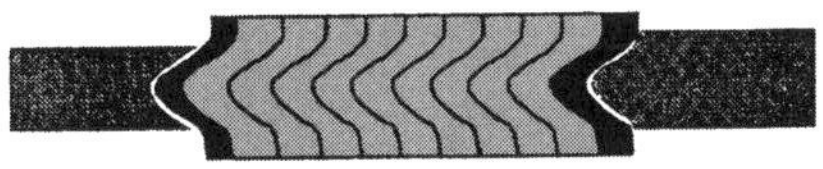

FIGURE 28.16 With inner and outer metal rings.

TABLE 28.32 Color Coding and Abbreviations for Spiral-Wound Gasket Metal Winding Materials

Material	Abbreviation	Color code
Carbon steel	CRS	Silver
304 SS	304	Yellow
304L SS	304L	No color
309 SS	309	No color
310 SS	310	No color
316L SS	316L	Green
347 SS	347	Blue
321 SS	321	Turquoise
430 SS	430	No color
Ni-Cu Grade 400	MON (Monel)	Orange
Nickel 200	NI	Red
Titanium	TI	Purple
20 Cb-3 alloy	A-20	Black
Ni-Mo Grade B2	Hast B (Hastelloy B)	Brown
Ni-Mo-Cr Grade C-276	Hast C (Hastelloy C)	Beige
Ni-Cr-Fe, Grade 600	INC 600 (Inconel 600)	Gold
Ni-Cr-Fe-Cb, Grade 625	INC 625 (Inconel 625)	Gold
Ni-Cr-Fe-Ti, Grade X-750	INX (InconelX-750)	No color
Ni-Fe-Cr, Grade 800	INC 800 (Incoloy 800)	White
Ni-Fe-Cr-Mo-Cu, Grade 825	INC 825 (Incoloy 825)	White
Zirconium	ZIRC	No color

TABLE 28.33 Color Coding and Abbreviations for Spiral-Wound Gasket Nonmetallic Filler Materials

Material	Abbreviation	Color code
Flexible graphite	F.G.	Gray stripe
Polytetrafluoroethylene	PTFE	White stripe
Mica-graphite	Manufacturer's designation	Pink stripe
Thermiculite	Manufacturer's designation	Light blue stripe
Ceramic	CER	Light green stripe
Chrysotile asbestos	ASB	No stripe

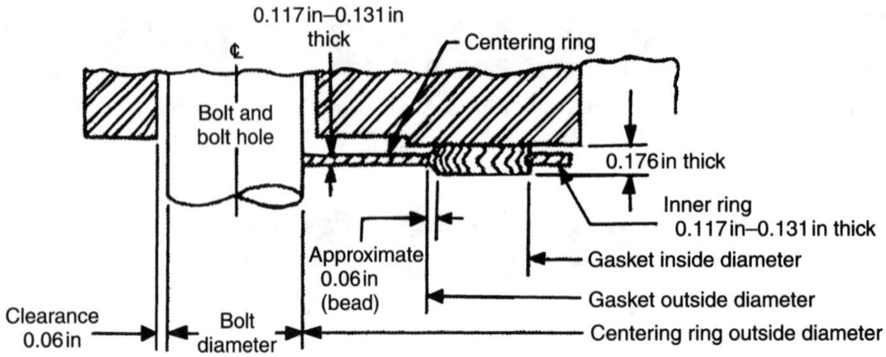

TABLE 28.34 Dimensions for Spiral-Wound Gaskets Used with ASME B16.5 Flanges*

Flange size (NPS)	Outside diameter of gaskets† Classes 150, 300, 400, 600	Outside diameter of gaskets† Classes 900, 1500, 2500	Inside diameter of gasket by class† 150	300	400	600	900	1500	2500	Outside diameter of centering ring by class† 150	300	400	600	900	1500	2500
½	1.25	1.25	0.75	0.75		0.75		0.75	0.75	1.88	2.13		2.13		2.50	2.75
¾	1.56	1.56	1.00	1.00		1.00		1.00	1.00	2.25	2.63		2.63		2.75	3.00
1	1.88	1.88	1.25	1.25		1.25		1.25	1.25	2.63	2.88		2.88		3.13	3.38
1¼	2.38	2.38	1.88	1.88		1.88		1.56	1.56	3.00	3.25		3.25		3.50	4.13
1½	2.75	2.75	2.13	2.13		2.13		1.88	1.88	3.38	3.75		3.75		3.88	4.63
2	3.38	3.38	2.75	2.75		2.75		2.31	2.31	4.13	4.38		4.38		5.63	5.75
2½	3.88	3.88	3.25	3.25		3.25		2.75	2.75	4.88	5.13		5.13		6.50	6.63
3	4.75	4.75	4.00	4.00		4.00	3.75	3.63	3.63	5.38	5.88		5.88	6.63	6.88	7.75
4	5.88	5.88	5.00	5.00	4.75	4.75	4.75	4.63	4.63	6.88	7.13	7.00	7.63	8.13	8.25	9.25
5	7.00	7.00	6.13	6.13	5.81	5.81	5.81	5.63	5.63	7.75	8.50	8.38	9.50	9.75	10.00	11.00
6	8.25	8.25	7.19	7.19	6.88	6.88	6.88	6.75	6.75	8.75	9.88	9.75	10.50	11.38	11.13	12.50
8	10.38	10.13	9.19	9.19	8.88	8.88	8.75	8.50	8.50	11.00	12.13	12.00	12.63	14.13	13.88	15.25
10	12.50	12.25	11.31	11.31	10.81	10.81	10.88	10.50	10.63	13.38	14.25	14.13	15.75	17.13	17.13	18.75
12	14.75	14.50	13.38	13.38	12.88	12.88	12.75	12.75	12.50	16.13	16.63	16.50	18.00	19.63	20.50	21.63
14	16.00	15.75	14.63	14.63	14.25	14.25	14.00	14.25		17.75	19.13	19.00	19.38	20.50	22.75	
16	18.25	18.00	16.63	16.63	16.25	16.25	16.25	16.00		20.25	21.25	21.13	22.25	22.62	25.25	
18	20.75	20.50	18.69	18.69	18.50	18.50	18.25	18.25		21.63	23.50	23.38	24.13	25.13	27.75	
20	22.75	22.50	20.69	20.69	20.50	20.50	20.50	20.25		23.88	25.75	25.50	26.88	27.50	29.75	
24	27.00	26.75	24.75	24.75	24.75	24.75	24.75	24.25		28.25	30.50	30.25	31.13	33.00	35.50	

* Dimensions in inches.
† See Table 28.35 for tolerances.

TABLE 28.35 Tolerances for Spiral-Wound Gaskets for ASME Flanges (NPS ½–24) per ASME B16.20 (U.S. Customary Units)*

Nominal pipe size (NPS) flanges	Gasket inside diameter, ID	Gasket outside diameter, OD	Centering ring outside diameter	Gasket thickness†
8 and smaller, ASME B16.5	±0.016	±0.03	±0.03	±0.005
10 through 24, ASME B16.5	±0.03	+0.06, −0.03	±0.03	±0.005

* Dimensions in inches.
† Measured across the metallic portion of the gasket not including the filler, which may protrude slightly beyond the metal.

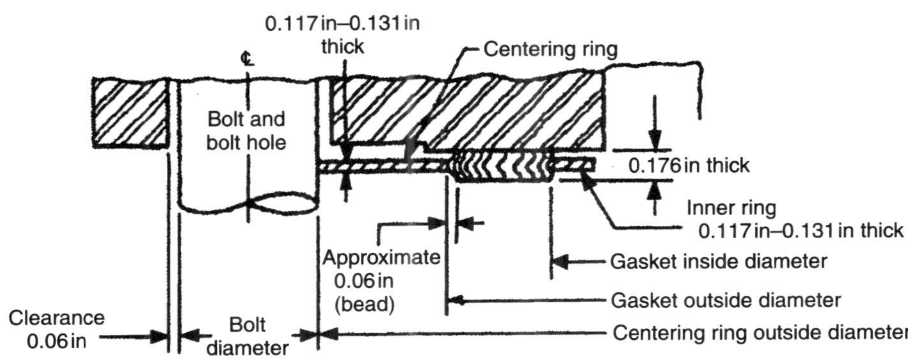

TABLE 28.36 Dimensions for Spiral-Wound Gaskets Used with ASME B16.47 Series A (MSS SP44) Flanges*

Flange size (NPS)	Class 150			Class 300			Class 400			Class 600			Class 900		
	Gasket		Centering ring outside diameter	Gasket		Centering ring outside diameter	Gasket		Centering ring outside diameter	Gasket		Centering ring outside diameter	Gasket		Centering ring outside diameter
	Inside	Outside		Inside	Outside		Inside	Outside		Inside	Outside		Inside	Outside	
26	26.50	27.75	30.50	27.00	29.00	32.88	27.00	29.00	32.75	27.00	29.00	34.13	27.00	29.00	34.75
28	28.50	29.75	32.75	29.00	31.00	35.58	29.00	31.00	35.13	29.00	31.00	36.00	29.00	31.00	37.25
30	30.50	31.75	34.75	31.25	33.25	37.50	31.25	33.25	37.25	31.25	33.25	38.25	31.25	33.25	39.75
32	32.50	33.88	37.00	33.50	35.50	39.63	33.50	35.50	39.50	33.50	35.50	40.25	33.50	35.50	42.25
34	34.50	35.88	39.00	35.50	37.50	41.63	35.50	37.50	41.50	35.50	37.50	42.25	35.50	37.50	44.75
36	36.50	38.13	41.25	37.63	39.63	44.00	37.63	39.63	44.00	37.63	39.63	44.50	37.75	39.75	47.25
38	38.50	40.13	43.75	38.50	40.00	41.50	38.25	40.25	42.25	39.00	41.00	43.50	40.75	42.75	47.25
40	40.50	42.13	45.75	40.25	42.13	43.88	40.38	42.38	44.38	41.25	43.25	45.50	43.25	45.25	49.25
42	42.50	44.25	48.00	42.25	44.13	45.88	42.38	44.38	46.38	43.50	45.50	48.00	45.25	47.25	51.25
44	44.50	46.38	50.25	44.50	46.50	48.00	44.50	46.50	48.50	45.75	47.75	50.00	47.50	49.50	53.88
46	46.50	48.38	52.25	46.38	48.38	50.13	47.00	49.00	50.75	47.75	49.75	52.25	50.00	52.00	56.50
48	48.50	50.38	54.50	48.63	50.63	52.13	49.00	51.00	53.00	50.00	52.00	54.75	52.00	54.00	58.50
50	50.50	52.50	56.50	51.00	53.00	54.25	51.00	53.00	55.25	52.00	54.00	57.00	†	†	†
52	52.50	54.50	58.75	53.00	55.00	56.25	53.00	55.00	57.25	54.00	56.00	59.00	†	†	†
54	54.50	56.50	61.00	55.25	57.25	58.75	55.25	57.25	59.75	56.25	58.25	61.25	†	†	†
56	56.50	58.50	63.25	57.25	59.25	60.75	57.25	59.25	61.75	58.25	60.25	63.50	†	†	†
58	58.50	60.50	65.50	59.50	61.50	62.75	59.25	61.25	63.75	60.50	62.50	65.50	†	†	†
60	60.50	62.50	67.50	61.50	63.50	64.75	61.75	63.75	66.25	62.75	64.75	68.25	†	†	†

* Dimensions in inches. See Table 28.37 for tolerances.
† Class 900 flanges are not manufactured in these sizes.

TABLE 28.37 Tolerances for Spiral-Wound Gaskets for ASME B16.47 Series A Flanges (NPS 26–60) per ASME B16.20 (U.S. Customary Units)*

Nominal pipe size (NPS) flanges	Gasket inside diameter, ID	Gasket outside diameter, OD	Centering ring outside diameter	Gasket thickness†
26 through 34, ASME B16.47 Series A	±0.03 in	±0.06 in	±0.03 in	±0.005 in
36 through 60, ASME B16.47 Series A	±0.05 in	±0.06 in	±0.03 in	±0.005 in

* Dimensions in inches.
† Measured across the metallic portion of the gasket not including the filler, which may protrude slightly beyond the metal.

TABLE 28.39 Tolerances for Spiral-Wound Gaskets for ASME B16.47 Series B Flanges (NPS 26–60) per ASME B16.20 (U.S. Customary Units)*

Nominal pipe size (NPS) flanges	Gasket inside diameter, ID	Gasket outside diameter, OD	Centering ring outside diameter	Gasket thickness†
26 through 34, ASME B16.47 Series A	±0.03 in	±0.06 in	±0.03 in	±0.005 in
36 through 60, ASME B16.47 Series A	±0.05 in	±0.06 in	±0.03 in	±0.005 in

* Dimensions in inches.
† Measured across the metallic portion of the gasket not including the filler, which may protrude slightly beyond the metal.

TABLE 28.40 Inner Ring Inside Dimensions for Spiral-Wound Gaskets for ASME B16.5 Flanges(1)

Flange size (NPS)	Pressure class						
	150	300	400	600	900	1500	2500
½	0.56	0.56	(2)	0.56	(3)	0.56	0.56
¾	0.81	0.81	(2)	0.81	(3)	0.81	0.81
1	1.06	1.06	(2)	1.06	(3)	1.06	1.06
1¼	1.50	1.50	(2)	1.50	(3)	1.31	1.31
1½	1.75	1.75	(2)	1.75	(3)	1.63	1.63
2	2.19	2.19	(2)	2.19	(3)	2.06	2.06
2½	2.62	2.62	(2)	2.62	(3)	2.50	2.50
3	3.19	3.19	(2)	3.19	3.19	3.19	3.19
4	4.19	4.19	4.19	4.19	4.19	4.19	4.19
5	5.19	5.19	5.19	5.19	5.19	5.19	5.19
6	6.19	6.19	6.19	6.19	6.19	6.19	6.19
8	8.50	8.50	8.25	8.25	8.25	8.12	7.88
10	10.56	10.56	10.25	10.25	10.25	10.15	9.75
12	12.50	12.50	12.50	12.50	12.38	12.38	11.50
14	13.75	13.75	13.75	13.75	13.50	13.38	(4)
16	15.75	15.75	15.75	15.75	15.50	15.25	(4)
18	17.69	17.69	17.69	17.69	17.50	17.25	(4)
20	19.69	19.69	19.69	19.69	19.50	19.25	(4)
24	23.75	23.75	23.75	23.75	23.75	22.75	(4)

Notes:
(1) Dimensions in inches. The inner ring thickness shall be 0.117–0.131 in. The inner ring inside diameter tolerances are:
NPS 1¼ through 3 ±0.03 in
NPS 4 and larger ±0.06 in
(2) There are no Class 400 flanges NPS ½ through 3. Use Class 600 flanges.
(3) There are no Class 900 flanges NPS ½ through 2½. Use Class 1500 flanges.
(4) There are no Class 2500 flanges NPS 14 through 24.

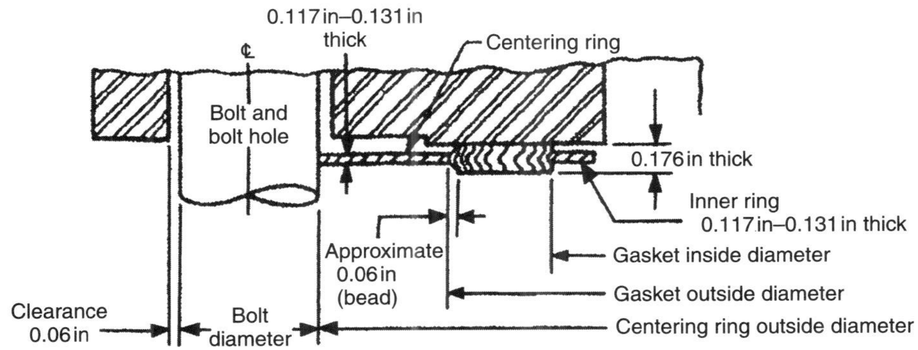

TABLE 28.38 Dimensions for Spiral-Wound Gaskets Used with ASME B16.47 Series B (API 605) Flanges*

Flange size (NPS)	Class 150			Class 300			Class 400			Class 600			Class 900		
	Gasket		Centering ring outside diameter	Gasket		Centering ring outside diameter	Gasket		Centering ring outside diameter	Gasket		Centering ring outside diameter	Gasket		Centering ring outside diameter
	Inside	Outside		Inside	Outside		Inside	Outside		Inside	Outside		Inside	Outside	
26	26.50	27.75	30.50	27.00	29.00	32.88	27.00	29.00	32.75	27.00	29.00	34.13	27.00	29.00	34.75
28	28.50	29.75	32.75	29.00	31.00	35.58	29.00	31.00	35.13	29.00	31.00	36.00	29.00	31.00	37.25
30	30.50	31.75	34.75	31.25	33.25	37.50	31.25	33.25	37.25	31.25	33.25	38.25	31.25	33.25	39.75
32	32.50	33.88	37.00	33.50	35.50	39.63	33.50	35.50	39.50	33.50	35.50	40.25	33.50	35.50	42.25
34	34.50	35.88	39.00	35.50	37.50	41.63	35.50	37.50	41.50	35.50	37.50	42.25	35.50	37.50	44.75
36	36.50	38.13	41.25	37.63	39.63	44.00	37.63	39.63	44.00	37.63	39.63	44.50	37.75	39.75	47.25
38	38.50	40.13	43.75	38.50	40.00	41.50	38.25	40.25	42.25	39.00	41.00	43.50	40.75	42.75	47.25
40	40.50	42.13	45.75	40.25	42.13	43.88	40.38	42.38	44.38	41.25	43.25	45.50	43.25	45.25	49.25
42	42.50	44.25	48.00	42.25	44.13	45.88	42.38	44.38	46.38	43.50	45.50	48.00	45.25	47.25	51.25
44	44.50	46.38	50.25	44.50	46.50	48.00	44.50	46.50	48.50	45.75	47.75	50.00	47.50	49.50	53.88
46	46.50	48.38	52.25	46.38	48.38	50.13	47.00	49.00	50.75	47.75	49.75	52.25	50.00	52.00	56.50
48	48.50	50.38	54.50	48.63	50.63	52.13	49.00	51.00	53.00	50.00	52.00	54.75	52.00	54.00	58.50
50	50.50	52.50	56.50	51.00	53.00	54.25	51.00	53.00	55.25	52.00	54.00	57.00	†	†	†
52	52.50	54.50	58.75	53.00	55.00	56.25	53.00	55.00	57.25	54.00	56.00	59.00	†	†	†
54	54.50	56.50	61.00	55.25	57.25	58.75	55.25	57.25	59.75	56.25	58.25	61.25	†	†	†
56	56.50	58.50	63.25	57.25	59.25	60.75	57.25	59.25	61.75	58.25	60.25	63.50	†	†	†
58	58.50	60.50	65.50	59.50	61.50	62.75	59.25	61.25	63.75	60.50	62.50	65.50	†	†	†
60	60.50	62.50	67.50	61.50	63.50	64.75	61.75	63.75	66.25	62.75	64.75	68.25	†	†	†

* Dimensions in inches. See Table 28.39 for tolerances.
† Class 900 flanges are not manufactured in these sizes.

TABLE 28.41 Inner Ring Inside Diameters for Spiral-Wound Gaskets Used between ASME B16.47 Series A Flanges (1)

	Pressure class				
Flange size (NPS)	150	300	400	600	900 (2)
26	25.75	25.75	26.00	25.50	26.25
28	27.75	27.75	28.00	27.50	28.00
30	29.75	29.75	29.75	29.75	30.50
32	31.75	31.75	32.00	32.00	32.00
34	33.75	33.75	34.00	34.00	34.00
36	35.75	35.75	36.13	36.13	36.25
38	37.75	37.50	37.50	37.50	39.75
40	39.75	39.50	39.38	39.75	41.75
42	41.75	41.50	41.38	42.00	43.75
44	43.75	43.50	43.50	43.75	45.50
46	45.75	45.38	46.00	45.75	48.00
48	47.75	47.63	47.50	48.00	50.00
50	49.75	49.00	49.50	50.00	(3)
52	51.75	52.00	51.50	52.00	(3)
54	53.50	53.25	53.25	54.25	(3)
56	55.50	55.25	55.25	56.25	(3)
58	57.50	57.00	57.25	58.00	(3)
60	59.50	60.00	59.75	60.25	(3)

Notes:

(1) Dimensions in inches. The inner ring thickness shall be 0.117–0.131 in. The inside diameter tolerance is ±0.12 in. These inner rings are suitable for use with pipe walls 0.38 in or thicker.

(2) Inner rings are required for Class 900.

(3) There are no Class 900 flanges NPS 50 and larger.

TABLE 28.42 Inner Ring Inside Diameters for Spiral-Wound Gaskets Used between ASME B16.47 Series B Flanges (1)

	Pressure class				
Flange size (NPS)	150	300	400	600	900 (2)
26	25.75	25.75	25.75	25.38	26.50
28	27.75	27.75	27.63	27.25	28.50
30	29.75	29.75	29.63	29.63	31.00
32	31.75	31.75	31.50	31.25	33.00
34	33.75	33.75	33.50	33.50	35.25
36	35.75	35.75	35.38	35.50	36.50
38	37.75	38.25	37.50	37.50	39.75
40	39.75	42.75	39.38	39.75	41.75
42	41.75	41.50	41.38	42.00	43.75
44	43.75	44.25	43.50	43.75	45.50
46	45.75	46.38	46.00	45.75	48.00
48	47.75	48.50	47.50	48.00	50.00
50	49.75	49.88	49.50	50.00	(3)
52	51.75	51.88	51.50	52.00	(3)
54	53.75	53.75	53.25	54.25	(3)
56	56.00	56.25	55.25	56.25	(3)
58	58.19	58.44	57.25	58.00	(3)
60	60.44	61.31	59.75	60.25	(3)

Notes:

(1) Dimensions in inches. The inner ring thickness shall be 0.117–0.131 in. The inside diameter tolerance is ±0.12 in. These inner rings are suitable for use with pipe walls 0.38 in or thicker.

(2) Inner rings are required for Class 900.

(3) There are no Class 900 flanges NPS 50 and larger.

Serrated Metal Core Gasket with Soft Conformable Facings

These gaskets are made from a solid serrated metal core faced on each side with a soft nonmetallic or soft metallic material. They are also known as a Kammprofile gaskets; the term is a derivative from the German words *kamm* and *profil*, meaning *comb profile*. Two profiles are commonly used: the DIN 2697 profile and a shallow profile. The shallow profile is similar to the DIN 2697 profile except that the groove depth is 0.5 mm (versus 0.75 mm for DIN 2697). The profile can be made from sheet metal or strip with a nominal thickness of 3 mm instead of a thickness of 4 mm for the DIN profile.

For the original German standard DIN 2697 profile see Fig. 28.17.

The most common facing used is flexible graphite. Other facings such as expanded or sintered PTFE and CAF or some soft metal, such as copper, can also be used. This gasket combines the strength and blowout and creep resistance of a metal core with a soft sealing material that conforms to the flange faces.

Standard NPS gaskets are available to suit ASME B16.5, BS 1560, and DIN series 2600 flanges. Gasket materials are listed in Table 28.43. The dimensions are shown in Tables 28.44 and 28.45.

These gaskets are available in three styles—PN, ZG, and ZA. Refer to Fig. 28.18. Style PN Flexpro gaskets are selected for use in confined locations, including male and female, tongue-and-groove, and recessed flange arrangements. A variation of the PN Flexpro, utilizing an integral outer locating ring for correct gasket positioning within the mating flange bolt circle, Style ZG Flexpro gaskets are recommended for use on standard

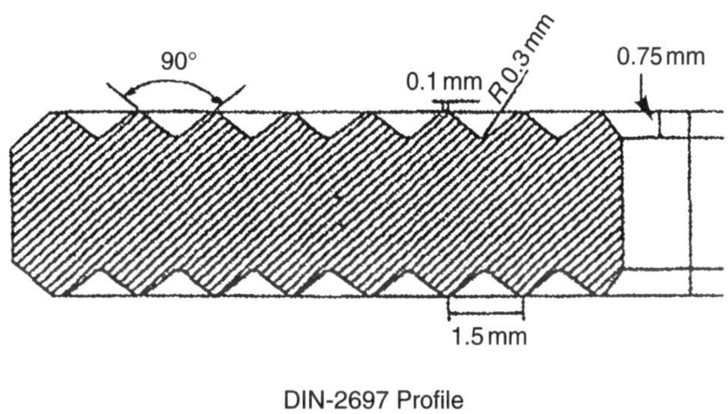

FIGURE 28.17 Serrated metal core gaskets with soft conformable facings.

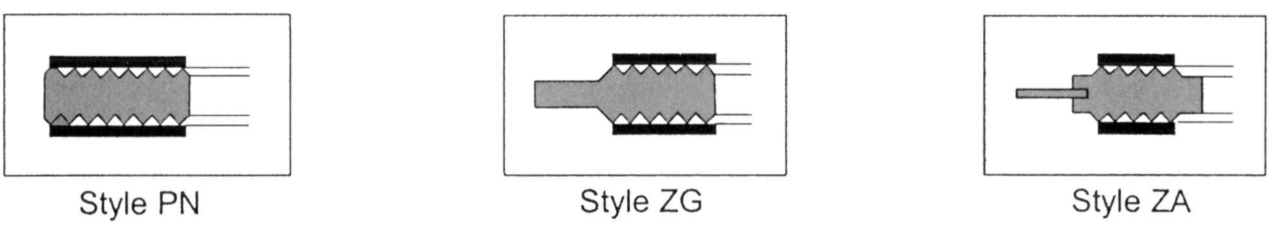

FIGURE 28.18 Styles of serrated metal core gaskets.

TABLE 28.43 Gasket Materials

Metallic core materials			Soft facing materials
Type 304 SS	Monel	Aluminum	Flexible graphite
Type 309 SS	Inconel 600	Copper	Thermiculite
Type 310 SS	Inconel 625	Brass	Nonasbestos
Type 316L SS	Inconel X-750	Nickel 200	PTFE
Type 321 SS	Incoloy 800	Alloy 20	PTFE + glass, silica,
Type 347 SS	Incoloy 825	Duplex	barium sulfate
Type 430 SS	Hastelloy B2	Titanium	Soft metals
Carbon steel	Hastelloy C276		

Style PN

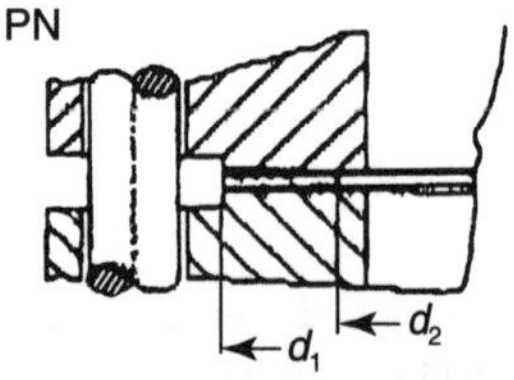

Style ZG & ZA

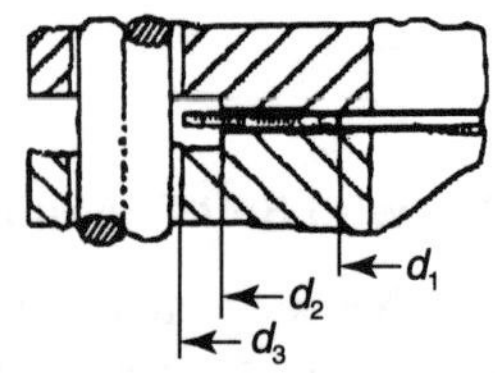

TABLE 28.44 Gasket Dimensions to Suit ASME B16.5 and BS 1560 Flanges, Class 150 to 2500*

			Style PN, ZG & ZA d_3						
NPS	d_1	d_2	150	300	400	600	900	1500	2500
$\frac{1}{2}$	$\frac{29}{32}$	$1\frac{5}{16}$	$1\frac{7}{8}$	$2\frac{1}{8}$	$2\frac{1}{8}$	$2\frac{1}{8}$	$2\frac{1}{2}$	$2\frac{1}{2}$	$2\frac{3}{4}$
$\frac{3}{4}$	$1\frac{1}{8}$	$1\frac{9}{16}$	$2\frac{1}{4}$	$2\frac{5}{8}$	$2\frac{5}{8}$	$2\frac{5}{8}$	$2\frac{3}{4}$	$2\frac{3}{4}$	3
1	$1\frac{7}{16}$	$1\frac{7}{8}$	$2\frac{5}{8}$	$2\frac{7}{8}$	$2\frac{7}{8}$	$2\frac{7}{8}$	$3\frac{1}{8}$	$3\frac{1}{8}$	$3\frac{3}{8}$
$1\frac{1}{4}$	$1\frac{3}{4}$	$2\frac{3}{8}$	3	$3\frac{1}{4}$	$3\frac{1}{4}$	$3\frac{1}{4}$	$3\frac{1}{2}$	$3\frac{1}{2}$	$4\frac{1}{8}$
$1\frac{1}{2}$	$2\frac{1}{16}$	$2\frac{3}{4}$	$3\frac{3}{8}$	$3\frac{3}{4}$	$3\frac{3}{4}$	$3\frac{3}{4}$	$3\frac{7}{8}$	$3\frac{7}{8}$	$4\frac{5}{8}$
2	$2\frac{3}{4}$	$3\frac{1}{2}$	$4\frac{1}{8}$	$4\frac{3}{8}$	$4\frac{3}{8}$	$4\frac{3}{8}$	$5\frac{5}{8}$	$5\frac{5}{8}$	$5\frac{3}{4}$
$2\frac{1}{2}$	$3\frac{1}{4}$	4	$4\frac{7}{8}$	$5\frac{1}{8}$	$5\frac{1}{8}$	$5\frac{1}{8}$	$6\frac{1}{2}$	$6\frac{1}{2}$	$6\frac{5}{8}$
3	$3\frac{7}{8}$	$4\frac{7}{8}$	$5\frac{3}{8}$	$5\frac{7}{8}$	$5\frac{7}{8}$	$5\frac{7}{8}$	$6\frac{5}{8}$	$6\frac{7}{8}$	$7\frac{3}{4}$
$3\frac{1}{2}$	$4\frac{3}{8}$	$5\frac{3}{8}$	$6\frac{3}{8}$	$6\frac{1}{2}$	$6\frac{3}{8}$	$6\frac{3}{8}$	$7\frac{1}{2}$	$7\frac{3}{8}$	—
4	$4\frac{7}{8}$	$6\frac{1}{16}$	$6\frac{7}{8}$	$7\frac{1}{8}$	7	$7\frac{5}{8}$	$8\frac{1}{8}$	$8\frac{1}{4}$	$9\frac{1}{4}$
5	$5\frac{15}{16}$	$7\frac{3}{16}$	$7\frac{3}{4}$	$8\frac{1}{2}$	$8\frac{3}{8}$	$9\frac{1}{2}$	$9\frac{3}{4}$	10	11
6	7	$8\frac{3}{8}$	$8\frac{3}{4}$	$9\frac{7}{8}$	$9\frac{3}{4}$	$10\frac{1}{2}$	$11\frac{3}{8}$	$11\frac{1}{8}$	$12\frac{1}{2}$
8	9	$10\frac{1}{2}$	11	$12\frac{1}{8}$	12	$12\frac{5}{8}$	$14\frac{1}{8}$	$13\frac{7}{8}$	$15\frac{1}{4}$
10	$11\frac{1}{8}$	$12\frac{5}{8}$	$13\frac{3}{8}$	$14\frac{1}{4}$	$14\frac{1}{8}$	$15\frac{3}{4}$	$17\frac{1}{8}$	$17\frac{1}{8}$	$18\frac{3}{4}$
12	$13\frac{3}{8}$	$14\frac{7}{8}$	$16\frac{1}{8}$	$16\frac{5}{8}$	$16\frac{1}{2}$	18	$19\frac{5}{8}$	$20\frac{1}{2}$	$21\frac{5}{8}$
14	$14\frac{5}{8}$	$16\frac{1}{8}$	$17\frac{3}{4}$	$19\frac{1}{8}$	19	$19\frac{3}{8}$	$20\frac{1}{2}$	$22\frac{3}{4}$	
16	$16\frac{5}{8}$	$18\frac{3}{8}$	$20\frac{1}{4}$	$21\frac{1}{4}$	$21\frac{1}{8}$	$22\frac{1}{4}$	$22\frac{5}{8}$	$25\frac{1}{4}$	
18	$18\frac{7}{8}$	$20\frac{7}{8}$	$21\frac{5}{8}$	$23\frac{1}{2}$	$23\frac{3}{8}$	$24\frac{1}{8}$	$25\frac{1}{8}$	$27\frac{3}{4}$	
20	$20\frac{7}{8}$	$22\frac{7}{8}$	$23\frac{7}{8}$	$25\frac{3}{4}$	$25\frac{1}{2}$	$26\frac{7}{8}$	$27\frac{1}{2}$	$29\frac{3}{4}$	
24	$24\frac{7}{8}$	$26\frac{7}{8}$	$28\frac{1}{4}$	$30\frac{1}{2}$	$30\frac{1}{4}$	$31\frac{1}{8}$	33	$35\frac{1}{2}$	

* Dimensions in inches.
d_1 = Inside diameter of gasket.
d_2 = Outside diameter of gasket.
d_3 = Diameter of locating ring.

TABLE 28.45 Gasket Dimensions to Suit DIN 2600 Series Flanges PN 64 to PN 400*

			Style PN, ZG & ZA per DIN 2697 d_3					
DN	d_1	d_2	64	100	160	250	320	400
10	22	40	56	56	56	67	67	67
15	25	45	61	61	61	72	72	77
25	36	68	82	82	82	82	92	103
40	50	88	102	102	102	108	118	135
50	62	102	112	118	118	123	133	150
65	74	122	137	143	143	153	170	192
80	90	138	147	153	153	170	190	207
100	115	162	173	180	180	202	229	256
125	142	188	210	217	217	242	274	301
150	165	218	247	257	257	284	311	348
(175)†	190	260	277	287	284	316	358	
200	214	285	309	324	324	358	398	442
250	264	345	364	391	388	442	488	
300	310	410	424	458	458			
350	340	465	486	512				
400	386	535	543					

* Dimensions in millimeters.
† Not commonly used size.
d_1 = Inside diameter of gasket.
d_2 = Outside diameter of gasket.
d_3 = Diameter of locating ring.

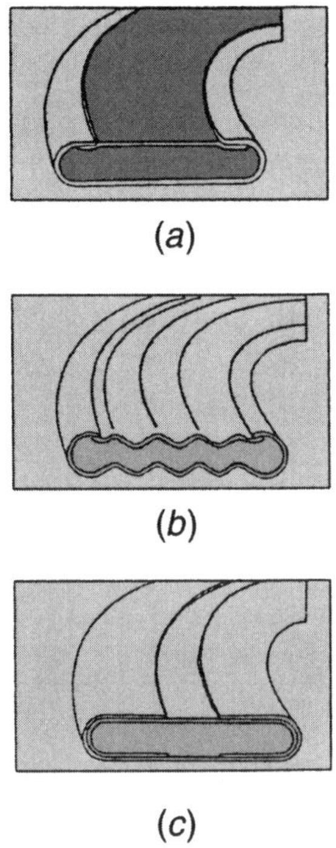

FIGURE 28.19 Metaljacketed gaskets: (*a*) smooth jacket, (*b*) corrugated jacket, (*c*) double jacket.

raised face and flat face flange assemblies. Style ZA Flexpro is a slight variation of Style ZG. The integral outer locating ring is replaced by a loose-fitting independent ring, which is preferred where flange differential radial thermal expansion may be encountered.

Metal-Jacketed Gaskets

Metal-jacketed gaskets, as the name suggests, consist of a metallic outer shell with either a metallic or nonmetallic filler. The filler material gives the gasket resilience, while the metal jacket protects the filler and resists pressure, temperature, and corrosion.

They are traditionally used for heat exchanger applications, pumps, and valves; however, the resilience and recovery properties of these gaskets are limited. Metal-jacketed gaskets require smooth flange surface finishes, high bolt loads, and flange flatness in order to seal effectively. A metal jacket formed from a corrugated jacket provides better resilience than one with a smooth jacket, since the corrugations form multiple seals across the flange sealing face. See Fig. 28.19*a* and *b*.

The most commonly used metal-jacketed gaskets are double-jacketed gaskets, in which the filler material is completely enclosed by a two-piece metal jacket, which covers both the inside and outside diameters and both contact surfaces (Fig. 28.19*c*). A double-shell jacketed gasket is constructed of two reversed wrap-round shells. This provides superior support and handleability and exhibits better resistance to high pressures.

Double-jacketed gaskets are frequently used on boilers and heat exchangers where ample bolting is available to correctly seat the gasket. They are designed for high pressure and temperature applications up to and inclusive of Class 900. The temperature limitation of the gasket is dictated by the combination of metallic and nonmetallic materials used in its construction. Gasket cross-sectional widths as narrow as $^5/_{16}$ in (8 mm) can be manufactured, depending on diameter. Very large gasket diameters can also be produced. Nominal gasket thickness is $^1/_8$ in (3.2 mm).

Tables 28.46 to 28.50 give dimensional specifications for metal-jacketed gaskets.

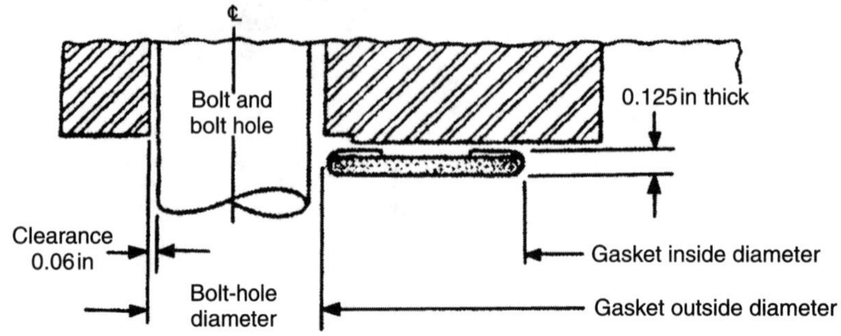

TABLE 28.46 Jacketed Gasket Dimensions for ASME B16.5 Flanges (U.S. Customary Units) (1)

		Gasket outside diameter by class (2)						
Flange size (NPS)	Gasket inside diameter	150	300	400	600	900	1500	2500
1/2	0.88	1.75	2.00	(3)	2.00	(3)	2.38	2.63
3/4	1.13	2.13	2.50	(3)	2.50	(3)	2.63	2.88
1	1.50	2.50	2.75	(3)	2.75	(3)	3.00	3.25
1 1/4	1.88	2.88	3.13	(3)	3.13	(3)	3.38	4.00
1 1/2	2.13	3.25	3.63	(3)	3.63	(3)	3.75	4.50
2	2.88	4.00	4.25	(3)	4.25	(3)	5.50	5.63
2 1/2	3.38	4.75	5.00	(3)	5.00	(3)	6.38	6.50
3	4.25	5.25	5.75	(3)	5.75	6.50	6.75	7.63
4	5.19	6.75	7.00	6.88	7.50	8.00	8.13	9.13
5	6.00	7.63	8.38	8.25	9.38	9.63	9.88	10.88
6	7.50	8.63	9.75	9.63	10.38	11.25	11.00	12.38
8	9.38	10.88	12.00	11.88	12.50	14.00	13.75	15.13
10	11.25	13.25	14.13	14.00	15.63	17.00	17.00	18.63
12	13.50	16.00	16.50	16.38	17.88	19.50	20.38	21.50
14	14.75	17.63	19.00	18.88	19.25	20.38	22.63	(3)
16	16.75	20.13	21.13	21.00	22.13	22.50	25.13	(3)
18	19.25	21.50	23.38	23.25	24.00	25.00	27.63	(3)
20	21.00	23.75	25.63	25.38	26.75	27.38	29.63	(3)
24	25.25	28.13	30.38	30.13	31.00	32.88	35.38	(3)

Notes:
(1) Dimensions in inches.
(2) See Table 28.47 for tolerances.
(3) There are no Class 400 flanges NPS 1/2 through NPS 3 (use Class 600), Class 900 flanges NPS 1/2 through NPS 2 1/2 (use Class 1500), or Class 2500 NPS 14 and larger.

TABLE 28.47 Tolerances for Metal-Jacketed Gaskets for ASME B16.5 Flanges (NPS 1/2–24) per ASME B16.20 (U.S. Customary Units)

Nominal pipe size (NPS) flanges	Gasket inside diameter, ID	Gasket outside diameter, OD	Gasket thickness
1/2 through 24	+0.06, −0.000 in	+0.06, −0.000 in	+0.03, −0.000 in

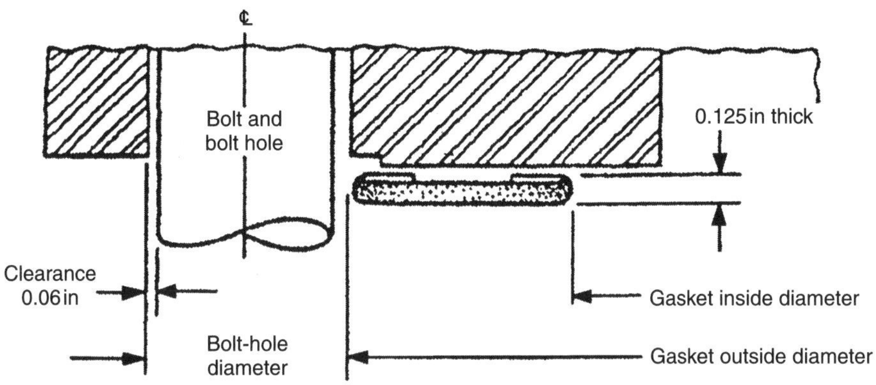

TABLE 28.48 Jacketed Gasket Dimensions for ASME B16.47 Series A Flanges (1)

Flange size (NPS)	Gasket inside diameter (2)	Gasket outside diameter by class (2)				
		150	300	400	600	900
26	26.50	30.38	32.75	32.63	34.00	34.63
28	28.50	32.63	35.25	35.00	35.88	37.13
30	30.50	34.63	37.38	37.13	38.13	39.63
32	32.50	36.88	39.50	39.38	40.13	42.13
34	34.50	38.88	41.50	41.38	42.13	44.63
36	36.50	41.13	43.88	43.88	44.38	47.13
38	38.50	43.63	41.38	42.13	43.38	47.13
40	40.50	45.63	43.75	44.25	45.38	49.13
42	42.50	47.88	45.75	46.25	47.88	51.13
44	44.50	50.13	47.88	48.38	49.88	53.75
46	46.50	52.13	50.00	50.63	52.13	56.38
48	48.50	54.38	52.00	52.88	54.63	58.38
50	50.50	56.38	54.13	55.13	56.88	(3)
52	52.50	58.63	56.13	57.13	58.88	(3)
54	54.50	60.88	58.63	59.63	61.13	(3)
56	56.50	63.13	60.63	61.63	63.13	(3)
58	58.50	65.38	62.63	63.63	65.38	(3)
60	60.50	67.38	64.63	66.13	68.13	(3)

Notes:
(1) Dimensions in inches.
(2) See Table 28.50 for tolerances.
(3) There are no Class 900 flanges NPS 50 and larger.

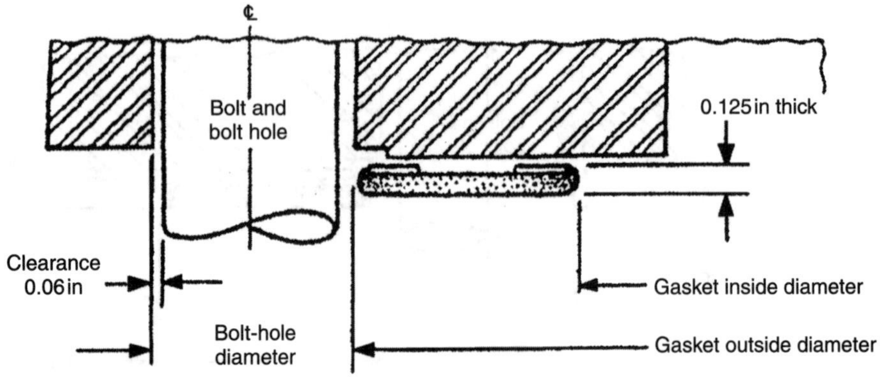

TABLE 28.49 Jacketed Gasket Dimensions for ASME B16.47 Series B Flanges (1)

Flange size (NPS)	Gasket inside diameter (2)	Gasket outside diameter by class (2)				
		150	300	400	600	900
26	26.50	28.44	30.25	29.25	30.00	32.88
28	28.50	30.44	32.38	31.38	32.13	35.38
30	30.50	32.44	34.75	33.63	34.50	37.63
32	32.50	34.56	36.88	35.75	36.63	39.88
34	34.50	36.69	39.00	37.75	39.13	42.13
36	36.50	38.75	41.13	40.13	41.13	44.13
38	38.50	41.00	43.13	42.13	43.38	47.13
40	40.50	43.00	45.13	44.25	45.38	49.13
42	42.50	45.00	47.13	46.25	47.88	51.13
44	44.50	47.00	49.13	48.38	49.88	53.75
46	46.50	49.31	51.75	50.63	52.13	56.38
48	48.50	51.31	53.75	52.88	54.63	58.38
50	50.50	53.31	55.75	55.31	56.88	(3)
52	52.50	55.31	57.75	57.13	58.88	(3)
54	54.50	57.50	60.13	59.63	61.13	(3)
56	56.50	59.50	62.63	61.63	63.13	(3)
58	58.50	62.06	65.06	63.63	65.38	(3)
60	60.50	64.06	67.06	66.13	68.13	(3)

Notes:
(1) Dimensions in inches.
(2) See Table 28.50 for tolerances.
(3) There are no Class 900 flanges NPS 50 and larger.

TABLE 28.50 Tolerances for Metal-Jacketed Gaskets for ASME B16.47 Series A and Series B Flanges (NPS 26–60) per ASME B16.20 (U.S. Customary Units)

Nominal pipe size (NPS) flanges	Gasket inside diameter, ID	Gasket outside diameter, OD	Gasket thickness
26 through 60	+0.13, −0.000 in	+0.13, −0.000 in	+0.03, −0.000 in

Tables in Chapter 28

Tables in Chapter 28 (Continued)

S · E · C · T · I · O · N · 9

RESOURCE LIST

CHAPTER 29
MANUFACTURERS, SUPPLIERS, AND FABRICATORS OF PIPING, VALVES, AND SPECIALTY ITEMS

TABLE 29.1 Sources for Steel Pipe

Company name and address	Product/services	Contact
Bethlehem Steel Corp. 1170 Eighth Ave. Bethlehem, PA 18016-7699 U.S.A.	Steel pipe	Telephone: (610) 694-2424 Email: webmaster@bethsteel.com http://www.bethsteel.com
The Punjab Steel Works B-38, Mayapuri Industrial Area Phase-I, New Delhi 110064 India	Steel pipe	Telephone: +91-11-5141318/5141309/5130960 Fax: +91-11-5144333/5137432 Email: pswforge@mantraonline.com http://www.pswforge.com
Umran Steel Pipe Inc. Meclisi Mebusan Cad. 85/5 80040 Salypazary Istanbul Turkey	Steel pipe	Telephone: +90 212 252 5280 Fax: +90 212 249 4349 Email: umran@umran.com http://www.umran.com
Florida Pipe & Steel, Inc. 13860 Wellington Trace No. 12 Suite 508 West Palm Beach, FL 33414 U.S.A.	Steel pipe	Telephone: (561) 642-1811 Fax: (561) 642-7431 Email: sales@floridapipe.com http://www.steel4sale.com
LaBarge Pipe & Steel Company 901 North Tenth St. St. Louis, MO 63101 U.S.A.	Steel pipe	Telephone: (314) 231-3400 Telephone: 1-800-LaBarge Fax: (314) 982-9395 Email: info@labargepipe.com http://www.labarge.com
LaBarge Pipe & Steel Company 100 Centerview Dr. Suite 100 Birmingham, AL 35216 U.S.A.	Steel pipe	Telephone: (205) 823-7881 Telephone: 1-800-LaBarge Fax: (205) 823-8388 Email: info@labargepipe.com http://www.labarge.com
LaBarge Pipe & Steel Company 500 East Morehead St. Charlotte, NC 28202 U.S.A.	Steel pipe	Telephone: (704) 333-1541 Telephone: 1-800-LaBarge Fax: (704) 331-9696 Email: info@labargepipe.com http://www.labarge.com
LaBarge Pipe & Steel Company 616 F.M. 1960 West Houston, TX 77090 U.S.A.	Steel pipe	Telephone: (281) 444-6910 Telephone: 1-800-LaBarge Fax: (281) 444-0842 Email: info@labargepipe.com http://www.labarge.com
Online Metals PO Box 19271 Seattle, WA 98109 U.S.A.	Steel pipe	Telephone: (800) 704-2157 Telephone: (206) 285-8603 Fax: (800) 533-6350 Email: sales@onlinemetals.com http://www.onlinemetals.com

TABLE 29.1 Sources for Steel Pipe (Continued)

Company name and address	Product/services	Contact
Online Metals PO Box 19271 Seattle, WA 98109 U.S.A.	Steel pipe	Telephone: (800) 704-2157 Telephone: (206) 285-8603 Fax: (800) 533-6350 Email: sales@onlinemetals.com http://www.onlinemetals.com
Jindal Group Corporate Office 56, Hanuman Rd. New Delhi 110 001 India	Steel pipe	Telephone: 91-11-3734816-20 Fax: 91-11-3343630 Email: jinpipe@giasdl01.vsnl.net.in http://www.202.54.15.254/jindal/main.htm
Steel City Corporation 2140 Edison Ave. San Leandro, CA 94577 U.S.A.	Steel pipe	Telephone: (510) 636-9351 Fax: (510) 636-9969 Email: sales@steelcitycorp.com http://www.steelcitycorp.com
Federal Steel Supply Inc. Executive Office 747 Goddard Ave. Chesterfield, MO 63005 U.S.A.	Steel pipe, tube, and fittings	Telephone: (800) 325-1584 Telephone: (636) 537-2393 Fax: (636) 537-8795 Email: info@fedsteel.com http://www.fedsteel.com
Glazer Steel & Aluminum 1556 Tchoupitoulas St. PO Box 50867 New Orleans, LA 70150 U.S.A.	Steel pipe	Telephone: (504) 529-2761 Telephone: (800) 824-1730 Fax: (504) 523-1839 Email: glazer@glazer.com http://www.glazer.com
Southern Pipe Industry (M) SDN BHD (7451-M) 4457, Mk. 15, Jalan Chain Ferry 12100 Butterworth, P.W. Malaysia	Steel pipe	Tel: 604-3317 393, 604-3317 741 Fax: 604-3319 435 http://www.spim.com.my
Reddy Pipe & Supply 17801 E. 11th St. Tulsa, OK 74108-5105 U.S.A.	Steel pipe and tube	Telephone: (918) 234-3651 Telephone: (918) 438-PIPE Fax: (918) 234-3701 http://www.reddypipe.com
Prime Pipe International Inc. 24900 Pitkin Rd. Suite 340 The Woodlands, TX 77386 U.S.A.	Steel pipe	Telephone: (281) 367-6608 Fax: (281) 367-9017 Email: bud@primepipe.com http://www.primepipe.com
Frichtl Steel and Welding 9550 East State Highway 33 Newton, IL 62448 U.S.A.	Steel pipe	Telephone: (618) 783-8323 Fax: (618) 783-3118 Email: sales@frichtlsteel.com http://www.frichtlsteel.com
Mandal Pipe PO Box 2566 Lilburn, GA 30048 U.S.A.	Steel pipe	Telephone: (770) 925-8885 Fax: (770) 925-8803 http://www.mandalpipe.com
Pipe Industries Corporation 5810 East 77th Ave. Commerce City, CO 80022 U.S.A.	Steel pipe	Telephone: (303) 287-1700 Fax: (303) 287-1720 Email: sales@pipeindustries.com http://www.pipeindustries.com
Texas Pipe Work PO Box 2937 Longview, TX 75606 U.S.A.	Steel pipe	Telephone: (903) 297-7833 Fax: (903) 297-9240 http://www.texaspipeworks.com
Ram Forge of Texas LLC PO Box 592 Industrial Park Rd. Navasota, TX 77868 U.S.A.	Steel pipe	Telephone: (936) 825-6501 Telephone: (713) 759-1953 Fax: (936) 825-7524 Email: ramf@ramforge.com http://www.ramforge.com

TABLE 29.1 Sources for Steel Pipe (Continued)

Company name and address	Product/services	Contact
Atlas Tube Inc. 200 Clark St. PO Box 970 Harrow, ON N0R 1G0 Canada	Steel pipe	Telephone: (800) 265-6912 Telephone: (519) 738-5000 Fax: (519) 738-3537 http://www.atlastube.com
Atlas Tube (U.S.) Inc. 13101 Eckles Rd. Plymouth, MI 48170 U.S.A.	Steel pipe	Telephone: (888) 785-8823 Telephone: (734) 738-5600 Fax: (734) 738-5604 http://www.atlastube.com
Welded Tubes, Inc. 135 Penniman Ave. Orwell, OH 44076 U.S.A.	Steel pipe	Telephone: (440) 437-5144 Fax: (440) 437-5180 http://www.weldedtubes.com
Kounan Steel Co., Ltd. No. 174, 275 Lane Shuenn-An Rd., Lu Jwu Shiang, Kaohsiung Shiann, Taiwan, R.O.C.	Steel pipe	Telephone: 886-7-6965818 Fax: 886-7-6960035 Email: kounan@kounan.com.tw http://www.kounan.com.tw
Am. Steel Pipe Supply Corp. PO Box 7 Pacific, MO 63069 U.S.A.	Steel pipe	Telephone: (800) 729-7473 Telephone: (800) SAY-PIPE Telephone: (636) 257-5151 Fax: (636) 257-3869 Email: info@steel-pipe.com http://www.steel-pipe.com
Commerce R US 18901 NW 11 St. Pembroke Pines, FL 33029 U.S.A.	Steel pipe	Telephone: (954) 436-8252 Fax: (954) 436-2128 Email: TUBE2@GATE.NET http://www.covesoft.com
Vision Metals Ann Arbor, MI U.S.A.	Steel pipe	Telephone: (800) 893-3650 Fax: (800) 893-3651 http://www.visionmetals.com
Bristol Metals, L.P. PO Box 1589 390 Bristol Metals Rd. Bristol, TN 37621 U.S.A.	Steel pipe	Telephone: (423) 989-4700 Fax: (423) 989-4790 Email: sales@brismet.com http://www.brismet.com
Piping & Equipment, Inc. 8781 Paul Starr Dr. Pensacola, FL 32514 U.S.A.	Steel pipe	Telephone: (850) 484-3994 Fax: (850) 474-0552 Email: piping@pipingequipment.com http://www.pipingequipment.com
Weldbend Corporation 6600 South Harlem Ave. Argo, IL 60501-1930 U.S.A.	Steel pipe	Telephone: (708) 594-1700 Fax: (708) 458-0106 Email: info@weldbend.com http://www.weldbend.com
Pacifictek & Supply Company PO Box 55271 7807 Long Point Suite 200 Houston, TX 77055 U.S.A.	Steel pipe	Telephone: (713) 956-1639 Fax: (713) 956-1611 Email: mails@pacifictek.com http://www.pacifictek.com
Dial Tube and Steel Company Dial Tube–Chicago 5133 West 70th Place Bedford Park, IL 60638 U.S.A.	Steel pipe	Telephone: (800) 323-2728 Fax: (708) 594-1086 Email: info@dialtube.com http://www.dialtube.com
Dial Tube and Steel Company Dial Tube–Los Angeles 1540 Date St. Montebello, CA 90640 U.S.A.	Steel pipe	Telephone: (800) 423-4363 Email: info@dialtube.com http://www.dialtube.com

TABLE 29.1 Sources for Steel Pipe (Continued)

Company name and address	Product/services	Contact
Surya Global Limited Padma Tower-I, Rajendra Place New Delhi 110008 India	Steel pipe	Tel: +91 11 5740237, 5750579 Fax: +91 11 5763997, 5753955 Email: sglobal@del2.vsnl.net.in http://www.suryaglobal.com
Bihar Tubes Ltd. 37, Hargonind Enclave Vikas marg Delhi 110 092 India	Steel pipe and tube	Tel: +91-11-2158876, 2163437 Fax: +91-11-2163537 Email: btltd@nda.vsnl.net.in http://www.bihartubes.com
Tubacex Europe, B.V. Stationsplein 4 3331 LL Zwijndrecht Holland	Steel pipe	Telephone: 31-78-620.5333 Fax: 31-78-620.5330 Email: sales@tubacex.nl http://www.tubacex.com
Tubacex, Inc. 12707 North Freeway Suite 100 Houston, TX 77060 U.S.A.	Steel pipe	Telephone: (281) 875-8660 Fax: (281) 875-8665 Email: houston@tubacex.com http://www.tubacex.com
Tubacex Canada, Inc. 1351 Matheson Blvd. East Suite 24 Mississauga, Ontario L4W 2A1 Canada	Steel pipe	Telephone: (905) 629-2011 Fax: (905) 629-4468 Email: tubacex@tornet.nexsys.ca http://www.tubacex.com
Tubacex Italia S.A. Piazza IV Novembre, 4 20124 Milano Italy	Steel pipe	Telephone: (3902) 667-15105 Fax: (3902) 667-12964 Email: tubacex@tin.it http://www.tubacex.com
Tubacex France Cedex 78 64170 Serres Sainte Marie France	Steel pipe	Telephone: 335.59.83.23.85 Fax: 335.59.83.39.11 Email: tti.france@wanadoo.fr http://www.tubacex.com
Chitrakoot Speciality Tubes Limited 306, Nirmal Towers, Dwarakapuri Colony Punjagutta, Hyderabad 500082 Andhra Pradesh India	Steel pipe	Tel: +91-40-335 4759 / 335 3116 Fax: +91-40-335 5086 Email: info@cstubes.com http://www.cstubes.com
Greenville Tube Corporation PO Box 30 Greenville, PA 16125 U.S.A.	Steel pipe	Telephone: (724) 588-6300 Fax: (724) 588-1492 Email: info@greenvilletube.com http://www.greenvilletube.com
Ocean International Suppliers, Inc. 5300 S. Florida Avenue Suite G-7 Lakeland, FL 33813 U.S.A.	Steel pipe	Telephone: (863) 648-9799 Fax: (863) 647-3070 Email: info@oceanint.com http://www.oceanintl.com
Stupp Corporation 12555 Ronaldson Road Baton Rouge, LA 70807-1503 U.S.A.	Steel pipe	Telephone: (800) 535-9999 Telephone: (225) 775-8800 Fax: (225) 775-7610 Email: stuppcorp@stuppcorp.com http://www.stuppcorp.com
H & H Tube Manufacturing Co. PO Box 455 108 Garfield St. Vanderbilt, MI 49795 U.S.A.	Steel pipe	Telephone: (517) 983-2800 Fax: (517) 983-4530 Email: info@h-htube.com http://www.h-htube.com

TABLE 29.1 Sources for Steel Pipe (Continued)

Company name and address	Product/services	Contact
Douglas Bros 475 Riverside Parkway Portland, ME 04103 U.S.A.	Steel pipe	Telephone: (207) 797-6671 Fax: (207) 797-8385 Email: douglas@douglasbrothers.com http://www.douglasbrothers.com
Douglas Bros 49 Artley Rd. Savannah, GA 31408 U.S.A.	Steel pipe	Telephone: (912) 748-7175 Fax: (912) 748-7197 Email: douglas@douglasbrothers.com http://www.douglasbrothers.com
USA Interpipe 24026 County Rd. 3118 Winona, TX 75792 U.S.A.	Steel pipe	Telephone: (800) 987-9312 Telephone: (903) 983-2999 Fax: (903) 948-4733 http://www.usainterpipe-steel-pipe.com
USA Interpipe R.R. 3, 3320 Miles Rd. Mount Hope, Ontario L0R 1W0 Canada	Steel pipe	Telephone: (905) 679-6999 Telephone: (877) 468-PIPE Fax: (905) 679-6544 http://www.usainterpipe-steel-pipe.com
USA Interpipe 805 Iere Avenue Ville Ste-Catherine, Quebec J0L 1E0 Canada	Steel pipe	Telephone: (514) 638-3320 Telephone: (888) 514-0040 Fax: (514) 638-3340 http://www.usainterpipe-steel-pipe.com
USA Interpipe 307-6th Street Calgary, Alberta T0M 0G0 Canada	Steel pipe	Telephone: (403) 947-2730 Telephone: (800) 434-8014 Telephone: (888) 923-3395 Fax: (403) 947-8919 http://www.usainterpipe-steel-pipe.com
Tachia Yung Ho Machine Industry Co. Ltd. No. 69, Yu Shih Road Tachia, Taichung Hsien Taiwan	Steel pipe	Telephone: 886-4-6815496-9 Fax: 886-4-6815456 Email: yhmco@yhmco.com.tw http://www.yhmco.com.tw
U.S. Steel Tubular Products 600 Grant St., Room 2001 Pittsburgh, PA 15219-2749 U.S.A.	Steel pipe	Telephone: (800) 553-3850 Telephone: (800) 553-3963 Fax: (412) 433-3878 Email: usstubular@uss.com http://www.usstubular.com
U.S. Steel International 600 Grant St., Room 2421 Pittsburgh, PA 15219-2749 U.S.A.	Steel pipe	Telephone: (412) 433-4531 Fax: (412) 433-4527 Email: ussi@uss.com http://www.usstubular.com
U.S. Steel International–Canada Noracor Metals and Materials, Inc. 89 Queensway West Suite 212 Mississauga, Ontario L5B 2V2 Canada	Steel pipe	Telephone: (905) 848-0909 Fax: (905) 848-0771 Email: ytrembley@uss.com http://www.usstubular.com
Kee Klamp Kee Industrial Products, Inc. 100 Stradtman Street Buffalo, NY 14206 U.S.A.	Steel pipe and fittings	Telephone: (800) 851-5181 Telephone: (716) 896-4949 Fax: (716) 896-5696 Email: info@keeklamp.com http://www.keeklamp.com
Kee Klamp Limited 10 Worton Dr. Worton Grange Reading RG2 0TQ United Kingdom	Steel pipe and fittings	Telephone: +44 (0) 118 931 1022 Fax: +44 (0) 118 931 1146 Email: sales@keeklamp.com http://www.keeklamp.com

TABLE 29.1 Sources for Steel Pipe (Continued)

Company name and address	Product/services	Contact
Kee Industrial Products, Ltd. 219 Connie Crescent, Unit 9 Concord, Ontario L4K 1L4 Canada	Steel pipe and fittings	Telephone: (800) 851-5181 Telephone: (905) 669-1494 Fax: (905) 669-4347 Email: info@keeklamp.com http://www.keeklamp.com
Kee Klamp GmbH Voltenseestrasse 22 D-60388 Frankfurt/Main Germany	Steel pipe	Telephone: +49 (0) 6109 32081 Fax: +49 (0) 6109 36402 Email: vertrieb@keeklamp.com http://www.keeklamp.com
Kee Klamp 30 Boulevard Pasteur 75015 Paris France	Steel pipe	Telephone: +(33) 1535 81426 Fax: +(33) 1535 81405 Email: vente@keeklamp.com http://www.keeklamp.com
Tubular Steel, Inc. 1031 Executive Parkway Dr. St. Louis, MO 63141-6351 U.S.A.	Steel pipe	Telephone: (314) 851-9200 Fax: (314) 851-9336 Email: info@tubularsteel.com http://www.tubularsteel.com
Tubular Steel, Inc. 519 North Cass Ave. Suite 509 Westmont, IL 60559-1514 U.S.A.	Steel pipe	Telephone: (630) 515-5000 Fax: (630) 515-5005 Email: illinois.sales@tubularsteel.com http://www.tubularsteel.com
Tubular Steel, Inc. 15810 Park Ten Place Suite 350 Houston, TX 77084-5193 U.S.A.	Steel pipe	Telephone: (281) 579-4200 Fax: (281) 579-4204 Email: texas.sales@tubularsteel.com http://www.tubularsteel.com
Tubular Steel, Inc. 100 Canal Rd. Fairless Hills, PA 19030-4304 U.S.A.	Steel pipe	Telephone: (215) 337-7000 Fax: (215) 337-7023 Email: pennsylvania.sales@tubularsteel.com http://www.tubularsteel.com
Tubular Steel, Inc. 2750 North Locust Avenue Rialto, CA 92376-4108 U.S.A.	Steel pipe	Telephone: (909) 429-6900 Fax: (909) 429-6939 Email: california.sales@tubularsteel.com http://www.tubularsteel.com
Dalian Haneda Steel Tube Co., Ltd. No. 382, Wuchang St. Zhongshan District Dalian China	Steel pipe	Telephone: 0086-411-2644726 Fax: 0086-411-2809932 Email: info@chinahaneda.com http://www.chinahaneda.com
Apex Tubes Pvt. Ltd Regd.Off. 1/6B Asaf Ali Rd. New Delhi, 110 002 India	Steel pipe	Telephone: 91-11-3234770, 3235201 Fax: 91-11-3236408, 3238991 Email: apextubes@eth.net http://www.apexpipes.com
Steel & Pipe Supply PO Box 1688 555 Poyntz Ave. Manhattan, KS 66505 U.S.A.	Steel pipe	Telephone: (800) 521-2345 Fax: (785) 587-5176 Email: sales@spsci.com http://www.spsci.com
Bonney Forge US Rt. 522 South Mt. Union, PA 17066 U.S.A.	Steel pipe	Telephone: (800) 345-7546 Telephone: (814) 542-2545 Fax: (814) 542-9877 Email: sleone@bonneyforge.com http://www.bonneyforge.com

TABLE 29.1 Sources for Steel Pipe (Continued)

Company name and address	Product/services	Contact
Bull Moose Tube Company 1819 Clarkson Rd. Chesterfield, MO 63017 U.S.A.	Carbon steel pipe and fittings	Telephone: (800) 325-4467 http://www.bullmoosetube.com
Friedman Industries, Inc. 1211 Judson Rd. Longview, TX 75601 U.S.A.	Carbon steel pipe and fittings	Telephone: (800) 527-8671 Fax: (903) 758-2265 http://www.friedmanindustries.com
Texas Pipe and Supply Co., Inc. 2330 Holmes Rd. Houston, TX 77051-1098 U.S.A.	Carbon steel pipe and fittings	Telephone: (713) 799-9235 Fax: (713) 799-8701 Email: tps@texaspipe.com http://www.texaspipe.com
Weldbend Corp. 6600 South Harlem Ave. Argo, IL 60501-1930 U.S.A.	Carbon steel pipe and fittings	Telephone: (708) 594-1700 Fax: (708) 458-0106 http://www.weldbend.com
Sen Fung Rollform Machinery Corp. 20-1 Tingtienliao,Tamshui Town Taipei Hsien Taiwan	Carbon steel pipe and fittings	Telephone: 886-2-26220000 Fax: 886-2-26252086 Email: senfung@ms2.hinet.net http://www.senfung.com
Total Alloys 16 Arvida St. Malaga Western Australia 6090	Carbon steel pipe and fittings	Telephone: (08) 9249 9660 Fax: (08) 9249 9661 Int'l Fax: + 61 8 9249 9661 Email: sales@totalalloys.com.au http://www.totalalloys.com.au
Norca Corporation 185 Great Neck Rd. Great Neck, NY 11022 U.S.A.	Carbon steel pipe and fittings	Telephone: (516) 466-9500 Fax: (516) 482-3367 http://www.norca.com
Van Leeuwen Pipe and Tube 10 Nealy Blvd. Trainer, PA 19061 U.S.A.	Carbon steel pipe and fittings	Telephone: (610) 485-5411 Fax: (610) 485-7353 http://www.vanleeuwen.com
Van Leeuwen Pipe and Tube 15333 Hempstead Rd. PO Box 40904 Houston, TX 77240-0904 U.S.A.	Carbon steel pipe and fittings	Telephone: (713) 466-9966 Fax: (713) 466-7423 http://www.vanleeuwen.com

TABLE 29.2 Sources for Cast Iron Pipe

Company name and address	Product/services	Contact
Hup Aik Huat Group 543 Yishun Industrial Park A Hup Aik Huat Hi-Tech Building Singapore 768765	Cast iron pipe and fittings	Telephone: (65) 755-1551 Fax: (65) 755-6556 http://www.hah.com.sg
AB&I Foundry 7825 San Leandro St. Oakland, CA 94621 U.S.A.	Cast iron pipe and fittings	Telephone: (800) GOT-IRON Telephone: (510) 632-3467 Fax: (510) 632-8035 http://www.abifoundry.com
Fengtai (Handan) Alloy Casting Co., Ltd. Beizhangzhuang town Handan, Hebei P.R. China	Cast iron pipe and fittings	Telephone: 0086-310-6091097 Fax: 0086-310-6092377 Email: zft@fengtaicasting.com http://www.fengtaicasting.com
N & W Construction Co. PO Box 3665 Victoria, TX 79703-3665 U.S.A.	Cast iron pipe and fittings	Telephone: (361) 582-4132 Fax: (361) 582-4170 Email: vnorrell@tisd.net http://www.tisd.net
S.P.T. METAL CORP. 19-1 Da Yeh St., Da Liao Hsiang Kaohsiung Hsien Taiwan	Cast iron pipe and fittings	Telephone: 886-7-7810525 Fax: 886-7-7821533 Email: spt@ksts.seed.net.tw http://www.spt.com.tw

TABLE 29.3 Sources for Ductile Iron Pipe

Company name and address	Product/services	Contact
United States Pipe & Foundry Comp. General Office PO Box 10406 Birmingham, AL 35202 U.S.A.	Ductile iron pipe, fittings, valves, and hydrants	Telephone: (205) 254-7000 http://www.uspipe.com
EBAA Iron Sales, Inc. PO Box 857 Eastland, TX 76448 U.S.A.	Ductile iron pipe and PVC pipe	Telephone: (254) 629-1731 Telephone: (800) 433-1716 Fax: (254) 629-8931 http://www.ebaa.com
The Ford Meter Box Company, Inc. PO Box 443 775 Manchester Avenue Wabash, IN 46992-0443 U.S.A.	Ductile iron pipe and fittings	Telephone: (219) 563-3171 Fax: (800) 826-3487 International Fax: (219) 563-0167 http://www.uniflange.com
The American Cast Iron Pipe Company III River Chase Office Plaza Suite 226 Birmingham, AL 35244 U.S.A.	Ductile iron pipe and fittings	Telephone: (800) 442-2347 http://www.acipco.com
Stanton plc Lows Lane, Stanton-by-Dale Ilkeston, Derbyshire DE7 4QU United Kingdom	Ductile iron pipe and fittings	Telephone: +44 (0) 115 930 5000 Fax: +44 (0) 115 932 9513 Email: sales@stanton.plc.uk http://www.stanton.plc.uk

TABLE 29.4 Sources for Aluminum Pipe

Company name and address	Product/services	Contact
Online Metals PO Box 19271 Seattle, WA 98109 U.S.A.	Pipe	Telephone: (800) 704-2157 Telephone: (206) 285-8603 Fax: (800) 533-6350 Email: sales@onlinemetals.com
M.C. Welding Services Ltd. 5506-238th St. Langley, BC V3A 7N6 Canada	Pipe and fittings	Telephone: (604)-533-4400 Email: info@mcwelding.com
D.P. Nicoli, Inc. 19600 SW Cipole Rd. Tualatin, OR 97062 U.S.A.	Pipe and fittings	Telephone: (503) 692-6080 Fax: (503) 692-1799 Email: or@dpnicoli.com
Metalcraft Enterprises, Inc. 202 Industrial Dr. New Haven, MO 63068 U.S.A.	Pipe and fittings	Telephone: (800) 325-0890 Fax: (573) 237-2330

TABLE 29.5 Sources for Titanium Pipe

Company name and address	Product/services	Contact
Custom Alloy Corp. 40 Seascape Village Aptos, CA 95003 U.S.A.	Titanium pipe and fittings	Telephone: (408) 685-6510 Fax: (408) 685-6514 Email: 104645.3254@compuserve.com http://ourworld.compuserve.com
Penn Machine 100 Bethel Rd. Aston, PA 19014-3485 U.S.A.	Titanium pipe and fittings	Telephone: (800) PENN-USA Int. Tel: (001) 1-610-497-3300 Fax: (001) 610-497-3325 http://usdropforge.com
Penn Machine 315 North Wayside Dr. Houston, TX 77020-7514 U.S.A.	Titanium pipe and fittings	Telephone: (800) PENN-USA Int. Tel: (001) 1-713-679-7900 Fax: (001) 713-679-7920 http://usdropforge.com
Valley Metals 13125 Gregg St. Poway, CA 92064 U.S.A.	Titanium pipe and fittings	Telephone: (858) 513-1300 Fax: (858) 513-6600 Email: sales@valleymetals.com http://www.valley-metals.com

TABLE 29.6 Sources for Copper Pipe

Company name and address	Product/services	Contact
Wolverine Tube, Inc. U.S./International Sales Office 2100 Market St. PO Box 2202 Decatur, AL 35609-2202 U.S.A.	Copper pipe	Telephone: (800) 874-6072 Telephone: (256) 353-1310 Fax: (256) 580-3825 http://www.wlv.com
H & H Tube & Manufacturing Co. PO Box 455 108 Garfield St. Vanderbilt, MI 49795 U.S.A.	Copper pipe and tube	Telephone: (517) 983-2800 Fax: (517) 983-4530 Email: info@h-htube.com http://www.h-htube.com
National Copper Products 415 East Prairie Ronde St. Dowagiac, MI 49047 U.S.A.	Copper pipe and tube	Telephone: (616) 782-2141 Fax: (616) 782-8538 http://www.nationaltube.com
Cerro Copper Products Co. PO Box 66800 St. Louis, MO 63166 U.S.A.	Copper pipe and tube	Telephone: (618) 337-6000 http://www.cerrocu.com
Northwest Copper Works, Inc. 1303 N. River St. Portland, OR 97227 U.S.A.	Copper pipe and fittings	Telephone: (503) 284-2191 Fax: (503) 284-3506 Email: sales@nwcopper.com http://www.nwcopper.com
1-Stop Piping and Tubing Supply Inc. San Diego, California U.S.A.	Copper pipe and tube	Email: Tubecoinc@aol.com http://www.piping-tubing.com
Phillips & Johnston, Inc. 21W179 Hill Ave. Glen Ellyn, IL 60137 U.S.A.	Copper pipe and tube	Telephone: (630) 469-8150 Fax: (630) 469-8048 Email: SalesDept@Phillips-Johnston.com http://www.phillips-johnston.com
United Industrial Piping, Inc. 9740 Near Dr. Cincinnati, OH 45246 U.S.A.	Copper pipe	Telephone: (800) 633-9690 Telephone: (513) 874-2004 http://www.unitedpiping.com
Ryerson Tull, Incorporated 2621 West 15th Place Chicago, IL 60608 U.S.A.	Copper pipe	Telephone: (773) 762-2121 Email: questions@ryersontull.com http://www.ryersontull.com
Glazer Steel & Aluminum 1556 Tchoupitoulas St. PO Box 50867 New Orleans, LA 70150 U.S.A.	Copper pipe	Telephone: (504) 529-2761 Telephone: (800) 824-1730 Fax: (504) 523-1839 Email: glazer@glazer.com tttp://www.glazer.com
Simonelli Trafilerie SpA 25068 Sarezzo (Brescia) Via Seradello, 213 Italy UE	Copper pipe	Tel. int++39 030 89191 Fax: int++39 030 8911025 http://simonellitrafilerie.com
Farmer's Copper & Industrial Supply Inc. 202 37th St. Galveston, TX 77553 U.S.A.	Copper pipe and fittings	Telephone: (800) 231-9450 Telephone: (409) 765-9003 Fax: (409) 765-7115 Email: sales@farmers-copper.com
Nibco Inc. 1516 Middleburg St. Elkhart, IN 46516-4714 U.S.A.	Copper pipe and fittings	Telephone: (800) 234-0227 Telephone: (888) 446-4226 Fax: (800) 234-0557 Fax: (219) 295-3221/3445 http://www.nibco.com

TABLE 29.7 Sources for Pipe Fabricators

Company name and address	Product/service	Contact
The Spool Group Enrique Foster Sur 39 Piso 4 Las Condes Santiago Chile	Pipe fabricator	Telephone: (511) 451-5012 Fax: (511) 562-2905
Boccard 2500 Galveston Rd. Houston, TX 77017-1925 U.S.A.	Pipe fabricator	Telephone: (713) 643-0681 Fax: (713) 643-4939 http://www.boccard.co.
Bristol Piping Systems 390 Bristol Metals Rd. Bristol, TN 37620 U.S.A.	Pipe fabricator	Telephone: (281) 444-8906 Fax: (281) 580-5915
Team Industries, Inc. 4002 Century Lane Chadds Ford, PA 19317 U.S.A.	Pipe fabricator	Telephone: (610) 558-9571 Fax: (610) 558-9572
BendTec, Inc. 366 Garfield Ave. Duluth, MN 55802 U.S.A.	Pipe fabricator	Telephone: (218) 722-0205 Fax: (218) 722-6598 http://www.bendtec.com
Hanjung 838 Yoksam-Dong Kannam-Ku Seoul Korea	Pipe fabricator	Telephone: 82 2 513 6196 6236 Fax: 82 2 513 6080 6076
International Piping Systems, Ltd 1700 S. Westport Dr. Port Allen, LA 70767 U.S.A.	Pipe fabricator	Telephone: (713) 995-6822 Fax: (713) 995-0234 http://www.turner-industries.com
PipeWorks, Inc. 469 West 19th St. Reserve, LA 70084-0545 U.S.A.	Pipe fabricator	Telephone: (800) 256-3142 Telephone: (504) 536-3142 Fax: (504) 536-8074
Scott Process Systems, Inc. 1160 Sunnyside St. Hartville, OH 44632 U.S.A.	Pipe fabricator	Telephone: (330) 877-2350 Fax: (330) 877-1524 http://www.processsystems.com
B. F. Shaw Inc. 8545 United Plaza Blvd. Baton Rouge, LA 70809 U.S.A.	Pipe fabricator	Telephone: (225) 932-2500 Fax: (225) 932-2661 http://www.shawgrp.com
B. F. Shaw, Inc. 150 Old Airport Rd. Lauren, SC 29360 U.S.A.	Pipe fabricator	Telephone: (864) 682-4000 Fax: (864) 682-4000 http://www.shawgrp.com
Sunland Fabricators, Inc. 30103 Sunland Dr. Walker, LA 70785 U.S.A	Pipe fabricator	Telephone: (225) 667-1000 Fax: (225) 667-1019 http://www.shawgrp.com
Process Fabricators, Inc. 800 New Natchitoches Rd. West Monroe, LA 71291 U.S.A	Pipe fabricator	Telephone: (318) 387-0212 Fax: (318) 387-9011 http://www.shawgrp.com

TABLE 29.7 Sources for Pipe Fabricators (Continued)

Company name and address	Product/service	Contact
Manufacturas Shaw South America, C.A. Zona Industrial Av. Principal Maracaibo, Venezuela	Pipe fabricator	Telephone: 11-5861-360-153 Fax: 11-5861-360-160 http://www.shawgrp.com
Belelli Energy srf via G. Taliercio Mantova 46100 Italy	Pipe fabricator	Telephone: 39 0376 333 327 Fax: 39 0376 333 818
Cimtas Enka 2 Binasi, Balmumcu 80780 Istanbul Turkey	Fabrication, erection, structural Engineering	Telephone: (90212) 275 2900 Fax: (90212) 275 2909 http://www.cimtas.com.te
Hisentepe Mah Gemsaz Merkii, PK92 Gemlik 16600 Bursa Turkey	Fabrication, erection, structural engineering	Telephone: (90224) 519 0250 Fax: (90224) 519 0256

TABLE 29.8 Sources for PVC and CPVC Pipe

Company name and address	Product/service	Contact
Corzan Industrial Systems 1271 William Robert Dr. Marietta, GA 30060 U.S.A.	PVC/CPVC pipe and fittings	Telephone: (800) 331-1144 Telephone: (770) 421-8564 Fax: (770) 421-1163
Harvel Plastics, Inc. Kuebler Rd. PO Box 757 Easton, PA 18044	PVC/CPVC pipe and fittings	Telephone: (610) 252-7355 Fax: (610) 253-4436 Email: info@harvel.com http://www.harvel.com
Spears Manufacturing Company 543 Industrial Dr. Lewisberry (Harrisburg) PA 17339 U.S.A.	PVC/CPVC pipe and fittings	Telephone: (800) 233-0275 Telephone: (717) 938-8844 Fax: (717) 938-6547 http://www.spearsmanufacturing.com
Vanguard Piping Systems, Inc. 901 N. Vanguard St. McPherson, KS 67460 U.S.A.	PVC/CPVC pipe and fittings	Telephone: (800) 775-5039 Int. Tel: 1 (316) 241-6369 Fax: (800) 775-4068 Email: service@vanguardpipe.com http://www.vanguardpipe.com
Nibco Inc. 1516 Middleburg St. Elkhart, IN 46516-4714 U.S.A.	PVC/CPVC pipe and fittings	Telephone: (800) 234-0227 Telephone: (888) 446-4226 Fax: (800) 234-0557 Fax: (219) 295-3221/3445 http://www.nibco.com

TABLE 29.9 Sources for Polyethylene Pipe

Company name and address	Product/service	Contact
Endot Industries, Inc, 60 Green Pond Rd. Rockaway, NJ 07866 U.S.A.	Polyethylene pipe and tube	Telephone: (800) 443-6368 Fax: (973) 625-4087 Email: endot@worldnet.att.net http://www.endot.com
CPP Engineering LLC PO Box 648 1180 West 5th St. Marysville, OH 43040 U.S.A.	Polyethylene pipe and fittings	Telephone: (937) 642-0669 Fax: (937) 642-0907 http://www.cpp-engineering.com
Plastics Pipe Institute 1825 Connecticut Ave. NW Washington, DC 20009 U.S.A.	Polyethylene pipe	Telephone: (800) 510-2772 Fax: (202) 462-9779 http://www.plasticpipe.org
ADS 3300 Riverside Dr. Columbus, OH 43221 U.S.A.	Polyethylene pipe	Telephone: (614) 457-3051 Fax: (614) 538-5204
Big "O" Inc. 254 Thames Rd. East Exeter, Ontario N0M 1S3 Canada	Polyethylene pipe	Telephone: (519) 235-0870 Fax: (519) 235-3238
Bruce Tile Inc. R.R. 3 Walkerton, Ontario N0G 2V0 Canada	Polyethylene pipe	Telephone: (519) 368-7015 Fax: (519) 368-7017
Crumpler Plastic Pipe PO Box 2068 Roseboro, NC 28382 U.S.A.	Polyethylene pipe	Telephone: (800) 334-5071 Telephone: (910) 525-4046 Fax: (910) 525-5801
Diller Tile Co., Inc. PO Box 727 111 E. Pine St. Chatsworth, IL 60921 U.S.A.	Polyethylene pipe	Telephone: (815) 635-3131 Fax: (815) 635-3133
Francesville Drain Tile Corp. PO Box 368 Francesville, IN 47946 U.S.A.	Polyethylene pipe	Telephone: (219) 567-9133 Fax: (219) 567-9296
Hancor, Inc. PO Box 1047 Findlay, OH 45839 U.S.A.	Polyethylene pipe	Telephone: (800) 537-9520 Telephone: (419) 422-6521 Fax: (419) 424-8302
Ideal Drain Tile Ltd. 8 Weldon Ave. Arva, Ontario N0M 1C1 Canada	Polyethylene pipe	Telephone: (519) 473-2669 Fax: (519) 641-2524
Lane Enterprises 34 Strohm Rd. Shippensburg, PA 17257 U.S.A.	Polyethylene pipe	Telephone: (717) 532-5959 Fax: (717) 532-7997
Plastic Tubing Inc. PO Box 878 Roseboro, NC 28382 U.S.A.	Polyethylene pipe	Telephone: (910) 525-5121 Fax: (910) 525-4934
Plastic Tubing Industries, Inc. Vulcan Rd. PO Box 607356 Orlando, FL 32860 U.S.A.	Polyethylene pipe	Telephone: (407) 298-5121 Fax: (407) 578-9393

TABLE 29.9 Sources for Polyethylene Pipe (Continued)

Company name and address	Product/service	Contact
Prinsco Inc. Box 265 Prinsburg, MN 56281 U.S.A.	Polyethylene pipe	Telephone: (800) 992-1725 Telephone: (320) 978-4116 Fax: (320) 978-8602
Quail Piping Products, Inc. PO Box 1470 Magnolia, AR 71753 U.S.A.	Polyethylene pipe	Telephone: (888) 919-7473 Fax: (870) 234-5121
Quality Culvert 406 Winton St. PO Box 1045 Wausau, WI 54402-1045 U.S.A.	Polyethylene pipe	Telephone: (715) 845-8559 Telephone: (888) 588-0054 Fax: (715) 849-2081
Springfield Plastics, Inc. Rt. 1, Box 171 Auburn, IL 62615 U.S.A.	Polyethylene pipe	Telephone: (800) 252-3361 Fax: (217) 438-6949
ISCO Industries 926 Baxter Ave. Louisville, KY 40204 U.S.A.	Polyethylene pipe	Telephone: (502) 583-6591 Telephone: (800) 345-4726 Fax: (502) 584-9713 http://www.isco-pipe.com
DRISCOPIPE 2929 N Central Expwy, #300 Richardson, TX 75080 U.S.A.	Polyethylene pipe	Telephone: (800) 527-0662 Fax: (972) 783-2647 http://www.driscopipe.com
Vanguard Piping Systems (Canada) Inc. 4641 Byrne Road Burnaby, BC V5J 3H6 Canada	Polyethylene pipe	Telephone: (888) 747-3739 Fax: (604) 431-5029 Email: sales@vanguard.ca http://www.vanguardpipe.ca
Vanguard Industries, Inc. 901 N. Vanguard St. McPherson, KS 67460 U.S.A.	Polyethylene pipe	Telephone: (800) 775-5039 Outside USA 1 (316) 241-6369 Fax: (800) 775-4068 Email: service@vanguardpipe.com http://www.vanguardpipe.com
KEFCO PO Box 126 Perkasie, PA 18944 U.S.A.	Polyethylene pipe	Telephone: (800) 457-0440 Email: sales@kefco.com http://www.kefco.com
CSR POLYPIPE PO Box 390 Gainesville, TX 76240 U.S.A.	Polyethylene pipe	Telephone: (800) 433-5632 Fax: (940) 668-2704 Email: PolyPipe@csra.com http://www.csra.com
Plexco General Office 1050 Route 83 Suite 200 Bensenville, IL 60106-1048 U.S.A.	Polyethylene pipe	Telephone: (630) 350-3700 Fax: (630) 350-2704 Email: info@plexo.com http://www.plexco.com
Plexco PO Box 608 217 Old Calhoun Falls Rd. Abbeville, SC 29620-0249 U.S.A.	Polyethylene pipe	Telephone: (864) 446-2136 Fax: (864) 446-2139 http://www.plexco.com
Plexco 1806 West Stone Ave. Fairfield, IA 52556-2151 U.S.A.	Polyethylene pipe	Telephone: (515) 472-3137 Fax: (515) 472-1658 http://www.plexco.com
Plexco 1601 West 287 Bypass Waxahachie, TX 75165-4709 U.S.A.	Polyethylene pipe	Telephone: (972) 937-0852 Fax: (972) 937-4530 http://www.plexco.com

TABLE 29.9 Sources for Polyethylene Pipe (Continued)

Company name and address	Product/service	Contact
Plexco R.R. 6 Industrial Park Bloomfield, IA 52537-0157 U.S.A.	Polyethylene pipe	Telephone: (515) 664-1286 Fax: (515) 664-1298 http://www.plexco.com
Plexco PO Box 23520 10420 Lexington Dr. Knoxville, TN 37933-1530 U.S.A.	Polyethylene pipe	Telephone: (865) 966-5822 Fax: (865) 675-9212 http://www.plexco.com
Plexco 1280 Jefferson Lane Colton, CA 92324-3530 U.S.A.	Polyethylene pipe	Telephone: (909) 420-5500 Fax: (909) 370-3841 http://www.plexco.com
Plexco 14381 Lear Blvd. Reno, NV U.S.A.	Polyethylene pipe	Telephone: (702) 677-1700 Fax: (702) 677-1712 http://www.plexco.com
Plexco De Mexico, S.A. de C.V. La Noria #103 Parque Industrial Queretaro Santa Rosa Jauregui C.P. 76220 Queretaro, Queretaro Mexico	Polyethylene pipe	Telephone-FAX, From USA Only Telephone: 011 52 42 16 2009 Fax: 011 52 42 16 1989 http://www.plexco.com
KWH Pipe 350 Lexington Ave. New York, NY 10016 U.S.A.	Polyethylene pipe and fittings	Telephone: (212) 687-8670 Fax: (212) 687-6798 http://www.kwhpipe.ca
KWH Pipe 6507 Mississauga Rd. Mississauga, ON L5N 1A6 Canada	Polyethylene pipe and fittings	Telephone: (905) 858-0206 Fax: (905) 858-0208 http://www.kwhpipe.ca
KWH Pipe 17918-55 Ave. Unit 4 Cloverdale, BC V3S 6C8 Canada	Polyethylene pipe and fittings	Telephone: (800) 668-1892 Fax: (604) 574-7073 http://www.kwhpipe.ca
KWH Pipe 7333 Place des Roseraies Suite 101 Anjou, PQ H1M 2X6 Canada	Polyethylene pipe and fittings	Telephone: (514) 352-3540 Fax: (514) 352-3290 http://www.kwhpipe.ca
KWH Pipe 37 Centre St. North Huntsville, ON P1H 2K8 Canada	Polyethylene pipe and fittings	Telephone: (705) 789-2396 Fax: (705) 789-7003 http://www.kwhpipe.ca
KWH Pipe 16800 Devonshire St. Suite 320 Granada Hills, CA 91344 U.S.A.	Polyethylene pipe and fittings	Telephone: (818) 831-8787 Fax: (818) 831-2999 http://www.kwhpipe.ca
ERS, Inc. 10154 Toebben Dr. Suite D Independence, KY 41051 U.S.A.	Polyethylene pipe and fittings	Telephone: (606) 525-6768 Fax: (606) 525-6763 http://www.hdpe-pipe.com
Superlon Plastics 2116 Taylor Way Tacoma, WA 98421 U.S.A.	Polyethylene pipe	Telephone: (253) 383-5877 Fax: (253) 383-5870 Email: info@superlon.com http://www.superlon.com

TABLE 29.10 Sources for Crossed-Linked Polyethelene (PEX) Pipe

Company name and address	Product/service	Contact
Vanguard Piping Systems, Inc. 901 N. Vanguard St. McPherson, KS 67460 U.S.A.	Cross-linked polyethylene (PEX) pipe	Telephone: (800) 775-5039 Int. Tel: 1 (316) 241-6369 Fax: (800) 775-4068 Email: service@vanguardpipe.com http://www.vanguardpipe.com
Vanguard Piping Sys. (Canada) Inc. 4641 Byrne Rd. Burnaby, BC V5J 3H6 Canada	Cross-linked polyethylene (PEX) pipe	Telephone: 1 (888) 747-3739 Fax: 1 (604) 431-5029 http://www.vanguard.ca.com Email: sales@vanguard.ca Email: technical@vanguard.ca
B & H Plumbing Co., Ltd. 805# Summit Building, No. 4 Shanghai Rd. Dalian China	Cross-linked polyethylene (PEX) pipe	Telephone: 86-411-2677991 Fax: 86-411-2591796 Email: info@bhplumbing.com http://www.bhplumbing.com

TABLE 29.11 Sources for Miscellaneous Plastic Pipe

Company name and address	Product/service	Contact
Orion Fittings, Inc. PO Box 17-1580 Kansas City, Kansas 66117 U.S.A.	Polypropylene pipe and fittings	Telephone: (913) 342-1653 Fax: (800) 777-1653 Email: Orion@orionfittings.com http://www.orionfittings.com
Wenzhou Zhaoflon Co., LTD. 66-14 Sanpaifang Wenzhou, Zhejiang P.R. China 325000	Fluoroplastic pipe	Telephone: 0086-577-8225050 Fax: 0086 577-8247734 Email: zhaoflon@mail.wzptt.zj.cn http://www.zhaoflon.com

TABLE 29.12 Sources for Plastic-Lined Pipe

Company name and address	Product/service	Contact
Plastic Lined Piping Systems PO Box 126 206 N. Sixth St. Perkasie, PA 18944 U.S.A.	Plastic-lined pipe and fittings	Telephone: (800) 637-9569 Fax: (800) 637-9573 Email: info@linedsteel.com http://www.linedsteel.com
Fusibond Piping Systems, Inc. 2615 W. Curtiss Downers Grove, IL 60515 U.S.A.	Plastic-lined pipe and fittings	Telephone: (630) 969-4488 Fax: (630) 969-2355 Email: sales@fusibond.com http://www.fusibond.com
Resistoflex Marion Manufacturing and Sales One Quality Way Marion, NC 28752 U.S.A.	Plastic-lined pipe, fittings, and expansion joints	Telephone: (828) 724-4000 Fax: (828) 724-4783 http://www.resistoflex.com
Resistoflex Bay City Manufacturing 4675 E. Wilder Rd. Bay City, MI 48706 U.S.A.	Plastic-lined pipe, fittings, and expansion joints	Telephone: (517) 671-3100 Fax: (517) 671-3173 http://www.resistoflex.com
Resistoflex Asia-Pacific Manufacturing and Sales 16 Gul Link Singapore 629484	Plastic-lined pipe, fittings, and expansion joints	Telephone: (65) 863-1559 Fax: (517) 863-1560 http://www.resistoflex.com

TABLE 29.13 Sources for Rubber Lining and Lined Pipe

Company name and address	Product/service	Contact
Kefco PO Box 126 Perkasie, PA 18944 U.S.A.	Rubber-lined pipe, fittings, valves, and tanks	Telephone: (800) 457-0440 sales@kefco.com http://www.kefco.com
Lined Process Systems Pty. Ltd. 30 Peachtree Rd. Penrith NSW 2750 Sydney Australia	Rubber-lined pipe and valves	Telephone: 61(02) 4721 7333 Fax: 61(02) 4721 7404 Email: staff@linpro.com.au http://www.linpro.com.au
Blair Rubber Company 1252 Mina Ave. Akron, OH 44321 U.S.A	Rubber lining	Telephone: (800) 666-9334 Fax: (330) 321 5583

TABLE 29.14 Sources for Fiberglass Pipe

Company name and address	Product/service	Contact
Conley Corp. 2795 East 91st St. Tulsa, OK 74137 U.S.A.	Fiberglass pipe	Telephone: (800) 331-5502 Telephone: (918) 299-5051 Fax: (918) 299-5907 Email: webmaster@conleyfrp.com http://www.conleyfrp.com
Tubular Fiberglass Corporation 11811 Proctor Rd. Houston, TX 77038 U.S.A.	Fiberglass pipe and fittings	Telephone: (281) 847-2987 Fax: (281) 847-1931 Email: TFC@TubularFiberglass.com http://www.tubularfiberglass.com
Dover Industrial PO Box 5010 111 MacNab Ave. Strathroy, Ontario, N7G 3J3 Canada	Fiberglass pipe and fittings	Telephone: (800) 410-4802 Telephone: (519) 245-6100 Fax: (519) 245-6078 Email: sales@doverindustrial.com http://www.doverindustrial.com
ERSHIGS 742 Marine Dr. PO Box 1707 Bellingham, WA 98227 U.S.A.	Fiberglass pipe, vessels, and valves	Telephone: (360) 733-2620 Fax: (360) 733-2628 Email: webmaster@ershigs.com http://www.ershigs.com
Eastern States Associates, Inc. 21 Wilbraham St. Suite 211 Palmer, MA 01069 U.S.A.	Fiberglass pipe and tanks	Telephone: (413) 283-4412 Fax: (413) 283-4269 Email: sales@easternstates.com http://www.easternstates.com
Star Fiberglass 2425 S.W. 36th St. PO Box 37389 San Antonio, TX 78237 U.S.A.	Fiberglass pipe, tube, and fittings	Telephone: (210) 434-5043 Fax: (210) 434-7543 Email: starfgs@texas.net http://www.onr.com
Non Metallic Resources, Inc. 8375 Zeigler Blvd. Mobile, AL 36608 U.S.A.	Fiberglass pipe and fittings	Telephone: (334) 633-3303 Fax: (334) 633-3318 http://www.nmr-sealtech.com
L. F. Manufacturing, Inc. PO Box 578 Hwy 290 East Giddings, TX 78942 U.S.A.	Fiberglass pipe and tanks	Telephone: (800) 237-5791 Telephone: (409) 542-8027 Fax: (409) 542-0911 Email: lfmfg@bluebon.net http://www.lfm-frp.com
Pipe Wrap Para Fijar Enterprises, Inc. PO Box 270190 Houston, TX 77277-0190 U.S.A.	Fiberglass pipe	Telephone: (713) 522-4467 Fax: (713) 522-5114 http://www.piperepair.net
Fibercast Co. PO Box 968 Sand Springs, OK 74063 U.S.A.	Fiberglass pipe	Telephone: (918) 245-6651 Telephone: (800) 331-4406 Fax: (918) 241-1143 http://www.fibercast.com
Prolite Plastics Ltd. 1620 Kingsway Ave. Port Coquitlam, BC V3C 3Y9 Canada	Fiberglass pipe	Telephone: (604) 944-6260 Fax: (604) 944-8854
Corrosion Prod. de Mexico S.A. de C.V. Nayarit 204 Int. 2 Col. Guadalupe Tampico, Tamps. 89120 Mexico	Fiberglass pipe	Telephone: 52-12-170335 Fax: 52-12-170435

TABLE 29.14 Sources for Fiberglass Pipe (Continued)

Company name and address	Product/service	Contact
Fiberdur-Vanck GmbH Industriepark Emil Mayrisch D-52457 Aldenhoven-Siersdorf Germany	Fiberglass pipe	Telephone: 49-2464-972-230 Fax: 49-2464-972-115
Flexachem Mfg. Ltd. Donnybrook Commercial Centre Donnybrook, Douglas Cork Ireland	Fiberglass pipe	Telephone: 353-21-363742 Fax: 353-21-891297
Flexachem U.K. Ltd. 1, Minster Close, Hucknall Nottingham NG 15 8DR United Kingdom	Fiberglass pipe	Telephone: 44-115-964-2099 Fax: 44-115-964-2093
AAL & ASSOCIATES c/o M.A. Ibrahin Office 10 Midan Abou Lithamin, Mohendessien Giza, Cairo Egypt 12411	Fiberglass pipe	Telephone: 202-302-4488 Fax: 202-347-6176 Email: aibrahim@acs.auc.eun.eg
Kaytec Ltd—Sales Rep. PO Box 1065 26102 Kiryat Motzkin Israel	Fiberglass pipe	Telephone: 972-4-872-1564 Fax: 972-4-872-1294
Champion Taiwan Co., Ltd. PO Box 57–58 6th Fl., No. 159, Hsin-An St. Taipei 110 Taiwan ROC	Fiberglass pipe	Telephone: 886-2-732-1519 Fax: 886-2-732-0765 Email: champ@ms16ihinet.net
Cosmo Trading Co., Ltd. 9-211 HO, 984 Sihungdong, Kumchun-Ku Seoul Korea	Fiberglass pipe	Telephone: 82-2-893-2443 Fax: 82-2-893-2442 Email: cosmocho@netsgo.com
Abacus Corporation Pty. Ltd. (A) 1 Defu Lane 10 03-571 Singapore 539182	Fiberglass pipe	Telephone: 65-285-6100 Fax: 65-281-7761 Email: abetrad@singnet.com.sg
Ameron International Corporation 245 South Los Robles Ave. Pasadena, CA 91101-2894 U.S.A.	Fiberglass pipe	Telephone: (626) 683-4000 Fax: (626) 683-4060 http://www.ameron.com
Fiberglass Pipe Group 5300 Hollister Rd. Suite 111 Houston, TX 77040 U.S.A.	Fiberglass pipe	Telephone: (713) 690-7777 Fax: (713) 690-2842
Win Fung Fiberglass SDN BHD No. 224B, Jalan 4D Kampung Baru Subang 40000 Shah Alam Selangor Darul Ehsan Malaysia	Fiberglass pipe	Telephone: (603) 747-1828 Fax: (603) 746-8137 Email: enquiry@winfung-frp.com.my http://www.winfung-frp.com.my
Flowtite Botswana (Pty.) Ltd. Plot 43180 PO Box 1110 Phakalane Industrial Site Gaborone Botswana	Fiberglass pipe	Telephone: 267.309 746 Fax: 267.308 332

TABLE 29.14 Sources for Fiberglass Pipe (Continued)

Company name and address	Product/service	Contact
Flowtite Africa (Private Ltd.) 10 Floor, Megawatt House Samora Machel/Leopold Takawira PO Box 4710, Harare Zimbabwe	Fiberglass pipe	Telephone: 263.4-700 131 263.4-731 603 Fax: 263.4-708 538 263.4-720 632
Flowtite Southern Africa	Fiberglass pipe	Telephone: 27.21.850 27 42 Fax: 27.20.850 31 61
Flowtite Greece	Fiberglass pipe	Telephone: 30.1.80 37 100 Fax: 30.1.80 37 101
Engineered Pipe Systems (Flowtite) Chaussee de la Hulpe, 178 B-1170 Brussels Belgium	Fiberglass pipe	Telephone: 32.2.674 82 25 Fax: 32.2.674 82 23
Owens-Corning Eternit (Flowtite) Rohre Gmbh Gewerbegebiet Am Fuchsloch D-04720 Grosssteinbach Germany	Fiberglass pipe	Telephone: 49.3431-71 82 0 Fax: 49.3431-70 23 24
Flowtite Pipe & Tank A/S PO Box 2059 3239 Sandefjord Norway	Fiberglass pipe	Telephone: 47.33.44.9121 Fax: 47.33.47.4220
Flowtite Iberica, S.A. Edificio Valrealty Basauri, 17 28023 Madrid Spain	Fiberglass pipe	Telephone: 34.91.372 81 43 Fax: 34.91.372 95 71
Owens-Corning Yapi Merkezi Boru Sanayi ve Cemil Topuzlu Cad. Is Bankasi C Blok, No. 6 Ciftehavuzlar, Istanbul Turkey	Fiberglass pipe	Telephone: 90.216-467 0850 Fax: 90.216-467 00 32
Amiantit Fiberglass Egypt 9 Menies St., El Korba Heliopolis Cairo Egypt	Fiberglass pipe	Telephone: 20.2.416 08 00 Telephone: 20.2.416 08 59 Fax: 20.2.290 43 40
Amiantit Fiberglass Industries Ltd. PO Box 589 Dammam 31421 Saudi Arabia	Fiberglass pipe	Telephone: 966.3.857 11 60 Fax: 966.3.857 90 39
Gulf Eternit Industries Co. PO Box 1371 Dubai United Arab Emirates	Fiberglass pipe	Telephone: 971.4-857 256 Fax: 971.4-851 935

TABLE 29.14 Sources for Fiberglass Pipe (Continued)

Company name and address	Product/service	Contact
Engineered Pipe Systems Owens-Corning Science & Technology Center 2790 Columbus Road Route 16 Granville, OH 43023-1200 U.S.A.	Fiberglass pipe	Telephone: 1.740.321.5530 Fax: 1.740.321.7433
Flowtite Argentina S.A. Cerrito 836 Piso 9 Frente Buenos Aires (1010) Argentina	Fiberglass pipe	Telephone: 54.11.4 816 88 58 Fax: 54.11.4 816 84 22
Flowtite Argentina S.A. Casilla de Correo No. 1, Suc. 22 5022 Cordoba Argentina	Fiberglass pipe	Telephone: 54.351.4 65 30 60 Fax: 54.351.4 65 43 79
Flowtite Andercol S.A. Cra. 33 No. 5G-73 Medellin Colombia	Fiberglass pipe	Telephone: 57.4.312 72 22 Fax: 57.4.311 82 68
Hobias Pipe 1413 Richey Rd. Houston, TX 77073-3058 U.S.A.	Fiberglass pipe	Telephone: (281) 821-2200 Fax: (281) 821-7715 Email: info@hobiaspipeusa.com

TABLE 29.15 Sources for Glass Pipe

Company name and address	Product/service	Contact
Starr Fiber Glass 2425 S.W. 36th St. PO Box 37389 San Antonio, TX 78237 U.S.A.	Glass pipe	Telephone: (210) 434-5043 Fax: (210) 434-7543 Email: starfgs@texas.net
Hobas Pipe USA 1413 Richey Rd. Houston, TX 78237 U.S.A.	Glass pipe	Telephone: (281) 821-2200 Fax: (281) 821-7715 Email: info@hobaspipeusa.com

TABLE 29.16 Sources for Vitrified Clay Pipe

Company name and address	Product/service	Contact
GBH Goh Ban Huat Berhad 238 Jalan Segambut 51200 Kuala Lumpur Malaysia	Vitrified clay pipe	Email: misweb@gbhgroup.com.my http://www.gbhgroup.com.my
Asahi Danke 555-2 Daiba Kamui-cho Asahikawa Hokkaido 070 Japan	Vitrified clay pipe	Telephone: 0011 81 166 62 0852 Fax: 0015 81 166 61 0864
Rockle Bros. Ltd. 42 Chang An East Rd. Taipei 104 Taiwan R.O.C.	Vitrified clay pipe	Telephone: 0011 886 2 2531 5999 Fax: 0015 886 2 2511 1597
Rich Building Materials Supplies Ltd. Flat C 1 7/F Universal Industrial Centre 19-25 Shan Mei St. Fo Tan, N.T. Hong Kong	Vitrified clay pipe	Telephone: 852 2606 1206 Fax: 852 2691 3534
Stewarts Building Products Ltd. PO Box 2079, 15 Hynds Rd. Tauranga New Zealand	Vitrified clay pipe	Telephone: 0011 64 7 578 7117 Fax: 0015 64 7 578 2112
Pyung An Brick 291-29 Yangjae-Dong, Seocho-Gu Seoul Korea	Vitrified clay pipe	Telephone: 82 2 577 4249 Fax: 82 2 577 5723

TABLE 29.17 Sources for Concrete Pipe

Company name and address	Product/service	Contact
Hanson Pipe and Products, Inc. PO Box 569470 Dallas, TX 75356-9470 U.S.A.	Concrete pipe and fittings	Telephone: (972) 262-3600 Fax: (972) 264-6236 Email: GHA@onramp.net http://www.ghainc.com
The Hawkeye Group Highway 61 South Mediapolis, IA 52637 U.S.A.	Concrete pipe and fittings	Telephone: (319) 394-3197 Fax: (319) 394-3977 Email: info@hawkeye-pipe.com http://www.hawkeye-pipe.com
A-LOK Products, Inc. PO Box 1647 697 Main St. Tullytown, PA 19007 U.S.A.	Concrete pipe and fittings	Telephone: (800) 822-2565 Telephone: (215) 547-3366 Fax: (215) 547-5260 Email: info@a-lok.com http://www.a-lok.com
Besser Company 801 Johnson St. Alpena, MI 49707 U.S.A.	Concrete pipe and fittings	Telephone: (517) 354-4111 Telephone: (888) 8-237737 Fax: (517) 354-3120 Email: IntPipe@aol.com http://www.home.aol.com/IntPipe
Sherman Dixie Concrete Industries, Inc. 3641 Central Pike Nashville, TN 37076 U.S.A.	Concrete pipe and fittings	Telephone: (615) 889-0700 Fax: (615) 889-7492 http://www.shermandixie.com
Permatile Concrete Products Company PO Box 2049 Bristol, VA 24203 U.S.A.	Concrete pipe and fittings	Telephone: (800) 662-5332 Fax: (540) 669-2120 Email: mail@permatile.com http://www.permatile.com
Price Brothers Company Pipesite PO Box 825 Dayton, OH 45401-0825 U.S.A.	Concrete pipe and fittings	Telephone: (937) 226-8700 Fax: (937) 226-8711 Email: marketing@pipesite.com http://www.pipesite.com
CSR Hydro Conduit 16701 Greenspoint Park Dr. Suite 350 Houston, TX 77060 U.S.A.	Concrete pipe and fittings	Telephone: (281) 872-3500 Fax: (281) 872-5709 Email: hydro@csra.com http://www.csra.com
Irving N. Loomis and Sons, Inc. PO Box 129 (Route 706) Montrose, PA 18801 U.S.A.	Concrete pipe and fittings	Telephone: (800) 278-1375 Fax: (570) 278-2886 Email: inloomis@epix.net http://www.concretebyloomis.com
Underground Specialties Cambridge Inc. 131 Shearson Cambridge, ON N1T 1J3 Canada	Concrete pipe and fittings	Telephone: (519) 662-3185 Fax: (519) 662-4666 Email: markusi@golden.net http://www.undergroundspec.com
Ozkanlar Hydraulics & Machine Ind. Trade Limited Co. Sincan Organize Sanayi Bolgesi Buyuk Selcuklu Sk. No. 1 Sincan/Ankara Turkey	Concrete pipe and fittings	Telephone: (90-312) 267 18 13-15 Fax: (90-312) 267 04 05 Email: ozkanlar@virtual.tr-net.net.tr http://www.ozkanlarmakine.com.tr
Carder Concrete Products 8311 West Carder Court Littleton, CO 80125 U.S.A.	Concrete pipe and fittings	Telephone: (303) 791-1600 Fax: (303) 790-1710 http://www.carderconcrete.com
Discount Drainage Supplies 2600 South Arlington Rd. Akron, OH U.S.A.	Concrete pipe and fittings	Telephone: (800) 407-0011 Email: dd-sales@neo.rr.com http://www.discountdrainage.com

TABLE 29.17 Sources for Concrete Pipe (Continued)

Company name and address	Product/service	Contact
Con Cast Pipe RR #3 299 Brock Road Guelph, ON 1H 6H9 Canada	Concrete pipe and fittings	Telephone: (800) 668-7473 Email: sales@concastpipe.com http://www.concastpipe.com
Choctaw Alabama Region Main Office 3750 Western Bypass Montgomery, AL 36108 U.S.A.	Concrete pipe and fittings	Telephone: (334) 281-0820 Fax: (334) 281-0831 http://www.choctawpipe.com
Choctaw Central Region Main Office 501 East Jefferson St. West Memphis, AR 72301 U.S.A.	Concrete pipe and fittings	Telephone: (501) 735-5514 Fax: (501) 735-0331 http://www.choctawpipe.com
Choctaw New Orleans Cement Products Gulf Coast Region Main Office 13201 Old Gentilly Rd. New Orleans, LA 70129 U.S.A.	Concrete pipe and fittings	Telephone: (800) 238-5294 Telephone: (901) 346-1400 http://www.choctawpipe.com
Choctaw Southern Region Main Office 2840 W. Northside Dr. Jackson, MS 39213-4654 U.S.A.	Concrete pipe and fittings	Telephone: (601) 982-1100 Fax: (601) 366-0655 http://www.choctawpipe.com
Ameron Concrete & Steel Pipe Group 10681 Foothill Blvd. Rancho Cucamonga, CA 91730 U.S.A.	Concrete pipe and fittings	Telephone: (909) 944-4100 Fax: (909) 944-4112 http://www.ameron.com

TABLE 29.18 Sources for Pipe Supports

Company name and address	Product/service	Contact
Bergen-Power Pipe Supports 225 Merrimac St. Woburn, MA 01801 U.S.A.	Pipe supports	Telephone: (617) 935-9550 Fax: (617) 938-0026
Bergen-Power Pipe Supports 810 Pleasent Hill Rd. Wallingford, PA 19086 U.S.A.	Pipe supports	Telephone: (610) 872-5576 Fax: (610) 872-3420
Bergen-Power Pipe Supports 829 Beaver Ave. Pittsburgh, PA 15233 U.S.A.	Pipe supports	Telephone: (412) 323-1545 Fax: (412) 323-1904
Basic Engineers Inc. Seventh & Blaw Ave. PO Box 38010 Pittsburgh, PA 15238 U.S.A.	Pipe supports	Telephone: (412) 826-1900 Fax: (412) 826-0173
Power Pipe Supports, Inc. PO Box 461 Donora Industrial Park Eighth Street Gate Donora, PA 15033 U.S.A.	Pipe supports	Telephone: (412) 379-5212 Fax: (412) 379-9363
Mercer Engineers, Inc. 17440 North Dallas Parkway Suite 139 Dallas, TX 75287 U.S.A.	Pipe supports	Telephone: (214) 732-6166 Fax: (214) 732-7563
Tompkins & Kinner Co., Inc. PO Box 7627 Shawnee Mission, KS 66207-0627 U.S.A.	Pipe supports	Telephone: (913) 345-1066 Fax: (913) 345-8905
Robert K. Griffith & Associates PO Box 111 Pearl Ave. Putman, CT 06260 U.S.A.	Pipe supports	Telephone: (203) 928-7971 Fax: (203) 928-9302
Power Sales 1137 E. Sandpiper Dr., No. 104 Tempe, AZ 85283 U.S.A.	Pipe supports	Telephone: (602) 839-5624
Power Piping International, bv PO Box 161 5050 Ads Goirle The Netherlands	Pipe supports	Telephone: 31-013-348515 Fax: 31-013-347925 Telex: 52750
Dave Copeman Dave Copeman Sales 1220 Bryn Mawr Lane RR 3 Campbellville Ontario L0P 1B0 Canada	Pipe supports	Telephone: (416) 332-8501 Fax: (416) 332-8501
Specialty Support Systems Inc. PO Box 226 Morrisville, PA 19067 U.S.A.	Pipe supports	Telephone: (215) 945-1033 Fax: (215) 945-3701
Piping Technology & Products, Inc. PO Box 34506 Houston, TX 77234-4506 U.S.A.	Pipe supports	Telephone: (713) 731-0030 Fax: (713) 731-8640 Email: info@pipingtech.com http://www.pipingtech.com/

TABLE 29.18 Sources for Pipe Supports (Continued)

Company name and address	Product/service	Contact
Lisega Inc. 375 West Main St. Newport, TN 37821 U.S.A.	Pipe supports	Telephone: (423) 625-2000 Fax: (423) 625-9009
PHS Industries, Inc. 434 Latigue Rd. Waggaman, LA 70094 U.S.A.	Pipe supports	Telephone: (504) 431-7722 Fax: (504) 431-7900
Rilco Manufacturing Company 11435 Brittmoore Park Dr. Houston, TX 77041 U.S.A.	Preinsulated hot and cold pipe supports	Telephone: (713) 466-4777 Fax: (713) 466-6547
Anvil International, Inc. Precision Park 160 Frenchtown Rd. North Kingstown, RI 02852 U.S.A.	Pipe supports (licensed to manufacture Grinnell hangers)	Telephone: (401) 886-3040 Fax: (401) 886-3053
Pipe Supports Ltd Salwarpe Rd. Droitwich, Worcestershire WR9 9BH United Kingdom	Pipe supports	Telephone: +44.1905.795500 Fax: +44.1905.794126
Pipe Supports USA, Inc. 2579 Lester St. Harvey, LA 70058 U.S.A.	Pipe supports	Telephone: (504) 367-7473 Fax: (504) 367-7484
Pipe Supports India Ltd Plot No. 29 Industrial Estate Perungudi, Chennai 600 086 India	Pipe supports	Telephone: +91.44.4961003 Fax: +91.44.4961395
Pipe Supports Asia Ltd 35/2 Bangna-Trad Rd. Km. 11.5 Bangplee, Samut Prakam 10540 Thailand	Pipe supports	Telephone: +662.316.3912-3 Fax: +662.316.2432
Binder Engineering Group 34 Clune St. Bayswater Western Australia 6053 Australia	Pipe supports	Telephone: +61 (08) 9377-2900 Fax: +61 (08) 9377-1855 Email: binder.sales@cbits.com.au
Binder Engineering Group 28 Treforest Dr. Clayton Victoria 3168 Australia	Pipe supports	Telephone: (03) 9543-9433 Fax: (03) 9543-9633 Email: bindervic@bigpond.com.au
Binder Engineering Group 59 Railway Parade Rocklea Queensland 4106 Australia	Pipe supports	Telephone: (07) 3274-3230 Fax: (07) 3274-4596 Email: binderqld@ecn.net.au
Binder Engineering Group 39 Brodie St. Rydalmere New South Wales 2116 Australia	Pipe Support	Telephone: (02) 9898-0822 Fax: (02) 9898-0844 Email: bindernsw@bigpond.com.au
Binder Engineering Group 27B Hamilton Rd. Singapore 209196	Pipe supports	Telephone: +65 291 5589 Fax: +65 297 3921 Email: bindasia@pacific.net.sg
Binder Engineering Group Jalan Pangeran Diponegoro 108 Bekasi Tambun Indonesia	Pipe supports	Telephone: +62 (021) 882 2150 Fax: +62 (021) 880 7334 Email: binder_indo@ibm.net

TABLE 29.19 Sources for Expansion Joints

Company name and address	Product/service	Contact
EJMA, Inc. 25 North Broadway Tarrytown, NY 10591 U.S.A.	Expansion joints, standards	Telephone: (914) 332-0040 Fax: (914) 332-1541
American BOA, Inc. PO Box 1301 Cumming, GA 30028 U.S.A.	Expansion joints	Telephone: (800) 856-4580 Fax: (770) 889-0661
Badger Industries, Inc. 100 Badger Dr. Zelienople, PA 16063 U.S.A.	Expansion joints	Telephone: (724) 452-4500 Fax: (724) 452-0802
Expansion Joint Systems, Inc. 10035 Prospect Ave., Suite 202 Santee, CA 92071 U.S.A.	Expansion joints	Telephone: (619) 562-6083 Fax: (619) 562-0636
Flexider Sp.A. C. so Romania 501/24 Torino Italy	Expansion joints	Telephone: 39-011-2627111 Fax: 39-011-2620904
Hyspan Precision Products, Inc. 1685 Brandywine Ave. Chula Vista, CA 91911 U.S.A.	Expansion joints	Telephone: (618) 421-1702 Fax: (619) 421-1702
Pathway Bellows, Inc. PO Box 3027 115 Franklin Rd. Oak Ridge, TN 37831 U.S.A.	Expansion joints	Telephone: (615) 483-7444 Fax: (615) 482-5600
Senior Flexonics Inc. 2400 Longhorn Industrial Dr. New Braunfels, TX 78130 U.S.A.	Expansion joints	Telephone: (800) 847-5746 Fax: (830) 629-6899
Wahlco Engineered Products, Inc. 29 Lexington St. Lewiston, ME 04240 U.S.A.	Expansion joints	Telephone: (207) 784-2338 Fax: (207) 753-1338
Proco Products, Inc. 2431 North Wigwam Dr. PO Box 590 Stockton, CA 95201-0590 U.S.A.	Expansion joints	Telephone: (209) 943-6088 Fax: (209) 943-0242 Email: sales@procoproducts.com http://www.procoproducts.com
Unaflex, Inc. 2056 North Dixie Highway Fort Lauderdale, FL 33305 U.S.A.	Expansion joints	Telephone: (800) 327-1286 Telephone: (305) 561-0500 Fax: (305) 565-8473

TABLE 29.20 Sources for Steam Traps

Company name and address	Product/service	Contact
Yarway Corp. 480 Norristown Rd. Blue Bell, PA 19422 U.S.A.	Steam traps	Telephone: (610) 832-2266 Fax: (610) 825-6679
Collin's Pipe and Supply 11 Thompson Rd. East Windsor, CT 06088 U.S.A.	Steam traps	Telephone: (800) 876-5406 Fax: (860) 292-5358
Mueller Steam Specialty 1491 NC Highway 20 West St. Pauls, NC 28384 U.S.A.	Steam traps	Telephone: (910) 865-8241 Fax: (910) 865-8245
Plenty Steam Products Unit 1, Lakeside Business Park Pinfold Road, Thurmaston Leicester UK	Steam traps	Telephone: 0116-26-5172 Fax: 0116-269-3695 http://www.plenty.co.uk
Plenty Steam Products 6630 Roxburgh Dr. Suite 175 Houston, TX 77041 U.S.A.	Steam traps	Telephone: (713) 937-3863 Fax: (713) 937-0968
Spirax-Sarco 1150 Northpoint Blvd. Blythewood, SC 29016 U.S.A.	Steam traps	Telephone: (803) 714-2000 Fax: (803) 714-2200 http://www.spiraxsarco-usa.com
Armstrong-Yoshitake, Inc. Three Rivers, MI 49093 U.S.A.	Steam traps	Telephone: (616) 279-3600 Fax: (616) 273-8656 http://www.armstrong-intl.com

TABLE 29.21 Sources for Strainers

Company name and address	Product/service	Contact
Collin's Pipe and Supply 11 Thompson Rd. East Windsor, CT 06088 U.S.A.	Strainers	Telephone: (800) 876-5406 Fax: (860) 292-5358
Plenty, Inc. 6630 Roxburgh Dr. Suite 175 Houston, TX 77041 U.S.A.	Strainers	Telephone: (713) 937-3863 Fax: (713) 937-0968
Hayward Industrial Products 900 Fairmount Ave. Elizabeth, NJ 07207 U.S.A.	Strainers	Telephone: (908) 351-5400 Fax: (908) 351-7893
Mueller Steam Specialty 1491 N.C. Hwy. 20 West St. Pauls, NC 28384 U.S.A.	Strainers	Telephone: (910) 865-8241 Fax: (910) 865-8245
Fabrotech Industries, Inc. 2089 Ninth Ave. Ronkonkoma, NY 11779 U.S.A.	Strainers	Telephone: (516) 467-2040 Fax: (516) 467-2196

TABLE 29.22 Sources for Pipe Fittings

Company name and address	Product/service	Contact
Alloy Stainless Products Co. Inc. 611 Union Blvd. Totowa, NJ 07512 U.S.A.	Pipe fittings	Telephone: (973) 256-1616 Telephone: (800) 631-8372 Fax: (973) 256-5256 Email: aspaba@aol.com
Amcast Industrial Corp. Elkhart Products Corp. PO Box 1008 Elkhart, IN 46515 U.S.A.	Pipe fittings	Telephone: (219) 264-3181 Fax: (219) 264-4835
American Cast Iron Pipe Co. PO Box 2727 Birmingham, AL 35202 U.S.A.	Pipe fittings	Telephone: (205) 325-7856 Telephone: (205) 325-7862 http://www.acipco.com
Anvil International, Inc. Research & Development Center 20 Thurber Blvd. Smithfield, RI 02917 U.S.A.	Pipe fittings	Telephone: (401) 233-3927 Fax: (401) 781-9912
BendTec, Inc. 366 Garfield Ave. Duluth, MN 55802 U.S.A.	Pipe fittings	Telephone: (218) 722-0205 Fax: (218) 722-6598 Email: employeesname@bendtec.com
Bestweld, Inc. 40 Robinson St. Pottstown, PA 19464 U.S.A.	Pipe fittings	Telephone: (610) 718-9700 Fax: (610) 718-9800 Email: www.fittings@bestweld.com http://www.bestweld.com
Bonney Forge Corporation Cedar & Meadows St. PO Box 359 Allentown, PA 18105 U.S.A.	Pipe fittings	Telephone: (610) 435-9611 Fax: (610) 435-6973 http://www.bonneyforge.com
Capitol Mfg./Camco Ftgs. Co. A Harsco Company 1125 Capitol Ave. Crowley, LA 70526 U.S.A.	Pipe fittings	Telephone: (337) 783-8626 Fax: (337) 783-5360
Dresser Industries DMD Division 41 Fisher Ave. Bradford, PA 16701 U.S.A.	Pipe fittings	Telephone: (814) 362-9200 Fax: (814) 362-9388 http://www.dresser.com/dmd
Dynamic Products, Inc. 16520 Peninsula Blvd. Houston, TX 77015 U.S.A.	Pipe fittings	Telephone: (281) 457-3500 Fax: (281) 457-2879 Email: dpi@flash.net
Edward Vogt Valve Co. Vogt Valve Operations 5171 Maritime Rd. Jeffersonville, IN 47130 U.S.A.	Pipe fittings	Telephone: (812) 218-7700 Fax: (812) 218-7777
Flowline Division Markovitz Enterprises Inc. PO Box 7027 New Castle, PA 16107-7027 U.S.A.	Pipe fittings	Telephone: (724) 658-3711 Fax: (724) 658-6117
Flo-Mac, Inc. 1846 East 60th St. Los Angeles, CA 90001 U.S.A.	Pipe fittings	Telephone: (323) 583-8751

TABLE 29.22 Sources for Pipe Fittings (Continued)

Company name and address	Product/service	Contact
Gerlin, Inc. 170 Tubeway Dr. Carol Stream, IL 60188 U.S.A.	Pipe fittings	Telephone: (630) 690-7000 Fax: (630) 690-9701 Email: JB@gerlin.com
GR Piping Components, Inc. 4600 Allen St. Houston, TX 77077-3464 U.S.A.	Pipe fittings	Telephone: (713) 864-0100 Fax: (713) 869-6938
JCM Industries 200 Old Boston Rd. PO Box 1220 Nash, TX 75569-1220 U.S.A.	Pipe fittings	Telephone: (903) 832-2581 Fax: (903) 838-6260
Kedco Mfg. Ltd. 645 Keddco St. PO Box 999 Sarnia, Ontario N7T 7K6 Canada,	Pipe fittings	Telephone: (519) 336-2960 Fax: (519) 334-3181
Kingdom Precision Castings 011-886-7771-3121 c/o Alliance Metals Group, LLC 2700 Freeway Blvd., Suite 500 Minneapolis, MN 55430 U.S.A.	Pipe fittings	Telephone: (612) 585-1387 Fax: (612) 561-8558 http://www.super.com.tw
M.P. Industries, Inc. (MPI) 102 N. Cool Springs Rd. PO Box 190 O'Fallon, MO 63366 U.S.A.	Pipe fittings	Telephone: (636) 240-6196 Fax: (636) 272-3555 http://www.mpfittings.com
Mueller Co. (Headquarters) 500 West Eldorado St. Decatur, IL 62525 U.S.A.	Pipe fittings	Telephone: (217)-423-4471 Fax: (217) 425-7245 http://www.muellerflow.com
Mueller Industries Mueller Copper Fittings LP 3400 Mueller Brass Rd. Covington, TN 38019 U.S.A.	Pipe fittings	Telephone: (901) 476-5858 Fax: (901) 475-1334
NIBCO Inc. 1516 Middlebury St. PO Box 1167 Elkhart, IN 46515-1167 U.S.A.	Pipe fittings	Telephone: (219) 295-3000 Fax: (219) 523-3611 http://www.nibco.com
Pennsylvania Machine Works 100 Bethel Rd. Aston, PA 19014 U.S.A.	Pipe fittings	Telephone: (610) 497-3300 Fax: (610) 497-3325 http://www.pennusa.com
The Phoenix Forge Group 800 Front St. Catasauqua, PA 18032 U.S.A.	Pipe fittings	Telephone: (610) 264-2861 Fax: (610) 268-0530 http://www.phoenixforge.com
Premier Copper & Brass/Atlantic, Inc. 3797 Progress Rd. Norfork, VA 23502 U.S.A.	Pipe fittings	Telephone: (757) 857-7030 Fax: (757) 853-3843 Email: premier@series2000.com

TABLE 29.22 Sources for Pipe Fittings (Continued)

Company name and address	Product/service	Contact
Romac Industries, Inc. 21919 20th Ave. SE Suite 100 Bothell, WA 98021-4404 U.S.A.	Pipe fittings	Telephone: (206) 624-6491 Fax: (206) 624-1971 http://www.romacindustries.com
Shaw NAPTech, Inc. North AM Piping Technologies 851 South Freeport Indust Pkwy. Clearfield, UT 84015 U.S.A.	Pipe fittings	Telephone: (801) 773-7300 Fax: (801) 773-6185
Smith-Blair, Inc. 30 Globe Ave. Texarkana, AR 71854 U.S.A.	Pipe fittings	Telephone: (870) 773-5127 Fax: (870) 773-5212
Swagelok Companies 29500 Solon Rd. Solon, OH 44139 U.S.A.	Pipe fittings	Telephone: (440) 248-4600 Fax: (440) 349-5806 http://www.swagelok.com
Thomas & Betts Corporation 8155 T&B Blvd. Memphis, TN 38125 U.S.A.	Pipe fittings	Telephone: (901) 682-8221 Fax: (901) 252-1333 http://www.TNB.com
Trinity Industries Fittings and Flange Group PO Box 568887 Dallas, TX 75357-8887 U.S.A.	Pipe fittings	Telephone: (214) 589-7856 Fax: (214) 589-7594 http://www.tffg.net
Tyco Flow Control Research and Development Center 1467 Elmwood Ave. Cranston, RI 02910 U.S.A.	Pipe fittings	Telephone: (401) 781-1551 Fax: (401) 781-7653
Victaulic Co. of America 4901 Kesslersville Rd. PO Box 31 Easton, PA 18044 U.S.A.	Pipe fittings	Telephone: (610) 559-3300 Fax: (610) 250-8817 Email: victaulic@victaulic.com http://www.victaulic.com
Ward Manufacturing, Inc. 115 Gullic St. PO Box 9 Blossburg, PA 16912 U.S.A.	Pipe fittings	Telephone: (570) 638-2131 Fax: (570) 638-3358
Weldbend Corporation 6600 South Harlem Ave. Argo, IL 60501 U.S.A.	Pipe fittings	Telephone: (708) 594-1700 Fax: (708) 458-0106
WFI International, Inc. PO Box 7303 Houston, TX 77248 U.S.A.	Pipe fittings	Telephone: (713) 695-3633 Fax: (713) 695-3528 http://www.wfi-intl.com
T. D. Williamson, Inc. PO Box 2299 Tulsa, OK 74101-2299 U.S.A.	Pipe fittings	Telephone: (918) 254-9400 Telephone: (918) 446-1941 Fax: (918) 446-0398 http://www.tdwilliamson.com

TABLE 29.23 Sources for Pipe Flanges

Company name and address	Product/service	Contact
Alloy Stainless Products Co. Inc. 611 Union Blvd. Totowa, NJ 07512 U.S.A.	Pipe flanges	Telephone: (973) 256-1616 Telephone: (800) 631-8372 Fax: (973) 256-5256 Email: aspaba@aol.com
American Cast Iron Pipe Co. PO Box 2727 Birmingham, AL 35202 U.S.A.	Pipe flanges	Telephone: (205) 325-7856 Telephone: (205) 325-7862 http://www.acipco.com
Anvil International, Inc. Research & Development Center 20 Thurber Blvd. Smithfield, RI 02917 U.S.A.	Pipe flanges	Telephone: (401) 233-3927 Fax: (401) 781-9912
BendTec, Inc. 366 Garfield Ave. Duluth, MN 55802 U.S.A.	Pipe flanges	Telephone: (218) 722-0205 Fax: (218) 722-6598 Email: employeesname@bendtec.com
Bestweld, Inc. 40 Robinson St. Pottstown, PA 19464 U.S.A.	Pipe flanges	Telephone: (610) 718-9700 Fax: (610) 718-9800 Email: fittings@bestweld.com http://www.bestweld.com
Coffer Corp. 4444 Center St. Houston, TX 77007 U.S.A.	Pipe flanges	Telephone: (713) 868-4421 Telephone: (800) 231-0131 Fax: (713) 862-8445 http://www.coffercorp.com
Conbraco Industries PO Box 247 Matthews, NC 28106 U.S.A.	Pipe flanges	Telephone: (704) 841-6092 Fax: (704) 841-6021 http://www.conbraco.com
Gerlin, Inc. 170 Tubeway Dr. Carol Stream, IL 60188 U.S.A.	Pipe flanges	Telephone: (630) 690-7000 Fax: (630) 690-9701 Email: JB@gerlin.com
GR Piping Components, Inc. 4600 Allen St. Houston, TX 77077-3464 U.S.A.	Pipe flanges	Telephone: (713) 864-0100 Fax: (713) 869-6938
M.P. Industries, Inc. (MPI) 102 N. Cool Springs Road PO Box 190 O'Fallon, MO 63366 U.S.A.	Pipe flanges	Telephone: (636) 240-6196 Fax: (636) 272-3555 http://www.mpfittings.com
Nibco Inc. 1516 Middlebury St. PO Box 1167 Elkhart, IN 46515-1167 U.S.A.	Pipe flanges	Telephone: (219) 295-3000 Fax: (219) 523-3611 http://www.nibco.com
Premier Copper & Brass/Atlantic, Inc. 3797 Progress Rd. Norfork, VA 23502 U.S.A.	Pipe flanges	Telephone: (757) 857-7030 Fax: (757) 853-3843 Email: premier@series2000.com
Romac Industries, Inc. 21919 20th Ave. SE Suite 100 Bothell, WA 98021-4404 U.S.A.	Pipe flanges	Telephone: (206) 624-6491 Fax: (206) 624-1971 http://www.romacindustries.com

TABLE 29.23 Sources for Pipe Flanges (Continued)

Company name and address	Product/service	Contact
Smith-Blair, Inc. 30 Globe Ave. Texarkana, AR 71854 U.S.A.	Pipe flanges	Telephone: (870) 773-5127 Fax: (870) 773-5212
Trinity Industries Fittings and Flange Group PO Box 568887 Dallas, TX 75357-8887 U.S.A.	Pipe flanges	Telephone: (214) 589-7856 Fax: (214) 589-7594 http://www.tffg.net
Tyco Flow Control Research and Development Center 1467 Elmwood Ave. Cranston, RI 02910 U.S.A.	Pipe flanges	Telephone: (401) 781-1551 Fax: (401) 781-7653
Velan, Inc. 2125 Ward Ave. Montreal, Quebec H4M 1T8 Canada	Pipe flanges	Telephone: (514) 748-7743 Fax: (514) 341-3032 http://www.velan.com
Ward Manufacturing, Inc. 115 Gullic St. PO Box 9 Blossburg, PA 16912 U.S.A.	Pipe flanges	Telephone: (570) 638-2131 Fax: (570) 638-3358
Weldbend Corp. 6600 South Harlem Ave. Argo, IL 60501 U.S.A.	Pipe flanges	Telephone: (708) 594-1700 Fax: (708) 458-0106
WFI International, Inc. PO Box 7303 Houston, TX 77248 U.S.A.	Pipe flanges	Telephone: (713) 695-3633 Fax: (713) 695-3528 http://www.wfi-intl.com

TABLE 29.24 Sources for Valves

Company name and address	Product/service	Contact
American Valve PO Box 35229 Greensboro, NC 27425 U.S.A.	Ball, gate, globe, check, butterfly, gas service, radiation, low-pressure, plumbing, and industrial valves	Telephone: (800) 645-0101 Telephone: (336) 668-0554 Fax: (800) 876-0109 Fax: (336) 668-4070 http://www.americanvalve.com
Anderson Greenwood 3950 Greenbriar Stafford, TX 77477 U.S.A.	Pressure relief, tank blanketing products, instrumental valves, and manifolds	Telephone: (281) 274-4400 Fax: (281) 240-1800 http://www.andersongreenwood.com
Anderson Greenwood/Crosby 43 Kendrick St. Wrentham, MA 02093-0308 U.S.A.	Pressure relief, tank blanketing products, instrumental valves, and manifolds	Telephone: (508) 384-3121 Fax: (508) 384-8675 http://www.andersongreenwood.com
Anderson Greenwood Corrie Way Bredbury Industrial Estate Stockport, SK6 2ST United Kingdom	Pressure relief, tank blanketing products, instrumental valves, and manifolds	Telephone: 44-161-494-5363 Fax: 44-161-494-5672 http://www.andersongreenwood.com
Anderson Greenwood/Crosby Market Harborough Leics., LE16 9EE United Kingdom	Pressure relief, tank blanketing products, instrumental valves, and manifolds	Telephone: 44-185-846-7281 Fax: 44-185-843-4728 http://www.andersongreenwood.com
Anderson Greenwood/Crosby 45 Tuas Ave. 9 Singapore 63189	Pressure relief, tank blanketing products, instrumental valves, and manifolds	Telephone: 65-861-1655 Fax: 65-861-6997 http://www.andersongreenwood.com
Crosby Valve & Engineering Co. Ltd. Crosby Rd. Market Harborough LE169EE United Kingdom	Thermal relief, safety, assisted safety, and pilot valves. Specialty products: changeover valves, test benches, and in situ set pressure verification device	Telephone: 01858-467281 Fax: 01858-434728 http://www.energyweb.net
Atwood & Morrill Co., Inc. 285 Canal St. Salem, MA 01970-4595 U.S.A.	Free-flow reverse current valves, main steam isolation valves, and Tricentric® metal seated butterfly valves	Telephone: (978) 744-5690 Fax: (978) 741-3626 Email: sales@atmor.com http://www.weir-valve.com
Weir Valve Operations PO Box B27 Birkby, Huddersfield West Yorkshire HD2 2UR United Kingdom	Free-flow reverse current valves, main steam isolation valves, and Tricentric® metal seated butterfly valves	Telephone: +44 (0) 1484 820820 Fax: +44 (0) 1484 820484 Email: info@weir-valve.com http://www.weir-valve.com
Hopkinsons Ltd. P.O. Box 17154 Jebel Ali Free Zone Dubai UAE	Free-flow reverse current valves, main steam isolation valves, and Tricentric® metal seated butterfly valves	Telephone: +971 (4) 881 9868 Fax: +971 (4) 881 9870 Email: dubai@hopkinsons.co.uk http://www.weir-valve.com
Hopkinsons Ltd. Yang Ming Plaza A-14D No. 10 Xiao Ying Rd. Asian Games Village Chao Yang District Bejing 100101 PR China	Free-flow reverse current valves, main steam isolation valves, and Tricentric® metal seated butterfly valves	Telephone: +86 (10) 6492 6555,6525 Fax: +86 (10) 6492 8204 Email: china@hopkinsons.co.uk http://www.weir-valve.com
Hopkinsons Ltd. c/o Samyang Engineering Co Ltd 302 Cheungkwun Building 612 Sinsa-Dong Kangnam-Gu Seoul Korea	Free-flow reverse current valves, main steam isolation valves, and Tricentric® metal seated butterfly valves	Telephone: +82 (2) 515 2887 Fax: +82 (2) 516 6581 Email: korea@hopkinsons.co.uk http://www.weir-valve.com

TABLE 29.24 Sources for Valves (Continued)

Company name and address	Product/service	Contact
Hopkinsons Ltd. Unit 8-12-12 Menara Mutiara Bangsar Jalan Liku Off Jala Riong 59100 Kuala Lumpur Malaysia	Free-flow reverse current valves, main steam isolation valves, and Tricentric® metal seated butterfly valves	Telephone: +60 (3) 287 3560 Fax: +60 (3) 287 3561 Email: malaysia@hopkinsons.co.uk http://www.weir-valve.com
SEBIM 21 de la Palunette F 13220 Chateauneuf les Martigues France	Free-flow reverse current valves, main steam isolation valves, and Tricentric® metal seated butterfly valves	Telephone: +33 (0) 4 42 07 00 95 Fax: +33 (0) 4 42 07 11 77 Email: sebimgroup@compuserve.com http://www.weir-valve.com
RSBD Rue du Mouton F 18108 Vierzon Cedex France	Free-flow reverse current valves, main steam isolation valves, and Tricentric® metal seated butterfly valves	Telephone: +33 (0) 2 48 71 32 33 Fax: +33 (0) 2 48 71 09 45 Email: RSBD@compuserve.com http://www.weir-valve.com
Sarasin Rue Jean Baptiste Grison 21 du Bois Rigault F 62880 Vendin Le Vieil France	Free-flow reverse current valves, main steam isolation valves, and Tricentric® metal seated butterfly valves	Telephone: +33 (0) 3 21 79 54 50 Fax: +33 (0) 3 21 28 62 00 Email: sarasin@compuserve.com http://www.weir-valve.com
SEBIM Lilac Cottage Benover Rd. Yalding, Kent ME18 6AS United Kingdom	Free-flow reverse current valves, main steam isolation valves, and Tricentric® metal seated butterfly valves	Tel/Fax +44 (0) 892 730 668 Email: abradford@hopkinsons.com.uk http://www.weir-valve.com
Flowguard Limited Watford Bridge New Mills, High Peak Derbyshire, SK22 4HJ United Kingdom	Free-flow reverse current valves, main steam isolation valves, and Tricentric® metal seated butterfly valves	Telephone: +44 (0) 1663 745976 Fax: +44 (0) 1663 742788 Email: sales@flowguard.com http://www.weir-valve.com
Batley Valve Longlands Industrial Estate Ossett West Yorkshire, WF5 9JF United Kingdom	Free-flow reverse current valves, main steam isolation valves, and Tricentric® metal seated butterfly valves	Telephone: +44 (0) 1924 275931 Fax: +44 (0) 1924 280026 Email: batley@hopkinsons.co.uk http://www.weir-valve.com
AUMA USA 4 Zesta Drive Pittsburgh, PA 15205 U.S.A.	Electric multiturn valve actuators, electric quarter-turn valve actuators, electric linear actuators, electric damper drives, electric actuator controls, and gearboxes	Telephone: (412) 787-1340 Fax: (412) 787-1223 http://www.auma-usa.com
Emerson 8000 West Florissant Ave. P.O. Box 4100 St. Louis, MO 63136 U.S.A.	Fluid control and fluid power applications, specialized solenoid valves, pilot control valves, dribble control valves, and angle body piston valves	Telephone: (314) 553-2000 http://www.emersonelectric.com
Babbitt Steam Specialty Co. PO Box 51208 Mount Pleasant St. New Bedford, MA 02745 U.S.A.	Ball, gate, globe, check, butterfly, high-pressure, plug, pressure reducing, instrument relief, solenoid, and steam valves	Telephone: (800) 804-7563 Telephone: (508) 995-9533 Fax: (508) 955-2701 Email: tbssco@aol.com http://www.babbitt.thomasregional.com
Edward Vogt Valve Co. 1900 South Saunders St. Raleigh, NC 27603 U.S.A.	Gate, globe, and check valves and the Edward Vogt line of fittings and unions	Telephone: (919) 832-0525 Fax: (919) 831-3369 Email: eddie.vogt@btrinc.com http://www.edwardvalves.com
Worcester Controls 33 Locke Dr. PO Box 538 Marlborough, MA 01752 U.S.A.	Wafer, flanged, Series 45, Series 151/301, Series 51/52, Series 59, high-pressure/high-temp., and high-purity/high-vacuum valves	Telephone: (508) 481-4800 http://www.worcestercc.com

TABLE 29.24 Sources for Valves (Continued)

Company name and address	Product/service	Contact
Worcester Controls 20 Mid-Dominion Acres Scarborough, Ontario M1S 4A5 Canada	Wafer, flanged, Series 45, Series 151/301, Series 51/52, Series 59, high-pressure/high-temp., and high-purity/high-vacuum valves	Telephone: (416) 298-1671 http://www.worcestercc.com
Cashco, Inc. 607 W. 15th St. Ellsworth, KS 67439 U.S.A.	Pressure regulators and control valves	Telephone: (785) 472-4461 Fax: (785) 472-3539 Email: webmaster@cashco.com http://www.cashco.com
CCI World Headquarters 22591 Avenida Empresa Rancho Santa Margarita, CA 92688 U.S.A.	Pressurizer safety and relief valves, main steam isolation valves, safety control valves, relief control valves, feed water isolation, and damped feed water check valves	Telephone: (949) 858-1877 Fax: (949) 858-1878 Email: info@ccivalve.com http://www.ccivalve.com
Check-All Valve Manufacturing Co. 1800 Fuller Rd. West Des Moines, IA 50265 U.S.A.	Assorted valves	Telephone: (515) 224-2301 Fax: (515) 224-2326 Email: sales@checkall.com http://www.checkall.com
Conval 265 Fields Rd. Somers, CT 06071 U.S.A.	Globe valves, gate valves, actuated valves, bellows valves, quad-flow valves, throttling valves, WhisperJet valves, and AutoSeal valves	Telephone: (860) 749-0761 Fax: (860) 763-3557 Email: sales@conval.com http://www.conval.com
Cooper Cameron Valves 16500 South Main Missouri City, TX 77489-1300 U.S.A.	ANSI class valves and related systems used in the production, transmission, and refining of oil and gas products	Telephone: (800) 323-9160 Telephone: (281) 499-8511 Fax: (281) 499-5211 http://www.ccvalve.com
Copes-Vulcan Inc. 9843 Martin Ave. PO Box 577 Lake City, PA 16423-0577 U.S.A.	Severe-duty control valves, general-service control valves, trim/valve internals, actuators, and accessories	Telephone: (814) 744-1500 Fax: (814) 774-1681 Email: inquiry@copes.com http://www.copes.com
Copes-Vulcan Inc. Road Two, Industrial Estate Winsford, Cheshire CW7 3QL England	Severe-duty control valves, general-service control valves, trim/valve internals, actuators, and accessories	Telephone: 44-01606 552041 Fax: 44-01606 558275 Email: sales@cv1.co.uk http://www.copes.com
Stockham 2129 3rd Ave., SE Cullman, AL 35055 U.S.A.	Bronze, iron, ball, butterfly, and commercial bronze valves	Telephone: (800) stockham Telephone: (256) 775-3800 Fax: (256) 775-3860 http://www.stockham.com
Curtisss-Wright Flow Control Corporation 1966E Broadhollow Rd. E. Farmingdale, NY 11735 U.S.A.	Pressure relief valves, safety relief valves, solenoid valves, control valves, and specialty valves	Telephone: (516) 293-3800 Fax: (516) 293-3404 http://www.cwfc.com
DeZURIK 9843 Martin Ave. P.O. Box 577 Lake City, PA 16423 U.S.A.	Rotary control valves, V-port valves, globe control valves, butterfly valves, plug valves, knife gate valves, and metal seated ball valves	Telephone: (814) 774-1500 Fax: (814) 774-1681 Email: info@dezurikcopesvulcan.com http://www.dezurikcopesvulcan.com
DeZURIK Riverside Ave. North Sartell, MN 56377 U.S.A.	Rotary control valves, V-port valves, globe control valves, butterfly valves, plug valves, knife gate valves, and metal seated ball valves	Telephone: (320) 259-2000 Fax: (320) 259-2227 Email: info@dezurikcopesvulcan.com http://www.dezruikcopesvulcan.com
Durabla Fluid Technology PO Box 566 Exton, PA 19341-0566 U.S.A.	WLC® check, SCV check, DLC® check, Ecalibur® check, PDC® check, GLC® check, GPV® check, and basic check valves	Telephone: (610) 363-8903 Fax: (610) 524-9242 http://www.durabla-checkvalves.com

TABLE 29.24 Sources for Valves (Continued)

Company name and address	Product/service	Contact
DynaQuip Controls 10 Harris Industrial Park St. Clair, MO 63077 U.S.A.	Ball valves, valve actuators, and ball valve automation	Telephone: (800) 545-3636 Fax: (314) 629-5528 Email: info@dynaquip.com http://www.dynaquip.com
Flo-Tork, Inc. 1701 North Main St. PO Box 68 Orville, OH 44667-0068 U.S.A.	Hydraulic rotary actuators for nuclear-powered submarines and other naval vessels	Telephone: (330) 682-0010 Fax: (330) 682-6857 Email: executive@flo-tork.com http://www.flo-tork.com
Flow Safe, Inc. S-3865 Taylor Rd. Orchard Park, NY 14127 U.S.A.	Spring-operated safety valves, micro safety relief valves, liquid pressure relief valves, unloader valves, and pilot-operated pressure and safety relief valves	Telephone: (716) 667-3641 Telephone: (800) 828-1036 Fax: (716) 667-3642 Fax: (716) 662-2580 Email: info@flowsafe.com http://www.flowsafe.com
Flowserve-Corporation 222 W. Las Colinas Blvd. Suite 1500 Irving, TX 75039 U.S.A.	Automatic control valves, quarter-turn valves, nuclear gate and check valves, intelligence systems for valves, actuators, and valve automation systems	Telephone: (972) 443-6500 Fax: (972) 443-6800 http://www.flowserve.com
Flow-Tek, Inc. 7404 Fairfield Rd. Columbia, SC 29203 U.S.A.	Ball valves	Telephone: (803) 754-8201 Fax: (803) 754-2501 Email: flotek@flow-tek.com http://www.flow-tek.com
Groth Corporation 1202 Hahlo St. PO Box 15293 Houston, TX 77220-5293 U.S.A.	Pilot-operated relief valves, blanket gas regulators, and pressure/vacuum relief valves	Telephone: (713) 675-6151 Fax: (713) 675-3714 Email: gpg@grothcorp.com http://www.grothcorp.com
High Pressure Equipment Co. PO Box 8248 1222 Linden Ave. Erie, PA 16505 U.S.A.	Taper seal valves, high-pressure valves, specialty valves, medium-pressure valves, and ultrahigh-pressure valves	Telephone: (814) 838-2028 Telephone: (800) 289-7447 Fax: (814) 838-6075 http://www.highpressure.com
Hunt Valve Company, Inc. 1913 E. State St. Salem, OH 44460 U.S.A.	Mega-Flo descale spray valves, combo directional control valves, Mod-Stack valves, poppet valves, bypass valves, industrial air valves, pneumatic valve couplings, and safety lockout valves	Telephone: (330) 337-9535 Fax: (330) 337-3754 Email: sales@huntvalve.com http://www.huntvalve.com
Hydroseal 1610 U.S. Hwy. 259`N Kilgore, TX U.S.A.	Ball and seat pressure relief valves, relief valves, and hydro safety and relief valves	Telephone: (903) 984-8574 Telephone: (800) 256-8574 Fax: (903) 984-6307 Email: info@hydroseal.com http://www.hydroseal.com
Jamesbury Inc. Headquarters 640 Lincoln St. Box 15004 Worcester, MA 01615-0004 U.S.A.	Value-Line ball valves, high-performance valves, special service ball valves, ball valve options and accessories, high-performance butterfly valves, and special service butterfly valve options and accessories	Telephone: (508) 852-0200 Fax: (508) 852-8172 http://www.jamesbury.com
Process Development & Control 1075 Montour West Industrial Park Coraopolis, PA 15108 U.S.A.	Econ-o-trol Series 09, medium-duty Series 01, thread-end Series 07, flanged-end Series 39, Excel Series 40, reduced port Series 15, elastomer-lined Series 27, and combustion valve Series 22 valves	Telephone: (724) 695-3400 Fax: (724) 695-8635 Email: pdcvalve@worldnet.att.net http://www.pdcvalve.com

TABLE 29.24 Sources for Valves (Continued)

Company name and address	Product/service	Contact
Kerotest Manufacturing Corp. 5500 Second Ave. Pittsburgh, PA 15207 U.S.A.	Gate valves, ball valves, needle and gage valves, and globe valves	Telephone: (412) 521-7681 Telephone: (412) 521-7263 Fax: (412) 521-5990 Email: kerotest@kerotest.com http://www.kerotest.com
MCF Valves, Inc. 3409 Brinkman Houston, TX 77018 U.S.A.	Manual and automated valves, including lever, hi-rise, and locking type handles and accessories	Telephone: (713) 869-3100 Fax: (713) 869-5511 http://www.mcfvalves.com
Jordan Valve 3170 Wasson Rd. Cincinnati, OH 45209 U.S.A.	Pneumatic control valves, electric control valves, precision low-flow control valves, sanitary valves, pinch valves, and disk valves	Telephone: (513) 533-5600 Fax: (513) 871-0105 Email: info@richardsind.com http://www.jordanvalve.com
Colesco Inc. 6060 Interstate Circle Cincinnati, OH 45242 U.S.A.	Jordan Valve representative; pneumatic control valves, electric control valves, precision low-flow control valves, sanitary valves, pinch valves, and disk valves	Telephone: (513) 489-1122 Fax: (513) 489-9062 http://www.jordanvalve.com
Gibson Sales, Inc. P.O. Box 450 Chagrin Falls, OH 44022 U.S.A.	Jordan Valve representative; pneumatic control valves, electric control valves, precision low-flow control valves, sanitary valves, pinch valves, and disk valves	Telephone: (440) 543-0300 Fax: (440) 543-1230 http://www.jordanvalve.com
Tri-State Technical Corp. 3601 Commerce Dr. Bldg. 109 Baltimore, MD 21217 U.S.A.	Jordan Valve representative; pneumatic control valves, electric control valves, precision low-flow control valves, sanitary valves, pinch valves, and disk valves	Telephone: (410) 536-4242 Fax: (410) 536-0583 http://www.jordanvalve.com
Samson Controls Inc. 4111 Cedar Blvd. Baytown, TX 77520-8588 U.S.A.	Control valves, actuators, positioners, controllers, process control equipment, lined globe valves, lined butterfly valves, and eccentric disk valves	Telephone: (281) 383-3677 Fax: (281) 383-3690 Email: samson@samson-usa.com http://www.samson-usa.com
Nu-Torque, Inc. 31206 NE Cherry Valley Rd. Duvall, WA 98019-7732 U.S.A.	Marine valve automation specialists	Telephone: (800) 798-5026 Telephone: (425) 788-6275 Fax: (425) 788-7948 http://www.nu-torque.com
PBM, Inc. 1070 Sandy Hill Rd. Irwin, PA 15642 U.S.A.	Ball valves, rotor valves, and actuators for controlling and automating process lines	Telephone: (800) 967-4PBM Telephone: (724) 863-0550 Fax: (724) 864-9255 http://www.pbmvalve.com
Powell Valves 2503 Spring Grove Ave. P.O. Box 145434 Cincinnati, OH 45250-5434 U.S.A.	Bronze valves, iron valves, bolted bonnet cast steel valves, pressure seal cast steel valves, corrosion-resistant steel valves, and cryogenic valves	Telephone: (513) 852-2000 Fax: (513) 852-2997 Email: wmpowell@msn.com http://www.powellvalves.com
Varec Vapor Control Inc. 10800 Valley View St. Cypress, CA 90630-5016 U.S.A.	Relief valves and back-pressure check valves	Telephone: (714) 761-1300 Fax: (714) 952-2701 http://www.varec.com
H.O. Trerice Co. 12950 W. Eight Mile Rd. Oak Park, MI 48237-3288 U.S.A.	Globe valves, ball valves, rotary control valves, E-ball rotary control valves, and temperature and pressure instrumentation and controls	Telephone: (248) 399-8000 Fax: (248) 399-7246 http://www.hotrericeco.com

TABLE 29.24 Sources for Valves (Continued)

Company name and address	Product/service	Contact
Townley Manufacturing Company, Inc. P.O. Box 221 Candler, FL 32111 U.S.A.	Stainless steel, rubber-seated, Towniprene® lined and ductile iron knife gate valves, wafer-style, lug-style, and super slurry butterfly valves	Telephone: (800) 342-9920 Fax: (352) 687-1862 Email: townley@townley.net http://www.townley.net
Warren Controls Corporation 2157 Rt. 57 Broadway, NJ 08808 U.S.A.	Products for commercial, industrial, and military markets	Telephone: (908) 689-1400 Fax: (908) 689-6839 http://www.warrencontrols.com
Walworth 8383 Commerce Park Dr. Suite 608 Houston, TX 77036 U.S.A.	Cast steel, forged steel, and bronze and iron gate globe, and check valves, lubricated plug valves, pressure and safety relief valves, industrial ball and butterfly valves, and waterworks valves	Telephone: (800) 735-6007 Telephone: (713) 777-7788 Fax: (800) 736-4392 Fax: (713) 981-1246 Email: walworth@walworthusa.com http://www.walworthvalves.com
Walworth de Mexico S.A. de C.A. Avenida de la Industria Lote 16 Fraccionamiento Industrial El Trebol Tepotzotlan 54600 Estado de Mexico Mexico	Cast steel, forged steel, and bronze and iron gate, globe, and check valves, lubricated plug valves, pressure and safety relief valves, industrial ball and butterfly valves, and waterworks valves	Telephone: (5) 899-1700 and (5) 876-0100 Fax: (5) 899-1782 and (5) 899-1703 Email: walworth@mail.internet.com.mx http://www.walworthvalves.com
Yarway PO Box 350 480 Norristown Rd. Blue Bell, PA 19422-0760 U.S.A.	Blow off valves, throttling valves, globe valves, water and liquid level gages, and electronic level detection systems	Telephone: (610) 825-2100 Fax: (610) 825-1235 http://www.yarway.com
Val-Matic Valve & Manufacturing Corp. 905 Riverside Dr. Elmhurst, IL 60126 U.S.A.	Air valves, check valves, eccentric plug valves, and foot valves for positive pump prime on negative suction systems	Telephone: (630) 941-7600 Fax: (630) 941-8042 http://www.val-matic.com
The Valv-Trol Co. 1340 Commerce Dr. PO Box 2259 Stow, OH 44224-1000 U.S.A.	High-pressure water valves, accessories, and services	Telephone: (800) 497-2136 Telephone: (330) 686-2800 Fax: (330) 686-2820 Email: sales@valv-trol.com http://www.valv-trol.com
Victaulic World Headquarters PO Box 31 4901 Kesslersville Rd. Easton, PA 18044-0031 U.S.A.	Butterfly, ball, circuit balancing, check, swing check, and plug valves	Telephone: (610) 559-3300 Fax: (610) 250-8817 http://www.victaulic.com
Valvetechnologies, Inc. 5904 Bingle Rd. Houston, TX 77092 U.S.A.	Flanged ball valves, weld end ball valves, Xactrol control valves, electronic relief valves, parallel slide valves, and specialty valves	Telephone: (713) 860-0400 Fax: (713) 860-0499 Email: sales@valv.com http://www.valv.com
Valcor Engineering Corporation 2 Lawrence Rd. Springfield, NJ 07081 U.S.A.	Dri-Solenoid valves, Teflon® solenoid valves, multipurpose solenoid valves, cryogenic valves, and inert instrument-class valves	Telephone: (973) 467-8400 Fax: (973) 467-8382/(973) 467-4905 Email: aerospace@valcor.com http://www.valcor.com
Nibco International 1516 Middlebury St. PO Box 1167 Elkhart, IN 46515-1167 U.S.A.	Ball, butterfly, gate, globe, angle, and check valves, plumbing and heating valves, flanged butterfly valves, bronze ball valves, carbon and stainless steel ball valves	Telephone: (219) 295-3221 Fax: (219) 523-3632 http://www.nibco.com

TABLE 29.24 Sources for Valves (Continued)

Company name and address	Product/service	Contact
Nibco Sp. Z o.o. UL Rokicinska 299-301 92-614 Lodz Poland	Ball, butterfly, gate, globe, angle, and check valves, plumbing and heating valves, flanged butterfly valves, bronze ball valves, carbon and stainless steel ball valves	Telephone: +48/42 6488192 Fax: +48/42 6488208 Email: nibco@nibco.com.pl http://www.nibco.com
PA Ventiler AB Elementvagen 1 437 36 Lindome Sweden	Ball, butterfly, gate, globe, angle, and check valves, plumbing and heating valves, flanged butterfly valves, bronze ball valves, carbon and stainless steel ball valves	Telephone: +46/31 992500 Fax: +46/31 992503 http://www.nibco.com
NCI Marketing 280A Industrial Parkway South Aurora, Ontario L4G 3T9 Canada	Ball, butterfly, gate, globe, angle, and check valves, plumbing and heating valves, flanged butterfly valves, bronze ball valves, carbon and stainless steel ball valves	Telephone: (905) 727-5545 Fax: (905) 727-4088 Email: admin@nci-marketing.com http://www.nibco.com
Nibco Pacific (Flow Controls) Ltd. Unit A, 16 Floor Empireland Commercial Bldg. 81-85 Lockhart Rd. Wanchai Hong Kong	Ball, butterfly, gate, globe, angle, and check valves, plumbing and heating valves, flanged butterfly valves, bronze ball valves, carbon and stainless steel ball valves	Telephone: +852/2512 8398 Fax: +852/2570 9428 Email: sales@nibcofc.com.hk http://www.nibco.com
Nibco de Mexico Alfonso Reyes 4080 NTE Monterrey, N.L. 64500 Mexico	Ball, butterfly, gate, globe, angle, and check valves, plumbing and heating valves, flanged butterfly valves, bronze ball valves, carbon and stainless steel ball valves	Telephone: +528/351-6308 Fax: +528/351-3180 Email: nibcomexico@infosel.net.mx http://www.nibco.com
Engineered Valves 33 Centerville Road PO Box 6164 Lancaster, PA 17603 U.S.A.	Diaphragm valves, ball valves, knife-gate, slide gate, and custom fabricated valves, hygienic diaphragm valves, plastic-lined valves, and burner safety shutoff valves	Telephone: (800)-366-1111 Telephone: (717) 509-2200 Fax: (717) 509-2336 http://www.engvalves.com
Engineered Valves 725 Cochran St., Unit E Simi Valley, CA 93065 U.S.A.	Diaphragm valves, ball valves, knife-gate, slide gate, and custom fabricated valves, hygienic diaphragm valves, plastic-lined valves, and burner safety shutoff valves	Telephone: (800) 926-884 Telephone: (805) 520-7200 Fax: (805) 520-7205 http://www.engvalves.com
Engineered Valves Av. 11 de Septiembre #1363 Of. 1403 Providenica Santiago Chile	Diaphragm valves, ball valves, knife-gate, slide gate, and custom-fabricated valves, hygienic diaphragm valves, plastic-lined valves, and burner safety shutoff valves	Telephone: +56-2-264-9491 Telephone: +56-2-264-9637 Fax: +56-2-236-1799 http://www.engvalves.com
Engineered Valves Urb. Valle San Luis #188 Calle Via de la Colina Caguas, PR 00725 Puerto Rico	Diaphragm valves, ball valves, knife-gate, slide gate, and custom-fabricated valves, hygienic diaphragm valves, plastic-lined valves, and burner safety shutoff valves	Ph./Fax: (787) 703-0402 http://www.engvalves.com
Engineered Valves Insurgentes SUR No. 670–7° P. Col. Del Valle C.P. 03100 Mexico D.F. Mexico	Diaphragm valves, ball valves, knife-gate, slide gate, and custom-fabricated valves, hygienic diaphragm valves, plastic-lined valves, and burner safety shutoff valves	Telephone: +52-5-669-5002 Fax: +52-5-669-5289 http://www.engvalves.com

TABLE 29.24 Sources for Valves (Continued)

Company name and address	Product/service	Contact
Engineered Valves Richards Street, Kirkham Lancashire PR4 2HU United Kingdom	Diaphragm valves, ball valves, knife-gate, slide gate, and custom-fabricated valves, hygienic diaphragm valves, plastic-lined valves, and burner safety shutoff valves	Telephone: +44-1772-682696 Fax: +44-1772-686006 http://www.engvalves.com
Engineered Valves Units 1903-04 CRE Centre 928-930 Cheung Sha Wan Road Hong Kong	Diaphragm valves, ball valves, knife-gate, slide gate, and custom-fabricated valves, hygienic diaphragm valves, plastic-lined valves, and burner safety shutoff valves	Telephone: +852-2741-6302 Fax: +852-2741-6605 http://www.engvalves.com
Nil-Cor®-Operations PO Box 2058 Alliance, OH 44601 U.S.A.	Ball valves, butterfly valves, check valves, ceramic valves, and automated/control valves	Telephone: (800) 321-9865 Telephone: (330) 823-0500 Fax: (330) 821-6722 Email: info@nilcor.com http://www.nilcor.com
J. R. Clendenning Co. PO Box 50 Royersford, PA 19468 U.S.A.	Backflow preventers, automated control valves, check and foot valves	Telephone: (800) 860-5726 Telephone: (610) 495-5600 Fax: (800) 495-3072 Fax: (610) 495-3072 Email: jrclen@aol.com http://www.flowmatic.com
Flowmatic Valves 145 Murray St. Glens Falls, NY 12801-4424 U.S.A.	Backflow preventers, automated control valves, check and foot valves	Telephone: (800) 833-2040 Telephone: (518) 761-9797 Fax: (800) 314-3155 Fax: (518) 761-9798 Email: flowmatic@flowmatic.com http://www.flowmatic.com
Indelac Controls, Inc. 6810 Powerline Dr. Florence, KY 41042 U.S.A.	Rotary electric actuators, electric spring return and electric fail-safe devices	Telephone: (800) 662-9424 Email: indelac@one.net http://www.indelac.com
Limitorque PO Box 11318 5114 Woodall Rd. Lynchburg, VA 24506-1318 U.S.A.	Electric valve actuators	Telephone: (804) 528-4400 Fax: (804) 845-9736 http://www.limitorque.com
Limitorque Corp. Trinity House, Kennet Side Newbury Berkshire, RG14 5EH England	Electric valve actuators	Telephone: 44-1-635-46999 Fax: 44-1-635-36034 http://www.limitorque.com
Limitorque Nippon Gear Co., Ltd. Asahi-Seimei Bldg. 9F 1-11-11 Kita-Saiwai, Nishi-Ku Yokohama-Shi (220-0004) Japan	Electric valve actuators	Telephone: 045-326-2065 Fax: 045-320-5962 http://www.limitorque.com
Limitorque Corp. 8901 Jameel Rd. Suite 180 Houston, TX 77040 U.S.A.	Electric valve actuators	Telephone: (713) 690-1960 Fax: (713) 690-1277 http://www.limitorque.com
Limitorque India, Ltd. 302, Mansarovar 90 Nehru Place New Delhi 110019 India	Electric valve actuators	Telephone: 91-11-6431-748 Fax: 91-11-6432-749 http://www.limitorque.com

TABLE 29.24 Sources for Valves (Continued)

Company name and address	Product/service	Contact
Limitorque Australia, Ltd. Division of Invensys Flow Control Australasia Pty., Ltd. 17 Scoresby Road Bayswater, Victoria 3153 Australia	Electric valve actuators	Telephone: 613-9729-0555 Fax: 613-9729-8225 http://www.limitorque.com
Limitorque Asia Pty., Ltd. 48A Loyang Way Singapore 508741	Electric valve actuators	Telephone: 65-214-2120 Fax: 65-214-2123 http://www.limitorque.com
Equipment Group, Inc. 85 Bodwell St. Avon, MA 02322-1190 U.S.A.	Globe-style valves, angle valves, special application valves, control valves	Telephone: (508) 586-4600 Fax: (508) 427-8971 http://www.halliburton.com
Dresser Italia S.p.A Via Italo Betto, 11 27058 Voghera PV Italy	Globe-style valves, angle valves, special application valves, control valves	Telephone: +39 0383 6911 Fax: +39 0383 367 166 Email: marketing@dvcd.com http://www.power-technology.com
Dresser Equipment Group, Inc. 11100 West Airport Blvd. Stafford, TX 77477 U.S.A.	Globe-style valves, angle valves, special application valves, control valves	Telephone: (281) 568-2211 Fax: (281) 568-1414 http://www.power-technology.com
Dresser Valve Division Far East Headquarters Business Suite 19A-9-1, Level 9 UOA Centre, No. 19 Jalan Pinang 50450 Kuala Lumpur West Malaysia	Globe-style valves, angle valves, special application valves, control valves	Telephone: +60-3-2163-2322 Fax: +60-3-2163-6312 http://www.power-technology.com
Dresser Valve Divison Colombres 29 10° A Buenos Aires Argentina	Globe-style valves, angle valves, special application valves, control valves	Tel/Fax: +54-11-4982-7378 http://www.power-technology.com
Dresser Valve Europe Hans Kudlich-Strasse 35A2100 Korneuburg (b. Wien) Austria	Globe-style valves, angle valves, special application valves, control valves	http://www.power-technology.com
Rotork Controls (Canada) Ltd. No. 1-2, 5650 Tomken Rd. Mississauga, ON L4W 4PI Canada	Electric and fluid power valve actuators and associated control systems	Telephone: (905) 602-5665 Fax: (905) 602-5669 http://www.rotork.com
Ontario Pollution Control Equip. Assoc. Box 137, Midhurst Ontario L0L 1X0 Canada	Electric and fluid power valve actuators and associated control systems	Telephone: (705) 725-0917 Fax: (705) 725-1068 Email: opcea@opcea.com http://www.opcea.com
Bettis 18703 GH Circle PO Box 508 Waller, TX 77484 U.S.A.	Actuators and controls	Telephone: (281) 463-5100 Fax: (281) 463-5103 http://www.bettis.com
Velan Proquip 835 Fourth Line Oakville, Ontario L6L 5B8 Canada	Wafer check valves	Telephone: (905) 842-1721 Fax: (905) 849-0923 http://www.proquipvalve.com
Mueller Co. 500 West Eldorado St. PO Box 671 Decatur, IL 62525 U.S.A.	Water distribution products, gas distribution products, Hersey water meters, gate valves, and butterfly valves	Telephone: (800) 423-1323 http://www.muellerflo.com

TABLE 29.24 Sources for Valves (Continued)

Company name and address	Product/service	Contact
Spirax Sarco, Inc. 1125 S. Cedar Crest Blvd. Suite 101 Allentown, PA 18103 U.S.A.	Hygienic ball valves	Telephone: (800) 251-7676 Fax: (800) 966-3232 http://www.spiraxsarco-usa.com
M. R. Franceschini, Inc. San Alberto Building 605 Condado Ave. Suite 307 San Juan, PR 00907-3823 Puerto Rico	Hygienic ball valves	Telephone: (787) 725-7080 Fax: (787) 721-5762 http://www.spiraxsarco-usa.com
Tyco International Inc. One Tyco Park Exeter, NH 03833 U.S.A.	Butterfly valves, hygienic butterfly valves, check valves, knife-gate valves, high-performance butterfly valves	Telephone: (603) 778-9700 http://www.tyco.com
Tyco Valves & Controls Distribution Tollaan 105 A/B 1932 St. Stevens Woluwe Belgium	Butterfly valves, hygienic butterfly valves, check valves, knife-gate valves, high-performance butterfly valves	Telephone: +32 2 725 0090 Email: tvcbrussels@Tyco-Valves.com http://www.tyco.com
Tyco Valves & Controls Distribution Na Petrinach 55 16100 Praha 6 Czech Republic	Butterfly valves, hygienic butterfly valves, check valves, knife-gate valves, high-performance butterfly valves	Telephone: +44 220 610379 Email: tvceasteur@Tyco-Valves.com http://www.tyco.com
K. S. Kaalund A/S Int. Bregnerodvej 102 3460 Birkerod Denmark	Butterfly valves, hygienic butterfly valves, check valves, knife-gate valves, high-performance butterfly valves	Telephone: +45 45 816511 Email: kaalund@Tyco-Valves.com http://www.tyco.com
Keyfin Oy Paalukatu 1 53500 Lappeenranta Finland	Butterfly valves, hygienic butterfly valves, check valves, knife-gate valves, high-performance butterfly valves	Telephone: +358 5 6146400 http://www.tyco.com
Tyco Valves & Controls Distribution 26 bis, Avenue de Paris 95230 Soisy-sous-Montmorency France	Butterfly valves, hygienic butterfly valves, check valves, knife-gate valves, high-performance butterfly valves	Telephone: +33 139 343536 Email: tvcparis@Tyco-Valves.com http://www.tyco.com
Tyco Valves & Controls Distribution Nobelstrasse 14 41189 Monchengladbach Germany	Butterfly valves, hygienic butterfly valves, check valves, knife-gate valves, high-performance butterfly valves	Telephone: +49 2166 9550 Email: tvcmgb@Tyco-Valves.com http://www.tyco.com
Tyco Valves & Controls Distribution Vaci Ut 202 II/201 1138 Budapest Hungary	Butterfly valves, hygienic butterfly valves, check valves, knife-gate valves, high-performance butterfly valves	Telephone: +36 1 3502966 Email: tvceasteur@Tyco-Valves.com http://www.tyco.com
Tyco Valves & Controls Distribution Via Piacenze 29018 Lugagnano Val d'Arda Piacenza Italy	Butterfly valves, hygienic butterfly valves, check valves, knife-gate valves, high-performance butterfly valves	Telephone: +39 052 3890201 Email: tvclugagn@Tyco-Valves.com http://www.tyco.com
Tyco Valves & Controls Distribution PO Box 22745 Sharjah United Arab Emirates	Butterfly valves, hygienic butterfly valves, check valves, knife-gate valves, high-performance butterfly valves	Telephone: +971 6 555 3699 Email: tvcuae@Tyco-Valves.com http://www.tyco.com
Tyco Valves & Controls Distribution PO Box 1017 31431 Dammam Saudi Arabia	Butterfly valves, hygienic butterfly valves, check valves, knife-gate valves, high-performance butterfly valves	Telephone: +966 3 882 6365 Email: tvcsarabia@Tyco-Valves.com http://www.tyco.com
Tyco Valves & Controls Distribution Mijkenbroek 22 4824 AB Breda Netherlands	Butterfly valves, hygienic butterfly valves, check valves, knife-gate valves, high-performance butterfly valves	Telephone: +31 76 5434100 Email: tvcbreda@Tyco-Valves.com http://www.tyco.com

TABLE 29.24 Sources for Valves (Continued)

Company name and address	Product/service	Contact
Tyco Valves & Controls Distribution Ringeriksveien 241 1313 Skui (Oslo) Norway	Butterfly valves, hygienic butterfly valves, check valves, knife-gate valves, high-performance butterfly valves	Telephone: +47 67 151500 Email: tvcoslo@Tyco-Valves.com http://www.tyco.com
Tyco Valves & Controls Distribution Oddzial w Polsce Al Jerozolimskie 202 02-486 Warsaw Poland	Butterfly valves, hygienic butterfly valves, check valves, knife-gate valves, high-performance butterfly valves	Telephone: +48 22 8740186 Email: tvceasteur@Tyco-Valves.com http://www.tyco.com
Industra Estrada de Albarraque 5 Centro Empresarial Sintra-Estoril Bloco A3 Linho 2710-297 Sintra, Portugal	Butterfly valves, hygienic butterfly valves, check valves, knife-gate valves, high-performance butterfly valves	Telephone: +351 21 9109500 Email: industra@Tyco-Valves.com http://www.tyco.com
Tyco Valves & Controls Distribution ul. Maroseika 17/6 101000 Moscow Center Russia	Butterfly valves, hygienic butterfly valves, check valves, knife-gate valves, high-performance butterfly valves	Telephone: +7 095 956 3830 Satellite: +7 503 956 3830 Email: tvcmoscow@Tyco-Valves.com http://www.tyco.com
Measure Tech (Pty.) Ltd. Ground Floor, Helvetia House Greenvale Road, Wilbart 1610 Germiston Edenvale South Africa	Butterfly valves, hygienic butterfly valves, check valves, knife-gate valves, high-performance butterfly valves	Telephone: +27 11 455 5710 Email: measuretech@Tyco-Valves.com http://www.tyco.com
Tyco Valves & Controls Distribution Ctra. Sabadell-Rubi, km. 13 Poligon Can Roses, Nau 70 08191 Rubi Barcelona, Spain	Butterfly valves, hygienic butterfly valves, check valves, knife-gate valves, high-performance butterfly valves	Telephone: +34 93 5881676 Email: tvcrubi@Tyco-Valves.com http://www.tyco.com
Erichs Armatur Travbanegatan 8 20039 Malmo Sweden	Butterfly valves, hygienic butterfly valves, check valves, knife-gate valves, high-performance butterfly valves	Telephone: +46 4031 1550 Email: erichs@Tyco-Valves.com http://www.tyco.com
Neotecha Ag Etzelstrasse 37-39 8634 Hombrechtikon Switzerland	Butterfly valves, hygienic butterfly valves, check valves, knife-gate valves, high-performance butterfly valves	Telephone: +41 55 254 4141 Email: neotecha@Tyco-Valves.com http://www.tyco.com
Tyco Valves & Controls Distribution UK Unit 9 Acorn Business Centre Heaton Lane Stockport, SK4 1AS Cheshire United Kingdom	Butterfly valves, hygienic butterfly valves, check valves, knife-gate valves, high-performance butterfly valves	Telephone: +44 161 477 5524 Email: tvcmachester@Tyco-Valves.com http://www.tyco.com
DynaTorque 1934 E. Sherman Blvd. Muskegon, MI 49444 U.S.A.	Industrial quarter-turn gear operators used with ball valves, high-performance butterfly valves, and plug valves, usually designed as ANSI 150 valves and higher	Telephone: (231) 739-1431 Fax: (231) 739-4514 http://www.dynatorque.com
Velan Valve Corp. 7007 Cote de Liesse Montreal, QC H4T 1G2 Canada	Forged steel and cast steel gate, globe, and check valves, bellows seal valves, knife-gate valves, butterfly valves, general-purpose ball valves, cryogenic valves	Telephone: (514) 748-7743 Fax: (514) 748-8635 Email: info@velan.com http://www.velan.com
Velan Valve Corp. 208 Lindbergh Ave. Wilmington, DE 19804-3316 U.S.A.	Forged steel and cast steel gate, globe, and check valves, bellows seal valves, knife-gate valves, butterfly valves, general-purpose ball valves, cryogenic valves	Telephone: (302) 994-1176 Fax: (302) 994-6156 http://www.velan.com

TABLE 29.24 Sources for Valves (Continued)

Company name and address	Product/service	Contact
Velan Valve Corp. 289 Antioch Rd. Lake Lure, NC 28746 U.S.A.	Forged steel and cast steel gate, globe, and check valves, bellows seal valves, knife-gate valves, butterfly valves, general-purpose ball valves, cryogenic valves	Telephone: (828) 625-9441 Fax: (828) 625-9437 http://www.velan.com
Velan Valve Corp. 45 W. Franklin St. Bellbrook, OH 45305 U.S.A.	Forged steel and cast steel gate, globe, and check valves, bellows seal valves, knife-gate valves, butterfly valves, general-purpose ball valves, cryogenic valves	Telephone: (937) 848-2011 Fax: (937) 848-2144 http://www.velan.com
Velan Valve Corp. 1343 Paddock Place Bartlett, IL 60103 U.S.A.	Forged steel and cast steel gate, globe, and check valves, bellows seal valves, knife-gate valves, butterfly valves, general-purpose ball valves, cryogenic valves	Telephone: (630) 736-9395 Fax: (630) 736-9397 http://www.velan.com
Velan Valve Corp. 21931 Harborbreeze Lane Huntington Beach, CA 92646 U.S.A.	Forged steel and cast steel gate, globe, and check valves, bellows seal valves, knife-gate valves, butterfly valves, general-purpose ball valves, cryogenic valves	Telephone: (714) 965-6277 Fax: (714) 965-6737 http://www.velan.com
Velan Valve Corp. 8825 51st Ave. Edmonton, AB T6E 5H1 Canada	Forged steel and cast steel gate, globe, and check valves, bellows seal valves, knife-gate valves, butterfly valves, general-purpose ball valves, cryogenic valves	Telephone: (780) 465-1122 Fax: (780) 465-0403 http://www.velan.com
Velan Valves Ltd. Cambridge Rd. Whetstone, Leicester LE8 6LH United Kingdom	Forged steel and cast steel gate, globe, and check valves, bellows seal valves, knife-gate valves, butterfly valves, general-purpose ball valves, cryogenic valves	Telephone: 44-116-275-0206 Fax: 44-116-275-0224 http://www.velan.com
Velan S.A. 90 rue Challemel Lacour 69367, Lyon Cedex 7 France	Forged steel and cast steel gate, globe, and check valves, bellows seal valves, knife-gate valves, butterfly valves, general-purpose ball valves, cryogenic valves	Telephone: (33) 4 78 61 67 00 Fax: (33) 4 78 72 12 18 http://www.velan.com
Velan GmbH Daimlerstrasse 8 Willich, D-47877 Germany	Forged steel and cast steel gate, globe, and check valves, bellows seal valves, knife-gate valves, butterfly valves, general-purpose ball valves, cryogenic valves	Telephone: (49) 2154-4938-0 Fax: (49) 2154-4938-99 http://www.velan.com
Velan Valve PO Box 61 5400 AB Uden The Netherlands	Forged steel and cast steel gate, globe, and check valves, bellows seal valves, knife-gate valves, butterfly valves, general-purpose ball valves, cryogenic valves	Telephone: +31 62952 1802 Fax: +31 41325 3321 http://www.velan.com
Velan Valve Room 2906 of Newton No. 3 Building No. 88 Jianguo Road Chaoyang District Bejing 100022 P.R. China	Forged steel and cast steel gate, globe, and check valves, bellows seal valves, knife-gate valves, butterfly valves, general-purpose ball valves, cryogenic valves	Telephone: 86-10-65665088, 65665089 Fax: 86-10-65675744 http://www.velan.com

TABLE 29.24 Sources for Valves (Continued)

Company name and address	Product/service	Contact
Velan Valve Block A-1, Flat 604 Glaxo Apartments Mayur Vihar-PH-1, Delhi-110091 India	Forged steel and cast steel gate, globe, and check valves, bellows seal valves, knife-gate valves, butterfly valves, general-purpose ball valves, cryogenic valves	Telephone: 91-11-271-2196 Fax: 91-11-225-3533 http://www.velan.com
Velan Valve 4-31-3 Kamimeguro Meguro-ku, Tokyo 153-0051 Japan	Forged steel and cast steel gate, globe, and check valves, bellows seal valves, knife-gate valves, butterfly valves, general-purpose ball valves, cryogenic valves	Telephone: 81-3-3792-1891 Fax: 81-3-3792-1891 http://www.velan.com
Velan Valve 18 Eastwood Place Singapore 486543	Forged steel and cast steel gate, globe, and check valves, bellows seal valves, knife-gate valves, butterfly valves, general-purpose ball valves, cryogenic valves	Telephone: (65) 246-1108 Fax: (65) 244-9791 http://www.velan.com
Velan Valve Corp. 730 N. Post Oak Rd. Suite 311 Houston, TX 77024 U.S.A.	Forged steel and cast steel gate, globe, and check valves, bellows seal valves, knife-gate valves, butterfly valves, general-purpose ball valves, cryogenic valves	Telephone: (713) 682-1837 Fax: (713) 682-6071 http://www.velan.com
Engineered Piping Products, Inc. 6317 Arundel Cove Rd. Baltimore, MD 21226 U.S.A.	PVC, CPVC, Polypro, PVDF valves, Asahi PVC, CPVC, Polypro, and Teflon valves	Telephone: (888) 355-8254 Telephone: (410) 355-8250 Fax: (410) 355-8274
Engineered Piping Products, Inc. 247 Bucheimer Rd. Frederick, MD 21701 U.S.A.	PVC, CPVC, Polypro, PVDF valves, Asahi PVC, CPVC, Polypro, and Teflon valves	Telephone: (877) 492-6422 Telephone: (301) 668-4315 Fax: (301) 668-4318
Spartan Controls Ltd. 305-27 St. Calgary, Alberta T2A 7V2 Canada	Control valves, regulators, instrumentation, and performance services to the process control industries	Telephone: (403) 207-0700 Fax: (403) 207-0874 http://www.frco.com
Norpac Controls Ltd. 30 Gostick Place North Vancouver British Columbia V7M 3G3 Canada	Control valves, regulators, instrumentation, and performance services to the process control industries	Telephone: (604) 980-5701 Fax: (604) 986-0483 http://www.frco.com
Atlantic Controls Ltd. PO Box 1250 32 Mcllveen Drive Saint John, New Brunswick E2L 4G7 Canada	Control valves, regulators, instrumentation, and performance services to the process control industries	Telephone: (506) 634-7437 Fax: (506) 634-0859 http://www.frco.com
Atlantic Controls Ltd. 102 Chain Lake Dr. Suite 200 Halifax, Nova Scotia B3S 1A7 Canada	Control valves, regulators, instrumentation, and performance services to the process control industries	Telephone: (902) 450-1488 Fax: (902) 450-2398 http://www.frco.com
Laurentide Controls Ltd. 18000 Trans Canada Hwy Kirkland, Quebec H9J 4A1 Canada	Control valves, regulators, instrumentation, and performance services to the process control industries	Telephone: (514) 697-9230 Fax: (514) 697-9335 http://www.frco.com
Septre Controls Ltd. 498 Henderson Drive Regina, Saskatchewan S4N 6E3 Canada	Control valves, regulators, instrumentation, and performance services to the process control industries	Telephone: (306) 721-6925 Fax: (306) 721-6936 http://www.frco.com

TABLE 29.24 Sources for Valves (Continued)

Company name and address	Product/service	Contact
Controltech, Inc. 7 Sylvan Way Winnipeg, Manitoba R2R 2B9 Canada	Control valves, regulators, instrumentation, and performance services to the process control industries	Telephone: (204) 633-9197 Fax: (204) 632-9608 http://www.frco.com
Fisher-Rosemount Systems, Inc. 320 Pinebush Road, Unit #12 Cambridge, Ontario N3C 2Z3 Canada	Control valves, regulators, instrumentation, and performance services to the process control industries	Telephone: (519) 624-2800 Fax: (519) 624-2801 http://www.frco.com
Control Southern Inc. 2200 Riverchase Center Suite 708 Birmingham, AL 35244 U.S.A.	Control valves, regulators, instrumentation, and performance services to the process control industries	Telephone: (205) 988-5700 Fax: (205) 988-4206 http://www.frco.com
PCE Pacific, Inc. 3710 Woodland Drive #2000 Anchorage, AK 99517 U.S.A.	Control valves, regulators, instrumentation, and performance services to the process control industries	Telephone: (907) 243-3833 Fax: (425) 487-1114 http://www.frco.com
Associated Process Controls 7150 Koll Center Parkway Pleasanton, CA 94566 U.S.A.	Control valves, regulators, instrumentation, and performance services to the process control industries	Telephone: (925) 846-9000 Fax: (925) 846-0396 http://www.frco.com
Proconex, Inc. PO Box 10696 Baltimore, MD 21265-0696 U.S.A.	Control valves, regulators, instrumentation, and performance services to the process control industries	Telephone: (410) 597-9000 Fax: (410) 265-8370 http://www.frco.com
Applied Control Equipment 7000 South Potomac St. Englewood, CO 80122 U.S.A.	Control valves, regulators, instrumentation, and performance services to the process control industries	Telephone: (303) 799-9300 Fax: (303) 799-1912 http://www.frco.com
Key Controls Inc. 8416 Laurel Fair Circle Suite 112 Tampa, FL 33610 U.S.A.	Control valves, regulators, instrumentation, and performance services to the process control industries	Telephone: (813) 626-1126 Fax: (813) 621-3134 http://www.frco.com
Process Controls 99-1295 Waiua Place Aiea, HI 96701 U.S.A.	Control valves, regulators, instrumentation, and performance services to the process control industries	Telephone: (808) 487-7717 Fax: (808) 488-3343 http://www.frco.com
Novaspect, Inc. 1776 Commerce Dr. Elk Grove Village, IL 60007-2120 U.S.A.	Control valves, regulators, instrumentation, and performance services to the process control industries	Telephone: (847) 956-8020 Fax: (847) 956-8588 http://www.frco.com
Applied Control Equipment 919 Nutter Blvd. Billings, MT 59105 U.S.A.	Control valves, regulators, instrumentation, and performance services to the process control industries	Telephone: (406) 656-5601 Fax: (406) 252-1945 http://www.frco.com
Fisher-Rosemount Systems, Inc. 8301 Cameron Rd. Austin, TX 78754 U.S.A.	Control valves, regulators, instrumentation, and performance services to the process control industries	Telephone: (512) 835-2190 Fax: (512) 832-3443 http://www.frco.com
Northeast Controls Inc. 3 Badertscher Dr. Clifton Park, NY 12065 U.S.A.	Control valves, regulators, instrumentation, and performance services to the process control industries	Telephone: (518) 664-6600 Fax: (518) 664-8500 http://www.frco.com
Noble Alloy 10945 Day Rd. Houston, TX 77043 U.S.A.	Ball valves, check valves, and multiport ball valves	Telephone: (800) 688-4893 Fax: (713) 984-0099 http://www.noblealloy.com

TABLE 29.24 Sources for Valves (Continued)

Company name and address	Product/service	Contact
Seeley & Jones 408 Woodmont Rd. Milford, CT 06460 U.S.A.	Diaphragm valves, ball valves, cryogenic valves, and general-purpose valves	Telephone: (203) 877-4425 Fax: (203) 877-4260 http://www.mccannainc.com
Tri-State Technical Sales 3601 Commerce Dr., Bldg. 109 Baltimore, MD 21227 U.S.A.	Diaphragm valves, ball valves, cryogenic valves, and general-purpose valves	Telephone: (410) 536-4242 Fax: (410) 536-0583 http://www.mccannainc.com
Benchmark Controls 356 Manchaug Manchaug, MA 01526 U.S.A.	Diaphragm valves, ball valves, cryogenic valves, and general-purpose valves	Telephone: (508) 476-1122 Fax: (508) 476-1133 http://www.mccannainc.com
Tri-State Tech. Sales 4 Central Ave. West Orange, NJ 07052 U.S.A.	Diaphragm valves, ball valves, cryogenic valves, and general-purpose valves	Telephone: (201) 672-9100 Fax: (201) 672-5660 http://www.mccannainc.com
Tri-State Tech. Sales 382 Lancaster Rd. Malvern, PA 19355 U.S.A.	Diaphragm valves, ball valves, cryogenic valves, and general-purpose valves	Telephone: (610) 647-5700 Fax: (610) 647-3905 http://www.mccannainc.com
Asco 50-60 Hanover Rd. Florham Park, NJ 07932 U.S.A.	Specialized solenoid valves, pilot control valves, dribble control valves, and angle body piston valves	Telephone: (973) 966-2000 Fax: (973) 966-2628 http://www.asco.com
Crane 860 Remington Ave. Bolingbrook, IL 60440 U.S.A.	Bronze, iron, steel, and stainless steel gate, globe, and check valves	Telephone: (800) 323-3679 Fax: (630) 226-2641 http://www.cranevalve.com
Center Line & Flowseal 3201 Walnut Ave. Long Beach, CA 90807 U.S.A.	Bronze, iron, steel, and stainless steel gate, globe, and check valves	Telephone: (562) 426-2531 Fax: (562) 426-6797 http://www.cranevalve.com
Jenkins 254 Henry St. Brantford, Ontario N3T 5TZ Canada	Bronze, iron, steel, and stainless steel gate, globe, and check valves	Telephone: (519) 759-3911 Fax: (519) 759-7970 http://www.cranevalve.com
Xomox 4444 Cooper Rd. Cincinatti, OH 45242 U.S.A.	Quarter-turn valves and actuators for process industry applications	Telephone: (513) 745-6000 http://www.xomox.com
Norstrom Valves 1511 Jefferson St. Sulphur Springs, TX 75482 U.S.A.	Tapered plug valves, polyethylene valves, steel ball valves, butterfly valves, and multiport valves	Telephone: (903) 885-4693 Fax: (903) 439-3411 http://www.nordstromvalves.com
Neles Controls, Inc. 42 Bowditch Shrewsbury, MA 01545 U.S.A.	Control valves, rotary plug valves, globe and angle valves, modular ball valves, and high-performance butterfly valves	Telephone: (508) 852-3567 Fax: (508) 856-8762 http://www.nelescontrols.com
EIM Valve Controls 13840 Pike Rd. Missouri, TX 77489 U.S.A.	Shipboard valve control networks, electric actuators, electric controls, and manual operators	Telephone: (713) 499-1561 Fax: (713) 499-8445 http://www.eim-co.com
Adams Valves Inc. 10649 Haddington Dr., No. 160 Houston, TX 77043 U.S.A.	Valves	Telephone: (713) 973-2490 Fax: (713) 973-2788 http://www.adams-valves.com

TABLE 29.24 Sources for Valves (Continued)

Company name and address	Product/service	Contact
American AVK Co. 5286 East Home Fresno, CA 93927 U.S.A.	Valves	Telephone: (559) 452-4300 Fax: (559) 251-0426
American Cast Iron Pipe Co. PO Box 2727 Birmingham, AL 35202 U.S.A.	Resilient seated gate valves and check valves	Telephone: (205) 325-7856 Telephone: (205) 325-7862 http://www.acipco.com
Anvil International, Inc. Research & Development Center 20 Thurber Blvd. Smithfield, RI 02917 U.S.A.	Butterfly valves	Telephone: (401) 233-3927 Fax: (401) 781-9912
BendTec, Inc. 366 Garfield Ave., Duluth, MN 55802 U.S.A.	Valves	Telephone: (218) 722-0205 Fax: (218) 722-6598 Email: employeesname@bendtec.com http://www.bendtec.com
Cesare Bonetti SPA Cesare Bonetti, 17-20024 Garbagnate Milanese Italy	Bellow valves, globe valves, gate valves, and high-temperature ball valves	Telephone: 011+39(2) 9956462 Fax: 011+39(2) 9952483 http://www.cesare-bonetti.it
Bonney Forge Corp. Cedar and Meadows St. PO Box 359 Allentown, PA 18105 U.S.A.	Forged steel valves and cast steel valves	Telephone: (610) 435-9611 Fax: (610) 435-6973 http://www.bonneyforge.com
Clow Valve Co. 902 South Second St. Oskaloosa, IA 52577 U.S.A.	AWWA valves and hydrants	Telephone: (641) 673-8611 Fax: (641) 673-8289 http://www.clowvalves.com
Conbraco Industries PO Box 247 Matthews, NC 28106 U.S.A.	Ball valves, check valves, forged gate, globe, and check valves	Telephone: (704) 841-6092 Fax: (704) 841-6021 http://www.conbraco.com
EKK Eagle American, Inc. 33 Plan Way, Bldg. 5 Warwick, RI 02886 U.S.A.	Valves	Telephone: (401) 732-0333 Fax: (401) 732-2201
Energovalves 4647 Pine Timbers Suite 115 Houston, TX 77041 U.S.A.	Gate, globe, swing check, and ball valves	Telephone: (713) 895-9599 Fax: (713) 895-7585 http://www.energovalves.com
East Jordan Iron Works PO Box 439 5000 Airport Rd. East Jordan, MI 49727 U.S.A.	AWWA C500, C515 gate valve lines	Telephone: (231) 536-2261 Fax: (231) 536-4486 http://www.energovalves.com
ITT Industries, Inc. 33 Centerville Rd. PO Box 6164 Lancaster, PA 17603-2064 U.S.A.	Diaphragm valves and actuators, ball valves, butterfly valves, knife-gate valves, slurry gate valves, and check valves	Telephone: (717) 509-2200 Fax: (717) 509-2349
Kennedy Valve 1021 East Water St. Elmira, NY 14902-1516 U.S.A.	AWWA and UL/FM hydrants and valves	Telephone: (607) 734-2211 Fax: (607) 734-3288 http://www.Kennedyvalve.com

TABLE 29.24 Sources for Valves (Continued)

Company name and address	Product/service	Contact
Kingdom Precision Castings Co., Ltd. c/o Alliance Metals Group, LLC 2700 Greway Blvd., Suite 500 Minneapolis, MN 55430 U.S.A.	Valves	Telephone: (612) 585-1387 Fax: (612) 561-8558
Kitz Corporation of America 10750 Corporate Dr. Stafford, TX 77477-4008 U.S.A.	Gate, ball, butterfly, direct seal, and HF-acid valves in brass, bronze, iron, stainless, and high alloy	Telephone: (281) 491-7333 Telephone: (800) 772-0073 Fax: (281) 491-9402 http://www.kitz.com
M&H Valve Company A Division of McWane, Inc. PO Box 2088 Anniston, AL 36202 U.S.A.	Gate valves, butterfly valves, check valves, hydrants, mud valves, eccentric plug valves, flap valves, and pressure relief valves	Telephone: (256) 237-3521 Fax: (256) 237-8630 http://www.mh-valve.com
Milliken Valve Co. Inc. 1995 Highland Ave. Suite 500 Bethlehm, PA 18020 U.S.A.	AWWA C500, C515 gate valve lines	Telephone: (610) 861-8803 Fax: (610) 861-8094 Email: millikenvalve@nni.com
Milwaukee Valve Co. 2375 South Burrall St. Milwaukee, WI 53207 U.S.A.	Bronze valves, butterfly valves, iron valves, commercial ball valves, industrial ball valves, cast steel valves, dry bulk valves, transformer valves, fire protection valves, actuation controls, actuators	Telephone: (414) 744-5240 Fax: (414) 744-5840 http://www.milwaukeevalve.com
MSA a.s. Hlucinska 41 747 22 Dolnl Benesov Czech Republic	Valves	Telephone: 011+420-653-541-142 Fax: 011+420-653-541-360
Neles Controls, Inc. 42 Bowditch Dr. Shrewsbury, MA 01545-8004 U.S.A.	Valves	Telephone: (508) 852-3567 Fax: (508) 595-5019
Orbit Valve 7200 Interstate Dr. Little Rock, AR 72209 U.S.A.	Valves	Telephone: (501) 570-5656 Fax: (501) 570-5765
Pacific Valves 3201 Walnut Ave. Long Beach, CA 90807 U.S.A.	Class 150 through 2500 flanged and butt weld end gate, T-globe, Y-globe, and swing check valves	Telephone: (562) 426-2531 Fax: (562) 426-0210 http://www.cranevalve.com
Parker Hannifin Corporation 6035 Parkland Blvd. Cleveland, OH 44124-4141 U.S.A.	Valves	Telephone: (216) 531-3000 Fax: (216) 896-4022
Perfection Corporation 222 Lake St. PO Box 10 Madison, OH 44057 U.S.A.	Valves	Telephone: (440) 428-1171 Fax: (440) 428-7325 http://www.perfecorp.com
Romac Industries, Inc. 21919 20th Ave. SE Suite 100 Bothell, WA 98021-4404 U.S.A.	Valve actuators	Telephone: (206) 624-6491 Fax: (206) 624-1971 http://www.romacindustries.com

TABLE 29.24 Sources for Valves (Continued)

Company name and address	Product/service	Contact
Swagelok Companies 29500 Solon Rd. Solon, OH 44139 U.S.A.	Needle, ball, check, plug, and relief valves	Telephone: (440) 248-4600 Fax: (440) 349-5806 http://www.swagelok.com
Trinity Industries Fittings and Flange Group PO Box 568887 Dallas, TX 75357-8887 U.S.A.	Valves	Telephone: (214) 589-7856 Fax: (214) 589-7594 http://www.tffg.net
UMAC, Inc. 120 South Ship Rd. Exton, PA 19341 U.S.A.	Excess-flow valves	Telephone: (610) 524-6800 Fax: (610) 524-6806
U.S. Pipe & Foundry Co. PO Box 10406 Birmingham, AL 35202 U.S.A.	Valves	Telephone: (205) 254-7299 Fax: (205) 254-7405
United Valve 707 Michigan South Houston, TX 77587 U.S.A.	Valves	Telephone: (713) 944-9852 Fax: (713) 944-5964 http://www.unitedvalve.com
The Walworth Company 8383 Commerce Park Dr. Houston, TX 77036-7425 U.S.A.	Cast, forged, and bronze gate, globe, and check valves, lubricated plug valves, safety relief valves, ball valves, and butterfly valves	Telephone: (713) 777-7788 Fax: (713) 981-1246 Email: walworth@walworthusa.com http://www.walworth.com
Waterous Company 125 Hardman Ave. S South St. Paul, MN 55075-1191 U.S.A.	Fire pump valves	Telephone: (651) 450-5000 Fax: (651) 450-5090 http://www.waterous.com
Zy-Tec Global Corp. 10600 Corporate Dr. Stafford, TX 77477 U.S.A.	Cast steel, low-temperature and chrome alloy, forged steel, floating and trunnion-mounted ball, pressure seal, and other valve types	Telephone: (281) 565-1010 Telephone: (800) 231-3530 Fax: (281) 565-3171 http://www.zy-tech.com

INDEX